T0230677

MARITIME TRANSPORTATION AND HARVESTING OF SEA RESOURCES

PROCEEDINGS OF IMAM 2017, 17TH INTERNATIONAL CONGRESS OF THE INTERNATIONAL MARITIME ASSOCIATION OF THE MEDITERRANEAN, LISBON, PORTUGAL, 9–11 OCTOBER 2017

Maritime Transportation and Harvesting of Sea Resources

Editors

C. Guedes Soares & Ângelo P. Teixeira
Instituto Superior Técnico, Universidade de Lisboa, Portugal

VOLUME 1

CRC Press
Taylor & Francis Group
Boca Raton London New York Leiden

CRC Press is an imprint of the
Taylor & Francis Group, an **informa** business

A BALKEMA BOOK

CRC Press/Balkema is an imprint of the Taylor & Francis Group, an informa business

© 2018 Taylor & Francis Group, London, UK

Typeset by V Publishing Solutions Pvt Ltd., Chennai, India
Printed and bound in Great Britain by CPI Group (UK) Ltd, Croydon, CR0 4 YY

All rights reserved. No part of this publication or the information contained herein may be reproduced, stored in a retrieval system, or transmitted in any form or by any means, electronic, mechanical, by photocopying, recording or otherwise, without written prior permission from the publisher.

Although all care is taken to ensure integrity and the quality of this publication and the information herein, no responsibility is assumed by the publishers nor the author for any damage to the property or persons as a result of operation or use of this publication and/or the information contained herein.

Published by: CRC Press/Balkema
   P.O. Box 11320, 2301 EH Leiden, The Netherlands
   e-mail: Pub.NL@taylorandfrancis.com
   www.crcpress.com – www.taylorandfrancis.com

ISBN: 978-0-8153-7993-5 (Hardback, set of 2 volumes + CD-ROM)
ISBN: 978-1-351-21450-6 (eBook, set of 2 volumes)
ISBN: 978-0-8153-7611-8 (Hardback, Volume 1)
ISBN: 978-0-8153-7990-4 (Hardback, Volume 2)

Maritime Transportation and Harvesting of Sea Resources – Guedes Soares & Teixeira (Eds)
© 2018 Taylor & Francis Group, London, ISBN 978-0-8153-7993-5

Table of contents

Maritime safety and human factors

Hydrodynamics

Hydrodynamics – foils

Hydrodynamics – resistance

Hydrodynamics – seakeeping

Hydrodynamics – CFD

Hydrodynamics – manoeuvring

Ship structures

VOLUME 2

Composite materials

Shipbuilding and repair

Ship design

Ship lifecycle

Machinery and control

Marine environment

Wave modelling

Offshore oil and gas

Offshore renewable energy

Fisheries

Preface

This book presents the Proceedings of the 17th International Maritime Association of the Mediterranean Congress held in Lisbon, Portugal between 9 and 11 October 2017, under the theme "Maritime Transportation and Harvesting of Sea Resources".

The International Congress of International Maritime Association of the Mediterranean has nearly a 40-year history since the first Congress of the International Maritime Association of East Mediterranean (IMAEM) held in Istanbul in 1978. It is a voluntary organization, established in 1974, by institutions originating from Bulgaria, Egypt, Greece, Italy, Yugoslavia and Turkey, and was progressively enlarged to other Mediterranean countries and neighbouring regions.

After 1990, the name of the Association, and of the Congresses, was changed to the International Maritime Association of the Mediterranean (IMAM). Since then, it has remained unchanged, being an association of academic and professional institutions who aim to contribute to the efficient, economic and safe operation of ships, offshore and coastal structures and port infrastructures in order to promote the advancement of the use of the seas and coastal waters for a sustainable development of the societies.

For the second time in its history, this biannual Congress is organized in Lisbon, Portugal, where the Centre for Marine Technology and Ocean Engineering (CENTEC) will be taking the responsibility of the preparation of the Conference Proceedings as well as of the logistic organization of the Congress.

The Congress is addressed to individuals from industry, research organizations, universities, government agencies, certifying authorities as well as designers, operators and owners who contribute to improved knowledge about the marine environment, ship and offshore design, building and maintenance technology, maritime transportation and port operation and exploitation, offshore oil and gas exploitation, nautical activities and marinas, fisheries and aquaculture, maritime safety and protection of the environment.

Like previous Congresses the present edition of IMAM has a wide scope by including, in addition to ships and maritime transportation and ports, other topics of contemporary interest related with the sustainable exploitation of ocean and coastal resources and with renewable energy, areas that are believed to have become increasingly more important.

This year's Congress incorporates the International Workshop on Nautical Traffic Models, following the successful previous four workshops held in Shanghai (2012), Delft (2013), Wuhan (2014) and in Espoo (2015), that covers the general aspects of maritime risk and safety, maritime traffic simulation technologies, maritime traffic engineering, human factors and related fields. The Congress also includes special sessions in honour of Professors Peter Kolev, Vedran Žanić and Giovanni Benvenuto who have been active in IMAM and have retired recently.

About 300 abstracts have been submitted and after a review process around 160 papers have been accepted and included in this book. Amongst the topics covered in this year's event are the most updated aspects and developments concerning the design and operation of ships, namely, hydrodynamics, seakeeping and manoeuvring, marine structures, machinery, resistance and propulsion systems, ship design, and shipyard technology. Similarly, papers reflecting a number of fundamental areas of the exploitation of ocean and coastal resources are included, such as marine environment, fisheries, maritime transportation and ports, maritime traffic modelling, coastal and offshore development, maritime safety, wave and wind modelling and renewable energies.

Thanks are due to the Scientific Committee who had most of the responsibility of reviewing the papers and to an equally numerous group of anonymous reviewers that have helped the authors in delivering better papers by providing them constructive comments to improve the papers. Unfortunately, several manuscripts that did not meet the quality criteria have not been included in the proceedings but we hope that this has contributed to a more consistent quality level of the papers included in these books.

Carlos Guedes Soares and Ângelo P. Teixeira

Maritime Transportation and Harvesting of Sea Resources – Guedes Soares & Teixeira (Eds)
© 2018 Taylor & Francis Group, London, ISBN 978-0-8153-7993-5

IMAM 2017 organisation

CONFERENCE CHAIRMAN

C. Guedes Soares, *IST, Universidade de Lisboa, Portugal*

INTERNATIONAL SCIENTIFIC COMMITTEE

Moustafa Abdel-Marksoud, *TU Harburg, Germany*
Ahmet Dursun Alkan, *Istanbul Technical University, Turkey*
Felice Arena, *University of Reggio Calabria, Italy*
Eugen Bârsan, *Constanta Maritime University, Romania*
Ermina Begovic, *University of Naples Federico II, Italy*
Kostas A. Belibassakis, *Technological Educational Inst. Athens, Greece*
Marco Biot, *University of Trieste, Italy*
Rui Carlos Botter, *Universidade de São Paulo, Brasil*
Emilio Campana, *INSEAN, Italy*
Francisco del Castillo, *Fundación Soermar, Spain*
Nastia Degiuli, *University of Zagreb, Croatia*
Roko Dejhalla, *University of Rijeka, Croatia*
Leonard Domnisoru, *University of Galati, Romania*
Cesar Ducruet, *Paris Geo, CNRS, France*
Selma Ergin, *Istanbul Technical University, Turkey*
R. Cengiz Ertekin, *University of Hawaii at Manoa, USA*
Stefano Ferraris, *FINCANTIERI, Italy*
Massimo Figari, *University of Genova, Italy*
Juana Fortes, *LNEC, Portugal*
Yordan Garbatov, *IST, Universidade de Lisboa, Portugal*
Petar Georgiev, *Technical University of Varna, Bulgaria*
José M. Gordo, *IST, Universidade de Lisboa, Portugal*
Omer Goren, *Istanbul Technical University, Turkey*
Gregory Grigoropoulos, *National Technical University of Athens, Greece*
Stein Haugen, *NTNU, Norway*
Dimitrios Konovessis, *Nanyang Technical University, Singapore*
Pentti Kujala, *Aalto University, Finland*
Jean-Marc Laurens, *ENSTA Bretagne, France*
Heba Leheta, *Alexandria University, Egypt*
Ould El Moctar, *University of Duisburg, Germnay*
Lorenzo Moro, *Memorial University, Canada*
Nikitas Nikitakos, *University of the Aegean, Greece*
Luís R. Nuñez Rivas, *ETSIN, Universidad Politecnica de Madrid, Spain*
Andrea Orlandi, *Consorzio LaMMA, Florence, Italy*
Joško Parunov, *University of Zagreb, Croatia*
Fernando López Peña, *University A Coruña, Spain*
Francisco Taveira Pinto, *University of Porto, Portugal*
Floriano C.M. Pires, Jr., *Universidade Federal do Rio de Janeiro, Brasil*
Jasna Prpić-Oršić, *University of Rijeka, Croatia*
Vicent Rey, *University of Toulon, France*

Cesare Rizzo, *University of Genoa, Italy*
Enrico Rizzuto, *University of Genova, Italy*
Germán Rodríguez, *U. Las Palmas de Gran Canaria, Spain*
Eugen Rusu, *Dunarea de Jos University of Galati, Romania*
Antonello Sala, *ISMAR, Italy*
Antonio Scamardella, *Parthenope University (Naples), Italy*
Kostas Spyrou, *National Technical Univ. Athens, Greece*
Nikolaos Ventikos, *National Technical Univ. Athens, Greece*
Michele Viviani, *University of Genova, Italy*
K. Visser, *Delft Technical University, The Netherlands*
Xinping Yan, *Wuhan University of Technology, China*
Ermal Xhelilaj, *University of Vlora, Albania*
Panos Yannoulis, *OCEAN KING, Greece*

NATIONAL SUPPORT COMMITTEE

Ângelo P. Teixeira, *IST, Universidade de Lisboa, Portugal (Chair)*
Dina Dimas, *Arsenal do Alfeite, Portugal*
Miguel Morgado, *Bureau Veritas, Portugal*
António Oliveira, *Transinsular, Portugal*
Paulo Parracho, *DGRM, Portugal*
Pedro Ponte, *APSS, Portugal*
Tiago A. Santos, *IST, Universidade de Lisboa, Portugal*
Abel Simões, *ENIDH, Portugal*
Paulo Viana, *DNV GL, Portugal*

TECHNICAL PROGRAMME SECRETARIAT

Maria de Fátima Pina, *IST, Universidade de Lisboa, Portugal*
Sandra Ponce, *IST, Universidade de Lisboa, Portugal*
Bruna Covelas, *IST, Universidade de Lisboa, Portugal*

Maritime Transportation and Harvesting of Sea Resources – Guedes Soares & Teixeira (Eds)
© 2018 Taylor & Francis Group, London, ISBN 978-0-8153-7993-5

IMAM organisation

IMAM EXECUTIVE COMMITTEE

Prof. Marco Altosole, *University of Genova, Italy*
Dr. Sergey Baskakov, *Odessa State Maritime University, Ukraine*
Dr. Elena-Felicia Beznea, *University "Dunarea de Jos" of Galati, Romania*
Prof. Selma Ergin, *Istanbul Technical University, Turkey*
Prof. Petar Georgiev, *Technical University of Varna, Bulgaria*
Prof. Carlos Guedes Soares, *Instituto Superior Técnico, Portugal (Vice-President)*
Prof. Heba W. Leheta, *Alexandria University, Egypt*
Prof. Fernando López Peña, *University of A Coruña, Spain (Past President)*
Prof. Jasna Prpic Orsic, *University of Rijeka, Croatia (President)*
Prof. Fabien Remy, *École Sup. d'Ingenieurs Marseille, France*
Prof. Constantinos (Kostas) Spyrou, *Hellenic Institute of Marine Technology, Greece*
Dr. Ermal Xhelilaj, *Vlora University, Albania*

IMAM TECHNICAL COMMITTEES

Hydrodynamics

Stefano Brizzolara, *Virginia Tech, USA*
Dario Bruzzone, *University of Genoa, Italy*
Pierre Ferrant, *École Centrale Nantes, France*
Omer Goren, *Istanbul Technical University, Turkey*
Gregory Grigoropoulos, *National Technical University of Athens, Greece*
Atilla Incecik, *University of Strathclyde, United Kingdom*
Fernando López Peña, *University of A Coruña, Spain*
Adolfo Marón, *CEHIPAR-Madrid, Spain*
Touvia Miloh, *Tel Aviv University, Israel*
Bernard Molin, *École Centrale Marseille, France*
Marcelo Neves, *Federal University of Rio de Janeiro, Brazil*
Jasna Prpić Oršić, *University of Rijeka, Croatia*
Constantinos Spyrou, *National Technical Univ. Athens, Greece*
Pandeli Temarel, *University of Southampton, United Kingdom*
Leszek Wilczynski, *CTO – Gdańsk, Poland*

Marine Structures

Dino Cervetto, *RINA, Italy*
Matteo Codda, *CETENA-Genoa, Italy*
Ionel Chirica, *University of Galati, Romania*
Leonard Domnisoru, *University of Galati, Romania*
Yordan Garbatov, *Instituto Superior Técnico-Lisbon, Portugal*
Mohammad Reza Khedmati, *Amirkabir University of Technology, Iran*
Heba W. Leheta, *Alexandria University, Egypt*
Mario Maestro, *University of Trieste, Italy*
Joško Parunov, *University of Zagreb, Croatia*
Cesare Rizzo, *University of Genoa, Italy*

Emmanuel Samuelides, *National Technical University of Athens, Greece*
Ajit Shenoi, *University of Southampton, United Kingdom*

Machinery & Control

Giovanni Benvenuto, *University of Genoa, Italy*
Andrea Cogliolo, *RINA, Italy*
Christos Frangopoulos, *National Technical University of Athens, Greece*
Antonio Paciolla, *University of Naples (Federico II), Italy*
George Palambrou, *National Technical University of Athens, Greece*
Luca Sebastiani, *CETENA-Genoa, Italy*

Shipyard Technologies

Ashutosh Sinha, *SSA, United Kingdom*
Niksa Fafandjel, *University of Rijeka, Croatia*
Luigi Mor, *Nuovi Cantieri Apuania, Italy*
Kalman Ziha, *University of Zagreb, Croatia*

Design of Marine Systems

Ahmed Alkan, *Yildiz Technical University – Istanbul, Turkey*
Ernesto Fasano, *University of Naples, Italy*
Miguel Ángel Herreros, *UPM-ETSIN, Spain*
Kristofor Lapa, *University of Vlora, Albania*
Apostolos Papanikolau, *National Technical University of Athens, Greece*
Panos Yannoulis, *OCEAN KING, Greece*
Vedran Zanic, *University of Zagreb, Croatia*

Safety of Marine Systems

Eugen Barsan, *Constanta Maritime University, Romania*
Gianfranco Damilano, *ATENA, Italy*
Alberto Francescutto, *University of Trieste, Italy*
Paola Gualeni, *University of Genoa, Italy*
Heba Leheta, *Alexandria University, Egypt*
Enrico Rizzuto, *University of Genoa, Italy*
Ângelo Teixeira, *Instituto Superior Técnico – Lisbon, Portugal*

Marine Environment

Jose Antunes do Carmo, *University of Coimbra, Portugal*
Felice Arena, *University of Reggio Calabria, Italy*
K.A. Belibassakis, *Technological Educational Inst. Athens, Greece*
Juana Fortes, *LNEC, Portugal*
German R. Rodriguez, *University of Las Palmas, Spain*
Eugen Rusu, *University of Galati, Romania*
Agustin Sanchez-Arcilla, *UPC-Barcelona, Spain*
Lev Shemer, *Tel Aviv University, Israel*

Protection of the Environment

Ruggero Dambra, *CETENA-Genoa, Italy*
Selma Ergin, *Istanbul Technical University, Turkey*
Corrado Schenone, *University of Genoa, Italy*
Massimo Figari, *University of Genoa, Italy*

Ports & Transports Systems

Makoto Arai, *Yokhohama University, Japan*
Rui Carlos Botter, *University of S.Paulo, Brazil*
Dimitrios Lyridis, *National Technical University of Athens, Greece*
Eden Mamut, *Ovidius University of Constantza, Romania*
Nikitas Nikitakos, *Aegean University, Greece*
Harilaos Psaraftis, *Technical University of Denmark, Denmark*
Giovanni Solari, *University of Genoa, Italy*

Offshore & Coastal Development

Francisco Taveira Pinto, *University of Porto, Portugal*
Mohamed Chagdali, *University Ben M'Sik Casablanca, Morocco*
Inigo Losada, *University of Cantabria, Spain*
Spyros Mavrakos, *National Technical University of Athens, Greece*
Vicent Rey, *University of Toulon, France*
Leonardo Brunori, *RINA, Italy*

Aquaculture & Fishing

Aida Campos, *IPMA, Portugal*
Teresa Dinis, *Algarve University, Portugal*
Rajko Grubisic, *University of Zagreb, Croatia*
Barry O'Neil, *MARLAB, United Kingdom*
Daniel Priour, *IFREMER, France*
Antonello Sala, *ISMAR, Italy*
Emma Tomaselli, *RINA, Italy*

Small & Pleasure Crafts

Carlo Bertorello, *University of Naples, Italy*
Dario Boote, *University of Genoa, Italy*
Izvor Grubisic, *University of Zagreb, Croatia*
Massimo Musio-Sale, *University of Genoa, Italy*
Lorenzo Pollicardo, *Federagenti Yacht, Italy*
Antonio Scamardella, *Parthenope University (Naples), Italy*

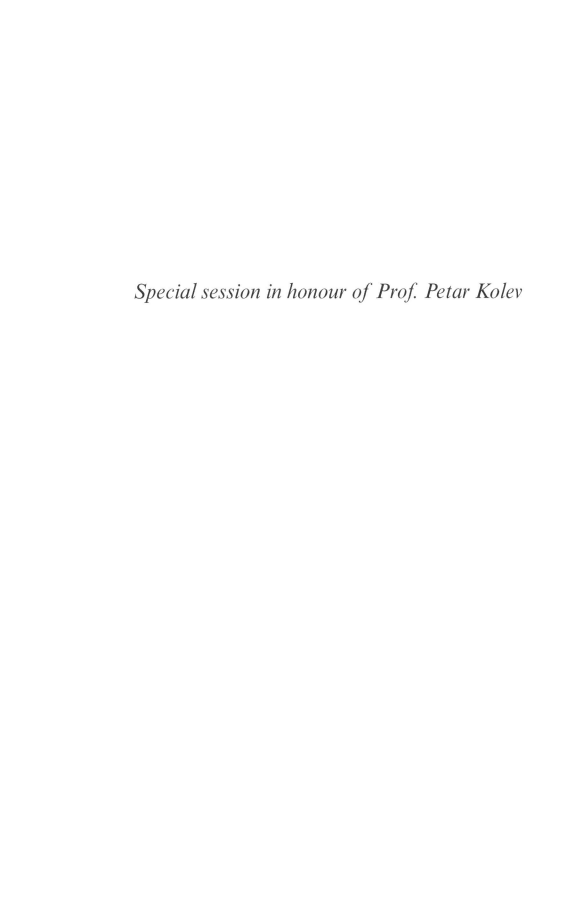

Special session in honour of Prof. Petar Kolev

Maritime Transportation and Harvesting of Sea Resources – Guedes Soares & Teixeira (Eds)
© 2018 Taylor & Francis Group, London, ISBN 978-0-8153-7993-5

Ship design, decades of development

T. Damyanliev & P. Georgiev
Technical University of Varna, Varna, Bulgaria

ABSTRACT: The principal objective of this work is to review the developments in the field of ship design in Bulgaria in the last four decades, driven by the leadership of Professor Kolev and his group of associate researchers. A very advanced research has been performed on different aspects of ship design, including optimization and on-board computer systems proven to be a good solution in many national and international research projects and practical applications. The paper also reviews the state of the art of ship design, in particular, the inverse and conceptual ship design, computer system for conceptual and robust design, sensitivity analysis of the optimum solution and different developments and implementations in the ship operation employing on-board loading software systems.

1 INTRODUCTION

The knowledge of any complex technical system, such as the ship, is gained through the results of systematic research in individual scientific fields (subject sciences), which are determined by the subsystems and their qualities.

Following the leading shipbuilding school, the Bulgarian academic education related to the ship design, construction and exploitation has been divided into the following fields:

– Ship theory;
– Structural mechanics and ship structures;
– Ship propulsion system;
– Marine engineering;
– Vessel devices, equipment and systems;
– Electrical equipment and ship automation;
– Shipbuilding and shipyard technology;
– Transportation and fleet exploitation;

These fields evaluate the design solution accounting for the ship performance, ship strength, necessary ship propulsion, devices and equipment and constrained by shipbuilding conditions and existing shipbuilding technologies and port and transportation limitations. Regardless of the fact that these field cover a very deep knowledge about the ship as a system, that can be transformed into a process of defining theoretical model(s) that can solve the "forward" problems, as can be seen in Fig. 1. To define the ship main dimensions, hull form and its coefficients, propulsion power system, a deep analysis based on the ship design theory needs to be performed and it is usually defined as a "reverse" problem (see Figure 1).

From a historical point of view, the work of Henri-Louis Duhamel du Monceau (1700–1782),

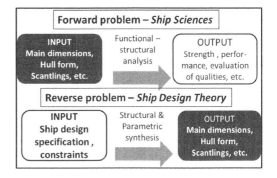

Figure 1. Forward and reverse design problem.

Euler L. (1707–1783) and Swedish admiral Fredrik Henrik af Chapman (1721–1808) can be accepted as the founders of the Ship Design Theory. They clearly formulated the principal objective of the ship design and drawn the first formulation of the ship mathematical model including the mass equilibrium and stability.

The principal development in the field of ship design in Bulgaria has been driven by Professor Petar Kolev and his research associates and several applications were reported in (Kolev, 1979, 1987).

The process of ship design starts from an idea to build a ship satisfying the regulations and accounting for the existing constraints, as can be seen in Figure 2, which presents the stages of ship design and associated scientific fields and corresponding tools as well.

The *Fleet composition* stage precedes the ship design and it is associated with the "*External task*" (Pashin, 1983). During this stage and based on the analysis of the transportation of goods and

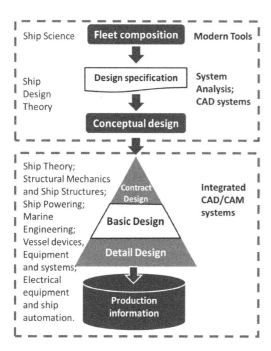

Figure 2. Ship design process.

commercial activities, the number and main characteristics of the ship, including the deadweight, propulsion system, cargo capacity, speed, range distance, crew number etc. are defined. All characteristics are included in the ship design specification

During the following conceptual design a reverse problem is solved. Taking into account the ship design specification, constraints and considering the ship as a complex engineering system, the basic characteristics, transforming it as a physical object (dimensions, architectural type, weights, volumes, etc.) are determined. This design solution should be optimal with respect to a predetermined assessment criterion. This stage is defined as *Conceptual design*.

Typically, the next two stages in the design are defined as *Contract* and *Basic design*. Based on conceptual design solution, or on existing design data in the case of a standard or family ship, negotiations with the ship owner take place leading to a decision being made regarding whether the project will go ahead or whether modifications are needed.

During the *Basic design* stage, the major equipment selection, general arrangements, systems design, space allocation and structural design and stability calculations are given a final approval by the Classification Societies and ship owner. This phase finishes with all drawings, material estimates, equipment lists, weld lengths and weight and centres of gravity reports. The preliminary structural definition developed in this phase is used for the detailed design and preparation of the ship building information.

The basic design is supplemented by the detailed design, which provides an additional information relating to the design of details, structural details as well as to subsystems such as electrical, navigation and piping arrangements and marine engineering.

As a rule, during these stages the reverse design problem is solved by modern integrated CAD/CAM systems. They are used in the preparation of the documentation related to ship construction. In Europe, the commercial CAD systems of companies as Aveva (http://www.aveva.com), SENER (http://www. ingenieriayconstruccion.sener), Autoship Systems Corporation (http://www. autoship. com/), Dassault Systemes (https://www.3ds.com) and others have become popular.

2   METHODOLIGAL BASIS OF CONCEPTUAL SHIP DESIGN

The exploration of the theoretical and applied aspects of conceptual ship design was the priority task in the work of Prof. P. Kolev in the 1970 s. In a number of publications and research projects (Kolev, 1972, 1972a, 1977), issues related to system analysis as a methodological base, optimization methods as a tool for finding the optimal design solution, were developed and analysed.

In general, the system analysis of the complex engineering systems includes three main components—decomposition, analysis and synthesis (Figure 3).

The most common strategies for decomposing the technical systems and processes that are associated with the ship can be formulated as:

– *Functional decomposition*, i.e. decomposition based on the analysis of system functions. The rationale for the separation of "functional

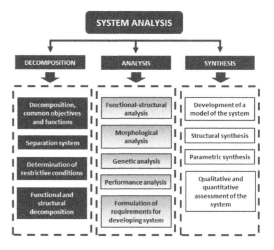

Figure 3.   Components of the system analysis.

subsystems" is the equivalence of the functions performed by the group of elements constituting the subsystem;

– *Life cycle decomposition* – decomposes any system where, at different stages of its operation, the subsystems change the basic laws of their operation;
– *Decomposition on subsystems* (structural decomposition) – an example of this may be the volume of generated and used information as a part of the overall database.

The analysis of the system includes:

– *Functional-structural analysis* – specification of the composition and operation laws of elements; the operating algorithm and the interoperability of the subsystems; formulation of design variables and uncontrollable parameters; the constraints that formulate the acceptable area of the project solution; analysis of the integrity of the system; formulation of the general requirements for the created system;
– *Morphological analysis* – interconnections between the components of the system;
– *Genetic analysis* – analysis of the background and the causes of developments of such systems;
– *Performance analysis* – formation of performance indicators, justification of efficiency criteria, immediate evaluation and analysis of the results obtained.

In the application of the system, analysis of complex engineering systems and their automated design is associated with the structural and parametric synthesis and assessment of the system.

Parametric synthesis defines the parameters of elements of a system with a particular structure that ensures its functionality. In case of existing mathematical model the parametric synthesis is accomplished solving an optimization problem.

The process of realizing a *quantitative system analysis* applicable to the conceptual design of a ship is illustrated by the algorithm proposed by Wagner, (1972). Independent on the type of the subject, the system analysis includes several stages as can be seen in Figure 4.

The design specification is related to the formulation of the basic requirements for the ship in terms of its core functions and characteristics defined by the higher-level system analysis. For the ship, this higher level could be the task of fleet composition.

The *mathematical model* of the ship needs to give information about:

– Mathematical and geometrical description of the object, defining the ship as a real physical object;
– Predicting the quality of the ship resulting from its functions, meeting the requirements

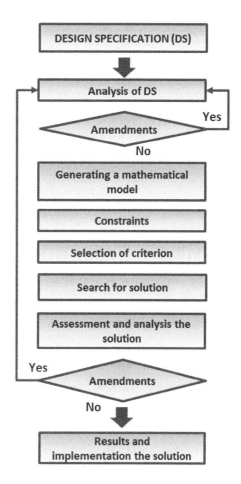

Figure 4.   Stages of system analysis.

of the Classification Societies, conventions and others;
– Indicators to assess the effectiveness of the project solution, including economic indicators as CAPEX, OPEX.

The requirements and *constraints* are determined by the parameters included in the design specification, model applicability constraints and other functional requirements to ensure the quality of the ship.

The implementation of the system analysis requires *selection of criterion*. Finding the solution is performed through an optimization procedure or by variant calculations.

The *assessment and analysis of the project solution* include a study of the robustness and sensitivity of the obtained optimal solution.

Applying a system analysis in the 1970 s, a number of authors developed computer systems for conceptual ship design. As for an example Nowacki (2010) described the details of many of these studies. Initially, the search for an optimal design solution, including main dimensions or their relations was based on variant calculations,

Later, the solution was defined by using a non-linear programming (Novacki et al., 1970) using an optimization algorithm known as *Sequential Unconstrained Minimization Technique*-SUMT (Fiacco et al., 1972).

The optimization task is defined as:

$$F(\mathbf{X^*}) = \min F(\mathbf{X},\mathbf{Q}), \mathbf{X} \in \mathbf{R} \tag{1}$$

Subject to:

$$G\{g_j(\mathbf{X},\mathbf{Q})\} \leq 0 \ , j = 1,2,...,m \tag{2}$$

$$H\{h_i(\mathbf{X},\mathbf{Q})\} = 0 \ , i = 1,2,...,p \tag{3}$$

where: $F(\mathbf{X})$ is the objective function to be minimized (or maximized) over the variables $(\mathbf{X}, \mathbf{Q})$; G is the inequality constraints; H is the equality constraints; $\mathbf{X} = \{x_k\}$, $k = 1, 2,...,n$ is the vector of design variables; $\mathbf{Q} = \{q_l\}, l = 1, 2,..., s$ is the vector of uncontrollable parameters.

The solution to this task is based on sequential unconstrained minimization of the transformed objective functions in the following form:

$$P(\mathbf{X},\mathbf{Q},r_k) = F(\mathbf{X},\mathbf{Q}) + 1/r_k \sum \left\{\min[0; g_j(\mathbf{X})]\right\}^2 +$$
$$+ 1/r_k \sum [h_i(\mathbf{X})]^2 \tag{4}$$

$$F(\mathbf{X^*}) = \lim \left\{\min P(\mathbf{X},\mathbf{Q},r_k)\right\}, r_k \to 0 \tag{5}$$

where r_k is the penalty parameter ($r_k > 0$)

This optimization procedure, as was proposed by Gallin (1973), was implemented to the linear mathematical model without intermediate checks of the compatibility of the design solution with the constraints.

3 SENSITIVITY ANALYSIS OF OPTIMAL SOLUTIONS

3.1 *Parametric studies of optimal solutions*

The use of an optimization procedure in the initial stages of ship design is of a great importance, but a more valuable information could be gained through the parametric studies of the obtained optimal solution. The theoretical background of these studies at early stages of ship design was presented by Kolev, (1987) and Kolev et al. (2005).

Immediate relation to the task of the nonlinear programming is the Lagrange function.

$$L(\mathbf{X},\mathbf{u},\mathbf{w}) = F(\mathbf{X}) + \sum_{j=1}^{m} u_j g_j(\mathbf{X}) + \sum_{i=1}^{p} w_i h_i(\mathbf{X}) \tag{6}$$

where $\mathbf{u} \{u_j\}$, $\mathbf{w}\{w_i\}$ is the Lagrange multipliers.

For an optimum solution $\mathbf{X^*}$ the Kuhn-Tucker conditions has to be satisfied:

$$g_j(\mathbf{X^*}) \leq 0, \quad j = 1,2,...,m \tag{7}$$

$$h_i(\mathbf{X^*}) = 0, \quad i = 1,2,...,p \tag{8}$$

$$u_j^* g_j(\mathbf{X^*}) = 0, \quad j = 1,2,...,m \ ; \ u_j^* \geq 0 \ \ j = 1,2,...,m \tag{9}$$

$$\nabla L(\mathbf{X^*}) = \nabla F(\mathbf{X^*}) + \sum_{j=1}^{m} u_j \nabla g_j(\mathbf{X^*}) + \sum_{i=1}^{p} w_i \nabla h_i(\mathbf{X}) \tag{10}$$

Kuhn-Tucker conditions permit finding the Lagrange multipliers $\{u_j^*\}$, $\{w_i^*\}$ if the first and the second derivatives of the objective function and the constraints could be calculated at the optimum point $\mathbf{X^*}$. This is the reason to use quadratic regression approximations for the objective function and constraints. These approximations are acceptable because of the smooth changes of the objective function and constraints in the limited area around the optimum point. The regression equation for the objective function is:

$$F(Z,Y) = a_0 + \sum_{k=1}^{n} a_k z_k + \sum_{k=1}^{n}\sum_{\leq l=1}^{s} a_{kl} z_k z_l + \sum_{l=1}^{s} d_l y_l +$$
$$+ \sum_{k=1}^{n}\sum_{\leq l=1}^{s} d_{kl} z_k y_l \tag{11}$$

where z_k, y_l are the relative deviations from the optimum values of the design variables and uncontrollable parameters.

$$z_k = \frac{x_k - x_k^*}{\delta x_k^*} \ , k = 1,2,...,n; y_l = \frac{q_l - q_l^*}{\delta q_l^*}, l = 1,2,...,s \tag{12}$$

The regression equations for the constraints are similar to the one presented by Eqn (11).

3.2 *Metamodeling technique in sensitivity analysis*

The idea of approximation of the surface around the found optimal solution was further extended by Kolev et al. (2005) and Georgiev (2008). These studies presented a methodology for implementing the metamodels for the optimization and sensitivity analysis. The framework of the methodology is shown in Figure 5, where Step 1 comprises the metamodel fitted around the found optimum based on the Response Surface Methodology (RSM) (Mayers & Montgomery, 2002). A polynomial regression is chosen due to its transparency and simplicity and the low order of the nonlinearity of the objective function. Space-filling design, suitable for the computer experiments is used. A comparison of different types of experimental designs has been reported by Georgiev & Damyanliev (2005).

Step 2 includes a linearization of the non-linear problem. Since the methods for the linear programming are well established, a simple approach to solve a nonlinear optimization problem is to linearize it and then use these methods to find the perspective direction for the truth optimum.

In Step 3, the linear programming problem is solved. The objective function and constraints are polynomials and any suitable algorithm may be used in Step 4 and Step 5 includes the calculation of the Lagrange multipliers if the optimizer does not calculate them.

The optimization problem, using the Lagrange function and Kuhn-Tucker conditions for optimality, is defined as:

$$\nabla f(\mathbf{X}^*) + \sum_{j \in A} u_j \nabla g_j(\mathbf{X}^*) = 0 \qquad (13)$$

where u_j are the Lagrange multipliers for a set of active constraints A and $\mathbf{X}^*$ is the found optimum.

Equation (13) for n design variables and m active constraints in matrix form is:

$$\left[\nabla \mathbf{g}(\mathbf{x}^*) \right]_{n \times m} \left[\mathbf{u} \right]_m = \left[-\nabla f(\mathbf{x}^*) \right]_n \qquad (14)$$

The system, presented by Eqn (14), has a unique solution when $n = m$, but in this case, there is not a place for an optimization. Usually $n > m$ and a least square solution can be used.

Step 6 includes the sensitivity analysis, where the effect of small changes in the right-hand sides (rhs) of the constraints of the optimum solution; the sensitivity of the optimal solution to changes of the design variables $\mathbf{X}^*$ and to changes of uncontrollable parameters $\mathbf{Q}$ is defined.

A methodology was proposed in (Georgiev, 2008) to demonstrate the conceptual design of 45000 tDW bulk carrier. The objective function was to minimize Required Freight Rate (RFR) and design variables are main dimensions and block coefficient of the ship (L, B, D, d, C_B).

Uncontrollable parameters are the fuel oil price and the price of unit hull structure. The active constraints are: g_1–is the cargo hold capacity; g_2–is the freeboard according to ICLL66; g_3–is the initial metacentric height; g_4–is the required deadweight. Based on the fitted metamodels a new better (by about 8%) design solution for the objective function was found with 3 active constraints. Figure 6 shows the constraints and objective function.

From the sensitive analysis can be concluded that small changes in right-hand, i.e. decreasing the freeboard and required deadweight leads to decreasing of the objective function. These conclusions agree with the facts that the freeboard type B-60 increases the efficiency of the ship and a bigger ship (with greater deadweight) is more effective.

Based on the normalized sensitivity the relative importance of L and C_B is higher and B practically does not influence on the objective function. Increasing the length and draught and decreasing C_B has a positive influence on RFR.

The influence of the fuel oil price is twice bigger than the cost of hull structures.

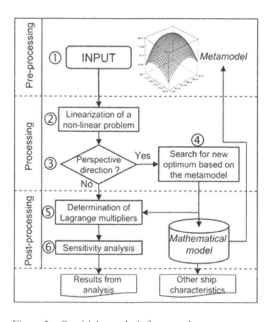

Figure 5.   Sensitivity analysis framework.

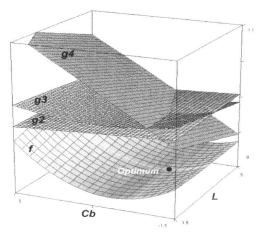

Figure 6.   The objective function and active constraints.

3.3 Dual response problem

The modern design process requires obtaining high quality products at low costs, which is an economical and technological challenge for the designer. Defined more than 50 years ago, the Taguchi Robust Parameter Design (RPD) is a tool for a proper choice of the levels of design (control) variables to achieve the robustness in the change in the uncontrollable (noise) factors. The Taguchi's RPD permits a simultaneous optimization of the mean and standard deviation responses. Georgiev & Pentschew (2002) and Georgiev (2004) demonstrated the applicability of the Taguchi's approach to a ship design hull form improvement.

Combined implementations of the Taguchi approach for RPD and Response Surface Methodology (RSM) (Mayers & Montgomery, 2002) lead to the formulation of the Dual Response System (DRS) (Koksoy & Doganaksoy, 2003).

Georgiev (2006, 2008) proposed an approach for solving DRS in the conceptual ship design by simultaneously optimizing the primary and secondary responses subjected to functional and design variables constraints.

The dual response problem may be treated as a special case of the multi-objective problem, i.e. as a bi-objective problem. In contrast to the single-objective optimisation, in the multi-objective optimization there is no a single global solution and it is often necessary to determine a set of points that all fit the constraints. In this case, all feasible points define the Pareto frontier, which is a curve for two object functions, surface for three and hyper surface for more objectives.

The metamodel of the object function presents a response surface that contains both n control ($\mathbf{X}$) and l noise ($\mathbf{Z}$) variables. Thus, the estimated model is:

$$y(\mathbf{X},\mathbf{Z}) = b_0 + \mathbf{X}^T\mathbf{b} + \mathbf{X}^T\mathbf{B}\mathbf{X} + \mathbf{Z}^T\mathbf{c} + \mathbf{X}^T\Delta\mathbf{Z} \qquad (15)$$

In this model, it is assumed the linear effects in $\mathbf{X}$ and $\mathbf{Z}$, a two-factor interaction and pure quadratic terms in $\mathbf{X}$ (the term $\mathbf{X}^T\mathbf{B}\mathbf{X}$) and two-factor interactions involving the control and noise variables (the term $\mathbf{X}^T\Delta\mathbf{Z}$).

The above model involves the control and noise variables as

$$y(\mathbf{X},\mathbf{Z}) = f(\mathbf{X}) + h(\mathbf{X},\mathbf{Z}) + \varepsilon \qquad (16)$$

where $f(\mathbf{X})$ is the portion that accounts only the control factors, interactions between them and second-order terms and h($\mathbf{X}$, $\mathbf{Z}$) are the terms involving the noise factors and interactions between the controllable and noise factors, where $h(\mathbf{X},\mathbf{Z})$ is:

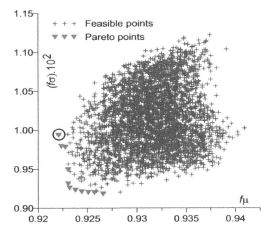

Figure 7. Feasible and Pareto points for DRS of ship design.

$$h(\mathbf{X},\mathbf{Z}) = \sum_{i=1}^{l}\gamma_i z_i + \sum_{i=1}^{l}\sum_{j=1}^{n}\delta_{ij}x_i z_j \qquad (17)$$

If it is assumed that the noise variables have a mean zero, variance σ_z^2, and covariance zero, and that the noise variables and the random error ε have a zero covariance, then the mean model and variance of the response are:

$$f_\mu : E_z[y(\mathbf{X},\mathbf{Z})] = f(\mathbf{X}) \qquad (18)$$

$$f_\sigma : Var_z[y(\mathbf{X},\mathbf{Z})] = \sigma_z^2 \sum_{i=1}^{l}\left[\frac{\partial y(\mathbf{X},\mathbf{Z})}{\partial z_i}\right]^2 + \sigma^2 \qquad (19)$$

Figure 7 shows the results of the DRS of 45000 tDW bulk carrier. The sampling includes 65536 points with 15850 feasible, where 11 of them are the Pareto points.

4 SOFTWARE TOOLS FOR CONCEPTUAL SHIP DESIGN

4.1 Computer aided preliminary design system PROCONS

The computer aided preliminary design system PROCONS has been reported in (Kolev et al, 1987a, Kolev & Damyanliev, 1991). The software tool involves a system analysis when formulating the ship design. According to the authors, this can be done on three levels—compatibility, optimization and investigation as can be seen in Figure 8.

The aim of the first level, *compatibility*, is to build a simplified mathematical model by analysing the ship subsystems.

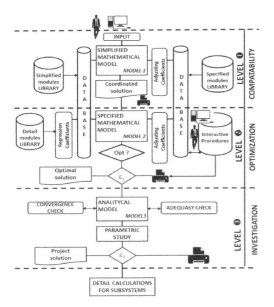

Figure 8. Scheme of computer system PROCONS.

At the second level, *optimization*, in an interactive mode, the model is further expanded and adjusted to the closest prototype. An optimal solution is obtained based on the model and defined constraints.

To prevent a reduction the confidence of calculation due to the use of simplified models, the so-called *adjusting* coefficients in the following form were implemented:

$$V_r(\mathbf{X}) = u_r w_r V_{or}(\mathbf{X}) + c_r \qquad (20)$$

where: $V_r(\mathbf{X})$ is the adjusted value of rth design variable; $V_{or}(\mathbf{X})$ is the variable value obtained by simplified model; u_r is the adjusting coefficient obtained by the system; w_r is the adjusting coefficient given by the designer; c_r is the adjusting constant given by the designer.

$$u_r = \frac{V_{sr}(\mathbf{X})}{V_{or}(\mathbf{X})} \qquad (21)$$

where $V_{sr}(\mathbf{X})$ is the variable value obtained by a specified module.

The analysis of the design solution is realized in the third level, *investigation*. To provide a subsequent analysis of the design solution, an analytical model based on Design of Experiments (DoE) was used (Kolev, 1993).

By the so-called "*project solutions* " it is possible to proceed to the next steps of the design process defined by the authors as *Detail Calculations for Subsystems*.

4.2 Conceptual ship design system "Expert"

The basic principles used in the PROCONS system have been employed in the development of the *Expert* system later on in (Damyanliev & Boev 1998, Damyanliev 2002).

The advantages of this system are seen as:

- Versatility with respect to the object being analysed—design of different ship types;
- Versatility with regard to the type of design task—an interactive formulation of the initial data and requirements and constraints
- Possibilities for updating the existing and generating new modules to the system libraries ensuring its continuous extension;
- "User-friendly" environment, working in the interactive mode.

Figure 9 shows a schematic diagram of the system.

The application software includes a library of modules of the ship mathematical model and the computational procedures associated with them.

The mathematical models of the ship are defined by 31 subsystems describing the ship as a physical object (geometry, volume and weight characteristics), its qualities (stability, cargo capacity, performance, manoeuvrability, etc.), cost estimates—CAPEX OPEX and criteria for the assessment of the optimal design solution.

For each of the subsystems, it is possible to generate, by a specialized external program for the system, unlimited number of modules specific to a particular vessel type or applied engineering models. The structure of the modules is unified.

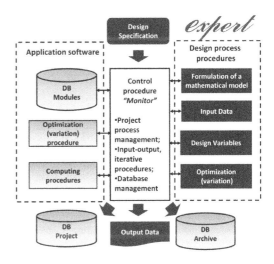

Figure 9. Scheme of conceptual ship design system *Expert*.

The formulation of the terms of the specific design task is carried out in several steps. Initially, the user selects the active subsystems and the corresponding modules that construct the mathematical model. The choice depends on the type of ship and the logic of the design task. For each specific model the system automatically generates the necessary input data. The design variables, constraints and objective functions are selected in an interactive mode depending on the logic of the design.

The vector of the design variables $\mathbf{X} = \{x_k\}$, k = 1, 2,..., n is usually formed by main dimensions and hull form coefficients. This vector may include other variables for an example the ratio of the amount of ballast water and displacement in the design of container vessel.

Very often the optimal solution lies on the boundary defined by the constraints $\mathbf{G}:\{g_j(\mathbf{X}, \mathbf{Q})\} \leq 0$, $j = 1,2,...,m$). These constrains are connected with some of the required qualities of the ship. By defining the upper and lower bounds, the allowable area for design variable is formed.

The selection of the objective function $F(\mathbf{X}, \mathbf{Q})$ depends on the ship type and the goal of the design task. Most often this is an economic indicator such as the Required Freight Rate or the coefficient of utilization of the deadweight ($\eta_{DW} = DW/\Delta$). The objective function can be formed by one or more criteria (single or composite) or multi-criteria such as ranked estimates of more varied parameters.

The system has two main functions: 1) ship synthesis and search for optimal solution and 2) variant calculations by one or more modules, available from the library.

The ability of the framework was recently demonstrated by solving four design tasks, including optimization of the fleet composition; conceptual design of bulk carrier; impact of the limited draught of the design solution; influence of stowage factor on the optimal solution (Damyanliev et al. 2017).

5 ON BOARD COMPUTER SYSTEMS

For more than two decades, a group of research associated led by Prof. Kolev has been developing and installing an Auto Loading Computer On-board System (ALCOS) on various types of ships (Kolev et al.1994). The system fulfils the requirements for a Loading Computer System as a computer based system that includes a loading computer (hardware) and a calculation program (software), capable easily and quickly to ascertain any ballast or loading condition that:

– the longitudinal and local strength will not exceed the permissible ones, and
– the stability complies with the stability requirements applicable to the ship.

Such loading instrument is mandatory for bulk carriers with a length of 150 m and above. Recently, the MARPOL Annex I, Ch.4, the IBC/BCH Code and IGC/GC Code have been amended, requiring tankers to be equipped with a stability instrument, capable of handling both intact and damage stability. The requirement is applicable from 1 January 2016 (https://www.dnvgl.com/).

ALCOS is designed as a complete object-oriented application system, including all typical elements of the interface. The different modules are included as DLL (Dynamic Link Libraries) that's why it is possible to "change" ALCOS by extending it.

Besides the required modules for the longitudinal strength and stability assessment, there are modules for solving additional tasks such as Draught survey; Ullage Survey; Loading/Unloading Sequences plan; Lashings of containers; EDIFACT/ Baplie Transfer files; Longitudinal strength after flooding of bulk carriers. The software has been approved by the leading Classification Societies and installed on more than 100 ships.

A screen with container cargo operations on a 9800 tDW multipurpose vessel is shown in Figure 10. The results from stability and longitudinal strength evaluation of a bulk carrier are presented in Figure 11.

The software is successfully used as a simulation tool for defining the metamodels for safety assessment of ships (Georgiev 2010, 2011).

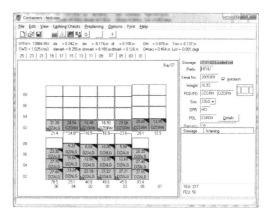

Figure 10.   Container cargo operations of loading condition defined by EDIVACT/Baplie format.

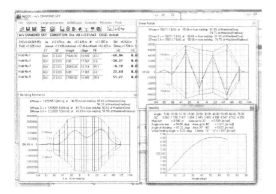

Figure 11. Longitudinal strength and stability evaluation of a bulk carrier.

5 CONCLUSIONS

This work presented here reviewed the most important developments in the field of ship design performed by Professor Kolev and his research associates, including different software systems that have been installed in many ships in the last four decades.

ACKNOWLEDGEMENTS

The authors express their gratitude to the organizing committee of the IMAM 2017 Congress for dedicating a Special Session in honour of Professor Petar Kolev of Technical University of Varna, Bulgaria for his contributions for several decades in different topics in the field of ship design and letting this article to be published.

REFERENCES

Ashik, V.V. 1975. *Ship design*, Sudostroenie, 1975 (in Russian).

Damyanliev, T. 2002. Program Environment for Decision Making Support System. *MEET/MARIND*, Vol.5.

Damyanliev, T., Georgiev, P. & Garbatov, Y. 2017. Conceptual ship design framework for designing new commercial ships. In: Guedes Soares, C. & Garbatov, C. (eds.) *Progress in the Analysis and Design of Marine Structures*. London: Taylor & Francis Group, pp 183–189.

Damyanliev, T.P., Boev S.M. 1998. Conceptual design system—EXPERT., SAC Shipbuilding '98, ISSN 1310–8573 (in Bulgarian).

Fiacco A.V., McCormick G.P. 1972. *Nonlinear Programming: Sequential Unconstrained Minimization Techniques*, 1972, Moscow (in Russian).

Gallin, C.1973. Which way computer aided preliminary ship de-sign and optimization, *ICCAS* – Papers.

Gaykovich, A.I. 2001. *Fundamentals of theory for design of complicated technical systems*, Morinteh, Sankt-Petersburg, (in Russian).

Georgiev, P. & Damyanliev, T. 2005. Metamodels in ship de-sign. *Proceedings of the IInd International Congress on Mechanical and Electrical Engineering and Marine Industry* (MEEMI), Vol. 2: 268–278.

Georgiev, P. & Pentchew, P. 2002. Parameter Ship Hull Design based on the Taguchi method. *Shiffbauforschung*, 41, Heft Nr. 3/ 4: 19–28.

Georgiev, P. 2004. Experimental design for ship hull form improvement. *Proceedings of the XVIIth International Conference on Marine Science and Technology*, Black Sea, Vol. 1:185–191.

Georgiev, P. 2006. Dual response system in the early stages of ship design. *Proceedings of the VIIIth International Conference on Marine Science and Technology*, Black Sea, vol. 1:107–112.

Georgiev, P. 2008. Implementation of metamodels in ship design. Maritime Industry, *Ocean Engineering and Coastal Resources—Guedes Soares* & Kolev (eds), 2008, Taylor & Francis Group, London, pp 419–427.

Georgiev, P. 2010. Probabilistic presentation of the bending moments of bulk carriers using metamodels. *Proceedings of the IXth International Conference on Marine Science and Technology*, Black Sea, pp 82–8.

Georgiev, P. 2011. Safety analyses for bulk carriers using metamodels of still water loads. *Advances in Marine Structures* - Guedes Soares & Fricke (eds), Taylor & Francis Group, London, UK, pp. 669–677.

Koksoy, O, & Doganaksoy, N. 2003. Joint Optimization of Mean and Standard Deviation Using Response Surface Methods, *Journal of Quality Technology*, Vol. 35, No 3: 239–252.

Kolev, P.N. 1972a. Determination of optimal ship characteristics, *National Scientific Conference*.

Kolev, P.N. 1987. *Sensitivity Studies of the Optimum Decisions in Pre-Contracted Ship Design*. D.Sc. thesis, Varna, 1987 (in Bulgarian).

Kolev, P.N. 1992. Parametric Studies of the Optimum Designs, *Proceeding of IMAM*, pp. 17–21.

Kolev, P.N., Damyanliev T.P. 1991. *CAD Systems in Ship Design*, Technical University of Varna, (in Bulgarian).

Kolev, P.N., Damyanliev, T.P., Georgiev, P.G. 2005. Optimization and robust investigations in ship design. *Maritime Transportation and Exploitation of Ocean and Coastal Resources* – Guedes Soares, Garbatov & Fonseca (eds), Taylor & Francis Group, London, pp. 875–881.

Kolev, P.N., Simeonov, I.S., Abaddjiev, K. 1987a. PROCONS—A computer aided preliminary design system, *IMAEM'87*, vol. 2.

Kolev, P.N.1972. Methods for optimization at early stages of ship design, *Shipbuilding and Shipping*, No 7, (in Bulgarian).

Kolev P.N. 1977. Optimization Methods in Pre-Contracted Ship Design, *Chalmers Technical University*, Gothenburg, 1977.

Kolev P.N. 1979. *Design and architecture of ships*, part I, II, Technical University of Varna, 1979 (in Bulgarian).

Kolev P.N., Petrova Z. Georgiev P. and Kolev N. 1994. Main Features and Possibilities of Auto Loading System ALCOS, *Black Sea*, pp.133–135.

Myers, R.H. & Montgomery, D.C. 2002. Response surface methodology, *Process and product optimization using de-signed experiments*, New York, John Wiley & Sons, Inc, 2002.

Nogid, L.M. 1955. *Ship design theory*, Sydpromgiz, 1955 (in Russian)

Nowacki, H.2010. Five decades of Computer-Aided Ship Design, Computer-Aided Design 42, 956–969.

Nowacki H, Brusis F, Swift P.M. 1970. Tanker preliminary design—an optimization problem with constraints. *Transactions of SNAME*;78.

Pashin V.M. 1983. *Ship optimization (system approach-mathematical models)*, Leningrad, 1983.

Wagner H.M.1972. *Principles of Operation Research*, Mir, 1972, Moscow. (in Russian).

Maritime Transportation and Harvesting of Sea Resources – Guedes Soares & Teixeira (Eds)
© 2018 Taylor & Francis Group, London, ISBN 978-0-8153-7993-5

Challenges in stability assessment of offshore floating structures

T.M. Hendriks & J. Mendonça Santos
GustoMSC, Schiedam, The Netherlands

ABSTRACT: Stability plays a key role in the design of floating structures. Its impact on safety and operational aspects render it a fundamental consideration in the preliminary design stages. Stability assessment of floating structures intended to work offshore, here generally referred to as offshore floating structures, present several specific challenges. These arise from diverse sources: from the departure of classical ship hydrostatic behavior due to different shapes and sizes, to the multitude of operational modes and specific loads and loading conditions inherent to the many different missions these units are intended for. This paper revisits several of such challenges faced in the stability calculations of offshore floating structures as well as its approval by regulatory authorities. Particular attention is paid to the one concerning the identification of the "most critical axis" for GZ curve calculation and a free-twist method is discussed as a preferred method for stability calculations of freely floating structures.

1 INTRODUCTION

Stability is one of the pillars of naval architecture, interlinked, and thus affecting and being affected by, many other disciplines concurring in the naval architecture compromise. Stability plays a key role in the design of floating structures. Its impact on safety and operational aspects render it a fundamental consideration in the preliminary design stages.

Stability assessment of floating structures intended to work offshore, here generally referred to as offshore floating structures, present several specific challenges. These arise from diverse sources: from the departure of classical ship hydrostatic behavior due to different shapes and sizes, to the multitude of operational modes and specific loads and loading conditions inherent to the many different missions these units are intended for. This paper revisits several of such challenges faced in the

stability calculations of offshore floating structures as well as its approval by regulatory authorities.

GustoMSC is known as a reputable independent design and engineering company in the offshore market, providing designs for mobile offshore units and associated equipment. Over its long history GustoMSC has designed offshore units of different type and serving different market sectors of the offshore industry. Pictured in figure 1 are examples of jack-ups of GustoMSC design: a) drilling jack-up used in the offshore oil & gas exploration market; b) wind turbine installation jack-up used in the offshore wind/renewable energy market; c) accommodation jack-up used supporting different offshore operations. Examples of GustoMSC designed semi-submersible units are shown in figure 2: a) drilling unit; b) heavy-lift unit used in the offshore construction market; c) accommodation unit. Figure 3 provides examples of ship-shaped offshore units (vessels) built to GustoMSC

Jack-ups: a) drilling;   b) wind turbine installation/construction;   c) accommodation

Figure 1. Jack-ups: a) drilling; b) wind turbine installation/construction; c) accommodation.

Semi-submersibles: a) drilling;   b) heavy-lift/construction;   c) accommodation

Figure 2.   Semi-submersibles: a) drilling; b) heavy-lift/construction; c) accommodation.

Vessels: a) drillship;   b) heavy-lift/construction;   c) pipe-layer/construction

Figure 3.   Vessels: a) drillship; b) heavy-lift/construction; c) pipe-layer/construction.

design: a) drillship; b) heavy-lift/construction unit; c) pipe-laying vessel.

The diverse missions and operational profiles of the different offshore units imply different challenges concerning the assessment of stability performance. For drilling units, one must cater for various drilling equipment loads playing a significant role in stability, these in many different loading conditions pertaining to the large number of operations and operational modes the units have. Similar considerations are also valid for the other mentioned units.

One feature common to all offshore units, irrespective of being a jack-up, semi-submersible or ship-shaped, is the very large wind exposed area due to the amount of equipment and structures that are required to execute their mission. Thus it is of utmost importance reckoning that the assessment of stability performance is very much dominated by the evaluating the capacity of a unit to withstand a wind overturning moment in several different scenarios and often in combination with numerous operational loads. Mendonça Santos & Alves (2016) noted this fact for drillships and proposed a wind heeling moment curve based on a realistic approach making use of basic naval architectural stability principles and also based on wind

tunnel test data. One of their motivations was the realization of the degree of conservatism introduced by the approach used in the calculation of wind loads and wind heeling moment curves set by regulations and generally adopted. One factor contributing to this is a certain ambiguity in the rules in the form of worded prescription not supported by a universally understandable mathematical formulation. This method has already been employed in recent projects and been accepted by classification societies in approval in principle exercises. It is the author's opinion that the curve proposed by Mendonça Santos & Alves (2016) may be applicable to other offshore ship-shaped units.

Challenges with respect to wind considerations in stability calculations have been addressed by many authors; reviewing and referencing such work goes beyond the scope of this paper. It is however noteworthy to mention recent developments in this subject, in particular recent work triggered by work to update the SNAME T&R Bulletin 5–4 Guidance on Wind Tunnel Testing on MODUs covering both the long standing empirical calculation methods and new techniques (e.g. CFD) alike. One of the first references following this work is the recently proposed (Breuer et al. 2017) power law formulation to replace the step function customarily used (and prescribed by regulations) in describing wind profiles for use in stability matters.

Other challenges particular to offshore units relate to methods and equipment used for station keeping. For units with dynamic positioning system the effects of thruster forces need to be dealt with in stability calculations. In turn stability calculations of moored units need to cater for the effects of mooring lines. Takarada et al. (1986) have pointed the significance of mooring lines effects in the stability of semi-submersibles and Nishimoto et al. (1991) presented an extensive study on the influence of such effects in relation to stability criteria. Many authors have revisited this matter after that, and this subject is still very much in debate noting recent works such as Luxcey et al. (2017).

HSE (2006) is a good reference giving an all-encompassing overview of stability issues specific to semi-submersibles and still relevant to date. Breuer & Rousseau (2009) review questions to classification society regarding stability issues of jack-ups. This is an informative and interesting publication reckoning that many stability challenges are associated with the interpretation of rules and regulations.

The evaluation of stability is performed by assessing compliance with regulatory requirements in the form of stability criteria. Many authorities have nowadays adopted almost the same requirements. These requirements lean on the application of wind heeling moment, equilibrium angles, down-flooding and range of stability.

One of the most pertinent challenge in assessing stability of offshore units (not ship-shaped) is the identification of the "most critical axis" for GZ curve calculation. An issue addressed early on by Vassalos et al. (1985) and Santen (1986).

A discussion of the various ways to calculate the righting arm curves is given, together with problems met in practice in calculations of stability parameters as well as in the interpretation of the results. A common practice is to calculate righting arms using free trim method with a fixed heeling axis direction. In this paper, a free twist method (varying axis direction) is discussed as a preferred method for stability calculations of freely floating structures, looking at the increase in potential energy to identify the critical axis directions. This issue will be further looked in more detail in this paper.

A recent addition to the world of offshore floating structures are the floating wind turbines, figure 4 depicts the GustoMSC designed Tri-floater, a semi-submersible type floating wind turbine.

The novelty factor of such structures poses the greatest challenges in the assessment of their

Figure 4. Floating wind turbine—GustoMSC Tri-Floater.

stability performance. Collu et al. (2014) mention that on of most difficult tasks lays with selecting the appropriate criteria to assess stability of these structures. In fact, only recently have classification societies published their first set of requirements concerning these specific structures.

Huijs (2015) discusses the influence of mooring system in the stability of a semi-submersible floating wind turbine. Despite the parallel with the same issue in other aforementioned semi-submersibles, floating wind turbines introduce fresh challenges as their size and loads impose a departure from current accepted practice with respect to catering for mooring systems in stability calculations. The order of magnitude of maximum operational wind overturning moment is comparable an often larger than for survival conditions.

2 PROBLEMS MET IN STABILITY CALCULATIONS OF OFFSHORE RIGS

In the assessment of the safety of a floating offshore rig, hydrostatic stability plays a crucial role. Initially, the requirements focused on the stability parameters like initial stability (GM) and ratio between area under the righting and wind overturning arm. Additional insight obtained from actual experience and accidents led to changes to the requirements such that they became more specific.

Initially, each classification society had its own set of rules. In the past years, the criteria of the various regulations attained a certain degree of similarity though still differing in details; the common ground being the IMO MODU code.

In general, these requirements were derived from the requirements for ships. However, the form, scale and configurations of offshore structures are increasingly different from that of ships. Thus, transverse stability is not sufficient to establish the safety of floating rigs. Classification societies address this issue by prescribing the righting arms to be calculated for any heeling axis orientation. From the obtained results, the most critical axis direction is to be selected.

This approach applies to all offshore units including jack-ups, semi-submersibles and almost any other vessel dedicated to the offshore oil industry.

Experience in the past years has shown that the actual calculation of the righting arm curve is not as straightforward as it looks. It seems that the availability of powerful computer programs has reduced the attention of the user to getting reliable results. Note that the maximum loading conditions of the units and thus their commercial value depends on it. Therefore, unrealistic results of stability calculations must be avoided.

3 SUMMARY REVIEW OF REQUIREMENTS

Below, a schematic overview is given of the requirements as found in the IMO MODU code rules. It is certainly not meant to provide a detailed overview of the requirements. For details, reference is made to the specific requirements set by the various regulatory bodies (IACS, IMO MODU, Classification societies MODU rules, amongst others). General stability requirements for offshore units (IMO MODU Code) are presented in figures 5, 6 and 7.

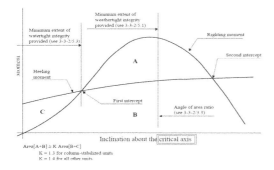

Figure 5.   Intact stability requirements.

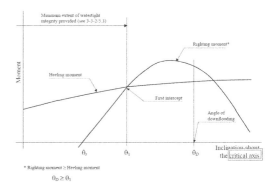

Figure 6.   Damage stability requirements—collision.

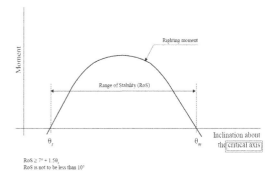

Figure 7.   Residual damage stability requirements—flooding any compartment.

Quite often additional clarification is needed regarding the proper interpretation of the requirements. For instance, the definition of the "critical axis" noted in the figures above is not clear. It is accepted that the critical axis is defined as the axis of inclination about which the maximum KG (vertical center of gravity) has its lowest value. (UK HSE 2005).

Also, details on the calculation procedure are missing. In most cases classification societies addresses free trim but there is no mention thereof in the rules.

4 STABILITY CALCULATIONS

There are various ways to perform a stability analysis. The simplest and failsafe approach is where the axis is fixed and the unit is heeled around this axis with zero trim. Various axis directions are to be investigated to arrive at the most critical one.

Similar as with ships, free trim can be introduced as a means to obtain the lowest righting arm curve. For a given axis direction and heel angle, free trim results in the lowest potential energy contained in the floating unit. The reason being that by setting the unit free to trim, potential energy is released which would otherwise be present due to forcing the unit to have a non-zero trim moment. But this is only true when the unit is stable in trim. A visualization of the difference in the fixed trim vs. free trim GZ curves and the resultant trim for a ship-shaped and jack-up unit are shown on the figures 8 and 9.

For ship shaped units, the longitudinal axis coincides with the axis for which the water-plane is minimal. Thus, for ship-shaped units heel around this axis combined with free-trim is a sensible way to do the calculations. Would we perform stability calculations around a different axis direction (as required for offshore structures) strange things may occur. For instance, if we select a near transverse axis direction, the ship would experience large trim angles in order to achieve zero trim moment. For 80 degrees axis direction (figure 10 b), the righting arm curve does not continue beyond about 5 degrees. For a proper continuation, heel has even to be reduced. Details can be found in Santen (2009a, b) and Santen (2013).

Similar problem appears with wrong selection of inclination axis for offshore structures. The first indication of this problem manifested when the GZ calculations failed to yield results for the full range of angles of inclination specified for the calculation run.

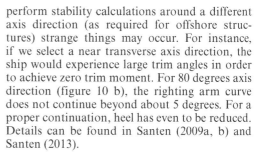

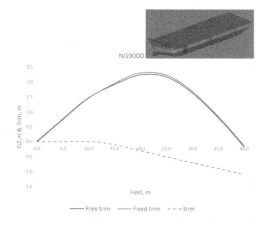

Figure 8. GZ curve—Free-trim vs. Fixed-trim (ship-shaped hull).

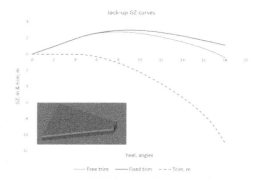

Figure 9. GZ curve—Free-trim vs. Fixed-trim (jack-up).

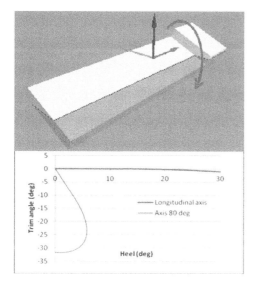

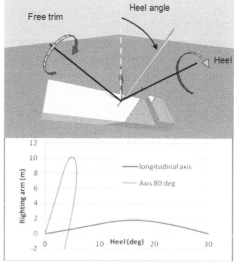

Figure 10. GZ curves and trim angles for two axes of inclination.

What happens when pushing the heel angle be-yond the angle for which the calculation fails (5 deg in figure 10) depends on the software implementation. Some programs will just stop presenting results. But is has also seen that results are presented which are in fact false. Further examination revealed that beyond a heel angle for which failure occurs, there are no trim angles for which zero trimming moment could be achieved. When forcing structure to such a heel angle, the structure changes position in an abrupt manner which would not yield a proper continuation of the righting arm curve. In fact, the structure switches to a new right arm curve. It is quite simple to find these undesirable steps by comparing the gain in potential energy with the change in area under the GZ curve. They should be almost identical.

The results, while being plausible, can lead to unexpected and wrong conclusions. Even well tested intact hull forms, were found to be not adequate under the range of stability criteria.

Questioning the adequacy of the conventional methods, the investigation led to a return to first principles of engineering. The conventional approach to calculating the energy to heel the vessel is to look at the area under the righting arm moment curve (van Santen 1986, see also Breuer & Sjölund 2006).

The work needed to incline (further referenced as energy to incline) corresponds to the increase in potential energy (EP) and can be determined by a rigorous method. This procedure requires an accurate de-termination of the position of VCB (center of buoyancy). For a constant displacement, the energy to incline equals the change in potential energy from the equilibrium position to the inclined position. The increase is proportional to the change in vertical distance between the center of gravity and the center of buoyancy:

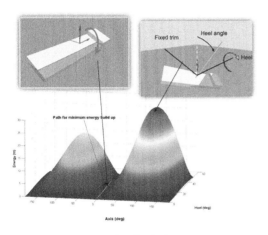

Figure 11. Energy surface for the barge.

$$E_{P\varphi} - E_{P0} = \Delta \cdot (VCB_\varphi - VCB_0) = \int_0^\varphi M(\varphi)d\varphi$$

(1)

where: $M(\varphi)$ is the righting moment.

$$\frac{E_{P\varphi} - E_{P0}}{\Delta} = (VCB_\varphi - VCB_0) = \int_0^\varphi GZ(\varphi)d\varphi$$

(2)

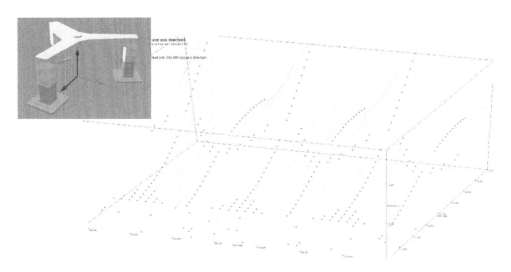

Figure 12. Energy plot for intact stability of Tri-floater.

The most critical axis can be viewed as the axis direction for which a given heel angle is reached with the least effort. In this way, the least energy is to be fed into the system in order to reach that particular heel angle, i.e. where the increase of VCB is minimum.

Based on this principle, the free twist method (also called variable axis direction) has been developed within GustoMSC (Santen, 1986 and Santen, 2009a). In this method, the direction of the heeling axis is varied for each heel angle such that the moment around the initial vertical axis is zero. By doing this, the trimming moment (being the horizontal component of the moment around the inclined twist axis) is zero whilst the trim angle is nil. In general, it results in the lowest energy build up in the system, as noted by Santen (2009a).

As an example, figure 11 shows the amount of energy fed into a barge depending on axis direction and heel angle. In this case, heel around an almost longitudinal axis (0 deg) results in the lowest build-up of potential energy.

It is important to aware that in the free twist method trim is always zero. Hence the heel angle is always equal to the steepest slope of the initial horizontal plane.

Figure 12 gives examples of energy plots for intact stability of a more complicated hull shape of Tri-floater. The green points mark the path where the system will have minimum potential energy for each heel angle. The red points show the path with maximum potential energy.

5 ISSUES IN CLASS APPROVAL

Not being aware of the problems mentioned above and the differences in the software are often the reason for long discussions with classification societies about the approval of stability results.

Small wind turbine installation jack-ups (like figure 1 b) are classed as jack-ups but their length over width ratio is such that they resemble conventional ships. For operational reasons, large center compartments are present. When such a compartment is damaged the range of positive stability has to be larger than a specific value being $7 + 1.5 \times$ static angle (see figure 7). For the hull shown in figure 13 with a damaged forward center compartment, a free trim calculation results in a range of stability of about 13.1 degrees, see figure 14. A problem with the results is that for a heel angle exceeding 11 degrees, the vessel becomes unstable in trim. So, though the calculation indicates a range of positive stability of 13 degrees this would not be confirmed by actual testing as the vessel would capsize when heeled beyond about 11 degrees. The result of a free twist calculation

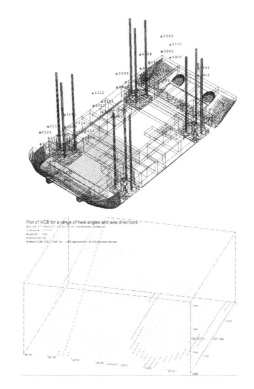

Figure 13.   Model and the energy plot in intact condition of turbine installation jack-up.

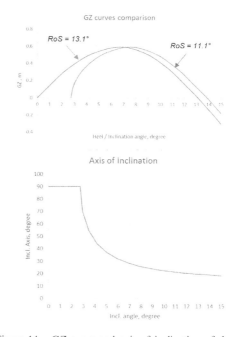

Figure 14.   GZ curve and axis of inclination of damaged turbine installation jack-up.

19

shows a much smaller range of positive stability of about 11.1 degrees. For a proper interpretation, it is necessary that in addition to the righting arm, the results must indicate if the vessel is stable in trim (or twist).

The reason for this is also seen in the energy plot for small heel angles (fig 13) and in the plot of the heeling axis direction (fig 14, bottom). There is a sudden change when heeled from the equilibrium position (around 2.7 degrees heel) to larger heel angles.

6 CONCLUSIONS

Experience with stability calculations shows that extending the free trim method as used for ships to offshore structures in combination with varying heeling axis direction may lead to severe interpretation problems. Awareness of these problems is not widespread amongst the industry. This requires more attention.

The main stability properties can be determined by analysis of the energy plots. Applying the free twist method for GZ curve calculation appears to be more rational than the conventional free-trim method.

The calculation procedures given by class—containing descriptions like "most critical axis" and "free to trim" method—are in urgent need of a clarification note explaining how to do the calculations and how to handle various problems.

It is important to look carefully at the results from the computer programs with good understanding of the calculation method and the validity of the results. In order to do this, the stability programs should have options to show information on trim angles, axis of inclination (if variable), VCB value and stability (in trim) of the position at any heel angle. The program should also identify possible discontinuities in the GZ curve based on the change in VCB versus the gain in area under the GZ curve.

The stability of offshore units is typically governed by criteria related to the ability to withstand a wind heeling moment. Rendering the assessment of wind loads and heeling moments is essential for the design and operation.

The presented method for wind heeling moment curve for drillships is simple, based on a realistic approach and shows a better fit to wind tunnel tests results than the other available calculation methods. Therefore, it is suitable to be applied to new drillships designs throughout the design spiral. The method results in gains in allowable KG which would allow increasing capacity/loads and establishing loading conditions with more favorable GM in terms of roll motion.

ACKNOWLEDGEMENTS

The authors wish to acknowledge the assistance provided by GustoMSC and the support from Joost van Santen in the preparation of this paper.

We would like to express our special thanks to Professor Peter Kolev from the Technical University of Varna for sharing his passion and knowledge in naval architecture with his students for more than four decades. He learned us about the fine line between "too difficult" and "just challenging enough". In this way he enabled us pushing the boundary and exerting additional efforts in our daily work. In doing so he helped us growing both as specialists in our fields and as a person.

T.M. Hendriks would like to express her appreciation and gratitude to Prof. Kolev for his guidance, patience and encouragement throughout her student and researcher times at the Technical University of Varna.

J. Mendonça Santos would like to thank Prof. Kolev, who was instrumental in facilitating my stay at the Technical University of Varna conducting research for my graduation.

REFERENCES

ABS. 2016. Rules for Building and Classing Mobile Offshore Drilling Units. Houston: American Bureau of Shipping.

Breuer, J.A. & Sjölund, K.G., 2006. Orthogonal Tipping in Conventional Offshore Stability Evaluations. Proceedings of the 9th International conference on Stability of Ships and Ocean Vehicles, 25–29 September, 2006 Rio de Janeiro, Brazil, pp. 112.

Breuer, J.A. & Sjölund, K.G., 2009. Steepest descent method. Resolving an old problem. 10th International Conference on Stability of Ships and Ocean Vehicles, 22–26 June 2009, St. Petersburg, Russia, pp.87–100.

Breuer, J.A. & Rousseau, J.H., 2009. Stability of jack-ups: top ten questions from industry to class. *15th International Conference: The Jack-Up Platform - Design, Construction and Operation*: City University London.

Breuer, J.A., Cheater, B., Chapman, C., Bhaumik, T., Masciola, M., Peak, A., Wang, T. & Brekke, J.M., 2017. A power law formulation equivalent to the step function for use in wind tunnel testing. *Proceedings of the 22nd Off-shore Symposium, February 2017, Houston, US.* SNAME.

Collu, M., Maggi, A., Gualeni, P., Rizzo, C.M. & Brennan, F., 2014. Stability requirements for floating offshore wind turbines (FOWT) during assembly and temporary phases: overview and application. *Ocean Engineering 84*: 164–175.

Huijs, F., 2015. The influence of the mooring system on the motions and stability of a semi-submersible floating wind turbine. *34th International Conference on Offshore Mechanics & Arctic Engineering; 31 May - 5 June 2015; St. John's, Canada*, pp. 41497. ASME.

HSE. 2005 *Research Report 387 - Stability*. Norwich: Crown/Health & Safety Executive.

HSE. 2006. Research Report 473 - Review of issues associated with the stability of semi-submersibles. Norwich: Crown/Health & Safety Executive.

IMO. 2009. Code for the Construction and Equipment of Mobile Offshore Drilling Units (2009 MODU Code). London: IMO Publishing.

Luxcey, N., Johanenssen, Ø. & Fouques, S., 2017, Static stability of floating units in operational condition: a physics driven approach. *36th International Conference on Offshore Mechanics & Arctic Engineering; 31 May - 5 June 2015; Trondheim, Norway*, pp. 62489. ASME

Mendonça Santos, J. & Alves, C., 2016. A wind heeling moment curve for ship-shaped MODU early design stability considerations, *Maritime Technology and Engineering 3*, Chapter: 118. Publisher: Taylor & Francis Group, Editors: C. Guedes Soares, T.A. Santos, pp.8.

Nishimoto, K., Brunozi, P.F. & Babadopulos, J.L., 1991. Analysis of mooring lines and riser effects on the stability of semisubmersibles. *10th International Conference on Offshore Mechanics & Arctic Engineering; 23–28 June 1991; Stavanger, Norway*, pp. 687. ASME.

NMD. 2011. Regulations of 20 December 1991 No. 878 on stability, watertight subdivision and watertight/ weathertight means of closure on mobile offshore units. Haugesund: Norwegian Maritime Authority.

Santen, J.A. van, 1986. Stability calculations for jack-ups and semi-submersibles. Conference on computer Aided. Design, Manufacturer and Operation in the Marine and Offshore Industries, September 1986, Washington D.C., US. Editors: G.A. Keramidas & T.K.S. Murthy, WIT Press.

Santen, J.A. van, 2009a. The use of energy build up to identify the most critical heeling axis direction for stability calculations for floating offshore structures. *10th International Conference on Stability of Ships and Ocean Vehicles, 22–26 June 2009, St. Petersburg, Russia*, pp.65–76.

Santen, J.A. van, 2009b. The use of energy build up to identify the most critical heeling axis direction for stability calculations for floating offshore structures, review of various methods. *15th International Conference: The Jack-Up Platform - Design, Construction and Operation*: City University London.

Santen, J.A. van, 2013. Problems Met in Stability Calculations of Offshore Rigs and How to Deal with Them. *13th International Ship Stability Workshop, 23–26 September 2013, Brest, France*, pp.9–17.

Takarada, N., Nakajima, T. & Inoue, R., 1986. A phenomenon of large steady tilt of a semisubmersible platform in combined environmental loadings. *3rd International Conference on Stability of Ships and Ocean Vehicles, 22–26 September 1986, Gdansk, Poland*, pp.225–238.

Vassalos, D., Konstantopoulos, G., Kuo, C., & Welaya, Y., 1985. A realistic approach to semisubmersible stability. *SNAME Transactions, Vol.93*, pp. 95–128.

Maritime Transportation and Harvesting of Sea Resources – Guedes Soares & Teixeira (Eds)
© 2018 Taylor & Francis Group, London, ISBN 978-0-8153-7993-5

Parametric modelling of tanker internal compartment layout for survivability improvement within the framework of regulations

H. Jafaryeganeh, M. Ventura & C. Guedes Soares
Centre for Marine Technology and Ocean Engineering (CENTEC), Instituto Superior Técnico, Universidade de Lisboa, Lisbon, Portugal

ABSTRACT: This work deals with the categorization of regulatory limits for oil tankers internal layout design after generating a parametric model with proper design variables and the capability of evaluating regulations restrictions. First, the regulations relevant for the internal layout type and dimensions will be categorized. Then, a method for generating a parametric model is presented taking into consideration the regulatory limitations. An adaptive procedure will be used to evaluate the provided model with respect to the damage stability regulations. Finally, the importance of design variables will be discussed and potential merit functions will be identified for future optimization studies.

1 INTRODUCTION

The internal layout of the ship's hull is established in the initial stage of ship design. The estimation of internal compartment size and arrangement depends on the type of compartment, its content and usage. On the other hand, survivability regulations impose limitations in the subdivision arrangement of the ship, which complicates the process of internal compartment design, especially when the main hull dimensions have to be kept constant (Santos & Guedes Soares, 2011). The variety of designs are possible in the framework of the obligatory regulations, thus the internal layout design can be investigated with optimization methods to identify the optimum design among the possible solutions (Nowacki, 2010). So, to define the feasible domain of the problem, an investigation is necessary to identify the effective regulatory criteria on the design of the internal layout. Additionally, the modelling method of the internal compartments should have the ability to provide different designs based on these regulations. The mentioned earliest steps are discussed in the present work for oil tankers, which can be applied for future improvement of internal layouts.

Decades ago, responding to some ship accidents, the International Maritime Organisation (IMO) introduced some first deterministic requirements for ships' watertight subdivision. The specified requirements included the number and arrangement of watertight bulkheads and presented criteria for the assessment of the ships' stability after damage, which was laid down in the international convention of Safety Of Life At Sea (SOLAS). Later, with the development of the probabilistic damage stability assessment concept, IMO addressed the damage stability of dry cargo ships by the probabilistic method. For passenger and dry cargo ships, damaged stability regulations were harmonized in a unique probabilistic one (Papanikolaou & Eliopoulou 2008; Santos & Guedes Soares, 2010). For liquid cargo ships, the mandatory regulations are defined in the International Convention for the Prevention of Pollution from Ships (MARPOL), which still applies the combination of the determinist and probabilistic approaches. These requirements are not only used as criteria for subdivision of the oil tankers but also, applied as merit functions to improve the tankers safety.

Santos & Guedes Soares, (2005a) applied a multi-objective optimization of fast ferry watertight subdivision, which demonstrated an approach similar to the one that is now developed for tankers. This is complementary to using Monte Carlo methods to assess damaged ship survivability (Santos & Guedes Soares, 2005b).

Papanikolaou et al. (2010) applied an optimization procedure for the arrangement of the cargo area of an Aframax class tanker with aiming of minimization of "accidental oil outflow" and improvement of economic competitiveness. Also, Harries et al. (2011) combined the simulation of key measures of merit in the early phase of ship design for an Aframax tanker. The approach considered payload, steel weight, strength, oil outflow, stability and hydrodynamics simultaneously.

A compatible parametric model is required for implementing automatic criteria and merit function evaluation. In fact, parametric modelling has been used for an increasing number of aspects of the design such as the hull form (Harries, 2003), the hull structures configuration (Roh & Lee, 2007) and the hull compartment layout (Koelman, 2012).

Regarding the internal layout, modelling the ship compartments introduce a challenge for defining the spaces as volumetric entities in the CAD software. The spaces can be defined either by their boundaries without referring to the realistic plane or by a series of sections alongside the compartments (Lee et al. 2009). In the first approach, the duality between spaces and planes is not addressed while, in the second one, the relationship between spaces and planes is defined.

However, a more complex approach was applied for the definition of fully duality between the planes and spaces in Koningh et al. (2011). They used the method of "binary space partitioning" to generate internal subdivision with minimum input by the designer. This method can be applied to create alternative topologies in the early stage of ship design, while, the topology of the internal layout is often very similar in a particular class of ships. For the ship safety evaluation with the fixed topology, Lee et al. (2004) provided a three-dimensional geometric modeler, which builds the model of damaged ship rapidly and efficiently.

The present work is focused on the mandatory regulations for internal layout of oil tankers and the study of their effect on the parametric modelling of the main internal subdivision. A classification of the regulatory obligations is done based on the implicit or explicit limits for internal layouts dimensions. A method of parametric modelling is presented to evaluate the mentioned restrictions. The study is focused on providing a model for oil tankers assuming a given fixed hull form, thus, the importance of the design variables is discussed with some samples.

2 REGULATIONS RELEVANT TO THE INTERNAL LAYOUT OF OIL TANKERS

The regulations define limitations for the internal layout of oil tankers. These regulations can be divided on explicit and implicit restrictions on the dimensions and type of internal layout. Some of the requirements specify minimum value for dimensions of some compartment directly, such as minimum values for the wing tank or double bottom. On the other hand, other requirements define the limits for the hypothetical values, such as "accidental oil flow parameters", which is derived from the dimensions of the internal layout.

2.1 Explicit limitation for the internal layouts

Some of the regulations indicate minimum or maximum values for the dimensions of the internal layout. The following categorization is based on these upper or lower limit values.

2.1.1 SOLAS requirements for the bulkheads

According to SOLAS, all ships shall have at least the following transverse watertight bulkheads: a) one collision bulkhead b) one aft peak bulkhead c) one bulkhead at each end of the engine room.

SOLAS 74/78, chapter II, part B-2, regulation 12, indicates the limitations for the minimum and maximum position of collision bulkhead from the forward perpendicular.

2.1.2 Explicit MARPOL limitation for internal layout of tankers

MARPOL 73/78 regulations define the requirement for the cargo area of oil tankers. Fig. 1 outlines a categorization of all types of regulation of MARPOL 73/78, Annex I, Prevention of Pollution by Oil, for the tankers, this category is based on the regulation consequence on the internal layout definition.

As it is shown, the tanker deadweight changes the category of the minimum requirements effectively.

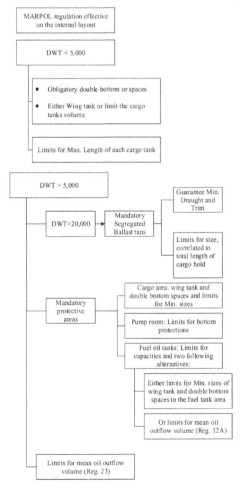

Figure 1. Overview of MARPOL regulations affecting on the internal layout of oil tankers.

24

As a preliminary requirement for the tank arrangement, the entire cargo tank length shall be protected by ballast tanks or spaces other than tanks that carry oil. Also, the deadweight of the vessel specifies the type of required spaces in the internal layout arrangement. The double bottom tanks are obligatory for all oil tankers. However, the wing tanks are obligatory for the tankers that have deadweight greater than 5,000 tonnes, though, they could be replaced with the limiting the size of cargo tanks for the tankers with deadweight lesser than 5,000 tonnes. The minimum dimensions of the mentioned tanks are specified in Regulation 19.

On the other hand, every crude oil tanker of 20,000 tonnes deadweight and above shall be provided with segregated ballast tanks (Regulation 18). Although these tanks may coincide with the specified tanks into the Regulation 19, their minimum requirement should follow different relations. The capacity of segregated ballast tanks shall be at least such that, the ship's draughts and trim can meet the specified requirements in the Regulation 18. Concisely, the regulation defines correlations among the capacity of the segregated ballast tank, the main dimensions of the ship, the total length of cargo tanks and oil outflow parameter (in the case of exceeding 200,000 tonnes deadweight).

Adequate protection against oil pollution in the event of collision or stranding is provided by limiting the maximum length of cargo tanks (for DWT < 5,000) and mean outflow parameter (for DWT ≥ 5,000) in the Regulation 23. Thus, if the deadweight is less than 5,000 tonnes, the internal arrangement affects the upper limit of the cargo tanks length according to the regulations. In this case, the maximum dimension of the cargo tanks length is depended on the number of longitudinal bulkheads (Regulation 23.3.2).

2.2 Implicit limitation of internal layout in the regulation

The following categorization is based on the implicit effect of MARPOL regulations on the internal layout of oil tankers.

2.2.1 Outflow parameter
For oil tankers of 5,000 tonnes deadweight and above, the regulations define the limits for the parameter of the mean oil outflow. The mean oil outflow shall be calculated independently for side damage and for bottom damage and then combined into the non-dimensional oil outflow parameter. Both of them are calculated by the probabilistic approach. The probability of penetrating cargo tanks is calculated by using the compartment boundaries, as shown in Figure 2.

All in all, the internal layout boundary dimensions and cargo tank volumes are correlated with

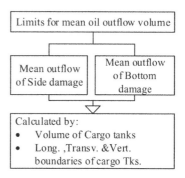

Figure 2. Overview of oil outflow limitation affecting on the internal layout of oil tankers.

Table 1. Damage locations along the length of the ship.

Ship length (L) (m)	Damage positions
L > 225	anywhere in the ship's length
150 < L < 225	anywhere in the ship length except machinery space, which shall be treated as single floodable compartment
100 < L < 150	anywhere in the ship's length except machinery space
L < 100	administration may allow relaxation form the damage requirements

evaluation of the probability of cargo tank damage, outflow parameters. Consequently, variation in these dimensions leads to modify the mean outflow parameter, which is assumed as hypothetical safety evaluation for the tankers.

2.2.2 Intact stability
There are no specified requirements of intact stability for the internal layout of tankers. However, any variation in the in internal layout dimensions or arrangement may result in changes in the cargo and ballast capacity and centre of gravity, consequently, the metacentric height and righting lever is affected. Intact stability regulation defines some criteria for the minimum initial metacentric height and specifications of the righting lever curve. Thus, the results of internal designs should be checked with the intact stability criteria.

2.2.3 Damage stability
According to MARPOL, every oil tanker shall comply with the subdivision and damage stability criteria after the assumed specified side or bottom damage, for any operating draught reflecting actual partial or full load conditions consistent with trim and strength of the ship as well as relative densities of the cargo.

The main damage criteria, which affect the internal arrangement and compartment sizes, are the limitations for the final draught, heel, trim and righting lever curve. These conditions should be checked for any possible damage case, which is due to specified side or bottom damage extents in the presented location in Table 1. Thus, any dimension and arrangement of ship internal layout can be affected by the results of such analysis.

3 PARAMETRIC MODEL OF THE INTERNAL LAYOUT

The ship hull shape is assumed to be fixed in this study. Furthermore, the compartment arrangement can be assumed constant in a layout to generate the parametric model. For example, the location of the main engine room usually is considered forward of the aft peak bulkhead in typical ship tankers, though, the regulations do not have obligation for such arrangement and the engine room position is depended on the type of propulsion system, capacity of the required power, shaft length and so on.

By subtracting the fixed spaces from the tanker ship hull, the available cargo area can be subdivided parametrically. A variety of layout configurations can be considered for the cargo area of oil tankers. For example, the different number of transverse and longitudinal bulkheads leads to various configurations. This diversity can be presented by introducing the proper parameters in the model.

3.1 Model parameters

A set of parameters control the detail of arrangement in the internal layout, as listed in the bellow:

- Number of transverse bulkheads in the cargo area (N_T)
- Number of longitudinal bulkheads in the cargo hold area (N_L)
- The total length of available cargo tanks (L_t)
- The positions of the transverse bulkheads alongside the ship $(l_i, i=1: N_{T-1})$
- The position of the collision bulkhead (L_{CBD})
- The positions of longitudinal bulkheads $(T_i, i=1: N_L)$
- Widths of the wing tanks for each longitudinal subdivision $(w_i, i=1: N_T)$
- Heights of double bottom tanks for each longitudinal subdivision $(h_i, i=1: N_T)$

Different types of configurations can be produced by changing the mentioned parameters, for example, the configuration of the inclined double bottom can be produced by defining two different height for parameters of two ends of the double bottom tanks, while, the stepped arrangement is generated by keeping the mentioned parameters equally. Similarly, the configurations with parallel (or inclined) walls of wing tanks can be achieved by

considering two equal (or non-equal) widths of two ends of wing tank in a subdivision arrangement.

The design variables can be defined by choosing the important parameters, which are depending on the optimization problem definition. For example, the total length of cargo holds can be eliminated from the list of design variables in the case of study of fixed hull shape because the available space of cargo hold is calculated by subtracting the total length of the tanker from the length of the engine room and collision bulkheads. The length of the engine room is almost constant as described before, and the collision bulkhead position can only have small variations relative to the ship length according to the regulations. Thus, the total length of available cargo tanks does not vary significantly for each design.

3.2 Method of generating the tanks and compartments

Once the available cargo space is specified for each configuration, the longitudinal subdivisions are generated by the boundary boxes. Each longitudinal boundary box contains the space between two consecutive transverse bulkheads, as shown in Figure 3. Then, the wing and double bottom tanks are generated for each longitudinal subdivision, which is presented schematically in Figure 4.

Accordingly, the remaining central space will be considered as cargo tanks, which can be

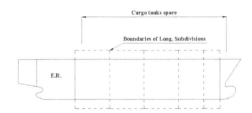

Figure 3. Schematic longitudinal subdivisions of cargo tanks space.

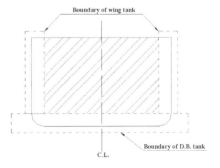

Figure 4. Sectional view of wing and double bottom tanks generation process.

divided based on the number of longitudinal bulkheads. A schematic sectional view is shown in Figure 5 for the case of two longitudinal bulkhead definitions. This kind of arrangement is based on the mandatory regulation for having the double bottom and wing tanks as the protective spaces. Also, the separated marginal spaces can be classified in the category of segregated ballast tanks, which is required to comply with the defined minimums, as described in the regulatory limits section.

The sizes of the wing and double bottom tanks can be defined with non-equal parameters for two consecutive longitudinal subdivisions. Although, the mentioned parameters are usually equal for the wall-sided part of the hull in the typical tanker. Their minimum requirement can be controlled by generating proper sectional adaptive lines as described in Jafaryeganeh et al. (2016).

In this method, the internal compartment generation is based on four steps:

- Creation of the solid domain boxes for boundaries of the tanks or compartments
- Boolean operations to combine or subtract the spaces
- Calibration of the provided spaces according to the ship hull shape
- Generation of a set of sections alongside compartment for hydrostatic calculations

Once the complete internal layout is generated, the actual dimensions of the compartments are determined. Then, the intact stability analysis is performed and the outflow parameter is calculated. Afterward, the results are verified against the regulation limitations. Figure 6 presents the process of generating the internal layout and criteria check in the regulatory framework.

At the end, the parameters or results of internal model calculations should be checked with the imposed constraints by the criteria of the regulations, as described previously.

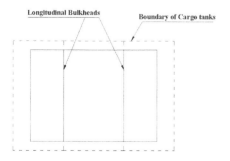

Figure 5. Sectional view of cargo oil generation based on the number of longitudinal bulkheads.

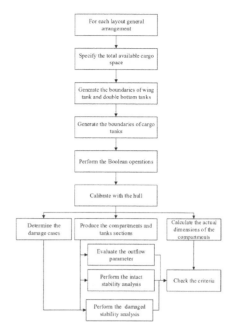

Figure 6. Providing of generic model of tanker internal layout.

3.3 *Providing the damage cases for damage stability evaluation in parametric model*

The possible damage cases need to be identified to evaluate the ship performance after the damage. The regulations defined a damaged area, which is located in the position described in Table 1.

Two types of side and bottom damage accidents are predicted according to the MARPOL regulation.

The damage extents are presented in Table 2 and Table 3 for side and bottom damage cases respectively.

Because different dimensions and positions of internal layout members lead to different damage scenarios, a technique is applied for the identification of all possible damage cases for each generated parametric model. Figure 7 presents the flowchart of this process.

The hypothetical damaged areas are simulated with boundary boxes, which can be located anywhere alongside the tanker. Then, an intersecting space is considered between the damaged boundary boxes and the provided internal layout. The damaged compartments or tanks are identified for each damage case.

Additionally, if any damage of a lesser extent than the maximum value specified in Table 2 and Table 3 would result in a more severe condition, such damage shall be considered according to the regulation. The existence of such damage cases

Table 2. Side damage extent dimensions.

Dimensions	Value
Longitudinal	min[$L^{2/3}/3$;14.5]
Transverse	min[$B/3$;11.5]
Vertical	without limit

Table 3. Bottom damage extent dimensions.

DWT	<20,000		>20,000
Dimensions	X_L from F.P	other parts	anywhere
Longitudinal	min[$L^{2/3}/3$;14.5]	min[$L^{2/3}/3$;5]	Note[1]
Transverse	min[$B/6$;10]	min[$B/6$;5]	$B/3$
Vertical	min[$B/15$;6]	min[$B/15$;6]	breach of outer hull

Note[1]: The same values of the less than 20,000 tonnage deadweight is applied

Deadweight in Tonnes	X_L
DWT < 20,000	0.3L
20000 < DWT < 75,000	0.4 L
DWT > 750,000	0.6L

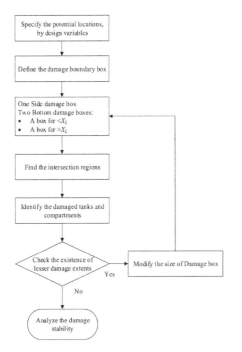

Figure 7. Identification of possible damage cases for each internal layout.

is checked by comparing the size of the damage box with the internal layout parameters. Then, the lesser damage cases are included by changing the

Figure 8. A schematic view of similar damage cases due to different locations of damage boundary boxes.

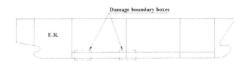

Figure 9. A schematic view of dissimilar damage cases due to locate damage boundary box in proper position.

damage box dimensions and evaluating the intersection space with the internal layout.

It should be noted that, if the position of the hypothetical damage box is located anywhere on the ship as mentioned in the regulations, its intersection with the internal layout may result in a lot of damage cases with the similar damaged compartments. Figure 8 presents a sample of different damage locations with similar damage case results. These repetitive damage scenarios would increase the run time for evaluation of internal layouts during the optimization iterations. Thus, the potential points of generating dissimilar damage scenarios are identified in the parametric model of internal layout. Those points include the intersection of longitudinal, transversal and vertical intersection of the watertight members. Afterwards, the damage box is positioned in two situations relative to each point, first, it should cover the maximum extent of the damage after each point and second, it should include each intersection point. Figure 9 shows a schematic longitudinal view of the generation of only dissimilar damage cases. In this sample, the longitudinal positions of damage boundary boxes are chosen using the relevant design variables of transverse bulkheads locations.

Therefore, the runtime of evaluation damage cases is reduced to dissimilar damage cases.

4 CASE STUDIES

A shuttle tanker has been considered as a case study and its main particulars are listed in Table 4. In the reference design, the main subdivision consists of eight transverse bulkheads and a central longitudinal bulkhead. Figure 10 shows the initial general arrangement of the taker.

In the reference design, the engine room is located forward. However, various layout arrangements can be generated by changing the position of the engine room alongside the ship.

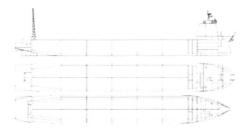

Figure 10. Initial general arrangement of the shuttle tanker.

Table 4. Main particulars of the Shuttle Tanker.

Particular	Value	Units
Length between perpendiculars	234.48	m
Breadth	32.0	m
Height	18.5	m
Draft	12.0	m
Deadweight	63,000	ton
Block Coefficient	0.881	–

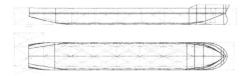

Figure 11. Generation of the tanks and compartments boundary.

Figure 12. Calibrated tanks and compartments.

Considering the ship's deadweight, mandatory compartments are the segregated ballast tanks, double bottom tanks, and wing tanks. Thus, the internal layout is generated with the described method. Figure presents the generation of the tanks and compartments boundaries and Figure 12 shows the calibrated internal compartments of the ship for the Alternative 1, as described in below.

Table 5 presents two alternative designs among the vast number of possible ones. Regardless of the regulatory limitations, some of the design variables should be restricted to generate a feasible design. For example, the height of double bottom and the width of the wing tanks are considered to be constant for each longitudinal subdivision.

The total cargo area length can only be considered as an effective design variable if the main

Table 5. Parameters and values used in the two case studies.

Parameters	Alternative 1	Alternative 2
NT	8	8
NL	1	1
Lt	222.48	224.12
$LCBD$	12	10.36
$l1$	27.6	20.6
$l2$	27.84	32.92
$l3$	27.84	17.84
$l4$	27.84	32.84
$l5$	27.84	32.84
$l6$	27.84	23.84
$l7$	27.84	30.14
$l8$	27.84	33.1
$wi; i=1: NT$	3	4.2
$hi; i=1: NT$	2.5	1.9

particulars of the ship are also assumed as design variables. A major variation of the cargo length requires variation of the total length of the ship, consequently, the breadth, height and deadweight of the ship would be affected. In this study, the main particulars of the ship are kept constant. Also, the engine room space is assumed to be constant for any alternative design, and thus, the total cargo hold length can only be increased by reduction of the collision bulkhead to the forward perpendicular, as applied in the alternative 2.

The minimum distance of two consecutive transverse bulkheads is controlled by the limitation defined implicitly in the damage stability analysis because, if the distance between these bulkheads decreases, the damage boundary box may include several compartments, and consequently the survival damage criteria would not pass. Considering the tanker deadweight, the maximum distance between two consecutive transverse bulkheads is controlled implicitly by the limits defined for the outflow parameter.

Because wing tanks and double bottom tanks are assumed to be the only segregated ballast tanks in the presented internal model, the minimum values of double bottom height and wing tanks width are not only controlled by the explicit limits of the regulations, but also, are affected by the minimum requirement of the total ballast tanks volume.

The damage cases are identified automatically for each alternative. The number of side damage cases is 58 for each alternative case. The numbers of bottom damage cases are 15 and 23 for alternative 1 and alternative 2 respectively. Since the heights of double bottom tanks are larger than the specified vertical extent of the bottom damage in alternative 1, some of the cargo tanks do not breach according to regulations and the number of damage cases is reduced relative to the alternative 2. Similarly, the number of side damage cases would be reduced, if the wing tanks widths are larger than the horizontal extent of the side damage. However, such variation in the number of damage cases needs more studies to realize its consequence in the internal layout dimensions and arrangements for all possible cases in the feasible domain of design variables.

5 CONCLUSIONS

The mandatory regulations from IMO are categorized for internal layout of the oil tankers. In these regulations, implicit and explicit limits are classified mostly based on the main characteristics of the ship, such as deadweight, length and breadth. These characteristics are also effective on the arrangement of the internal layouts. A method was developed for the parametric modelling of the internal layout of oil tankers based on the IMO regulations. Also, a technique was applied to the identification of all dissimilar damage cases for the provided model. These methods can be applied for the comparison of alternative designs of the internal spaces during the conceptual design stage. The evaluation of the stability and the compliance with the criteria of the regulations can be performed automatically.

The most important parameters required to produce models of tankers with realistic layout topologies are identified. An automatic process is developed to generate different layout and evaluation of safety criteria with available commercial software. CAESES/FRIENDSHIP-Framework® produces the geometric model and MAXSURF® computes the stability. Those criteria, which specify the implicit limitations for the internal layout, can also be used as merit functions for the improvement of the ship survivability after the damage.

The method presented is intended to be a component of a ship synthesis model to be used in optimization studies, where the size and arrangement of internal layout members are being decided.

ACKNOWLEDGEMENTS

This work was performed within the Strategic Research Plan of the Centre for Marine Technology and Ocean Engineering (CENTEC), which is financed by Portuguese Foundation for Science and Technology (Fundação para a Ciência e Tecnologia-FCT).

REFERENCES

FRIENDSHIP-SYSTEMS. FRIENDSHIP-Framework, *www.CAESES.com.*

Harries, S. (2003). Geometric Modelling and Optimization. OPTIMISTIC- *Optimization in Marine Design, 39th WEGEMT Summer School, Berlin, Germany:* 11–33.

Harries, S; Tillig, F; Wilken, M; Zaraphonitis, G. (2011). An Integrated approach for Simulation in the Early Ship Design of a Tanker. *10th international conference on computer and IT application in the maritime industries, Hamburg.* ISBN: 978-3-89220-649-1

IMO (2006). MARPOL, International Convention for the Prevention of Pollution from Ship.

IMO (2012). SOLAS, the International Convention for the Safety of Life at Sea.

Jafaryeganeh, H; Ventura, M; Guedes Soares, C. (2016). Parametric modelling for adaptive internal compartment design of container ships, *Maritime Technology and Engineering,* Guedes Soares, C. & Santos T.A., (Eds.), Taylor & Francis Group, London, UK: 655–661.

Koelman, H.J. (2012). An approach to modelling internal shapes of ships to support collaborative development. *Proceedings of TMCE, Karlsruhe, Germanyr,* 685–696.

Koningh, D; Koelman, H; Hopman, H. (2011). A novel ship subdivision method and its application in constraint management of ship layout design. *Journal of Ship Production and Design,* 27(3): 137–145.

Lee, D; Lee, S.S; Park, B.J. (2004). 3-D geometric modeler for rapid ship safety assessment. *Ocean Engineering.* 31(4):1219–1230.

Lee, S.U; Roh, M.I; Cha, J.H; Lee, K.Y. (2009). Ship compartment modelling based on a non-manifold polyhedron modelling kernel. *Advances in Engineering Software.* 40: 378–388.

Nowacki, H. (2010). Five decades of Computer-Aided Ship Design, *Computer-Aided Design,* 42: 956–969.

Papanikolaou. A; Eliopoulou. E (2008). On the development of the new harmonised damage stability regulations for dry cargo and passenger ships, *Reliability Engineering and System Safety,* 93:1305–1316.

Papanikolaou. A; Zaraphonitis. G; Boulougouris. E; Langbecker, U (2008). Multi-objective optimization of oil tanker design, *J Mar Sci Technol,* 15:359–373.

Roh, M.I; Lee, K.Y. (2007). Generation of the 3D CAD model of the hull structure at the initial ship design stage and its application. *Computers in Industry,* 58: 539–557.

Santos, T.A. and Guedes Soares, C. (2005a) Multi-Objective Optimization of Fast Ferry Watertight Subdivision. Guedes Soares, C. Garbatov Y. & Fonseca N., (Eds.). *Maritime Transportation and Exploitation of Ocean and Coastal Resources.* London, U.K.: Taylor & Francis Group; pp. 893–900.

Santos, T.A. and Guedes Soares, C. (2005b) Monte-Carlo Simulation of Damaged Ship Survivability. *Journal of Engineering for the Maritime Environment.* 219, Part M: 25–35.

Santos, T.A. and Guedes Soares, C. (2010) Probabilistic Approach to Damage Stability. Guedes Soares, C. & Parunov J., (Eds.). *Advanced Ship Design for Pollution Prevention.* 1 ed. London, U.K.: Taylor & Francis Group; pp. 227–242.

Santos, T.A. and Guedes Soares, C. (2011) Deterministic and Probabilistic Methods Applied to Damage Stability. Guedes Soares, C. Garbatov Y. Fonseca N. & Teixeira A.P., (Eds.). *Marine Technology and Engineering.* London, UK: Taylor & Francis Group; pp. 1297–1312.

Maritime Transportation and Harvesting of Sea Resources – Guedes Soares & Teixeira (Eds)
© *2018 Taylor & Francis Group, London, ISBN 978-0-8153-7993-5*

Offshore sulfide power plant for the Black Sea

L. Stoev
Keppel FELS Baltech, Varna, Bulgaria

P. Georgiev
Technical University of Varna, Varna, Bulgaria

Y. Garbatov
Centre for Marine Technology and Ocean Engineering (CENTEC), Instituto Superior Técnico,
Universidade de Lisboa, Lisbon, Portugal

ABSTRACT: Hydrogen sulfide—a highly poisonous, explosive and corrosive gas has been identified as a renewable energy source from the Black Sea with a great potential. There were several methods suggested for hydrogen sulfide exploration and utilization—from hydrogen dissociation, storage and usage as energy fuel, to power generation in hydrogen sulfide 'fuel cells'. The main objective of this article is to present a concept for an Offshore Sulfide Power Plant (OSPP) of a commercial significance. In this study, different designs of Ocean Thermal Energy Conversion (OTEC) floating, spar or semi-submersible design concepts are discussed. A study was made on an octagonal semi-submersible and spar type sulfide power plant that apart from hydrogen sulfide provides an option for utilization of multi-renewable energy sources, such as OTEC, solar and wind. For the Concept Assessment purpose, the stability, motions, structural response, hull configuration, lightship weight, deep-water pipe disconnection, mooring, etc. issues are identified and discussed. In the present study, a concept structural design and general arrangement of OSPP is presented. The basic evaluation, limited to the weight and cost of the floating structure, is performed, which is going to be used for a further and more thorough feasibility study.

1 HYDROGEN SULFIDE AS A RENEWABLE ENERGY SOURCE

The offshore technologies are traditionally referred to Oil & Gas industry. However, since year 2014 the hydrocarbon discovery and exploration is not a priority of the marine & offshore engineering anymore. There are other disruptive technologies related to renewable energies, seabed mining and subsea engineering.

Renewable Energy Sources (RES)-based energy industry is the most dynamic and perspective sector of the global energy industry, which solves a number of problems of depletion of traditional organic energy resources, but also contributes to environmental safety. The renewable offshore energy sources to be considered are: Wind energy; Solar energy; Geothermal and OTEC energy; Hydro energy—wave, tidal, river and sea currents; Energy based on hydrogen sulfide (H_2S) dissolved in the seawater or being in the gas phase of mud volcanoes;

Besides already known offshore renewable sources of energy, the Black Sea offers another, unique one—hydrogen sulfide, which is a disruptive threat and a disruptive RES at the same time. The Black Sea, a highly-isolated inland sea, is the largest anoxic zone in the world. The solubility of H_2S gas in the water is really high and according to Henry's law (SOEST, 2013) it depends on temperature, salinity, pH, pressure, and nature of H_2S and water molecules (see Figure 1, Naman, 2008).

Researchers allocated considerable amounts of dissolved hydrogen sulfide and other sulfides (H_2S, HS^- and S^{2-}) with significant concentration at a depth of 150 m and more. Further research outlined three main sources of hydrogen sulfide (Azarenkov, 2014):

– sulfides are a product of the metabolism of the sulfate-reducing bacteria, living in the greater Black sea depths, and reducing the natural sea water salts—sulfates (SO_4^{2-})—to sulfides.
– sulfides in the anaerobic layers of sediments at the sea bottom are about 1–2 cm thick.
– sulfides are released by some groundwater tectonic cracks at the sea bottom.

A high content of organic matter, with maximum processes of bacterial sulfate reduction is the major source of this hydrogen sulfide zone.

Hydrogen sulfide is one of the most poisonous gases, but it has a great economic value to obtain hydrogen via dissociated into hydrogen and

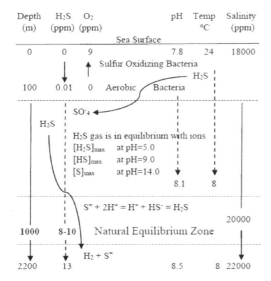

Depth (m)	H₂S (ppm)	O₂ (ppm)		pH	Temp °C	Salinity (ppm)

$$Depth\ (m)\quad H_2S\ (ppm)\quad O_2\ (ppm)\qquad\qquad pH\quad Temp\ ^\circ C\quad Salinity\ (ppm)$$

Sea Surface

0   0   9   7.8   24   18000

Sulfur Oxidizing Bacteria

100   0.01   0   Aerobic / Bacteria   H₂S

H₂S   SO⁻₄ ←

H₂S gas is in equilibrium with ions
[H₂S]$_{max}$ at pH=5.0
[HS]$_{max}$ at pH=9.0
[S]$_{max}$ at pH=14.0

8.1   8

$S^= + 2H^+ = H^+ + HS^- = H_2S$

20000

1000   8-10   Natural Equilibrium Zone

$H_2 + S^=$

2200   13   8.5   8   22000

Figure 1.   Natural equilibrium of the Black Sea.

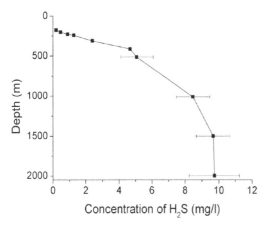

Figure 2.   Hydrogen Sulphide concentrations in the Black Sea.

Sulphur. Thus, the Black Sea has not only a serious environmental contamination, but it is also a potential source of hydrogen energy, if a decomposition process can be developed.

The daily production of H₂S gas by the sulphur reducing bacteria in the Black Sea is about 10,000 tons. If to assume that an annual increase of H₂S of the Black Sea is not less than 4–9 million of tones, then the total of its reserves could be estimated on 28–63 billion of tones (Azarenkov, 2014). The interface between anaerobic and aerobic water layers is situated at a depth of about 200 m. Under 150–200 m, H₂S concentration increases gradually until 1 km, and finally reaches a nearly constant value of 9. 5 mg/l at a depth around 1.5 km, Figure 2.

2   OFFSHORE SULFIDE POWER PLANT (OSPP)

2.1   Sulfide fuel

The hydrogen produced from H₂S decomposition could be used for power plant fuel. There are several methods for H₂S decomposition, including thermal, thermochemical, electrochemical, photochemical and plasm-chemical techniques. However, applying any of these methods for H₂S decomposition to H₂ and S is a costly process, e.g. the Clauss process (Sassi, 2008) that requires high temperatures or energy. There are also issues about the hydrogen storage.

H₂S dissolved in the Black Sea waters could be processed (oxidized) in "fuel cell" in order energy to be generated (Beschkov, 2013).

All the technologies that utilize H₂S from the Black Sea waters as an energy source directly or indirectly require a huge mass of water—90 m³/s to be pumped up from about 1000 meters depth, where the H₂S concentration is reaching 8 mg/l (Beschkov, 2013).

Similar are the requirements for OTEC (Ocean Thermal Energy Conversion) technology. For generating 10 MW in OTEC plant, the water mass flow rate of 30 m³/s (Aldo, 2005) is required to be lifted through a cold-water pipe (CWP) from over 600 m depths in the tropical seas, (see Figure 3).

For the first 50 m or so, near the surface, the turbulence maintains the temperature uniform at some 25°C. It then falls rapidly reaching 4 or 5°C in deep places. The temperature difference between the upper (warm) and bottom (cold) sea water layers' ranges from 10°C to 25°C, with the higher values found in the equatorial waters. This implies that there are two enormous reservoirs providing a heat source and the heat basin required for a heat engine. A practical application is found in the

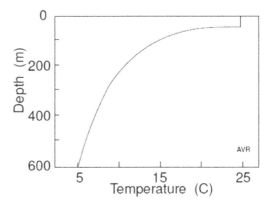

Figure 3.   Typical ocean temperature profile in the tropics.

OTEC system (heat engine) designed to transform the thermal energy into electricity.

2.2 OSPP basis of design

The OSPP design, discussed here, is based on the OTEC floating platform concept that integrates a 6-meter diameter 1000 m long deep-water pipe (DWP). The sulfide power plant equipment is envisaged to operate on a floating platform, positioned beyond the continental shelf at a distance of 10 to 90 km from the shore, depending on the water depth. The platform is provided with a DWP, lifting sea water to the sea surface from about a 1000 m depth. The dissolved in the water sulfides are processed thru fuel cells where the energy is generated. The treated sulfate water is returned back through return pipes to the natural anaerobic zones (to a 150 m depth and below) where the saprophyte bacteria regenerates the sulfates to sulfides. This represents a closed process, taking place in an oxygen-free environment, which does not admit propagation of anaerobic pathogenic bacteria. The floating power platform will be connected to an onshore or floating transformer station by high-voltage cables, which are not discussed in this study.

2.3 OSPP concept design

An Offshore Power Plant could be floating or fixed offshore. It could be designed as a Multi-Renewable Energy Plant (MREP) that utilizes deep water (H_2S, thermal), wind and solar energies. MREP is to be based on proven gas & oil offshore technologies and the newest OTEC concepts. There are certain issues that need to be considered—poisonous and corrosive H_2S, gas lift effect and deep pipe installation.

A floating platform is generally the preferred configuration, to minimize the total length of the Cold-Water Pipe (CWP). Heat exchangers, turbines, pumps, and generators are above and just under the surface. The CWP extends down to the cold water layer located at a 1000 m depth, and carries the cold water to the surface. The industry showed a strong ambition to have the first commercial OTEC plant, which can be seen in Figure 4 (Miller (2014)) in the water before 2020.

The 2.5 MW Mini-Spar Pilot Plant (Lockheed, 2011) is a lower cost OTEC pilot plant relative to the 10-MW semisubmersible baseline design developed by Lockheed Martin. An analysis of alternatives with the objective of reducing the baseline cost resulted in a Floating Option and a Land-Based Option. The Floating Option—Mini-Spar comprises a complete OTEC system, albeit at a smaller 2.5-MW capacity relative to the baseline Top-Level Requirements.

Figure 4.   Lockheed 10-MW OTEC Semi-sub concept (Miller (2014)).

The 2.5-MW Mini-Spar, Figure 5 below, has a cell spar hull structure of four steel cylinders to support the OTEC equipment, seawater ducting, CWP, mooring, and subsea power export cable. The most of the OTEC process equipment is submerged and mounted externally to the hull cylinders, and the ammonia turbine-generator is mounted within the topsides with other standard marine auxiliary equipment. The spar's deep draft results in excellent stability; even the 100-year storm event (18-meter waves high) creates pitch motions that are small enough to allow a fixed connection between the vessel and the CWP. The deep hulls also direct the seawater discharge plume to beneath the euphotic zone, simulating the ocean dynamics expected by future large OTEC plants. The 2.5-MW Mini-Spar's function is to transition OTEC technology from the development stage to commercial operation.

Multiple criteria are considered in addition to the Top-Level Requirements and include the ability to demonstrate system performance; obtain requisite environmental measurements for commercial applications; validate the cost and schedule data; establish the operation and maintenance requirements; scale to larger utility capacities with minimal risk perceptions; show long-term survivability at sea and to provide a system-level integration experience.

It is anticipated the final design, procurement, fabrication, and the installation will take three years.

The rough order magnitude (ROM) capital cost for the Mini-Spar is $107M. The critical factor in

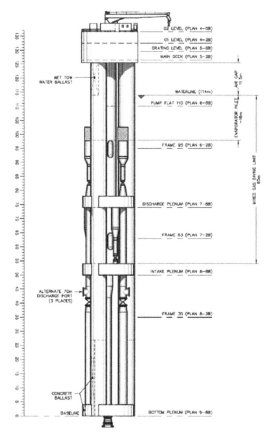

Figure 5. Lockheed 2.5-MW OTEC Mini-Spar Pilot Plant (Lockheed, 2011).

commercial Ocean Renewable Deep-Water Energy Plants is the DWP (CWP in OTEC) and its integration to the floating hull. In the Lockheed's 2.5MW OTEC spar concept, 2.5-meter diameter CWP, made of High density polyethylene (HDPE) is applied. For 10-MW/50-MW or bigger OTEC floating plants, 4 m to 10 m diameter CWP made of Fiber Reinforced Plastic (FRP) is proposed.

Sulfide energy technology also requires H_2S rich (8 mg/l) water mass of 50–90 m^3/s to be lifted though a big diameter (6–10 m) pipe, (Beschkov, 2013).

2.4 Deep water pipe—purpose and integration

Lifting the sulfide water solutions from a 1000 m depth is a high energy consuming process, for which big, heavy and costly equipment is required. Instead, gas-lift—a method used in the Gas & Oil industry could also be considered.

However, gas-lift effect is difficult to be modelled in such large scale for CFD analysis. The

gas-lift effect could facilitate a deep-water lift, but also could have a negative impact on the upstream flow—high turbulence, gas phase that needs to be handled, etc.

Four-meter diameter CWP of fiber reinforced plastic (FRP) is envisaged for 10-MW OTEC application as can be seen in Figure 6.

Innovative method for CWP installation and handling by use of "Gripper", Figure 7 is proposed by Lockheed.

For a deep-water pipe installation, also the jack-up or "jack-down" technology could be considered.

An integration of CWP with the floating hull is shown in Figure 8.

Although it is developed for a mini spar hull, the integration philosophy is applicable also for a bigger spar structure. However, the operational and on-field environment loads have to be analysed and applied accordingly. Also, DWP installation and mooring philosophy is an important part of the solution.

Figure 6. Lockheed four-meter CWP core, (Miller (2014)).

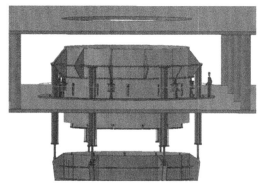

Figure 7. Cold Water Pipe handling system (Gripper), (Miller (2014)).

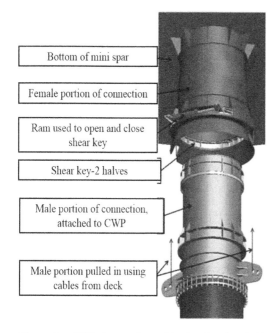

Bottom of mini spar

Female portion of connection

Ram used to open and close shear key

Shear key-2 halves

Male portion of connection, attached to CWP

Male portion pulled in using cables from deck

Figure 8. CWP integration with the floating hull, (Miller (2014)).

2.5 Mooring

The mooring system, as presented in Figure 9 (Lockheed, 2011) and considered in the present analysis is a catenary mooring made up of a platform chain, sheathed spiral strand and anchor chain. All components are a nominal 65 mm diameter, and the water depth is assumed constant 1100 m. The actual seafloor topography and the precise anchor positions must be accounted for in the final global response analysis, which would be conducted in the preliminary design phase.

2.6 Motion

Motions and accelerations are computed along the centreline of the spar at the mean water line. A summary of the extreme motions is given in Table 1.

Accelerations of the 2.5-MW cell spar for the 100-year condition are summarized in Table 2.

The following conclusions can be made for the 2.5MW cell spar:

Mooring: The nine-line catenary mooring system satisfies API and ABS mooring criterion for the 100-year environments. This is the appropriate load case for a 20-year permanent mooring. The controlling condition is the 100-year storm with one line missing.

Platform Motions and Accelerations: Maximum platform responses in the 100-year event are:

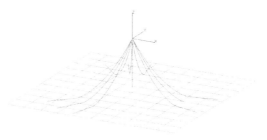

Figure 9. Cell spar coupled analysis model (Lockheed, 2011).

Table 1. Extreme motion responses—intact mooring, (Lockheed, 2011).

	Surge (m)	Sway (m)	Heave (m)	Roll (deg)	Pitch (deg)	Yaw (deg)
Max	220.17	195.21	0.21	5.21	6.47	6.78
Min	−168.45	−0.01	−2.28	−1.51	−6.93	−5.99
Std Dev	10.26	6.25	0.21	0.85	1.17	1.00

Table 2. Extreme acceleration responses—intact mooring, (Lockheed, 2011).

	Surge (m/s²)	Sway (m/s²)	Heave (m/s²)	Roll (deg/s²)	Pitch (deg/s²)	Yaw (deg/s²)
Max	1.61	0.07	0.18	0.05	0.81	0.33
Min	−1.57	−0.06	−0.15	−0.05	−0.77	−0.30
Std Dev	0.43	0.02	0.04	0.01	0.20	0.09

Platform Offset:	237 m,
Bottom of CWP Offset:	304 m,
Heave:	−2.4 to + 0.25 m,
Pitch/Roll:	+/− 6.5 deg,
Horizontal Acceleration:	0.27 g,
Vertical Acceleration:	0.02 g max.

For the final design the following must be considered: the bottom of the CWP must have sufficient clearance from the seafloor to prevent contact under the motions computed and given the local bathymetry; the configuration/design of the power cable must accommodate the computed platform motions from the global response analysis.

2.7 OTEC/OSEPP structure

After considering different designs—Semi-submersible, SEVAN-type, Octa-buoy, an octagonal spar is discussed as can be seen in Figures 10 and 11, as a commercial 10-MW concept in the present study.

The submerged water pumps are replaced with bigger capacity pumps placed in pump rooms in the octagonal hull structure. A 6-meter diameter

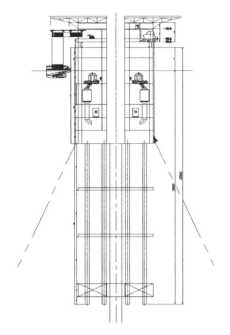

Figure 10.   Elevation view of the OTEC/OSEPP spar.

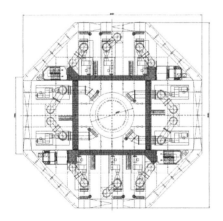

Figure 11.   Pump room deck of the OTEC/OSEPP spar.

deep-water pipe servicing a 10-MW power plant is envisaged to be made of FRP material.

Beside the OSEPP structure configuration and general arrangement, also stability, motion, mooring, installation and maintenance will be evaluated and proposed. Special attention will be paid on the deep-water pipe and its integration with the floating hull.

3   CONCLUSIONS

Answering the main challenge that both Ocean Thermal Energy Conversion and H$_2$S technology

are facing—the lifting of huge amount of water mass (flow over 30 t/s) from the sea depths of 1,000 meters, requires a technically feasible and economically liable concept of a floating platform to be developed. It shall service over 10-MW power Sulfide or Ocean Thermal Energy Conversion power plant. The spar type structure proposed is based on proven oil & gas technology and is designed so pumps and power equipment chosen for a deep water renewable energy power plant of commercial scale to be installed.

As the use of HDPE solution is not available for CWP with a diameter bigger than 2.5 m, FRP material and technology is chosen for the 6-meter diameter deep water pipe. The pipe integration with the floating hull depends on the pipe installation technology, operational and environment loads.

Despite of numerous prototypes developed during the last 5 years, there is still not any commercial size OTEC or other deep water energy floating platform built. It seems that the concept of OSEP may contribute to the efforts for having a Deep Ocean Renewable Energy floating plant of commercial scale soon.

REFERENCES

Aldo V. da Rosa, 2005, Fundamentals of Renewable Energy Processes, Elsevier Inc., ISBN 13: 978-0-12-088510-7.

Azarenkov N., Borts B., Tkachenko V, 2014, Hydrogen sulfide energy of the Black Sea, East European Journal of Physics. Vol.1 No.3 pp. 4–21.

Beschkov V., Christov V., 2013, Hydro sulfide energy from the Black Sea - Sulfide Power Plants (SPP), International Scientific Applied Conference "New Technologies in the Offshore Industry", Varna, pp. 2–5.

Lockheed Martin, 2011, Configuration Report and Development Plan Volume 4 - Site Specific OTEC Pilot Plant, 2.5 MW OTEC Mini-Spar Pilot Plant, OTEC-2011-001-4.

Miller A., Rosario T., Ascari M., 2012, Selection and validation of a minimum-cost cold water pipe material, configuration, and fabrication method for ocean thermal energy conversion (OTEC) systems, Proceedings of SAMPE, Baltimore, MD, p. 27.

Naman S., Ture I., Veziroglu T., 2008, Industrial extraction pilot plant for stripping H$_2$S gas from Black Sea Water, International Journal of Hydrogen Energy, Vol. 33, pp. 6577–6585.

Sassi M, Gupta A., 2008, Sulfur Recovery from Acid Gas Using the Claus Process and High Temperature Air Combustion (HiTAC) Technology, American Journal of Environmental Sciences 4 (5): pp. 502–511.

SOEST, 2013, Dissolved Gases other than Carbon Dioxide in Seawater, OCN 623 – Chemical Oceanography, Libes, Chapter 6 – pp. 147–158.

Special session in honour of Prof. Vedran Zanic

Maritime Transportation and Harvesting of Sea Resources – Guedes Soares & Teixeira (Eds)
© 2018 Taylor & Francis Group, London, ISBN 978-0-8153-7993-5

Multi-objective scantling optimization of a passenger ship structure

J. Andric, P. Prebeg & T. Stipic
Faculty of Mechanical Engineering and Naval Architecture, University of Zagreb, Zagreb, Croatia

ABSTRACT: Structural optimization aspects of cruise ships has been reviewed and discussed through this paper. Multi-objective scantling optimization (with Weight & VCG as objectives) of post-Panamax cruise ship has been presented as an example. Structural design software MAESTRO and in-house developed, framework for the design support system DeMak—OCTOPUS Designer was used to generate a set of non-dominated designs. Two Pareto optimal designs marked as *minWGT* and *minVCG* have been compared and differences in structural response of both variants and prototype structure were discussed.

1 INTRODUCTION

The structural complexity of cruise ships has been increased over last few decades. Continuous tendency to increase the size and the capacity of cruise ships (Levander, 2006) makes structural design as a very important objective and one of the most important issues in rational technical design of those sophisticated and complex objects. Suitable structural design procedures and techniques have to be implemented to support design process through all phases, in most efficient way. However, it is of the highest importance to improve structural design process at the earliest stage of design where the basic concept and all strategic decisions have been made (ISSC, 2003). Those demands have created a need to use the state of the art decision support methods and techniques in connection to structural design methods with the goal to find most rational design solutions and improve overall cruise ship design. Also, those possibilities enable more space for general designer creativity and implementation of new and innovative feasible concepts.

Through this paper structural optimization aspects of cruise ships will be reviewed and discussed in more detail. Multi-objective scantling optimization (with Weight & VCG as objectives) of post-Panamax cruise ship has been presented as an example.

1.1 *The global structural response aspects of a cruise ships*

The structural design methodology is particularly complex for the modern multi-deck ships (cruise ships, passenger ships, ferries, RoPax ships, etc.), that are characterized by the extensive superstructure and its interaction with the lower hull.

Today, in a modern cruise ship the volume of superstructure is even bigger than the volume of the lower hull. Very useful contribution, still actual, in the clarification of the most important aspects of a modern cruise ship structural design can be found in Gudmunsen (1995), while in Levander (2006) most important aspects related to general cruise ship design has been discussed. For cruise ships, only the full ship 3D FE analysis is considered sufficient by ISSC (1997, 2003) for the correct assessment of the global structural response.

Correct modelling of large side openings is essential for this kind of ships due to the fact that reduced shear stiffness of the superstructure sides has a strong influence on the longitudinal hull girder stress distribution over a ship cross section height. This issue is particularly important for the new cruise ships due to fact that over 70% of all cabins are balcony cabins (Levander, 2006). As a consequence, the simple beam theory cannot be used for preliminary considerations and the influence of the superstructure on the primary strength has to be taken into account starting from the concept design phase.

The main disadvantage of the required 3D full ship FE model is a large amount of work needed for the model preparation and the evaluation of its feasibility. Therefore, it cannot be used efficiently for the exploration of different concepts, but mostly for the verification purposes of the final structural configuration. The research task for the structural design in a concept phase would be to establish simple, fast and flexible, but sufficiently accurate design oriented analysis model enable to modelling the complex structural response.

Following ISSC (2003) recommendations Andric and Zanic (2010) proposed *generic* global 3D FE model for the concept design phase. Authors declared that the proposed model ensures rapid generation of different structural topological concepts, correct structural assessment (compared to full ship FE models) and easily inclusions in formal structural optimization procedure. The generic models have been tested and verified by comparing the global structural response on several complex examples (cruise ship, livestock carrier, etc.)

1.2 *Structural optimization of a cruise ships*

Very limited number of papers w.r.t to structural optimization of cruise/passenger ships can be found.

Regarding the synthesis tasks using 3D FEM, the paper by Hughes et al. (1980), and the book Hughes (1983) presented an analysis model for design of the 3D thin-walled structure, using combination of 3D FEM analytical model and the sequential LP-based synthesis model. In Zanic and Jancijev (1986) a somewhat simpler combination of 2D FE membrane model of ship projected onto the centerline plane and the partial 3D FE model around midship section was applied to simulate the global structural response of large passenger ferry and to perform weight optimization of the amidships part of the structure. In Jancijev et al. (2000) for the structural optimization of Ro-Pax ship the analysis models were a ship segment modeled by the FE stiffened membrane elements and a girder system based on bracketed beams.

For scantling structural optimization 2D model of midship section or partial FE or analytical based models with the predefined superstructure efficiency are still in use, see Zanic et al (2007), Richier at al.(2007a), Richier at al.(2007b) and Caprace et al (2010). But, as Romanoff et al. (2013) identified these approaches are limited w.r.t structural response calculations due to several reasons:

The calculation of deck efficiency coefficients (used for correction of normal strain distribution along hull girder) has been done using separate 3D FE analysis, or data from previous designs.

During the scantling optimizations if scantlings change a lot, deck efficiency coefficients has been also changed and need to be updated. So, approach is limited to scantling optimization with the small scantling changes compare to prototype, if the deck efficiency coefficient stay fixed during the design process.

This approach cannot be used for the efficiently investigation of different structural topology/ geometry concepts, because superstructure efficiency are unknown and therefore separate 3D FE analysis has to be done previously.

Romanoff et al. (2011) presents the interaction between the hull and the superstructure in optimized large cruise ship. Optimization is carried out with respect to weight and vertical centre of gravity (VCG) and set of Pareto optimal solutions has been generated. The used structural model was simplified 3D model of the whole ship based on *Coupled Beams* theory (Naar et al., 2004). Authors underline the necessity to use structural model of the whole ship due to complex vertical and shear coupling over model length which cannot be adequate represent using 2D cross section models.

The opportunity to make the methodological step forward has been identified during the last decade based on the fact that the advances in the fast optimization methods and tools, due to the increased computer speed, have not been followed by the improvements in the design oriented analysis models coupled with them. It is of particular importance for the concept design phase where approximately 70–80% of assets have to be determined.

To perform formal structural optimization of cruise ship several characteristics in design problem definition have to be taken into account:

- Along the ship, the topology and geometry are not monotonously comparing to other ship types (tanker, LNG, bulk carrier, etc.) → several sub-models may be needed for different zone along the ship.
- Functionality of the ship w.r.t to her mission (cruising and fulfilling passenger expectations) is dominant goal and that fact reduces possible structural solutions leading to situation where the ship topology and geometry are mainly fixed by general designer/architect → topology optimization has to be started as soon as possible in cooperation with general designer who is crucial in definition of possible topology variants (see Andric et al., 2017).
- Position of VCG is very important due to stability demands. It means that mass reductions in upper parts of superstructure can be main design objective, Romanoff et al. (2013).
- Possible inclusion of different sandwich structure concepts of superstructure decks combine with laser welding technology are under investigation (see Romanoff, 2011-TWS).
- Minimum plate thickness are generally limited to 5 mm because of the uncertainties related to production, fatigue, buckling, and vibration issues (see Lillemäe et al. 2017).
- Alternative material to be use in upper part of superstructure can be reconsider → in implementation of different non-steel materials (aluminium alloys, composites, etc.) rational approach have to be taken w.r.t possible increase of production cost. Bacicchi et al. (2000) give brief review regarding use of alternative materials in superstructure of cruise ship.

2 OVERALL DESIGN FRAMEWORK FOR OPTIMAL STRUCTURAL DESIGN SOLUTIONS

A design framework to performed optimal structural design solution can be split in two steps:

STEP-I: Investigation on different topological/geometrical concepts of superstructure.

STEP-II: Multi-objective scantling optimization of chosen topological variant(s).

STEP-I has been presented in more detail in Andric et al. (2017). Authors presented methodology how to perform fast investigation of main topological parameters that influenced optimal structural solution of simplified passenger ship example.

This paper presents in more detail STEP-II of the procedure where for the fixed topological solution multi-objective scantling optimization has been performed for post-Panamax cruise ship design. For more information about multi-objective scantling optimization that are not given in this paper see, Zanic (2013) and Zanic et al. (2013).

2.1 Design support system OCTOPUS

Due to the complexity of the design process in the real practical applications it was necessary to develop the design environment which enables a flexible execution of the described process. In this section DeMak—OCTOPUS Designer, framework for the decision support problem manipulation with components *DeMak* (Δ and Σ modules) and *DeView* (Visualization Γ modules), will be presented. The work on *DeMak* started back in 1990 (see Zanic et al., 1992) but recently it has been redefined to enable easy implementation of all new developments, together with the flexible graphical user interface which enables easy problem definition and problem solving as well as visualization of the output and the final design selection.

The whole framework is programmed in VS2008 in .NET 2.0 technologies using several computer languages (C#, FORTRAN and C++).

The main components of OCTOPUS Designer are DeMakGUI (see Fig.1), DeMakMain and DeModel. DeMakGUI and DeMakMain are problem (i.e. analysis model) independent. DeModel component wraps given User Model computation component Ψ (e.g. full MAESTRO code for structural problems). It gives prescribed interface from the input in modules (**Φ, ε**), for structure and loads, to the calculation modules in Ψ. This enables communication between User Model and User Model Independent components. DeMak-Main is the main component that encapsulates

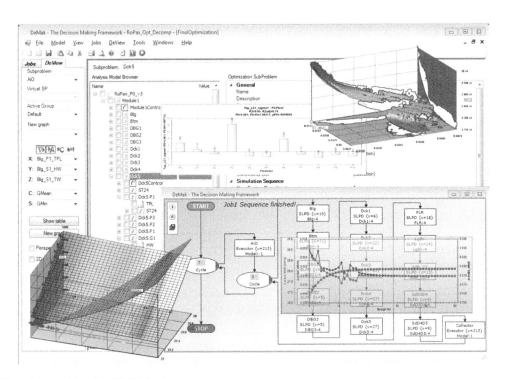

Figure 1.   DeMak—OCTOPUS Designer.

functionalities necessary for solving DSP Problem. The corresponding classes in DeMakMain (see Fig.2) are: *DeMakProblem* class which contains all jobs (*DeMakJob*) which the designer has to perform during design of a certain model based on modules Φ - Ω implemented using prescribed *IAnalysisModel* interface. Currently designer can perform two types of jobs, which are defined in the classes *DOEBasedJob*, and *OptimizationJob*. *OptimizationJob* can be *SimpleOptimi-zationJob* or *SystemOptimizationJob*. *SimpleOptimization-Job* is intended for simple concept phase optimizations without decomposition and hybrid solvers. Basically it contains one instance of *Optimiza-tionSubProblem* class (explained in sequel), while *SystemOptimizationJob* class contains the definition of all fields and methods necessary to enable optimization of the real ship/shipping problems which demand decomposition, hybrid solvers and some advanced logic. More elaborate description of *SystemOptimizationJob* implementation can be found in (Prebeg et al., 2009).

DOEBasedJob, where DOE stands for Design of Experiments, is an abstract class which contains fields and methods necessary for all types of design experiments jobs. It contains an instance of selected Design Experiments plan from appropriate DOE class. *RobustnessJob* was the first active class that inherits *DOEBasedJob*. Its purpose is robustness calculation for criteria (e.g. life cycle cost, yard costs), based on changes in uncontrollable parameters (see (Zanic et al., 2010)).

Surrogator class executes simulations prepared by appropriate DOE plan, and saves all the data

necessary for generation of surrogate of each selected criteria by the appropriate surrogate modeling technique. This includes observed parameters and criteria values for each executed experiment, as well as all mathematical manipulations like normalization and denormalization which needs to be done in order to generate efficient surrogate models. Methods *CreateSurrogateModel( )* or *UpdateSurrogateModel( )* if the surrogate models were already prepared for the previous set of experiments data, generates/updates surrogate model based on the experiments data. *Predict*() method predicts approximated criteria value on the non-observed location using previously created surrogate models of that criteria. It is important to mention that the *Surrogator* besides the fact that it is a *DeMakJob*, it also implements the *IAnalysisModel* interface, which enables treatment of a *Surrogator* instance as any other analysis model in Optimization Jobs.

In order to enable parallel (concurrent) execution of solution sequences, the new OCTOPUS Designer environment includes asynchronous parallel task manager (DeCluster) based on .NET Remoting (see (Prebeg, 2011)), replacing Meiko Computing surface application of parallel algorithm on 40 transputers (Zanic et al., 1993).

3 STRUCTURAL MODEL, LOADS AND STRUCTURAL CONSTRAINTS

Post-Panamax cruise ship, already used in study done by Romanoff et al. (2013) has been chosen

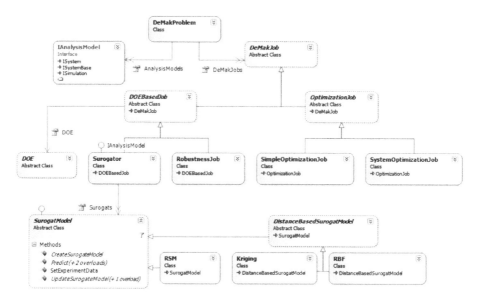

Figure 2. DeMakMain basic classes.

42

to demonstrated multi-objective scantling opti-mization procedure. Ship has following main par-ticulars: L = 275 m, B = 42.6 m and H = 43.75 m. The web frame spacing is w = 2730 mm with pil-lar on each web frame and every fourth web frame in recess area. The loads components (hull girder bending moments and local pressures) have been determined according DNV Rules (2007).

3.1 Generic 3D FEM model

Generic FE model is presented in Fig. 3 and can be easily generated. The model length should extend approximately over the full length and depth of the ship, while the full or half-breadth modelling depends upon the degree of structural symmetry. All the effective longitudinal material was included in the coarse mesh model using orthotropic panel approach length. Similarly transverse structure (i.e. web frames, watertight and fire bulkheads) are represented in the model. Plated areas such as decks, outer shell and bulkheads were repre-sented by the special stiffened shell macroelements. MAESTRO stiffened macroelement (MAESTRO, 2016) uses the NASTRAN type QUAD4 4-node shell elements enhanced with stiffeners in their proper geometrical position regarding axial and bending energy absorption. Primary transverse deck frames/girders and pillars were modelled with beam elements.

The primary objective of using generic model is to provide structural designers with the fast but accurate enough calculation of primary hull girder stress distributions over ship height by including all relevant 3D effects and also for bottom/deck gril-lage calculation and determination of deck trans-verse/longitudinal girders and pillars. In general, generic models on the topological level can ensure rapid exploration of different concepts by allow-ing changes of previously identified topological parameters (eg. size of side openings, size/stiffness of ventilation tubes, the geometry of superstruc-ture, etc.) with the goal to simultaneously generate acceptable structural layout and feasible structural design.

3.2 Loads

Deck loading has been implemented along generic model with deck pressure as specified in Fig.4.

Two global load cases have been examined:

- LC1-Combination of rule hogging wave bend-ing moment and the maximum still water bend-ing moment that gives maximal longitudinal and shear stresses ($M_{Hogg, tot}$ = 8 420 230 kNm).
- LC2-combination of rule sagging wave bending moments and minimum hogging still water bend-ing moment that can result in potential buckling problems in superstructure decks ($M_{Sagg, tot}$ = -4 427 070 kNm).

FE model was balanced in the quasi-static position (by changing heave, trim and roll using

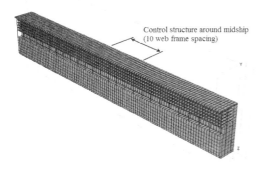

Figure 3. Full ship/half breadth generic 3D FEM model.

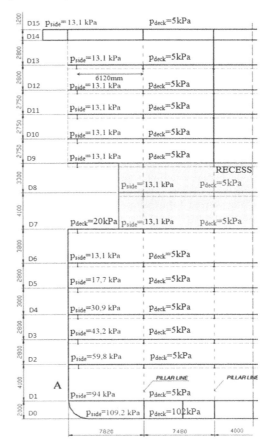

Figure 4. Midship geometry with decks loads.

Table 1. MAESTRO structural constraints.

Item	Limit state	Description	Safety factor, γ_{DNV}
1	PCSF	Panel Collapse—Stiffener Flexure	1.04
2	PCCB	Panel Collapse—Combined Buckling	1.04
3	PCMY	Panel Collapse—Membrane Yield	1.23
4	PCSB	Panel Collapse—Stiffener Buckling	1.04
5	PYTF	Panel Yield—Tension Flange	1.04
6	PYTP	Panel Yield—Tension Plate	1.23
7	PYCF	Panel Yield—Compression Flange	1.04
8	PYCP	Panel Yield—Compression Plate	1.23
9	PSPB	Panel Serviceability—Plate Bending	1.00
10	PFLB	Panel Failure—Local Buckling	1.00

MAESTRO hydrostatic balancing option) to achieve minimum reaction in the artificial supports.

3.3 Structural constraints

Structural constraints that include collapse and serviceability constraints (buckling and yield) are in-built in MAESTRO software and adjusted to fit DNV rules are given in Table 1. Their general form reads:

- $g_i(\mathbf{x}^{Total}) \geq 0$, $i = 1,\ldots,n$ constraints

For the purpose of the presentation of results, whose normalize form of g(x) is needed, the strength ratio R is defined as follows:

$$R(\mathbf{x}) = Q(\mathbf{x}) / Q_L(\mathbf{x});$$

where: $\mathbf{x}$ is vector of current values of structural descriptors including scantlings, $Q(\mathbf{x})$ is load effect and, $Q_L(\mathbf{x})$ is its limit value for particular failure mode. Failure criterion is given by:

$\gamma R(\mathbf{x}) < 1$ where: γ is prescribed safety factor or in normalized form using 'adequacy parameter' the constraint now reads g(R(x)), where

$g(R(\mathbf{x})) = (1 - \gamma R) / (1 + \gamma R)$; note: $-1 < g(R(\mathbf{x})) < 1$.

The structural adequacy parameters have to be equal or greater than zero, or $g(R(\mathbf{x})) \geq 0$, to satisfy necessary structural strength requirement. Constraints regarding structural strength were used in design procedure via safety factors defined according to DNV Rules.

4 STRUCTURAL OPTIMIZATION SETUP

Multiobjective structural optimization is performed with the minimization of weight and vertical centre of gravity position as objectives. In-house design support environment OCTOPUS Designer—DeMak has been used for the task with an integrated MESTRO structural analysis for stress calculation and strength adequacy checks.

MOPSO Hypercube multiobjective algorithm was used for the Pareto front generation.

Design variables were plate thickness and scantlings of longitudinal stiffeners. Web frame and stiffeners spacing were keep fixed during optimization process.

MAESTRO criteria combined with DNV safety factors, as specified in Table 1, were used as structural constraints. Based on calculated stiffened panel adequacy values two global deterministic safety measures are calculated for usage in optimization process. The first is the cumulative sum of the all adequacy criteria that are less than 0.05 (abbreviation used for this measure in optimization diagrams is GMean):

$$g_{<0.05} = \sum_{i=1}^{n_p} \sum_{j=1}^{n_c} g_{ij}\Big|_{<0.05} = \text{GMean}; \quad (1)$$

The second is average value of the lowest (worst) 5% of adequacy criteria (abbreviation used for this measure in optimization diagrams is GMin):

$$g_{min\,5\%} = \frac{\sum_{i=1}^{n_{5\%}} g_{ij}\Big|_{5\%}}{n_{5\%}} = \text{Gmin}; \quad (2)$$

5 STRUCTURAL OPTIMIZATION RESULTS

Pareto frontier with 22 non-dominated projects was generated. On Fig.5 2D Pareto frontier is presented with the marked prototype (P0), minimum weight (minWGT) and vertical center of gravity (minVCG) projects, while on Fig.6 3D Pareto frontier is presented with overall structural safety measure GMIN presented on vertical axis. Structural mass is presented as mass of control structure of 10 web frame spacing amidship with transverse structure included.

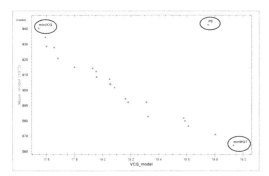

Figure 5. 2D Pareto frontier with non-dominated variants.

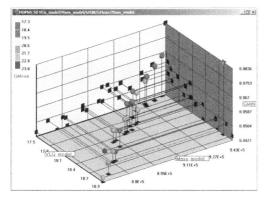

Figure 6. 3D Pareto frontier with non-dominated variants.

It can be identified that both extreme projects on Pareto frontier *minWGT* and *minVCG* have smaller mass then prototype (*P0*). For the *minVCG* structural mass has been just slightly reduced 0.15 t/m, while for the *minWGT* reduction is significantly higher, about 3.26 t/m or 9.3%. Vertical centre of gravity (VCG) for *minWGT* variant is 0.2 m higher (1%) compare to prototype, while for *minVCG* is reduced for 1.2 m or 6.5% compare to *P0*.

MinWGT variant has about 8.5% lower mass and 7% higher VCG compare to *minVCG* variant.

Summary of obtained output results of examined three variants are presented in Table 2.

On Fig.7 plate thickness has been presented for all three examined variants. It can be seen that larger differences are in lower hull structure while in superstructure relative similar scantlings have been obtained for all three variants.

MinWGT variant has more uniform distribution of structural mass along ship height which resulted in much higher VCG, while *minVCG* variant obtained feasible structure solution with increased structural thicknesses of double bottom

Table 2. Summary of results for three examined variants.

Variants	VCG (m)	Mass of control structure (t)	Unit mass (t/m)
P0	18.6	945	34.61
minVCG	17.4	941	34.46
minWGT	18.8	865	31.35

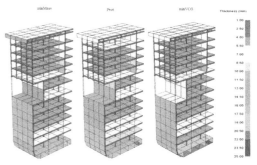

Figure 7. Comparison of the thickness of plating for the three relevant designs.

and lower decks which in parallel reduces VCG but increase mass.

6 STRUCTURAL RESPONSE COMPARISON

Three examined structural variants have been compared w.r.t. 1) stiffness through obtained vertical displacements, 2) primary stress distribution over ship height and 3) structural feasibility of stiffened panels.

This study confirmed results obtained by Romanoff et al. (2013) that decks bend with different curvature, so the different vertical displacements have been identified for each deck as presented on Fig.8 for *minWGT* variant. Similar behaviour is present for each variant. Prototype structure has higher stiffness, while *minWGT* are most flexible (*minVCG* is between) as presented on Fig.9. *minWGT* has in average 15% higher displacement compare to prototype structure(*P0*).

Primary stress distribution over ship height is presented on Fig.10 for LC1 and points around CL for all examined variants.

MinWGT variant expressed about 10% higher stresses in double bottom structure compare to *minVCG*. In recess area all variants expressed shear leg effect with reduction of stresses side shell to CL.

Plot of compliment with structural feasibility criteria has been presented on Fig.11 for LC2 (Sagg

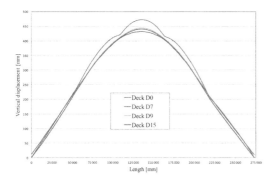

Figure 8. Vertical displacement of different decks for *minWGT* variant.

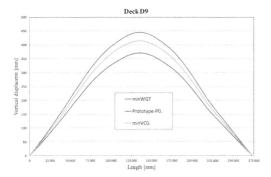

Figure 9. Vertical displacement of Deck 9 for different design variants.

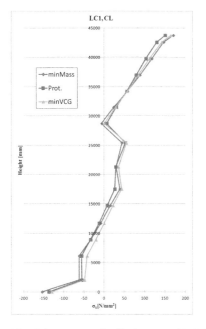

Figure 10. Primary stress distribution over ship height.

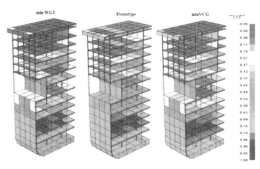

Figure 11. Structural feasibility criteria in LC2.

case). Upper decks are satisfied w.r.t buckling criteria just over minimum allowable level (g > 0).

7 CONCLUSIONS

Multi-objective scantling optimization (with Weight & VCG as objectives) of post-Panamax cruise ship has been presented as an example. Two Pareto optimal designs marked as *minWGT* and *minVCG* have been compared and differences in structural response of both variants and prototype structure were discussed.

ACKNOWLEDGEMENTS

Thanks are due to prof. Vedran Zanic for his long-term support, enthusiasm and efforts to introduce decision support techniques to a ship structural design practice.

REFERENCES

Andric, J. and Zanic, V. (2010), The global structural response model for multi-deck ships in concept design phase, *Ocean Engineering* 37 (2010): 688–704.

Andric, J., Prebeg, P., Piric, K. (2017), Influence of different topological variants on optimized structural scantlings of passenger ship, Progress in the Analysis and Design of Marine Structures (MARSTRUCT 2017) Guedes Soares & Garbatov (Eds), Vol.1. pp. 351–358, Lisbon, Portugal.

Bacicchi, G., Nevierov, A. (2000), Use of Alternative Structural Materials in the Superstructure of Cruise Vessels, International Conference on Ship and Shipping Research- NAV 2000, Venice, Italy.

Caprace, J.D., Bair, F. & Rigo, P. (2010), Multi-Criteria Scantling Optimization of Cruise Ships, Ship Technology Research, Schifftechnik, 57(3): 210–220.

Det Norske Veritas, (2007), Direct Strength Analysis of Hull Structures in Passenger Ships. Classification Notes No.31.8., October 2007.

Gudmunsen, M.J., (1995), Some Aspects of Modern Cruise Ship Structural Design. Cruise Shipbuilding in the USA, Lloyd's Register of Shipping, May 1995.

Hughes, O.F. (1983). Ship Structural Design: A Rationally-Based, Computer-Aided, Optimization Approach. John Wiley&Sons, NY.

Hughes, O.F, Mistree, F, Zanic, V. (1980), A Practical Method for the Rational Design of Ship Structures. Journal of Ship Research, 24(2):101–113.

ISSC (2003),Technical Committee II.1, Quasi-Static Response, Proceedings of the 15th International Ship and Offshore Structures Congress, Vol.1, San Diego, USA.

ISSC (1997), Technical Committee II.1. Quasi-Static Response. Proc. of the 13th International Ship and Offshore Structures Congress Trondheim, Norway.

Jancijev, T., Zanic, V., Andric, J. (2000), Mathematical Models for Analysis and Optimization in Concept and Preliminary Ship Structural Design, Proceedings of IMAM 2000 Congress, pp.15–22, Napoli, Italy.

Levander, K. (2006), The Limits to Growth in Cruise Ship Size, International Marine Design Conference, IMDC 2006, pp. 207–230, Ann Arbor, USA.

Lillemäe-Avi, I., Remes, H., Dong, Y., Garbatov, Y., Quéméner, Y., Eggert, L., Sheng, Q., Yue, J. Benchmark study on considering welding-induced distortion in structural stress analysis of thin-plate structures, Progress in the Analysis and Design of Marine Structures (MARSTRUCT 2017) Guedes Soares & Garbatov (Eds),, Vol.1. pp.387–394, Lisbon, Portugal.

MAESTRO Version 11.3. (2016), Program Documentation, Advanced Marine Technology Center, DRS Technologies Inc.

Naar, H., Varst, P., Kujala, P. (2004), A theory of coupled beams for strength assessment of passenger ships. Marine Structure, 17(8): 590–611.

Prebeg, P., 2011, "Multi-criteria design of complex thin-walled structures," Ph.D Doctoral dissertation, University of Zagreb, Zagreb.

Prebeg, P., Zanic, V., and Kitarovic, S. (2009), "The Design Methodology with the Sequencer for Efficient Design Synthesis of Complex Engineering Systems," 50th AIAA/ASME/ASCE/AHS/ASC Structures, Structural Dynamics, and Materials Conference, USA

Richir, T., Caprace, J-D., Losseau, N., Bay, M., Parsons, M.G., Patay, S., Rigo, P.(2007a), Multicriterion Scantling Optimization of the Midship Section of a Passenger Vessel Considering IACS Requirements. Proceedings of PRADS 2007 Conference, pp.758–763, Huston, USA.

Richir, T., Losseau, N., Pircalabu, E., Toderan, C. &Rigo, P.(2007b), "Least Cost Optimization of Large Passenger Vessel", Proceedings of the 1st International-Conference on Advancements in Marine Structures, 2007, pp. 483–488, Glasgow, UK.

Romanoff, J. 2011. Interaction between laser-welded web-core sandwich deck plate and girder under bending loads. Thin-Walled Structures 49: 772–781.

Romanoff, J., Remes, H., Varsta, P., Jelovica, J., Klanac, A., Niemela, A., Bralic, S., Naar, H. (2013), Hull-superstructure interaction in optimized passenger ships. Ships and Offshore Structures, 8: 612–620.

Zanic, V. (2013), Methods and concepts for the multi-criteria synthesis of ship structures, Ships and Offshore Structures, 8(3–4): 225–244.

Zanic, V., Andric, J., Prebeg, P. (2013), Design Synthesis of Complex Ship Structures, Ships and Offshore Structures, 8(3–4): 383–403.

Zanic, V., Andric, J., Prebeg, P. (2007), Decision Support Methodology for Concept Design of Multi-Deck Ship Structures. Proceedings of PRADS 2007 Conference, pp.468–476, Huston, USA.

Zanic, V., Das, P.K., Pu, Y., and Faulkner, D. (1993), "Multiple criteria synthesis techniques applied to reliability based design of SWATH ship structure," Integrity of Offshore Structures, D. Faulkner, Incecik, Cowling, ed., EMAS Scientific Publications, Glasgow, pp. 387–415.

Zanic, V., Jancijev, T. (1986), Structural design and analysis of night ferry 'Amorella' with 2200 passengers and 600 cars. Technical report for BRODOSPLIT Shipyard, University of Zagreb.

Zanic, V., Kitarovic, S., and Prebeg, P. (2010), "Safety as Objective in Multicriterial Structural Optimization," 29th International Conference on Ocean, Offshore and Arctic Engineering, ASME, Shanghai, China, pp. 899–910.

Parametric equations for the design of a logistic support platform

T.S. Hallak & M. Ventura
Centre for Marine Technology and Ocean Engineering (CENTEC), Instituto Superior Técnico, Universidade de Lisboa, Lisbon, Portugal

C. Guedes Soares
Centre for Marine Technology and Ocean Engineering (CENTEC), Instituto Superior Técnico, Universidade de Lisboa, Lisbon, Portugal. Also, Subsea Technology Laboratory, Ocean Engineering Department, COPPE, Federal University of Rio de Janeiro, Rio de Janeiro, Brazil

ABSTRACT: The discoveries of Brazil's pre-salt layer brought to light the logistic problem of staff transfer to fields considerably far from the coast. Several additional fields—distant up to 300 km from the coast—will be explored, thus a solution for the problem proves necessary. This paper analyzes the alternatives for a logistic support platform, to serve as a hub in the staff transfer process. Since accommodation units are usually used in the offshore industry as an appendix of production/drilling rigs, the hub constitutes an innovative design, and introduces new challenges in the people transfer interfaces. A market analysis of accommodation units was performed in order to evaluate the tradeoffs between constructing, upgrading and renting a platform, and estimate expenditures for further financial analysis. Extensive data gathering on accommodation units in the various configurations has also been conducted, allowing statistical analyses and the obtainment of indices and regressions relating the design parameters.

1 INTRODUCTION

Due to the elevated oil prices' recovery during the post-crisis, i.e. from 2011 until 2014 and to the continuous draining of oil resources located next to shore, an increase interest on deep waters' exploration and production (E&P) units has been observed recently.

Contemporarily, E&P advances over Brazil's pre-salt layer oil reservoirs and the amount of offshore units operating considerably far from the coast there (up to 300 km) is expected to increase significantly in the next years. The time for a watercraft to reach a production site is considerably higher when compared with a helicopter. As a matter of fact, staff transfer between coast and production rigs is nowadays performed in Brazil mainly via helicopters. Most of the personnel working in Bacia de Campos units (responsible for around 80% of Brazil's oil production) fly from Macaé airport, located by the coast in Rio de Janeiro's state.

This brings to light the logistic problem of staff transfer to distant production rigs, because helicopters have limited autonomy and the people transfer interfaces between watercrafts and production rigs (e.g. offshore basket transfer) allow only low-flow people transfer and correspondent smaller time windows for operation due to environmental conditions.

At the same time, Brazilian offshore industry is already one of the fittest in what concerns the commissioning of state-of-the-art accommodation units (or flotels)—about ten different flotels in monohull or semi-submersible configurations had operated in Brazilian waters, mainly in Bacia de Campos, starting from 2007. These units however were commissioned for particular purposes not related to the staff transfer logistics. An innovative and efficient solution for the logistic support function started only more recently justifiably being sought. A feasible design solution for the logistic support function, currently being studied by several authors, is a logistic support platform, very much like a flotel, located at an intermediate point between the coast and the production fields, having room for several helidecks—for helicopters will transfer people between the logistic support platform and the production rigs—and likely capability of connecting itself to High Speed Vessels (HSVs)—since vessels may transfer people between the coast and the platform. Another possibility is the use of helicopters during all people transfer process and the logistic support platform as an intermediate site for the refueling of helicopters.

Innovative configurations that seek to fulfill all mentioned logistic requirements were initially proposed by (Oliveira et al., 2013) and (Malta et al., 2014). The proposed engineering solutions differ from consolidated designs especially on the people transfer interface between the platform and HSVs. The prior proposed a monocolumn design solution, similar to a spar platform, but characterized by a small draft, big diameter and an internal moonpool connected to the sea via an opening at the free surface's draught. The latter thereon proposed another ambitious design, that of a semi-submersible with an internal dock, i.e. an uncovered space between the pontoons. Both design solutions were motivated by the inspiration of sheltering waters within the flotel's extent in order to accommodate the HSVs, and thus allow on demand staff transfer in and out the platform. The monocolumn—without wave suppressing devices—proved to be an unfeasible solution (unless considerably modified), amplifying wave amplitude inside the moonpool instead of mitigating it. The semi-submersible with internal dock on the other hand was numerically and experimentally analyzed and a couple of solutions, characterized by different wave suppressing devices, were pointed as effective solutions. Anyway, the implementation of any of these design solutions still waits for deeper studies on the hydrodynamic characteristics of the platform.

In what concerns the methods for designing such units, several authors have been interested on the subject, but this will be further explained in detail in chapter 3. As a general rule, published research neither enters deeply on the study of offshore accommodation units' market, nor seek to find relevant qualitative correlations between flotels' parameters—which may be relations between platform's geometry, costs, weights, capabilities, etc. This work thus seeks to clarify better the qualitative advantages and disadvantages of every flotel configuration for the offshore industry, as well as offer quantitative correlations between flotels' design parameters; but also discuss the behavior of the offshore accommodation units' market. The final goal is to consolidate a basis for the design of a logistic support platform which solves the staff transfer logistical problem.

2 MARKET ANALYSIS

In the last 10 years, oil prices changed abruptly in what can be seen as two cycles of rise and fall. While in the middle of 2008 oil prices reached the historical record of 155 USD per barrel, a sudden drop was observed due to the economic crisis, and half a year later, the price fell below 50 USD per barrel. Afterwards oil prices recovered relatively fast. From the beginning of 2011 until the middle of 2014, prices remained at a plateau of around 100 USD per barrel. A sudden drop was observed right afterwards, with prices reaching at some

points less than 30 USD per barrel. In the last year (from mid-2016 to mid-2017), prices have remained somewhat stable at around 50 USD per barrel.

It is important to understand the dynamics of the oil prices and the relations between such prices and the offshore accommodation units' market. The first of the abovementioned price drops happened mainly due to the 2008 economic crisis, for oil demand is known to be price-inelastic and somewhat demand-elastic, thus economies' slowdown considerably affected the oil prices. After the recovery of oil prices, investments on the E&P industry led the oil market to an excess of supply. Since oil production countries, including OPEC members, initially were mostly not willing to reduce their output, a new drop in oil prices was observed. OPEC's late 2016 agreement (171st Meeting held in Vienna) was extended half a year later (172nd Meeting held in Vienna) in order to maintain the price's stability; the production adjustments will so be ruling at least until the end of March 2018.

The demand for offshore accommodation, however, depends basically on offshore investment and production, not directly on oil prices, for the accommodation units are connected to drilling and production rigs—sometimes only for maintenance or upgrading purposes. Thus, the utilization rate of accommodation units followed a trend similar to the variation of oil prices, but with a time lag of between 1 and 2 years, which is explained by the fact that commissioning contracts are signed with such an anticipation time before operation begins. Indeed, utilization rates dropped considerably around 2011 and again around 2016, recently causing offshore accommodation companies to scrap a few units in order to stabilize flotels' supply and demand. The total amount of offshore accommodation units is still expected to decrease in the next years.

Contract values, on the other hand, follow a trend more similar to the variation of production rates—also with a time lag of 1 to 2 years—but fluctuating considerably. Thus, contract values have dropped only slightly after the 2008 economic crisis, but not right after the latter drop in oil prices, after all offshore output took more time to suffer any significant change—there being cases, in countries like Brazil, in which the output never actually decreased from one year to the other. The current oil prices, plateaued at a lower level, are still not expected to increase. This is causing offshore accommodation companies to reduce OPEX, thus creating the expectation that contract values will decrease in the next couple of years.

Contract values are usually firmed in PDPR (per day *pro rata*), according to a time charter or a bareboat contract; and may have options after the initial period of operation. Contract values fluctuate considerably and depend strongly on the operation site, flotel's capabilities and amount of time operating. For a set of DP semi-submersibles studied, while contracts for operation in the North Sea are firmed at around 300 kUSD per day (Norway) and between 250 and 300 kUSD (UK), contracts for similar flotels are around 150 kUSD in Brazil—all time charts. Similar flotels will be commissioned for 60 to 80 kUSD in Mexico, but according to bareboat practice. The highest contract values observed within the period covered by the data compiled were: 395 kUSD per day for a state-of-the-art semi-submersible in Brazil; and 357 kUSD per day for a DP semi-submersible in Norway, a contract comprising only 2 weeks of operation.

No proportionality rule can be derived when comparing the abovementioned contract values and the OPEX for the same flotels. Anyway, OPEX values were around 90, 85 and 70 kUSD per day in Norway, Brazil and UK, respectively, for the same DP semi-submersibles back in 2012. These values are now decreasing considerably due to OPEX cuts in order to stabilize the market in the current period of low oil prices. OPEX depend strongly on the existence of DPS, since oil consumption is considerably higher in those cases. As a matter of fact, back in 2009 when oil prices oscillated between 50 and 90 USD per barrel, OPEX for DP semi-submersibles were 10 to 15 kUSD (per day) higher when compared to non-DP units. In 2012, oil prices were somewhat stable at around 100 USD per barrel and the difference was about 20 kUSD per day. All this indicates that an index of about 200 times the barrel price seems quite reasonable as a daily operational expenditure related to the usage of a DPS in a semi-submersible.

On the other hand, information on CAPEX values for offshore accommodation units is extremely hard to find, so roughly a dozen values regarding total construction costs could be obtained. Regarding loans, bonds and refurbishment, only another dozen of values were available. Nowadays, construction costs for a DP semi-submersible are normally between 250 and 350 Mill USD.

During the compilation of data it has also been observed that no flotels actually have a designed capability for connection with HSVs. Moreover, only one of the flotels analyzed could operate with two helicopters simultaneously—a semi-submersible. Thus, no flotels currently in the market are capable of solving high demanding people transfer process for the exploration of Brazil's pre-salt layer. Moreover, in order to comply with the standards of offshore aviation, it is most unlikely that mono-hulls could operate with three or more helicopters. Barges, on the other hand, have no sufficient seakeeping performance for the Brazilian waters. Taking all this into account, it is clearly unavoidable that the logistic support platform must be designed

from scratch, and it turns out that semi-submersibles are indeed the most likely to comply with the design requirements. The connection to HSVs remains a critical issue, but the semi-submersible with internal dock conceived by Malta et al. (2014) is currently the most advanced proposed solution and also an inspiration.

Anyway, the design of an accommodation unit from scratch is not justified in most cases, especially in a period where oil prices are low and not expected to increase, for investments in the E&P industry decreases considerably and the high CAPEX associated to constructing (or even upgrading) an accommodation unit makes unlikely that investors will put money on such a business. A look into the utilization rate of flotels is another relevant insight when one is deciding if a new flotel should be constructed or not. Utilization rates nowadays are indeed low, pointing to a high risk of having the platform laid up.

Lastly, as a general rule companies that hold accommodation units do not hold the production/drilling units that are connected to them, so the operation of accommodation units is usually done according to commissioning contracts between different companies and for a period of time that usually goes only from a few months to a few years. Thus, if a company needs to commission a flotel only for a specific task, the risks of constructing it and after sell it or handle to rent it out for several years are hardly ever justified.

3 LOGISTIC SUPPORT PLATFORM DESIGN

In order to design offshore accommodation units, the first step is to compile all possible relevant information related to them, in terms of parametric relations between design variables, as no open source database exists concerning such data.

Starting from construction and operation site, it has been verified that most of the offshore accommodation units were built in China or Singapore, independently of vessel's configuration, but also in other Asian countries (Japan, Indonesia, and South Korea). The Swedish shipyard GVA also built a few units in the 80 s.

Flotel's possible operation sites are strongly dependent on flotel's configuration. In order to operate in harsh environments, for instance, it is necessary to have great seakeeping performance, thus only semi-submersibles are usually operating in North Sea's Norwegian and British sectors. A few monohulls however have also operated in the North Sea (Danish sector and a state-of-the-art monohull for 600 POB in UK). Other regions more kin to semi-submersibles are the Gulf of Mexico and Australia due to recurrent hurricanes and cyclones.

Brazilian waters are considered benign, thus both semi-submersibles and monohulls may operate there. The only CSS in the market had also operated in Brazil—Bacia de Campos. Lastly, barges do not have sufficient seakeeping performance to operate in any of the abovementioned seas and are usually not equipped with DPS, thus they only operate in mild seas or close to the shore (Malaysia, Caspian Sea and West Africa). On the other hand, due to their relative low cost, they are actually the preferable solution for these operation sites.

Recent research done on the design of an accommodation unit was published starting from Pardo & Fernandez (2012), where a preliminary and strictly qualitative study on flotels was performed. It includes even accommodation units not designed for the offshore industry and coastels, i.e. accommodation units located by the coast and therefore built in accordance with different requirements, having to operate under different regulations, sometimes having also to meet luxury standards.

For what concerns effectively the design phases of an accommodation unit, several authors approached the subject, e.g. (Melo et al. 2015) and (Moyano, 2016)—the former concerning a monohull accommodation for the Brazilian waters and the latter concerning an accommodation rig that seeks to solve the already mentioned logistic complication of staff transfer process from land to production rigs considerably far from the coast. Research on the dimensioning of offshore platforms has also been performed, e.g. by (Sharma et al., 2010), where different optimization methods are discussed for the dimensioning of a semi-submersible platform. Their approach however is a general one and it does not consider the particularities of accommodation units.

An example of such a method is thereon given by (Gallala, 2013), where a non-linear optimizer was developed on Excel® to determine the main dimensions of a four-leg semi-submersible platform, again not an accommodation unit. Whilst the spreadsheet offers a straightforward way to solve the problem, the optimized geometries resulted from the method proved to be feasible and cost-efficient. A set of geometric characteristics are kept variable and independent; the optimizator then seeks to minimize the structural weight of the platform (related to construction costs) whereas other objectives (variable deck load, open deck area and motion characteristics) are handled via constraints and inputs.

The referenced bibliography serves as a starting point for the effective design of the logistic support platform. But it was only the compilation of data that allowed high fidelity comparisons between different configurations, and offered quantitative relations between platform's parameters in a way

to be included in methods such as the one proposed by Gallala (2013), thus permitting the effective design of the unit.

4 DEVELOPMENT OF PARAMETRIC RELATIONS

An extensive research on web, magazines, brochures and catalogues has been performed and datasheets were developed in order to further analyze all the compiled information. It was possible to look into a total amount of 72 floating accommodation units, 27 of them semi-submersibles—including 1 non-conventional compact semi-submersible (CSS), 27 barges, 17 monohulls.

Database includes all relevant information that could be found, which includes:

- Vessels' particulars (IMO, construction and operation sites, year of construction/update, etc.)
- POB and amount of crew members
- Main dimensions (total length, breadth, draught, depth, pontoons/columns' dimensions, etc.)
- Operational and survival air gap for semi-subs
- Weights and volumes (gross tonnage, displacement, deadweight, lightweight, etc.)
- Capacities (water and fuel storage, emergency power, generation power, etc.)
- Superstructure's inner areas (accommodation areas, recreational, hospital, workshop, etc.)
- Open area on deck, and open storage area
- Information related to helidecks, cranes, gangways, Dynamic Positioning System (DPS) and other systems
- Prices and costs (construction cost and capital expenditures, operational expenditures, contract values, etc.)

The scarcity of information regarding prices and costs and vessels' lightweight and operational displacement are the most noteworthy.

Available deck plans are also scarce. Only 10 out of the 72 units could be analyzed in this level of detail. Anyway, several relevant data points could be obtained only via deck plans' analysis, such as superstructures' inner areas, mess room capacity and elements' dimensions for semi-submersibles' hulls (pontoon and columns' dimensions).

The compilation of data allowed both qualitative and quantitative analyses and comparisons. The scarcity on prices and costs' information allowed only a more qualitative market analysis, as already explicated in chapter 2, but also some quantitative relations. In this chapter, however, significant regression curves were traced in order to correlate relevant flotel's parameters (dimensions, weights, capacities, etc.). These correlations introduce new quantitative information that may further be included in methods for the effective dimensioning of an accommodation unit. Thus, several plots were drawn. Figures 1 to 13 are graphical representations of these plots; most of the data was fitted to a linear model.

Since one of the basic requirements for an accommodation unit is the capacity of accommodation, it is important to relate vessel's parameters with POB capacity. It was found however that for both barges and semi-submersibles there's no strong correlation between vessels' dimensions and POB. Figure 1 for instance shows that semi-submersibles' total available area for topside is independent of POB capacity. Because the critical issue on a semi-sub is its operability, they are usually over dimensioned for the POB capacity and therefore include a large number of beds (at least 270, according to the database). Figure 2, on the other hand, shows that the same is not valid for monohulls, whereas monohulls' dimensions indeed increase with POB capacity.

Whilst vessels' operational displacement and lightweight are usually hard to find information, it is also interesting to related vessels' capacities with gross tonnage (GT), for this will be the best of the available estimates of vessels' size. Figures 3 and 4 thereon present the correlation between GT and

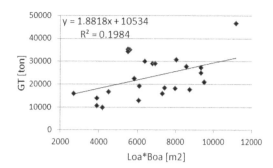

Figure 3. Regression for platform's gross tonnage estimate (only semi-submersibles).

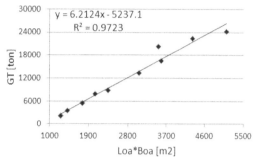

Figure 4. Regression for vessel's gross tonnage estimate (only monohulls).

parameter Loa*Boa (total length times moulded breadth) for semi-submersibles and monohulls, respectively. Whilst GT may be only roughly estimated for a semi-sub, the regression for monohulls is acceptable.

Offshore accommodation units' main purpose is to serve as accommodation in the strict sense; however they are normally operating side by side with another offshore unit (drilling or production rigs), connected via gangway, sometimes for maintenance/upgrading reasons in the latter. Thus, accommodation units are usually well equipped on the main deck with cranes, gangway, workshop, etc. A good estimation for the open deck area may be performed with aid of Figure 5, where a regression is traced for semi-subs and monohulls as a function of vessels' gross tonnage. Data analyzed from barges show that for this configuration, open area on deck is actually not correlated with vessel's dimensions, and may reach up to 70% of total available area on topside (Loa*Boa).

Figure 6 presents a graph for estimating the total generation power based on vessel's gross tonnage. Regression is quite significant for DP semi-submersibles and DP monohulls. Points related to barges on the other hand are sparse, anyway it is

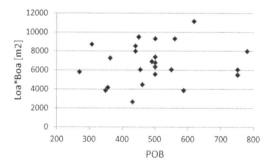

Figure 1. No correlation between platform's dimensions and maximum people on board (only semi-submersibles).

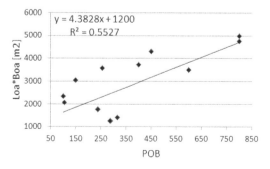

Figure 2. Correlation between vessel's dimensions and maximum people on board (only monohulls).

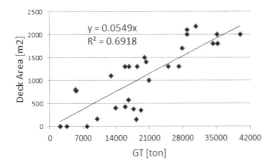

Figure 5.   Regression for vessel's open deck area estimate.

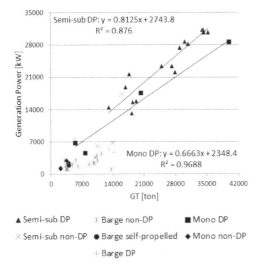

Figure 6.   Total power for every kind of accommodation unit.

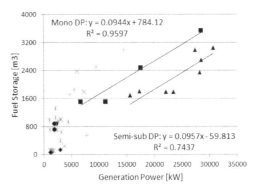

Figure 7.   Fuel storage for every kind of accommodation unit.

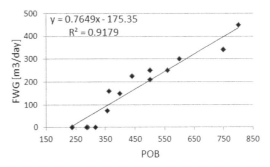

Figure 8.   Regression for fresh water generator's capacity.

now clear that barges require a far lower generation power, thus have relative lower operational costs due to to energy consumption. Figure 7 on its turn presents a graph for the fuel storage tank volume based on the total power. Again, regression lines were traced for DP semi-submersibles and DP monohulls.

Other vessel's capacities such as fresh water generation (FWG) and water storage may be estimated as function of POB, since water is crucial basically for the living quarters. Figure 8 also suggests that the cost of buying a fresh water generator is only justified for more than 350 people on board.

Although barges are usually not equipped with DPS, connection via gangway between semi-submersibles or monohulls and other offshore structures requires a good station keeping during operation. Thus, DPS of 2nd or 3rd generation are usually present in those configurations; this however does not exclude the usage of a mooring system, common to all offshore accommodation units.

DPS maximum power is well correlated with vessel's total power, as can be observed in Figure 10. An index of 75% (max DPS power divided by max generation power) is a good estimate for semi-submersibles. Monohulls, due to their hydrodynamic advantages, require lower DPS power, an index of 31% seems also reasonable for a first estimate.

Emergency power may also be estimated based on total power—it is usually between 3% and 6% for both semi-submersibles and monohulls—see Figure 11. An index of 5% is acceptable for a first estimate; however, values may reach up to 15% in some cases where generation power is low.

Finally, even though vessels' operational displacement is usually not an available information, a regression considering all configurations was traced and show reasonable linearity between displacement and gross tonnage (Figure 12). Vessel's deadweight, on the other hand, may be only roughly estimated as a linear function of gross tonnage (Figure 13).

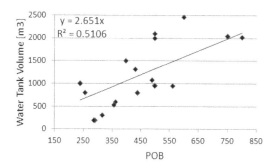

Figure 9. Regression for fresh water storage's maximum capacity.

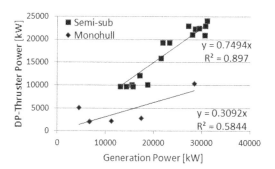

Figure 10. Correlation between dynamic positioning system's power and total power.

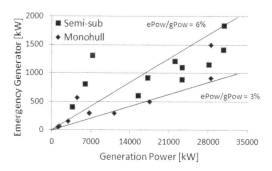

Figure 11. Correlation between emergency power and total power.

4.1 Superstructure's areas

From the available deck plans it was possible to separate inner superstructure areas as belonging to different kinds of facilities. The proposed division is into 9 kinds:

- Working Areas (e.g. offices, bridge)
- Machinery (generators room, A/C room)
- Accommodation (including private WCs)
- Bathrooms (restrooms, changing rooms)
- Accesses (corridors, stairs and lifts)

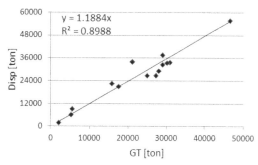

Figure 12. Regression for vessel's displacement estimate.

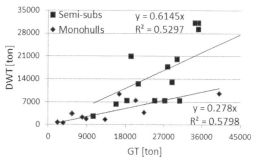

Figure 13. Regressions for vessels' deadweight estimates.

- Mess (including galley)
- Common (recreational areas, lobby)
- Services (laundry, storage)
- Hospital (or clinic)

For barges and semi-submersibles, it has been observed that some facilities may be allocated under the main deck, thus not belonging to the superstructure. This is mainly the case for the machinery area and some services areas.

Of course, most of the facilities are designed for the POB requirement, or similar requirement, e.g. the mess room that is dimensioned for the expected number of people eating simultaneously—usually around half of the POB requirement and area of ~1.7 m² for each eater. Total accommodation area on its turn is dimensioned based on the POB requirement with night stay; and rooms' average dimensions/capacity. It has been observed that single rooms are about 8 m², double rooms 12 m² and double bunk bed rooms 15 m². Moreover, linear regression analyses show that 1.12 m²/POB and 0.16 m²/POB are good indexes for services and hospital areas, respectively.

Machinery areas on the other hand depend strongly on vessel's mechanical/electrical apparatus, thus are better correlated with total electrical generation power instead of POB requirement. Figure 14 shows a significant trend for the machinery area estimation.

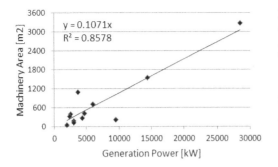

Figure 14.   Regression for machinery area estimate.

Some areas related to the specific kinds of facilities may be estimated as function of the total superstructure's inner area, this is mainly the case for accesses areas and services areas. The former has an index (access area / total superstructure's inner area) between 15% and 20%, and the latter around 10%. Surprisingly however, hospital area was the most precise empirical index, around 1.2%.

Data obtained also allowed the evaluation of the correlation coefficients regarding different kinds of facilities. It has been verified that the most correlated couples are mess room/common areas, common/accommodation areas and services/accesses areas. All correlations were negative, showing that usually there's a tradeoff between, e.g. more spacious accommodation rooms or bigger amount of recreational facilities.

A first estimation for total superstructure's inner area may also be convenient during the design process. Multivariate regression based on POB requirement and available area for topside has been performed, a useful formula for barges and semi-submersibles is:

$$SE_{area}\,[m^2] = 5.36 \cdot POB + 2.62 \cdot L_{OA}[m]B_{OA}[m] - 5760$$

where SE_{area} = total inner area of superstructure; POB = Maximum people on board (including crew); L_{OA} = topside's total length; and B_{OA} = topside's total breadth.

5   CONCLUSIONS

Compilation of available relevant data regarding offshore accommodation units has been performed. This allowed the study of the offshore accommodation units' market, including the description of the dynamics of such a market. It allowed also the execution of several analyses for obtaining significant regressions relating flotels' dimensions, capabilities, weights and other parameters mostly in the form of linear regressions and indexes. Superstructures' inner areas were also analyzed in order to perceive how the available area for each kind of facility is dimensioned. During the execution of this work, the greatest hindrance found was the scarcity of data, especially regarding weights and expenditures.

The design of a logistic support platform for the Brazilian waters is currently a study focus. All analyses performed allowed the comparison between different flotels' configurations, indicating which configurations are capable to operate there in compliance with the limiting design requirements of such a platform. Lastly, the quantitative analyses performed serve as basis for the effective design of such a platform, its hull and superstructure, as well as assign its necessary capabilities for proper operation, and estimate its costs.

ACKNOWLEDGEMENTS

This work has been supported by EMBRAPII-COPPE Unit—Subsea Technology, within the project "Subsea Systems", which is conducted in cooperation with COPPE (UFRJ) and is financed by PETROGAL Brazil. The third author holds a visiting position at the Ocean Engineering Department, COPPE, Federal University of Rio de Janeiro, which is financed by the program "Ciências sem Fronteiras" of Conselho Nacional de Pesquisa of Brazil (CNPq).

REFERENCES

Gallala J.R. (2013). Hull Dimensions of a Semi-Submersible Rig – A Parametric Optimization Approach. Master's thesis, NTNU, Trondheim, Norway.

Malta E.B., Ruggeri F., de Mello P.C., Vilameá E.M., de Oliveira A.C. & Nishimoto K. (2014). Semi-submersible Hub Platform with an Internal Dock Model Testing. Proc. 33rd International Conference on Ocean, Offshore and Arctic Engineering. OMAE2014-24418.

Melo J.V., Motta R.F. & Martins Filho P.D. (2015). Monohull flotel (Embarcação Flotel Monocasco in portuguese). Undergrad project, UFRJ, Rio de Janeiro, Brazil.

Moyano S.F.M. (2016). Design of a Logistic Hub Platform for Oil & Gas Production Fields. Master thesis in Naval Architecture & Marine Engineering, IST, Lisbon, Portugal.

de Oliveira A.C., Vilameá E.M., Figueiredo S.R. & Malta E.B. (2013). Hydrodynamic Moonpool Behavior in a Monocolumn Hub Platform With an Internal Dock. Proc. 32nd International Conference on Ocean, Offshore and Arctic Engineering. OMAE2013-10709.

Pardo M.L. & Fernandez R.P. (2012) Accommodation Vessels. Journal of Shipping and Ocean Engineering 2, pp. 327–339.

Sharma R., Kim T.W., Sha O.P. & Misra S.C. (2010). Issues in Offshore Platform Research – Part 1: Semi-submersibles. International Journal of Naval Architecture & Ocean Engineering 2, pp. 155–170.

Maritime Transportation and Harvesting of Sea Resources – Guedes Soares & Teixeira (Eds)
© *2018 Taylor & Francis Group, London, ISBN 978-0-8153-7993-5*

Practical ship hull structural scantling optimization using ABS-HSNC criteria

T. McNatt, M. Ma, J. Shaughnessy & K. Stone
Advanced Marine Technology Center, DRS Technologies Inc., Stevensville, USA

ABSTRACT: Naval ships require reduced structural weight while meeting safety and service life require-ments. These objectives can be met using state-of-art structural design and optimization procedures that integrate the complexities of naval ship structural arrangements, loading, and performance criteria. Ship structural optimization is a mixed discrete-continuous design problem constrained by buckling of struc-tural elements and material strength, and involves the optimization of a large number of variables (con-tinuous/discrete) such as plate thickness, scantlings of stiffeners and frames, and the (discrete) number of stiffeners and frames. This paper presents a method to optimize global hull girder scantlings using three objectives: structural weight, vertical center of gravity, and structural safety. The mid-body of a 150-meter notional frigate is optimized to meet High Speed Naval Craft (HSNC-ABS) criteria. The results demon-strate that the method effectively performs limit-state based ship structural optimization with multiple objectives, using an integrated set of tools and computer based methods.

1 INTRODUCTION

Naval ship structural design, like many real world problems, involves simultaneous consideration of several objectives, such as weight, safety, vertical center of gravity, manufacturing cost, and life-cycle performance. These objective functions are non-commensurable and often conflicting. Because of the uncertainty of environmental loads, and the complexity of the mission requirements and struc-tural configuration, naval ship structural scant-lings are rarely programmatically or systematically optimized for weight, safety and vertical center of gravity. In practice, once the hull envelope and structural arrangements are developed to meet the operational requirements, the structural scantlings tend to be started by adapting existing designs and sized by using local design criteria. A 3D finite ele-ment model is then constructed only for checking the structural scantlings based on the local design criteria. For structural components which sat-isfy the design criteria (for example, *ABS's High Speed Naval Craft* buckling criteria), the struc-tural scantlings are often left as they are, and for structural parts which do not meet design criteria, the structural scantlings are simply scaled up from a local perspective, by undertaking manual design iterations. This piecemeal approach can lead to the following problems: (1) The resulting overall structural design being sub-optimal in terms of limit states, weight, vertical center of gravity, cost to manufacture, and a poor compromise of design

objectives; (2) A costly manual set of iterative changes being required in the design stage, which are not well integrated on a global ship structure level.

In order to resolve these problems, using opti-mization algorithms to automatically select better structural scantlings and topology becomes a natu-ral choice. The programmatic or systematic optimi-zation approach does not rely on time consuming manual design iterations, and enables designers to focus on high level engineering decisions.

Optimization algorithms implement numerous methods and approaches, each designed to target a specific type of problem. Optimization algorithms can be viewed from their respective mathematical approaches (e.g. deterministic or gradient) and/ or from their heuristic/stochastic approaches. Recently, the heuristic and evolutionary proce-dures, including concepts inspired by natural bio-logical systems such as Genetic Algorithm (GA) and Simulated Annealing (SA), have been growing in popularity as more and more researchers dis-cover the benefits of these approaches (Zanic, et. al., 2013, Sekulski, 2014). While the heuristic and evolutionary based optimization algorithms are very powerful and straightforward to implement, one of the drawbacks is that a large population of candidate points needs to be evaluated frequently. For ship and offshore structural design, it is often required to use a finite element analysis to obtain high fidelity structural limits. For a large complex system, a finite element analysis can be computa-

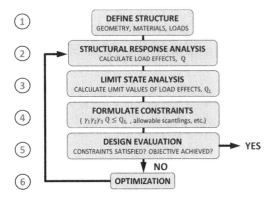

Figure 1. Rationally-based design procedure.

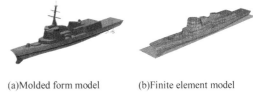

(a)Molded form model   (b)Finite element model

Figure 2.   A notional frigate.

Table 1.   Main particulars of the notional frigate.

Length Overall	150 m
Length Between Perpendiculars	135 m
Breadth	17 m
Waterline Breadth at LCG	15 m
Depth	9.4 m
Draught	4 m
Displacement	4000 ton

tionally demanding, and despite steady advances in computing power, the expense of repeatedly running analysis codes remains nontrivial. Therefore, it is imperative to limit the number of FEM response calculations in the optimization process.

A limit state based ship structural optimization, using a two-step iterative procedure and a multi-objective Simulated Annealing method, was proposed by Ma, Hughes and Paik, (Ma, Hughes and Paik, 2013a). This procedure is predicated on the rationally-based ship structural design procedure, first introduced by Professor Owen Hughes in 1977 (Hughes, 1983), Figure 1. Professor Hughes developed the MAESTRO computer code (MAES-TRO 2017) to implement a practical method for the application of this rational procedure to ship structural design.

The implemented optimization strategy, which is analogous to Collaborative Optimization (CO) of the Multidisciplinary Design Optimization (MDO) field, decomposes a large computationally expensive problem into a number of sub-problems. With the use of local subspace optimizers, each design cluster is given control over its own set of local design variables and is charged with satisfying its own limit state constraints, while a system-level optimizer provides coordination, global-level optimization, and ensures that local design decisions are fully integrated at the global system level.

2   HIGH SPEED NAVAL CRAFT (HSNC) LIMIT STATES

A limit state is defined as the condition beyond which a structural member or entire structure fails to perform its designated function. It is now well recognized that the limit state approach is a better basis for design since it is difficult to determine the real safety margin of any structure using linear elastic methods. Four types of limit states are

usually considered: serviceability limit state (SLS), ultimate limit state (ULS) or ultimate strength, fatigue limit state (FLS), and accidental limit state (ALS). The limit state type used in this paper is the ultimate limit state (ULS) for stiffened panels, which are the basic strength members in ships and offshore structures. ULS represents the collapse of a structure due to a loss of structural capacity in terms of stiffness and strength that typically arises from the buckling and plastic collapse of structural components. There are a number of closed-form ULS solutions, including formulations proposed by class societies. *For naval ships, American Bureau of Shipping (ABS) has published Guide for Building and Classing High Speed Naval Craft (HSNC). In this paper, the HSNC limit states, which include both buckling and yielding criteria, and local design criteria, are systematically applied to a full ship finite element model for* analysis, evaluation *and structural scantling optimization. The notional frigate finite element model was automatically* created from a NAPA-Steel molded form structural model (Figure 2(a)) using the NAPA-Steel/MAESTRO interface. The model has over 61,000 nodes and 125,000 elements, as shown in Figure 2(b). In addition, the weight distribution, tank loads and floating condition defined in the NAPA hydrostatic module are also translated into MAESTRO. The main particulars of the model are listed in Table 1.

In order to evaluate HSNC limit states, limit state evaluation panels have to be prepared first. An evaluation panel is a collection of finite elements, and is often flat (or slightly curved), rectangular and supported by bulkheads, frames and girders. In Figure 3, each colored rectangle

Figure 3.   Limit state evaluation panels.

Figure 4.   Notional model HSNC Design Pressure Region.

represents an evaluation panel, which is a plate, a stiffener, or a stiffened panel. The limit state panels are automatically prepared from MAESTRO.

To evaluate the limit states, a strength ratio is computed for each panel in the model and for each load case defined for each limit state. Using only the strength ratio to evaluate the structure can be difficult to visualize or use in optimization as it can range from zero to infinity. Instead, MAES-TRO uses an adequacy parameter. The adequacy parameter, g, is defined as:

$$g = \frac{1-r}{1+r}$$

The adequacy parameter ranges from -1 to +1. An adequacy parameter of zero indicates that the structure, under the defined loads, just satisfies the limit state, including the safety factors. Negative values indicate that the structure's response, with the user defined safety factors, exceeds the limit state.

To calculate HSNC design pressure, minimum plate thickness and beam minimum required section modulus, groups of the plating and beams are defined from a finite element model. Each of these groups is assigned a location category which is used to determine the applicable rules (e.g. bottom, side, deckhouse front-1st tier, etc.). Tank boundaries are automatically detected/evaluated based on the defined volume groups and the density of their assigned contents. Figure 4 illustrates the HSNC location groups of the notional model.

When evaluating the HSNC rules, MAESTRO will first compute the design pressures for each patch of plate based on its location category and its physical location in the ship. Geometric data (i.e. aspect ratio, location, flare angle, etc.) is extracted directly from the model automatically. If multiple design pressures apply, MAESTRO will select the more conservative value (e.g., a tank boundary that is also part of a watertight bulkhead; or side shell plating near the bow where Slamming Pressure and Fore End pressure both apply). Minimum section modulus is computed for all beam elements

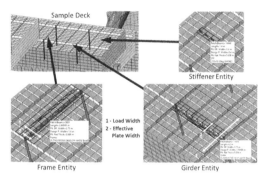

Figure 5.   Beam Entity Examples for Section Modulus Evaluation.

contained in the HSNC groups. In the MAESTRO properties, part of the beam element definition is whether it is to be used as a stiffener, frame or girder. Stiffener, frame and girder evaluation entities are defined using the true span of the beam with the following order of priority:

Stiffeners are broken by frames, girders and bulkheads

- Frames are broken by other frames, girders, stanchions and bulkheads
- Girders are broken by other girders, stanchions and bulkheads.

This hierarchy is used to determine both the beam length and spacing (load width) used in computing the minimum section modulus. Figure 5 presents samples of each beam type under a deck and their associated supported structure. Both the load width (used for computing the required section modulus) and the effective plate width (used for computing the actual section modulus) are shown graphically in the figure. The beam length used in the section modulus evaluation tends to be conservative in that the full length of the beam is used since bracketed end connections or radiuses in the corners for frames are not modeled.

Figure 6 shows HSNC local design pressure, and Figure 7 shows the structures not meeting HSNC

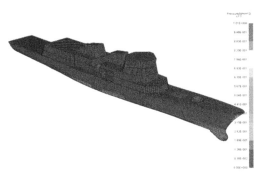

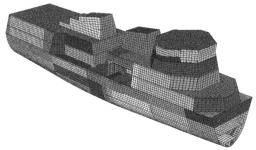

Figure 8.   Structural Design Clusters.

Figure 6.   Notional Model HSNC Design Pressures (N/mm2).

Figure 7. Structures not meeting HSNC minimum scantling criteria.

minimum plate thickness and minimum section modulus of the initial design model.

3   DESIGN CLUSTERS

In order to develop a realistic model with reasonable production characteristics, in which each individual evaluation patch, or stiffened panel, isn't assigned different plate and/or stiffener properties during the optimization, MAESTRO uses a group of structural elements called a "design cluster" during the optimization. The purpose of the design cluster is to group particular areas of structure in which the scantling properties will be the same for all of the elements in the design cluster at the completion of the optimization. For example, a deck within a module could be defined as a design cluster consisting of the deck plating and longitudinal stiffeners, and a second design cluster consisting of the deck's transverse frames. Each design cluster can also have its own set of constraints such as minimum or maximum plating thickness, web height, or stiffener spacing. The MAESTRO user is responsible for defining the design clusters, and can define the design clusters in coordination with the shipyard's production processes to further reduce manufacturing costs for the vessel. The con-

cept of a "design cluster" is a natural extension of the optimization's *design variable linking* approach. In Figure 8, each color represents a structure for which the plate thickness, stiffener dimensions and stiffener spacing are all uniform.

4   LIMIT STATE BASED OPTIMIZATION

Structural design optimization methodologies and techniques have been around for many years. Recently, the heuristic and evolutionary procedures have been growing in popularity. In this paper, a genetic algorithm (GA) is used for optimization. There have been many examples of using heuristic based optimization methods to optimize ship structures. Most of them use simple empirical constraints, such as scantling rules and stress limits specified by class societies, to size the structural scantlings (Sekulski, 2014). They usually do not include a high fidelity calculation of actual stresses, which in most cases requires a 3D finite element analysis. To accurately evaluate the limit states of a stiffened panel, the actual stresses must be used. For a single panel where the applied load is often to be assumed constant throughout the design iterations, the normal heuristic based optimization procedure can be used. However, to optimize a system of stiffened panels, the applied load can no longer be assumed constant. The loads of each individual panel are functions of the system's structural scantlings. As panel scantlings get updated in each perturbation, the corresponding panel loads are also changed. The optimization for a system of panels becomes an iterative process. While evaluating limit states for each individual panel is relatively efficient, the panel loads must be obtained by finite element analysis of the entire structure.

A single panel is defined as a plate, a stiffener, or a plate with stiffeners. At the single panel level, there are two objective functions, Weight (Wp) and Safety of the structural panel (ηp). The definition of the single panel weight was given in the authors' previous studies (Ma, et al, 2013a and 2015, Hughes, et. al, 2014). The panel safety is defined as the ratio of the applied loads to the panel ultimate limit states, which includes yield and instability. The applied loads of a stiffened panel are the results of hull girder primary, secondary and tertiary loads from all load cases.The aggregated fitness function f can be expressed as:

$$f = \cfrac{1}{\lambda_w \cfrac{w_p}{w_{p0}} + \lambda_\eta \cfrac{\eta_{p0}}{\eta_p} + \sum_{i=1}^{m} c_i g_i} \quad (1)$$

where f_w and f_n are the normalized individual fitness functions of weight (W_p) and safety measure (η_p). W_{p0} and η_{p0} are the nominal initial design value of a stiffened panel weight and safety measure respectively, g_i is the constraint penalty function, and c_i is the coefficient of the penalty function. c_i is 0 if the design variables satisfy the constraints, and is 1 if they violate the constraints.

For example, g_i can be a ratio of a stiffener flange width to the stiffener web height.

The optimization of a system of stiffened panels, such as a ship structure, requires an iterative procedure that reintegrates the system of panels with updated scantlings, and reanalyzes the panel loads using a global finite element model and global loading conditions. The multi-level collaborative optimization process employed is illustrated in Figure 9.

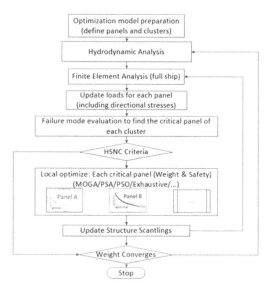

Figure 9. Multi-level collaborative optimization procedure.

In the most inner loop (first level), the multi-objective optimization is performed at the single stiffened panel level, assuming the applied load is constant. The HSNC limit state criteria, along with the HSNC minimum plate thickness and minimum sectional modulus criteria, are used to optimize the panel's weight and safety. At this level, the number of possible solutions can reach into the millions for a single stiffened panel optimization. When all stiffened panel local optimizations are completed for the current macro design cycle, the updated scantlings are reintegrated into the revised global structure, and this updated structure is reanalyzed and evaluated against the comprehensive set of load cases and invoked limit state criteria as shown by the 'second loop' of Figure 9. The applied loads of stiffened panels are then updated. The finite element analysis iteration is the second level of the optimization, which serves to integrate and reevaluate all stiffened panels' optimal scantlings. In addition, other global level optimization objectives, such as reducing the number of different plates and stiffeners, matching preferred plate and stiffener combinations, optimizing hull girder cross section moment of inertia, and minimizing the vertical center of gravity of the structure, are optimized at this level.

5 DESIGN LOADS DEVELOPMENT PROCESS

The notional frigate's hydrodynamic motions and forces arising from wave loading were calculated using MAESTRO-Wave, a potential flow based computational tool that predicts the motions and wave loads of any type of vessel. Due to the complexities of the hull geometry and limitations of strip theory, MAESTRO-Wave's 3-dimensional panel method was used for the analysis. MAESTRO-Wave leverages the high fidelity inertial structural model and automatically extracts a hydrodynamic mesh from the finite element model. In order to compute the motions and loads, MAESTRO-Wave first calculates the velocity potential, source strengths, and flow velocities at

Figure 10. Notional Model Hydrodynamic Panels.

the centroids of the hydrodynamic panels for each speed, heading, and wave frequency and then maps the source strength directly to the structural panels. This allows the equations of motion to be formulated using the structural mesh, rather than the hydrodynamic mesh, which ensures equilibrium for the structural model (i.e. bending moments, shear forces, and torsional moments are automatically in closure). A roll damping model which accounts for potential flow and viscous damping effects was used. A typical critical roll damping coefficient, which is the ratio of the actual (potential plus viscous) damping to the critical damping, of 0.05 was used. This coefficient can be calibrated using model test or CFD results for future analyses.

Due to the level of detail in ship finite element models and the computational power required to solve the source strength for each hydrodynamic panel, MAESTRO-Wave offers different levels of discretizing the finite element mesh into the hydrodynamic mesh. One method is to compute the loads on a one-to-one ratio between the finite element and hydrodynamic meshes. The notional model has 13,761 wetted finite elements. To provide a relative measure of the number of wetted panels, one common industry accepted 3D panel code limits models to only 5,000 hydrodynamic panels. MAESTRO can also use the evaluation patches automatically defined and shown in Figure 3 as the hydrodynamic panels. This technique reduces the number of hydrodynamic panels to 4,846.

The number of hydrodynamic panels can be further reduced by using MAESTRO's built-in panel merging algorithm which automatically merges adjacent evaluation patches into larger panels, while maintaining a user-specified aspect ratio. Figure 10 shows the hydrodynamic panels used in the MAESTRO-Wave runs.

Table 2. Notional Combatant Operating Profile.

Speed	Heading	0~5 m	5~10 m	10~16 m
5	180	0.01250	0.02500	0.00000
5	135	0.02500	0.37500	0.80800
5	90	0.02500	0.05000	0.00000
5	45	0.01250	0.05000	0.04200
5	0	0.08750	0.02500	0.00000
15	180	0.17500	0.02300	0.00000
15	135	0.17500	0.33750	0.14200
15	90	0.17500	0.00000	0.00000
15	45	0.17500	0.04450	0.00800
15	0	0.08750	0.02250	0.00000
25	180	0.02500	0.02250	0.00000
25	135	0.05000	0.03750	0.00000
25	90	0.05000	0.00000	0.00000
25	45	0.05000	0.0500	0.00000
25	0	0.02500	0.00250	0.00000

Table 3. Wave Scatter Diagram of Combined N. Pacific Region.

	Tp=3.2(s)	Tp=4.8(s)	Tp=6.3(s)	Tp=7.5(s)	Tp=8.6(s)	Tp=9.7(s)	Tp=10.9(s)	Tp=12.4(s)	Tp=13.8(s)	Tp=15(s)	Tp=16.4(s)	Tp=18(s)	Tp=20(s)	Tp=22.5(s)
Hs=0.01m	201.9	403.05	1276.7	1174.63	1195.64	1096.57	1017.76	502.88	325.74	216.91	93.82			
Hs=0.5m	4895.8	9773.4	30958.2	28483	28992.6	26590.2	24679.2	12194	7898.8	5259.8	2275			
Hs=1.5m	3635.55	7298.03	72791.8	42468.6	46346.5	42468.6	21220.8	18177.8	7863.56	3904.85	1696.59	1427.29		
Hs=2.5m		9250.74		43096.9	47012.9	32327.6	21558.4	18461.5	9950.04	9950.04	4315.68	2157.84	1178.82	539.46
Hs=3.5m				3770.91	32844.2	37619	22574	12908.6	6936.93	6936.93	2406.69	1505.79	823.68	373.23
Hs=4.5m					2254.92	15452.3	30913.4	17663.5	9509.04	4754.56	4116.54	1372.18	839.04	515.66
Hs=5.5m						1661.08	12434.5	21316.3	7655.64	3825.2	3311.68	1105.64	681.2	413.96
Hs=6.5m							1658.16	14229.3	7662.41	3829.56	3319.61	1105.44	681.03	414.54
Hs=7.5m								4407.79	9488.67	4745.39	1232.24	685.75	280.63	257.42
Hs=9m								481.95	1553.58	10361	4488.75	1122.66	612.36	279.72

6 DESIGN LOADS BASED ON FREQUENCY DOMAIN ANALYSIS

A MAESTRO-Wave frequency domain analysis was performed on the notional frigate model at Full Load using the following combinations of speeds, headings, and wave frequencies:

Speeds: 5, 15, 25 knots
Headings: 0, 45, 90,135, 180 degrees
Wave Frequencies: 0.2 rad/s to 1.8 rad/s in increments of 0.05 rad/s

The frequency domain analysis produced a database of regular unit wave Response Amplitude Operators (RAOs). The RAO database was combined with a notional combatant operating profile (Table 2), a composite wave scatter diagram (Table 3), and used a probability of exceedance of 1.0E-8 to generate lifetime expected loads. The following dominant load parameters (DLPs) were to be evaluated for the structural optimization:

- Vertical bending moment
- Vertical shear force at ¼ and ¾ of the vessel length
- Vertical acceleration at bow
- Relative vertical velocity at bow
- Longitudinal torsional moment
- Horizontal bending moment
- Horizontal shear force

Based on the linear frequency domain hydrodynamic analysis, the maximum wave bending moment (less stillwater) is 364,845 kN-m occurring at sea state 6, 15 knots and head seas. The maximum combined sagging moment due to waves, 277,877 kN-m, is obtained by superposing the still water bending moment on the wave bending moment.

7 HSNC SLAMMING BENDING MOMENT

Section 3–2-1 of the HSNC rules provides equations for wave bending moments amidships, still water bending moment, and slamming induced bending moment. The largest of these vertical bending moments is used to determine the minimum required section modulus over the midships 0.4 L and is used as input for the vertical bending moment DLP. The rule based slamming induced bending moment is given as follows,

$$M_{sl} = C_3 \Delta \left(1 + \eta_{cg}\right)\left(L - l_s\right) kN - m \qquad (2)$$

where

$$l_s = \frac{A_R}{B_{wl}} = \frac{\dfrac{0.697\Delta}{d}}{B_{wl}}$$

Table 4. Input Data for Slamming Bending Moment.

Variable	Value	Units
C_3	1.25	
Δ	4000	MT
l_s	46.47	M
A_R	697.00	m²
B_{wl}	15	M
η_{cg}	0.0687	g's
d	4	M
L	135	M
N_2	0.0078	

Using the equation (2) and the inputs from Table 4, the HSNC Slamming Induced Bending Moment is 473,077 kN-m. The slamming amplification factor is estimated as the ratio of the HSNC slamming bending moment to the wave sagging moment.

For the notional frigate model, the slamming amplification factor is 1.7.

8 DESIGN LOADS BASED ON TIME DOMAIN ANALYSIS

The ABS Guidance Note Subsection 6/1 requires that a non-linear seakeeping analysis is to be performed to establish the instantaneous design loads at a specific time when each DLP reaches its maximum. The frequency domain extreme load analysis results were used to identify the magnitude of each DLP and the speed, heading and sea state that it occurs at. Time-domain simulations associated with each DLP's critical speed, heading, and sea state were run to capture the nonlinear effects. Seven different one-hour weakly nonlinear time domain simulations were performed for the Full Load condition. Each time domain simulation uses a unique speed, heading, and sea state combination. The panel pressures, point forces, accelerations, and hull girder loads were calculated at 0.05 second intervals, and recorded at every 0.5 second. The peak load of a DLP from the time domain simulation is extracted as a design load. For bow down sagging events, the slamming amplification factor is applied. Figure 11 shows the vertical bending moment comparison of the HSNC rule based envelope to all design loads. The maximum sagging and hogging moments are 524 kN-m, 499 kN-m and 516,204 kN-m respectively. Figure 12 illustrates the wetted panel pressure distribution under sagging and hogging conditions.

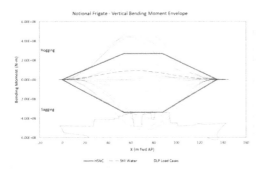

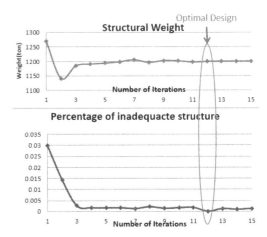

Figure 11. Vertical Bending Moment Envelope Comparison.

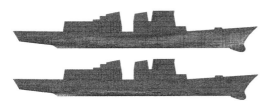

Figure 14. Design history of the full ship.

Figure 12. Pressure Distribution of Sagging and Hogging Conditions.

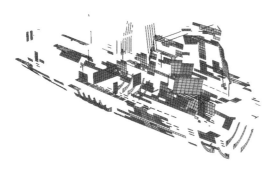

Figure 13. Structures with negative adequacy parameter for all load cases.

9 FULL SHIP ANALYSIS, EVALUATION AND OPTIMIZATION

A large portion of the notional model, shown in Figure 8, is selected for structural optimization. The section weighs 1,306 tons (65% of the total structural weight) and is subdivided to 177 design clusters. 14 load cases are applied to the global model. All panels were evaluated using HSNC limit state criteria. Within the section, 39.4 tons of the structure, shown in Figure 13, did not meet HSNC buckling and yielding crite-

ria in the initial design. The minimum adequacy parameter was −0.78, which indicated some panels were severely under designed for the given loads.

To correct the inadequate design while minimizing the structural weight, a hierarchical multi-level, multi-objective heuristics based optimization, proposed in this paper, is used. The values of upper and lower bound constraints for scantlings are:

$$\begin{cases} 5mm \leq t_p \leq 50mm \\ 60mm \leq b_{fs} \leq 400mm \\ 5mm \leq t_{fs} \leq 30mm \\ 20mm \leq h_{ws} \leq 300mm \\ 5mm \leq t_{ws} \leq 30mm \end{cases} \quad (3)$$

In spite of this relatively weak initial design, the optimization converged in 15 design cycles, Figure 14. The optimization not only corrected all of the inadequate structure, it was also able to save 5.5% or 71.82 tons of structural weight.

10 CONCLUSIONS

In this paper, a hierarchical, multi-level collaborative optimization procedure was successfully used to optimize a large and complex structure on the basis of weight and safety. A set of seakeeping loads are generated using weakly nonlinear time domain potential flow based panel code, MAESTRO-Wave. The slamming effect is considered by applying ABS HSNC rules. In each optimization design cycle, the finite element method was used to accurately calculate the stresses throughout the structure. The examples show that even for an inadequate initial design, the method can find an optimal design within a short time, and is thus a valuable and attractive tool for naval vessel structural optimization.

REFERENCES

ABS Rules for Building and Classing High-Speed Naval Craft, 2015.

Hughes, O., Ship Structural Design: A Rationally-Based, Computer-Aided, Optimization Approach 1st edition, New York (NY): John Wiley & Sons, 1983

Hughes, O.F., Ma, M. and Paik, J.K.(2014) "Applications of Vector Evaluated Genetic Algorithm (VEGA) in Ultimate Limit State Based Ship Structural Design",, Proceedings of the ASME 2014 33th International Conference on Ocean, Offshore and Arctic Engineering, OMAE2014, June 9–14, 2014, San Francisco, USA

ISSC. (2012), 18th International Ship and Offshore Structures Congress, Committee III.1, Ultimate Strength.

Ma, M., and Hughes, O. Paik, J.K. (2013a), Ultimate strength based stiffened panel design using multiobjective optimization methods and its application to ship structures, Proceedings of the PRADS2013

Ma, M., Zhao, C.B. and Hughes, O. (2013b), A practical method to apply hull girder sectional loads to full-ship 3D finite-element models using quadratic programming, Ships and Offshore Structures, 04 April 2013

Ma, M., Hughes, O., McNatt, T., (2015) "Ultimate limit state based ship structural design using multi-objective discrete particle swarm optimization", Proceedings of the OMAE 2015, St. Johns, Canada

MAESTRO Version 11.4.9 (2017). Program documentation, Advanced Marine Technology Center, DRS Technologies Inc., http://www.maestromarine.com.

Sekulski, Z., (2014), "Ship Hull Structural Multiobjective Optimization by Evolutionary Algorithm", Journal of Ship Research, Vol 58, No 2., pp 45–69

Zanic, V., Andric, J., & Prebeg, P., (2013), "Design synthesis of complex ship structures", Ships and Offshore Structures, Vol 8, pp 383–403

Maritime Transportation and Harvesting of Sea Resources – Guedes Soares & Teixeira (Eds)
© 2018 Taylor & Francis Group, London, ISBN 978-0-8153-7993-5

Framework for risk-based salvage operation of damaged oil tanker after collision in the Adriatic Sea

J. Parunov, S. Rudan & M. Ćorak
Faculty of Mechanical Engineering and Naval Architecture, University of Zagreb, Zagreb, Croatia

B. Bužančić Primorac
Faculty of Electrical Engineering, Mechanical Engineering and Naval Architecture, University of Split, Split, Croatia

M. Katalinić
Faculty of Maritime Studies, University of Split, Split, Croatia

ABSTRACT: The study describes the framework for risk calculation of post-accidental salvage operation of damaged double-hull tanker in the Adriatic Sea. The principal failure mode driving the presented risk assessment is the ship hull girder collapse, representing one of the most unfavorable outcomes of a ship accident. Risk of such an event depends on the hull girder probability of failure and the cost of its consequences. Risk assessment is performed by employing short-term ship structural reliability analysis for the failure probability calculation and available models for estimating cost of an oil spill, as the most important consequence of ship hull girder collapse. The risk of the hull girder failure of damaged tanker, expressed in monetary units is compared with the initial cost of the oil spill, which may occur after a collision accident in which the inner hull is damaged. It was found that the risk of hull girder collapse of damaged ship during towing in the Adriatic Sea is negligible, despite huge economic loss that could occur. That is the consequence of the mild sea environment where hull girder failure probability of damaged ship is rather low.

1 INTRODUCTION

Collisions involving tankers are major type of accidents yielding to large scale oil spills, sometimes with huge economic consequences including environmental disasters (Kim et al. 2015). Collision can even lead to the total loss of the ships involved, as a most serious outcome of a marine accident (e.g. Jia and Moan, 2008., Hussein and Guedes Soares 2009., Faisal et al. 2017., Parunov et al. 2017a). Such scenario is particularly dangerous in ecologically sensitive regions as well as in countries heavily depending on tourism (Klanac et al. 2013, Tabri et al. 2015). Despite constant improvements in ship design and operation aiming to reduce the number of marine accidents, ship collisions and associated consequences could not be completely prevented. It is therefore of the utmost importance to react fast and properly in case of a collision accident in order to minimize post-accidental consequences to the damaged ship safety and to the environment.

The present study describes the framework for calculation of the post-accidental salvage operation risks for a damaged double-hull tanker in the Adriatic Sea. The principal failure mode driving risk assessment is the ship hull girder collapse, representing one of the most unfavorable outcomes of a ship accident. Risk of such an event is a function of the hull girder probability of failure and the cost of its consequences (Dong and Frangopol 2015). Risk assessment is in the present study performed by employing short-term ship structural reliability analysis for the failure probability calculation and available models for estimating cost of oil spill, as the most important consequence of the ship hull girder collapse.

The principal aim of the study is to compare the risk of a hull girder failure of damaged tanker, expressed in monetary units, with the initial cost of the oil spill, which may occur after collision accident in which the inner hull is ruptured. Such comparison enables to identify the efficiency of the risk control measures in order to prevent post-accidental hull girder collapse of damaged tanker. If risk control measures considerably reduce the total cost of oil spill, they can be proposed for further investigation while otherwise, measures are rather inefficient. The main risk control measure considered in the present paper is the selection of the route along which the ship can be towed to a safe location.

The paper is organized as follows. In Section 2 of the paper collision scenario for oil tanker in the

Adriatic is described as well as the numerical simulation of such a scenario.

As a result of the numerical simulations, credible damage of the double-hull oil tanker is selected as representative for the risk calculation. Assessment of the volume of oil that is spilled to the environment immediately after the accident and an evaluation of the associated cost is performed in Section 3. Damage consequences on ship structural strength and post-accidental behavior of the damaged ship is described in the Section 4, while possible towing routes for assumed accident location are described in Section 5. Section 6 deals with wind-generated waves in the Adriatic and associated hull-girder vertical wave bending moments. Structural reliability calculations of the damaged ship are presented in Section 7 while risk analysis for the damaged ship in tow is performed in Section 8. Paper ends with concluding section based on numerical results obtained.

2 COLLISION SCENARIO AND NUMERICAL SIMULATION

A collision between a tanker and a ferry is considered to be an accident with high probability of occurrence in the Adriatic Sea and the numerical simulation of such an accident is performed by non-linear FEM analysis using LS-Dyna software (Parunov et al. 2017a). The situation during the impact may vary but it is reasonable to assume that the worst case scenario considers the bow of the ferry hitting the tanker orthogonally.

The Aframax class tanker is assumed as the struck ship and a ferry as the striking. The main particulars for both the struck and the striking ship are listed in Table 1.

Because a very fine mesh is required in the collision zone, only a portion of the struck ship is modelled, consisting of the half of three cargo holds. The striking ship bow is modelled in detail and the rest of the ship, i.e. the ferry hull, is modelled appropriately by beam finite elements. Reference collision scenario set-up is presented on Figure 1.

Destruction of the structure and the plastic strain in the damaged area for the reference model

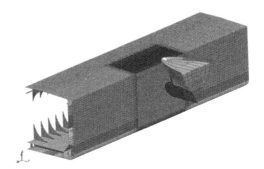

Figure 1. Reference model collision set-up.

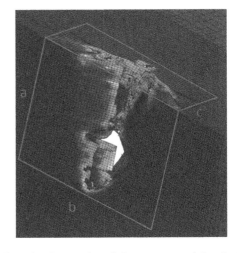

Figure 2. Destruction of the structure and the plastic strain.

is presented on Figure 2. A number of FEM analyses were then performed with main parameters (striking ship speed and displacement, collision angle and location) systematically varied as described in previous work (Galletta 2015, Parunov et al. 2017a, 2017b).

Struck ship structural damage consists of bulb induced rupture and fore peak induced damage.

This defines the volume of structural damage and the quantification of that volume is not straightforward. It is assumed, however, that total width b, height a and depth c of the plastic deformation zone define the "damage box" of the structure.

Monte Carlo (MC) simulation of damage size for random collision scenarios in the Adriatic is performed by Parunov et al. (2017a) based on the numerical simulation of the ship-ship collision. It was found that damage size specified in Table 2 is realistic and occurs with rather high probability of 8.4% as a result of MC simulation:

Table 1. Main struck and striking ship particulars.

	Struck ship (tanker)	Striking ship (ferry)
Lpp (m)	236	120
B (m)	42	19.6
D (m)	21	12.2
T (m)	15.1	5.3
Δ (t)	133000	6889

Table 2. Parameters of damage size (m).

Damage height a	7
Damage length b	6
Damage depth c	2

Damage depth c specified in Table 2 is such to cause rupture of the inner hull and it is credible for risk analysis performed in the present study. Therefore, damage size parameters specified in Table 2 are assumed in all further calculations. It is assumed that damage extents from the main deck downwards. The adequacy of these damage size parameters is verified by detailed comparative analysis of the numerical simulations and damage size parameters provided by IMO (2003), as described by Parunov et al. 2017b.

3 WEIGHT AND COST OF OIL SPILL

In the case of collision damage, oil spill occurs in the case when the inner hull is breached. To calculate the amount of oil spilled to the environment is not an easy task and models of different complexity exist. The simplest approach is to assume that all cargo from damaged tanks would be released (Kim et al. 2015). The more realistic approach would be to calculate amount of spilled oil analytically taking into account both unidirectional flow due to hydrostatic imbalance and also bi-directional flow caused by different densities of oil and sea water (Tavakoli et al. 2012., Kollo et al. 2017).

The simplified approach is used in the present study, where the amount of oil spill is determined from the damage stability calculation as difference between cargo run-off weight and water ingress in damaged cargo tanks.

Thus, if one cargo tank in the midship region is damaged, then cargo run-off weight in the damaged cargo tank reads 9660 t. The weight of water ingress in damaged tank reads 7560 t. Therefore, in the hydrostatic equilibrium, difference between run-off weight and sea water ingress would read 2100 t. If two tanks in the midship region are damaged, then difference between run-off weight and flooded sea water reads 3734 t. Although the volume of tanks in the midship region is the same, due to different trim and heel at equilibrium, oil spill for two damaged tanks is not twice than for the case of one tank damaged.

There are several models, available from the literature, used to estimate the total cost of the oil spills, as reviewed by Kim et al. (2015). The cost of oil spill represents the cost of the environmental clean-up (i.e., the removal of oil) and the claims paid for compensation (i.e., property damage of

economic users such as fisheries, tourism and recreational users). Large differences in the total costs of oil spills estimated by different models for the same amount of spilled oil are found by Kim et al. (2015), where the most conservative model resulting in the largest cost is the one by Ventikos and Sotiropoulos (2014), while the lowest predicted cost results from the model provided by Yamada (2009).

Non-linear regression model for the cost of oil spill is generally provided by the following equation:

$$y = Ax^B \tag{1}$$

where y is the total cost of oil spill in US$, x is the weight in tones of spilled oil, while A and B are model-dependent parameters. Thus, for the model of Ventikos and Sotiropoulos (2014), A and B read 24,020 and 0.8447 respectively, while for the model proposed by Yamada (2009), A and B read 38,905 and 0.66 respectively.

Therefore, the total cost of oil spill reads 15,380,000 US$ and 25,000,000 US$ for one and two cargo tank ruptures respectively, according to the model of Ventikos and Sotiropoulos (2014). While according to the model of Yamada (2009), cost of oil spill is much lower and reads 6,064,000 US$ and 8,864,000 US$ for one and two cargo tank damages respectively.

4 POST-ACCIDENTAL BEHAVIOR OF DAMAGED SHIP

After a collision event, struck ships may capsize rapidly, sink slowly or stay afloat. If the ship still stays afloat after collision, hull-girder collapse may occur when the hull's maximum residual load-carrying capacity is insufficient to sustain the corresponding hull girder loads applied. The collapse of a hull girder may lead to large oil spill in addition to the one occurred immediately after collision. Such additional cost is large, but it happens only with a certain probability of the damaged hull girder collapse. Therefore, such potential additional cost may be expressed in terms of risk, i.e. as a product of the probability of a hull girder collapse and associated cost.

The collision affects the ultimate hull girder capacity in the damaged region, the still water bending moment (SWBM) distribution along the vessel as well as the vertical wave bending moments (VWBM). For tankers in full load condition, damage in the midship region is typically the worst situation, leading to considerable increase of the sagging SWBM. VWBM is usually much lower compared to the intact ship, but it depends on the weather conditions during damaged ship towing to the protected harbor and also on the exposure period.

The progressive collapse analysis (PCA) method initially proposed by Smith represents nowadays the most frequently used method for ultimate strength assessment. Based on the PCA results, design curves are developed for the studied oil tanker showing dependency of the residual ultimate strength in sagging on the damage extent, Figure 3. Curves in Figure 3 represent the ratio of the ultimate bending moment capacity in damaged (M_{ud}) and intact condition (M_{u0}). It is conservatively assumed in PCA that collision damage always starts from the main deck. Upper curve in Figure 3 is used for the case when only outer shell is damaged, while the lower curve is used when both outer and inner shells are breached. The latter curve includes also the influence of the rotation of neutral axis on the ultimate strength of damaged ship in sagging (Parunov et al. 2017b). For damage height given in the Table 2 ($a = 7$ m), $a/D = 0.33$, so the reduction of the ultimate bending moment capacity M_{ud}/M_{u0} reads 0.85. As the ultimate bending moment capacity in sagging of the intact ship reads 8470 MNm, the residual ultimate bending moment capacity in sagging may be calculated as 7200 MNm.

In the present case, for damage in the midship region, SWBM in damage condition is determined based on the damage stability calculations. Calculation details are presented by Bužančić Primorac et al. (2015) where increase of the SWBM in damaged, compared to the intact condition, is approximately by factor 1.35 for one cargo tank damage and 1.5 for two cargo tank damaged. As the midship SWBM in sagging for the intact ship reads 1556 MNm, SWBM of damaged ship would read 2100 MNm and 2334 MNm for one and two tank damages respectively.

The wave-induced bending moment should be determined based on the actual wave conditions (e.g. significant wave height and wave length, etc.) at the moment of accident at sea together with the vessel speed in association with a short-term response during the rescue operation. In the present study, ship is assumed to be damaged in the Adriatic and specific wave-induced response is considered. In order to have efficient procedure for post-accidental safety assessment of damaged ship, fast calculation of global wave loads is required. Therefore, charts for estimation of VWBM at amidships on damaged Aframax oil tanker in the Adriatic Seas are developed by Parunov and Ćorak (2015). Seakeeping assessment of damaged ship was performed by a 3D panel method. It was assumed that the mass of flooded seawater becomes an integral part of the ship mass and moves with the ship. The spectral analysis is performed using Tabain's wave spectrum, developed specifically for the Adriatic Sea (Tabain 1997).

Standard deviation of the VWBM at amidships is calculated for three different ship speeds (0, 5 and 10 knots) as well as for four different heading angles (180° – head seas, 135° – bow seas, 45° – quartering seas and 0° – following seas). Design charts are presented in Parunov and Ćorak (2015) for different significant wave heights, but only result for $H_S = 4$ are reproduced in Figure 4.

It is also possible to estimate the number of cycles n in a storm of certain duration T in seconds as:

$$n = T / T_Z \tag{2}$$

where T_Z is the mean response period, which may be estimated from Figure 5 (Parunov and Ćorak 2015).

The most probable extreme VWBM M_w^* in short-term sea state with n number of cycles is given as:

$$M_w^* = \sigma\sqrt{2\ln n} \tag{3}$$

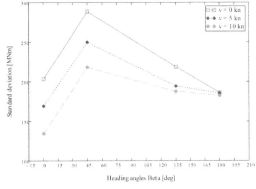

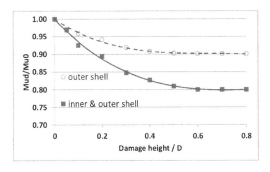

Figure 3. Reduction of ultimate longitudinal strength of damaged Aframax oil tanker (Parunov et al. 2017b).

Figure 4. Standard deviations of VWBM for different ship speeds and heading angles $H_S = 4$ m (Parunov and Ćorak, 2015).

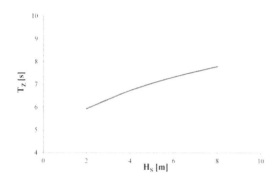

Figure 5. Mean zero crossing period T_z as a function of the significant wave height H_s for Tabain's wave spectrum (Parunov and Ćorak 2015).

where σ is the standard deviation of VWBM given in Figure 4.

The mean value $\overline{M}_{we}$ of the random extreme VWBM in sea state with n response cycles is given as:

$$\overline{M}_{we} = M_w^* + \frac{\sigma \cdot 0.5772}{\sqrt{2\ln n}} \qquad (4)$$

While the standard deviation of the random extreme VWBM reads:

$$\sigma_e = \frac{\pi}{\sqrt{6}} \frac{\sigma}{\sqrt{2\ln n}} \qquad (5)$$

Equations (4) and (5) represent parameters of Gumbel distribution as the extreme value probability distribution function to be used in damaged ship structural reliability analysis.

5 TOWING ROUTES OF DAMAGED SHIP

As described in previous sections, Aframax oil tanker is assumed to be damaged in midship area in collision accident with a ferry in the Adriatic Sea. Damage parameters are such to cause oil spill from the cargo tanks. It is further assumed that accident occurred near the island of Palagruža, in the central part of the Adriatic. That area is identified as the region with increased risk of collision, according the study of Zec et al. (2009).

After such accident, there are several options for towing of damaged oil tanker. Three potential towing routes assumed in the present study are shown in Figure 6.

First option, named Towing route 1, is from the accident location, to the Italian cost. This is the shortest towing path and reads about 35 nautical miles (nm). By assuming towing speed of 5 knots, towing duration would be 7 hours.

Figure 6. Adriatic Sea with potential towing routes.

The second option, Towing route 2, is from Palagruža to the Croatian island of Korčula, where a shipyard is located. Distance from Palagruža to Korčula reads about 55 nm, corresponding to the towing duration of about 11 hours.

The third option assumed in the present paper is named Towing route 3, from Palagruža to the Bay of Saldun, close to the city of Trogir, where also shipyard exists. Towing distance in that case reads 70 nm, while towing duration is about 14 hours.

6 WIND-GENERATED WAVES IN THE ADRIATIC SEAS

Wind-generated waves are the dominant type of surface waves in the Adriatic Sea due to its fetch limiting three-sided enclosed boundaries and surrounding orography which dominates wind speed and direction. Wind, and consequent wave characteristic, of the basin are well described in literature (Katalinic 2014). Two most often storm level winds are identified. *Bura* family winds, arriving from the north-east quadrant, gaining speed down the Dinaric mountain range along the east coast reaching maximum gale speeds up to 68 m/s with direction transversally across the basin limiting fetch to about 100 km, and resulting in maximum wave height of more than 6 meters. The other, *sirocco* (*jugo*) family winds arrive from the south-east quadrant, blowing along the basin with more than 750 km of fetch, usually lasting several days with top speeds of about 30 m/s, these winds create the highest known waves in the Adriatic (in its northern part) with the highest recorder wave of 10.8 meters.

There are numerous approaches to evaluate a certain sea condition differing in available tools and input parameters. Although numeric wave models (e.g. SWAN *"Simulating WAves nearshore"*, Booij

1999; or ECMWF *"European Center for Medium range Weather Forecast"* WAM model results available online via national meteorological center) are considered state-of-the-art Reikard (2011) showed that their result inferior to statistical models in short-term response.

Accurate statistical models (regression and neural network) are being developed for the Adriatic Sea (Farkas 2016, Mudronja 2017) based on a satellite calibrated database of wave (and wind) parameter measurements across 39 locations, at 6-hour intervals, from 1992.–2016. obtain from OCEANOR.

Furthermore, both of these approaches can be validated and improved in the Adriatic by means of available in-situ measurements obtained within the Italian RON project taken by buoys located along the western coast in front of Venezia, Ancona, Ortona and Monopoli (Liberti 2013).

For a given, post-accidental decision making scenario, computational efficiency is of the utmost importance for wave condition evaluation and subsequent VWBM calculation. Thus, a relatively simple and straightforward software tool, cgWindWaves (RUNET 2016) is chosen. The Adriatic water region is defined for the purpose of fetch calculation at a certain location (e.g. near island Palagruža—central Adriatic) as presented in Figure 7.

For the purpose of this study it is assumed that weather conditions in the time of accident are such that waves with significant wave height of 4 m are coming from south-east (*sirocco/jugo wind*). This would correspond to the upper level of sea-state 5 (or lower level of sea-state 6) according to the Douglas scale which is standard to seafarers and in meteo-reports delivered to them. Significant wave height of 4 meters, with significant period of 8 seconds, near Palagruža would correspond to southeast wind speed of 16.3 m/s. Significant wave height was evaluated by Bretschenider's method and the directional spectrum effects were included by the effective fetch method. A spectral description by the JONSWAP formulation of the given sea state is shown in Figure 8.

If the given wave spectrum is given a random phase seed and an inverse discrete FFT (fast Fourier transform) is performed, a representation of possible time-series wave elevation can be acquired which may be of use to other professionals involved in the salvage operation as it is straightforward and intuitive for sea state description (Figure 9).

Therefore, during towing to the Italian coast (Towing route 1), beams seas condition will dominate as well as during Towing route 2. When towing the damaged ship along Towing route 3, stern quartering seas (45 degrees in Figure 4) may be assumed.

Based on these information and using Figure 4, probabilistic model of VWBM during towing period may be calculated. Moments (mean value μ_e and standard deviation σ_e) of the extreme (Gumbel) distribution are determined for each of the towing routes and specified in Table 3.

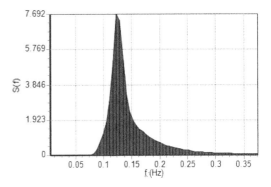

Figure 8. JONSWAP wave spectrum—Palagruža $Hs = 4$ m.

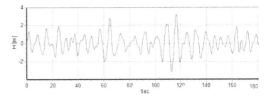

Figure 9. $Hs = 4$ m, time series representation.

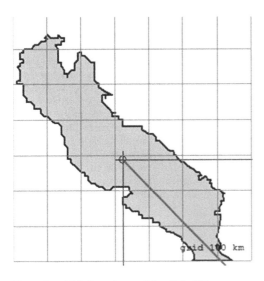

Figure 7. Adriatic region model- cgWindWaves.

Table 3. Moments of Gumbel distribution for VWBM on different towing routes.

Towing route	μ_e	σ_e
1	946	71
2	970	69
3	1092	76

Table 4. Summary of stochastic model adopted.

Variable	Distribution	Mean	COV
M_u (MNm)	Deterministic	7200	
M_{sw} (MNm) (one tank damaged)	Deterministic	2100	
M_{sw} (MNm) (two tank damaged)	Deterministic	2334	
M_w (MNm) (Route 1)	Gumbel	946	0.08
M_w (MNm) (Route 2)	Gumbel	970	0.07
M_w (MNm) (Route 3)	Gumbel	1092	0.07
χ_{sw}	Gaussian	1.0	0.05
χ_u	Log-normal	1.1	0.12
χ_w	Gaussian	1.0	0.1
χ_{nl}	Gaussian	1.03	0.15

7 STRUCTURAL RELIABILITY OF DAMAGED SHIP IN TOW

Structural reliability analysis is performed for the assumed damage scenario for three towing routes. Following limit-state function is used:

$$\chi_u M_u - \left(\chi_{sw} M_{SW} + \chi_w \chi_{nl} M_w\right) < 0 \qquad (6)$$

where M_u is the deterministic residual ultimate hull-girder bending capacity in sagging; M_{sw} the deterministic still-water bending moment of damaged ship in sagging; M_w the random variable extreme vertical wave bending moment for the towing route and corresponding period; $\chi_u, \chi_w, \chi_{nl}, \chi_{sw}$ the random variables representing the modelling uncertainty of ultimate strength, linear wave load, non-linearity of wave load and still water load respectively. Summary of the stochastic model used in calculation is presented in Table 4.

Safety indices and failure probabilities for three towing routes are calculated by FORM and specified in Tables 5 and 6 for one and two cargo tank damages respectively.

Structural reliability analysis performed so far refers to the "as-built" ship, i.e. unaffected by the corrosion degradation. Corrosion degradation reduces considerably ultimate longitudinal strength. In this particular case of Aframax oil tanker, the ultimate vertical bending moment

Table 5. Safety indices and hull-girder failure probabilities for one tank damage.

Towing route	β	P_f
1	6.59	2.27E-11
2	6.61	1.94E-11
3	6.23	2.32E-10

Table 6. Safety indices and hull-girder failure probabilities for two tank damage.

Towing route	β	P_f
1	6.22	2.55E-10
2	6.21	2.58E-10
3	5.86	2.30E-09

Table 7. Safety indices and hull-girder failure probabilities for one tank damage of "corroded" ship.

Towing route	β	P_f
1	5.37	3.94E-08
2	5.34	4.65E-08
3	4.99	3.02E-07

Table 8. Safety indices and hull-girder failure probabilities for two tank damage of "corroded" ship.

Towing route	β	P_f
1	4.93	4.11E-07
2	4.90	4.79E-07
3	4.57	2.44E-06

capacity is reduced because of the corrosion deduction thickness according to ICAS (2014) by 17%. Therefore, ultimate bending moment capacity in corroded state reads 5976 MNm. Safety indices and failure probabilities of corroded ship are provided in Tables 7 and 8. By comparing results from Table 5 and 6 with those from Table 7 and 8, one can observe increase of the failure probability of corroded ship for about three orders of magnitude.

8 RISK OF SHIP IN TOW

Risk is defined as the product of consequences of an unwanted event and its probability of occurrence. In the present paper, unwanted event is the hull girder collapse while the consequences are expressed in terms of monetary value. Based on previous studies, it is assumed that the major contribution to the total economic loss related to the ship hull girder collapse belongs to the environmental costs (Dong and Frangopol, 2015). Therefore, risk is in the present study defined as:

$$R = P_{H-G\,failure}C(H-G\,failure) \qquad (7)$$

where $P_{H-G\,failure}$ is actually hull-girder failure probability P_f calculated in the previous section, while $C(H-G\,failure)$ is additional cost because of the hull-girder failure. That cost is calculated as the cost of the additional oil spill resulting from the total loss of ship. Therefore, it is assumed that all crude oil remaining on board is spilled to the environment. As subject Aframax oil tanker in full load condition carries about 108000 t of crude oil, difference between this value and initial oil spill resulting directly from damage represents the amount of the crude oil at risk.

For the case of rupture of one cargo tank, crude oil at risk reads 105900 t, while for the case of damage of two cargo tanks, weight of crude oil at risk reads 104266 t. Only cost model proposed by Ventikos and Sotiropoulos (2014) is used in this analysis. Therefore, by applying equation (1), total cost of oil spill reads 421,800,000 US\$ and 416,300,000 US\$ in the case of one tank damage and two tank damages respectively.

Risk of hull-girder collapse of "as-built" ship is calculated by multiplying failure probabilities specified in Table 5 and 6 with associated costs of oil spill, while risk of "corroded" ship hull is obtained using failure probabilities given in Table 7 and 8. Results are presented in Table 9 and Table 10 for "as-built" and "corroded" hull states respectively.

It may be seen that the risk involved with hull girder collapse of a double hull oil tanker is neg-

Table 9. Risk of hull-girder failure of "as-built" damaged oil tanker (US\$).

Towing route	1 Tank rupture	2 Tank rupture
1	0.009	0.106
2	0.008	0.106
3	0.098	0.957

Table 10. Risk of hull-girder failure of "corroded" damaged oil tanker (US\$).

Towing route	1 Tank rupture	2 Tank rupture
1	17	171
2	20	200
3	127	1020

ligible. This is primarily the consequence of the low level of wave loads in the Adriatic. Therefore, according to the present study, it is not necessary to consider risk of hull girder collapse when towing damaged ship to the safe location. Risk involved in such an unfavorable event is much lower compared to the cost of oil spill occurring as a consequence of inner hull damage in collision.

9 CONCLUSIONS

The study describes framework for calculation of the risk of post-accidental salvage operation of damaged double-hull tanker in the Adriatic Sea. The principal failure mode driving the risk assessment is the ship hull girder collapse, representing one of the most unfavourable outcomes of ship accident. Risk of such an event depends on the hull girder probability of failure and the cost of its consequences. Risk assessment is performed by employing short-term ship structural reliability analysis for the failure probability calculation and available models for estimating the cost of oil spill, as the most important consequence of ship hull girder collapse.

It was found that the risk of hull girder collapse of damaged ship during towing is negligible, despite huge economic loss that could occur. That is the consequence of the rather low hull girder failure probability, since the hull girder wave bending moments in the Adriatic Sea are small. For other wave environment, the conclusion could be different.

Therefore, it is of the outmost importance to minimize damage of the initial spill oil that can happen immediately after collision of an oil tanker. The cost of the initial spill in the Adriatic Sea is surpassingly larger than the risk of the additional cost that may occur as the consequence of hull girder collapse.

ACKNOWLEDGMENTS

This work has been supported in part by Croatian Science Foundation under the project 8658 and by Research Support Grant of University of Zagreb.

REFERENCES

Booij, N., Ris, R.C., Holdthuijsen, L.H., 1999. A third generation wave model for coastal regions: model description and validation. J. Geophys. Res. 104 (C4), 7649–7666.

Bužančić Primorac, B., Ćorak, M., Parunov, J. 2015. Statistics of still water bending moment of damaged ship. Analysis and Design of Marine Structures. Guedes

Soares, C., Shenoi, A. R. (ed.). London: Taylor & Francis Group, 2015. 491–497.

Dong, Y., Frangopol D.M., 2015. Probabilistic ship collision risk and sustainability assessment considering risk attitudes, Structural Safety 53, 75–84.

Faisal, M., Noh, S.H., Kawsar, R.U., Youssef, S.A.M., Seo, Y.K., Ha, Y.C. & Paik, J.K., 2017. Rapid hull collapse strength calculations of double hull oil tankers after collisions, Ships and Offshore Structures, 12:5, 624–639.

Farkas A, Parunov J, Katalinić M. 2016. Wave Statistics for the Middle Adriatic Sea. Pomorski zbornik. 13;52(1): 33–47.

Galletta M., 2015. Parametric analysis of collision between a tanker and a ferry, University of Genova, Master Thesis, course: Naval Architecture, supervisors: Cesare Mario Rizzo and Smiljko Rudan.

Hussein, A.W., Guedes Soares, C., 2009, Reliability and Residual Strength of Double Hull Tankers Designed According to the new IACS Common Structural Rules, Ocean Engineering, Vol. 36, pp. 1446–1459.

IACS, 2014, InternationalAssociation of Classification Societies. Common structural rules for bulk carriers and oil tankers. London: IACS.

IMO, 2003. Revised Interim Guidelines for the approval of alternative methods of design and construction of oil tankers under regulation 13F(5) of Annex I of MARPOL 73/78, Resolution MEPC.110(49), International Maritime Organization.

Jia, H., Moan, T., 2008. Reliability Analysis of Oil Tankers with Collision Damage, 27th International Conference on Offshore Mechanics and Arctic Engineering, Estoril, Portugal, Paper number; 57102.

Katalinić M, Ćorak M, Parunov J. 2014. Analysis of wave heights and wind speeds in the Adriatic Sea. In 3rd International Conference on Maritime Technology and Engineering (MARTECH), Lisabon, Portugal, (pp. 5–7).

Kim, Y.S., Youssef, S.A.M., Ince, S.T., Kim, S.J., Seo, J.K., Kim, B.J., Ha, Y.C., Paik, J.K. 2015. Environmental consequences associated with collisions involving double hull oil tankers, Ships and Offshore Structures, 10(5): 479–487.

Klanac, A., Duletić, T., Erceg, S., Ehlers, S., Goerlandt, F., Frank, D. 2013. Environmental risk of collision of enclosed seas: Gulf of Finland, Adriatic and implications to ship design, 5th International Conference on Collision and Grounding of Ships, Espoo, Finland, pp. 55–65.

Kollo, M., Laanearub, J., Tabri, K., 2017. Hydraulic modelling of oil spill through submerged orifices in damaged ship hulls, Ocean Engineering 130, 385–397.

Liberti L, Carillo A, Sannino G., 2013. Wave energy resource assessment in the Mediterranean, the Italian perspective. Renewable Energy. 28; 50:938–49.

Mudronja L, Matić P, Katalinić M., 2017. Data-Based Modelling of Significant Wave Height in the Adriatic Sea. Transactions on Maritime Science. 20; 6(01):5–13.

Parunov, J., Rudan, S., Ćorak, M. 2017a. Ultimate hull-girder-strength-based reliability of a double-hull oil tanker after collision in the Adriatic Sea. Ships and Offshore Structures. 12, S1; S55–S67.

Parunov, J., Ćorak, M., 2015, Design charts for quick estimation of wave loads on damaged oil tanker in the Adriatic Sea, Towards Green Marine Technology and Transport, Guedes Soares C., Dejhalla R., Pavletic D. (Eds.). London, UK: Taylor & Francis Group, pp. 389–395.

Parunov, J., Ćorak, M., Rudan, S. 2017b. Correlation analysis of IMO collision damage parameters, Progress in the Analysis and Design of Marine Structures/Guedes Soares, C., Garbatov, Y. (eds.). Taylor & Francis Group, 477–485.

Reikard G., Pinson P., Bidlot JR., 2011. Forecasting ocean wave energy: the ECMWF wave model and time series methods. Ocean engineering. 31;38(10):1089–99.

RUNET software & expert systems, 2016. cgWindWaves —Forecasting of wind generated waves v4.03, User's Manual, Ioannina, Greece.

Sormunen O-V, Ehlers S and Kujala P. 2013. Collision consequence estimation model for chemical tankers, Proc IMechE, Part M;, Journal of Engineering for the Maritime Environment; 227 (2); 98–106.

Tabain, T., 1997. Standard wind wave spectrum for the Adriatic Sea revisited. *Brodogradnja*, 45 (4), 303–313.

Tabri, K., Aps, R., Mazaheri, A., Heinvee, M., Jönsson, A., Fetissov, M., 2015. Modelling of structural damage and environmental consequences of tanker grounding. In: Analysis and Design of Marine Structures V: 5th International Conference on Marine Structures, 25-27.03.2015, Southampton UK. Ed. C. Guedes Soares and R. Ajit Shenoi. Taylor & Francis, 703–710.

Tavakoli, M.T., Amdahl, J., Leira, B.J., 2012. Analytical and numerical modelling of oil spill from a side tank with collision damage. Ships Offshore Struct. 7 (1), 73–86.

Ventikos NP, Sotiropoulos FS. 2014. Disutility analysis of oil spills: graphs and trends. Mar Pollut Bull. 81:116–123.

Yamada Y. 2009. The cost of oil spill from tankers in relation toweight of spilled oil. Mar Technol. 46:219–228.

Zec D., Maglić L., Šimić Hlača M., 2009. Maritime transport and possible accidents in the Adriatic Sea. 17th Annual Conference of the European Environment and Sustainable Development Advisory Councils EEAC; Dubrovnik.

Special session in honour of Prof. Giovanni Benvenuto

Maritime Transportation and Harvesting of Sea Resources – Guedes Soares & Teixeira (Eds)
© 2018 Taylor & Francis Group, London, ISBN 978-0-8153-7993-5

Waste heat recovery from dual-fuel marine engines

M. Altosole, M. Laviola & R. Zaccone
Department of Electrical, Electronic, Telecommunications Engineering and Naval Architecture,
University of Genova, Italy

U. Campora
Department of Mechanical, Energy, Management and Transportation Engineering, University of Genova, Italy

ABSTRACT: A steam plant layout for the Waste Heat Recovery (WHR) of a marine four-stroke Dual Fuel (DF) engine is analysed by numerical simulation. The different values of exhaust gas mass flow and temperature between the two fuel modes involve different Rankine cycle thermodynamic characteristics, and therefore different heat exchangers sizes in order to obtain the greater waste heath recovery from the engine in both its operating modes. The two developed WHR solutions, optimized for each fuel type, are compared for the normal working conditions of the engine. In particular, the comparison is focused on the maximum plant efficiency and on the size and weight of the Heat Recovery Steam Generator (HRSG) components. Finally the best WHR compromise solution is indicated and tested for different load conditions and fuel types of the considered DF engine.

1 INTRODUCTION

The themes of the pollutant emissions and fuel costs reduction are becoming an increasingly crucial issue in the shipping field. The first topic is due mainly to the increasingly stringent legislation adopted by the Marine Environment Protection Committee (MEPC) regarding sulphur oxides (SO_x) and nitrogen oxides (NO_x) ships emissions. These two pollutant substances are subject to strict limits in the Emission Control Areas (ECA), currently located in the European North Sea, Baltic Sea, English Channel and in both east and west coasts of North America. Moreover ECA limits will be extended to further water areas.

The carbon dioxide (CO_2) emission from ships is also considered with increasing attention in order to control the greenhouse effect on the planet. The reduction of CO_2 and other pollutants can be achieved by using more efficient engines and/ or alternative less polluting fuels, as in the case of Natural Gas (NG), in ships but also in smaller applications (Altosole et al. 2014). About this, the liquefied natural gas (LNG) fueled fleet increased in the 2010–2014 period, and a further increase is expected in the next future (Pedersen 2015). LNG engines reduce strongly the pollutant emissions typical from diesel engines fueled by Heavy Fuel Oil (HFO). In detail: $-25 \div -30\%$ CO_2; -25% CO; -85% NO_x; $-98 \div 100\%$ SO_x; $-90 \div 99\%$ particulate matter (Altosole et al. 2017).

A further emissions reduction can be achieved by the Waste Heat Recovery (WHR) from the ship propulsion engines (about 25% of the energy delivered by the fuel combustion), for the production of electric energy through a steam turbine, reducing the onboard diesel generators power and their related emissions. In addition, the simultaneous use of Dual-Fuel (DF) engines with WHR steam plants increase the onboard energy efficiency, with consequent reduction of fuel cost and pollutants.

Several studies on WHR systems, fed by marine diesel engines exhaust gases, are proposed in literature (Ioannidis 1984, Ito & Akagi 2007, Dzida. 2009a and b, Akiliu & Gilani 2010, Dimopopulos & Kakalis 2010, Grimmelius et al. 2010, Chul & Min 2013, Altosole et al. 2015, Benvenuto et al. 2014, 2015) but as far as authors' knowledge is concerned, only Livanos et al. (2014) refer to marine DF engines combined with WHR steam plants.

It is known that marine diesel oil characteristics (i.e. lower heating value, in cylinders air fuel ratio values, sulphur content, ignition start modality), in diesel cycle (compression ignition), are different in comparison with NG, burned in the same engine according to Otto cycle (spark-ignition). Therefore, DF engines exhaust gases, depending on the fuel typology, present different properties as: mass flow, temperature, chemical compositions, sulphur oxide presence (diesel cycle) or near absence (in Otto cycle, during the NG mode, about 99% of the fuel is NG and 1% is diesel oil for the pilot ignition). These differences affect the optimization procedure of the WHR steam cycle thermodynamics, and therefore some characteristics of the WHR plant components, (e.g. heat exchanges surfaces).

Starting from an original WHR single pressure steam plant layout, this paper investigates the different Rankine cycle characteristics and heat exchangers geometry for a marine DF WHR plant. Each WHR system design is optimized in order to obtain the maximum heat recovery, as far as possible, in both the engine fuel modes. All the optimized WHR plants are able to produce electric power and saturated steam for onboard utilities.

Finally, the best compromise solution will be addressed for the considered DF engine.

2 WHR SYSTEM MODELLING

The considered WHR steam plant layout for saturated steam production is illustrated in Figure 1.

The WHR system recovers the waste heat of the DF engine through the Heat Recovery Steam Generator (HRSG), composed by an Economizer (E) and an Evaporator (EV). The produced saturated steam is delivered to a Steam Turbine (ST)

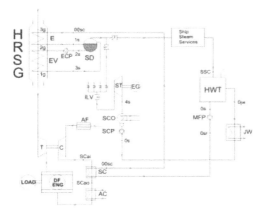

Figure 1. WHR plant layout, where 0jw = Jacket Water (JW) outlet; 0s = Heat Water Tank (HWT) water outlet; 0sr = Scavenger (SC) water inlet; 00sc = SC water outlet and economizer (E) water inlet; 00sr = economizer water inlet; 1g = Evaporator (EV) gas inlet; 1s = economizer water outlet; 2g = economizer gas inlet; 2s = EV water outlet; 3g = HRSG gas outlet; 3s = EV steam outlet; 4s = Steam Turbine (ST) outlet; AC = Air Cooler; AF = Air Filter; C = turbocharger Compressor; DF ENG = Dual Fuel Engine; ECP = Evaporator Circulating Pump; EG = Electric Generator; ILV = isenthalpic Lamination Valve; MFP = Main Feed Pump; SCai = SC air inlet; SCao = SC air outlet; SCO = Steam Condenser; SCP = Scavenge Condensing Pump; SD = Steam Drum; SSC = Steam Service Condenses; T = turbocharger Turbine.

driving an Electric Generator (EG). The exhaust steam condenses inside a Steam Condenser (SCO). The saturated water is extracted from the SCO by a Steam Condensing Pump (SCP): it is preheated by the Jacket Water (JW) before to be delivered to the Heat Water Tank (HWT), from which the Main Feed Pump (MFP) moves to the HRSG economizer, after to be warmed by the turbocharger compressor outlet hot air in the engine Scavenger (SC). A part of the saturated steam is taken from the Steam Drum (SD) to satisfy the steam service of the ship.

A numerical code is developed in MATLAB ® environment for the preliminary design of the WHR steam plant components, whose size, weight and performance are evaluated. The detailed procedure for the WHR steam plant Rankine cycle characteristics definition, components design and overall performance optimization is reported in Benvenuto et al. (2014) therefore only a brief description is described in the following.

2.1 Basic equations

The steady state continuity and energy equations are the main equations of the developed design procedure for the several components:

$$\sum (M_i - M_o) = 0 \tag{1}$$

$$\sum (M_i h_i - M_o h_o) = Q' - P \tag{2}$$

where M_i = inlet mass flow rate; M_o = outlet mass flow rate; h_i = inlet specific enthalpy; h_o = outlet specific enthalpy; Q' = heat flow; and P = power.

The HRSG considered in this study is characterized by a cross-counter flow configuration with horizontal finned tubes. The finned pipes wall overall heat exchange coefficient K (for economizer and evaporator) is calculated as:

$$\frac{1}{K} = \frac{1}{h_e} + R_e + \frac{s}{k} \frac{A_e}{A_{ml}} + R_i \frac{A_e}{A_i} + \frac{1}{h_i} \frac{A_e}{A_i} \tag{3}$$

where A_e = pipe wall external area; A_i = pipe wall internal area; A_{ml} = pipe wall logarithmic mean area; h_e = external pipe convective heat transfer coefficient; h_i = internal pipe convective heat transfer coefficient; k = pipe wall thermal conductivity; R_e = pipe external fouling thermal resistance; R_i = pipe internal fouling thermal resistance; s = pipe wall thickness.

Convective heat transfer coefficients are determined by using finned tube correlations reported in Benvenuto et al. (2011).

For each HRSG heat exchangers, the pertinent heat exchange area A, necessary for the pipes length evaluation, is determined by means of the balance equation applied to the gas-wall and wall-steam heat exchange (Benvenuto et al. 2012):

$$Q' = K \, A \, \Delta T \tag{4}$$

where ΔT is the logarithmic mean temperature difference between hot gas and steam-water. From the thermal exchange area A, it is possible to evaluate the total length of the pipes and the steam pressure losses.

The design of SCO, JW and SC heat exchangers is carried out through a very similar procedure. The steam turbine mathematical model is based on the typical steady state non-dimensional mass flow map (Cohen et al. 1987), while the HRSG steam drum and heat water tank are modelled as simple fluid mixing tanks. The mathematical model was validated by comparing simulation results with data of similar existing plants. A very good agreement, especially in design conditions, was observed not only for physics and thermodynamic parameters, but also for the main characteristics of the components (i.e. heat exchangers surface and HRSG pipes number).

The difference between calculated parameters and reference data is generally less than 1% in design conditions, while is less than 4% in off-design working conditions. Further details on the whole validation process can be found in Benvenuto et al. (2012), Benvenuto et al. (2014) and Benvenuto et al. (2015).

3 CASE STUDY

The examined WHR steam plant scheme is applied to an existing cruise ferry of the GNV shipping line (overall length = 211 m; displacement = 49257 tons; cruising speed = 22 knots). The propulsion system consists of two controllable pitch propellers, each driven by two diesel engines Wartsila 16V46C, delivering 16800 KW at 500 rpm each one. In the normal sea going condition, each engine runs at 72÷80% of the Maximum Continuous Rating (MCR). This ferry travels every day between two main Italian ports, sailing six days a week in wintertime and all days in summertime. To fulfill the electric users, four diesel generators Wärtsilä 6R32LNE (2430 kWe at 720 rpm) are installed.

The thermal power required by the ship services is satisfied by 1.15 kg/s of saturated steam at least 7 bar. In this study, it is considered that the services steam is produced by the proposed WHR steam plant (one for each engine) and delivered to a single steam turbine, while the originally

four–stroke diesel engines are substituted by an equal number of MAN 51/60 18V DF four-stroke engines, whose MCR power is 17550 KW at 500 rpm. The ship engine room is able to contain the new engines equipped with WHR systems, satisfying the mentioned required thermal power at 75% MCR load (about 13160 KW) of the engines.

The engine manufacturer data of Figure 2 and Figure 3 show the remarkable difference between the two possible fuels (HFO or NG) in terms of exhaust gas mass flow rate and temperature at the turbocharger turbine outlet. It is important to consider that in NG mode, the HFO fuel tank heating is not necessary and then the ship services thermal power is reduced by 70%: therefore only 0.35 kg/s of saturated steam, at least 2 bar, is sufficient in this condition.

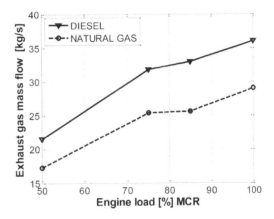

Figure 2. Exhaust gas mass flow in different fuel modes (MAN 51/60DF Project Guide 2014).

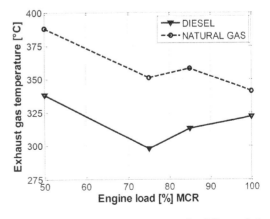

Figure 3. Exhaust gas temperature in different fuel modes (MAN 51/60DF Project Guide 2014).

4 WHR OPTIMIZATION

Once defined the engine working conditions, the Rankine cycle parameters (i.e.: pressures, temperatures, water/steam mass flow rate) in the several WHR plant components (HRSG, steam turbine, condenser, etc.) are thermodynamically optimized. To this aim, the optimization goal is the maximum efficiency η_{CC} of DF engine—WHR combined plant (DF-WHR):

$$\eta_{CC} = \frac{P_{DF\,E} + P_{ST} - P_{MFP} - P_{SCP}}{M_f\,LHV_f} \qquad (5)$$

P_{MFP} = power of the main feed pump; P_{SCP} = power of the steam condensing pump; P_{ST} = steam turbine power.

In equation(5) the fuel mass flow and its lower heating value are referred to HFO, in the case of diesel engine cycle, while 99% NG plus 1% Marine gas Oil (MGO) is considered for NG mode. In accordance with manufacturer indications, the following lower heating values are adopted: 42700 kJ/kg for HFO and MGO, and 49000 kJ/kg for NG.

As explained in Benvenuto et al. (2014), starting from a series of "fixed parameters", reported in Table 1, the WHR steam plants for both fuel modes of the engines are optimized by the variation of the values shown in Table 2.

The optimization procedure takes in to account some constrains, mainly indicated in Hansen (1990). One of the most important is the stack temperature limited to 160 °C, as minimum value, as suggested by several marine engine manufacturers. Prudently, in this research, the limit is raised to 170 °C in the case of HFO fuel. On the contrary, this limit is not considered for NG, due to the substantial absence of sulfur in this fuel type (the eventually sulfur content in the diesel oil of the pilot ignition is negligible, since the fuel oil injected in each cycle is about 1% of NG mass).

Table 1. Constant parameters (75% MCR power).

Fixed parameters	NG	HFO
Exhaust gas mass flow rate [kg/s]	26	32
Exhaust gas temperature [°C]	352	299
Condenser pressure [bar]	0.08	0.08

Table 2. Constant parameters (75% MCR power).

Optimization parameters	Range	Variation step
SD pressure [bar]	8÷20	0.5
Pinch point temperature difference ΔT_{pp} [°C]	5÷20	1

A further constrain in the HRSG design procedure is the back pressure after the engine turbocharger. The MAN society suggests 35 mbar of maximum backpressure. The engine specific fuel consumption (sfc) increase, due to the back pressure, is considered as suggested by Wartsila: 0.3 g/KWh for each 100 mm H_2O of back pressure increase.

As reported above, when the engine burns HFO as fuel, the onboard services require saturated steam at least 7 bar. This represents a further important constrain, not existing in NG fuel mode. For this reason, the evaporator steam drum pressure minimum value, adopted in the variation range of Table 2, is 8 bar, to consider the pressure loss due to the ship steam service pipes. In order to not differentiate excessively the two WHR systems, the same value is supposed in the case of NG fuel.

For the HRSG (vertical type) design, more detailed are reported in Benvenuto et al. (2011).

The main results of the WHR steam plant optimization procedure are shown in Table 3, where the results are pertinent to both the fuel modes of the engines.

The gas and water/steam temperature distributions are reported for both the NG engine—WHR Combined Plant (named NG-WHR CP) and the Diesel Engine—WHR Combined Plant (named DE-WHR CP) in Figure 4 and Figure 5, respectively.

Due to the minor NG-WHR HRSG pinch point temperature differences (ΔT_{pp} in Table 3), and also due to a greater number of the NG-WHR HRSG

Table 3. DF-WHR CP geometrical and plant performance parameters.

Optimization parameters	NG	HFO
SD pressure [bar]	8	8
Pinch point temp. difference ΔT_{pp} [°C]	5	13
HRSG efficiency [%]	58	43.9
HRSG outlet temperature (3 g) [°C]	162	178
HRSG gas side pressure loss [mbar]	30.1	34.6
+ E_{sfc} (pressure loss) [g/kWh]	0.92	1.06
HRSG height [m]	4.3	3.6
HRSG weight [t]	28.6	21.7
HRSG economizer pipes number	7	3
HRSG evaporator pipes number	34	28
η_E [%]	47.3	45.7
η_{CC} [%]	50.3	47.8
η_{COG} [%]	55.2	52.5
Engine power (75% MCR) [kW]	13163	13163
ST power [kW]	1718	1199
ST efficiency [%]	70.9	70.9
ST mass flow rate [kg/s]	3.54	2.47
SSS mass flow rate [kg/s]	0.35	1.15

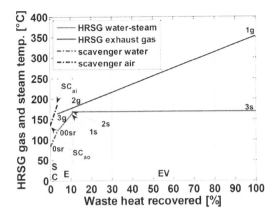

Figure 4.   HRSG temperature profiles (NG-WHR CP).

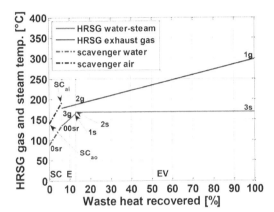

Figure 5.   HRSG temperature profiles (DE-WHR CP).

economizer pipes, the HRSG outlet section exhaust gas temperature (i.e. 3$_g$ in the scheme of Figure 1) is lower in the case of NG engine, in comparison with the HFO (162°C and 178 °C, respectively).

As for HRSG, the data reported in Table 2 show a higher efficiency (η_{HRSG}) when the engine fuel is NG:

$$\eta_{HRSG} = \frac{T_{1g} - T_{4g}}{T_{1g} - T_{amb}} \qquad (6)$$

where $T1_g$ = EV inlet gas temperature; $T4g$ = HRSG outlet gas temperature; and $Tamb$ = ambient temperature.

The exhaust gas temperature at the outlet section is 170 °C for the HFO and lower for NG. The economizer pipes number is greatly different for the two fuel types, while a minor difference is for EV pipes number and HRSG height and weight (Table 3).

In the NG case, the increase of the efficiency η_{CC}, in comparison with the engine efficiency η_E (i.e. without WHR system), is about 3%; while the same efficiency difference is about 1% less than HFO fuel. This is mainly due to the steam turbine power and the minor ship steam service mass flow rate required in the case of engine running with NG fuel, as reported in Table 3. In the same table, the DF-WHR plant cogeneration efficiency (η_{COG}) is determined by:

$$\eta_{COG} = \frac{P_{DF\,E} + P_{ST} + Q'_{SSS} - P_{MFP} - P_{SCP}}{M_f\,LHV_f} \qquad (7)$$

where Q'_{sss} = heat flow of the ship steam services; and the other terms are the same defined in equation (5).

5   OFF-DESIGN CONDITIONS

A second numerical code has been developed to assess the DF engine—WHR steam plant performance in off-design load conditions. Once estimated the main characteristics of the steam plant components (through the design procedure previously described), the steady-state performance of the WHR system is evaluated by using the same basic equations.

For validation purposes, the same engine working condition considered for the WHR steam plant design (i.e. 75% MCR power at 488 rpm) is reproduced by the off-design simulator. The results are compared in Table 4, denoting small discrepancies with data reported in Table 3. In detail, the fourth column of Table 4 shows the investigation of the final considered solution, represented by the WHR plant optimized for the NG engine. In this case, it is necessary the partial closing of the ILV valve, located at the steam turbine inlet, in order to maintain the HRSG outlet temperature over the limit of 170 °C. This valve reduces its outlet pressure, resulting unchanged the steam enthalpy (a 30% of steam pressure drop is considered, as shown in Table 4) and involving an increase of the HRSG steam drum pressure up to 8.4 bar (necessary for the ship steam service supply in the case of HFO use).

The engine power values considered for the off-design conditions of the WHR system (optimized for the NG fuel mode of the engine) are:

 – 100% MCR power at 500 rpm;
 – 85% MCR power at 500 rpm;
 – 75% MCR power at 488 rpm;
 – 50% MCR power at 462 rpm;

The computation results are reported in Table 5 (NG mode) and Table 6 (HFO mode), while

Table 4. DF-WHR CP off design simulators results at 75% MCR engine power and 488 rpm speed.

WHR plant parameters	WHR NG -NG fuel-	WHR HFO -HFO fuel-	WHR NG -HFO fuel-
SD pressure [bar]	7.9	8	8.4
ST [% of pSD]	100	100	70
ΔTpp [°C]	5.1	12.9	70
HRSG parameters			
$HRSG_{eff}$[%]	58.3	43.8	46.0
Gas temperature (3g) [°C]	164	178	171
HRSG pr. loss [mbar]	30	34.5	43.0
+ Esfc [g/kWh]	0.92	1.05	1.32
HRSG height [m]	4.3	3.6	4.3
HRSG weight [t]	28.6	21.7	21.7
HRSG E pipes number	7	3	7
HRSG EV pipes number	34	28	34
Engine & WHR parameters			
η_E [%]	47.3	45.7	45.6
η_{CC} [%]	50.2	47.8	47.7
η_{COG} [%]	55.1	52.5	52.4
75% MCR Engine [kW]	13163	13163	13163
ST power [kW]	1702	1203	1274
ST efficiency [%]	71.1	71.0	70.8
ST mass flow rate [kg/s]	3.52	2.47	2.61
SSS mass flow rate [kg/s]	0.35	0.35	1.15

Table 5. NG engine—WHR combine plant off design simulator results for different engine power.

Engine & WHR plant parameters	Engine power [MCR%] – NG			
	50	75	85	100
Engine speed [rpm]	462	488	500	500
SD pressure [bar]	6.7	7.9	8.3	8.7
ST [% of p_{SD}]	100	100	100	100
ΔT_{pp} [°C]	2.6	5.1	5.7	6.2
η_E [%]	46.3	47.3	47.3	47.5
η_{CC} [%]	50.0	50.2	50.1	50.0
η_{COG} [%]	52.7	52.3	51.9	51.6
ST power [kW]	1384	1702	1801	1807
ST efficiency [%]	71.4	71.1	69.9	69.8
ST mass flow rate [kg/s]	3.01	3.52	3.87	3.80
SSS mass flow rate [kg/s]	0.35	0.35	0.35	0.35
$HRSG_{eff}$[%]	68.3	58.0	57.1	53.2
Gas temperature (3g) [°C]	148	164	169	174
HRSG pr. loss [mbar]	14.1	30.0	31.2	39.8

efficiency, pressure and temperature values are illustrated from Figure 6 to Figure 10 for a better understanding.

The engine types and DF-WHR CP efficiencies are visualized in Figure 6. As typically found, the NG engine efficiency is always 1÷2% higher than HFO one but it is always lower than DE-WHR CP efficiency.

The increase of DF-WHR CP efficiency, in comparison with single NG engine and diesel engine, is illustrated in Figure 7. The figure highlights the WHR system efficiency increase at low engine powers for NG fuel, in comparison with

Table 6. DE engine—WHR combine plant off design simulator results for different engine power.

Engine & WHR plant parameters	Engine power [MCR%] – DE			
	50	75	85	100
Engine speed [rpm]	462	488	500	500
SD pressure [bar]	11.8	8.4	8.5	9.1
ST [% of pSD]	40	70	90	100
ΔT_{pp} [°C]	3.6	7.0	7.1	7.6
η_E [%]	44.9	45.6	46.5	46.0
η_{CC} [%]	47.6	47.8	48.9	48.6
η_{COG} [%]	54.6	52.5	53.2	52.1
ST power [kW]	1075	1274	1616	1981
ST efficiency [%]	69.8	70.8	70.8	70.5
ST mass flow rate [kg/s]	2.07	2.61	3.31	3.98
SSS mass flow rate [kg/s]	1.15	1.15	1.15	1.15
$HRSG_{eff}$[%]	53.0	46.0	49.1	49.0
Gas temperature (3g) [°C]	171	171	170	176
HRSG pr. loss [mbar]	21.0	43.0	48.2	58.0

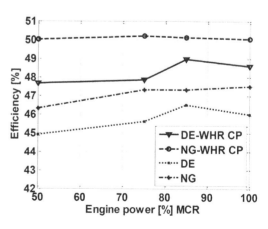

Figure 6. Engine and DF-WHR CP efficiency vs engine power.

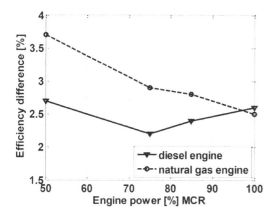

Figure 7. DF-WHR CP efficiency increase compared to the single diesel engine and NG engine.

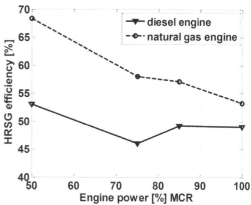

Figure 9. HRSG efficiency vs engine power.

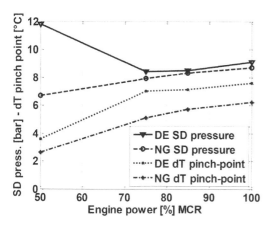

Figure 8. SD pressure and ΔT_{pp} vs engine power.

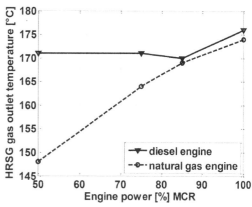

Figure 10. HRSG gas outlet temperature vs engine power.

HFO. This difference decreases up to zero around MCR power.

Figure 8 shows that the HRSG steam drum pressure always decreases together with the NG engine power reduction, since the ILV valve is always full open in NG mode. On the contrary, when the engine is HFO fueled, the progressive ILV valve closure in accordance with engine power decrease (to satisfy the above mentioned constrains: SD 8 bar minimum pressure and HRSG exhaust gas exit temperature not less to 170°C) leads to a slow SD pressure reduction, while the same pressure increases when the engine power drops below 75% MCR value. The same figure reports that the pinch point temperature differences decrease approximately in the same way for the two engine fuels, and that the ΔT_{pp} is always higher in HFO engine mode.

Figure 9 shows that the HRSG efficiency is always better in NG mode and it increases while the engine power decreases. This behavior is due to the contemporary decrease of the HRSG outlet gas temperature, as illustrated in Figure 10.

6 CONCLUSIONS

In DF marine engines the exhaust gas mass flow and temperature values are significantly different, depending on the engine fuel mode (HFO or NG). As a consequence, the choice of the best WHR characteristics is rather complex.

Through a numerical model, developed for the geometry definition of the main WHR components, the steady state performance of the proposed WHR plant has been investigated for both engine fuels, at several engine delivered powers. Although the proposed solution is optimized for NG fuel use, the WHR efficiency is similar for both fuel modes at high power values, while the behavior is different by reducing propulsion engine power. The HFO constrains (170°C for the HRSG outlet gas temperature and 8 bar for the minimum pressure of the ship steam services) involve the steam pressure lamination at the steam turbine inlet through the ILV valve. In NG mode, these constrains do not exist, thus HRSG steam drum sliding pressure solution can be adopted and an ever greater HRSG efficiency, in comparison with HFO mode, is possible at low engine loads. In addition, by using NG fuel, HFO heating is no longer necessary and therefore less steam mass flow is required for the ship service, resulting into a further improvement of the WHR system overall efficiency.

REFERENCES

Akiliu, B.T. & Gilani, S.I. 2010. Mathematical Modelling and Simulation of a Cogeneration Plant. *Applied Thermal Engeneering*, No. 30: 2545–2554.

Altosole, M., Benvenuto, G., Campora, U., Laviola, M., Zaccone, R. 2017. Simulation and performance comparison between diesel and natural gas engines for marine applications. *Journal of Engineering for the Maritime Environment*, 231 (2): 690–704.

Altosole, M., Buglioni, G., Figari, M. 2014. Alternative propulsion technologies for fishing vessels: A case study. *International Review of Mechanical Engineering*, 8 (2): 296–301.

Altosole, M., Laviola, M., Trucco, A., Sabattini, A. 2015. Waste Heat Recovery systems from marine diesel engines: comparison between new design and retrofitting solutions. In Guedes Soares & Santos (eds), Maritime Technology and Engineering; Proc. intern. symp., Lisboa, 14–17 October 2014.

Benvento, G., Bertetta, D., U. Campora, Carollo, F. 2012. COGAS Plant as Possible Future Alternative to the Diesel Engine for the Propulsion of Large Ships. In Rizzuto & Guedes soares (eds), 14th International Conference on Maritime Association of Mediterranean, Genova, Italy, 13–16 September 2011.

Benvenuto, G., Campora, U., Laviola, M., Zaccone, R. 2015. Comparison of Waste Heat Recovery Systems for the Refitting of a Cruise Ferry. In Altosole & Francescutto (eds), 18th International Conference on Ships and Shipping Research, Lecco, Italy, 24–26 June 2015.

Benvenuto, G., Campora, U., Trucco, A. 2014. Optimization of Waste Heat Recovery from the Exhaust Gas of Marine Diesel Engines. *Journal of Engineering for the Maritime Environment*, (online version: June 9, 2014, pp. 1–12).

Chul, C.B. & Min, K.Y. 2013. Thermodynamic analysis of a dual loop heat recovery system with trilateral cycle applied to exhaust gases of internal combustion engine for propulsion of the 6800 TEU container ship. *Energy*, 58: 404–416.

Dimopopulos, G.G. & Kakalis, N.M.P. 2010. An Integrated Modelling framework for the Design Operation and Control of Marine Energy Systems. Proc. CIMAC Congress, Bergen, Norway, Paper No. 15.

Dzida, M. 2009a. On the Possible Increasing of Efficiency of Ship Power Plant with the System Combined of marine Diesel Engine, Gas Turbine and Steam Turbine, at the Main Engine-Steam Turbine Mode of Cooperation. *Polish Maritime Research* 1(59), Vol. 16: 40–44.

Dzida, M. 2009b. On the Possible Increasing of Efficiency of Ship Power Plant with the System Combined of marine Diesel Engine, Gas Turbine and Steam Turbine, at the Main Engine-Steam Turbine Mode of Cooperation. *Polish Maritime Research* 2(60), Vol. 16: 47–52.

Grimmelius, H., Boonen, E.J., Nicolai, H., Stapersma, D. 2010. The integration of Mean value First Principle Diesel Engine models in Dynamic Waste Heat and Cooling Load Analysis. Proc. CIMAC Congress, Bergen, Norway, Paper No. 280.

Ioannidi, J. 1984. Thermo Efficiency System (TES) for reduction of Fuel Consumption and CO2 Emission. *Publ. No.: P3339161*. Copenhagen (Denmark).

Ito, K. & S. Akagi. 2007. An optimal planning method for a marine heat and power generation plant by considering its operational problem. *Energy Research*, Vol. 10 (1): 75–85.

Livanos, G.A., Theotokatos, G., Pagvonis, D.N. 2014. Technoeconomic investigation of alternative propulsion plants for Ferries and RoRo ships. *Energy Conversion and Management* 79: 640–651.

Maritime Transportation and Harvesting of Sea Resources – Guedes Soares & Teixeira (Eds)
© 2018 Taylor & Francis Group, London, ISBN 978-0-8153-7993-5

Numerical and experimental investigation for the performance assessment of full electric marine propulsion plant

M. Martelli & M. Figari

Department of Electrical, Electronic, Telecommunications, Naval Architecture and Marine Engineering (DITEN),
Polytechnic School of Genoa University, Genova, Italy

ABSTRACT: The advantage of hybrid diesel-electric propulsion is well known in terms of fuel consumption, especially for ships with variegated speed profiles or with different propulsion loads. Nowadays several hybrid configurations are available and in the preliminary design phase, the selection of the best configuration among several options could be a time-consuming task. Therefore, in this paper, a methodology is proposed, based on the main propulsion design parameters, to identify the global ship energy efficiency in function of the operational profile, considering both design and off-design conditions. In order to validate the method, a sea trial campaign has been done and the records have been analyzed. The developed tool helps, during the design phase, to choose the optimal propulsion configuration, number, type, and size, of the main engines and generators. The methodology is used to compare two design solutions, a conventional diesel propulsion and a proposed full electric plant using a 'father and son' power generation.

1 INTRODUCTION

Since a long time, the energy efficiency is one of the most important aspects, together with the safety during navigation, for the ship owners.

Nowadays several strategies could be adopted, to improve the propulsion ship performance in terms of fuel consumption, for example, optimizing from an hydrodynamics point of view (Nelson et al. 2013), both the hull and propeller geometry; this solution is only feasible for a new ship.

Fluoropolymer painting increases the energy efficiency of existing hulls, by providing a low-friction, ultra-smooth surface on which organisms have great difficulty settling (Candries et al. 2000).

Another solution is to replace fixed pitch propellers with controllable pitch propellers (Altosole et al. 2012, 2014) or ducted propellers; this last device is suitable for heavily loaded propellers, such as trawlers and tugboats, where high thrust is needed at low vessel speed (Martelli et al. 2016).

Some solutions recently proposed to increase the thermal efficiency of diesel engines, i.e. magnetic devices for fuel condition, were tested and no improvement was assessed (Gabiña et al. 2016).

In the case of an existing boat, the efficiency could be improved replacing the main propulsion components with more efficient and newer ones, or it is possible to redesign the whole propulsion plant.

The latter solution will be deep analyzed in this paper.

Energy optimization requires engineering studies and real data feedback. Engineering studies require technical data and drawings of existing vessels, generally not available. This drawback can be overcome by the availability of real field data, in particular, to better understand which kind of improvements could be done it is necessary to define the energy profile of the vessel. This kind of data come from the Energy Audit, an engineering test for the monitoring of energy usage during normal fishing activities (Buglioni et al. 2012, Notti & Sala 2014).

To analyze the huge amount of data, and to help the designer to develop a new power architecture is needed rigorous mathematical models and a computer-based tool.

Therefore, in this paper, a methodology is proposed, based on the main propulsion design parameters, to identify the global ship energy efficiency in function of the operational profile, considering both design and off-design conditions.

The developed tool helps, during the design phase, to choose the optimal propulsion configuration, number, type, and size, of the main engines and generators. For validation purposes, the propulsion plant of an existing oceanographic vessel, equipped with a four-stroke diesel engine that drives a ducted controllable pitch propeller is modeled.

To confirm the method reliability a sea trial campaign has been done and the data have been analyzed (Notti & Sala 2012). The proposed methodology was also used to design a new propulsion configuration, with electric prime movers and considering an asymmetrical power generation with a battery pack. This configuration gives a great flexibility to the propulsion plant and it is optimized for all the different missions that oceanographic vessels have to deal with.

At the end of the paper, a comparison between the two design solutions, existing and the new one is carried out; the saving in term of fuel consumption during several real operative situations is highlighted.

2 EFFICIENCY EVALUATION

The global propulsive efficiency is more than the combination of the energy efficiency of the isolated machinery or elements that compose the propulsion plant. All the propulsion elements (engine, gearbox, bearing, propulsor) interact, affecting each other. In this view, a mathematical model, static or dynamic, is needed first to assess the single elements behavior, secondary to catch the mutual interactions. Since in design phase, thousands of combinations should be studied, to obtain reliable results in a reasonable time, a steady state approach has been used. In the following, the methods adopted to model the engine, the propeller, the transmission line and the mutual interaction are presented.

The first element to be modeled is the thermal engines (both prime mover/s and diesel generator/s). A great number of propulsion plants deal with several operation profiles that differ in terms of ship speed, propulsor loads, boundary conditions and constraints. Due to economical (Castles et al. 2009), environmental (Eyring et al. 2005, Larsen et al. 2015), and legislative (IMO 2009a, b) constraints, the knowledge of the fuel consumption on the whole set of engine working points is a crucial aspect. Based on previous motivations, the standard data, often available from engine manufacturers, only on a cubic power request, are not enough. The specific fuel consumptions q_s, and consequently the engine efficiency η_{Eng}, is assessed using a polynomial surface, its form is reported in following:

$$q_s(N,P_b) = \sum_{i=1}^{4}\sum_{j=1}^{4} P_{ij} N^i P_b^j \tag{1}$$

where P_{ij} are the coefficients of the polynomial, obtain through the analysis of several fuel consumption data related to different four stroke diesel engines.

The fuel map obtained is function of both of engine speed N, and of brake power P_B, and used as response surface (Altosole & Figari 2011).

The propeller performances are evaluated using open water characteristics, thanks to which it is possible to evaluate the non-dimensional thrust coefficient K_T and torque coefficient K_Q depending on both the non-dimensional advance coefficient J and the propeller pitch angle φ.

$$T = K_T(J,\varphi)\rho n^2 D^4$$
$$Q_o = K_Q(J,\varphi)\rho n^2 D^5 \tag{2}$$

where ρ is the seawater density, D is the propeller diameter and n is the propeller revolution regime. Among the several methods that could be used to evaluate the propeller characteristics, the systematic series approach (Kuiper 1992) results more suitable for this application, due to its low computational cost.

The ship drag and the propulsive coefficients can be modeled using three different methods: towing tank tests, statistical regression (Holtrop & Mennen 1982, Von Oortmerssen 1971) or the systematic series.

Once both hull and propeller performance, together with their mutual interaction, are known, it is possible to obtain the equilibrium point in term of shaft line revolution and required power, using the well know engine-propeller matching procedure. The drag-thrust equilibrium problem is solved using the non-dimensional factor, K_T/J^2, from which the advance coefficient J, the propeller rotational regime n and the required propeller power P_B are obtained, for each velocity and for each propeller pitch.

After this, to evaluate the propulsive energy consumption of the vessels, first, the overall propulsive coefficient (OPC) has to be calculated (17th ITTC 1984) as follows:

$$O.P.C. = \frac{P_E}{P_B} = \frac{1-t}{1-w}\eta_o\eta_r\eta_m \tag{3}$$

where P_E is the effective power; P_B is the brake power; η_o is the propeller open water efficiency; η_r is the relative rotative efficiency; and η_m is the mechanical efficiency defined as the product between the gear and bearings efficiencies.

The overall propulsive coefficient it is not sufficient to identify the global ship efficiency because does not take into account the efficiency of the prime mover. In fact, an optimum working point from the hydrodynamic point of view not always matches with a good performing engine working point. In order to define an holistic assessment of the propulsion energy efficiency, the mass fuel flow rate $\dot{m}_f$ has been introduced, depending on specific fuel consumption q_s and the delivered power P_B, see Equation 4. The latter is then used to define the global propulsive energy index η_{TOT} as the ratio between the effective power P_E and the chemical fuel power $\dot{m}_f H_i$, as shown in Equation 5, where H_i is the lower heating value of the fuel. By substitution, it was possible to obtain η_{TOT} (Martelli et al. 2016) as shown in Equation 6.

$$\dot{m}_f = q_s P_B \tag{4}$$

Figure 1.   R/V "G. Dallaporta".

$$\eta_{TOT} = \frac{P_E}{\dot{m}_f H_i} \qquad (5)$$

$$\eta_{TOT} = \frac{OPC}{q_s H_i} \qquad (6)$$

The 'propulsion global energy efficiency' could be now assessed for several speeds, under different operational conditions, and in case that the propulsion plant has two degrees of freedom, for all the possible equilibrium pair (n, φ) as shown in Figure 5.

In order to have a complete overview of the energy production and demand, it will be mandatory to assess also the 'ship global efficiency'; in authors' opinion this means to take into account also the energy consumption, or the power required by the auxiliary systems, P_{Aux}.

$$\eta_{Ship} = \frac{P_E + P_{Aux}}{\sum_{j=1}^{n} \dot{m}_{f_j} H_{i_j}} \qquad (7)$$

This last formula expresses the efficiency of the whole ship during operations, taking into account the total amount of the fuel burned onboard. Since the lack of experimental data, this aspect will be the target of further analysis in future.

3   CASE STUDY

To validate the proposed methodology, the performance of the Italian National Research Council (CNR) ship "G. Dallaporta" are analyzed. A numerical model has been developed and the performance and energy efficiency has been evaluated.

Analyzing the historical navigation data, an operational profile made up by four different missions and the harbor stops has been identified. The four missions are briefly described in the following.

Sea Water Sampling: The ship sails from the harbor to the sea area to be sampled at cruising speed. Once the sampling area is reached, the ship stops at the first sampling point, collects the water and moves to the next point, usually about four miles away. About 20 minutes, sailing at cruising speed, are needed to this transfer phase. The duration of this type of mission is not easily defined because it depends on the number of stations sampled.

Offshore Platform Monitoring: The pattern of this task is similar to the previous one. The difference is that the sampling takes place at four points around the offshore platform, about 1.5 miles far away from it, plus four additional points near the platform along which the ship moves. Such sampling takes about 3 hours and a half, including one and a half hour needed for stations near the platform.

Acoustic Survey: The campaign aims to associate the acoustic survey of the pelagic fishery with the actual amount of biomass sampled by fishing. The ship, in the sampling area, carries out a serpentine pathway to cover a large sea area. This path is carried out at a speed not exceeding nine knots, in order to avoid disturbances to the acoustic equipment. At constant intervals, the ship shoots and tows the fishing gear for about 30 minutes at a speed of 4 knots.

Fishing gear testing: In this activity, the performance and the behaviour of standard and innovative fishing gears, are tested; the ship tows the fishing gear, with a high propeller load, with a speed between three and four knots for about one hour, five times a day, three days a week.

The ship operates overall more than 200 days per years, and the summary of the previously described activities are shown in Table 2.

The step forward is the calibration of both ship and towing instruments. A comparison of the ship drag obtained with two different methods (Holtrop & Mennen 1982, Von Oortmerssen 1971), and towing tank data are presented in Figure 2. Von Oortemerssen results present a hump not feasible for a displacement hull. This strange behavior was due to some typos in the original paper (Helmore 2008). Due to these uncertainties, all the next evaluations were performed by using the towing tank data, because available.

Table 1.   Main   characteristics   of   the   R/V   "G. Dallaporta"

Year of construction	2001
Length overall (LOA)	35.0 m
Breadth	7.67 m
Draft	3.0 m
Gross tonnage	286 GT
Displacement	312 tons
Main engine	Wärtsilä 810 kW
Auxiliary Engine	Cat170 kVA
Propeller	CPP in Nozzle
Crew	7 + 11 researchers

Table 2. "G. Dallaporta" operational profile.

	Days per year	Total hours	Percentage
	[gg]	[h]	[%]
Platform monitoring & Water Sampling	63	530	31,2%
Acoustic Survey	26	273	12,9%
Fishing gear testing	113	1300	55,9%

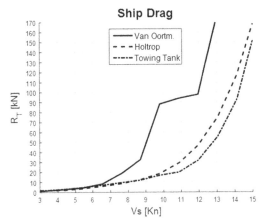

Figure 2. Ship drag comparison.

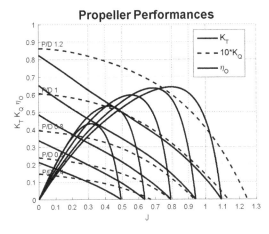

Figure 3. Propeller open water characteristics.

The propeller performance is modeled using the ducted propeller series published by (Kuiper 1992). The results are shown in Figure 3 as function of the propeller pitch angle.

The application of Equation 1, to assess the fuel consumption in the whole engine envelope, leads to the results shown in Figure 4. In the Figure, the

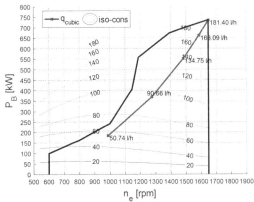

Figure 4. Engine fuel consumption map.

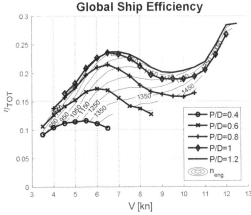

Figure 5. Propulsion energy efficiency.

calculated fuel consumption [*l/h*], is compared with the data coming from the engine manufacturer, available only on the nominal propeller 'cubic' curve. The discrepancy between data is less than 4%, in line with the measurements tolerance.

The global propulsive efficiency is calculated for all the operational profiles, for several ship speed and for different propeller pitch angle. For sake of shortness, only the results concerning the navigation condition are shown in Figure 5. This figure suggests also the optimum cruising speed, referred to the propulsion energy index.

4 SEA TRIALS

Sea trials campaign is fully described in (Capasso et al. 2016). It has been carried out during regular sampling and sailing activities from 2011 to 2015 aiming at the definition of the energy profile of the vessel (Sala et al. 2011, Notti & Sala 2012).

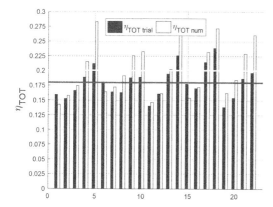

Figure 6.   Energy efficiency: Predicted vs. Measured.

The collected records range from 20 up to 40 hours of continuous monitoring of the propulsion system.

Vessel's position, course, and speed are monitored with a digital GPS. The vessel is also equipped with an log to measure the speed through water.

The propulsion system is equipped with a torque meters, for the assessment of the delivered torque. Two flow meter are installed on the fuel feed line and a on the return line. An optical encoder operated as rotation counter for the calculation of the shaft rotational speed. Each channel has been sampled every two seconds.

Figure 6 shows a comparison of all the propulsive efficiencies corresponding to the 22 different tests done in navigation mode, varying both engine speed and propeller pitch angle.

The results are compared with the results of the sea trials. The average of the global propulsive energy efficiency is about 0.18 for experiments and about 0.2 using the numerical tool. An average difference of 8% between the real and the forecasted values is experienced.

The difference could be due to the meteo-marine conditions (i.e. current), not monitored during the sea trials and object of future work. In the authors' opinion, the results are reliable enough to be used for the design of a new propulsion plant using the proposed methodology.

5   PROPOSED HYBRID ARCHITECTURE

By using the developed numerical code, several feasible plant architecture have been evaluated, and after a preliminary selection, taking into account the structural constrains, the promising solutions is to use an unique electric motor as prime mover. Therefore, the attention will be focused on the power generation. To the correct assessment of the energy production, and of the energy storage

capacity, is not sufficient the study only of the ship mission profile, a second step is needed. For this reason, every mission (Acoustic Survey, Water Sampling, Platform Monitoring and Fishing gear testing) is divided into the sub-missions. The sub-missions are defined as the part of the activities where the propulsion power requirement maintains similar magnitude. For an oceanographic vessel, five sub-missions have been identified: Navigation full load, Navigation half load, Towing, Manoeuvring, Water Sampling.

Using the propeller-engine equilibrium procedure, and assessing the optimum propeller pitch, the propulsion power requirement for each task is evaluated, and the results shown in the next figure.

Once known the required power for every condition, it is possible to choose the diesel generators number and size. Due to space constraints, the number of a diesel generator is set to two.

As shown in Figure 7, the best solution is to adopt a *"Father & Son"* configuration. This means that the two diesel generators have different power levels. The use of asymmetric power generation allows the engines to run near their optimum working point, in almost all conditions.

An additional feature of the proposed full electric propulsion plant is the use of the energy storage in accumulators. This solution leads to navigate in so-called *"ZEM"* mode (zero emission mode) where all the thermal engines are switched off. Several advantages of *ZEM* are present: the silent navigation, avoiding disturbances coming from the noise and vibrations of internal combustion engines; the possibility to navigate in marine protected area; the possibility to increase the overall efficiency recharging the batteries when the engine works at partial load (of course the batteries could be recharged with a shore connection, during harbor stops). Actually, the energy density of batteries does not allow a

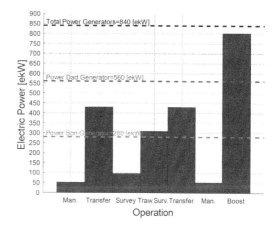

Figure 7.   Power request for each task during the acoustic survey.

91

long range, but when batteries technology becomes mature, this bottleneck could be overtaken.

Therefore, to design the battery pack it is useful to express the total energy spent to perform each subtask, as shown in Figure 8. This figure support the ship designer to choose both the number and the capacity of the batteries. In fact, having in mind which task should be accomplished in "*ZEM*" mode, the correct sizing of the battery pack is possible.

Eventually, after all the previous considerations, the proposed propulsive architecture is shown in Figure 9. With the new propulsive configuration the ship would sail in ten different propulsive configurations, each of which is optimized to perform every single sub-task with a maximum energy efficiency:

- Batteries (ZEM)
- "Son" diesel generator
- "Son" diesel generator + Batteries IN
- "Son" diesel generator + Recharge Batteries
- "Father" diesel generator
- "Father" diesel generator + Batteries IN
- "Father" diesel generator + Recharge Batteries

- "Father" diesel generator + "Son" diesel generator
- "Father" diesel generator + "Son" diesel generator + Batteries IN
- "Father" diesel generator + "Son" diesel generator + Recharge Batteries

6 CONCLUSIONS AND RECCOMENDATIONS

The paper presented a methodology to assess, in an original way, the energy propulsion efficiency. The methodology was applied to a real case study for the validation through dedicated sea trials. By using the presented methodology, a new hybrid propulsion system has been designed with the aim to improve the ship's energy efficiency.

The proposed design, after the careful analysis of the different operating profiles of the ship, lead to a particular configuration of the generation system, "*Father & Son*" which refers to an asymmetrical sizing of the diesel generators, whence the name. This solution gives a great flexibility to the plant and allows using the thermal engines near their optimum working point.

The main idea behind this work is to develop a tool that can help the designer in the early design phase. The correct use of this tool helps to reduce the environmental impact of a propulsion ship system, by reducing both emissions and fuel consumption.

As shown in Figure 10, installing the new hybrid propulsion system, design by using the developed numerical code, the annual fuel saving could be around 11% and consequently an increase in energy efficiency up to 17%, compared to the actual propulsion system. These values could decrease because actually, the losses due to the electrical transmission are not taken into account (from literature they should be 2–4%)

The study carried out is not only applicable to the specific case of "G. Dallaporta", but to a wide range of working and research ships.

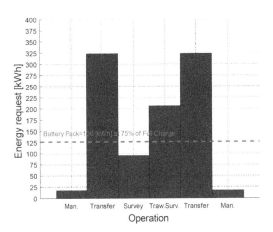

Figure 8. Acoustic survey: Energy request.

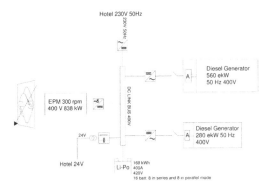

Figure 9. Layout of the full electric proposed propulsion plant.

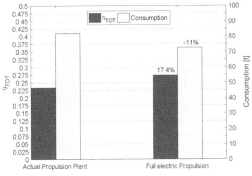

Figure 10. Saving, in terms of fuel and energy, of the proposed propulsion plant.

When the propulsion system deals with different design speeds or very different propulsive loads, such as harbor and escort tug, supply vessels, anchor handling, etc., the choice of an asymmetrical power generation could give the best advantages.

Further studies will address the auxiliary and deck systems in order to include them in the energy assessment methodology. In addition, the development of a suitable power management system will be a crucial task to manage the correct machinery switch between the different propulsive modes.

ACKNOWLEDGEMENTS

The authors wish to thank the CNR-ISMAR of Ancona, and in particular Eng. Gabriele Buglioni and Eng. Emilio Notti, for their support and valuable contribution to the research.

REFERENCES

17th ITTC 1984. International Towing Tank Conference Proceedings. Swedish Maritime Research Center SSPA, Goteborg, Sweden.

Altosole, M. & Figari, M. 2011. Effective simple methods for numerical modeling of marine engines in ship propulsion control systems design. Journal of Naval Architecture and Marine Engineering, December 2011, pages 129–147.

Altosole, M., Martelli, M. & Vignolo, S. 2012. A mathematical model of the propeller pitch change mechanism for the marine propulsion control design. Sustainable Maritime Transportation and Exploitation of Sea Resources—Proceedings of the 14th International Congress of the International Maritime Association of the Mediterranean, IMAM 2011, 2, pp. 649–656.

Altosole, M., Buglioni, G. & Figari, M. 2014. Alternative propulsion technologies for fishing vessels: A case study. International Review of Mechanical Engineering, Vol. 8 (2): pages 296–301.

Buglioni, G., Notti, E. & Sala, A. 2012. E-Audit: Energy use in Italian fishing vessels. In Rizzuto & Guedes Soares (eds) Sustainable maritime transportation and exploitation of sea resources. Taylor & Francis Group, London.

Candries, M., Atlar, M. & Anderson, C.D. 2000. Considering the use of alternative antifoulings: the advantages of foul-release systems.

Capasso, C., Veneri, O., Notti, E., Sala, A., Figari, M. & Martelli M. 2016. Preliminary Design of the hybrid propulsion architecture for the research vessel "G. Dallaporta". Proceeding of the 4th International Conference on Electrical System for Aircraft, Railway, Ship propulsion and Road Vehicles & International Transportation Electrification Conference.

Castles, G., Reed, G., Bendre, A. & Pitsch R. 2009. Economic benefits of hybrid drive propulsion for naval ships. IEEE Electric Ship Technologies Symposium, Baltimore, MD, 2009, pp. 515–520.

Eyring, V., Köhler, H.W., Lauer, A. & Lemper B. 2005. Emissions from international shipping: The impact of future technologies on scenarios until 2050. J. Geophys. Res., 110–117.

Gabiña, G., Basurko, O.C., Notti, E., Sala, A., Aldekoa, S., Clemente, M. & Uriondo, Z. 2016. Energy efficiency in fishing: Are magnetic devices useful for use in fishing vessels?. Applied Thermal Engineering, 94: 670–678

Helmore, P.J. 2008. Update on Von Oortmerssen's Resistance Prediction. Royal Institution of Naval Architects—International Maritime Conference, Pacific 2008, pp. 437–448.

Holtrop, J. & Mennen, G.J. 1982. An approximate power prediction method. International Shipbuilding Progress, Vol. 29, pp. 166–170.

International Maritime Organization—Part a (IMO). 2009. Interim guidelines for voluntary verification of the energy efficiency design index. MEPC.1/Circ.682, 17 August 2009. London: IMO.

International Maritime Organization—Part b (IMO). 2009. Guidelines for voluntary use of the ship energy efficiency operational indicator. MEPC.1/Circ. 684, 17 August 2009. London: IMO.

Kuiper, G. 1992. The Wageningen propeller series. Delft: Marin.

Larsen, U., Pierobon, L., Baldi, F., Haglind, F. & Ivarsson, A. Development of a model for the prediction of the fuel consumption and nitrogen oxides emission trade-off for large ships. Energy, Volume 80, 2015, Pages 545–555.

Martelli, M., Vernengo, G., Bruzzone, D. & Notti, E. 2016. Overall Efficiency Assessment of a Trawler Propulsion System Based on Hydrodynamic Performance Computations. Proceeding of the 26th International Ocean and Polar Engineering Conference (ISOPE 2016), Volume 4, pp- 875–882, 2016, Rhodes, Greece, June 26-July 2, 2016.

Nelson, M., Temple, D.W., Hwang, J.T., Young, Y.L., Martins, J.R.R.A. & Collette M. 2013. Simultaneous optimization of propeller–hull systems to minimize lifetime fuel consumption. Applied Ocean Research, Volume 43, 2013, Pages 46–52.

Notti, E. & Sala, A. 2012. On the opportunity of improving propulsion system efficiency for Italian fishing vessels. Proc of the Second International Symposium on Fishing Vessel Energy Efficiency (E-Fishing2012), Vigo, Spain, May, 22–24, 2012.

Notti, E. & Sala, A. 2014. Propulsion system improvement for trawlers. Developments in Maritime Transportation and Exploitation of Sea Resources—Proceedings of IMAM 2013, 15th International Congress of the International Maritime Association of the Mediterranean, 2, pp. 1085–1090.

Sala, A., De Carlo, F., Buglioni, G. & Lucchetti, A. 2011. Energy performance evaluation of fishing vessels by fuel mass flow measuring system. Ocean Engineering, Volume 38: pp.804–809.

Von Oortmerssen G. 1971. A power prediction method and its application to small ship. International Shipbuilding progress, Volume 12 n° 207, pp. 397–415.

Maritime Transportation and Harvesting of Sea Resources – Guedes Soares & Teixeira (Eds)
© *2018 Taylor & Francis Group, London, ISBN 978-0-8153-7993-5*

Manoeuvring model and simulation of the non-linear dynamic interaction between tethered ship and tug during escort

B. Piaggio, M. Martelli, M. Viviani & M. Figari
Department of Electrical, Electronic, Telecommunications, Naval Architecture and Marine Engineering (DITEN), Scuola Politecnica—Università di Genova, Genova, Italy

ABSTRACT: When dealing with towing and Escort operations—indissolubly—a wider and complete modelling of the involved dynamics becomes fundamental, on both ship and tug side, even more so making unavoidable the need of correctly ponder the cables constraining coupling. With the aim to better understand the peculiarities of tug-ship interaction, a 6-DOF time domain simulator comprehensive of the propulsion dynamics has been developed, across the inter-connection imposed by the towing-line. Particular attention is focused on discovering the operational capabilities of the tug in exerting a force on the assisted ship during Escort, and the subsequent great influence that the reaction makes on the tug handling and effectiveness. A conceptual and critical discussion stands out, preparing the ground to future design strategies and hinting different solutions to be investigated, directly facing the operative profile of the vessel.

1 INTRODUCTION

A whichever ocean-going vessel requires the assistance of tugs whereas its manoeuvring capabilities could result particularly reduced and limited or if—some way—certain geographical constraints in the seaway to be traced or meteorological circumstances would impose it. In fact, the abilities in governing the ship result immediately decayed especially at least in two scenarios: or in the case the navigation velocity to be sustained has to be so far reduced to turn the vessel unstable or uncontrollable, *i.e.* in many scenarios of access to harbours, channels, mooring and docking operations, or else in the case the ship incurs in a whichever sort of mechanical fault in the propulsion or in the means of govern.

In such a broad panoply of scenarios, the towing service principally concerns the partial or total handling of the ship, whether in confined waters, or open seas and oceans, either on-line or off-line. In function of the velocity of the ship, fundamentally two different methodologies of towing are discerned according to the means of propulsive allocation logics: the Direct and Indirect Towing.

The Direct Towing Mode involves tugs with high manoeuvring capabilities at low speed, characterized by relatively small sizes, handyness and minimal appendix configuration. Those are enabled to provide push and pull flexibly, for an all-around 360 degrees, particularly in confined waters, in function of the design specifications of bollard-pull, by aligning the tow-line with the thrust. Their main performance characteristic therefore is the installed power on board (Hensen 2003).

The Indirect Towing—instead—is targeted to aid sailing ships during the navigation in an intermediate range of velocities in-between the harbour manoeuvres and the normal design cruise speed (*i.e.* around 6–10 knots) in order to ensure operational safety and—therefore—reasonably reduce any involved risk along the route: the tug must be capable of braking and steering the running tethered ship in open seas, straits, channels, in the optic of safe assist in the trace-keeping the vessel along particular sea-ways (*i.e.* Gray & Reynolds (2001)). In such situations, the forces and the dynamics involved definitely result superior of many orders of magnitude if compared to the scenarios at lower speeds, and therefore particular hull and appendages geometries combined with specific propulsive solutions are required (Allan & Molyneux 2004, The Glosten Associates 2004), in order to be able of taking advantage of the hull's hydrodynamic forces to generate a towing force. Purposely designed with increased projected lateral area, those tugs in order to generate pull, operate forcing the hull to drift: the forces so born, indeed, in analogy with the theory of slender lifting bodies, are exploited by orienting the bow of the tug across non-zero incidence angle relatively to the flow, with resultants in general +70–80% superior to the classic bollard pull (Figs. 1–2 from Capt. Henk Hensen (2012) and Canadian and Environmental Assessment Registry (2004)).

While the direct towing mode is nowadys widely known and historically consolidated in the normal practice, the potentiality of the latter sowed the particular interest of the present investigation, in the effort of well understanding the challenging governing principles: across a thorough study and modelling of the dynamics, indeed, the aim of the study is addressed to investigate the phenomena, so that its conveniences and disadvantages stand out, by directly facing all the implicitly hidden issues head on.

1.1 Escort manoeuvring

The fundamental mission of an Escort Tug is to tether, *i.e.* "escort", ships with weak manoeuvring capabilities in the transit of zones of particular interest and hazard, with the intent of intervening, in case of emergency, in their seaway-route control, so that the risk of groundings or collisions is reduced.

In function of the vessel's attitude and relative positioning with respect to the tethered ship, basically the escort tug is capable of transmitting to the assisted ship's stern a pull force, either across the combination of braking and steering components or in a pure mode, by imposing a constrained drift and a desired towing angle. In that way the propulsive power is just entrusted of the only duty of maintaining the non-zero relative angle, and therefore is not directly implied in the generation of the hydrodynamic forces useful in the exploit of the towing reaction (Fig. 2). For this reason the pro-

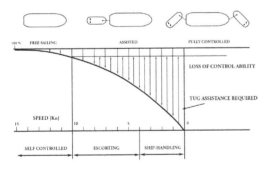

Figure 1. Definition of towing operations.

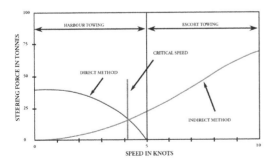

Figure 2. Direct and indirect towing comparison of a 40 t Tug.

pulsion must be much more flexible and responsive in the thrust vectorialization: proper propulsive solutions have to be selected (Allan 2000, Robert Allan, and B. Smoker 2013). Besides the pure braking and steering abilities, then, many are the combined escort manoeuvres in emergency: schematically three primary assistance procedures can be distinguished in the yaw-check aid, by retarding, assisting or opposing the turn, in function of which is the need.

In the practice, the tug behaves similarly to an external appended airfoil rudder free to displace astern the tethered ship, and in order to fulfil such a scope the typical hull geometries consist of large skegs or equivalent lateral exposed surfaces, so that the hydrodynamic lifting performances in enhanced when inclined to the flow. Subsequently—the need of modelling the tug's manoeuvring at high speed and at large drift angle arises, to be subject later on to the towing cable constraint.

1.2 The state of the art

Consequently to various accidents occurred in the last decades, raising pressures pointing to an improvement in terms of safety of the maritime transportations pushed to the introduction of the Escorting service, since the common cause of such casualties was ascribable to the lost of the propulsion and/or the steering in confined waters near the shore. Collaterally to the economic damages, indeed, the consequences of those led to disasters of environmental nature (see Exxon Valdez oil spill 1989 in Prince William Sound, Alaska), that yielded the USA to draw up and draft the Oil Pollution Act in the 1990 — OPA'90 33 CFR 168 August 1994 (United States Congress 101st 1990). This is a statutory regulation according to which the escort by means of two suitable tugs of support turned to be mandatory for determined classes of tankers and chemical-carriers approaching the national coastal areas.

Following up—the whole classification societies of the shipping world (*e.g.* American Bureaux of Shipping (2016)) moved on, adjusting and defining indexes and standards of performance, editing rules *ad hoc* and introducing the special "Escort" class notation with particular added care to the safety side. Indeed transversal dynamic stability is a first order issue and the attention of the classification societies hence moved forward sensibly, deepening some dynamic aspects about the phenomena (see AHTS Bourbon Dolphin capsize accident). The whole definitely led to the requirement of proper sea trials, or substitutive suitable numeric calculations, in order to accomplish a measure of the performances.

The present study proposes to move the topic in the target, moving the spotlight upstream the problem, with the aim of gathering all the technical aspects of the tug design with the real final operative needs, in terms of desirable specific capabilities, with the scope of enclosing the whole picture in a unique tool. On the one hand many aspects arise about the design stage, such as propulsive solution selection (Kooren, Quadvlieg, & Aalbers 2000, SAFETUG 2015, Jrgens & Palm 2015) and the optimization of the static and dynamic performances of the hull (*e.g.* Pure Drift CFD computations (Smoker 2009, Allan & Molyneux 2004, Molyneux & Earle 2001)) And again both the towing cable (Papoulias & Bernitsas 1988, Bernitsas & Kekridis 1986) and the winch selection (R. Allan 2012, Couce, Couce, & Formoso 2015, der Laanand Kees Kraaijeveld 2013, Griffin 2004) become matter of interest in the dynamic coupling in-between the vessels. On the other hand, likewise operative consequences descended, in terms of pull performances (steering and braking), safety and dynamic stability.

With such an aim the whole study is addressed to a close-up analysis and rationalization of the complete dynamics (Waclawek & Molyneux 2000, Li & Calisal 2005, Merrick 2002, Frans Quadvileg 2006) in terms of equilibria finding and stability assessment (Bernitsas & Kekridis 1986), so that the development of the aforementioned tool be means of a proactive investigation in both the directions, *i.e.* optimization of the tug concept in hull geometry, propulsive means and towing variables across its direct repercussion on the operability and controllability, and, *vice-versa*, the opposite, identifying eventual lacks and difficulties on the latter side resorting again to the design stage. In order to put the groundwork to this long-term project, here—in the first instance—the basic concept design of the mathematical model is laid.

2 MATHEMATICAL MODEL

The developed towing simulator consists of a set of differential equations, algebraic equations and parametric tables that represent all the parts contributing to the overall dynamics: the whole model is subdivided in a Ship Block (*i.e.* Single Screw Merchant Ship), a Tug (single or expansible to multi-tug ops) Block and a Towing Cable Block, playing the role of the dynamic coupling of the system kinematics.

The complex is considered to be manoeuvring in an undisturbed, homogeneous, isotropic environment—the still water—without any dependency from the sea position or depth. No environmental forces are considered in this preliminary stage: those will be object of further investigation. Then, in order to satisfy the four quadrant demand, the physical system ship has been decomposed in different logical blocks in which all the mechanical and hydrodynamic phenomena involved are studied in their mutual interaction, undertaken the peculiarities of the case.

2.1 *Vessel dynamics*

The kinematic state of a vessel is completely determined by assessing the time-evolution of the six classical generalized coordinates $\eta = [x, y, z, \theta, \psi]^T$ in the earth fixed $n - frame$: according to Fossen (2011)—therefore—the well known equations of motion in vectorial form in terms of the $b - frame$ (body-fixed) velocities $v = [u, v, w, p, q, r]^T$ can be expressed by applying the classical total force and moment equations for rigid bodies across the operator $J_\Theta(\varphi, \theta, \psi)$.

The kinematic modelling of the vessel motions adopts six degrees of freedom particularly because

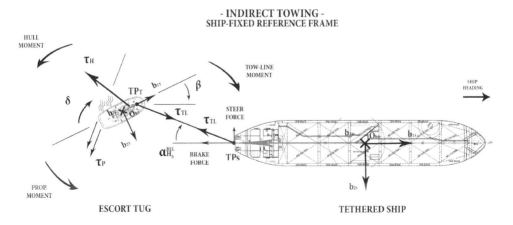

- INDIRECT TOWING -
SHIP-FIXED REFERENCE FRAME

Figure 3. The escort dynamics.

of the non-negligible roll motion effect that exists during the towing operations. In fact due to possible important towing angles, prominent tensions, and heavy allocation logics of the propulsive forces, many transient scenarios in the dynamic could result critical for the operability and stability of the tug in particular.

When dealing with towing operations, either in the direct mode or in the indirect mode, the need of describing the involved dynamics regards wider domains than in the classical navigation modelling in which small drift angles are considered ($u > >v$). In fact in both the cases conditions far from the straight one are involved. On the one hand, where the full range $0-180\ deg$ of angles of incidence of the water on the hull drifting is involved at low speed (harbour manoeuvring), on the other hand particular large drift scenarios are encountered at high speed during escort (Molyneux & Bose 2007). Identically the modelling of the propulsive forces in oblique flow becomes essential.

Since the description of the forces acting on a classical merchant vessel is well known in literature, here the attention is focused on the tug modelling, observing anyway that with the exception of the means of manoeuvring, the modelling criteria almost overlap.

2.2 Propulsion dynamics

While in the case of a classical slow merchant ship configuration the steering capability is entrusted to the hydrodynamic performances of a conventional rudder-propeller configuration, the tug here considered relieves on a twin Azimuthal Stern Drive-system (ASD), in order to enhance the steering capabilities of the vessel.

2.2.1 ASD Propulsion

The vectorizing of the thrust is attained across the pure rotation of the ducted fixed propeller (FP) $K_a 4-7019A$ belonging to the Wageningen Series (Oosterveld 1970) locally subject to the perturbed water inflow. In order to correctly accomplish the evaluation of the propulsive forces, thus, a series of intermediate kinematic variables of the fluid has to be computed, so that the local flow field around the propeller position is well considered consequently to the hull interaction. Indeed, the flow at the azimuthal propeller is evaluated in terms of magnitude and direction, considering the relative tug motions, the propeller orientation and including wake and straightening effects (V. Ankudinov, Kaplan, & Jacobsen 1993).

In principle, consequently to the misaligned condition to the flow induced whether by the local drift or the ASD angle in the behind hull condition, all the classical coefficients of the propulsion

chain have to be re-considered from a local propeller-fixed standpoint, in order to catch the right axial inflow component of the water given the effective ASD angle. This will result,—therefore, in the dependency of the entire propulsion from the effective local axial component of velocity at the propeller disc. Unfortunately, a clear lack of experimental data exist from this point of view. Therefore, some simplistic assumptions have been made, imagining that both the straightening and propulsive coefficients $\gamma v, \gamma p, \gamma r, (1-t), (1-w)\ and\ \eta_R$ will (more or less smoothly) vary between their normal straight values (valid in the range $\beta_R \cong 0-30\ deg$) to 1 as a threshold angle ($\beta_R \cong 60\ deg$) is approached. This implicitly assumes that the hull will not affect the fluid when a higher drift angle are encountered.

Once the local flow field is computed on each turned ASD, the propeller hydrodynamic forces in the axial body-fixed reference are hence evaluated according to the open water four quadrant performance of thrust and torque, C_T and $10C_Q$, respectively (FREng 2011, Oosterveld 1970).

Simplistically, since propeller functioning in oblique flow, Cross-Flow Drag on the ducts, Loading asymmetries, Propeller-Propeller interactions and deeper Propeller-Hull interactions (Nienhuis 1992, Brandner & Renilson 1998) still are not included in the present level-zero modelling. These aspects will be analysed in future by means of experimental campaigns and, eventually, through direct numerical calculations, because of their great relevancy and impact on manoeuvring. Finally, a first order actuating angular velocity on the ASD angle variation is included, in order to match and tune the steering gear functioning.

2.2.2 Shaft-line dynamics

The propulsion plant dynamics are described through-out the dynamical coupling between the engine and the propeller across the shaft-line once reached the equilibrium between the propeller request and the engine provided torque. Hence the engine's brake torque Q_B reduced downstream the gear with ratio i, at the slow shaft-end, will be $Q_s = i\eta_G\ Q_B$ and the propeller's requested one will be Q_D, up to the frictional efficiency of the shaft-line, which is assessed standalone in the term $Q_{fric.} = (1-\eta_s)Q_s$. Finally the balancing equation is obtained, in which, the positiveness or negativity of the terms, is implicitly assigned. The overall polar inertia I reduced to the slow shaft is considered.

2.2.3 Engine dynamics

Performance maps in terms of power, consumptions and revolutions are exploited all over the working area of the engines in order to correctly reflect the behaviour in both design and off-design conditions, allowing, therefore, the study of the transient steps in an acceptable approximate way (Altosole, Figari, & Martelli 2012). The engine control is based on properly independent designed governors on the revolution speed and propeller pitch across the combinator law, relying on the most basic Proportional-Integral (PI) controller on the fuel flow $\dot{q}_f$ for sake of simplicity. A continuous monitoring of the error between the required set-point $n_{ref.}$ and the actual value of RPMs $n_{eng.}$ of the engine's crankshaft is assessed, and suitably the percentage of fuel flow to be fed in order to make zero the error is assigned and allocated.

2.3 Hull forces

The force break-down starts from the Oltmann & Sharma's 3-DOF Model (Oltmann & Sharma 1984, T. Jiang & Sharma 1998) of which all the needed coefficients are provided at slow speed, either for the tanker, either for the tug, and, then, proceeds extending the dynamics to all the 6-DOF (Martelli, Viviani, Altosole, Figari, & S.Vignolo 2014) Hence, the planar Ideal Fluid effect τ_I, the Hull Lifting effect τ_{HL}, the Cross-Flow Drag effect τ_{HC} are evaluated according to the generalized irrotation potential theory (Imlay 1961), to the Prandtl & Tietjens theory of slender lifting bodies in inclined flow, and the strip theory modelling, respectively. The ordinary hull resistance τ_{Rt} to pure longitudinal advance is then evaluated.

$$\tau_H = \tau_I + \tau_{HL} + \tau_{HC} + \tau_{Rt} + \tau_G \qquad (1)$$

Then—following the same logic—the model is extended to 6-DOF across the additional added mass coefficients, the damping forces acting out of the horizontal plane and the coupling in-between the horizontal and vertical modes. In particular, additionally to the classic coupling interpretation a further relevant effect should be introduced, linking again the vertical and horizontal motions: according to Kijima & Furukawa (1998) in fact, since non-zero heeling angle could generate asymmetries in the immersed hull when the ship rolls, additional hull lift and drag forces would arise even at zero incidence angles such as in the case of cambered foil with dihedral angle effect (Hoerner, F., & Borst 1965, Hoerner, F., & Borst 1975)), making the full picture attractive in the case of the drifting escort configuration. Concerning the Hydrostatic Restoring Forces τ_G, in function of the hull geometries those are derived around the equilibrium attitude.

Anyway—on the one hand—it is opportune to be reminded that the simulation outcomes here reported will just be correctly be interpretable and comparable within the reasonable limits of the tug which is adopted: the Sharma's tug, in fact, is a large size oceanic tug, not intended to carry out such kind of manoeuvres, but—on the other hand—all its abilities or inabilities in bearing those are designated to spark considerations, either in the limits of the modelling, either in the limits of the selected tug in exploiting his steering and braking capabilities.

By the way, on the long term side, the hull force architecture modelling can be thought such as a customizable parametric performance block, easily identifiable across opportune model testing either in tank, or numerically (Bruzzone, Ruscelli, Villa, & Vivani 2015), once the hull geometries are varied. To pursue such a scope eventual brute force systematic variations or wiser optimization problems subject to constraints could be later faced rigorously, making possible to have direct and bilateral sight on the final operative and control side, once the any design solution is undertaken. This part, together with the propulsion modelling effectively constitutes the core of the dual constrained optimization concept behind the motivations of the study here begun.

2.4 Tow-line dynamics

With the purpose of modelling the cable several approaches stand out, each one with different levels of accuracy and faithfulness to reality. Among these arise Finite Element Methods (FEM), Finite Difference Methods (FDM), Catenary Equations, Lump-Mass-Spring Formulations (LMS) and Finite Segment Approaches (FSA). All these methods are based on a particular and generalized mathematical schemes of the problem whom formulation easily can be extended to describe different configurations of cables and wires, amongst mooring cables, towing lines and suspended cables. Here the catenary method is adopted, as first simplified approach in a quasi-static standpoint, and in particular the elastic theory is treated, considering a linear elastic limit around +20% of elongation before yielding.

2.4.1 The general elastic catenary
The word catenary is derived from the latin word catena, referring to the shape of a chain or wire hanging between two points under its own weight. Let be a cable of length L attached to a point TP_s and a point TP_T, standing for towing point at ship and tug-end, respectively: those points will define the cable supports and they are fixed on board the ships following the motions (Fig. 2). In time they will differ horizontally and vertically about the double quantities $[l(t), h(t)]$, whom combined with the cable's unstretched length L (eventually $L(t)$ if controllable) define the triplet of the boundary conditions of the problem. It is assumed the cable always belongs to a vertical terrestrial plane passing through the end-points, and therefore its planar representation completely defines the problem.

The equations are therefore obtained using a Lagrangian approach: starting from support TP_s and moving along the cable profile, each point on the cable is identified by the Lagrangian coordinate s along the line with respect to the origin. Whereas the cable is extensible and strainable, when stretched an additional coordinate p has to be introduced in order to describe the deformed configuration. In order to follow the plane in his time evolution, it is convenient to introduce a new reference frame, the $c-frame$, in which the 2-D catenary analytically could be written with origin fixed to the cable's ship-end: that is the $\mathbf{c} = \{O_c, \underline{c}_1, \underline{c}_2, \underline{c}_3\}$ reference, with $\underline{c}_1$ pointing towards the tug-end, $\underline{c}_2$ rightwards, and $\underline{c}_3 = \underline{n}_3$ vertical (Fig. 5).

The cable is assumed to be homogeneous, subject to an uniform weight. The bending stiffness is ignored, and hence a perfectly flexible cable is assumed Highly tensioned cables in matter, that are taut, require in particular an elastic model.

The non-linear general elastic catenary equations (2) hence are solved simultaneously in time in function of the unstretched length $L(t)$ at each step—given the pair $[l(t), h(t)]$—in order to find the horizontal and vertical towing reaction components at the extremes of the line, $T_H(t)$ and $T_V(t)$, respectively. W is the weight per unit of length, E is the Young's Modulus and A is the resistant area.

$$
\begin{cases}
l = \dfrac{T_H L}{EA} + \dfrac{T_H L}{W}\left[asinh\left(\dfrac{T_V}{T_H}\right) - asinh\left(\dfrac{T_V - W}{T_H}\right)\right] \\
h = \dfrac{WA}{EA}\left(\dfrac{T_V}{W} - \dfrac{L}{2}\right) + L \\
\& + \dfrac{T_H L}{W}\left\{\left[1 + \left(\dfrac{T_V}{T_H}\right)^2\right]^{1/2} + \left[1 + \left(\dfrac{T_V - W}{T_H}\right)^2\right]^{1/2}\right\}
\end{cases}
\tag{2}
$$

Figure 4. Escort stand-by operation.

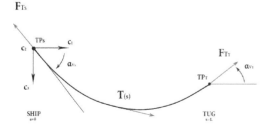

Figure 5. Catenary plane.

Then, once known the geometry of the cable and once the reaction vectors are well identified in space at the ship and tug-end (magnitude T and orientation across the pair of horizontal and vertical angles $[\alpha_H, \alpha_V]$, straight-forward is the projection (3) of each reaction switching from the $\mathbf{c}-frame$ to the respective $\mathbf{b}-frame$.. The description of the towing dynamics therefore is accomplished.

$$
\mathbf{F_T}\big|_{b-Ref} = \left[\mathbf{R}_b^n(\varphi, \theta, \psi)\right]\left[\mathbf{R}_n^c(\alpha_H)\mathbf{R}_n^c(\alpha_V)\right]\mathbf{F_T}
\tag{3}
$$

As result the cable dynamics is thought such as sequence of quasi-static scenarios in the tensioning of the cable, and, when completely straightened, it will behave almost like a spring segment (kinematic bond), and not even remotely intends to describe the real accelerated and impulsive dynamic of a cable in a multiphase fluid with wind and waves: in this way a future development could definitely be a dynamic Lagrangian lumped mass modelling (Nakayama, Yasukawa Hirata, & Hata 2012).

2.4.2 Towing parameters and solver

The towing dynamics mainly depend on all the mechanical characteristics of the cable (*i.e.* unstretched length L and Young Modulus E), playing on the toughness of the response in slacking, but also rely on the positioning of the towing points on board the vessels ($\mathbf{TP_s} - \mathbf{O_{b_s}}$) and ($\mathbf{TP_T} - \mathbf{O_{b_T}}$), conditioning the levers available to generate the moments. Then immediate extensions could be faced in function of the third boundary condition, the unstretched length, when logics of dynamic load control of the towing winch would become subject of study when dealing with peak loading. Non-linear constitutive laws in stretching-tensioning further can be implemented.

Eventually, in function of the kind of line chosen, nominally the mechanical properties, the minimum tensile strength $\sigma_{tens.}$ is considered, i.e. the maximum tension before breaking, or equivalently the maximum elongation ε_{break}. The consitutive thresholds are checked constantly along the time evolution and in the case of overtake the simulation will be by default interrupted, but if it was of interest the following dynamic of the unleashed ship or tug (*i.e.* dangerous roll angles of tug, prediction of the unleashed route of the ship given environmental constraints in the surrounding) it can be continued voiding the dynamical coupling.

2.4.3 Line's tension and hawser's control

Having the knowledge of the towing line's dynamic behaviour all around his possible domain of existence reveals very useful in tracing and checking the variables involved in the towing operations, with consciousness of the evolution in time of those quantities. In a control logic for the towing operations, those variables immediately would gain main importance on board the tug in terms of optimal allocation and safety across the towing hawser: the check of tensioning of the line by the use of extensometers or load cells on the cable's winches is clearly possible and normal practice, making that variable extremely interesting for further developments on winch logics.

3 METHOD OF INVESTIGATION AND SIMULATIONS

The typical initial condition scenario to be simulated regards the tug following the aided ship, holding an in-line standby position, in which the cable is maintained in the limit between the unloaded condition and the loaded one, keeping ready in case of any request of support from the mother ship. In event of sudden need—in fact—the tug must be prompt in assisting partially or intervene totally in the handling and manoeuvring of the ship. Once the eventual failure recognition is aknowledged, then, the switch from the standby to the operative condition must be as fast as possible, but never losing of sight the safety and stability's perspective according to which the chase of a smart and smooth manoeuvre is decisive.

Even more so, the loading of the cable and the rate in changing position are of fundamental importance and criticality: the high drift angles attainable, with subsequent high levels of tensioning and transmitted pull to the ship's stern, are enabled by the propeller's power and angle, and any uncontrolled transient or perturbation of the final equilibrium configuration could make the tug oscillate and traverse. Furthermore, also the way in which the allocation is raised, reduced or removed, highly affects the procedure along the manoeuvre: in fact the assisted ship loses speed and gradually the tug must adapt his attitude consequently while reducing speed.

In order to separate and distinguish all the matters mentioned, schematically, a simplified approach is carried out first, so that the tug's behaviour could be studied detached from the dynamics of the ship and from the time variant control logics. In this way the attention can be focused exclusively on the pure tug in a partial captive way, in a partial scenario of the entire dynamics.

3.1 Escort mimic

The ideal scenario concerns the ship sailing straight ahead inexorably with constant speed, fully restrained on the degrees of freedom and therefore insensitive of the whichever towing reaction is exploited at the line's end. By letting vary the angle and/or the number of revolutions of the ASD propellers the identification

of all the final configurations of equilibrium of the tug is made possible measuring the vessel positioning and attitude and the cable's tensioning and orientation: as result, the regime's behaviour and the fictitious steering and braking pull are individuated. Each one of those final pictures of equilibrium would practically represent instant by instant what during the generic manoeuvre would happen if the inertial effects were neglected, such as that was thought as a sequence of quasi-static frames at regime. Obviously this is far away from the reality in which the accelerations and the impulsive tensioning play crucial roles, but it is considered a fundamental step in order to understand the pure behaviour and the abilities of the partially-isolated dynamical system of the tug "hull + propulsion" under the tow-line's force, purging out any other undesired artefact. This—in a decoupling effect modelling criteria—by fact is aimed to distinguish the intrinsic behaviour of the inclined hull to the flow such as an ideal limit scenario, eliminating any effect of the tug's dynamic control and any misleading transient. This approach, indeed, effectively reflects the testing logic adopted by the towing tanks in evaluating the escort capabilities at the design stage besides the classical PMM manoeuvring programme, with the effort of mimicking the real phenomena.

As an example, the critical scenario with initial speed at 10 kn is considered (Fig. 6). Starting from that, the engine power is gradually increased (up to the maximum), spanning time by time the thrusters angle domain. The set of simulations is intended to show how the equilibrium configurations will vary, by gradually taking advantage of all the available power, in order to reveal which the possible maximum steering and braking capabilities are if compared to the bollard pull in the boosted condition, T_S^* and T_B^*, respectively. Subsequently, the associated maximum breaking and steering abilities are hence individuated at 105 *deg* the first and 135 *deg* of ASD angle the latter. Once again must be reminded that the present vessel is not intended to carry out such manoeuvres in an outright escort manner, neither in the size nor in the hull shape, and therefore different performances from the canonical are clearly expectable, but—anyway—it offers an interesting first standpointfrom which start for future considerations and hull implementations.

3.2 Escort full DOF

When the fully released DOF scenario has to be faced, immediately the spectra of dynamics becomes wider: whereas some kinematic configurations where previously achievable dynamically, now by accounting the complete dynamical interaction and the progressive loss of speed opportune control of the tug has to be considered necessarily in order to pursue the sliding equilibria configurations and

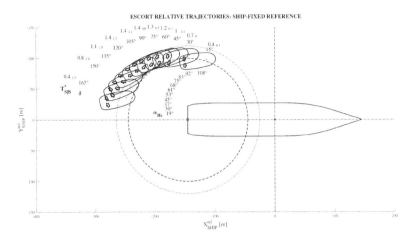

ESCORT RELATIVE TRAJECTORIES: SHIP-FIXED REFERENCE

Figure 6.   Escort: Tanker 10 kn—Tug 10 kn (179.4 rpm).

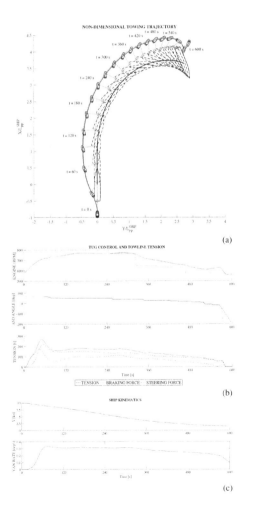

(a)

(b)

(c)

Figure 7.   Escort Full DOF Turning.

avert the instability threats. The manoeuvres to be faced in the reality could be ideally distinguished in partial or total yaw-check support (*i.e. route-keeping*, either on-line or off-line, in the event of failures or bad weather conditions) and forced arrest in order to evade ahead obstacles (*i.e. pure stop* or *turning* according to the amount of lateral clearances). As an example, an induced turning manoeuvre in the case of the complete loss of propulsion by the tanker is reported (Fig. 5).

Starting at equilibrium from the speed of 10 knots, once the propulsion and control means are lost, the tug engages the manoeuvre rising up the engines and moving portside astern the ship so that the steering force is transmitted (Fig. 5a): the position then is held making turning the ship and slowing her down until being arrested safely. Shaft revolutions and propeller angles are progressively adapted and decreased in time in order to maintain the relative positioning and the effectiveness of the manoeuvre (Fig. 5b). The induced kinematics of the ship in terms of speed loss and yaw rate are illustrated in Figure 5c.

4   CONCLUSIONS AND RECOMMENDATIONS

The operative performance specification for an escort tug is represented by the maximum steering and breaking force at the maximum speed of the escort service, *i.e.* 10 knots. This, from the operator and commercial standpoint, regards the achievement of the maximum transmittable force to the ship's stern in the most efficient way given the specific tug size: the indirect towing concept immediately pushes the dissertation in the necessity of the right representation of the dynamics of the coupled system *ship-tug* subject to the tow-line constraint. With a view to lay the groundwork for future research analysis in such a field, the preliminary concept design of the dynamics modelling has been laid, with the forthcoming intent of moulding it to reality across opportune model and numerical identification of the enclosed parameters. The rationalization of the whole consist in the outright comprehension of the complete coupled dynamics especially in terms of the hydrodynamic force generation of an inclined hull run over the flow at high speed up to very large drift angles and the relative ability to transmit it to the assisted ship across the towing line: suitable devised tank trials necessarily descend.

Indissolubly either the hull geometries or the propulsive choice affect the ability of the tug to achieve a desired configuration, pointing out the need of considering the problem sequentially at different levels of accuracy: the intrinsic hydrodynamic behaviour of tug's hull in fact cannot disregard the propulsive allocation means and so *vice-versa*, playing crucially both on the manoeuvrability and handling of the vessel. The resulting problem becomes very complex: Naval Architecture, Propulsion, Manoeuvrability, Experience and Control thoroughly merge in the ideation and realization of the optimal modalities of towing, always keeping in mind the mandatory safety issues.

In this sense the ideal tug to be adopted should be able to offer to the pilot the wider realizable domain of operability with great readiness in the response and handling: the great forces exploitable on the tethered ship indeed rely on the combination of the steering and braking components and the control of the manoeuvre is left to the experience and feel of the helmsman. At least two are therefore the governing kinematic variables emerging looking at the steady scenarios: the first is the tunableamount of implied hydrodynamic force across the drift angle and the second is the selectivity in the orientation of the resultant on the ship's stern, across the horizontal cable's angle, the whole bounded by a sector of annulus of admissible kinematic configurations identified by the un-stretched condition of the line and an upper elongated length safely selected below the minimum tensile limit. Dangerous uncontrollable and unstable scenarios in the tensioning of the line that could make the tug divert from the operation and potentially endanger it must be avoided in terms of roll motions, route-keeping, handling and efficiency in transmitting the pull force.

Given the coupled configuration, the tensioning of the line highly affects the tug manoeuvrability, undermining the desirable performances empirically or numerically predictable: the length, the elasticity of the cable and the positioning of the towing staple on board the tug become a matter of interest, but even more so the right combination of hull shape and propulsive solution becomes crucial,

shrinking or enlarging the operability of the tug already at the early conception of the design stage. Inparticular lateral exposed area and its longitudinal distribution along the hull drifting, together with thrust magnitude are the parameters which the designer has to carefully determine in order to optimise the tug's capabilities in terms of balancing and counter-balancing available levers in yaw with the scope of maintaining the tug misaligned to the flow. Indeed, by correctly designing the hull shapes, properly tuning the thrust allocation and vectorizing the equilibria with the hydrodynamic forces and the towing reaction, the domain of equilibria could be spanned completely or not. Then the time variant equilibria finding during the speed decaying manoeuvre and the relative stability assessment is left to the helmsman or to opportune control laws across the contemplation of suitable logics, *i.e.* definition of adequate objective functions to be optimized and pursued.

Once defined the tug's characteristics and the subtended hydrodynamic characteristics derived from model tests or numerical calculations, the resultant time-domain simulator stands out congregating the whole shades of the phenomena in an unique tool. This—in a long term expectation—does not just disclose to this day an immediate useful analytic and predictive instrument intended to be hypothetical aid whether to the planning, training or emulation of the operations of real units undertaking dangerous routes, but still further it reveals a powerful mean—in the future—to be encompassed and to be explored with the aim of concurrent design optimization in the hull geometries, propulsive configurations and towing parameters. The whole, with the final challenging target of individuation of smart governing logics and consequent synthesis of automatic controllers, able to impose and keep the assisted ship along a predefined sea-way safely, in event of failures or atmospheric disturbances.

REFERENCES

Allan, R. & D. Molyneux (2004). Escort tug design alternatives and a comparison of their hydrodynamic performance. *Transactions of the Society of Naval Architects 112*, 191–205.

Allan, R.G. (2000). The evolution of escort tug technology... fulfilling a promise. *SNAME Transactions, Vol. 108, 2000, pp. 99–122 108*, 99–122.

Allan R. (2012). Escort winch, towline, and tether system analysis. Technical report, Prince William Sound Regional Citizens' Advisory Council, Anchorage, AK.

Altosole, M., M. Figari, & M. Martelli (2012). Time-domain simulation for marine propulsion applications. Volume 44, pp. 36–43.

American Bureaux of Shipping (2016). Guide for Building and Classing—Offshore Support Vessels.

Ankudinov, V., P. Kaplan, & B. Jacobsen (1993). Assessment and principal structure of the modular mathematical model for ship maneuverability predictions and real-time maneuvering simulation. *MARSIM93, international conference on marine simulation and ship manoeuvrability, International Marine Simulator Forum (IMSF), St. Johns, NL, Canada*, pp.4051.

Bernitsas, M. & N. Kekridis (1986). Nonlinear stability analysis of ship towed by elastic rope. *Journal of Ship Research* 30 No. 2, 136–146.

Brandner, P. & M. Renilson (1998). Interaction between two closely spaced azimuthing thrusters. *Journal of Ship Research* 42 No.1, 15–32.

Bruzzone, D., D. Ruscelli, D. Villa, & M. Vivani (2015). Numerical prediction of hull force for low velocity manoeuvring. In *18th International Conference on Ships and Shipping Research, NAV 2015*, pp. 284–295.

Canadian Environmental Assessment Registry (2004). Tethered escort of tankers. Technical report, State of Washington: Department of Ecology Lacey, Washington.

Capt. Henk Hensen (2012). The art of tugnology: the ultimate balance between safe ship assist and safe escorting duties. Technical report.

Couce, L.C., J.C.C. Couce, & J.F. Formoso (2015). Operation and handling in escort tugboat manoeuvres with the aid of automatic towing winch systems. *The Journal of Navigation 68*, 71–88.

der Laanand Kees Kraaijeveld, M.V. (2013). Safewinch tackles slack wires and peak loads. *Tugnology Day 2*, No. 2.

Fossen, T.I. (2011). *Handbook of Marine Craft Hydrodynamics and Motion Control*. Norwegian University of Science and Technology—Trondheim, Norway.

Frans Quadvileg, a. S.K. (2006). Development of a calculation program for escort forces of stern drive tug boats. *International Tug & Salvage Convention.*

FREng, J.S.C. (2010/2011). *Marine Propellers and Propulsion*. City University London—Institute of Marine Engineering, Science and Technology 2010/11.

Gray, D.L. & E. Reynolds (2001). Eng. methodologies used in the preparation of escort requirements for the ports of sanfrancisco and los angeles. *Marine Technology 38 No. 1*, 51–64.

Griffin, B. (2004). Ship assist and escort winches for dynamic seas. *International Tug & Salvage Convention.*

Hensen, C.H. (2003). *Tug Use In Port*. The Nautical Institute.

Hoerner, S.F., & H. Borst (1965). Fluid dynamic drag: Theoretical, experimental and statistical information. *Hoerner Fluid Dynamics.*

Hoerner, S.F., & H. Borst (1975). Fluid-dynamic lift: Information on lift and its derivatives in air and in water dynamics. *Hoerner Fluid Dynamics.*

Imlay, F.H. (1961). The complete expressions for added mass of a rigid body moving in an ideal fluid. *Technical Report DTMB 1528. David Taylor Model Basin. Washington D.C..*

Jiang, T., R.H. & S.D. Sharma (1998). Dynamic behavior of a tow system under an autopilot on the tug. *International Symposium and Workshop on Forces Acting on a Manoeuvring Vessel Manoeuvring Vessel, Val de Reuil, France.*

Jrgens, D. & M. Palm (2015). Voith water tractor improved manoeuvrability and seakeeping behaviour. *Tugnology.*

Kijima, K. & Y. Furukawa (1998). Effect of roll motion on manoeuvrability of ship. *International Symposium and Workshop on Force Acting on a Manoeuvring Vessel, Val de Reuil, France—Dep. of Naval Architecture and Marines Systems Engineering, Kyushu University.*

Kooren, T., F. Quadvlieg, & A. Aalbers (2000). Rotor tugnology. *The 16th International Tug and Salvage Convention.*

Li, Y. & S.M. Calisal (2005). Numerical simulation of ships manoeuvrability in wind and current. *Marine Technology 42 No. 3*, 159–176.

Martelli, M., M. Viviani, M. Altosole, M. Figari, & S.Vignolo (2014). Numerical modelling of propulsion, control and ship motions in 6 degrees of freedom. *Proceedings of the Institution of Mechanical Engineers Part M: Journal of Engineering for the Maritime Environment 228*(4), 373–397.

Merrick, J.R.W. (2002). Evaluation of tug escort schemes using simulation of drifting tankers. *Simulation 78 Issue 6*, 380–388.

Molyneux, D. & N. Bose (2007). Escort tug at large yaw angle: Comparison of cfd predicitions with experimental data.

Molyneux, W. & G. Earle (2001). A comparison of forces generated by a hull and three different skegs for an escort tug design. *Research Report, Fort St. John. National Research Council Canada Institute of Marine Dynamics.*

Nakayama, Y., H. Yasukawa, N. Hirata, & H. Hata (2012). Time domain simulation of wave-induced motions of a towed ship in head seas. *Proceedings of the Twenty-second International Offshore and Polar Engineering Conference Rhodes, Greece—Graduate school of Engineering, Hiroshima University, Japan.*

Nienhuis, U. (1992). Analysis of thruster effectivity for dynamic positioning and low speed manoeuvring. *PhD thesis.*

Oltmann, P. & S. Sharma (1984). Simulation of combined engine and rudder manoeuvres using an improved model of hull-propeller-rudder interactions. *Hamburg, Technical University of Hamburg.*

Oosterveld, M.W.C. (1970). *Wake Adapted Ducted Propellers*. H. Veenman & Zonen N.V. - Wageningen.

Papoulias, F.A. & M.M. Bernitsas (1988). Autonomous oscillations, bifurcations, and chaotic response of moored vessels. *Journal of Ship Research 32 No. 2*, 220–228.

Robert Allan, and B. Smoker (2013). A review of best available technology in tanker escort tugs. Technical report, Prince William Sound Regional Citizens Advisory Council, An-chorage, AK.

SAFETUG, J.I.P. (2015). Ship assist in fully exposed conditions. *Tugnology.*

Smoker, B. (2009). Escort tug performance prediction: A cfd method. Technical report, B. Eng, University of Victoria.

The Glosten Associates (2004). Study of tug escorts in puget sound. Technical report, State of Washington: Department of Ecology Lacey, Washington.

United States Congress 101st, P.W.B. (1990). Oil pollution act (opa). Technical Report (101 H.R.1465, P.L. 101–380).

Waclawek, P. & D. Molyneux (2000). Prediction the performance of a tug and tanker during escort operatioins using computer simulations and model tests. *SNAME Transactions 108*, 21–43.

Maritime Transportation and Harvesting of Sea Resources – Guedes Soares & Teixeira (Eds)
© 2018 Taylor & Francis Group, London, ISBN 978-0-8153-7993-5

Surrogate models of the performance and exhaust emissions of marine diesel engines for ship conceptual design

M. Tadros, M. Ventura & C. Guedes Soares

Centre for Marine Technology and Ocean Engineering (CENTEC), Instituto Superior Técnico, Universidade de Lisboa, Lisbon, Portugal

ABSTRACT: The estimation of the performance and the exhaust emissions of a marine engine is an important step during the ship early design stage in order to choose the efficient ship propeller and to verify the compliance with the EEDI requirements and the exhaust emissions limitations defined by the International Maritime Organization. In this paper, a surrogate model is developed based on the response surface methodology to estimate the performance and the exhaust emissions of two four-stroke marine turbocharged diesel engines for different speeds and loads. The response surface is obtained based on data generated from 1D engine simulation software. A polynomial regression model is used to generate equations of brake specific fuel consumption, CO_2 and NO_x emissions of the selected engine as functions of the engine speed and load. These generated equations can be further used to predict the behavior of the selected engines or similar engines.

1 INTRODUCTION

Complex and large systems are used nowadays frequently in many different fields, but with the aid of advanced computers, modeling and simulation of these systems becomes easier. However, they require a large number of numerical iterations with different analysis and need some experts to identify the optimal solution.

A ship is a very complex system and its design requires different optimization methods in order to control the huge amount of variables involved. These variables vary with the different types and sizes of the ships, the requirements of the owner and the international regulations.

Ship design deals with many fields (Watson, 1998) and each one has its characteristics and calculation methods. The ship's hull, the ship resistance and the propulsion system are controlled using different methods as Taylor, series 60, British Ship Research Association (BSRA), Guldhammer and Harvald (1970) and Holtrop and Mennen (Carlton, 2011). These methods are based on large amount of real data collected from existing ships. In order to simplify the calculations of the ships, these methods are presented either in the form of chart or equations. They are considered as approximation models that help the ship designer to model standard ships and to optimize a new series of ships able to achieve the specific requirements.

Diesel engines, used either as a source of mechanical or of electrical power, are very important components of the ship. However, it is a multi-disciplinary application field deals with the four major academic disciplines in mechanical systems: design methodology and interactive graphics; dynamic systems and control; machine dynamics; and tribology. It requires an optimization approach, system dynamics approach, dynamic motion analysis and friction analysis (Xin, 2011).

Figari and Altosole (2007) and Altosole and Figari (2011) used control methods to calculate the dynamic behavior of a propulsion system in which they recommend the thermodynamic calculation for a better accuracy. Different 1D commercial software packages of cycle simulation are used to compute the engine performance which required more details and effort to calibrate and simulate each engine separately as in Cordiner et al. (2007) and Theotokatos et al. (2016) in which the dual fuel engine is simulated and in Gharahbaghi et al. (2006), Mosbach et al. (2006), Lee and Mastorakos (2007) and Morcinkowski et al. (2012) where the homogeneous charge compression ignition (HCCI) is considered. Also, different 0D codes using Matlab/Simulink were developed by Benvenuto and Campora (2002), Maftei et al. (2009), Benvenuto et al. (2013), Tadros et al. (2015), Benvenuto et al. (2016) and Altosole et al. (2017) to compute the behavior of spark and compression ignition engines. These models are validated using real engine data to verify the limitations of the exhaust emissions as shown in Figure 1.

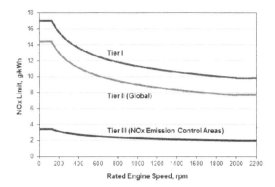

Figure 1. MARPOL Annex VI NOx emission limits (www.dieselnet.com).

The International Maritime Organization (IMO) (http://www.imo.org) has applied another restriction to achieve the most efficient design of the ship which is the energy efficient design index (EEDI) as presented in the following equation. This index helps to maximize the dead weight of the ship and to minimize the fuel consumption and thus the CO_2 emissions.

$$EEDI = \frac{CO_2\ emissions}{transport\ work} \qquad (1)$$

where the CO_2 emissions represents the total amount of CO_2 from all internal combustion engines on the ship and the transport work depend on the ship dead weight and ship design speed.

Different theories, methods and software are updated to verify the market requirement for both engines and ship. However, it takes long time to get the results.

Approximation models (surrogate models) are one of the methods used to reduce the time consumed doing simulations and analysis which leads to an economy of calculations. They are constructed by understanding the relationship between design parameters and output performance over the entire design space. Queipo et al. (2005) discussed the concepts, methods and techniques of different surrogate models and their effects on the analysis and optimization of different systems. This paper focuses on the development of a surrogate model of the engine performance for a future use instead of a detailed modeling of the engine. A fourth order polynomial regression model is used to generate equations of brake specific fuel consumption (BSFC), CO_2 and NOx emissions of two four-stroke marine turbocharged diesel engines as functions of speed and load.

This regression model depends on response surface methodology (RSM) implemented in Matlab in which the response surface is based on

the generated data from two models developed and simulated using 1D engine simulation software (Ricardo Wave Software, 2016). Figure 2 presents a flow chart for developing a surrogate model coupling Ricardo Wave® and Matlab®.

These generated equations can be used during the ship early design stage to achieve the required power of the ship and to verify the compliance with the EEDI requirements and the limitation of exhaust emissions applied by IMO. They can also be connected to ship routing code to determine the behavior of the ship during sailing and to analysis the amount of fuel consumption and exhaust emis-

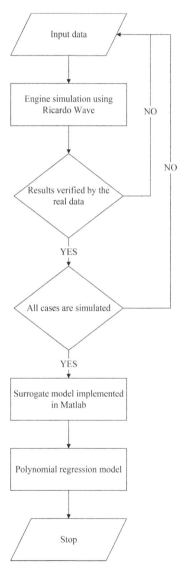

Figure 2. Flowchart for surrogate model via Ricardo Wave/Matlab coupling.

sions during the voyage (Vettor and Guedes Soares, 2016, Vettor et al., 2016, Perera and Mo, 2017).

The rest of the paper is organized as follows: section 2 presents the main technical data of the diesel engine used during the simulation; section 3 gives an overview of the methods and theories used to simulate the engine and the surrogate model; the simulation and results are presented in section 4, while conclusion and final remarks are discussed in section 5.

2 GENERAL PRESENTATION OF THE ENGINES

2.1 MAN R6-730 engine

The first engine chosen for simulation is the MAN R6-730 marine turbocharged diesel engine, a 4-stroke, 6 cylinder in-line, with a speed range of 1,000–2,100 RPM (http://www.engines.man.eu/). It is designed for the propulsion of small boats like yachts and patrol boats. The maximum power of the selected engine is 537 kW at 2,300 RPM and the following table shows the main characteristics of the engine.

The simulated data to compute the performance of the engine is validated using real data from the manufacturer. The BSFC calculated from Ricardo Wave are compared with the data from the manufacturer for different cases on the theoretical propeller curve and on the maximum load limit. A comparison between the calculated BSFC from Ricardo Wave and the real one is presented in Figure 3 and Figure 4. A 5% maximum error is achieved for the different cases between the simulated case and the manufacturer data.

2.2 MAN 18V32/44CR engine

The second engine chosen is the MAN 18V32/44CR. It is a large marine four-stroke tur-

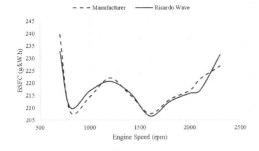

Figure 3. Comparison between the values of BSFC at the theoretical propeller curve of the manufacturer and Ricardo Wave for MAN R6-730.

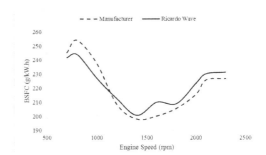

Figure 4. Comparison between the values of BSFC at the maximum load of the manufacturer and Ricardo Wave for MAN R6-730.

Table 1. Technical data of MAN R6-730.

Bore (mm)	128
Stroke (mm)	166
No. of cylinders	6
Displacement (liter)	12.82
BMEP (bar)	21.9
Piston speed (m/s)	10.5
Rated speed (RPM)	2,300
Maximum torque (Nm)	2,512
Engine speed range (RPM)	1,000–2,100
Specific fuel consumption (g/kW.h)	225
Power-to-weight ratio (kW/kg)	0.41
Exhaust gas status	IMO Tier 2

Table 2. Technical data of MAN 18V32-44CR.

Bore (mm)	320
Stroke (mm)	440
No. of cylinders	18
Displacement (liter)	640
BMEP (bar)	23.06
Piston speed (m/s)	11
Rated speed (RPM)	750
Maximum torque (Nm)	116,886
Engine speed range (RPM)	450–750
Specific fuel consumption (g/kW.h)	190
Power-to-weight ratio (kW/kg)	0.1
Exhaust gas status	IMO Tier 2

bocharged diesel engine. It may be applied for mechanical or diesel-electric propulsion drive in Bulker, container vessel, general cargo vessel and ferry. The maximum power of this engine is 9,180 at 750 RPM and Table 2 shows the main characteristics of the engine.

The model developed in Ricardo Wave is also calibrated and validated using real data from the manufacturer attached to a fixed pitch propeller (FPP) as presented in Figure 5 and then the per-

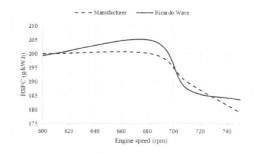

Figure 5. Comparison between the values of BSFC at the maximum load of the manufacturer and Ricardo Wave for MAN 18V32/44CR.

formance of the operating range is estimated for different engine speed and loads.

3 ENGINE SIMULATION MODEL AND RESPONSE SURFACE METHODOLOGY

3.1 *Engine simulation model*

The two marine turbocharged diesel engines considered are modeled using 1D engine simulation software to calculate the performance of each engine for different speeds and loads. The simulated data are compared and validated by the data presented from the manufacturer as described above. The model helps to minimize the BSFC and achieve the limitation of NOx emissions according to the IMO restrictions. The turbocharger, intercooler and intake and exhaust manifolds are modeled and simulated according to the methods presented in Watson and Janota (1982). The different processes inside the cylinders are calculated according to the first law of thermodynamics (Heywood, 1988) taking into account the coefficient of the equation of the heat transfer suggested by Woschni (1967). The heat release rate (HRR) of the combustion process is modeled using the Wiebe function expressed in Watson et al. (1980) considering the effect of injection timing on the engine performance (Tadros et al., 2016). The CO2 emissions are estimated as function of BSFC using the emission factor while the NOx emissions are calculated using the extended Zeldovich mechanism using two zones of combustion (Heywood, 1988).

3.2 *Response surface methodology*

The RSM is a type of surrogate model that employs a statistical procedure, as suggested by Box and Wilson (1951) to obtain a surface response using a polynomial model to describe the relationship between the inputs and the outputs. A good RSM

allows estimating the coefficients of the model with low uncertainty which is measured by narrow error bars of the model's regression coefficients. The RSM is widely used in different fields with different polynomial equations according to the complexity of the system. It is applied to both experimental test and simulation where, the computer simulation is faster and more economic than the experimental test, however, a large amount of accurate data simulated and validated with the experimental one provide a good accuracy to the polynomial regression model.

In case of internal combustion engines, Xin (2011) supports the use of second-degree polynomial model which show good results to generate the surface response in comparison with the experimental data as presented in (Ganapathy et al., 2011) by equation (2).

$$Y = \beta_o + \sum_{i=1}^{n} \beta_i X_i + \sum_i \sum_j \beta_{ij} X_i X_j + \sum_{i=1}^{n} \beta_{ii} X_i^2 + \varepsilon \quad (2)$$

where Y is the response, Xi are numeric values of the factors, terms βo, βi, βii and βij are regression coefficients, i and j are linear and quadratic coefficients and ε is the experimental error.

RSM is used to optimize different engine parameters in order to obtain the desired combustion and performance characteristics as described in (Win et al., 2005, Atmanlı et al., 2015, Hirkude and Padalkar, 2014). Different mathematical models are built according to the generated surface and linked to an optimization code to minimize or maximize the functional target.

4 SIMULATION AND RESULTS

The performance of the two four-stroke marine turbocharged diesel engine is estimated using 1D engine simulation software. The BSFC, CO2 and NOx emissions are the main results calculated functions of speeds and loads as shown from Figure 6 to Figure 11. The BSFC is validated using the data provided from the manufacturer as presented above and the NOx emissions are verified by the limitation applied by the IMO. The RSM is then applied as a type of surrogate model in which the response surface is generated according to the simulated data. A fourth polynomial regression model shows a good fitting with the simulated data than the lower-order model. It is used to generate equations of BSFC, CO2 and NOx emissions function of speed and load and is presented by the following equation:

$$\begin{aligned}
f(x,y) &= p00 + p10 \times x + p01 \times y + p20 \times x^2 \\
&+ p11 \times xy + p02 \times y^2 + p30 \times x^3 + p21 \times x^2 y \\
&+ p12 \times xy^2 + p03 \times y^3 + p40 \times x^4 + p31 \times x^3 y \\
&+ p22 \times x^2 y^2 + p13 \times xy^3 + p04 \times y^4
\end{aligned} \quad (3)$$

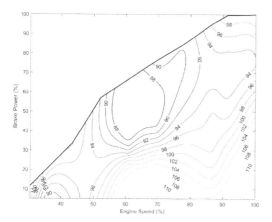

Figure 6. Variation of BSFC of engine operating range.

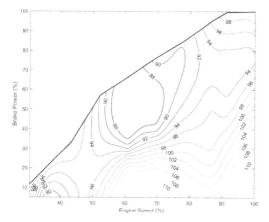

Figure 7. Variation of CO_2 emissions of engine operating range.

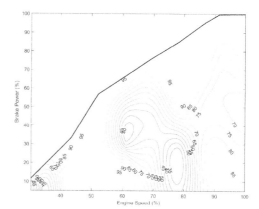

Figure 8. Variation of NOx emissions of engine operating range.

Table 3. Value of equation coefficients of BSFC and NOx emissions MAN R6-730.

Coefficient	BSFC	NO_x emissions
p00	4.213	−6.243
p10	−26.840	52.240
p01	11.190	−11.760
p20	76.150	−137.900
p11	−57.170	66.340
p02	7.610	−10.840
p30	−85.240	157.500
p21	72.450	−145.600
p12	3.181	93.200
p03	−9.721	−32.550
p40	32.490	−61.650
p31	−23.590	67.170
p22	−18.670	−23.160
p13	17.220	−27.110
p04	−2.274	21.340

where x is the engine speed and y is the brake power and both of them are non-dimensional.

Engine speed and load are expressed as percentages from the rated speed and the maximum load respectively. The BSFC equation is also calculated as a percentage of BSFC at the rated speed, while the equation of NOx emissions is a percentage of the NOx limit at the rated speed as suggested by the IMO. The CO_2 emissions is calculated using the BSFC multiplied by the emission factor which is 3.17 as presented by Kristensen (2012). Equation (4) shows the relation between the CO_2 emissions and BSFC.

$$CO_2 \ emissions = BSFC * EF \qquad (4)$$

where EF is the emission factor.

The performance of the small engine MAN R6-730 is shown from Figure 6 to Figure 8. The curve of the load limit is presented in equation (5) to define the limitation of the brake power for each engine speed.

$$y = \begin{cases} 1.6394x - 0.3853 & 0.3 > x \geq 0.43 \\ 2.7028x - 0.8341 & 0.43 > x \geq 0.52 \\ 1.1019x - 0.0041 & 0.52 > x \geq 0.91 \\ 1 & 0.91 > x \geq 1 \end{cases} \qquad (5)$$

The generated surface of BSFC and NOx emissions show a goodness of fit with R-square equals to 0.8983 and 0.5505 respectively and the coefficients of the equation are presented in Table 3. Due to low value of R-square of NOx emissions, the sum of squares due to error (SSE) and the root mean squared error (RMSE) are 0.5034 and 0.1896 respectively where the two values are closer to 0 which indicates a fit that is more useful for prediction.

The performance of the large engine MAN 18V32/44CR is shown from Figure 9 to Figure 11 and the maximum load limit for the operating range using FPP are presented in equation (6).

$$y = \begin{cases} 0.901x^3 + 0.4792x^2 & 0.5 > x \geq 0.9 \\ \quad - 0.1547x - 0.0049 & \\ 0.9918x + 0.0082 & 0.9 > x \geq 1 \quad (6) \\ 1 & 1 > x \geq 1.06 \end{cases}$$

The generated surface of BSFC and NO_x emissions using the regression model show a goodness of fit with R-square equals to 0.9340 and 0.8168

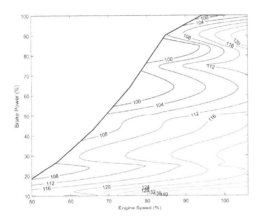

Figure 9. Variation of BSFC of engine operating range.

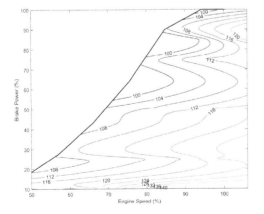

Figure 10. Variation of CO_2 emissions of engine operating range.

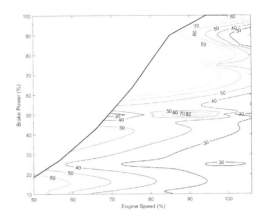

Figure 11. Variation of NOx emissions of engine operating range.

Table 4. Value of equation coefficients of BSFC and NOx emissions for MAN 18V32/44CR.

Coefficient	BSFC	NO_x emissions
p00	5.404	17.950
p10	−24.460	−89.420
p01	10.820	24.380
p20	48.630	179.600
p11	−32.440	−156.800
p02	−6.538	77.720
p30	−38.050	−161.500
p21	17.600	228.800
p12	32.890	−98.680
p03	−7.723	−45.860
p40	10.150	53.580
p31	0.869	−97.070
p22	−19.190	26.840
p13	0.710	36.270
p04	2.356	5.021

respectively and the coefficients of the equation are presented in the following table:

5 CONCLUSIONS

In this paper, a fourth-order polynomial regression model implemented in Matlab to represent a RSM is used to generate equations of BSFC, CO_2 and NO_x emissions of two four-stroke marine turbocharged diesel engine function of speed and load. The response surface is based on results for each engine from 1D engine simulation software. These models help to estimate the fuel consumption and exhaust emissions of the selected or similar engine during the ship early design stage to choose the efficient ship propeller and to verify the compliance with the EEDI requirements and the exhaust emissions limitations applied by IMO. It can also be coupled to a routing code to estimate the amount of fuel consumption and exhaust emissions during the ship voyage.

ACKNOWLEDGEMENTS

This work was performed within the scope of the Strategic Research Plan of the Centre for Marine Technology and Ocean Engineering (CENTEC), which is financed by the Portuguese Foundation for Science and Technology (Fundação para a Ciência e Tecnologia—FCT).

REFERENCES

Altosole, M. & Figari, M. 2011. Effective simple methods for numerical modelling of marine engines in ship propulsion control systems design. *2011*, 8: 19.

Altosole, M., Benvenuto, G., Campora, U., Laviola, M. & Zaccone, R. 2017. Simulation and performance comparison between diesel and natural gas engines for marine applications. *Journal of Engineering for the Maritime Environment*, 231: 690–704.

Atmanlı, A., Yüksel, B., İleri, E. & Deniz Karaoglan, A. 2015. Response surface methodology based optimization of diesel–n-butanol—cotton oil ternary blend ratios to improve engine performance and exhaust emission characteristics. *Energy Conversion and Management*, 90: 383–394.

Benvenuto, G. & Campora, U. 2002. Dynamic simulation of a high-performance sequentially turbocharged marine diesel engine. *International Journal of Engine Research*, 3: 115–125.

Benvenuto, G., Laviola, M. & Campora, U. 2013. Simulation model of a methane-fuelled four stroke marine engine for studies on low emission propulsion systems. Guedes Soares, C. & Lopez Pena F., (Eds.). *Developments in Maritime Transportation and Exploitation of Sea Resources.* Taylor & Francis Group London, UK; 591–597.

Benvenuto, G., Laviola, M., Zaccone, R. & Campora, U. 2016. Comparison of a natural gas engine with a diesel engine for marine propulsion. Guedes Soares, C. & Santos T.A., (Eds.) *Maritime Technology and Engineering III.* London, UK: Taylor & Francis Group 725–734.

Box, G.E.P. & Wilson, K.B. 1951. On the Experimental Attainment of Optimum Conditions. *Journal of the Royal Statistical Society. Series B (Methodological),* 13: 1–45.

Carlton, J. 2011. *Marine Propellers and Propulsion.* Elsevier Science.

Cordiner, S., Rocco, V., Scarcelli, R., Gambino, M. & Iannaccone, S. 2007. Experiments and Multi-Dimensional Simulation of Dual-Fuel Diesel/Natural Gas Engines. Consiglio Nazionale delle Ricerche.

Figari, M. & Altosole, M. 2007. Dynamic behaviour and stability of marine propulsion systems. *Journal of Engineering for the Maritime Environment*, 221: 187–205.

Ganapathy, T., Gakkhar, R.P. & Murugesan, K. 2011. Optimization of performance parameters of diesel engine with Jatropha biodiesel using response surface methodology. *International Journal of Sustainable Energy*, 30: S76-S90.

Gharahbaghi, S., Wilson, T.S., Xu, H., Cryan, S., Richardson, S., Wyszynski, M.L. & Misztal, J. 2006. Modelling and Experimental Investigations of Supercharged HCCI Engines. SAE Technical Paper 2006–01-0634.

Guldhammer, H.E. & Harvald, S.A. 1970. *Ship Resistance: Effect of Form and Principal Dimensions.* Akademisk Forlag.

Heywood, J.B. 1988. Internal combustion engine fundamentals. McGraw-Hill.

Hirkude, J.B. & Padalkar, A.S. 2014. Performance optimization of CI engine fuelled with waste fried oil methyl ester-diesel blend using response surface methodology. *Fuel*, 119: 266–273.

Kristensen, H.O. 2012. Energy demand and exhaust gas emissions of marine engines. Technical University of Denmark.

Lee, C.-W. & Mastorakos, E. 2007. Numerical simulations of homogeneous charge compression ignition engines with high levels of residual gas. *International Journal of Engine Research*, 8: 63–78.

Maftei, C., Moreira, L. & Guedes Soares, C. 2009. Simulation of the dynamics of a marine diesel engine. *Journal of Marine Engineering & Technology*, 8: 29–43.

Morcinkowski, B., Brassat, A., Pischinger, S., Adomeit, P., Ewald, J., Albin, T. & Abel, D. 2012. Detailed Simulation of Cycle-to-Cycle Fluctuations at Gasoline Auto Ignition Engines and Derivation of a Synchronous Simulation and Control Approach*. *IFAC Proceedings Volumes*, 45: 154–161.

Mosbach, S., Kraft, M., Bhave, A., Mauss, F., Mack, J.H. & Dibble, R.W. 2006. Simulating a Homogeneous Charge Compression Ignition Engine Fuelled with a DEE/EtOH Blend. SAE International.

Perera, L.P. & Mo, B. 2017. Marine Engine-Centered Data Analytics for Ship Performance Monitoring. *Journal of Offshore Mechanics and Arctic Engineering*, 139: 021301-021301-8.

Queipo, N.V., Haftka, R.T., Shyy, W., Goel, T., Vaidyanathan, R. & Kevin Tucker, P. 2005. Surrogate-based analysis and optimization. *Progress in Aerospace Sciences*, 41: 1–28.

Ricardo Wave Software 2016. WAVE 2016.1 Help System.

Tadros, M., Ventura, M. & Guedes Soares, C. 2015. Numerical simulation of a two-stroke marine diesel engine. *In:* Guedes Soares, Dejhalla & Pavleti (eds.), *Towards Green Marine Technology and Transport.* Taylor & Francis Group, London: 609–617.

Tadros, M., Ventura, M. & Guedes Soares, C. 2016. Assessment of the performance and the exhaust emissions of a marine diesel engine for different start angles of combustion. *In:* Guedes Soares & Santos (eds.), *Maritime Technology and Engineering 3.* Taylor & Francis Group, London: 769–775.

Theotokatos, G., Stoumpos, S., Lazakis, I. & Livanos, G. 2016. Numerical study of a marine dual-fuel four-stroke engine. *In:* Guedes Soares & Santos (eds.), *Maritime Technology and Engineering 3.* Taylor & Francis Group, London: 777–783.

Vettor, R. & Guedes Soares, C. 2016. Development of a ship weather routing system. *Ocean Engineering,* 123: 1–14.

Vettor, R., Tadros, M., Ventura, M. & Guedes Soares, C. 2016. Route planning of a fishing vessel in coastal waters with fuel consumption restraint. *In:* Guedes Soares & Santos (eds.), *Maritime Technology and Engineering 3.* Taylor & Francis Group, London: 167–173.

Watson, D.G.M. 1998. *Practical Ship Design.* Elsevier.

Watson, N. & Janota, M.S. 1982. Turbocharging the Internal Combustion Engine. Wiley.

Watson, N., Pilley, A.D. & Marzouk, M. 1980. A Combustion Correlation for Diesel Engine Simulation. SAE Paper 800029.

Win, Z., Gakkhar, R.P., Jain, S.C. & Bhattacharya, M. 2005. Parameter optimization of a diesel engine to reduce noise, fuel consumption, and exhaust emissions using response surface methodology. *Proceedings of the Institution of Mechanical Engineers, Part D: Journal of Automobile Engineering,* 219: 1181–1192.

Woschni, G. 1967. A Universally Applicable Equation for the Instantaneous Heat Transfer Coefficient in the Internal Combustion Engine. SAE Technical Paper 670931.

Xin, Q. 2011. Diesel Engine System Design. Woodhead.

Ports

Maritime Transportation and Harvesting of Sea Resources – Guedes Soares & Teixeira (Eds)
© 2018 Taylor & Francis Group, London, ISBN 978-0-8153-7993-5

Sustainable urban logistics for efficient and environmentally friendly port-cities

A. Anagnostopoulou
Center for Research and Technology Hellas—Hellenic Institute of Transport, Athens, Greece

E. Sdoukopoulos & M. Boile
Department of Maritime Studies, University of Piraeus, Piraeus, Greece
Center for Research and Technology Hellas—Hellenic Institute of Transport, Athens, Greece

ABSTRACT: This paper presents a methodology aiming to support local and regional authorities of port-cities and regions in identifying appropriate measures to improve efficiency and sustainability of urban freight transport and distribution. To this end, it undertakes a thorough review of many successful but often disparate relevant initiatives, implemented in Europe and the US, assessing their transferability and the potential benefits to be realized once implemented in a port-city environment. This process can provide a solid basis for developing a sustainable urban logistics plan setting short and long-term targets. A case study in the city of Piraeus demonstrated the applicability of the proposed approach as well as its successful transferability potential to port cities in Europe and beyond. This approach may be incorporated in the urban logistics planning process, to support decision making.

1 INTRODUCTION

1.1 Background

The high degree of urbanization characterizing the world today (54% of global population lives in urban areas), which is expected to further grow in the near future (a 12% increase is being anticipated until 2050) (United Nations, 2015), has resulted into a considerable increase in the size of urban settlements as well as in the density of their activities especially with regard to urban freight transport and distribution services, in an effort to successfully meet the requirements of what turns to be a rapidly increasing consumer-oriented world. Urban freight transport systems are often plagued with major inefficiencies stemming from competing interaction with passenger transport activities, limited available capacity, lack of coordination among stakeholders, and extended peak hours of operation, which significantly contribute towards higher congestion levels, delays, energy consumption and resulting environmental implications. While passenger mobility is being viewed by the general public as a necessity, urban freight transport is typically seen as the 'bad neighbor'. With the aim to address inefficiencies, reduce the associated externalities and enhance the public image of urban freight transport, several initiatives have been undertaken worldwide at a local, national, regional or even international level. Their scope and focus includes policy formation (e.g. Sustainable Urban Logistics Plans—SULPs, etc.) and the introduction of appropriate regulatory measures, which may be infrastructure, facility, time or vehicle-related (e.g. access restrictions, pricing policies, etc.); efficient information exchange (e.g. Information and Communication Technologies, etc.); infrastructure developments (e.g. Urban Consolidation Centres, freight un/loading spaces, etc.); vehicle technologies and exploitation of alternative energy sources (e.g. electric or hybrid vans, trucks, etc.); and new business processes and models. The successful investigation, selection and implementation of appropriate measures, tailored to the conditions and characteristics of each local context is of utmost importance for moving towards a more efficient, productive and sustainable urban freight transport system, driving local economic growth and prosperity. Such a process proves to be increasingly complex and multi-parametric and becomes more challenging in port-city environments, where consideration should be given to the operations of the port in relation to the urban area, which increase the density of urban freight transport activities and the number of actors involved.

Globalization and world trade evolution have altered the intricate relationship between ports and cities with the latter becoming focal points of inland transport networks with increased job opportunities for covering the freight transport needs of both inland and urban areas and serving as major commercial centers. Ports and inland terminals play a vital role as well, and have been effectively integrated into global supply chains going beyond their traditional transshipment function and strengthening their role in logistics. The so-called 'terminalization' of supply chains implies that ports and inland terminals are increasingly acting as the main buffer in supply chains with supply chain managers often adopting a hybrid inventory approach effectively coupling 'inventory in transit' and 'inventory at terminal' strategies in an effort to minimize warehousing capacities at distribution centres (Rodrigue and Notteboom, 2009).

Such an increased interaction of ports with inland transport terminals has often resulted into the development of several large-scale logistics poles within the respective zones of port-cities. Port-related operations, therefore, add a new dimension and different dynamics and implications to the urban freight transport system. With very little room for land and facility expansion and considering new market related trends and developments (e.g. introduction of mega vessels), ports and their surrounding urban areas are often struggling with rapidly increasing freight volumes and heavy-truck flows, taking into consideration the modal share of most ports. The urban freight distribution problem in port-cities should therefore be approached from a different perspective compared to other urban environments, considering all additional parameters that should be taken into account for resolving extra problems that may be created but also exploiting the new opportunities that may arise due to the port activity.

Congestion is regarded as the most important challenge that modern supply chains are facing today since, in many cases, the available transport infrastructure is limited, impeding the efficient planning and accessibility of urban freight transport services. The importance of the associated environmental concerns and management issues has also risen considerably attracting the increased interest of both public and private stakeholders. To this end, the implementation of energy efficient and environmentally-friendly urban freight transport measures can provide a common ground for dialogue between private and public sector stakeholders (Taniguchi et al., 2001) meeting, respectively, profit maximization and social inclusion targets.

Motivated by the aforementioned existing trends and observations, this paper aims to present an approach to be considered in urban freight transport planning in port cities. The approach was adopted in the city of Piraeus in Greece, for determining the most appropriate, to the respective local context, set of actions, efficiently combining successful but often disparate urban freight trans-

port initiatives, and investigating their implementation potential for improving the current energy and environmental performance of the urban freight transport system. The analysis of the latter, taking into consideration the operation of Greece's largest port which the city of Piraeus surrounds, resulted in the identification of major inefficiencies and local disturbances that need to be properly addressed for achieving both operational and environmental cost reductions.

The paper adds to the existing literature on urban freight transport research by discussing and analyzing the critical factors that influence decision-making with regard to urban freight transport planning, and utilizing the respective research perspectives for urban freight transport modelling and analysis with emphasis on port-cities. To this end, the remainder of the paper is structured as follows: Section 2 provides an overview of the existing literature on urban freight transport research, providing the corresponding theoretical background and summarizing the international experiences gained with regard to successfully implemented relevant policies and measures. More specifically, pertinent perspectives from Europe and the U.S. are being presented, highlighting common challenges and urban freight transport planning processes and coordination strategies that can be adopted considering different port structures and freight movements, as encountered in practice. The key observations drawn serve as background for proposing, in section 3, a rational urban freight transport planning approach, efficiently incorporating diverse experiences and practices. The approach has been implemented in the city of Piraeus, where the largest port in Greece and one of the largest in the Mediterranean is located. Through an empirical analysis, useful knowledge and insights were generated regarding the applicability and effectiveness of the proposed approach as well as of selected measures, as presented in section 4. The last section concludes the paper by providing targeted recommendations for efficiently integrating urban freight transport planning into the overall process of developing sustainable urban mobility action plans with focus on port-cities.

2 LITERATURE REVIEW AND INTERNATIONAL EXPERIENCES

2.1 *Urban freight transport systems*

Several different frameworks are available for supporting efficient urban freight transport planning facilitating the proper selection, and combination, of targeted measures for improving the overall performance and efficiency of urban freight transport systems, as described in the relevant literature (Ogden, 1984) (Lindhom & Behrends, 2010) (Rosso & Comi, 2011) (Suksri et al. 2012) (Balm et al., 2014) (Imanishi & Taniguch, 2016). The implementation of urban freight distribution (or consolidation) centers is among the most extensively studied measures worldwide, considering also their efficient integration and combination with other relevant measures and solutions (Boile et al., 2008). Along this, a number of coordination strategies have been investigated even among competing companies, realizing the important benefits that can be generated, at both ends, from the development of such synergies (i.e. reduced operational costs, traffic congestion, pollutant emissions, empty vehicle-kms, etc.).

Thomson and Taniguchi (2001) effectively summarize and present a number of different models and strategies that can be adopted for ensuring efficient urban freight transport operations improving profitability, optimizing traffic flows and enhancing freight accessibility into city centers. Ambrosini and Routhier (2004) classify urban freight transport models into 'operational' considering economic improvements and 'systemic' focusing on environmental impacts. They study a number of real-life cases in Europe, Asia and the US and identify two critical factors for ensuring an efficient and effective urban freight transport system i.e. cooperation among all different stakeholders and ICT utilization. Taniguchi and Tamagawa (2005) analyze the behavior of different urban freight transport stakeholders and consider two popular measures in city logistics (i.e. truck access restrictions and pricing) for evaluating their respective implementation costs, benefits and stakeholders' acceptance. An increased concern on the environmental implications of urban freight transport operations characterizes more recent literature with Giuliano et al. (2013) presenting the most effective urban freight transport strategies that can be implemented, in terms of both economic and environmental benefits, considering the freight flows and particular characteristics of each urban area.

Overall, urban freight transport research confirms the heavy contribution of relevant operations in congestion increase, pollution and public health deterioration highlighting a number of critical points that should be carefully considered such as increased data availability, better understanding of the needs for urban freight supply and demand, etc. for moving towards energy efficient and sustainable urban freight transport systems meeting, to the greatest extent possible, the local needs of each urban area. The latter necessitates the commitment of all relevant stakeholders for implementing tailored solutions that can effectively preserve a city's heritage and social cohesion. This is being widely perceived as a major challenge in several cities (Transportation Research Board, 2013).

2.2 *Urban freight in port cities*

In several cases, inland transport terminals serve as a buffer of urban distribution centers forming, in that way, a storage and inventory cluster. Such an approach offers better flexibility and can ensure back-loading minimizing consequently 'empty' kilometers. These new concepts of 'agile' ports and 'extended warehouses' are not fully applied yet in practice mainly due to the business practices in which industry stakeholders are currently used to as well as due to the absence of proper policy frameworks for undertaking and sustaining such new business models. To this end, the urban environment of port-cities presents another unique dimension, creating new challenges and opportunities, which should be carefully considered for successfully implementing new energy efficient and environmentally friendly urban freight transport measures.

Port-cities have experienced a rapid growth in the freight volumes required to be handled, while, however relevant environmental impact mitigation efforts are scarce and often limited to diverting part of the port traffic along peripheral to the port-city routes. Port-cities are confronted with the difficulty of establishing and sustaining an efficient collaboration mechanism between municipalities and port authorities, due to different and in some cases contradictory interests and perspectives, resulting in several cases in unplanned activities imposing significant negative externalities to urban community stakeholders (Blanco, 2014). Such a cooperation scheme constitutes the main basis for adopting a coherent urban freight transport strategy that can effectively contribute towards improving the overall performance and efficiency of the urban freight transport system and provide substantial benefits especially to increasingly stressed-out areas (BESTUFS, 2007a). Considering the extra burden of port-generated traffic that such systems are coupled with as well as the existence of various sectors of the economy that are tightly related to port activities, the potential benefit of even marginal improvements in such urban areas can be multiplied with regard to other city environments.

2.3 *US and EU perspectives*

In modern societies, urban freight transport has been a major issue and strategies for freight movements within and outside urban areas are developed based on systematic analysis and modeling (Taniguchi & Russell, 2014). A major effort in this field was concentrated on "freight-focused" solution approaches for urban goods movement in the U.S. developing frameworks for action, and in the EU on "integrated" approaches (BESTUFS, 2007b; SUGAR, 2012) based on best practices and successful measures for freight movements and passengers' mobility. Although in the US the private sector plays a key role in urban freight, a "focused" planning is followed involving issues regarding operations management, infrastructure, pricing, regulation, safety and environmental protection. Federal Highway Administration (FHWA) is responsible to develop and maintain processes of the urban logistics developing customized action frameworks for urban areas in an attempt to mitigate the related current and future public health and environmental impacts (Federal Highway Administration, 2010). The "freight-focused" action frameworks adopted in the US and designed to address a large scale area are more stable and long lasting enabling needs determination, infrastructure developments, education, land-use decisions, accountability, funding and other critical issues. However, recent observations reveal that this "freight-focused" planning in an extended urban area underlies risks of the current urban freight flows determination in a micro level since detailed data about truck movements and the types of vehicles used are not available. To this end, Metropolitan Planning Organizations (MPOs) have begun to collect urban freight data regarding the truck movements in an attempt to control the large number of freight flows as well as the corresponding inefficiencies occurred (Giuliano et al., 2013).

On the other hand, current EU "integrated" attempts are mainly addressing passenger mobility and only a limited range of initiatives exists for urban freight distribution such as the ENCLOSE (2014) project that builds up a suitable and usable framework for the definition of SULPs for small/mid-sized historic towns, the ELBA (2014) project which investigates, implements and pilot tests a number of "intermediate" and "flexible" transport and logistics schemes, operated by eco-friendly vehicles and integrated in the broader context of mobility measures and more recently, the SMILE (2015) project which focuses on the development and implementation of innovative strategies and measures on energy efficient urban freight solutions for smart cities utilizing available technologies and building upon previous experiences and current Sustainable Urban Mobility Plans (SUMPs) of the participating cities. In addition, the trend toward small scale implementations of successful measures and policies (mostly in the city centers) adopted in many EU Member States during the last decade emerges as the need to understand the interactions among various measures for developing sustainable urban freight plans embedded in the general SUMPs of the areas, since some measures

are supporting each other; some are conditional to each other; some may deter or impede each other's effectiveness; some measures taken together may have a multiplier effect of their impacts, others taken together may produce less than the sum of their potential individual impact.

Furthermore, in both the US and EU, port authorities have established various operational strategies in mitigating the adverse environmental impacts of port related traffic, reduce congestion and energy consumption within the port area and the near port network with emphasis given on gate strategies and truck drayage operations. A gate appointment system, extended gate hours, traffic streamlining, off-peak deliveries and possibly off port marshalling of trucks (Boile et al., 2008; Dougherty, 2010) are some of the solutions developed to extent current capacity, efficiently accommodate the increasing freight volumes, address current inefficiencies and improve the overall performance of the system (Maguire et al., 2010).

3 TOWARDS A RATIONAL PLANNING APPROACH INCORPORATING DIVERSE EXPERIENCE

3.1 Urban planning for port cities

Port cities have peculiarities of their own as regards freight movements, and face the complex challenge of balancing the demand for additional space for both port and city activities. Based on the matrix of port-city relations (Figure 1) proposed by Ducruet

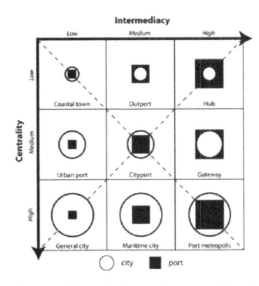

Figure 1. Relationship between the port and the city (Ducruet and Sung-Woo, 2006).

and Sung-Woo (2006), we can understand the nature of the relations between port activities and urban freight. It is observed that the importance of the port in economic activities and the importance of the city in port operation activities are imbalanced in almost all cases while the balanced "cityport" is a theoretical base that does not exist in practice. Although urban planning could provide the basis to overcome the difficulties derived from the interaction of port activities and urban growth, the main players are not always willing to exchange information and facilitate productive communication, or broadly implement actions, which can improve energy efficiency of urban distribution in port cities and ultimately, creating new wealth and building stronger local economies. The evolution of urban freight flows in port cities indicates the active involvement of relevant stakeholders and a deep transformation and reorganization of current procedures and operations.

The policy challenges for ports and cities depend on the local conditions and are usually different (Figure 2). In case where their priorities are likely to come into conflict, a mix of policies and commonly acceptable plans that mitigate negative impacts can ensure success of the territory (Merk, 2014). A better understanding of the local context can provide the basis for effective measures and sustainable development. Hence, a shared vision of future development needs for the port and the city is required to combine objectives of both port and city. Stronger relations and cooperation between port authorities and city stakeholders are recognized as the answer to rationalize the use of urban land and transportation.

The considered actions and best practices both in micro- and macro-level (Figure 3) should have the support of the relevant groups of interest of the urban areas in order to facilitate the implementation process and avoid conflict of interests among the business and social stakeholders. In particular, the industry focuses on the minimization

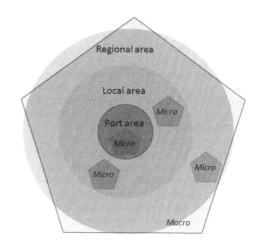

Figure 3.   Level of planning.

of the operational costs in order to support their competitiveness and their economic activity and on the other hand, the public authorities aim to improve the quality of life improving public places, safety, air and noise pollution. Especially, in port cities the relationship among city and port activities is more sensitive and long term outcomes in energy efficiency could be the catalyst to build the dialogue between industry and public authorities opening the road for synergies and expanding the cooperation network. Even if the relevant stakeholders have different objectives and priorities, an efficient urban freight system offers added value accommodating anticipated economic growth in the urban area and mitigating the related environmental impacts. In this way, sustainability awareness could evolve and get strengthened over the years, becoming the core of the strategic objectives that drive public authorities and industry.

Given the fact that different types of urban areas have different characteristics and different sets of issues to address, a rational approach is proposed to determine the best mix of strategies and initiatives. For the purpose of achieving the highest benefit in port cities, the proposed approach aims to develop proper SULPs that will interact successfully with current SUMPs in order to improve operations of freight distribution in the urban area close to the port facilities and achieve energy efficiency and reductions in environmental impacts. The concept of the proposed approach is that many successful and often disparate urban freight initiatives, that have been undertaken, can be combined to improve and make more efficient transport movements in urban areas. This concept stems from the experience of many promising initiatives that did not manage to achieve their full potential

Area of interest	Port	City	Port-city
Economic	Port volumes	Value added, diversification	Smart port growth strategies Maritime clusters
Transportation	Freight	Passengers	Dedicated freight corridors or smart coexistence of freight and passenger traffic
Labour	Efficiency	Employment	High value added port-related employment
Environment	Limit impacts	Quality of life	Green growth
Land use	Cargo handling, industry	Urban waterfront as opportunities for housing	Mixed developments, with a role for port functions
Structural logic	Closed industrial cluster	Open networks with pure agglomeration effects	Mix

Figure 2.   Policy challenges for ports and cities (OECD, 2014).

and in many cases did not prove to be sustainable beyond a pilot period or over a long period of time. This is often due to reasons that do not have to do with the potential of the initiative in itself, but with reasons such as not being communicated and implemented properly. The active involvement of all required parties is critical to achieve the broader acceptance and fit the initiative properly within a broader mobility planning framework.

This is further supported by previous facts, which have shown that no single measure, no matter how efficient it is, can be considered as a panacea to solve the problems of urban freight distribution and in general, measures taken independently may not achieve their full potential. Strategic and institutional planning should be implemented at a micro and macro level, so that disparate and fragmented, yet promising, efforts are integrated in a holistic and coordinated micro/macro port-city related strategy for reducing energy consumption of freight related activities in a short and long term horizon.

Cities should incorporate new selected measures into their SULPs and SUMPs based on commonly acceptable objectives of both port and city authorities, and gain acceptance from the involved stakeholders exploiting any existing actions in the current operational and strategic plans. Figure 4 summarizes the crucial aspects in terms of geographical and operational level that should be taken into consideration to establish efficient planning policies.

3.1.1 *Basket of measures and policies*

Typically, the private sector objectives relate to cost reduction and profit maximization, while public sector initiatives primarily concern the social benefit. At the crossroad of the interests of both public and private sector is the goal for energy reduction, which is translated into cost reduction for the private sector and to reduced emissions for the public sector. The planned measures should be designed in a short- and long-term horizon and cover the extended area of both city and port, so as to assess whether the energy savings may be achieved. The particular measures adopted by each port city had to be derived

by a "basket of measures and policies" generated according to the current conditions, needs and specific objectives of each area as depicted in Figure 5. This proposed concept focuses on the planning of actions that provide the means to facilitate improvements and cover the different objectives of the area (i.e. operational, environmental, energy, social, etc.). Therefore, these actions should be planned based on the classification of the potential measures with their associated efficacy called "association matrix". The "association matrix" matches the priorities and objectives identified in a port city with the solutions offered by the state-of-practice which shape the "basket of measures and policies" that are anticipated to contribute to the area improvement and flexibility for future extensions.

Furthermore, the candidate measures require further specification and adjustment while each port city has to produce and/or re-assess its own urban logistics plan. The urban logistics plans should encourage the efficient transportation of goods, the good functioning of the land-side of the sea port and the harmonious interaction between the city and the port. This has to be done in consultation and cooperation with the port authority and other stakeholders, considering the unique elements, challenges and opportunities presented in the urban area. For those cities with established urban logistics plans, the re-assessment consists of bringing them up-to-date regarding any significant changes, new data that has become available since their adoption, or changing needs and priorities. On the other hand, for those cities that do not have an established freight mobility and logistics plan, the re-assessment consists of bringing together all existing city structural and transport/mobility plans and assessing how any significant changes or new data that has become available since their adoption may have altered their direction. For these cities the analysis and assessment of the existing structural and transport/mobility plans could lead to a definition of broad guidelines for the future specification of an urban logistics plan, or at least the specifications required making freight movement needs explicit in existing general mobility plans.

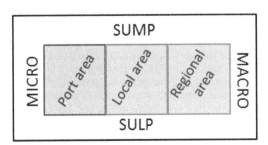

Figure 4. Planning aspects of a port city.

Figure 5. Scheme for selection and implementation.

4 CASE STUDY

4.1 *The city of Piraeus*

The city of Piraeus has a population of 175.000 people and it is home to Greece's main port, one of the largest in the Mediterranean. The port comprises two container terminals from where almost 70% of the imported containers (approximately 358,000 TEUs) carry freight intended to be consumed locally in the metro area. The central district area is composed of a network of narrow streets running sharply uphill, several of them with stone staircases. The commercial units, which are clustered in the city center, face an increasing demand of being supplied with products efficiently, at low cost and in a timely manner. Daily traffic congestion constitutes a major problem in the area, which is exacerbated by the high port related business activity. Urban freight logistics activity in the Piraeus city center, which is close to the port, is characterized by a large number of deliveries in small roads, with frequent trips, resulting in many vehicle kilometers travelled. Limited capacity, lack of coordination and extended peak hours of operation result in increased congestion, delays, energy consumption and other environmental impacts such as noise and air pollution.

The proposed methodology provides a systemic guidance for the preparation, implementation and assessment of the strategic urban logistics plan developed in Piraeus city. The strategic vision is to build a "smart city" offering high quality of life and interoperable systems. In terms of urban freight distribution, the city aims to reduce the generated pollutant emissions improving the environmental conditions of the local area and also reduce energy consumption and other adverse impacts. Additionally, the city authorities attempt to support the territorial sustainability of the region as well as to enhance the environmental consciousness of the citizens ensuring also future development and viability. In an effort to achieve the engagement and the dialogue with the industry at a local level, a Memorandum of Cooperation has been signed between the Municipality of Piraeus and the Piraeus Commercial Association. The state-of-practice was used as the building block of the association matrix. The basket of measures was generated based on the associated efficacy of the possible measures/policies that deal with the needs and the priorities of the area as defined by port and city representatives. The selected measures presented the highest applicability and the broader energy impact.

A set of potentially effective actions aligned with the corresponding association matrix have been identified and included in the official "Strategic Plan for Urban Development" of the Piraeus city with the aim to enhance competitiveness of the greater area. Specifically, the proposed areas of intervention include: (i) Investment in research and innovation, (ii) Development of maritime grid and blue economy (with emphasis on tourism, culture and creative economy), (iii) Revitalizing urban areas, (iv) Promotion of sustainable urban development and multimodal mobility (supporting the role of Piraeus Port in the southeast Mediterranean cruise market), (v) Connection and cooperation between the port and the city (spreading information widely, supporting engagement of relevant clusters and stakeholders), (vi) Support education and training, (vii) Improvement of urban environment, (viii) Development of smart infrastructure solutions, communication services and advanced technological applications (supporting port-city interface).

In the context of the sustainable urban mobility as well as the development of smart infrastructure solutions and advanced technological applications, a pilot activity has been initiated and it is indicatively presented. The pilot activity has been implemented in the city centre and it is focused on the development of on the road parking bays for loading/unloading supported by ICT tools for real time information and management. On the basis of the pilot results, the total delays occurred due to urban logistics or in other words, the total idle time of the vehicles in the road network close to delivery parking spaces was reduced by up to 67.83%. Furthermore, the related generated CO_2 emissions were reduced by 56% minimizing the air pollutant emissions and improving the air quality of the city. Overall, the pilot action was evaluated as a promising solution and was recommended for a large scale implementation.

5 CONCLUSIONS

The proposed selection scheme is transferable and could assist local and regional authorities in improving sustainability in the urban freight sector while reducing the energy consumption and other adverse impacts. Sustainability awareness has evolved and strengthened over the years, becoming the core of the strategic objectives that drive public authorities and industry. The commitment to sustainable development could be reflected in the integration of economic decisions with their related social and environmental impacts. One of the main challenges is to develop proper actions that are able to satisfy the participating stakeholders of both the port and the city since they usually have competing interests, desires and needs. Resources and synergistic strategies broadly implemented helped streamline and follow a systemic approach to decision-making in selecting and implementing best strategies to improve energy efficiency of freight distribution in urban areas creating new wealth and building stronger local economies.

REFERENCES

Ambrosini, C., & Routhier, J. 2004. Objectives, methods and results of surveys carried out in the field of urban freight transport: An international comparison. Transport Reviews 24(1): 57–77.

Balm, S., Browne, M., Leonardi, J. & Quak, H. 2014. Developing an evaluation framework for innovative urban and interurban freight transport solutions. Procedia – Social and Behavioral Sciences 125: 386–397.

BESTUFS. BESTUFS Policy and Research Recommendations III. 2007a. Port cities and innovative urban freight solutions. Managing urban freight transport by companies and local authorities. European Commission, 6th Framework Programme for Research and Technological Demonstration, http://www.bestufs.net/bestufs2_policy.html.

BESTUFS. 2007b. Good Practice Guide on Urban Freight Transport. European Commission, 6th Framework Programme for Research and Technological Demonstration, http://www.bestufs.net/.

Blanco, E.E. Urban Freight and Port Cities. 2014. World Bank, Washington, DC. © World Bank, https://openknowledge.worldbank.org/handle/10986/17835 License: CC BY 3.0 IGO.

Boile, M., Golias, M.M. & Theofanis, S. 2008 Evaluating Roadside Impact of Different Gate Operation Strategies at a Container Terminal Using Microsimulation. Proceedings of the 49th Annual Transportation Research Forum, Texas A&M University, College Station, Texas.

Boile M., Theofanis S. Strauss-Wieder A. (2008) Feasibility of Freight Villages in the NYMTC Region. Task 3 Report. http://www.nymtc.org/project/freight_planning/freight_village.html

Brown Jr., E.G., Kelly B.P. & Dougherty, M. 2014. California Freight Mobility Plan. California State Transportation Agency, http://www.dot.ca.gov/hq/tpp/offices/ogm/cfmp.html.

Dougherty, P. 2010. Evaluating the Impact of Gate Strategies on a Container Terminal's Roadside Network using Micro-simulation: The Port of Newark/Elizabeth Study case study. Ph.D. Dissertation, New Brunswick, NJ.

Ducruet, C. & Sung-Woo, L. 2006. Frontline soldiers of globalization: Port-city evolution and regional competition. Geojournal 67(2): 107–122.

ENCLOSE, ENergy efficiency in City LOgistics Services for small and mid-sized European Historic Towns. 2014. European Union, Intelligent Energy Europe Programme, http://www.enclose.eu/content.php?lang=en.

ELBA, Integrated Eco-friendly Mobility Services for People and Goods in Small Islands. 2014. European Union, LIFE+ program, Environment Policy and Governance, http://www.elba-lifeplus.eu/

FHWA, Federal Highway Administration. 2010. Urban Freight Case Studies, http://ops.fhwa.dot.gov/freight/technology/urban_goods/index.htm.

Giuliano, G., O'Brien, T., Dablanc, L. & Holliday. 2013. NCFPR Report 23: Synthesis of freight research in urban transportation planning, NCFRP Project 36 (05).

Imanishi, Y. & Taniguch, E. 2016. Framework of the urban road freight transport—Lessons learnt from case studies. Transportation Research Procedia 12: 627–633.

Lindhom, M. & Behrends, S. 2010. A holistic approach to challenges in urban freight transport planning. 12th World Conference on Transport Research Society, Lisbon, Portugal.

Maguire, A., Ivey, S., Lipinski, M.E. & Golias, M.M. 2010. Relieving Congestion at Intermodal Marine Container Terminals: Review of Tactical/Operational Strategies. 51st Annual Transportation Research Forum, Arlington, Virginia.

Merk, O. 2014. The effectiveness of port-city governance. In Port-City Governance: 279–291.

OECD, Organisation for Economic Co-operation and Development. 2014. The Competitiveness of Global Port-Cities, http://www.oecd-ilibrary.org/urban-rural-and-regional-development/the-competitiveness-of-global-port-cities_9789264205277-en.

Ogden, K.W. 1984. A framework for urban freight policy analysis. Transportation Planning and Technology 8(4): 253–265.

Rodrigue, J-P. & Notteboom, T. 2009. The terminalization of supply chains: reassessing the role of terminals in port/hinterland logistical relationships. Maritime Policy & Management 36(2): 165–183.

Russo, F. & Comi, A. 2011. Urban freight transport measures: environmental evidences from the cities. 1st World Sustainability Forum, Electronic conference.

SMILE, SMart green Innovative urban Logistics for Energy Efficient Mediterranean cities. 2015. European Union, MED Programme, http://www.programmemed.eu/en/the-projects/project-database/results/view/single.html?no_cache=1&idProject=73.

SUGAR, Sustainable Urban Goods Logistics Achieved by Regional and Local Policies. 2012. City Logistics Best Practices: a Handbook for Authorities. European Union, INTERREG IVC programme, http://www.cei.int/SUGAR.

Suksri, J., Raicu, R. & Yue, W.L. 2012. Towards sustainable urban freight distribution—a proposed evaluation framework. Proceedings of the Australasian Transport Research Forum, Perth, Australia.

Taniguchi, E. & Russell, T.G. 2014. City logistics: Mapping the future. CRC Press.

Taniguchi, E., & Tamagawa, D. 2005. Evaluating city logistics measures considering the behavior of several stakeholders. Journal of the Eastern Asia Society for Transportation Studies 6: 3062–3076.

Taniguchi, E., Thompson, R., G., Yamada, T., & Van Duin, R. 2001. City Logistics—Network modelling and intelligent transport systems. Oxford: Pergamon.

Thomson, R.G., & Taniguchi, E. (ed.) 2001. Modeling City Logistics and Transportation. Handbook of Logistics and Supply-Chain Management: 393–405, Elsevier: Amsterdam.

Transportation Research Board. 2013. City Logistics Research: A Transatlantic Perspective, Summary of the First EU-U.S. Transportation Research Symposium.

United Nations. Department of Economic and Social Affairs, Population Division. 2015. World urbanization prospects: The 2014 revision. ST/ESA/SER.A/366.

Maritime Transportation and Harvesting of Sea Resources – Guedes Soares & Teixeira (Eds)
© 2018 Taylor & Francis Group, London, ISBN 978-0-8153-7993-5

A conceptual framework for port assessment

X. Bellsolà Olba, W. Daamen & S.P. Hoogendoorn
Department of Transport and Planning, Faculty of Civil Engineering and Geosciences, Delft University of Technology, Delft, The Netherlands

T. Vellinga
Department of Hydraulic Engineering, Faculty of Civil Engineering and Geosciences, Delft University of Technology, Delft, The Netherlands

ABSTRACT: Port authorities want to attract and serve the highest possible number of vessels. Assessing the impacts of different vessel demands is crucial to enhance port efficiency and, for the port planning phase, to choose the most efficient design. In this paper we introduce a conceptual framework for the assessment of port efficiency as a whole. The different elements of the assessment methodology are presented and their relations are detailed. Our suggested conceptual framework is the basis for a tool to structure the assessment of a port and support the choice for specific port expansions or the application of new vessel traffic management strategies.

1 INTRODUCTION

Maritime transportation has become a key part of the supply chain network. Following the world economic growth, maritime transport has continuously been increasing and gaining more and more importance over time (UNCTAD 2016). Ports need to develop and improve to be able to handle the increase of vessel flows and sizes for keeping their competitiveness. The nautical infrastructures, such as waterways and channels, have a high level of complexity and low flexibility and adaptability. Moreover, there are many different stakeholders involved within the operations in a port, such as shareholders, business customers (terminal operators, transport companies, shipping lines, etc.), the government (state and local) and the community (environment and education groups, port visitors, local population, media, etc.). Port authorities are focused on maintaining or increasing port efficiency over time with their decisions. Assessing the impacts of vessel traffic is crucial to enhance port efficiency and, for the port planning phase, to choose the most suitable design. However, each stakeholder has its own interests, and decisions might have different effects depending on the role assumed for each stakeholder. Therefore, a need to choose the indicators and methods that allow port authorities to assess the impact of their decisions on port efficiency exists. In order to improve port efficiency, changes can be made in the port infrastructure or the port operations, such

as port expansions or setting up traffic new traffic regulations or restrictions.

Extensive research exists on either capacity and efficiency, risks or environmental port assessment, which are identified as crucial elements of port assessment. In relation to capacity, different level of detail have been developed, such as terminal, waterway or port capacity (Bellsolà Olba et al. 2017, Dekker 2005, Kia et al. 2002, Piccoli 2014). Safety and risks in ports have been addressed at an individual level (collisions) (Goerlandt & Montewka 2015), as well as at a more aggregated level (port accidents) (Yip 2008). Other research focused on the environmental efficiency (Chang 2013) and economic efficiency for port expansion (Dekker and Verhaeghe 2008). However, an overall assessment methodology for ports, that integrates a trade-off between several indicators, e.g. port capacity, risks, environmental issues and costs, has not yet been developed. In this research, we aim to provide a conceptual framework that structures the factors influencing port efficiency, their connections and how to derive them, which are needed by port authorities for a comprehensive port assessment. The use of a tool that is implemented based on this framework will allow decision makers to include more factors and their relative importance to support their choices for port assessment.

The remainder of this paper is structured into three main sections. Section 2 provides general information about the structure of an assessment methodology. In Section 3, the conceptual framework for port assessment is developed and their different parts are explained in detail with their dependencies and the information required for each of them. Section 4 concludes the paper, focusing on the benefits of this conceptual framework.

2 BACKGROUND INFORMATION

Port assessment is sophisticated due to the large number of stakeholders involved in running a port and the complex port processes. As described in the previous section, extensive research has been done in assessing specific issues in ports with different aims, e.g. safety and capacity. In this research, we want to combine all these individual elements and come to a comprehensive assessment of port efficiency as a whole. Thus, its main parts and its structure need to be identified.

In Figure 1, the general steps necessary to build our assessment methodology are presented. First, the scope of the assessment and the target group for whom the methodology is developed should be defined. Since ports depend on a wide variety of dependent issues, the information needed to build a multidisciplinary assessment is directly dependant on the scope and purpose. The combination of previous research and the analysis of needs and interests should help to identify the assessment criteria and indicators to be used in each case. In the case of ports, as previously mentioned, a large number of stakeholders are involved, such as port authorities, terminal operators, shipping companies, who have an influence on the indicators used. Thus, different weights for the stakeholders should be considered to represent their influence in the assessment. Moreover, each port has specific characteristics that require the methodology to be flexible. The input data and the calculation process include the use of all the parameters necessary to recreate reality and calculate the indicators required. Finally, the desired output results can be used to assess the current situation or to compare

Purpose
Scope & target group
Assessment criteria, Indicators & Stakeholder Analysis
Input data & Calculation process
Output results & Assessment

Figure 1.   Assessment methodology steps.

between different scenarios identifying the best solution.

3 CONCEPTUAL FRAMEWORK

After identifying and describing the main steps of a generic assessment methodology, this section presents the conceptual framework that describes all the elements needed for a specific port assessment in a structured way.

Figure 2 presents this framework, where the different steps identified in Figure 1 are elaborated, and their relations are identified. A comprehensive explanation describing the different steps is presented below.

3.1   Scope and target group

The port assessment begins by defining its *scope*. In this research the scope is to assess ports from a navigational and operational perspective, where the nautical infrastructure and traffic demand are the main elements to be assessed. Therefore, we aim to combine different fields of research, such as port capacity and risks, in order to support long term decision making.

The *target group*, for whom this methodology is developed, is port authorities, who are the stakeholders in charge of long term decisions in a port, including improving operations and port planning. Since terminals involve different stakeholders with different roles and interests, they should be treated as separate processes. Thus, in this research, we assume that the increase/decrease in nautical traffic can be accommodated by the terminal operations.

3.2   Purpose and assessment criteria

The *purpose* of the assessment is to support port authorities in their decision making. They aim to keep their port efficiency at an optimal level while considering a wide variety of multidisciplinary issues. They want to be prepared for the future and identify the consequences of their long term decisions.

Based on the target group and the purpose of the assessment, a set of suitable *assessment criteria* can be chosen. As introduced in Section 1, ports have been previously assessed considering different factors, but the combination of them as an overall assessment has not been developed yet. Therefore, the main criteria to be used in the assessment of the nautical infrastructure and operations of a port are the capacity, the risks and the environment. The combination of these 3 elements allows port authorities to have an aggregated picture of the performance of a port, while considering their trade-offs.

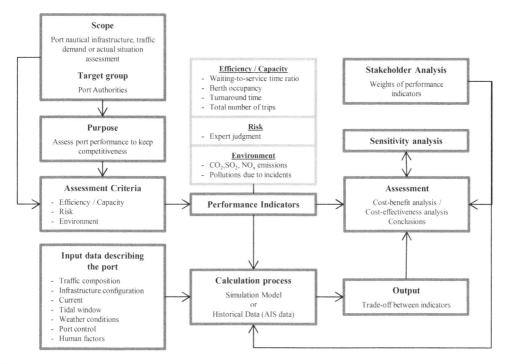

Figure 2. Conceptual Framework for Port Assessment.

The choice of the assessment criteria is subjective, and directly dependent on the scope, purpose and target group. Even having the same purpose, the choice of the indicators could be different. Thus, the choice should be discussed in detail and supported by the results.

3.3 *Performance indicators*

There is a vast number of *performance indicators* previously used for different port assessment purposes, which can be identified for each of the assessment criteria necessary for this research. In the next paragraphs, the relevant indicators for each of the assessment criteria are reviewed and described.

Capacity and efficiency have been quantified with many different indicators in previous research. Usually terminal capacity has been measured with the waiting-to-service time ratio and berth occupancy indicators (Layaa & Dullaert 2014). The same indicators have been proven to be useful when looking to the nautical part of the port (Bellsolà Olba et al. 2017, Tang et al. 2016), since they give an indication of how the port utilization is, although berth occupancy may not be sufficient to identify specific bottlenecks by itself. Another useful efficiency indicator that have been used to support the previous ones is the turnaround time

(Layaa & Dullaert 2014), which is the time spent by a vessel in the port.

Finally, the total number of vessel trips in the port, considering both vessel movements between terminals or the vessel trips when entering or leaving the port, also indicates the capacity of the port network to handle specific traffic demands, and can be used to quantify the throughput of the port (Bellsolà Olba et al. 2017). By using all these indicators, and not just one, different traffic situations or port designs can be compared and the bottlenecks can be identified.

Risk is a more subjective topic and it is difficult to quantify. A wide variety of quantitative maritime risk assessment research calculated the risk values in a data-driven or probabilistic way (Goerlandt & Montewka 2015, Li et al. 2012), where the probabilities of occurance and the probable causes where used to calculate risks. However, in case of not having historical data, these approaches could not be applied. Different research have proven that expert judgment is a useful way to identify and quantify risks, based on their risk perception (Debnath & Chin 2009, Pak et al. 2015). Pak et al. (2015) developed a macroscopic port model and calculated the navigational risk of different ports based on vessel flows. For our purposes, a similar approach could be developed by using the risk contribution of different factors affecting navigation and its consequences, e.g. the infrastructure design, the environmental conditions or the traffic conditions. Based on expert judgement, the weighting of these factors that are related to risks and consequences in navigation can be derived. By combining all these factors, a value of risk can be calculated for different scenarios and assessed within a risk scale defined by the target group.

Environmental port efficiency has been addressed considering the terminal dimensions, cargo handled, vessels per terminal and the CO_2 emissions (Chang 2013). Other research has also considered SO2 and NO_X emissions (De Meyer et al. 2008) or has proven that vessel speed is directly related to the emissions (Psaraftis et al. 2009), thus different values for emissions might be associated to different speeds and vessel types. For this port assessment, the nautical environmental efficiency could be evaluated by considering the CO_2, SO_2 and NO_X emissions, according to the vessel types and dimensions, and combine these with their sailing time, as done in previous recent research.

3.4 Calculation process

In order to obtain the performance indicators, a *calculation process* is needed and it is related to the data processing that leads to the desired output. There are two main possibilities, either historical analysis or forecasting expected scenarios. For the first purpose, both data-driven or simulation-based analysis would work, while forecasting future scenarios with new port designs should be developed based on a simulation model that includes the performance indicators for the specific assessment purposes. The simulation model has to fit the scope and purpose of the assessment method. Hence, in this research, a simulation model that includes nautical infrastructure and traffic characteristics, as well as, environmental and port control factors, that influence navigation, should be considered to replicate properly vessel navigation and port operations. However, the level of detail of the model would not need to be microscopic, since the assessment is on an aggregated level. To simulate port operations, event-based simulation models have been proven to be useful, even though other types of models could be used. The level of detail of simulation models could be macroscopic. However, when using latest simulation models that include vessel interactions and crew behaviour (Shu et al. 2016), the human behaviour influence could be considered and analysed as a source of risk and uncertainty. In the case of having historical data available and the need to assess an existing design in the future, a data-driven model could be developed, with the difficulties of dealing with incoherent and even incorrect data, and the limitations of a single port design.

3.5 Input data describing the port

The method chosen for the calculation process requires the necessary *input data* that allows the representation of all the port nautical operations in a realistic and complete way, which level might change depending on the purpose, the assessment criteria and performance indicators chosen for the assessment. The different components for a safe and efficient vessel navigation in waterways have been comprehensively summarized by Froese (2015), and the same components can be used to represent vessel navigation in ports for our purposes. According to Froese (2015), a total of 7 main components should be covered and they are summarized in Table 1. The traffic composition includes the information related to the vessel flows inside the port. Since vessels have different sailing requirements and restrictions in different layouts, the infrastructure configuration should describe the different port areas to represent different navigational situations within a port. The current, tidal window and weather conditions are directly linked with the infrastructure and the traffic composition, because depending on the first, certain traffic restrictions might occur in specific areas of a port. The port control includes the international laws and regulations, as well as specific regulations which are needed for this certain port. Moreover, the need of pilotage and tugs, and the current situation of the Vessel Traffic Assistance (VTS) is specified. Although, human factors are difficult to include in navigation, and, as previously mentioned, the simulation model to be used could include this uncertainty in the different processes, different levels of communication and assistance between vessels and between vessels and the port authorities could be defined for a specific port.

3.6 Output & stakeholder analysis

After the calculation process, the *output* can be obtained. This step aims to calculate the different performance indicators previously defined. These indicators can be combined in different ways, and they are dependent on the stakeholders that are relevant for port assessment. Thus, a *stakeholder analysis* allows to identify which are the relevant stakeholders involved in port navigation and operations, and their specific influence. Depending on the outcome from the analysis, the performance indicators might need to be weighted according to the influence of each of the stakeholders.

In order to compare the different indicators, there is the possibility to make an economic evaluation of all of them, in order to find out the best trade-off between them for different scenarios. In economic terms, there are several approaches

Table 1. Input data for port assessment.

Main component	Sub-component
Traffic composition	Type (container, bulk, tanker, etc.)
	Dimensions (length and beam)
	Origin-Destination flows
	Itinerary (tugs/pilot requirements)
Infrastructure configuration	Type (straight, bend, intersection, etc.)
	Location
	Depth
	Width
	Speed regulations
	Other regulations (air draft allowance, etc.)
Current	
Tidal window	
Weather conditions	Wind
	Visibility
	Time of the day
Port control	Laws and regulations
	Specific port regulations
	Pilotage / tugs
	Vessel Traffic Services (VTS)
Human factors	Level of communication/assistance

that have been used. In order to support strategic planning of port expansions, Dekker (2005) developed a methodology focused on the viability of the investments and mainly using TEU/year as capacity indicator and the costs associated per TEU. Another economic study proved that there is a positive relation between the economic performance and the environmental performance of a port (Cheon et al. 2017). Thus, having less environmental polluting situations would lead to higher environmental efficiency. Since the economic evaluation of risks might be difficult, port authorities should define acceptable risk levels to be able to compare them all.

3.7 *Assessment and sensitivity analysis*

Finally, the output provided allows to perform the *assessment* of the port. If the performance indicators have been translated into monetary units, these indicators can be compared and assessed using a cost-benefit analysis (CBA) approach. However, without translating all the performance indicators into economical terms, they can also be assessed using a cost-effectiveness analysis (CEA) approach. Both methods are tools to compare between different scenarios. Port authorities should define a set of decision criteria based on their interests and the stakeholders influence, which might change in each specific case. By changing the input data and repeating this process several times, different

scenarios can be built and compared between them to find out the best solution to the specific objective proposed.

A *sensitivity analysis* needs to be performed to validate the influence of certain changes in traffic or infrastructure design over each of the performance indicators calculated, and, thus, assure the validity of their assessment. Moreover, the importance of the different criteria considered, as well as the stakeholders influence on the final results can be analysed. This sensitivity analysis also allows to identify how much the results depend on certain stakeholder weight for decision making.

4 SUMMARY AND FUTURE RESEARCH

The conceptual framework presented in this paper intends to guide decision-makers in the process of port assessment. In this research, we presented the generic steps for developing an assessment methodology and afterwards a conceptual framework with the steps and relations between each part of the assessment methodology with an application to the assessment of port designs and nautical operations.

The purpose of the assessment determines the assessment criteria to be used in each case, which will condition the choice for specific performance indicators for the assessment of a port. For our purposes efficiency, risks and environmental performance are the criteria chosen. From these the most suitable performance indicators can be selected. Depending on the calculation process chosen, either data-driven or simulation modeling can be performed in order to calculate the desired output and, finally, assess a port based on the combined results.

The comparison of different scenarios for future strategic decisions, such as Vessel Traffic Management (VTM) measures, development of new terminals or infrastructure modifications, will help port authorities to take the decision based on the key factors that determine their future efficiency and performance.

Future research should further develop this comprehensive framework and provide more detailed insights about the port assessment process, in-depth guidelines, and the results from the application in real case studies.

REFERENCES

Bellsolà Olba, X., Daamen, W., Vellinga, T. & Hoogendoorn, S.P. 2017. Network Capacity Estimation of Vessel Traffic: An Approach for Port Planning. *J. Waterw. Port, Coast. Ocean Eng.* ASCE 143(5).

Chang, Y.T. 2013. Environmental efficiency of ports: A Data Envelopment Analysis approach. *Marit. Policy Manag.* 40, 467–478.

Cheon, S., Maltz, A. & Dooley, K. 2017. The link between economic and environmental performance of the top 10 U.S. ports. *Marit. Policy Manag.* 44, 227–247.

De Meyer, P., Maes, F. & Volckaert, A. 2008. Emissions from international shipping in the Belgian part of the North Sea and the Belgian seaports. *Atmos. Environ.* 42, 196–206.

Debnath, A. & Chin, H. 2009. Hierarchical Modeling of Perceived Collision Risks in Port Fairways. *Transp. Res. Rec. J. Transp. Res. Board* 2100, 68–75.

Dekker, S. 2005. Port Investment Towards an Integrated Planning of Port Capacity. Delft University of Technology.

Dekker, S. & Verhaeghe, R.J. 2008. Development of a Strategy for Port Expansion: An Optimal Control Approach. *Marit. Econ. Logist.* 10, 258–274.

Froese, J. 2015. Safe and Efficient Port Approach by Vessel Traffic Management in Waterways, in: Ocampo-Martinez, C., Negenborn, R.R. (Ed.), Transport of Water versus Transport over Water. *Operations Research/Computer Science Interfaces Series.* Springer, pp. 281–296.

Goerlandt, F. & Montewka, J. 2015. A framework for risk analysis of maritime transportation systems: A case study for oil spill from tankers in a ship–ship collision. *Saf. Sci.* 76, 42–66.

Kia, M., Shayan, E. & Ghotb, F. 2002. Investigation of port capacity under a new approach by computer simulation. *Comput. Ind. Eng.* 42, 533–540.

Layaa, J. & Dullaert, W. 2014. Measuring and analysing terminal capacity in East Africa: The case of the seaport of Dar es Salaam. *Marit. Econ. Logist.* 16, 141–164.

Li, S., Meng, Q. & Qu, X. 2012. An overview of maritime waterway quantitative risk assessment models. *Risk Anal.* 32, 496–512.

Pak, J.Y., Yeo, G.T., Oh, S.W. & Yang, Z. 2015. Port safety evaluation from a captain's perspective: The Korean experience. *Saf. Sci.* 72, 172–181.

Piccoli, C. 2014. Assessment of port marine operations performance by means of simulation. Delft Univeristy of Technology.

Psaraftis, H.N., Kontovas, C.A. & Kakalis, N.M.P. 2009. Speed Reduction As An Emissions Reduction Measure For Fast Ships, in: *10th International Conference on Fast Sea Transportation FAST.* pp. 1–125.

Shu, Y., Daamen, W., Ligteringen, H. & Hoogendoorn, S.P. 2016. Verification of Route Choice Model and Operational Model of Vessel Traffic. *Transp. Res. Rec. J. Transp. Res. Board* 2549, 86–92.

Tang, G., Wang, W., Song, X., Guo, Z., Yu, X. & Qiao, F. 2016. Effect of entrance channel dimensions on berth occupancy of container terminals. *Ocean Eng.* 117, 174–187.

UNCTAD 2016. Review of Maritime Transport (UNCTAD/RMT/2016).

Yip, T.L. 2008. Port traffic risks—A study of accidents in Hong Kong waters. *Transp. Res. Part E Logist. Transp. Rev.* 44, 921–931.

Maritime Transportation and Harvesting of Sea Resources – Guedes Soares & Teixeira (Eds)
© 2018 Taylor & Francis Group, London, ISBN 978-0-8153-7993-5

Liner service operational differentiation in container port terminals

A.M.P. Carreira & C. Guedes Soares
Centre for Marine Technology and Ocean Engineering (CENTEC), Instituto Superior Técnico, Universidade de Lisboa, Lisbon, Portugal

ABSTRACT: A model based on data from calls of regular liner services at a major Portuguese container port terminal shows that call size explains quay service time to a large extent. However, by including additional categorical variables, the model exposes the dependency of service time on the services themselves, with some services doing better than others. Quay flow, the ratio between call size and quay service time, however, explains operational differentiation better than service time. Full understanding of this phenomenon, which may be caused by a number of reasons, including commercial policies of the terminal operator, requires additional data and modelling, not covered by this model.

1 INTRODUCTION

In the pursuit of the maximization of their financial indicators, namely revenue and profit, container port terminals may discriminate among their clients, the regular shipping services calling them. Such discrimination, or differentiation, may apply to tariffs, usually negotiated and subject to contracts between port terminals and shipping services (Bichou, 2013; Stopford, 2009) and, therefore, easily regarded as a market modulator, apparently privileging frequent, high cargo volume customers.

This phenomenon of price discrimination, is widely studied in business and economics (e.g., Varian, 2010). Usually, trade quantities or call frequency are discrimination criteria, namely by means of volume discounts. However, client differentiation may also encompass operations, namely quay cargo handling, because, along with cost, efficiency is a major factor of choice of port terminals by shippers and shipping companies (Hoshino, 2010).

Efficiency encompasses several components, namely handling time, which is nowadays a major competitive factor between ports at the terminal level (Notteboom & de Langen, 2015), and service time, which is a meaningful indicator of terminal operations performance (e.g. Bichou, 2013). Service time is defined as the elapsed time from the start to the end of the unloading or loading of the cargo of a ship. Other terminal performance indicators include berth waiting time, ship dwell time, or, in a more synthesized view of port performance, total time in port or ship turnaround time (De Langen et al. 2007; Ducruet et al. 2014).

Service time is the main performance indicator, even if, in some instances, it is not the longest part of turnaround time, like, for instance, when the call size, i.e., the amount of cargo unloaded and loaded during the port call, is small. In such a case, port navigation and mooring/unmooring times may take longer than quay service time. In general, however, service time is the major component of time in port. This is the case in major port terminals, and it is rather clear from the data used in this research, which also reveals a strong variability in call sizes and service times.

This paper analyses the relation between call sizes and service times, and the factors that explain that relation, using regression methods, and how they describe terminal performance on a client basis.

2 METHODS AND RESULTS

The analysis of the relation between quay service time and call size was performed for the largest container port terminal in Portugal, Terminal XXI of Porto de Sines, using data supplied by the Port Authority on all ship calls in the period between January 2013 and September 2015.

In this period, the terminal handled 3 042 483.5 TEUs (some call records show fractions of TEU) from 2 498 calls of 23 shipping services.

Raw data, per call, includes

- Liner Service name
- Port entry Date/Time
- Operation start Date/Time
- Operation end Date/Time
- Port leaving Date/Time

Table 1. Summary of Terminal data.

	Cargo	QST	QF	TIP	PF
	TEU	hr	TEU/hr	hr	TEU/hr
Total	3 042 483.50	31 046.6	N/A	53 745.6	N/A
Average	1 217.97	12.4	103.2	21.5	66.1
σ	894.32	18.1	43.6	30.5	37.0
Max (call)	6 770	785.6	282.4	1 009.4	231.8
Min (call)	36	1.0	1.0	3.7	0.9
Median	1 000.50	10.2	97.9	16.3	62.7
Mode	549	9.1	76.3	15.1	82.6
Range	6 734.00	787.4	281.4	1 005.7	230.9

- TEU Loaded
- TEU Unloaded

From this data, the following was obtained

- Liner Services coded as single letters (A to W)
- Quay service time (QST, total and per call)
- Time in Port (TIP, total and per call)
- Cargo (TEU Loaded + TEU Unloaded, total and per call)
- Quay flow (QF = Cargo/QST)
- Port flow (PF = cargo/Time in Port)
- Season of the year

Table 1 summarizes the data.

2.1 Data preparation

The raw QST data summarized in Table 1 exhibits a long tail. The corresponding extremely long calls, which are only a few, have very long service times that are unrelated to quay operations.

For instance, the call with the longest service time, 785.6 hours (close to 33 days), with a cargo batch of 769 TEU, could have been handled in less than 8 hours (quay flow average of 103.2 TEU/hr.). Moreover, other calls were handled during the duration of this call. It looks like non-operational reasons are a better explanation to such a long quay time. Discarding the far outliers, 39 in total, leads to the base data used for the analysis.

A drawback of this data set is that several liner services have a very low number of calls (Table 2), which represent sample sizes that are unsuitable for proper analysis. Only liner services with sample sizes larger than 50 are considered for the analysis. This leaves out services B, E, K, L, Q, R, T and W, reducing the number of liner services in the analysis from 23 to 17. The final sample size (calls) is 2276. Table 3 summarizes the data to analyze.

The histograms of the quay service times and call sizes of the final data set are depicted in Figure 1 and Figure 2, respectively.

Table 2. Number of calls per liner service.

Liner_Svcs	Number_Calls	Lin_Svcs	Number_Calls
A	84	M	138
B	13	N	119
C	136	O	136
D	204	P	126
E	37	Q	43
F	209	R	32
G	125	S	163
H	28	T	13
I	198	U	244
J	262	V	132
K	6	W	5
L	6		

Table 3. Summary of data for analysis.

	Cargo (TEU)	QST (hr)	QF (TEU/hr)
Total	2 717 491.75	26085.8	N/A
Average	1193.98	11.5	103.0
σ	852.88	6.3	42.4
Max (call)	5 112	35.5	282.4
Min (call)	36	1.0	7.7
Median	1009	10.2	97.5
Mode	690	9.1	70.4
Range	5 076	34.5	274.7

2.2 Quay service time as a function of independent variables

Quay service time is fitted by regression models that start by taking into account call sizes only, and then add new categorical variables corresponding to the liner services.

2.2.1 Effect of call size (cargo)
Call size explains 0.58 of the variance of QST. The corresponding simple linear regression model obeys the following function

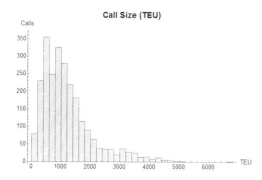

Figure 1. Quay service time histogram after the purging of the far outliers and liner services with sample sizes lower than 50.

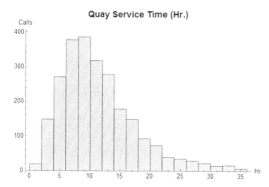

Figure 2. Call size histogram after the purging of the far outliers and liner services with sample sizes lower than 50.

$$QST(TEU) = 4.769 + 0.00560 \times TEU,$$

where TEU is call size. Statistics for this model are depicted in Figure 3.

2.2.2 Effect of liner services (clients)

Quay service time behaviour is different for each liner service.

This can be seen in the box diagrams of Figure 4, each liner service QST has its own median, and its own distribution, as confirmed by a Kruskal-Wallis test, with *statistic* = 966.843 and *P-Value* = 7.8 10^{-245}.

By adding categorical variables *Clients*, corresponding to the liner services, the regression model is multi-line, of the form

$$QST_C(TEU) = \alpha_0 + \alpha_C + \beta\, TEU,$$

where *TEU* is call size, α_0 is a constant that is added to α_C, the coefficient specific to Client C, to obtain the intersection, and β is the slope. The new model,

R^2_{adj}	0.584833				
		Estimate	Standard Error	t-Statistic	P-Value
Parameters	1	4.7691	0.145242	32.8355	6.99286×10^{-194}
	cg	0.00560491	0.0000989931	56.6191	$1.100356408647 \times 10^{-436}$
		DF	SS	MS	F-Statistic P-Value
ANOVA	cg	1	51986.5	51986.5	3205.73 $1.100356408645 \times 10^{-436}$
	Error	2274	36876.9	16.2167	
	Total	2275	88863.3		

Figure 3. Statistics of simple call size-based regression model.

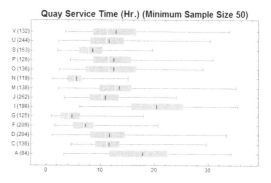

Figure 4. The box plots of the QST of the Clients (liner services) don't have the same mean nor median. Liner services are identified by their code letters, on the left, and sample sizes in brackets.

with additional variables Clients, explains 0.64 of the variance of QST, a small increase of 0.06 (10% of the final R^2_{adj}) over the simpler cargo-based model. The plots of both models are shown superimposed on the data in Figure 5, which provides visual insight into what is already suggested by the model parameters of Figure 6: each liner service is handled differently at the container terminal.

Figure 6 includes the tables of the model parameters, ANOVA and corresponding statistics

Only n-1 coefficients α_C are estimated because n categorical variables represent n-1 degrees of freedom. The Client variable corresponding to the smallest sample size (liner service A) was the one chosen to be left out.

2.2.3 Analysis of model fitting

The tables in Figure 6 show that not all parameters have the same confidence level.

Parameters c, f, g, p, and u have p- values above 0.05, which means the null hypothesis—that quay service time is explained by pure randomness and not by these parameters—cannot be discarded. The p-values obtained from the analysis of variance (ANOVA) also leads to not discarding the hypothesis that parameters p, and u do not explain the variance of quay service time. In the case of

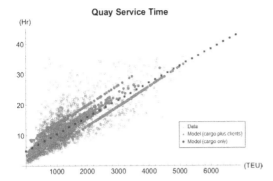

Figure 5. Categorical variables Clients allow for a better explanation of the data than just call size.

R^2_{adj}	0.64416			

		Estimate	Standard Error	t-Statistic	P-Value
Parameters	1	3.95443	0.205343	19.2577	$1.2*10^{-76}$
	cg	0.00650412	0.000162456	40.0363	$2.08308*10^{-265}$
	c	-0.52615	0.307548	-1.71079	0.0872569
	d	-1.02912	0.257006	-4.00426	0.0000642179
	f	0.405507	0.278641	1.45531	0.145724
	g	-0.691825	0.353275	-1.95832	0.0503153
	i	-3.87955	0.409382	-9.47661	$6.38535*10^{21}$
	j	-1.46538	0.229425	-6.38719	$2.0446*10^{10}$
	m	3.62117	0.307719	11.7678	$4.51979*10^{31}$
	n	-1.35851	0.346477	-3.92094	0.0000908304
	o	0.979712	0.307095	3.19026	0.00144094
	p	-0.180928	0.319407	-0.566449	0.571145
	s	0.545546	0.2962	1.84182	0.065633
	u	0.183044	0.235706	0.776574	0.437491
	v	0.77201	0.311649	2.47717	0.0133153

		DF	SS	MS	F-Statistic	P-Value
ANOVA	cg	1	51986.5	51986.5	3740.19	$1.165099282497*10^{481}$
	c	1	519.879	519.879	37.403	$1.12869*10^9$
	d	1	388.093	388.093	27.9216	$1.38446*10^7$
	f	1	79.9382	79.9382	5.7512	0.016558
	g	1	483.385	483.385	34.7774	$4.25146*10^9$
	i	1	1000.55	1000.55	71.9855	$3.85904*10^{-17}$
	j	1	338.08	338.08	24.3234	$8.73697*10^{-7}$
	m	1	2149.54	2149.54	154.65	$2.16598*10^{-34}$
	n	1	165.584	165.584	11.9131	0.000567655
	o	1	176.942	176.942	12.7302	0.000367292
	p	1	0.411871	0.411871	0.0296323	0.863343
	s	1	57.3568	57.3568	4.12656	0.0423317
	u	1	5.26283	5.26283	0.378637	0.538396
	v	1	85.2922	85.2922	6.13639	0.0133153
	Error	2261	31426.5	13.8994		
	Total	2275	88863.3			

Figure 6. The model with the categorical variables Clients, explains the data better than the simple linear model, which only accounts for call size.

variable p it can be stated that this variable does not explain quay service time almost at all.

2.3 Quay service time of individual liner services

Each liner service can as well be the subject of an individual analysis whose outcomes can be compared with the previous model, to eventually obtain more insight on the hypothesis of quay service performance differentiation.

The call data of each client can be fitted to a simple linear model that is a function of call size. This model can then be compared with the full model above. This modelling approach is carried out on the same data set of the full model. The call size and quay service time histograms for each liner service are shown on Figure 7 and Figure 8, respectively.

The individual models have the general form $QST_{LS}(TEU) = \alpha_{LS} + \beta_{LS}\, TEU_{LS}$, and their parameters are listed on Table 4.

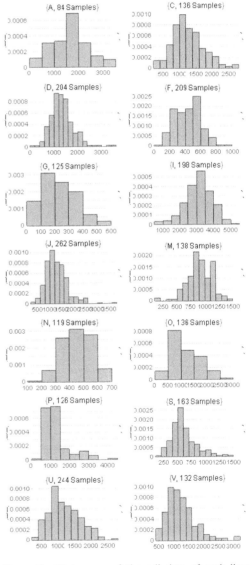

Figure 7. Histograms of the call sizes of each liner service.

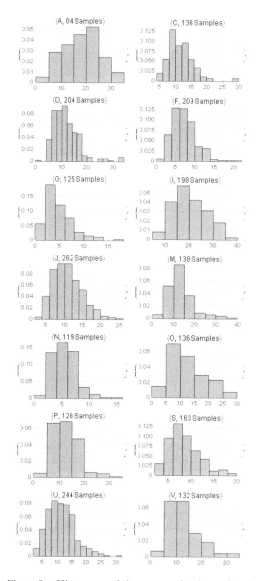

Figure 8. Histograms of the quay service times of each liner service.

The individual linear models and the corresponding data are sketched in Figure 9. The SnDispersion of each liner service's data set is also included.

SnDispersion is an unbiased measure of data dispersion, and the higher its value the lower the dispersion of the data set (Rousseeuw & Croux, 1993).

However, not all models explain the behaviour of the QST with call size in the same way for all the terminal clients, and, as with the full model, not all individual models have the same confidence level.

Table 4. Liner Services QST individual models.

Clt	Sample		Model params.		P-Values		
	Size	R^2adj	α_{LS}	β_{LS}	α_{LS}	β_{LS}	Anova
A	84	0.6212	4.503	0.0080	0.0005	0	0
C	136	0.2077	6.827	0.0038	0.0000	0	0
D	204	0.4371	2.096	0.0071	0.0103	0	0
F	209	0.5028	1.274	0.0136	0.0037	0	0
G	125	0.5847	0.436	0.0188	0.2363	0	0
I	198	0.5996	1.051	0.0062	0.3655	0	0
J	262	0.3106	4.762	0.0047	0.0000	0	0
M	138	0.1144	5.046	0.0093	0.0131	0	0
N	119	0.3385	0.039	0.0122	0.9573	0	0
O	136	0.6768	1.674	0.0093	0.0194	0	0
P	126	0.4014	6.271	0.0046	0.0000	0	0
S	163	0.3760	2.950	0.0090	0.0000	0	0
U	244	0.3795	3.932	0.0064	0.0000	0	0
V	132	0.3928	3.511	0.0075	0.0015	0	0

2.3.1 Analysis of the individual models

A first result is that the amount of variation explained by the individual models is, in general, quite low, as expressed by the corresponding value of the R^2_{adj}.

Only the models of liner services A, F, G, I and O have a R^2_{adj} value above 0.5. For the model of service M, this value is very low (0.11).

These single variable linear models are poor representations of the quay service times of the liner services, except for the five cases mentioned before. The intersect parameters α_{LS} of the models of liner services G, I and N are outside the 95% confidence level (p-value ≥ 0.05), and, in the case of service N, the null hypothesis that the value of this parameter is the outcome of pure randomness, should be accepted.

The p-values of the slope parameter and of the ANOVA of all models are well below 0.05, i.e., call size is an explanatory variable of quay service time variance (as was expectable and in line with the outcomes of the full model).

2.4 Performance analysis

Quay service time is generally considered as the main indicator of container terminal performance, but the previous analysis illustrates that such a conclusion is not easily derived from the call data.

Although call size is an explanatory variable of quay service time,it leaves unexplained a sizeable amount of the variance of this performance indicator.

Including categorical variables that represent the port terminal clients adds little explanatory value to the resulting models.

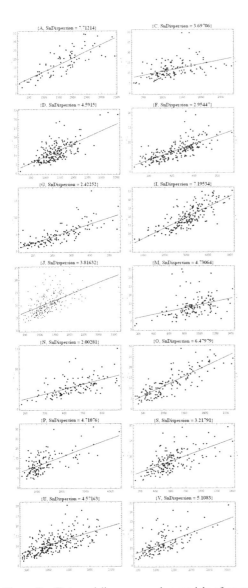

Figure 9. Data and linear regression models of quay service time as a function of call size, for each liner service. The *SnDispersion* of each liner service data set is included.

#	Quay Service Time Models	R^2_{adj}
1	Single Continuous variable cargo	0.5808
2	Continuous variable cargo, and aditional categorical variable seasons	0.5863
3	Continuous variable cargo and categorical variable clients (liner services)	0.6440
4	Continuous variables cargo, and clients' time mismatch with previous client call T/T0m	0.5833
5	Continuous variables cargo, and clients' call size maximum, average and dispersion (standard deviation and range) avgcg, σcg, maxcg, rgcg	0.5918
6	Continuous variables cargo, categorical variables seasons and clients (liner services)	0.6466
7	Continuous variables cargo, categorical variables seasons and clients (liner services), clients' time mismatch with previous client call T/T0m	0.6484
8	Continuous variables cargo, categorical variables seasons and clients (liner services), and clients' call size average avgcg	0.6463
9	Continuous variables cargo, categorical variables seasons and clients (liner services), clients' time mismatch with previous client call T/T0m, and clients' call size maximum, average and dispersison (standard deviation and range) avgvg, σcg, maxcg, rgcg	0.6476
10	Continuous variables cargo and average takt time avgT	0.5872
11	Continuous variables cargo, categorical variables clients (liner services) and crossed-effects (cargo × clients)	0.6624
12	Continuous variables cargo, categorical variables clients (liner services), crossed-effects (cargo × clients) and clients' call size average avgcg	0.6622
13	Continuous variables cargo, categorical variables clients (liner services), crossed-effects (cargo × clients) and average takt time avgT	0.6622
14	Continuous variables cargo, categorical variables clients (liner services), crossed-effects (cargo × clients) and maximum takt time maxT	0.6622
15	Continuous variables cargo, categorical variables clients (liner services), crossed-effects (cargo × clients) and range of takt times rgT	0.6622
16	Continuous variables cargo, categorical variables clients (liner services) and crossed-effects (avgT × clients)	0.6302

Figure 10. Quay Service Time Regression Models. Takt time, in this context, is the time between two successive port calls by the same liner service (liner service inter-arrival time).

In fact, adding more variables, obtainable from the raw call data, does not improve the explanatory properties of the regression models, as highlighted in the analysis preliminary to the work presented here, whose results are summarized on the table of Figure 10. This analysis was performed on the outlier-cleaned data, with no sample size restrictions, a caveat that does not impact the main outcome, which is the relative explanatory power of models with increasing number of variables.

The mentioned limitations of service time modeling, based on the data from liner services calls, motivate other approaches to the terminal service performance related to each individual client. One such approach relies on flows instead of times. Flows (cargo-to-time ratios) are service rates and may express performance better than service times.

2.4.1 *Quay flow (QF)*

Quay flow is already part of the working data. Table 5 lists, for each liner service, the averages of call sizes, quay service time and quay flow.

Table 5. Liner services sorted by average quay flow.

Spl Clt	Size	SnDisp	CallSz Avg(TEU)	QST Avg(Hr)	QF Avg TEU/Hr	Param α_C
G	125	2.423	229.9	4.76	48.32	−0.692
F	209	2.954	432.6	7.17	60.30	0.406
M	138	4.731	908.8	13.49	67.39	3.621
S	163	3.218	610.7	8.47	72.08	0.546
N	119	2.003	451.3	5.53	81.59	−1.359
O	136	6.480	1155.5	12.45	92.81	0.980
A	84	7.712	1668.6	17.80	93.75	0.000
V	132	5.108	1237.9	12.78	96.88	0.772
U	244	4.572	1189.8	11.51	103.37	−0.183
P	126	4.711	1329.0	12.42	107.03	−0.181
C	136	3.697	1261.9	11.64	108.45	−0.526
D	204	4.592	1348.4	11.70	115.29	−1.029
J	262	3.816	1290.8	10.88	118.59	−1.465
I	198	7.195	3112.7	20.32	153.18	−3.880

Table 6. Correlation matrix.

	SplSz	SnDisp	CallSz Avg	QST Avg	QF Avg	Param α_C
SplSz	1	−0.142	0.166	−0.026	0.359	−0.316
SnDisp	−0.142	1	0.790	0.928	0.557	−0.065
CallSz_Avg	0.166	0.790	1	0.914	0.891	−0.523
QST_Avg	−0.026	0.928	0.914	1	0.712	−0.168
QF_Avg	0.359	0.557	0.891	0.712	1	−0.636
Param	−0.316	−0.065	−0.523	−0.168	−0.636	1

The table also includes sample sizes on column 2, SnDispersion values on column 3 and the parameters α_C of the full model on the last column. The table is sorted by increasing values of average quay flow, i.e., increased service performance.

Looking at the Param α_C column gives the idea that the negative α_C values of the liner services at the bottom half of the table, apparently corresponding to better average QF's, should as well correspond to better average service times (QST). But the correspondence, if any, seems to be with average call sizes, which show some consistency with the corresponding QST's. Table 6, which is the correlation matrix of the six variables in Table 5, helps explaining that consistency, with a high correlation value between average call size and average QST (0.914). Average call size is also highly correlated with average quay flow (0.891). In turn, average QF is the variable that has the most negative correlation coefficient with the full model parameters α_C (−0.636), Although this is not a strongly significant value, it points to the idea that the lower the liner service coefficient in the model, the lower the corresponding service time and the higher the flow.

2.4.2 Other effects

The inspection of Table 6 gives some more insights. The effect of individual liner services sample sizes is not relevant, for the range of sizes used here (≥50). Sample dispersion, measured by SnDispersion, is highly correlated with average QST, strongly correlated with average call size and somehow correlated with quay flow.

Parameters α_C correlate negatively with all other variables, and, beyond QF, only have some correlation with average call size

3 DISCUSSION

The differences between the client ranking obtained from the client parameters of the full model and the client ranking obtained from the client quay flows legitimizes the doubt that those parameters describe client differentiation. Quay flow looks like being a more robust indicator of client differentiation.

These rankings, however, allow concluding that there is operational client differentiation at this port terminal. No reasons for this differentiation are proposed because both the terminal and the liner services may contribute to this phenomenon. For instance, a large ship whose length allows for the operation of, say, four or five quay cranes, is probably serviced at a faster rate than a ship than can only be serviced by a single crane, assuming that the cranes have no operational differences and that they face the same quay to yard and yard-specific capabilities. A shipping service operating large ships may, therefore, be serviced faster, on a TEU basis, then a shipping service operating small ships. Other factors, for instance stowage plans, may also contribute to observable quay service differentiation.

The container port terminal may also engage in commercial policies. Such policies can be regarded as a form of product differentiation (Varian, 2010), where liner services may be served differently, considering the existence or lack of alternative port terminals suitable to handle their cargo flows (port terminals have some monopolistic power at least over some liner services that have no suitable alternatives to serve their hinterland). This topic is outside the scope of this research. Additional data on client-supplier relationships in the context of maritime container transportation might help getting insight into this matter.

Nevertheless, by making the correspondence between the liner services letter codes and the shipping companies they belong to, it is possible to understand how the terminal operator and the shipping companies value their business relationships. For instance, the four services at the bottom of Table 5 (best performers) are MSC services (the top performer is a Far East bound service using VLCC ships). MSC is a carrier with a special business relationship with Sines Container Terminal XXI, whose operator is PSA, instantiating the relationship that these two companies have been building in the last years Additional examples of this relationship are the MSC PSA European terminal in Antwerp and the MSC PSA Asia terminal in Singapore.

4 CONCLUSIONS

Operational service differentiation at container port terminals can be spotted in the parameters of categorical variables, corresponding to client liner services, in linear regression models. However, in a real case with a sizeable data set of liner services calls at the major Portuguese container port terminal, those parameters correlate poorly with service time, which is generally accepted as the main terminal performance indicator.

Quay flow, however, looks like a more adequate service performance indicator, which correlates much better with the categorical variables model parameters and plays the role of a benchmark reference in the context of finding other suitable client differentiation indicators.

Although based on a single empirical case, this analysis allows concluding that it is possible to obtain container port terminals client differentiation indicators from call data. Additional data might reveal other aspects of terminal servicing of their clients, including commercial policies taking place along with operational differentiation. Criteria for obtaining that data are outside the scope of this study.

ACKNOWLEDGEMENTS

This work was performed within the Strategic Research Plan of the Centre for Marine Technology and Ocean Engineering, which is financed by Portuguese Foundation for Science and Technology (Fundação para a Ciência e Tecnologia-FCT).

REFERENCES

Bichou, K. (2013). Port Operations, Planning and Logistics. Lloyd's Pactical Shipping Guides. Informa Law from Routledge.

De Langen, P., Nijdam, M., & Van Der Horst, M. (2007). New Iindicators To Measure Port Performance. *Journal of Maritime Research*, *IV*(1), 23–36.

Ducruet, C., Itoh, H., & Merk, O. (2014). Time Efficiency at World Container Ports. International Transport Forum Discussion Papers.

Hoshino, H. (2010). Competition and Collaboration among Container Ports. *The Asian Journal of Shipping and Logistics*, *26*(1), 31–47. http://doi.org/10.1016/S2092-5212(10)80010-0

Notteboom, T.E., & de Langen, P.W. (2015). Container Port Competition in Europe. In Handbook of Ocean Container Transport Logistics – Making Global Supply Chain Effective, International Series in Operations Research & Management Science (Vol. 220, pp. 75–95). http://doi.org/10.1007/978-3-319-11891-8_3

Rousseeuw, P.J., & Croux, C. (1993). Alternatives to the Median Absolute Deviation. *Journal of the American Statistical Association*, *88*(424), 1273–1283. http://doi.org/10.1080/01621459.1993.10476408

Stopford, M. (2009). *Maritime Economics* (Third Edit). Routledge.

Varian, H.R. (2010). *Intermediate Microeconomics, A Modern Approach 8th edition* (8th ed.). W.W. Norton & Company.

Maritime Transportation and Harvesting of Sea Resources – Guedes Soares & Teixeira (Eds)
© 2018 Taylor & Francis Group, London, ISBN 978-0-8153-7993-5

Commercial maritime ports with innovative mooring technology

E. Díaz-Ruiz-Navamuel, A. Ortega Piris & C.A. Pérez-Labajos
Members of Ocean and Coastal Planning and Management R&D Group, School of Maritime Engineering,
Department Sciences and Techniques of Navigation and Shipbuilding, University of Cantabria,
Santander, Spain

L. Sanchez Ruiz & B. Blanco Rojo
Members of Bank and Finances of the Company I+D+i Group, University of Cantabria, Santander, Spain

ABSTRACT: This article is a fundamental part of a doctoral thesis that investigates the best system of automatic tie to install to tie automatically in the port of the Santander in Cantabria (Spain) (Diaz 2016). For the development, the doctoral thesis was necessary to do an exhaustive analysis of the systems that existed on the market to know who was the suitable one to be installed in the port of the study. This presentation is a synthesis of the study realized of the presence in the world of the selected AMS. In this article, a study has been made of the different ports in which the automatic mooring system using vacuum suction cups is installed (AMS). Currently there are 31 ports in which different models of this mooring system are operating. The reasons they have been installed are different but the results are equally good in all cases.

1 INTRODUCTION AND BACKGROUND

For thousands of years, the maritime world has used ropes, cables and lines to moor its vessels. These systems have been reliable and dependable, but now it seems that they are out of synch with the latest developments in the maritime industry, with its continuous improvements in productivity and efficiency.

In the case of container ship and bulk transport terminals, the most common problems are the need for docks with a greater draft, length and storage space, which means that these often have to be constructed in locations which are not sheltered from high tides and strong winds. In order to offset the movements caused by these adverse meteorological conditions, it is cheaper and easier to install these automatic mooring systems using vacuum cups than to construct seawalls

Innovation has taken place in the maritime sector in all areas, including that of mooring systems, though less intensely than in other areas since the conventional system is still used in most ports. At present, some innovative automatic mooring systems are being developed and installed in some terminals (Caro 2014, Díaz et al., 2016).

Until the Second World War, the exploitation and organization of maritime traffic did not change substantially. The loading and unloading operations were slow and laborious. Thus, in the post-war era, with the expansion of the market and the rapid rise in labor costs, the system was put under tremendous stress. Congestion in ports increased and the ingenuity had to be sharpened to seek through innovation in both technology and in processes a response to such problems. (Caro 2014).

The maritime industry responded to the new challenges with two "revolutions" in the two sub-sectors of maritime transport:

– In non-scheduled traffic through the development of integrated transport systems (bulk carriers)
– In regular lines by grouping together the general cargo by means of the phenomenon of containerization.

All of which led to profound changes being made in commercial seaports in order to respond to these challenges. (Camarero Orive et al. 2011).

As a consequence of these revolutions, new innovations have taken place in the technological as well as the organizational sphere, with the appearance of new means of traffic, new equipment and new methods.

In particular the increase in the production of cars and the incessant need to transport them from the factories to their different points of sale has led to different forms of regular traffic, such as combined train-boat-truck transport, and the need to transport and deliver goods in the shortest possible time and with the lowest costs in order to maximize profits. Subsequently, the possibility arose for using these types of vessels with aft or side ramps not only for the transport of cars but also for Ro-Ro goods and for goods on platform. Hence, it became necessary to construct larger vessels in order to be able to load greater volumes of merchandise and to make other routes.

This innovative process, though it is in advanced stages in the major industrial countries, cannot yet be said to have concluded. (Natarajan, Ganapathy 1997, Natarajan, Ganapathy 1995, Nakamura et al., 2007).

The "AMS" system closed a great technological gap between the vessels and the installations. Prior to it, vessel mooring systems lagged behind the great technological advances made in the marine terminals (Fang, Blanke & Leira 2015).

This AMS is a good solution for RoRo and ferry terminals with RoRo freight, because today the sea plays an increasingly important role in reducing road traffic jams. Good examples are the maritime lines specialized in car transport, which form an integral part of logistics within the global automobile manufacturing chain, and passenger ferries that are mostly employed on short-haul routes where the reliability of the schedules and rapid handling are of vital importance.

As with container ships, RoRo ships and RoRo freight shipping on regular routes have also led to the adaptation of different ports in order to handle this type of traffic. With the increase in the size of vessels, the most important challenge is often safety, both environmental and personal, when the dock space for the ropes is restricted. The AMS is very effective in shortening the stay of ships in port, and allows the terminals to operate in adverse environmental conditions that are sometimes incompatible with the traditional mooring system. (Fang, Blanke 2011).

Container terminals have become a crucial link in today's global economy. They are often the main logistical hub of a broad geographic region that ensures a fluid exchange of consumer goods, raw materials and industrial products. The gains in efficiency and productivity that can be achieved by using automatic mooring systems both on the ship and on land are potentially significant for streamlining the logistics chain. This can have a major impact on the commercial success of container terminal operators and shipping (Jin et al., 2014).

2 AIMS, HYPOTHESES AND METHODOLOGY

2.1 General and specific aims

As mentioned above, the conventional mooring system is traditionally used to perform the mooring maneuvers of merchant ships, but a new method of automatic mooring by means of vacuum suction cups (AMS) for merchant ships is currently being developed and installed in different ports.

In this context, the general aim of this paper is to analyze the degree of implantation of these systems and their location in the world.

2.2 Hypotheses

The proposed aims are to be reached under the following hypotheses:

- Hypothesis 1: The AMS is increasingly present in ports around the world.
- Hypothesis 2: The AMS has been installed in different types of terminals covering, in each case, quite different needs.

2.3 Methodology

To assess the implantation of the automatic mooring system, a complete compilation of general and local information is made to clearly define the objectives of the study (González 2006, Moyano Retamero, 2002). In order to reach the proposed objectives, the following working methodology was proposed in which two parts can be distinguished. (Camarero Orive et al., 2011).

In the first part, the characteristics of the different types of automatic mooring system by vacuum suction cups that can be found on the market are analyzed (Ortloff et al., 1986).

In the second, a study is made of the presence of these types of AMS in the world.

Over the years, several different patents have been developed for automatic mooring systems for merchant and recreational vessels, but the focus of our study is a system with vacuum suction cups whose first registered patent dates from the year 2001 (Hadcroft, J. and PJ Montgomery, 2001), although the system was first used in 1998.

The paper begins by describing the different models of this AMS that are on the market; it then goes on to identify the different types of terminals and docks in which it is installed; and it finsihes by analyzing the data obtained.

3 MODELS OF AMS

The AMS has several different models, depending on the use to be made of it, the power of retention required, the vessels, frequency of use, docks, etc.

These systems can be installed on the ship itself or on the dock.

The Table 1 shows the images of the different models of the Quay Sailor range and the monitoring system.

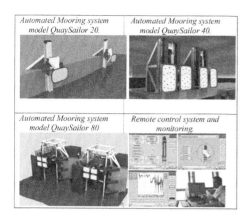

| Automated Mooring system model QuaySailor 20. | Automated Mooring system model QuaySailor 40. |
| Automated Mooring system model QuaySailor 80 | Remote control system and monitoring. |

Table 1. Illustrations of the different models of AMS. *Source. http://www.cavotec.com.*

4 DOCKS WITH AMS AND VESSELS THAT USE IT

The system can be used in any type of dock and for any type of vessel. It is currently in use in container terminals, RoRo and passenger and bulk liquids and solids terminals.

4.1 *RoRo and passenger terminals*

Terminals of this type that are already operating with "AMS" allow the docking of larger vessels, without the need for expensive dock extensions, or berthing dolphins or similar. This system is installed in the following terminals:

Port of Picton, First System, New Zealand (1998)
The first AMS ever was used in this port in 1998. The first model that came onto the market was the AMS i-400, and it was installed in the hull of the passenger ferry "Aratere", a ship with a length of 150 meters and 12,000 tons of displacement. The system was used with a frequency of three times per day and made the regular route from Picton to Wellington.

The system consisted of the installation on the hull of four suction cups, with two cups at bow and two at stern, with a power of 20 tons of retention per unit.

The system operated from 1998 to 2009, but since 2003 it has been used simultaneously with a prototype that was installed on the dock, whose description can be found in the patent of Montgomery, P. J. and B. J. Rositer. (2002).

Port of Picton, prototype AMS 400, New Zealand (2003)
In 2003, a prototype of the AMS, was installed in the dock of Picton, New Zealand. In this case,

a mooring robot of 400 KN was installed on the dock.

The system was also used by the ship, the "Aratere", with a frequency of three times a day. In this case, the operator was Kiwirail Ltd who installed the prototype in 2003 and used it until 2005 when the next model was installed.

Port of Melbourne & Davenport, Australia (2003)
The AMS was installed in these two docks (Melbourne and Davenport) on the edge of the dock. The retention capacity is produced by 4 robots × 400 KN = 160 tons and the frequency of use is of one berth per day.

The AMS technology was first installed in the port of Melbourne for RoRo vessels such as the 149-meter long Searoad Tamar, with 13,697 tons of displacement and the 118-meter Mersey Searoad, with 7,928 tons of displacement.

The units are placed in pairs, one at bow and the other at stern.

Port of Picton, New Zealand (2005)
In the year 2005, the prototype was replaced by the AMS TM 400 model, which is described in the patents of Montgomery, P. J. and B. J. Rositer. (2003), Montgomery, P.J. and B.J. Rositer. (2005). This model is the one that has been installed in the port of Melbourne, Australia, since 2003. In the dock at Picton, two robots with a power of 400 Kn were installed, making a total of 80 tons of power of retention.

Dover, UK (2005)
In June 2005, for the purpose of testing and trialling, installed the world's most powerful automatic mooring unit, the AMS with a holding capacity of 80 tons per unit, in a newly built dock in the port of Dover, in the United Kingdom.

It was installed in the busiest ferry port in northern Europe: Dover Quay 8 is used by Ro-Pax ferries of up to 185 meters in total length.

The high frequency of use of the docks and the conditions of strong winds of up to 60 knots and the height of the waves of up to one meter made it necessary to install an automatic mooring system that could withstand such extreme conditions.

During testing, this AMS successfully completed some 750 mooring operations, demonstrating the efficiency of the AMS automatic mooring system even with large variations in tidal height and extreme environmental conditions.

Ports of Hov and Sælving, Denmark (2009)
In the ports of Hov and Sving in Denmark, the AMS has been installed since 2009. It is also installed on the dock and operates below the edge. The retention capacity produced by the 2

robots × 400 KN. It is used on a dock where the 91-meter long Kanhave ship does the regular Hov-Sælving (Samsø) route.

These ports saw the need to reduce the times used in the maneuvers in order to maximize the regular passenger line between Hov on the mainland of Denmark and Saelvig on the island of Samsø.

The two AMS are installed on the two ports of this line, making maneuvering safer and reducing the number of crew members required for docking or departure, even in adverse weather conditions. In twenty-five seconds and through a remote control system located on the ship's bridge, the vessel is moored or released.

Ports of Spodsbjerg and Tårs, Denmark (2012)

The model installed in these ports is also the AMS, installed on the edge of the dock, with a retention capacity of 400 KN = 40 tons, and is used with a frequency of 36 moorings per day. The vessels that use these berths are ferries of 99 meters of length, the "Lolland" and the "Langeland", which cover the regular line between Spodsbjerg and Tårs.

These AMS are installed in the two ports of call of the regular line, reducing costs of crew, time and maneuvering materials (ropes and lines), and equipment maintenance.

Port of Wellington, New Zealand (2012)

The AMS was installed in the Port of Wellington in the year 2012 in order to have the same system as in the Port of Picton, and thus have the same maneuverability conditions in the two ports of the regular line, which has a frequency of three times per day.

As in the Port of Picton, the vessel that uses it is the ferry of 180 meters in length, the "Aratere", later substituted by the vessel the "Kaitaki" de 181 meters in length.

Housælvig/Samsø Municipality, Denmark (2013)

In 2013, the port authority of Hou and Svlig in Denmark decided to install and put into operation two units of the AMS, similar to those operating on other docks of these two ports since 2009. The plan was to install a robot in each port of the regular ferry line between Hou in Jutland and Svlig on the west coast of the island of Samsø.

The problem that these two ports have is the strong wind of more than 40 knots, and the variations in the height of the tide. Hence, they needed to speed up the maneuvers and to increase security during the time the ship was docked.

The frequency of use is seven berths per day in each port and the vessel that makes the trips is a ferry of 99 meters.

Port of Helsinki, Finland (2014)

The Port of Helsinki granted a license to the company for the installation in the Länsisatama dock of six AMS 15 units. This equipment has been in operation since the end of the year 2015. The retention capacity of each of these devices is 40 tons per unit for the passenger ferries of 186 meters in length. The ferries cover the route between Helsinki and Tallin with a frequency of six times per day.

Den Helder/Teso, Holanda (2015)

The Teso company runs a high-speed ferry route between Den Helder and Texel Island with 2 passenger ferries. Following increasing pressure from municipalities to improve the environmental quality of air in port areas, Teso decided to change the ferry docking procedures for ferries of between 110 and 130 meters in length, which kept their engines running on idle all the time they were in the dock. This was made possible by the installation of two AMS units that ensured the secure mooring just by pressing the remote control button on the radio control on board the ferry.

The two units are able to exert a clamping force of 400 kN, and are attached to a floating steel structure (pontoon) in the terminal.

Teso is benefiting not only from the very fast and secure mooring of its ferries, but it also saves a lot of fuel. In addition, the saving in time has allowed the ferry line to better maintain the schedule of about 16 daily calls.

Lavik & Oppedal/Norled, Norway (2015)

In 2015, the AMS was installed for a renewable energy ferry line in Norled in Western Norway. This service is the first in the world to operate with fully battery operated ferries. The AMS has an electrical system that allows the recharge of the batteries of the ferry during its stay in the dock: the ship's batteries are connected to the mooring system by means of a connector that they have called AMP. (Murray et al., 2009).

The vessel that uses it has a length of 86 meters and a capacity of 120 cars and 360 passengers, and the frequency of use is 17 berths per day.

Port of Ballen & Kalundborg, Denmark (2015)

Færgen A / S purchased its first AMS system in 2008 when they won the Hou-Samsø route operation. In 2014, having lost the renewal of the concession, they decided to remove the two AMS units, restore them and install them in another nearby site, and did so on the route between Ballen and Kalundborg. The route is covered by a ferry of 91 m in length with a frequency of five times a day.

4.2 Container terminals

Container ships are becoming larger and larger and in many ports the docking of these large ships with ropes and lines can easily take longer than 40 minutes. AMS can guarantee the docking of these large ships in a matter of seconds, allowing port staff to have faster access to the vessel to begin loading operations. In the ports that have this system, an increase in the effectiveness of the cranes has been observed, as AMS provides a stable docking, with few movements of the ship, providing greater security in the handling of the containers. This increases loading and unloading rates, thus shortening the stay of the vessels in port (Sakakibara, Kubo, 2007). The container terminals where the system is installed are:

Dock N° 6, Salalah, Omán (2006)

In this dock, the model installed since 2016 is the AMS. This system is also installed on the dock and operates under the edge of the quay. The retention capacity is 600 KN times 4 units, the frequency of use of the dock is three or four times per week and the vessels that use it are of 350 meters in length and 130,000 GTs.

Dock N° 1, Salalah, Oman (2009)

In dock N° 1 of the Port of Salalah, Oman, 12 robots of 200 KN are installed on the dock, though they work below it. The dock is used three or four times per week by different container ships of around 362 meters in length.

Beirut, Lebanon (2014)

Since 2014, the AMS has been operating in the Port of Beirut in the container ship terminal for vessels of up to 350 meters in length, made up of 42 robots of 20 tons of retention each.

Economically, it was not feasible to extend the existing breakwater to protect a new dock extension of 500 meters, and the Port of Beirut sought a solution to protect ships from the wave-induced movements on that dock (Lee, Hong & Lee, 2007).

The docking frequency on this quay is 5 to 8 berths per week and it is operated by the owner of the facility, which is the Port of Beirut itself.

Port of Salalah, Oman (2015)

In 2014, 8 units of 400 Kn each were installed in the Port of Salalah in another of its six container ship docks. At the same time, the retention capacity of the 2009 facility of Dock No. 1 was increased by 2,400 tons and that of Dock 6 was raised to 3,200 tons. The increase was requested by the Port to ensure that more ships could benefit from the system where the ship's structure prevented all units from being used.

The frequency of use of this dock is 3 to 4 times per week and it is used by vessels up to 350 meters in length.

Port de Ngqura, South Africa (2015)

In 2013, the Transnet Port Authority installed the automatic mooring system in the Ngqura Port in the container terminal. They installed 26 units of the AMS for the four container docks of this port on the east coast of South Africa. Each unit has a retention power of 200 Kn.

The units were installed to moor container ships from 1,500 TEU up to 13,000 TEU and up to 366 meters in length.

The main reason for the introduction of the system was to cushion on board the effects of long waves and strong winds, especially during the winter, which cause excessive movement in the vessels along the dock wall and interfere with the crane operations.

4.3 Bulk load terminals (solids and liquids)

The transportation of bulk cargo, both solids and liquids, accounts for the largest volume of trade and transportation worldwide, shipping by sea being the best option for this kind of goods.

The vessels are not usually used on fixed routes, but rather are usually chartered for a single trip. The terminals in which they operate normally have to compensate the ship-owner if there is a loss of time for unjustified delays in the loading or unloading terminals.

The AMS makes the ship's stay more predictable and stable, reducing downtime due to adverse conditions in the port.

These systems are installed in the following ports:

Bulk liquid and fuel jetty, Parker Point, Dampier, Australia (2011)

The model installed in this jetty is the AMS, positioned on the jetty and made up of 8 units of 200 KN each. The jetty is used with a frequency of one time per week and is used by tankers of up to 60,000 displacement tons (dwt). These units can resist winds of up to 45 knots and tides of up to one meter of wave height.

Port de Utah Point, Port Hedland, Australia (2012)

The model used in this port is the AMS, which is installed on the face of the dock, as if it were a defense. The Hedland dock is in the west of Australia.

Its retention capacity is produced by 14 robots of 200 Kn each. The frequency of use of the system is once every two days and it is used by vessels that do not cover a regular line. This berth can house bulkcarrier vessels of up to 295 meters in length.

Dock Nº 7, Gerald ton, Australia (2012)
In this dock, the model used is the AMS, installed on the edge of the dock. It consists of 12 units of 200 Kn each. The dock is used with a frequency of three or four times a week, in which ships of different sizes dock, generally Panamax bulkcarriers (294 m. in length).

In the past, this port used to suffer from many delays in the cargo operations due to the continuous swells, and on some occasions the ships had to leave the dock and wait at anchor for the sea to calm (Banfield, Flory, 2010).

Jan de Nul, Brisbane, Australia (2013)
In 2013, the company Jan de Nul was hired to run the dredging operations required to obtain the necessary landfill to build a second runway at Brisbane Airport. This process consisted of the dredging of seafloor sand and its unloading in a fixed location. This dredging and unloading cycle had to be undertaken in 12 hours but during the last stages of the planning of the project, a complication arose concerning the mooring of the dredger/hopper, "Charles Darwin" for the unloading of the filler material, as required by the Brisbane Port Authority. This made it impossible to meet the 12 hour cycle time requirement using conventional methods. Hence, Jan de Nul could not meet the landfill delivery obligations for Brisbane Airport on time.

The use of the AMS with 20-ton retention per unit from a previous customer, thus shortening the delivery, installation and start-up times. was pro-

Table 2. Mooring system installed on pier.

Year	Country	Nº Robot	Cap Tons/unit	USE	Vessel	Length (m)	USE year
1998	N Zealand	1	20	3 per day	Ferry	150	1080
2003	N Zealand	2	20	3 per day	Ferry	150	1080
2003	Australia	4	40	1 per day	Ro-Ro	150	360
2005	N Zealand	2	40	3 per day	Ferry	150	1080
2005	UK	1	80	2 per day	Ro-ro	185	720
2006	Oman	4	60	4 per week	Bulk	350	208
2007	Canada	4	20	15 per day	All	225	5400
2009	Oman	12	20	7 per week	Cont.	362	364
2009	Denmark	2	40	7 per week	Ferry	91	364
2009	Denmark	2	40	7 per week	Ferry	91	364
2011	Australia	8	40	1 per week	Tanker	300	52
2012	Australia	14	20	3 per week	Bulk	295	156
2012	Denmark	1	40	36 per day	Ferry	99	12960
2012	Denmark	1	40	36 per day	Ferry	99	12960
2012	N Zealand	1	40	3 per day	Ferry	180	144
2012	Australia	12	20	4 per week	Bulk	294	208
2013	Denmark	1	40	7 per day	Ferry	99	364
2013	Denmark	1	40	7 per day	Ferry	99	364
2013	Australia	2	20	2 per day	Dredge	183	720
2014	Lebanon	42	20	7 per week	Cont.	300	364
2014	Finland	6	40	6 per day	Ferry	186	2160
2015	Norway	18	20	3 per week	Bulk	305	156
2015	Holland	2	40	16 per day	Ferry	130	5760
2015	Holland	2	40	16 per day	Ferry	130	5760
2015	Norway	1	20	17 per day	Ferry	86	6120
2015	Norway	1	20	17 per day	Ferry	86	6120
2015	Oman	8	40	4 per week	Cont.	350	208
2015	Denmark	2	40	5 per day	Ferry	91	1800
2015	Denmark	2	40	5 per day	Ferry	91	1800
2015	S Africa	26	20	2 per day	Cont.	366	720
31		191					69916

Source: Author.

posed o solve the problem and to reduce mooring times from two hours to approxiately 30 seconds.

This AMS with eight AMS has a frequency of use of several berths per day with a dredge of 183 m. LKAB Narvik, Norway (2015).

Since 2015, a mooring system made up of 18 units of the AMS of 200 Kn each have been installed in the dock of Narvik, Norway, located on the new iron mineral dock of LKAB.

In this dock, bulk carriers of up to 185,000 dead-weight and 305 meters in length can berth. This is the first AMS installation on a solid-bulk dock in Europe, and the first on the Arctic Circle. The frequency of use of this dock is 2 to 4 times per week.

4.4 *Installations in locks y prototypes*

St. Lawrence Seaway Dock, Great Lakes, Canada (2007)

The model installed on this dock is the AMS consisting of 4 units of 200 Kn each. The first installations were carried out in Welland Canal Lock No. 7 with 2 different versions of the AMS.

The dock is used with a frequency of several times per day, between 5 and 15 times, by different vessels of up to 225 meters in length.

Since 2007, AMS has worked on the development of the best possible AMS system for use in the locks and the result was a new system. It was installed in 2013 in Lock # 4 of Beauharnois, with a change of level of 13.5 million liters of water.

The AMS is currently being installed in the remaining 13 locks.

5 DISCUSSION

This automatic mooring system was present in 31 Ports in the year 2016, according to the information compiled, (Cavotec, 2015), and they are currently scheduled to install it in a further three ports.

Today, AMS is a widely accepted technology that has made more than 40,000 mooring operations, with a security ratio of one hundred percent, and is installed on all types of docks and used by different types of ships such as ferries, bulk carriers, and RoRo and container carriers.

The total number of maneuvers currently carried out with this type of automatic mooring system is 69,916. The year 2015saw the highest number of installations undertaken, nine in total, and also the largest increase in the number of maneuvers, reaching 28,444.

In 2012, the increase in the number of maneuvers (26,428) was almost the same with just 5 installations, the explanation for this being that the installations made in 2015, although they were

Table 3. Maneuver with AMS installed on dock each year.

Year	N° of installation per year	N° Robot installed per year	Increased of maneuver per year	Sum of maneuvers
1998	1	1	1080	1080
2003	2	6	1440	2520
2005	2	3	1800	4320
2006	1	4	208	4528
2007	1	4	5400	9928
2009	3	16	1092	11020
2011	1	8	52	11072
2012	5	29	26428	37500
2013	3	4	1448	38948
2014	3	54	2524	41472
2015	9	62	28444	69916
Total	31	191	69916	

Source: Author.

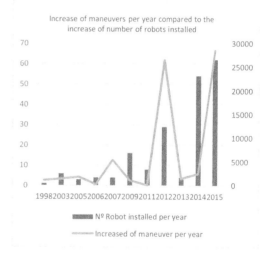

Graph 1. Increase of maneuvers per year compared to the increase of number of robots installed. Source: Author.

mostly in ferry terminals, were in terminals with fewer daily stops.

As can be observed in the graph 1, which compares the number of maneuvers performed with the increase in the number of robots installed each year, a greater number of maneuvers does not correlate to a greater number of robots installed. In the years 2006, 2007 and 2013, 4 robots were installed and 2008, 5,400 and 1,448 maneuvers were performed, respectively.

There is no direct relation that indicates that the more robots installed per year, the more maneu-

Relationship between sum of maneuvers and number of facilities per year

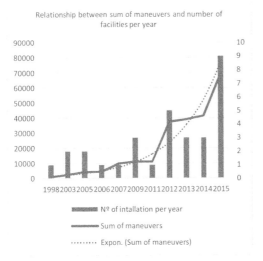

Graph 2. Difference between N° of installation per year with sum of maneuvers and N° of robots. Source: Author.

Table 4. Terminals where the AMS is installed and its use.

Terminal Type	Number of terminals	Number of Robot	Maneuvers per year	Maneuvers per day
Pax-RoRo	19	35	61360	170
Container	4	88	1656	5
Bulk Load	5	56	780	2
Others	3	12	6120	17

Source: Author.

vers performed. This is because each port requires an independent study to verify the need for the installation, the feasibility of its use and the determination of the most suitable model for the environmental conditions, type of dock and type of vessels using the berth (Baan, J.1983).

As can be seen in Graph 2, the number of maneuvers performed is always on the rise, but not at a continuous rate.

The same occurs in the relation between the number of ports in which the system is installed each year and the number of robots installed in these ports. It can be observed that the exponential tendency line of the number of ports in which the system is installed is upward.

The table 4 shows the relation between the different types of terminals in which this type of automatic mooring system is installed with the number of docks, number of robots and number of maneuvers per year and per day.

Relationship betweeen terminal types, robots and maneuvers per year

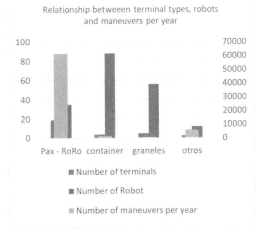

Graph 3. Difference between terminal types, N° of robots and N° of maneuvers per year. Source: Author.

Table 5. Pictures of some models of the ports on which it is operating. Source: Author.

The graph 3 shows the relation between the different types of terminals, the number of installations in these ports and the number of robots and maneuvers performed per year in each type of terminal

6 CONCLUSIONS

It can be observed that this type of installation is more commonly found in Roro and passenger terminals, but the highest number of robots are installed in container terminals, the system being installed in four docks with 88 between them.

In view of the results obtained from the research described above, the following conclusions have been reached:

1. The use of the automatic mooring system allows the safety margins to be increased during the stay of the docked vessel. The ship can remain safely docked with more virulent winds than with the conventional system, without the need to resort to tugs.
2. This type of installation is more commonly found in Roro and passenger terminals.
3. The highest number of robots are installed in the container terminals.
4. The number of robots installed in each dock depends on the size of the vessels, the type of dock and the meteorology of the place, not on to the number of stops.
5. With the automatic mooring system, the cost of the mooring service is reduced.
6. The number of maneuvers performed with this system is always on the rise, but not at a continuous pace.
7. Nowadays engineers continue to develop AMS automatic mooring systems and are perfecting new forms of technology which may be used to improve safety, operational efficiency and infrastructure savings.
8. The utilization of any system of automatic tie is beneficial for the whole personnel involved in the maritime business.
9. In Ferry and RORO terminals, the main objective is the optimization of the time scale: with the installation of these systems, the time required to perform the maneuvers is reduced and safety during the maneuvers and the stay in port increases

We can assure that both raised hypotheses are fulfilled.

REFERENCES

Baan, J., 1983 A rigid arm single point mooring system for vessels., DE FR GB IT NL SE edn, B63B 27/34, B63B 22/02, (NL.
Banfield, S. & Flory, J. 2010, "Effects of fiber rope complex stiffness behavior on mooring line tensions with large vessels moored in waves", OCEANS 2010IEEE, pp. 1.
Burges, A., 1978 A Mathematical model for the calculation of the harbour entrance manouevre., Cogress edn, 7° International Harbour Congress of Amberes, Amberes.
Camarero Orive, A., López Ansorena, I., González Cancelas, N. & Pery Paredes, P. 2011, "Análisis de las diversas metodologías para el estudio del sistema de atraque en terminales portuarias".
Caro, R.V. 2014, "Revolución en los sistemas de amarre de los buques", Revista general de marina, vol. 266, no. 3, pp. 475–490.
Cavotec 2015, 12/01/2016-last update, Cavotec [Homepage of Cavotec S.A.], [Online]. Available: http://www.cavotec.com/ [2016, 12/01/2016].
Diaz, E, 2016, "Innovation in the mooring system of the maritime commercial ports", Universidad de cantabria
Díaz, E., Ortega, A., Pérez, C., Blanco, B., Ruiz, L. & Oria, J. 2016, "Empirical analysis of the implantation of an automatic mooring system in a commercial port. Application to the Port of Santander (Spain)".
Fang, S. & Blanke, M. 2011, "Fault monitoring and fault recovery control for position-moored vessels", International Journal of Applied Mathematics and Computer Science, vol. 21, no. 3, pp. 467–478.
Fang, S., Blanke, M. & Leira, B.J. 2015, "Mooring system diagnosis and structural reliability control for position moored vessels", Control Engineering Practice, vol. 36, pp. 12–26.
Goekdeniz, N. & Deniz, U. 2006, "Dynamics of ships and fenders during berthing in a time domain", Ocean Engineering, vol. 33, pp. 14–15.
González, D.D.C. 2006, Estudio paramétrico de las fuerzas en sistemas de amarre para buques amarrados en Port., Universidade do Porto.
Hadcroft, J. & Montgomery, P.J., 2001Mooring device.
Jin, H., Su, X., Yu, A. & Lin, F. 2014, "Design of automatic mooring positioning system based on mooring line switch", Dianji yu Kongzhi Xuebao/Electric Machines and Control, vol. 18, no. 5, pp. 93–98.
Lee, D., Hong, S. & Lee, G. 2007, "Theoretical and experimental study on dynamic behavior of a damaged ship in waves", Ocean Engineering, vol. 34, no. 1, pp. 21–31.
Montgomery, P.J. & Rositer, B.J., 2003 Ship based mooring device.
Montgomery, P.J. & Rositer, B.J., 2005 Mooring robot.
Montgomery, P.J. & Rositer, B.J., 2002 Mooring robot.
Moyano Retamero, J. & Losada Rodríguez, M.Á 2002, Fiabilidad y riesgo en sistemas de atraque, amarre y defensa, Universidad de Granada, Grupo de Ports y Costas, Granada.
Murray, J., Gupta, A., Seng, F.K., Mortensen, A. & Tung, W.T. 2009, "Disconnectable Mooring System for Ice Class Floaters", ASME 2009 28th International Conference on Ocean, Offshore and Arctic Engineering American Society of Mechanical Engineers, pp. 169.
Nakamura, M., Kajiwara, H., Inada, M., Hara, S., Hoshino, K. & Kuroda, T. 2007, "Automatic mooring system for ship", pp. 1039.

Natarajan, R. & Ganapathy, C. 1997, "Model experiments on moored ships", Ocean Engineering, vol. 24, no. 7, pp. 665–676.

Natarajan, R. & Ganapathy, C. 1995, "Analysis of moorings of a berthed ship", Marine Structures, vol. 8, no. 5, pp. 481–499.

Ortloff, J.E., Ziarnik, A.P., Filson, J.J. & Gadbois, J.F. 1986, Vessel mooring system,.

Sakakibara, S. & Kubo, M. 2007, "Ship berthing and mooring monitoring system by pneumatic-type fenders", Ocean Engineering, vol. 34, no. 8–9, pp. 1174–1181.

Schelfn, T.E. & Östergaard, C. 1995, "The vessel in port: Mooring problems", Marine Structures, vol. 8, no. 5, pp. 451–479.

Shang, G., Zhen, H. & PingRen, H. 2011, "Automatic multi-point mooring vessel monitoring system", pp. 316.

Maritime Transportation and Harvesting of Sea Resources – Guedes Soares & Teixeira (Eds)
© 2018 Taylor & Francis Group, London, ISBN 978-0-8153-7993-5

A green port—case study of Port of Koper

E. Twrdy
Faculty of Maritime Studies and Transport, University of Ljubljana, Portorož, Slovenia

E. Hämäläinen
Brahea Centre, University of Turku, Turku, Finland

ABSTRACT: European transport market has changed in the last twenty years. With the establishment of production facilities (automobile industry) in the countries of Central and Eastern Europe, cargo flows from Adriatic ports are growing. The strategic goal of Adriatic ports is to become one of the most important gateway regions in Europe and to develop from a handling port into a logistic centre. Port of Koper (as one of Adriatic ports) is the most successful port with the highest throughput of containers and cars. Increase of traffic goes along with the concerns for the environment and new investments are part of the strategy to be a green port. One of the strategic orientations of port until 2030 is to be institutional stakeholder of sustainable development of natural and social environment. In this paper we will present the investments in terminals and capacities that will allow this port to be green.

1 INTRODUCTION

Global trade growth, increasing sizes of vessel and the need to modernize port facilities are the key reasons for urgent investments in ports, if ports don't want to loss cargo, throughput and competitive position on market. Port development can have a negative impact on the surrounding area, which is why sustainable or green port development is important. With increased concerns about the environmental impact due to the development of ports, a research in the field of ecological issues in ports and port management policies in relation to green port development (Peris-Mora et al., 2005; Gupta et al., 2005; Darbra et al., 2009, Nebot et al., 2017) has been done. They appointed that green ports became an important issue and a critical part of sustainable supply chain management research and that there is a need for innovative solutions for sustainable port development. New green ports must be in harmony with the ecosystem, but also prepared to change for new solutions.

In Sustainable Ports—A Guide for Port Authorities (PIANC, 2014) green port is defined as a port in which the port authority and port users pro-actively and responsibly develop and operate, based on an economic green growth strategy. Lam and Notteboom (2012) indicate impacts caused by ports in three groups: air pollution, water pollution and the maintenance and upgrading of port infrastructure, causing a high impact on marine ecosystems due to dredging and civil works.

To achieve the concept of development for Green Port it is necessary to integrate environmen-tally friendly method of port activities, port operations and management. In the paper we present ways that Port of Koper uses to define measures for establishment of ecological/green sea port.

2 PORT OF KOPER

Development of transport and logistics sector in Slovenia depends mostly of the Port of Koper as one of the most important creators of traffic flows. The economic effects of port activity are reflected in direct surroundings and wider environment, as port presents a generator for economic development. Port of Koper is designed for (the) handling various types of goods. The port consists of twelve specialised and highly efficient terminals, i.e.: container and ro-ro terminal, car terminal, general cargo terminal, livestock terminal, fruit terminal, timber terminal, terminal for minerals, terminal for cereals and fodder, European energy terminal, alumina terminal, liquid cargoes terminal and passenger terminal.

The port of Koper presents a natural waterway that penetrates deep into the European continent. It is located in Mediterranean, at the northernmost part of Adriatic Sea. The good location of the port provides the cheaper naval route from the Far East via Suez to Europe. The difference is some 2,000 nautical miles—the port of Koper is closer to destinations east of Suez than the ports of Northern Europe are. Port of Koper is connected to all major world ports with regular and reliable shipping lines. Land transport, from Koper by road and by railway to the main industrial centres in Central

Europe, is approximately 500 km shorter than from North European ports. The port of Koper is also a transit port, serving more than 70 percent to non-domestic hinterland markets and less than 30 percent to the Slovenian market. The port therefore has a wider European role and value.

Port of Koper is, together with Trieste, Venice and Rijeka, a member of NAPA—North Adriatic Ports Association. The geographical position of this ports is crucial since they are on the Adriatic Baltic Corridor and Mediterranean Corridor. This are two of the most important European Corridors that connected Europe from North to East and from West to East (Figure 1).

Around 60 percent of ports throughput is transported to and from the port by rail, which means that more than 650 wagons arrive (to) and leave the port on a daily basis. The trains are connecting the port (in maximum 2 days) with all main hinterland markets in Slovenia, Austria, Hungary, Slovakia, Germany, Czech Republic, Italy, Croatia and Poland.

The entire area of port of Koper extends over 1,600 hectares. A big part of this area is dedicated to cars and containers. Investments in technological modernisation in connection with the further development of container services and preparations for the extension of container terminal (Figure 2) on Pier I are now in process (enabling total transshipment of 1.000.000 TEUs).

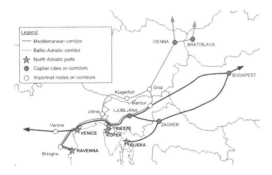

Figure 1. Geographical position of NAPA ports.

Figure 2. Port of Koper.
Source: Port of Koper.

The traffics in the port are growing fast, overpassing the total of 22 million tonnes in 2016 and with a main focus on containers (almost 845.000 TEU) and new vehicles (more than 749.000). This throughput shows that Port of Koper is the biggest container terminal in the Adriatic and one of the biggest car terminals in the Mediterranean. Port of Koper has ambitious ongoing investment plans estimated at an amount up to 300 million EUR. The existing port capacities are well utilized and with new investment they plan to enlarge some vital parts of the port.

3 WHY IS PORT OF KOPER A GREEN PORT?

For the application of "Green Port" concept it is necessary to include the term "green" growth in the further development. Therefore, there is a need of the implementation of policies relevant for reduction of the emissions of harmful substances in the atmosphere and in the sea, the appropriate landscape design with trees that absorb the noise and pollution, use of renewable energy for port operations and activities (solar, wind, energy from the sea), recycling... In Port of Koper are concerned for the wellbeing of the people that live and work in this region. They aim to ensure the preservation of the environment and to operate in symbiosis with the community. Slovenia's National Spatial Plan includes the plan for the expansion and spatial organization of the Port of Koper in accordance with legislative provision and stringent environmental standards.

3.1 Development strategy

Development of the "Green Port" is based on the strategy where all activities of work and planning of the port include the field of environmental protection, energy conservation and environmental care. Port of Koper has a detailed development strategy based on this issue, which is presented at their web site (http://www.zivetispristaniscem.si—Figure 3). In 2014 Port of Koper has been awarded by ESPO in recognition of its work in creating a sustainable future for the port and its surroundings (www.luka-kp.si). ESPO—The European Sea Ports Organization have a regular survey in all EU ports to check if they obtain the environmental priorities. Their development strategy on environmental management is based on measures and solutions that should be introduced to meet legislative demands, with the use of best available technology, to insure as little as possible impacts on the environment.

To protect the environment and natural habitats, they follow some main guidelines that consist of

Figure 3. Plans for green port.
Source: http://www.zivetispristaniscem.si.

the introduction of modern energy-efficient technology, monitoring and result reporting of emissions into the environment, prompt and efficient responses in emergency situations that can happen in the port, and with constant improvement of the environment management system. In this way, they ensure that the development of port operations is in balance with the environmental, social and economic demands.

3.2 "Green" activities in Port of Koper

Port of Koper organized its entire operations in accordance with international quality standards for the protection of the environment, health and safety and food safety management. They constantly improve the ecology of the port zone by introduction of vegetation, and the creation of new habitats in port surroundings. More than 2,000 trees, including 200 olive trees, have been planted in the last ten years.

In the port's institute TOC, they develop some new technological solutions for environmental protection, prepare research in areas involving technology and ecology, with particular emphasis on renewable energy sources, natural resources, waste management solutions, and the preservation of marine and coastal ecosystems.

One of the activity that can increase environmental problem in port is dredging. In the Port of Koper removed every year around 80.000 m3 of sediments. They try to use dredging materials as row material for civil engineering application in the port area. With appropriate treatment this material can be used also as a building composite.

The biggest environmental problem presents a European Energy Terminal, where they handle with coal and iron ore. The area where the port is located is exposed to a strong wind—bora that blew over 100 km/h. To prevent dusting in the area, in the beginning they used water spray system. In the last years they developed a paper mill sludge; the cellulose pellets that are a waste by-product of the pulp and paper industry. The dust-suppression system involves mixing the pelletized paper mill sludge with water and then spraying the resultant solution on the stockpiles of coal and ore. When the mixture dries, short cellulose fibres cover coal and ore with a crust that prevents dusting (Figure 4).

In the Waste Management Centre, they sort and collect majority of waste generated at the Port of Koper for further processing, using appropriate recycling methods. All bio-degradable waste has a prior to conversion into compost (Figure 5). They provide this processing services also for other enterprises and concerns in the region.

Other activities that are planned in Port of Koper are:

– use of solar energy, where the roofs of warehouses shall be equipped with photovoltaic cells,
– creation of biofuels from processing waste engendered by port operations,

Figure 4. Iron ore and coal terminal.
Source: www.luka-kp.si.

Figure 5. Centre for waste collection.
Source: http://www.luka-inpo.si.

- use of 'grass-carpets' on the roofs of semi-open storage have a positive ecologic effects on the buildings themselves and on the wider environment,
- establishment of Technology park with an exposition of typical machinery used in operations. This presentation will provide an overview, displaying the history of Koper, the development of the seaport and the new expansion projects, including the new sports and recreational facilities. A viewing platform will offer an impressive panorama of the port and its broader surroundings.

3.3 Example of green investments—container terminal

Container throughput in the Port of Koper has grown very fast in the last years and the existing container terminal is at the limit of the capacity. With such an intense growth of TEUs transhipped in Koper it is necessary to start with the construction on a new container terminal and reconstruction and extension of the current container terminal (Twrdy et al 2013).

The extension of Pier I for 100 metres and new warehousing surfaces for containers are in line with the estimated growth of traffic as well as with the exploitation of present and future terminal capacities. New projects and potential investments are important steps for the development of the Port of Koper, enhancing its performance and increasing its market share. According to this plans Port of Koper purchased two Super Post Panamax cranes, 22 electrical RTG cranes and two RMG electrical cranes.

All new investments in transport infrastructure equipment on container terminal are environmentally friendly and present a new step for the development of the port.

Figure 6.   Container terminal.
Source: www.luka-kp.si.

4   CONCLUSIONS

Ports are complex environmental systems The Port of Koper intends to increase their cargo operations from the current 22 million tons to 30–40 million tons in five to ten years, almost doubling the cargo capacity, and potential number of vessels calling. Further development of Port of Koper will depend on the rate of investment in increasing capacity. With the sufficient draft (15 m) and a longer berth (with additional length of 100 m) port will be able to accept the ships with capacity of 20.000 TEU. In the long term strategy there is in the plans the construction of the third pier with the capacity of 1 mio TEUs. New projects and potential investments are important steps within the development of the Port of Koper enhancing its performance and increasing the market share.

A sustainable development of Port of Koper is focused on the balance between environmental issue and economic benefits. Enlargement of port area is in accordance with Slovenia's National Spatial Plan, with legislative provision and environmental standards. All new investments in infrastructure could present a positive effect on welfare and ecological protection of air. The green port concept anticipate less carbon emission and new equipment is in compliance with these standards.

REFERENCES

Darbra et al., 2009, Survey on environmental monitoring requirements of European ports, J. Environ. Manag. 90 (3).
Lam, J.S.L., Notteboom, T., 2012. In: The Green Port Toolbox: a Comparison of Port Management Tools Used by Leading Ports in Asis and Europe. International Association of Maritime Economists (IAME) Conference, Taipei, Taiwan, 2012.
Nebot et al., 2017, Challenes for the future ports. What can be learnt from the Spanish Mediterranean ports?, Ocean & Costal Management 137.
Peris-Mora et al., 2005, Development a system of indicators for sustainable port management, Mar. Pollou. Bull. 50.
Twrdy et al., 2013, Limitation and Restrictions on the Admission of Postpanamax Container Ships in the Port of Koper, Shipbuilding: Theory and Practice of Naval Architecture, Marine Engineering and Ocean Engineering, Vol. 64 No. 4.
"Sustainable Ports-A Guide for Port Authorities" (PIANC, 2014) (www.pianc.org)
www.luka-kp.si
http://www.zivetispristaniscem.si

Maritime Transportation and Harvesting of Sea Resources – Guedes Soares & Teixeira (Eds)
© 2018 Taylor & Francis Group, London, ISBN 978-0-8153-7993-5

Challenges for implementation of the green corridor in Brazil

D.A. Moura
Federal University of ABC, Santo André, Brazil

R.C. Botter & J.F. Netto
University of São Paulo, São Paulo, Brazil

ABSTRACT: This paper will report on the challenges of establishing a green sea corridor in Brazil. It will deal with the legal aspects, the public policies related to the Brazilian maritime navigation sector, the integration of logistic modalities, the technical and political obstacles for a continental extension country. The focus of this project is to study European green corridors and to analyze the feasibility of adopting something similar for Brazil, as a focus on multimodality, especially in the use of short sea shipping transport, taking advantage of the extensive coastline that Brazil possesses. The specific objectives are to identify sustainable transportation in the green corridor, clean technologies applicable to multimodal corridors with emphasis on short sea shipping.

1 INTRODUCTION

The integration of transportation modals in Brazil cannot be considered a relevant point for the national logistics system. There are several infrastructure problems that interfere with the flow of transportation of goods, which contributes to high logistics costs and loss of competitiveness in the national and international scenario.

The world soybean market is a clear example. We are the second largest exporter of soybeans in the world. The world harvest in 2015/2016 was 312,362 million tons. The United States had a production of 106,934 million tons and Brazil was the second largest producer, with a harvest of 95,631 million tons.

The state of Mato Grosso is the largest soybean producer in Brazil. It is located in the central region and produces 30,514 million tons. However, Mato Grosso is located far from port areas to flow its production abroad, mainly to the Asian market, which is Brazil's largest soybean consumer.

In addition to transporting soybeans for export, short sea shipping on the Brazilian coast is another example of a major challenge for the green corridor.

This paper intends to approach some legal and current aspects of logistics infrastructure in Brazil for the implementation of a sustainable logistics corridor.

2 RELEVANT ASPECTS OF A SUSTAINABLE LOGISTICS SYSTEM

A sustainable logistics system must meet the needs of society, aiming to reduce or eliminate negative environmental impacts such as emission of polluting gasses, congested roads or noise pollution. It should, on the contrary, optimize the use of resources, reduce operating costs and waste, use energy in an intelligent and renewable way, and create economic value for society (Psaraftis, Panagakos, 2012; Lee, et al., 2016; Jafarzadeh et al., 2017).

In order to meet the needs of society, there must be a participation of public and private institutions with the same objective. Suiting infrastructure to promote the best performance of the logistics system in each transport corridor is an essential condition for the success of the project. Committed companies, research centers, universities, municipal, state and federal public agencies and the community are key factors to initiate a sustainable transport engagement.

Promoting sustainable growth is associated with the use of efficient, preferably green, resources that are economically competitive and do not bring disruption to the environment and society. The focus should be on the use of renewable energy in all transport operations between origin and destination, seeking low carbon economy, innovation in transportation, storage and transshipment operations, regardless of the modalities used to transport the products (Psaraftis, Panagakos, 2012; Jafarzadeh et al., 2017; Eng-Larsson and Kohn, 2012; Wang et al., 2015).

The use of information and communication technology is an ally for the application of sustainable transport systems. It can contribute immensely to monitoring transportation processes to eliminate or minimize congestion bottlenecks and eliminating bureaucratic documentation in operations, to simplify the system, and to support staff in transportation operations on highways, urban transport, waterways, railways, ports, maritime and air transport (Psaraftis and Kontovas, 2010; Chapman et al. 2003; Perego et al., 2011; Wang et al., 2015).

Eliminating or reducing paperwork along the transport chain is the essential role of the area of information technology and communication, which should support the transport of goods in a green corridor. Only then, one can think of optimizing the operations and helping in the decision making that contributes to an integrated logistics system (Presbitero et al., 2017; Perego et al., 2011; Wang et al., 2015).

Along with the use of information and communication technology that eliminates or minimizes paper bureaucracy throughout the freight transport chain, the focus is on the use of the best modalities at each stage of the transportation of merchandise. By optimizing modals using clean energy, the benefits will be great for sustainable transportation (Perego et al., 2011; Eng-Larsson and Kohn, 2012; Wang et al., 2015).

Thus, it is important the commitment of all actors in the sustainable transport corridors to invest in innovative technologies, regardless of the mode of transportation used in logistics operations.

The type of fuel used in transportation operations directly influences the sustainability of operations. Using renewable energy is an important differential for green corridors. Using modal which minimizes or eliminates polluting gasses is also essential (Psaraftis, Panagakos, 2012).

In the waterway modal, it is essential to limit or reduce the emission of gaseous pollutants using a parameter that is defined by the International Maritime Organization (IMO). The same is true on roads to reduce or eliminate CO_2, with the use of electric motors in vehicles, or with the use of renewable energies such as solar, for example (Psaraftis, Panagakos, 2012; Lee, et al., 2016; Sys et al. 2016; Jafarzadeh et al., 2017; Eng-Larsson and Kohn, 2012). It is also applicable for the case of NOx nitrogen oxide gas and for particulate matter (PM). The concept also applies to locomotive engines in the use of the rail modal and the whole transshipment system. Avoiding or eliminating the use of fossil fuel is a relevant condition in the sustainable transport process (Psaraftis, Panagakos, 2012; Sys et al. 2016; Liljestrand, 2016; Zis and Psaraftis, 2017; Pålsson, and Kovács, 2014; Amaya et al., 2016).

Conventional transportation must be rethought and innovated. New alternatives should be used, such as hybrid system, use of renewable energy, liquefied petroleum gas, biomethane, compressed natural gas etc. For urban areas, electric motors or compressed natural gas could be used in transport vehicles. The compressed natural gas in vehicles could be used on highways. However, it would be essential to have several points of supply for the trucks along the way. The power grid in port operations could be used in river and maritime transport, for example, the use of wind to feed ports in sustainable logistics operations. Hydrogen gas for vehicles engines and LNG gas for ships are two other possibilities (Zis and Psaraftis, 2017; Jafarzadeh et al., 2017).

Attention should be given to the emission of organic compounds, oils used in transport operations, tires, batteries, etc.

The ballast water treatment of ships corroborates for sustainable transport. All attention is essential to avoid the transfer of pathogenic aquatic organisms.

Another essential factor for the right development of sustainable logistics system is to avoid or minimize congestion in transportation operations. Monitoring all operations to avoid any kind of congestion will contribute to the sustainable logistics system and for this, information technology and communications systems are important for success in implementing green sustainable corridors. Thus, it is possible to optimize the logistics operations by optimizing the routes and the operations of transport, transshipment, movement and storage. Limiting the speed of vehicles used in transport modes is also an interesting and important measure for the entire logistics system (Liljestrand, 2016; Chapman et al. 2003; Presbitero et al., 2017; Perego et al., 2011; Wang et al., 2015).

3 METHODOLOGY

The methodology consisted of the qualitative type research. It was carried through by means of personal interviews, with entrepreneurs, directors and managers of the transportation industry. The criterion used for selection of the companies in the qualitative research was based on the importance of the company inside its segment. However, other data had been collected personally in the other actors of the national transportation industry.

The methodological procedures adopted were based on the opinion of experts. This type of research design can be used to answer questions about relationships, including those of cause. Thus, the questioning of the participants happened through questionnaires.

Regarding the questionnaire, the survey method involves structured questions that the respondents answered and which was carried out to describe the current stage of the national transportation industry.

It was accomplished through personal interviews, in loco, with entrepreneurs, presidents, directors and managers in the Brazilian maritime industry (short sea shipping), road transportation and railway transportation industry. It was included 30 companies of the three segments of transportation. The total of the companies working in transportation systems, almost 90% of the total has been interviewed.

4 CHALLENGES FOR IMPLEMENTATION OF GREEN CORRIDORS FOR BRAZIL

The main types of cargo in the Short Sea Shipping are the liquid bulk of Transpetro, subsidiary of Petrobras, which transports oils through Brazil. Second are solid bulks, for example, bauxite and, subsequently, the general cargo. The general cargo transport has been increasing in percentage in the Short Sea Shipping. Short Sea Shipping with containers has grown by an average of 8% in recent years.

The maritime short-sea transportation can be analyzed according to the type of cargo. They can be: dry bulk, wet bulk, general cargo, containerized cargo.

The transport of dry bulk is basically: wheat, salt, alumina, corn, manganese, iron ore, limestone and bauxite.

The transport of wet bulk and gas involves light and dark oil derivatives, soya oil, solvent and aromatics, alcohol, acetate monomer, acids, gasoline, styrene and Liquefied petroleum gas (LPG).

The general cargo transport and containerized Cargo is usually: rice, paper, construction material, electronics, PET resin and dry chemicals.

4.1 Logistic corridor for soybean exportation

In general, the lack of integration between modes is a negative point in the transportation of goods throughout Brazil. Transportation of soybeans for export can be used as an example. The largest production is located in the Center-West region of Brazil, in the state of Mato Grosso (MT). The second largest exporter is the state of Rio Grande do Sul (RS), located in the southern region of the country and the third largest producer is the state of Paraná (PR), also located in that region, according to Figure 1.

The state of Mato Grosso is the largest exporter of soybeans in Brazil, responsible for 29.18% of all Brazilian production, approximately 30,514 million tons, followed by the state of Paraná with 18.68%, approximately 19,534 million tons and the state of Rio Grande do Sul with 17.89%, approximately 18,714 million tons. Those three states together correspond to approximately 65.75% of all soybeans exported.

Mato Grosso (MT) – 1st
Rio Grande do Sul (RS) – 2nd
Paraná (PR) – 3rd

Figure 1. Three largest exporters of soybeans in Brazil.

Especially analyzing the state of Mato Grosso do Sul in the Central-West region of Brazil, which is the largest exporter of soybeans, it is distant from the region of ports and the largest port in Brazil, the port of Santos, located in the Southeast region, in the State of São Paulo.

The port of Santos, in São Paulo, handled approximately 29% of the whole national soybean load. From the total amount of soybean exported from the state of Mato Grosso, approximately 51% passes through the port of Santos. Therefore, the port of Santos concentrates the greatest demand for soybeans for export from the state of Mato Grosso.

The extension of the soybean transportation corridor between the state of Mato Grosso (in the Center-West region) and the Port of Santos (Southeast region), is approximately 2,116 km. In that corridor the following modals of transportation are used:

– the road is used in the first part of the route, which corresponds approximately to 819 km. In the second part, the railway is used, being 1,296.6 km away from the Port of Santos, on the coast of São Paulo, to be exported to Asia, mainly to China. The average speed of the rail mode is around 27.3 to 24.1 km/h, but when the train arrives in the urban area, the speed is around 12 km/h. As for the road modal the average speed is 30 km/h, because of roads of very precarious conditions for the transport.

When the trucks arrive to supply the silos, which store the soybeans to be loaded by the rail mode, they find huge queues to release the goods. This is a hindrance to a green corridor in soy transport. The highways between the producing areas and the storage silos are badly conserved, in general, causing a reduction of the speed of the truck and, consequently, increasing the travel time, logistical costs, emission of polluting gasses, etc. There is not, either, an information and communication technology system to manage transportation operations, to provide logistical support to workers and drivers, and to manage the operations of queuing and unloading of soy in silos.

The fleet used in road transport is more than 10 years old. It does not use any technological system or innovation to reduce polluting gasses. In addition, there has not a type of fuel that collaborates to reduce the emission of these polluting gasses emitted by the trucks yet.

There are no appropriate locations to support truck drivers, with an adequate space and infrastructure for rest, to contribute to their good physical and mental health.

As for the rail modal, the infrastructure is not appropriate. There is a lack of public and private sector investment to improve those conditions.

Investments in Logistics in Brazil are still very incipient, as the public sector does not have the financial resources to invest in infrastructure and also finds barriers to obtaining financial resources from the private sector, given the legal problems and rules that make it difficult to attract private financial resources. There are constant changes of regulations and political turbulence, besides the great corruption that Brazil has experienced along many years.

Brazil falls short of the expected when the performance indicators (KPIs) are applied, according to the literature, referring to the green corridors implemented in Europe.

The indicator of logistics costs related to freight is high in Brazil, due to infrastructure problems, labor agreements, transshipment costs, storage, etc. This indicator weighs negatively when one assesses the feasibility of a green corridor for Brazil.

Both road and rail modals have a poor quality indicator in services. Transportation time is very high for soybeans from the source to the port of Santos. Another indicator of evaluation is reliability in time. Brazil is precarious on this item too, which hinders logistical operations and brings obstacles to the system.

The use and application of information technology in the transport of soybeans between Mato Grosso and the port of Santos is practically non-existent and requires a lot of investment, mainly from the private sector to improve logistics operations. The frequency of the service is also not a positive item in the system. Cargo safety is precarious, especially in the road modal, due to the conditions of the roads and the vehicles used, besides the ability of the drivers to carry out operations.

4.2 North-South sort sea shipping logistic corridor

Part of the soybean produced in the state of Mato Grosso is transported by waterway and exported through the North of the Country. However, the lack of infrastructure in the terminal, located in Porto Velho, in the state of Rondônia, hinders the implementation of a logistically sustainable green corridor. After Porto Velho, the soybean is transported by waterway along the Madeira River to the port of Itacoatiara in Manaus (Amazonas) or to the port of Santarém (Pará), both in the Northern Region of Brazil. However, one can find problems with the Madeira River depth, which during drought season is around 2 meters deep, as well as finding a lack of adequate signaling. Both problems are surely obstacles to more intense and effective use of this route as an alternative for sustainable logistic transport.

In addition to the transport of soybeans, it is possible to mention short sea shipping along the Brazilian coast, which could be a future example of a green corridor to interconnect the North, Northeast, Southeast and South regions. There is a huge flow of goods between those regions of Brazil. However, road transport predominates having, approximately, 60% of the national transport matrix.

The equivalent of a distance of more than 5,000 km between origin and destination is transported in Brazil via highway. The highways are often precarious and lack public sector investment. The regulation to have the private sector participating in investments in infrastructure is still incipient and bureaucratic. The concession of public services to the private initiative is low, slow and bureaucratic too.

The lack of rules in the medium and long term hinders the acceptance of national and international investors.

Improving logistics between the Southern and Northern regions of Brazil with Short Sea Shipping might be a way to start a green logistics corridor. There is potential for implementation. However, there is a need for government support at federal, state and municipal levels. According to Figure 2, Brazil has enormous potential to explore the use of maritime transport of goods, allowing the main Brazilian ports to be interconnected and intermodal, linking rail, road and waterway modalities to supply the entire national market, since about 80% of the population live on average 200 km from the country's coast.

Once the use of maritime transport predominates over the road, the emission of greenhouse

Figure 2.   Short sea shipping in Brazil.

gasses (GHG) may reduce, as well as the emission of carbon dioxide (CO_2), sulfur oxide (SO_x), nitrogen oxide (NO_2) And Atmospheric Particulate Matter (Liljestrand, 2016; Zis and Psaraftis, 2017; Jafarzadeh et al., 2017; Pålsson, and Kovács, 2014).

Public policies focused on the development of Short Sea Shipping are essential. Today this type of transportation does not significantly attract shipping companies to operate in Brazil. It is also necessary to reduce bureaucracy and documentation required to carry out this type of transport and promote intermodality, using only electronic and integrated documents.

Therefore, if the international rules that limit the Nitrogen Oxide (NO_x) emissions of the new maritime diesel engines are applied, the contribution of that factor to sustainable transport would be relevant and would corroborate a green corridor in Brazil. Limiting the values for the emission of harmful gasses from the new engines of locomotives and waterway vessels would be an important step for Brazil to enter the era of sustainable transport (Liljestrand, 2016; Zis and Psaraftis, 2017; Jafarzadeh et al., 2017).

5   CONCLUSIONS

Brazil demands structure to implement a green corridor system, with a focus on short sea shipping. That demand is explained by the size of the country itself and by having the largest part of the population living around 200 km close to the coast. Integrating modals and using them optimally is the most feasible and economical solution to reduce the emission of pollutant gasses from the means of transportation.

There is a need for public policies with partnerships between companies and public institutions and the private sector to increase investments in this transportation segment, building ships with technology and innovation in the use of more efficient fuels, in order to eliminate or reduce polluting gasses (GHG—Greenhouse Gas). Using engines with new technologies, investing in information and communication technology, and empowering people with new technologies of innovation in the sector, can insert Brazil among nations that use sustainable transport systems and contribute to a cleaner planet.

Several structural problems hinder the rapid development of the use of short sea shipping in Brazil. In general, companies that transport their products have many complaints about operational issues. They cite, for example, the high shipping time in the maritime modal (Short Sea Shipping) and the low supply of ships for the routes interconnecting the Southern and Northern regions of Brazil; as well as little reliability in deadlines.

Companies intend to increase the use of Short Sea Shipping with personal hygiene products, cleaning products, cosmetics and pharmaceuticals.

In recent years, the ports of Suape (Pernambuco—Northeast of Brazil) and Vila do Conde (Pará—in the North) presented the highest percentage increase in tonnage handled by short sea shipping. On the other hand, Manaus (Amazonas—North Region), Santos (São Paulo—Southeastern Region) and Paranaguá (Paraná—South Region) are identified as the ports with the highest potential for sending cargo. Manaus, Santos, and Suape are those with the highest potential for receiving short sea shipping cargo. The route Manaus (North)—Santos (Southeast) has the highest growth potential for Brazilian short sea shipping in the near future.

Some favorable factors of the short sea shipping are lower unit cost, low loss ratio, reduction of damages, accidents reduction, lower fuel consumption and, consequently, the reduction of CO_2 emissions.

Short sea shipping companies expect greater reliability, consistency, security of supply and frequency of ships. Shipowners, however, expect an increasing offer of competitive costs and greater flexibility for urgent shipments.

According to the shipowners, the main obstacles to the Short Sea Shipping are the high cost of bunker and bureaucracy. They call for a tax treatment of fuels similar to that given to long-haul ships. As far as bureaucracy is concerned, it is known that the cargo of short sea shipping is managed in much the same way as a foreign trade commodity.

The merchant navy does not encourage investment in productive capacity, the supply of new ships, or improvement of port infrastructure. The provision of exclusive terminals so that the short sea shipping would not compete with that of the international vessels within the port terminals would, for example, be a very reasonable improvement.

ACKNOWLEDGEMENTS

Research supported by FINEP and CNPQ.

REFERENCES

Amaya, A.F.C., Torres, A.G.D., Acosta, D.A. 2016. Control of emissions in an internal combustion engine: first approach for sustainable design. *International Journal on Interactive Design and Manufacturing*: 10, 275–289.

Chapman, R. l., Soosay, C., Kandampully, J. 2003. Innovation in logistic services and the new business model — A conceptual framework. *International Journal of Physical Distribution & Logistics Management*: Vol. 33, No. 7.

Eng-Larsson, F., Kohn, C. 2012. Modal shift for greener logistics – the shipper's perspective. *International Journal of Physical Distribution & Logistics Management*: Vol. 42, No. 1, 36–59.

Jafarzadeh, S., Paltrinieri, N., Utne, I.B., Ellingsen, H. 2017. LNG-fuelled fishing vessels: A systems engineering approach. *Transportation Research Part D* 50: 202–222.

Lee, T.-C, Lam, J.S.L., Lee, P.T.-W. 2016. Asian economic integration and maritime CO2 emissions. *Transportation Research Part D* 43: 226–237.

Liljestrand, K. 2016. Improvement actions for reducing transport's impact on climate: A shipper's perspective. *Transportation Research Part D* 48: 393–407.

Perego, A., Perotti, S., Mangiaracina, R. 2011. ICT for logistics and freight transportation: a literature review and research agenda. *International Journal of Physical Distribution & Logistics Management*: Vol. 41, No. 5, 457–483.

Presbitero, A., Roxas, B, Chadee, D. 2017. Sustaining innovation of information technology service providers - Focus on the role of organisational collectivism. *International Journal of Physical Distribution & Logistics Management*: Vol. 47, No. 2/3, 56–174.

Psaraftis, H.N.; Kontovas, C.A. 2010. Balancing the economic and environmental performance of maritime transportation. *Transportation Research Part D*: *Transport and Environment*, V. 15, 458–462.

Psaraftis, H.N.; Panagakos, G. 2012. Green Corridors in European surface freight logistics and the SuperGreen project. Procedia – *Social and Behavioral Sciences*, v. 48, 1723–1732.

Pålsson, H., Kovács, G. 2014. Reducing transportation emissions—A reaction to stakeholder pressure or a strategy to increase competitive advantage. *International Journal of Physical Distribution & Logistics Management*: Vol. 44, No. 4, 283–304.

Sys, C., Vanelslander T., Adriaenssens, M., Rillaer, I.V. 2016. International emission regulation in sea transport: Economic feasibility and impact. *Transportation Research Part D* 45: 139–151.

Wang, Y., Rodrigues, V.S., Evans, L. 2015. The use of ICT in road freight transport for CO2 reduction – an exploratory study of UK's grocery retail industry. *The International Journal of Logistics Management*: Vol. 26, No. 1, 2–29.

Zis, T., Psaraftis, H.N. 2017. The implications of the new sulphur limits on the European Ro-Ro sector. *Transportation Research Part D* 52: 185–201.

Maritime Transportation and Harvesting of Sea Resources – Guedes Soares & Teixeira (Eds)
© 2018 Taylor & Francis Group, London, ISBN 978-0-8153-7993-5

Methodology for the identification of the potential hinterland of container terminals

T.A. Santos & C. Guedes Soares

Centre for Marine Technology and Ocean Engineering (CENTEC), Instituto Superior Técnico, Universidade de Lisboa, Lisbon, Portugal

ABSTRACT: A methodology is presented for determining the hinterland of container terminals, which takes in consideration the geography of the road and rail networks, the presence and location of intermodal terminals and seaport terminals. Unimodal and intermodal freight transportation solutions are considered in the calculation of the total transportation and handling costs across the geographic region surrounding the terminals. Costs comprise road haulage from load centers to ports, rail transportation in intermodal cases and handling of containers in intermodal and seaport terminals. An application of this methodology to the multi-port system in Southern Portugal is carried out, allowing the identification of hinterlands and competition margins for the different container terminals.

1 INTRODUCTION

The containerized cargo segment of the shipping industry has been growing consistently on a global basis for many years now. Although it is questionable if this trend is to continue in the future at the same pace, it certainly has made a lot of pressure on the land side part of logistic chains, including port container terminals, road and rail networks and intermodal terminals. This has also been felt in Portugal and the Portuguese government has accordingly promoted the development of the existing terminals and more recently has been encouraging the construction of new greenfield terminals.

Consequently, container terminals operate within a highly competitive industry and, in addition, are often situated in multi-port systems where significant competition between terminals exists. Many different factors have influence on the competitiveness of container terminals in a given geographic region, most notably the cost factor, which largely determines the extent of the hinterland of each terminal.

Portugal possesses a multi-port gateway system, known as the Portuguese range, which consists mainly of Leixões (serving the north of the country) and Lisboa, Setúbal and Sines (serving the south). As the country is by itself a limited hinterland, ports have been long promoting the penetration in the Spanish market, especially in the regions next to the Portuguese border which are landlocked and for which Spanish ports are difficult to access. Santos & Guedes Soares (2017) have studied ongoing developments in this area and point out that although ports have not used such terminology, what they are promoting is a regionalization strategy, as shown in Figure 1,

Figure 1. Discontinuous hinterland captured through a successful port regionalization.

along the general theoretical lines put forward by Notteboom & Rodrigue (2005).

Such strategies require the improvement, expansion and development of container terminals in ports in order to handle the enlarged container volumes expected. In this context, a long and difficult discussion has been going on concerning the best location for new terminals. The Portuguese economy ministry has presented in 2014, (ME 2014), a plan to improve transportation infrastructures, which includes the addition of new terminals and improvements in the existing ones.

However, in the south sector (Lisboa, Setúbal and Sines), as these ports are closely spaced, an intense competition exists for cargos. This means that each port is pursuing a strategy of development of its own capacity in this market segment and all three agree that regionalization towards Spain is necessary. The government is in turn slowly making progress in the improvement of the railway connection to Spain, which will eventually bring additional cargo to these ports.

The best options regarding development of container terminals involve a detailed knowledge of the cost and time advantages of each existing terminal and proposed location taking in consideration the geographical distribution of containerized cargo across the entire region. That is, the actual throughput of each terminal will result from the logistic costs implied by each one and the said geographic dispersion of containers. It may be said that the transportation costs largely determine what is called here the potential hinterland. However, as is well known, in addition to cost and time issues (or generalized cost), there are other parameters that influence decision making such as range of destinations offered by the terminal, quality of service and reliability of service. The throughput of each terminal will be a consequence of transportation costs in first place also of and these other factors.

Upon review of the existing literature, it was found that the cost and time relative advantages of different container terminals in ports in the south sector of the Portuguese range are not known. Similarly, no such assessment has been done for the new proposed locations. The present paper presents a numerical model of transportation costs coupled to a model of transportation networks, specially developed to support the cost calculations. These models attempt to cover the mentioned gap in the literature and in the technical knowledge required to support decisions on such a difficult matter.

The paper is organized in the following manner. Section 2 presents a literature review on models and techniques used to evaluate costs and transit time in logistic chains across a geographical region, including different modes of transportation and intermodal transportation solutions. Section 3 presents the numerical model adopted in this study to evaluate transportation costs. Section 4 presents the transportation networks model developed for this study which comprises the road and rail network along the Portuguese coast serving the three mentioned ports. The results of the numerical calculations carried out over these transportation networks are shown in section 5, allowing the definition of the potential hinterland of different container terminals. Section 6 puts forward the main conclusions of this study.

2 LITERATURE REVIEW

The main aim of this paper is to present a method to determine the natural hinterland of container terminals of ports which are integrated in a multi-port system. The literature on terminal hinterlands is rather extensive, but the work of Notteboom (2008) is of special interest.

According with Rodrigue *et al.* (2017), following the classical definition of Slack (1993), this natural hinterland is the summation of a main hinterland and a competition margin. The main hinterland represents the geographic region across which the terminal has a dominant or even exclusive share of the market. The competition margin comprises the geographic region across which the terminal is engaged in competition with other terminals. In this case, factors such as accessibility, costs, quality and reliability of services will determine the split of cargo between different terminals in competition. In small or medium sizes regions which are split in various sub-regions by geographic features such as rivers or mountains, terminal natural hinterland may actually be clearly distinguishable and it is an objective of this paper to present a method which is capable of revealing this underlying structure of space.

The natural hinterland may be determined by calculating the cost and time necessary to reach individual geographic locations, allowing the determination of which port is in advantage for that specific location. In certain locations, several ports will have relatively similar costs and times of access and competition will be fierce. Another approach is to analyze the cargo throughput statistics of ports and in particular their geographical distribution across the hinterland and determine for which geographical locations a particular port is in advantage. This is the approach taken in Santos *et al.* (2015). It is then possible to compare the potential natural hinterland of a port with its de facto natural hinterland, evaluate the differences and analyze possible reasons for the discrepancies.

An example of transportation networks applied to the identification of port hinterlands is the study of Zondag *et al.* (2010). These authors present a mathematical model that may be used for example by port administrations for the purpose of evaluating the extension of port hinterlands. An application is made to the assessment of the hinterland characteristics of different ports in the Northern European range (Antwerp to Bremen). The model comprises an assessment of transportation costs and logistic costs in hinterland distribution and in maritime transportation, coupled to a forecasting model for freight flows and a multinomial logit model for freight flow allocation.

It is also worth mentioning Lim & Thill (2008) and Thill & Lim (2010), who study the accessibility from multiple locations in the US to different port ranges (organized by coast). This work builds upon considerable literature on accessibility measurements in transportation networks based on such seminal works as Morris *et al.* (1979). Thill

& Lim (2010) have systematically calculated the accessibility of ZIP code locations across the US to different port ranges (in the various US coasts) based on transportation costs and applying a gravimetric model. The road and rail transportation networks are modelled in detail using nodes and links, including also interconnections between sub-networks at intermodal terminals, allowing for various possible combinations of means of transportation for the same route.

Concerning studies of the hinterland of Portuguese ports, there are very few comprehensive studies in the open literature. APL (2002a) presents some limited data on the geographical origin of cargos (all types) flowing through the port of Lisboa, split between imports and exports and NUTS 3 region (Portugal). Another example of hinterland studies is provided in Macário (2009), which contains a study dedicated to the Portuguese scenario in which the hinterlands of ports and, in special, logistic platforms situated across the country, are identified, based on its accessibility using the road network. APL (2002) had also previously provided a rough definition of the hinterland of the port of Lisboa. More recently, Baptista (2012) presents a study of the hinterland of the port of Leixões using the data contained in the Portuguese Customs database for the cargos which go through the port of Leixões. This author also combines this data with social and economic indicators to improve the overall results. Santos et al. (2015) have also used the Portuguese Customs database.

Another approach is that of Reis & Macário (2008) and Reis (2014), which present an agent based approach which allows the estimation of the performance in terms of cost and transit time of the Portuguese transportation system (road and rail networks) while distributing containers coming from Portuguese ports (Sines) across the hinterland. Various scenarios of growth of demand are considered and the performance of the transportation system is evaluated in these scenarios.

More recently, ACP (2016) published a study on the development of new deep water container terminals in Portugal which contains some definitions of the port hinterlands in Portugal, based again on road access time but also on Voronoi diagrams which consider the road haulage quoted price. These analyses were carried out for the whole of the Iberian Peninsula, with a crude spatial definition but considering all major Iberian ports.

This literature review allows the general conclusion that port hinterlands in Portugal are insufficiently studied, especially concerning the identification of the potential hinterland that result from cost advantages of specific ports in each geographical region.

3 METHODOLOGY FOR POTENTIAL HINTERLAND IDENTIFICATION

The identification of the hinterland of different container terminals is carried out by calculating the minimum transportation costs from each individual origin of cargo to each container terminal. For that purpose, a set of possible routes is defined according with the characteristics and lay-out of the transportation networks and the one with the minimum cost is determined. The routes include unimodal road based routes from the cargo origin to the seaport container terminal or intermodal routes. Several variants of unimodal routes may exist taking advantage of the comprehensive road network. The intermodal routes comprise an initial road based pre-haulage connecting the cargo origin with some intermodal terminal and then the transportation of cargo using a high capacity mode of transportation such as rail or inland waterway until the seaport container terminal. In the case of intermodal routes, additional costs related with the handling of cargo units arise in the intermodal terminals.

The set of possible routes between a certain origin and a seaport terminal is defined as R and r as the route under evaluation. The network of transportation in the region under analysis comprises N nodes and L links, with n and l denoting individual nodes and links. Each link is characterized by a type (road, rail, inland waterway), a length d_l (distance) and a specific cost of utilization c_l. Each node is characterized by specific costs of utilization, which might be zero or might assume non-zero values of c_{nu} and c_{nl}. These values represent costs of unloading and loading units of cargo at those nodes which represent intermodal terminals. Finally, also the costs associated with handling of the containers in the seaport terminal are considered, these being c_{su} and c_{sl}.

Considering the above, the total transportation cost across a given route r may be calculated as follows:

$$C_r = \sum_{l=1}^{L}(\delta_{rl} \cdot C_l \cdot d_l) + \sum_{n=1}^{N}[\delta_{rn} \cdot (C_{nu} + C_{nl})] + C_{su} + C_{sl}$$

(1)

where:

Δ_{rl} is a binary variable to consider whether the link is used in route r or not,

Δ_{rn} is a binary variable to consider whether the node is used in route r or not.

It is important to have in mind that the specific cost of utilization is the same for all types of road links, but its value varies with the total distance covered by road, that is, it depends on the sum of the lengths of all road links used in each route r. This specific cost decreases as larger distances are covered, as will be seen in Figure 6. The same occurs for rail links, as will be seen in Figure 7.

Having calculated the total transportation costs associated with each route r between an origin and a seaport terminal, it is possible to find the route with minimum costs:

$$C_{rmin} = \min_{r \in R}(C_r) \qquad (2)$$

It is also possible to determine, for each origin, the seaport (out of the set of seaports available, S) which has the minimum transportation costs:

$$CrminS = \min_{s \in S}(C_{rmin}) \qquad (3)$$

This allows the assignment of origins (which in this study will be municipalities) to the natural hinterland of seaports. It is also possible to determine which seaports will be the second, third or fourth option for each origin.

The cost C_r thus includes all transportation and handling costs up to the moment the container is loaded in the ship. It is acknowledged that logistic costs such as stock-out costs and inventory carrying costs (storage costs, obsolescence costs, depreciation costs, insurance costs, and opportunity costs of capital) are not included in this model. It is also assumed that transitions between modes of transportation are smooth, implying that cargo does not have to wait significant amounts of time and, therefore, no intermodal or seaport storage costs exist. This is further supported by the fact that such terminals generally do not charge storage for some days (demurrage time, free time).

This calculation method has been implemented in a dedicated Fortran code. This code reads the data on the transportation networks (nodes and links definition) from a specially developed database. Cost parameters for the different modes of transportation and handling operations are also included in this same database. At this stage, no restrictions are included in terms of maximum capacities of links (roads and railway lines), container terminals and intermodal terminals. Therefore, in this paper, only the potential hinterlands of seaport container terminals are evaluated.

4 APPLICATION TO CONTAINER TERMINALS IN SOUTHERN PORTUGAL

4.1 The multi-port system in southern Portugal

The main commercial ports that constitute the Portuguese range are divided in two groups: the northern group, with Leixões and Aveiro; the southern group, with Lisboa, Setúbal and Sines. The range comprises three core European network ports: Leixões, Lisboa and Sines. A total of five port authorities exist, managing each one two (in one case only one) geographically close ports. In these ports, only some are handling containerized cargo, namely Leixões, Lisboa, Setúbal and Sines. The port of Leixões possesses a container terminal (TCL), which serves mainly the north and center of Portugal and is engaged primarily in Short Sea Shipping (SSS). As it is separated by more than 300km from Lisboa (and even more distant from the other southern ports), it is considered in this study that its hinterland is entirely separate from the hinterland of the southern ports.

The southern ports mentioned above are spread along the coast throughout a distance of not more than 150km. Lisboa possesses three container terminals (Liscont, Sotagus, Operlis), which, together with Sadoport container terminal in Setúbal, serve the Lisboa metropolitan area and the southern part of the country and are engaged primarily in SSS. PSA Sines (Terminal XXI), in Sines, is engaged mainly in transshipment operations. Terminal XXI is the terminal with the largest handling capacity, followed by the two terminals in Lisboa (Liscont and Sotagus). These two container terminals have been challenged since 2002 and 2004 by, respectively, Sadoport and PSA Sines. Sadoport is the most recent terminal, somewhat under equipped, considering the available area and quay length. Operlis is a small terminal, operating only geared feeder vessels engaged in the Portuguese Atlantic islands trades. This terminal is located also in Lisboa, immediately adjacent to Sotagus terminal, and in this study it will be considered that its hinterland is identical to that of Sotagus.

Figure 2 shows the number of TEUs handled throughout the last 15 years. The most significant ports, according with their TEU throughput, are Sines and Leixões. However, in 2015 around 78% of the containers handled in Sines were transshipped containers.

As mentioned, this paper will restrict its hinterland analysis to the ports in the southern sector of the Portuguese range, since these are the ones

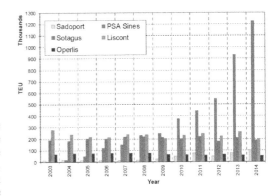

Figure 2. TEUs handled in the Portuguese ports.

located closer in geographic terms. Figure 3 assists in defining the geographic scope of the area which should be considered in the analysis of the port's hinterlands for containerized cargo. This figure shows the export and import cargos through the port of Lisboa, in 2000, split between NUTS 3 regions of origin or destination (in percentage). The regions NUTS 3 located closer to Lisboa along the coast represent 97.6% of exports and 98.1% of imports of this port. This figure includes all types of cargo and it is probable that containerized cargo is somewhat more spread through Portugal, as the study of Santos *et al.* (2015) indicates.

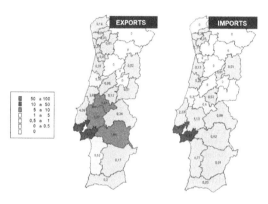

Figure 3. Exports and imports through the port of Lisboa (adapted from APL (2002a)).

Overall, it is assumed that the NUTS 3 regions indicated in Figure 3 with letters A-H are the ones relevant for the port of Lisboa and other ports in southern Portugal (Setúbal and Sines) and the model of transportation networks developed in this study covers these NUTS 3 regions. The region indicated as H is only partially considered in this study by including its westernmost part in the model, basically to provide geographical continuity between regions E, D and G.

4.2 *Transportation networks model*

Figure 4 shows the road and rail networks across the geographical area considered in this study. Regarding the road network, most of the links included in the model represent motorways (shown in the Figure as A10, A8, etc), but some national roads and short urban roads are also included, as necessary. The Figure also shows the railway lines, but only the ones used by freight trains have been modelled, basically the line coming from Entroncamento in the north and giving access to the ports of Lisboa, Setúbal and Sines.

A database has been developed which includes the definition of the nodes and links used to model the transportation networks. Links are characterized by type, length, nodes connected and typical average speeds. Nodes are characterized by costs of use of the nodes, in case the nodes represent

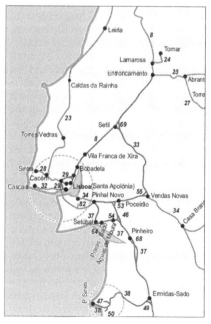

Figure 4. Road (left) and rail (right) networks (adapted respectively from GoogleMaps and Infraestruturas de Portugal).

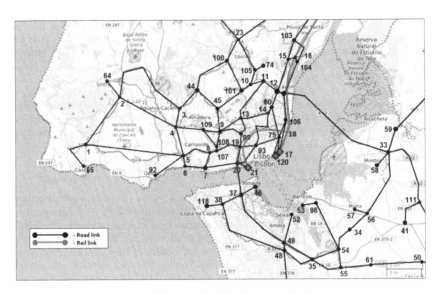

Figure 5. Transportation networks model (fragment showing the core region).

intermodal terminals or container terminals in seaports.

Figure 5 shows part of the model which has been built for representing the rail, road and inland waterway networks in the region of Lisboa. Links between nodes are represented by straight lines but their length has been included in the model in real value taking in consideration the layout of the road or railway line. The road network is modeled in a simplified manner, as necessary to connect the different municipalities (considered as load centers) to the different container terminals in an efficient way.

Speeds in various types of links are set according with the values in Table 1. Rail speed has been considered with the same value as considered in Reis (2014). It is however recognized that different speeds are possible in different links within the rail sub-network, as this might be included in the model in the future. Speed in inland waterways is considered as being 18 km/h, which is approximately 10 knots, considered to be the service speed of barge-tugs currently planned to be used in river Tagus.

Table 1 also includes the speed used over road links, which may be seen to be lower in urban roads than in other road types. The speeds have been defined so that they are 10–20 km/h inferior to maximum legal limits. A special case has been considered for the links representing the bridges over Tagus, which have the characteristics of motorways, but are typically congested. For these specific links, a speed of 60 km/h has been considered. For purpose of comparison, SPC (2012) uses 65 km/h in its simulator as a pre-defined value but this may be changed by the user. Peneda (2009) considered a speed of 80 km/h in motorways and 45 km/h in roads.

Table 1. Speed in different types of links.

Road	60.0 km/h
Urban road	40.0 km/h
Motorway	80.0 km/h
Bridges over Tagus	60.0 km/h
Rail	37.0 km/h
Inland waterway	18.0 km/h

4.3 Transportation costs

Several studies and research projects have analyzed transportation costs in the EU. RECORDIT (2003) reports costs resulting from a survey carried out for three long-distance corridors in the EU, indicating that road haulage (short distance) may cost between 0.62 and 1.89 €/TEU.km. Road haulage for large distances costs 0.29–0.69 €/TEU.km, rail costs 0.23–0.68 €/TEU.km and inland waterways 0.13 €/TEU.km. Short sea shipping (mainly short ferry crossings in the case of this study) costs 0.27–0.84 €/TEU.km. The ranges of costs are wide with smaller values applicable for south Europe. Notteboom (2008) states that the transportation costs by container ship from East Asia to Northern Europe were about 0.12 €/FEU.km (around 0.06 Euro/TEU.km). Road haulage from Northern Europe ports was about 1.5–4 €/FEU.km and inland waterways costed 0.5–1.5 €/FEU.km. SPC (2012) indicates a cost of road haulage of 1.1 €/km for a load of 18 tons. Kreutzberger (2010) indicates costs for rail transportation of 19–23 €/train.km leading to 0.23–0.34 €/TEU.km (for trains of 700m and 585m respectively). The general conclusion is that these costs vary according with a number of parameters: mode of transportation, length of route, size of the ship or train used. On top of this, there will be market fluctuations of prices and inflation.

Concerning the Portuguese case, values have been obtained from CEGE (2014), Reis (2014) and Martins (2013). Figure 6 shows the road transportation costs as function of distance. The costs are reported in CEGE (2014) and Reis (2014) and it may be observed that there is a large increase in costs per TEU.km for short distances, which are those prevailing in the current study. Martins (2013) reports values for road transportation and these conform with the values reported in Figure 6.

Figure 7 shows the rail transportation costs as function of distance, also based in CEGE (2014) and Reis (2014). The costs show significant dispersion, allowing only the conclusion that they are higher at shorter distances (below 100 km).

These figures also show the cost models adopted in this study (black line) which basically reproduces the reduction in cost per km for larger distances and takes asymptotic values of 1.6 €/TEU.km and 0.45 €/TEU.km for road and rail transportation (medium haul). These values are in the medium-high range of costs reported in RECORDIT (2003) for road haulage and rail. The studies mentioned above for the Portuguese do not consider the cost of inland waterway transportation since containers are not carried using barges in Tagus. However,

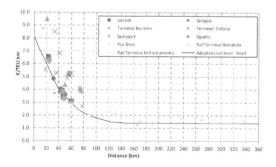

Figure 6.   Road transportation costs.

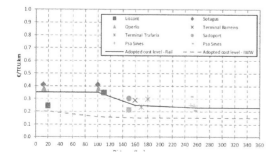

Figure 7.   Rail transportation costs.

Table 2.   Cargo unit handling costs (values taken from various tariff regulations of container terminals).

Type	Terminal	Unload(€)	Load (€)
Seaport	Sotagus	23.0	103.0
Seaport	Liscont	30.0	118.0
Seaport	Operlis	29.0	105.0
Seaport	PSA Sines	27.0	115.5
Seaport	Sadoport	29.5	152.0
Intermodal (Rail)	All	12.5	12.5
Intermodal (IWW)	All	12.5	12.5

in APL (2002b) estimated values are shown within the context of a study of a river transportation service. A value of 0.16 €/TEU.km is estimated for the specific cost of inland waterway transportation of containers, a value not significantly different from the one mentioned in RECORDIT (2003) but significantly lower than the values reported in Notteboom (2008).

4.4   *Handling costs*

The model also includes cargo handling costs which are incurred when the cargo units need to be unloaded/loaded in a different mode of transportation. The costs of the initial loading of the container in the road truck is considered included in the freight charge of the road haulage company. However, in intermodal terminals (road-rail or road-inland waterway) and seaport container terminals (road-ship or rail-ship) a fee is charged for unloading and then for loading the cargo unit. The costs implied by these fees have been taken from the tariff regulations of the different seaport and intermodal terminals, namely PSA (2016), TSA (2016), Liscont (2017), Sotagus (2017) and Infraestruturas de Portugal (2017). It is known that commercial discounts over the table of tariffs are offered to good customers by terminal operators, but these values are not publicly disclosed. Therefore, the values in these references have been used in this study.

Table 2 shows the values adopted for each terminal for unloading and loading of cargo units. For intermodal terminals, values where only available for the rail terminal in Bobadela, so these were taken also for the other intermodal terminals (road-rail and road-inland waterway).

5   RESULTS FOR POTENTIAL HINTERLANDS OF CONTAINER TERMINALS

This section presents the results of numerical calculations of transportation costs carried out using the transportation network model described above. For each municipality covered by the model, numbering a total of 67 municipalities, transportation costs have been calculated for various possible routes to each seaport container terminal, including both intermodal and unimodal routes. The minimum transportation cost between each municipality and each seaport terminal has then been identified. Finally, the seaport terminal with minimum transportation costs is also identified. Municipalities are thus assigned to the main hinterland of one seaport terminal if the minimum transportation cost occurs for that specific terminal and this corresponds to "Area 1" in Figures 10–13. First, the main hinterlands of the seaport terminals in Southern Portugal are shown ("Area 1" for each municipality) in Figure 8, based on minimum distances from each municipality (represented by the geographical location of its main town) to the seaport terminals. The distance is calculated in each case along the road network. Figure 9 presents also the preferred seaport terminal, but based on minimum transportation costs from each municipality.

Comparing these sets of results based on different criteria, it may be seen that when distance only is considered, Liscont terminal shows a more comprehensive main hinterland located in the municipalities to the west of Lisboa and that Sadoport terminal shows a more comprehensive main hinterland extending to the east and north of Setúbal.

It may also be seen that the main hinterland of Sotagus extends north of Lisboa but also to the east, across river Tagus. These hinterlands arise naturally from geographic conditions with some influence of road network characteristics. Therefore, when transportation costs are considered, the hinterlands of Liscont and Sadoport are consider-

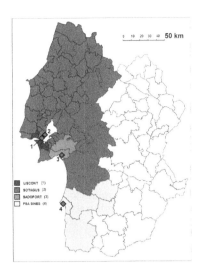

Figure 9. Main hinterland of ports in Southern Portugal based on transportation cost.

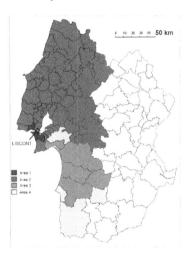

Figure 10. Main hinterland and competition margins of Liscont terminal (port of Lisboa).

ably reduced. The reason for the later observation will be made clear below.

Until this point only the main hinterland of seaport terminals has been considered. However, the model also allows an assessment of the competition margins of terminals. In this study, the competition margins of each terminal are composed of the municipalities for which the transportation costs are larger than for the "preferred" terminal. Therefore, each municipality will have a second, third and fourth preference, according with increasing transportation costs, corresponding to "Area 2", "Area 3" and "Area 4" in Figures 10–13. Figures 10 and 13 show the main hinterland and competition margins of the two terminals in the

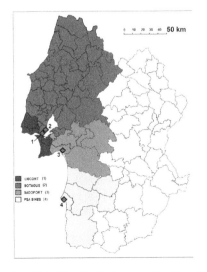

Figure 8. Main hinterland of ports in Southern Portugal based on road distance.

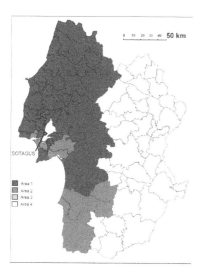

Figure 11. Main hinterland and competition margins of Sotagus terminal (port of Lisboa).

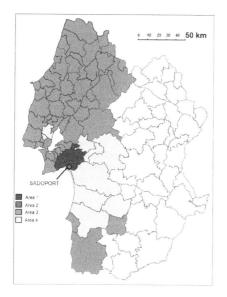

Figure 12. Main hinterland and competition margins of Sadoport terminal (port of Setúbal).

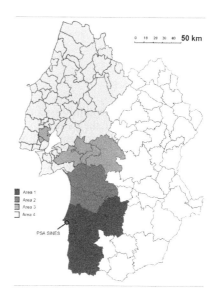

Figure 13. Main hinterland and competition margins of PSA Sines (port of Sines).

in Sines, due to its location in the extreme south of the region considered in this study, ends up having a very clear cut hinterland: some municipalities in the south form its main hinterland and in all other municipalities it is the least preferable terminal, considering transportation costs.

6 CONCLUSIONS

This paper has presented a methodology for identifying the potential hinterland of seaport container terminals, based on calculating the transportation costs between municipalities and the different seaport container terminals, using the existing road and rail networks.

Transportation costs include also handling costs in intermodal terminals and seaport terminals up to the moment when the containers are loaded in the ship. Only export cargo has been considered, but the methodology is also applicable to import cargos. An application to container terminals in southern Portugal has been shown.

The methodology has allowed the identification of the main hinterland of container terminals in seaports. It was confirmed that the model predicts slightly different main hinterlands if distance is used instead of transportation costs as criterion in the definition of hinterlands. The competition margins were also identified for the different container terminals. This allows container terminals to characterize municipalities regarding their own competitive position vis-à-vis their competitors

port of Lisboa. The municipalities in which Sotagus and Liscont are the first or second preference are the same, with the main hinterland of Sotagus being much more comprehensive, since the terminal is in general easier to reach from the inland parts of the country.

Figures 12 and 13 show the main hinterland and competition margins for Sadoport and PSA Sines. It may be seen that Sadoport is the third preference in the main hinterland of the Lisboa terminals (mainly north of river Tagus). The terminal and assess in which ones they are nevertheless competitive and it is worth taking measures to attempt to capture those markets.

ACKOWLEDGEMENTS

This work was performed within the scope of the Strategic Research Plan of the Centre for Marine Technology and Ocean Engineering (CENTEC), which is financed by the Portuguese Foundation for Science and Technology (Fundação para a Ciência e Tecnologia—FCT).

REFERENCES

ACP (2016), Port terminals and logistic infrastructures in Portugal: analysis of the feasibility of developing a new deep water terminal (in Portuguese), Porto, Portugal.

APL (2002a), Development of a strategic plan for the port of Lisbon—Synthesis report (in Portuguese), Lisboa, Portugal.

APL (2002b), Study of the logistics of the port of Lisbon (in Portuguese), Lisboa, Portugal.

Baptista, H. (2012), Definition of a port hinterland: case study for the port of Leixões (in Portuguese), MSc Thesis, University of Porto, Portugal.

CEGE (2014), Container terminals in the region of Lisbon-Setúbal—a Comparative Analysis (in Portuguese), coordinated by A. Felício, ISEG—School of Economics and Management, Lisboa, Portugal.

Infraestruturas de Portugal (2017), Regulation on access and tariffs of the rail freight terminals of Leixões and Bobadela (in Portuguese), Lisboa, Portugal.

Kreutzberger, E. (2010) Lowest Cost Intermodal Rail Freight Transport Bundling Networks: Conceptual Structuring and Identification. European Journal of Transport and Infrastructure Research, Issue 10(2), June, pp. 158–180.

Lim, H., Thill, J. (2008), Intermodal freight transportation and regional accessibility in the United States, Environment and Planning A, Vol 40, Issue 8.

Liscont (2017), Tariff Regulations (in Portuguese), Lisboa, Portugal.

Macário, R. (2009), Paradigms and Development Strategies: "Portugal Logistico", 15th Congress of APDR, 6–11 July.

Martins, S. (2013), Economic evaluation of the project "Rail freight axis Lisbon-Germany" developed by DB Schenker (in Portuguese), Faculdade de Ciências, University of Lisbon, Portugal.

ME, 2014. Strategic Plan for Transports and Infrastructures – 2014-2020 Horizon, Lisboa, Portugal.

Morris, J.M., Dumble, P.L., Wigan, M.R. (1979), Accessibility indicators for transport planning. Transportation Research Part A, Vol. 13, Issue 2, pp. 91–109.

Notteboom, T. (2008), The relationship between seaports and the intermodal hinterland in light of global supply chains: European challenges. Discussion paper 2008-10. OECD-ITF Round Table on Seaport Competition and Hinterland Connections, Paris, 10–110 April.

Notteboom, T., Rodrigue, J. (2005), Port regionalization: towards a new phase in port development, Maritime Policy and Management, Vol. 32, N°3, pp. 297–313.

Peneda, M. (2009), Spatial analysis of the competitive potential of the Logistic Portugal plan (in Portuguese), MSc thesis, Instituto Superior Técnico, Universidade de Lisboa, Portugal.

PSA (2016), Tariff Regulations (in Portuguese), Sines, Portugal.

Reis, V. (2014), Analysis of mode choice variables in short-distance intermodal freight transport using an agent-based model, Transportation Research Part A, Vol.61, pp.100–120.

Reis, V., Macário, R. (2008), Fitness: an agent based modelling approach to freight intermodal chains, Proceedings of European Transport Conference.

Rodrigue, J.-P., Comtois C., Slack, B. (2017), The geography of transport systems, Fourth Edition, New York: Routledge.

Sadoport (2017), Tariff Regulations for Zone 2 (in Portuguese), Setúbal, Portugal.

Santos, T.A., Guedes Soares, C. (2016), Modeling of Transportation Demand in Short Sea Shipping, Maritime Economics and Logistics, accepted for publication.

Santos, T.A., Guedes Soares, C. (2017a), Quantitative assessment of connectivity provided by liner services and ports, submitted for publication.

Santos, T.A., Guedes Soares, C. (2017b), Development dynamics of the Portuguese range as a multi-port gateway system, Journal of Transport Geography, 60, pp. 178–188.

Santos, T.A., Santos, A.M.P., Guedes Soares, C. (2015), Competition dynamics of ports in the Portuguese range, in Towards Green Marine Technology and Transport, Edited by Guedes Soares, C. Dejhalla, R. Pavletic, D. (Eds), Taylor & Francis Group, London, pp. 705–716.

Slack, B. (1993), Pawns in the game: ports in a global transportation system. Growth and Change, Vol. 24, pp. 379–388.

Sotagus (2017), Tariff Regulations (in Portuguese), Lisboa, Portugal.

SPC (2012), The supply chain simulator of the Spanish SSS Promotion Center (in Spanish). ANAVE Bulletin, May, pp. 14–16.

Thill, J., Lim, H. (2010), Intermodal containerized shipping in foreign trade and regional accessibility advantages, Journal of Transport Geography, Vol. 18, pp. 530–547.

TSA (2017), Tariff Regulations (in Portuguese), Lisboa, Portugal.

Zondag, B., Bucci, P., Gützkow, P., de Jong, G. (2010), Port competition modeling including maritime, port, and hinterland characteristics, Maritime Policy & Management, Vol. 37, Issue 3.

Maritime Transportation and Harvesting of Sea Resources – Guedes Soares & Teixeira (Eds)
© 2018 Taylor & Francis Group, London, ISBN 978-0-8153-7993-5

Technological processes inside a ro-ro ferry terminal

V. Stupalo, N. Kavran & M. Bukljaš Skočibušić
Faculty of Transport and Traffic Sciences, University of Zagreb, Zagreb, Croatia

ABSTRACT: The aim of this paper is to contribute to a better understanding of technological processes within the ro-ro ferry terminal regarding passengers and road vehicles. Good understanding of these technological processes is a basis for analysis and improvement of quality of service towards these two types of users. An analysis of technological processes in departure and arrival in ro-ro ferry terminals, described in PIANC guide Port facilities for ferries, was conducted. Based on these findings technological processes in departure and arrival were modelled by using Event-driven Process Chain diagram (EPC diagram), separately for passengers and road vehicles. This research provides a framework for analysis, evaluation and improvement of quality of service within the ro-ro ferry terminal from the perspective of passengers and road vehicles.

1 INTRODUCTION

Guide *Port facilities for ferries* (PIANC 1995) gave narrative description of technological processes in departure and arrival for passengers within ferry port. However, it didn't model the sequence of processes through which port service users such as passengers and road vehicles go within ferry port area. To understand and have sufficient familiarity with technological processes is crucial to obtain complete insight into the maritime passenger port in order to be able to analyze it, evaluate it and, finally, improve it.

In order to identify main technological processes to be considered when evaluating quality of service in ro-ro ferry terminal, analysis of guide *Port facilities for ferries* (PIANC 1995) will be conducted. As the result of this analyses technological processes in departure and arrival in ro-ro ferry terminal, separately for passengers and for road vehicles, will be outlined. Their sequence will be modelled by using ordered graph of events and activities (EPC diagram).

Focus of this paper will be on passengers and road vehicles as users of ro-ro ferry terminal.

This paper is structured in five chapters. In the first chapter, the research topic is explained and the approach described. Relevant services and processes involving passengers and road vehicles within the ro-ro ferry terminal will be analyzed in the second chapter. In the third chapter, technological processes will be modelled. Applicability of these modelled technological processes will be discussed in fourth chapter. Conclusions and suggestions for further research will be part of the last fifth chapter.

2 RO-RO FERRY TERMINAL

The ro-ro transport system was first applied to merchant ships after the Second World War in the late 1940s and early 1950s, and was encouraged by technical developments on land and sea. Today ro-ro ships are popular on many shipping routes. In maritime passenger traffic, they are particularly popular on short-sea ferry routes (IMO 2016).

Configuration and dimension of maritime passenger port significantly varies from port to port. Maritime passenger port can be a simple object along the coast dedicated exclusively for embarkation and disembarkation of passengers, which as such, mostly serves small number of passengers and small number of smaller ships. Likewise, it can be a port that serves a significantly larger number of passengers and larger ships, and, in addition to embarkation and disembarkation services, offers a variety of supplementary services to all transportation participants in the port (passengers, ship owners, operators, concessionaires, etc.). Obviously, the latter port requires significantly more complex organization and configuration as well as greater dimension of the port area than the former (Stupalo 2015, p. 29).

Ro-ro ferry terminals, as opposed to purely passenger ferry ports, in addition to service of embarkation and disembarkation of passengers, provide service of embarkation and disembarkation of vehicle, i.e. ro-ro units. In ro-ro ferry terminals primarily road transport vehicles are handled (e.g., bicycles, motorcycles, cars, caravans, lorries, semitrailers, trailers, buses, road trains), while rail vehicles are less presented. Rail vehicles are not handled in none of the Croatian ro-ro ferry terminals, but are handled in some Italian ports (e.g. port of Villa San Giovanni in region Calabria and port of Messina in region and island of Sicily). These Italian ports offer embarkation and disembarkation of passenger train (Rail.cc 2016) thus allowing the passenger to travel from Rome to Palermo by single train. Similar service is also available in German ports (e.g. port of Puttgarden) and Danish ports (e.g. port of Rødby).

One of the main purposes of ro-ro ferry terminals is to ensure quick and safe embarkation and disembarkation of passengers and vehicles to/from the ship. Identifying and removing bottlenecks, increasing reliability, level of safety, mobility of passengers and vehicles are some of the measures that can contribute to fulfilment of port main purposes and to provision of service at satisfactory level.

In order to evaluate service quality in a given transport facility, it is necessary to identify existing services and processes that involve users from whose perspective the service will be analyzed and evaluated.

The users of ro-ro ferry terminals are passengers, ro-ro units, shipping companies (ship owners) and other port concessionaires (operators, logistics service providers, etc.). Focus of this research will be on passengers and road vehicles.

Port services available within European maritime passenger ports were identified in a study (Pallis & Vaggelas 2006, Vaggelas & Pallis 2010) that was conducted in 20 major European maritime passenger ports, for ferries and cruise ships. The study identified a total of 70 elements of port services that ports provide to its users. All identified elements were not available in every port, since each element was identified in at least one of the total 20 ports analyzed, so the number of this elements varied from port to port.

Wide variety of components needs to be taken into consideration while analysing quality of service within ro-ro ferry terminal. Different port areas have different characteristic and purposes requiring it to be analysed separately. Therefore, in the following chapter, possible division of the maritime port land area is further researched.

3 ANALYSIS OF POSSIBLE SCENARIOS OF TECHNOLOGICAL PROCESSES IN RO-RO FERRY TERMINAL

While researching for adequate subdivision of ro-ro ferry terminal area, no generally accepted area subdivisions were found. From the perspective of the port functional elements, following subdivisions of maritime ferry port area were identified in the literature:

– marshalling yard, passenger facilities and ferry berth(s) with three main elements: fendering system, ramps(s) and scour protection (Agerschou et al. 2004, pp. 291–297),
– landside facilities, dockside facilities and enroute (TCRP Report 165 2013, p. 9–28),
– terminal forecourt or landside, terminal or wetside, and buildings (PIANC 1995, pp. 33–38).

Since focus of this research is to identify technological processes within the land port area, and by the term land port area all port infrastructure and superstructure, from the edge of the coastline until the final land border of the port area, is implied, the last listed subdivision was accepted. Ro-ro ferry terminal area was therefore subdivided into three areas:

1. terminal forecourt (landside),
2. passenger terminal building(s) (PTB) and
3. terminal (wetside).

PIANC guide *Port facilities for ferries* was analysed in order to identify technological processes within this area. Possible scenarios for the organization of technological processes in the ro-ro ferry terminals, described in this literature, were modelled using *Event-driven Process Chain diagram*, hereinafter: EPC diagram.

EPC diagram describes the process initiated by an event, which starts and ends with event(s). It is used to describe business processes and workflows in many industries. A major strength of EPC diagram is its simplicity and easy-to-understand notation. It is also one of the methods used by *Business Process Management* (BPM) *methodology* (For more: Božić et al. 2016, p. 509).

Technological processes in departure and arrival for passengers and road vehicles were modelled separately. In following diagrams (Figs. 1–5), numbers and letters in brackets are related to area illustrated in figures of PIANC Guide (PIANC 1995, pp. 34–35)—figures entitled *Ferry terminal prototype* and *Typical layout of terminal forecourt*. Diagrams were modelled using Microsoft Visio 2013.

In order to capture various possible scenarios for organization of technological processes in the ro-ro ferry terminal within one diagram, e.g. for departing passengers, application of EPC diagram rules was not fully complied, i.e. while splitting and consolidating paths use of possible connectors and triggering combinations was not fully complied with.

There are special rules for logical operator connections between events and activities. It is not possible to have a single event connect to multiple activities with OR or XOR logical operator. This is because an event represents a status of state, not

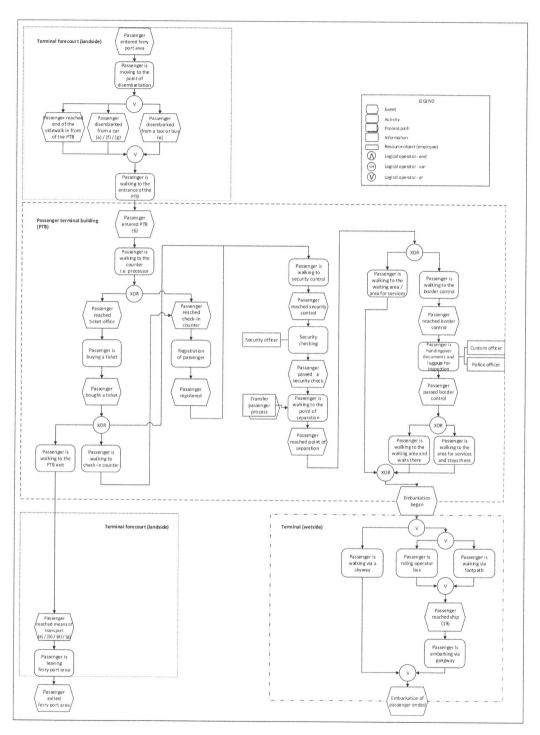

Figure 1. EPC diagram of various technological processes in ro-ro ferry terminal for departing (outbound) passengers.
Note: Numbers and letters in brackets are related to area illustrated in figures of PIANC Guide (PIANC 1995, pp. 34–35).
Source: Modelled by Authors using data from (PIANC 1995).

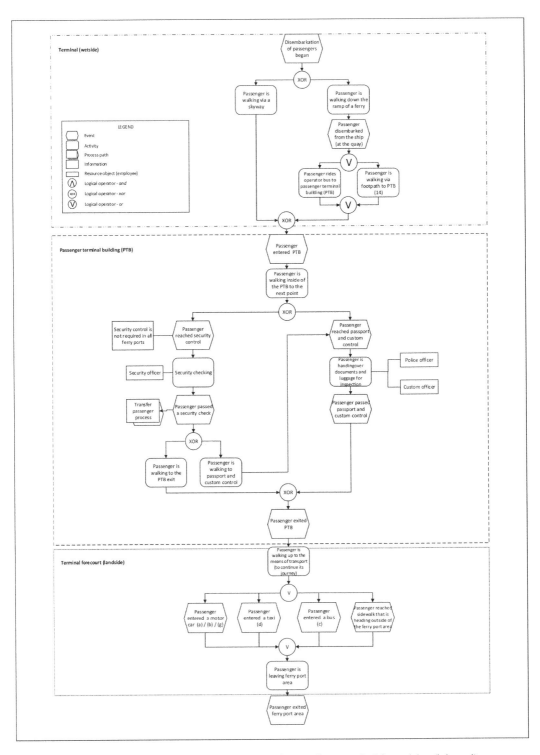

Figure 2. EPC diagram of various technological processes in ro-ro ferry terminal for arriving (inbound) passengers.
Note: Numbers and letters in brackets are related to area illustrated in figures of PIANC Guide (PIANC 1995, pp. 34–35).
Source: Modelled by Authors using data from (PIANC 1995).

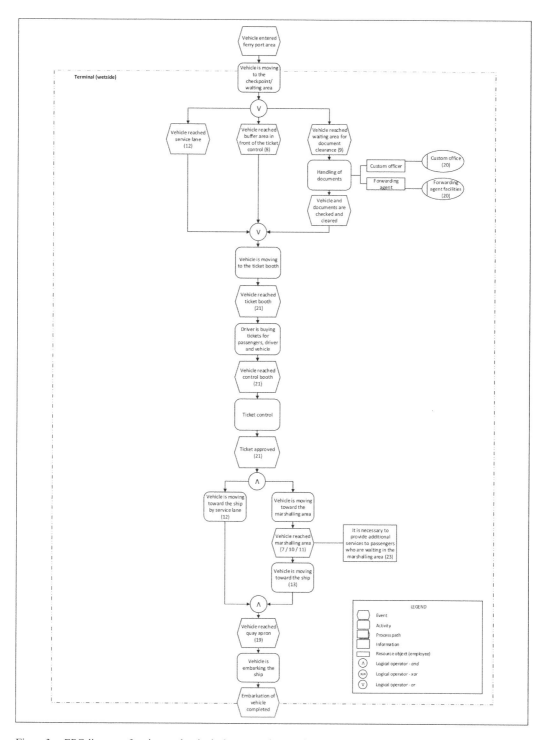

Figure 3. EPC diagram of various technological processes in ro-ro ferry terminal for departing (outbound) road vehicles.
Note: Numbers and letters in brackets are related to area illustrated in figures of PIANC Guide (PIANC 1995, pp. 34–35).
Source: Modelled by Authors using data from (PIANC 1995).

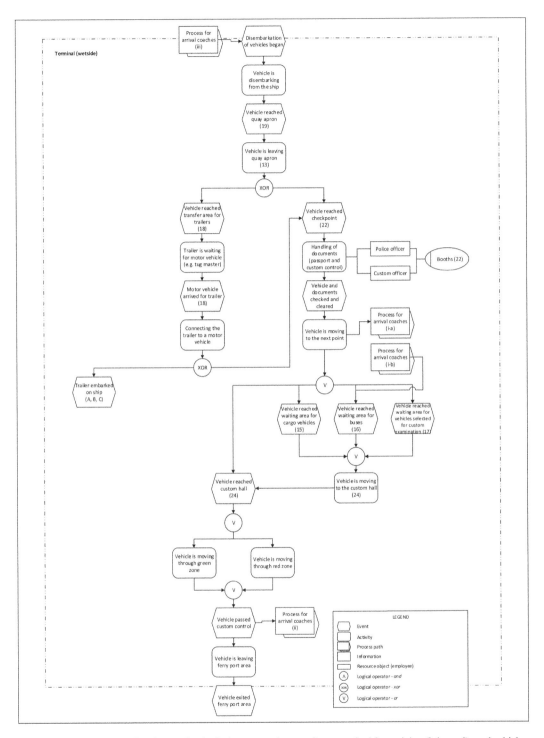

Figure 4. EPC diagram of various technological processes in ro-ro ferry terminal for arriving (inbound) road vehicles.
Note: Numbers and letters in brackets are related to area illustrated in figures of PIANC Guide (PIANC 1995, pp. 34–35).
Source: Modelled by Authors using data from (PIANC 1995).

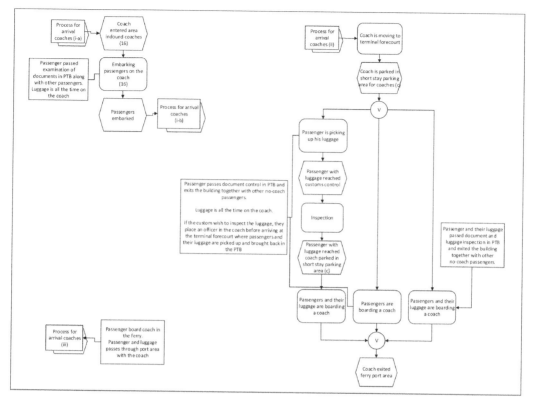

Figure 5. EPC diagram of various technological processes in ro-ro ferry terminal for arriving (inbound) coaches. Note: See note and legend in Diagram 4. Source: Modelled by Authors using data from (PIANC 1995).

a status of decision, and in this case, decision is required in order to decide what path of multiple outcomes should be followed (Monk & Wagner 2013).

4 APPLICABILITY OF MODELLED TECHNOLOGICAL PROCESSES

Technological processes in previous chapter were modelled using EPC diagrams (Figures 1 to 5). Processes shown in the diagrams above are described in the PIANC guide (PIANC 1995, pp. 28–38), numbers and letters in brackets are related to areas illustrated in figures of PIANC guide. Focus of this paper is visualization of these descriptive data using graphical method of EPC diagrams, so they are not further explained in this paper.

Visualisation of these data is used as a tool for identification of services and processes that involves passengers and road vehicles inside a ro-ro ferry terminal. Diagrams are used as tools for identification of a relations inside a ro-ro ferry terminal

area due to its ability to visualise such relationship, helping the researcher to understand how different areas are connected.

These diagrams can also be used as a tool in construction of an event driven simulation model since EPC diagrams illustrate event driven processes. Different simulation software's can be used for construction of this model, for example ARIS simulation module which is based on EPC modelling language, or ARENA simulation software (Kelton et al. 2015) which is based on SIMAN simulation language.

Simulation models can be further used as a tool in evaluation of existing or new technological processes inside an existing or a new ro-ro ferry terminal and as a tool in defining alternative technological processes inside an existing ro-ro ferry terminal, since diagrams above illustrate different possible scenarios of processes inside a ro-ro ferry terminal.

Technological process inside a specific ro-ro ferry terminal is defined by wide range of conditions, e.g. type of traffic in the port (domestic/ international), share of different categories of traffic units (bicycle, car, lorry, trailer, bus, etc.), customs and passport control regulations, security systems, type of check-in, terminal lay-out and function, cargo handling equipment, additional services within the port, etc.

Connections of these conditions and different possible scenarios of processes inside a ro-ro ferry terminal are illustrated in the diagrams (Figures 1 to 5). These conditions influence how specific processes are performed in a specific ro-ro ferry terminal, and depending on traffic volume, influence on the level of service provided to the users in a given object. These conditions affect construction of the simulation model preventing the construction of singular simulation model valid for all ro-ro ferry terminals. Nevertheless, EPC diagrams above (Figures 1 to 5) provides insight into the prototype of ro-ro ferry terminal enabling researcher to better understand technological processes inside a ro-ro ferry terminal in order to be able to analyze technological processes inside a specific ro-ro ferry terminal and to construct simulation model which can be used as a tool to evaluate specific ro-ro ferry terminal.

5 CONCLUSIONS

Ro-ro ferry terminals are important links between land and maritime transport. They act as technical support of maritime transport and, together with shipping routes, constitute its main elements. Both shipping routes and ro-ro ferry terminals should be taken into account while evaluating overall quality of maritime transport service, since both are essential to providing transport service to passengers and thus influence passenger overall service satisfaction. In this paper, technological processes of traffic inside ro-ro ferry terminals, involving passengers and road vehicles, were modelled by EPC diagrams. Familiarity with technological processes is needed in order to describe, measure and evaluate activities and processes in a given transport facility, compare it with other transport facilities, identify problems and propose possible solutions.

Diagrams display different scenarios of technological processes that may be applied in the ro-ro ferry terminal. Reasons for existence of different scenarios within different ports are due to the physical conditions of a port area, rules of the country, local custom routine, type of traffic inside of the port (road vehicle, domestic/international traffic), etc. Outcome of each of these processes affects how passenger perceives quality of transport service inside the port.

Further research resulting in a model of ro-ro ferry terminal for analysis of functionality and connectivity of certain port facilities as well as evaluation of quality of service is recommended. Also as a further research, construction of EPC diagrams of technological processes for passengers on board the ship is recommended, as this would enable better evaluation of overall maritime passenger transport service.

REFERENCES

Agerschou, H., Dand, I., Ernst, T., Ghoos, H., Jensen, O.J., Korsgaard, J., Land, J.M., McKay, T., Oumeraci, H., Petersen, J.B., Runge-Schmidt, L. & Svendsen, H.L. 2004. *Planning and design of ports and marine terminals, 2nd ed.* London, UK: Thomas Telford.

Božić, D., Stanković, R. & Rogić, K. 2016. Possibility of applying business process management methodology in logistic processes optimization. *Promet-Traffic & Transportation* 26(6): 507–516.

IMO 2016. http://www.imo.org/en/OurWork/Safety/Regulation s/Pages/RO-ROFerries.aspx, July 5, 2016.

Kelton, W.D., Sadowski, R.P., Zupick, N.B. 2015. *Simulation with Arena, 6ed.* New York, USA: McGraw-Hill Education

Monk, E.F. & Wagner, B.J. 2013. *Concepts in enterprise resource planning, 4ed.* Boston, USA: Course Technology, Cengage Learning.

Pallis, A.A. & Vaggelas, G.K. 2006. Benefits from port services provision in passenger ports: a constructive categorization. *Paper presented at the 3rd International Conference on Maritime Transport (MT), 16–19 May 2006:* pp. 1–10. Barcelona, Spain.

Permanent International Association of Navigation Congresses (PIANC) 1995. *Port facilities for ferries, Practical Guide.* Brussels, Belgium.

Rail.cc 2016. https://en.rail.cc/rome-messina-palermo/intercity notte-1959–1961/night-trains-city/10684/236 #rail, September 5, 2016.

Stupalo, V. 2015. *Methods for evaluation of capacity and level of service in ferry port.* Zagreb: Faculty of Transport and Traffic Sciences of University of Zagreb. (doctoral thesis).

TCRP Report 165 2013. *Transit capacity and quality of service manual, 3rd ed.* Washington, DC, USA: Transportation Research Board of the National Academies. Available at: http: //www.trb.org/main/blurbs/169437. aspx, April 20, 2014.

Vaggelas, G.K. & Pallis, A.A. 2010. Passenger ports: services provision and their benefits, *Maritime Policy & Management: The flagship journal of international shipping and port research,* 37(1): 73–89.

Maritime transportation

Maritime Transportation and Harvesting of Sea Resources – Guedes Soares & Teixeira (Eds)
© 2018 Taylor & Francis Group, London, ISBN 978-0-8153-7993-5

Integrated transportation system under seasonality effects

J. Milanez Zampirolli, T. Novaes Mathias & R.C. Botter
Department of Naval and Ocean Engineering, Polytechnic School of University of São Paulo, São Paulo, Brazil

ABSTRACT: This study presents an analysis of a soybean transportation system in Brazil using discrete event simulation model to measure the fleet necessary to transport the demand from farms to the export terminal. The system is composed of different transport modals, where grains are transported by truck from the farms to the storages. After that, barges on waterways transport the product to the final terminal, where ships are loaded for export at the end of this process. The current study brings a series of comparisons between scenarios for evaluation of existent bottlenecks in the intermodal transport system, making it possible to evaluate interferences of water level and seasonally of soybean production in the system.

1 INTRODUCTION

The agribusiness of a nation is an extremely important area for the economic growth. On the last decades, Brazilian soybean cultivation had an explosive growth: about 49% of planting grain areas of the country. This increase in productivity is highly associated to the technological progress and the efficient handling of grain by producers and transporters.

However, Brazilian transport system presents high costs if compared to great grain exporters, being almost two times more expensive than USA. The transportation of soy in the Center-West, due to the condition of roads conservation, had significant losses of competitiveness. The logistic cost in the country increased expressively, overcoming in average, in 83% that from the United States and in 94% that from Argentina which are the major competitors of Brazil in the soy sector according to Jank et al. (2004/05).

The problem of the transportation of soybean grains from Center-West region of Brazil to ports, in the east coast, has great importance, since the perspective of growth of production of grains is to increase it due to the potential of productive land available to plant grains and also to the great investments in technology. Therefore, the main idea of this problem is to evaluate the impact of dimensioning fleets combined with the evaluation of port capacity for grain exportation, as well as the necessity of establishing intermediate stocks throughout the chain, considering two seasonal factors in this system. This system is composed by road, waterway and maritime modals, where the simulation model approaches from the farms where grains are produced and then transported

by truck to intermediate storages, where they are loaded into fluvial barges with destination to the port for their final exportation. The current study will present a series of comparisons among scenarios for the evaluation of bottlenecks that could exist in the intermodal transport system, specially evaluating scenarios without seasonal restrictions, where the fleet and necessary stocks are substantially different if the seasonality is not considered. Thus, it was considered two different fluctuations in this system. First the soy bean production, which changes during the year, and the water level in the rivers, that affects the capacity of transportation.

The innovation in this work can be justified since any simulation research with high complexity, and considering the combination of different seasonality types, was not found in the literature.

2 LITERATURE REVIEW

One of the most important steps in a paper is the literature review. This is the stage of the project as it theoretically underlies the problem and how the author will treat it. Through the analysis of the existing literature, the author of the research should make a conceptual structure that will support the development of labor (Silva & Menezes, 2001).

When talking about transport of goods, the first thing to do is to define how they will be transported. In better words, to define which mode of transportation would be better to transport the cargo from a one place to another.

According to Kopytov and Abramov (2013), solving problems with optimal cargo transportation is a complex and complicated task. To choose the best modal transportation we need to look for all cargo characteristics such as:

✓ Transportation routes;
✓ Types of transport means and their combinations;
✓ Types of cargo packaging;
✓ Transportation technologies and others.

In order to choose the less costly way to transport goods, Zhao et. al. (2005) listed the main factors when shippers and carriers are choosing the transportation mode based on: Cargo value; Cargo Volume; Distance; Cost; Time; Contract; Reliability; Obtainable factor; Transportation frequency; and Accessibility.

Although every kind of freight has its own suitable transport mode, it is possible to make a better decision to reduce the transportation costs by following those cargo characteristics.

The systems of decision support, according to Junior et al. (2006), are computational models that aim to systematize and support decision processes of enterprises.

That said, they give tools of modeling and analysis with the objective of enabling users to solve problems in an integral form, giving information acquired from several sources of data from the company. The major characteristics of support systems, according to Buosi (2004), are:

✓ Focus on decisions;
✓ Emphasis at flexibility, adaptability and capacity of giving fast answers;
✓ Initiated and controlled by the user;
✓ Support for personal styles of decision-making.

Lane (2000) uses Churchman's theory of systems approach as a modeling approach by collecting three characteristics. First, is the concept of information feedback loops, second is the computer simulation and third is the need to engage with mental models. Respectively, the first one involves the collection of information about the state of the system followed by actions that changes this state. Then, the capacity of human brain to deduce the consequential dynamic behavior, without a computational tool. Thirdly is more about quantitative and subjective information. Those characteristics are the bases for organizational decision-making.

To Pegden et al. (1990) simulation is a way to represent a real system through a computational model that enables the understanding of all actions that happen in a system in the exact sequence that they happened and the evaluation of alternatives that can bring improvements to the operation.

Harrington et al. (1992) highlights the financial impacts at logistics and marketing by rational decisions using simulation techniques in a scenario characterized by the increase of external competition, economic uncertainty and fluctuations of the financial market.

Chowdhury (1989) proposes the simulation as the ideal tool to study and evaluate great and complex systems where it occurs interactions between activities. So, the author shows that using average values to analyze transport systems can induce wrong conclusions, since there are fluctuations and random variations on real systems, and arguments the importance of testing the influence of several phenomena of random characters.

Botterat et al. (1988) highlights the importance of using the technique of discrete events simulation for the measurement of integrated transport systems to the waterway Tietê-Paraná to transport alcohol and diesel oil.

Cochran and Lin (1989) illustrate the employability of the probabilistic simulation technique to help the development of a master plan of cargo transport in Arizona (USA). Due to the dynamic and complex character of the problem, authors used this simulation. Hameri & Paatela (1995) used the simulation to evaluate a supply chain system that encompasses several characteristics such as the demand of final clients, economic performance, and physical restrictions at material flow, amongst others. Thereby, authors used the implemented simulation model to evaluate alternative operation scenarios of the system.

Crainic & Laporte (1997) showed a com-pilation of mathematical models for the planning in a tactical, operational or strategic level, where the simulation is a key tool for evaluation and measurement of a transport service. So, the importance of simulation is highlighted by the possibility of making "what-if" analysis for questions that appeared during the strategic planning, and also can be the tool that will give subsides for the elaboration of a transport master plan, i.e., it will indicate policies that will guide the daily operation.

An important aspect of maritime navigation logistics, according to Stanford (2008), is how one can optimize the matrix of logistic transport. Thereby, four variables were proposed in maritime logistics model: distance to be covered, type of vessel, size and speed.

The distance is a crucial aspect, since it directly affects the cost and time of travel. The size of the vessel is important since it can obtain economy of scale, and this happens because the bigger the vessel, the more it is possible to be transported, causing the cost by unit to be less. However, restrictions of size can exist in some ports due to the average depth of draught of each port. Speed determines the cost of the travel, i.e., the fuel consumption and costs associated to the movement of the ship. Last, the type of vessel will determine its efficiency and which cargo should be transported on it.

To Stanford (2008) the maritime transportation of cargo is a low cost and high volume business, presenting worries of infrastructure at ports to receive bigger ships, where it is necessary a higher draught depth and, consequently, of the berth where the ship will be more likely to load and/or unload.

3 METHODOLOGY

According to Prodanov and Freitas (2013) the methodology of a report has a proposal to validate the techniques that are used to build the knowledge.

The simulation method implies in modeling processes or systems that it is capable of reproducing answers that could be given by the real system in a series of events during a determined period of time (Schriber, 1974).

To Chwif and Medina (2006) the simulation is an indispensable methodology to the solution of engineering of projects and administration problems. Through a simulation model it is possible to capture the dynamic and random nature of a real model with more precision and reproduce it in a computational model. To the development of a simulation model, authors suggest three steps, as shown in Figure 1.

Thereby, every project starts in the conception step, where the objectives are outlined and when discoveries happen. This step should be discussed exhaustively and all the team involved should have their expectations well aligned. After all the discussion, having the project clarified, objectives are established and the conceptual model is created, it should be validated by specialists of the process so input data can be collected and adjusted for a probability distribution that represents them. Authors emphasize that it is important to use all the time necessary to the construction of the conceptual model, because it will guide the sequence of the project.

With a conceptual model that is validated, documented and with input data ready to feed the computational model, the step of implementation starts. In this part, the conceptual model will be implemented in a computational model by a simulation language. This new model should be compared to the conceptual model with the purpose of checking if its operation attends to what was established at the conception step. After that, the model goes through two important steps: validation and verification, where validation consists in the acceptance or not of the model as an acceptable representation of reality, while verification is to prove that the conceptual model was correctly implemented.

In addition, when analyzing results, it can be experimented proposed scenarios and redefined new ones. In this step it is possible to verify the bottlenecks the system will generate for each simulated scenario. To Netto (2012) it is important to give the necessary treatment to the analysis of available data, because it plays a function of extreme influence at the output results of the model.

Besides the methodology presented, it was developed a discrete simulation model to evaluate the capacity of intermodal system of transportation located at middle-west to the north of Brazil as shown in the Figure 2.

To develop the computational model of this system, it was divided in 3 big sub-systems: 1) Roadway System; 2) Waterway System and 3) Seaway System.

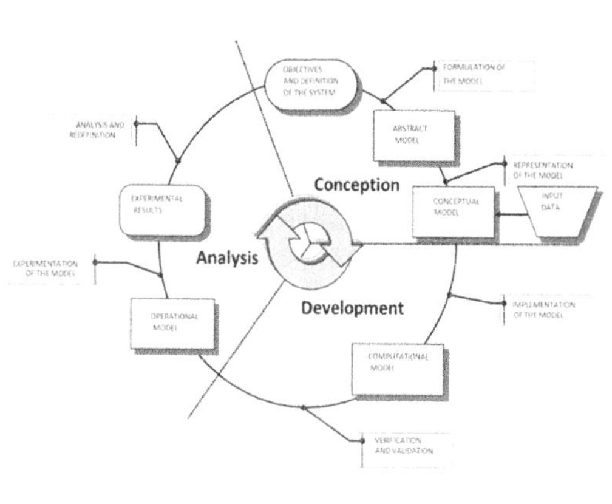

Figure 1. Steps of the simulation project. Source: Chwif and Medina, 2010.

Figure 2. System geo-located. Source: Authors.

Figure 3. Soy bean exportation system. Source: Authors.

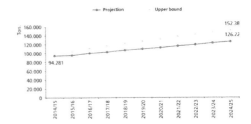

Figure 4. Soybean Growth Projection. Source: Ministério da Agricultura, Pecuária e Abastecimento (2015).

All the soybean is created in the 3 loading points that are defined by a center of gravity and are transported by trucks to the silos. Then, the trucks unload the soybean into the silos to load the barges.

After that, the barges go through waterways until the export terminal. By this point, the export terminal receives the soybean from the barges to load the ships that are going to the final customer.

Also, the trucks can go straight to the export terminal without using the waterways as shown in the Figure 3.

For the transportation system developed in this simulation model, it was necessary inputs as truck, barge and ship fleets, the distances between each places, the velocity of each kind of vehicle and the simulation time. In addition, the demand was needed to make the whole system begin.

Also, to make sure we are representing a real system, every data such as uncertainties on the demand, capacities and speeds of vehicles, equipment production rates, etc. is set according to a probability density function. Every probability function was taken and analyzed from real data systems.

For instance, speeds of truck were taken from their board computer and we considered the probability function that fits better to this case.

As outputs, this model calculates vehicles cycle time, storage levels, average time in queue, length of the queue, number of elements in queue and the quantity of soybean that was exported at the end of the simulation time.

All the data analysis will allow the manager to make better decisions once it is possible to see the whole system behavior.

4 RESULTS

The consumption of soybean around the world is growing not only because the productivity is getting better where grain is planted, but also the area where is planted and because the population is getting bigger, as shown in Figure 4.

To validate the model developed, it was made a comparison between the amount of soybean that was transported by the real system and the one the simulation model proposed.

Table 1. Real × Simulated cycle time.

	Cycle time	
	Calculated (hours)	Simulated (hours)
Madeira River	935,7	945,4
Tapajós River	439,3	485,0
Tocantins River	231,6	349,2

Source: Authors.

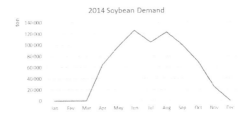

Figure 5. Graphic of 2014 soybean demand. Source: Authors.

Table 1 shows the differences between cycle times that were calculated using the distances, velocity and rates of loading and unloading grain, and the simulated output from this model.

Considering the planned demand during 2014 as shown in Figure 5, it is possible to realize not only the behavior of all intermediate storages used in this system, but also the level of utilization of all equipment that was necessary to handle the cargo.

Although all storages have their own pattern, only Marabá and Miritituba reach the maximum level, while Porto Velho and the export port of Vila do Conde keep their capacity low (Figure 5). This happens because there is only one barge in each waterway that has to transport all grains to the final port, so the storage gets full occupied in consequence of barges cycle time.

The equipment level of utilization is helpful to understand how much of each equipment is used during the simulation time. In the Table 2 it is demonstrated that the system capacity is not used in its total usage rate, it means that this system can handle an expansion of its capacity.

Figure 6. Graphic of 2014 soybean demand. Source: Authors.

Table 2. Equipment rate.

Equipment	Utilization
Barge Loader in Vila do Conde	9%
Barge Unloader in Vila do Conde	12%
Truck Unloader in Marabá	6%
Truck Unloader in Praia Norte	0%
Truck Unloader in Vila do Conde	0%
Truck Unloader in Porto Velho	2%
Truck Unloader in Miritituba	10%
Berth Utilization Rate	92%
Barge Loader in Marabá	55%
Barge Loader in Praia Norte	0%
Barge Loader in Vila do Conde	0%
Barge Loader in Porto Velho	47%
Barge Loader in Miritituba	42%

Source: Authors.

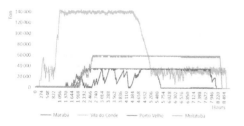

Figure 7. Graphic of 6,000,000 demand. Source: Authors.

Testing a scenario with 6,000,000 ton of grain, the storages follow a different pattern than the one tested to validate. In this case, the utilization of most of the silos kept their level close to the maximum capacity of each storage.

Therefore, following the same idea of the validation scenario, the utilization rates of resources got increased, allowing the analyst to verify where it is possible to make changes in the system that can result in an increase in its capacity.

The rates of utilization increase because the volume transported also was increased; even with 6,000,000 ton of grain, the system is still not in 100% of its capacity.

It is possible to verify that if the manager increases the equipment efficiency, the shape of

Table 3. Equipment rate.

Equipment	Utilization
Barge Loader in Vila do Conde	76%
Barge Unloader in Vila do Conde	89%
Truck Unloader in Marabá	25%
Truck Unloader in Praia Norte	0%
Truck Unloader in Vila do Conde	25%
Truck Unloader in Porto Velho	25%
Truck Unloader in Miritituba	76%
Berth Utilization Rate	90%
Barge Loader in Marabá	67%
Barge Loader in Praia Norte	0%
Barge Loader in Vila do Conde	0%
Barge Loader in Porto Velho	45%
Barge Loader in Miritituba	74%

Source: Authors.

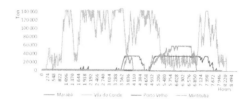

Figure 8. Graphic of 6,000,000 demand and increasing the rates of loading. Source: Authors.

Table 4. Equipment rate.

Equipment	Utilization
Barge Loader in Vila do Conde	35%
Barge Unloader in Vila do Conde	73%
Truck Unloader in Marabá	17%
Truck Unloader in Praia Norte	0%
Truck Unloader in Vila do Conde	17%
Truck Unloader in Porto Velho	17%
Truck Unloader in Miritituba	51%
Berth Utilization Rate	54%
Barge Loader in Marabá	90%
Barge Loader in Praia Norte	0%
Barge Loader in Vila do Conde	0%
Barge Loader in Porto Velho	69%
Barge Loader in Miritituba	85%

Source: Authors.

storages levels change (Figure 8) and the rates of equipment (Table 4) in the whole system decrease.

This happens because when the equipment is more efficient and the ship loading is faster, the system behavior is more likely to be idle than highly utilized. In other words, when the equipment gets free faster it is possible to have more cargo to transport if it is available, when you do not have this cargo available, the system become less efficient.

5 CONCLUSIONS AND CONSIDERATIONS

Despite all the results and analysis, it is possible to understand that in the first scenario, the amount of cargo that will be transported by the fluvial port has to wait many hours until it completes one barge and goes through the waterway to the export terminal. This happens because in this case, the soybean that will load the barges was transported from the farm to the fluvial port by truck and it was necessary a large quantity of trucks to do so.

When the model is processed with 6,000,000 ton (Figure 5), the average waiting time decreases substantially. On the other hand, the fleet number is increased. Then, when the volume of cargo is higher, the loading of the barges is faster and the average waiting time to complete one barge is shorter. By this point, the system is strongly dependent of the equipment instead of the fleet, as shown in the previous scenario (Figure 7).

To conclude, understanding a complex and detailed system (seasonality, transportation fleet, cargo variability, etc.) is only possible using simulation methodology that enables the manager to make the best decision possible regarding all constraints in computational model that imitate the system in real life.

In this study it was not directly considered the perishability, however the storage capacity is low and the flow of transportation fast that is reasonable not to consider this factor.

REFERENCES

Barnhart, C.; Laporte, G. (2007) Handbook in OR & MS, Vol 14. Elsevier B.V.

Belton, V., Stewart, J.T. (2001) Multiple criteria decision analysis—An integrated approach. Kluwer Academic Publishers, London.

Botter, R.C., Brinati, M.A., Roque, J.R. (1988) Dimensionamento de um Sistema integrado de transporte de álcool e óleo diesel na região de influência da hidrovia Tietê-Paraná. XII Congresso Nacional de Transportes Marítimos e Construção Naval, Rio de Janeiro, RJ. XXI Simpósio Brasileiro de Pesquisa Operacional em Conjunto com o VI Congresso Latino-Ibero-Americano de Pesquisa Operacional e Engenharia de Sistemas, Engenharia Naval e Oceânica, p. 13–44.

Buosi, T. (2004) Sistemas computacionais de suporte à medição de desempenho: Proposição de critérios para Análise, comparação e aquisição de sistemas. Dissertação (Mestrado) - Escola de Engenharia de São Carlos—Universidade de São Paulo, São Carlos.

Chowdhury, K.H. (1989) Simulation in marine transportation system. Marine Technology, Vol. 26, No. 1, 74–85.

Churchman, C.W. (1968) The Systems Approach, Delacorte Press, New York.

Chwif, L.; Medina, A.C. (2006) Modelagem e Simulação de Eventos Discretos. 1. ed. São Paulo.

Cochran, J.K. e Lin, L. (1989) Application of computer simulation to freight transport systems. Journal of Operational Research Society, Vol. 40, 433–441.

Crainic, T.G. e Laporte, G. (1997) Planning models for freight transportation. European Journal of Operations Research, Vol. 97, 408–439.

Hameri, A.P. e Paatela, A. (1995) Multidimensional simulation as a tool for strategic logistics planning. Computers in Industry, Vol. 27, 273–285.

Harrell, C., Ghosh, B.K., Bowden JR, R.O. (2004) Simulation Using Promodel. McGraw Hill, Nova Iorque, EUA.

Harrington, T.C.; Lambert, D.M.; Sterling, J.U. (1992) Simulating the financial impact of marketing and logistics decisions.International Journal of Physical Distribution & Logistics Management, Vol. 22 No. 7, 3–12.

Hijjar, M.F. (2004) Logística, soja e comércio internacional. Centro de Estudo em Logística. COPPEAD, UFRJ, Rio de Janeiro.

Hoff, R.K., O'Kray, C. (2014) 2014/15 Soybean Production Forecast at Record 97 Million Metric Tons (MMT); Exports Forecast at Record 50 MMT. USDA—United StatesDepartmentofAgriculture.

Jank, M.S.; Nassar, A.M.; Tachinardi, M.H. (2004/05) Agronegócio e Comércio Exterior Brasileiro. Revista USP, São Paulo, n. 64, p. 14–27, dez/fev.

Jank, M.S.; Nassar, A.M.; Tachinardi, M.H. (2004/05) Agronegócio e Comércio Exterior Brasileiro. Revista USP, São Paulo, n. 64, p. 14–27, dez/fev.

Junior, R.F.T.; Fernandes, F.C.F. Pereira, N.A. (2006), Sistemas de apoio à decisão para programação de produção em fundição de mercado. Gestão&Produção, v.13, n.2.

Kelton, W.D., Sandowski, R.P., Sandowski, D.A. (1998) Simulation with Arena. McGraw Hill, Boston, MA.

Kopytov, E.; Abramov, D. Multiple-criteria choice of transportation alternatives in freight transport system for different types of cargo. Proceedings of the 13th International Conference "Reliability and Statistics in Transportation and Communication", 2013.

Lane, D.C. Should system dynamics be described as a "Hard" or "Deterministic" System Approach? System Research and Behavioral Science. Operational Research Department, London School of Economics and Political Science, University of London, London, UK. 2000.

Mankiw, N.G. (2006) Principles of Economics, 4th ed. South—Western College Pub.

Netto, J., F.(2012) Modelo de simulação para dimensionamento da frota de contêineres movimentada por navios em rota dedicada. Dissertação (Mestrado) - Escola Politécnica da Universidade de São Paulo—USP, São Paulo. PEGDEN, C.D., SHANNON, R.E.; SADOWSKI, R.P. (1990) Introduction to Simulation Using SIMAN, 2nd ed., New York: McGraw-Hill.

Prodanov, C.C.; Freitas, E.C. Metodologia do trabalho científico: Métodos e Técnicas da Pesquisa e do Trabalho Acadêmico. 2. Ed. Universidade Feevale, 2013.

Schriber, T.J. (1974) Simulation Using GPSS. John Wiley & Sons, 1st Edition, New York, US.

Silva. E.L.; Menezes, E.M. Metodologia de pesquisa e elaboração de dissertação. 3. Ed. Florianópolis: LED/UFSC, 2001.

Stanford, M. (2008) Maritime Economics.3rded. New York: Routledge.

Maritime Transportation and Harvesting of Sea Resources – Guedes Soares & Teixeira (Eds)
© 2018 Taylor & Francis Group, London, ISBN 978-0-8153-7993-5

A case study on the capacity of a Brazilian iron ore port distribution center in Asia

J.F. Netto & R.C. Botter
University of São Paulo, São Paulo, Brazil

ABSTRACT: This case study presents a Brazilian company, the world leader in the production of iron ore, which intends to carry a large amount of its product to meet an existing demand in some Asian countries. Thinking about the logistics gains of its operation, the company invested in the construction of a large maritime terminal in Asia to operate as an Iron Ore Distribution Center for the operation to take place with economies of scale from the use of larger ships in transport from Brazil and redistribution to Asian ports through smaller ships and lots. The use of larger ships would reduce the amount of long distance travel and would lead the company to reduce its transport costs, enabling investments in quality of operation. In addition, it is important that the terminal operates at its maximum capacity and to make it happen the maintenance actions and equipment stops should be well planned. This work shows that the discrete events simulation is a suitable tool to show the capacity of the DC (Distribution Center) and its level of services when operating with different fleet profiles for import ships.

1   INTRODUCTION

This paper presents a methodology for dimensioning and evaluating the capacity of port terminals using simulation as the major tool to verify service levels offered by the terminal. The concept of capacity associated to service level is something that can be used along with simulation in several situations that involve logistic and commercial planning of organizations, because it enables a practical and direct vision of real behavior of a system, facilitating the comprehension of existent bottlenecks and being a way to see provided services from the point of view of who use those services.

A discrete events simulation model that represents the operations at the terminal is built, in which the systems are divided according to the characteristic of the tasks. This analysis considers the application of systemic approach.

The terminal that is the object of this study is a distribution center, built in Asia to improve the exportation of iron ore produced in Brazil, through the use of large vessels (400,000 DWT), achieving economy of scale. In this DC also is made the blending to meet the needs of Asian customers. The simulation will be used to give data necessary for the decision making in analysis that will consider the necessity of promoting efficient services and that guarantee better efficiency for resources.

For the application of the methodology shall be considered the discharge of two types of iron ore produced in Brazil, in this article identified

as Northern Ore (NO) and Southern Ore (SO), their storage and blending to production of two types of final products, which are stored in different stockyards where they wait to be loaded onto smaller vessels and distributed to final buyers. The objective of this study is to consider the logics of routering and blocking analysis with the simulation model of the Distribution Center allowing a better evaluation of the impacts that the operational restrictions represent for the terminal.

1.1   The terminal

The distribution center studied in this work, as previously mentioned, receives the Brazilian ore in ships with capacity of 400,000 DWT and 220,000 DWT, unloading these vessels in a single import berth, equipped with three ship unloaders (SU), whose effective discharge capacity is 5.159 t/h. For the system of discharge, it is considered the incidence of preventive maintenance of 69 hours with intervals of 730 hours, corrective maintenance of 474 hours per year, distributed throughout this period, and 671 hours of operational stoppages per year.

The ore discharged in this DC goes to three stockyards, that have capacity for 1.01 million tons of iron ore (NO and SO) and are served by a stacker, a reclaimer and a stacker-reclaimer (SR). These quantities of ore are used to produce two different types of blend: B1, which corresponds to 19.2% of exportations and is composed by 70%

NO and 30% of SO, and B2, corresponding to 80.8% of the total produced and composed by 60% of NO and 40% of SO.

The products B1 and B2 are stored in two other stockyards (D and E), with capacity for 716,000 tons of blended and where operate a reclaimer and two SR's. There is an operational blocking in the operation of these two SR's that will be better detailed later, due to the position of these equipment are along the positioning rails.

The blending operations occur from the recovery of the production inputs in stockyards A, B and C, according to the proportions presented. Therefore, the effective blending rates are obtained as a function of the productivity of the reclaimers, which are 5,750 t/h per equipment, with corrective maintenance of 69 hours with intervals of 730 hours, 900 hours of corrective maintenance and 500 hours of operational stoppage.

Finally, products B1 and B2 are reclaimed from stockyards D and E to be loaded on Capesize vessel types (170,000 DWT) and be brought to customers. The operation of reclaim occur at an effective rate of 5,750 t/h, with corrective maintenance of 69 hours at intervals of 730 hours, 840 hours of corrective maintenance and 720 hours of operational stoppages, which affect the operations of the shiploader.

2 LITERATURE REVIEW

Due to the increased use of maritime shipping to transport goods between countries geographically distant it became necessary to rely on port operations most efficient in order to maximize the capacity of port terminals across the planet leading to the existence of many works and researches whose main objective is measure the levels of service and ways to evaluate the operational capacity of a port.

Lima et al. (2015) evaluate port operations in accordance with the times that ships wait in queue at the port terminals emphasizing the need to control and manage the operations at the terminals. The authors constructed a model to maximize the quantity of delivered goods, using the queuing theory and simulation (using the Simian software) as tools to support in obtaining results. However, the results obtained were related more to routes to be followed than the infrastructure improvements. Lee and Choo (2015) dealt with the sea freight as a set of networks to be optimized while keeping the balance of mass between the nodes of these networks and seeking to maximize the quantity transported.

Zehendner et al. (2015) studied a way to increase the efficiency of available port equipment at a terminal by reducing delays in loading and discharge and also reducing the time of wait in queue of attended ships reinforcing the time on queue as the object of efforts to increase the offered level of service.

Another highlight in this review is the large amount of iron ore produced each year in Brazil, making this product an object of many studies related to its supply chain and alternatives to make it a more sustainable industry, such as the study Macedo et al. (2002).

Finally, authors such as Gualda (1995) and Chwif and Medina (2007) were studied to a better understanding of the methodology used in this study, to be presented afterward.

3 METHODOLOGY

The used approach for the determination of capacity of the terminal, with a new layout, consider three important steps. Firstly, it was used the so-called systems approach, in order to conduct the analysis with accuracy and more detailed characteristics of the system. The port terminal is divided into subsystems according to the characteristics of the operations performed.

Following, a simulation model of discrete events is built to represent the reality of operations of subsystems that compose the terminal, starting from the collection of premises and the creation of a conceptual model that should list all characteristics to be represented. This model should be validated according to real and existent data to gain reliability and to enable analysis of terminal expansion in a direct way, through the construction of scenarios.

Finally, using the validated model and the constructed scenarios, it is applied the concept of capacity associated with the service level, to determine the actual capacity of the terminal in accordance with the improvements (expansion) and to investigate which of the subsystems will be the system bottleneck.

3.1 Systemic approach

Before to evaluate the capacity of exportation on port terminal, associated to pre-established service levels, a systemic approach of the terminal after expansion should be applied so that by a discrete events simulation model may be possible to identify the bottleneck of the project and the actions to be taken in order to ensure that the subsystems operate at maximum operational capacity.

The systemic approach is applied in a project or masterplan of a terminal of transport because allows a more detailed analysis of the existing components and subsystems and determining the system that limits the capacity of the terminal, its "bottleneck".

According to Gualda (1995), is necessary to apply a suitable method for the treatment of a problem with systemic approach. To detail the subsystems of the terminal of transports enables to solve a problem neatly, and the systemic approach is a method based on premises of scientific methodology to analyze a system by the build of a model.

Thus, for the application of the systemic approach on a terminal of transport, its subsystems and components must be identified through the links between its objectives in order to facilitate the modelling and analysis of the obtained results afterward. So the systemic approach method will be used in solve of the proposed problem and therefore the terminal will be divided into three subsystems: discharge system, storage and blending system and loading system. The Figure 1 presents the layout of the terminal divided into subsystems according to what is described.

The discharge subsystem is composed by the arrival of ships fulllfield with iron ore and their discharge through shipunloaders that operate with discharge lines for stacking in the yards through stackers and stacking function of stackers-reclaimers, in what can be called interface between the discharge subsystem and the storage subsystem.

After its stacking, the inputs of production are stored until when there is space in the stockyards D and E and availability of equipment for the production of blends B1 and B2, with priority for the one that must be exported in greater quantity. These products are then stored in the terminal waiting for ships to be loaded. This is, therefore, the DC storage and blending subsystem. Finally, reclaimers take off the stored material that is directed by the loading line until loading subsystem, composed by ship loaders and mooring berth, with limited capacity to certain types of ships.

It must be noted that this systemic approach can be considered for any kind of port terminal, since all of them can be divided into three types described above.

3.2 Simulation model

The simulation of the expansion scenario of the ore terminal will be executed through a model validated with current operations and service level, obtained through analysis of data and collection of information together with mining company. The validation of the discrete events simulation model is a really important stage on the process of analysis of the capacity of the new terminal, from which the simulation model will may be considered a reliable tool to answer the questions of project.

The conceptual model is the enumeration of premises and rules to be considered at implementing the computational model. With the computational model ready it can be possible the search for the validation through input data that represent the performance and characteristics of simulated system (validation) which can generate a review of conceptual model and reimplementation of computational logic. Figure 2, adapted from Chwif and Medina (2007), represents this interactive process, executed until the model is validated and able to be used to represent the system with its improvements and expansion.

With a simulation model representing expansion scenarios, it is possible to determine the capacity of the terminal using the concept of capacity associated with the service level for operations at the terminal.

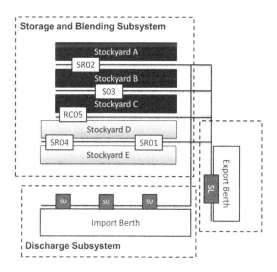

Figure 1. Layout of expansion of ore terminal divided into subsystems.

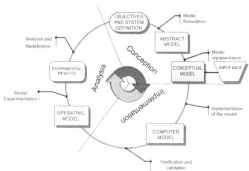

Figure 2. Representation of steps followed in a simulation project (Source: Chwif and Medina 2007).

3.3 Capacity associated with the level service

According to Gualda (1995), modeling of transport terminals should help the planning and decision making about solutions to be adopted at planning, conception or expansion of a terminal. Still, the capacity of a transport terminal is obtained through the determination of maximum level of the demand imposed to the terminal in a pre-defined period of time, without any criteria of stipulated service level is violated.

Service level is a rate of performance "offered by a component, subsystem or system, that aims to translate the quality of the service offered by the terminal and that can be measured" (Gualda 1995). The capacity of the terminal, therefore, is given by the lower capacity of components and subsystem, that is to say, by the bottleneck. This offered service level can be a competitive differential that is so important as the discount at prices, advertisement or favorable payment conditions.

With the simulation model, it will be possible to obtain several rates and service level obtained at operation of constructed scenarios and, from those results, major service levels of the terminal will be highlighted. Those who are not attended before will be verified, limiting, this way, the capacity of movement and exportation of iron ore.

Such methodology will be used for all defined subsystems and will allow a definition of what are bottlenecks of the operation. More than that, which subsystem will limit the capacity of exportation of this terminal.

4 CONSTRUCTION OF MODEL AND ANALYSIS

As described above, a simulation model that represents the operations on subsystems of the port terminal was built and, from the level of services to be considered, the expected answers about the capacity of the terminal will be obtained. It is important, however, that logics incorporated by the model are detailed, as well as expected service levels, by who use the services of the port terminal.

4.1 Discrete events simulation model

The Figure 3 shows the layout of discrete events simulation model developed with all the considered logics (discharge, storage/blending and loading).

The discrete events simulation model was built in software ARENA® and it was included the modeling of three subsystems, interconnected. The model size is 11 MB and its processing speed depends on the exported demand, lasting 2 or 3 minutes per replication (simulations were done with 10 replications after tests that proved to be a reasonable number).

4.2 Conceptual mode: Characteristics of subsystems

In a conceptual model, all logics and premises to be incorporated at computational model should be listed. Considering that the project uses systemic approach, a particular analysis of subsystems that compose the terminal, characteristics of each subsystem are numbered. The discrete events simulation model can be used with a few alterations to represent other terminals, since it aims to increase its coverage ability. That is to say, represent several terminals from a generic logic for port terminals.

- Discharge subsystem: the main characteristics of this subsystem are related to the ships and facilities used for discharge operations, which result in the operation of stacking of iron ore used as production input in stockyards. The cargo to be discharged from ships is generated in the simulation model from the characteristics informed by the user, guaranteeing the maintenance of the mass balance, since all the products that arrive at the terminal must be stacked and, in the end, serve as product for export or be held in stock.

The user of the simulation model must fill in the information regarding the quantity of each imported product and the percentages of the cargo of each product to be transported by each type of ship.

Some input data for discharge operations:

- Berthing Maneuver: 1.86 hour
- Unberthing Maneuver: 0.85 hour
- Pre-Operation Time: 1.05 hour
- Post-Operation Time: 1.24 hour

- Storage and blending subsystem: the storage of iron ore starts at the moment when the products are unloaded and stacked at the input stockyards, from where they are reclaimed to produce blends B1 and B2 to be stored again as products of export.

The model differentiates the way in which the blending to production of B1 and B2 is done, prioritizing the production of the second product, since its demand is significantly higher and thus a better balance between supply and demand is sought. When B1 and B2 are produced, they are stacked in stockyards D and E where they remain until the moment when the ship is berthed and should be loaded with such products. In another words, the subsystem includes from the stacking of the inputs to the reclaim of the blended ones.

Figure 4 shows the capacity of the five existing stockyards and the product to be stored in each of them.

In the model, if a stack is occupied with a stacking operation, it is locked for reclaim, and vice versa. About stockyard equipment:

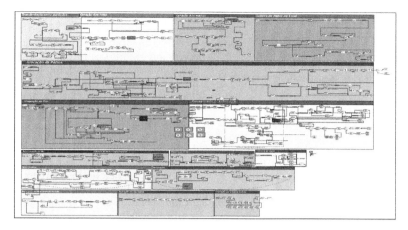

Figure 3. Layout of the discrete events simulation model developed in ARENA®.

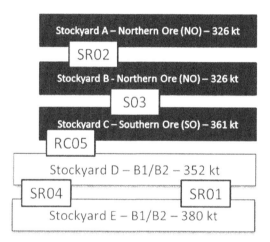

Figure 4. Layout and capacity of the stockyards.

The reclaimers have an effective rate of 5,750 t/h, with preventive maintenance of 69 hours every 730 hours, 900 hours of corrective maintenance and 500 hours of operational shutdowns distributed throughout each year. These values also apply to the SR's reclaim function;

SR's can operate as stackers or as reclaimers but should privilege the mode of operation according to the desire of the model user;

Shipunloaders (SU's) have an effective rate of 5,159 t/h, with preventive maintenance of 69 hours every 730 hours, 474 hours of corrective maintenance and 671 hours of operational shutdowns distributed throughout each year;

The stacking rate will be a consequence of the productivity rates of the shipunloaders, since the stacking capacity of the stackers and SR's are higher than the productivity rate of the discharge system.

Loading subsystem: the main features of this system include mooring berths, ship loaders (SL) and the loading line which directs the iron ore reclaimed from stockyards to the loading of the ships. There is a SL that operates with a capacity resulting from the operation of the reclaimers used in stockyards D and E.

The vessels used for the export and distribution of B1 and B2 are Capesize type with a capacity of 170,000 DWT. For operation at the export berth, the following operating times are considered:

- Berthing Maneuver for 170,000 DWT: 1.24 hour
- Unberthing Maneuver for 170,000 DWT: 0.59 hour
- Pre-Operation Time: 1.34 hour
- Post-Operation Time: 0.83 hour

In order to generate the vessels at simulation model, the amount of them required to meet the demand to be exported is calculated considering the fleet profile mentioned. With the amount, the interval between ships is obtained: average interval is calculated ensuring that these vessels arrive at port all within a year and it is used at an exponential distribution (EXP) to represent the arrivals of ships.

As mentioned previously, it is necessary to verify the operational restrictions between the equipment that operate in the DC to model the available routes. One of the restrictions that could occur would be due to the limitation at the operation of SR 01 and SR 04, which need to maintain a distance of approximately 100 meters from each other due to their format. So in this work, the distance between SR 01 and SR 04 will be measured in stockyards spaces with storage capacity

Figure 5. Arrangement of stockpiles at stockyards D and E.

of 4,000 t of ore, and the space limitation is called fence ("x" spaces, for example). For preliminary understanding, it should be noted that stockyards D and E have a "useful" length for 3 stockpiles of 124.5 meters and 131.8 meters respectively. It means that, depending on how an SR is working at the middle stack, it can completely block the another one. Figure 5 shows how the stockpiles are arranged in stockyards D and E.

It was defined that the fence should be of 32 spaces for 4,000 t of iron ore, which are equivalent to the 130 meters of restriction existing between SR 01 and SR 04, but this value can be the subject of further analysis if necessary.

4.3 Considered service levels

The company of the maritime terminal established some service levels considered limiting to prevent the performance of the terminal to fail and, this way, to prevent the generation of unnecessary costs due to the payment of fines stipulated at contracts with navigation companies. Such fines generally take into consideration the time that ships wait on queue what is called demurrage, charged when the waiting is higher than the one stipulated at the contract.

To minimize queues, the terminal should operate with maximum efficiency. All subsystems should operate together to result in high productivity rates for ore loading so that the ships will keep moored for an acceptable time, minimizing the time that other ships wait on queue. Loading rates generally are measured in tons per hour and depend directly of reclaiming rates of ore at storing yards, at the interface between storage and loading subsystems. This way, it is considered the average time that ships wait on queue as an indicator that represents the service level of the terminal.

Other indicator that is verified by professionals responsible for the operation at the port is the occupation of installed equipment at the terminal. This occupation is the relation between time when equipment are operating and hours calendar of the period which the analysis is done, generally all 8,760 hours of an year. To mooring berths, the occupation is the relation between total time during the period considered for the berth to be unavailable for mooring of a ship because it is already being used by another vessel and it includes, also,

the period in which ships are executing maneuvers of mooring and unmooring of berths.

The occupation of the equipment and the berth is very important because can result on an increase of the possibility of a ship wait a time bigger than the acceptable. Furthermore, very high occupations can compromise the terminal in case of necessity of preventive maintenance on the components of all the subsystems.

Therefore, the average waiting time of ships on the queue and the occupation of equipment and berths are the two main indicators that represent the level of service on the port terminal and the considered values are pre-set by the responsible for the operation of the terminal. In this article is considered the occupation of the berths and the acceptable value is 90%.

The occupation of resources of 90% is considered because, according to the company that commands the operation of the terminal, is a value that corresponds to the limit to prevent risks of stopping operations. This would cause loss of efficiency and, consequently, increase of queues.

5 SIMULATION RESULTS

As described by Chwif and Medina (2007), in studies of simulation, it is very common to evaluate possibilities for improvements of existing systems when there are performance indicators to be analyzed. This process is called validation in which it is analyzed whether the responses obtained with the model and the input data are similar to the results obtained in real operations.

However, for systems in development such as the Distribution Center presented in this work, it is not possible to validate the model, but only get its verification, which is the process by which are obtained the expected responses of a model, for example an increase in the waiting time due to increase of the demand or increase on operational capacity due to the increase in available resources. For this case, a scenario is defined from which tests and analyzes are made, called the Base Scenario. This base scenario should consider all operational assumptions and input data expected as real and it is the scenario whose results are taken as indicators to be sought in actual system operations.

From this base scenario are constructed the sensitivity analyzes to obtain responses of system according to changes in some of its elements and parameters.

5.1 Base scenario

The base scenario of the simulation of operation of DC considers the assumptions already presented in the conceptual model besides considering that all imported ore from Brazil, NO or SO, is carried in ships of 400,000 DWT.

Table 1 shows the results obtained in the base scenario simulation.

It is observed from results that the capacity of the terminal turns around the 28.9 Mtpy (millions of tons per year) considering all the evaluated parameters with occupation of 86.9% in the discharge berth and 89.6% in the loading berth. It possible note that the bottleneck of system is the occupancy of loading/export berth (89.6%).

Table 1. Results of simulation of Base Scenario.

% NO imported by Ships of 400,000 DWT	**100%**
% NO imported by Ships of 220,000 DWT	**0%**
% SO imported by Ships of 400,000 DWT	**100%**
% SO imported by Ships of 220,000 DWT	**0%**
Total demand loaded on ships (t)	28,918,652
Total demand loaded on ships (Mtpy)	28.9
Total demand unloaded from ships (t) – Import	28,875,569
Total demand NO unloaded (t) – Import	17,864,638
Total demand SO unloaded (t) – Import	11,010,932
Total demand B1 loaded on ships (t) – Export	5,694,643
Total demand B2 loaded on ships (t) – Export	23,224,009
Total produced blend B1 (t)	5,699,043
Total produced blend B2 (t)	23,182,137
Total blended (t)	28,881,180
Number of vessels for exportation generated	175.5
Number of vessels for exportation attended	169.5
Total time of berthing (h)	348.9
Total time of unberthing (h)	163.5
Total occupation time (including maneuvers) for discharge (h)	7,593.6
Total occupation time (including maneuvers) for loading (h)	7,845.6
Number of vessels for importation attended during simulation	75
Total occupation of shiploaders (%)	55.1%
Total occupation discharge/import berth (%)	86.9%
Total occupation of loading/export berth (%)	89.6%
Average rate of loading of export berth (t/h)	3,683.9
Average occupation of stacking at SR 02	10.3%
Average occupation of stacking at S 03	65.2%
Average occupation of stacking at SR 04	14.4%
Average occupation of stacking at SR 01	30.7%
Average occupation of reclaiming at SR 02	45.1%
Average occupation of reclaiming at RC 05	45.1%

(Continued)

Table 1. *(Continued)*.

Average occupation of reclaiming at SR 04	41.6%
Average occupation of reclaiming at SR 01	30.9%
Time SR 02 stacking (h)	898.7
Time S 03 stacking (h)	5,708.8
Time SR 04 stacking (h)	1,259.7
Time SR 01 stacking (h)	2,691.5
Time SR 02 reclaiming (h)	3,951.2
Time RC 05 reclaiming (h)	3,951.2
Time SR 04 reclaiming (h)	3,646.6
Time SR 01 reclaiming (h)	2,708.6
Average rate of SR 04 stacking (t/h)	4,372.5
Average rate of SR 01 stacking (t/h)	4,369.7
Average rate of SR 02 reclaiming (t/h)	4,560.4
Average rate of RC 05 reclaiming (t/h)	4,536.7
Total cargo stacked with SR 04 (t)	3,929,788
Total cargo stacked with SR 01 (t)	4,945,781
Total cargo reclaimed with SR 02 (t)	16,630,353
Total cargo reclaimed with RC 05 (t)	12,288,299
Blending rate with SR 04 stacking (t/h)	7,322
Blending rate with SR 01 stacking (t/h)	7,303
Blending rate with SR 02 reclaiming (t/h)	4,530
Blending rate with RC 05 reclaiming (t/h)	2,780
Import vessels 220,000 DWT attended	0
Import vessels 400,000 DWT attended	75
Average time (h) export vessels waited for product	52.1
Number of EXPORT ships that waited for lack of products in the stockyards	59
Average time (h) that B1 vessels waited for product	77.0
Average time (h) that B2 vessels waited for product	46.0
Number of B1 ships that waited for lack of products in the stockyards	12
Number of B2 ships that waited for lack of products in the stockyards	47
Time (hours) of shutdown with corrective maintenance and failures subsystem discharge	1,113.0
Time (hours) of shutdown with corrective maintenance and failures blending	1,402.0
Time (hours) of shutdown with corrective maintenance and failures subsystem loading	1,559.0

From this base scenario is done the sensitivity analysis that evaluates the impact of changing the import fleet profile of the SO.

5.2 *Sensitivity analysis: Profile of the Southern Ore imports fleet (SO)*

In order to evaluate the impact of the alteration of the fleet profile used in the SO importation, three new profiles are established:

- Scenario 1: 80% vessels of 400,000 DWT - 20% vessels of 220,000 DWT

Table 2. Results of sensibility analysis.

% NO imported by Ships of 400,000 DWT	100%	100%	100%	100%
% NO imported by Ships of 220,000 DWT	0%	0%	0%	0%
% SO imported by Ships of 400,000 DWT	100%	80%	60%	40%
% SO imported by Ships of 220,000 DWT	0%	20%	40%	60%
Total demand loaded on ships (Mtpy)	28.9	29.0	29.1	29.2
Total demand unloaded from ships (t) – Import	28.9	30.0	29.0	29.1
Total blended (t)	28.9	29.0	29.0	29.1
Number of vessels for exportation generated during simulation	176	176	175	175
Number of vessels for exportation attended during simulation	170	170	171	171
Total time of berthing (h)	349	358	366	377
Total time of unberthing (h)	164	168	171	176
Number of vessels for importation attended during simulation	75	79	83	89
Total occupation of shiploaders (%)	55,1	55,6	56,0	56,4
Total occupation of discharge/import berth (%)	86,9	88,6	90,7	93,2
Total occupation of loading/export berth (%)	89,6	89,8	90,1	90,2

- Scenario 2: 60% vessels of 400,000 DWT – 40% vessels of 220,000 DWT
- Scenario 3: 40% vessels of 400,000 DWT – 60% vessels of 220,000 DWT

From changing the input data of the model are obtained the results of these sensitivity analysis, shown in Table 2.

It is possible to evaluate from the results of Table 2 that the import of SO with ships of 220,000 DWT is bad for terminal operations due to the fact that occupancy of berths increases and exceeds the limit of 90% considered, specially the import/discharge berth.

6 CONCLUSIONS

This paper had the objective of, using concepts of systemic approach and capacity associated to service levels, propose an analysis of results obtained through a simulation model and determine the capacity of movement of products in a port terminal.

The first step after the construction of the simulation model was the definition of a Base Scenario

that made possible verify that the DC capacity is 28.9 Mtpy.

It is noteworthy that the occupation of export berths is at high levels due to the fact that the simulations considered a data entry of 30 Mtpy and the terminal cannot attend this demand. The occupation of the import berth shows that the class of ships has considerable impact directly on the operation.

Smaller ships, in addition to operating at a lower discharge rate, still cause the occurrence of more time for maneuvers and pre-operation times, directly impacting at availability of the berths.

The concept of capacity associated to the service level is, so, a really important and useful tool to execute projects that considerate implantation and improvements or expansion of terminals, such the studied ore port, allied to systemic approach that considerate different subsystems that compose a terminal, making it possible to determine the maximum capacity of movement of products and also what should be the next step of this project. This is because we know what is the bottleneck of the system, making it possible to look for an increase of capacity through other actions executed directly to the identified restriction.

As recommendation for future works is proposed that an analysis of the operation described and modeled in this paper be done from modeling the handling of the iron ore at the port continuously (using loading & unloading rates) and the arrivals and departures of ships, discretely, adopting a combined simulation approach.

REFERENCES

Chwif, L., and A.C. Medina. 2007. *Modelagem e Simulação de Eventos Discretos—Teoria e Aplicações.* São Paulo: Editora do Autor.

Gualda, N.D.F. "Terminais de Transportes: Contribuição ao Planejamento e ao Dimensionamento Operacional" Ph.D. thesis, Department of Transportation, Polytechnic School of University of São Paulo, São Paulo.

Lee, H., and S. Choo. 2015. "Optimal Decision Making Process of Transportation Service Providers in Maritime Freight Networks." *KSCE Journal of Civil Engineering* 60:1–11.

Lima, A.D.P., F.N. Mascarenhas, and E.M. Frazzon. 2015. "Simulation-based Planning and Control of Transport Flows in Port Logistic Systems." *Hindawi Publishing Corporation* 2015. http://dx.doi.org/10.1155/2015/862635.

Macedo, A.B., Freire, D.J.A.M., and Akimoto, H. 2003. "Environmental Management in the Brazilian Non-Metallic Small-Scale Mining Sector." *Journal of Cleaner Production* 11: 197–206.

Zehendner, E., G.R. Verjan, N. Absi, S.D. Péres, and D. Feillet. 2015. "Optimized Allocation of Straddle Carriers to Reduce Overall Delays at Multimodal Container Terminals." *Flexible Services and Manufacturing Journal* 27: 300–330.

Maritime Transportation and Harvesting of Sea Resources – Guedes Soares & Teixeira (Eds)
© 2018 Taylor & Francis Group, London, ISBN 978-0-8153-7993-5

A comparative analysis of Brazilian maritime transport by cabotage between 2010 and 2015 using network theory

C.C. Ribeiro Santos
SENAI CIMATEC, Salvador, Brazil

M. do Vale Cunha
Instituto Federal da Bahia, Barreiras and SENAI CIMATEC, Salvador, Brazil

H. Borges de Barros Pereira
Universidade do Estado da Bahia and SENAI CIMATEC, Salvador, Brazil

ABSTRACT: In Brazil, the excessive concentration of freight transportation by the road modal brought to the surface the need to use more efficient and competitive modes for the transportation of finished products. With an extensive coastline of more than 7000 km and more 230 port structures divided between public and private, maritime transport, especially cabotage, has been consolidating in Brazil as an alternative capable of transporting goods safely and economically. Based on this context, this work presents a comparative analysis of the maritime movement of ships by cabotage between the years 2010 and 2015, using network theory. These networks allow, from the established relations, to identify important properties and topologies essential for the understanding of the relationships existing in the maritime movement by cabotage on the Brazilian coast. The results show how the main indexes of networks vary between the years 2010 and 2015, for the network as a whole and for specific vertices, through their centralities of degree and intermediation. This ensures a more detailed monitoring of the main ports (i.e. Santos, Suape, Rio Grande and Vitória) and howthese influence the Brazilian maritime network.

1 INTRODUCTION

Organizations operate in a highly competitive, dynamic, complex and unstable global environment. These characteristics denote scenarios of unpredictability, where all of its operations and activities need to be viewed and revised continuously. In this business context, it is of fundamental importance that organizations seek effective strategies for their operations—planning, marketing, finance, production, quality and logistics—because they are increasingly important and because they are directly related to end-user activities of production systems.

It seeks, therefore, to reduce costs throughout the value chain and provide customer satisfaction. From this perspective it is possible to understand why logistics is evident in the current market environment. The fact that this is supported by the pillars: transportation, inventory management, order processing, support activities and aim to provide the customer with the levels of satisfaction desired by them, makes logistics assume a key role for organizations: To be a component capable of generating a fundamental competitive advantage, that is, to provide goods and services in a correct way,

in exact times and places, in the desired condition and at the lowest possible cost.

However, when it comes to improving operational efficiency in physical distribution, it is not enough to consider only the most used means of transportation in Brazil—the road; It is necessary to analyze all available transport matrix to achieve a service capable of satisfying the sales channel. This vision considers each step of the transportation process, always seeking to identify the possible alternatives, often discarded or poorly explored.

The five main basic modalities are: road, rail, waterway, pipeline and air, and their relative importance must be measured in terms of system mileage, traffic volume, revenue and composition nature. Precisely because of these issues, the waterway model is an important alternative for the Brazilian transport matrix, especially when it comes to cabotage, due to the geographic characteristics of Brazil.

2 MARITIME NETWORKS

Recent studies by Clark (2004) on the geography of the world economy point to the constant reduction of trade barriers around the world, with the consequent reduction of transport costs in general. Glaeser & Kohlhase (2004) argue that the world market is experiencing a new context in which political, economic and technological transformations have promoted the globalization and regionalization of market processes. In this new configuration of markets, the use of maritime transport becomes a competitive path for modern organizations.

Bird (1984) points out that, in this new global competitive environment, organizations have come to value and scientifically analyze routes and seaports around the world in search of competitive advantages. Thus came the Global Maritime Network.

Changes in the economic organization of transport lines and the requirements imposed by shippers are reflected in the new geographic organization of maritime networks. The structure of liner networks evolves over time, so the position of the ports as nodes in the network also changes over time. Understanding these changes is critical to analyzing the competition and growth prospects of container ports (Ducruet, 2016).

Despite the finding of the importance of these maritime networks to increase levels of competitiveness through their analysis, these maritime networks have not yet received as much attention as the terrestrial transport networks in which ports are also embedded. This is because maritime networks are increasingly integrated with other transport networks: the "new paradigm" proposed by Robinson (2002):

Although maritime transport accounts for about 90% of world trade volumes, it still has not attracted as much attention as other transport systems from a graphical point of view. As a result, the relative status and evolution of seaports in the networks are still not well understood. Robinson (2002).

Ducruet (2016) points to a recent lack of technical and scientific studies on the subject of maritime networks. For these authors, the use of maritime structures by the modern organizations of a country, city or region is one of the main ways to achieve competitive business objectives. Countries such as China and England grow substantially in competitive advantages for the development of their internal and external maritime structures, mediated by important analyzes of maritime networks.

For Ducruet (2016) maritime networks are among the oldest forms of spatial interaction. So-called port hierarchies and the already established spatial pattern of sea routes can be considered as examples of regionalization and globalization of trade patterns and business cycles between organizations, cities, regions or countries. Ducruet, Zaidi and Inria (2012) argue that a detailed analysis of the maritime network geography of a region provides

competitive information of relevance in the logistic decision making.

In this sense, Frémont (2007) explains that maritime networks around the world, after decades of adaptation, have succeeded in establishing a global network for the transport of maritime containers, as a consequence of a technological revolution in containerization that has gradually produced new forms of relations between countries, Port regions and cities, supported by continued pressure on transport costs and an increasing power of transport alliances and large operators.

Thus, it is in this competitive scenario of maritime networks that the object of this study is presented. It is known that an aquatic system is divided according to the geographical feature of the navigation route. Inland waterway transport uses navigable rivers and maritime transport encompasses the circulation of the seas and oceans. Because these are two different systems, the analysis of maritime transport is subdivided into four types of navigation: long haul, cabotage, maritime support and port support. The objective of this scientific research will be the maritime networks of the Brazilian cabotage of containers.

3 METHODOLOGY

3.1 *Background*

Network theory has been prominent in research seeking to understand the structure and dynamics of systems that contain connected elements. Within this context, a network is the abstraction of these elements, called vertices, and their mutual relationships, called edges or arcs.

These sets can be represented by a graph $G(t) = \{V,E\}_t$, where V is the set of vertices containing n elements, and E, with m elements, is the set of edges, i.e., of pairs of vertices that are related through some pre-established criteria, and t is the instant of time that the graph lies.

Networks may contain directed edges (arcs) or not. In other words, the edge emanating from a vertex origin and focuses on target vertex.

A network can be considered as a complex network, depending on the values of its indexes, and classified according to classical topologies (Carneiro, Rios & Rios, 2016, p. 95):

- **Regular**, if all vertices have the same number of connections;
- **Random**, if its degree distribution follows a normal distribution (Erdös and Rényi Prize, 1960);
- **Small-world** network, if the connections between the vertices favor short distances between any two vertices in the network, making the network more efficient in terms of information transmission. (Milgram, 1967; Watts & Strogatz, 1998);

- **Scale-free** network, if its degree distribution follows a power law, i.e., $P(k) \sim k^\gamma$, which favors the existence of hubs, which are vertices that concentrate many connections. In the theoretical model, the growth of the network is based on the preferential adhesion of vertices (Barabási and Albert, 1999).

In addition to revealing the structure of the network, the network metrics allow us to highlight vertices of high prestige that concentrate or intermediate many connections.

3.2 Used metrics

To characterize the entire network we will use the following indicators (Barabási, 2016):

- **Number of vertices** (*n*): represents the total vertices of the network, or:

$$n = |V| \tag{1}$$

- **Number of edges** (*m*): represents the total of edges of network, or:

$$m = |E| \tag{2}$$

- **Average degree** (*<k>*): calculates the average number of connections of the vertices, or:

$$\langle k \rangle = \frac{2m}{n} \tag{3}$$

with

$$2m = \sum_{i=1 \in V}^{n} k_i \tag{4}$$

And k_i the degree of *i*, i.e. number of connections of *i*.

- **Degree distribution** (*P(k)*): is the probability distribution of degrees of all vertices over the whole network.
- **Average minimal path length** (*<ℓ>*): The average of the shortest paths of network, or:

$$\langle \ell \rangle = \frac{1}{n(n-1)} \sum_{i \neq j} \ell_{ij} \tag{5}$$

Where ℓ_{ij}, represents the shortest distance (minimum path), in terms of edges, between vertices *i* and *j*.

- **Diameter** (*D*): represents the largest minimum path among the vertex pairs in the network.
- **Density** (*Δ*): is the total of existing edges (*m*), divided by the maximum possible number of edges (*N(N − 1)/2*), for non-directed networks, i.e.:

$$\Delta = \frac{2m}{n(n-1)} \tag{6}$$

- **Average clustering coefficient** (*<C>*): Mean clustering coefficient of the vertices of the network,

$$\langle C \rangle = \frac{1}{n} \sum_{i=1}^{n} C_i \tag{7}$$

C_i measures the clustering coefficient of a vertex *i*, which means the proportion of existing edges between neighbors of vertex *i*, E_i and the maximum possible number of edges, $C_i = 2E_i/[k_i(k_i-1)]$.

To specifically characterize each vertex of the network, we use the measures of centrality:

- **In-degree** (k_i^{in}): reports the number of edges incident on the vertex *i*.
- **Out-degree** (k_i^{out}): reports the number of edges emanating from the vertex *i*.
- **Degree** (k_i): as stated above, it reports the total number of edges m_i incident or emanating from a vertex *i*, i.e.:

$$k_i = k_i^{in} + k_i^{out} \tag{8}$$

- **Beetweness** (B_u): Measure how much the vertex u intermediates connections between other vertices, that is:

$$B_u = \sum_{\substack{i,j \in V \\ u \neq j \neq i}} \frac{\sigma_{i,j}(u)}{\sigma_{i,j}} \tag{9}$$

$\sigma_{i,j}(u)$ is the number of minimum paths between vertices *i* and *j* that must pass through *u*, and $\sigma_{i,j}$ is the total of minimum paths between vertices *i* and *j*.

Metrics can be calculated for the entire network or its largest component. A component is a connected subgraph, where all pairs of vertices are connected by a path

If weighted or multiple edges are considered in the network, the in-degree and out-degree will change according to the weights of the edges or the number of repeated edges. Consequently, all measures that depend on the in- and out-degrees will undergo the same adjustment.

3.3 Data set

From the Statistical System of Waterways of ANTAQ, we obtained the following information for cabotage on the Brazilian coast:

- all types of cargo moved via cabotage from 2010 to 2015;
- all types of cargo moved via cabotage monthly from 2010 to 2015;

- the cargo's city and state of origin;
- the cargo's city and state of destination;
- the cargo's maritime installation of origin (port or private use terminal-PUT);
- the cargo's destination maritime installation (port or private use terminal-PUT);
- the total value of the cargo transported via cabotage from 2010 to 2015.

3.4 Building cabotage networks

With the data in hand, the system was modeled, for each year, in the form of a directed network, where:

$$V = \{ p_1, p_2 ..., p_i, ..., p_n \} \qquad (10)$$

Where the set of the vertices V is the set ports/PUTs and p_i corresponds to some port and,

$$E = \{ (p_i, p_j) \}; (p_i, p_j) \neq (p_j, p_i) \qquad (11)$$

Where the set of the edges E represents the trips originating in p_i and destination in p_j.

Figure 1 shows the static network that considers all the ports and trips in the period between 2010 and 2015 of cabotage transport on the brazilian coast. The vertices are distributed according to their latitude and longitude, drawn on the map of Brazil for a better visualization. Nodes of different grades are proportional to their sizes and follow a gray scale, according to legend. The edges have thickness proportional to the number of trips made in the period. It's important to say that some trips of the cabotage enter by inland waterways.

4 RESULTS AND DISCUSSION

4.1 *Network indexes*

Figure 2 shows the evolution of the Brazilian cabotage network between the years 2010 and 2015. The vertices are distributed according to their latitude and longitude, adjusted to the map of Brazil for a better visualization. Nodes of different grades are proportional to their sizes and follow a gray scale, according to legend. The edges do not have differentiated weights and are not multiple edges. Table 01 shows the network indexes for the network in each year, both disregarding multiple edges. In addition, Table 01 shows the indices for the largest connected grid of the network, with no directed edges, for the purpose of network topology analysis. The results reveal that:

- There are few changes from year to year;
- The average grade shows that there are between three to four different trips per port each year. This number is from eight to nine different trips in the largest connected network (which corresponds, on average, to 67% of the entire network);
- On the other hand, densities are very low, averaging 3%, rising to 11% on average in the largest connected grid. This means that, in theory, this waterway mode allows the number of different routes to be increased;
- The diameters are high, on average 6, which means that there are still very distant ports in terms of intermediate stops. Of course, the high size of the coast contributes to this;
- Meanwhile, the average of the routes between two ports is between 2 and 3. In other words, one or two stops on the average between two ports
- Clustering coefficient reach 22% in the original network and 63% in the largest connected

Figure 1. Static network of cabotage shipping in the brazilian coast from 2010 to 2015.

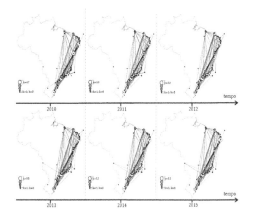

Figure 2. Dynamic network of brazilian cabotage maritime network, from 2010 to 2015.

network. This, coupled with low minimum paths, reflects the degree of efficiency of the network, since most of the neighboring ports, in terms of edges, are also connected to each other.

4.2 Networks topologies over time

In order to evaluate the networks topologies, according to the classification proposed in Section 3.1, it was necessary to make the following adjustments for each network:

- Make $(p_i, p_j) = (p_j, p_i)$, that is, transform all edges into non-directed;
- eliminate multiple edges, i.e., repeated edges;
- make the edges with the same weight;
- to Filter the largest component of the network and consider it only if it contains more than 50% of the vertices of the network.

Once the necessary adjustments have been made, the metrics are computed for the network each year, see Table 1, in the lines that correspond to the largest component.

The next sections evaluate each one of the topologies presented in Section 3.1, to verify if they characterize the Brazilian maritime cabotage networks.

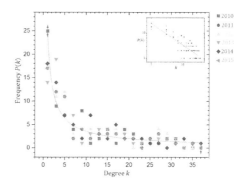

Figure 3. Distribution of degrees for the largest components of the brazilian cabotage maritime network, between 2010 and 2015. In inbox, the same data in logarithmic scale. In red, the average point adjustment ($\gamma = 0,66 \pm 0,08$, with $R^2 = 0,81$).

4.2.1 Random and regular topologies

The grid distribution of a network gives us clues about its topology. Figure 3 show that the frequency distributions of degrees in all years do not follow a normal curve. This means that the average degree is not representative and the network has no regular or random topologies.

4.2.2 Scale-free topology

The distributions of Figure 3 reveal the low frequency of ports that have many connections. Vertices in this distribution condition are called Hubs.

However, networks cannot be considered scale-free topology in any of the networks due to $\gamma < 2$, which was already expected for the few vertices.

The fact that the distributions resemble more a power law than a normal curve leads us to assume that new routes to be created are more likely to be in ports that are considered Hubs than in ports with few routes. Section 4.3 presents the ports that are considered Hubs in this study.

4.2.3 Small-world topology

According to Watts and Strogatz (1998), a non-directed network, which has more than 50% of its vertices connected (largest component), will be small-world if the mean clustering of its vertices is much larger than the equivalent random network's vertex clustering, with its vertices mean minimum path close to that of the random network's equivalent.

Being the small world maritime network in this period of time, we conclude that it is efficient for the transport of loads. This shows, on average, that most routes are direct from one port to another or have one to two intermediaries. In addition, due to the clustering, each port in media has 60% of its neighbors connected to each other.

Table 1. Network indices corresponding to each year of the brazilian cabotage network, 2010 to 2015, without multiple edges.

Indices	2010	2011	2012	2013	2014	2015
n	114	114	114	114	114	114
m	400	407	404	431	451	446
<k>	3.50	3.57	3.54	3.78	3.96	3.91
Δ	0.047	0.032	0.031	0.033	0.035	0.035
D	6	6	6	7	5	5
<ℓ>	2.42	2.24	2.34	2.44	3.34	2.46
<C>	0.19	0.22	0.19	0.21	0.2	0.22
Largest component*	66% 2010	61% 2011	63% 2012	66% 2013	67% 2014	79% 2015
n	75	70	72	75	76	90
m	303	312	313	330	345	347
<k>	8,08	8,91	8,69	8,80	9,08	7,71
Δ	0,109	0,129	0,122	0,119	0,121	0,087
D	7	6	5	5	5	7
<ℓ>	2,67	2,48	2,47	2,39	2,34	2,63
<C>	0,54	0,63	0,52	0,50	0,49	0,51
<ℓ>rand**	2,20	2,24	2,20	2,16	2,13	2,45
<C>rand**	0,14	0,16	0,10	0,11	0,13	0,11

*Network indices of largest component with non-directed edges.
**Corresponding to random network (even n and <k>).

4.3 Highlight vertices

We highlight here the main ports in each year, regarding the number of connections with other ports (high grade k) and the number of routes between two ports.

4.3.1 Degree Centralities over time

Figure 4 shows the ports that stood out for the value of their degree centralities in each year. That is, by the number of routes to different ports.

Ports that have stood out in terms of connections with others are the main areas of port influence in Brazil, such as Santos-SP (2011 to 2013 and 2015). We can consider the ports in this condition as being *hubs*.

This result may be interesting for a manager who wants to route a large load to as many different places with as few stops as possible. Just forward to one of the hubs and from there you will have direct access to many destinations. However, because of the flow in these places, there is a high chance of overriding, resulting in the undesirable demurrage.

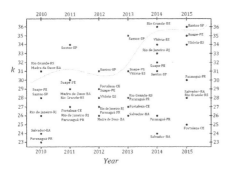

Figure 4. The seven ports that most stood out in terms of the centrality of degree in the Brazilian maritime cabotage network in each year, from 2010 to 2015. The ports above the dotted line have the largest centralities in the corresponding year.

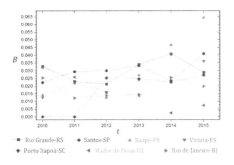

Figure 5. The seven ports that stood out most in terms of the beetweness in the Brazilian maritime cabotage network and its evolution between 2010 and 2015.

4.4 Beetweness Centralities over time

Figure 5 shows the evolution of the values of these centralities for the ports that had on average the highest values of beetweness.

A port that has a high value of beetweness, is between most of the paths between two ports. Suape-PE in 2014 and 2015, obtained the highest value of beetweness, nevertheless in 2014 this port was the fourth of greater degree (Figure 4).

Thus, a cargo that was transported from one port to another in this period has a high probability of passing through the port of Suape-PE in an intermediate stop.

However, the Suape-PE port was not always on top of the beetweeness. In 2010, it was the fifth most intermediate port.

5 CONCLUSIONS

The modeling of the waterway of Brazilian cabotage through a network allowed us to obtain relevant information to know the modal in more depth.

We noticed that the networks vary little from year to year, which suggests a robust network, which is interesting for times of economic crisis, but not for times of non-crisis.

The densities are low, revealing the existence of many possibilities of routes between ports not yet explored.

However, the networks are well grouped. For direct routes between a port A and a port B, if an intermediate stop is required in a port C, connected to A, there is a 60% chance of occurring, that is, of B and C also having routes between them.

This, coupled with the fact that the average minimum paths are low, reveals that the networks show small-world topologies, which makes them efficient, without many stops in cargo transportation. But the low density and the high diameter shows that it is still possible to increase considerably the number of direct routes between ports.

The analysis of the centralities allows us to know which ports have more different routes (hubs) and the ones that allow stops between routes of any other two ports of the network.

Finally, we suggest that future studies use this model for comparisons between different modes of transport (air and land). We believe that the information obtained from the dynamics of the evolution processes in a maritime network would help in the decision making for logistics managers.

REFERENCES

Barabási, A.L.; Albert, R. Emergence of scaling in random networks. Science, v. 286, 1999, p. 509–512.

Barabási, A. L.; Network science. Cambridge University Press, 2016.

Bird, J.; Seaport development: some questions of scale. Seaport Systems and Spatial Change, Chichester: Wiley, p. 21–41, 1984.

Clark, X.; Dollar, D. and Micco, A.; Port efficiency, maritime transport costs, and bilateral trade. Journal of development economics, v. 75, n. 2, p. 417–450, 2004.

Ducruet, C. and Zaidi, F.; Maritime constellations: a complex network approach to shipping and ports. Maritime Policy & Management, v. 39, n. 2, p. 151–168, 2012.

Ducruet, C.; Ports in proximity: Competition and coordination among adjacent seaports. Routledge, 2016.

Erdos, P. and Rényi, A.; On the evolution of random graphs. Publ. Math. Inst. Hung. Acad. Sci, v. 5, n. 1, p. 17–60, 1960.

Fremont, A.; Global maritime networks: The case of Maersk. Journal of Transport Geography, v. 15, n. 6, p. 431–442, 2007.

Glaeser, E.L. and Kohlhase, J.E.; Cities, regions and the decline of transport costs. Papers in regional Science, v. 83, n. 1, p. 197–228, 2004.

Milgram, S.; The small world problem. Psychology today, New York, v. 2, n. 1, p. 60–67, 1967.

Pereira, H.B.B.; Santos, C.C.R.; Cunha, M.V. and Lopes, C.R.S. Network Teoria das redes para estudos na área de saúde [Theory for healthcarestudies]. In Carneiro, T.K.G.; Rios, J.A.; Souza, C.R.B. (Eds.). Tecnologias aplicadas à saúde [Technologies appliedtohealth]. Salvador: Edifba, 2016. chp. 5, p. 89–115.

Robinson, R.; Ports as elements in value-driven chain systems: the new paradigm. Maritime Policy & Management, v. 29, n. 3, p. 241–255, 2002.

Watts, D.J. and Strogatz, S.H. Collective dynamics of small-world networks. Nature, v. 393, n. 4, 1998, p. 440–442.

Maritime Transportation and Harvesting of Sea Resources – Guedes Soares & Teixeira (Eds)
© 2018 Taylor & Francis Group, London, ISBN 978-0-8153-7993-5

The impact of the Maritime Labour Convention in the maritime transport

S. Santos
Shore Excursions Business, Unit Manager Buzz DMC, Lisboa, Portugal

A. Simões
Escola Superior Náutica (ENIDH), Paço de Arcos, Portugal

ABSTRACT: The MLC (Maritime Labour Convention, 2006), was created to provide decent work conditions to the seafarers and fair competition for the ship owners. It became the fourth pillar of the international regulation together with SOLAS, STCW and MARPOL. Through the analysis of interviews, carried out to different maritime players, and questionnaires, carried out exclusively to seafarers on duty, we realise that there are still several requirements that are not properly implemented and some improvements can already be identified. The fact that the seafarers regard the MLC, 2006 as a compilation of existing conventions, regulations and laws, that crucial motivational factors are not mentioned on the Convention and a certain distrust on its inspections, disserve the importance of this Convention. Its moderate impact would benefit from a proper dissemination of its contents, goals and boundaries.

1 INTRODUCTION

It was the circumstances of Titanic's sinking and the great loss of life that accompanied it which prompted the international community to seriously revise their laws concerning the safety of life at sea. In 1914 SOLAS (Safety Of Life At Sea) Convention, generally regarded as the most important of all international treaties concerning the safety of merchant ships, had its first version in response to the Titanic disaster. It's latest version dates of 1974, but every year many amendments enter in force. Further regulations where implemented in order to cover all aspects of the maritime regulation such as STCW (International Convention on Standards of Training, Certification and Watchkeeping for Seafarers) and MARPOL (International Convention for the Prevention of Pollution from Ships). However, regarding the seafarers working conditions, many conventions, laws and standards had not been ratified. States interpreted the conventions differently and the ruled for ship owners and seafarers differed from country to country. Various seafarer's organisations, ship owners and governments started working together, in 2001, to create new maritime labour rules.

The MLC (Maritime Labour Convention, 2006), was created to provide decent work conditions to the seafarers and fair competition for the ship-owners. It became the fourth pillar of the international regulation together with SOLAS, STCW and MARPOL.

By 15 June 2017, from the 185 ILO states, 84 have ratified this Convention. These 84 states represent 91% of the world's gross tonnage. Despite the success of its ratification what was indeed—its real impact on the improvement of the seafarer's working conditions?

Through the analysis of interviews, carried out to different maritime players, and questionnaires, carried out exclusively to seafarers on duty, we realise that there are still several requirements, that are not properly implemented, and some improvements can already be identified. The fact that the seafarers regard the MLC, 2006, as a compilation of existing conventions, regulations and laws, that crucial motivational factors are not mentioned on the Convention and a certain distrust on its inspections, disservice the importance of this Convention. Its moderate impact would benefit from a proper dissemination of its contents, goals and boundaries.

2 MLC BACKGROUND AND IMPLEMENTATION

"The MLC means that (…) consumer goods that we take for granted in our daily lives, will have been transported across the seas (…) by seafarers who

will be entitled to decent working conditions. It also means that the next time you may be lucky enough to be on a cruise ship or commercial yacht for the holidays of a life time, your crew will have a safe and healthy working environment. (...)". This statement from Cleopatra Doumbia-Henry, former Director of the International Labour Standards Department of the International Labour Organization (ILO) and the "Mother of MLC", summarizes the aim of this Convention to provide decent work conditions to the seafarers and fair competition for the ship owners.

It was the circumstances of Titanic's sinking and the great loss of life that accompanied it which prompted the international community to seriously revise their laws concerning the safety of life at sea. In 1914 SOLAS Convention, generally regarded as the most important of all international treaties concerning the safety of merchant ships, had its first version in response to the Titanic disaster. It's latest version dates of 1974 and further regulations where implemented in order to cover all aspects of the maritime regulation such as STCW and MARPOL. However, in regards to the seafarers working conditions, many conventions, laws and standards had not been ratified. States interpreted the conventions differently and the ruled for ship-owners and seafarers differed from country to country.

The Maritime Labour Convention is constituted by upgrading existing international conventions such as:

– Freedom of Association and Protection of the Right to Organise Convention, 1948 (No. 87)
– Right to Organise and Collective Bargaining Convention, 1949 (No. 98)
– Forced Labour Convention, 1930 (No. 29)
– Abolition of Forced Labour Convention, 1957 (No. 105)
– Minimum Age Convention, 1973 (No. 138)
– Worst Forms of Child Labour Convention, 1999 (No. 182)
– Equal Remuneration Convention, 1951 (No. 100)
– Discrimination (Employment and Occupation) Convention, 1958 (No. 111)

The ILO's Governing Body has also designated another four conventions as "priority" instruments, thereby encouraging member states to ratify them because of their importance for the functioning of the international labour standards system. The four governance Conventions are:

1. Labour Inspection Convention, 1947 (No. 81)
2. Employment Policy Convention, 1964 (No. 122)
3. Labour Inspection (Agriculture) Convention, 1969 (No. 129)
4. Tripartite Consultation (International Labour Standards) Convention, 1976 (No. 144)

In 2001, the ship owners and seafarers in the Joint Maritime Commission jointly proposed a new approach and preferred solutions, known by the "Geneva Accord", for the existing maritime labour standards in order to better achieve two goals: seafarer's right to decent work and help ensure fair competition for ship owners. This was the first step towards the creation of a comprehensive Convention. After 5 years of intense negotiations, between the involved parties, desiring to create a single coherent instrument embodying, as far as possible, all up-to-date standards of existing International Maritime Labour Conventions and Recommendations, as well as the fundamental principles to be found in other international labour Conventions, in February of 2006, the MLC was adopted in Geneva on the 94th ILC session.

The MLC, 2006 was structured based on the desire to create a single, coherent instrument embodying as far as possible all up-to-date standards of existing international maritime labour Conventions and Recommendations, as well as the fundamental principles to be found in other international labour Conventions, in particular:

– Freedom of Association and Protection of the Right to Organise Convention, 1948 (No. 87)
– Right to Organise and Collective Bargaining Convention, 1949 (No. 98)
– Forced Labour Convention, 1930 (No. 29)
– Abolition of Forced Labour Convention, 1957 (No. 105)
– Minimum Age Convention, 1973 (No. 138)
– Worst Forms of Child Labour Convention, 1999 (No. 182)
– Equal Remuneration Convention, 1951 (No. 100)
– Discrimination (Employment and Occupation) Convention, 1958 (No. 111)

The implementation of the MLC 2006, has 5 main objectives:

• Updating and consolidating existing ILO conventions;
• Recruiting, developing, motivating and retaining qualified staff;
• Preventing poor working and living conditions being used to create a competitive advantage;
• Creating a level playing field;
• Establishing the MLC, 2006 as the 4th pillar in international maritime regulations

The Convention intends to minimize the non-conformities verified during the inspections and that are part of the following Conventions:

– Safety of Life at Sea, 1974;
– Preventing Collisions at Sea, 1972;
– Standards of Training, Certification and Watch keeping for Seafarers, 1978;

- Marpol 73/78;
- United Nations Convention on the Law of the Sea, 1982.

The Convention entered into force on 20 August 2013 for the first 30 members with registered ratification, and a combined world GT of nearly 60%, becoming the fourth pillar of the international regulation together with SOLAS, STCW and MARPOL.

There are five primary Titles in the Convention, each of which serves a distinct purpose. Each Title is composed of regulations, standards (Part A), and guidelines (Part B). The regulations and standards are mandatory, while the guidelines are more particularized suggestions for implementation that may or may not be followed at the signing party's discretion.

The first Title – *Minimum Requirements for Seafarers to Work on a Ship* – establishes a minimum age of sixteen for seafarers, mandates recruitment and training procedures, and requires each seafarer to produce a medical certificate verifying his good health before he is employed.

The second Title – *Conditions of Employment* – requires all seafarers to enter into a written employment agreement with the ship owner. Calls for payment on at least a monthly basis, establishes that the standard work day will be eight hours and overtime pay must be at least 25 percent greater than the standard rate, and mandates rights to both shore leave and repatriation.

The third Title – *Accommodation, Recreational Facilities, Food and Catering* – establishes specific standards for the size and furnishings of living quarters, and mandates that all religious and cultural food requirements be respected and accommodated by the ship owner.

The fourth Title – *Health Protection, Medical Care, Welfare and Social Security Protection* – requires that onboard medical care be given to seafarers in need at the ship owner's cost, ensures that sick or injured seafarers will be paid as long as they remain on board the ship, and provides for occupational safety standards.

The fifth and final Title – *Compliance and Enforcement* – requires that ships carry a maritime Labour certificate which certifies compliance. Provides that each individual nation will be responsible for enforcing these provisions over all ships that sail under their flag, grants certain protections to whistleblowers, and allows member nations to perform inspections of ships from other member nations that enter their ports to ensure compliance.

Compliance with the Convention's regulations and standards will be policed via:

Article XIII – establishes a committee that will oversee compliance and continually review the Convention for amendment. This committee will consist of representatives from each ratifying nation, as well as seafarer and ship owner representatives, to ensure that all interests are protected;

Article V – provides for in-port ship inspections by ratifying nations and declares that member nations will be forbidden from favoring ships that fly the flag of non-ratifying states.

The MLC, 2006, defines seafarers as "all persons who are employed or are engaged or work in any capacity on board a ship to which the Convention applies". This includes not just the crew involved in navigating or operating the ship but also, for example, persons working in hotel positions that provide a range of services for passengers on cruise ships or yachts. The MLC, 2006 caters for all the seafarers working on board ships that fly the flag of countries that have ratified the Convention.

The Convention applies to a wide range of ships operating on international and national or domestic voyages. It covers all ships other than those which navigate exclusively in inland waters or waters within, or closely adjacent to sheltered waters or areas where port regulations apply. The Convention applies to all those ships, whether publicly or privately owned, that are ordinarily engaged in commercial activities, except for ships engaged in fishing or in similar pursuits, ships of traditional build such as dhows and junks and warships or naval auxiliaries.

3 MLC IMPACT ON SEAFARERS

Despite the success of its ratification, what was indeed its real impact on the improvement of the seafarer's working conditions?

Questionnaires were carried out exclusively to seafarers on duty 2 years after the Convention enter into force in order to measure its impact in their working environment. Questionnaires had both closed and open ended questions.

We receive answers of 97 participants, mainly males (82%), from 27 different nationalities. The seafarers that participated in this questionnaire, disseminated online, are embarked mainly on oil tankers but also in cargo ships, cruise vessels and containers.

The principal results extracted for the answers demonstrated that:

- 77% of the participants were aware of the existence of the MLC and were made aware of it by the ship owners;
- 48.5% of the seafarers stated they have never been confronted with a situation of non-compliance;
- 48.5% states that their ship owner has showed willingness to comply with the MLC's provisions;
- 71.1% believes that the MLC will bring them more confidence claiming for their rights.

From the open-ended questions, we understood further seafarer's concerns. Sharing the professional and personal space, it's a characteristic of the maritime labour. Therefore, seafarer's concerns are of a broader dimension and need to be equally catered for.

Two years after the enter into force of the Convention was still possible to observe disparities of treatment and on board conditions depending not only on the ship owner but also the maritime sector of the seafarer. Seafarers shared their dissatisfaction with the Convention exclusions that, from their perspective, assists evading the compliance of the basic working and living conditions and creates an unfair labour differentiation between seafarers.

The low rate of ratifications, on the Asian and South American continents, it's also seen as a concern for the seafarers as it not only allows ships under convenience flags to benefit from inspections gaps as the use of accredited recruitment agencies in countries that have not ratified the Convention may be delayed or ignored. Seafarers claim the urge of using accredited recruitment agencies, standardize, and concentrated hiring processes to improve safety on board and to assist disseminating the Convention.

Another concern lays in the fact that seafarers regard the MLC as a compilation of existing conventions, regulations and laws which all together have had little impact in their welfare as they suffered from low ratification, were too detailed, which allowed different interpretations, and were outdated due to slow amends processes. The reinforcement of its provisions inspections, its determinant to obtain a successful implementation of the Convention and consequently increase the seafarer's welfare and trust. However, a certain distrust on the PSC (Port State Control) inspector's qualifications to run proper inspections on the Convention's provisions, since they are mainly qualified to inspect technical and maritime matters, may disserve it's importance.

Delays on salaries payment, salary not equivalent to effort and hours worked, lack of equity regarding progression on the career and increase on salary. Lack of adequate food, suppression or inadequate time for meals, insufficient potable water, safety and neglected hygiene, denial of access to health care and non-compliance of repatriation are the aspects more often violated as per the seafarers.

Seafarer's, and some authors, also claim that despite common belief, MLC key objectives will not be achieved with the convention itself as the MLC does not properly address key motivation factors. Seafarer's mentioned as motivation factors the establishment of minimum salary, long term guarantee contract, guaranteed maximum period of stay on-board and minimum period of ashore staying to properly address fatigue factors and establishment of minimum safe manning in relation to ship type, size and trade.

However, some are able to point improvements since the entry into force of the MLC such as electronic recording of hours of work with overtime payment, annual wage increases, training provided and paid by the ship owner, recruitment through credible and certified agencies, improved access to health care, improvement of recreation and accommodation on board, better balance between contract length and holidays, wiliness on clarifying crew doubts regarding MLC and creation of means of anonymous reporting non-compliance.

This different perception of the MLC's impact on their welfare is undoubtfully caused by the different timings of its implementation by the involved maritime players.

Interviews to maritime players were also carried out so they could contribute, with their opinion, to identify implementation challenges and suggest improvements.

From these interviews, we understand that a correct mind set from the ship owners and regular inspections on the Convention compliance will determinate its successful implementation. The confidence that seafarers will feel on reporting abusive situations will also be determinant and this factor, may be conditioned by fear and labour insecurity, reinforce the need of a stronger HR (Human Resources) policy on the maritime sector.

The social related provisions of the Convention, larger living quarters, on board hygiene, decent food, access to health care and recreational activities, cap on working hours, assistance to repatriation and social security payment, were considered the most relevant for the seafarer's welfare by the interviewed.

A widely and proper implementation of the Convention is regarded as priority over amends, such not only because of the early stage of its entry into force but also of the costs implications to the sector and the number of seafarers affected. Nevertheless, the interviewed mentioned that introducing other social aspects into the Convention provisions such as a minimum of social security payments being established worldwide, an insure to provide compensation to seafarers and their families in the case of abandonment and a retirement entitlement for seafarers, would be of relevance. The interviewers also mentioned the urge of an effective solution concerning the hours of work vs rest hours.

4 CRITICAL ANALISYS OF RESULTS

It's unanimous that the unification of the existent maritime conventions was a necessity. Increasing the on board professional and living standards will turn this professional sector more attractive and consequently more will be motivated to pursue a career at sea. In addition, the rate of drop out will decrease and the sector will be able to keep its good assets for longer. This will inevitably lead to an increase of the maritime safety, improve the service, increase the ship owners profit, decrease a non-favorable competition and these combined factors will ultimately benefit the final customer.

Despite it's obvious benefits the MLC's impact on the seafarer's welfare was, at the time of the study, moderate. The fact that seafarers regard the MLC as a compilation of existing conventions, regulations and laws which previously have had little impact in their welfare, that key motivational factors are not properly address and its different timings of implementation by the involved maritime players are some of the causes of this moderate impact.

The MLC depends on a set of factors to achieve its goal being one of the most important its implementation monitoring. Seafarers question the capability and knowledge of the Port State Control Officers to properly monitor its entry into force, as they are mainly qualified for technical and maritime matters. However, it is important to know that, as per ILO's guidelines for Port State Control the officers carrying out inspections under the Maritime Labour Convention 2006, needs to have adequate qualifications and must be trained for this purpose.

The PSC must have a number of officers trained to carry out inspections under the MLC, which, in most cases, will involve personnel that are already qualified under the existing international PSC arrangements, but it can, as well, in some countries, be carried out by an authorized officer that are not necessarily qualified as a PSC Officer for other purposes. For example, a maritime labour inspector.

While the proper training of the PSC Officer for the MLC inspections it's still of major importance, the fact that other qualified officer can carry out inspections increases its effectiveness.

Another important factor is that some of the key motivational factors pointed out by the Seafarers, not well addressed in the Convention, are not all its competence and should not diminish its importance. Factors such as minimum wage are of difficult resolution due to the different pay levels for different nationalities based on the different worldwide cost of living as well. These kind of factors, although extremely important, will need to be discussed in a different level, as too detailed Conventions, as observed in the past, tend to have a low ratification and a consequently low impact.

Its moderate impact would benefit from a proper dissemination, throughout the maritime industry, of its contents, goals, and boundaries to achieve a proper implementation and consequently a stronger impact. As mentioned by the maritime players during the interviews, more important than considering amends, it's a proper implementation. Implementing a Convention of this magnitude and with so many players involved, take considerable time. This difference it's perceived by the seafarers. As mentioned previously and showed by the results, however we should understand this as a normal situation considering that the study was carried out in an early stage of its implementation. Being able to already classify its impact as moderate, it's a sign of success as it is a rapid ratification.

Despite the need of a solid implementation, the MLC has recently suffered an amendment. Seafarers are now protected in case of crew abandonment, with the entry into force of new provisions regarding this matter, and can rest assure that the Convention is willing to improve and to not suffer from slow amends as seen on previous. This is also an important step on increasing its importance among the seafarers.

5 CONCLUSIONS

Seafarers have always faced limited rights. However the situation is about to improve based on the entry into force of the MLC 2006 and its subsequent ratifications by different countries.

The MLC, 2006 faced a number of initial problems in its implementation, at the level of the Flag States, especially because of its heterogeneity and differences in the economic and social development of the ratifying States. Moreover, the South American and Asian continent still have a low rate of ratification which demonstrates that the 2006 MLC is still not widespread and that there is still a great way to go until its complete acceptance.

There is no doubt that the MLC 2006 represents a success and a significant positive step for the plight of the seafarers as regards their rights. There are inherent weaknesses that have been recognized which need further considerations so as to make the instrument achieve its full purpose and bring about fairness and equity to the concerns of the seafarers.

Being at its early stage of implementation, the interpretation of some items and compliance it's difficult. Without experience and when many doubts arise, the Flag States are following the good procedures and experiences of other Flags. Due to this, the initial expectations as to the implementation and immediate effect of the Convention have been drastically reduced. The MLC 2006 has however not been without criticisms with respect to certain omissions and neglect of certain issues that concern seafarers.

These gaps in ratification allow ship owners to move records of their vessels to countries where MLC is not yet implemented, especially in Flags of Convenience, allowing the existence of totally unprotected trade arenas, as the clause of "no more favorable treatment" it's not applicable there yet.

Such conditions of implementation and consequent lack of uniformity hamper to a certain extent the confidence that the seafarers place in the MLC, 2006. However, and although it still does not fully comply with its objectives, we cannot deny that, in its short implementation period, the MLC 2006, has made a significant contribution to the improvement of working conditions and the habitability of seafarers. Since its entry into force, the number of vessels arrested by the Paris MoU authorities, associated with the non-compliance of the Convention, has increased.

As verified by the survey carried out in this project, although the impact of the Convention on the seafarer's day to day is only moderate, they are more aware of their rights and more confident in their claims since its entry into force. The fact that the MLC 2006 it's mentioned when they place a complaint and it's an excellent example of this. Investing on the training of inspectors responsible for monitoring its implementation will certainly turn the Convention more credible in the eyes of the most skeptical seafarers and will help ensure that the Convention's provisions are met, thus ensuring their long-term sustainability.

It is important to clarify the objectives and limits of the Convention's intervention so its objectives and goals are not misunderstand, diminishing its importance. The Convention would greatly benefit from clarification sessions where its competences would be explained clearly to the seafarers and would assist them understanding that certain claims should be addressed to other organisms such as ITF (International Transport Workers' Federation), Flag States and Unions.

During this project both the seafarers and the maritime players that participated in interviews and questionnaires, had a conservative position regarding amendments to the Convention. Although they had a few suggestions, they all stated it was better to first implement the Convention correctly and worldwide spread it instead of amending it. However, the recent introduction of the amendment related to financial security of seafarers in cases of abandonment, and contractual claims for compensation in the event of a seafarer's death or long-term disability due to an operational injury, illness or hazard, was welcome within the maritime community.

Since this project was realized in an early stage of its implementation, it would be of relevance to verify its impact nowadays, 4 years after its entry into force.

REFERENCES

American Bureau of Shipping, 2010, Guidance Notes On The ILO Maritime Labour Convention 2006, New York.

Australian Maritime Safety Authority, 2013, *A Guide to the Implementation of the MLC in Australia,* Australian Government.

DNV (2009), "Voluntary declaration of maritime labour compliance—Part II of 'Seacrown'," Ship Management Inc.

ILO (2006a), *Labour standards: Maritime Labour Convention,* Geneva, Switherland.

ILO (2006b), *MLC Maritime Labour Convention,* 2006, Geneva, Switherland.

ILO (2006c), *Resolution concerning recruitment and retention of seafarers* (Vol. XI, 94th (Maritime) Session), Geneva, Switherland.

ILO (2009a), *2006,* Geneva, Switherland: International Labour Organization.

ILO (2009b), *Maritime Labour Convention 2006,* Geneva, Switherland.

ILO (2014), Rules of the Game, A brief introductionto International Labour Standards Geneva, Switherland.

ILO. (1976), I*LO Convention 147 concerning minimum standards in merchant ships,* Geneva, Switherland.

IMO (1972), *The international regulations for preventing collisions at sea,* London, U.K. International Maritime Organization.

IMO (1973), *The international convention for the prevention of pollution from ships,* London, U.K. International Maritime Organization.

IMO (1978a), *The international convention for the safety of life at sea,* London, U.K. International Maritime Organization.

IMO (1978b), *The international convention on standards of training, certification and watchkeeping for seafarers* (2010 ed.), London, U.K. International Maritime Organization.

IMO (1995), *Procedure for port state control, Resolution A, 787(19),* London, U.K. International Maritime Organization.

IMO (1996), *International safety management code* (2010 ed.), London, U.K. International Maritime Organization.

IMO (1999), *Amendments to the procedures for port state control, Resolution A, 882(21),* London, U.K. International Maritime Organization.

IMO (2002), *The international code for the security of ships and of port facilities,* London, U.K. International Maritime Organization.

International Labour Organization, 2006.

Susana Santos, 2015, Critério de análise do impacto da implementação da Convenção sobre o Trabalho Marítimo (MLC 2006), Master Thesis, ENIDH, Portugal.

Weston, C., 2014, *Maritime Labour Convention Deficiencies/Detentions,* Australian Maritime SafetyAuthority, Canberra.

http://ilo.org/global/standards/maritime-labour-convention/lang--en/index.htm.

Maritime traffic models

Maritime Transportation and Harvesting of Sea Resources – Guedes Soares & Teixeira (Eds)
© 2018 Taylor & Francis Group, London, ISBN 978-0-8153-7993-5

Mapping of the ship collision probability in the Strait of Istanbul based on AIS data

Y.C. Altan
Department of Civil Engineering, Bahcesehir University, Turkey

E.N. Otay
Department of Civil Engineering, Bogazici University, Turkey

ABSTRACT: Ships in congested waterways are more prone to collision than those in open sea conditions. The overall statistics of maritime accidents are known in most waterways, yet their spatial distribution in detailed maps is not common. The collision probability is mostly determined on route-based methods, however it distorts the spatial traffic distribution. Hereby, a collision model is developed based on molecular collision diameter which uses long-term AIS data to calculate the geometric collision probability. To analyze the maritime accidents, the Strait of Istanbul (SOI) is divided into sectors and further subdivided into cells of each the geometric collision probability is calculated using one-year continuous AIS data. Model results show that the collision probability increases in the narrow passages and in the sharp turns of the SOI. The spatial distribution of accidents in high resolution maps obtained from the sector based approach will be beneficial for stakeholders to take necessary precautions.

1 INTRODUCTION

Ship to ship collision is one of the most encountered accident types in congested waterways. First studies of ship collision started in 1950s (Nichols 1950, Sadler 1957) where they rely solely on deterministic formulas and therefore come short in reflecting the real sea conditions (Wylie 1962). In 1970s, two pioneering works applied the molecular collision theory for the maritime accidents (Fujii & Shiobara 1971, Macduff 1974) and introduced the first collision probability theory based on collision diameter. Pedersen (1995) later quantified Fujiii and Shiobaras (1971) collision diameter to calculate the probability of causation factor. Montewka et al. (2010) improved the collision diameter as the minimum distance to collision (MDTC) defined as the distance where the collision is unavoidable by any maneuver. Ylitalo (2010) used ship tracks from AIS data to create accident probability map of a waterway in the form of passage histograms.

An alternative to molecular collision theory is the ship domain approach where the approach of two ships within a pre-defined distance is defined as a critical situation (Goodwin 1975) to predict the number of collisions (Goerlandt & Kujala 2011, Merrick, Van Dorp, Blackford, Shaw, Harrald, & Mazzuchi 2003, Wang & Chin 2016). Although the ship domain approach has been successfully used in waterways with steady congestion levels, its variable domain size causes uncertainties in the SOI whose

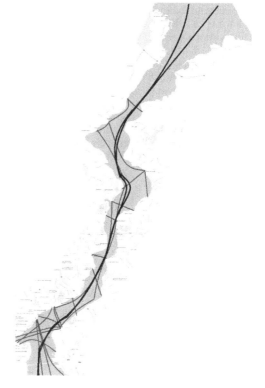

Figure 1. Representative sketch of transit and local traffic in the SOI.

congestion levels show spatial and temporal variations. Therefore, the molecular collision approach is used for determining the collision map of the SOI.

Spatial distribution of maritime collision probability is studied using molecular collision theory. In the following sections, the application of collision theory on a spatial distribution, use of long-term AIS data as input in a congested waterway and the comparison of model results with the past maritime collision records are discussed. The model is applied to the Strait of Istanbul (SOI) which is known as one of the most complex waterways in the world. In addition to sharp turns and strong currents, the dense local traffic makes the navigation more difficult (Altan & Otay 2017). The model accounts both for the scheduled (with pre-determined routes shown in Figure 1) and the more significant unscheduled local traffic which is accessed through long-term AIS data.

2 METHODOLOGY

Molecular collision theory has been widely used for predicting the maritime collision probability. Past results in the literature confirmed its reliability. Basically, the molecular collision theory states that when the center-to-center distance of two particles moving within a domain are equal or less than a certain value there will be a collision. This distance is called as the collision diameter and it varies with the size and velocity of the molecules.

For the maritime application, molecules are assumed as ships whose velocities and dimensions are used to calculate the collision diameters. Collision diameter of two ships that are approaching to each other, is defined as the projection of the ship dimensions to the line which is perpendicular to the relative velocity vector. According to Pedersen (1995), the collision diameter (schematically shown in Figure 2) can be calculated as:

$$D_{ij} = \frac{L_i V_j + L_j V_i}{V_{ij}} \sin\theta + B_j \left(1 - \left(\sin\theta \frac{V_i}{V_{ij}}\right)^2\right)^{\frac{1}{2}}$$
$$+ B_i \left(1 - \left(\sin\theta \frac{V_j}{V_{ij}}\right)^2\right)^{\frac{1}{2}} \tag{1}$$

where,

D_{ij}, is the collision diameter between the ships i and j.

L_i and L_j, are the overall lengths (LOA) of ships i and j, respectively.

V_i and V_j, are the velocity of the ships i and j, respectively.

V_{ij}, is the relative velocity of the ships

B_i and B_j, are the width of the ships i and j, respectively.

θ, is the encounter angle of the ships i and j.

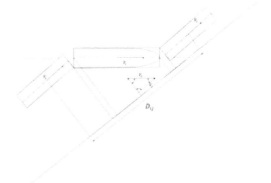

Figure 2. Collision diameter representation.

Equation (1) gives reasonable results when the encounter angle is greater than 10° or smaller than 170°; such encounters are considered as crossing. Possible encounter types and angles are as follows:

- Head-on ($0° \leq \theta \leq 10°$)
- Crossing ($0° \leq \theta \leq 170°$)
- Take-over ($170° \leq \theta \leq 180°$)

The number of collisions for a given ship, at a given time period is given as the number of ship centers located inside the area swept by the collision diameter. The swept area is the product of collision diameter, relative velocity of the particles and the time period. The number of collisions for a given ship can be formulized as:

$$D\rho_j \| \vec{V}_i - \vec{V}_j \| \Delta t \tag{2}$$

where,

D, is the collision diameter.

$\| \vec{V}_i - \vec{V}_j \|$, is the relative velocity of the ships i and j.

Δt, is the time period for which the collision probability is calculated.

ρ_j, is the ship density, defined as the number of ships per unit area, per time and calculated as in Equation (3):

$$\rho_i = \frac{Q_i}{v_i \Delta t} \tag{3}$$

where,

Q_i, is the number of ships travelling in the ith direction.

v_i, is the velocity of ship travelling in the ith direction.

The ship density formula in Equation (3) not only considers the number of ships in ith direction in a given area, it also considers the travel speed of these ships. If the ships in a certain direction are slower, they would stay a longer time inside the considered area and consequently, would have a higher density then the faster ones.

The geometric collision formula in Equation (2) gives the number of collisions for a given ship in the considered area; however, it does not indicate which area of a certain domain is more exposed to collision. Therefore, spatial variability cannot be found when the molecular collision theory is applied directly. In order to get the spatial variability, the theory is applied in sectors along the waterway (Figure 3). The question remains is the size of the sectors. In order to get a meaningful distribution of collisions, the navigation characteristics within a particular sector should remain constant. The characteristics include, course over ground (COG), speed over ground (SOG) and ship density. For a congested waterway, setting up sectors with steady-state characteristics is very difficult. Therefore, sectors are subdivided into smaller cells (Figure 3) to which the geometric collision probability formula is applied individually. The cell sizes are selected to be larger than the collision diameter; Otherwise the collision diameter may exceed the cell border and create artificial collision. The formula applied cell by cell is given in Equation (4).

$$N_c(ik, jk) = \sum_s \sum_i \sum_j \sum_{ik(s)} \sum_{jk(s)}$$
$$\frac{\rho_i(ik(s),s)\rho_j(jk(s),s)V_{ij}(s)}{\Delta l(ik,s)\Delta l(jk,s)\Delta t} \qquad (4)$$

where,

N_c, is the number of collisions
s, represents the sector
i & j, represent the entrance (N, S, E, W) of the sector
ik & jk, represent the cell inside the sector according to the entrance
ρ, is ship density
V_{ij}, is relative velocity
D_{ij}, is collision diameter
Δl, travel distance
Δt, considered time

Although the parameters inside the cells are uniformly distributed throughout the solution process of Equation (4), variation of parameters among the cells makes the equation inherently probabilistic. Therefore, an extra probability function is not added to the equation. According to the formula, the main factors determining the number of collisions in a given area are the density of the ships, the collision diameter and the relative velocity of the ships. Since ship collision is a rare event, more than one collision at the same time is not considered in the study.

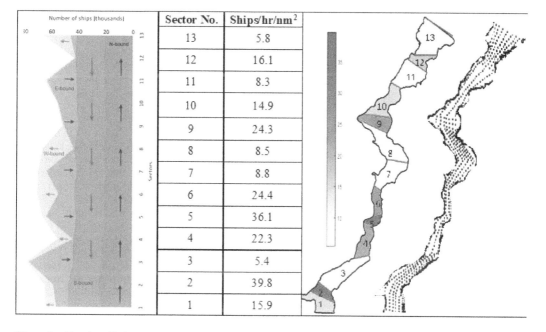

Sector No.	Ships/hr/nm^2
13	5.8
12	16.1
11	8.3
10	14.9
9	24.3
8	8.5
7	8.8
6	24.4
5	36.1
4	22.3
3	5.4
2	39.8
1	15.9

Figure 3.   Vessel traffic in sectors and cells along the strait of Istanbul.

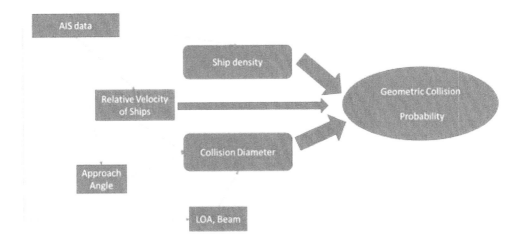

Figure 4. Schematic representation of the geometric collision probability parameters.

3 APPLICATION

The spatial distribution of the geometric collision probability within the Strait of Istanbul is determined using long-term AIS data for maritime traffic information. A detailed discussion of the maritime traffic data and the navigational characteristics of the SOI can be found in Altan & Otay (2017). Analysis of one year of continuous AIS data in the SOI, the number of ship passages are determined in all four directions (left most column in Figure 3). The traffic densities in number of ships per hour per nautical square meter are shown in sectors together with the cells within each cell are shown right most column in Figure 3.

A schematic representation of the parameters used in Equation (4) during the solution procedure is shown in Figure 4. The required maritime traffic data for Equation (4) are analyzed at the entrances of 13 sectors and the ship densities are updated for each of the 235 cells in the SOI.

4 RESULTS AND DISCUSSION

In the SOI, the most critical parameter which controls the collision probability in Equation (4) is the ship traffic density. The density map has been analyzed based on long-term AIS data in the SOI and given in Altan and Otay (2017). The second most important parameter is the collision diameter, whose details can be found in Altan (2016). The final outcome of this study is the spatial distribution of geometric collision probability in the SOI (Figure 5). The collision probabilities distributed over the 235 cells in the SOI increases towards the central cells along the waterway. This is mainly

due to the transit vessels which set a course close to the centerline of the channel to stay within the officially enforced boundaries of the Traffic Separation Scheme (TSS). As a result, this leads to an increasing ship density at central cells.

The spatial distribution of geometric collision probability among the 235 cells help to identify the accident hotspots along the SOI. From the navigational point of view, the maritime risk is related to the collision density per unit surface area. However, the cell based probabilities in Figure 5 are biased with cell size. These are divided by the cell area to obtain the probability density per unit area given at each cell center in units of number of collisions per square nautical mile per year. According to the collision probability density, the most dangerous waters for each nautical mile square are located in Sector 10 (Figure 3). Whereas the least dangerous waters in terms of collision probability is in Sector 12. Sector 4, 5, 2 and 9 are above the average value.

From the administrative point of view, sectoral collision probabilities are more important than the probability density per unit area in the SOI. Especially for accident response planning, it is important which sectors are more likely for a maritime accident to take place. For that purpose, the cell probabilities within each sector are summed up to calculate the sectoral collision probabilities.

The present model results (Figure 6a) are compared to two past studies, an earlier model Figure 6 b) and the actual accident logs (Figure 6c) of collisions in the SOI. Although the time periods are different, the locations of the actual ship collisions recorded by the Ministry of Transportation in the SOI between 1982 and 2003 (Akten 2004) are captured well with the present 2D model results based on one-year AIS data from 2015.

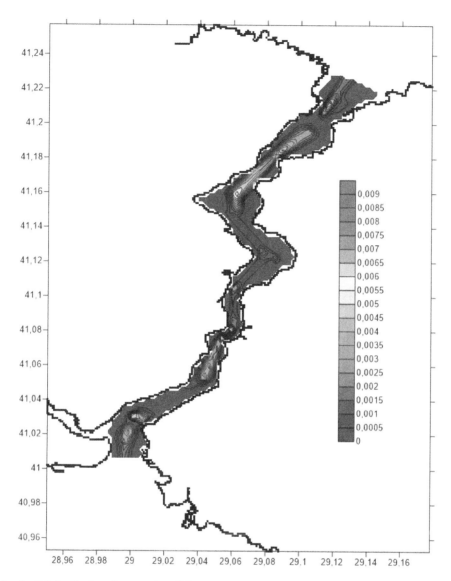

Figure 5. Spatial distribution of geometric collision probability in the SOI.

On the other hand, the collision probabilities predicted with a one-dimensional ship motion model (Otay & Özkan 2003) could not capture some of the observed collision sites. The present model shows improvements over the 1D collision model developed by (Otay & Özkan 2003) in identifying various collision hotspots in all sectors as in the past accident logs. The advantage of the present study over the accident logs is that it would show the collision probability before the accident happens since the present analysis is based on AIS readings which are continuously available for any time period since 2014 onwards.

5 CONCLUSIONS

In this study, a framework for the geometric collision theory is proposed, implemented to the Strait of Istanbul (SOI) and results are mapped. The results show the spatial distribution of the geometric collision probability throughout 235 cells in the SOI so the captains and the stakeholders can quantify and quickly assess the navigation risks. Another advantage of mapping the collision probabilities is that, captains can plan their routes through the SOI at minimum long-term risk. In addition to the theories in the literature, the missing parts of the geometric

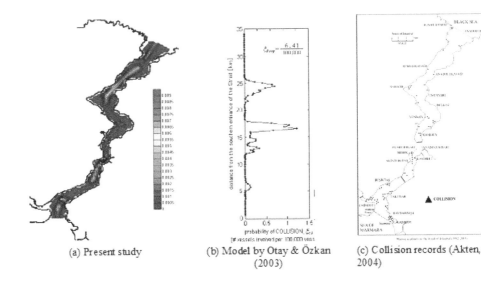

| (a) Present study | (b) Model by Otay & Özkan (2003) | (c) Collision records (Akten, 2004) |

Figure 6.   Comparison of geometric collision results with the past studies.

collision theory is improved in terms of spatial distribution. Geometric collision probabilities show some important hotpots along the Strait. Results are congruent with the past studies and accident locations. Another improvement is the two-dimensional positioning of the geometric collision probability map. So the captains can decide their lateral position in the sector for minimizing the collision probability. According to the probability density per nm^2, Sector 4 is on average the most dangerous sector in the Strait. The highest probability cell is observed at Sector 5. In general, the sectors with double bends have the highest collision probabilities. The proposed framework can be applied to all congested narrow waterways around the world. Improvements can be done by applying the system to calculate the dynamic collision probability. By using real-time maritime traffic data, dynamically changing collision probabilities can be calculated and mapped for live short-term predictions. Predicting near future will need a bigger database and behavioral algorithms to analyze the captain. From the entrance pattern, navigation pattern of a ship throughout the SOI can be predicted. Collision probabilities can be predicted before entering the relevant sector.

REFERENCES

Akten, N. (2004). Analysis of shipping casualties in the bosphorus. *Journal of Navigation 57*(03), 345–356.

Altan, Y.C. (2016). *Analysis and Modeling of Maritime Traffic and Ship Collision in the Strait of Istanbul Based on Automatic Vessel Tracking System*. Ph. D. thesis, Bogazici University.

Altan, Y.C. & E.N. Otay (2017). Maritime traffic analysis of the strait of istanbul based on ais data. *Journal of Navigation* –(–), –.

Fujii, Y. & R. Shiobara (1971). The analysis of traffic accidents. *Journal of Navigation 24*(04), 534–543.

Goerlandt, F. & P. Kujala (2011). Traffic simulation based ship collision probability modeling. *Reliability Engineering & System Safety 96*(1), 91–107.

Goodwin, E.M. (1975). A statistical study of ship domains. *Journal of Navigation 28*(03), 328–344.

Macduff, T. (1974). The probability of vessel collisions. *Ocean Industry 9*(9).

Merrick, J.R., J.R. Van Dorp, J.P. Blackford, G.L. Shaw, J. Harrald, & T.A. Mazzuchi (2003). A traffic density analysis of proposed ferry service expansion in san francisco bay using a maritime simulation model. *Reliability Engineering & System Safety 81*(2), 119–132.

Montewka, J., T. Hinz, P. Kujala, & J. Matusiak (2010). Probability modelling of vessel collisions. *Reliability Engineering & System Safety 95*(5), 573–589.

Nichols, C. (1950). Lessons from some collisions and groundings at sea. *Journal of Navigation 3*(02), 166–182.

Otay, E.N. & S. Özkan (2003). Stochastic prediction of maritime accidents in the strait of istanbul. In *Proceedings of the 3rd International Conference on Oil Spills in the Mediterranean and Black Sea regions*, pp. 92–104.

Pedersen, P.T. (1995). Collision and grounding mechanics. *Proceedings of WEMT 95* (1995), 125–157.

Sadler, D.H. (1957). The mathematics of collision avoidance in two dimensions. *Journal of Navigation 10*(04), 306–319.

Wang, Y. & H.-C. Chin (2016). An empirically-calibrated ship domain as a safety criterion for navigation in confined waters. *Journal of Navigation 69*(02), 257–276.

Wylie, F. (1962). Mathematics and the collision regulations. *The Journal of Navigation 15*(1), 104–112.

Ylitalo, J. (2010). *Modelling marine accident frequency*. Ph. D. thesis, Aalto University.

Maritime Transportation and Harvesting of Sea Resources – Guedes Soares & Teixeira (Eds)
© 2018 Taylor & Francis Group, London, ISBN 978-0-8153-7993-5

Risk assessment methods for ship collision in estuarine waters using AIS and historical accident data

P. Chen
Faculty of Technology, Policy and Management, Delft University of Technology, The Netherlands
School of Navigation, Wuhan University of Technology, China

J. Mou
School of Navigation, Wuhan University of Technology, China
Hubei Key Laboratory of Inland Shipping Technology, School of Navigation, Wuhan University
of Technology, Wuhan, China

P.H.A.J.M. van Gelder
Faculty of Technology, Policy and Management, Delft University of Technology, The Netherlands

ABSTRACT: Maritime transport adds significant contributions to the world trade and economy. Estuarine waters, due to the heavy congestion of inbound and outbound—and complicated vessel traffic, witnesses high risk of maritime accidents especially collision accidents. To quantify such risk, to analyze its characteristics, and to facilitate maritime safety management, this paper established a risk assessment method for ship collision using AIS and historical accident data. First, a collision candidate detection model based on Fuzzy Quaternions Ship Domain theory was established. By introducing Bayesian networks, an accident causational probability model was built based on historical accident data and previous research. Western Shenzhen waters are chosen as research area to conduct a case study. The preliminary result of the case study indicates that method this paper proposed can be applied to analyze risk of ship collision and its characteristics, as reference for maritime safety management in the region.

1 INTRODUCTION

Maritime transportation makes significant contributions to the development of world economy and cargo transportation, around 80% of the cargo are conducted by ship (Asariotis et al., 2016). Estuarine waters, as vital corridors, play a significant role in connecting inland and open sea shipping. With the rapid development of China's economy and shipping industry, the trends of ships' maximization and speed's increase become more tangible, resulting in increasing traffic density and complicating the navigation environment in such areas. In the meantime, maritime accidents especially ship collisions are posing risk to the safety of the public's life, property and environment. To improve navigation safety level and economic development, it is desirable to conduct researches on risk of ship collision accident.

Many efforts have been put into researches on risk from various perspectives. Some researchers focus on the fundamental definitions and perspectives of risk (Aven et al., 2011; Aven, 2012; Goerlandt & Montewka, 2015; Kaplan, 1997).

Although such ideas vary among different application fields, the idea that risk should contain elements of "scenario, likelihood and consequence" is well accepted among academia. When it comes to engineering practical, risk is denoted as product of probability of accident and its consequence (Kristiansen, 2005). In maritime field, compared with research on accident consequence, probability analysis approach received increasing interests from scholars due to its cost-effective nature in reducing accident risk (Pedersen, 2010).

The common approach for probability estimation of ship collision was originated from research of Fujii (1971) and Macdoff (1974) as Eq. 1 indicates:

$$P = N_A \cdot P_C \tag{1}$$

According to such method, risk of ship collision consists of two parts; the number of collision candidates N_A and the causational probability P_C. Collision candidate indicates ships in an encounter course where collision would happen if no evasive maneuver was performed. In the meantime, causational probability describes the possibility of collision candidate resulting in collision by environmental factors or human errors etc. Such method provides a concise and simple structure to perform probability analysis, which is popular in risk analysis of ship collision accident.

To obtain number of collision candidates, three categories of methods were proposed according to literature. Pedersen and et al (2010), as the representative of the first category, designed a statistical probability estimation model taking vessel traffic flow and its characteristics (volume, speed, spatial distribution etc.) into consideration. Similar work was also conducted by Fowler & Sørgård, and other researchers (Ylitalo, 2010). Application of ship domain theory is the second approach of collision candidate analysis. Qu and et al (2011) analyzed collision risk in Singapore strait, within which overlap of ship domain was utilized to identify collision candidate. Szlapczynski, et al (2016) incorporated domain violation and time to it as an alternative of distance at closest point of approach (DCPA) and time to the closest point of approach DCPA, which are classical indices to measure risk of ship collision. The third category, compared with approaches aforementioned, determines candidate with ship maneuverability into consideration. Monteweka, et al (2010) introduced minimum distance to collision (MDTC) to replace the collision diameter as new determination of collision candidate by measuring the smallest distance between ships with which collision can be avoided by taking maneuvers. The wide application of ship automatic identification system (AIS) has also offered new opportunity on ship collision risk analysis. AIS data contains ship static and dynamic information (e.g. positon, speed, course, ship length, width etc.), which can be utilized to evaluate safety level of certain water areas and determine collision candidate by analyzing actual behavior of ship. Mou et al (2010) conducted pilot study on introducing AIS data into risk analysis of ship collision in busy waters. Silveira, et al (2013) designed an algorithm to determine collision candidate by comparing future distance and collision diameter of encounter ships using AIS data and analyzed traffic pattern in coast of Portugal. Zhang et al (2015) developed a new method to evaluate near miss of encounters by taking distance, relative speed and difference of bearing of ships based on AIS data, which is also applicable to evaluate collision candidate.

Causational probability reflects the possibility of ships falling into collision accident. Historical accident statics and their analysis were initially applied into research on causational probability of maritime accident (e.g. Kujala et al., 2009). By introducing analytical methods such as Fault Tree Analysis (FTA), mechanism of accident and causal relationship between factors which could lead to accident can be obtained (e.g.Martins & Maturana, 2010). With the development of research on maritime transport system, the idea that maritime transport system is a complicated social-technical system and human factors are main contributor of accident have been well accepted. Traditional accident analysis methods such as historical data analysis and FTA may be insufficient under new context. Bayesian network, as a graphical probability inference method, is widely applicated in research on causational probability due to its advantage on dealing with complicated causal relationships between factors and capability in two-way inference and update with new evidence (Zhang & Thai, 2016). Martins, et al (2013) established a Bayesian network model to evaluate risk of oil tanks collision and identify critical activity in ship operation. Sotiralis, et al (2016) obtained ship collision probability due to human factors and most influential factors to human performance and ship collision by integrating Bayesian network and Technique for Retrospective and Predictive Analysis of Cognitive Errors (TRACEr). Based on empirical data such as AIS data and historical accident records, this paper proposes a risk assessment method for ship collision in estuarine waters which can provide insights such as spatial-temporal distribution of collision candidate of risk for the maritime safety management to facilitate their decision-making process. The main contents of this paper are structured as follows: Section 3 explains the methodology, framework and details of models which consists the risk assessment method this paper proposed. In section 4, a case study on western Shenzhen waterways are conducted and results are shown. Section 5 gives conclusions and future for this research.

2 MODELS FOR SHIP COLLISION RISK

The objective of risk assessment is to identify the scenario which could result in undesired event and estimate its possibility and consequence and to support decision making process. To obtain a comprehensive analysis of ship collision risk and to offer decision-making support to the safety management participators, a risk assessment method should address the following questions properly, which are the primary goal of this paper:

- What is the definition of risk this method adopted? How to measure it?
- From management perspective, what are the characteristics of collision risk that can be utilized to mitigate risk and improve navigational safety? How to obtain them?

This paper accepts that risk is possibility of collision accidents' occurrence in certain waterways during certain time. The applied methodology is based on the common approach that risk is product of geometry collision probability (Number of collision candidates) and causational probability. By introducing empirical data of AIS data and historical collision accident records, this paper designed theoretical calculation model to quantitively assess collision risk and analyze its characteristics. The risk assessment framework is shown in Figure 1.

To establish such theoretical model, a ship collision candidate detection model was established based on Fuzzy Quaternion Ship Domain (FQSD) theory, as the approach for obtaining the number of collision candidates. A causal probability model for ship collision was then proposed based on Bayesian network theory.

2.1 Collision candidate detection model

2.1.1 Fuzzy quaternion ship domain

In this paper, Fuzzy Quaternion Ship Domain theory (FQSD) (Wang, 2010) was introduced to determine if collision risk exists between encounter ships. Compared with traditional ship domain, FQSD not only takes ship maneuvering characteristics into consideration, but also includes reflection of seamanship of officer on watch from COLREG by defining different radii in fore, aft, starboard and port side of own ship by introducing factors such as ship speed, length, and concepts in international convention of collision avoidance at sea (COLREG). The formulation of FQSD are as follows:

$$FQSD_k(r) = \left\{ (x,y) \middle| f_k(x,y;Q(r)) \le 1, k \ge 1 \right\} \quad (2)$$

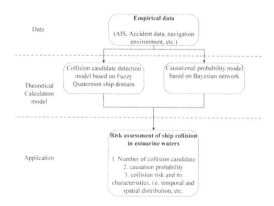

Figure 1. Framework of risk assessment method for ship collision in estuarine waters.

$$f_k(x,y;Q(r))$$
$$= \left(\frac{2x}{(1+\operatorname{sgn}x)R_{fore}(r)+(1-\operatorname{sgn}x)R_{aft}(r)} \right)^k$$
$$+ \left(\frac{2y}{(1+\operatorname{sgn}y)R_{starb}(r)+(1-\operatorname{sgn}y)R_{port}(r)} \right)^k \quad (3)$$

$$Q(r) = \left\{ R_{fore}(r), R_{aft}(r), R_{starb}(r), R_{port}(r) \right\} \quad (4)$$

$$R_i(r) = \left(\frac{\ln\frac{1}{r}}{\ln 2} \right)^{\frac{1}{k}} R_i, \quad i \in \{fore, aft, starb, port\} \quad (5)$$

$$\begin{cases} R_{fore} = (1+1.34\sqrt{k_{AD}^2+(\frac{k_{DT}}{2})^2})L \\ R_{aft} = (1+0.67\sqrt{k_{AD}^2+(\frac{k_{DT}}{2})^2})L \\ R_{starb} = (0.2+k_{DT})L \\ R_{port} = (0.2+0.75k_{DT})L \end{cases} \quad (6)$$

$$\begin{cases} k_{AD} = \frac{A_D}{L} = 10^{0.3591 \lg V_{own}+0.0952} \\ k_{DT} = \frac{D_T}{L} = 10^{0.5441 \lg V_{own}-0.0795} \end{cases} \quad (7)$$

Equation 2 gives a mathematical description of FQSD based on the analytical framework of Wang (2009). (x, y) is relative position of target ship. k is factor which defines the shape of domain. In this paper k = 2 was introduced to reflect the elliptical shape. sgn$x$, sgn$y$ are functions determining which radii should be applied (sgn ≥ 0 if x or $y \ge 0$). $Q(r)$ is the set of radii of ship domain, which consists of radii of ship domain in four directions respectively $(R_{fore}(r), R_{aft}(r), R_{starb}(r), R_{port}(r))$. r is the fuzzy factor which determines the scale of ship domain. The value of radii decreases with the increment of r. V, L are speed and length of own ship respectively. By introducing factors of advance distance (kAD) and tactical diameter (kDT) which can be obtained based on Eq. 7, boundary of ship domain can vary dynamically according to ship's speed and length. Radii of FQSD can be rewritten into Eq. 8 according to Eq.7.

$$\begin{cases} R_{fore} = (1+1.34\sqrt{1.55v^{0.72}+.017v^{1.09}})L \\ R_{aft} = (1+0.67\sqrt{1.55v^{0.72}+0.17v^{1.09}})L \\ R_{starb} = (0.2+0.83v^{0.5441})L \\ R_{port} = (0.2+0.62v^{0.5441})L \end{cases} \quad (8)$$

2.1.2 Collision candidate detection model

Collision candidate is encountered ships which are in a situation that they would collide with each other if no evasive maneuver was performed. It indicates ships with geometrical possibility of collision. By detecting whether the domain of ship was violated by another ship and their tendency of movement, this paper established a collision candidate detection model using FQSD and relative motion of encountered ships based on AIS data. The flow chart of this model is shown in Figure 2.

The specific procedure is as follows:

Step 1. Data sampling

Sample ship AIS data within research area into dataset at time interval Δt. The time interval is introduced from literature (Qu et al., 2011), which is set as 30 min.

Step 2. Domain violation determination

For dataset $D_T(P_1, P_2, ..., P_i)$ at time T, $P_i(x_i, x_y, type_i, L_i, B_i, COG_i, SOG_i$ denotes an AIS data record of ship i. To determine whether there is domain violation, ship i is chose as own ship and the distance of other ships to ship i within D_T was utilized to determine if they are within the ship domain of own ship or not using FQSD model.

Step 3. Relative motion determination

The violation of ship domain only provides spatial information about whether two ships are within close range. To further determine whether they

are on approaching course, the vectors of relative distance and speed of ships with domain violation are introduced using dot product method.

Step 4. Collision candidate record

If collision candidate is determined, the information of candidates (encounter time, position, speed, course over ground, ship length, width, type, etc.) will be recorded into text file for further analysis.

2.2 Causal probability model based on Bayesian networks

2.2.1 Definition of the variables

Maritime transport is a very complicated system and accidents such as ship collision are the results of coupling effect of multiple factors (Hänninen, 2014). Human factors have become major contributor due to the improvement in reliability of ship equipment and aids to navigations (Zhang & Thai, 2016). To model the causational probability of ship collision accident, this paper defined 23 influencing factors from "human, ship, environment" perspective based on the previous research by the author (Chen et al., 2015). The variables of Bayesian network are as follows: (1) Unused safety speed; (2) fail in duty, (3) Negligence in watchkeeping, (4) fatigue, (5) Improper lookout, (6) Detection failure, (7) Radar, (8) Uncontrolled situation, (9) Poor visibility, (10) Rough sea, (11) Maneuvering error, (12) Navigation error, (13) Steering failed, (14) Main engine failed, (15) Wrong assessment of situation, (16) Operation error, (17) improper emergency operation, (18) Improper steering control, (19) Avoiding error, (20) Uncoordinated avoidance operation, (21) equipment failure, (22) harsh environment and (23) illegal behavior leading to collision.

2.2.2 Structure development of the network

As a graphical tool to perform probability inference, the structure of Bayesian network reflects causal relationship between variables. Generally, the structure of Bayesian network was established by performing data learning, eliciting expert knowledge or the combination of them. Since the scale of accident data of this research is not sufficient to perform structure learning, this paper introduced a mapping technique proposed by Khakzad et al (2011) to translate a Fault Tree model (Chen et al., 2015) into Bayesian network. The method of translation is shown in Fig 3.

Based on such translation method, the fault tree model of the previous research is translated and modified into Bayesian network as shown in Figure 4.

2.2.3 Quantitative part of the network

The structure of Bayesian network reflects causational relationship between contributing factors

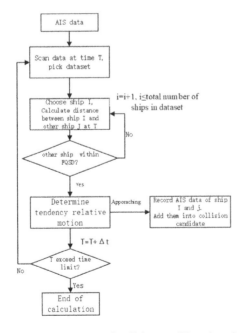

Figure 2. Flow chart of collision candidate detection model.

Fault Tree

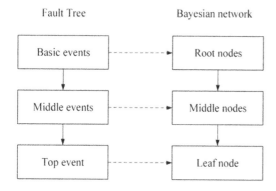

Figure 3. Translation of fault tree into Bayesian network (Khakzad et al., 2011).

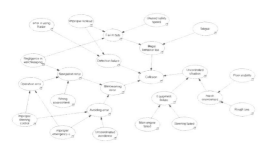

Figure 4. Structure of Bayesian network.

Table 1. CPT of Boolean logic "OR" of "Operation error".

Improper steering control	Yes		No	
improper emergency operation	Yes	No	Yes	No
Yes	1	1	1	0
No	0	0	0	1

Table 2. CPT of Boolean logic "AND" of "Detection error".

Improper lookout	Yes		No	
improper use of Radar	Yes	No	Yes	No
Yes	1	0	0	0
No	0	1	1	1

Table 3. Nodes with Boolean logic CPTs.

Boolean logic	Name of variables
AND	Illegal behavior leading to collision, Detection failure, Navigation error, Avoiding error, Uncontrolled situation, Harsh environment
OR	Fail in duty, Operation error, Maneuvering error, Equipment failure, Collision

of ship collision candidate. In the meantime, conditional probability tables (CPTs) of variables quantify the probability of states of variables. The number of causational probabilities of variables grows exponentially with the increment of the number of variables and their states (Montewka et al., 2014), which makes it very difficult to build model with multiple variables and states. To tackle with such challenge, methods on reducing probability inputs and parameter learning technique were proposed in the literature (e.g. Halpern & Sontag, 2013; Oniśko et al., 2001). As for the network this paper established, since it was mapped from a fault tree, the CPTs of middle nodes are based on Boolean logic, which simplify the complexity of CPTs and number of probability input to a large extent. Table 1 to 3 illustrate example CPTs for Boolean logic "AND", "OR" and nodes which are middle nodes of the network respectively.

As for the probability input for root nodes, Li (2014) conducted comprehensive research on probability of the input for Fault tree of ship collision accident in the research area. This paper introduced his work as input.

3 APPLICATION OF THE PROPOSED METHOD

To verify the efficacy of the proposed method, this section describes a case study where part of western Shenzhen waters is chose as research area. As for the research data, one-week ship AIS data (21st -27th April 2016) within the range of research area was collected and decoded for further process. Besides this, historical ship collision data from 2002 to 2012 was also collected.

3.1 Research area

Located in eastern bank of estuary of the Pearl River (Fig. 5), Shenzhen port is one of the most important corridors for maritime transportation of China. Connecting inland, Hong Kong and Macau, ship traffic in these areas are very active, among which western water of Shenzhen port are the busiest waterway, which results in a congested navigation environment with complicated ship traffic.

Figure 6 depicts the AIS data density distribution of one day in western Shenzhen waters, the color of dots indicates density of data.

As can be seen, composition of waterway here is very complicated and data clustered in some specific routes and area especially in the vicinity of western Shenzhen port (red frame).

3.2 Results

3.2.1 Results of collision candidate detection model

Collision candidate detection model is implemented using C# programming and applied to the AIS data within research area. Table 4 shows the results of detection within 7 days with fuzzy factors r = 0.2, 0.4, 0.6 and 0.8 respectively.

This paper chose 4 value of r (0.2, 0.4, 0.6, 0.8) to determine collision candidate with different risk of collision. According to Table 4, the number of

Figure 5. Location of research area (red frame).

Figure 6. AIS data density distribution in western Shenzhen waters.

Table 4. Number of collision candidates during time frame with different r values.

Day	r = 0.2	r = 0.4	r = 0.6	r = 0.8
21st April	23	12	7	4
22nd April	19	11	6	4
23rd April	36	21	16	11
24th April	33	23	16	8
25th April	42	27	20	10
26th April	98	73	54	27
27th April	98	68	42	27

collision candidates decreases with the increment of fuzzy factor r of FQSD. As developed by Wang (2010), $r \in (0,1)$ is a factor used to reflect collision risk of the corresponding fuzzy boundary of FQSD. This is reasonable since size of FQSD is determined by r when all other factors are the same (Eq. 2). The larger r is, the smaller FQSD will be. Goerlandt and et al (Goerlandt & Kujala, 2014) also pointed that domain with different r can be interpreted as areas navigator want to keep clear from other ship with different need. From this perspective, collision candidate with a larger r is closer to ship encountered which indicates more dangerous geometrical encounter situation. To estimate geometrical collision probability of the area, this paper adopts result of r = 0.4 as candidates with average level of risk to calculate annual number of collision candidates of 2016 based on Eq. 9, which is 12288.

$$N = \frac{\sum_{i=1}^{N_{day}} n_i}{N_{day}} \times N_{\text{day of the year}} \qquad (9)$$

where N is the estimated number of collision candidates per year; N_{day} is the number of days of dataset; n_i is the number of collision candidates in day i of data set; $N_{\text{day of the year}}$ is number of days of the year.

Figure 7 and 8 illustrate the spatial and temporal distribution of collision candidate within research area. Most of the collision candidate cluster in the vicinity of western public channels and northern waterways of Shenzhen port. This result is consistent with AIS data density distribution shown in Fig. 7 as ship traffic also congests in the same areas. As for temporal distribution of collision candidate, Figure 9 shows the average temporal distribution of collision candidate. Number

Figure 7. Spatial distribution of collision candidates (r = 0.4).

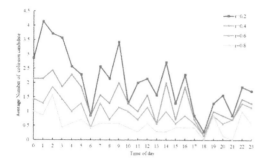

Figure 8. Temporal distribution of collision candidates.

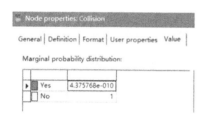

Figure 9. Causational probability of ship collision.

of candidate fluctuates throughout the day and reaches peak value (4.1) and minimal value (0.3) at 1 AM and 6 PM. Among all the candidate, about 26% of them are dangerous encounter($r = 0.8$), and number of them during 19 o'clock and 5 o'clock are relatively higher than that during day time (6–18 o'clock). Such finding indicates that night time in western Shenzhen waters needs to be focused on due to its higher number of dangerous encounter.

3.2.2 Causational probability and risk of collision

For the nodes which are translated from "And" and "Or" gate in Fault Tree, their CPTs are determined by Boolean logic, which reduces the probability input for Bayesian network to a large extent. To obtain the causational probability of ship collision in western Shenzhen waters, probability input for root nodes of the model are estimated on the basis of historical ship collision records from 2002 to 2012 and research of Li (2014). Probability for root nodes are shown in Table 5.

Probability inference of the causational probability model of ship collision is performed by using update function of genie software when all probability inputs are prepared. Due to the fact that values of probability inputs are so small that result of the model cannot be illustrated correctly in bar chart of genie interface, only result of causational probability of node "collision" is shown in Figure 9.

The result of causational probability of ship collision is 4.38×10^{-10} per vessel transit Based on Eq. 1, risk of ship collision in the research area is estimate to be 5.38×10^{-6} which means that 0.538 collision accident occurs per 100000 of vessel transits within research area.

To get more insights about causational probability of ship collision, this paper performs sensitivity analysis on the Bayesian network to obtain sensible variables. Generally, there are three methods to perform sensitivity analysis: "1). *difference in probability of collision given variable's state correspondingly; 2) sensitivity value; 3) mutual information*" (Hänninen & Kujala, 2012). This paper introduced the first one to perform sensitivity analysis and result is shown in Figure 10.

The value indicated by Figure 10 is the difference of the probability of node "collision" under two states of each analyzed variables. The difference of collision probability can indicate the maximum change a variable can produce, which can be used as an indication of sensitivity of node "collision" to the variables (Hänninen & Kujala, 2012). A shorter bar in the figure indicates collision is more sensitive to this variable. Nodes such as "fatigue", "improper lookout", "negligence in watchkeeping" and "unused safe speed" have larger influence on probability of ship collision while variables such as

Table 5. Probability input of root nodes (Li, 2014).

Variable	Probability
Fatigue	1.10E-05
Negligence in watchkeeping	2.87E-05
Unused of safe speed	6.24E-06
Improper lookout	2.38E-05
Error in using Radar	1.15E-05
Improper emergency operation	5.95E-05
Wrong assessment of situation	1.18E-05
Improper steering control	9.93E-06
Uncoordinated avoidance	9.69E-06
Rough sea	8.27E-06
Poor visibility	8.81E-06
Steering engine failed	6.70E-06
Main engineering failed	6.70E-06

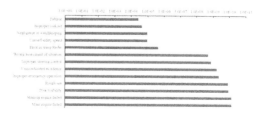

Figure 10. Sensitivity analysis of the network.

"steering engine failed" and "main engine failed" influence the causation probability the least. Such finding points out which aspects should be paid attention by maritime safety administration of the research area, in the meantime, it also validated the argument that human relevant factors are the major contributor of accident.

3.3 Discussions

The case study quantified the risk of ship collision in western Shenzhen waters and its characteristics such as spatial-temporal distribution of collision candidate and sensitive contributing factors of causational probability. According to the maritime safety situation report of 2011and 2012, which are adopted by Shenzhen MSA, maritime accident happened between 0 AM and 8 AM take up about 50% of the total amount of accident.

Figure 11 shows the spatial distribution of ship collision accident during 2003–2012 which has explicit location information (GPS coordinates). Compared with Fig.11, spatial distribution of collision candidate identified areas with high occurrence of accident, which is indicated by red circle in Fig. 11, while some discrepancies still exist. Collision candidates also congest in the public southern waterway which, however, is inconsistent with pattern in Fig 11.

As for the causational probability, the result of Bayesian network in this paper is significantly smaller than results of some existing literature (Hänninen & Kujala, 2010; Pedersen, 2010). This is because that the causational probability the model calculated is a frequency based value. The probability inputs are based on the occurrence frequency of those variables (Li, 2014) which takes traffic volume of the area into account. Therefore, the result of causational probability and risk of collision is also based on traffic volume of research area at that year.

Figure 11. Spatial distribution of collision accidents with explicit location data during 2003–2012.

4 CONCLUSIONS

Based on ship AIS data and historical collision records, this paper performed a case study in the estuarine waters of Shenzhen, China. on collision risk by applying FQSD and Bayesian network. To obtain such probability, a collision candidate detection model and causational probability model are designed and established respectively. A case study on western Shenzhen waters was conducted by implementing the method this paper proposed and the risk of ship collision, which is the estimated frequency of ship collision accident, is obtained. In the meantime, characteristics of collision risk such as the spatial-temporal distribution of collision candidates, sensitive contributing factors of causational probability are also analyzed. The findings indicate that such method can be applied for risk assessment of ship collision accidents as a decision support tool for maritime safety administration to improve navigation safety.

However, some deficiencies need to be improved in further research: 1) The time interval of candidate detection is introduced from literature (Qu et al., 2011) fixed throughout the calculation, which can be region dependent. However, the time for ships finishing an encounter situation should be taken into consideration and quantified; 2) The result of collision candidate detection shows some discrepancies from the historical spatial distribution of collision candidates. Although the incompleteness of the accident analysis is one factor causing such discrepancy, the accuracy of the detection model needs to be improved to reduce the misdetection rate; 3). Structure of Bayesian network is based on a fault tree of previous research, such method of establishing a network restrains its capability in dealing with uncertainty of the causal relationship between variables and modelling variables with multiple states other than 2 states. In future work, thorough research taking methods such as expert elicitation should be applied to improve the causation probability model.

ACKNOWLEDGEMENTS

The work presented in this paper is financially supported by China Scholarship Council (Grant number 201606950005) and National Natural Science Foundation of China (Grant number 51579201).

REFERENCES

Asariotis, R., Benamara, H., Hoffmann, J., Premti, A., Valentine, V., Youssef, F., 2016. Review of Maritime Transport, 2016.

Aven, T., 2012. The risk concept—historical and recent development trends. Reliability Engineering & System Safety 99, 33–44.

Aven, T., Renn, O., Rosa, E.A., 2011. On the ontological status of the concept of risk. Safety Science 49, 1074–1079.

Chen, P., Mou, J., Li, Y., 2015. Risk analysis of maritime accidents in an estuary: a case study of Shenzhen Waters. Scientific Journals of the Maritime University of Szczecin 42 (114), 54–62.

Goerlandt, F., Montewka, J., 2015. Maritime transportation risk analysis: Review and analysis in light of some foundational issues. Reliability Engineering & System Safety 138, 115–134.

Goerlandt, F., Kujala, P., 2014. On the reliability and validity of ship–ship collision risk analysis in light of different perspectives on risk. Safety Science 62, 348–365.

Halpern, Y., Sontag, D., 2013. Unsupervised Learning of Noisy-Or Bayesian Networks.

Hänninen, M., 2014. Bayesian networks for maritime traffic accident prevention: Benefits and challenges. Accident Analysis & Prevention 73, 305–312.

Hänninen, M., Kujala, P., 2010. The Effects of Causation Probability on the Ship Collision Statistics in the Gulf of Finland. TransNav, the International Journal on Marine Navigation and Safety of Sea Transportation 4, 79–84.

Hänninen, M., Kujala, P., 2012. Influences of variables on ship collision probability in a Bayesian belief network model. Reliability Engineering & System Safety102, 27–40.

Kaplan, S., 1997. The Words of Risk Analysis. Risk Analysis 17, 407–417.

Khakzad, N., Khan, F., Amyotte, P., 2011. Safety analysis in process facilities: Comparison of fault tree and Bayesian network approaches. Reliability Engineering & System Safety 96, 925–932.

Kujala, P., Hänninen, M., Arola, T., Ylitalo, J., 2009. Analysis of the marine traffic safety in the Gulf of Finland. Reliability Engineering & System Safety 94, 1349–1357.

Li, Y., 2014. Study of Ship Collision Accident Pattern in Estuary based on Data Mining. vol. Wuhan University of Technology.

Martins, M.R., Maturana, M.C., 2010. Human Error Contribution in Collision and Grounding of Oil Tankers. Risk Analysis 30, 674–698.

Martins, M.R., Maturana, M.C., 2013. Application of Bayesian Belief networks to the human reliability analysis of an oil tanker operation focusing on collision accidents. Reliability Engineering & System Safety 110, 89–109.

Montewka, J., Ehlers, S., Goerlandt, F., Hinz, T., Tabri, K., Kujala, P., 2014. A framework for risk assessment for maritime transportation systems—A case study for open sea collisions involving RoPax vessels. Reliability Engineering & System Safety 124, 142–157.

Montewka, J., Hinz, T., Kujala, P., Matusiak, J., 2010. Probability modelling of vessel collisions. Reliability Engineering & System Safety 95, 573–589.

Mou, J.M., Tak, C.V.D., Ligteringen, H., 2010. Study on collision avoidance in busy waterways by using AIS data. Ocean Engineering 37, 483–490.

Oniśko, A., Druzdzel, M.J., Wasyluk, H., 2001. Learning Bayesian network parameters from small data sets: application of Noisy-OR gates. 27, 165–182.

Pedersen, P.T., 2010. Review and application of ship collision and grounding analysis procedures. Marine Structures 23, 241–262.

Qu, X., Meng, Q., Suyi, L., 2011. Ship collision risk assessment for the Singapore Strait. Accident Analysis & Prevention 43, 2030–2036.

Silveira, P.A.M., Teixeira, A.P., Soares, C.G., 2013. Use of AIS Data to Characterise Marine Traffic Patterns and Ship Collision Risk off the Coast of Portugal. Journal of Navigation 66, 879–898.

Sotiralis, P., Ventikos, N.P., Hamann, R., Golyshev, P., Teixeira, A.P., 2016. Incorporation of human factors into ship collision risk models focusing on human centred design aspects. Reliability Engineering & System Safety 156, 210–227.

Szlapczynski, R., Szlapczynska, J., 2016. An analysis of domain-based ship collision risk parameters. Ocean Engineering 126, 47–56.

Wang, N., 2010. An Intelligent Spatial Collision Risk Based on the Quaternion Ship Domain. Journal of Navigation 63, 733–749.

Wang, N., Meng, X., Xu, Q., Wang, Z., 2009. A Unified Analytical Framework for Ship Domains. Journal of Navigation 62, 643.

Ylitalo, J., 2010. Modelling marine accident frequency. vol. Aalto University.

Zhang, G., Thai, V.V., 2016a. Expert elicitation and Bayesian Network modeling for shipping accidents: A literature review. Safety Science 87, 53–62.

Zhang, W., Goerlandt, F., Montewka, J., Kujala, P., 2015. A method for detecting possible near miss ship collisions from AIS data. Ocean Engineering 107, 60–69.

Maritime Transportation and Harvesting of Sea Resources – Guedes Soares & Teixeira (Eds)
© 2018 Taylor & Francis Group, London, ISBN 978-0-8153-7993-5

Risk analysis for maritime traffic in the Strait of Gibraltar and improvement proposal

N. Endrina & J.C. Rasero
Department of Nautical Sciences, University of Cádiz, Spain

J. Montewka
Finnish Geospatial Research Institute, Masala, Finland
Department of Transport and Logistics, Gdynia Maritime University, Gdynia, Poland

ABSTRACT: The Strait of Gibraltar (SoG) is one of the major navigation areas in the world. The maritime traffic registered in the area is approximately 110,000 ship movements per year. Unfortunately the area is facing numerous accidents involving collisions. The aim of this paper is to carry out a risk analysis for maritime traffic in the SoG based on the accidents statistics, AIS data and dedicated software such as IWRAP. On the basis of the outcomes, we investigate two risk control options (RCOs) focused in changes in the current TSS. Finally, we found that the level of risk collision in this area could be effectively reduced with the application of one of the RCOs investigated.

1 INTRODUCTION

The Strait of Gibraltar (SoG) is a natural sea passage located between Spain and Morocco, linking the Mediterranean Sea with the Atlantic Ocean. It accommodates very high volume of traffic, being one of the main shipping areas, with approximately 110000 ship movements per year.

The European coast, limited by Cape Trafalgar and Europe Point, is 55 nautical miles long, whereas the African coast from Cape Spartel to Punta Almina is 42 NM long. The width of the Strait varies between 24 NM to 7.4 NM close to Tarifa Island, leaving therefore very little freedom to ship navigation, however the depth of the Strait poses no problems to navigation, since it falls in the range of 300–1200m. Moreover, strong eastern or western winds and frequent fogs characterize the area.

Its geographic situation involves a privileged location, and for that is upmost important the control of the water by coastal States (Morocco, Spain and the Government of Gibraltar). Currently, the jurisdictional limits of each State are not known with certainty, resulting in several conflicts in the area.

In order to enhance navigational safety, the International Maritime Organization (IMO) adopted different measures to control and organize the maritime traffic. These are: traffic separation scheme (TSS), precautionary areas, mandatory ship reporting systems and vessel traffic systems. The TSS in the Strait the Gibraltar came into force in 1970 by Resolution A.161 (IMCO 1968). Over the years, this scheme has undergone slight modifications, such as the incorporation of inshore traffic zones (IMO 1994) or easier access to the anchorage areas of Tanger Mediterraneo port, (IMO 2014). However, the main modification was established in 2006 - (IMO 2006), when two precautionary areas (PA) were introduced to enhance the maritime traffic flows. First PA was established on the eastern part side of the TSS; and the second PA was established off the Moroccan port area in the TSS, see Fig. 1. These modifications altered maritime traffic significantly but the collisions continue to occur, however in different locations, see Fig. 3. Introduction of those safety measures intended to increase the safety in the analyzed area, solved some problems but maybe created others, eventually leaving the level of safety in the area unknown.

Figure 1. Gibraltar TSS adopted at 1st July 2007. COLREG.2/Circ.58.

Studies relating to waterway risk assessment have received growing interest in the last years. A number of risk assessments in particular areas can be found in the literature (Cucinotta et al. 2017, Valdez et al. 2016, Ventikos & Rakas 2015, Zaman et al. 2015, Xiaobo et al. 2011, Kujala et al. 2009, Uluscu et al. 2009). These assessments are very helpful for the review and development of new regulations in order to improve maritime safety.

Therefore the aim of this study is two-fold. First, we carry out a risk analysis of maritime transportation focusing on the most common accidents to ships in the area of the SoG. These are collision and groundings. To this end we apply dedicated professional software called IWRAP that provides the spatial picture of the risk level in the area. As input data to this tool we use AIS data to determine the distribution of maritime traffic in the area and to determine the number of ships on potentially accidental courses. Second, based on the obtained level of risk, risk control options (RCOs) are proposed and evaluated for the current traffic organization. Finally, the study provides a basis for further research and decision-making process.

The remainder of this paper is organized as follows. Section 2 gives a background of some commonly found elements of risk perspective and the approach taken in this paper. The collision model and the methodology that is used to develop it are presented in section 3. Section 4 presents the results and two new proposals to reduce the risk of collision in the area under study. The model and its results are discussed in section 5. Section 6 concludes the paper.

2 CHALENGES IN MARITIME TRANSPORTATION RISK ANALYSIS

In risk analysis as a scientific discipline numerous the whole spectrum of scientific approaches exists. On one end of it sits strong realist approach, where risk perspectives consist exclusively of probabilistic risk measures. The evidences for these probabilities are based on data or models. It means probabilities of accident occurrences are calculated from observed frequencies or through probabilities models. On the other end of spectrum sits strong constructivists view on risk, according to which risk is nothing but observer's perception about a given situation, and varies between observers, (Goerlandt & Montewka 2015a).

The recent review of models pertaining to the field of maritime transportation risk analysis reveals that the field is dominated by the realist view, see for example (Goerlandt & Montewka 2015a, Mazaheri et al. 2014, Li et al. 2012,

Pedersen 2010). This can be attributed to relatively rich databases about accidents at sea, which in turn allows for probability, or rather frequency, estimation related to a given accident. However, the large databases may also be misleading, especially when the concept of probability is misused. This is the case, if the database covering large number of various accidents is treated as if the data refer to the same transportation system operating under the same or similar conditions. This is called the exchangeability of the events, which is a prerequisite for the frequency assessment based on data. However, each chain of events leading to accident is different. The ships involved, crew, external and internal conditions, the situational context—in each accident these are different. This means that the exchangeability of the events is not met, (Apostolakis 2014), leading to the uncertainty that is difficult to evaluate and address.

The uncertainties arise mainly from the paucity in data—the number of accidents in the analysed area is low and those accidents are of various nature. These gaps in background knowledge lower the level of accuracy of a model, thus it shall not be used to seek an accurate estimates of the measure of the risk level. The main challenge arise when the numbers obtained in the course of databases review are considered "the ultimate truth" about the probability of an accident in the analysed area. This, when taken further, leads to another challenge related to the economical quantification of benefits of risk control options (RCOs). Therefore in this study we use the accident statistics as a secondary source of data to update the results obtained from the modelling process, which is seen as the primary source of information for further analysis. Moreover, the obtained numbers are treated as indicators of risk level rather than "the true risk", and the effect of RCOs is expressed as relative changed of risk level compared to some base-line value calculated for a given setting (Montewka et al. 2016). We can also assume that by using relative differences rather than the absolute numbers we can—to some extent—eliminate the effect of the existing uncertainty in the selection of the RCOs.

By adopting such approach it is feasible to differentiate among various designs of traffic flow in the analysed area, based on the criteria selected, which in this case is risk level. By doing that we produce a message to decision makers that is affected to lesser extent by the gaps in background knowledge.

In this paper we take moderate realist's perspective on risk, meaning that we use available databases and modelling techniques in order to quantify the risk. The latter is measured with the use of the concept of probability, understood here as the empirical probability or relative frequency.

It is defined as the ratio of the number of outcomes in which a specified event occurs to the total number of trials, however not in a theoretical sample space but in an actual experiment. By doing that we attempt to reflect on the chances for an accident to happen (A), as perceived by a risk analyst, given the background knowledge (BK) i.e. accidents' databases, traffic flow data.

Thus the description of risk as adopted in this paper can be given as follows, where "~" signifies "is described by":

$$R \sim P(A|BK) \tag{1}$$

To this end utilize the existing data and evidence to model maritime traffic and the resulting collisions; to some extent we attempt to deal with the uncertainties The latter is reflected in the way how we use the obtained results. We see calculated risk values as indicators for the purpose of relative comparison between various options rather than absolute numbers, referring to the unknown but "true" risk.

3 METHODS AND DATA

3.1 Modeling principles

To calculate the P(A|BK) we used commercial software package called IALA Waterway Risk Assessment program (IWRAP), which was developed by IALA together with Canadian Coast Guard and Technical Universities of Denmark and Wismar during 1990s. Later, the commercial version IWRAP Mk2 was realized in 2009 supported by GateHouse, (IALA 2009). IWRAP is a tool for estimating the annual a number of collisions and groundings in the modelled waterway. Also, it allows comparisons of various navigational route layouts by assessing the frequencies of collisions. IWRAP and tools alike are commonly used to diagnose the risk level in maritime transportation systems and to evaluate the effectiveness of proposed risk control options, see for example (Guze et al. 2016, Goerlandt & Kujala 2011, Gucma 2006, Montewka et al. 2010, Sormunen et al. 2013, Merrick et al. 2003, Rong et al. 2015, Przywarty et al. 2015).

In IWRAP the Pedersen's model (Pedersen, 1995) is adopted to calculate the probability of collision—P_{coll}—in waterways, as follows:

$$P_{coll} = N_G P_C \tag{2}$$

where N_G stands for geometric number of collision candidates; and P_C means causation factor. To calculate the N_G value it is necessary to know the lateral traffic distribution (identifies where ships move on the leg) and the traffic volume/composition, as well as ship dimensions. IWRAP

Mk2 allows performing these operations through the import of data provided by the Automatic Identification System (AIS), which is known as very good source of shipping related information, (Rong et al. 2015; Silveira et al. 2013; Mou et al. 2010; Weibin et al. 2015; Van Iperen 2015). In addition, the program estimates the N_G value according to the different types of collision. Head-on, overtaking and crossing collisions as depicted in Fig. 2. As an example, N_G value for head-on collision is calculated as follows:

$$N_G^{head-on} = L_W \sum_{i,i} P_{Gi,j}(head-on)\frac{V_{ij}}{V_i^{(1)}V_j^{(2)}}$$
$$\times \left(Q_i^{(1)} Q_j^{(2)}\right) \tag{2}$$

where L_w is the length of the waterway; $P_{Gi,j}$ is the probability that two ships of ship classes i and j collide in a head-on meeting situation if no aversive manoeuvres are made; V_{ij} is the relative speed between the vessels; $V\alpha(\beta)$ is the speed of the ship class α moving in direction β 1; and $Q\alpha(\beta)$ is the number of passages per time unit for ship class α moving in direction β.

Regarding to P_C, IWRAP allows adjusting this factor for the different types of collisions in a specific area, as well as for the different types of ships. However, the program proposes default causation factor values based on the literature and Bayesian Networks models as presented in Table 1. However those choices do not necessarily reflect the local

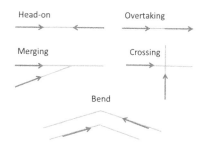

Figure 2. Types of collisions in IWRAP Mk2.

Table 1. Default causation factor of IWRAP Mk2 program, (Friis-Hensen et al. 2008).

Condition	Causation factor
Head on collision	0.5×10^{-4}
Overtaking collision	1.1×10^{-4}
Crossing collision	1.3×10^{-4}
Collision in bend	1.3×10^{-4}
Collision in merging	1.3×10^{-4}

conditions, since the values proposed result in different spatial resolution of the accidents. Therefore the numbers taken from the literature need to be tailored to the analyzed area accounting for the existing background knowledge. This is due to existing uncertainty due to the validity of those numbers, whether they hold globally or just reflect some local conditions and specific ship type.

3.2 Background knowledge

3.2.1 Accident data

Casualty analysis has been carried out on historical data for the period 2000–2015. These data have been obtained by the casualties module of the IHS Fairplay Sea-web™ database, known earlier as *Lloyds Register Fairplay*. The fleet statistics for the same period was obtained by the Spanish Maritime Safety Agency.

For this study, only those collisions, which occurred when both ships were underway, will be considered, excluding those occurred during berthing operations. Table 2 shows the number of collisions registered in the area for the period 2000–2015. A total of 10 collisions happened while the vessels were sailing, with 20 involved vessels. It is important to highlight that there are 8 RoPax ships involved, which do regular line between the ports in the area.

The frequency of collision calculated for the period 2000–2015 is 0.625 collisions per year, meaning one collision every 1.6 years. The frequency of collisions per movement[1] year calculated has been estimated in 1.3369E-05.

In Figure 3 the locations are shown of the collisions occurred for the period 2000–2015. The red

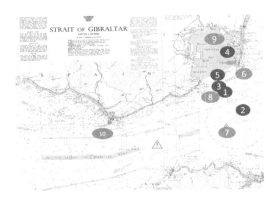

Figure 3. Map of the collision registered for the period 2000–2015. Admiralty Chart 142: Strait of Gibraltar. UK Hydrographic Office.

spots represent those collisions occurred before the entry in force of the TSS modification at 2007, while the green spots represent the collisions occurred after. Except an isolated case, all collisions occurred in the eastern part of the SoG, and none of the accidents occurred in the TSS area. This data could indicate that the establishment of the TSS has positively contributed to minimize the risk of accidents in the area. However, we cannot say the same about the eastern side. While the amendment in 2007 established a precautionary area in the eastern part of the Gibraltar TSS; the fact is accidents are still happening. This area, between Algeciras Bay and Ceuta port, could be called like "black area" for maritime navigation; characterized by a high density of traffic, in which the traffic flow is not well defined. For example, in the northern part of the precautionary area, near Carnero Point, 4 collisions were recorded (with numbers 1,3,5 and 8 in the Fig. 3).

3.2.2 Traffic data

All data related to lateral traffic distribution and traffic composition (number, type and size of vessels, average speed, etc.) are calculated based on AIS data. The AIS data is referred to the SoG area for the months of December 2012, March and July 2013. The sum of these months (90 days) is sufficient to calculate the annual frequency of collision by the program. The selection of these months is based on traffic statistical data and climatic data (winds and mists). The above months correspond to a greater number of traffic crossings and more adverse weather conditions in the area. Thus, the estimation of the frequencies of collision is calculated for the worst of the possible scenarios.

Regarding to the causal factor, the model was performed with the default values given by the IWRAP software.

Table 2. Collisions registered in SoG for the period 2000–2015. IHS Fairplay Sea-web™ database.

N°	Ships involved	Date
1	Ciudad de Ceuta—Ciudad de Tanger	16/07/2000
2	Mar Rocio—SKS Trinity	16/08/2000
3	Indalo—Al Mansour	08/05/2001
4	Spetses—Van Gogh	26/09/2004
5	Atlas—Avemar	28/11/2006
6	Torm Gertrud—New Flame	12/08/2007
7	New Glory—Milleniun II	13/01/2012
8	Le Sheng—Cap Med	30/05/2014
9	Wisby Argan—Celsius Mumbai	11/10/2014
10	María Dolores—Rinconcillo	20/07/2015

[1]Frequency per movement year is referred to the total number of involved ships between the fleet at risk. The fleet at risk corresponds to the number of records of ship movements in the area during the period 2000–2015.

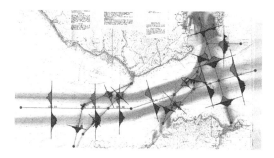

Figure 4. Traffic flow model by IWRAP Mk2.

Table 3. The probability of collisions per year according to the type of collision.

Type of collision	P (collisions)
Head on	0.0525
Overtaking	0.0715
Crossing	0.0489
Bend	0.2526
Merging	0.0759
Collision total	0.5017

Based on the imported data from AIS, IWRAP is capable of visualizing the graphical representation of the traffic distribution (density plot). The program uses a color code where blue corresponds to the areas with the highest traffic density, followed by red and yellow zones. Then, the routes were modelled through the density plot. Each route is defined by the position of two waypoints (A, B), the length of a leg (|AB|) and the width of the area covered by all the hypothetical paths leading from A to B. Width is set manually according to the distribution of the traffic obtained from AIS. In Figure 4 the following are depicted: density plot, the defined legs and the histograms of the traffic distribution for each leg.

4 RESULTS

4.1 *Risk level in the analyzed sea area*

The modeling process resulted in the definition of the area as a graph consisting of nodes and arcs. The nodes refer to the waypoints, whereas the arcs to the traffic lanes joining those waypoints. The lateral distribution of traffic along the arcs and traffic type (i.e. waterways that are crossing, parallel or merging) is obtained from AIS data. The layout of the graph is presented in Figure 5.

Based on the graph the probability of collision between two ships in the analyzed area is obtained. Further the probability of specific type of collision is derived (while overtaking, crossing, head-on, merging or at a bent), which allows assigning specific probability to a given leg, as presented in Table 5.

The probability for a collision between two ships underway in the area within one year yields 0.50, which can be translated into 1 collision every two years, according to the definition of probability adopted in Section 2. Based on accidents statistics collected for the period 2000–2015, the frequency of collision calculated is 0.625 collisions per year, i.e. 1 collision each 1.6 year. The result obtained in the course of modelling showed rather good

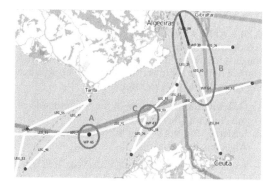

Figure 5. Results for the SOG model. Color code such that blue corresponds to the areas with the highest frequency of collision, followed by red and yellow zones.

agreement with the statistical data. For this study, the collision frequency values are not of particular importance; what matter is the relative distribution of risk across the analyzed legs. For this reason, default causal factors have not been modified.

In general, the zones with the highest probability of collision are as follows—see Figure 5: A (waypoint 46); B (leg 68, waypoints 30 and 64); and C (waypoint 47).

Zone A (waypoint 46) has the higher frequency of collision to bend type with 0.20 incidents per year. This type of collision occurs when two ships moving in opposite directions on the same leg collide in the curvature of the same. In waypoint 46, there is a change of course for both directions of the TSS—east and west—due to the curvature that the scheme presents by the orography of the zone. However, collisions have not been registered in this area during the period of study. This fact could be because ships sailing on this fairway reinforce the watching to follow the compliance of the TSS.

Zone B is characterized by the high frequency of collision to head on and overtaking type in leg 68. In addition, waypoint 30, 64 and 63 show high probability of collisions of the following types: crossing, bend and merging. In this area is difficult to determinate precisely the traffic flow because the flow is mixed with ships entering/leaving in the Algeciras and vessels doing bunkering or provisions in the proximity of the Bay.

Finally, zone C (*waypoint 47*) accommodates a new junction of routes in the central part of the TSS, which includes the incorporation of a precautionary area to permit to the ships entry or leave to Tanger-Med port.

The eastern part of the SoG (zone B and C) could be considered as the highest risk area. This shows good agreement with the accident statistics, which are demonstrated that the majority of collisions occurred here—see Figure 3.

4.1.1 Risk control options

Accounting for the high level of risk of collision in the eastern part of the SoG we introduce two risk control options for lowering the risk level, addressing zones B and C as follows:

1. Ordering the traffic flow in the western side of the Algeciras Bay entrance.
2. Ordering the traffic flow in the eastern side of Algeciras Bay entrance.

To measure the effectiveness of those RCOs we updated the already developed risk model with the use of IWRAP. The composition of the new model with the number of ships passing in each leg is shown in Table 4 and illustrated in Figure 6.

Prior to the introduction of the proposals the probability of collisions in the area per year is 0.017; where the highest probabilities are attributed to *leg 68* and *waypoint 30*, with 0.06 and 0.09 respectively.

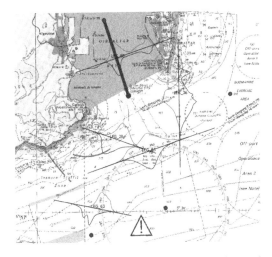

Figure 6. Eastern zone model in IWRAP. The legs and lateral traffic distribution histograms are showed; where leg 68 and waypoint 30 (blue color) have got the highest collision frequency values.

Table 4. Model composition for the eastern part of SOG.

Id.	Denomination	Traffic Direction	N° of passing
Leg 68	Algeciras Bay	North	31864
		South	31698
Leg 28	W Entrance Bay	North	6987
		South	14160
Leg 63	Precautionary area	North	10380
		South	7242
Leg 62	Off TSS—N traffic line	North	6398
		South	228
Leg 36	E Entrance Bay	West	7524
		East	3469
Wp 30	Entrance Bay	–	–
Wp 31	Off TSS—N traffic line	–	–

4.1.2 RCO 1: ordering the traffic flow in the western side of the Algeciras Bay entrance

The goal is to order the traffic flow along legs 28 and 62 to reduce the head-on collisions in Carnero Point. Thus, ships departing from the Algeciras Bay towards the TSS or departing from the TSS towards the Bay would not meet each other. Therefore, an extension toward east (approximate 600 m) of the TSS—central separation zone is proposed. This measure involves the following changes:

- Change over lateral traffic distribution in legs 62, 28 and 68 and
- Waypoint 31 will be shifted eastward.

4.1.3 RCO2: ordering the traffic flow in the eastern side of Algeciras Bay entrance

The goal is to order the traffic flow along *legs 36* to reduce the head-on situations at Europa point. Thus, ships departing from the Algeciras Bay towards Mediterranean Sea or leaving from Mediterranean towards the Bay would not meet each other. Therefore, a new routing system is proposed in Europa Point to reduce the head on collision frequency. This measure involves the following change:

- Change over lateral traffic distribution in *legs 36* and *68*.

However, this measure alone does not prevent the collision risk near to Carnero Point (where 4 collision have been collected). Hence it should be complementary to the RCO1

4.2 RCOs' effectiveness assessment

The effectiveness of the RCOs in terms of the relative reduction of the risk levels, with the reference to the level prior to the introduction of the RCOs, is depicted in figures 7 and 8, as well as it is elaborated in this section.

For the RCO 1, the main results are:

- The risk level over the eastern SoG is reduced by 7.6%.
- For collision type, the most significant change occurs for the head-on where the risk is reduced by 42.8%. However, the risk of the overtaking collision increases by 22.4%.
- For legs and waypoints, the most significant change occurs for the leg 68 and 28, where the risk is reduced by 19% and 26% respectively. The above values are directly related to the head-on collisions. For this type of collision, the risk is reduced by 58% and 100% in leg 68 and 28 respectively.

For the RCO 2, the main results are:

- The risk level in the eastern SoG is reduced by 10.8%.
- For collision type, the most significant change occurs for the head-on where the risk is reduced by 50.5%. However, the risk of the overtaking collision increases to 15.9%.
- For legs and waypoints, the most significant change occurs for the leg 36, where the risk of collision is reduced by 14%. In addition, the head-on collision risk in this leg is reduced by 99%.

Therefore, results show that the application of both options reduces the risk level of head-on collisions in the areas of Carnero and Europa Point, where the risk would become negligible

5 DISCUSSION

For assessing the strength of evidence for each type of evidence, namely data and models, a set of evidential qualities has been used according to Goerlandt & Montewka (2015b), and depicted in Table 5.

We consider that data used in the study have low number of errors and high reliability and accuracy of recording since these data are provided by AIS system. However, the amount of data available is limited. Furthermore, the outputs obtained in the course of modelling showed rather good agreement with the statistical data. In addition, the tool is strongly recommended and used to diagnose the risk level in maritime transportation systems.

The causation probabilities used in the tool have been derived from the literature and BNs, and do not take into account specific characteristics of the area under study, e.g. the presence of the TSS and the traffic monitoring. In the future, a research to define these causation probability values in this area would be beneficial to carry out.

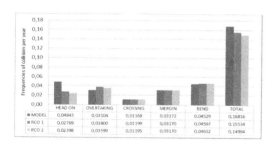

Figure 7. Collision frequencies according to the collision type in the Model and RCOs.

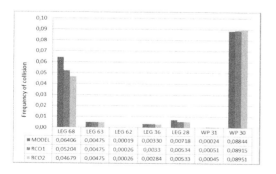

Figure 8. Collision frequencies distributed by legs and waypoints in Model and RCOs.

Table 5. Assessment scheme for evidence qualities based on (Goerlandt & Montewka 2015b).

Evidence type	Strong evidential	Weak evidential
Data		
Quality	Low number of errors	High number of errors
	High accuracy of recording	Low accuracy of recording
	High reliability of data resource	Low reliability of data resource
Amount	Much relevant data available	Little data available
Model		
Empirical validation	Many different experimental test performed	No or little experimental confirmation available
	Existing experimental test agree well with model output	Existing experimental test show large discrepancy with model output
Theoretical viability	Model expected to lead to good predictions	Model expected to lead to poor predictions

Conventionally the effect of RCOs on risk is measured through the cost-benefit analysis. However, it may be challenging to conduct, since the obtained risk numbers are claimed to be merely indicators, not the absolute numbers. Therefore, the numbers make sense when used for comparative purposes as presented here, with respect to some predefined base-line level. Any other use of the results may be misleading.

With regard to the above, it is known that different maritime waterway risk models can give different results and therefore, to compromise their reliability (Goerlandt & Kujala 2014). Hence, the development of a quantitative uncertainty analysis to explore the reliability of our research could be a good way in the future. To do so, we could use the approach that Talavera et al. (2013) propose to manage the uncertainties given in IWRAP using the Demspter–Shafer theory.

6 CONCLUSIONS

The aim of this study is to carry out risk analysis with respect to collisions at sea for the Strait of Gibraltar to learn about the existing risk level. Based on the results obtained we proposed two RCOs suitable for the area, that may reduce the level of collision risk therein.

To achieve the above we utilize professional tool called IWRAP and available accident statistics as well as the AIS data to describe the traffic.

To talk about risk we adopted moderate realists' view on risk, admitting that the obtained numbers do not refer to an absolute truth, and shall be seen as a kind of indicators, that when used for the comparative purposes may provide meaningful results, as long as no economical quantification of the RCOs is involved.

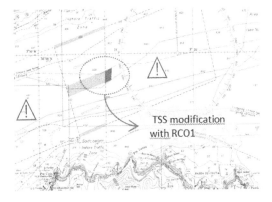

Figure 10. Proposal of amendment to TSS.

Finally, following RCOs is expected to reduce the collision risk level in the easter side of SoG by 11%:

- Extension toward east of the TSS central separation zone in 0.32 nautical mile (600m).

In this way, the amended separation zone would be defined as "A separation zone, half a mile wide, is centred upon the following geographical positions: (1) 35°59'.10N, 005° 25'.37W (2) 35°58'.36 N 005°28'.19W", see Figure 10.

- New routing system southern of Europa Point. Traffic flow could be ordered through separation lines or two-way route according to Res.A.572(14) (IMO 1985).

ACKNOWLEDGEMENTS

Dr. D. Per Christian Engberg from Gatehouse is thanked for his help with IWRAP program.

The authors wish to thank the Spanish Maritime Safety Agency for providing the necessary AIS data used in the analysis.

REFERENCES

Apostolakis, G.E. 2014. Beware of the assumptions: Decision making and statistics. In *Probabilistic Safety Assessment and Management—PSAM12*, Honolulu, 22–27 June 2014.

Cucinotta, F., Guglielmino, E. & Sfravara, F. 2017. Frequency of collisons in the Strait of Messina through regulaoty and environmental constraints asessment. *The Journal of Navigation*, pp. 1–21.

Friss-Hanses, P., Sonne, E. & Engberg, P. 2009. *The Bassy Toolbox Baltic sea safety.* Technical University of Denmark.

Goerlandt, F. & Kujala, P. 2011.Traffic simulation based ship collision probability modeling. *Reliab. Eng. Syst. Saf.* 96(1): 91–107.

Goerlandt, F. & Kujala, P. 2014. On the reliability and validity of ship-ship collision risk analysis in light of different perspectives on risk. *Safety Science* 62: 348–365.

Goerlandt, F. & Montewka, J. 2015a. Maritime transport risk analysis: Review and analysis in light of some foundational issues. *Reliab. Eng. Syst. Saf.* 138: 115–134.

Goerlandt, F. & Montewka, J. 2015b. A framework for risk analysis of maritime transport systems: A case of study for oil spill from tankers in a ship-ship. *Safety Science* 76: 42–66.

Gucma, L. 2006. The method of navigational risk assessment on waterways based on generalised real time simulation data. *Int. Conference on Marine Simulation and Ship Manoeuvrability*, Terschelling, The Netherlands.

Guze, S., Smolarek, L. & Weintrit, A. 2016. The area-dynamic approach to the assessment of the risks of ship collision in the restricted water. *Sci. Journals Marit. Univ. Szczecin* 45(117): 88–98.

IALA 2009. IALA Recommendation O-134 on the IALA risk management tool for ports and restricted waterways. International Association of Marine Aids to Navigation and Lighthouse Authorities, May–2009.

IMCO 1968. Resolution A.161(ES.IV) Recommendation on establishing traffic separation schemes and areas to be avoided by ships of certain classes. Inter-Governmental Maritime Consultative Organization. London, 20 November 1968.

IMO 1994. COLREG.2/Circ.40 New and amended traffic separation schemes. International Maritime Organization. London, 9 June 1994.

IMO 2006. COLREG.2/Circ.58 New and amended existing traffic separation schemes. International Maritime Organization. London, 11 December 2006.

IMO 2014. COLREG.2/Circ.66 Amended traffic separation schemes. International Maritime Organization. London, 11 December 2006.

Kujala, P., Hanninen, M., Arola, T. & Ylitalo, J. 2009. Analysis of the marine traffic safety in the Gulf of Finland. *Reliab. Eng. Syst. Saf.* 194: 1349–1357.

Li, S., Meng, Q. & Qu, X. 2012. An overview of maritime waterway quantitative risk assessment models. *Risk Anal.* 32(3): 496–512.

Mazaheri, A., Montewka, J. & Kujala, P. 2014. Modeling the risk of ship grounding—a literature review from a risk management perspective. *WMU J. Marit. Aff.* 13(2): 269–297.

Merrick, J.R.M., Van Dorp, J.R., Blackford, J.P., Shaw, G.L., Harrald, J. & Mazzuchi, T.A. 2003. A traffic density analysis of proposed ferry service expansion in San Francisco Bay using a maritime simulation model. *Reliab. Eng. Syst. Saf.* 81(2): 119–132.

Montewka, J., Hinz, T., Kujala, P. & Matusiak, J. 2010. Probability modelling of vessel collisions. *Reliab. Eng. Syst. Saf.* 95(5): 573–589.

Montewka, J., Goerlandt, F., Innes-Jones, G., Owen, D., Hifi, Y. & Puisa, R. 2016. Enhancing human performance in ship operations by modifying global design factors at the design stage. *Reliab. Eng. Syst. Saf.* 159: 283–300.

Mou, J.M., van der Tak, C. & Ligteringen, H., 2010. Study on collision avoidance in busy waterways by using AIS data. *Ocean Engineering*, 37, pp. 483–490.

Pedersen, P.T. 1995. Collision and grounding mechanics. In Proceeding of the Danish Society of Naval Architect and Marine Engineers 125: 157.

Pedersen, P.T. 2010. Review and application of ship collision and grounding analysis procedures. *Mar. Struct.* 23(3): 241–262.

Przywarty, M., Lucjan Gucma, P.D., Krzysztof A. & Andrzej P.D. 2015. Risk analysis of collision between passenger ferry and chemical tanker in the western zone of the Baltic Sea," *Polish Marit. Res.* 2(22): 3–8.

Rong, H., Teixeira, A.P. & Guedes Soares, C. 2015. Simulation and analysis of maritime traffic in the Tagus river estuary using AIS data. In *Maritime Technology and Engineering*, C. Guedes Soares and T. A. Santos (eds.).London, UK: Taylor & Francis Group, 185–193.

Silveira, P.A.M., Teixeira, A.P. & Guedes Soares, C. 2013. Use of AIS Data to Characterise Marine Traffic Patterns and Ship Collision Risk off the Coast of Portugal. *Journal of Navigation*, 66(6), pp. 879–898.

Sormunen, O-V. E., Ehlers, S. & Kujala, P. 2013. Collision consequence estimation model for chemical tankers. *Proc. Inst. Mech. Eng. Part M J. Eng. Marit. Environ.* 227(2): 98–106.

Talavera, A., Aguasca, R., Galvan, B., Cacereno, A. 2013. Application of Dempster-Shafer theory for the quantification and propagation of the uncertainty caused by the use of AIS data. *Reliab. Eng. Syst. Saf.* 111: 95–105.

Ulusçu, Ö.S., Özbaş, B., Altıok, T. & Or, İ. 2009. Risk analysis of the vessel traffic in the strait of Istanbul. *Risk Analysis*, 29(10): 1454–1472.

Valdez Banda, O.A., Goerlandt, F., Kuzmin V., Kujala P. & Montewka, J. 2016. Risk management model of winter navigation operations. *Marine Pollution Bulletin* 108(1–2): 242–262.

Van Iperen, W.H. 2015. Classyfing ship encounters to monitor traffic safety on the North Sea from AIS data. TransNav, the International Journal on Maritime Navigation and Safety of Sea Transportation, 9(1): 53–60.

Ventikos, N.P. & Rakas, D.K. 2015. Avoiding collisions, enhancing marine safety—A simplified model for the Aegean Sea. *Scientific Journal of the Maritime University of Szczecin* 42(114): 78–85.

Webin, Z., Goerlandt, F., Montewka, J. & Kujala, P. 2015. A method for detecting possible near miss ship collision from AIS data. *Ocean Engineering* 107: 60–69.

Xiaobo, Q., Qiang, M. & Suyi, L. 2011. Ship collision risk assessment for the Singapore Strait. *Accident Analysis and Prevention* 43: 2030–2036.

Zaman, M.B., Kobayashi, E., Wakabayashi N. & Maimun, A. 2015. Risk of navigation for amrine traffic in the Malacca Strait using AIS. *Procedia Earth and Planetary Science* 14: 33–40.

Maritime Transportation and Harvesting of Sea Resources – Guedes Soares & Teixeira (Eds)
© 2018 Taylor & Francis Group, London, ISBN 978-0-8153-7993-5

Hierarchical control for ship navigation under winds: A case in the Three Gorges Reservoir area

D. Jiang

School of Shipping and Naval Architecture, Chongqing Jiaotong University, China

ABSTRACT: In the Three Gorges Reservoir area of China, the frequent wind has become one of the important factors that affects the safety of ship navigation. According to the requirements of the ship stability in the "statutory inspection rules for inland river ships of China", the theory of ship motion is deployed to analyze the relationship between vessel stability and anti wind grade, and the wind load for typical ships in the Three Gorges river area is calculated based on the navigating zone. On the other hand, the calculation formula of allowed navigation speed is put forwarded under wind conditions. In line with the structure characteristics and operating traits of each ship type, the hierarchical control strategy for navigation under the condition of winds is proposed on the basis of wind resistance and safe speed of various ship types which aims improving the safety level of navigation and the shipping efficiency under winds weather in the Three Gorges river.

1 INTRODUCTION

The Three Gorges Hydro hub is the world's largest water conservancy project which aims to exploit and govern the the Yangtze river. The Three gorges reach lies in the throat part of the water transportation main of China which is also the control segment for passing the Chuan River. The total length of the Three Gorges hub reach is 45.6 km which locates in the connection piece of the middle and upper reach of the Yangtze river. The map of the channel of the Three Gorges area is shown in Figure 1.

The navigation conditions is improved significantly after the impoundment of the Three Gorges project while the freight capacity in ship lock is growing rapidly. However, frequent occurrence of high wind and other bad weather cause serious jam of ships in the Three Gorges Reservoir area. In the middle of November in the year 2010, frequent severe weather such as wind and fog caused the backlog of more than 500 ships which were not evacuated

entirely until the middle of January in the year 2011. A surplus of ships for a long time will cause huge economic, social and environmental damage.

At present, the studies of safety of ship navigation in stormy waves are mainly concentrated in safe navigation method, navigation safety assessment, classification of vessel sustainability in winds and seas and etc. The main research methods include qualitative analysis, statistical analysis, fuzzy comprehensive evaluation, checking and fault tree, reliability analysis method and etc (Wang 1997, Zhao 1998, Liu et al. 2003, Foxwell 1996, Wang & Jia 1998). The management measures for ships against wind weather are temporary and limited. There is lack of research on navigation techniques for ships under the wind condition. Therefore, study on management strategy of navigation under the high winds in the Three Gorges Reservoir area, adopting hierarchical navigation control technology, to ease the shipping backlog in the Three Gorges hub area and to protect the water traffic safety of the Three Gorges hub river is very necessary for reducing the economic and social loss.

2 NAVIGATION STATUS IN THE THREE GORGES RIVER UNDER WIND WEATHER

2.1 *Characteristics of ships in the lock approach waters*

The upper approach channel of the Three Gorges ship lock is narrow where ship maneuvering is

Figure 1. Channel map of the Three Gorges area.

constrained. In winds, downriver ships is sailing harder than upriver ships. The descending ships go through the approach water in queue, and the following ship need to wait until the proceeding ship berth completely. When ship run before the wind, the reverse effect is poor and the speed is low, thus ship's ability to resist wind and water is weak which cause ship drift and may drift to no-navigable zone or trapped to the separation levee. Light ship, cruise ship, and container ship are more significantly affected by wind.

2.2   Ship navigation features in other waters

Ship navigating between the mouth of Xiling Gorge and the upper approach channel of the Three Gorges ship lock should comply with "the separation navigation rules of the Three Gorges" and the driver should choose route based on the characteristics of this vessel and the external factors. Ships with strong power, good stability, high freeboard and well wind resistance ability should up run against the wind and flow, and down run downwind and downstream.

From upstream to downstream in the left bank, there are branch creeks such as Miao river, Duanfang creek, Baisui creek, Taiping creek, providing shelter and berth for ships with power shortage or weak wind resistance ability.

There are mainly wharf, anchorages, shipyards, gas stations at waters along the right bank. In this area, ships are docking intensively and sailing in and out frequently, where drift speed of ship is fast affected by wind. Among them, 95% of the downstream dangerous goods carriers are empty vessels which are affected by wind obviously. Moreover, external wind of Lanling creek is stronger, ships should turn in the dangerous goods anchorage to adjust along the hull and sail into the Lanling creek with bow inclined upwards.

The No. 1 to No. 10 breasting dolphin in Shawan anchorage are used for container ship berthing. Affected by strong wind, the container ship is not easy to control, part of which choose to anchor in the dolphin outside the gate.

3   WIND-ANTI GRADE OF THE SHIPS

3.1   The basic requirements for the ship stability

In various performances of the ship, stability is of special significance. The condition of stability directly affect the buoyancy, seaworthiness, quickness and maneuverability, the economy and safety of operation of the ship.

Prescribed in "the law of legal inspection for inland river ships", all kinds of ships with calculated load condition, the stability criterion should meet the relevant requirements. The Three Gorges hubs river is divided into the level C area, the wind pressure stability criterion for ship sailing in class C area should at the same time satisfy the following conditions:

$$K_f = \frac{9.81\Delta GM}{M_f} \times \frac{F}{B} \geq 1, \text{ when } \frac{F}{B}$$

$$\geq 0.125 \text{ let } \frac{F}{B} = 0.125; \tag{1}$$

$$K_f = \frac{9.81\Delta GM}{M_f} \times \frac{d}{B} \geq 1, \text{ when } \frac{d}{B}$$

$$\geq 0.125, \text{ let } \frac{d}{B} = 0.125; \tag{2}$$

where, Δ represents the displacement of the ship in calculated load cases, GM represents the initial stability height with free surface correction accounting load cases, B is the molded breadth, d represents the moulded draft in calculated load cases, F represents the minimum freeboard along the direction of the heading accounting load cases.

3.2   Derivation of wind load formula

1. According to fluid mechanics theory, formulas of wind pressure and wind speed are shown as follow:

$$p = C_m \times \frac{1}{2} \times \rho \times V^2 \tag{1}$$

By referring to the code of "ships wind pressure calculation (CB/Z801-80)", wind pressure calculation formula for river boat is:

$$C_m = 0.453 + 1.48\frac{Z \times L}{A}$$

where, ρ represents air density (under 20°C and normal atmospheric pressure, which $\rho = 1.205$ kg/m³), V represents the speed corresponding to the centroid of the lateral projection area, L is the length overall, A represents the windage area of the ship under calculated load cases, C_m is the wind pressure of calculated unit.

2. Vessels navigating in the J segment of level C area, the common function of wind heeling moment and the water roll torque are considered to determine the requirements of wind resistance, of which the wind pressure stability criterion should be:

a. $M_f + M_j \leq \frac{F}{B} \times 9.81\Delta GM$ \hfill (2)

234

b. $M_f + M_j \leq \dfrac{d}{B} \times 9.81 \Delta GM$ (3)

where, Δ represents the displacement of the ship in calculated load cases, GM represents the initial stability height with free surface correction accounting load cases, B is the molded breadth, d d represents the molded draft in calculated load cases, F represents the minimum freeboard along the direction of the heading accounting load cases. M_f represents wind heeling moment, M_j represents water roll torque (Yang & Liu (2014)).

$$M_f = C_m \times \frac{1}{2} \times \rho \times V^2 \times A \times Z \times 10^{-3}$$ (4)

According to (2) and (4), there is,

$$M_f \leq \frac{F}{B} \times 9.81 \times \Delta \times GM - M_j$$

$$V_{a0} = \sqrt{\frac{2 \times (9.81 \times \Delta \times GM \times F - B \times M_j) \times 10^3}{(0.453 \times A + 1.48 \times Z \times L) \times \rho \times Z \times B}}$$ (5)

According (3) and (4), there is,

$$M_f \leq \frac{d}{B} \times 9.81 \Delta GM - M_j$$

$$V_{b0} = \sqrt{\frac{2 \times (9.81 \times \Delta \times GM \times d - B \times M_j) \times 10^3}{(0.453 \times A + 1.48 \times Z \times L) \times \rho \times Z \times B}}$$ (6)

According to the "Rules of legal inspection for inland river ship", water roll torque M_j can be calculated as follow,

$$M_j = 9.81 C_j L_s d (KG - a_1 d)$$

where, L_s represents waterline length of the ship in calculated load cases, d represents the molded draft in calculated load cases, KG represents the vertical height of the ship's center of gravity to the baseline in calculated load cases, a_1 is coefficient of which the value can be selected from Table 1 according to the value of B_s/d, C_j is the torrent coefficient of which the value can be selected from Table 2 on the basis of f.

Table 1. Relation between value B_s/d and a_1.

B_s/d	Less than 4.5	5.0	5.5	6.0	6.5
a_1	0.500	0.495	0.475	0.440	0.405

B_s/d	7.0	7.5	8.0	8.5	Greater than 9.0
a_1	0.350	0.285	0.225	0.475	0.440

Table 2. Relation between value f and C_j.

f	Less than 1	2	3	4	5
C_j	0.255	0.279	0.301	0.326	0.346
f	6	7	8	Greater than 9.0	
C_j	0.358	0.365	0.372	0.440	

where, B_s represents the maximum line width in calculated load cases, f is coefficient, $f = 0.013 V_J \Delta / L_s$, V_J is the computation speed, for ship sailing in the J_1 segment, if the ship's maximum speed $V_m \leq 5.83$ m/s, take $V_J = V_m$; when the ship's maximum speed $V_m > 5.83$ m/s, take $V_J = 5.83$ m/s. For ship sailing in the J_2 segment, when the ship's maximum speed $V_m \leq 4.44$ m/s, take $V_J = V_m$, when $V_m > 4.44$ m/s, let $V_J = 4.44$ m/s.

3. Transition of the calculated wind speed and the Beaufort scale

a. Ratio of instantaneous wind speed with the average wind speed in 10 minutes is called sudden wind rate. According to the definition of sudden wind rate (G), for safety and convenience of calculation, selecting average wind speed in 1 second as instantaneous wind speed, thus the wind speed can be written as,

$$V_a = 0.9 \times \sqrt{\frac{(9.81 \times \Delta \times GM \times F - B \times M_j) \times 10^3}{(0.453 \times A + 1.48 \times Z \times L) \times \rho \times Z \times B}}$$ (7)

$$V_b = 0.9 \times \sqrt{\frac{(9.81 \times \Delta \times GM \times d - B \times M_j) \times 10^3}{(0.453 \times A + 1.48 \times Z \times L) \times \rho \times Z \times B}}$$ (8)

b. The scale of wind is measured with wind speed while wind speed along the vertical direction changes. Quoting wind velocity gradient in the "ship design practical handbook", calculated wind speed is converted into Beaufort scale wind speed in 10 meters high.

$$V_{10} = V \times \left(\frac{10}{Z}\right)^{1/7}$$ (9)

Take 1 m when Z is less than 1 m.

3.3 Instance of typical ship in the three gorges reservoir area

According to the current situation of the Three Gorges shipping, the tonnage of mainstream ships passing through the ship lock is between 3000 to 3500. With the development of the shipping economy in upper Yangtze river and the improvement of navigation condition in upstream, according to the standard ship principal dimensions in the Three

Table 3. The designation of the representative ships.

Ship type	Tonnage	Principal dimensions (length × width × draft, m)
Bulk carrier	5000t	110 × 19.2 × 4.3
Liquid cargo ship	3000t	95 × 16.2 × 3.6
Container ship	350TEU	105 × 19.2 × 4.1
Ro-ro	110 stall	108 × 16.8 × 2.6

Table 4. Calculation for wind resistance velocity.

Ship type	V_a(m/s)	V_{a10}(m/s)	V_b(m/s)	V_{b10}(m/s)
Bulk carrier	13.9	16.2	15.2	18.4
Liquid cargo ship	11.2	13.5	13.6	15.8
Container ship	8.6	10.4	10.5	16.3
Ro-ro	16.3	19.5	19.1	23.9

Gorges Reservoir area, selection of representative ship types and dimensions are as shown in Table 3.

Since beam wind has greatest influence on ship stability, thus the horizontal windage area of the ship is computed considering beam wind. The wind resistance velocities of each representative ship are shown in Table 4.

By analyzing the calculation results in Table, the minimum wind resistance velocity of Beaufort scale in 10 meters high for 5000 tons bulk carrier with full load is 16.2 m/s, of which the wind load is Beaufort scale 7. For 110 car ro-ro, the minimum wind resistance velocity of Beaufort scale in 10 meters high is 13.5 m/s and the wind load is Beaufort scale 6. For 350 TEU container ship, the wind load is Beaufort scale 5. For 3000 tons liquid cargo ship, the minimum wind resistance velocity of Beaufort scale is 19.5 m/s and the wind load is Beaufort scale 8.

4 HIERARCHICAL CONTROL STRATEGY FOR SHIP UNDER STRONG WIND

4.1 Control standard for ship navigation

Taking reference of the calculation method of safety speed in the stormy waves at sea, the computation formula of safety speed for inland ship under wind is put forwarded as follow.

$$V_s = V_0\left[1 - (\frac{m}{L} + N)\right] \tag{10}$$

where, V_s represents the allowed speed in stormy waves, V_0 is the design speed for ship, L represents the overall length, m and N are parameters of which the value refers to Table 5.

The allowed speeds of the ship under different wind conditions are calculated according to Eq.10. Select the allowed speed under the most unfavorable conditions as navigation safety limit value. The wind speed restrictions for ship navigation and the allowed safe navigational speed are shown in Table 6.

4.2 Hierarchical control strategy for ship navigation

On the basis of structure characteristics and load stations of each ship type, classified control for ships navigation in the Three Gorges River is conducted. The navigation strategies for each type of ship are shown in Table 7.

Via analysis of the results in Table 7, it is indicated that under the influence of cross wind, the ratio of windage area and ship load has evident effect on the speed limit. The greater the ratio is,

Table 5. Values for parameters m and N.

Beaufort scale	Upwind		Inclined against the wind	
	m	N	m	N
5	9	0.02	7	0.02
6	13	0.06	10	0.05
7	21	0.11	14	0.08
8	36	0.18	23	0.12

Beaufort scale	Crosswind		Downwind	
	m	N	m	N
5	3.5	0.01	1	0
6	5.0	0.03	2	0.01
7	7.0	0.05	4	0.02
8	10.0	0.07	7	0.03

Table 6. Navigable limit standard for various ship.

Ship type	Tonnage	Wind speed Limit/(m/s)	Wind grade	Safe speed/(kn)
Bulk	5000 tons	16.2	7	6.8
carrier	3000 tons	11.5	6	7.8
Liquid	3500 tons	19.5	8	6.6
cargo ship	1000 tons	14.3	7	5.9
Container	350TEU	9.7	5	9.7
ship	200TEU	8.4	5	9.5
Ro-ro	Passenger ship	12.4	6	11.2
	cargo ship	11.3	7	9.3
Cruiser	330 seats	11.3	6	8.8
	460 seats	9.6	5	11.2

Table 7. The navigation limit strategy for ships under wind condition.

Ship type	Tonnage	Beaufort force 5	Beaufort force 6	Beaufort force 7
Bulk carrier	5000 tons	8.7 kn	8.0 kn	6.8 kn
	3000 tons	8.6 kn	7.8 kn	ban
	1000 tons	8.2 kn	7.3 kn	ban
Liquid cargo ship	3500 tons	8.6 kn	7.9 kn	6.6 kn
	1000 tons	8.3 kn	7.4 kn	5.9 kn
	500 tons	8 kn	6.9 kn	ban
Container ship	350TEU	9.7 kn	ban	ban
	200TEU	9.5 kn	ban	ban
	100TEU	ban	ban	ban
Ro-ro	passenger ship	11.2 kn	ban	ban
	cargo ship	10.1 kn	9.3 kn	ban
Cruiser	330 seats	12.8 kn	ban	ban
	460 seats	12.6 kn	11.2 kn	ban
	670 seats	10.5 kn	8.8 kn	ban

the smaller the speed limit value, such as liquid cargo ship. If the ratio is smaller, the speed limit value will be higher, such as full loaded container ship and cruiser (Liu & Zhang (2007)).

In conclusion, management strategies for ship voyage under wind environment is enacted as follow,

1. When the wind scale reaches level 5, navigation of 100 TEU container ship is forbidden.
2. When the wind scale reaches level 6, navigation of container ship, ro-ro passenger ship and cruiser with 670 seats are prohibited.
3. When the wind scale reaches level 7, navigation of bulk carrier less than 3000 tons and 500 tons liquid cargo ship are prohibited. And ban all types of ro-ro, container ship and cruiser sailing.
4. When the wind scale reaches level 8, navigation of all kinds of ships are banned. Navigation speed limit of representative ships under other wind scale refer to Table 6.
5. When the wind scale is less than 5,all types of ship can navigate freely. While the wind scale reaches 8, navigation is prohibited.

5 DISCUSSION

In this paper, the wind resistance classification for typical ship types in the Three Gorges reach is determined in terms of the characteristics of the ship stability, and the navigation control standards and hierarchical navigation control concept are proposed. Since the speed limitation are deduced

from the wind pressure, however there are many other factors such as water flow, visibility, etc. which may have influence on ship navigation. Thus, the navigable limit standard proposed in this paper can only be taken as reference while wind takes a main role of the impact on ship navigation. The pilot should pay attention to the conditions such as ship load, weather, water and so on when making reference of the navigable limit standards under wind weather.

REFERENCES

Foxwell D. 1996. Structures in Question. Safety at Sea 1:35–37.
Liu, D.G., Zheng, Z.Y. & Wu, Z.L. 2003. Review on safety assessment for ships under heavy sea 1:114–118.
Liu, H.Q. & Zhang, W.H. 2007. Research on classification controlling criterion and strategies for safety of expressways in fog area. Highway 10:134–137.
Wang, F.W. & Jia, C.Y. 1998. The study of vessel sustainability in winds and seas. Journal of Dalian Maritime University 1: 61–63.
Wang, F.C. 1997. Estimate of the safety of the voyage in rough. Proceedings of maritime navigation domain of Chinese institute of navigation during 1995–1997, 52–56.
Yang, M.J. & Liu, Q.L. 2014. Calculation method of wind anti grade for inland ships. China water transport 11:16–17.
Zhao, H.S. 1998. Avoidance of dangerous wind zone in typhoon. Proceedings of symposium on navigation safety and typhoon prevention experience of maritime navigation committee of Chinese institute of navigation, 169–174.

Maritime Transportation and Harvesting of Sea Resources – Guedes Soares & Teixeira (Eds)
© 2018 Taylor & Francis Group, London, ISBN 978-0-8153-7993-5

Dynamic risk assessment for nautical traffic

M. Li, J. Mou, Y. He, F. Ning & Y. Xiong
School of Navigation, Wuhan University of Technology, China

P. Chen
Faculty of Technology, Policy and Management, Delft University of Technology, The Netherlands
School of Navigation, Wuhan University of Technology, China

ABSTRACT: In some busy port and waterways, the risk of vessel traffic accidents is increasing. From the view point of nautical traffic, the risk is closely linked with ship behaviors and hydrodynamic. To substantially improve navigation safety, it is demanding to model dynamic risk of ship collision or grounding accidents. This paper presents a method combining Nomoto model and Monte Carlo simulation to determine the risk. The stochastic phenomenon of traffic behaviors is described by the Nomoto model under random rudder angle, heading and speed. The dynamic risk is derived from the final distribution of the ship pitch and the span in the fairway at the next moment. Taking a local segment of the Yangtze River in China as an example, the study shows that the method is novel and effective to assess the ship's dynamic risk and beneficial for vessel traffic management and channel design.

1 INTRODUCTION

In some hub ports and busy waterways, the trends of ships' maximization and increasing speed become more tangible, and the increasing traffic density make the navigation environment more complicating. Vessel traffic in the vicinity of busy waterways and large ports pose high risk to both private and public sectors. Safety and risk management related to the vessel traffic has been becoming a very important issue in maritime industry (Mou et al. 2016).

Many efforts have been put into researches on maritime risk from various perspectives. Kaplan (1997) defined risk as a complete set of triplets which consists of "scenario, likelihood and consequence". All of these triplets represent the three questions about risk: "What can go wrong? How likely it is? And what is the consequence?" Aven (2012) classified the definitions of risk into nine categories. Goerlandt (2014, 2015) reviewed the concepts in risk assessment of maritime transportation and proposed that, in the application fields, risk is strongly tied to probability and expected values; Kristiansen (2005) denoted risk as product of probability of accident and its consequence. According to investigation, collision accidents account for almost half of the total accidents and percentage still remains high. Therefore, many studies will be the risk of colli-

sion as a research focus. Zhang (2011) established the risk assessment model of ship collision in the Yangtze River based on the grid technology. By choosing the ship collision risk as the state variable and selecting the ship encounter mode, wind speed and visibility as the control variables, Wen (2012) fitted the collision risk state equation, based on the order Probit model. Mou (2010) assessed the ship dynamic collision risk based on the SAMSON model using AIS (Automatic Identification System) data and took the DCPA (Distance to Closest Point of Approach) and TCPA (Time to Closest Point of Approach) into consideration. Szlapczynski (2006) developed a uniform ship collision risk calculation model based on ship domain. Wang (2010, 2012) assessed the spatial collision risk level at different encounter situation based on fuzzy quaternion ship domain. Xiang (2013) identified the ship collision risk using DBSCAN algorithm based on ship domain. Tam (2010) proposed a ship collision risk assess method by changing the perspective of own ship and target ship based the COLREGS (International Regulations for Preventing Collisions at Sea); by choosing the ship collision risk as the state variable and selecting the ship encounter mode, wind speed and visibility as the control variables. In addition to the risk of collision, narrow waters also need to study on grounding risk of the ship. Many scholars (e.g. Fujii et al. 1974;

Pedersen 1995; Simonsen1997; Amrozowicz 1996; Amrozowicz et al. 1997; Aven 2013) had modeled this type of accident in order to predict the likelihood of the accident given certain criteria.

Generally, it is important to conduct studies on ship collision and grounding risk, respectively, based on DCPA and TCPA, ship domain, and the collision risk calculation still sequential order Probit model. Because, it is very difficult to describe dynamic risk. AIS data consists of information of ship movement in the geospatial space, reflecting the behavioral characteristics with interaction between ship and environment, AIS data is promising for risk assessment.

Ship collision and grounding are the main accidents. If it is not only causing significant casualties and property damage, but also lead to the transport of goods and coastal residents living a huge negative impact. Based on AIS data, this paper presents a method combining Nomoto model and Monte Carlo simulation to determine the risk. The stochastic phenomenon of traffic behaviors is described by the Nomoto model under random rudder angle, heading and speed. The dynamic risk is derived from the final distribution of the ship pitch and the span in the fairway at the next moment. The maneuverability indexes of Nomoto model are identified by using AIS data. The distributions of rudder angles, heading, and speed are obtained by investigation and the statistical of AIS data. Taking a local segment of the Yangtze River in China as an example, a hundred thousand simulation has been conducted for vessel traffic for a series of very short time span. The probability of grounding can be calculated by the total occurrences of ship beyond fairway. While the probability of risky encounter, which is very important concept of collision avoidance, has been determined by the violation of ship domains. The risk of grounding is determined by grounding probability, and the risk of collision is determined by probability of risky encounter. The results of aforementioned are defined as measurement of grounding and collision risk, respectively.

This paper presents a method combining Nomoto model and Monte Carlo simulation to assess the risk of collision and grounding. The results show the dynamic risk of a ship navigating along channel under a certain rudder. The risk of grounding is acceptable with level of 10^{-4} at the beginning. While the risk of own ship conflicting the domain of target ship, is very much simplified from real collision and cannot be validated in this stage. Generally, the method is novel and effective to assess the ship's dynamic risk and beneficial for vessel traffic management and channel design.

2 METHODOLOGY

It is well-known that risk of vessel traffic is closely linked with ship behaviors, and ship behaviors are the reflections of ship performance, human factors and environmental conditions. Due to the fluctuation of above mentioned factors, the behaviors are stochastic. Monte Carlo methods (or Monte Carlo experiments) are a broad class of computational algorithms that rely on repeated random sampling to obtain numerical results. The main idea is using randomness to solve engineering issues, which are influenced by coupled factors. Monte Carlo methods are powerful to tackle three kind problems: optimization, numerical integration, and generating draws from a probability distribution (Kroese 2014).

In this paper, the ship behavior responding to random rudder angle is calculated on a certain ship control model named Nomoto model (Eq. (1)). The longitudinal and lateral distance X and Y from the present position, i.e., the pitch and span of ship maneuvering, are predicted (Eq. (2)) for a short time.

The distributions of rudder angles, heading, and speed are obtained by investigation and the statistical of AIS data. During simulation, the inputs are rudder angle, heading and speed, while the outputs are ship track points. After carrying out a hundred thousand simulation for vessel traffic in a series of very short time span, the probability of grounding can be calculated by the total occurrences of ship beyond fairway. While the probability of risky encounter, which is very important concept of collision avoidance, can be determined by the violation of ship domains (Fig. 1). It is assumed that the risk of collision is equal to percentage of own ship simulated which lies in target domain. In this paper, the design of the simulation is shown in Fig. 2.

$$T\ddot{\psi} + \dot{\psi} = K\delta \qquad (1)$$

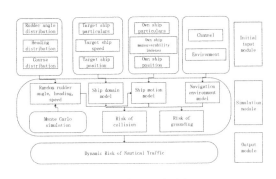

Figure 1. Framework of dynamic risk assessment.

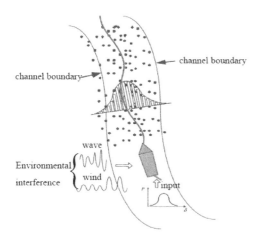

Figure 2. Design of the simulation.

where: K(turning ability index), T(turning lag index) is the usual maneuverability indexes, which can be calculated by the data of the actual Zig-Zag test; $\ddot{\psi}$ ship heading angle; $\dot{\psi}$ for the ship's first angular speed (ROT), δ is the rudder angle

$$\begin{cases} X(t) = \int_0^t 0.9v_0 \cos\left[K\delta_0(t - T + Te^{-\frac{t}{T}}) \right] dt \\ Y(t) = \int_0^t 0.9v_0 \sin\left[K\delta_0(t - T + Te^{-\frac{t}{T}}) \right] dt \end{cases} \quad (2)$$

Where: t is the time of ship's movement; v_0 is the initial speed of the ship; X_t and Y_t are the longitudinal and lateral distance of the ship at time t.

The proposed dynamic risk model uses inputs from a wide range of parameters, e.g. channel, and target ship domain, own ship parameters. In general, the dynamic information regards characteristics of an object or conditions changing over time.

For simulation, this paper analyzes the parameters of environmental parameters and ship parameters respectively. The environmental parameters include the channel parameters, target ship domain parameters. Channel parameters mainly consider the width of the channel; and ship domain parameters will be calculated by using AIS data. Own ship's parameters include ship course, speed, rudder angle. These parameters can directly affect the trajectory of the ship, the final position and time of the ship movement is an important indicator of dynamic risk.

2.1 Channel parameters

The inland waterway is a general term for rivers, lakes, reservoirs, and canals. The navigable conditions include current, stage, meteorological conditions, channel layout and boundary. Channel layout includes: water depth, width and bending radius.

In this paper, a straight and narrow channel is demonstrated. It is assumed that the ship will be ground outside the fairway and therefore the water depth outside of the channel will not be taken into account. The channel width will be used as a channel parameter. In this simulation, the horizontal distance of the narrowest part of the local segment is taken as the channel width.

2.2 Target ship domain

Ship domain plays a very important role in risk assessment. It was first presented in early 1970s by Fujii for a narrow channel of Japanese Sea. Later, Goodwin established a domain model in open sea. In 1980s, Goldwell studied the domain for hand-on and overtaking encounters in the restricted waters. There is still without a universal definition for the concept of ship domain. In the past four decades, many researchers have presented various ship domains with different shapes and sizes. Reviewed from the factors affecting ship domain, the typical ship domain could be roughly distinguished into circle, elliptical and polygonal shapes. Meanwhile, to meet the current development of the marine traffics, the existing models of ship domain needed to be improved. For a more systematic and flexible ship domain model, the shape and size depend on a number of variables, such as fairway condition, ship style, and density of traffic flow and so on.

Based on the AIS data, for every vessel in the survey area, its relative bearing and distances from all other vessels can be plotted every certain interval. Looking the maximum density of vessel traffic around the own ship, the boundary of ship domain can be determined. According to ship's type, speed, density of traffic flow, day or nighttime, and channel condition, the ship domain can be deliberately categorized and presented. Therefore, AIS data is used to calculate the ship domain.

2.3 Own ship maneuverability indexes

The Nomoto model with K, T indexes which can be expressed by the linear hydrodynamic derivatives. And the indexes K and T are called ship maneuverability indexes. Ship maneuverability indexes K and T are important indexes with clear physical meaning. According to maneuverability indexes K, T, the advance and lateral distance from steering point are calculated.

2.4 Ship heading

The distribution of the ship heading at a certain is subject to ship performance human factors, moment and external factors. AIS data can be used to analyze the distribution of ship heading. After the ship steering, the ship will respond, so the position of the ship at the next moment is related to the heading of the ship.

2.5 Ship speed

The distribution of the ship speed at the channel also can be analyzed by AIS data. The main factor affecting the position of the ship at the next moment is the speed, speed is the great impact on the pitch and span of the ship maneuvering.

2.6 Ship rudder angle

In the case of steering operations, crews will take different rudder angles during different situations for the safety of navigation. According to onboard observation, the distribution of the rudder angle can be revealed. Especially, for a fixed ship type, the distribution is obvious. According to the knowledge and experience of pilots, ship is steered with a rudder angle from 5° to 15°. Most crews will choose 5° rudder angle, and a very small number of people will choose 10° or even 15° rudder angle. Therefore, a normal distribution is assumed for the ruder.

3 APPLICATION

3.1 Research area

The Yangtze River is the longest river in Asia and the third longest in the world. In the lower reaches of the Yangtze River, the ship has a higher density, and some of the channels are narrow, and grounding and collision is frequently experienced. In this paper, the narrowest straight part of the Yangtze River is employed to simulate and analyze. The study area is the channel between Changqing sand and Minzhu sand, the latitude range of 032° 02' N-032° 04' N and longitude range of 120° 31 'E-120° 33' E, as shown in Fig. 3.

3.2 Initial condition

The assumption in this paper is no wind and no current, any own ship and target ship are in the middle of the channel. Target ship starts at a distance of 610 meters from own ship. Own ship taking a random rudder angle, keep the steering angle unchanged and the ship continues to navigate for a while. Target ship keeps a speed of 6 knots. Theo-

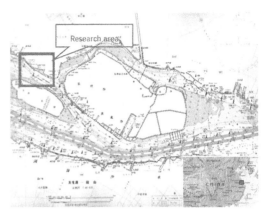

Figure 3. Research area (in red rectangle).

retically, the location of own ship into the target ship domain are recorded as a risk of collision, and ship outside the channel as a grounding risk.

1. Channel parameters

The channel width will be used as the channel parameter, and the channel width is the horizontal distance between the sides of the channel and perpendicular to the centerline of the channel. According to Yangtze River waterway chart, the narrowest channel width is about 200 meters. In this paper, it is assumed that the segment is a straight channel, and the channel boundary is a line. In this environmental model, the channel is divided into upstream and downstream, two-way navigation, set the main channel with direction of 305° or 125°, and width of 200 meters.

2. Target ship domain

Regarding risk of collision, it is assumed that target ship is stand-on ship, keeping original speed and course. The ship domain model used in this paper can be referred to Ren (2014). In the simulation, set the ship's long axis to 2 times of ship length, the span is 2.5 times of ship width. And the dimension of target ship is 150 meters long, 30 meters wide.

3. Own ship maneuverability indexes

The simulated ship prototype is from uses Huang (2016), with identified ship's K, T indexes (Table 1).

4. Rudder angle

As mentioned at section 3.6, when the ship is maneuvering in a certain water, the rudder angle is subject to a normal distribution. Based upon the experience of crews, the means and standard deviation values of the normal distribution are 2° and 5° respectively.

5. Ship speed

When determining the speed of the ship, the statistics based on AIS data is conducted. The speed

Table 1. prototype ship and K, T.

MMSI	Length/m	width/m	K/s−1	T/s
413201760	95	15	0.1053	27.077

distribution of cargo ship, as shown in Fig. 4, is tested by Gaussian and Weibull destruction, and the confidence level is 95%. The results show that the Weibull distribution is more acceptable.

6. Ship heading

According to the AIS data statistical analysis of the ship heading, the ship's heading distribution is concentrated between 300°-310° and 120°-130°. The distribution was tested performed by Gaussian and Weibull distribution, and the confidence level was 95%. The results show that the Gaussian distribution is better for the interpretation of ship heading (Fig. 5.)

3.3 Results

In order to obtain the dynamic risk of the ship during the navigation, the simulation were conducted. After carrying out a hundred thousand simulation for vessel traffic in a series of very short time span, and the relative position of the ship is shown in Figs. 6–10, within 20s, 30s, 40s and 50s after the steering. In every figure, the ellipse stands for target ship domain, and blue clusters stand for trajectory of own ship. Three vertical lines stand for left, right channel boundary and center line of channel. Grounding is recognized when own ship run out the boundary of the channel. And collision is defined when own ship lies in target domain.

Figs. 6–8 demonstrate all of the ship trajectories are concentrated in the interior of the channel, and own ship does not enter the target ship domain at the beginning, so there is no risk of collision and grounding. In Fig. 9, part of trajectories are out the channel boundary, while still no trajectory is coming into the domain of the target

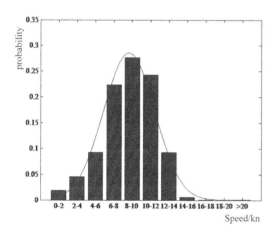

Figure 4. Cargo ship speed distribution.

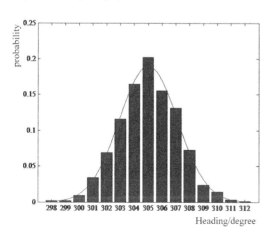

Figure 5. Ship heading distribution.

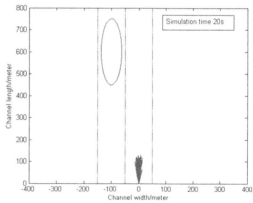

Figure 6. Results of 20s simulation time output.

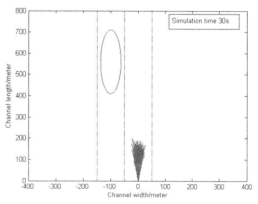

Figure 7. Results of 30s simulation time output.

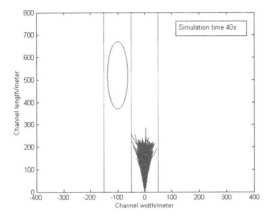

Figure 8.   Results of 40s simulation time output.

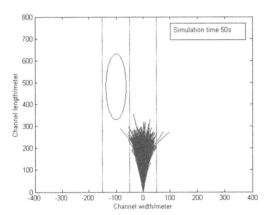

Figure 9.   Results of 50s simulation time output.

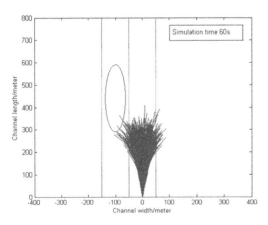

Figure 10.   Results of 60s simulation time output.

ship, so there is very small risk of grounding and zero risk of collision. Fig. 10 indicates some trajectories of ships are outside the fairway boundary,

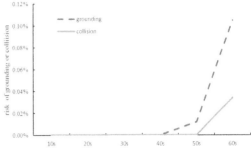

Figure 11.   The risk of grounding or collision.

and a small number of trajectories enter the ship domain, the risk of collision and grounding is perceptible. Moreover, as time goes by, the risk of the ship climbs dramatically (Fig.11).

3.4   *Discussion*

The method is derived from point view of nautical operation, and the simulation can be taken as a part of nautical traffic model. Compared with other studies (Huang et al. 2017; Zhang et al.2011), the result of grounding is acceptable. However, during simulation an assumption, that the risk of collision is equal to percentage of own ship conflicting the domain of target ship, is very much simplified from real situation. It is imposable to validate the risk of collision in this stage.

The results show the dynamic risk is linked with ship location and time. As time goes by, the risk of the ship climbs dramatically (Fig.11). Generally, there is no risk of collision in the first 40s of the ship. After then, the risk of ship is emerging and increasing. The risk of grounding increases faster than the risk of collision, but the latter one is very much influenced by the relative distance of target ship. It is recommended that crews should timely take proper rudder angle to adjust the ship position in the channel.

The three dynamic inputs (speed/heading/rudder angle) are the key parameters to affect the ship location at the next moment. Although, the simulation based on relatively ideal environment such as no wind and no current, it can be easily applied to the probability of grounding and risky encounter in more complicated situation.

4   CONCLUSIONS

In some busy port and waterways, the risk of vessel traffic accidents is increasing. From the view point of nautical traffic, the risk is closely linked with ship behaviors and hydrodynamic. This paper presents a method combining Nomoto model and Monte Carlo simulation to determine the dynamic risk of ship grounding and collision. The study shows that the method is novel and effective to assess the ship's dynamic risk and beneficial for vessel traffic management and channel design.

A case study on the Yangtze River was conducted by the method, which used to estimate the probability of grounding or risky encounter, considering the distribution of parameters such as ship speed, heading, and rudder angle. In future work, this model shall be improved to study on dynamic risk in a more complicated environment. Different distributions of ship speed during different time periods (such as day and night) will be incorporated into the model. The actual topography of channel and depth, which are important to grounding analysis, will be considered to improve the model's efficacy.

ACKNOWLEDGEMENTS

The work presented in this paper is financially supported by National Natural Fund of China (Grant number 51579201), Independent innovation fund for graduate students of Wuhan University of Technology (Grant number 175212004)and the Open Foundation of Nation Water Transportation Safety Engineering Technical Center (Grant number: 17KF02).

REFERENCES

Amrozowicz M. 1996a. The need for a probabilistic risk assessment of the oil tanker industry and a qualitative assessment of oil tanker groundings. Massachusetts Institute of Technology, Cambridge.

Amrozowicz M. 1996b. The quantitative risk of oil tanker groundings. Massachusetts Institute of Technology, Cambridge

Aven, T. 2012. The risk concept—historical and recent development trends. Reliability Engineering & System Safety99, 33–44.

Aven T. 2013. A conceptual framework for linking risk and the elements of the data–information–knowledge–wisdom (dikw) hierarchy. Reliability Engineering System Safety, 111, 30–36.

Aven T, Reniers G. 2013. How to define and interpret a probability in a risk and safety setting. Safety Science, 51(1):223–231.

Chen, P., Mou, J., Li, Y. 2015. Risk analysis of maritime accidents in an estuary: a case study of Shenzhen Waters. Scientific Journals of the Maritime University of Szczecin42 (114), 54–62.

Daidola J.1995. Tanker structure behavior during collision and grounding. Marine Technology, 32(1):20–32.

Fujii Y. 1974. The Effect of Darkness on the Probability of Collision and Stranding. Journal of Navigation, 27(2):243–247.

Fujii Y., Shiobara R.1971. The Analysis of Traffic Accidents. Journal of Navigation, 24(4):534–543.

Goerlandt, F., and Montewka J.2015. Maritime transportation risk analysis: Review and analysis in light of some foundational issues. Reliability Engineering & System Safety, 138: 115–134.

Goerlandt, F. 2015. Risk analysis in maritime transportation: principles frameworks and evaluation. Ph. D dissertation of Aalto University.

Goerlandt, F., and Montewka, J. 2014. Review of risk concepts and perspectives in risk assessment of maritime transportation. Safety and Reliability: Methodology and Application 2014: 1547–1554.

Greidanus H, Alvarez M, Eriksen T, et al.2013. Basin-Wide Maritime Awareness From Multi-Source Ship Reporting Data. Transnav International Journal on Marine Navigation & Safety of Sea Transportation, 7(2):185–192.

Huang, X. 2016. Prediction on ship behaviors in curve channel of Yangtz River based on AIS data. Wuhan University of Technology.

Huang C, Chen W., Chai T.2017. IWRAP MK II Software Application in Ship Collision and Grounding Risk assement in Yangtze River Estuary Area. Navigation of China, 40(1): 79–82+118.

Kaplan, S., 1997. The Words of Risk Analysis. Risk Analysis17, 407–417.

Kroese D P, Brereton T, Taimre T, et al. 2014. Why the Monte Carlo method is so important today. Wiley Interdisciplinary Reviews Computational Statistics, 6(6):386–392.

Kristiansen S.2005. Maritime Transportation: Safety Management and Risk Analysis. Marine Safety, 36(2):250–251.

Kujala P, Montewka J. Modelling Marine Accident Frequency.

Li H X.2014. Study on the Data-Mining Based Mechanism for the Occurrence of Ship Collision Accidents Caused by Human Factors. Applied Mechanics & Materials, 686:321–326.

Minorsky V U. 1959. An analysis of ship collision with reference to protection of nuclear power plants. Journal of Ship Re-search, 3(2):1–4.

Montewka, J., Hinz, T., Kujala, P., et al.2010. Probability modelling of vessel collisions. Reliability Engineering & System Safety, 95(5): 573–589.

Mou, J.M., Tak, C.V.D., Ligteringen, H., 2010. Study on collision avoidance in busy waterways by using AIS data. Ocean Engineering37, 483–490.

Mou, J., Chen, P., He, Y., Yip, T., Li, W., Tang, J., & Zhang, H. 2016. Vessel traffic safety in busy waterways: a case study of accidents in western shenzhen port. Accident Analysis & Prevention.

Pedersen P.1995. Collision and grounding mechanics.

Pedersen P.2010. Review and application of ship collision and grounding analysis procedures. Marine Structures, 23(3):241–262.

Ren Y. Mou J., Du Y., Zhu J.2012. Study on Ship Domain Using AIS Data. The International Workshop on Next Generation of Nautical Traffic Model.

Szlapczynski R. 2006. A Unified Measure Of Collision Risk Derived From The Concept Of A Ship Domain. Journal of Navigation, 59(3):477–490.

Tam C., Bucknall R. 2010. Collision risk assessment for ships. Journal of Marine Science and Technology, 15(3):257–270.

Wang N. 2010. An Intelligent Spatial Collision Risk Based on the Quaternion Ship Domain. Journal of Navigation, 63(4):733–749.

Wang N, Meng X, Xu Q, et al. 2009. A Unified Analytical Framework for Ship Domains. Journal of Navigation, 62(62):643–655.

Wen Y Q. 2012. Risk Degree Analysis of Ship Collision Risk in Channel Based on Ordered Probit Model. China Safety Science Journal, 22(2):134–139.

Weng J, Meng Q, Qu X. 2012. Vessel Collision Frequency Estimation in the Singapore Strait. Journal of Navigation, 65(2):207–221.

Xiang Z, Hu Q, Shi C. 2013. A Clustering Analysis for Identifying Areas of Collision Risk in Restricted Waters. Transnav International Journal on Marine Navigation & Safety of Sea Transportation, 7(2):101–105.

Zhang G, Thai V V. 2016. Expert elicitation and Bayesian Network modeling for shipping accidents: A literature review. Safety Science, 87:53–62.

Zhang D, Yan X P, Liu J X.2011. Study of grid-based collision risk assessment model for main route of the Yangtze River. Navigation of China, 34(1):44–43.

Zhang D.2011. Navigational Risk Assessment in the Dry Season of Yangtze River. Wuhan University of Technology.

Maritime Transportation and Harvesting of Sea Resources – Guedes Soares & Teixeira (Eds)
© 2018 Taylor & Francis Group, London, ISBN 978-0-8153-7993-5

Challenges in modelling characteristics of maritime traffic in winter conditions and new solution proposal

J. Montewka & R. Guinness
Finnish Geospatial Research Institute, Masala, Finland

L. Kuuliala
Finnish Transport Safety Agency, Helsinki, Finland

F. Goerlandt & P. Kujala
Department of Mechanical Engineering, Aalto University, Espoo, Finland

M. Lensu
Finnish Meteorological Institute, Helsinki, Finland

ABSTRACT: Modelling maritime traffic in ice-covered waters is known for numerous challenges pertaining to proper characterization of ships navigating in ice. In this paper we discuss strength, weaknesses and potential application areas of the existing approaches to modelling ship performance in ice. These approaches can be divided into two categories: engineering-based approaches and data-driven approaches i.e. based on long-term observations of winter traffic. Finally, by combining these two approaches, we propose a hybrid model, in order to utilize the strength of each modelling approach to the maximum. We assign a given approach to our modelling space based on environmental and operational conditions. The obtained results reveal that the hybrid model is capable of providing reliable information about ship performance in a wide range of conditions, accounting for existing operational conditions. In principle, it can be used for the purpose of route optimization in ice and maritime traffic modelling.

1 INTRODUCTION

Modelling maritime traffic in ice-covered waters is gaining global attention, mainly due to new prospective routes through the Arctic Ocean (Somanathan, Flynn and Szymanski, 2009; Ho, 2010; Bergström et al., 2014; Bergström, Erikstad and Ehlers, 2016; Aksenov et al., 2017). However there are numerous challenges in proper characterization of ships navigating in ice. These can be divided into two groups: environmental and operational. The former refer to the effects of the ice environment on the behaviour of ships, primarily their speed. The latter means the effect of the winter navigation system (e.g. ice-breakers) on ships' speed. Conventionally, the effect of the ice environment on ship performance has been estimated with the use of engineering models, accounting for the ice-breaking process in relatively simple ice conditions. For recent developments see, for example, (Su, Riska and Moan, 2010; Külaots et al., 2013; Tomac et al., 2014; Zhou, Peng and Qiu, 2016; Kuuliala et al., 2017). The effect of operational features, however, has not been taken into account,

adding therefore another element to the total uncertainty budget related to the assessment of ship performance in ice (Choi et al., 2015).

Therefore, in this paper we discuss the challenges related to appropriate modelling of ship speed in ice for the purpose of maritime traffic model and route finding in ice (Kotovirta et al., 2009; Guinness et al., 2014). We discuss the strengths, weaknesses, and potential application areas of the existing approaches. We classify existing approaches as either classical, engineering-based approaches, or as data-driven approaches. The latter are based on long-term observations of winter traffic. Finally, by combining these two approaches, we propose a solution in a form of a hybrid model. The main intention of the hybrid model is to utilize the strength of each modelling approach to the maximum by assigning a given approach to a modelling space based on certain conditions (i.e. environmental and operational).

The obtained results reveal that the hybrid model is capable of providing reliable information about the performance of a ship in a wide range of conditions, accounting for environmental variability and existing operational conditions. The hybrid model is suitable for the purpose of maritime traffic simulation, as well as safe route planning in ice for a single ship or group of similar ships, accounting for the economy and safety of a voyage.

The paper consists of six sections, including the introduction, and is structured as follows: Section 2 elaborates on existing modelling approaches and discusses the related challenges. Section 3 describes the concept of a hybrid model and a developed example. The example hybrid model is then applied in a routing tool, which is described in Section 4. The obtained results are shown and discussed in Section 5, and Section 6 presents our conclusions.

2 CHALLENGES OF EXISTING SOLUTIONS

The main aim of this section is to discuss the approaches to estimating ship performance in ice-covered waters, suitable for modelling maritime traffic and route finding in ice. To this end, we first show the strengths of the existing approaches. Secondly we point to the area where the modelling approaches are falling short.

2.1 *Overview of engineering models*

With the use of engineering models, ship transit in ice is modelled by solving equations of motion of a rigid body for one or more degrees of freedom. The majority of models falling into this category are quantity-oriented, where the focus is on detailed quantification of the forces acting on a ship as continuous functions, thus allowing one to compute her speed. There are two major force components acting on a ship: resistance due to the ship moving in ice and net thrust provided by the ship's propulsion system. Resistance varies greatly depending on ice conditions, such as level ice, ridged ice, ice channel, or presence of ice compression. Level ice resistance can be broken down into three components: breaking, crushing and submergence (Lindqvist, 1989). Ridged ice resistance can be divided into components caused by breaking of level ice and the consolidated layer, displacing ridge keel mass at the bow, and the sliding of keel mass along the parallel section of the ship hull (La Prairie, Wilhelmson and Riska, 1995). Compression resistance follows from the increased contact forces on the sides of a ship due to transversal compressive strain of the ice cover (Riska et al., 1995). However the more complex ship motion model, more degrees of freedom (DoF) are concerned, the less complex ice conditions are considered in the overarching ship-ice interaction model.

The main reason for that is the computational efficiency of such a model.

For example, in models adopting one degree of freedom, where only the surge direction is considered, at least two types of ice are considered: level ice and ridged ice (Kotovirta et al., 2009; Kuuliala et al., 2017). Sometimes also ice compression is accounted for (Kaups, 2012; Külaots et al., 2013). However, the majority of 3- and 6-DoF models consider only level ice (Su, Riska and Moan, 2010; Tan et al., 2013). In the case of 3-DOF models, sway and yaw are added (Su, Riska and Moan, 2010). For 6-DoF models, an additional three directions of motion are accounted for: roll, pitch and heave (Tan et al., 2013).

To set up a meaningful set of equations of motions one needs to possess detailed data describing the ship and surrounding ice cover. To define thrust generated by a ship detailed information about her propulsion system are required. To determine the resistance components detailed data on ship hull and ice cover (including mechanical properties of the ice) are needed.

Once the data are at hand the detailed assessment of the ship performance in given ice conditions can be made. Generally the aim of this kind of assessment is to find ways to decrease the resistance of the ship. This in turn allows optimal design of a ship for an intended purpose, within predefined boundary conditions. Alternatively maximum operational conditions can be specified, within which the ship can safely operate.

2.2 *Overview of data-driven models*

Another group of models evaluating ship performance in ice is based on empirical data. This group can be further divided into quantity-oriented models (with both inputs and outputs defined as continuous functions) and event-oriented models (with inputs or outputs discretized into different classes). The former, similarly to engineering models, tend to predict a ship's speed based on data obtained in the course of sea-trials and dedicated campaigns (Keinonen, 1996) or model tests (Kujala and Arughadhoss, 2012; Suominen and Kujala, 2014; Cho and Lee, 2015; Jeong et al., 2017) with respect to specific ship type, e.g. an icebreaker or strong, specialized ice-going vessel. Event-oriented models, on the other hand, focus on specific events that occur in time (e.g.: ship's speed falling within a certain speed range or ship besetting in ice) and on the causality relationship between observable factors affecting the event. See, for example, (ENFOTEC, GeoInfo and McCallum, 1996; Tunik, 2000; Montewka et al., 2015). This type of modelling, unlike the majority of quantity-oriented approach, is able to describe the joint effect of various ice features on a ship's speed. These can include also operational aspects of the winter navigation system, like the presence of IBs, existence of ice channels or specificity of winter navigation operations carried out in a given area. All these may have a prominent effect on the speed of ships navigating in the area, the probability of besetting and waiting time while beset, which are not accounted for in quantity-oriented models. These may be included in the event-oriented models, but without generating insight into the physics of the process of ice breaking. This makes the event-oriented models of little use for the ship hull design process. They can, however, be used for prediction the vessels performance in strategic (long term planning, risk assessment of transits in ice covered waters) or operational settings (route finding). See, for example, (Kujala, Montewka and Filipovic, 2014; Montewka et al., 2015; Fu, Zhang, Montewka, Yan, et al., 2016).

For instance, an event-oriented model for a specific ship or a group of ships or can be developed with the use of a combined set of data about an ice field, observed or obtained from a numerical ice model, and a set describing ship speed obtained from the Automatic Identification System (AIS). If appropriate modelling techniques are applied (e.g. machine learning) they lead to a model which is computationally fast to deliver its output (even though the process of learning the model from the data may be time consuming) easy to validate and can be updated if new knowledge about the conditions/inputs is gained.

2.3 Environmental challenges related to traffic modelling in ice-covered waters

We list here two environmental challenges, which refer to the effect of the environment on a ship's speed.

The first challenge relates to the ice thickness description assumed by the ship performance models. These models often limit the description to level ice and ice channels (Naegle, 1980; Lindqvist, 1989; LaPrairie, Wilhelmson and Riska, 1995; Mulherin et al., 1996; Riska et al., 1997; Kotovirta et al., 2009; Su, Riska and Moan, 2010; Lubbad and Løset, 2011). In some cases, information about ice ridges is incorporated (Riska et al., 1997; Juva and Riska, 2002). In reality, level ice prevails in relatively narrow time windows, usually in the beginning of the ice season and its relative contribution decreases strongly throughout the ice season. The ice is increasingly forced into motion by wind and currents, breaks up and deforms as the season proceeds (Löptien and Dietze, 2014). Close to the coasts, the ice usually appears as fast ice, while the ice is adrift elsewhere. When ice cover

stresses generated by wind and ocean currents build up it finally fails and accumulates to form ridges (Löptien et al., 2013; Kubat, Fowler and Sayed, 2015). In case of the Baltic Sea, the ice ridges can become 5–15 m thick and account for 10–50% of the total sea ice volume (Leppäranta and Myrberg, 2009). Such a massive pile of ice, when encountered by a ship may introduce serious impediments to her progress. The ice resistance dramatically increases and the ship speed drops. The ship needs to ram the ice or, if not able to back any more, she becomes beset.

The ice cover stresses may also manifest as local compression event, which is the second environmental challenge for winter navigation. In compression event ship channel closes and the ice exerts pressure against ship hull. The ship experiences significant resistance increase, and the whole engine power is needed to proceed. Even with full power, often ships get stopped (Rosenblad, 2007; Kubat, 2012; Kubat, Babael and Sayed, 2013; Montewka, Goerlandt and Kujala, 2014; Fu, Zhang, Montewka, Zio, et al., 2016). The compression events are often localized in space and time and depend on local floe geometry (Leisti and Riska, 2010). In a severe compression event the ice may start to fail and pile up against the ship's parallel section of the hull. If a ship is caught in such a situation, pressures on the hull will be high with the loads of magnitude of 1 MN/m (Riska et al., 1995). This can cause damages to a ship, moreover smaller vessels can be completely lifted onto the ice (Kubat, Fowler and Sayed, 2015).

Presently, the required joint effect of all relevant ice features on ship speed, including the effect of ridging and ice compression lacks its proper description, despite numerous efforts taken up-to-date, (Külaots et al., 2013; Tomac et al., 2014). Moreover, suggestions have been made to move from the deterministic, quantity-oriented models towards probabilistic and event-oriented models, e.g. (Kotovirta et al., 2009). This would also allow modelling the uncertainty associated with ship performance in ice.

2.4 Operational challenges related to traffic modelling in ice-covered waters

Operational challenges refer to the effect of existing services (e.g. icebreakers—IBs) and traffic organization (e.g. presence of ice channels, advised safe speed) on ships' speed, besetting probability, and waiting time.

Engineering-based models for ship performance in ice provide valid results for areas of high ice concentration, accounting for a wide range of ice conditions, but fail to account for the presence of ice channels and IBs assistance (Kuuliala et al.,

2017). The latter is usually available in such conditions and affect significantly attainable speed and transit time (Liikennevirasto, 2017). Moreover, engineering-based models can estimate ship performance in a wide range of ice conditions, comparing ice resistance and available thrust, providing the maximum attainable speed. The latter, however, may not always correspond to the safe speed.

On the other hand, data-driven models provide valid information for the already existing range of ice conditions but fail to predict beyond that. Data-driven models inform the users about factual performance of ships navigating in the area, however, there are numerous hidden variables affecting the performance that has been attained (Aldous *et al.*, 2015), and only rough estimates of the relationship between ice conditions, expressed in simplified manner (ice concentration or equivalent ice thickness), and ship performance can be established.

3 A HYBRID MODEL

3.1 *A concept of hybrid model*

A hybrid model results from applying two or more related but different models and then synthesizing the results into a single score or spread in order to improve the accuracy of predictive analytics.

In predictive modelling and other types of data analytics, a single model based on one data set can have biases, high variability, overfitting, or outright inaccuracies that affect the reliability of its analytical findings. By combining different models or analysing multiple data sets, one can reduce the effects of such limitations and provide better generalization.

In a hybrid model, several sub-models cover the modelling space, and the choice of the coverage area for a given sub-model is made based on the accuracy achieved by the sub-model in the given area. Therefore, the areas of low accuracy for one sub-model are covered by another which scores better in those areas. Therefore, a logical condition needs to be introduced that switches between sub-models employed in the overarching hybrid model. This needs to be dependent upon some observable and preferably unambiguous factor or set of factors. These, in the case of maritime transportation in ice-covered waters, could be as follows: ice concentration and presence of IBs. The details of our developed hybrid model are presented in the following section.

3.2 *Construction of the hybrid model*

The hybrid model (HM) presented here is tailored for specific operational conditions prevailing in the Northern Baltic Sea, as well as for a specific

ship (1A super bulker). The HM encompasses two sub-models, as follows: 1) sub-model A', 2) sub-model B. The sub-model A' is obtained from an engineering-based model, which is stochastic by nature and is introduced in our earlier work (Kuuliala *et al.*, 2017). A' is developed for level and ridged ice conditions and ice concentration close to 100%. It calculates ship speed for a 20 km transit trough simulated ice cover. To cover various combinations of ice conditions and the stochastic variation present in ridged ice cover A' is executed 20.000 times in a loop. The response variable of this sub-model are: ship speed, v, and the probability of besetting in ice, $p(v=0)$, for a given set of ice conditions. The ship speed is taken as an average obtained in the course of simulations, where series of two hundred ship transits is conducted for each set of ice conditions. Each series exhibits certain speed variation due to the stochastic generation process of the ridged ice cover. The probability of besetting in ice is calculated as a fraction of cases where the ship got stuck in ice during a simulation to the total number of simulated transits in given ice conditions, and refers to 20 km transit. In the simulation a ship is considered beset when she is not able to ram the ice or the ramming is not successful after four attempts. The explanatory variables are: ice concentration, level ice thickness and equivalent ice thickness. The latter concept utilizes the geometric method, where all the ice forms in some certain area are taken into account and the result is given by one simple thickness value. The sub-model A' takes a form of a two-dimensional look-up tables defined in terms of the two thicknesses, as depicted in Figure 1.

The sub-model B is obtained in the course of the extended analysis of winter traffic as presented in our earlier work (Lensu and Kokkonen, 2017). It utilizes a database that combines the AIS messages archived by the Finnish Traffic Administration, ice information products of Finnish Meteorological Institute, and a ship particulars database. The database covers nine ice seasons 2007–2016. The ice information consists of ice chart data (covering the whole period) and ice model data (covering selected years). The basic AIS-retrieved position

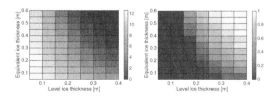

Figure 1. The results obtained from the engineering sub-model A' – independent navigation for a specific 1A Super class ship.

reports include ship identifier, time, location, speed, course, and heading. Each report is linked with ice parameters found at the time and location of broadcasting the report. Most results are for independently navigating IA Super ships. B takes the form of look-up tables showing the speed and variation of speed as a function of thickness from ice charts and ice concentration for the studied region and ship type, as presented in Figure 2. B does not consider the probability of getting stuck resulting from increasing ice resistance, since such events can be distinguished from events of deliberate waiting only by detailed analyses of individual voyages. l. It is evident from the results that modelling of the Baltic wintertime navigation system (WNS) as a collection of ships experiencing speed reduction due to physical ice resistance is a narrow picture. Independent ice transit is not the rule and cases of independent icebreaking transit by merchant vessels are rare exceptions. For the majority of the fleet, the indirect effects of ice conditions and operational conditions dominate: the location, age and condition of the channels, navigating in convoys and the proximity of other ships in general, scheduled meeting times and other icebreaker assistance practices. The patterns that emerge from the analysis give insight into how the winter navigation should be approached by more holistic, system-oriented methods.

3.3 Form of a hybrid model for the Baltic Sea

The HM is constructed in a manner that allows the model to cover wide range of operations performed within the Baltic wintertime navigation system. The sub-model A' covers well the area of independent navigation; however it fails to reflect the reality in harsh conditions, where the assistance of IBs is available. For example, if the thrust delivered by the ship propulsion system does not exceed the added resistance of ice, that is calculated based on the physics of the ice breaking process, the ship is unable to proceed independently—this is reflected by sub-model A'. However, the existence of IBs

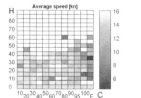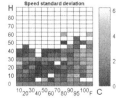

Figure 2. The results obtained from the empirical sub-model B—independent navigation for a specific IA Super class ship. *H* denotes ice thickness in [m], *C* means concentration in [%].

makes the non-independent navigation possible for those conditions, thus the factual, observable ship's speed becomes non-zero; this is reflected by the sub-model, B. However, in such situations the speed of the ship is dependent on the speed of other assisting ships, such as an icebreaker or other ships in convoy. Therefore, speed is not dependent only on surrounding ice conditions, rather operational characteristics of the BWNS affect prominently the performance of the system. For that reason the speed of a ship navigating in harsh conditions obtained with the use of the theoretic sub-model A' should be supplanted by the empirical sub-model B, which enables more realistic estimation of ship speed under such conditions.

The response variable of the HM is ship speed and the probability of besetting in ice. Whereas the explanatory variables are level ice thickness, equivalent ice thickness and ice concentration. Thus, the HM takes the advantage of both sub-models to cover the following: the operational conditions of independent navigation, steaming assisted by an icebreaker (both in the presence of high ice concentration) and operational practices when navigating in medium and low ice concentration conditions. The HM takes the following form:

$$HM(X,c) = \begin{cases} A'(X,c), & c \geq c_t \wedge uA' \neq 0 \\ B(X,c), & c < c_t \\ B(X,c), & c \geq c_t \wedge uA' = 0 \end{cases} \quad (1)$$

Where X stands for the set of explanatory variables, which are level ice thickness (h_i) and equivalent ice thickness (h_{eq}); $v_{A'}$ stands for the speed of ship obtained with the use of meta-model A', c denotes ice concentration. The following logical condition is used, governing the selection of a sub-model: if the ice concentration exceeds adopted threshold, the sub-model A' is selected to yield ship's speed given ice conditions. Then, if the speed is zero then the response of the sub-model A' is cross-checked with the sub model B. In case both sub-models yield null speed, the HM delivers null speed as well. However, if B provides non-zero speed, then the HM returns this value as the outcome speed of a ship given ice conditions. If the ice concentration is lower than the adopted concentration threshold c_t, B is the source of speed for HM. The HM seems well suited for route optimization in realistic, complex ice conditions for a specific ship or a group of similar ships. This is demonstrated in section 4. The description of a ship for which the HM is tailored is given in the following sub-section.

3.4 Ship-related features covered by HM

The HM is developed for a specific ship, which is a medium-sized bulk carrier having the ice class equivalent to IA Super, according to Finnish-Swedish Ice Class rules (TRAFI, 2010). Ships of this class are defined as follows: "Ships whose structural strength in essential areas affecting their ability to navigate in ice essentially exceeds the requirements of ice class IA and which as regards hull form and engine output are capable of navigation under difficult ice conditions" (TRAFI, 2010). This means that an IA Super ship has such structure, engine output, and other properties, which make her capable of navigating in difficult ice conditions without the assistance of IBs. The design requirement for this ice class is a minimum speed of 5 knots in 1 m thick brash ice channels with a 0.1 m thick consolidated layer of ice on top (TRAFI, 2010). Moreover, to ease the process of ice breaking, the ship is equipped with an "ice knife" at the bow. Her particulars are presented in Table 1.

However, one important parameter, which describes a ship's inertia and her ability to break the ice, meaning mass of the ship (displacement) cannot be determined accurately from the available data sets as it depends on loading condition. This is a significant source of uncertainty in case of engineering-based models. Neither is this parameter is known for the data-driven model. Since the data-driven model covers a 10-year period and thousands of AIS points describing ship parameters while transiting the Baltic Sea, it represents a range of loading conditions for all those transits. All the recorded tracks of the analysed IA Super type ships that were used to develop sub-model B are depicted in Figure 3. Based on our background knowledge about the routes of the ships, type of trades they made, and harbours visited, we made an assumption about half-load conditions. Since in case of sub-model B the speed of a ship is taken as an average over thousands of entries, such an assumption on half-loaded condition may be quite reasonable. In case of sub-model A', the speed is calculated following the above-mentioned assumption, thus in some cases the model may overestimate the results (if the real ship transits ice in ballast condition) or underestimate it when

Figure 3. The AIS-retrieved ice transit tracks of the analysed ship during 2007–2016.

the real ship proceeds through the ice fully loaded. However, those discrepancies are minor for the conditions of low ice concentration. In case of high ice concentration, the serious underestimation of A' (null speed) is compensated by B, as indicated by Eq. 1.

4 APPLICATION OF THE HYBRID MODEL IN ROUTE FINDING IN ICE ALGORITHM

In this section we demonstrate the effect that the HM has on the route selection and transit time calculation of 1A super bulk carrier. The HM is implemented into a routing tool based on our earlier work, presented in (Guinness et al., 2014). The tool utilizes the A* path finding algorithm to find the optimal path between two points over a gridded map of the sea area in question. The size of a grid cell corresponds to the ice model resolution of 1×1 NM. The A* algorithm searches for the path through the grid cells, which has the lowest cost function. The cost function is tantamount to travel time between point of origin and destination. Each grid cell of navigable waters has assigned two parameters that affect the transit time: attainable speed and the probability of getting stuck. The navigable waters for a given ship are defined through the ship draught with safety margin and sea bathymetry. The former is provided by the user, whereas the latter is obtained from GEBCO open service (www.gebco.net). Sea ice information is obtained from the HELMI ice model (Haapala, Lönnroth and Stössel, 2005), and three parameters affecting ship performance in ice are considered: level ice thickness, ridged ice thickness and ice concentration. To evaluate ship performance in ice two models are used for the purpose of

Table 1. Main characteristics of the analysed ship.

Parameter	Value
Ice class	1AS
DWT	21353 t
Length	149.3 m
Breadth	24.6 m
Draught	9.4 m
Power	9720 kW
Max speed in open water	15.5 kn

comparison. First, we adopt sub-model A', which is a typical, engineering type of model, suitable for the given purpose, adopted in earlier route finding tools (Kotovirta et al., 2009). Second, we adopt our hybrid model, accounting for the presence of dirways and icebreaker waypoints, as a part of operational routine in the area affecting travel time. Finally, for each ship performance model, the tool delivers the optimal (fastest) path's waypoints, expected speed along the path, and travel time.

Subsequently, the obtained routes are benchmarked with the recorded speed of the same ship that navigated independently between the same

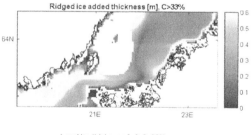

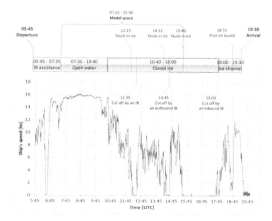

Figure 4. Annotated time history of the 1A super bulk carrier in her journey in the Northern Baltic Sea during severe ice conditions of 6th of March 2011, (Montewka et al., 2015).

Ridged ice added thickness [m], C>33%

Level ice thickness [m], C>33%

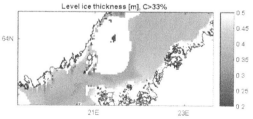

Figure 5. Ice conditions for the analysed journey as obtained from: HELMI ice model (upper panes) and satellite picture (lower pane).

points in the past (6th of March 2011). Part of her voyage was through the open water, however, upon entering the ice field her performance decreases and she was brought to a halt three times, as can be seen from her speed profile depicted in Figure 4. A detailed description of this voyage is given in our earlier work (Montewka et al., 2015). The relevant ice parameters obtained from the HELMI model for the time of voyage are shown in Figure 5. The ridged ice thickness in the figure does not describe the actual size of ridges; it corresponds rather to the concentration of the ice ridges in the given grid cell.

5 RESULTS

The results that are obtained are depicted in Figure 6 and in Table 2. Therein, three routes are shown, two obtained from the routing tool and a real track of a ship as recorded in the Automatic Identification System (AIS), referred to as benchmark path. The routes' lengths and travel time are shown in Table 2. Analysing the picture, one sees that the geographical location of the routes obtained with the use of routing tool differ from the real path. We can define two reasons for that. First is the difference between actual and modelled ice conditions, especially in the location of the ice edge. This becomes evident when comparing

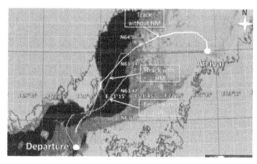

Figure 6. Comparison of routing results: route with HM adopted, route without HM and the real track followed by a ship as recorded in the AIS.

Table 2. Parameters of the compared routes.

Route	Length [NM]	Transit time [hrs]
Benchmark path from AIS—green	~112	~11
Optimized without HM—white	~88	~18
Optimized with HM—red	~90	~10

the ice picture as obtained from HELMI with the satellite image (Figure 5). Due to strong western wind, the ice mass was moved eastwards, and apparently this has not been fully captured by HELMI which shows low but nonzero concentration in the actually ice-free coastal lead. The second reason is rooted in the operational pattern of IBs, which avoid navigating too close to the landmasses and chose their routes based on the location and number of ships that need assistance. For example, two IBs escorting the departing bulk carrier were heading towards the Swedish coast to take several ships that required assistance towards Finnish harbors. Thus, the IBs chose the route for the bulk carrier taking into account the other operations as well.

When analysing the routes lengths, it is obvious that the optimized routes are shorter than the AIS track. However this comes from the operational practices of IBs operating in the area on that particular day. The track of the ship as recorded by AIS was to a large extent determined by the IBs, which escorted the ship to the edge of the ice while departing, and then indicated her entrance to the ice field while approaching her destination. In fact, the entrance position is the same for two optimized routes, and 10 NM North with respect to the actual track of the ship.

When it comes to transit time, the route computed with the use of HM is very close to the actual travel time. Whereas the route where sub-model A' was used delivers significantly higher travel time, which comes from the fact, that in heavily ridged area close to the destination, ship speed is very low. This could be true, if there was no assistance available. However, with the presence of IBs, the ship can proceed with higher speed and reach her destination faster.

6 CONCLUSIONS

In this paper we discussed the challenges related to the existing approaches for modelling ship performance in ice. Two major types of models are discussed, engineering and data-driven. Each modelling approach has its pros and cons, which pertain to accuracy, scope, proper reflection of actual conditions, and computational time. To address some of those challenges, we propose a hybrid model that combines both types of model and is found suitable for ice routing, as well as trafficability assessment in ice covered waters. With the hybrid model we expand the modelling space, by including the operational aspects of winter navigation, which are not covered by either of the models alone. Moreover, the computational time is reduced to minimum, since the HM takes the form

of look-up tables. This can be further improved, for instance, by fitting an appropriate mathematical function to the data contained in the tables

Accommodating the HM in the ice routing tool, the following can be obtained: 1) better knowledge on the fastest route through the ice accounting for the operational practices existing in the area; 2) better knowledge on the areas that are difficult for a ship to proceed independently, where the assistance of IBs may be required; 3) better knowledge on the effectiveness of IBs services in the area.

The main sources of uncertainty of the proposed solution pertains to: 1) lack of information about loading conditions of a ship, which affects the icebreaking process, 2) assumption that the speed reduction is due to ice conditions, whereas there may be other reasons, like scheduling in the harbor operations, and 3) quality and accuracy of ice information (especially in temporal dimension) used to develop B sub-model.

Since the HM is ship specific, the ships that would benefit from it the most are those which are capable of independent navigation in ice. The other, less capable ships usually rely fully on icebreaker assistance (e.g. towing, escorting) and do not need any ice routing information.

ACKNOWLEDGEMENTS

The research presented in this paper has been conducted in the context of the "Strategic and Operational Risk Management for Wintertime Maritime Transportation System" (BONUS STORMWINDS) project. This has received funding from BONUS, the joint Baltic Sea research and development programme (Art 185), funded jointly from the European Union's Seventh Programme for research, technological development and demonstration, and from Baltic Sea national funding institutions: the Academy of Finland. The financial support is acknowledged.

REFERENCES

Aksenov, Y., Popova, E.E., Yool, A., Nurser, A.J.G., Williams, T.D., Bertino, L. and Bergh, J. (2017) 'On the future navigability of Arctic sea routes: High-resolution projections of the Arctic Ocean and sea ice', *Marine Policy*, 75, pp. 300–317. doi: 10.1016/j.marpol.2015.12.027.

Aldous, L., Smith, T., Bucknall, R. and Thompson, P. (2015) 'Uncertainty analysis in ship performance monitoring', *Ocean Engineering*, 110, pp. 29–38. doi: 10.1016/j.oceaneng.2015.05.043.

Bergström, M., Ehlers, S., Erikstad, S.O., Erceg, S. and Bambulyak, A. (2014) 'Development of an Approach Towards Mission-Based Design of Arctic Maritime

Transport Systems', in *Volume 10: Polar and Arctic Science and Technology*. ASME, p. V010T07A029. doi: 10.1115/OMAE2014-23848.

Bergström, M., Erikstad, S.O. and Ehlers, S. (2016) 'A simulation-based probabilistic design method for arctic sea transport systems', *Journal of Marine Science and Application*, 15(4), pp. 349–369. doi: 10.1007/s11804-016-1379-1.

Cho, S.-R. and Lee, S. (2015) 'A prediction method of ice breaking resistance using a multiple regression analysis', *International Journal of Naval Architecture and Ocean Engineering*, 7(4), pp. 708–719. doi: 10.1515/ijnaoe-2015-0050.

Choi, M., Chung, H., Yamaguchi, H. and Nagakawa, K. (2015) 'Arctic sea route path planning based on an uncertain ice prediction model', *Cold Regions Science and Technology*, 109, pp. 61–69. doi: 10.1016/j.coldregions.2014.10.001.

ENFOTEC, GeoInfo and McCallum, J. (1996) Safe speed in ice: an analysis of transit speed and ice decision numerals. Ontario, Canada.

Fu, S., Zhang, D., Montewka, J., Yan, X. and Zio, E. (2016) 'Towards a probabilistic model for predicting ship besetting in ice in Arctic waters', *Reliability Engineering & System Safety*, 155, pp. 124–136. doi: http://dx.doi.org/10.1016/j.ress.2016.06.010.

Fu, S., Zhang, D., Montewka, J., Zio, E. and Yan, X. (2016) 'A Fuzzy Event Tree Model for Accident Scenario Analysis of Ship Stuck in Ice in Arctic Waters', in *Volume 8: Polar and Arctic Sciences and Technology; Petroleum Technology*. ASME. doi: 10.1115/OMAE2016-54882.

Guinness, R.E., Saarimaki, J., Ruotsalainen, L., Kuusniemi, H., Goerlandt, F., Montewka, J., Berglund, R. and Kotovirta, V. (2014) 'A method for ice-aware maritime route optimization', in *IEEE/ION PLANS 2014*. Monterey, CA: IEEE, pp. 1371–1378. doi: 10.1109/PLANS.2014.6851512.

Haapala, J., Lönnroth, N. and Stössel, A. (2005) 'A numerical study of open water formation in sea ice', *Journal of Geophysical Research: Oceans*, 110(C09011), pp. 1–17. doi: 10.1029/2003JC002200.

Ho, J. (2010) The implications of Arctic sea ice decline on shipping, Marine Policy. doi: 10.1016/j.marpol.2009.10.009.

Jeong, S.-Y., Choi, K., Kang, K.-J. and Ha, J.-S. (2017) 'Prediction of ship resistance in level ice based on empirical approach', *International Journal of Naval Architecture and Ocean Engineering*. doi: 10.1016/j.ijnaoe.2017.03.007.

Juva, M. and Riska, K. (2002) On the power requirement in the Finnish-Swedish ice class rules. techreport. Espoo, Finland.

Kaups, K. (2012) *Modeling of Ship Resistance in Compressive Ice*. phdthesis. Aalto University.

Keinonen, A. (1996) Icebreaker characteristics synthesis. Ottawa.

Kotovirta, V., Jalonen, R., Axell, L., Riska, K. and Berglund, R. (2009) 'A system for route optimization in ice-covered waters', *Cold Regions Science and Technology*, 55(1), pp. 52–62. doi: 10.1016/j.coldregions.2008.07.003.

Kubat, I. (2012) 'Quantifying Ice Pressure Conditions and Predicting the Risk of Ship Besetting', in *International Conference and Exhibition on Performance of Ships and Structures in Ice 2012 (ICETECH 2012)*. Society of Naval Architects and Marine Engineers (SNAME).

Kubat, I., Babael, H. and Sayed, M. (2013) 'Analysis of Besetting Incidents in Frobisher Bay during 2012 Shipping Season', in *Proceeding of International Conference on Port and Ocean Engineering undere Arctic Conditions (POAC)*. Espoo, Finland.

Kubat, I., Fowler, C.D. and Sayed, M. (2015) *Snow and Ice-Related Hazards, Risks and Disasters, Snow and Ice-Related Hazards, Risks and Disasters*. Elsevier. doi: 10.1016/B978-0-12-394849-6.00018-4.

Kujala, P. and Arughadhoss, S. (2012) 'Statistical analysis of ice crushing pressures on a ship's hull during hull–ice interaction', *Cold Regions Science and Technology*, 70, pp. 1–11. doi: 10.1016/j.coldregions.2011.09.009.

Kujala, P., Montewka, J. and Filipovic, A. (2014) 'Risk management for Arctic shipping', in *Proceedings of the 21th Symposium on Theory and Practice of Shipbuilding*. Baska, Croatia.

Külaots, R., Kujala, P., von Bock und Polach, R. and Montewka, J. (2013) 'Modelling of ship resistance in compressive ice channels', in *Proceedings of the 22nd International Conference on Port and Ocean Engineering under Arctic Condition*. Espoo, Finland.

Kuuliala, L., Kujala, P., Suominen, M. and Montewka, J. (2017) 'Estimating operability of ships in ridged ice fields', *Cold Regions Science and Technology*, 135, pp. 51–61. doi: 10.1016/j.coldregions.2016.12.003.

LaPrairie, D., Wilhelmson, M. and Riska, K. (1995) 'Transit simulation model for ships in Baltic ice conditions. Documentation of the calculated routine.'

Leisti, H. and Riska, K. (2010) Description of ice compression. Deliverable 4.1. SAFEWIN project FP7-RTD-233884. Espoo, Finland.

Lensu, M. and Kokkonen, I. (2017) Inventory of ice performance for Baltic Sea IA Super traffic 2007–2016. N/A. Helsinki.

Leppäranta, M. and Myrberg, K. (2009) *Physical oceanography of the Baltic Sea*. Springer/Praxis Pub.

Liikennevirasto (2017) Finland's winter navigation. Instructions for winter navigation operators. Helsinki, Finland.

Lindqvist, G. (1989) 'A straightforward method for calculation of ice resistance of ships', in. Luleå University of Technology, pp. 722–735.

Löptien, U. and Dietze, H. (2014) 'Sea ice in the Baltic Sea—revisiting BASIS ice, a historical data set covering the period', *Earth Syst. Sci. Data*, 6, pp. 367–374. doi: 10.5194/essd-6-367-2014.

Löptien, U., Mårtensson, S., Meier, H.E.M. and Höglund, A. (2013) 'Long-term characteristics of simulated ice deformation in the Baltic Sea (1962–2007)', *J. Geophys. Res. Oceans*, 118, pp. 801–815. doi: 10.1002/jgrc.20089.

Lubbad, R. and Løset, S. (2011) 'A numerical model for real-time simulation of ship-ice interaction', *Cold Regions Science and Technology*, 65(2), pp. 111–127. doi: 10.1016/j.coldregions.2010.09.004.

Montewka, J., Goerlandt, F. and Kujala, P. (2014) 'Empirical, probabilistic models of ship besetting in ice incidents in the Northern Baltic Sea', in *International Workshop on Next Generation Nautical Traffic Models*. Wuhan.

Montewka, J., Goerlandt, F., Kujala, P. and Lensu, M. (2015) 'Towards probabilistic models for the prediction of a ship performance in dynamic ice', *Cold Regions Science and Technology*, 112, pp. 14–28. doi: 10.1016/j.coldregions.2014.12.009.

Mulherin, N.D., Eppler, D.T., Proshutinsky, T.O., Proshutinsky, A.Y., Farmer, D. and Smith, O.P. (1996) 'Development and results of a Northern Sea Route transit model'. {US} Army Corps of Enigineers, {CRREL}.

Naegle, J.N. (1980) Ice-resistance prediction and motion simulation for ships operating in the continuous model of icebreaking. phdthesis. The University of Michigan.

La Prairie, D., Wilhelmson, M. and Riska, K. (1995) *A transit simulation model for ships in Baltic Ice conditions*. M 200. Espoo, Finland: Helsinki University of Technology.

Riska, K., Kujala, P., Goldstein, R., Danilenko, V. and Osipienko, N. (1995) 'Application of results from the research project "A ship in compressive ice" to ship operability', in *POAC '95*. Murmansk.

Riska, K., Wilhelmson, M., Englund, K. and Leiviskä, T. (1997) 'Performance of merchant vessels in ice in the Baltic'. Winter Navigation Research Board, Helsinki.

Rosenblad, M. (2007) Increasing the safety of icebound shipping—WP4 Operative Environment (icebreaker operations). Espoo, Finland: Helsinki University of Technology.

Somanathan, S., Flynn, P. and Szymanski, J. (2009) 'The Northwest Passage: A simulation', *Transportation Research Part A: Policy and Practice*, 43(2), pp. 127–135. doi: 10.1016/j.tra.2008.08.001.

Su, B., Riska, K. and Moan, T. (2010) 'A numerical method for the prediction of ship performance in level ice', *Cold Regions Science and Technology*, 60(3), pp. 177–188. doi: 10.1016/j.coldregions.2009.11.006.

Suominen, M. and Kujala, P. (2014) 'Variation in short-term ice-induced load amplitudes on a ship's hull and related probability distributions', *Cold Regions Science and Technology*, 106, pp. 131–140. doi: 10.1016/j.coldregions.2014.07.001.

Tan, X., Su, B., Riska, K. and Moan, T. (2013) 'A six-degrees-of-freedom numerical model for level ice–ship interaction', *Cold Regions Science and Technology*, 92, pp. 1–16. doi: 10.1016/j.coldregions.2013.03.006.

Tomac, T., Klanac, A., Katalinić, M., Ehlers, S., Von Bock Und Polach, R., Suominen, M. and Montewka, J. (2014) 'Numerical simulations of ship resistance in model ice', in *Developments in Maritime Transportation and Exploitation of Sea Resources—Proceedings of IMAM 2013, 15th International Congress of the International Maritime Association of the Mediterranean*, pp. 847–851.

TRAFI, T.S.A. (2010) 'Maritime safety regulation. Ice class regulations and the application thereof'. Transport Safety Agency, Helsinki, Finland.

Tunik, A. (2000) 'Safe speeds of navigation in ice as criteria of operational risk', *International Journal of Offshore and Polar Engineering*, 10(4).

Zhou, Q., Peng, H. and Qiu, W. (2016) 'Numerical investigations of ship–ice interaction and maneuvering performance in level ice', *Cold Regions Science and Technology*, 122, pp. 36–49. doi: 10.1016/j.coldregions.2015.10.015.

Maritime Transportation and Harvesting of Sea Resources – Guedes Soares & Teixeira (Eds)
© 2018 Taylor & Francis Group, London, ISBN 978-0-8153-7993-5

A model for predicting ship destination routes based on AIS data

H. Rong, A.P. Teixeira & C. Guedes Soares
Centre for Marine Technology and Ocean Engineering (CENTEC), Instituto Superior Técnico, Universidade de Lisboa, Lisbon, Portugal

ABSTRACT: The paper presents a data mining method on Automatic Identification System data. The approach automatically identifies maritime traffic junctions and applies a logistic regression model for predicting the ships' destination routes based on a set of characteristics of the ships and of their behaviour at a particular junction. In the context of maritime traffic "junction" is a place at which two or more routes diverge. In this area, changes in the course over ground of ships with different final destination routes are consistently observed. The approach consists of detecting turning points based on the Douglas & Peucker algorithm and clustering them based on the density-based spatial clustering of applications with noise algorithm. Then, a multinomial logistic regression model is developed and applied for predicting the most probable destination route of the ships according to their current state in the junction. The proposed method is applied to the southbound traffic leaving the traffic separation scheme off Cape Roca toward the ports of Lisbon, Setubal and Sines.

1 INTRODUCTION

Maritime transportation is responsible for approximately 90% of the world's trade and more than 50,000 ships sail in ocean each day. Surveillance of maritime traffic is therefore of great importance for maritime safety and security.

The need to ensure the safety of navigation has led to the implementation of Automatic Identification System (AIS) base stations, which receive and maintain records of messages with dynamic and static information transmitted by ships. Although several types of errors in AIS messages have been identified (Norris, 2007), it is obvious that the system provides plenty of data for maritime traffic analysis, modelling and monitoring tasks. Mou et al. (2010) performed a statistical analysis for collision involved ships in a Traffic Separation Scheme (TSS) off Rotterdam Port by using AIS data. Qu et al. (2011) proposed three indices, including speed dispersion, degree of acceleration and deceleration and number of fuzzy ship domain overlaps based on the analysis of AIS data to assess ship collision risk in the Singapore Strait. Silveira et al. (2013) proposed a method to determine the number of collision candidates based on the available AIS data and the concept of collision diameter. Rong et al. (2016) used the concept of "ship domain" to derive near collision scenarios, and a detailed statistical analysis was performed to evaluate and characterize near ship collision scenarios off the coast of Portugal.

Understanding maritime traffic patterns is key to maritime surveillance that relies on an efficient description of the traffic routes. The ship motion data provided by AIS can be grouped based on the similarity behaviour of the ships' trajectories to provide an overview of the general motion patterns and can be used to model the ship routes.

Aarsæther & Moan (2009) applied computer vision techniques to automatically split trajectories into manoeuvre sequences and grouped ship trajectories according to their similarity. From a statistical study of manoeuvre sequences, the voyage plans of ship groups were derived. Etienne et al (2010) applied data mining techniques on ship trajectories to characterize the maritime traffic in the Iroise Sea. In this study, maritime traffic network is represented by a zone graph and spatial-temporal patterns of main routes are analysed. Pallotta et al. (2014) presented a cluster-based method to detect waypoints in an unsupervised way. In this study, ship routes are modelled by a series of straight lanes and turning sections, and path prediction can be performed based on the learned maritime traffic pattern.

Based on statistical analysis of ship trajectories following the same route, spatial and temporal outliers can be detected. Guillarme & Lerouvreur (2013) presented a simple Markov chain approach to represent ship transitions and junctions are learned based on a clustering technique.

In these studies, straight routes connecting turning sections have proven to be suitable for modelling the motion patterns of maritime traffic. However, ship motion patterns in a maritime junction, where several turning sections overlap have not been fully discussed. The knowledge of ship motion patterns in maritime junctions is important to improve the understanding of ships' behaviour and to assist operators in identifying ships' destination routes.

The learning of ships' motion patterns in a maritime junction and the prediction of ships' destination, according to its current state, is a classification task. In the last decade, Multinomial Logistic Regression (MLR) has proven to be a powerful tool for classification problems in several maritime studies. Vander Hoorn & Knapp (2013) proposed a multi-layered risk assessment framework, which can estimate various types and degrees of seriousness of incidents by means of binary logistic regression. Li et al. (2014) also used a binary logistic regression model to develop a novel quantitative safety index for worldwide sea-going vessels based on information on their condition and safety records. Psarros et al. (2011) found that pirate success attack rate decreases with vessel size while increases with the improvement of pirate capability by estimating the probability of pirates attack through logistic regression modelling.

In this paper, an unsupervised methodology for identifying maritime junctions with few assumptions about the behaviour of the maritime traffic is presented. Then, ship motion patterns in a maritime junction are modelled and a multinomial logistic regression model is applied for estimating the most probable destination route of the ships. The approach is applied to the southbound traffic leaving the Traffic Separation Scheme off Cape Roca toward the ports of Lisbon, Setubal and Sines.

2 METHODOLOGY

The prediction of the ships' final destination routes at a maritime junction includes to following tasks: *i*) maritime junction detection; *ii*) characterization of the ships' features at the junction; *iii*) prediction of the ships' final destination routes.

Turning point detection and clustering techniques are applied to detect maritime junctions. Turning points of ship trajectories are detected based on Douglas & Peucker (DP) algorithm that allows detecting ship directional changes at the trajectory level based on spatial information (Douglas & Peucker, 1973), which is more reliable than the use of the Course Over Ground (COG) for turning point detection. A turning section corresponds to an area

where a significant density of turning points is observed. In addition, a junction area is defined in this study as an aggregation of several overlapping turning sections. For this purpose, the Density-based spatial clustering of applications with noise (DBSCAN) algorithm is used (Ester et al. 1999), which is a density-based clustering algorithm adequate to specific maritime applications (Guillarme & Lerouvreur, 2013; Pallotta et al., 2013). Then, the maritime traffic features at the junction are collected and used as explanatory attributes for classifying ships in terms of their destination routes. The ship destination route prediction is achieved by using a MLR model trained by appropriate AIS data.

2.1 *Maritime junction identification*

Maritime junction is automatically identified based on clustering turning points detected in ship trajectories. A turning point in ship trajectory is where a significant directional change is observed. The Douglas & Peucker (DP) algorithm typically used for ship trajectory compression, can also be used for turning point detection. The DP algorithm is an approach for reducing the number of points in a curve that is approximated by a series of points (Douglas & Peucker, 1973). In the field of moving object analysis, this algorithm has been used to compress trajectories data (Etienne et al., 2010; Cao et al., 2006). This algorithm, together with a suitable threshold parameter can be used to identify the relevant turning points along the trajectory.

According to the DP algorithm the first and last points of a given trajectory are kept automatically. Then, the farthest point P_{max} from the line defined by the first and last points of the trajectory is found. If the transverse distance of the point P_{max} to this line calculated by Eq. (1) exceeds the threshold ε_{DP}, then the point is retained, the trajectory is then split at that position P_{max} and the algorithm is recursively applied to both sub-trajectories. Otherwise, any point between first and last points of the trajectory (or sub-trajectories) can be discarded. See Figure 1 for an illustration of the DP algorithm. In the figure the solid lines represent the original trajectory, and the dashed lines represent a DP simplified trajectory. In this study, the threshold ε_{DP} is set to 500 meters. P_1 is the furthest point to the line defined by P_{start} and P_{end}. The distance, $\text{dist}\big((P_{start}, P_{end}), P_1\big)$, exceeds the threshold ε_{DP} meaning that P_1 will be retained. However, $\text{dist}\big((P_{start}, P_1), P_2\big)$ and $\text{dist}\big((P_1, P_{end}), P_3\big)$ are smaller than the threshold ε_{DP}, and therefore, the remainder points of both original sub-trajectories are discarded.

$$\text{dist}\big((P_a,P_b),p\big)=$$
$$=\frac{\left|\big(y_b-y_a\big)*x_p-\big(x_b-x_a\big)*y_p+x_by_a-y_bx_a\right|}{\sqrt{\big(y_b-y_a\big)^2+\big(x_b-x_a\big)^2}}\quad(1)$$

where $\text{dist}\big((P_a,P_b),p\big)$ is the distance from point $p=\big(x_p,y_p\big)$ to the line defined by two points $P_a=\big(x_a,y_a\big)$ and $P_b=\big(x_b,y_b\big)$.

In the context of maritime traffic, the retained points detected by the DP algorithm can be considered as turning points, and the clustering of those turning points define the turning sections along the ship route. Turning points clustering is based on DBSCAN algorithm (Ester et al. 1996). This algorithm forms clusters of elements on the basis of the density of points in their neighbourhood. In the clustering procedure, the points are classified as core points, density-reachable points and noise. A point p is defined as core point if at least *min Pts* points are within the distance ε of it. Such neighbourhood points are density-reachable from p and belong to the same cluster. Then, those points that are not density-reachable from other points are considered as noise. It should be mentioned that the DBSCAN technique does not require the specification of the number of clusters in the data a priori. However, the setting of parameters *min Pts* and ε requires some knowledge about the traffic characteristics in the study area. The detection of turning sections are based on clustering turning points of each trajectory determined by DP algorithm. A conceptual flowchart of the process is shown in Figure 2. Finally, a junction is defined as an aggregation of overlapping turning sections of different routes.

At the junction, the maritime traffic is analysed so as to characterize a set of relevant attributes, e.g. time of the day, ship speed and heading, turning rates, ship type among others, which are then used for training the MLR model used to predict ships' destination routes.

2.2 MLR model for predicting ship destination route

In areas of complex traffic patterns such as junction areas where the marine traffic converges or diverges into different routes, ship route prediction is of importance for understanding the ships' behaviour as well as for traffic surveillance. However, static information, such as Estimated Time of Arrival (ETA) and destination are manually set and not fully reliable (Last, 2014). Therefore, a method for predicting the ships' destination is developed based on a Multinomial Logistic Regression (MLR) model.

The MLR is an extension of the binary logistic regression that has proved to be a powerful modelling tool to predict categorical placement or the probability of category membership on a dependent variable based on multiple independent variables. There are several studies in the maritime domain using logistic regression models (Bergantina & Marlow, 1998; Li et al., 2014). In this context, the goal is to construct a MLR model that explains the relationship between a set of explanatory attributes (independent variables) and the destination of the ships (dependent variable), so that the destination of a "new ship" can be predicted given its current state.

In a junction, the ship trajectories are classified by destination routes $\{R_1,...,R_K\}$. The MLR is then used to predict the destination that one particular ship belongs to from $i=1$ to K.

The dependent variable is the ship destination index ($i=1$ to n) and the independent variables correspond to a set of attributes β that are relevant to predict the ship destination. The probability of each category (i) is calculated by:

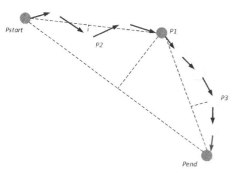

Figure 1. Illustration of Douglas & Peucker (DP) algorithm.

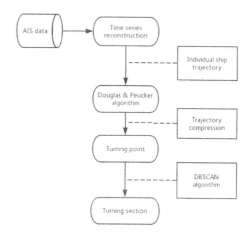

Figure 2. Turning sections identification based on AIS data.

$$\Pr(y_i = k \mid x_i, \beta) = \frac{\exp(\beta(k)^T x_i)}{\displaystyle\sum_{k=1}^{K} \exp(\beta(k)^T x_i)} \qquad (2)$$

where $\beta(k)$ is the set of MLR parameters for class k, $\beta \equiv \{\beta(1),...,\beta(K)\}$ and $\beta(1)$ is always set to zero as the probability for 1st class can be directly calculated by subtracting the sum of other classes. x_i is a set of explanatory variables associated with observation i. The index corresponding to the maximum value of $\Pr(i)$ is taken as the predicted ship destination.

3 CASE STUDY

The maritime areas off the continental coast of Portugal are crossed a complex network of routes where routes connecting northern Europe and the Mediterranean Sea meet vessels bound to and leaving from national ports. Two Traffic Separation Schemes (TSSs) located off Cape Roca and off Cape San Vicente organize the Northbound and Southbound Traffic of dangerous and non-dangerous cargo off the cost of Portugal and influence significantly the routes of the ships entering and leaving the national ports, of Lisbon, Setubal and Sines, as shown in Figure 3.

In this paper, the suggested approach to classify ship trajectories according to the ships' final destination routes is applied to the southbound maritime traffic data from the Traffic Separation Scheme off Cape Roca to the ports of Lisbon, Setubal, Sines over a period of three months (from 1st October to 31th December 2015). Figure 3 shows the ensemble of ship trajectories from the southbound traffic off Cape Roca to the ports of Lisbon, Setubal, Sines.

3.1 Maritime traffic groups

First the southbound maritime traffic that leaves the Traffic Separation Scheme off Cape Roca is grouped into traffic-groups or main routes defined by the ship trajectories corresponding to the different destination routes. In particular, the maritime traffic in this area is divided into four traffic-groups which are: 1) southbound ship traffic that maintain the traffic lane toward the TSS off Cape San Vicente; and the ship traffic to the ports of 2) Lisbon; 3) Setubal and 4) Sines. The traffic-groups generally show a certain geographic clustering of the ship trajectories, as the individual ships follow a similar path to their destination routes. Figure 4 shows the southbound maritime traffic from the Traffic Separation Scheme off Cape Roca grouped in terms of the different destination

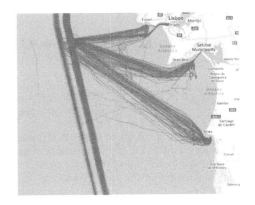

Figure 3. Southbound ship traffic in TSSs off Cape Roca.

routes. The number of ship trajectories in each group is shown in Table 1. One can see that most of the southbound traffic follows the main traffic-lane to the TSS off Cape San Vincente. Moreover, the number of ship trajectories approaching to the three ports is similar.

3.2 Turning points, sections and junction

Knowing where a ship changes its heading is important to characterize the normal behaviour of the traffic along the route and, therefore, the DP algorithm described before is used for ship trajectory compression and turning point detection.

The turning points detected along the ship routes from the TTSs to three ports are shown in Figure 5a (black points). Then, turning sections are identified by the spatial distribution of the coordinates of the turning points detected. As a consequence, a junction corresponds to a spatial area that aggregates several overlapped turning sections.

Figure 5c shows the resulting junction area defined by the outline of clustered turning points (represented by yellow dots in Figure 5c).

3.3 Characteristics of the maritime traffic in the junction

A statistical analysis of the maritime traffic shows that the southbound traffic of TSSs off Cape Roca consists mainly of cargo ships and tankers, which account for 67.8% and 27.2% respectively.

Different routes are used by different ship types. Most of the ships that visit the ports of Lisbon and Setubal are cargo ships, 75% and 92% respectively, whereas 53% of the ships toward the port of Sines are tankers, as shown in Table 2. Passenger ships only navigate in routes to the port of Lisbon and to the TSSs off Cape San Vincente. On the other

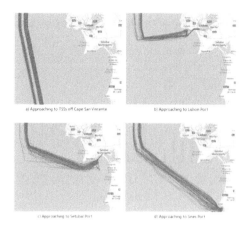

a) Approaching to TSSs off Cape San Vincente

b) Approaching to Lisbon Port

c) Approaching to Setubal Port

d) Approaching to Sines Port

Figure 4.   Traffic-groups by destination.

Table 1.   Number of trajectories in each traffic-group.

Traffic group	No. of trajectories
TSSs off Cape San Vincente	1396
Lisbon Port	56
Setubal Port	66
Sines Port	70
Sum	1588

hand, excluding those navigating to TSS off Cape San Vincente, 78.7% of the tankers sailed to Sines Port. The characteristics of the ship type distributions of each traffic group imply that ship type can be used as an indicator in the prediction of the ships' destination.

Figure 6 shows the ship speed distribution of each traffic group in the junction. It can be seen that there is not a significant difference between the ship speed distributions in the junction. However, the average speed of ships navigating to the TSS off Cape San Vincente is the highest among the four traffic groups, as shown in Table 3 and Figure 6.

Directional changes and deceleration happen under the condition that ships are about to change route. Turning rate of ships is calculated from COG of AIS data, so a negative value of turning rate means that ships turn counter clockwise. According to the average turning rate of each traffic group, ships to the port of Lisbon Port turn faster, followed by those to ports of Setubal and Sines. Accordingly, ship speed and turning rate in the junction can be considered as indicators for predicting the ships' destination.

Figure 7 shows COG distributions of each traffic group in the junction. Obviously, there are significant differences between each traffic group. One may notice four peaks in the histogram, which

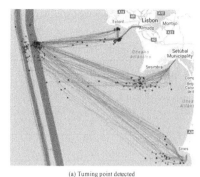

(a) Turning point detected

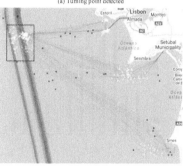

(b) Clustering of turning point

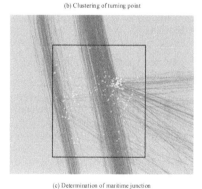

(c) Determination of maritime junction

Figure 5.   Turning sections and maritime junction detected.

correspond to the four destination routes of the traffic groups.

3.4   Prediction of ships' final destination

At this stage a MLR model is developed to predict at the junction the ships' final destination. A set of 5 independent variables consisting of the time of day, speed, ship type, heading, turning rate have been selected to construct the model.

Initially it is considered that all southbound traffic at TSS of Cape Roca is directed to TSSs off Cape San Vincente (in lane). However, when the ships leave southbound traffic lane to the ports of

Table 2. Ship type distribution of traffic groups in the junction.

Destination	Cargo	Tanker	Passenger	Other
TSSs off Cape San Vincente	941	385	6	64
To Lisbon	42	8	3	3
To Setubal	61	2	0	3
To Sines	33	37	0	0

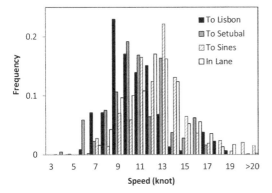

Figure 6. Histograms of ship speed for each traffic group in the junction.

Table 3. Average speed and turning rate of traffic groups in the junction.

Destination	Average speed (knot)	Average turning rate (degree/min)
TSSs off Cape San Vincente	12.326	−0.021
To Lisbon	10.504	−2.617
To Setubal	10.256	−1.871
To Sines	10.325	−1.235

Lisbon, Setubal and Sines their turning rate will change. Therefore, the turning rate of ships can be useful to predict the ships' destination. Additionally, the ship heading taken by the ships when following a specific trajectory also helps to distinguish to which route the ship belongs to.

In order to facilitate statistical analysis of the MLR model, the values of heading, turning rate, ship speed and ship type are discretized and normalized to 0~1. Considering the actual situation of navigation, heading is divided in 180 bins. Turning rate and ship speed is discretized into 4 bins, which are: low, medium-low, medium and high, as shown in Table 4. In this study, three ship types are considered which are passenger ship, cargo and tanker, also normalized from 0 to 1, as shown in Table 5.

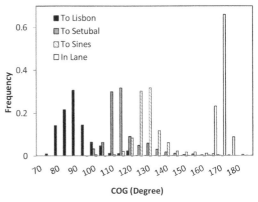

Figure 7. Histograms of COG for each traffic group in the junction.

Table 6 shows that the training sample is divided into four categories (final destination routes), and the trajectories maintaining the southbound traffic lane are set to 0 as the reference category in MLR model (40.5% training sample).

Table 7 shows the MLR model coefficients for the various destination routes and the Wald test on the model coefficients. Wald test is used for assessing the contribution of individual predictors in a given model. The Wald test (W_j) is the ratio of the square of the regression coefficient β_j^2 to the square of the standard error $SE_{\beta_j}^2$ of the coefficient (j), as shown in

$$W_j = \frac{\beta_j^2}{SE_{\beta_j}^2} \qquad (3)$$

According to the value of the Wald test, heading, turning rate and ship type are the most important variables for predicting the ships' destination among the 5 selected independent variables. The Wald test also confirms the findings of the characterization of the maritime traffic in the junction.

It can be seen that the COG distributions of 4 traffic groups are clearly different, which indicates that heading is the variable that most contributes to ships' destination prediction. However, ship speed distributions do not change significantly between the 4 traffic groups in the junction, and therefore, its role for predicting the ships' final destination routes is limited.

For the purpose of illustrating the prediction model, two ship trajectories from the TSS Cape of Roca to the ports of Lisbon and Setubal are tested. The tested trajectories are shown in Figure 8 by black dots. In Figure 8a, the ship enters the junction area and is initially moving southwards on the traffic lane for about 13 minutes. After these 13 minutes the MLR model correctly predicts the motion as approaching to the port of Lisbon.

Table 4. Normalization of ship speed and turning rate.

Speed (knot)	Turning rate (degree/min)	Discrete value	Normalized value
0~5	0~5	Low	0.25
5~10	5~10	Medium-low	0.50
10~15	10~20	Medium	0.75
>15	>20	High	1.00

Table 5. Normalization of ship type.

Ship type	Normalized value
Other	0.25
Passenger	0.50
Cargo	0.75
Tanker	1

Table 6. Train data set summary.

Destination	Categories	Frequencies	Percentage (%)
TSSs off Cape San Vincente	0	1500	40.5
To Lisbon	1	761	20.5
To Setubal	2	611	16.5
To Sines	3	830	22.4

Table 7. MLR model parameters.

Category	Independent variables	Coefficient value	Wald test (w_j)
1 - Port of Lisbon	Intercept	72.122	
	Time of day	0.292	0.573
	Speed	−0.916	1.643
	Ship Type	−7.669	58.473
	Heading	−93.931	901.398
	Turning Rate	10.906	97.235
2 - Port of Setubal	Intercept	53.756	
	Time of day	0.500	2.018
	Speed	−1.117	2.887
	Ship Type	−7.077	68.216
	Heading	−62.057	514.878
	Turning Rate	8.244	60.234
3 - Port of Sines	Intercept	28.492	
	Time of day	0.209	0.457
	Speed	0.636	1.118
	Ship Type	5.587	47.466
	Heading	−40.726	327.665
	Turning Rate	5.946	35.326

However, the ship bound for port of Setubal is firstly predicted as approaching to Sines Port at (minute 5) and then the MLR model correctly predicts the destination of the ship to the port of Setubal at minute 13, as shown in Figure 8b.

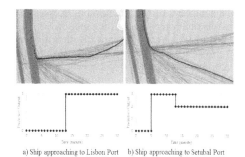

a) Ship approaching to Lisbon Port b) Ship approaching to Setubal Port

Figure 8. Route prediction applied to AIS data.

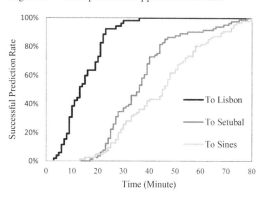

Figure 9. Successful prediction rates as a function of time.

Figure 9 shows the performance of MLR model in predicting the ship destination in terms of the successful prediction rate as a function of time (min). Most of the ships are classified correctly by the developed MLR model after 60 minutes navigating in the junction (100% to Lisbon Port, 90.1% to Setubal Port and 81.4% to Sines Port). 98.1% of the ships bound for port of Lisbon are successfully classified in 30 minutes, which indicates that these ships manoeuvre earlier than those approaching to Setubal and Sines ports. It should be mentioned that the destination prediction is conducted continuously when ships are navigating in the junction area, and therefore, fluctuations on the predicted final destination can occur, as shown in Figure 8b. Therefore, successful predictions are recorded only when the predicted destination corresponds to the corrected one and does not change. The times at which these correct predictions are achieved are recorded and used to assess the performance of the MLR model, as shown in Figure 9.

4 CONCLUSIONS

The proposed data mining method provides a relatively straight forward and unsupervised approach to determine maritime junction and to predict the ships' destination routes at the junction.

The junction is an automatically identified using ship trajectory compression and turning point detection techniques and clustering methods. In particular, the trajectory compression DP algorithm is adopted for turning point detection and a density-based clustering algorithm is applied for clustering these dispersed points into clusters. The clusters of turning points forms turning sections of the route, which benefits greatly from the previous compression step. Moreover, a junction at which two or more routes diverge is defined to aggregate the turning sections to the different routes.

A MLR model is developed to predict the ships' destination at the junction based on 5 attributes: time of day, ship type, speed, ship heading and turn rate. The model is applied to ship trajectories from southbound traffic of the TSS off Cape Roca to the ports of Lisbon, Setubal and Sines.

It is shown that the ship heading, the turning rate and ship type are most important variables in predicting ships' destination routes. The results obtained show that most of the ships are classified correctly by the developed MLR model after 60 minutes navigating in the junction. However, successful prediction rate of the model depends on the ships' destination routes.

This work is just the first step for maritime route modelling and there are many challenging issues. The MLR model only uses static and dynamic information provide by AIS data. Since the motion pattern of ships are related to their schedules, more accurate predictions may be obtained if this information is taken into account. The characterization of ships' destination routes and, consequently, the route that the ships are expected to follow is useful for real-time applications such as maritime situational awareness and abnormal behaviour detection.

ACKNOWLEDGEMENTS

This work was performed within the Strategic Research Plan of the Centre for Marine Technology and Ocean Engineering, which is financed by Portuguese Foundation for Science and Technology (Fundação para a Ciência e Tecnologia-FCT).

REFERENCES

Aarsæther, K.G and Moan, T., 2009. Estimating Navigation Patterns from AIS. *The Journal of Navigation*, 62(4), 587.

Bergantina, A. and Marlow, P., 1998. Factors influencing the choice of flag: empirical evidence. *Maritime Policy & Management*, 25 (2), 157–174.

Cao, H., Wolfson, O. and Trajcevski, G., 2006. Spatio-temporal data reduction with deterministic error bounds. *VLDB Journal*, 15(3), 211–228.

Douglas, D.H. and Peucker, T.K., 1973. Algorithms for the reduction of the number of points required to represent a digitized line or its caricature. *Cartographica: The International Journal for Geographic Information and Geovisualization*, 10 (2), 112–122.

Ester, M., Kriegel, H.-P., Sander, J. and Xu, X., 1996. A density-based algorithm for discovering clusters in large spatial databases with noise. *Second International Conference on Knowledge Discovery and Data Mining*, E. Simoudis, J. Han, & M.U. Fayyad (Eds.), pp. 226–231, AAAI Press.

Etienne, L., Devogele, T. and Bouju, A., 2010. Spatio-Temporal Trajectory Analysis of Mobile Objects Following the Same Itinerary. *Remote Sensing and Spatial Information Sciences*, 38(II), 86–91.

Last, P., Bahlke, C., Hering-bertram, M. and Linsen, L., 2014. Comprehensive Analysis of Automatic Identification System (AIS) Data in Regard to Vessel Movement Prediction. *The Journal of Navigation*, 67, 791–809.

Le Guillarme, N. and Lerouvreur, X., 2013. Unsupervised extraction of knowledge from S-AIS data for maritime situational awareness. *Information Fusion* 16th, 2025–2032.

Li, K.X., Yin, J. and Fan, L., 2014. Ship safety index. *Transportation Research Part A: Policy and Practice*, 66(1), 75–87.

Mou, JM., Tak, C. and Ligteringen, H., 2010. Study on collision avoidance in busy waterways by using AIS data. *Ocean Engineering* 37 (5–6): 483–490.

Norris, A. (2007). AIS implementation – success or failure. *The Journal of Navigation*, 60, 373–389.

Pallotta, G., Vespe, M. and Bryan, K., 2013. Vessel pattern knowledge discovery from AIS data: A framework for anomaly detection and route prediction. *Entropy*, 15(6), 2218–2245.

Psarros, G.A., Christiansen, A.F., Skjong, R. and Gravir, G., 2011. On the success rates of maritime piracy attacks. *Journal of Transportation Security*, 4(4), 309–335.

Qu, X., Meng, Q. and Suyi, L., 2011. Ship collision risk assessment for the Singapore Strait. *Accident Analysis and Prevention*, 43(6), 2030–2036.

Rong, H., Teixeira, A.P. and Guedes Soares, C., 2016. Assessment and characterization of near ship collision scenarios off the coast of Portugal. *Maritime Technology and Engineering 3*, Guedes Soares, C. & Santos T.A., (Eds.), Taylor & Francis Group. London. 871–878.

Silveira, P.A.M., Teixeira, A.P. and Guedes Soares, C., 2013. Use of AIS Data to Characterise Marine Traffic Patterns and Ship Collision Risk off the Coast of Portugal. *The Journal of Navigation*, 66, 879–898.

Vander Hoorn, S. and Knapp, S., 2015. A multi-layered risk exposure assessment approach for the shipping industry. *Transportation Research Part A: Policy and Practice*, 78, 21–33.

Maritime Transportation and Harvesting of Sea Resources – Guedes Soares & Teixeira (Eds)
© 2018 Taylor & Francis Group, London, ISBN 978-0-8153-7993-5

AIS data analysis for the impacts of wind and current on ship behavior in straight waterways

Y. Zhou & T. Vellinga
Department of Hydraulic Engineering, Delft University of Technology, Delft, The Netherlands

W. Daamen & S.P. Hoogendoorn
Department of Transport and Planning, Delft University of Technology, Delft, The Netherlands

ABSTRACT: Due to the increasing ship traffic flow in ports, maritime traffic safety has attracted much attention. In addition to traffic flow, the ship safety is influenced by external navigational factors (visibility, wind and current), etc. In this paper, we investigate the effect of navigational factors on ship behavior (speed and path) by collecting the raw AIS data and locally measured visibility, wind, and current data in the port of Rotterdam. The results reveal that the wind mainly affects the paths of ships by the force of cross-wind, while the current impacts the speed over ground of ships when the current is with or against the heading of ship. The impacts on different sizes of ships are different as well. The analysis results could assist the port authority in predicting ship traffic in different situations, and be used in the development of a new maritime traffic model to simulate ship behavior while considering the external navigational factors.

1 INTRODUCTION

Waterborne transport has been an important means of cargo transportation as the most economical method. More than 80% of world merchandised trade by volume are carried by sea (EuropeanCommission, 2013). The understanding and effective management of maritime traffic will benefit the overall performance of the sea ports and inland waterways. Due to the increasing ship traffic flow in hub ports, e.g. the port of Rotterdam, the maritime traffic safety is an important and sensitive issue. Unlike the large space for ship maneuvering at sea, ports and inland waterways are restricted areas. In such areas, the impacts of external navigational factors may lead to serious consequences, such as grounding or collision with vast loss of life and property. The understanding of ship behavior in real-life situations is of theoretical and practical significance.

In current studies of maritime traffic, various models are developed for risk assessment (Goerlandt and Kujala, 2011) (Montewka et al., 2011) (Park et al., 2016) (Fernandes et al., 2016) and capacity analysis (Özkan et al., 2016). However, most of these models include few external factors or make the external impacts as an assumption. This is partly due to a lack of insight into the relations between the observed ship behavior and the external conditions.

In order to investigate ship behavior, Automatic Identification System (AIS) data have proven to be a valuable source. AIS has been installed on all passenger ships and sea-going ships larger than 300 Gross Tonnage (GT), according to the requirement of International Maritime Organization (IMO). Many papers present analyses of ship behavior patterns based on AIS data (De Boer, 2010) (Shu et al., 2013) (Zhou et al., 2015) (Rong et al., 2015, Xiao et al., 2015). Combining AIS data with some meteorological data, the general impacts of visibility, wind and current on ship behavior are also presented (Shu et al., 2017). However, due to a lack of detailed hydrological information and ship behavioral attributes in the collected data, the impacts of other factors cannot be eliminated and the impact of wind and current from different directions is not fully investigated.

In this paper, the impacts of the wind and current on ship behavior are systematically analyzed based on raw AIS data and meteorological and hydrological data in a straight waterway in the port of Rotterdam for the whole year 2014. Using the actual ship heading, the direction of wind and current are defined into four directions relative to the ship movement. With a comparative analysis of ship behavior (indicated by path and speed over ground (SOG)) in different situations, the impacts of different wind and current conditions are revealed. The results will help researchers to simulate ship behavior in different external conditions and provide the port authority with an insight into relations between ship behavior and external factors.

In Section 2, the collected data set is introduced. Section 3 explains the proposed methodology for data analysis. The impact analysis results for wind and current are presented in Sections 4 and 5, respectively. Section 6 concludes the paper with discussion and recommendations for further research.

2 DATA DESCRIPTION

The research area is a nearly straight waterway, Nieuwe Waterweg, located at the entrance of the port, as shown in Figure 1. The length of the research area is about 2.3 km. By choosing a straight waterway, the impact of waterway intersection on ship behavior is eliminated.

2.1 AIS data

The AIS data are collected from the port authority of Rotterdam, covering the whole year of 2014, including 2,299,842 messages. Every sea-going ship, even under the GT limit of IMO's regulation, has installed AIS and used it in all voyages. For the inland ships, both commercial and recreational ships, and sailing vessels longer than 20 meters are mandatory to use AIS since Dec. 1st, 2014 according to Central Commission for the Navigation of the Rhine. The regulation applies to most inland vessels in the Netherlands. The year 2014 is thus a transition year, during which more and more inland vessels are recorded by AIS. Thus, the majority of vessels in this research are seagoing ships.

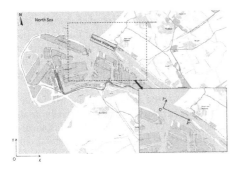

Figure 1. Layout and coordinate system of the research area in port of Rotterdam. (The X-Y coordinate system is the Dutch geographical coordinate system, Rijksdriehoeksmeting (RD system). In the cut out area, the transposed system is indicated, so the inbound ships sail in the X'-direction, while the lateral deviations from the straight path are visible in Y'-direction.)

In the collected AIS data, the cargo ships (993,566 messages, 43.2%), tankers (522,614 messages, 22.7%) and passenger ships (77,724 messages, 3.4%) are selected as the research objects. Other ships, such as pilot ship, tug, dredger, are not included in the analysis because the behavior of such ships in working and non-working status is different, while their working status is not indicated in the AIS messages. As for the cargo ships, since there is no secondary categorization of ship type in the collected AIS data, these ships cannot be identified as container, general cargo ship or bulker exactly. Thus, the impact of ship type on behavior is not specified.

To some extent, the ship size determines the windage area and the volume under water, which are relevant to the impacts of wind and current on ship behavior. In this paper, the ships are classified using beam as the criterion. In this research, the minimum beam in the data set is 6 meters, while the maximum beam is 79 meters. To make the proposed methodology generic, the beam intervals of four ship classes are determined as: (1) beam < 10 m, (2) 10 m ≤ beam < 23 m, (3) 23 m ≤ beam <33 m, (4) beam ≥ 33 m.

The collected AIS data contain three types of information:

- Static information: Maritime Mobile Service Identity number, type, length, beam, sensor type.
- Dynamic information: utc time, X-position, Y-position, SOG, course over ground (COG), heading, navigation status, etc.
- Voyage-related information: draught.

The attributes to describe ship behavior (position, SOG, COG and heading) are illustrated in Figure 2.

The AIS messages describe the dynamic position of ship (X and Y) in the RD system. In order to explicitly compare the ship behavior, the coordinate system is transposed, as shown in Figure 1. The origin lies at the west corner of research area. Thus, the ship position is described by the distance

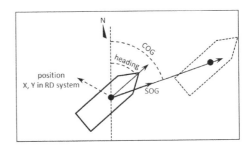

Figure 2. Illustration of behavioral attributes in AIS data.

to the northwest boundary of research area (the Y'-axis) and the lateral distance to the dam (the X'-axis).

2.2 Meteorological and hydrological data

The meteorological condition refers to wind and visibility. This is locally measured data in 2014, collected from the port authority of Rotterdam. The wind velocity data are at an interval of 5 minutes, while the visibility is measured every minute.

The hydrological condition refers to the current velocity. The data are also from the port authority. Unlike wind, the locally measured current velocity is not representative for the whole area, due to the propagation of flow and the difference through the water-depth. Thus, the data of current velocity are calculated using the SIMONA model (Vollebregt et al., 2003) with the measured water level from eight stations around the port as input. The collected data describe the current in 41×7 orthogonal curvilinear grids with a resolution about 85 meters. The current velocity in each grid cell is presented by 10 layers with an average depth at an interval of 15 minutes.

3 DATA ANALYSIS METHODOLOGY

Since the research area is a nearly straight waterway, the observed COG of the ships are always parallel to the bank. Thus, only the impacts of wind and current on ship path and SOG are analyzed. To compare the ship behavior when passing the same location, a set of cross-sections are developed parallel to Y'-axis. The ship behavior data are interpolated to the cross-sections by the last message before and the first message after the cross-section. By calculating the proceeded distance of ships between two adjacent AIS messages in data set, the interval distance between cross-sections is determined as 65 meters, with 35 cross-sections in total. This value guarantees that there is at least one AIS message in between two adjacent cross-sections for 75 percent of the data.

The proposed research methodology is illustrated in Figure 4. To eliminate the impact of ship encounters, the processed data set excludes ships with an encounter to other ships during the voyage in the research area. According to the International Regulations for Preventing Collisions at Sea, the conducts of ships at sea are regulated in two situations, being in sight of another ship and in restricted visibility. Preliminary analyses of ship behavior in the port of Rotterdam show that the ships behave differently when the locally measured visibility range is less than 2000 meters. To

eliminate the impact of visibility, the situation with visibility larger than 2000 meters are chosen.

Based on a preliminary analysis of ship behavior in different external conditions, some thresholds are used to set up the situations for impact analysis. For both wind and current, there are three situations, being weak, average, and strong. When the wind speed is less than 8 m/s, it is deemed as weak wind with little impact on ship behavior. Wind speed larger than 13.7 m/s is classified as strong wind. The impact of such wind on ship behavior is not analyzed, due to a lack of data in these rare conditions. As for current velocity, only the surface velocity and the depth-averaged velocity are known in a real-life situation. In the research area, the depth-average current speed is smaller than the surface current speed due to the reverse flow near the bottom. Thus, the surface current is identified as the indicator to represent the current condition. The current speed less than 0.37 m/s is deemed as weak current, while the current speed larger than 1.45 m/s is strong. The impact of strong current is not investigated either.

To analyze the impacts of wind and current on ship behavior, four relative directions to ship behavior are defined as secondary situations rather than the original wind/ current directions. The four relative directions are with the wind/ current, against the wind/ current, wind/ current from the port side and wind/ current from the starboard side, as shown in Figure 3. The directions are determined by the angle between wind/ current direction and heading of the ship. In this way, the wind and current situation is linked to the dynamic ship motion, which would better reveal the impacts than using the original geographical directions. The ship behavior in each secondary situation for both wind and current are compared to the ship behavior in the unhindered situation using five descriptive statistics, being 1st, 25th, 50th, 75th, 99th percentile. The statistical test of t-test is also performed to each pair of comparison. In this paper, p-value at 0.05 is taken as the criterion to

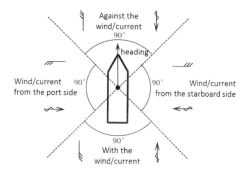

Figure 3. Four relative directions of wind and current.

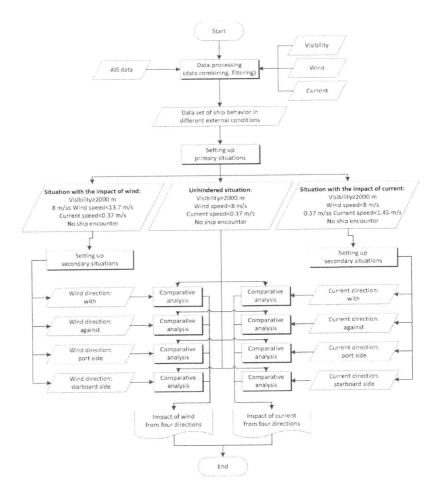

Figure 4. Flow diagram of the impact analysis based on AIS data. (The secondary situations are determined by the angle between wind/current direction and heading of ship. With the wind/current means the direction of wind/current is in the range from right ahead to 45 degrees on either side of the ship, while against means the direction is in the range from right aft to 45 degrees abaft the beam. Port side means the direction is in the range from 45 degrees afore to 45 degrees abaft the beam on the port side, while starboard side means the same range on the starboard side of ship.)

decide the significance level. The null hypothesis of the t-test is that 'the unhindered and hindered ship behavior are from the same distribution'. In the result of t-test, if H is equal to 1, it means the null hypothesis is rejected. In this way, the situation with an impact on ship behavior is recognized.

4 IMPACT OF WIND

The impact of wind on ship path and SOG is discussed in this section. The data set of both inbound and outbound ships are analyzed. As an example, the results of inbound ships (North Sea-Nieuwe Waterweg) with medium beam (10m≤ beam < 23m) are presented in details. The results

of outbound ships (Nieuwe Waterweg-North Sea) are similar. When the impacts on different size of ships are different, a comparison between different classes of ships is also given.

4.1 Path

In the situations 'with the wind' and 'against the wind', the paths of ships are similar to the paths in the unhindered situation. The statistical test result also proves that the lateral distances to the starboard bank in these two situations are not significantly different to the unhindered paths. The statistical analysis result is presented in Table 1.

However, the cross-wind does influence the paths of the ships, as shown in Figure 5. Since the transposed coordinate system is orthogonal while the dam (the starboard bank) is with a slight bend, the lateral distance to the starboard bank increases with larger distance to the entrance. With the wind from the port side, the paths are closer to the starboard bank. On the contrary, the paths are closer to the port bank when the wind is from the starboard side. Both results show that the wind force will push the ship to the other side, which is as we expected. The statistical results also indicate that the hypothesis is rejected that ship paths in crosswind situations are equal to the unhindered paths.

The deviation direction of all the ships in the same crosswind situation are the same. However, the extent of the impacts varies among different ship sizes. The differences between the mean values on each cross-section compared to the unhindered situation is shown in Figure 6. It reveals that with port wind, small ships (beam < 10m) sail closer to the starboard bank than large ships (beam ≥ 10m). With the wind force to the starboard bank, large ships appear to keep more distance to the bank to prevent collision due to their large inertia and possible shallow water near the bank. Meanwhile, starboard-wind impacts is larger for large ships

Table 1. T-test results between paths in wind-hindered and unhindered situations.

Situation	No. of cross-sections with H = 1*	Average p-value
With the wind	2	0.2917
Against the wind	0	0.6668
Wind from the port side	35	5.1172×10^{-4}
Wind from the starboard side	35	4.9795×10^{-7}

* The total number of cross-sections is 35.

than for small ships. In both crosswind situations, large ships bear larger wind force than small ships. However, the wind from the starboard side pushes ships to the portside bank, which also implies that they sail closer to the centerline of the waterway with sufficient water-depth and room for ship maneuvering.

4.2 SOG

The differences of SOG in the situations 'with the wind', 'wind from the port side', and 'wind from the starboard side' are quite small. The t-test results also show that in these three situations, ship's SOGs are not significantly different from the unhindered situation. However, when ships sail against the wind, there is a decrease in SOG of ships, as shown in Figure 7. The t-test results also indicate the difference of SOG between the situation 'against the wind' and the unhindered situation. For all the ships, it will increase the fuel consumption to maintain a same SOG as in the unhindered situation, when the wind is from ahead. It is neither economical for the ship owner, nor environment-friendly for the port. It is observed that the impacts of wind on SOG are similar to different size of ships. The reason is that even the wind force is large, the ships would increase the level of engine operation to maintain a speed avoiding maneuvering failure of rudder effect in any circumstances.

5 IMPACT OF CURRENT

5.1 Path

There is no significant difference between paths in the situations 'against current' and 'current from 0ort side', which is supported by the t-test results. However, in the situations 'with current' and 'cur-

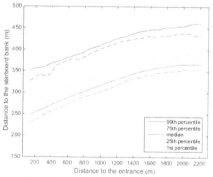

(a) Situation: wind from the port side (dashed lines) and unhindered situation (solid lines).

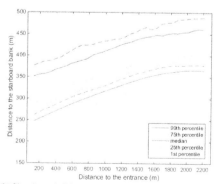

(b) Situation: wind from the starboard side (dashed lines) and unhindered situation (solid lines).

Figure 5. Ship path as a function of wind conditions.

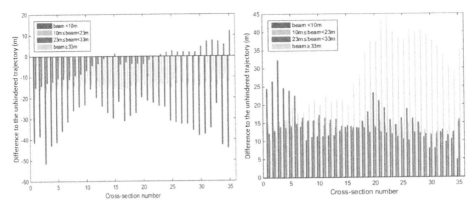

Figure 6. Difference to the unhindered paths in cross-wind situations (positive value means hindered path closer to the port bank, while negative value means hindered path closer to the starboard bank).

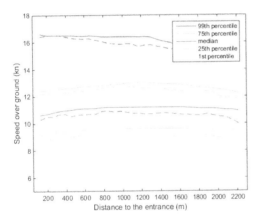

Figure 7. Ship SOG in the situation 'against the wind'.

rent from the starboard side', ships sail further to the starboard bank, as shown in Figure 8. In the t-test results, for 24 cross-sections the hypothesis that the paths in the situation 'with current' are equal to the unhindered paths is rejected, with an average p-value of 0.0488, which is close to the acceptance value 0.05. It means the ships sail further to the starboard bank when the current pushes them forward, but the distance difference is small as can be seen from Figure 8. In the situation 'current from starboard side', a significant difference is observed in all cross-sections. The ships sail further to the starboard bank, but the distance deviation varies a lot. The reason of such behavior variation in the starboard-current needs to be further investigated.

5.2 SOG

Figure 9 presents the impacts of current on the SOG of ships. The current from the port side does not have a significant impact on SOG, as also indicated by the t-test results. The t-test result also indicates that the hypothesis that SOG in the situation 'current from the starboard side' is equal to the unhindered situation is not accepted, which suggests this direction of current does impact SOG. However, the hindered SOG fluctuates a lot. For most ships, the SOG increases, especially for 1st percentile value. The ships maneuver frequently under the bank effect when the current is from the starboard side.

It can also be observed that the SOG of ships increases in the situation of 'with current' and decreases in the situation of 'against current', which follows our expectations. In the statistical results for both situations, the hypothesis that the SOG of ships is equal to the unhindered SOG is rejected for all cross-sections. However, the impacts of current among different ship sizes are different, as shown in Figure 10. The impact of current on small ships (beam < 10m) is the largest. Then, the impact decreases with an increase of the size of the ships ($10m \leq$ beam <33m). However, the impact on large ships (beam $\geq$ 33m) increases again. When the ship size increases, the gravity of ships also increases. Thus the frictional resistance effect of current on ships decreases. However for large ships (beam $\geq$ 33m), the draught is bigger than smaller ships, which means the ship is impacted by more layers of current. The current force on large ships is also larger. Furthermore, the extent of SOG increase in the situation 'with current' is larger than the impact of SOG decrease in the situation 'against current'. It is because the ships would increase the level of engine operation to maintain a proper speed for ship maneuvering.

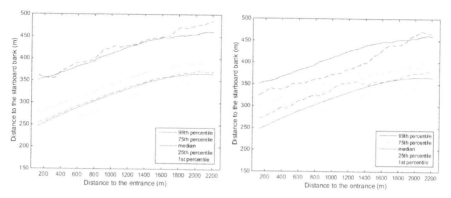

Figure 8. Ship path as a function of current conditions.

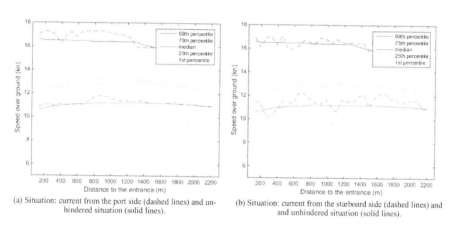

(a) Situation: current from the port side (dashed lines) and unhindered situation (solid lines).

(b) Situation: current from the starboard side (dashed lines) and
and unhindered situation (solid lines).

Figure 9. Ship SOG as a function of current conditions.

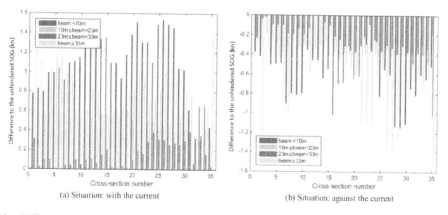

(a) Situation: with the current

(b) Situation: against the current

Figure 10. Difference to the unhindered SOG in situations 'with the current' and 'against the current' (positive value
means an increase of SOG, while negative value means a decrease of SOG).

6 CONCLUSIONS AND DISCUSSION

This paper investigates the impacts of wind and
current on ship behavior (indicated by path and
SOG) using AIS data from a straight waterway in
the port of Rotterdam, the Netherlands.

It is shown that the cross-wind influences
the paths of ships, while the paths present no

271

significant difference in the situations 'with the wind' or 'against the wind' compared to unhindered situation. The wind from the port side appears to 'push' the ships towards the starboard side of the bank, and vice versa. The port side wind has a larger impact on small ships than on large ships. On the contrary, the impact of wind from the starboard side is larger for large ships than for small ships. The reason is the insufficient waterdepth and room for maneuvering on the starboard side of ships. The impact of wind of SOG is only observed in the situation 'against the wind' with a decrease of SOG for all ship sizes.

In the situations 'against current' and 'port side current', no impact on the path of ships is revealed. However, the paths in the situations 'with the current' and 'starboard side current' showed a larger distance to the starboard bank compared to the unhindered situation. In the situation 'with current' and 'against current', SOG of ships increases and decreases respectively. The impact is larger for small ships than large ships, and least on medium ships. The impact in the situation 'with current' is larger than 'against current'. The port-side current does not influence SOG, while the starboard current influences SOG, due to the bank effect.

It can be concluded that paths of ships are easier to be impacted by wind than current, especially by cross-winds. The SOG of ships will decrease when sailing against either wind or current, but increase only when sailing with current. In addition, the impact of current from starboard side is possibly due to the bank effect. But further research on the impact of bank effect on ship behavior is required.

The analysis result could benefit both researchers and port authority. For the researcher, a detailed insight into the impact of wind and current helps to predict ship behavior in different external conditions. For the port authority, the revealed impact pattern of wind and current can be used to predict ship behavior in forecasted external factors, which helps the ship traffic management and risk control in port.

Within this paper, only a straight waterway is investigated. In a later stage, a port area with more complex layout will be analyzed to identify the impact on all ship behavioral attributes, being path, SOG, COG and heading, in particular for cases where the course does not correspond to the heading.

ACKNOWLEDGEMENTS

This work is financially supported by China Scholarship Council and Delft University of Technology with the Grant number 201306950015. The authors would like to thank the department of Data Management in the port of Rotterdam during the data collection, and appreciate Frank Cremer for accessing AIS data, Cor Mooiman for providing wind and visibility data, and Bob van Hell for simulating current data.

REFERENCES

De Boer, T. (2010) An analysis of vessel behaviour based on AIS data. TU Delft, Delft University of Technology.

Europeancommission (2013) Ports: an engine for growth. Brussels.

Fernandes, R., Braunschweig, F., Lourenço, F. & Neves, R. (2016) Combining operational models and data into a dynamic vessel risk assessment tool for coastal regions. *Ocean Science*, 12, 285.

Goerlandt, F. & Kujala, P. (2011) Traffic simulation based ship collision probability modeling. *Reliability Engineering & System Safety*, 96, 91–107.

Montewka, J., Krata, P., Goerlandt, F., Mazaheri, A. & Kujala, P. (2011) Marine traffic risk modelling–an innovative approach and a case study. *Proceedings of the Institution of Mechanical Engineers, Part O: Journal of Risk and Reliability*, 225, 307–322.

Özkan, E.D., Nas, S. & Güler, N. (2016) Capacity Analysis of Ro-Ro Terminals by Using Simulation Modeling Method. *The Asian Journal of Shipping and Logistics*, 32, 139–147.

Park, J., Han, J., Kim, J. & Son, N.-S. (2016) Probabilistic quantification of ship collision risk considering trajectory uncertainties. *IFAC-PapersOnLine*, 49, 109–114.

Rong, H., Ap, T. & Guedes Soares, C. (2015) Simulation and analysis of maritime traffic in the Tagus River Estuary using AIS data, Taylor & Francis Group. London.

Shu, Y., Daamen, W., Ligteringen, H. & Hoogendoorn, S. (2013) Vessel Speed, Course, and Path Analysis in the Botlek Area of the Port of Rotterdam, Netherlands. *Transportation Research Record: Journal of the Transportation Research Board*, 63–72.

Shu, Y., Daamen, W., Ligteringen, H. & Hoogendoorn, S.P. (2017) Influence of external conditions and vessel encounters on vessel behavior in ports and waterways using Automatic Identification System data. *Ocean Engineering*, 131, 1–14.

Vollebregt, E.A., Roest, M. & Lander, J. (2003) Large scale computing at Rijkswaterstaat. *Parallel Computing*, 29, 1–20.

Xiao, F., Ligteringen, H., Van Gulijk, C. & Ale, B. (2015) Comparison study on AIS data of ship traffic behavior. *Ocean Engineering*, 95, 84–93.

Zhou, Y., Daamen, W., Vellinga, T. & Hoogendoorn, S. (2015) Vessel classification method based on vessel behavior in the port of Rotterdam. *Scientific Journals of the Maritime University of Szczecin*, 114, 86–92.

Maritime safety and human factors

Maritime Transportation and Harvesting of Sea Resources – Guedes Soares & Teixeira (Eds)
© 2018 Taylor & Francis Group, London, ISBN 978-0-8153-7993-5

Human factor influence on ship handling errors in restricted waters— stress susceptibility study

T. Abramowicz-Gerigk & A. Hejmlich
Department of Ship Operation, Gdynia Maritime University, Gdynia, Poland

ABSTRACT: The paper presents investigations on the human factor influence on ship handling errors with respect to human susceptibility to stress including a comprehensive research on the relations between human personalities, attitude towards risk and ship handling errors during navigation in restricted waters. The research had been carried out with a group of Ship Masters, maritime pilots and students from Gdynia Maritime University. The research consisted of physical trials performed as a single-man watch on the Full Mission Ship-handling Simulator and single operator of a man-manned physical model and psychological surveys carried out to determine personality and attitude towards risk of the research participants. On the basis of test results a model of stress susceptibility was proposed for each personality. The statistical analysis of the results along with the assessment of the conditional probability of a navigational error allowed determining a degree of stress susceptibility.

1 INTRODUCTION

The operational decision making process of bridge personnel is related to the time stress, simultaneous multiple tasks requirements, constantly changing information from multiple sources and uncertainty (EMSA, 2014, Soares Guedes & Teixeira, 2001). It is essential for the seafarers they are convinced of their performance in difficult situations (Battacharya, 2012; Berg, 2013; Gerigk, 2015).

The paper presents preliminary results of the research on correlations between personality characteristics of a seafarer, his attitude towards risk, stress vulnerability level and ship handling errors in restricted waters. The research participants were students of Gdynia Maritime University, experienced Ship Masters and experienced Sea Masters working as harbor pilots. The participants took part in the survey research, expressing their opinions in psychological questionnaires. The psychological analysis of the questionnaires allowed defining their psychological profiles—personality, stress vulnerability level and attitude towards risk.

The next stage of the investigations was a physical ship handling test performed on Full Mission Ship Handling Simulator—SimFlex Navigator 4.0. The main objective of the research was to prove that the accuracy of manoeuvring task realisation, dependent on personal skills and experience, is to a large extent dependent on the personal psychological profile of a seafarer.

The research participants were divided in two groups: first group composed of 22 students and second group composed of 32 Ship Masters.

There were three components taken into account in the assessment of human psychological profile: personality characteristics, attitude towards risk and stress level.

Personality characteristics according to the Big Five (Big 5) personality traits are the five broad domains which define human personality and account for individual differences (Costa & McCrae, 2003):

– Openness – tendency for positive thinking, tolerance for new experiences and cognitive curiosity
– Extraversion – reflects a level of activity, energy, positive emotions
– Conscientiousness – shows a level of perseverance and motivation for goal achievement
– Neuroticism – tendency for feeling fear, confusion, anger, guilt
– Agreeableness – attitude towards other people, altruism versus antagonism.

The attitude towards risk is related to the instrumental and stimulating risk:

– Instrumental risk – means the risk is undertaken to perform the task in possible best way, considering the results of failure (Makarowski, 2012)
– Stimulating risk – risk undertaken for self-pleasure, self-excitement regardless of consequences

The stress level assessment is related to the following elements (Plopa & Makarowski, 2010):

– Emotional tension – comes from anxiety, excessive nervousness resulting with low energy, fatigue, developing tendency to quit task before starting it
– External stress – comes from unfair judgment by others, frustration, feeling that burden have been put on us will overgrow our possibilities
– Intrapsychic stress – internal tension from bad memories which induces anxiety, resignation, and pessimism

Each component has doubtlessly influence for a human being behavior and tasks performance.

The analysis of the questionnaires gives as a result human psychic expressed in numbers. The obtained results can be input into mathematical models and possible correlations with committed errors can be determined.

2 RESEARCH TOOLS AND METHODOLOGY

The research tools and methodology used in the presented psychological tests were as follows:

– Stressors ranking questionnaire (Abramowicz-Gerigk & Hejmlich, 2015; Abramowicz-Gerigk et al. 2015)
– Questionnaire of stress sense (Plopa & Makarowski, 2010)
– Questionnaire of Stimulation and instrumental risks (Makarowski, 2012)
– Questionnaire NEO-FFI (Costa & McCare, 2003)

The physical tests of ship handling skills were carried out on Full Mission Bridge simulator Sim-Flex Navigator 4.0 and man manned physical ship models The area used for maneuvering exercises on the Full Mission Simulator was area of Port of Amsterdam (Fig. 1).

The ship selected for the tests was a container ship "Dona-C". Her main dimensions—length overall and breadth were accordingly equal to LOA = 318.24 m and B = 42,8 m. The container ship "Dona-C" entering Noordelijk Sluiseiland lock of 50 m breadth is presented in Figure 2.

The exercise task was composed of the following sequential manoeuvres:

– Entering the harbor trough outer breakwaters heads,
– Approach the lock,
– Passing through the lock,
– Transiting Ijmouiden Canal
– Berthing starboard side alongside the container terminal.

The exercise scenario included hydro-meteorological conditions and hydrodynamic effects acting on the ship:

– Wind
– Current
– Shallow water and bank-wall effects
– Interaction forces between the own ship and overtaken or passed by ships.

The external conditions and initial ship parameters were exactly the same for each research participant. Each exercise was monitored and recorded in the data base along with the identified errors.

The following maneuvering errors have been considered (Abramowicz-Gerigk, 2012):

– Deviation from the planned route
– Not sufficient or exaggerated settings of main engine, rudders and thrusters
– Wrong coordination of maneuver in time

For the assessment of the quality of maneuvers the following penalty scale has been introduced (Fig. 3–Fig. 5):

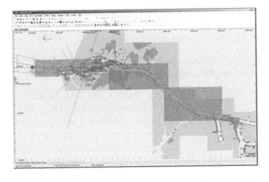

Figure 1. Exercise area for ship handling skills test on Full Mission Ship Handling Simulator—Port of Amsterdam.

Figure 2. Container ship "Dona-C" entering Noordelijk Sluiseiland lock—Full Mission Ship Handling Simulator—Simplex Navigator 4.0.

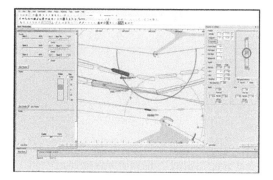

Figure 3. Exceeding 6 knots speed limit in Port Entrance—1 point penalty.

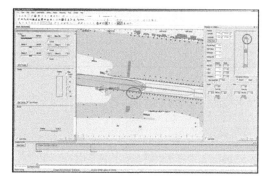

Figure 4. Allison with the lock's wall with the lateral speed of 0.1 m/s—2 points penalty.

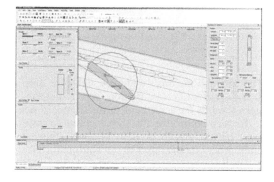

Figure 5. Example of grounding—10 points penalty.

- Speed exceeding, deviation from planned track, excess of main engine power – 1 point
- Hitting the lock wall or berth with sideway speed less than 0,15 m/s – 2 points
- Hitting the lock wall or berth with sideway speed higher than 0,15 m/s – 5 points
- Grounding, collision, allision – 10 points

3 RESULTS

The maneuvers performed in the environment of Full Mission Simulator allowed to test the hypotheses which link the number of ship handling errors with risk, stress and the Big 5 traits.

Psychological questionnaires created and decoded by a professional psychologist have delivered a numerical value of contents of ten examined factors: neuroticism, extraversion, openness, conscientiousness, agreeableness, emotional tension, external stress, intrapsychic stress, stimulating risk and instrumental risk in a psychological profile of each participant. The numerical values of each factor were mathematically analysed by Pearson's linear correlation formula with the number of manoeuvring errors committed by each participant. The results are presented in Tables 1, 2 and 3.

The Pearson's linear correlation coefficients were determined for the variables analysed in two groups of participants. The mean and standard deviation values, and sten scores for the analysed variables are presented in Table 1.

The sten scores indicate an approximate position of each participant with respect to other people in that population. They are calculated by a linear transformation: standardized, multiplied by the desired standard deviation of 2 and then the mean of 5.5 is added. The resulting decimal value is rounded to an integer.

The Pearson's correlation coefficients for the analysed variables in the group of students are presented in Table 2.

The Pearson's correlation coefficients for the analysed variables in the group of ship masters are presented in Table 3.

Table 1. Mean and standard deviation and sten score values for the analysed variables.

Variable	Students			Masters		
	M	SD	StSc	M	SD	StSc
Neuroticism	20.05	8.83	5	16.00	6.37	**3**
Extraversion	31.02	9.74	7	30.41	7.43	7
Openness	25.32	5.40	4	25.81	5.83	7
Conscientiousness	36.41	6.69	**8**	40.53	6.26	**8**
Agreeableness	24.14	6.13	4	29.13	4.67	5
Emotional tension	16.91	6.72	3	16.72	5.02	3
External stress	15.86	3.62	3	13.25	4.15	2
Intrapsychic stress	14.36	5.18	3	13.59	4.25	2
Stimulating risk	11.86	4.73	4	9.25	4.90	3
Instrumental risk	10.82	3.63	5	12.88	2.27	6

M – mean, SD – standard deviation, StSc – sten score.

Table 2. The Pearson's correlation coefficient "r" for the variables in the group of students.

Variables	1	2	3	4	5	6	7	8	9	10	11	12	13
Stimulating risk	1.00												
Instrumental risk	−0.29	1.00											
SR-IR	0.87	−0.72	1.00										
Emotional tension	−0.32	0.08	−0.27	1.00									
External stress	0.02	−0.25	0.14	0.33	1.00								
Intrapsychic stress	0.09	−0.19	0.16	0.74	0.22	1.00							
ET+IS	−0.06	−0.20	0.06	0.40	0.18	0.22	1.00						
Extraversion	0.30	−0.05	0.24	−0.37	0.26	−0.41	−0.42	1.00					
Neuroticism	0.04	−0.22	0.14	0.81	0.49	0.75	0.84	−0.23	1.00				
Openness	0.03	0.19	−0.07	−0.24	0.11	−0.39	−0.33	0.25	−0.26	1.00			
Conscientiousness	0.28	0.40	0.00	−0.61	−0.32	−0.63	−0.66	0.54	−0.57	0.48	1.00		
Agreeableness	−0.10	−0.08	−0.03	−0.09	−0.22	0.09	−0.01	0.12	−0.01	0.20	0.23	1.00	
Number of errors	−0.01	0.17	−0.10	−0.42	−0.48	−0.58	−0.52	−0.11	−0.57	0.23	0.38	−0.33	1.00

Table 3. The Pearson's correlation coefficient "r" for the analyzed variables in the group of ship masters.

Variables	1	2	3	4	5	6	7	8	9	10	11	12	13	14	15
1 Stimulating risk	1.00														
2 Instrumental risk	−0.38	1.00													
3 SR-IR	0.94	−0.67	1.00												
4 Emotional tension	0.26	−0.29	0.32	1.00											
5 External stress	0.22	−0.29	0.29	0.68	1.00										
6 Intrapsychic stress	0.34	−0.33	0.39	0.85	0.66	1.00									
7 ET+IS	0.31	−0.32	0.17	0.97	0.69	0.96	1.00								
8 Extraversion	0.40	0.07	0.29	−0.27	0.07	−0.23	−0.27	1.00							
9 Neuroticism	0.20	−0.26	0.25	0.74	0.53	0.75	0.77	−0.10	1.00						
10 Openness	0.20	0.04	0.14	−0.06	0.07	−0.02	−0.04	0.30	0.03	1.00					
11 Conscientiousness	−0.11	0.35	−0.22	−0.45	−0.32	−0.66	−0.57	0.29	−0.49	0.12	1.00				
12 Agreeableness	−0.07	0.28	−0.16	−0.49	−0.21	−0.41	−0.47	0.44	−0.26	−0.09	0.23	1.00			
13 Number of errors	0.19	−0.29	0.26	0.16	0.30	0.16	0.17	0.17	0.16	0.16	0.22	0.03	1.00		
14 Collision, grounding	0.28	−0.34	0.35	0.28	0.44	0.28	0.29	0.13	0.17	0.14	0.14	0.06	0.88	1.00	
15 Grounding at control point	0.58	−0.23	0.55	0.31	0.19	0.35	0.34	0.19	0.16	0.03	−0.08	−0.16	0.38	0.50	1.00

4 DISCUSSION OF THE RESULTS

In both groups of research participants a week positive correlation between number of errors and personality characteristic conscientiousness was observed. The results of trials carried out by the students show no correlation with risk approach and negative correlation with stress level. For the Ship Masters being a reference group a week (positive) correlation of ship handling errors number with conscientiousness r = 0.22, stress level r = 0.35 and medium positive correlation with approach towards risk r = 0.55 were observed.

The main reason of the differences is the low accumulation of professional experiences among students being in phase so called "character crystallization".

For the Ship Masters an isolated additional testing point defined as "control point" was introduced. There is a phenomenon of a very strong bank effect at a certain place of Ijmouiden canal.

The interaction forces suddenly push the ship's bow to port and create suction force at the ship's stern towards the canal bank at the starboard side.

To keep control of the ship a countermeasure must be undertaken by Ship Master within seconds, otherwise the ship will run a ground.

The best countermeasure to bring the ship back to track is a "force maneuver"—maximum engine power and starboard full rudder in a short period of time.

This situation is doubtlessly stressful and allows observing Ship Master's reaction to cope with a kind of emergency. A few seconds for decision making and action are vital for success or failure. Not all the participants decided to use this force manoeuvre and the ship grounded in many cases.

The last row of Table 3 "grounding in control point" shows the Pearson's correlation of the failure and three tested components of human psychological profile. The attitude towards risk correlates moderately with the value of r = 0.55

(p < 0.001), stress level correlates weakly with value of r = 0.34 (p < 0.001).

There was no correlation observed for any personalities.

The attitude towards risk factor—mathematical subtraction of decoded questionnaires values SR-IR, SR positive correlation, IR negative correlation with number of errors for all participants is presented in Figure 6.

The results can be summarised as follows:

- 7 out of 32 participants with factor SR-IR > 0:5 run aground
- 3 out of 32 participants with factor SR-IR = 0:1 run aground
- 22 out of 32 participants with factor SR-IR < 0:2 run aground

The distribution of participants with respect to approach towards risk factor value and number of groundings is presented in Table 4.

The stress level factor ET+IS (mathematical addition of decoded questionnaires values ET+IS, both ET and IS show positive correlation with number off errors) for all participants is presented in Figure 7:

- 22 out of 32 participants with factor ET+IS < 35.5: 3 run aground
- 10 out of 32 with factor ET+IS > 35.5:5 run aground

The distribution of participants with respect to the stress level value and number of groundings are presented in Table 5.

The assessment of total probability of groundings due to the risk and stress level factors obtained using Bayesian networks are presented in Figures 8 and 9.

The list of the stressors presented below was ranked by the participants as per their personal feelings and analysed statistically. The highest

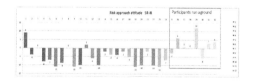

Figure 6. Approach towards risk factor—correlation SR-IR for all participants.

Table 4. Distribution of participants with respect to the approach towards risk factor and number of groundings—risk factor.

Risk factor SR-IR	No.Part.	No. GR.	Prob.	Part. Distr.
	–	–	–	%
0	22	2	0.09	68.75
= 0	3	1	0.33	9.38
>0	7	5	0.71	21.88

No. Part. – No. of Participants, No. GR – No. of groundings, Prob. – probability, Part. Distr. – participant distribution.

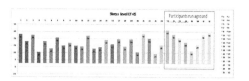

Figure 7. Stress level factor ET+IS for all participants.

Table 5. Distribution of participants with respect to the risk factor value and number of groundings—stress level factor.

Stress level	No. Part.	No. GR.	Prob.	Part. Distr.
ET+IS	–	–	–	%
35.5	22	3	0.14	68.75
>35.5	10	5	0.50	31.25

No. Part. – No. of Participants, No. GR – No. of groundings, Prob. – probability, Part. Distr. – participant distribution.

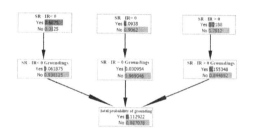

Figure 8. Bayesian network for the probability assessment of grounding due to risk factor.

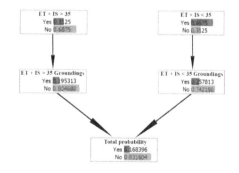

Figure 9. Bayesian network for probability assessment of grounding due to stress factor.

score of 87.69% belongs to the stressor No. 14—continuous, various inspections when in port, the lowest score of 49.23% belongs to stressor No. 12—lack of Internet access.

The list of the stressors presented ranked by participants

1. Continuous wariness about ship's safety in aspect of fire, collision and grounding

2. Wariness about ship's safety in difficult weather conditions: storm, fog, navigation in ice
3. Continuous wariness about possible failure of ship's equipment: main engine, diesel generators, steering equipment
4. Tiresomeness of navigation in the dense traffic areas
5. Frequency of in and out maneuvers
6. Port maneuvers in bad Weather conditions—strong wind and current
7. Maneuvers in restricted areas (small ports)
8. Prolonged manoeuvring: Panama Canal, Suez Canal, Great Lakes Seaway, Antwerp, Hamburg, Buenos Aires
9. Continuous wariness about safety of the crew
10. Possibilities of accident at work, loss of life, conflicts among the crew members
11. Low competency of crew members
12. Lack of Internet access
13. Various visitors on board when in port. Lack of time to rest
14. Continuous, various inspections when in port
15. Shortage of crew members on board
16. Time stress, continuous acting in haste
17. Pressure from ship's owner (work more and faster)
18. Pressure from Charterers
19. Lack of competency of shore based office—people at Owners office
20. Conflicts ship-shore office
21. Limited relax possibilities—short and busy port stay
22. Work at night
23. Paper work overload
24. Sleeping trouble due to frequent time zones changing

At least half of these stressors are related to the "external pressings" exerted on ships and could be easily reduced or even removed by administrative and organizational corrective action.

The new view theory says that human being is not a weak link in a safety chain. Human being is very often forced to take an action even against his common sense. This is typical conflict of interest between safety and economy.

Ship Master on board a ship has to create safe working environment to meet the company and charterers expectations. It is a conflict of safety culture and financial success culture. New view theory says that human errors come not from his bad performance but they come from controversies between different goals to be achieve by all the parties involved in the business for whom unfortunately the safety is not a priority (Morawski, 2005).

5 CONCLUSIONS

The results of the research presented in the paper have confirmed the hypothesis of correlations between stress level, attitude towards risk and number of errors leading to serious accident.

A probability of grounding due to high stress level reads 0.17. Probability of grounding due to risk approach attitude comes to 0.11.

The hypothesis of personality characteristic influence on human behaviour has been confirmed partly. Only personality defined as "Conscientiousness" shows week correlations with number of errors committed.

It is worthy to emphasise that 13 out of 32 (41%) of participants have the maximum level of conscientiousness in their profile.

REFERENCES

Abramowicz-Gerigk, T. 2012. *Safety of critical manoeuvres of ships in Motorway of the Sea transportation system.* Publishing House of Warsaw University of Technology. Warsaw 2012, in Polish.

Abramowicz-Gerigk, T., Hejmlich, A. 2015. Human factor modelling in the risk assessment of port manoeuvers. *TransNav - The International Journal on Marine Navigation and Safety of Sea Transportation.* Vol. 9, z. 3, 427–433.

Abramowicz-Gerigk, T., Burciu, Z., & Hejmlich, A. 2015. The concept of a method for the assessment of human factor influence on the risk of accident in maritime transport. *Logistyka* No. 4/2015, 17–23, in Polish.

Battacharya, S. 2012. The effectiveness of the ISM: a qualitative enquiry. *Marine Policy* No. 36, 2012.

Berg, H. 2013. Human Factors and Safety Culture in Maritime Safety (revised). *The International Journal on Marine Navigation and Safety of Sea Transportation*, Vol. 7 No. 3, 343–352.

Costa, P. & McCrae, R. 2003. A contemplated revision of the NEO Five-Factor Inventory. Baltimore, Maryland, USA: Department of Health and Human Services, National Institute on Aging, NIH.

EMSA, 2014. Report No. PP092663/1-1/2, R. 2EMSA/OP/10/2013 Risk Level and Acceptance Criteria for Passenger Ships. First interim report, part 2: Risk Acceptance Criteria. European Maritime Safety Agency. 2014.

Gerigk, M. 2015. Modeling of event trees for the rapid scenario development. *Safety and Reliability: Methodology and Applications.* Nowakowski et al. (Eds), Taylor & Francis Group, London 2015, 275–280.

Guedes Soares, C., Teixeira, A.P. 2001. Risk assessment in maritime transportation. *Reliability engineering and System Safety* 74, 299–309.

Hejmlich, A. 2014. Human factor in safety assessment of port manoeuvres. *Journal of Faculty of Navigation Gdynia Maritime University,* Vol. 29, 81–88.

Hejmlich, A. 2016. Stress influenceon the safety of manouvering in restreicted waters. Scientific Journals of Gdynia Maritime University. Vol. 97/2016, 25–36.

Makarowski, R. 2012. Stimulation and instrumental risk questionnaire not for sportsmen only. *Psychological Journal,* Vol 18 No. 2, in Polish.

Morawski, J.M. 2005. Man and technology. Secrets of mutual conditioning. Publishing House ASPRA-JR, Pułtusk.

Plopa, M., Makarowski, R. 2010. Stress feeling questionnaire—book for students. Visia Press & IT, Warsaw.

Maritime Transportation and Harvesting of Sea Resources – Guedes Soares & Teixeira (Eds)
© *2018 Taylor & Francis Group, London, ISBN 978-0-8153-7993-5*

Maritime search, rescue and assistance

A. Anechitoae, A.A. Cîmpanu & M.S. Matasa
University "Ovidius" Constanta, Romania

ABSTRACT: The institution of maritime rescue encourages maritime commercial activities by the fact that, thus, there are governed such clear rights and obligations saving life at sea and shipping goods. The maritime rescue, as any other legal institution related to maritime events—collision, crash, etc.—has its distinctive features. The maritime rescue may be considered as the operation that arises from maritime collision, because, while collision stems from a breach of a negative duty necessary in maritime navigation, i.e. of not harming the other, the maritime rescue is the implementation of positive obligations required to vessel captains by the material requirements of the marine life that adds to the elements of the legal concept which can be summed up as follows: to go to the aid of a vessel in danger, provided that the vessel does not expose itself, through this action, to a serious danger.

1 INTRODUCTION

Theoretically, "assistance" means the aid provided to a ship in danger in order to get out from a situation, and "rescue" – the help rendered to a vessel in danger, which, because it lost its manoeuvre ability, cannot cooperate with the opportunity of the aid it receives. We speak of "assistance" when the help comes in time in order to avoid danger, i.e. before sinking, and of "rescue" when the act does not occur until after the shipwreck has already begun.

In practice there were used the concepts of "assistance" in order to refer to the help rendered to vessels in danger, with the purpose of avoiding a more serious accident, and "rescue" in order to consider the contrary, when the aid is provided under serious conditions or when a first accident has already occurred.

Historically, the aim of rescue rules was to counteract the temptation of the saviour to acquire goods from stranded ships. In modern times, the aim was actually the desire to provide an incentive for the rescue efforts and, therefore, to maintain property values.

1.2 *Terms and definitions*

The search activity is an operation normally coordinated by a coordination centre of rescue operations or by a subsidiary of the coordination centre of rescue operations, using the available personnel and resources to locate persons in distress.

In its turn, rescue is an operation to retrieve persons in distress, provide for their initial medical or other needs, and deliver them to a place of safety.

The search and rescue service is the performance of distress monitoring, communication, coordination and search and rescue functions, including provision of medical advice, initial medical assistance, or medical evacuation, through the use of public and private resources including co-operating aircraft, vessels and other craft and installations.

Search and rescue facility represents any mobile resource, including designated search and rescue units, used to conduct search and rescue operations.

Noting the great importance attached in several conventions to the rendering of assistance to persons in distress at sea and to the establishment by every coastal State of adequate and effective arrangements for coast watching and for search and rescue services, Having considered Recommendation 40 adopted by the International Conference on Safety of Life at Sea, 1960, which recognizes the desirability of co-ordinating activities regarding safety on and over the sea among a number of inter-governmental organizations, Desiring to develop and promote these activities by establishing an international maritime search and rescue plan responsible to the needs of maritime traffic for the rescue of persons in distress at sea, Wishing to promote co-operation among search and rescue organizations around the world and among those participating in search and rescue operations at sea„ International Maritime Organization concluded, at Hamburg on 27 April 1979 the "1979 International Convention on Maritime Search and Rescue".

2 THE DISTINCTION BETWEEN MARITIME RESCUE AND MARITIME ASSISTANCE

The Convention for the unification of certain rules of law relating to assistance and rescue (Brussels, September 23, 1910) makes no distinction between maritime rescue and maritime assistance.

The Rescue International Convention, adopted in London on April 28, 1989shows that the rescue operation means any act or activity undertaken to assist a vessel or any other property in danger, in navigable waters or in any waters.

The obligation to assist is imposed by Art. 10, section 1 of the Convention: "Every master is bound, to the extent that he can do it, without endangering his ship and the persons on board, to assist any person in danger of being lost at sea".

In practice, the term "assistance" is used to refer to the help provided to a vessel in distress, in order to avoid a more serious event, and the notion of "rescue"—to designate the help given after the production of maritime distress.

The assistance always involves the existence of a state of real danger.

The rescue is conceivable only if the vessel, being in the open sea, is in such danger that, without immediate help, it would be completely lost; the assistance consists of preventive measures in order to avoid major damage to both the body of the vessel and the goods.

The maritime assistance service is provided to a maritime vessel by another vessel or by specialized intervention services in a timely manner so that an imminent danger can be avoided, i.e. before the shipwreck occurs. The maritime rescue service is the service provided by another maritime vessel or by specialized services that are called and that intervene only after the imminent danger has occurred, i.e. it intervenes after the shipwreck has already begun.

However, the legal rules of international law do not make a distinction specific to the rescue action or to the maritime assistance action. Most national laws are circumscribed to a single conceptual significance, i.e. providing aid to a vessel in real, potential, serious and imminent danger.

If the vessel is not in such a situation, the aid granted will not be legally qualified as an act of assistance, as a rescue act or as an act of maritime assistance and rescue, which may lead to the request of a remuneration right in relation to the applicable legal regulations in maritime law.

3 THE STIMULATING FACTOR BY REWARD

The stimulating factor is particularly evident in the rules for calculating the rescue reward. If the rescue effort was successful, the reward has to clearly exceed the normal remuneration for such services. The assessment must recognize the danger associated with the rescue operation—i.e., the danger faced by the saviour, the risk to fail, or the risk to damage the saved party—and the value of the property saved. The rescue reward, which is from 5 to 10 percent of the total value of the property saved is not an unusual percentage, which means that a reward for a general cargo ship loaded to its maximum capacity may be a considerable amount. It goes without saying that the perspective of such compensation is a strong incentive for any potential saviour.

The stimulating factor is also a reason for another rescue law, namely, the so-called "no cure, no pay" principle (if the rescue operation fails, the reward is not granted). This means that no reward can be asked unless the rescue operation was successful. If the attempt fails, the rescuer cannot make any claim, not even for the direct reimbursement of expenditures. The principle of "no cure, no pay" encourages rescuers to do everything they can, hoping to win a generous reward if their efforts result in success.

Another incentive is provided for the establishment of a maritime lien on the ship and cargo in order to ensure the claims after rescue. Thus, rescuers can be relatively sure that the rescue reward will be paid.

4 PRINCIPLES OF MARITIME RESCUE

The philosophy of the rescue service was epitomized in maritime law literature as follows:

"The maritime rescue principles are based on the simple premise that everyone who helps in order to save marine property is entitled to reasonable compensation for their efforts, and those who have benefited from these efforts should contribute to a reward in proportion to the value of the property saved. This, of course, led to the famous triumvirate of danger, volunteering and success that, with few exceptions, must exist in order to be able to talk about a rescue service" (Gold).

Thus, in order to reward a rescue service, three main factors should exist:

• Danger
• Volunteering
• Success.

According to article 87 of O.G. 42/1997 (r) on maritime transport and on inland waterways, when the commander/leader of the vessel flying the Romanian flag receives a message indicating that there is a vessel in danger, is obliged, to the extent that he does not endanger his own vessel, crew, passengers and/or cargo, to move with all possible speed to that vessel in order to provide it the necessary assistance and to save the people in danger on board of that ship.

The vessel commander/leader is obliged to give, after collision, support to the other vessel, crew and passengers and, where possible, to indicate to the other vessel the name of his own vessel, its port of registry and the nearest port to which it will get [Article 88, O.G. 42/1997 (r)].

The vessel commander/leader has no obligation to provide assistance and rescue if the master/leader of the vessel in distress expressly refuses help and if he receives information that help is no longer necessary [Art. 88, O.G. 42/1997 (r)].

The grounds for not granting the aid, due to refusal will be recorded in the logbook.

Co-operation between States [article 3 (3.1–3.1.2), International Convention on Maritime Search and Rescue, (SAR 1979), Hamburg, 27 April 1979]. Parties shall co-ordinate their search and rescue organizations and should, whenever necessary, co-ordinate search and rescue operations with those of neighbouring States. Unless otherwise agreed between the States concerned, a Party should authorize, subject to applicable national laws, rules and regulations, immediate entry into or over its territorial sea or territory of rescue units of other Parties solely for the purpose of searching for the position of maritime casualties and rescuing the survivors of such casualties. In such cases, search and rescue operations shall, as far as practicable, be co-ordinated by the appropriate rescue co-ordination centre of the Party which has authorized entry, or such other authority as has been designated by that Party.

5 THE EXISTENCE OF DANGER

The situation threatening the vessel which needs help is an objective and necessary circumstance qualifying an aid as rescue. It has a rather vague nature, especially since there is no legal definition in the internal and international regulations, so that it inherently raises the question on the nature or the type of danger affecting the vessel, so that the effective intervention of the rescuer determines a royalty, or the question related to the lower limit of possible dangers.

The first question that arises is: what can be called "danger"?

The answer is that, basically, there should be a risk of physical and extensive damage of the ship and cargo. Thus, if there is a risk of total loss of ship or cargo, we can clearly speak of a rescue situation. An unfortunate example is the case of Costa Concordia cruise ship with 4,229 people on board which, on January 13, 2012, struck the rocks and partly sank near the Giglio Island in the Tyrrhenian Sea. The supreme value to defend, i.e. the life

and integrity of the passengers, was threatened by documents containing false information about the true state of danger, which would have required urgent rescue actions.

6 VOLUNTEERING

Concerning volunteering, the wrecker should not be under the pre-existing duty to provide salvage services. The Law of Salvage does not apply where there is an official duty of the coastal guard, the Navy etc. to provide help.

The claim of the wrecker to a reward depends on the successful outcome. This premise is reflected in the traditional phrase "no cure, no pay". Until recently, the matter of the success of the salvation effort did not have any difficulties. If a vessel was in danger, it should have been immediately removed from the state of danger to safety. When the vessel had difficulties due to engine problems, it had to be towed into a port where it could benefit at least from temporary repairs. As a consequence, the wrecker was rewarded for salvage of the property.

Thus, a wrecker who had managed to tow a vessel loaded with explosive substances which had a fire on board, and thus prevented damage to the shore, would not enjoy any kind of reward if the ship exploded and sank. A similar situation existed when the wrecker has saved the ship-owner from liability on the saved property.

Major disasters involving oil carriage proved serious problems with the rules of property salvage.

Imagine a tank loaded to maximum capacity which is out of control and heading to the shore, having the potential trajectory to break and cause to a significant environmental damage. The prospect that the wrecker who prevented the oil pollution was not entitled to a reward may not be satisfactory if the ship sank during the rescue operation. No need to say that such an operation has saved the ship-owner from liability for damage caused by pollution

7 CONCLUSIONS

The maritime rescue is a maritime event with distinctive features that are based on the spirit of solidarity and on mutual human aid, which must govern the activity at sea.

The legal rules established in this regard give expression to this spirit. But they cannot offer the possibility to obviously frame the maritime rescue in a pre-existing legal typology.

Except where action is mandatory, all the other aid at sea is based on legal rescue contracts. These contracts will entitle the rescuer to a reward according to the efforts and risks assumed but also to the payment of the possible expenses related to the damage caused by the partial rescue or by the intervention that triggered more damages than those that would have occurred if the rescuer had not intervened.

Although the obligation of ships to go to the assistance of vessels in distress was enshrined both in tradition and in international treaties (such as the International Convention for the Safety of Life at Sea (SOLAS), 1974), there was, until the adoption of the International Convention on Maritime Search and Rescue (SAR), no international system covering search and rescue operations. In some areas there was a well-established organization able to provide assistance promptly and efficiently, in others there was nothing at all.

The technical requirements of the International Convention on Maritime Search and Rescue (SAR) are contained in an Annex, which was divided into five Chapters. Parties to the Convention are required to ensure that arrangements are made for the provision of adequate SAR services in their coastal waters. Parties are encouraged to enter into SAR agreements with neighbouring States involving the establishment of SAR regions, the pooling of facilities, establishment of common procedures, training and liaison visits. The Convention states that Parties should take measures to expedite entry into its territorial waters of rescue units from other Parties.

REFERENCES

Alexandrescu, A. (1997). *Accidentele de navigaţie maritimă: prevenire si cercetare*. Constanţa: Andrei Şaguna.

Anechitoae C. & Marinescu, C. 2012. *Maritime rescue.* In Galaţi: 7th International Conference on European Integration—Realities and Perspectives Danubius University May.

Anechitoae C. 2008. *Introducere în drept maritime.* Bucureşti: Bren.

Anechitoae C. 2008. *Maritime Systems Geopolitics.* Bucharest: Military Publishing House.

Anechitoae C. 2009. *Drept maritim şi portuar. Maritime and inland water law. Droit maritime et fluvial—Bibliografie selectivă. Selectiv bibliography*. Bibliographie sélective. Vol. I. Ediţie trilingvă. Bucharest: Academia Română.

Anechitoae C. 2009. *Introduction to Harbor Law.* Bucharest: Bren Publishing House.

Anechitoae C. 2015. *Legislaţie în domeniul asistenţei şi salvării pe mare. Curs universitar.* Bucureşti: Ed. Pro Universitaria.

Anechitoae C. 2015. *Politici şi geostrategii în transporturile navale. Curs universitar.* Bucureşti: Ed. Pro Universitaria.

Anechitoae C., Marinescu, C. (2012), *Maritime rescue,* în 7th International Conference on European Integration—Realities and Perspectives Danubius University May 18, 2012 – May 19, Galaţi, 2012, EIRP Proceedings, Vol. 7 (2012), pp. 328–331.

Anechitoae, C. 2005. *Convenţii internaţionale maritime. Legislaţie Maritimă. Vol. I, Vol. II.* Bucureşti: Bren.

Anechitoae, C. 2013. *Maritime and inland water law. Selective bibliography*. Germany: Lambert Academic Publishing.

Brasoveanu F. (2011), Dreptul international intre stiinte, diplomatie si politica, Bucuresti, Universul Juridic, 2011,

Brasoveanu F., (2013), *Drept internatioanl public I*, Editura Editura ProUniversitari, Bucuresti

Braşoveanu F., (2013). *Dreptul european al mediului,* Editura Pro Universitaria, Bucureşti.

Brasoveanu F., (2014), *Drept internatioanl public II*, Editura Editura ProUniversitari, Bucuresti.

Brasoveanu F., (2014). Dreptul mediului, Editura ProUniversitari, Bucuresti.

Brasoveanu F., (2016), *Institutii fundamentale ale dreptului international public*, Editura Sitech, Craiova.

Brasoveanu F., (2016). Dictionar de Dreptul mediului, Editura Sitech, Craiova.

Bureau International du Travail. [1996]. *Prevention des accidents a bord des navires en mer et dans les ports.* [2e edition].—Geneve: Bureau International du Travail.

Convention on the International Regulations for Preventing Collisions at Sea. 1972. (London, 20 October 1972).

International Safety Management (ISM) Code was adopted by the IMO through Resolution A.741 (18) on the 4th of November 1993. It is given legal force on May 1994 by the IMO's Maritime Safety Committee when it was included in the new Chapter IX of SOLAS 1974.

Manolache, O. 2001. *Contractul de salvare maritimă.* Bucureşti: ALL Beck.

Maraloi, C. 2003. *Căutarea şi salvarea pe mare.* Constanţa: Ed. Academia Navală "Mircea cel Bătrân".

Marin M., (2010) *Trading Partners in Conflict. Mediation, Arbitration, Court—the Current Settlement Possibilities*, lucrare publicata in Ovidius University Annals Economic Sciences Series, partea I, Volume X.

Mihai G.M. (2005) *Arbitraj Maritim si Comercial International—monografie.* Ed. Dobrogea, Constanta.

Mihai G.M. (2006). Dreptul Comertului International—curs universitar, Ed. Ex Ponto, Constanta.

Pandele, A. L., (2006) *Conosamentul.* Bucureşti, Ed. C. H. Beck.

Pandele, A. L., (2010). *Contracte de transport maritim* Caiet de studiu individual IFR, Ovidius University Press, Constanţa.

Pandele, A. L., (2010). *Drept maritim. Caiet de studiu individual IFR*, Ovidius University Press, Constanţa.

The law of the sea. 1991. New York: United Nations.

Tiţa—Călin, I. 2014. *Cine îi apără pe lupii de mare? Vol. I.* Constanata: Dobrogea.

Tiţa—Călin, I. 2014. *Cine îi apără pe lupii de mare? Vol. II.* Constanata: Dobrogea.

http://eur-lex.europa.eu/legal-content/RO/TXT/?uri=CELEX:52009XC0611%2801%29.

http://www.imo.org/en/About/Conventions/ListOfConventions/Pages/International-Convention-on-Maritime-Search-and-Rescue-(SAR).aspx.

Maritime Transportation and Harvesting of Sea Resources – Guedes Soares & Teixeira (Eds)
© 2018 Taylor & Francis Group, London, ISBN 978-0-8153-7993-5

An approach for an integrated assessment of maritime accidental oil spill risk and response effectiveness for the Northern Baltic Sea

F. Goerlandt, L. Lu & O.A. Valdez Banda
Marine Technology, Research Group on Maritime Risk and Safety, Aalto University, Espoo, Finland

J. Rytkönen
Finnish Environment Institute, Helsinki, Finland

ABSTRACT: Despite significant decreases in maritime accident rates and volumes of accidentally spilled oil over the past decades, major maritime oil spills remain high on the list of concerns of environmental protection agencies of coastal states, because of which appropriate response capabilities are maintained. In the Northern Baltic Sea, winter conditions are characterized by a periodical sea ice cover. Resulting navigational operations correspond to particular oil spill risks. This paper presents an approach to develop a tool for assessing accidental oil spill risks from shipping, and the effectiveness of oil spill response in different conditions in an integrated manner. As the tool itself is currently is under development, focus is on the underlying conceptual basis and the overall layout of the model. The main purpose of the tool is to support strategic decision making concerning the required capacity for spill response in sea ice conditions in different areas of the Northern Baltic, and to identify conditions for which additional risk mitigating measures should be prioritized. This article provides the background of the tool development, its underlying conceptual and methodological basis, and a general overview of the layout of the elements included in the model.

1 INTRODUCTION

Maritime oil spills are a major concern to coastal communities, as these have the potential to cause serious damage to the marine environment and lead to significant economic losses. Risks of operational and accidental spills are mitigated through specific technical and operational measures implemented through international, regional and national policy instruments (Hassler 2011).

Notwithstanding a remarkable decrease in major maritime oil pollution accidents, effective response remains an essential pillar of environmental protection. In the Baltic Sea Region, national authorities are responsible for maintaining response capacity, where cooperation is strengthened on a regional level through the Baltic Marine Environmental Protection Commission (HELCOM) and on a European level through the European Maritime Safety Agency (EMSA). Cooperation e.g. concerns information and experience exchange, establishment of common operational procedures and response strategies, and periodic training (EMSA 2004). An important aspect of ensuring adequate capacity is to assess the oil spill risks, and several guidelines for risk assessment and response plan-

ning have been published (IMO 2010, IPIECA and IOGP 2013).

Several approaches for and applications of maritime oil spill risk assessment have also been presented in the literature, often making use of spatial analyses with dedicated Geographic Information System (GIS) tools (Pelot and Plummer 2008). Methodological approaches for accidental maritime oil spill risk assessment include system simulation (Merrick et al. 2002), Bayesian Networks (Klemola et al. 2009), traffic flow analysis (Akhtar et al. 2012), fuzzy systems (Ventikos et al. 2013), and qualitative risk matrices (Lee & Jung 2013).

In the Baltic Sea, relatively much work has been performed on assessing maritime oil spill risks and related ecological impacts, see e.g. ICES (2017) for an overview. For the specific case of the maritime accident risk in the Northern Baltic Sea in winter conditions, a comprehensive sub-regional risk and response preparedness assessment has been performed (COWI 2011). However, the adopted traffic flow-based methodology does not account for the specific nature of wintertime operations. A central aspect to ensure the safety of navigation in the Northern Baltic Sea is the Finnish-Swedish Winter Navigation System, which leads to particular operations each with different risks involved. A general traffic flow-based methodology is unrealistic as it does not account for the specific nature of different operations. Moreover, it has been found that also in open water conditions, the methods closely related to the one applied in COWI (2011) leads to unreliable results (Goerlandt & Kujala 2014). Other work includes a high-level risk analysis of winter navigation based on accident data and expert judgment (Valdez Banda et al. 2015) and a more comprehensive risk management model of oil transportation in the Gulf of Finland, focusing on the effectiveness of preventive risk mitigating options (Valdez Banda et al. 2016).

This article extends the existing literature on maritime risk assessment in the Baltic Sea area by presenting ongoing work for developing a tool for assessing accidental oil spill risk and the related and response effectiveness in an integrated manner. After outlining the aims and scope (Section 2), focus is on the conceptual foundations underlying the tool, including a brief outline of the generic methodology applied in the tool development (Section 3) and the on the integrated tool's main elements and their interrelations (Section 4). A short discussion and conclusion section (Section 5) briefly reflects on the approach taken and outlines future planned work.

2 AIMS AND SCOPE

National states must balance different motives, such as protecting their marine areas and shorelines from environmental and economic damages caused by accidental spills, while minimizing the costs of investing in building and maintaining response capacity (Hassler 2011). International collaboration is an effective means to share the financial burdens of maintaining a certain response capacity. Different stakeholders consider a rationalization of the decisions about the required capacity beneficial, for which a common understanding of the oil spill risks is considered essential (COWI 2011). Similarly, it is important to gain insight into how effective the response fleet is under different oil spill and environmental conditions, which can e.g. be useful to consider specific risk mitigating actions for such conditions (Juntunen et al. 2005).

In light of the above, the aim of the tool under development is to provide comprehensive insights in the maritime oil spill risks in the Northern Baltic Sea, as well as in the effectiveness of the response fleet for handling the spill.

While the suggested approach has wider application potential, the tool under development focuses on accidental oil spill risk of tankers in the

Northern Baltic Sea area under wintertime conditions. The Northern Baltic Sea area is delineated here in line with the HELCOM response areas 'Gulf of Bothnia' and 'Gulf of Finland' (HELCOM 2015), see Figure 1, whereas winter conditions are taken here as the period during which the area is normally characterized by the potential presence of sea ice. Based on information in Jalonen et al. (2005), this is the period early November to late April. Both conditions with and without sea ice present are considered. With the aim of supporting strategical planning, the tool considers the effect of climate change to the prevailing ice conditions under different ice climate scenarios. According to HELCOM (2013), a reduction of the sea ice extent is one of the effects of climate change in the Baltic Sea area. In light of this, accounting for future response needs in ice conditions is a relevant policy issue.

The spill risk is considered as the amount of oil flowing out of a damaged ship hull in different time periods, and its subsequent drift in the sea area. The oil spill response effectiveness concerns the amount of oil which can be recovered, as well as the time in which this can be achieved.

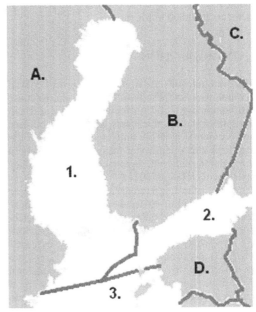

1. Bothnian Bay	A. Sweden
2. Gulf of Finland	B. Finland
3. Baltic Proper	C. Russian Federation
	D. Estonia

Figure 1. Northern Baltic Sea, based on HELCOM (2015).

3 CONCEPTUAL-METHODOLOGICAL BASIS

3.1 Risk concept and perspective

When developing a risk assessment tool, it is important to provide clarity about the risk-conceptual basis, because risk means different things to different people at different times (Meyer & Reniers 2016). In particular, there is a need for aligning the risk concept (what risk is understood to be) and the risk perspective (how risk is described) with appropriate risk analysis methods and measurement tools. Clarity on a risk-theoretical basis is important for practical reasons such as communicating risk analysis results to end-users and stakeholders (Veland & Aven 2013), and for setting the expectations of the tool's capabilities at an appropriate level. Conceptual clarity is also essential for defining how the risk assessment tool is intended to be used, which in turn relates to the essential question to what extent the tool meets its envisaged usability, i.e. to test its pragmatic validity (Goerlandt et al. 2017).

In the presented approach to oil spill risk and response effectiveness assessment, the understanding of the risk concept is in line with the recommendations in (SRA 2015, p.3): risk is the occurrence of some specified consequences of the activity and associated uncertainties. Hence, the focus is not on the frequency or probability of occurrence of specific event and on accurate estimation, as in some earlier presented oil spill risk models (COWI 2011, Akhtar et al. 2012), but more widely on the uncertainties, the evidence on which the expressed uncertainties are based, and on the potential for surprises. Correspondingly, based also on SRA (2015), the risk perspective can be summarized as follows:

$$R \sim (C, Q, SoE, D \mid BK) \qquad (1)$$

where C are the consequences of the activity, Q a quantified measure of the uncertainties associated with this consequence, SoE a qualitative description of the strength of evidence for the quantified uncertainty measure, and D a qualitative description of the deviations compared to the quantified risk description (corresponding to the potential for surprises). BK is the background knowledge, to which the risk description is conditional.

Correspondingly, the risk assessment tool under development is primarily intended to serve as input stakeholder deliberations, as a basis for reflection and discussion. The model serves to provide insights about the system under study as well as the assumptions upon which the tool is based, which have been proposed as appropriate model uses in contexts in which accurate prediction is not possible (Hodges 1991).

3.2 BNs as a risk modeling tool in light of available evidence

Understanding risk descriptions essentially as expressions of uncertainty of an assessor or a group of assessors in light of the available background knowledge, the developed tool for supporting risk assessment applies Bayesian Networks (BNs) as a modelling tool. Understanding probabilities as an expression of uncertainty of an assessor in light of available evidence as in Lindley (2006), BNs allow a coherent treatment of uncertainties about the specified consequences, as a type of expression of argument presented in a quantified form (Watson 1994). For risk modeling, BNs have a number of favorable characteristics: including i) the ability to contextualize the occurrence of specific events through situational factors (observable aspects of the studied system), ii) integration of different types of evidence through knowledge-based probabilities, iii) relatively simple incorporation of alternative hypotheses about phenomena which are not well understood, and iv) ready availability of tools for applying sensitivity analysis.

In mathematical terms, BNs represent a class of probabilistic graphical models, defined as a pair $\Delta = \{G(V,A), P\}$. (Koller & Friedman, 2009), where $G(V,A)$ is the graphical component and P the probabilistic component of the model. $G(V,A)$ is in the form of a directed acyclic graph (DAG), where the nodes represent the variables $V = \{V_1, \dots, V_n\}$ and the arcs (A) represent the conditional (in) dependence relationships between these. P consists of a set of conditional probability tables (CPTs) $P(V_i \mid Pa(V_i))$ for each variable V_i, $i = 1, \dots, n$ in the network. $Pa(V_i)$ signifies the set of parents of V_i in G: $Pa(V_i) = \{Y \in V \mid (Y, V_i)\}$. A BN encodes a factorization of the joint probability distribution (JDP) over all variables in V:

$$P(V) = \prod_{i=1}^{n} P(V_i \mid Pa(V_i)) \qquad (1)$$

BNs are relatively widely used tools for risk modeling (Fenton & Neil, 2012). Nevertheless, care must be taken that there is evidence for the causal nature of the connections between model variables, which is due to the complexity of complex systems is not a straightforward issue. This is also a reason why an uncertainty-based risk perspective, which apart from the quantified uncertainty measures also focuses on the strength of evidence and the potential for surprises, see Section 3.1, is appropriate for activities in complex systems (Johansen & Rausand 2014), such as oil spills from accidents in maritime operations, and the response to these. Considering that the risk description is an expression of uncertainty of an assessor or a group of assessors, the evidence for the existence of causal connections between system aspects (and correspondingly, model elements) is essential for the validity of expert judgments (Rae & Alexander 2017).

3.3 Tools for expressing the strength of evidence and the potential for surprises

The risk perspective of Section 3.1 also requires tools to qualitatively assess the strength of evidence and the deviations compared to the risk quantifications, and communicate these to stakeholders and decision makers.

Concerning the assessment of the strength of evidence, several proposals have been made both on how to rate and visually present these, see e.g. Goerlandt & Reniers (2016) for a state of the art review and Berner & Flage (2016) for another recent contribution. For the current approach, the strength of evidence (SoE) assessment scheme by Goerlandt and Reniers (2016) is adopted, see Table 1 to Table 4.

Concerning the qualitative assessment of the deviations compared to the quantified risk description, a concept proposed by Aven (2013) is applied. The 'assumption deviation risk' is a qualitative assessment of the possible deviation from the quantified results under the base case assumptions. In the practical application of the presented approach, a scheme proposed by Goerlandt & Montewka (2015) is applied. This scheme considers following aspects:

1. Magnitude of deviation
2. Direction of deviation
3. Consequence range where deviation has effect
4. Strength of justification

Table 1. Criteria for SoE: data.

Evidential characteristic	Strong	Weak
Quality	Low number of errors	High number of errors
	High recording accuracy	Low recording accuracy
	Data source very reliable	Data source not very reliable
Amount	Much relevant data	Little data available

Table 2. Criteria for SoE: models.

Evidential	Strong	Weak characteristic
Empirical validation	Many experimental tests performed; Existing experimental tests agree well with model output	No/little experimental confirmation available; Existing experimental tests show large discrepancy with model output
Theoretical viability	Model expected to lead to good predictions	Model expected to lead to poor predictions

Table 3. Criteria for SoE: judgments.

Strong	Broad intersubjectivity: more than 75% of the peers support the judgment
Medium	Moderate intersubjectivity: between 25% and 75% of peers support the judgment
Weak	Predominantly subjective: less than 25% of the peers support the judgment

Table 4. Criteria for SoE: assumptions.

Agreement among peers	
Strong	Many (more that 75%) would have made the same assumption
Medium	Several (between 25% and 75%) would have made the same assumption
Weak	Few (less than 25%) would have made the same assumption
Influence on results	
Strong	The assumption has only local influence
Medium	The assumption has wider influence in the analysis
Weak	The assumption greatly determines the results of the analysis

4 OVERVIEW OF THE INTEGRATED TOOL

4.1 General overview

With the overall aims of assessing the oil spill risk from tanker traffic in the Northern Baltic Sea under winter conditions, and the effectiveness of the response to these spills, see Section 2, the main elements of the model are schematically shown in Figure 2.

The model provides the oil spill risk and response effectiveness for different locations in the Northern Baltic Sea during winter conditions, for response areas defined in Section 2. It includes information about sea ice conditions under different climate

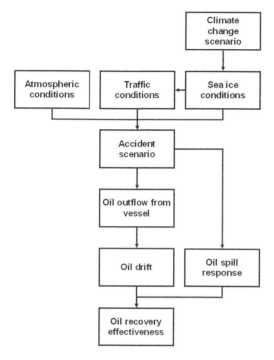

Figure 2. Main elements of the overall integrated oil spill risk and response effectiveness tool.

change scenarios. Together with traffic conditions and atmospheric conditions, these form the basis of defining accident scenarios. For these scenarios, the oil outflow from the vessels is assessed, and how the oil drifts in the sea area. Together with an assessment of the oil spill response, which depends also on the accident scenario (e.g. the spill location, and atmospheric and sea ice conditions), the oil recovery effectiveness is assessed.

4.2 Evidence for the different model elements

In this section, a very brief overview is given for the evidence which is available or which is being gathered for developing the integrated model. As the main intention is to provide a broad overview of the evidence underlying the model development, brevity is sought, For details, the reader is referred to appropriate references.

4.2.1 Climate change and sea ice scenarios
According to HELCOM (2013), the largest changes in the temperature of the sea surface are projected to occur in the Bothnian Bay and Bothnian Sea during summer, and in the Gulf of Finland in the spring. Near the end of the 21st century, summer sea surface temperature increases between 2 °C and 4 °C are projected, with the latter more likely

for the Northern Baltic Sea areas. Furthermore, both the extent of sea ice cover and the length of the ice season are expected to diminish.

As evidence for the strategic oil spill risk and response model, different representative climate change scenarios are simulated in a coupled ice-ocean model system called Nemo-Nordic, see Höglund et al. (2017). These simulations provide insight in the projected long-term monthly and decadal annual ice extent in the Baltic Sea, and give distributions of sea ice thickness and concentration.

4.2.2 Traffic conditions
Traffic conditions concern both the general traffic flows, considering which vessels operate in which sea areas, as well as the operational characteristics of the ship navigation in different sea areas and ice conditions. The former issue concerns spatial information concerning the vessels (types, sizes, cargos), whereas the latter relates to which operational mode is selected (independent navigation, icebreaker assistance such as convoy, escort, or towing). Different vessel types and sizes pose different accidental oil spill risks (McAllister et al. 2003), and also different winter navigation operations involve different risk levels (Valdez Banda et al. 2016).

Traffic conditions occurring in the wintertime maritime traffic system are investigated based on an elaborate database which integrates data from the Automatic Information System (AIS) and sea ice data from the (Lensu & Kokkonen 2017). Ongoing analyses focus on which operation types are found in different parts of the Northern Baltic Sea, and on the vessel types and sizes in these operations.

4.2.3 Atmospheric conditions and accident scenarios
The circumstances under which the accident occur can have an important influence on the effectiveness of the response operations (Kang et al. 2016).

For the model under development, the atmospheric conditions of the accident scenarios are defined based on an accident analysis of winter navigation accidents in the Northern Baltic Sea (Goerlandt et al. 2017). This analysis integrates data from accident data bases with data from the Automatic Information System, sea-ice data, atmospheric data and other situational data to provide insights into the context under which the accidents have occurred. This includes the atmospheric conditions during the accidents, and the operational characteristics.

This analysis of historic accident cases is the most comprehensive source of evidence for describing the accident scenarios. Compared to open water conditions, significantly more accidents occur in winter conditions, see Valdez Banda et al. (2015), so patterns can be identified in the typical contexts under which these occur. Nevertheless, maritime accidents are rare events and historic accidents are limited as a sources. Hence, the accident scenarios considered in the developed model will also consider information from the traffic system analysis (Section 4.2.2) and expert judgment.

4.2.4 Oil outflow from vessel

A central issue in the modeling effort is the amount of oil outflow from the damaged vessel. In state-of-the-art models, the influence of the ice conditions either not accounted for, or included based on a number of simple assumptions. Similarly, the dynamics of the oil outflow (i.e. the period during which the oil outflow takes place) is not accounted for, and a simple assumption is made that all oil is spilled instantaneously (Valdez Banda et al. 2016).

Due to the specific nature of collisions and groundings in, and the relevance of the dynamics of the oil outflow for response operations, the model under development improves the state of the art in oil outflow modeling on these two issues. First, the effect of ice conditions on the dynamics of ship-ship impact and the associated damage to the ship hull is modeled (Nelis et al. 2015, Tabri et al. 2016). Second, the dynamics of the oil outflow from hull damages is modeled (Kollo et al. 2017). Integrating these models with a database of tanker designs, similarly as in Goerlandt et al. (2015), provides the evidence for the oil outflow for the considered accident scenarios (Section 4.2.3).

4.2.5 Oil drift

Once the oil flows out from the damaged vessel, it drifts in the sea, and various processes take place by with the oil spreads, weathers and undergoes physical change.

For assessing the fate of the oil in the sea area, the model under development applies Seatrack Web, the HELCOM oil drift forecasting system. This tool is used as an operational forecasting spill for use in response operations, for detecting the vessel which has caused the pollution, and as a tool for simulating oil spill scenarios for response planning (Kostianoy et al. 2014). The model also has a basic functionality for assessing the drift in ice conditions, which is however improved especially in relation to how the oil is trapped in deformed ice components due to sea ice drift.

The evidence for the tool development is based on a large number of simulated oil spill scenarios, defined based on the accident scenarios (Section 4.2.3) and the models for oil outflow from the vessel (Section 4.2.4).

4.2.6 Oil spill response and recovery effectiveness

The oil spill response and recovery effectiveness is modeled based on information about the vessels and their equipment, expert information concerning the operational aspects of the response operations, and evidence concerning the trafficability of the response vessels in different ice conditions. Insights are also obtained from literature sources, e.g. IPIECA and IOGP (2013) and Kang et al. (2016).

A similar approach is adopted as in Juntunen et al. (2005), where the recovery effectiveness of individual response vessels is modeled conditional to key aspects of the oil spill (oil type, amount of spilled oil,…), the response vessel (speed and time to reach the accident location, equipment, storage capacity,…), and the environmental conditions under which the response operation takes place.

5 DISCUSSION AND CONCLUSIONS

In this paper, a broad overview is given of a tool under development to assess the oil spill risk and response effectiveness in the Northern Baltic Sea. Focus is on oil spills from tankers in winter conditions, but the tool design is flexible in the sense that it can be extended to other conditions as well.

Apart from providing an overview of the tool and the evidence available or under development for its construction, attention is also given to its risk-conceptual and methodological basis. In particular, attention is given to tools for assessing the strength of evidence underlying the tool construction, and tools for assessing the potential for surprises compared to the model results. Clarity about the underlying approach to risk analysis is important as presently many approaches to risk analysis co-exist, Lack of clarity can lead to misguided expectations about what kind of decision support the tool offers, which can also lead to challenges in risk communication.

The tool is developed as a stand-alone model, but it is built such that it can in principle be linked with software facilitating easy updating of the evidence underlying the model. For instance, traffic analysis and oil drift tools could in the future be coupled to the tool through specific development work.

The first aim of this endeavor is to provide response authorities and other national and international stakeholders tools for assessing the oil spill risks from maritime transportation in wintertime conditions, based on best available evidence. Such an assessment on a national and regional level can provide input for decisions related to the required oil spill response capacity for these conditions. Significantly, whereas state-of-the-art analyses only provide estimates of the total accidental outflow, based on conservative assumptions, the developed tool more accurately accounts for the dynamics of the oil outflow, providing insight in the relation between the size and duration of the spill. Also, given that climate change is expected to have significant impacts on the extent of sea ice in the Northern Baltic Sea, the tool accounts for different climate change scenarios based on which the required response capacity in sea ice conditions can be assessed.

The second aim of the model is providing insights into the oil recovery effectiveness under different environmental conditions. This information shows how well the response system is expected to work in case of a spill. Importantly, insights can be obtained about the limitations of recovery. Identifying contexts in which recovery is not very effective, may be used to consider specific risk mitigation measures under these conditions, such as imposition of traffic restrictions or ice-breaker assistance.

The presented tool under development goes beyond the state-of-art in many respects. Nevertheless, once the integrated tool is developed, it is important to test the tool and evaluate its functionality. This concerns both the design of the tool and the plausibility of the outcomes of specific scenarios, as well as its performance in actual user and stakeholder contexts. These aspects are left for future research.

ACKNOWLEDGEMENTS

The work in this paper has been performed as part of the BONUS STORMWINDS project. Funding has been received from BONUS (Art. 185), funded jointly from the European Union's Seventh Programme for research, technological development and demonstration, and from the Academy of Finland.

REFERENCES

Akhtar J., Bjørnskau T., Jean-Hansen V. 2012. Oil spill risk analysis of routeing heavy ship traffic in Norwegian waters. *WMU Journal of Maritime Affairs* 11:233–247.

Aven T. 2013. Practical implications of the new risk perspectives. *Reliability Engineering and System Safety* 115:136–145.

Berner C.L., Flage R. 2016. Comparing and integrating the NUSAP notational scheme with an uncertainty based risk perspective. *Reliability Engineering and System Safety* 156:185–194.

COWI. 2011. BRISK: Sub-regional risk of spill of oil and hazardous substances in the Baltic Sea. COWI A/S, Kongens Lyngby, Denmark, Project no. P-070618

EMSA. 2004. Action Plan for Oil Pollution Preparedness and Response. European Maritime Safety Agency, Lisbon, Portugal, 67 p.

Fenton, N., and Neil, M. 2012. Risk assessment and decision analysis with Bayesian networks. CRC Press.

Goerlandt F., Khakzad N., Reniers G. 2017. Validity and validation of safety-related quantitative risk analysis: A review. *Safety Science*, doi: 10.1016/j.ssci.2016.08.023.

Goerlandt F., Kujala P. 2014. On the reliability and validity of ship-ship collision risk analysis in light of different perspectives on risk. *Safety Science* 62:348–365.

Goerlandt F., Montewka J. 2015. A framework for risk analysis of maritime transportation systems: A case study for oil spill from tankers in a ship-ship collision. *Safety Science* 76:42–66.

Goerlandt F., Reniers G. 2016. On the assessment of uncertainty in risk diagrams. *Safety Science* 84:67–77.

Hassler B. 2011. Accidental versus operational oil spills from shipping in the Baltic Sea: risk governance and management strategies. *Ambio* 40(2):170–178.

HELCOM. 2013. Climate change in the Baltic Sea Area HELCOM thematic assessment in 2013. Baltic Sea Environment Proceedings No. 137. Baltic Marine Environment Protection Commission (Helsinki Commission).

HELCOM. 2015. HELCOM Response Sub-regions. Baltic Marine Environment Protection Commission (Helsinki Commission), Response Group.

Hodges J. S. 1991. Six (or so) things you can do with a bad model. *Operations Research* 39(3):355–365.

Höglund A., Pemberton P., Hordoir R., Schimanke S. 2017. Ice conditions for maritime traffic in the Baltic Sea in future climate. *Boreal Environment Research*, submitted.

ICES. 2017. Interim report on the Working Group on Risks of Maritime Activities in the Baltic Sea (WGMABS), 7–11 November 2016, ICES Headquarters, Copenhagen, Denmark. ICES CM 2016/SSGEPI:14. 12 pp.

IMO. 2010. Manual on Oil Spill Risk Evaluation and Assessment of Response Preparedness. IMO Publishing.

IPIECA, IOGP. 2013. Oil spill risk assessment and response planning for offshore installations. The global oil and gas industry association for environmental and social issues and International Association of Oil & Gas Producers Oil Spill Response Joint Industry Project.

Jalonen R., Riska K., Hänninen S. 2005. A Preliminary Risk Analysis of Winter Navigation in the Baltic Sea. *Winter Navigation Research Board, Report No. 57*.

Johansen I.J., Rausand M. 2014. Defining complexity for risk assessment of sociotechnical systems: A conceptual framework. *Proceedings of the Institution of Mechanical Engineers, Part O: Journal of Risk and Reliability* 228(3):272–290.

Juntunen T., Rosqvist T., Rytkönen J., Kuikka S. 2005. How to Model the Oil Combating Technologies and Their Impacts on Ecosystem: A Bayesian Networks Application in the Baltic Sea; *ICES Council Meeting documents*; International Council for the Exploration of the Sea, Copenhagen, Denmark.

Kang J., Zhang J., Bai Y. 2016. Modeling and evaluation of the oil-spill emergency response capability based on linguistic variables. *Marine Pollution Bulletin* 113(1–2):293–301.

Klemola E., Kuronen J., Kalli J., Arola T., Hänninen M., Lehikoinen A., Kuikka S., Kujala P., Tapaninen U. 2009. A cross-disciplinary approach to minimising the risks of maritime transport in the Gulf of Finland. *World Review of Intermodal Transportation Research* 2(4):343–363.

Koller D., Friedman N. 2009. Probabilistic graphical models: principles and techniques (1st ed.). The MIT Press.

Kollo M., Laanearu J., Tabri K. 2017. Hydraulic modelling of oil spill through submerged orifices in damaged ship hulls. *Ocean Engineering* 130:358–397.

Lee M., Jung J-Y. 2013. Risk assessment and national measure plan for oil and HNS spill accidents near Korea. *Marine Pollution Bulletin* 73(1):339–344.

Lensu M., Kokkonen I. 2017. Inventory of ice performance for Baltic IA super traffic 2007–2016. *Winter Navigation Research Board Research Report*, 54p. Finnish Transport Safety Agency and Swedish Maritime Administration.

Lindley D.V. 2006. Understanding uncertainty. John Wiley & Sons, Inc., Hoboken, New Jersey, USA.

McAllister T., Rekart C., Michel K. 2003. Evaluation of accidental oil spills from bunker tanks (Phase I). *Technical Report No. SSC-424*. Ship Structures Committee, Washington DC, USA.

Meyer T., Reniers G. 2016. Engineering risk management. De Gruyter Graduate (2nd ed.), Berlin, Germany.

Merrick J.R.W., van Dorp J.R., Mazzuchi T., Harraled J.R., Spahn J.E., Grabowski M. 2002. *Interfaces* 32(6):25–40.

Nelis S., Tabri K., Kujala P. 2015. Interaction of ice force in ship-ship collision. Proceedings of the ASME 34th International Conference on Ocean, Offshore and Arctic Engineering OMAE2015, 31.05-05.06.2015, OMAE2015-41351.

Pelot R., Plummer L. 2008. Spatial analysis of traffic and risks in the coastal zone. *Journal of Coastal Conservation* 11:201–207.

Rae A., Alexander R. 2017. Forecasts or fortune-telling: When are expert judgments of safety risk valid? *Safety Science*, doi: 10.1016/j.ssci.2017.02.018

Tabri K., Goerlandt F., Kujala P. 2016. Influence of compressive ice force in bow-to-aft ship collision. Proceedings of the 7th International Conference on Collision and Grounding of Ships, Ulsan, South-Korea, p. 131–137.

Valdez Banda O.A., Goerlandt F., Montewka J., Kujala P. 2015. A risk analysis of winter navigation in Finnish sea areas. *Accident Analysis and Prevention* 79:100–116.

Valdez Banda O.A., Goerlandt F., Kuzmin V., Kujala P., Montewka J. 2016. Risk management model of winter navigation operations. *Marine Pollution Bulletin* 108(1–2):242–262.

Veland H., Aven T. 2013. Risk communication in the light of different risk perspectives. *Reliability Engineering & System Safety* 110:34–40.

Ventikos N.P., Louzis K, Koimtzoglou A. 2013. The shipwrecks in Greece are going fuzzy: A study for the potential of oil pollution from shipwrecks in Greek waters. *Human and Ecological Risk Assessment: An International Journal*, 19(2):462–491.

Watson S.R. 1994. The meaning of probability in probabilistic safety analysis. *Reliability Engineering & System Safety* 45(3):261–269.

Maritime Transportation and Harvesting of Sea Resources – Guedes Soares & Teixeira (Eds)
© 2018 Taylor & Francis Group, London, ISBN 978-0-8153-7993-5

Causation analysis of ship collision accidents using a Bayesian Network approach

S. Gong, J. Mou & X. Zhang
School of Navigation, Wuhan University of Technology, Hubei, China

ABSTRACT: Water traffic accidents not only cause serious economic losses, but also lead to environmental pollution and personal casualties. According to the report of EMSA (European Maritime Safety Agency), collision is one of the prominent casualties of marine incidents, and about 67% of the accidents are related to human errors (EMSA, 2015). To probe the mechanism of ship collision, this paper analyzed 81 typical ship collision accident reports during the past 10 years in China. Firstly, the Bayesian Network (BN) model on ship collision is developed based on the accidents data. Secondly, the significant types of human error are analyzed by performing a sensitivity analysis. The validity of the model is demonstrated by applying practical accidents data later. And, last of all, the most probable causation chain leading to ship collision are gained by the backward inference of present BN model and some relevant recommendations have also been put forward.

1 INTRODUCTIONS

Maritime transport is the backbone of the international trade and global economy. Around 80% of global trade by volume are carried by sea and are handled by ports worldwide (United Nations Conference on Trade and Development, UNCTAD, 2014). However, the rapid development of the maritime trade and the continuous growth of vessel traffic have also lead to the increase of marine accidents, and most of the accidents are due to human errors (Fortland,2014; Hetherington et al.,2006; Corovic et al.,2013). IMO (International Maritime Organization) whose aim is to create a regulatory framework for safety and security of crew and ship, struggles to decrease human errors on-board ships by requiring a set of rules and activities (IMO, 2013). However, marine accidents caused by human error are still on-going and has not reduced to desired level yet (Gaonkar et al., 2011; Noroozi et al., 2014). And according to the annual overview report of EMSA, collision is one of the prominent casualties of marine incidents (EMSA, 2015). Moreover, most collision accidents are mainly low probability-high consequence in nature (Zhang and Thai, 2016). It is therefore of great importance to probe the mechanism of collision accidents and find out the most probable causation chain of the accidents, especially the human erroneous actions related, which can inform countermeasure design to prevent and control such occurrences.

Many researchers have dedicated great effort to understanding how and why accidents happen.

As a result, lots of theories and models of accidents causation have been developed over the past decades.

The Heinrich Domino Model describes an accident as a linear one-by-one progression that occurs in a fixed and logical pattern (Weaver, 1971). The Swiss Cheese Model, as an accident investigation model, involves the sequences of decisions and actions that resulted in the accident and shows against each step, the possible recommendations from investigations (Kletz, 2001). The Loss Causation Model is organized in such a way that it establishes a hierarchy of events relative to their respective precursor conditions (Kujath et al., 2010). The SHIPP Model is aimed to detect hazards, assess them, forecast, avert their occurrences, and continue monitoring the occurrences. The model relies on process history, accident precursor information, and accident causation modeling (Rathnayaka et al., 2011). The conceptual model is designed to analyze marine accidents (Mullai et al., 2011). The Fuzzy Fault Tree Analysis (FFTA) model is applied to analysis the root cause of Arctic Marine accidents from 1993 to 2011(Kum et al., 2015). A new method to analyze accidents has been proposed by combining the accident analysis methodology of CASMET with the Technique for the Retrospective and predictive Analysis of Cognitive Errors (TRACEr) to identify the main task errors, technical equipment and cognitive domains involved in the accidents (Graziano et al., 2016). The evidence-based and expert-supported approach is proposed to structure a model assessing the probability of ship-grounding accidents (Mazaheri et al., 2016).

Other popular models of accident causation include the SHEL Model, the CFAC, and MORT (Lehto and Salvendy, 1991). While these accident models are detailed, they are complex and take so much time to build (Afenyo et al., 2017). As a first step to decision making, simpler and time-efficient methodologies are required. The proposed paper focused on presenting a methodology that is simple and easy to execute.

The proposed method aims at probing the mechanism of collision accidents and find out the most probable causation chain of the accidents from past accident data using a Bayesian Network (BN) based methodology. In this methodology, the prior probabilities can be updated as new information becomes available. Potential contributory factors can be identified and subsequently controlled with relevant safety measures. The use of a BN provides provides the researchers with a great tool to represent multivariate state of causal factors compared to binary states in a tool like the Fault Tree (Afenyo et al.,2017).

The rest of this paper is organized as described following. Section 2 gives a brief introduction to Bayesian Network. Section 3 presents the proposed Bayesian Network modelling on ship collision. The validation and application of the proposed BN model are described in Section 4. A discussion of the results is presented in Section 5 and conclusions in Section 6.

2 BAYESIAN NETWORK

The Bayesian Network (BN) is a probabilistic graphical based network, mainly for describing knowledge uncertainty (Martin et al., 2009; Jensen et al., 2009). BN follows a Direct Acyclic Graph (DAG) structure and is made up of nodes and arrows. The node is representative of random variables while the arrows are the probabilistic relationships between these variables. The relationships in the BN model describes dependency among the variables. In its simplest form, it is represented as two nodes which depict the random variables. These nodes are connected by directed arrows. A line from y to x depicts dependence between the two variables. A simple interpretation of this connection is that the variable y has an impact on x. x is called the child-node of y, while y is the parent-node of x.

The DAG is basically the qualitative description of the BN. The quantitative relationship is described using the conditional probability table (CPT) for discrete random variables. The basis of the Bayesian Network is the Bayes theory, which is expressed as:

$$P(A|E) = \frac{P(E \mid A)P(A)}{P(E)} \tag{1}$$

Where P(A|E) is referred to the posterior probability when given an evidence of E, P(E|A) is the likelihood which represents how likely the evidence is true, P(A) is the probability of A before observing the evidence of E, and P(E) is normalization factor (Zhang and Thai,2016).

To describe this mathematically, a BN, designated as B here, can be defined as a DAG that depicts a joint probability distribution (JPD), over the variables V. B is defined by the pair <G, Θ>. G is the DAG with nodes Y_1, Y_2, Y_3............Y_n with the arrows representing the dependency between the variables. Θ describes the set of parameters of the network. The set is made up of the parameter $\theta_{(Y_i\pi_i)} = P_B(Y_i\pi_i)$, that is for realizing each of y_i of Y_i conditioned on π_i, which are the parameters of Y_i in G. Therefore, B defines a special Joint Probability Distribution (JPD) over V (Ben-Gal, 2007). This relationship is shown as Eq. (2).

$$P_B = (Y_1, Y_2, Y_3......Y_n) = $$
$$\prod_{i=1}^{n} P_B(Y_i\pi_i) = \prod_{i=1}^{n} \theta_{(Y_i\pi_i)} \tag{2}$$

3 BAYESIAN NETWORK MODELING ON SHIP COLLISION

The mechanism of ship collision may be very complicated, and many different factors are responsible for the causes of accidents. However, almost most of the accidents are due to human errors (Fortland, 2014; Hetherington et al., 2006; Corovic et al., 2013). The focus of the present study is to probe the mechanism of collision accidents and find out the most probable causation chain of the accidents, especially the human errors related, by the proposed BN model. Fig.1 represents the framework of the present methodology and the detailed modelling process is described from Section 3.1–3.3.

3.1 Determination of network nodes

When a vessel is involved in a collision situation with another vessel, the detailed collision avoidance decision process must be very complex. However, almost all the process includes the following three stages (Fig.2).

According to the above process of collision avoidance decision, and after reading and analyzing all the collected accidents reports, the main type of human errors is obtained, which means the network nodes are also determined.

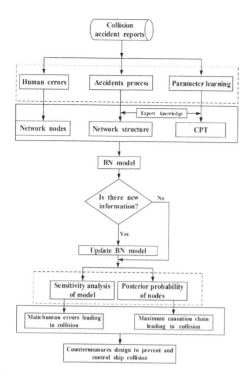

Figure 1.   The present framework of methodology.

Figure 2.   The process of collision avoidance decision.

For convenience of modelling, we use different alphanumeric characters, as network nodes, to represent different types of human errors (see following).

Each node of the network has two states: "True" or "False". "True" indicates the incident represented by the node occurred, while "False" is on the contrary, indicates the incident does not happen.

Main types of human errors in different stages:
Information
Perception:

1. Improper lookout-----S1
2. Improper duty-----S2
3. Improper use of Radar and ARPA-----S3
4. Improper signal exhibition-----S4
5. Improper use of VHF/AIS-----S5

Decision and Judgement:

1. Assumption made on scanty information-----D1
2. The inadequate risk assessment-----D2
3. Failure to assess the collision risk-----D3
4. Fail to detect target ship at early time-----D4
5. Uncertain with the situation-----D5

Take Action:

1. Not proceed at safe speed-----A1
2. Fail to take measures in time-----A2
3. Improper action of avoidance-----A3

3.2   Determination of network structure

After the network nodes are determined, the network structure should be developed. In the present study, this process is conducted by two steps: (1) Analyzing the causation chain of every single accident; (2) Integrate all the available causation chain of every accident. For instance, the causation chain of accident A may be seen in Fig.3, and the causation chain of accident B may be seen in Fig.4.

When the causation chains of accident A and B are integrated, the following network structure are obtained (see Fig.5).

And the present BN network structure can be determined after integrating all the single causation chain of 81 typical collision accidents (see Fig.6).

Figure 3.   The causation chain of accident A.

Figure 4.   The causation chain of accident B.

Figure 5.   The network structure obtained by integrating causation chain of accident A and B.

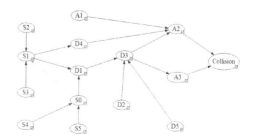

Figure 6.   The network structure of the present BN model.

3.3 Determination of conditional probability tables(CPT)

To determine the CPT of network nodes means to obtain the prior probability of nodes. In general, there are roughly two approaches to determine the CPT of the network nodes: Expert knowledge or literature-based and machine learning-based approach. In the present study, we prefer to combine the machine learning-based approach with the expert knowledge-based approach to obtain the CPT from the data sets. The state matrix of the basic nodes in every single accident is obtained by the statistic of the accidents data firstly; Then, the CPT of each network node is automatically determined by the execution of the parameter learning-based algorithm.

4 VALIDATION AND APPLICATION OF THE PROPOSED BN MODEL

In the above section, the detailed Bayesian Network is developed. In this section, the validation of the proposed BN model is demonstrated by practical accidents firstly; Then, the significant human errors leading to collision is distinguished by sensitivity analysis and the most probable causation chain of the accidents, especially the human errors related, is derived by the application of the present BN model later.

4.1 Validation of the BN model

To verify the validation of the present BN model, 20 typical and detailed collision accidents are selected and applied in the BN model.

Here we take an accident happened in the Strait of Dover on October 29th in 2008 as an example.

After reading and analyzing the accident report. First, we recognized the main human errors leading to the accident, such as "Improper information communication", "Improper duty", "Improper lookout", "Improper use of Radar and ARPA", "Improper use of AIS", "The inadequate risk assessment", "Failure to assess the collision risk", "Fail to detect target ship at early time" and "Fail to take measures in time".

Then we set all the state of the network nodes corresponded with the above human errors to be "True", and check the posterior probability of the collision occurrence. As seen in Fig.7, when all the mentioned human errors occurred, the probability of the collision occurrence will be 96% which means there is little chance that the collision can be avoided. After analyzing another 19 accidents reports by using the present BN model, it is found that when all the human errors derived from the

practical reports occurred, there are 6 accidents with a probability above 96% to happen, 12 with the probability between 90% and 95%, and the probability of the other accidents will happen is all between 85 and 90%, which demonstrates the present model can be used to model the causation chain of ship collision with considerable reliability.

4.2 Sensitivity analysis

As discussed in the Section 3, the sensitivity analysis is performed to identify the most critical variables or human errors leading to collision in the present BN model. In our study, we use the basis for measure the degree of sensitivity is the change ratio of parameters, which can be described in Eq. (3).

$$\left| \frac{\text{Posterior Probability-Prior Probability}}{\text{Prior Probability}} \right| \quad (3)$$

According to Eq. (3), the results of the sensitivity analysis are shown in Fig. 8. The probabilities of the most sensitive factors are higher relative to others. Only the most significant changes are shown in Fig.8.

4.3 Causation analysis of collision

In Section 4.1 and Section 4.2, the validation of the present BN model has been demonstrated, and the most significant human errors leading to collision have also been identified. In this section, the most probable causation chain will be derived by

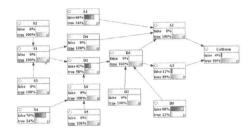

Figure 7. The validation of the present model.

Figure 8. Change ratio of the critical human errors.

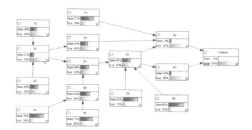

Figure 9. The inference graph of the present BN model.

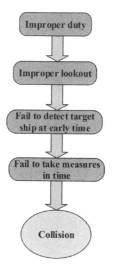

Figure 10. The most probable causation chain of the ship collision.

the back reasoning of the proposed BN model, which can be interpreted as the process to identify the child-nodes with maximum posterior probability when the top event is set to 100% (see Fig.9).

It can be easily found that when the probability of the top event(Collision) is set to 100%, the child- node of the father-node with the maximum probability is "A2". And when "A2" is determined, the child-node of "A2" with the maximum probability is also obtained (see "D4").

In this way, the most probable causation chain of ship collision derived from the present BN model can be abstracted to: S2→S1→D4→A2→Collision. When interpreted as semantic information, the most probable causation chain can be described in Fig.10

5 DISCUSSION

Sensitivity analysis is technique that can help validate the probability parameters of a Bayesian network. This is done by investigating the effect of small changes in numerical parameters (i.e., probabilities) on the output parameters (e.g., posterior probabilities). Highly sensitive parameters affect the reasoning results more significantly. Identifying them allows for a directed allocation of effort to obtain accurate results of a Bayesian network model.

Fig.8 shows the results of the sensitivity analysis for ship collision. The variables with significant changes include fail to take measures in time, improper action of avoidance, the inadequate risk assessment, assumption made on scanty information, not proceed at safe speed, failure to assess the collision risk. The changes of other variables may be relative minor; however, it does not mean these variables are of little significance. In fact, almost all the collisions begin with these variables, such as improper duty and improper signal exhibition, which can be seen in Fig.10. When the ship is sailing at the sea, almost every single error, especially human error related, may have a Domino effect on ship collision. Despite the diversity and complexity of the causes leading to ship collision, however, generally doing the following may reduce the probability of occurrence of the accident and hence the associated consequences:(1) full understand and adherence to the navigational rules and regulations, (2) giving the crew good training on the comprehensive quality, (3) encourage and strengthen the BRM (Bridge Resource Manage). It should be noted that while these recommendations are general, specific precautions need to be taken to reduce the probability of the occurrence of accidents.

6 CONCLUSIONS

In this paper, Bayesian Network (BN) based methodology for the causation chain analysis of ship collision has been presented. The use of Bayesian Network offers analysts the opportunity to model interdependencies among the casual factors, which is not possible in conventional methods like the Fault Tree. To develop the present BN model, the network node and network structure are determined by analyzing and abstracting the 81 typical and detailed collision accident reports firstly. Secondly, the CPT (conditional probability table) is determined by the combination of parameter learning based approach and expert knowledge-based approach. The validation of the present model is demonstrated by 20 practical accident reports later. And by sensitivity analysis, the critical human errors leading to ship collision are obtained, which include fail to take measures in time, improper action of avoidance, the inadequate risk assessment, assumption made on scanty information, not proceed at safe speed, failure to assess the collision risk, improper duty and improper signal exhibition. In addition, the most probable causation chain is derived from the present BN model by backward inference, which can be seen in Fig.10. Last of all, to reduce the probability of occurrence of marine ship collision, some general recommendations have been proposed.

However, there are still many aspects which need to be improved about this paper. Firstly, the network structure of the present BN model is determined by the expert knowledge in the field. This might cause bias during the procedure. Secondly, due to the small scale of the accident data, there may be some uncertainty in the process of BN modelling. In the further study, much work will be done in these aspects to improve the present BN model.

ACKNOWLEDGEMENTS

The work presented in this paper is financially supported by the National Natural Science Fund of China (Grant No.51579201).

The models described in this paper were created using the GeNIe Modeler, available free of charge for academic research and teaching use from BayesFusion, LLC, http://www.bayesfusion.com/.

REFERENCES

Afenyo M, Khan F, Veitch B, et al. (2017) Arctic shipping accident scenario analysis using Bayesian Network approach [J]. Ocean Engineering, 133:224–230.

Ben-Gal, I., (2007) Bayesian Networks. In: Ruggeri, F., Faltin, F., Kenett, R. (Eds.), Encyclopedia of Statistics in Quality and Reliability. John Wiley and Sons, New York.

Corovic, B., Djurovic, P. (2013) Research of marine accidents through the prism of human factors. Promet Traffic Transport. 25 (4), 369–377.

EMSA, 2015. Annual Overview of Marine Casual ties and Incidents 2015.

Fotland, H., (2004) The International Maritime Hu man Element Bulletin, No: 3, April 2004. Published by Nautical Institute, London, UK.

Gaonkar R S P, Xie M, Ng K M, et al. (2011) Subjective operational reliability assessment of maritime transportation system [J]. Expert Systems with Applications, 38(11):13835–13846.

Graziano A, Teixeira A P, Soares C G.(2016) Classifica tion of human errors in grounding and collision accidents using the TRACEr taxonomy [J]. Safety Science, 86:245–257.

Hetherington C, Flin R, Mearns K. (2006) Safety in ship ping: the human element [J]. Journal of Safety Research, 37(4):401.

IMO, 2013. ISM Code and Guidelines on Implemen tation of the ISM Code 2010. <www.imo.org>.

Jensen, J., Soares, C.G., Papanikolaou, A., (2009). Methods and tools. In: Papanikolaou, A. (Ed.), Risk-Based Ship Design. Methods, Tools and Applications. Springer, Berlin, 195–301.

Kletz, T., (2001) Learning from Accidents Third edi tion. Gulf Professional Publishing, Oxford.

Kujath M F, Amyotte P R, Khan F I. (2010) A conceptual offshore oil and gas process accident model. [J]. Journal of Loss Prevention in the Process Industries, 23(2):323–330.

Kum S, Sahin B. (2015) A root cause analysis for Arctic Marine accidents from 1993 to 2011 [J]. Safety Science, 74:206–220.

Lehto M, Salvendy G. (1991) Models of accident causation and their application: Review and reappraisal [J]. Journal of Engineering & Technology Management, 8(2):173–205.

Martín J E, Rivas T, Matías J M, et al. (2009) A Bayesian network analysis of workplace accidents caused by falls from a height. [J]. Safety Science, 47(2):206–214.

Mazaheri A, Montewka J, Kujala P. (2016) Towards an evidence-based probabilistic risk model for ship-grounding accidents [J]. Safety Science, 86:195–210.

Mullai A, Paulsson U. (2011) A grounded theory model for analysis of marine accidents [J]. Accident; analysis and prevention,, 43(4):1590–603.

Noroozi A, Khan F, Mackinnon S, et al. (2014) Determina tion of human error probabilities in mainte nance procedures of a pump [J]. Process Safety & Environmental Protection, 92(2):131–141.

Rathnayaka S, Khan F, Amyotte P. SHIPP method ology: Predictive accident modeling approach. Part I: Methodology and model description [J]. Process Safety & Environmental Protection, 2011, 89(3):151–164.

Ren J, Jenkinson I, Wang J, et al. (2008) A methodology to model causal relationships on offshore safety assessment focusing on human and organizational factors. [J]. Journal of Safety Research, 39(1):87–100.

UNCTAD, 2014. Review of Maritime Transport 2014. http://unctad.org/en/pages/PublicationWebflyer. aspx?publicationid = 1068.

Weaver D A. Symptoms of operational Error [J]. Professional Safety, 2006 (April).

Zhang G, Thai (2016) V V. Expert elicitation and Bayesian Network modeling for shipping accidents: A literature review [J]. Safety Science, 87:53–62.

Maritime Transportation and Harvesting of Sea Resources – Guedes Soares & Teixeira (Eds)
© 2018 Taylor & Francis Group, London, ISBN 978-0-8153-7993-5

Decision support tool for modelling security scenarios onboard passenger ships

I. Gypa, E. Boulougouris & D. Vassalos
University of Strathclyde, Glasgow, UK

ABSTRACT: The vast growth in the global cruise industry, coupled with a number of major maritime casualties has brought about emphasis on the importance of ensuring precise and effective ship evacuation. The use of simulation software tools is the current-state-of-the art, in order to improve the existing evacuation procedures. At the same time, growing concern over maritime security issues has given rise to a new requirement for these simulation tools, particularly regarding the security of large passenger ships. In response to the aforementioned concerns, this paper aims to address a number of scenarios pertinent to the maritime security problem. Using the evacuation simulation software tool EVI and a large cruise vessel as a test case, a series of Monte Carlo base simulations have been conducted with focus placed on scenarios relating to important aspects of the Ship Security Plan. This includes procedures such as the evacuation of large spaces and the movement of all passengers to their cabins.

1 INTRODUCTION

Recent decades have witnessed a vast growth in the global cruise industry with cruise travel increasing by more than 500% since 1990, in terms of passenger carrying capacity (Brida & Zapata-Aguirre, 2009). In order to accommodate this rising demand the industry has responded by building increasingly larger and higher capacity vessels. Harmony of the Seas, now the largest cruise ship in the world, was delivered to Royal Caribbean Cruises on 12th May 2016 and has a gross tonnage of 226,000 GT and is able to carry 5500 passengers and 2400 crew members (Tinsley, 2016). Passenger capacities of this magnitude have never before been witnessed, raising new questions regarding the safety of such vessels, in particular regarding their safe and timely evacuation. This in combination with a number of recent maritime casualties has brought the subject of evacuation to the attention of the global maritime industry (Gypa et al., 2017).

Moreover, in recent years there is a growing concern over maritime security issues. There are many potential threats facing passenger vessels, such as hostage scenarios, attempts to sink the ship using small vessels as waterborne improvised explosive devices and waterside attacks. This concern has given rise to a new requirement for computer based simulations, particularly regarding the security of large passenger ships (Gypa et al., 2017).

The challenge addressed in the present work is the feasibility of the development of a Decision Support Tool (DST) capable of addressing Maritime Security Scenarios onboard large passenger ships. The security problem is existing and it has been addressed by IMO's International Ship and Port Facility Security (ISPS) code and the addition of Chapter XI-2 in the 1974 SOLAS Convention (IMO, 2002). Unfortunately, most of the analysis is based on experts' judgement and the results are qualitative in nature. The aim of the present paper is to prove that it is feasible to develop a tool capable of modelling security scenarios onboard large passenger ships.

2 EVACUATION ANALYSIS

The evacuation of a large passenger vessel is a highly complex event, presenting a number of difficulties at system, procedural and behavioural levels (IMO, 2016). The current state-of-the-art on evacuation of large passenger ships is provided by advanced computer-based simulation models, which are able to represent the vessel's internal arrangement in detail and take into consideration the interactions between the passengers, crew members and ship's layout (Boulougouris & Papanikolaou, 2002). In the following sections, a series of modelling techniques are briefly described, as well as some evacuation simulation software tools, which are based on these techniques.

2.1 Modelling techniques

In the Social Forces approach, the interaction between an agent and an obstacle is modelled in detail (Helbing & Molnár, 1995). At each time t, the movement of each pedestrian is determined, by using a force-based system, which is controlled by the distances between the agent and the obstacle (Sicuro et al., 2016, Vassalos, 2005).

The Cellular Automata Models are discrete dynamical systems in time and space and they simplify the problem of obtaining information from the surrounding locality by dividing the space into a grid. These techniques consider both homogeneous and heterogeneous groups of individuals (Sicuro et al., 2016).

One of the most predominant techniques is that of the Agent-Based Models (ABM). The ABM technique is a computational model class that simulates the interactions between the agents for the evaluation of the effects on the system as a whole (Ginnis et al., 2010).

Other modelling techniques include the Lattice Gas Modelling, the Game Theoretical Modelling, the Fluid-dynamics models and the models based on Animals (Sicuro et al., 2016).

2.2 Evacuation simulation software tools

Based on the above mentioned modelling techniques, a series of evacuation simulation tools have been developed, some of the most important are shown in Table 1.

In this study, the evacuation simulation tool EVI is used (Vassalos, 2005). EVI is based on multi-agent modelling techniques, which have been proved to be the most suitable for the passenger evacuation simulations. Each passenger and crew member is modelled as an individual agent with unique characteristics and behaviours. EVI uses a mesoscopic—hybrid approach, which is a multi-level planning structure, combining the macroscopic and microscopic models. The macroscopic model relates to the flow model, which is described by IMO in the Annex 2 of the Revised Guidelines for evacuation analysis, while the microscopic model is associated with human behaviour models (Vassalos, 2005).

For all the evacuation simulation software tools, the path-planning process is of great importance, due to the complexity of the vessel's layout. In EVI, when an agent has a specific location in a specific space onboard the vessel, the distance information between

Table 1. Evacuation simulation software tools (Sicuro et al., 2016, Ginnis et al., 2010).

Simulation tool	Modelling technique
AENEAS	ABM
maritimeEXODUS	ABM
EVI	Mesoscopic Approach
MonteDEM	Cellular Automata Models
IMEX	ABM
VELOS	Virtual Reality

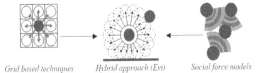

Grid based techniques   Hybrid approach (Evi)   Social force models

Figure 1. Space modelling techniques (Vassalos et al., 2003).

the agent and all the doors of the space is obtained and the agent chooses the shortest route. The shortest route is calculated by Dijkstra's shortest path algorithm (Dijkstra, 1959). EVI combines grid-based techniques and social forces models, see Figure 1, so that an agent chooses the direction of movement, when other agents or obstacles are present.

EVI is capable of modelling very complex evacuation scenarios, with a series of different parameters in a realistic sea environment (Sicuro et al., 2016, Vassalos et al., 2003, Vassalos et al., 2002).

3 MARITIME SECURITY

A growing concern over maritime security issues has given rise to a new requirement for computer based simulations, particularly regarding the security of large passenger ships. This is a problem that has been addressed by IMO through the ISPS code and the addition of Chapter XI-2 to the 1974 SOLAS Convention (IMO, 2002). ISPS aims at the detection of security threats and implementation of security measures, the establishment of roles and responsibilities concerning maritime security for governments, local administrations, ship and port industries at the national and international level, the collection and dissemination of security-related information and the provision of a methodology for security assessments, in order to react to changing security levels.

The European Union has also attempted to address this issue with the adoption of Directive 2005/65/EC (EC, 2005) on enhancing port security along with the revised Commission Regulation (EC) No. 324/2008 (EC, 2008) aimed at monitoring the implementation of the directive and Regulation EC725/2004 (EC, 2004), which describes procedures for conducting inspections aimed at monitoring the application of the directive.

The US government has expressed a certain degree of trepidation regarding the terrorist threat to passenger vessels and in response the US Coast Guard (USCG) has developed, with the assistance of the Department of Homeland Security (DHS), a Maritime Security Risk Analysis Model (MSRAM) aimed at mitigating the risks associated with terrorist attacks on vessels operating within US waters (GAO, 2010). While there may not be any definitive evidence that cruise ships are being specifically

Table 2. Maritime security incidents 1960–2017 (Gypa et al., 2017).

Year	Ship's name	Ship's type	Incident's type
1961	Santa Maria	Cruise Ship	Hijacking
1963	Anzoategui	Cargo Ship	Hijacking
1970	SS Columbia Eagle	Cargo Ship	Hijacking
1985	MS Achille Lauro	Cruise Ship	Hijacking
1996	MV Avrasya	Ferry	Hijacking
2000	USS Cole	Navy Ship	Waterside Attack
2002	SS Limburg	Oil Tanker	Waterside Attack
2004	SuperFerry 14	Ferry	Explosives onboard ship
2010	M/V M. Star	VLCC Oil Tanker	Explosives onboard ship
2017	Saudi Arabian Ship	Navy Ship	Waterside Attack

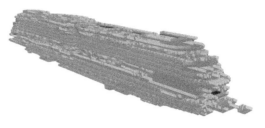

Figure 2. Model of the cruise ship in EVI.

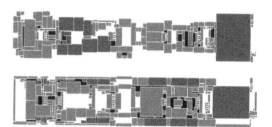

Figure 3. Assembly station in Decks 5&6.

targeted, maritime intelligence officials have identified the presence of terrorist groups that have the ability to carry out such attacks and thus the cause for concern. This has been reported both by USCG and RAND Corporation (Chalk, 2008). The potential threats that passenger vessels include face range from hostage scenarios to attempts at sinking the vessels through the use of small craft as waterborne improvised explosive devices (DHS, 2008). The Cruise Lines International Association (CLIA) has also reported that the greatest security concern for such ships is a waterside attack. In 2006, RAND reported that despite the implementation of security measures such as screening of passengers, crew and baggage, if militants were successful in bypassing any of these barriers, they would be able to carry out a wide range of attack scenarios.

Some of the most significant maritime security incidents are shown in Table 2. There are several types of such incidents, like hijacking (Goossens, 2014, Desert Sun, 1963, Johns, 2014, Goossens, 2013, Pallardy, 2010), waterside attacks (House Armed Services Committe Staff, 2001, Reuters, 2017) and explosives onboard the vessels (BBC, 2004, BBC, 2010).

4 CASE STUDY

The vessel chosen for this case study is a 13-deck passenger cruise ship (see Figure 2). Such a vessel is the perfect candidate for security-based simulations due to its large passenger capacity and therefore large potential loss of life. There are in total 23 assembly stations, of which 17 are on Deck 5 and 6 on Deck 6, as shown in Figure 3.

Two scenarios have been selected for further investigation, where the complete number of passengers and crew (3001 passengers & 801 crew)

are used in the simulations. The demographic characteristics have been taken into consideration, according to IMO MSC.1/Circ.1533–Day Case. This describes the make-up of the population in terms of age, gender, physical attributes and response durations. More specifically, there are 50% female and 50% male agents onboard the vessel, the age of the agents follows a uniform random distribution, the response duration a truncated logarithmic normal distribution and the speed is modelled as a statistical uniform distribution having minimum and maximum values.

4.1 Scenario No. 1

The first scenario investigated, deals with the evacuation of a large density populated area within the vessel. Such a scenario is of particular interest with regards to security, as it is often the case that terrorists target large crowded public spaces, where they stand to inflict the most harm. The space selected in this case is the restaurant located in Firezone 6 of Deck 4, where 500 passengers and 20 crew members are initially placed, as shown in Figure 4. The remaining passengers and crew are allocated randomly throughout the vessel's decks. The evacuation of the specific place is done in conjunction with the mustering of the rest of the ship, in order to study the potential congestions. It is also interesting to see how the evacuation of a large density populated area affects the evacuation of each deck separately.

4.2 Scenario No. 2

The second scenario investigates the time required for the agents to move from their initial positions

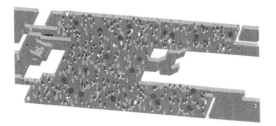

Figure 4. Restaurant Deck 4–Scenario No. 1.

Figure 5. Cabins Deck 8–Scenario No. 2.

to their cabins, see Figure 5. This may be the case, when the crew are alerted to a potential security threat and subsequently inform all passengers to return to their cabins, as a precautionary measure. The foreseeable problems with such a procedure could involve the congestion of stairwells and corridors, all of which are accounted for and assessed within the simulations. The 3800 agents (passengers and crew) are allocated randomly throughout the vessel's decks, a percentage of whom are initially in their cabins.

5 RESULTS AND DISCUSSION

In this section, all the results for the two scenarios are presented. For both scenarios, the times for the deck clearance and the mustering of the agents are calculated, while for the first scenario also the space clearance time is calculated and for the second one the return-to-cabin time is calculated. Each scenario has been simulated 100 times, where the walking speed, reaction time and initial location of agents are randomly generated each time.

5.1 Scenario No. 1

The space's mean evacuation time is approximately 5 minutes (303.8 seconds) with a standard deviation of 21 seconds, from the moment the agents initially realise the situation, until the space is completely empty. It is shown in Figure 6 that the simulation data follow an approximately normal distribution.

As depicted from Table 3 and Figure 7, Deck 7 has the slowest evacuation time with the highest occurrence (80/100 simulations), where as Deck00 5.1 has the fastest evacuation time with the highest occurrence (99/100 simulations). These conclusions are realistic, as in deck 6 six large assembly stations are positioned and thus a high percentage of agents passes through deck 7 in order to reach those assembly stations. Moreover, Deck00 5.1 is the vessel's lowest deck with a few crew agents, who evacuate the space fast.

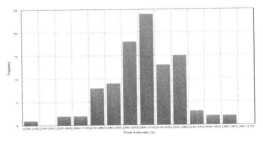

Figure 6. Space clearance time histogram.

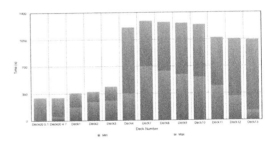

Figure 7. Deck clearance time vs Deck number—Scenario 1.

Table 3. Deck clearance statistics—Scenario 1.

	Max (s)	Min (s)	Max occurrence	Min occurrence
Deck00 5.1	298.5	16.5	0	80
Deck00 4.7	299.0	56.0	0	19
Deck1	358.5	175.0	0	0
Deck2	374.0	244.5	0	0
Deck3	437.0	260.0	0	0
Deck4	1205.5	349.5	1	0
Deck7	1289.0	697	99	0
Deck8	1275.5	640	0	0
Deck9	1263.5	594	0	0
Deck10	1244.0	559.5	0	0
Deck11	1071.0	442	0	0
Deck12	1055.5	311.5	0	0
Deck13	1043.0	124	0	1

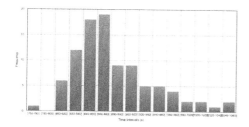

Figure 8. Mustering time histogram—Scenario 1.

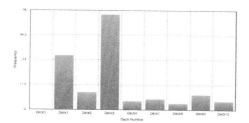

Figure 9. Return-To-Cabin maximum occurrence histogram.

The mean total mustering time is approximately 15 minutes with a standard deviation of 56 seconds. The histogram of the mustering time is presented in Figure 8.

5.2 Scenario No. 2

The Return-To-Cabin time is calculated as follows. For each cabin the minimum time of the cabin's full occupancy (each cabin can be fully occupied by 1, 2, or 3 agents) is calculated. For each simulation, the maximum time of the above times is kept and shows the moment that the last agent returned to their cabin. The average Return-To-Cabin time is approximately 1670 seconds. As depicted from Figure 9, there is a tendency for deck 3 to be the last one to be occupied (44/100 simulations). Delays have also been observed in deck 1 (25/100 simulations).

As shown in Table 4 and Figure 10, deck00 5.1 is the first one to be evacuated with the highest occurrence (92/100 simulations), whereas deck 4 is evacuated last (87/100 simulations).

Finally, the total mustering time lasts approximately 40 minutes. During the simulation, when the agents reach their cabin, they stay there for ten minutes before going to the assembly stations. It should be taken into consideration that while some agents have reached their cabins, there are other agents that have not reached their cabins yet. When the ten minutes pass and the first group of agents go to the assembly stations, there is counter-flow with the group of agents that have not reached

Table 4. Deck clearance statistics—Scenario 2.

	Max (s)	Min (s)	Max occurrence	Min occurrence
Deck00 5.1	286.5	17.0	0	92
Deck00 4.7	2152.5	1467.0	0	0
Deck1	2575.0	1920.0	0	0
Deck2	2682.5	2038.0	0	0
Deck3	2734.0	2048.0	0	0
Deck4	2811.5	2060.0	87	0
Deck7	2623.5	1873.0	13	0
Deck8	2601.5	1844.0	0	0
Deck9	2583.5	1825.5	0	0
Deck10	2529.5	1722.0	0	0
Deck11	817.5	482.0	0	0
Deck12	665.0	441.0	0	0
Deck13	326.0	96.5	0	8

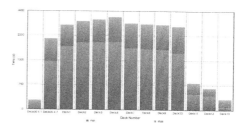

Figure 10. Deck clearance time vs Deck number—Scenario 2.

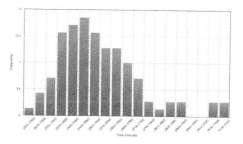

Figure 11. Mustering time histogram—Scenario 2.

their cabins. The histogram of the mustering time is presented in Figure 11.

6 CONCLUSIONS

This paper aims to prove that it is feasible to develop a tool capable of modelling security scenarios onboard large passenger ships. The results of the two case studies presented herein, have proved the feasibility and the vast potential of using the updated software in order to address security scenarios onboard. The analysis of the simulation statistics can provide very useful insight to the decision makers helping them to take well-informed, rational decisions in order to address security issues onboard passenger ships. The time required to evacuate a large restaurant due to a security threat or the time required for all passenger to return to their cabins in order to allow the crew to address any security threats onboard, can be of vital importance in particular situations.

The results of the present investigation are very promising for the future development of the tool. Future research steps include the investigation of security scenarios onboard RoPax ships and the challenges created by the large, open RoRo spaces, the development of scenarios, where explosives are placed in a specific space within the internal arrangement, in order to investigate the impact of the explosive types and mass, the awareness time, the arrangement and the room type to the probable loss of life (PLL), the estimation of the fatalities according to the distance of the agents relative to explosion blast and mustering time, the combined impact of a waterside attack to the ship which is berthed and the options for evacuating it given the constraints imposed by the environment, which might prohibit/limit the use of the existing life-saving appliances, the investigation of the combined flooding-evacuation-security problem with the use of time-domain flooding simulation tools such as Proteus coupled with EVI, the simulation of a moving threat onboard a passenger ship and the passengers are required to avoid different spaces as the threat changes place and the development of combined scenarios with security and hostile forces co-existing on a ship.

ACKNOWLEDGEMENTS

The authors wish to express their appreciation and gratitude to Luis Guarin, Yasmine Hifi and Eilidh McAdam from BBSaS for their support and advice during the process of this investigation and for supplying the evacuation analysis software tool EVI.

REFERENCES

BBC. 2004. *Bomb caused Philippine ferry fire* [Online]. Available: www.news.bbc.co.uk.
BBC. 2010. *Japan tanker was damaged in a terror attack* [Online].
Boulougouris, E. & Papanikolaou, A. 2002. Modeling & Simulation of the evacuation process of passenger ships. *IMAM.*
Brida, J., Gabriel & Zapata-Aguirre, S. 2009. Cruise Tourism: Economic, Socio-Cultural and Environmental Impacts. *International Journal of Leisure and Tourism Marketing,* 1, 205–226.

Chalk, P. 2008. The Maritime Dimension of International Security: Terrorism, Piracy, and Challenges for the United States. RAND Corporation.
Desert Sun. 1963. Hijacked Freighter in Venezuela Ship Race Spotted Today By U.S. Plane Now Tailing It.
DHS 2008. Small Vessel Security Strategy.
Dijkstra, E. 1959. A Note on Two Problems in Connexion with Graphs. *Numerische Mathematik,* 1, 269–271.
EC 2004. Regulation (EC) No 725/2004 of the European Parliament and of the Council of 31 March 2004 on enhancing ship and port facility security.
EC 2005. Directive 2005/65/EC of the European Parliament and of the Council of 26 October 2005 on enhancing port security.
EC 2008. Comission Regulation (EC) No 324/2008 of 9 April 2008 laying down revised procedures for conducting Commission inspections in the field of maritime security.
Gao 2010. Varied actions taken to enhance Cruise Ship Security, but some concerns remain.
Ginnis, A., Kostas, K., Politis, C. & Kaklis, P. 2010. Velos: a VR platform for ship-evacuation analysis. *Computer-Aided Design,* 42, 1045–1058.
Goossens, R. 2013. *Achille Lauro* [Online]. Available: www.ssmaritime.com.
Goossens, R. 2014. *TS Santa Maria—TS Vera Cruz* [Online]. Available: http://www.ssmaritime.com/.
Gypa, I., Boulougouris, E. & Vassalos, D. 2017. T-TRIG: Decision Support Tool for Modelling Security Scenarios Onboard Passenger Ships. Glasgow: University of Strathclyde.
Helbing, D. & Molnár, P. 1995. Social force model for pedestrian dynamics. *Physical Review E,* 51, 4282–4286.
House Armed Services Committe Staff 2001. The investigation into the attack on the U.S.S. Cole.
IMO 2002. International Code for the Security of Ships and Port Facilities (ISPS Code).
IMO 2016. MSC.1/Circ.1533: Revised Guidelines on Evacuation Analysis for New and Existing Passenger Ships.
Johns, S. 2014. *The SS Columbia Eagle mutiny, 1970* [Online]. Available: www.libcom.org.
Pallardy, R. 2010. Achille Lauro hijacking. *Britannica.*
Reuters 2017. Yemen's Houthis attack Saudi ship, launch ballistic missile.
Sicuro, D., Vasconcellos, J. & Vassalos, D. 2016. Representing Military Behavior in Naval Ship Evacuation Simulations including Flooding Damage Scenarios. *Design for Safety.* Hamburg, Germany.
Tinsley, D. 2016. Largest, most efficient yet for Royal Caribbean: Royal Caribbean has set the bar higher with 'Harmony of the Seas', drawing on elements from the previous Oasis-class ships and the Quantum generation. *The Motorship Magazine.*
Vassalos, D., Guarin, L., Vassalos, G., Bole, M., Kim, H. & Majumder, J. Advanced Evacuation Analysis—Testing the Ground on Ships. 2nd International Conference on Pedestrian and Evacuation Dynamics (PED), 2003.
Vassalos, D., Kim, H., Christiansen, G. & Majumder, J. 2002. A mesoscopic model for passenger evacuation simulation in a virtual ship-sea environment and performance-based evaluation. *1st International Conference on Pedestrian and Evacuation Dynamics.*
Vassalos, G. 2005. *Design for Ease of Passenger Evacuation in a Ship-Sea Environment.* Master of Philosophy, Universities of Glasgow and Strathclyde.

Maritime Transportation and Harvesting of Sea Resources – Guedes Soares & Teixeira (Eds)
© 2018 Taylor & Francis Group, London, ISBN 978-0-8153-7993-5

Scenario-based oil spill response model for Saimaa inland waters

J. Halonen & J. Kauppinen
South-Eastern Finland University of Applied Sciences, Kotka, Finland

ABSTRACT: As Lake Saimaa in the Eastern Finland is characterized by shallow waters, scattered formations of islands as well as narrow fairways and channels, the maritime safety is of the highest concern. Saimaa deep water route is proven difficult to navigate. Accident statistics indicate that the probability of grounding incident is relatively high in the Saimaa waterways. Several sensitive areas and habitats for endangered species situating in close proximity to the deep water route emphasize the significance of risk prevention. Inland oil spills have high potential to contaminate shorelines and to affect densely populated areas. In parallel with proactive measures, capability to deal with the consequences of an oil spill is needed. In order to improve the oil spill response capability, the regional response authorities initiated a joint project to develop an oil spill response management model for Saimaa inland waters. This paper describes the preliminary results of that development work.

1 INTRODUCTION

1.1 *Aim and scope*

The aim of this paper is to introduce an oil spill response management model for Lake Saimaa inland waters. Inland oil spill response model is based on the results of risk analysis on transportation and storage of oils in Lake Saimaa district, and the oil drift calculation models produced for identified high-risk locations. Modelling enables creation of scenario-based, site-specific contingency plans crucial for inland waters where response strategy is to be selected in a very short space of time. In Saimaa waterways where shores are near, window of opportunity to conduct an effective on-water response is further reduced due to the fast currents. In order to facilitate identification of vulnerable areas and establishing protection priorities and response strategy, management model includes easy-to-use decision support tools for response managers and on-scene response teams. Inland oil spill response management model is developed in close cooperation with the Finnish oil spill response authorities (See Chapter 8).

Responsibility in ship borne pollution response in Finland is divided between environmental and emergency authorities. In case of a major oil spill, the Finnish Environment Institute (SYKE) is the competent government pollution response authority. SYKE, accompanied with the Finnish Border Guard and the Finnish Defence Forces, conducts the oil spill response measures on the open sea. SYKE is also the nationally appointed authority that is empowered to request and give international assistance. The Regional Fire and Rescue Services (RFRS) are in charge of oil response operations in both inland and coastal waters. The Centres for Economic Development, Transport and the Environment (ELY Centres) assist the RFRSs in organising the oil spill response operation as well as approve the regional contingency plans of the RFRSs' (Hietala & Lampela 2007; SYKE 2017).

This paper focuses on the field of operation of RFRSs and ELY Centres and, particularly, on their area of responsibility in oil spill response on inland waterways of Lake Saimaa district. Secondly, this paper analyses the oil spill risk resulting from the vessel traffic. The risks originating from the land-based facilities, considered major sources of oil discharges in the region, are beyond the scope of this paper.

1.2 *Structure of the paper*

This paper provides an overview of the development of inland oil spill response model. Firstly, the geographical area, to which the model is targeted, is introduced to describe the unique, and rather challenging environment the response operation is ought to take place. Secondly, the risk assessment directing the contingency planning is presented and the main findings are highlighted, as well as some models to illustrate the impact areas of potential oil spills. Then the sensitiveness of environment and the principles behind the protection prioritisation are discussed in order to reason the map tools developed. In conclusion the outcomes are summarized and some future steps introduced.

2 SAIMAA LAKE AS UNIQUE OPERATING ENVIRONMENT

Lake Saimaa is the largest lake in Finland and the fourth largest lake system in Europe. Surface area of the waterbody is approximately 4400 square kilometers. Lake Saimaa is a labyrinthine lake system where in places there is more shoreline per unit of area than anywhere else in the world; the total length of the shoreline is about 15 000 kilometers. The number of islands within the area is about 14 000 (Finland's Environmental Administration 2015).

Waterways of Lake Saimaa are considered as part of the main transport corridor in the trans-European transport network (TEN-T). Lake Saimaa is connected to the Gulf of Finland and the Baltic Sea via Saimaa Canal by deep water route for merchant shipping with channel depth of 4,20–4,35 meters. The deep water route starts from the Gulf of Finland, goes through the Saimaa Canal and runs up to Siilinjärvi and Joensuu as illustrated in Figure 3. The total length of the Saimaa deep water route is about 760 kilometers, 255 kilometers of which (33,5%) is considered difficult to navigate. At Lake Saimaa the navigational challenges are due to the restricted waterways; shallow waters, straits, narrow channels and tight turns combined with the fast currents resulting in limited mistake margins (Finnish Transport Agency 2015).

These characteristics complicate also the spill response operations. In places, shallow waters do not allow recovery vessels to operate off the fairway zone. With narrow channels and canals as well as numerous islands and islets, the shoreline is in close proximity to the deep water route—often less than 100 meters away, in some places not more than 50 meters. In case of an oil spill, time for boom deployment to prevent oil washing ashore may be extremely short. Therefore it is noted, that in most areas near to Saimaa deep water route the risk for shoreline contamination is very high. As typical to inland water oil spills (Owens et al. 1993), spill is likely to affect populated areas and contaminate surface and groundwater water supplies.

In Saimaa region accessibility of threatened or contaminated shores may be limited, except for mainland where shoreline can be reached via small cottage roads. Most islands instead are difficult to reach as there are no roads, bridge connections, harbours or other kind of infrastructure and, in most cases, no space for helicopter landing due to dense forests.

The shallow waters and small water volume make the Lake Saimaa highly sensitive to oil pollution. Also cold winters and long periods of ice coverage slow the physical, chemical and biological decomposition of harmful substances (HELCOM 2010). In addition, several environmentally sensitive areas as well as habitats for endangered species are situated close the Saimaa deep-water route. For example, Saimaa is a sole habitat for critically endangered Saimaa ringed seal (*Pusa hispida saimensis*) (Toivola 2015). These characteristics necessitates a high-level preparedness with feasible action plans both for incident prevention, response and recovery.

3 RISK FOR SHIPBORNE OIL SPILL AT LAKE SAIMAA

3.1 *Estimated spill volume*

Due to the environmental sensitivity of the Lake Saimaa the transportation of hazardous liquid substances, including oil and oil products, is prohibited. However, as there are over 1500 vessels a year each of which carrying approximately 50 tonnes of bunker fuel oil, the total volume of oil transported annually exceeds 75 000 cubic meters. Most common bunker oils used are light fuel oils (MGO, MDO) with densities varying from 838,8 to 840,5 kg/m^3 (Heino et al. 2017). Results of the risk analysis indicated that volume of oil likely to spill, resulting from the damages typical to the incident scenarios of the area, is from 20 to 30 cubic meters, vessels in general carrying up to 100 cubic meters fuel oil each (Halonen et al. 2016).

3.2 *Probability of vessel accidents*

Risk assessment, accident probability and spatial distributions were based on the statistical data on incidents occurred between 1978 and 2014 on the Saimaa inland waterways. The nature of accident, position, vessel type, flag state, damages and environmental impacts, if any, were analysed from accident reports.

The results of the risk assessment are based on 116 cases considered relevant, excluding the occupational accidents and accidents of small crafts such as pleasure boats. Figure 1 shows the annual number of ship collisions and groundings within the period of study. As can be seen, accidents occur every year, the average accident frequency being five (5) incidents per year (Halonen et al. 2016).

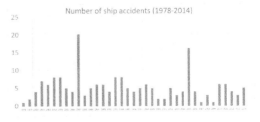

Figure 1. Number of ship accidents in Lake Saimaa between years 1978–2014 (Häkkinen 2016).

When the annual accident frequency was compared to the traffic density of the period, no clear correlation could be found. The causality might rather be derived from circumstantial factors, such as variations in water levels. According to the Finnish Maritime Administration (2007) two peaks in number of accidents seen in Figure 1, are explained by the altered procedures in accident reporting system, the increase of incidents being proportional to the improvement in reporting.

On Saimaa inland waterways, probability of a ship accident varies monthly (See Figure 2). Comparison of the number of incidents and the traffic volumes of each month indicates that the accidental risk is highest during autumn months, in November especially. At that specific time, primary causes reported were related to the prevailing conditions such as restricted visibility due to darkness or snowfall, or manoeuvring failures due to heavy ice conditions or drifted ice channels (Halonen et al. 2016).

The second highest accidental risk is apparent in April when the annual navigation season in Lake Saimaa and in the Saimaa Canal usually begins (Finnish Transport Agency 2015). At the time, the traffic is hampered by partial ice coverage and the maritime navigational aids displaced or damaged by drifting ice might cause uncertainty to the positioning of the vessel (Halonen et al. 2016). Figure 2 presents the monthly share of traffic volumes (with blue columns) and the monthly share of incidents (with orange columns). As can be seen, the traffic ceases for the most part during the winter months as the Saimaa Canal is closed.

In the accidents of Saimaa deep water route, dry cargo vessels represented 54,1% of the ship types involved. Majority (69,3%) of the accidents were groundings, the rest (30,7%) comprising of collisions. The main collisions types were identified as collisions with bridges (11,5%), collision with fixed

fairway structures and canal locks (7,7%) and collisions with another ship (3,8%). The distribution of the accident types have remained same over 40 years period of study, only the collisions with bridges indicating a subtle decreasing trend (Halonen et al. 2016).

Based on the severity of the damages and consequences reported, 11,1% of the accidents resulted in structural damages potential to cause spillage of fuel oil or lubricant oils. Thus, with the prevailing accident frequency, an accident resulting in an oil spill is likely to happen every 1,5 year. It is however worth of noting that no major oil spills have occurred in Saimaa inland waters during the period of the study (Halonen et al. 2016).

3.3 Accident locations

Geographical information of the incidents reported enabled identification of the most high-risk locations along the Saimaa deep water route. Distribution of incident positions revealed 104 fairway segments where the probability of incident was considered higher. These fairway segments were typically narrow channels or passages with strong current, canals, bridge openings or ferry crossings. Figure 3 provides an overview of the high risk areas

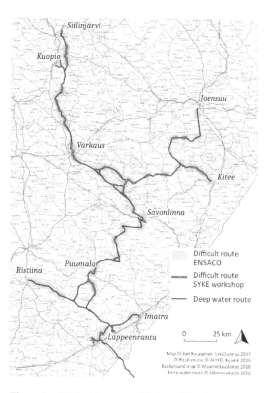

Figure 3. High risk areas of Saimaa inland waterways (Kauppinen 2017).

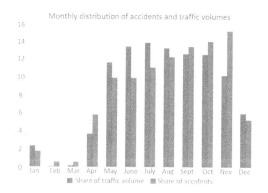

Figure 2. Monthly distribution of accidents in relation to the traffic volume (Häkkinen 2016).

of the Saimaa inland waterways. From a total of 116 incidents analysed, half (52,0%) occurred on these identified risk locations. Figure 4 provides a close-up of one of the high accidental risk locations.

When examining the number of incidents, most (35) occurred in Saimaa Canal in South Carelia, following the Kyrönsalmi strait (18) and Vekaransalmi strait (10) in South Savo. Regarding to the incident frequency in relation to the traffic volumes, the most significant risk locations were i) Ristiina fairway, ii) Konnus Canal, iii) Vihtakanta Canal and iv) Kyrönsalmi strait (Halonen et al. 2016).

Due to the low accident rate, relatively few incidents have occurred within the study period, though the timeframe was prolonged to cover the entire duration the statistics in question have been recorded. As the statistical data was limited, it was recognized that expert judgements were needed in order to derive valid conclusions. As Pedersen (2010) states, although the historical data would provide realistic figures, they nevertheless are inadequate to use for future predictions since they are relevant to the ship's structures and navigation equipment of the past, differing from those used today. Alongside with the technical development, the improvements in vessel traffic service, watch keeping and safety protocols, manning requirements etc. have altered, and will alter, the operational environment on inland waters. Thus, the expert advice was of great importance. It allowed identification of such risk locations where no

incidents have yet occurred, but which are considered to pose a potential risk.

4 OIL DRIFT MODELS FOR HIGH-RISK LOCATIONS

The route specific probabilities directed the studies on the impacts of a potential oil spill. These studies involve the modelling of the spreading of the oil with estimated spill volumes, and identification of the resources at risk within area likely to be impacted. The impact area was derived from the modelling outcome. Oil drift calculation models were produced for six most significant high-risk locations.

An example of the drift calculation model for shipborne spillage of light fuel oil in Vekaransalmi strait is presented in Figure 5, the colors representing the extent of the spreading in given time intervals up to 48 hours. This drift model demonstrated that an oil spill of 25 tonnes by volume is likely to affect eventually over 90 kilometers of shoreline including 100 small islands or islets. Within 48 hours with no intervening actions, oil contaminates 50 kilometers of shoreline (Halonen et al. 2017).

Drift models were of particular interest as prevailing fast currents are considered dominant

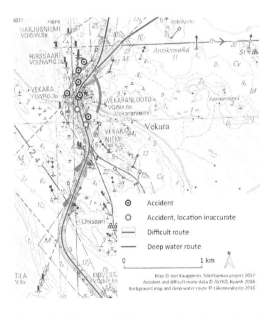

Figure 4. Close-up of Vekaransalmi strait considered as one of the high accidental risk locations in South-Savo region (Kauppinen 2017).

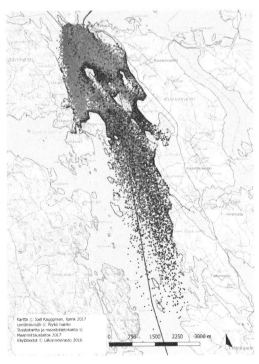

Figure 5. Oil drift model for Vekaransalmi strait calculated with 25 tonnes of light fuel oil (Kauppinen 2017).

factors in determining a response strategy. Together with demonstrating the effects of currents and winds on the oil trajectories (Häkkinen 2016), the drift modelling was used to display the effectiveness of specific, variously timed interventions. As result, the modelling calculations indicated, that the actions of the crew of the vessel involved are considered most effective, and in many cases, a single chance to restrict the oil from spreading. Especially in Lake Saimaa region where, due to long distances, it takes hours the RFRS reaches the incident scene, the response measures of the crew members are of highest importance (Heino et al. 2017). It is therefore recommended that in these particularly sensitive areas, the response capabilities of merchant vessels are improved to higher level than generally required by the SOPEP regulations (Shipboard Oil Pollution Emergency Plan, regulated by IMO).

Based on the drift modelling, amount of spilled oil has a minor effect to the extent of the impacted area. The spreading is more subject to the wind and currents as well as the duration of the leakage, than to the volume spilt. This finding was also supported by the results of modelling marine oil spills (Jolma 2009). Instead, the volume of spilled oil has an influence on the level of shoreline contamination (Jolma 2009; Kauppinen 2014). Figure 6 demonstrates the minor differences in the extent of impact areas resulting from oil spills of two different volume.

The modelling enables creation of site-specific, scenario-based contingency plans as the exposed areas and resources at risk can be analysed to direct the response resources accordingly. In addition to site-specific analysis on environmental impacts, the models were used for assessing the optimal access points and other logistical arrangements. These resulted in creation of separate tools, operational maps and maps introducing the vulnerable objects, to be discussed later in this paper.

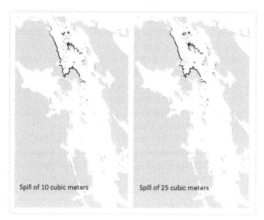

Figure 6.   Comparative drift models for 10 to 25 cubic meters oil spills with impact areas of similar extent (Kauppinen 2017).

The outcomes of the drift calculations are presented as oil drift scenarios and the data is included in the national situational awareness system (see Chapter 6) where it can be analysed connected with other spatial data, and used for site-specific planning.

5   PROTECTION PRIORITIES FOR NATURE AREAS AND SPECIES

The specific characteristics of the Lake Saimaa, the several environmental protection areas and numerous vulnerable species, makes the selection of response strategy difficult. One look at the sensitivity map (Toivola 2015) tells, that the entire area is covered with objects having a protection status of some sort. To facilitate the decision situation the response teams will be confronting, these objects, protected areas, sensitive ecosystems, critical habitats, endangered species and key resources considered sensitive to oil spills, are grouped into response priority ranks in order to facilitate establishing protection priorities and response strategy.

Protection prioritisation for the Saimaa nature areas and species is based on the methodology developed in Bothnian Bay oil spill development project (PÖK 2013). For the purpose of prioritisation, data on species and biotopes were mainly collected from the existing data sources, except the data on birds which was assembled from the information collected by the voluntary bird specialists. Saimaa prioritizing work is a synthesis of the efforts of ten environment specialists representing i.a. the ELY Centres and State Forest Enterprise (See Chapter 8).

Main objective of the prioritizing work was to produce viable, easy-to-use tool for oil spill responders. The protection prioritisation offers the response managers opportunity to take into account the vulnerable nature objects and areas when considering the response options. Prioritisation principles, jointly agreed prior an incident, helps resolving the issue of competing priorities in the event of limited protection and clean-up resources. Overall, the prioritisation improves both preparedness and spill response capability as decision making is based on profound analyses.

Prioritisation method bases on the objects' conservation status enhanced with experts judgements on how harmful the oil spill would be to the object in question. The conservation classification used is the International Union for Conservation of Nature (IUCN) scale (Figure 7). If the species is classified into Vulnerable (VU), Near Threatened (NT) and Least Concern (LC) classes, its regional conservation status or certain Natura-status raises the prioritisation a class higher.

The result of the prioritisation work is a simple two class model. Vulnerable objects are classified

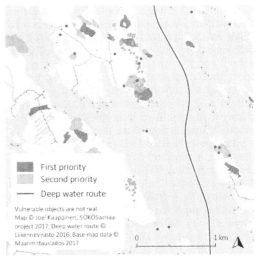

First priority
Second priority
Deep water route

Vulnerable objects are not real.
Map © Joel Kauppinen, SÖKÖSaimaa
project 2017; Deep water route ©
Liikennevirasto 2016; Base map data ©
Maanmittauslaitos 2017

0   1 km

Figure 7. IUCN conservation status and prioritizing method for Saimaa vulnerable nature objects (Kauppinen 2017).

Figure 8. Demonstration of the protection prioritisation classes for sensitive natural resources. Map sample is generated based on fictive data (Kauppinen 2017).

after their status into either first or second priority class and visualized accordingly. Figure 8 demonstrates the visualisation of the priority classes. Red objects are considered as first priority and orange objects as second priority. In case of an oil spill, this colour-coded system provides quick and easy visual determination of the protection priority class of the object or the area at risk. The prioritised sensitivity information is integrated into static map series as well as into an interactive web-based situational awareness system (See Chapter 6).

Protection prioritisation maps were generated as follows. The data was gathered on the area defined with a 1000 meter buffer from the shoreline in order to include vulnerable objects likely to be affected by onshore response operations. The size of the area ended up to be approximately 10 000 square kilometers.

Biotope data included 153 different biotope types in total, 76 types of which were classified as red and 77 types were classified as orange. Biotopes were categorized into groups according to their type and the prioritisation was ranked by groups. The largest groups of biotopes were groves (classified as red) and flood meadows (classified as orange).

Endangered species constituted of 309 prioritized species in total, 144 species of which were ranked into red class and 168 species into orange class. This number of species excludes all birds and Saimaa ringed seal that were analysed separately. In Saimaa area, the Siberian flying squirrel (*Pteromys volans*) is the most common endangered species with over 5700 observations. According to the IUCN-scale Siberian flying squirrel should, as an endangered species, belong to the red class. It was however lowered to the orange class for an oil spill does not affect to the squirrel habitat. A Spring pasqueflower (*Pulsatilla vernalis*) is the second common endangered species in Saimaa region with 225 observations. There are also several other

endangered species in the region with a few observations reported.

Saimaa ringed seal (*Pusa hispida saimensis*) was treated separately in the classification process because of its special status as one of the most endangered seal species. Data used for classification included the known nesting and resting spots of the seals. The Saimaa ringed seal is characterized by the breeding site fidelity, and thus the nesting spots are to be protected in order to ensure the nesting success. According to the specialist judgements, i.a. Sipilä (2016), contamination of nesting spots would have serious impacts for the survival of the species. In the classification process the criteria originally developed for analysing the adverse effects of human activity for nesting and resting spots of the seals was applied. In this criterion, most frequently used nesting areas receive the highest scores and rarely used resting areas situated away from nesting areas receive the lowest score. As the result, the scores are given from 1 to 10. All seal nesting or resting places that received four or more scores were classified into red protection prioritisation class, the rest of the places receiving the orange status.

The data on birds was collected from the voluntary bird specialists participating in counting the birds. The data consisted mostly of water birds and shorebirds (gulls) and information gaps were filled with data from Finnish important bird areas and Natura-datasets with less precise data quality. Bird datasets when incorporated into map formed large areas in the viewpoint of oil spill response. In protection prioritisation, these areas were left in orange class even though the area would have filled the criterion of the red class in order to maintain equality between objects.

As regards to the fish species, the Grayling (*Thymallus thymallus*) was represented in the data. It is classified as a near threatened species (NT, visualised as orange line) in Lake Saimaa (Rassi et al. 2010). Typical habitats for Graylings are shoals where they are known to live or go for spawn.

Analysis of the sensitive areas indicate that 2432shore segments, with shoreline length of 1584 kilometers in total, possess nature objects vulnerable to oil spills (not including birds and Grayling). Over 13% of these shores (327 segments with shoreline length of 206 kilometers) locate within one kilometer radius from the high-risk fairway sections. Regarding to the potential environmental impacts, severe consequences are more likely in Tappuvirta fairway where distinctly more nature objects are situated within one kilometer radius of the route.

Most challenging aspect with this straightforward method of prioritisation was to reach results equal between specific resources such as different species and biotopes. Other remarkable defect found was the data's reliability and lacking coverage in some areas. As directed by Finnish legislation, the data on vulnerable biotopes and species was classified as confidential official documents (Act on the Openness of Government Activities 621/1999).

6 MAP AND GIS TOOLS

The project produced three series of static maps. The first map set illustrates ship accidents on high risk routes (See Figure 4) inserted on nautical chart including fairway zones, aids to navigation and water depths facilitating the analyzation of route-dependent reasons making each route risky. These *Risk Maps* are used for planning and training purposes, as well as for risk prevention.

The second set of maps created are called *Operational Maps* providing shoreline segmentation and logistics data, like locations of temporary storage sites for oily wastes. Map sets for each of four RFRS regions are produced on terrain maps and on nautical charts with similar operational elements. Map set covering the whole area includes 588 map sheets, index pages and legend, accompanied with data sheets for 157 transport and storage points with detailed information, pictures and aerial photographs. These maps are meant for use during an incident response as well as for contingency planning.

The third set of maps identifies sensitive environmental resources, vulnerable species and nature areas. These *Prioritisation Maps* provide all objects within the area with protection priority class together with similar shoreline information than the operational maps. Prioritisation map set

includes 178 map sheets, index and legend pages. These maps are used by the response managers to define the priorities for protection and to select response tactics.

Response tools developed are compatible with the existing operational systems of response authorities to ensure the feasibility and durability of the outputs. Produced GIS-data is available through national situation awareness system BORIS 2.0 developed by SYKE. BORIS, acronym from words Baltic Oil Response Information System, is a web based geographic information system for spill response planning and data storage (SYKE 2017). The system allows user to combine several data layers to e.g. study relations between high-risk fairways, spill trajectories and vulnerable resources, and offers an access to additional, more specific information than the static maps can provide.

7 CONCLUSIONS

Lake Saimaa is a unique combination of sensitive freshwater environment, heavy vessel traffic and shallow waters with seasonal variations such as cold climate and ice coverage. Risk for shoreline oil contamination is high and several high-risk fairway sections situate near shores with vulnerable nature objects causing significant risk for the environment. Saimaa deep water route is proven difficult to navigate due to its narrowness and fast currents. It is recognized that the probability of ship accidents is relatively high in the Saimaa watercourse. Ship incidents are reported every year, yet no large scale oil spills have occurred. Since the risk cannot be entirely eliminated it is important to get prepared for response operation to ensure the consequences are minimized. As large parts of the Saimaa water body are too shallow for oil spill response vessels to operate and long distances hamper the effectiveness of immediate response measures, abilities of merchant vessel crews to limit the spreading of the oil should be supported. Self-sufficiency of merchant fleet, by providing oil booms and sorbent material onboard, is recommended in order to widen the window of opportunity for effective response.

The project results, described in this paper, serve as practical tools contributing to the scenario-based site-specific contingency planning. Easy-to-use decision support tools, such as response priority ranks represented with colour codes and guidelines introducing preferred response methods for various freshwater habitats and shoreline types, facilitate establishing response strategy. Results are compiled into an inland oil spill response manual, which will be published as a web based application.

As the project is underway, there is still many subjects to develop and further studies are needed.

As the risk analysis presented in this paper is based on estimations derived from statistical data and expert knowledge, there is need to elaborate these assessments with mathematical modelling. Also the studies on environmental impacts of an oil spill need to be extended to cover socio-economic impacts.

ACKNOWLEDGEMENTS

This study is executed under the project SÖKÖ-Saimaa (2016–2018) financed by the Finnish Oil Spill Compensation Fund of the Finnish Ministry of the Environment, South-Eastern Finland University of Applied Sciences Xamk and the Rescue Services of Southern Carelia, Northern Carelia, Southern Savonia and Northern Savonia.

The inland oil spill response management model is a joint effort of the project working group comprised of the designated spill response specialists representing the above mentioned rescue services together with the Finnish Transport Agency, the Emergency Service College, the Finnish Environment Institute and the Centres for Economic Development, Transport and the Environment of South Savo, North Savo, North Karelia and Southeast Finland (ELY Centres).

Special acknowledgment is made to the contribution of specialists from ELY Centres, State Forest Enterprise and the voluntary specialists to the prioritizing work, and to our co-workers at Xamk; Emmi Rantavuo, Mikko Pitkäaho, Krista Surakka, Vuokko Malk and Jouni-Juhani Häkkinen.

REFERENCES

Act on the Openness of Government Activities 621/1999, 24§ (14). Available at <http://finlex.fi/en/laki/kaannokset/1999/ en19990621.pdf> [Ref. 27.4.2017].

Finland's Environmental Administration 2015. State of the surface waters. Available at <www.ymparisto.fi/fi-FI/Vesi/Pintavesien_tila/Pintavesien_tilan_seuranta/Saimaa_Haukiselka(33512)> [Ref. 26.4.2017]

Finnish Maritime Administration 2007. Alusonnettomuusana-lyysi 2001–2005. Merenkulkulaitoksen julkaisuja 5/2007. Helsinki. ISBN: 978-951-49-2128-5. p. 7

Finnish Transport Agency 2015. Saimaa VTS Master's Guide. Available at <www.liikennevirasto.fi/web/en/merchant-shipping/vts/saimaa> [Ref. 26.4.2017]

Halonen, J., Malk, V. & Kauppinen, J. 2017. Alusöljyvahingon jätelogistiikka. Article in Itä-Suomen maa-alueiden ja Saimaan vesistöalueen öljyn—ja vaarallisten aineiden varastoinnin ja kuljetusten ympäristöriskien älykäs minimointi ja torjunta. Malk V. (Ed)., In: Xamk Kehittää 3, Kaakkois-Suomen ammattikorkeakoulu. ISBN: 978-952-344-007-4. p. 299.

Halonen J., Häkkinen J., & Kauppinen J., 2016. Alusliikenteen riskialueet Saimaa syväväylällä alusöljyvahingon näkökulmasta. *Tutkimusraportti Älykö-hankkeen*

vesiliikenteen riskikohteiden kartoituksesta. Kymenlaakson ammattikorkeakoulun julkaisuja Sarja B. Tutkimuksia ja raportteja nro 160, Kymenlaakson ammattikorkeakoulu. ISBN: 978-952-306-174-3. p. 31, 37–38.

Heino, H.; Voroshilin, D.; Heikkilä, H. Halonen, J. & Häkkinen, J. 2017. Haverialuksen miehistön ensitoimenpiteet alusöljyvahingossa. Article in *Itä-Suomen maa-alueiden ja Saimaan vesistöalueen öljyn- ja vaarallisten aineiden varastoinnin ja kuljetusten ympäristöriskien älykäs minimointi ja torjunta*. Malk V. (Ed)., In: Xamk Kehittää 3, Kaakkois-Suomen ammattikorkeakoulu. ISBN: 978-952-344-007-4. p. 127–128, 144, 146.

HELCOM 2010. Hazardous substances in the Baltic Sea. An integrated thematic assessment of hazardous substances in the Baltic Sea. Baltic Sea Environment Proceedings no 120B. Helsinki Commission. ISSN 0357-2994. p. 68–69.

Hietala, M. & Lampela, K. 2007. Öljyntorjuntavalmius merellä—työryhmän loppuraportti. Suomen ympäristö 41/2007. Suomen ympäristökeskus. Edita Prima Oy, Helsinki. ISBN (PDF) 978-952-11-2913-1. p. 21.

Häkkinen, J. 2016. Saimaan vesistön öljyvahinkoskenaarioiden mallintaminen, Tutkimusraportti Älykö-hankkeessa tehdyis-tä Saimaan alueen öljyn leviämismallinnuksista vedessä. Kymenlaakson ammattikorkeakoulun julkaisuja Sarja B. Tutkimuksia ja raportteja nro 158, Kymenlaakson ammattikorkeakoulu: ISBN: 978-952-306-168-2.

Jolma, K. 2009. Kokonaisselvitys valtion ja kuntien öljyn-torjuntavalmiuden kehittämisestä 2009–2018. Suomen ympäristökeskus, Helsinki. p. 10.

Kauppinen, J. 2014. Suuren alusöljyvahingon sosioekonomiset vaikutukset Merenkurkun ja Perämeren matkailualueilla. Master's thesis, University of Oulu, Department of Geography. p. 36.

Owens, E.H.; Taylor, E.; Marty, R. & Little, D.I. 1993. An Inland Oil Spill Response Manual to Minimize Adverse Environmental Impacts. Article in Oil Spill Conference Proceedings 1993. p. 105.

Pedersen, Terndrup P. 2010. Review and application of ship collision and grounding analysis procedures. Article in *Marine Structures* no 23. p. 242.

PÖK 2013. Ensisijaisesti suojeltavat kohteet. Perämeren öljyntorjunnan loppuraportit, työryhmä 3. Unpublished report. Economic Development, Transport and the Environment of North Ostrobothnia, Oulu. p. 15.

Rassi, P.; Hyvärinen, E.; Juslen, A & Mannerkoski I. 2010. The 2010 Red List of Finnish species. Ministry of the Environment, Helsinki. p. 343.

Sipilä, T. 2016. State Forest Enterprise. Expert judgement. SÖKÖSaimaa workshop on sensitivity mapping. Mikkeli, South-Savo. 20.4.2016.

SYKE 2017. Environmental emergency response in Finland. Suomen ympäristökeskus, Finnish Environment Institute. Document available <http://www.ymparisto.fi/en-US/Waters/Environmental_emergency_response_in_Finland> [Ref. 26.4.2017]

Toivola, V. 2015. Saimaan syväväylän alueen alusöljy—ja aluekemikaalivahinkojen torjunnan yhteistoimintasuunni-telma. Etelä-Savon elinkeino-, liikenne ja ympäristökeskus. Raportteja 39/2015. ISBN: 978-952-314-255-8. p. 54.

Maritime Transportation and Harvesting of Sea Resources – Guedes Soares & Teixeira (Eds)
© 2018 Taylor & Francis Group, London, ISBN 978-0-8153-7993-5

Trends and needs for research in maritime risk

S. Haugen
Department of Marine Technology, Norwegian University of Science and Technology, Trondheim, Norway

N.P. Ventikos
Laboratory for Maritime Transport, School of Naval Architecture and Marine Engineering,
National Technical University of Athens, Greece

A.P. Teixeira
Centre for Marine Technology and Ocean Engineering (CENTEC), Instituto Superior Técnico,
Universidade de Lisboa, Lisbon, Portugal

J. Montewka
Department of Navigation and Positioning, Finnish Geospatial Research Institute, Masala, Finland
Department of Transport and Logistics, Faculty of Navigation, Gdynia Maritime University, Gdynia, Poland

ABSTRACT: This paper describes the scope of the recently created Maritime Accident Risk Network (MARNet) that aims at providing the maritime industry with new knowledge on prevention, control and mitigation of serious accidents. The paper starts by presenting the objectives, focus areas and research topics addressed by the network MARNet. Then, the present position, trends and research needs in maritime risk are discussed. The emphasis is on the main focus areas of the Maritime Accident Risk Network, which include: *i*) Learning from accidents, *ii*) Understanding and communicating the risk picture, *iii*) Effective response to accidents and *iv*) Risk management. The three first research topics are all part of a general risk management process, but address different aspects of the process, while the last one aims to provide the framework that integrates the other focus areas. Although this paper presents an overview on the topic of maritime risk, the contents are mainly focused on the knowledge and contribution of the authors to the developments in this area and, therefore, it does not claim to be an extensive general bibliographic review of the subject matter.

1   INTRODUCTION

The maritime sector is very important globally for promoting trade and economic growth. Both in terms of the volume of the transported goods but also in its contribution to socio-economic growth, the shipping industry holds a leading role worldwide. Therefore, monitoring of risk level and efforts to enhance the safety performance of the shipping industry is of great interest and involves a plethora of stakeholders. Even though great progress has been done in this field over the years, unwanted events still regularly occur and is a cause for concern, both from the point of view of loss of life, environmental impact and economic costs.

Together, the four authors of this paper have established a group of researchers called MAR-Net, for Maritime Accident Risk Network. The focus areas of MARNet are: *i*) Learning from accidents, *ii*) Understanding and communicating

the risk picture, *iii*) Effective response to accidents and *iv*) (Life Cycle) Risk management.

In the following, we briefly address the four focus areas of MARNet, and illustratively, some specific topics where we see a need for improvement compared to present practices in the industry.

2   LEARNING FROM ACCIDENTS

At least as for safety, the shipping industry is largely reactive in nature, despite the risk-based instruments that have been introduced in the last decades in the regulatory framework for maritime safety (Guedes Soares & Teixeira, 2001; Guedes Soares et al. 2010). Highly indicative of this reactive nature is the fact that the safety regulations have significantly evolved under the impact of accidents. Such accidents, with severe consequences in terms of loss of human life and environmental damage, were for example those of the ferry Estonia in 1994 and the tankers Erika in 1999 and Prestige in 2002. Estonia sank in the Baltic Sea after losing her bow visor under heavy weather resulting in the loss of 852 lives. This accident, which is recorded as one of the worst maritime disasters of the 20th century, led to the development of new requirements for enhanced stability of passenger ships which were introduced in SOLAS90. They improved the damage stability criteria and ensured that roll on/roll off ships could survive with 50 cm of water on the vehicle decks; the accident also played a catalytic role in the development of requirements for Voyage Data Recorders (VDR). Erika sank in rough weather off the coast of Brittany causing an oil spill of 20.000 tonnes and Prestige sank off north-west Galicia spilling about 66.000 tonnes of heavy fuel oil. After the sinking of Erika, besides a series of amendments adopted on IMO level, on EU level legislative packages ERIKA I, ERIKA II and ERIKA III were adopted (the latter after the Prestige incident). Furthermore, Erika and Prestige significantly contributed to the acceleration of single-hull tankers phase-out.

The results of several studies have shown that a large proportion of ship accidents are due to human and organizational factors and, therefore, it is necessary to understand better how the operational risk levels depend on the human performance. In correspondence with the strong cultural attraction to reacting to historical incidents a series of frameworks to enhance safety have been proposed. In this context, methodologies for accident analysis and codification that describe the sequence of accidental events and the influence of human and organizational factors in these events have been developed (Graziano et al. 2016; Guedes Soares et al. 2000). These methodologies, along with accident databases properly designed (Antão & Guedes Soares 2002), provide useful information to construct risk models that describe in probabilistic terms the main contributors to the risk (e.g. Sotiralis et al. 2016). Nevertheless, the Casualty Investigation Code adopted by IMO in 2008 clearly states that the objective of a safety investigation is to prevent marine casualties and similar incidents to occur in the future. In this context, it is much easier to understand the characteristics of the dominant perspective on safety, i.e. Safety-I as described by Hollnagel (2014). Safety-I is defined by its opposite, by the lack of safety; it is perceived as a condition where the number of adverse outcomes (accidents/incidents/near misses) is as low as possible. Consequently, through this concept of safety, we try to increase the level of safety by studying situations where there was a recorded lack of it. An inherent assumption with respect to the Safety-I perspective is that systems are decomposable to less complex components and

that the functioning of those components can be described as bimodal: they are either considered to function correctly, or not and therefore fail. At the other end of the spectrum of this ratiocination by Hollnagel stands the description of Safety-II. Safety-II is defined by the presence of safety. In this case safety is perceived as a condition where the number of successful outcomes is as high as possible and is achieved by trying to make sure that things go right, rather than by preventing them from going wrong. Fundamental to Safety-II, and in direct contrast with the bimodality found in Safety-I, is the performance variability which is a key to organizational resilience. It is essential for individuals and organizations to adjust to the current conditions in order for things to work, an interesting insight into aspects of human performance can be found in (Ventikos et al. 2012) and (Ventikos et al. 2014). Also in the Safety-II concept it is proposed that it is more effective to learn from something that is much more frequent (successful operations) rather from something infrequent (accidents).

The maritime industry needs to move from a reactive to a proactive approach to safety; Safety-II is a more compatible way to achieve such a goal. A major factor that increases the need for this transition is the introduction of new technologies on ships (e.g. sensors, big data, and communications) and the increasingly growing interest in unmanned and automated ships.

Hence, the future picture of the industry will pose new hazards and risks. The methodologies that shaped the regulatory framework to date that is in a retrospective manner, will not be able to serve the industry as effectively as they have done. Formal Safety Assessment (FSA) adopted by IMO has been an effort for a more proactive approach to safety though the effectiveness of the method is highly dependent on the intellectual input process and the available data.

There is the need to consider new hypothetical hazards beyond the acceptance of evidence that came from real accidents (we also have to search for new black swans, possibly unknown-unknown in terms of probability and consequences, respectively). As the changes in the maritime industry are fast-paced in terms of technological innovations, standards and practices, we must move beyond by simply learning from accidents; we must build on that and further introduce parts and bits of a proactive way of thinking, designing and operating.

3 UNDERSTANDING AND COMMUNICATING THE RISK PICTURE

The amount of research in the field of maritime risk is significant and has been growing in the recent years. Three main lines of research can be distinguished therein: risk-based ship design, (Papanikolaou 2009; Puisa et al. 2014; Vanem et al. 2009) risk-based transportation system design (Montewka et al. 2014a; Goerlandt & Montewka 2015; Goerlandt 2015; Grabowski et al. 2000; Valdez Banda et al. 2015a, Sotiralis et al. 2016) and emergency response location and planning based on risk assessment of the anticipated accidents, (Aps et al. 2009; Haapasaari et al. 2014; Jolma et al. 2014; Lehikoinen et al. 2013). However, the view on risk adopted in the majority of those studies is rather uniform and dominated by a perspective postulated in a method called Formal Safety Assessment (FSA) by the International Maritime Organization (IMO), (IMO 2013). Therein risk (R) is defined as a combination of the probability (P) of an accident and its consequences (C), as follows:

$$R = f(P,C) \qquad (1)$$

As a risk metric, the FSA proposes risk index (*RI*), which is defined more explicitly as a product of the probability (*P*) and consequences (*C*):

$$\begin{aligned} RI &= P \times C \\ \log(RI) &= \log(P) + \log(C) \end{aligned} \qquad (2)$$

Within FSA, the *RI* serves as a crude indicator used for ranking various hazards and selecting the most relevant, which are then analysed in detail. However, it can lead to confusion, especially when comparing two situations A and B, where:

- A encompasses frequent events resulting in minor consequences—single or minor injuries and local equipment damage;
- B considers remote event of catastrophic consequence—multiple fatalities and total loss of a ship.

Even though the products of P and C in both cases are the same—following the FSA guidelines the risk indices are the same—for example, RI = 7 - these two situations differ substantially in its core. The available information about A is most likely better than in the case of B, as A occurs more frequently—is likely to occur once per month on one ship. B occurs rarely—once per year in a fleet of 1000ships, or likely to occur in the total life of several similar ships, for the adopted classification see (IMO 2013). The level of background knowledge about A and B is obviously different. This affects the level of uncertainty associated with the descriptions of situations A and B. Also the measures to control the risks in these two situations are probably different, as in the first case the focus might be given to the dimension of P, whereas in

the second case C might be subject to mitigation. Therefore, interpreting risk simply as a product of P and C leads to the misconception that risk is just a number, which is divorced from the scenarios of concern and available background knowledge. Applying this perspective, much of the relevant information needed for risk management is not properly reflected or even missing, (Aven 2013).

Therefore, a wider concept of risk should in our view be applied, allowing: 1) a systematic and hierarchical description of the risk associated with a given system; 2) reasoning about risk control options (RCO) in light of available background knowledge (BK); 3) reflection of the effect of BK on the evaluated risk and proposed RCOs (Montewka et al. 2014b). Fortunately, in risk analysis as a scientific discipline various scientific approaches exists, creating a kind of spectrum (Bradbury 1989; Rosa 1998; Shrader-Frechette 1991; Thompson and Dean 1996).

On one end of it sits strong realist approach, where risk perspectives consist exclusively of probabilistic risk measures. The evidences for these probabilities are based on data or models. It means probabilities of accident occurrences are calculated from observed frequencies or through probabilities models. On the other end of spectrum sits strong constructivists view on risk, according to which risk is nothing but observer's perception about a given situation, and varies between observers, (Goerlandt & Montewka 2015). Moreover, it would be beneficial to distinguish between the concept of risk and ways to describe risk, applying the following terminology: 1) the risk concept concerns what risk means in itself, what risk "is"; 2) a risk perspective is a way to describe risk, a systematic manner to analyse and make statements about risk; 3) a risk metric is the assignment of a numerical value to an aspect of risk according to a certain standard or rule (risk matrices address the likelihood an event occurrence or the consequence severity, or derivations such as expected values). Thus, we allow wider risk description, e.g. as a set of following elements: scenarios of concerns, their likelihoods and associated consequences, the level of background knowledge, expected uncertainties and biases. Depending on the level of BK, the risk could be measured differently. If there is enough reliable data to derive the probabilities and quantify consequences, we can attempt to do that and measure risk as the combination of those two. However, if the reliable data is missing, then the process of deriving probabilities from them is challenging, if possible. In such a case merging highly uncertain P with even more uncertainty C, just to follow the risk definition given by Eq.(1) may not lead to any meaningful and/or reliable results.

The field of maritime risk analysis is dominated by the realist view, where risk is mainly understood and measured as presented by Eq.(1), see for example (Goerlandt & Montewka 2015, Mazaheri et al. 2014). On one hand, this can be attributed to relatively large databases about accidents at sea, numerous models and tools to simulate maritime transportation systems and tools facilitating ship design. In the case of the latter the realist's view on risk may be justifiable, since the analysis addresses a system that is well-described and known in great detail with available large body of information and data, from the operation or simulation. However in case of maritime transportation systems, there are serious shortcomings related to the paucity and scatter of the available data, (Mazaheri & Montewka 2015; Valdez Banda et al. 2015b), underreporting of accidents and incidents, (Hassel et al. 2011; Psarros et al. 2010), and issues related to reliability and validity of the models, see for example (Goerlandt et al. 2016; Goerlandt & Kujala 2014).

These gaps in background knowledge lower the level of accuracy of a model, thus it shall not be used to seek an accurate estimates of the measure of the risk level. The main challenge arises when the numbers obtained in the course of databases review are considered "the ultimate truth" about the probability of an accident in the analysed area. This, when taken further, leads to another challenge related to the economical quantification of benefits of risk control options (RCOs). Instead, the obtained numbers should be rather treated as indicators of risk level rather than "the true risk", and the effect of RCOs is expressed as relative changes of risk level compared to some base-line value calculated for a given setting, see for example (Montewka et al. 2017). By using relative differences rather than the absolute numbers we can eliminate—to some extent—the effect of the existing uncertainty in the selection of the RCOs. By adopting such approach, it is feasible to differentiate among various designs of maritime transportation systems, based on the criteria selected, which in this case is risk level. By doing that we communicate to decision makers a message that is affected to lesser extent by the gaps in background knowledge.

The risk analysis is always performed for certain purpose in most cases to facilitate decision-making process. Risk analysis tends to combine the available BK on a subject of analysis to provide a platform for systematic elaboration on the effect of certain factors on the level of risk associated with a given activity. Since the BK may vary across the analysed system, it shall be reflected in the risk model. Moreover, this BK needs to be reflected in the effectiveness of the proposed RCOs. Since some of them will be more effective than the others.

Also, depending where on the time line of system development, the risk analysis is made; the level of BK will vary. If the risk analysis is conducted at the design stage, with numerous variables unknown, it would be more informative for decision-makers to work on relative changes in the probability of unwanted event (a single element) due to various RCOs, rather than "absolute" risk numbers, combining multiple uncertain elements. If the analysis is done in during the lifetime of the project, when more data is available, the risk measure can encompass more elements, if needed. Another relevant element of risk analysis is the quantification of the effect of alternative hypothesis on risk level. Since the variables and their relations in risk model are often not well defined, alternative paths are feasible. This needs to be addressed in the risk model, like the uncertainty assessment of the model elements, its outcome along with model sensitivity. Finally, after the risk model has been developed it must be validated, to demonstrate its fitness to the given purpose. This final process informs the decision-makers about the credibility of the risk model and obtained results. If the obtained results are not credible to the prospective stakeholders, the model is of limited use, unless it is further improved. However, if this improvement is not possible, some other method to evaluate and measure the safety shall be adopted.

Recent developments that have triggered a need for new and improved risk analysis methods are the development of autonomous ships. Autonomous ships will require continuously updated risk analysis to support automated decision-making and also to provide feedback to human operators that may intervene. In recent years, significant effort has been put into developing dynamic risk analysis (Haugen et al. 2015, Haugen et al. 2016b, Paltrinieri and Khan 2016, Wróbel et al. 2017) but most of this has been aimed at process industry and oil and gas industry, with their inspiration coming from the nuclear industry. Developing similar methods for the shipping industry is an important future task. This will also require new risk analysis methods, based e.g. on static and dynamic Bayesian networks (Haugen et al. 2016a, Sotiralis et al. 2016, Wróbel et al. 2016).

4 EFFECTIVE RESPONSE TO ACCIDENTS

The research on the various elements that influence an effective response to maritime accidents is vast, covering a wide range of specific problems such: the safety assessment of damaged ships (Teixeira & Guedes Soares 2010, Luís et al. 2009, Prestileo et al. 2013); the evacuation from large passenger ships under storm conditions, fires and progressive flooding; the models, tools and information systems to support maritime Search and Rescue (SAR) and oil combating operations, which are particularly important in sensitive an remote locations such as in the Arctic region, where, despite the cooperation between Arctic states, the response capacity is still very scarce and may not be sufficient (Takei 2013).

Maritime Search and Rescue (SAR) aims at assisting people in distress or danger at sea and involves activities such as assisting ships in difficulty, accident prevention, search and rescue, medical assistance and patient transport. The focus of the research in this topic has been on the searching stage of missing persons at sea and lost objects like drifting containers or wrecks, that can be decomposed into three steps (Kratzke et al. 2010). The first step consists of extracting all useful information about the missing object, including its last known position, wind and current fields, waves, among others (Vettor & Guedes Soares 2015). The second step aims at constructing the probability maps of the possible location of the missing object by synthesizing all the available information that influence the trajectory prediction of the drifting object (e.g. Zhang et al. 2017, Breivik & Allen 2008). The last step comprises the definition of optimal routes for the Search and Rescue Units (SRUs), which would maximize the probability of finding the object under the probability distribution maps (e.g. Kratzke et al. 2010, Zhang et al. 2016, Siljander et al. 2015), also when Unmanned Aerial Vehicles (UAVs) are used to assist maritime SAR missions (Ryan & Hedrick 2005).

The drift of an object at sea has many common features with the assessment of the drift of oil slicks, which have been modelled for some time ((Sebastião & Guedes Soares 2006), (Sebastião & Guedes Soares 2007), (Ventikos 2013)), which also has to track the motion of the oil slick in addition to its chemical changes (Sebastião & Guedes Soares 1995), as recently reviewed by (Spaulding 2017).

Much attention has been paid to the development of sophisticated multi-agent simulation tools for modeling emergency evacuation from passenger ships e.g. (Kim et al. 2004, Gwynne et al. 2003). Evacuation may be performed in various emergency situations, not necessarily associated with flooding and fires (e.g. Azzi et al. 2011, Galea et al. 2003). For example, the inability of a cruise ship to perform her mission (due to failure of power plants, grounding, among others) can also require evacuation. Almost all existing systems try to model passenger movements and to estimate influence of ship motions using velocities—or accelerations-based models (Balakhontceva et al. 2015) but not all of them explore the influence of

ship dynamics (irregular waves) on the evacuation process (Balakhontceva et al. 2016).

Typically, fire simulations in ships have been based on Computational Fluid Dynamics (CFD), often referred to as field models, coupled one way with evacuation simulation. Thus, changes in the geometry created by the evacuating population, e.g., opening and closing of doors, do not impact the fire dynamics. This is due to the extremely long run times associated with the CFD fire simulations, which is particularly a problem in the early stage of the ship design. A novel approach to reduce the run times was developed in the EU FP7 project FIRE-PROOF known as the HYBRID approach (Burton et al. 2007) that replaces the field model with a zone model in appropriate parts of the geometry leading to a considerable saving in run time, thus allowing more realistic evacuation simulations.

Less effort has been directed towards including such evacuation models in a holistic framework for risk assessment, which involves the definitions of a set of relevant evacuation scenarios used to estimate the overall risk associated with a specific passenger ship (Vanem & Skjong 2006).

Concerning flooding simulation, Virtual Reality based Decision Support Systems to assist the coordination of damage control teams and to take the appropriate counter-measures have also been proposed (Varela & Guedes Soares 2007). The tool was later improved to include sophisticated progressive flooding methods (Varela et al. 2014) and the inclusion of coupled flooding-ship motions under waves (Varela et al. 2015).

Despite the tools and models that have been developed to address individually specific aspects that influence an effective response to accidents, it is still necessary to develop integrated decision support systems for an effective fast response to ships in distress in the first hours following a casualty that can affect the vessel's chances of survival and the potential impact on the marine environment. This requires adequate communication and information exchange between vessel crew, SAR operators and other relevant parties in maritime distress situations (Nordström et al. 2016).

5 LIFE CYCLE RISK MANAGEMENT

All the topics that we are addressing in this paper are related to specific aspects of risk management, from understanding the risk picture, to effective response to learning lessons from what we experience. There are however also important aspects of risk management on a more general level where further development may be beneficial.

The introduction of FSA (IMO 2013) introduced important elements of more risk-informed thinking in the maritime industry. However, in recent years, the risk management standard ISO31000 (ISO 2009) has become popular and is being widely used. Further, the Risk Governance framework (IRGC 2012) has also gained a lot of recognition. It may therefore also be time to take some further steps in the maritime industry with respect to the subject matter.

One step is to increase the use of risk-based regulations. Introduction of new regulations is very often driven by a need to control negative effects of an activity or a product, including the risk associated with these. SOLAS and MARPOL are both examples of regulations that are introduced to reduce risk to people and the environment, respectively. Usually, this is done by prescriptive regulations, specifying in detail what should be done to control risk. There are clear disadvantages of this, most notably that the resulting risk level may vary significantly because "one size" will not fit all. Furthermore, prescriptive regulations can hinder innovation and introduction of safer solutions because they are not explicitly covered by the prescriptive regulations.

Risk-based regulations have been in use in some areas for some time, among others in the offshore oil and gas industry in some countries, e.g. in Norway, UK and Australia (Vinnem 2014). These regulations aim at resolving the issues mentioned above, giving a more consistent risk level and also allowing for new solutions to be introduced as long as risk criteria are met. On the other hand, this clearly also opens up for some problems. In particular, it is claimed that risk assessment is too uncertain to be a consistent tool that can be used, opening up for misuse of both data and models (ref Section 3 of this paper). The end result is that even if the risk assessments show acceptable and consistent risk results, the actual designs may not have improved safety.

In line with many other high-risk industries, further development of risk-based regulations in the maritime industry will be a step forward. This however needs to be done carefully, aiming at gaining the best from both prescriptive and risk-based regulations, while at the same time controlling the negative aspects. This requires careful consideration of where risk-based regulations are best suited (in line with the development of the Goal Based Standards GBS) for the building of ships). In areas where the technology and operations are well-established and stable and where the risk potential is limited, prescriptive regulations may be the best option, because it gives quick and simple answers to how to design and operate maritime systems. On the other hand, where things are developing rapidly and where risk is high, risk-based approaches is a better option.

Risk-based regulations need to be accompanied with strong guidance on how to do the risk management process, both in terms of identifying and analyzing risk and in terms of interpreting and using the results. This will ensure more consistent analysis. Certain minimum, prescriptive standards are also likely to be required, as a supplement to risk-based regulations, although this needs to be balanced against the flexibility that one wants to achieve using risk-based approaches.

Another aspect of risk management where significant improvement can be achieved is related to closer integration of risk management processes, in at least four dimensions.

First of all, a ship and its owner faces a variety of risks. Integrated management of these risks may be beneficial. Risks to people can often also represent risks to assets and to the environment, to reputation and to business in general. Understanding this and integrating management of these different dimensions of risk may save money and give a better result, with fewer losses.

Secondly, risk management needs to be integrated more closely into the whole process of designing and operating ships (in the sense of dealing with the entire life-cycle of the product, in our case of the ship—as shown below in this section). Risk should be one of the decision criteria that goes into this, in line with other criteria such as costs, flexibility of operation etc. If risk becomes an integrated part of the decision process, it increases focus and paves the way for more optimal decisions.

Thirdly, risk in the whole transportation system should be considered. For international bodies and governments, a holistic view on risk needs to be taken and risk should also be managed from this perspective.

Finally, the whole lifecycle of the transport system needs to be considered from the very beginning. Optimizing risk with respect to all phases, from construction through operation to decommissioning and scrapping should be the aim. Looking at just one phase at a time does not necessarily minimize risk overall.

Risk analysis is primarily used as a tool for achieving safe designs and to a lesser degree in operations. This optimizes risk "on average", but does not necessarily help us identify high-risk situations in operation, when risk is increased due to the particular circumstances. In the oil and gas and process industries, significant effort is now being put into developing methods for on-line risk assessment, to provide decision-support in operations. More and more data are available digitally and sensors are being increasingly used to give online status on systems, components and external and internal conditions. This can be used to tell us more about what the risk is right now and act as guidance in day-to-day risk management; the digitalization met in the shipping industry and the active introduction of big data highly support this perspective.

A modern topic related to risk management is resilience. This is a term that has been around for quite some time within the safety community, but there is still relatively limited practical research and applications (Lykos et al. 2016). Getting a better understanding of what resilience means in practice could be a first step forward to develop this into practical solutions. Integrated thinking is required also here, to develop resilient design, resilient operations, resilient procedures and resilient organizations.

6 CONCLUSIONS

The maritime industry is very important for trade and prosperity worldwide and in spite of significant developments in safety over the last decades, the industry still experiences serious accidents too often. In this paper, we have tried to highlight some of the areas where further research and development is required to improve safety even further. The focus has been specifically on four areas: *i*) Learning from accidents, *ii*) Understanding and communicating the risk picture, *iii*) Effective response to accidents and *iv*) (Life Cycle) Risk management. If further improvements could be made in these four areas, the maritime industry could improve its safety record even further.

MARNet has been developed and organized in a manner to accommodate and promote modern risk focused research and knowledge transfer for shipping and the offshore related industry aiming at the substantial strengthening of safety and the effective protection of the maritime and littoral environment.

REFERENCES

Antão, P. & Guedes Soares, C. (2002). "Organisation of Databases of Accident Data." *Risk Analysis III*, Wit-Press, UK, 395–403.

Aps, R., Fetissov, M., Herkül, K., Kotta, J., Leiger, R., Mander, Ü. & Suursaar, Ü. (2009). "Bayesian inference for predicting potential oil spill related ecological risk." *Safety and Security Engineering III*, M. Guaracio, C.A. Brebbia, and F. Garzia, eds., WIT Press, 149–159.

Aven, T. (2013). "Practical implications of the new risk perspectives." *Reliability Engineering and System Safety*, 115, 136–145.

Azzi, C., Pennycott, A., Mermiris, G. & Vassalos, D. (2011). "Evacuation Simulation of Shipboard Fire Scenarios." *Fire and Evacuation Modeling Technical Conference*.

Balakhontceva, M., Karbovskii, V., Rybokonenko, D. & Boukhanovsky, A. (2015). *Multi-agent Simulation of Passenger Evacuation Considering Ship Motions*. Procedia Computer Science, Elsevier Masson SAS.

Balakhontceva, M., Karbovskii, V., Sutulo, S. & Boukhanovsky, A. (2016). "Multi-agent simulation of passenger evacuation from a damaged ship under storm conditions." *Procedia Computer Science*, Elsevier Masson SAS, 80, 2455–2464.

Bradbury, J.A. (1989). "The policy implications of differing concepts of risk." *Science Technology Human Values*, 14(4), 380–399.

Breivik, Ø. & Allen, A.A. (2008). "An operational search and rescue model for the Norwegian Sea and the North Sea." *Journal of Marine Systems*, 69(1–2), 99–113.

Burton, D.J., Grandison, A.J., Patel, M.K., Galea, E.R. & Ewer, J.A. (2007). "Introducing a Hybrid Field/Zone Modelling Approach for Fire Simulation." *11th International Fire Science & Engineering Conference, Interflam 2007*, Royal Holloway College, University of London, UK, 1491–1497.

Galea, E.R., Lawrence, P., Gwynne, S., Filippinidis, L., Blackshields, D., Sharp, G., Hurst, N., Wang, Z. & Ewer, J. (2003). "Simulating Ship Evacuation Under Fire Conditions." *2nd International Pedestrian and Evacuation Dynamics Conference*, CMS Press, Greenwich, UK.

Goerlandt, F. (2015). "Risk analysis in maritime transportation: principles, frameworks and evaluation." Aalto University.

Goerlandt, F., Khakzad, N. & Reniers, G. (2016). "Validity and validation of safety-related quantitative risk analysis: A review." *Safety Science*.

Goerlandt, F. & Kujala, P. (2014). "On the reliability and validity of ship-ship collision risk analysis in light of different perspectives on risk." *Safety Science*, Elsevier Ltd, 62, 348–365.

Goerlandt, F. & Montewka, J. (2015). "A framework for risk analysis of maritime transportation systems: A case study for oil spill from tankers in a ship–ship collision." *Safety Science*, 76, 42–66.

Grabowski, M., Merrick, J.R.W., Harrold, J.R., Massuchi, T.A. & van Dorp, J.D. (2000). "Risk modeling in distributed, large-scale systems." *IEEE Transactions on Systems, Man and Cybernetics, Part A: Systems and Humans*, 30(6), 651–660.

Graziano, A., Teixeira, A.P. & Guedes Soares, C. (2016). "Classification of human errors in grounding and collision accidents using the TRACEr taxonomy." *Safety Science*, 86, 245–257.

Guedes Soares, C. & Teixeira, A.P. (2001). "Risk assessment in maritime transportation." *Reliability Engineering and System Safety*, 74(3), 299–309.

Guedes Soares, C., Teixeira, A.P. & Antão, P. (2000). "Accounting for Human Factors in the Analysis of Maritime accidents." *Foresight & Precaution*, M.P. Cottam, D.W. Harvey, R.P. Pape, and J. Tait, eds., Balkema, Rotterdam, 521–528.

Guedes Soares, C., Teixeira, A.P. & Antão, P. (2010). "Risk-based Approaches to Maritime Safety." *Safety and Reliability of Industrial Products, Systems and Structures*, Guedes Soares, ed., Taylor & Francis Group, London, 433–442.

Gwynne, S., Galea, E.R., Lyster, C. & Glen, I. (2003). "Analysing the evacuation procedures employed on a Thames passenger boat using the maritimeEXODUS evacuation model." *Fire Technology*, 39(3), 225–246.

Haapasaari, P., Dahlbo, K., Aps, R., Brunila, O.-P., Fransas, A., Goerlandt, F., Hanninen, M., Jonsson, A., Laurila-Pant, M., Lehikoinen, A., Mazaheri, A., Montewka, J., Nieminen, E., Nygren, P., Salokorpi, M., Tabri, K. & Viertola, J. (2014). *Minimizing risks of maritime oil transport by hoistic safety strategies.* Kotka.

Hassel, M., Asbjørnslett, B.E. & Hole, L.P. (2011). "Underreporting of maritime accidents to vessel accident databases." *Accident; analysis and prevention*, 43(6), 2053–63.

Haugen, S., Almklov, P.G., Nilsen, M. & Bye, R.J. (2016a). "Norwegian national ship risk model." *Maritime Technology and Engineering 3*, C. Guedes Soares and T.A. Santos, eds., Taylor & Francis Group, London, 831–838.

Haugen, S., Edwin, N.J., Vinnem, J.E. & Brautaset, O. (2016b). "Dynamic risk analysis for operational decision support." *Risk, Reliability and Safety: Innovating Theory and Practice,* L. Walls, M. Revie, and T. Bedford, eds., CRC Press, 776–772.

Haugen, S., Vinnem, J.E., Brautaset, O., Bye, R.J., Nyheim, O.M., Seljelid, J. & Wagnild, B.R. (2015). "Risk information for operational decision making in oil and gas operations." *Safety and Reliability of Complex Engineered Systems*, CRC Press, 405–413.

Hollnagel, E. (2014). Safety-I and Safety-II: The Past and Future of Safety Management. Ashgate Publishing, Limited., Surrey.

IMO. (2013). "Revised guidelines for formal safety assessment (FSA) for use in the imo rule-making process." *MSC-MPEC.2/Circ. 12.*

IRGC. (2012). *An introduction to the IRGC risk governance framework.* International Risk Governance Council (IRGC), Lausanne, Switzerland.

ISO. (2009). "Risk management - principles and guidelines on implementation." *International standard, ISO 31000*, International Organisation for Standardisation, Geneve.

Jolma, A., Lehikoinen, A., Helle, I. & Venesjärvi, R. (2014). "A software system for assessing the spatially distributed ecological risk posed by oil shipping." *Environmental Modelling & Software*, 61, 1–11.

Kim, H., Park, J.H., Lee, D. & Yang, Y.S. (2004). "Establishing the methodologies for human evacuation simulation in marine accidents." *Computers and Industrial Engineering*, 46(4 SPEC. ISS.), 725–740.

Kratzke, T.M., Stone, L.D. & Frost, J.R. (2010). "Search and Rescue Optimal Planning System." *13th International Conference on Information Fusion*, Edinburgh, Scotland, United Kingdom, 1–8.

Lehikoinen, A., Luoma, E., Mäntyniemi, S. & Kuikka, S. (2013). "Optimizing the recovery efficiency of Finnish oil combating vessels in the Gulf of Finland using Bayesian Networks." *Environmental science & technology*, American Chemical Society, 47(4), 1792–9.

Luis, R.M., Teixeira, A.P. & Guedes Soares, C. (2009). "Longitudinal strength reliability of a tanker hull accidentally grounded." *Structural Safety*, 31(3), 224–233.

Lykos, G.V., Ventikos, N.P. & Sotiralis, P. (2016). "Application of the FRAM for a Fire Onboard Focusing on Human Centered Design & Resilience Aspects." *Proceedings of the SEAHORSE Conference*, Glasgow, UK.

Mazaheri, A. & Montewka, J. (2015). "Usability of accident and incident reports for evidence-based risk modeling of ship grounding." *Safety and Reliability: Methodology and Applications* - Taylor & Francis Group, London, 75–82.

Mazaheri, A., Montewka, J. & Kujala, P. (2014). "Modeling the risk of ship grounding—a literature review from a risk management perspective." *WMU Journal of Maritime Affairs*, 13(2), 269–297.

Montewka, J., Ehlers, S., Goerlandt, F., Hinz, T., Tabri, K. & Kujala, P. (2014a). "A framework for risk assessment for maritime transportation systems - A case study for open sea collisions involving RoPax vessels." *Reliability Engineering and System Safety*, 124, 142–157.

Montewka, J., Goerlandt, F. & Kujala, P. (2014b). "On a systematic perspective on risk for Formal Safety Assessment (FSA)." *Reliability Engineering & System Safety*, 127(July), 77–85.

Montewka, J., Goerlandt, F., Innes-Jones, G., Owen, D., Hifi, Y. & Puisa, R. (2017). "Enhancing human performance in ship operations by modifying global design factors at the design stage." *Reliability Engineering and System Safety*, 159(February 2016), 283–300.

Nordström, J., Goerlandt, F., Sarsama, J., Leppänen, P., Nissilä, M., Ruponen, P., Lübcke, T. & Sonninen, S. (2016). "Vessel TRIAGE: A method for assessing and communicating the safety status of vessels in maritime distress situations." *Safety Science*, 85, 117–129.

Paltrinieri, N. & Khan, F. (2016). Dynamic Risk Analysis in the Chemical and Petroleum Industry: Evolution and Interaction with Parallel Disciplines in the Perspective of Industrial Application. Butterworth-Heinemann.

Papanikolaou, A. (2009). *Risk-Based Ship Design: Methods, Tools and Applications.* (A. Papanikolaou, ed.), Springer.

Prestileo, A., Rizzuto, E., Teixeira, A.P. & Guedes Soares, C. (2013). "Bottom damage scenarios for the hull girder structural assessment." *Marine Structures*, 33, 33–55.

Psarros, G., Skjong, R. & Eide, M.S. (2010). "Underreporting of maritime accidents." *Accident Analysis & Prevention*, 42(2), 619–625.

Puisa, R., Malazizi, L. & Gao, Q. (2014). *Risk models for aboard fires on cargo and passenger ships.* EC FP7 project FAROS (no 314817), FAROS Public Deliverable, Brookes Bell LLP.

Rosa, E.A. (1998). "Metatheoretical foundations for post-normal risk." *Journal of Risk Research*, 1(1), 15–44.

Ryan, A. & Hedrick, J.K. (2005). "A mode-switching path planner for UAV-assisted search and rescue." *Proceedings of the 44th IEEE Conference on Decision and Control, and the European Control Conference 2005*, Seville, Spain, 1471–1476.

Sebastião, P. & Guedes Soares, C. (1995). "Modeling the fate of oil spills at sea." *Spill Science & Technology Bulletin*, 2(2–3), 121–131.

Sebastião, P. & Guedes Soares, C. (2006). "Uncertainty in predictions of oil spill trajectories in a coastal zone." *Journal of Marine Systems*, 63, 257–269.

Sebastião, P. & Guedes Soares, C. (2007). "Uncertainty in predictions of oil spill trajectories in open sea." *Ocean Engineering*, 34(3–4), 576–584.

Shrader-Frechette, K.S. (1991). *Risk and rationality: philosophical foundations for populist reforms.* University of California Press, Berkeley.

Siljander, M., Venalainen, E., Goerlandt, F. & Pellikka, P. (2015). "GIS-based cost distance modelling to support strategic maritime search and rescue planning: A feasibility study." *Applied Geography*, 57, 54–70.

Sotiralis, P., Ventikos, N.P., Hamann, R., Golyshev, P. & Teixeira, A.P. (2016). "Incorporation of human factors into ship collision risk models focusing on human centred design aspects." *Reliability Engineering & System Safety*, 156, 210–227.

Spaulding, M.L. (2017). "State of the art review and future directions in oil spill modeling." *Marine Pollution Bulletin*, Elsevier Ltd, 115(1–2), 7–19.

Takei, Y. (2013). "Agreement on Cooperation on Aeronautical and Maritime Search and Rescue in the Arctic: an assessment." *Aegean Review of the Law of the Sea and Maritime Law*, 2(1), 81–109.

Teixeira, A.P. & Guedes Soares, C. (2010). "Reliability assessment of intact and damaged ship structures." *Advanced Ship Design for Pollution Prevention*, Guedes Soares and Parunov, eds., Taylor & Francis Group, London, 79–94.

Thompson, P.B. & Dean, W.R. (1996). "Competing conceptions of risk." *Risk: Health, Safety, Environment*, 7, 361–384.

Valdez Banda, O.A., Goerlandt, F., Montewka, J. & Kujala, P. (2015a). "A risk analysis of winter navigation in Finnish sea areas." *Accident Analysis and Prevention*, 79, 100–116.

Valdez Banda, O.A., Goerlandt, F., Montewka, J. &Kujala, P. (2015b). "A risk analysis of winter navigation in Finnish sea areas." *Accident Analysis & Prevention*, 79, 100–116.

Vanem, E., Puisa, R. & Skjong, R. (2009). "Standardized Risk Models for Formal Safety Assessment of Maritime Transportation." *ASME*, 51–61.

Vanem, E. & Skjong, R. (2006). "Designing for safety in passenger ships utilizing advanced evacuation analyses - A risk based approach." *Safety Science*, 44(2), 111–135.

Varela, J.M. & Guedes Soares, C. (2007). "A virtual environment for decision support in ship damage control." *IEEE Computer Graphics and Applications*, 27(4), 58–69.

Varela, J.M., Rodrigues, J.M. & Guedes Soares, C. (2014). "On-board decision support system for ship flooding emergency response." *Procedia Computer Science*, Elsevier Masson SAS, 29, 1688–1700.

Varela, J.M., Rodrigues, J.M. & Guedes Soares, C. (2015). "3D simulation of ship motions to support the planning of rescue operations on damaged ships." *Procedia Computer Science*, 51(1), 2397–2405.

Ventikos, N.P. (2013). "A case study for oil spillage in Greek waters." *Developments in Maritime Transportation and Exploitation of Sea Resources*, C. Guedes Soares and F. Peña López, eds., CRC Press, 853–860.

Ventikos, N.P., Lykos, G.V. & Padouva, I.I. (2012). "People may not change but perhaps their behavior can? Behavior based safety process in Short Sea Shipping safety." *Proceedings of International Research Conference on Short Sea Shipping (Short Sea Shipping 2012)*.

Ventikos, N.P., Lykos, G.V. & Padouva, I.I. (2014). "How to achieve an effective behavioral-based safety plan: the analysis of an attitude questionnaire for the maritime industry." *WMU Journal of Maritime Affairs*, 13(2), 207–230.

Vettor, R. & Guedes Soares, C. (2015). "Computational system for planning search and rescue operations at sea." *Procedia Computer Science*, Elsevier Masson SAS, 51, 2848–2853.

Vinnem, J.E. (2014). Offshore Risk Assessment (vol 1). Principles, Modelling and Applications of QRA Studies. Springer, London, UK.

Wróbel, K., Krata, P., Montewka, J. & Hinz, T. (2016). "Towards the Development of a Risk Model for Unmanned Vessels Design and Operations." *Transnav, the International Journal on Marine Navigation and Safety of Sea Transportation*, 10(2), 267–274.

Wróbel, K., Montewka, J. & Kujala, P. (2017). "Towards the assessment of potential impact of unmanned vessels on maritime transportation safety." *Reliability Engineering and System Safety*, Elsevier Ltd, 165, 155–169.

Zhang, J.F., Teixeira, A.P. & Guedes Soares, C. (2016). "Study on path planning strategies for search and rescue." *Maritime Technology and Engineering 3*, Guedes Soares & Santos, ed., Taylor & Francis Group, London, London, 937–942.

Zhang, J., Teixeira, A.P., Guedes Soares, C. & Yan, X. (2017). "Probabilistic modelling of the drifting trajectory of an object under the effect of wind and current for maritime search and rescue." *Ocean Engineering*, 129, 253–264.

Maritime Transportation and Harvesting of Sea Resources – Guedes Soares & Teixeira (Eds)
© 2018 Taylor & Francis Group, London, ISBN 978-0-8153-7993-5

Incorporation of safety barrier and HFCAS to human error analysis of major maritime accidents in China

B. Wu, L.K. Zong, X.P. Yan & Y. Wang

Intelligent Transport Systems Research Center (ITSC), National Engineering Research Center for Water Transport Safety (WTSC), Wuhan, P.R. China

ABSTRACT: Maritime transportation poses serious risk and major accidents sometimes occurred owing to sequential failure of safety barriers. This paper manages to discover the human error caused failure by using more than 10 major accidents report in China, where the human factors classification and analysis system is introduced and used to analysis the human error from a systematic perspective. The kernel of this proposed method is first to develop a framework for accident development, then utilizes the human factors classification and analysis system and safety barrier for analyzing the human error. The contributing factors of human error can be discovered and it will facilitate the understanding of major accident development in China by using "Eastern Star" accident as an example.

1 INTRODUCTION

The major accidents have received considerable attention in the recent decades. Many researchers have focused on this research in order to gain insights from major accidents. Okoh & Haugen (2013) focused on the maintenance-related major accidents, and the authors intended to investigate how maintenance impacts the occurrence of major accidents, moreover, they also managed to develop classification schemes for causes of maintenance-related major accidents by using safety barriers. Vinnem et al. (2016) discovered that maintenance of safety valves causes a few leaks in oil and gas industry. Khakzad et al. (2014) proposed a probabilistic risk assessment model by using event tree and hierarchical Bayesian analysis to analyze the major accidents, and the near miss data of offshore blowouts in the Gulf of Mexico is used for validation. Okoh & Haugen (2014) reviewed 183 detailed, major accident investigation and analysis reports from 2000 to 20111, and they intended to analyze the current and the trend of major accidents.

From the above analysis, the majority of the researches focused on the maintenance—based major accidents. In fact, Sarshar et al. (2015) concluded that the major accidents are characterized by a complex interaction of technical, human, organizational and environmental factors, and they used a number of theories of major accidents to analyze the contributing factors of offshore oil and gas operation. From the literature, there are more than three human reliability analysis methods. The first method is the human factors analysis and classification system (HFACS), this method views the human error from the retrospective perspective. Human

reliability analysis views the human error from both retrospective and prospective perspectives. The first generation technique is the technique for human error rate prediction (THERP), which assumes that the human errors are mainly caused by the inherent deficiencies of the human beings (Marseguerra et al., 2007). However, as the requirements on the working skills of the crews have changed from being manual skills to being knowledge intensive functions, the human reliability is assumed to be much more related to the contextual conditions rather than the characteristics of the tasks in engineering systems (Yang et al., 2013; Wu et al., 2017).

As this paper intends to analyze the human error of the major accidents that have been occurred in China, the HFACS method is selected. The remainder of this paper is organized as follows. Section 2 proposes the developed HFACS for human error analysis; Section 3 utilized HFACS to analysis major accidents in China, conclusions are drawn in Section 4.

2 DEVELOPED HFACS FOR HUMAN ERROR ANALYSIS

2.1 *Traditional HFACS*

The human factors analysis and classification system (HFACS) method is first derived from the aviation transportation (Wiegmann & Shappell 2001). Currently it is the most widely used method for human reliability analysis in different transportation modes. Based on this method, Celik & Cebi (2009) analyzed the human performance by incorporating the fuzzy analytical hierarchy process method and cognitive map for fire prevention (Soner et al., 2015). Moreover, Chauvin et al. (2013) and Chen et al. (2013) also carried out some researches by using HFACS. Compared with the above reviewed methods, HFACS method is so widely used because its explicit classification of the accident causation factors from a systemic perspective.

As shown in Fig. 1, the traditional HFACS views the accidents from a sequence of failures, including the organization influences, unsafe supervision, preconditions for unsafe acts, and unsafe acts. In order to simplify the modeling process, this is defined as the second level of HFACS, while the accident is assumed as the first level. Moreover, the second level can be divided into many influencing factors, and they are assumed as the third level. Take the organizational influences for example; it includes resource management, organizational climates and organizational process. The other factors in second level also can be classified by using this way, and they are shown in Fig. 1. The detailed description of these factors will be introduced in Subsection 2.3.

2.2 Proposed framework of HFACS

It can be seen that the traditional HFACS only gives the influencing factors and its descriptions. However, in practice, this method requires the operator should have good knowledge of both this method and also the domain knowledge of these influencing factors. In order to overcome this problem when applying this method to maritime transportation, the safety barrier is introduced in this paper. The proposed framework by incorporating HFACS and safety barrier is shown in Fig.2.

Safety barrier is also widely used for accident causation analysis. Xue (et al. 2013) introduced the safety barrier-based to analyze the offshore drilling blowouts, a typical process accident. The author treated the accident as a sequence of barriers failure in the process, by defining the function of each barrier, the consequences of the barrier failure were analyzed and discussed. This model is especially useful in the specific accident causation analysis, which was validated by using the Macondo accident in the case study. However, this model failed to define the scenario of the developing states in the process, which makes this model may have some weakness in analyzing the different types of accidents.

In this developed human reliability analysis model, the HFACS is divided into third level, second level, and finally the maritime accidents. Moreover, as the safety barriers may be used to prevent, control, and mitigate in the accident development process. It should be mentioned that the safety barriers can be classified into different types. Sklet (2006) thoroughly reviewed the definition, classification, and performance of the safety barrier. Since this method can intuitively demonstrate the function of accident prevention, it is widely used in the safety industry. In this paper, the author gives many different definitions of barrier types. However, as the maritime accident development is a dynamic process, the safety functions in a process model is introduced, which is shown in Fig.3.

2.3 Explanations of the influencing factors in HFACS

In order to facilitate the modeling process, the detailed descriptions of HFACS are shown in Table 1.

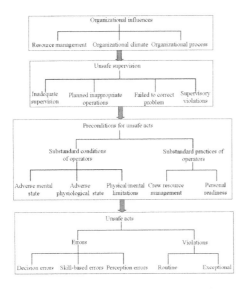

Figure 1. Framework of the traditional HFACS.

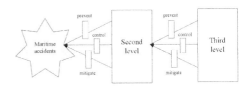

Figure 2. Developed human reliability analysis model.

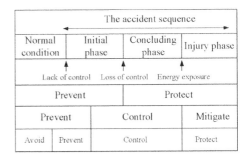

Figure 3. Safety function related to a process model.

Table 1. Detailed description of HFACS.

Second level	Third level	Explanations
Organizational influences	Resource management	Management, allocation, and maintenance of organizational resources, including human resource management (selection, training, staffing), monetary safety budgets, and equipment design.
	Organizational climate	The command structure (chain of command, delegation or authority and responsibility, communication channels, formal accountability for actions), policies (official guidelines that direct management decisions about hiring and firing, promotion, retention, raises, sick leave, drugs and alcohol, overtime, accident investigations, and the use of safety requirement), and culture (unofficial or unspoken rules, values, attitudes, and beliefs and customs of an organization).
	Organizational Process	Formal processes (operational tempo, time pressures, production quotas, incentive systems, schedules, etc.), procedures (performance standards, objectives, documentation, etc.), and oversight within the organization (organizational self-study, risk management, establishment and use of safety programs). In the shipping industry, this category includes the Safety Management System (SMS). According to the ISM code, every company should in fact develop, implement, and maintain a Safety Management System which includes the following functional requirements: (a) a safety and environmental protection policy, (b) instructions and procedures to ensure the safe operation of ships and protection of the environment in compliance with relevant international and flag State legislation; (c) defined levels of authority and lines of communication between, and amongst, shore and shipboard personnel; (d) procedures for reporting accidents and non-conformities with the provisions of this Code; (e) procedures to prepare for and respond to emergency situations; and (f) procedures for international audits and management reviews.
Unsafe Supervision	Inadequate leadership	Failure to provide guidance, failure to provide operational doctrine, failure to provide oversight, failure to provide training, failure to track qualifications, failure to track performance.
	Planned inappropriate operations	Improper or inappropriate crew scheduling and operational planning (crew paring, crew rest). The supervisor failed to provide correct data or to provide adequate briefing time. The mission is not defined in accordance with the rules (the speed, for example, may be excessive with regard to the environmental conditions). The manning may be insufficient regarding these conditions.
	Failure to correct known problems	Occurs when deficiencies among individuals, equipment, training, or other related areas are "known" to the supervisor, yet are allowed to go on uncorrected.
	Supervisory violations	Instances when existing rules, regulations, instructions, or standard operating procedures are willfully disregarded by supervisors during the course of their duties.
Preconditions for unsafe acts	Adverse mental states	Mental conditions that adversely affect performance: loss of situation awareness, mental fatigue, attention deficit, circadian dysrhythmia, complacency, and misplaced motivation.
	Adverse physiological states	Physiological, pharmacological, and medical abnormalities known to affect performance.
	Physical and/or mental limitations	This category includes instances when individuals do not have the knowledge, aptitude, skill, or time to deal safely with information.
	Ship Resource Mismanagement (SRM)	Failures of both inter—and intra-bridge communication as well as communication with or between pilots and with shore personnel. This category also includes those instances when crew members do not work together as a team or when individuals directly responsible for the conduct of operations fail to coordinate activities.
	Personal readiness failures	Occur when individuals fail to prepare physically or mentally for duty: violations of crew rest requirements, excessive physical training, self-medicating, and being under the influence of alcohol.
Unsafe acts	Routine violations	They occur every day as people regularly modify or do not strictly comply with work procedures, often because of poorly designed or defined work practices
	Exceptional violations	Behaviors that show willful disregard for the rules and regulations. For example, a merchant vessel that is navigating in the wrong direction in a Traffic Separation Scheme.
	Decision errors	Account for conscious, goal-intended behavior that proceeds as designed; yet, the plan proves inadequate or inappropriate for the situation.
	Skill-based errors	Occur with little or no conscious thought. They are related to automatic behavior. Onboard a vessel, an example could be an inadvertent activation of the rudder.
	Perceptual errors	Occur when sensory input is degraded. In collision cases, these errors are related to the fact that the crewmembers on board one vessel did not perceive the other vessel or perceived it very late.

Table 2. Requirements to barrier quality.

Quality/criterion	Specific requirement
Adequacy	Able to prevent all accidents within the design basis.
	Meet requirements set by appropriate standards and norms.
	Capacity must not be exceeded by changes to the primary system.
	If a barrier is inadequate, additional barriers must be established.
Availability, reliability	All necessary signals must be detectable when barrier activation is required.
	Active barriers must be fail-safe, and either self-testing or tested regularly.
	Passive barriers must be inspected routinely.
Robustness	Able to withstand extreme events, such as fire, flooding, etc.
	The barrier shall not be disabled by the activation of another barrier.
	Two barriers shall not be affected by a (single) common cause.
Specificity	The effects of activating the barrier must not lead to other accidents.
	The barrier shall not destroy that which it protects.

Majority of the explanations are adopted from (Chauvin et al., 2013), which also focuses on the maritime transportation. Moreover, some of the explanations are adopted from (Chen et al., 2013).

2.4 Descriptions of safety barriers

In order to identify the functions of safety barriers, the following questions should be considered to understand the function mechanism for maritime transportation.

1. Barriers that were in place and how they performed.
2. Barriers that were in place but not used.
3. Barriers that were not in place but were required.

Moreover, Hollnagel (2004) proposed the requirements of barrier quality from four criterions, which are adequacy, availability and reliability, robustness and specificity. The detailed descriptions of the requirements are shown in Table 2. In practice, when judging something whether it is a safety barrier, these four requirements can be used.

3 MAJOR ACCIDENTS ANALYSIS USING HFACS

3.1 Definitions of major accidents in China

In China, the maritime accidents are defined as negligible, minor, major, catastrophic, and the standard to define these classifications is shown in Table 3, which has been introduced by Zhang (et al., 2016) when analysis the maritime risk in the Tianjin Port.

It can be seen from Table 3 that the fatalities are all the same for the different types of ship tonnage. However, the economic losses differ from the different ship tonnage, therefore, when defining

the economic loss, the ship tonnage is taken into consideration, but the ship tonnage is not the parent node of the number of fatalities. Finally, the number of fatalities and economic loss can be used for defining the consequences of collision accidents, which is defined by the Table 3.

3.2 Data collection of major accidents

As shown in Table 3, the major and catastrophic may cause serious consequence; therefore, this paper collected the maritime accidents that caused major and catastrophic consequence for further research. The collected data are shown in Table 4. It can be seen this table that there are 11 catastrophic accidents occurred from the period of 1998 to 2016. However, it should be mentioned that since 2015, the standard of maritime consequence has changed. The major consequence has changed from 1–2 fatalities to 10 fatalities, while the catastrophic consequence has changed from over 2 fatalities to 30 fatalities. Therefore, if using this updated standard, some cases may be classified as different consequence. Moreover, some cases may also not meet the standard of major accident in the new standard. However, as the majority of the accidents occurred before 2015, the old standard is used in this paper but only the catastrophic accidents are considered in the old standard, and they are all assumed as major accidents.

3.3 Human error analysis using "eastern star" as an example

In 2015, the "eastern star" ship, when navigating in the midstream of Yangtze River, this ship was foundering owing to the heavy weather. This accident caused 445 fatalities after more than 5 days search and rescue. The Chinese Sate Council issued the investigation report after more than 7 months investigation. In this investigation, more than 1607

Table 3. Classification on maritime consequence from MoT.

Ship tonnage	Negligible	Minor	Major	Catastrophic
Ships over 3000 gross tonnage	Below minor accident	Serious injury, or economic loss between 500k and 3000k RMB	1–2 fatalities, or economic loss between 3000k and 5000k RMB	Over 2 fatalities, or economic loss over 5000k RMB
Ships between 500 and 3000 gross tonnage	Below minor accident	Serious injury, or economic loss between 200k and 500k RMB	1–2 fatalities, or economic loss between 500k and 3000k RMB	Over 2 fatalities, or economic loss over 3000k RMB
Ships below 500 gross tonnage	Below minor accident	Serious injury, or economic loss between 100k and 200k RMB	1–2 fatalities, or economic loss between 200k and 500k RMB	Over 2 fatalities, or economic loss over 500k RMB

Table 4. Collected data of major accidents in China.

Year	Consequence	Ship involved	Fatalities
1998	Catastrophic	Chuangjiangandu 16	5
1999	Catastrophic	Dashun	282
2000	Catastrophic	Rongjian	130
2001	Catastrophic	Bozhoudu 081	23
2001	Catastrophic	Yuhechuangke 00110	46
2002	Catastrophic	Changyun 1	40
2003	Catastrophic	Fuzhou 10	27
2012	Catastrophic	Mahedu 1004	11
2015	Catastrophic	Wanshenzhou 67	22
2015	Catastrophic	Eastern Star	442
2016	Catastrophic	Chuangguangyuanke 1008	15

copy of materials have been collected, and more than 500 000 words of inquiry records have also been collected. Moreover, the restore of ship trajectory was carried out by using AIS and GPS data, and the ship manoeuvring simulation is carried out to evaluate whether the ship handling is correct or not.

From the accident investigation report, it can be seen that the following information can be obtained.

1. The accidental ship has meet 5 ships in the accidental scene, and these 5 ships have been anchored after perceived the unsafe situation.
2. The ship captain and officer on water did not request for help and also did not broadcast the safety information to the passengers.
3. The ship condition is not good in the heavy sea after three times ship reconstruction but the abovementioned anchored ships is designed for the heavy weather.
4. The safety management system for the ship company is bad.

This information is useful to analyze the human error when applying the HFACS method. From Table 2, the factors in second level, including organizational influences, unsafe supervision, preconditions for unsafe acts, and unsafe acts, can be discovered

from the investigation report. This will not go in detail as the third level will be detailed explained.

The ship fail to detect the unsafe situation and it can be classified as perception errors, this is owing to the other 5 ships have anchored after perceiving the unsafe situation (in fact, the other 5 ships have better ship condition than the eastern star). From the perspective of safety barrier, it is the prevention barrier failed to work. The ship officer failed to request for help is assumed to be inadequate leadership, and this safety barrier is assumed to be control barrier. The ship condition is not good can be assumed to be resource management and perception errors, and the safety barrier is assumed to be prevention. The safety management system is assumed to be organizational process, and this is also assumed as prevention barrier.

For further research, more human and organizational factors can be discovered. First, in the Yangtze River, the SART and EPIRB is not installed in the ship, and the accident is discovered after the crews have escaped from the accidental scene, which is 2 or 3 hours later after foundering. This can be assumed as the resource management, and it is the prevention barrier. Second, the crews did not give suggestions to the caption or the caption did not adopt the suggestions when the ship was navigating in such heavy weather and in emergent situation, this is assumed to be ship resource management and fail to correct know problems; the safety barrier is prevention barrier. Third, the caption did not organize the crews and passengers to evacuate in the emergent situation, this is assumed as exceptional violations and the safety barrier is a mitigate barrier.

4 CONCLUSIONS

The main contribution of this paper is to propose the HFACS and safety barrier based method to analyze the human and organizational factors for major accidents in China. When only applying the HFACS method to human reliability analysis, the working mechanism of the safety barriers may be ignored in this modelling process. Moreover, if the safety barrier is not considered, the in-depth human and organizational factors also cannot be discovered.

In the above analysis, the importance of the different types of human and organizational factors is not considered. This can be done in the future, moreover, the different other maritime accidents can also be analyzed in the future to have a better understanding of the major accidents in China.

ACKNOWLEDGEMENTS

The research presented in this paper was sponsored by a grant from the Key Project in the National Science & Technology Pillar Program (Grant No.2015BAG20B05), grants from National Science Foundation of China (Grant No. 51609194), grants from the special funds of Hubei Technical Innovation Project (Grant No. 2016 AAA055) and the Fundamental Research Funds for the Central Universities (WUT: 2017IVA103).

REFERENCES

Celik, M., & Cebi, S. (2009). Analytical HFACS for investigating human errors in shipping accidents. Accident Analysis & Prevention, 41(1), 66–75.

Chauvin, C., Lardjane, S., Morel, G., Clostermann, J. P., & Langard, B. (2013). Human and organisational factors in maritime accidents: Analysis of collisions at sea using the HFACS. Accident Analysis & Prevention, 59, 26–37.

Chen, S. T., Wall, A., Davies, P., Yang, Z., Wang, J., & Chou, Y. H. (2013). A Human and Organisational Factors (HOFs) analysis method for marine casualties using HFACS-Maritime Accidents (HFACS-MA). Safety science, 60, 105–114.

Hollnagel, E. (2004). Barrier and accident prevention. Hampshire, UK: Ashgate.

Khakzad, N., Khakzad, S., & Khan, F. (2014). Probabilistic risk assessment of major accidents: application to offshore blowouts in the Gulf of Mexico. Natural hazards, 74(3), 1759–1771.

Marseguerra, M., Zio, E., & Librizzi, M. (2007). Human reliability analysis by fuzzy "CREAM". Risk Analysis, 27(1), 137–154.

Okoh, P., & Haugen, S. (2013). Maintenance-related major accidents: classification of causes and case study. Journal of Loss Prevention in the Process Industries, 26(6), 1060–1070.

Okoh, P., & Haugen, S. (2014). A study of maintenance-related major accident cases in the 21st century. Process Safety and Environmental Protection, 92(4), 346–356.

Sarshar, S., Haugen, S., & Skjerve, A. B. (2015). Factors in offshore planning that affect the risk for major accidents. Journal of Loss Prevention in the Process Industries, 33(33), 188–199.

Soner, O., Asan, U., & Celik, M. (2015). Use of HFACS–FCM in fire prevention modelling on board ships. Safety Science, 77, 25–41.

Sklet, S. (2006). Safety barriers: Definition, classification, and performance. Journal of loss prevention in the process industries, 19(5), 494–506.

Vinnem, J. E., Haugen, S., & Okoh, P. (2016). Maintenance of petroleum process plant systems as a source of major accidents?. Journal of Loss Prevention in the Process Industries, 40, 348–356.

Wiegmann, D. A., & Shappell, S. A. (2001). Human error analysis of commercial aviation accidents: Application of the Human Factors Analysis and Classification System (HFACS). Aviation, space, and environmental medicine, 72(11), 1006–1016.

Wu, B., Yan, X., Wang, Y., & Soares, C. G. (2017). An Evidential Reasoning—Based CREAM to Human Reliability Analysis in Maritime Accident Process. Risk Analysis.

Xue, L., Fan, J., Rausand, M., & Zhang, L. (2013). A safety barrier-based accident model for offshore drilling blowouts. Journal of loss prevention in the process industries, 26(1), 164–171.

Yang, Z. L., Bonsall, S., Wall, A., Wang, J., & Usman, M. (2013). A modified CREAM to human reliability quantification in marine engineering. Ocean Engineering, 58, 293–303.

Zhang, J., Teixeira, Â. P., Guedes Soares, C., Yan, X., & Liu, K. (2016). Maritime transportation risk assessment of Tianjin Port with Bayesian belief networks. Risk analysis, 36(6), 1171–1187.

Hydrodynamics

Maritime Transportation and Harvesting of Sea Resources – Guedes Soares & Teixeira (Eds)
© *2018 Taylor & Francis Group, London, ISBN 978-0-8153-7993-5*

On the influence of the parametric roll resonance on the principal wave-induced loads

M. Acanfora & T. Coppola

Industrial Engineering Department, Università degli studi di Napoli "Federico II", Italy

ABSTRACT: A safer and more reliable ship structure design needs continuous efforts in improving the estimation of the environmental critical loads. Full-scale observations and model tests have shown that the phenomenon of parametric roll resonance can occur in head, following and oblique seas. In this paper, we aim to qualitatively assess the variation of the internal loads in waves of a containership, in presence of parametric roll resonance. A hybrid non-linear simulation model in 6 degree of freedom will be used to evaluate ship dynamics in waves and to calculate shear force and structural moment distributions. In particular, the nonlinear restoring and non-linear Froude-Krylov action are implemented, together with non-linear inertial loads. Numerical time domain simulations will be carried out for a containership, in regular waves.

1 INTRODUCTION

The accurate prediction of wave-induced forces and moments on a hull operating in severe weather conditions plays an important role in assessing ship structural strength. Longitudinal strength calculation is a standard practice for evaluating the primary stresses on ship structures.

Internal loads in waves are usually obtained by linear strip-theory approach (Faltinsen 1990) that allows for spectral analysis in frequency domain, accounting for all the possible wave height—period combinations.

Nowadays, linear and non-linear time domain seakeeping codes represent the two main standards for direct calculation of ship motion and structural loads in waves (Beck & Reed 2001). Although linear analysis remains valuable for long term statistical analysis, it cannot model some motion instabilities such as the parametric roll resonance (Acanfora & Matusiak 2014). It is well known that this phenomenon is caused by restoring lever variation in waves (Paulling 2011). Full-scale observations and model tests have shown that the phenomenon of parametric roll resonance can occur in head, following and oblique seas. The occurrence of the phenomenon is mainly related to the submerged hull variation during the wave passage, that characterizes the new generation of containerships (Silva et al. 2010).

Despite several numerical and experimental studies have been carried out on parametric roll regarding ship motion and stability in waves (Hashimoto et al. 2008) (Silva et al. 2010; Lu et al. 2016; Galeazzi et al. 2013; Reed 2011), not sufficient attention has been given to its influence on induced internal loads for a ship.

In event of parametric roll, wave-induced loads could be influenced by larger ship motions. This concept is based on the evidence that the combination of all the ship rigid body motions could lead to larger loads on the hull, (such as inertial, restoring and Froude-Krylov loads), and thus to unexpected internal forces and moments on the hull structures.

Several research works focus on the non-linear effects of structural loads in head sea and oblique sea (Kukkanen & Matusiak 2014; Jensen et al. 2009; Fonseca & Guedes Soares 2004; Zhu & Moan 2014; Acanfora & Coppola 2014) for containership and ro-ro pax hulls. These type of vessels are particularly prone to non-linear effects in waves concerning the change of the immersed hull volume and the change of the shape of the waterplane. These non-linear effects are also the main reason for parametric roll development (Reed 2011).

In this paper a simulation model based on a panel discretization of the hull (Acanfora & Cirillo 2017b; Acanfora & Cirillo 2017a) and accounting for all the pertinent non-linearities, is enhanced by implementing a routine that calculates the hull girder loads in waves.

The applications of the implemented code will be carried out on the containership named C11, that has been extensively studied concerning parametric roll (Silva et al. 2010; Hashimoto et al. 2011; France et al. 2003; Acanfora et al. 2017b). The aim of this paper is to observe the effects of parametric roll on structural loads such as shear forces and moments (bending horizontal and torsional moments), in order to characterize the large openings in the deck of container-ship structures.

This is carried out by means of a qualitative analysis.

In the following section we introduce the numerical model implemented in Matlab-Simulink, focusing on the development of the routine for the calculation of the internal loads on hull girder. The model is then checked against numerical data obtained by an available linear strip theory code (Faltinsen 1990). Finally, the non-linearities of the internal loads are simulated and studied with particular attention to a condition where parametric roll develops. The present investigation refers only to regular waves.

2 THE NUMERICAL MODEL

The mathematical model presented in this section is meant to be a fast numerical simulation tool for ship motions in waves. It has been developed and applied in previous research works (Acanfora & Cirillo 2017b; Acanfora & Cirillo 2017a) showing a good agreements of the results with the available experimental data.

In the current paper we implement an additional routine, that discretizes all the forces applied on the intact ship, in hull girder loads and then estimates shear forces together with bending, horizontal and torsional moments.

In the following subsections a brief description of the ship dynamic model is provided, along with the equations for calculating internal loads acting on the rigid hull girder.

2.1 Model for the dynamics of the ship

The equations of motion of a ship in the time domain, for six degrees of freedom (6DoF), are presented in Equation 1 and 2. They refer to a body fixed co-ordinate system, centered at the initial center of gravity G of the vessel. The earth fixed axes, instead, present the XY plane coincident with the still water level. These two reference frames are shown in Fig. 1.

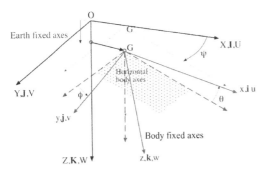

Figure 1. Reference frame for the ship model (Matusiak 2013).

The ship is assumed to be a rigid body, with a constant mass m, and matrix of inertia $\mathbf{I}$.

$$m[\dot{\mathbf{u}} + \boldsymbol{\omega} \times \mathbf{u}] = \mathbf{f}_{ext} + m\mathbf{g} \tag{1}$$

$$\mathbf{I}\dot{\boldsymbol{\omega}} + \boldsymbol{\omega} \times \mathbf{I}\boldsymbol{\omega} = \mathbf{m}_{ext} \tag{2}$$

The numerical integration is carried out in Matlab/Simulink by means of Runge-Kutta ODE45 algorithm. Ship linear and angular motions are solved from Equation 1 and 2 based on the conservation of momentum (Manderbacka et al. 2011). The velocity vector and the angular velocity vector, of the ship motions, are respectively $\mathbf{u}$ and $\boldsymbol{\omega}$. The external forces and moments, namely $\mathbf{f}_{ext}$ and $\mathbf{m}_{ext}$ consist of restoring, radiation and wave actions. The ship speed is achieved by applying a simple proportional-integral (PI) control law to keep it constant, since no resistance and no thrust are modeled.

The wave actions consist of diffraction and Froude-Krylov forces and moments. The simulation model works on a discrete representation of the hull, using triangular panels. Restoring and Froude-Krylov forces are evaluated by taking into account all the pertinent non-linearities (Matusiak 2013). The hydrostatic and the dynamic pressure are calculated assuming the actual wetted surface of the hull in wave, accounting for its motions in time.

The diffraction actions and the radiation forces, that are added mass and damping, are obtained from linear potential theory (Faltinsen 1990). More details for the external action numerical model (intact ship), can be found in the reference work (Matusiak 2013; Acanfora & Cirillo 2017a; Acanfora & Cirillo 2017b).

2.2 Sub-Model for internal loads

For any ship cross section, defined by the x-coordinate in the body frame of Figure 1, the two characteristic vectors (see Equation 3 and 4), can be introduced:

$$\mathbf{R}(x,t) = N(x,t)\mathbf{i}(t) + T_y(x,t)\mathbf{j}(t) + T_z(x,t)\mathbf{k}(t) \tag{3}$$

$$\mathbf{M}(x,t) = M_x(x,t)\mathbf{i}(t) + M_y(x,t)\mathbf{j}(t) + M_z(x,t)\mathbf{k}(t) \tag{4}$$

The axes-components of $\mathbf{R}(x,t)$ and $\mathbf{M}(x,t)$, are usually called the shear and compression forces and, respectively, the bending and torsional moments, relative to the considered x cross-section. For distributed load expressions, Equation 5 applies, where $G(x)$ is the center of each section coincident with the structural center of the main section.

$$\left|\begin{array}{l} N_x(x,t) = \int_0^x c_x(x,t)dx \\[2mm] T_y(x,t) = \int_0^x c_y(x,t)dx \\[2mm] T_z(x,t) = \int_0^x c_z(x,t)dx \\[2mm] M_x(x,t) = \int_0^x m_x(x,t)dx \\[2mm] M_y(x,t) = \int_0^x T_z(x,t) + m_y(x,t)dx \\[2mm] M_z(x,t) = \int_0^x \left(m_z(x,t) - T_y(x,t) \right)dx \end{array}\right. \qquad (5)$$

where:

$c(x,t) = \frac{\partial P(x,t)}{\partial x}$ is the linear density of the load, whose components are $[c_x(x,t), c_y(x,t), c_z(x,t)]$, (see Equation 6);

$m(x,t) = \frac{\partial MG(x)(x,t)}{\partial x}$ is the linear density of the load moment with $[m_x(x,t), m_y(x,t), m_z(x,t)]$ (see Equation 8);

$p(x,t)$ is the distributed load aft-ward the x-cross section;

$MG(x)(x,t)$ is the moment around $G(x)$, of the distributed load aft-ward the x-cross section.

In the body fixed frame (Fig. 1), the forces per unit length of a hull in waves can be expressed according to Equation 6. Where $m(x)$ is distributed ship mass, $\theta(t)$ and $\varphi(t)$ are respectively the pitch and roll angle of the ship, $f_{rest} = [f_{x,rest} \, f_{y,rest} \, f_{z,rest}]$ are the linear densities of the restoring forces, $f_{fk} = [f_{x,fk} \, f_{y,fk} \, f_{z,fk}]$ are the linear densities of the Froude-Krylov forces and the acceleration components (a_x, a_y, a_z) are calculated in the center of each x cross-section $G(x) = [x_c, y_c, z_c]$ according to the Equation 7.

radiation and diffraction actions. For these actions the linear density forces are obtained by means of an approximate uniform distribution having its center in the ship center of gravity G.

In the following we assume that the center of each cross section $G(x)$ lies on a longitudinal axis passing through the ship center of gravity rather than the structural center of the main section. It is always possible to obtain the internal moments with respect to a different center by the application of a proper transformation matrix.

We consider the exact Froude-Krylov and restoring linear density moments together with the approximate uniform distributions of radiation and diffraction moments. Thus the linear density of the load moments can be expressed as follows:

$$\left|\begin{array}{l} m_x(x,t) = m_{x,rest}(x,t) + m_{x,fk}(x,t) + m_{x,r}(x,t) - m(x)k_{xx}^2 \dot{p}(t) \\[2mm] m_y(x,t) = m_{y,rest}(x,t) + m_{y,fk}(x,t) + m_{y,r}(x,t) \\[2mm] m_z(x,t) = m_{z,rest}(x,t) + m_{z,fk}(x,t) + m_{z,r}(x,t) \end{array}\right.$$

$$(8)$$

For the moment around the x-axis we also need to introduce the inertial contribution $m(x)k_{xx}^2 \dot{p}(t)$ where k_{xx} is the roll radius of inertia assumed to remain constant along the ship length.

The linear density of the Froude-Krylov moments $m_{fk} = [m_{x,fk}, m_{y,fk}, m_{z,fk}]$ arise from the fact that the forces on each x-cross section do not reduce exactly in the center of gravity $G(x)$ of the cross-section. The same applies to the linear density function of the restoring moments $m_{rest} = [m_{x,rest}, m_{y,rest}, m_{z,rest}]$.

The uniform linear density of the radiation and diffraction moments $m_r = [m_{x,r}, m_{y,r}, m_{z,r}]$, that are disregarded by the assumption introduced on the corresponding forces f_r, is needed to balance ship moments.

$$\left|\begin{array}{l} c(x,t)i(t) = c_x(x,t) = -m(x)g\sin\theta(t) + f_{x,rest}(x,t) - m(x)a_x(x,t) + f_{x,fk}(x,t) + f_{x,r}(x,t) \\[2mm] c(x,t)j(t) = c_y(x,t) = m(x)g\cos\theta(t)\sin\varphi(t) + f_{y,rest}(x,t) - m(x)a_y(x,t) + f_{y,fk}(x,t) + f_{y,r}(x,t) \\[2mm] c(x,t)k(t) = c_z(x,t) = m(x)g\cos\theta(t)\cos\varphi(t) + f_{z,rest}(x,t) - m(x)a_z(x,t) + f_{z,fk}(x,t) + f_{z,r}(x,t) \end{array}\right. \qquad (6)$$

$$\left|\begin{array}{l} a_x = \dot{u} - vr + wq + \dot{q}z_c - \dot{r}y_c - (r^2 + q^2)x_c + prz_c + qpy_c \\[2mm] a_y = \dot{v} + ur - wq + rqz_c - \dot{r}x_c - (r^2 + p^2)y_c - \dot{p}z_c + qpx_c \\[2mm] a_z = \dot{w} - uq + vp + \dot{p}y_c - \dot{q}x_c - (q^2 + p^2)z_c + rpx_c + qry_c \end{array}\right. \qquad (7)$$

They take into account all the ship rigid body motions, for evaluating the linear density of the inertial loads.

All the above mentioned linear density forces are obtained with their exact value along the ship length, except for the $f_r = [f_{x,r}, f_{y,r}, f_{z,r}]$ that comprises

2.3 Evaluation of the linear density for restoring and Froude-Krylov forces and moments

The hull surface is discretized by means of panels, allowing for the pressure integration techniques on each panel of the wetted surface. For each panel the surface S and its normal vector $n = [n_x, n_y, n_z]$ are known. The numerical model calculates both the Froude-Krylov and restoring forces per unit length, by integrating the contribution of the all the panels across the x-cross section, indicated by i, divided by λ that is the distance between the

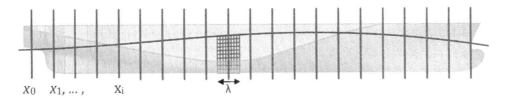

x_0 $x_1, ..., $ x_i λ

Figure 2. Example of the calculation of forces per unit length from panel discretization of the hull.

cross-sections (see Fig. 2). The Froude-Krylov and restoring moments per unit length, with respect to the center of gravity of each cross section $G(x)$, are calculated based on the same approach.

$$\begin{cases} f_{fk}(x_i) + f_{rest}(x_i) = \sum_i (p_i S_i n_i / \lambda) \\ m_{fk}(x_i) + m_{rest}(x_i) = \sum_i (p_i S_i (r_i \times n_i) / \lambda) \end{cases} \quad (9)$$

The stretched pressure method is used (Matusiak 2011).

$$p = \rho g[\zeta e^{-k(Z_C + \zeta)} + Z_C] \quad (10)$$

In Equation 10, ζ is the wave profile (see Equation 11), k is the wave number and Z_c is the local draft of any panel. All the coordinates $[X_c, Y_c, Z_c]$ of the center of each panel are evaluated in body fixed frame, accounting for the actual ship motions. It is important to underline that ω in Equation 11, is the wave frequency since the longitudinal coordinate X_c depends on ship forward speed V_S. The angle μ is the wave heading.

$$\zeta = A\cos[k(X_c\cos\mu - Y_c\sin\mu) - \omega t] \quad (11)$$

3 APPLICATIONS

The ship chosen for the application is the containership C11 whose main dimensions are reported in the Table 1. The C11 is a post-Panamax ship that has been extensively studied for being prone to parametric roll (France et al. 2003; Acanfora et al. 2017a; Silva et al. 2010).

The hull is discretized by 146196 triangular panels (see Fig. 2). It is found that more than the half of the panels has a length greater than 0.5 m while the largest panel has a length of almost 1.5 m.

The ship mass $m(x)$, has been distributed along the cross sections, in absence of reference data, according to the main standard for preliminary ship design, (see Fig. 3).

Table 1. Principal particulars of the C11 hull.

Dimension	Full scale
Length between perpendiculars, L (m)	262.0
Breadth, B (m)	40.0
Depth, D (m)	24.45
Draft molded on forward perpendicular, T_F (m)	11.72
Draft molded on aft perpendicular, T_A (m)	12.856
Displacement, Δ (tons)	76020
Center of gravity above keel, KG (m)	17.34
Transverse metacentric height, GM_T (m)	2.075
LCG measured from aft perpendicular (m)	122.78
Transverse radius of gyration in air, k_{XX} (m)	16.73
Longitudinal radius of gyration in air, k_{YY} (m)	62.55
Natural roll period, T_n (s)	25.2

It is assumed a constant interval between the frames of $\lambda = 13.1$ m that leads to 21 cross-sections from AP (aft perpendicular) to FP (forward perpendicular).

4 RESULTS

It is acknowledged the fact that the developed model for internal loads based on the panel discretization of the hull, presents some simplifications that need to be checked, prior using the code for further analysis. In this section the reliability of the implemented code is briefly tested by comparison of some results with the outcomes of a linear strip theory code. It is worth to recall that a linear strip theory code discretizes the ship into cross sections and it is based on potential theory equations that are solved on each cross section. The strip theory code used for the comparison of the results, is Seaload based on the equation presented in (Faltinsen 1990). The code is run by using 21 sections and $\lambda = 13.1$ m with the same mass distribution of Fig. 3.

4.1 Still water analysis

The first comparison concerns still water results by applying the developed panel method and the potential strip theory code.

In Fig. 4 the vertical shear force in still water is shown: it is positive according to the z-axis of Fig. 1, that is positive pointing down.

The bending moment in still water (hogging positive, sagging negative) is presented in Fig. 5. In both Fig. 4 and Fig. 5, the blue line (round dots) represents the results obtained from strip theory analysis, while the orange line (triangular dots) represents the results of the developed panel method. It is possible to observe somewhat small gaps in the shear and in the bending moment results, that are mainly due to the different hull discretizations of the two methods. Nevertheless, for the still water analysis there is a good match of the results, with an overall error, less than 1%.

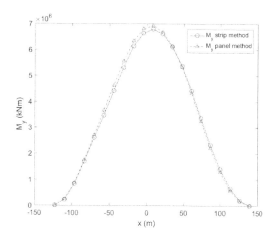

Figure 5. Bending moment distribution in still water for the developed method (panel) vs the linear strip theory method.

4.2 Small wave amplitude results

The simulation in wave by means of the non-linear panel method are carried out for a constant wave amplitude a = 0.5 m (that corresponds to a/L = 0.002) in order to compare the results with the linear strip theory results.

The hull is sailing in head sea with a constant speed of $V_S = 9.43$ m/s that corresponds to $F_N = 0.186$.

In Fig. 6 the shear force in regular waves is plotted for different wave frequencies. The reported shear force in wave, according to the linear analysis, represents the amplitude of the shear force oscillation around the still water value. The observed section is the number 15, that is at x = 73.723 m where the maximum absolute value of the vertical shear force is found (see Fig. 4).

In Fig. 6 it is possible to notice a good agreement of the results, although the panel method results do not exactly match the strip method results at the same frequencies.

A similar behavior can be observed for the bending moment. The amplitude of the bending moment oscillation in wave, around the still water value, at the mid-ship section, (that is the section number 10, where the maximum absolute bending moment in still water is found) is presented in Fig. 7. Also in this case there is not an exact match between the results of the two different methods along the wave frequencies, although the curves show a similar trend.

Except for the actions that accounts for all the pertinent non-linearities, that are weight, inertia, restoring and Froude-Krylov actions, the remaining actions that are diffraction and radiation are

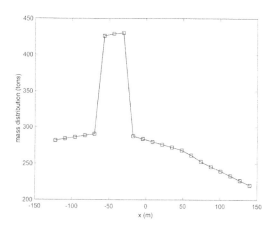

Figure 3. Weight distribution for the C11 hull. x = 0 corresponds to the ship longitudinal center of gravity G.

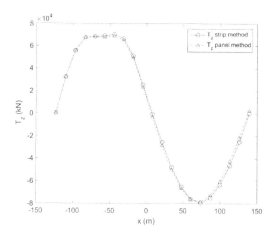

Figure 4. Vertical shear force distribution in still water for the developed method (panel) vs the linear strip theory method.

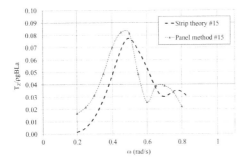

Figure 6. Shear force comparison for small wave amplitude between the strip theory code and the developed panel method code. Head sea, $F_N = 0.186$.

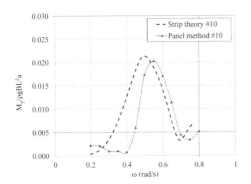

Figure 7. Bending moment comparison for small wave amplitude between the strip theory code and the developed panel method code. Head sea, $F_N = 0.186$.

assumed with a uniform distribution. This assumption was introduced for its simplicity, due to the unavailability of distributed sectional data output from the available code. Actually the assumption of uniform distribution of radiation and diffraction actions is retained far from a realistic distribution (as assumed by the strip theory code) and could explain the gap between the results. Nevertheless, the overall matching of the results is considered satisfactory despite the assumptions introduced in the developed panel method.

It is also worth to underline that the wave amplitude value chosen for the small amplitude application, that is 0.5 m, is close to the length of the majority of the triangles that constitute the panel hull discretization. The sensitivity of the developed panel method to all the geometrical and physical approximation with respect to the internal loads will be investigated in future works.

4.3 Non-linear behavior of the internal loads

In this subsection, the developed panel method is applied in order to disclose the effects of non-

linearities on the internal loads of the C11 hull. In Fig. 8 the maximum and the minimum peaks of the shear force are shown for different wave amplitudes. The analyzed condition refers to head sea navigation with $F_N = 0.186$ for two different wave amplitudes that are a/L = 0.002 and a/L = 0.015 (the latter corresponds to a = 4 m). The maximum and minimum peaks refer to the oscillation of the shear force in wave around the still water value of the shear force (that is subtracted from the plotted data).

It is possible to observe in Fig. 8 that the small wave amplitude results are symmetric. The large wave amplitude results instead show different values of the minimum and maximum peaks of the shear force. This asymmetric behavior is due to the effects of nonlinearities on the internal loads in wave.

It is possible to notice the same outcomes for the bending moment in Fig. 9. In particular, for the wave amplitude a/L = 0.015 the minimum peaks of the bending moment, that are sagging conditions, are more severe than the hogging conditions in wave, that are the maximum peaks. Similar out-

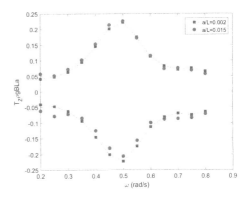

Figure 8. Maximum and minimum peaks of the shear force evaluated by the panel method code at different wave amplitudes. Head sea, $F_N = 0.186$.

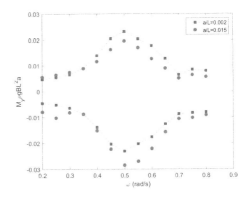

Figure 9. Maximum and minimum peaks of the bending moments evaluated by the panel method code at different wave amplitudes. Head sea, $F_N = 0.186$.

comes were found in (Kukkanen & Matusiak 2014) for a ro-ro pax hull, where a clear difference in sagging and hogging maximum values of internal loads was observed experimentally.

5 THE EFFECTS OF PARAMETRIC ROLL ON INTERNAL LOADS

The non-linear effects on ship dynamics and on ship internal loads are simulated in this section. The C11 hull is well known to be prone to parametric roll, thus it will observed how the development of the parametric resonance could influence the horizontal and the torsional moment. The latter has a peculiar relevance for containership structures, characterized by large openings in the deck that results in a reduced torsional resistance.

The application is conducted for the ship sailing at a speed of $V_s = 9.43$ m/s in head sea with wave amplitude $a = 4$ m (that is a/L = 0.015). The wave frequency that triggers out parametric roll is found equal to $\omega = 0.4$ rad/s. In the above mention condition, the ratio between the wave length λ_w and the ship length L is $\lambda_w/L = 1.47$ and the encounter frequency $\omega_e = 0.554$ rad/s.

In Fig. 10 a time history of the simulated case is shown. It is possible to observe a development of the parametric roll up to an amplitude of 40°; for the current application no viscous roll damping was accounted for. The reported forces and moment refers to section number 10 and they are expressed in the reference frame of Fig. 1. The first sub-plot shows the wave encounter by the ship that has $\omega_e = 0.554$ rad/s. The measured roll frequency is $\omega_\varphi = 0.275$ rad/s that gives a ratio $\omega_\varphi/\omega_e = 0.496$, that is expected in event of parametric resonance (Reed 2011). The time histories of the internal loads of section number 10 are plotted as well. The torsional and horizontal moments, together with horizontal shear force develop with the roll motion; it is also possible to observe a non-regular behavior of these time histories.

It is possible to evaluate the mean values of the internal loads, referring only to the time history where parametric roll is fully developed (see Fig. 10, for $t > 1700$ s). They are reported in Fig. 11 as continuous orange line. Due to the non-regular behavior of the time histories, it is not possible to define a regular amplitude of the internal loads around the mean value. It is thus evaluated the half distance between the maximum and minimum peaks of each internal load. In Fig. 11 the half distance between the max-min peaks, summed to the correspondent mean value, is plotted by dotted-line with blue marks. Although the horizontal shear load, the horizontal bending moment and the torsional moment shows an almost null mean values, they are characterized by peaks comparable with

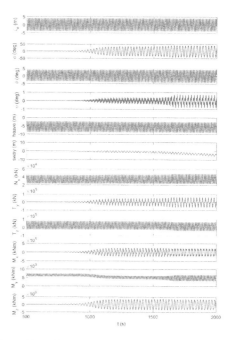

Figure 10. Time history of the ship with development of parametric roll. Internal loads refer to section #10 (mid-ship).

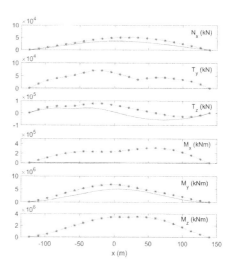

Figure 11. Internal loads in parametric roll condition (a = 4 m, $V_s = 9.43$ m/s, head sea, $\omega = 0.4$ rad/s) on each section. Mean values are in orange (continuous line); half distance between min and max peaks in blue marks (dotted marked line).

the vertical shear vertical and bending moment order of magnitude. The outcomes of this application disclose the need of future studies for assessing the severity of the consequences on ship structures in event of parametric roll.

6 CONCLUSIONS

In this paper a method for evaluating the effects of parametric roll on internal loads has been developed. Prior the commencement of the applications, the reliability of the results of the implemented method has been checked. This has been carried out by comparisons with linear strip theory results by assuming small waves amplitude. Despite the introduced approximations, it was found a fine match of the results. The non-linear behavior of the code was also observed: reasonable outcomes were found.

The implemented method, applied to an operational condition in head sea, prone to parametric roll, disclosed a modification in the vertical shear force and bending moment in wave. Moreover it resulted in the development of torsional moment and horizontal shear and bending moment. For a containership, characterized by structural sections with large openings, torsional loads could result in large stresses and deformations. The outcomes of the applications underlines the need for further investigations of parametric roll consequences on internal loads, including also oblique waves conditions. In future works, the numerical method will be enhanced and validated, allowing for a quantitative analysis.

REFERENCES

Acanfora, M. et al., 2017a. On the estimation of the design loads on container stacks due to excessive acceleration in adverse weather conditions. *Marine Structures*, 53, pp.105–123. Available at: http://dx.doi.org/10.1016/j.marstruc.2017.01.003.

Acanfora, M. et al., 2017b. Towards realistic estimation of ship excessive motions in heavy weather. A case study of a containership in the Pacific Ocean. *Ocean Engineering*, 138(March), pp. 140–150. Available at: http://dx.doi.org/10.1016/j.oceaneng.2017.04.025.

Acanfora, M. & Cirillo, A., 2017a. On the development of a fast modeling of floodwater effects on ship motions in waves. *Proceedings of the Institution of Mechanical Engineers, Part M: Journal of Engineering for the Maritime Environment*, p.147509021668743. Available at: http://journals.sagepub.com/doi/10.1177/1475090216687438.

Acanfora, M. & Cirillo, A., 2017b. On the intact stability of a ship in head and following sea: an analysis of the dynamic roll angle due to sudden heeling moments. *Journal of Marine Science and Technology*, 0(0), p.0. Available at: http://link.springer.com/10.1007/s00773-017-0446-x.

Acanfora, M. & Coppola, T., 2014. On the Geometrical Non-linearities of the Ship Load Expressions. In *Second International Symposium on Naval Architecture and Maritime, INT-NAM, 23–24 October 2014*. Istanbul.

Acanfora, M. & Matusiak, J., 2014. Quantitative Assessment of Ship Behaviour in Critical Stern Quartering Seas. *14th International Ship Stability Workshop*, (October), pp.19–27.

Beck, R.F. & Reed, A.M., 2001. Modern Computational Methods for Ships in a Seaway. *SNAME Transactions*, 109(September 2000), pp. 1–55.

Faltinsen, O.M., 1990. *Sea loads on ships and offshore structures*, Available at: https://books.google.com/books?hl=it&lr=&id=qZq4Rs2DZXoC&oi=fnd&pg=PR7&dq=falt insen+sea+loads+on+ships+and+offshore+ structures&ots=67ks1opXvp&sig=RhcUloA5klv8E7JQrRWODeC Y1K0.

Fonseca, N. & Guedes, S., 2004. Experimental investigation of the nonlinear effects on the vertical motions and loads of a containership in regular waves. *Journal of Ship Research*, 48(2), pp. 118–147. Available at: http://www.ingentaconnect.com/content/sname/jsr/2004/00000048/00000002/art00002.

France, W. et al., 2003. An investigation of head-sea parametric rolling and its influence on container lashing systems. *Marine Technology*, 40, pp. 1–19. Available at: http://www.ingentaconnect.com/content/sname/mt/2003/00000040/00000001/art00001.

Galeazzi, R., Blanke, M. & Poulsen, N.K., 2013. Early detection of parametric roll resonance on container ships. *IEEE Transactions on Control Systems Technology*, 21(2), pp. 489–503.

Hashimoto, H. et al., 2008. Prediction Methods For Parametric rolling with forward velocity and their validation—Final report of scape committee. In *The 6th Osaka Colloquium on seakeeping and stability of ships*.

Hashimoto, H., Umeda, N. & Sogawa, Y., 2011. Prediction of Parametric Rolling in Irregular Head Waves. *International Ship Stability Workshop*, pp. 213–218.

Jensen, J.J. et al., 2009. Wave induced extreme hull girder loads on containerships. *Trans. SNAME 116*, (V), pp. 128–151.

Kukkanen, T. & Matusiak, J., 2014. Nonlinear hull girder loads of a RoPax ship. *Ocean Engineering*, 75, pp.1–14. Available at: http://dx.doi.org/10.1016/j.oceaneng.2013.10.008.

Lu, J., Gu, M. & Umeda, N., 2016. Experimental and numerical study on several crucial elements for predicting parametric roll in regular head seas. *Journal of Marine Science and Technology*.

Manderbacka, T.L., Matusiak, J.E. & Ruponen, P., 2011. Ship motions caused by time-varying extra mass on board. In *12th International Ship Stability Workshop*. pp. 263–269.

Matusiak, J., 2013. *Dynamics of a Rigid Ship* SCIENZE +., Aalto University publication series.

Matusiak, J., 2011. On the non-linearities of ships restoring and the froude-krylov wave load part. *International Journal of Naval Architecture and Ocean Engineering*, 3(1), pp. 111–115.

Paulling, J.R., 2011. Parametric rolling of ships—then and now. *Fluid Mechanics and its Applications*, 97, pp. 347–360.

Reed, A.M., 2011. 26th ITTC Parametric Roll Benchmark Study. In *12th International Ship Stability Workshop*. pp. 195–204.

Silva, S.R.E. et al., 2010. Experimental Assessment of the Parametric Rolling on a C11 Class Containership. In *HYDRALAB III Joint User Meeting*. pp. 4–7.

Zhu, S. & Moan, T., 2014. Nonlinear effects from wave-induced maximum vertical bending moment on a flexible ultra-large containership model in severe head and oblique seas. *Marine Structures*, 35, pp.1–25. Available at: http://dx.doi.org/10.1016/j.marstruc.2013.06.007.

Maritime Transportation and Harvesting of Sea Resources – Guedes Soares & Teixeira (Eds)
© *2018 Taylor & Francis Group, London, ISBN 978-0-8153-7993-5*

Contributes to sailing yacht performance by foil hydrodynamic lift

C. Bertorello & E. Begovic

Marine Hydrodynamic Research Group, University of Naples Federico II, Naples, Italy

ABSTRACT: Since 2000 foils have been fitted to some racing sailing boats with significant results. Several multihulls as well as high performance dinghies use foils and take benefit from hydrodynamic lift. 2013 America's Cup was sailed almost entirely over the water by foiled catamarans. Recently foil application has been successfully proposed on mono hull ocean racers with low ballast ratio as IMOCA60 s. The foils used on both racing mono and multi-hulls allow benefits on both motion resistance and transversal righting moment through the exploitation of hydrodynamic lift; the vertical component of hydrodynamic force produced by leeward foil(s) takes place of hydrostatic buoyancy, but differently from Archimede's static force it increases with square speed. Larger righting moment and resistance reduction due to foils result in higher speed. The paper presents a general review of different foil applications on sailing craft and reports detailed characteristics of the most recent trends.

1 INTRODUCTION

Foil application to sailing craft dates from early sixties, when lifting surfaces have been fitted to very few experimental vessels. (The AYRS Members, 1970). At first, they got poor results. Later, from the earliest years of this century, l'Hydroptere (Fig. 1) astonishing performances as well as International Moth (Fig. 2) dinghies racing on foils

gave evidence to foil potential and became widely known, although considered as funny sailing freaks.

When 2013 America's Cup was raced mostly over the water by hydrofoil catamarans (Fig. 3) a general consideration of foils potential grow up and the idea to race or even to cruise foil borne came out from the dream's world.

Recently, the results of 2016 Vendee Globe (Fig. 4) and the proposals for next Transat 6.50 (Fig. 5) have shown foil advantages even on

Figure 1. Foiler Trimaran L'Hydroptere, photo ©Francis Demange.

Figure 3. ETNZ 2013 America's cup challenger.

Figure 2. International Moth sailing on foils.

Figure 4. Banque Populaire VIII, 2016 Vendee du Globe winner.

Figure 5. Seair 747, Mini650 on foils.

Figure 6. Quant 23 performance scow with foils.

Figure 7. Gunboat 37 First foiling catamaran with accommodation.

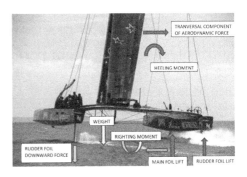

Figure 8. Forces by lifting surface and dynamic righting moment.

Figure 9. Enrico Forlanini hydrofoil, 1906.

ballasted oceangoing mono-hulls. Beside such widely known and documented racing applications, several attempts of high performance dinghies (Fig. 6) and cruising catamarans (Fig. 7) show wide interest in the exploitation of hydrodynamic forces on sailing vessels and lead to wonder if future performance cruisers will exploit hydrodynamic lift for drag reduction and righting moment increase.

The application of such concepts needs a critical evaluation of foil effects and the consideration of involved safety issues. Resistance breakdown and transversal stability have to be revised to take into account hydrodynamic forces produced by lifting surfaces as shown in Figure 8 (Heppel, 2015).

At first hydrodynamic lift provided by immersed appendages has been proposed for resistance reduction only, as exploited by hydrofoil power craft. The first successful foil borne sailing boats did not consider foil contribute to stability.

International Moth dinghies pioneered the use of foils to lift the hull out of the water, but got stability by shifting crew weight upwind (Fig. 2).

Further foil applications to racing catamarans take double benefit by hydrodynamic lift. The vertical component of hydrodynamic force produced by foils of leeward hull takes place of hydrostatic buoyancy within vertical and transversal equilibrium and increases with vessel square speed. The larger righting moment and the resistance reduction due to foils result in astonishingly higher speed (Paulin, 2015).

Ballasted mono-hulls are generally too heavy to get full hydrodynamic lift, nevertheless when Displacement/Length ratio is very low they can take advantage from a partially lifted hull provided by appropriate foil configuration as well as from dynamic righting moment as shown in the following by Figure 16.

Very recently few Mini6.50 that are light and largely canvassed mono-hulls have reached full hydrodynamic lift by twin forward foils and T shaped twin rudders (Fig. 5). Although foil sailing is, at present, almost totally related to high tech racing yachts, future wider application of this concept has to be considered. Performance and safety assessments of sailing vessels in partial or total hydrodynamic lift need specific consideration. This paper presents a general review of the most recent foil applications and highlights involved forces and transversal equilibrium specific aspects.

2 HYSTORICAL BACKGROUND

The early development of hydrofoils started over 100 years ago when Italian Enrico Forlanini achieved 36.9 knots with his 60 hp airscrew-driven boat in 1906 (Fig. 9). Several engineers took notice, among them the Wright Brothers and Alexander Graham Bell, both of whom experimented with foil-borne craft.

In mid 50ties Italian boatyard Rodriguez patented a foil system that was successfully applied to many HSC craft for passenger transportation. Hydrofoils up to 50 m Length over all (LOA) have been successfully built in several countries for commercial use. The concept although fascinating for performances and suitable to the mission profile was not accepted for pleasure craft with very few exceptions.

Sailing hydrofoils first attempts although taking benefit from previous experiences of powered hydrofoils were not successful. Their poor results were due to high structure weight and not efficient sails. Design was coherent and correct so that few attempts showed the path for the future development.

In 1955 *Monitor* (Fig. 10) reached 25 knots. She was designed by Gordon Baker and built by the Baker Manufacturing Company of Evansville, Wisconsin. The US Navy shared part of the cost of construction. In October 1956she was recorded at 30.4 knots and was later said to have sailed close to 40 knots.

In 1970 the first hydrofoil cruiser, David Keiper's *Williwaw* cruised throughout the South Pacific recording more than 20,000 miles.

During 70ties some vessels aimed at establish world speed record during Weymouth speed week tested foil configurations. In 1974 *Mayfly* a foiling cat established world record for A Class at 19.38 knots. In 1977 she set the bar higher at 23 knots. In 1976 *Icarus*, a modified foiling Tornado set a new world record in B class at 20.70 knots. By 1985 her designers and crew Grogono and Fowler raised the speed to 28.14 knots. These and other less successful attempts were very useful for present achievements.

They proposed lesson about foil configuration and behavior. Foil cavitation and ventilation were identified as key factors of foil performance.

What was not fully understood was the poor efficiency of the used rigs and sails. If laminated

sails had been available at that time, today's results had been achieved quite earlier.

In 1980 *Paul-Ricard* skippered by Eric Tabarly beat the schooner *Atlantic*'s transatlantic record by more than two days (Fig. 11). She was a trimaran with short and small outriggers fitted with large foils. The idea was to experience the amount of foil benefit on righting moment and to evaluate the potential of hydrodynamic lift highlighting concepts that had lead fourteen years later to a foilborne trimaran *L'Hydroptère*.

In 1990 *Hobie Trifoiler*, a twin-sail foiled trimaran with a mainsail on each outrigger, capable of 30 knots, was probably the fastest production sailboat in the world. In 1992 Russell Long broke his own world records for the fourth time in the *Trifoiler* at 43.55 knots.

In 1994 Alain Thébault's *L'Hydroptère* foiling trimaran was launched. In 2009 after several attempts set new outright world speed record over 500 m at 51.36 knots (Fig. 1). 2013 America's Cup was sailed in foiling AC72 (Fig. 8).

After a significant period during which foils were neglected except for the French program that led to L'Hydroptere, the racing catamarans 44 for the AC series experienced appreciable hydrodynamic lift due to dagger-board and rudder end plates. In 2004 the first 44 with both hulls out of the water were seen. Foils were soon adopted by all teams. While the first foil applications were limited by inefficient rigs and too heavy boats, America's Cup44 were so performing and efficient to fly easily over the water if adequate lift had been available. After these achievements, foils became fundamental for all types of catamarans related to America's Cup and the largest part of hydrodynamic design was focused upon them.

Beside foils aimed at developing full hydrodynamic lift several attempts to take benefit mainly of further transversal righting moment were performed on sailing vessels intended for displacement condition. On oceangoing multihulls leeward daggerboards shaped to produce a vertical force have been used since 90ties. The Dynamic Stability System (*DSS*) concept developed in early 10 s of this century to fit performance mono-hulls is intended

Figure 10.   In 1955 monitor.

Figure 11.   Paul Ricard.

Figure 12.   DSS concept.

Figure 14.   35 America's cup catamaran.

Figure 13.   Figaro 3 singlehanded one-design with foils.

to get higher transversal stability and to avoid bow diving when reaching at high speed (Fig. 12).

At Paris Boat show in December 2016 Beneteau unveiled renderings of the Figaro 3 (Fig. 13), the first production-built mono-hull with foils. Leading French design firm VPLP has drawn a contemporary looking race boat with foils sticking out of the sides like spiders' legs. These foils are designed to replace the traditional ballast tanks used on past Figaro models. Described as 'asymmetric tip foils' they work by creating vertical and lateral side forces with minimal drag. An important factor is also that they are able to retract within the boat's maximum beam.

The next generation Volvo Ocean Race boats, that will be used from 2019 edition, will be 65 ft foil-assisted mono-hulls. The race features mixed professional crews sailing around the world in a multi-stopover format.

Olympic Class Nacra catamaran has shaped daggerboard to get adequate hydrodynamic lift to sail most of time foil borne. Most sailors agree the foiling move it's a natural evolution for the sport and will be a fantastic addition for spectators.

2017 America's Cup is raced entirely on foils by catamarans able to tack and jibe without touching the water surface with their hulls that now have lost any hydrodynamic meaning (Fig. 14).

A few recent attempts of foiling yachts not intended for racing have to be mentioned.

The Q23 scow is a ballasted dinghy, which can also fly or a foiler which also is one of the fastest open sailing boat in the 7 m LOA range (Fig. 6).

Gunboat 37 (Fig. 7) was the first foiling catamaran with living spaces and inside accommodations. The performance were astonishing but a catastrophic capsize stopped the project highlighting that performance at sea is speed but also safety.

Finally it is worth to mention International Moth contribute to foil sailing and to foil control systems as one of the most coherent, constant and easy to identify. While Frank Raison's wooden scow Moth was the first foiling moth in 1974 it was in 1994 that Andy Patterson in the UK created the first modern narrow Moth. Although capable to sail foil-borne she was not successful around the buoys. It was Rohan Veal's persistence to apply the practice to the International Moth, that gave to the foil option an undisputable supremacy when Veal became the first to win a Moth world championship with foils in 2005.

Rohan understood that 15 to 20 degrees of windward heel makes a huge difference to how a Moth sails. It allows the main lifting foil to contribute lift to windward and complete unload the centreboard of all side loads, greatly reducing drag. There is also a slight righting moment gain from moving the small mass of the boat to windward of the center of lift.

3   SCIENTIFIC BACKGROUND

There is plenty of scientific and technical reference on sailing hydrofoil. Furthermore, several papers on powered hydrofoils constitute a strong and fundamental benchmark. Unfortunately, the most recent achievements on the matter have been got within America's Cup or top level racing teams so that they are still confidential. It is not possible, for space reasons, to present a complete report of so huge amount of published papers, but it is useful to separate them into four blocks. Significant papers for each block are reported in references.

1. Fundamental papers on powered hydrofoils. They are still a most significant contribute for the theoretical approach to foil concept,

342

configuration and equilibrium (Matveev, 2005, Reichel, 2007).

2. Papers relative to the first, sometime naïf attempts performed by sailing hydrofoil (The AYRS Members, 1970).

3. Technical reports of the few partly successful attempts carried on from seventies to the early years of this century. (Alexander, 1972, RINA, 1982).

4. Recent scientific papers aimed at considering and explaining the achievements of present sailing hydrofoils by theoretical and numerical approaches (Heppel, 2015, Paulin, 2015).

Most of papers from 60ties and 70ties have been carried on by tools available in that period when CFD codes where beyond any dream. That's why they are difficult to be considered by today's researcher, nevertheless they are still very important for the understanding of the phenomena connected to foils.

A general conclusion coming out from the chronological examination of scientific background is that at first the foil concept although exactly identified and correctly applied led to unsuccessful results on sailing craft due to inadequate materials and technologies available. After a few years, at the beginning of this century some racing catamarans within America's Cup preliminary racing circuit became so fast that particular daggerboard shapes gave so significant hydrodynamic lift to foresee foil-borne sailing.

The interest about the matter is very strong so that many comments and considerations useful for a deeper and further analysis based on a scientific approach are available (Loveday, 2006).

4 STATE OF THE ART

At present two main events in the sail racing field highlight foil importance and fundamental contribute, they are America's Cup and Vendee Globe. Following this path two important offshore races that are the Figaro and the Volvo race both sailed by one design boat have shifted to foil equipped yachts. Olympic catamaran Nacra has modified the daggerboards to take advantage from hydrodynamic lift.

America's Cup yachts, as always has been in yachting, are leading the technology advances. They are able to tack and jibe on foils and, even more important, they have developed hydraulic control systems of the foil attack angles. This allows longitudinal trim control and optimal tuning of the righting moment generated by foils. Recently, a significant improvement is given by the T shaped adjustable rudders in which through the inclination of rudder shaft the angle of attack of the tip foil is modified. Downward force produced by the upwind rudder allows extra righting moment and a longitudinal moment counteracting the bow down effect of the forward thrust applied in center of sail plan (Fig. 8). While foil trimming on America's Cup boats posed tough request in terms of human power due to the constrains imposed by racing rules, the easy availability of electric or thermal power on large cruising yachts allows to wonder about possible installation of controlled foils on future performance cruisers.

At present most of multihull foil configuration considers two aft foils fitted to the rudders and a main retractable foil protruding from the leeward hull (Fig. 8).

America's Cup catamarans consider two sets of main foils with different lifting area used according to wind (and craft) speed.

The un-ballasted mono-hulls sailing foil-borne as International Moth, or International 14' dinghies are fitted with two foils aligned with centerplane. The forward foil can be partially or totally trimmed to adjust vessel longitudinal trim and hull distance from water surface. The automatic system for foil flap setting can be considered one of the most valuable contribute to foil sailing development.

The very few ballasted mono-hulls able to get total hydrodynamic lift present a four foils configuration with twin foiled rudders and twin main foils close to the mast position (Fig. 5).

While the foils used for total hydrodynamic lift are very similar, in case of partial lift where the main foil aim is righting moment improvement, the foil configurations and even more the foil shapes can be very different. The necessity to retract the foils and to fit the transversal hull section have strong influence also. Any application of foils to sailing vessels is based on consolidated theoretical and experimental studies carried on in 60ties and 70ties. Further specific contributes on cavitation and ventilation are available also (Wu, 2016, Andrun et al., 2016). At present easy availability of CFD tools, together with renewed interest to the matter has encouraged new numerical researches (Lemini & Malathi, 2016). The interaction between aligned foils and the effects of the wakes of the forward foils on the after ones are among the most interesting research paths (Kinaci, 2015).

5 DRAG REDUCTION

Substantial higher speed obtained by foiled power craft is responsible of the initial foil diffusion within the marine field. This result was very important when light and high rpm diesel engines were limited in powering performances. That's why when better performing engines became available the hydrofoil concept loose the leadership among high speed passenger small craft in favor of simpler and cheaper catamaran hull-forms.

Foils on sailing boats produce a substantial drag reduction only when the hull is in total or almost total hydrodynamic lift. This happens on craft with very low Displacement/0,01 L³ ratio. Foils best behavior for drag reduction is not coupled with righting moment improvements. International Moth dinghies pioneered the use of total hydrodynamic lift by center line aligned small foils and got transversal stability by crew weight shifting.

Total hydrodynamic lift has been experienced only by very light un-ballasted vessels as racing catamarans whose stability rely on crew weight and hydrodynamic contributes. Very recently also ballasted mono-hulls with special features have been able to sail in full hydrodynamic lift (Fig. 5). In this case foil advantage is fully exploited for both drag reduction and righting moment increase. Nevertheless, the future application of foils on ballasted mono-hulls will be mostly concentrated on stability improvement and partial hydrodynamic lift that anyway will be beneficial for motion resistance.

In case of total hydrodynamic lift the advantages on motion resistance are relatively easy to assess. Resistance breakdown is relative to foils only and scientific references allows reliable preliminary predictions. The absence of wave pattern due to negligible foil volume not only reduces resistance absolute values but, more important, leads to almost quadratic resistance increase allowing speed values impossible to consider for conventional craft.

Much more difficult is to evaluate the advantages (if any) of foil application on resistance in case of partial hydrodynamic lift. There are counteracting effects, the intrinsic foil resistance, the reduction of canoe body wet surface, the shorter Length waterline (LWL) that leads to lower Length/Beam ratio. This last factor can influence negatively wave resistance. Unless total hydrodynamic lift would be achieved, the foil contribute to ballasted mono-hull performances is mainly limited to stability increment that is anyway quite significant to get higher speed. As regard drag reduction of sailing boats in partial hydrodynamic lift experimental and numerical dedicated studies are necessary, while lift contribute to righting moment is easy to evaluate as it will be shown in the following paragraph.

6 DYNAMIC TRANSVERSAL STABILITY

None among the peculiar characteristics of sailing vessels has pointed out technological resources or influenced hull forms such as transversal stability. In sailing craft the relation among stability, performances and safety is immediate and inescapable. Transversal stability is in most of cases the propulsion factor of sailing vessels.

Any mono or multi hull, both racer or cruiser rely on transversal righting moment to counteract the heeling moment due to wind force acting on the sails. The forward component of the resultant of the aerodynamic interaction between apparent wind and sail plan is the thrust that makes vessel moving forth. Larger righting moment allows larger sail area for a given wind speed and consequently larger aerodynamic resultant and larger thrust. For any vehicle, speed represents the even point between resistance and applied horsepower. Righting moment of sailing boats can be compared to rated horsepower of motor craft main engine.

The vessels considered for foil application are fast so that apparent wind angle is most of the time less than 90 degrees. At so small angles of incidence, the transversal component of aerodynamic resultant is quite large and results in high heeling moment. That's why transversal stability become most important, as in all high speed sailing craft. It is possible to increase sail area and consequently forward thrust, if transversal stability is adequate to keep the boat at a reasonable heeling angle. Reasonable, in this context, means safe, small enough to avoid too large resistance increment and finally suitable with on board ergonomics.

Ballasted mono-hulls sailing in partial hydrodynamic lift have to be considered separately from un-ballasted multihulls in full hydrodynamic lift. Foiling mono-hulls are generally fitted with canting keel and lateral tanks for water ballast as shown in Figure 16.

In this case transversal righting moment, in the static consideration of ship stability is due to Displacement (Δ) times Righting Arm (GZ), where GZ is the distance between Weight and Buoyancy directions (Fig. 16). Stability can be divided into two components, the so called Weight Stability related to the Center of Gravity (G) position and the Form Stability related to the position of the center of Buoyancy (B) of the inclined hull. Weight stability can be increased lowering or shifting offset upwind G. IMOCA60 s as several oceangoing racing mono-hulls use this last option through liquid ballast on the windward side and keel canted upwind (Fig. 16).

Form stability can be increased by the appropriate hull shape to get B far leeway offset when heeled and a consequent larger arm GZ of the righting couple. This generally implies resistance increment and leads to uncomfortable heeled sailing trim as in the scows or in modern offshore racers. Both weight and form stability are speed independent.

A further contribute to transversal righting moment can be provided by interaction between the water flow and a submerged foil or a streamlined cross section board. This is due to the vertical component of the hydrodynamic resultant applied to the foil (Fig. 15). It increases according to square speed so this contribution is generally called dynamic

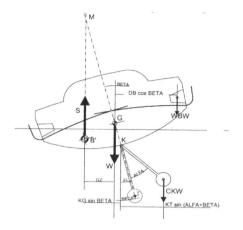

Figure 15. Static Righting Moments for IMOCA60 ballasted mono-hulls.

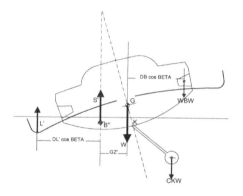

Figure 16. Static and dynamic righting Moment in partial Hydrodynamic lift for IMOCA.

stability (*DS*). The mentioned vertical component influences also hull vertical equilibrium and lifts the boat partially or totally out of the water with a strong reduction of motion resistance. In presence of hydrodynamic lift the Form Stability has to be considered according to new displaced volume and water plane. The lift can be treated as a weight removed from the ship bottom, recalling the stability model for a grounded ship. According to this modeling, the metacentric height reduces when increasing the lift. In the considered case the key factor to explain the large effect of hydrodynamic lift on stability is the foil geometry. It is possible to set the lift force far offset leeward than static buoyancy so that quite larger righting arm is obtained as shown in Figure 16. Differently from what happens with static Form Stability this leeward shift does not affect resistance significantly. When upward force that counteracts hull weight is due to contributes of both Buoyancy (*S*) and Hydrodynamic Lift (*L'*) it is very difficult to evaluate each of them (2). The Lift is the difference between ship Weight and Residual Buoyancy

force. The limit case in which Lift is equal to ship Weight is called Total Hydrodynamic Lift *L* (3).

Lift equations are:

$$\Delta = W = S \qquad (1)$$

$$L' + S' = W \qquad (2)$$

$$L = W \qquad (3)$$

Lift is function of square speed and of foil area S_F according to the expression (4)

$$L = \frac{1}{2}\rho S_F v^2 C_L \qquad (4)$$

IMOCA60's and Mini650's performances allow to consider future sailing in full hydrodynamic lift—on foils and T shaped rudders—as America's Cup catamarans or racing dinghies actually do.

Ship transversal equilibrium and righting moment are summarized by the following expressions, with reference to Figs. 15–16. The static righting moment *RM* at a finite angle β is given by three contributes:

– the static righting moment

$$RM = \Delta \cdot GZ \qquad (5)$$

– the water ballast moment *WBM*

$$WBM = WWB \cdot DB \cos \beta \qquad (6)$$

Where *WWB* is the weight of the water ballast
– the canting keel moment *CKM*

$$CKM = WCK \left(KG \sin \beta + KT \sin(\alpha + \beta) \right) \qquad (7)$$

WBM and CKM are not influenced by boat speed neither by the presence of foils.

In presence of lift component given by a leeward foil only, the expression of *RM* (5) changes in

$$RM' = (\Delta - L) \cdot GZ' + L \cdot DF \cos \beta \qquad (5)'$$

The first term $(\Delta - L) \cdot GZ'$ can be smaller than static one $\Delta \cdot GZ$, as (Δ-L) is smaller than ΔGZ' is different from *GZ*. *GZ* is the righting arm given by $GZ = (BM - BG)\sin \beta$ where *BM* is the transversal metacentric radius and *BG* is the difference between Metacenter and Center of Gravity heights. Metacentric radius *BM* is given by the expression $BM = I_x/V$ where Ix is moment of inertia of waterplane and V is hull volume. Due to the different dynamic hull volume and waterplane, it is not possible to say a priori that *GZ'* is smaller or greater than *GZ*. The second term $L \cdot DF \cos \beta$ can be quite large as DF depends on foil geometry and is limited by structural and practical reasons only.

In case (3), i.e. W = L, transversal righting moment, the main propulsion factor of a sailing boat, is given by $W \cdot d \cos \alpha$ where d is the distance of G from the point where dynamic force is considered applied. As this concept is applied to very light boats, the option to increase this moment above the limit given by W value seems interesting. This can be done by movable foils generally attached to the upwind rudder providing a downward force and consequently a further righting moment $L' \cdot d' \cos \alpha$. In this case the vertical equilibrium is modified as W = L-L'. Careful consideration of the induced resistance has to be done so that very seldom the upwind main foil is used to get extra righting moment leaving this task to small foils connected to the rudders. The downward force provided by the foiled upwind rudder provides also a longitudinal moment quite useful for longitudinal dynamic trim as shown in the following paragraph.

7 DYNAMIC LONGITUDINAL TRIM

First powered hydrofoil attempts highlighted longitudinal trim and potential instability as one of the critical points for the full exploitation of the concept. Designers of International Moth class successfully developed an automatic system that reading the distance from the water surface to lifted fore-body sets the main foil flap to avoiding excessive longitudinal trim. Larger and heavier boats on foils are intrinsically more stable.

When 2013 America's Cup was raced almost entirely on foils, the benefit given by foil angle attack adjustment was clearly identified. In present America's Cup edition all competitors are equipped with hydraulic servo equipment able to trim the main foil and foils set on the rudders. The first has always positive (lift up) trim. As regard the rudder with foils, the leeward is generally positive the upwind negative (downward thrust). This last contributes to transversal righting moment and origins a longitudinal moment counteracting the bow down tendency given by the considerable height of the point of application of the aerodynamic thrust (Fig. 8).

8 CONCLUSIONS AND FUTURE DEVELOPMENTS

The increasing foil diffusion and more in general the specific consideration of hydrodynamic lift in performance yacht design allows to assume even larger interest in the future. Although it is easy to predict wider foil application to racing yachts, it is more difficult to identify the future consideration of hydrodynamic lift in performance cruising yacht design. Future developments and researches pushed by the pressure of racing field and the increasing experience in foil safe sailing allow to forecast dedicated foil configurations suitable at least for performance cruising catamarans. Much more difficult is to consider foil application to small-medium size mono-hulls. Larger yachts present easier to solve layout problems, and could be more easily attracted by foil potential.

The application of *DSS* to Infiniti brand (Fig. 12) represents a significant step. While foils and substantial hydrodynamic lift are outside the possible consideration for performance cruisers, the benefits provided by a less invasive and easier to handle *DSS* board are considered interesting at present.

Beside sailing yachts, recent, symptoms of possible application of foils to power pleasure craft are visible. They are one of the fallout of the success of foils on racing sailing yachts. If this would develop as a trend, given the quite larger dimension of power craft market in respect to sailing yachts, the foil phenomenon could be one of the most significant technical issue in pleasure craft design, ever.

ACKNOWLEDGEMENTS

This work has been financially supported by University of Naples "Federico II" within the frame of the 2016–2017 research program.

REFERENCES

Andrun M., Saric B., Basic J., Blagojecic B., 2016. CFD Analysis of Surface—Piercing Hydrofoil Ventilation Inception. XXII Symposium Sorta.

Heppel P., 2015. Flight Dynamics of Sailing Foilers. 5th High Performance YD Conf., Auckland, 10–12 March, 2015.

Kinaci O., 2015. A numerical parametric study on hydrofoil interaction in tandem. Int. J. Nav. Archit. Ocean Eng 7: 25–40.

Lemini Y. et al., 2016. Design and Optimization of Hydrofoil using CFD and Structural Analysis, Int. J. of Sci. Eng. and Tech. Research, Vol. 05 Issue 44.

Loveday H., 2006. The Design of Hydrofoil System for Sailing Catamarans. Msc Thesis, University of Stellenbosh.

Matveev K., Duncan R., 2005. Development of the tool for predicting hydrofoil system performance and simulating motion of hydrofoil assisted boats. High Speed and High Performance Ship and Craft.

Paulin, A., Hansen, H., Hockirch K., Fisher, M., 2015. Performance assessment and optimization of a C-class Catamaran Hydrofoil configuration. 5th High Performance Yacht Design Conf., Auckland, 10–12 March, 2015.

Reichel M., Bednarek A., 2007. The experimental studies on hydrofoil resistance. Arch.Civil and Mech. Eng.,Vol. 7, No. 2.

RINA Small Craft Group 1982. Sailing Hydrofoils: Papers Presented at the International Conf., November 12, 1982.

The AYRS Members, 1970. Sailing Hydrofoils. The Amateur Yacht Research Society, UK.

Wu P., Chen J., 2016. Numerical study on cavitating flow due to a hydrofoil near a free surface. Journal of Ocean Engineering and Science I 238–245, 2016.

Maritime Transportation and Harvesting of Sea Resources – Guedes Soares & Teixeira (Eds)
© 2018 Taylor & Francis Group, London, ISBN 978-0-8153-7993-5

The effect of the stern shape and propeller location on the propulsive efficiency of a blunt ship

G. Bilici & U.O. Ünal
Istanbul Technical University, Turkey

ABSTRACT: A reliable powering prediction is of critical importance at the design stage of a ship and the evaluation of the hull/propeller interaction constitutes a fundamental part of the propulsion performance estimation. Whilst the hull form and propeller geometry directly affect the interaction characteristics, the axial distance between propeller and hull is also a critical factor. This paper investigates the effect of different propeller positions and stern shapes on the self-propulsion characteristics of Kriso Very Large Crude Carrier 2 (KVLCC2) tanker. Hydrodynamic performance of the reference propeller (KP458) was first examined in open-water conditions for validation purposes. Two different longitudinal positions of the propeller were considered in the self-propulsion cases. Section shifting method was used to generate two variations of the original stern form. The approach adopted was simply based on deforming the shape of the associated cross-sectional area curve. Negligible changes in the displacement and wetted surface area were basically aimed for the hull form modifications.

1 INTRODUCTION

Hydrodynamic resistance and propulsion efficiency are of critical importance in the hull form design of a ship. The latter is also directly related with the flow noise and hull vibration issues, while at the same time affecting the comfort level of the crew and passengers. Undoubtedly, a reliable powering prediction has also a great value at the design stage of a ship and the evaluation of the hull/propeller interaction constitutes a fundamental part of the propulsion performance estimation. The form of the stern region of a ship directly affects the hydrodynamic performance due to the strong adverse pressure gradients occurring in this area. Particularly for blunt ships, such as oil tankers, this characteristic can cause severe flow separation issues and hence can produce large power losses. Even in a hull design, which displays no flow separation, strong bilge vortices can be generated, which also influences the resistance and propulsion performance of the vessel. The stern region of the hull also interacts with the propeller and consequently, has a significant role in propulsive performance. Slight form modifications in this region can lead to high gains in the sense of powering efficiency. As is known, on the other hand, whilst the existence of the propeller can eliminate certain hydrodynamic problems such as flow separation, the required power to propel the ship increases due to the suction effect applied by the propeller around the stern. Undoubtedly, the location of the propeller should be selected with care to maximise the propulsive efficiency. Whilst the hull form and propeller geometry notably affect the interaction characteristics, the axial distance between propeller and hull is also a critical factor.

The propulsive performance is generally expressed with the quasi-propulsion coefficient, which excludes the mechanical loses. This coefficient is basically related with the efficiency of the propeller operating behind the ship and with the hull efficiency which includes the effect of both wake fraction and thrust deduction. Either experimental or computational techniques can be used to evaluate each component of the propulsive efficiency. Extensive research studies have been carried out so far, to investigate and improve the propulsion performance of the ships. Even the first experimental work of Taylor (1922) emphasised the importance of the geometry and position of the propeller concerning the propulsive efficiency. Kim et al. (2001), Choi et al. (2010), Castro et al. (2011) can be given as fundamental examples about this ongoing research area. In recent years, viscous flow computations by using Reynolds-Averaged-Navier-Stokes (RANS) solvers has become a widespread tool in the ship design process and satisfactory agreement with model-scale wake field measurements can be obtained. Conventional hull forms have been largely used in these studies to achieve reliable validations of the simulations by means of comparisons with the experimental data.

Kriso Very Large Crude Carrier 2 (KVLCC2) tanker of MOERI, a variant of KVLCC, is one of the most widely-known non-built hull forms, which was essentially designed for research purposes. The form has been extensively used for both experimental and computational studies in the literature (e.g. Uneo et al. 2009; Knutsson & Larssson 2011; Win 2014). Larsson et al. (2014), focusing on the state-of-the-art in computational fluid dynamics (CFD) for hydrodynamic applications, provides a comprehensive review on the studies involving KVLCC2 hull form. Although numerous studies have been performed about the manoeuvrability, hydrodynamic resistance etc. of this hull form, the investigations into the propulsive efficiency of KVLCC2 are still somewhat rare.

This paper investigates the effect of different stern shapes and propeller positions on the self-propulsion characteristics of KVLCC2 tanker by means of RANS simulations. The original propeller of KVLCC2 (KP458) was used in the study. Three variations of the original stern form were generated with negligible chances in the displacement and wetted surface area. By taking two different distances between the hull and the propeller into account, five different computational cases were systematically analysed at the self-propulsion point. The simulations, which also included the analyses of the non-propelled and open water propeller cases, provided detailed results about the components of the propulsive efficiency of this blunt hull form.

2 COMPUTATIONAL STUDY

2.1 Computational models and boundary conditions

As mentioned previously, the computations were performed with KVLCC2 hull and KP458 propeller at 1:58 model scale. The main particulars of the models are given in Tables 1 and 2.

The aft form variations were generated based on the section shifting method of Markov & Suzuki (2001). The modifications were applied within the last quarter portion of the hull form. A small amount of shifting for the selected cross sections was considered to minimally affect the main particulars of the original form. Consequently, the wetted area and displacement of KVLCC2 did not change more than ±0.15% in the new hull forms. The abbreviations of the generated forms, Var1 and Var2, which are stated throughout this paper, indicate the shifting of the cross-sections performed towards the bow and stern, respectively.

The comparative cross-sections and side views of the original and generated forms can be seen in Figure 1. Shown in Figure 2 are the sectional

Table 1. Main dimensions of KVLCC2 at 1/58 scale.

Particular	Dimension
L_{BP} [m]	5.51
L_{WL} [m]	5.61
B_{WL} [m]	1.00
D [m]	0.51
T [m]	0.35
∇ [m³]	1.60
S [m²]	8.08
C_B	0.81
LCB (%)	3.48
Fr	0.142
V [m/s]	1.047
Re	5.74×10^6

Table 2. Main dimension of KP458 propeller at 1/58 scale.

Particular	Dimension
Z	4
D [m]	0.17
P/D [0.7R]	0.721
Ae/A0	0.431
Rot. Dir.	Right
Hub Ratio	0.155

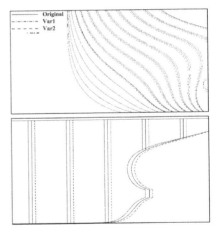

Figure 1. The comparative cross-sections and side views of the forms.

area curves of the three forms. As is seen, the differences between the curves are rather small. For the first three computational cases carried out, whilst above-mentioned hull forms were used, the distance between the stern of the hulls and the propeller (d) remained constant, allowing the original location of the propeller to change in accordance with the aft form. For the remaining cases, this distance slightly increased. One case was performed

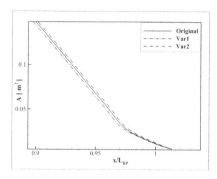

Figure 2. Sectional area curve of variation forms.

Table 3. Computational cases.

	Form	d
C1	KVLCC2	Original
C2	Var1	Original
C3	Var2	Original
C4	Var1	Increased
C5	KVLCC2	Increased

with the form Var1, whilst the propeller was kept at its original location. For the other case, on the other hand, the original KVLCC2 hull form was used, however the propeller slightly moved away from the stern. Table 3 summarises the above explained computational cases.

The comparative cross-sections and side views of the original and generated forms can be seen in Figure 1. Shown in Figure 2 are the sectional area curves of the three forms. As is seen, the differences between the curves are rather small. For the first three computational cases carried out, whilst above-mentioned hull forms were used, the distance between the stern of the hulls and the propeller (d) remained constant, allowing the original location of the propeller to change in accordance with the aft form. For the remaining cases, this distance slightly increased. One case was performed with the form Var1, whilst the propeller was kept at its original location. For the other case, on the other hand, the original KVLCC2 hull form was used, however the propeller slightly moved away from the stern. Table 3 summarises the above explained computational cases.

Bare hull, open water propeller and hull with propeller (self-propulsion) computational cases were separately considered in the study. A Cartesian coordinate system was adopted.

A medium sized rectangular prismatic solution domain was applied for the bare hull and self-propulsion cases. Accordingly, the computational boundaries at the upstream, downstream, sides and bottom were placed at a distance of 1.5 L, 3 L, 1.5 L and 1.5 L, respectively, from the centroid of the hull, where L implies the ship model length. Since the effect of the free surface was not considered in the simulations, the top boundary was simply placed at a height corresponding to that of the design waterline. A cylindrical inner domain with a diameter of 2D, where D implies the propeller diameter, was generated around the propeller. This provided a better numerical grid resolution control around the propeller and also allowed to use Multiple Reference Frame Model (MRF) method (Bhattacharyya et al. 2016) to simulate the propeller rotation.

For the open water propeller analyses, which were carried out for grid dependency and validation purposes, two nested cylindrical flow domains were generated. The inlet and outlet boundaries were placed at a distance of 1.5D and 5D, respectively, from the propeller plane. In the radial direction, the outer and inner boundaries were located at a distance of 4D and 2D, respectively.

As far as the boundary conditions applied are concerned, for the bare hull and self-propulsion cases, the flow velocity components and the turbulence properties were specified at the inlet boundaries. The velocity magnitude of 1.047 m/s was considered which corresponded to a Froude number (Fr) and Reynolds number of around 0.142 and 5.7×10^6, respectively. An insignificant turbulence intensity level of 0.5% along with the turbulence length scale of 0.05% were set at the inlet boundaries. At the outlet boundary, the atmospheric pressure was specified. A symmetry boundary condition, which can be interpreted as a slip wall, was used for the top and bottom boundaries. The intersection of the inner and outer domains was simply specified as an interface boundary to enable to fluid flow across it.

The boundary conditions applied for the open water propeller cases were basically similar to those of the other cases. The symmetry condition was set to the outer circular boundary in the radial direction.

Grid generation were performed by using a state-of-the-art mesh generation software package. The technique used is based on the hybrid type unstructured mesh generation which involves the use of quad-dominant surface meshing with hexahedral cell layers in the vicinity of the wall. The flow which is relatively distant from the body, on the other hand, is resolved with tetrahedral cell elements. The advantage of the mesh generation technique is the accuracy obtained in the near wall flow field due to the quadrilateral/hexagonal cell elements used in this region. Some views of the meshes generated around the hull and propeller are shown in Figure 3 and 4.

Care was taken to be able to accurately resolve the boundary layer as well as the viscous sub-layer

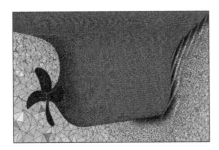

Figure 3. A view of the hybrid volume mesh around the hull.

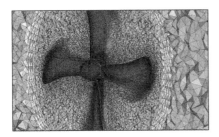

Figure 4. A view of the hybrid volume mesh around the propeller.

developing over the body. For his purpose, the y^+ value of the first cell adjacent to the surface was kept around 1 for most part of the hull and propeller.

2.2 Numerical method

Steady, incompressible Reynolds-Averaged-Navier-Stokes (RANS) equations were solved for the CFD simulations. The SST k-ω turbulence model (Menter 1994), which is based on the Bousinessq hypothesis (Tennekes & Lumley 1972), was used to calculate the turbulence field and hence to model the Reynolds stresses appearing in the RANS equations. The model can be directly integrated through the viscous sub-layer and therefore no special viscous correction for the near-wall was applied. A finite-volume method (Blazek 2001) was employed with a segregated algorithm to solve the RANS and turbulence transport equations. The standard pressure-correction procedure (SIMPLE) of Patankar (1980) were used for the pressure-velocity coupling. The spatial discretisation of the convective terms was achieved with a second-order-upwind scheme while the viscous terms were discretised with a second-order central difference scheme. The variations of the flow variables such as the drag and velocity in the wake field was systematically checked along with the scaled residuals in order to decide whether the

convergence was achieved. The effect of the free surface was ignored. MRF method (Mizzi et al. 2017) was adopted for the cases involving propeller. This approach allows a steady state calculation without moving/sliding the mesh around the propeller geometry.

2.3 Grid dependency and validation

Two separate grid dependency analyses were carried out for the bare hull (KVLCC2) and open water propeller (KP458) analyses to ensure that solutions were independent from the grid resolution. Three different mesh resolutions were considered for the calculations. The resolution of the meshes was systematically increased at each direction of the coordinate system by a factor of approximately $\sqrt[3]{2}$. This approach resulted in nearly doubled cell sizes for the refined mesh structures. Tables 4 and 5 summarise the main characteristics of the meshes used and basic parameters obtained from the computations. As seen from the table, the average y^+ value was kept nearly constant for all meshes in order to fully resolve the viscous sub-layer. The negligible differences obtained between the results of the mesh resolutions B and C for both grid dependence analyses pointed out that the B meshes generated were suitable to perform further simulations for the hull with propeller cases.

Table 4 also presents the experimental results found by Kim et al. (2001) who conducted detailed towing tank tests for a KVLCC2 model at 1:58 scale. The difference between the computed results and the experimental ones are not greater than 2% which validates the simulations performed. A comparison of the velocity contours found in this study with the experimental results of

Table 4. Grid dependence analysis for the bare hull.

	Total cells	Cells on the hull	Av. aspect ratio	1+k	C_T ($\times 10^3$)
A	5.8×10^6	69×10^3	7	1.183	4.21
B	12×10^6	133×10^3	5	1.179	4.17
C	27×10^6	259×10^3	4	1.176	4.16
Exp				1.160	4.09

Table 5. Grid dependence analysis for the propeller.

	Total cells	Av. aspect ratio	J	K_T	10 K_Q	η
A	5.8×10^6	10	0.55	0.112	0.143	0.677
B	12×10^6	9	0.55	0.113	0.144	0.678
C	27×10^6	7	0.55	0.114	0.145	0.679
Exp			0.55	0.116	0.153	0.661

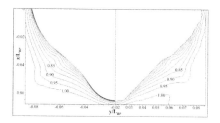

Figure 5. Comparison of the velocity contour at $x/L_{BP} = 0.85$; Present study and Lee et al. (2003) (from left to right).

Table 6. Open water results of the propeller.

J	CFD			Exp		
	K_T	$10 K_Q$	η	K_T	$10 K_Q$	η
0.45	0.153	0.180	0.608	0.156	0.189	0.590
0.50	0.133	0.162	0.653	0.136	0.172	0.631
0.55	0.113	0.144	0.687	0.116	0.153	0.660

Lee et al. (2003) is given in Figure 5. The general agreement of the contours is very good with the exception of the outermost contours. However, the shape of this contour level was found to be similar to that obtained in this study by the computational groups of Gotherburg 2010 Workshop (Larsson et al. 2014).

The experimental results of the open water tests of of URL-1 which were conducted for the KP458 propeller, are presented in Table 6 along with the results obtained in the present study. It can be seen that the agreement is good and the differences are within 2% of the experimental values.

3 RESULTS

The Var1 and Var2 bare hull forms were computationally analysed along with the original KVLCC2 bare hull form to extract the form effect on the propulsion efficiency. The exact grid resolution previously presented in the grid dependency analyses was used in the simulations to provide consistency. Small form modifications performed resulted in very similar pressure distributions on the hull. However, since a detailed form fairing study was not performed for the newly generated hull forms, due to the small inconsistencies around the transition zone between the original and modified portions, the pressure distribution altered in these regions and hence the form factor and total resistance coefficient (C_T) slightly increased for Var1 and Var2. Table 7 presents the calculated form factors and C_T values for the three cases. The general

structures of the flow on the hull were also similar for these cases. The associated limiting streamlines are comparatively shown in Figure 6. The original form produces a small separation zone near the stern profile. A streamline convergence is also visible below this zone which indicates a vortex-type separation around this region. The flow structure found is in good agreement with those found by Gothenburg 2010 Workshop groups, e.g. CHALMERS and IHRR (Larsson et al. 2014), who also conducted a numerical study with a RANS solver. The form modifications did not significantly affect this topology. However, whilst the inception of the streamline convergence moved a bit towards the stern, the separation zone slightly reduced, particularly for Var1.

Prior to comparing the propulsive efficiency of the hull forms, the self-propulsion point was determined for each computational case. This is basically an iterative process which aims to find the advance coefficient (J) where the thrust of the propeller (T) becomes equal to the hydrodynamic resistance (R_S) of the model. Three simulations, which were carried out at different advance coefficient values by changing the propeller rotation speed, were adequate for each case to accurately determine the self-propulsion point by means of a second order interpolation technique. Table 8 is

Table 7. Resistance and form factor coefficient for based and variation forms.

	1+k	Diff%	$C_T (\times 10^3)$	Diff.%
Original	1.18		4.17	
Var1	1.19	0.8	4.20	0.7
Var2	1.19	1.6	4.24	1.7

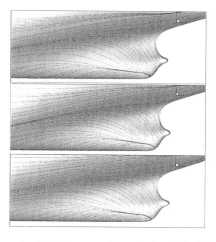

Figure 6. Limiting streamlines of the original form, Var1 and Var2 (from top to bottom).

Table 8. Self-Propulsion point, resistance, torque and moment values for Case–1.

J	N[rps]	R_s[N]	T [N]	ΔS[N]	Q[Nm]
0.493	12.48	23.390	25.7	−2.31	0.374
0.515	11.96	23.105	23.05	0.06	0.323
0.520	11.85	23.026	22.46	0.57	0.324

Table 9. Resistance, torque and moment values for all cases.

Case	J	R_T[N]	T[N]	Q_B[Nm]
C1	0.515	18.08	23.10	0.332
C2	0.516	18.20	22.82	0.331
C3	0.514	18.40	23.07	0.335
C4	0.516	18.20	22.40	0.324
C5	0.518	18.08	22.42	0.323

Table 10. Propulsion coefficients of KVLCC2 for cases.

Case	J	T	W	η_H	η_B	η_D
C1	0.515	0.218	0.245	1.036	0.730	0.756
C2	0.516	0.202	0.245	1.057	0.727	0.768
C3	0.514	0.203	0.241	1.050	0.726	0.762
C4	0.516	0.188	0.221	1.043	0.752	0.785
C5	0.518	0.194	0.227	1.042	0.750	0.782

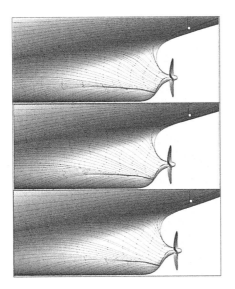

Figure 7. Limiting streamlines of C1, C2 and C3 (from top to bottom).

presented as an example of what was found as a result of this analysis. N, Δ_S and Q seen in the table imply propeller revolutions per second, tow force, and torque values, respectively.

The limiting streamlines obtained from the self-propulsion analyses with different stern forms can be observed in Figure 7. The main difference with respect to the bare hull cases is the complete absence of the flow separation due to the suction effect of the propeller. The three cases also produced very similar flow structures with a slight delay of streamline convergence apparent in C3.

Shown in Table 9 are the resistance, thrust and torque values obtained from all cases at the associated self-propulsion points. The J values presented are nearly identical, which implies an almost constant propeller rotation speed for all cases. As is seen, although the bare hull resistance values were increased for the two modified forms, Var1 and Var2, the required thrust force to propel the model for C2 seems to be slightly lower than that found for C1. On the other hand, the cases with a larger d value exhibited around 3% lower thrust value compared to the original model, C1. These findings underline the importance of the propeller-hull

interaction which will be quantitatively presented in the following paragraphs.

The propulsion coefficients for all cases investigated are listed in Table 10. The first three cases represent the propulsive differences solely due to the form variations. The thrust deduction (t) values of these cases indicate that the suction applied on the hull by the propeller was significantly decreased by about 7% with the effect of the form modifications. Figure 8 comparatively presents the pressure distributions on the aft section of the cases C1 and C2. The increase in the pressure distributions for C2 is apparent. The stern form modifications did not have a significant impact on the effective wake fraction (w). The three hulls displayed very similar effective wake distributions on the propeller plane as in Figure 9. Consequently, the w values obtained from these cases were roughly equal and hence, the reduction in the t values directly reflected to the hull efficiency (η_H), which increased around 1.5–2% for C2 and C3. The propeller efficiency behind the hull (η_B), on the other hand, which basically depends on the incoming flow distribution, remained almost constant.

The fact that two modified forms generated by means of section shifting in opposite directions, exhibited similar but higher hull efficiencies compared to the original form seems somewhat confusing. Since only two modified forms were analysed in this study, this finding should be validated with a similar systematic analysis with more form variations.

For the cases C4 and C5, since a larger d value was considered, the thrust deduction further reduced due to the effect of the weak propeller-hull interaction. The case C4, which was subjected to

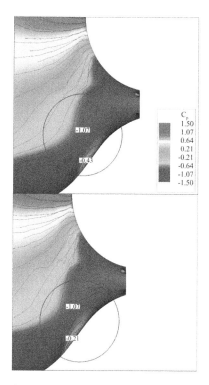

Figure 8. Pressure distributions on the hull for C1 (top) and C2.

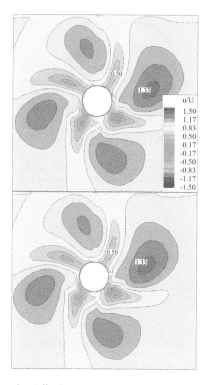

Figure 9. Effective wake distributions for C1 (top) and C3.

the additional influence of the form modification, displayed a dramatic reduction of around 14% in t. The relatively increased pressure distribution on the hull for the case C4 can be seen in Figure 10. Larger d value of the cases C4 and C5 resulted in a decrease in the wake fraction values for these cases. These arrangements also leaded to a significant decrease in the axial vorticity values at the propeller plane. As an example, the comparative vorticity distributions of the cases C2 and C4, which use the same hull form, Var1, are presented in Figure 11. Significantly smaller axial vorticity zones and intensities can be observed for C4. Since hull efficiency consist of the combination of thrust deduction and wake fraction, lower w values obtained from C4 and C5 did not allowed to achieve very high η_H values (Table 10). Still, the values found were slightly higher than that of C1. However, more efficient effective wake distributions of these arrangements leaded to around 3% higher η_B values with respect to that obtained with the original form.

The higher η_H values obtained due to the form modifications for C2 and C3, slightly increased the quasi-propulsion coefficient η_D compared to that found for C1. However, since the bare hull hydrodynamic resistance values were also higher

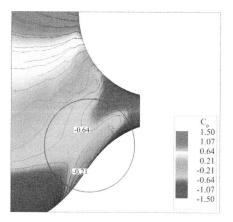

Figure 10. Pressure distributions on the hull for C4.

for C2 and C3, the thrust force required at the self-propulsion point remained almost constant. On the other hand, C4 and C5, with the influence of both form modifications and higher distance between the hull and propeller, displayed a notable increase of around 4% in η_D. The thrust force required to propel the model also reduced for these cases, which indicates a direct gain in powering.

353

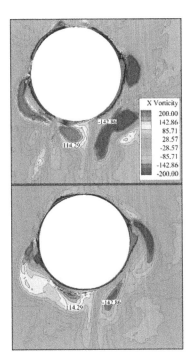

Figure 11. Axial vorticity distributions for C2 (top) and C4.

4 CONCLUDING REMARKS

The results of the RANS simulations to systematically analyse the quasi-propulsion efficiency of well-known tanker form KVLCC2 was presented. Section shifting method was used to generate two variations of the original stern form. Slight form modifications were considered without affecting the general form structure and characteristics. Different axial positions of the propeller were also considered in the self-propulsion cases. The main findings of the computational investigation are summarised in the following paragraphs.

The study revealed the importance of the aft form and axial propeller position regarding the propulsive efficiency. Even the insignificant modifications applied in the study provided rather high efficiency gains. Basically, the stern form modifications affected the thrust deduction but did not significantly modify the wake structure. A decrease of around 7% could be obtained in t with the new forms generated. Consequently, higher η_H and η_D values were achieved compared to those of the original form. Wake fraction and thrust deduction factors were strongly influenced by the axial position of the propeller. Increasing distance between the hull and propeller significantly reduced the t values particularly for C4. This arrangement also reduced the vor-

ticity levels at the propeller plane providing a more efficient wake structure. As a result of this, the efficiency of the propeller working behind the hull was improved about 3%. Combined effect of increasing values of both η_H and η_B provided a gain of around 4% regarding the quasi-propulsion efficiency.

This study encourages future works on the blunt ships involving systematic computational self-propulsion analyses of a wider range of hull modifications and propeller positions.

REFERENCES

Bhattacharyya, A., Krasilnikov, V., & Steen, S. (2016). A CFD-based scaling approach for ducted propellers. *Ocean Engineering, 123*, 116–130.

Blazek, J. 2001. Computational Fluid Dynamics: Principles and Applications. Computational Fluid Dynamics: Principles and Applications.

Castro, A.M., Carrica, P.M., & Stern, F. (2011). Full scale self-propulsion computations using discretized propeller for the KRISO container ship KCS. *Computers and Fluids, 51*(1), 35–47.

Choi, J.E., Min, K.-S., Kim, J.H., Lee, S.B., & Seo, H.W. (2010). Resistance and propulsion characteristics of various commercial ships based on CFD results. *Ocean Engineering, 37*(7), 549–566.

Kim, W.J., Van, S.H., & Kim, D.H. (2001). Measurement of flows around modern commercial ship models. *Experiments in Fluids, 31*(5), 567–578.

Knutsson, D. & Larssson, L. (2011). Large Area Propellers, *Second International Symposium on Marine Propulsors*, Hamburg, Germany.

Larsson, L., Stern, F. & Visonneau, M. (2010). Gothenburg 2010 a workshop on numerical ship hydrodynamics, *Proceedings of Gothenburg 2010 Workshop on Numerical Ship Hydrodynamics*. Gothenburg, Sweden, 2010.

Markov, N.E., & Suzuki, K. (2001). Hull Form Optimization by Shift and Deformation of Ship Sections, *Journal of Ship Research, 45*(3), 197–204.

Menter, F.R. (1994). Two-equation eddy-viscosity turbulence models for engineering applications. *AIAA Journal, 32*(8), 1598–1605.

Mizzi, K., Demirel, Y.K., Banks, C., Turan, O., Kaklis, P., & Atlar, M. (2017). Design optimisation of Propeller Boss Cap Fins for enhanced propeller performance. *Applied Ocean Research, 62*, 210–222.

Patankar, S.V. (1980). Numerical heat transfer and fluid flow, *CRC Press*.

Tennekes, H. & Lumley, J.L., (1972). A First Course in Turbulence, MIT Press, Cambridge, UK.

Ueno, M., Yoshimura, Y., Tsukada, Y., & Miyazaki, H. (2009). Circular motion tests and uncertainty analysis for ship maneuverability, *Journal of Marine Science and Technology, 14*, 469–484.

Win, Y.N. (2014). Computation of the Propeller-Hull and Propeller-Hull-Rud Interaction Using Simple Body-Force Distribution Model. Journal of Chemical Information and Modeling. (Doctoral dissertation). Osaka University.

URL-1. Retrieved May, 10, 2017 from http://www.simman2008.dk/KVLCC/KVLCC2/kvlcc2_geometry.html.

Maritime Transportation and Harvesting of Sea Resources – Guedes Soares & Teixeira (Eds)
© 2018 Taylor & Francis Group, London, ISBN 978-0-8153-7993-5

Estimating flow-induced noise of a circular cylinder using numerical and analytical acoustic methods

S. Ergin & S. Bulut

Faculty of Naval Architecture and Ocean Engineering, Istanbul Technical University, Istanbul, Turkey

ABSTRACT: In this paper, the flow-induced noise around the circular cylinder is investigated by using numerical and analytical methods. The steady and transient flow field data required for acoustic analysis are calculated by employing Computational Fluid Dynamics (CFD) using finite volume method. The turbulence is modelled by using the two-equation turbulence models. The flow-induced noise has been calculated by solving Ffowcs Williams and Hawkings (FW-H) equations. A k–ε sound model based on Proudman analogy is also employed to approximate the total sound power and flow noise generated by turbulent flow past over the circular cylinder, analytically. The results obtained using the FW-H and the k–ε sound methods are presented and compared with experimental and numerical data in the literature. The k–ε sound method can predict discrete values of sound pressure levels and has compatible harmonics into the broadband noise, especially. The agreement between the results are found to be good.

1 INTRODUCTION

The prediction of the flow-induced noise has been the target of those who have been researching in the field of acoustics for many years. Recently, flow noise around submerged bodies used in marine industries has become commercially and ecologically an important issue. In general, the sources of noise are defined as monopole, dipole and quadrupole sources. Monopole sources are produced by the transfer of mass or heat into a flow unsteadily. When the object moves through the flow, these sources occur as a result of displacement of the fluid. To estimate monopole terms, the geometrical properties and motion of the object should be known. Dipole sources occur when irregular flows interact with bodies or surfaces. The variable surface wall pressure and the surface shear force of the object give rise to the dipole terms (Lighthill, 1952). Velocity fluctuations in the boundary layer cause quadrupole noise sources. Quadrupole sources have a volumetric distribution of nonlinearities in the flow. According to Lighthill (1952), these nonlinearities are of two kinds. The first is that the local sound velocity is not constant and depends on the particle velocity, and the second is that the particle velocity near the body affects the sound velocity.

The Ffowcs Williams-Hawkings (1969) noise model which is developed based on the Lighthill (1952) analogy, is a very important study on the estimation of far field noise. This method evaluates the nonlinear pressure fluctuations at the surface of the sound source from the solution of the flow equations. The pressure changes in the far field are calculated by integrating these solutions. The noise of a moving object can be estimated using this method.

Turbulence-induced noise around submerged bodies has not been well understood because of the complexity of the problem. There is a limited number of studies in the literature regarding the estimation of the flow noise around the cylinder. For example, J.D. Revell (1997) conducted an experimental study to examine the relationship between the far field flow noise and the drag coefficient for cylinders. Experiments carried out at different Mach and Reynolds numbers revealed a strong relationship between the Sound Pressure Level (SPL) and the drag coefficient, C_d (J.D. Revell 1997). B. Cantwell and Donald Coles (1983) have experimentally examined the characterization of the flow of a cylinder in the near wake region with different Reynolds numbers (B. Cantwell and Donald Coles 1983). C. Nornberg (2002) and Watanabe (1996) investigated the relationship between the variable lift coefficient and the Reynolds number for the cylinder, experimentally. Y.T. Lu et al. (2008) calculated flow characteristics around the submarine and flow-induced noise using the FW-H method. Choi, W. et. al. (2016) performed noise analyses for a submerged cylinder using the FW-H method and the LES turbulence model without considering the quadrupole source terms.

In this study, the flow-induced noise around circular cylinder has been investigated by using numerical and analytical methods. The two-dimensional numerical solution of the flow has been obtained by using Finite Volume Method. The turbulence is modelled by using the k-ε and k-ω SST turbulence models. The flow-induced noise has been calculated by solving Ffowcs Williams and Hawkings (FW-H) equations. A k-ε sound model was also employed to approximate the total sound power and flow noise generated by turbulent flow past the circular cylinder. The method is based on Proudman analogy. The steady CFD results are used as an input for the k-ε sound model and the transient CFD results for the FW-H method. The flow with a Reynolds number of 90,000 is investigated for the mediums both water and air. The results have been compared with the available experimental and numerical results in the literature. The agreement between the results are found to be good.

2 METHODOLOGY

In the present work, the flow-induced noise around a circular cylinder is estimated by using various acoustic analogies. The characteristics of incompressible flow are first obtained by solving the governing equations of fluid using Computational Fluid Dynamics (CFD). Then, the noise generated at a certain distance from the circular cylinder is predicted using the calculated flow data as an input. The flow noise at the receiver points placed at the same locations is also calculated analytically using the k-ε noise method. The k-ε noise method based on the Proudman (1952) analogy accounts for both dipole and quadrupole noise sources. The results obtained using the FWH method and the k-ε method are presented and compared with experimental and numerical data.

2.1 Computation methodology of flow field

The simulations were conducted considering two-dimensional (2D) computational domain. Computational fluid dynamics (CFD) is used to obtain the unsteady and steady flow fields.

The realizable K-Epsilon two-layer model and the k–Omega SST (shear stress transport) model of Menter (1993 and 1994) are used to simulate the flow past over the circular cylinder. The numerical discretization scheme used to deal with the pressure-velocity coupling between the momentum and the continuity equations is SIMPLE algorithm. The convection discretization is defined by using the segregated flow solver with second-order accuracy. The time discretization is performed by using an implicit and second-order accurate scheme.

The diameter of cylinder, D is 0.019 m and the freestream Mach number is 0.2 corresponding to

Re ≈ 90.000, which is equal to the one used in the experiments of J.D. Revell et al. (1997). The size of computational domain in y-direction is equal to 29D. The inlet and the outlet in the numerical simulations are placed respectively, 3.4D and 15.4D from the cylinder as shown in Figure 1 (see, Liang 2006). The computational domain at the top and bottom boundaries are both located at 10.5D from the cylinder axis.

A structured mesh is employed with 166629 cells by locally refining the mesh in the near cylinder region and over the cylinder wake where the highest flow fluctuations are observed. A constant expansion of 1.2 is used in the radial direction away from the cylinder. Different representative cylinder lengths are used in the span-wise direction for two-dimensional simulations.

A spatial mesh resolution of Δy+≈1 is used on the cylinder wall to solve the near-wall flow. The smallest cell spacing in the radial direction $r_{min}/D = 7.45 \times 10^{-4}$. Figure 2 shows the computational mesh used in the two-dimensional simulations.

2.2 Acoustic computational methodology

2.2.1 The Ffowcs Williams and Hawkings method
The Ffowcs William and Hawkings method, based on the Lighthill analogy, allows prediction of a distant area loudness at some point (Ffowcs Williams-Hawkings 1969). Nonlinear pressure fluctuations on the sound source surface are obtained by solving the flow equations. The obtained solution is integrated

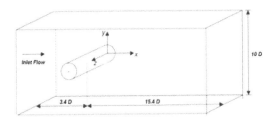

Figure 1.   Physical configuration of the cylinder flow.

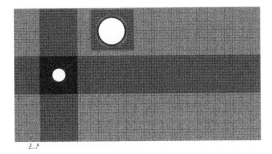

Figure 2.   Computational mesh in X–Y plane.

to calculate the pressure changes in the far field. The noises of moving objects can be estimated by this method. The FW-H Equation is defined as a non-homogeneous wave equation derived from the continuity equation and the Navier-Stokes equations.

Firstly, all near-field sources of sound are simulated using the RANS-based CFD solver (k-ε and SST k-ω). Then the Ffowcs Williams-Hawkings (FW-H) acoustic analogy gives the propagation of sound into far field using the near-field sound sources as the input. This aeroacoustic model is based on Farassat's Formulation 1 A of the FW-H analogy which uses an advanced-time formulation (or source time dominant algorithm) proposed by Casalino (D. Casalino, 2003). This yields the far-field acoustic pressure fluctuations computed using the FW-H formulations (Eqs. 1–4):

$$P'(\bar{x},t) = p'_T(\bar{x},t) + p'_L(\bar{x},t) + p'_Q(\bar{x},t) \tag{1}$$

where

$$p'_T(\bar{x},t) = \frac{1}{4\pi} \int_{(f=0)} \left[\frac{\rho_0(\dot{U}_n + U_{\dot{n}})}{r(1-M_r)^2} + \frac{\rho_0 U_n(r\dot{M}_r + a_0 M_r - a_0 M^2)}{r^2(1-M_r)^3} \right]_{ret} dS \tag{2}$$

$$p'_L(\bar{x},t) = \frac{1}{4\pi a_0} \int_{(f=0)} \left[\frac{\dot{L}_r}{r(1-M_r)^2} + \frac{L_r(r\bar{M}_r + a_0 M_r - a_0 M^2)}{r^2(1-M_r)^3} \right]_{ret} dS$$

$$+ \frac{1}{4\pi} \int_{(f=0)} \left[\frac{L_r - L_M}{r^2(1-M_r)^2} \right]_{ret} dS \tag{3}$$

$$p'_Q(\bar{x},t) = \frac{1}{4\pi} \left(\left(\frac{1}{c}\right)\left(\frac{\partial^2}{\partial t^2}\right) \int^t_{-\infty} \left[\int_{(f>0)} \frac{T_{rr}}{r} d\Omega \right] d\tau \right.$$

$$+ \left(\frac{\partial}{\partial t}\right) \int^t_{-\infty} \left[\int_{(f>0)} \frac{3T_{rr} - T_{ii}}{r^2} d\Omega \right] d\tau$$

$$\left. + \left(c \int^t_{-\infty} \left[\int_{(f>0)} \frac{3T_{rr} - T_{ii}}{r^3} d\Omega \right] d\tau \right) \right) \tag{4}$$

where

$$U_i = \frac{\rho u_i}{\rho_0} + v_i \left(1 - \frac{\rho}{\rho_0}\right) \tag{5}$$

$$L_i = \frac{P_{ij}}{n_i} + \rho u_i (u_n - v_n) \tag{6}$$

$$r = \left(x_{observer} - y_{face} \right) \tag{7}$$

where ρ_0 is the far-field reference density, u_i and v_i are, respectively, the fluid and surface velocity components in the x_i direction (Eq. 5). $P_{ij} = (p - p_0)\delta_{ij} - \sigma_{ij}$ is the compressive stress tensor, u_n and v_n are, respectively, the fluid and surface velocity components normal to the surface (f = 0) (Eq. 6). σ_{ij} is the viscous stress factor and δ_{ij} corresponds to Kronecker delta. a_0 represents the speed of sound in the far-field area and $M_i = a_0 v_i$ is surface Mach number vector. is the distance from a source point to the observer (Eq. 7). f < 0 defines a region inside S, f = 0 corresponds to the surface S and f > 0 describes an unbounded space outside S. A dot above a variable denotes the time derivative with respect to source time of that variable. $T_{rr} = T_{ij} r_i r_j$ describes the double contraction of T_{ij}.

T_{ij} is the Lighthill stress tensor, which is given by equation 8,

$$T_{ij} = \rho v_i v_j + \delta_{ij} \left[(p - p_0) - a_0^2(\rho - \rho_0) \right] - \sigma_{ij} \tag{8}$$

$p'_T(\bar{x},t)$ defines the noise generated by the displaced fluid due to the moving object. This is called thickness noise. The displacement of fluid as the body passes generates the monopole noise. $p'_L(\bar{x},t)$ refers to the normal stress on the surface due to the pressure distribution and the noise caused by the moving forces. This is called loading or lift noise. The unsteady motion of the force distribution on the body surface generates the dipole noise. $p'_Q(\bar{x},t)$ is called quadrupole noise, which accounts for nonlinearities in the flow.

2.2.2 The k-ε sound model

The k-ε sound method, based on the Proudman (1952) analogy, is used for the estimation of flow-induced noise. Both dipole and quadrupole noise sources are considered in acoustic calculations. The processes of noise generation and noise propagation are examined separately in this method. The flow field data required for acoustic analysis is obtained using computational fluid dynamics (CFD).

Flow noise sources are presented as the sum of surface and volumetric sources in equation 9;

$$P = \int_s P_s(y_b) ds + \int_v P_r(y) dv \tag{9}$$

Y_b and y represent the position vectors of the surface and volumetric source points, respectively. The surface and volume components of total acoustic power are calculated as given in equation 10 and equation 11 (Skvortsov et al. 2009, Proudman 1952).

$$P_s(y_b) = B\rho \left(\frac{k_0(y_b)}{c} \right)^3 \tag{10}$$

$$P_v(y) = A\rho\varepsilon(y)\left(\frac{\sqrt{k(y)}}{c}\right)^5 \quad (11)$$

C and ρ represent sound velocity and fluid density, respectively. $k(y)$ and $\varepsilon(y)$ correspond to turbulence kinetic energy and turbulence dissipation rate, respectively. $k_0(y_b)$ is the kinetic energy of the fluctuating component of the friction velocity at y_b. A and B are the coefficients obtained from different model calibrations (P. Croaker et al. 2011).

When $P_s(y_b)$ and $Pv(y)$ are considered to act as point sources, the acoustic sound intensity, I at a field point x is determined according to equation 12 as,

$$I(x) = I_s(x) + I_v(x) \quad (12)$$

$I_s(x)$ and $I_v(x)$ represent, respectively, the acoustic intensity that occurs due to surface sources and volumetric sources at the field point x as given in equation 13 and equation 14.

$$I_s(x) = \int_s \frac{P_s(y_b)}{4\pi R^2} ds \quad (13)$$

$$I_v(x) = \int_v \frac{P_v(y)}{4\pi R^2} dv \quad (14)$$

The spectrum for volumetric sources, f_v, is be determined by an approach such as in Equation 15, depending on the $\frac{w}{w_0}$ ratio.

$$f_v(w, q_v, r_v, C_v) \approx \begin{cases} \left(\dfrac{w}{w_0}\right)^{q_v}; & \dfrac{w}{w_0} \ll 1 \\[3mm] \left(\dfrac{w}{w_0}\right)^{q_v - r_v}; & \dfrac{w}{w_0} \ll 1 \end{cases} \quad (15)$$

Q_v is 8 for the volumetric sources because of the 8th power scaling law for acoustic power generated by quadrupole sources derived by Lighthill (1952). The acoustic spectra experience a power law decay of $-7/2$ and therefore $q_v - r_v = 23/2$.

For surface sources, a piecewise linear spectrum is used, expressed as in equation 16.

$$f_s(w, q_s, r_s, s_s, C_s) \approx \begin{cases} 0, 1^{r_s - q_s} \cdot C_s\left(\dfrac{w}{w_0}\right)^{q_s}; & \dfrac{w}{w_0} < 1 \\[3mm] C_s\left(\dfrac{w}{w_0}\right)^{r_s}; & 0,1 < \dfrac{w}{w_0} < 1 \\[3mm] C_s\left(\dfrac{w}{w_0}\right)^{s_s}; & \dfrac{w}{w_0} > 1 \end{cases} \quad (16)$$

Table 1. Calibration coefficients for volumetric and surface sources (P. Croaker et al., 2011).

Calibration coefficient	Volumetric	Surface
A	1.3E+3	–
B	–	6.6E-8
C	2.27/w_0	1/(3.61w_0)
q	8	1.2
r	23/2	−1.1
s	–	−3.5
F	2.0	0.09

where w_0 represents the characteristic frequency value, which is given below for surface sources and volume sources, respectively (Eqs. 17,18);

$$w_0 = F_v\left(\frac{\varepsilon}{k}\right) \quad (17)$$

$$w_0 = F_s\left(\frac{\varepsilon_0}{k_0}\right) \quad (18)$$

The correlation coefficients, A, B, C_v, C_s, F_v, F_s q_s, r_s, ve s_s obtained from the model calibration are given at Table 1 (P. Croaker et al., 2011).

The acoustic power and intensity spectra are determined by using the equation 19 and 20, respectively.

$$P(w) = P_s f_s(w, q_s, r_s, s_s, C_s) + P_v f_v(w, q_v, r_v, C_v) \quad (19)$$

$$I(w) = I_s f_s(w, q_s, r_s, s_s, C_s) + I_v f_v(w, q_v, r_v, C_v) \quad (20)$$

3 RESULTS AND DISCUSSION

Numerical and analytical studies are carried out to investigate the hydro-acoustic and aero-acoustic characteristics of a circular cylinder. The flow field results are given as an input data to the wave equations in order to attain the acoustic far-field. The acoustic and dynamic results are presented separately since the acoustic analogies are separated the flow field and acoustic computations.

3.1 Flow field analysis

The noise prediction is subject to the accuracy of the Computational Fluid Dynamics results. For a precise prediction of the acoustic far-field, first unsteady flow parameters are necessary to be measured accurately. Therefore, the non-dimensional coefficients are assessed by comparing them with existing experimental results. To examine the performance of the numerical simulations, here mean

Table 2. Comparisons of strouhal numbers and drag coefficients with the available results.

	Strouhal number		Mean Drag coefficient
Experiment (Park 2012)	0.195	Experiment (Cantwell & Coles 1983)	1.0–1.4
Experiment (Nornberg 2003)	0.180–0.191	Simulation, LES model 3D (Orselli et al. 2009)	1.08
Simulation, k-ϵ model 2D (Orselli et al. 2009)	0.282	Simulation, k-ϵ model 2D (Orselli et al. 2009)	0.479
Simulation, k-ω model 2D (Orselli et al. 2009)	0.247	Simulation, k-ω model 2D (Orselli et al. 2009)	0.944
Results, k-ϵ model 2D	0.286	Results, k-ϵ model 2D	0.679
Results, k-ω model 2D	0.254	Results, k-ω model 2D	1.229

drag coefficient, and Strouhal (1878) number are used as benchmark parameters.

Table 2 shows the values of mean drag coefficients and Strouhal numbers obtained from the two-dimensional computations. The available experimental and computed data are also presented. Regarding the turbulence models, the highest value of mean drag coefficient is attained by the k-ω SST model which agrees well with the experimental data. Strouhal (1878) number is critical in prediction of the vortex shedding, which shows the beginning of first harmony peak of noise. In comparison with the experimental data, all two-dimensional simulations present a slightly higher shedding frequency. It is expected that the 3D simulations will be agree better with the experiments than 2D simulations. As mentioned by Casalino (2003), the mean Reynolds stresses in 2D simulations are higher than 3D flows. It causes shorter mean recirculating regions resulting of the over prediction of the shedding frequency from 2D computation.

3.2 Flow-induced noise analysis

The flow field data is used as a sound source into the wave equations to predict the flow-induced noise. The first method used here is based on the Ffowcs Williams and Hawkings (1969) equations and its integral solution. The Ffowcs Williams and Hawkings (1969) equations can be obtained by manipulating the continuity and Navier Stokes equations and it is basically an inhomogeneous wave equation. The second method involves a k-ε acoustic model based on the Proudman's (1952) acoustical model to approximate the flow noise and extended to include the sound produced by a circular cylinder immersed in turbulent flow has been presented.

In all acoustic computations, the flow field is obtained by the k-ε and k-ω SST turbulence models, separately. The experimental data of Revell et al. (1997) is selected to be compared to the aero-acoustic numerical and analytical results.

Table 3. OASPL results for each representative length.

	$L_a =$ 3.16D	$L_a =$ 7.5D	$L_a =$ 10.0D	$L_a =$ 25.3D	Exp. Revel
OASPL (dB)	99.30	106.85	110.95	117.40	108.03

In the experiment, a cylinder diameter and span length of, respectively, D = 0.019 m and L = 25.3D is employed.

The acoustic results are expressed in terms of sound pressure level in the decibel scale (dB). The sound pressure level (SPL) is described as

$$SPL = 20 \log(P_e / P_{ref}) \tag{21}$$

where P_e is the effective sound pressure and P_{ref} indicates a reference pressure, which is used a value of 20 μPa in air and 1 μPa in water. The acoustic spectrum is figured considering quadrupole sources, dipole and monopole sources. Receiver point for the aero-acoustic analysis is placed perpendicular to the flow direction downward and 128 diameters away from the cylinder.

It is necessary to define a representative length in order to use the two-dimensional CFD results as an input data for the acoustic computations and consequently, flow data over the span direction can be used in the three-dimensional wave equation as the integration surfaces. The values of representative length $L_a = 3.16D$, 7.5D, 10.0D and 25.3D have been used in order to provide two-dimensional data from CFD simulations.

Regarding all noise amplitudes of the spectrum, an overall pressure sound level (OASPL) can be obtained as

$$OASPL = 20 \log \sqrt{\sum_i (10^{SPL_i / 20})^2} \tag{22}$$

The OASPL is calculated for each representative length and presented in Table 3. It can be seen from the Table 3 that the OASPL value for the length $L_a = 7.5D$ is the closest one to the experimental result.

Figures 3 and 4 show the acoustic spectrum obtained for each L_a by using k-ε turbulence model and k-ω SST turbulence model, respectively. They are compared with the corresponding experimental result of Revell et al. (1997). It is observed that the two-dimensional approach can predict discrete values of SPL associated with the fundamental frequency (Strouhal number) and its harmonics. The k-ε turbulence model presents a slightly higher shedding frequency whereas the k-ω SST turbulence model agrees well. Besides, the harmonics couples are well in timing when the k-ω SST model is used

Figures 3 and 4 present the frequency weighted sound pressure levels calculated using the k-ε acoustic model for different turbulence models. The results predicted using calibration parameters

(P. Croaker et al. 2011) are compared with the experimental data of Revel et al. (1997). The harmonics of the predicted SPL compares reasonably well with the experimental results; however, the SPL is underestimated over the acoustic spectrum. This method gives an estimation of the average sound pressure level and the broadband noise levels and it can be optimized in terms of calibration coefficients.

The hydro-acoustic numerical and analytical results are compared with the numerical results of W.S Choi et al. (2015) in Figures 5 and 6. The results are obtained with a cylinder diameter and span length of D = 0.02 m and L = 1.5D, respectively. $L_a = 1.5D$ is used as a representative length for 2D simulations. Receiver points for the hydro-acoustic analysis are placed perpendicular to the flow direction downward from the cylinder at distances of 100 mm, 200 mm and 400 mm. The comparisons

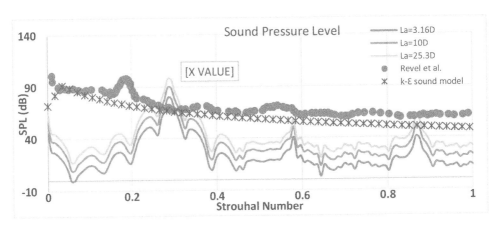

Figure 3. Comparison of sound pressure level spectra for two-dimensional approach (k–ε turbulence model, air).

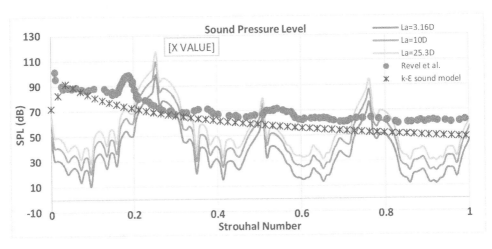

Figure 4. Comparison of sound pressure level spectra for two-dimensional approach (k–ω SST turbulence model, air).

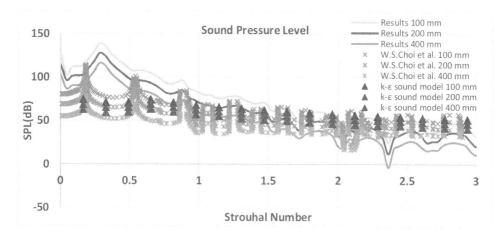

Figure 5. Comparison of sound pressure level spectra for two-dimensional approach (k–ε turbulence model, water).

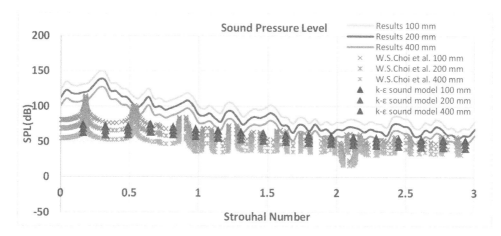

Figure 6. Comparison of sound pressure level spectra for two-dimensional approach (k–ω SST turbulence model, water).

of the predicted noise with the available numerical results at the receiver points around the cylinder in underwater environment for Re = 9×10^4 are shown in Figures 5 and 6.

Overall frequency characteristics of both the calculations using k-ε turbulence model and k-ω SST turbulence model show similar results throughout the receiver points with stationary peak frequencies. However, the results obtained by k-ω SST turbulence model is slightly higher compared with k-ε turbulence model. Analysis results in various receiver positions shows similar frequency characteristics with decrease in magnitude as distance increases.

The k-ε acoustic method can predict discrete values of SPL associated with the broadband noise and its harmonics, however, it can be optimized in terms of calibration coefficients.

Figure 7 shows sound pressure levels (SPL) which are obtained for various receiver positions and for the medium air. Receiver points for the aero-acoustic analysis are placed perpendicular to the flow direction downward from the cylinder at distances of 8D, 16D, 32D, 64D, 96D and 128D. Analysis results for various receiver positions show similar frequency characteristics with decrease in magnitude as distance increases. Comparing the SPL values at the vortex shedding frequency, there is a 6 dB decrease in SPL. This corresponds to 16D increase in the distance between circular cylinder and receiver point.

Figure 8 shows sound pressure levels which are obtained for various receiver positions and for the medium water. Receiver points for the hydro-acoustic analysis are placed perpendicular to the flow direction downward from the cylinder at

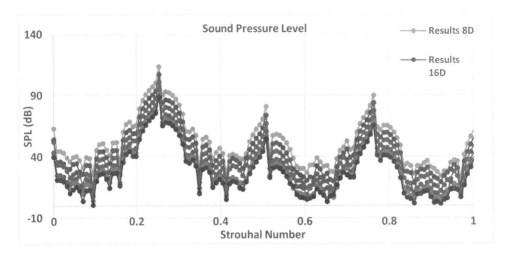

Figure 7. Sound pressure level spectra for various receiver positions (two-dimensional approach, k-ω SST turbulence model, air).

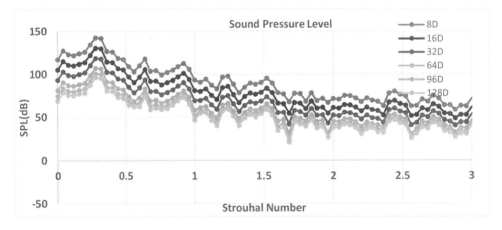

Figure 8. Sound pressure level spectra for various receiver positions (two-dimensional approach, k-ω SST turbulence model, water).

distances of 8D, 16D, 32D, 64D, 96D and 128D. The results for various receiver positions show similar frequency characteristics with decrease in magnitude as distance increases. Comparing the SPL values at the vortex shedding frequency, there is a 12 dB decrease in SPL. This corresponds to 16D increase in the distance between circular cylinder and receiver point.

4 CONCLUSIONS

The prediction of flow-induced noise around a circular cylinder in various environments is performed by two different methods: FW-H acoustic model and k-ε acoustic model. The noise sources

are computed from the computational fluid dynamics computations which are used as input data for these acoustic models. The noise sources is obtained from unsteady two-dimensional CFD computations by using k-ε turbulence model and k-ω SST turbulence model.

The k-ω SST turbulence model is found to give a better agreement with the available results in the literature. The span length is necessary for sound computation to provide flow information over the cylinder span.

A correlated flow over the cylinder span length is predicted using the 2D flow data. As a result, a discrete acoustic spectrum at the harmonic frequencies is calculated. A k-ε acoustic model based on Proudman's formula which is extended to comprise the sound induced by a circular cylinder immersed in turbulent flow is presented. This method can predict discrete values of SPL and has compatible harmonics into the broadband noise, especially.

Sound pressure levels are obtained at various receiver positions for the medium both air and water. The results at various receiver positions show similar frequency characteristics with decrease in magnitude as distance increases.

For future studies, three-dimensional simulations by using longer span lengths should be performed in order to obtain more accurate data for unsteady flow throughout the cylinder span. Besides, large eddy simulation should be taken into consideration to investigate the capability of this turbulence model to predict the flow-induced noise around the cylinder.

REFERENCES

Cantwell & D. Coles (1983). *An experimental study of entrainment and transport in the turbulent near wake of a circular cylinder*, J. Fluid Mech.

Casalino, D. & M. Jacob (2003). *Prediction of Aerodynamic Sound from Circular Rods via Spanwise Statiscal Modelling*, J. Sound and Vibration, Vol. 262, No. 4, pp. 815–844.

Choi, W.-S., Y. Choi, S.-Y. Hong, J.-H. Song, H.-W. Kwon & C.-M. Jung (2016). *Turbulence-induced Noise of a Submerged Cylinder Using a Permeable FW-H Method*, INJAOE, 2016, 1–8.

Choi W.-S. & S.-Y. Hong (2015). Turbulent-Induced Noise around a Circular Cylinder Using Permeable FW-H Method, Journal of Applied Mathematics and Physics.

Croaker P., A. Skvortsov & N. Kessissoglou (2011). *A Simple Approach to Estimate Flow-Induced Noise from Steady State CFD Data*, Proceedings of ACOUSTICS, s.54, Gold Coast, Australia.

Casalino D. (2003). *An advanced time approach for acoustic analogy predictions.* Journal of Sound and Vibration, vol 261, no. 4, pp. 583–612.

Liang C. & G. Papadakis, (2006). *Large eddy simulation of pulsating flow over a circular cylinder at subcritical*

Reynolds number, King's College London, Strand, London.

Lighthill, M.J. (1952). *On sound generated aerodynamically I: general theory.* Proceedings of the Royal Society, 564–587.

Lu Y.-T., H.-X. Zhang & X.-J. Pan (2008). *Numerical simulation of flow-field and flow-noise of a fully appendage submarine*, Shanghai Jiao Tong University.

Menter, F.R. (1993). "Zonal Two Equation k-ω Turbulence Models for Aerodynamic Flows", AIAA Paper 93–2906.

Menter, F.R. (1994), "Two-Equation Eddy-Viscosity Turbulence Models for Engineering Applications", AIAA Journal, vol. 32, no 8. pp. 1598–1605.

Norberg, C. (2002). *Fluctuating Lift on a Circular Cylinder: Review and New Measurements,* J. Fluids and Structures, Vol. 17, No. 1, pp. 57–96.

Orselli R.M., J.R. Meneghini & F. Saltara (2009). *Two and Three-Dimensional Simulation of Sound Generated by Flow around a Circular Cylinder,* 15th AIAA/CEAS Aeroacoustics Conference (30th AIAA Aeroacoustics Conference) Miami, Florida.

Park C.-W. & S.-J. Lee (2002). *Flow structure around a finite circular cylinder embedded in various atmospheric boundary layers*, The Japan Society of Fluid Mechanics and IOP Publishing Ltd.

Proudman, I. (1952). *Sound generation of noise by isotropic turbulence.* Proceedings of the Royal Society.

Revell, J.D., R.A. Prydz & A.P. Hays (1997). *Experimental Study of Airframe Noise vs. Drag Relationship for Circular Cylinders.* "Lockheed Report 28074, Final Report for NASA Contract NAS1–14403.

Skvortsov, A, K. Gaylor, C. Norwood, B. Anderson & L. Chen (2009). *Scaling laws for noise generated by the turbulent flow around a slender body*, Undersea Defence Technology: UDT Europe 2009, pp. 182–186.

Strouhal, V. (1878). *Ueber eine besondere Art der Tonerregung (On an unusual sort of sound excitation).* Annalen der Physik und Chemie, 3rd series, 5 (10), 216–251.

Watanabe, H., A. Ihara & H. Hashimoto (1996). *Hydraulic characteristics of a circular cylinder with a permeable Wall,* JSME International Journal, Series B 39.

Williams, J. E., & D. L. Hawkings (1969). *Sound Generation by Turbulence and Surfaces in Arbitrary Motion.* Philosophical Transactions of the Royal Society of London.

Maritime Transportation and Harvesting of Sea Resources – Guedes Soares & Teixeira (Eds)
© 2018 Taylor & Francis Group, London, ISBN 978-0-8153-7993-5

Modeling of performance of an AUV stealth vehicle. Design for operation

M.K. Gerigk
Department of Mechanical Engineering, Gdansk University of Technology, Gdansk, Poland

ABSTRACT: In the paper some results of research connected with modelling of performance and risk assessment of an AUV stealth vehicle are presented. A general approach to design of the stealth AUV autonomous underwater vehicle under consideration is introduced. The basic stealth characteristics of the AUV stealth vehicle are briefly described. The method of research is introduced. The AUV stealth vehicle concept is presented including the hull peculiarities. Between the results of research some basic stealth characteristics of the AUV stealth vehicle are presented. Some final remarks regarding the AUV stealth vehicle performance and its behaviour in the data operational conditions are given. The final conclusions are presented.

1 INTRODUCTION

Despite of the number and types of navy ships each of EU key navies should have a set of unmanned surface (USV – Unmanned Surface Vehicle) and underwater (UUV – Unmanned Underwater Vehicle) vehicles for the different missions to perform. Some of the vehicles should be partially or fully autonomous (AUV – Autonomous Underwater Vehicle).

The last decade it was the time when the research investigations towards implementing a new generation of the multi-task small navy ships and unmanned USV and UUV vehicles has been conducted. The small multi-task navy ships despite the patrol and typical combat missions may be the platforms for the unmanned air and maritime vehicles.

The discussion how many small multi-task navy ships and unmanned maritime vehicles of which type a navy needs seems to be over. The research teams are ready to implement the chosen solutions concerning both the small multi-task navy ships and USV, UUV and ASV vehicles.

Concerning the USV, UUV and ASV vehicles they may perform the typical patrol, reconnaissance or combat missions. Depending on the mission the vehicles may be equipped with either the sophisticated reconnaissance electromagnetic, hydro acoustic and IT-based equipment or conventional arms. The most advanced vehicles may be equipped with the small fast underwater missiles (Gerigk, 2014).

2 AN AUV VEHICLE CONCEPT

The primary aim of the research is to work out a functional model of the stealth AUV vehicle moving in the data operational conditions.

The novel solutions have been applied regarding so far the hull form, arrangement of internal spaces, materials and propulsion system. The final hull form is a combined stealth hull form. The arrangement of internal spaces has been designed according to the functional requirements and is very much affected by the sub-systems to be installed onboard.

The major sub-systems of the stealth AUV vehicle are as follows (Gerigk, 2015; Gerigk, 2016):

– ballast sub-system,
– energy supply sub-system (batteries),
– water-jet propulsion sub-system,
– T-foil stabilizing sub-system,
– steering sub-system,
– communication and navigation sub-system.
– dedicated sub-system.

The main parameters of the stealth AUV vehicle are as follows (Gerigk, 2015; Gerigk, 2016):

– overall length L – is equal to 2.2 meters (4.4 meters for the larger stealth AUV vehicle (LS-AUV),
– operational breadth B – is equal to 1.1 meters (2.2 meters for LS-AUV) without the additional equipment and appendages,
– height H – is equal to 0.35 meters (0.70 meters for LS-AUV) without the additional equipment and appendages,
– mass is equal to from 0.38 tons to 0.65 tons, depending on the mass of equipment installed onboard,
– averaged speed during the underwater mission for the submerged conditions (3 meters) v_{uw} – is equal to 1.0–2.0 meters per second.

The general visualizations of the hull form and arrangement of internal spaces of the stealth AUV vehicle are presented in Figure 1.

Figure 1. The general visualizations of the hull form and arrangement of internal spaces of the stealth AUV vehicle (Gerigk, 2014–2016).

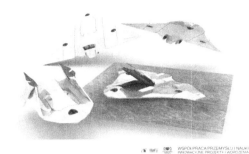

Figure 2. The visualization of the stealth triple-state FIST-RP vehicle (Gerigk, 2012–2016).

The general visualizations of the hull form and arrangement of internal spaces of the stealth AUV vehicle are presented in Figure 1.

3 THE CHARACTRISTICS

The new Polish solutions concerning the unmanned maritime vehicles (USV, UUV, ASV) have been worked out at the Gdańsk University of Technology by the team conducted by the author. The team work is devoted to the interdisciplinary research and application of advanced technologies. Some implementations concern development of innovative small multi-task navy ships and unmanned maritime vehicles (mainly UUV and ASV) (Gerigk, 2015; Gerigk, 2016).

The latest research and design concerns the stealth AUV vehicle. The research is sponsored by the National Centre for Research and Development (NCBiR) (Gerigk, 2015; Gerigk, 2016).

Intelligent and autonomous stealth AUV. Despite of the mission the stealth AUV vehicles should be applied with the coded communication and navigation subsystems to operate the vehicles above the water surface, on the water surface and under the water surface. The vehicles should be equipped with the energy supply subsystem (batteries) to perform up to several hours missions with the possibility to upload the batters using the submerged energy loading stands. According to the performed tasks some vehicles should be equipped with the autonomous intelligent (AI – Artificial Intelligence) steering and vehicle positioning subsystems enabling to use the data obtained from the vehicle sensor subsystems. The most advanced stealth AUV vehicles would be those which could independently communicate with the other vehicles (USV, UUV, ASV) and make decisions on their own.

Silent and invisible stealth AUV. Despite the above mentioned features the most important for the AUV vehicles is to design and manufactured using the most advanced STEALT technologies.

Nowadays, "stealth" does not only mean to avoid and/or absorb the radar radiation by the unique hull form and/or hull skin cover. "Stealth" concerns the propulsion system. The stealth AUV vehicles should be equipped with the silent both the electric engines and jet propellers. The noise and vibrations generated by a vehicle should on the lowest possible level. The subsystems and construction of the stealth AUV vehicle should protect the emission of heat. Moving under the water surface the stealth AUV vehicle should generate a small value boundary layer and wake to limit its own acoustic signal. This is why the innovative hull skin covers are of great importance.

A research is conduced to find a cover to make the hull skin invisible.

"Stealth" means to limit the probability to detect the AUV vehicle as much as possible using the well known and hardly known methods and means of reconnaissance.

The current investigations on the stealth technologies towards their application onboard the AUV vehicles are associated with the following problems:

- size and vehicle hull form,
- hull skin covers,
- minimizing the noise, vibrations and heat factors and impacts,
- minimizing own electromagnetic and acoustic signals,
- avoiding and absorbing the outside electromagnetic and acoustic signals,
- maximizing the invisibility.

4 A PERFORMANCE-ORIENTED RISK-BASED METHOD OF MODELING THE STEALTH CHARACTERISTICS

The research method for modeling the stealth characteristics, assessment of the stealth AUV vehicle performance and risk assessment is a kind of performance-oriented risk-based method which enables to assess the above mentioned at the design stage and in operation (Gerigk, 2010). The method takes into account the influence of design and operational factors on the performance and safety of the stealth AUV vehicle including many factors following from different sources. The holistic approach and system approach to safety have been applied.

For assessment of the stealth AUV vehicle performance the investigations using the physical models and computer simulation techniques have been applied. The performance assessment of the stealth AUV vehicle enables to identify the operational sequence of events for the conditions very close to reality (Gerigk, 2015; Gerigk, 2016: Gerigk, Wójtowicz, Zawistowski, 2015; Gerigk, Wójtowicz, 2015).

The risk assessment is based on application of the matrix type risk model which is prepared in such a way that it enables to consider almost all the scenarios of events. The criteria is to achieve an adequate level of risk using the risk acceptance criteria, risk matrix, (Gerigk, 2010). Providing a sufficient level of safety based on the risk assessment is the main objective. It is either the design, operational or organizational objective. Safety is the design objective between the other objectives. The measure of safety of the stealth AUV vehicle is the risk (level of risk).

The method itself is based on the following main steps:

– setting the requirements, criteria, limitations, safety objectives;
– defining the stealth AUV vehicle and operational environment;
– identifying the hazards and identifying the sequences of events (scenarios);
– assessing the stealth AUV vehicle performance;
– estimating the risk according to the event tree analysis ETA and matrix type risk model (risk is estimated separately for each scenario development);
– assessing the risk according to the risk acceptance criteria (risk matrix) and safety objectives;
– managing the risk according to the risk control options;
– selecting the design (or operational procedure) that meet the requirements, criteria, limitations, safety objectives;
– optimizing the design (or operational procedure);
– making the decisions on safety.

The structure of the method which combines the typical design/operational procedures with the risk assessment techniques is presented in Figure 3

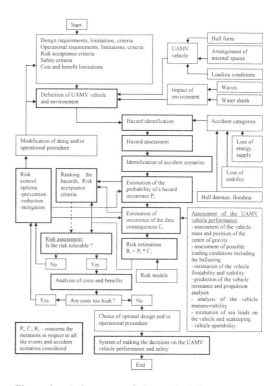

Figure 3. A Structure of the method for assessment of the UAMV vehicle performance and risk assessment (Gerigk, 2014–2016).

(AUVSI/ONR, 2007; Abramowicz-Gerigk, Burciu, 2014; Cwojdziński, Gerigk, 2014; Dudziak, 2008; Faltinsen, 1990; Faltinsen, 2005; Gerigk, 2010; Gerigk, 2015; Lamb, 2003; Szulist, Gerigk, 2015).

5 AN EXAMPLE OF INVESTIGATIONS OF THE STEALTH CHARACTERISTICS

As an example of investigations concerning predicting the influence of a stealth characteristic on the stealth AUV vehicle performance the impact of different hull skin covers (generated during the computer simulation) on the flow (boundary layer and wake) around the stealth AUV vehicle has been modeled.

During the computer simulation the mesh consisted of 3 275 000 elements. The numerical domain had the size of 5 meters in length, 1.5 meters of breadth and 1.2 meters of height (Kardaś, Tiutiurski, Gerigk, 2016). The domain is presented in Figure 4.

The water flow velocity was anticipated to be from 0.5 meters per second to 2.5 meters per second with the step 0.5 meters per second.

During the computer simulation the hull skin cover was generated by the skin roughness as follows:

Figure 4. A visualization of the numerical domain for simulating the impact of different hull skin covers on the flow around the stealth AUV vehicle.

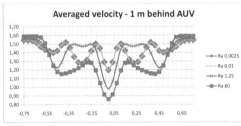

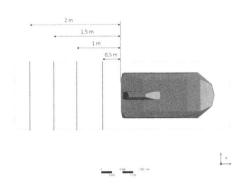

Figure 5. Some results of the impact of different hull skin covers on the flow behind the stealth AUV vehicle for the data skin roughness and flow velocity.

– Ra 80 – as a normal steel plate surface,
– Ra 1.25 – as a slightly polished steel plate surface,
– Ra 0.01 – as a polished steel plate surface,
– Ra 0.0025 – extremely polished steel plate surface (nano-surface).

The flow was estimated for example 0.5 meters, 1.0 meters, 1.5 meters and 2.0 meters behind the stealth AUV vehicle. Some results of the flow estimation for the data skin roughness and velocity are presented in Figure 5 (Kardaś, Tiutiurski, Gerigk, 2016).

6 CONCLUSIONS

In the paper some results connected with development of a functional model of the stealth AUV vehicle are presented.

Some data on the stealth AUV vehicle concept, major stealth characteristics, performance-oriented risk-based method of modelling the stealth characteristics are described in the paper.

As an example some results concerning predicting the influence of the different hull skin covers (generated during the computer simulation) on the boundary layer and wake are presented in the paper.

At the current stage of research an influence of combined stealth characteristics on the stealth AUV vehicle performance is investigated.

REFERENCES

Abramowicz-Gerigk T., Burciu Z., 2014. Safety assessment of maritime transport—Bayesian risk-based approach in different fields of maritime transport. *Developments in Maritime Transportation and Exploitation of Sea Resources, Guedes Soares & Lopez Pena (Eds)*, Volume 2, Taylor&Francis Group, London, UK, p. 699–703.

AUVSI/ONR, 2007. *Engineering Primer Document for the Autonomous Underwater Vehicle (AUV) Team Competition Association for Unmanned Vehicle Systems International (AUVSI) US Navy Office of Naval Research (ONR)*, Version 01 – July 2007.

Cwojdziński L., Gerigk M.K., 2014 The Polish innovative solutions concerning the maritime platforms and vehicles including the unmanned systems and vehicles (in Polish: Polskie innowacyjne rozwiązania w zakresie jednostek i obiektów morskich, w tym systemów bezzałogowych). *The New Military Technologies (in Polish: Nowa Technika Wojskowa)*, No. 11, 2014.

Dudziak J., 2008. The theory of ships (in Polish: Teoria okrętu), *The Foundation of Promotion of the Shipbuilding and Marine Economy (in Polish: Fundacja Promocji Przemysłu Okrętowego i Gospodarki Morskiej)*, Gdańsk 2008.

Faltinsen O.M.,1990. Sea Loads on Ships and Offshore Structures, *Cambridge University Press*, 1990.

Faltinsen O.M., 2005. Hydrodynamics of High-Speed Marine Vehicles, *Norwegian University of Science and Technology, Cambridge University Press*, 2005.

Gerigk M.K., 2010. A complex method for assessment of safety of ships in damage conditions using the risk analysis (in Polish: Kompleksowa metoda oceny bezpieczeństwa statku w stanie uszkodzonym z uwzględnieniem analizy ryzyka), *Monography No. 101 (in Polish: Monografie 101), Edited by the Gdańsk University of Technology (in Polish: Wydawnictwo Politechniki Gdańskiej)*, Gdańsk 2010.

Gerigk M., 2012. Assessment of safety of ships after the collision and during the ship salvage using the matrix type risk model and uncertainties. *Sustainable Maritime Transportation and Exploitation of Sea Resources. Guedes Soares & Rizzuto (Eds)*, Volume 2, pp. 715–719.

Gerigk M.K., Wójtowicz S., 2014. A model of the steering system of a small unmanned vehicle moving on the water surface (in Polish: Model systemu sterowania małego obiektu bezzałogowego poruszającego się na powierzchni wody), *The Logistics (in Polish: Logistyka)*, No. 6, 2014.

Gerigk M.K., 2015. The innovative multi-task ships and vehicles for the Polish Navy (in Polish: Innowacyjne wielozadaniowe jednostki i obiekty pływające dla komponentu morskiego sił zbrojnych RP), *The Manual, 11th International Conference & Exhibition "Advanced Technologies for Homeland Defense and Border Protection"*. Zarząd Targów Warszawskich S.A., Intercontinental Hotel, Warsaw, 14th May 2015.

Gerigk M.K., 2015. Modeling of performance and safety of a multi-task unmanned autonomous maritime vehicles. Modelowanie ruchu i bezpieczeństwa wielozadaniowego bezzałogowego autonomicznego pojazdu wodnego. *Journal of KONBIN, Safety and Reliability Systems*, No. 1 (33), Warszawa 2015.

Gerigk M.K., 2015. Innowacyjne rozwiązania w zakresie okrętów i obiektów pływających. *Logistyka, nr 3*, Poznań 2015.

Gerigk M.K., 2016. Challenges associated with the design of a small unmanned autonomous maritime vehicle. Scientific Journals of the Maritime University of Szczecin, *Zeszyty Naukowe Akademii Morskiej w Szczecinie*, nr 46 (118) 2016, 46 (118), 22–28 ISSN 1733-8670 (Printed) Received: 31.08.2015 ISSN 2392-0378 (Online) Accepted: 01.03.2016 DOI: 10.17402/113 Published: 27.06.2016.

Gerigk M.K., 2016. Konstrukcje bliskiej przyszłości. "PREZENTUJ BROŃ", *14th BALT-MILITARY-EXPO Baltic Military Fair*, Gdańsk, June 20–22, 2016.

Gerigk M.K., Wójtowicz S., Zawistowski M., 2015. A precise positioning stabilization system for the unmanned autonomous underwater vehicle for the special tasks (in Polish: Precyzyjny system stabilizacji pozycji autonomicznego pojazdu podwodnego do celów specjalnych), *The Logistics (in Polish: Logistyka)*, No. 3, 2015.

Gerigk M.K., Wójtowicz S., 2015. An Integrated Model of Motion, Steering, Positioning and Stabilization of an Unmanned Autonomous Maritime Vehicle. *TRANSNAV the International Journal on Marine Navigation and Safety of Sea Transportation*, Vol. 9, No. 4, December 2015, DOI: 10.12716/1001.09.04.18.

Kardaś D., Tiutiurski P., Gerigk M., 2016. Modelowanie opływu obiektu OWS. Model warstwy przyściennej. *Opracowanie nr 1/IMP, Projekt PBS3/A6/27/2017*, Politechnika Gdańska, 2016.

Lamb. G.R., 2003. High-speed, small naval vessel technology development plan, Total Ship Systems Directorate Technology Projection Report, NSWCCD-20-TR-2003/09, *Carderock Division, Naval Surface Warfare Center, Bethesda*, MD 20817-5700, May 2003.

Szulist N., Gerigk M.K., 2015. Metodyka nadawania cech stealth małym bezzałogowym pojazdom wodnym. *Logistyka, nr 4*, Poznań 2015.

Maritime Transportation and Harvesting of Sea Resources – Guedes Soares & Teixeira (Eds)
© 2018 Taylor & Francis Group, London, ISBN 978-0-8153-7993-5

Multibody dynamics of floaters during installation of a spar floating wind turbine

M.A.A.A. Hassan & C. Guedes Soares

Centre for Marine Technology and Ocean Engineering (CENTEC), Instituto Superior Técnico, Universidade de Lisboa, Lisbon, Portugal

ABSTRACT: Floating wind energy concepts have multiple challenges during installation. Float-over technologies have been developed and successfully applied to offshore installations of integrated topsides onto different fixed and floating platform substructures in the industry. This paper addresses numerical modelling and time-domain simulations during the float-over operation of a wind turbine on floating on a spar substructure. Time domain simulation results are presented for critical responses such as gripper forces, vessel and floating spar. A sensitivity study of the responses in different wave directions is also performed. For this type of marine operations, it is useful to evaluate the probability of acceptable weather conditions. The wave statistics for the North Sea was chosen for the operability and uptime analysis.

1 INTRODUCTION

The demands for renewable and reliable energy are requiring urgent actions due to global warming and the energy crisis. It is expected that 20% of the world's electricity to be generated by renewable energies by 2040 (IEA, 2014). Offshore wind accounted for 24% of total EU wind power installations in 2015, double the share of annual additions in 2014. This confirms the growing relevance of the offshore wind industry in the development of wind energy in the EU (EWEA, 2015).

However offshore wind is facing great challenges, as studies show that the capital cost of offshore wind power is more than twice of onshore projects (Mone et al., 2015). The turbines, although based on onshore designs, need to be designed with additional protection against corrosion and the harsh marine environment (Ciang et al., 2008). The more significant increase of cost offshore is due to increased investments in constructing expensive foundations at sea, transportation and installation of foundations, equipment and turbines, and operation and maintenance (van Kuik et al., 2016).

Offshore wind is a capital-intensive technology, and installation costs are an important cost component. But that is not the case for most of the deep offshore designs, which can be assembled onshore and then towed out to sea. This reduces the need to use vessels, cutting overall installation costs. It also reduces installation lead times as the deep offshore designs are less dependent on weather and sea conditions. The industry is, however, trying to minimize installation costs and

must continue to develop self-installing and shore assembled systems.

There are many different methods for offshore installation in the common practice. The methods for installing foundations depend on the foundation type. Monopiles (MP) is the most commonly used foundations for offshore wind turbines (OWT), but there is little work focusing on their installation phase. Monopiles are transported either on-board of a barge or an installation vessel or capped and wet towed (Herman, 2002). The size and weight of the monopiles are increasing, and it is expensive to use large installation vessels to carry out the operations. As an alternative, the wet tow method can be used. The wet tow of a single floating monopile has already been applied during installation of two wind farms (Npower Renewable, 2006; Ballast Nedam, 2011). Figure 1 shows the towing, ballasting and upending operation for a floating spar substructure of as an example. The transportation of more than one monopile per trip can be achieved using proper connection between the monopiles.

Another great challenge as the industry moves further offshore and into deeper waters, and the turbines and foundations are getting larger and heavier. Floating concepts are likely to be more cost-efficient for locations deeper than 100 [m] and three primary types are under consideration as shown in Figure 2. Up to now, there are only a few full-scale floating wind turbines installed for testing purposes, i.e., Hywind in Norway and WindFloat off the coast of Portugal (Statoil, 2012; Principle Power, 2011).

Figure 1. Towing, ballasting and upending of spar substructure.

Figure 2. Floating wind turbine concepts (Spar, semi-submersible and tension leg platform).

The type of the floating concepts in Hywind project is the floating spar. The spar-type floating foundation consists of a steel or concrete cylinder filled with a ballast of water and gravel to keep the center of gravity well below the center of buoyancy which ensures the wind turbine floats in the sea and stays upright since it creates a large right-ing moment arm and high inertial resistance to pitch and roll motions. The floater is ballasted by permanent solid iron or ballast, concrete or gravel from a chute. The draft of the floating foundation is usually larger than or at least equal to the hub height above the mean sea level for stability and to minimize heave motion. The spar-type floating wind turbine is usually kept in position by a taut or a catenary spread mooring system using anchor-chains, steel cables or synthetic fiber ropes.

2 PROBLEM DESCRIPTION

The available installation methods in common practice are all sensitive to weather conditions such as:

- Lifting the turbine using floating crane vessels
- Deploying and retrieving jack-ups legs

However, the operability of a jack-up vessel is limited by the water depth and the positioning process is time consuming and requires a low sea state. Floating vessels have more flexibility for offshore operations and will be effective in mass installations of a wind farm due to fast trans-portations between units. Lifting turbine nacelles and rotors at a large lift height need specific crane requirements.

Another way of installation is the floatover tech-nology. There are several substantial advantages to installing integrated topside onto a floating sub-structure using the floatover method, particularly for large topsides that exceed the single lift capacity of the available heavy lift fleet. These advantages include schedule and cost savings for the integra-tion and commissioning of modules on land rather than at sea. Uncoupling the top side fabrication schedules from the availability of heavy lift vessels is another advantage.

The floatover operation involves several major steps:

- Topside load-out onboard the installation vessel.
- Transportation the topside to the installation site.
- Approach the floating spar with the installation vessel
- Transfer the topside to the spar "mating operation"

There are some significant differences between installing a top structure on a fixed foundation ver-sus a floating foundation and of course between sheltered and open water installations. The most important difference is the length of time to per-form the load transfer between the transportation vessel and the floating structure.

For a fixed platform installations, jacks may be used to transfer most of the top structure load within a matter of one minute with jacking strokes on the order of 1.8 [m]. This means that the deck goes from a state of being clear of the mating points to being in synchronized contact over the course of about 6–12 wave periods. This minimizes the risk of excessive relative motions during the transfer. If this operation were attempted on a floating body, the 1.8 [m] jacking stroke would result in a depression of the floater (reduction in freeboard, or increase in the draft) while effectively transferring only a small fraction of the top side load, depending on the waterplane area of the floating substructure.

The performance of a successful floatover installation requires adequate design and analysis of each phase of the floatover installation and a sufficient weather window in which to perform each phase.

The analytical challenges related to floatover installation simulations are several and include multi-body hydrodynamics and prediction of relative motions and interface loads during the mating operations.

3 MODELLING OF THE COUPLED DYNAMIC SYSTEM

3.1 Installation methodology

This section describes the installation methodology and the way of hydrodynamic and mechanical coupling modelling. A floating installation vessel was chosen for the top structure "turbine" installation. The main dimensions of the vessel are presented in Table 1. The vessel is a monohull designed with a dynamic positioning system. The vessel is also equipped with a float-over skidding system for turbine installation. The system consists of hydraulic jacks used for lifting and lowering operations of the turbine. In addition, the vessel will be equipped with a motion compensation clamping system such as a gripper device to connect the floating spar to the vessel side during the skidding/jacking operation.

The overview of the installation method is shown in Figure 3. The floating sub-structure used in the model is to support a 6 MW offshore wind turbine and it is a spar type structure with main dimensions listed in

Table 1. Main parameters of the floating vessel.

Floating vessel		
Displacement	[MT]	31240
Length	[m]	161
Breadth		32.2
Depth		13.3
Draught		8.0

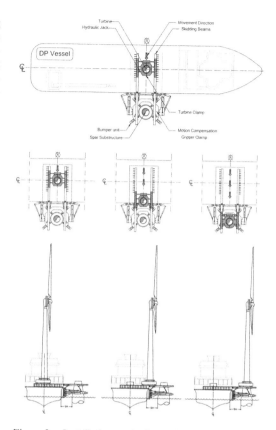

Figure 3. Installation method overview.

Table 2. Main parameters of the floating spar.

Floating spar		
Displacement	[MT]	11320
Beam @ W.L	[m]	9.5
Depth		92
Draught		83

Table 3. Main parameters of the turbine.

Turbine		
Weight	[MT]	970
Length	[m]	83.0
Diameter bottom		7.5
Diameter top		4.0

Table 2. The main dimensions of the "top structure-turbine" and its components are shown in Table 3.

The global coordinate system (GCS) is a right-handed coordinate system, with the following

373

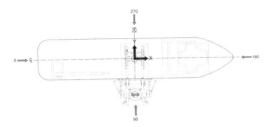

Figure 4. Global coordinate system and wave heading.

orientation used: X axis pointed towards the bow, Y axis towards the port side, and Z axis upwards. The origin is located at [mid-ship, center line, still water line] when the vessel is at rest. Figure 4 shows the definition of global coordinate system and wave directions applied in the current study. Beside the global coordinate system, the body fixed coordinate of each of the body is also defined.

The body-fixed coordinate moves with the body and is used to define the coupling points between bodies. Two body fixed coordinates are defined as follows:

a. The vessel-fixed coordinate system is overlapped with the global coordinate when the vessel is at rest;

b. The spar-fixed coordinate system is originated in the middle of the spar with an axis parallel to the global coordinate system in the initial condition.

3.2 Numerical model

3.2.1 Coupled system equation of motion

Every floating body has six degrees of freedom. Under the assumption that the response is linear and harmonic, the coupled differential equations of motion for the floating bodies can be written in the following form:

$$\sum_{j=1}^{18}[-\omega^2(M_{ij}+A_{ij})-i\omega B_{ij}+C_{ij}]\xi_j = F_i$$

for i = 1, 2... 18, where M_{ij} is the generalized mass matrix for the two floating bodies. A_{ij} and B_{ij} are the added mass and potential damping matrix respectively and C_{ij} is the restoring force matrix. ξ_j is the response motion amplitude in each of the six degrees of freedom for each barge. F_i is the complex amplitude of the wave exciting force for the two floaters.

3.2.2 Multibody simulations

It is effective to adopt the generalized mode concept to solve multi-body hydrodynamic interaction

problems. For a multi-body system of NB body units, the concept of generalized mode leads to 6XNB degrees of freedom assuming each body behaves as a rigid body.

In the case of multi-body which are hydrodynamically and mechanically coupled, the equations mentioned above need to be solved in a coupled matrix equation. In the case of a 3-body system considered in this paper, the system has 18 degrees of freedom. All the three bodies are subjected to wave-induced forces, hydrodynamic reaction forces and mechanical coupling effects. The equation of multi-body motion is given in Equation).

$$\begin{bmatrix} M^{11} & M^{12} & M^{13} \\ M^{21} & M^{22} & M^{23} \\ M^{31} & M^{32} & M^{33} \end{bmatrix} X \begin{bmatrix} \ddot{X}^1 \\ \ddot{X}^2 \\ \ddot{X}^3 \end{bmatrix} + D\dot{X}$$

$$+ \begin{bmatrix} \int_0^t H^{11}(t-\tau)d\tau & \int_0^t H^{12}(t-\tau)d\tau & \int_0^t H^{13}(t-\tau)d\tau \\ \int_0^t H^{21}(t-\tau)d\tau & \int_0^t H^{22}(t-\tau)d\tau & \int_0^t H^{23}(t-\tau)d\tau \\ \int_0^t H^{31}(t-\tau)d\tau & \int_0^t H^{32}(t-\tau)d\tau & \int_0^t H^{33}(t-\tau)d\tau \end{bmatrix} X \quad (1)$$

$$\begin{bmatrix} X^1 \\ \dot{X}^2 \\ X^3 \end{bmatrix} + \begin{bmatrix} C^{11} & C^{12} & C^{13} \\ C^{21} & C^{22} & C^{23} \\ C^{31} & C^{32} & C^{33} \end{bmatrix} X \begin{bmatrix} X^1 \\ X^2 \\ X^3 \end{bmatrix} = \begin{bmatrix} F^1 \\ F^2 \\ F^3 \end{bmatrix}$$

where:

M_{ij} = inertia and added inertia matrix of body i because of motion of body j.

D = damping matrix

H_{ij} = matrix of retardation functions of body i because of motions of body j.

C_{ij} = matrix of hydrostatic restoring forces of body i.

X_i = motion vector of body i.

F_i = vector of external forces on body i, including wave exciting forces and wave drift forces.

The inertia matrices, added inertia matrices, damping matrices, stiffness and first order wave loads are derived from diffraction analysis in the frequency domain. The retardation force matrix is excluded from the current study. This implies that the wave shielding of one body behind another body is not considered.

3.2.3 Wave forces on the vessel and floating spar

The potential added mass and damping coefficients, the hydrostatic stiffness and the first order wave excitation force transfer functions of the vessel are calculated in Ansys AQWA based on the 3D panel method. In the current analysis, waves are considered as the main effective factor. Current and wind forces are excluded from this study. The

374

external forces on the floating spar can be categorized to three main forces:

- Gravity forces
- Buoyancy forces
- Hydrodynamic wave forces

The hydrodynamic wave forces on the Spar are calculated by Morison's equation. Morison's equation is often used for slender structures with D/L ratio less than 0.2. Diffraction and radiation effects are considered insignificant for slender bodies. The Spar is divided to numbers of strips and the wave forces per unit length on each strip normal to the member can be determined from Morison's formula (Faltinsen, 1990):

$$F = F_I + F_D \qquad (1)$$

This equation assumes that the wave properties are unaffected by the presence of the structure. Therefore, the total wave load can be expressed as the sum of the inertia forces due to wave fluid acceleration and of the drag forces resulting from the wave fluid velocity. The differential form of the Morison equation is written as:

$$dF = C_m \rho dV \dot{U}_n + \frac{1}{2} C_d \rho dA |U_n| U_n \qquad (2)$$

where dF is the total wave force on the element of volume (dV) and the projected area (dA), $\dot{U}_n$ and U_n are the instantaneous wave fluid acceleration and velocities normal to the strip axis, ρ is the fluid density, and C_d and C_m are the drag and inertia coefficients. The modulus or absolute value sign in the drag force term is used to ensure the drag force is in the direction of the wave velocity. The inertia force consists of two parts, one proportional to fluid acceleration relative to earth (the Froude-Krylov component), and one proportional to fluid acceleration relative to the body (the added mass component). In this case, the inertia term can be re-written as follow:

$$F_I = \left(\Delta U_f + C_a \Delta U_r \right) \qquad (3)$$

where U_f is the fluid particle acceleration relative to earth and U_r is the fluid particle acceleration relative to the body, Δ is the mass of the fluid displaced by the body. Both C_d and C_m coefficients are chosen as per DNV-RP-C205.

3.2.4 Mechanical coupling
The mechanical coupling between the ship and the spar substructure is achieved by using the motion compensation gripper clamp. The physical model of the gripper clamp is a circular shape structure

with several contact elements in the inner diameter which act such as bumpers and/or motion damping elements. The motion compensation gripper clamp in the model is simplified by using three constraints. The main idea is to compensate surge, sway and heave motions of the spar substructure. Figure 5 shows an example the mechanical connections between floating spar and the float-over vessel. In the current study, two options will be investigated for the motion compensation feasibility regarding constraint the motions in x, y and z translations.

This is will be modelled using constraints object in time domain software. A constraint object comprises two coordinate systems, or frames of reference: the inner-frame and the outer-frame. In the current case study, the three translational DOFs are fixed, and the three rotational DOFs are free, which means that the outer-frame and inner-frame origins are co-incident, but the outer-frame can rotate independently from the inner-frame.

3.3 Eigenvalue analysis

The eigenvalue analysis is used to determine the eigen periods of the rigid body motions. The natural modes and natural periods are calculated by solving the following equation:

$$\left[-\omega^2 (M + A) + k \right].x = 0 \qquad (4)$$

where "M" is the mass matrix of the vessel and the Spar. A is the added mass matrix for the vessel. K is the stiffness matrix while x is the rigid body motion with 12 DOFs for the coupled system and 6 DOFs if only single body considered. In the current study, both the vessel and spar eigen modes are calculated. The natural periods and natural modes of the vessel alone and of the spar alone for heave, roll and pitch are shown in Figure 6 and

Figure 5. Motion compensation gripper clamp concept.

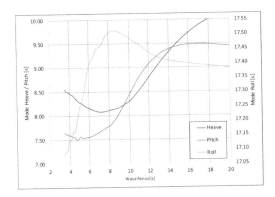

Figure 6. Eigen modes for heave, roll and pitch for vessel.

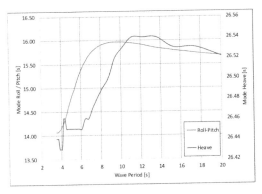

Figure 7. Eigen modes for heave, roll and pitch for spar.

Figure 7. The wave spectrum is also presented to show the different modes that are dominant for the response at different sea states.

3.4 Frequency domain and time domain

The main aim of the frequency domain calculations is to determine the wave excitation force, added mass, hydrostatic stiffness and wave damping for the installation vessel. The parameters in the 6 degrees of freedom are all dependent of wave frequency and wave heading. Calculations are made using AQWA which is based on the 3D-diffraction theory. The body surface is divided into several panels small enough to assume that the fluid pressure is constant over each element. All panels are covered with sources, sinks and dipoles to make sure that there is no flow perpendicular to each panel. In this way, a watertight shell is simulated. Since the panel is based on potential theory, the effect of flow separation is neglected. The method predicts waves and pressure differences correctly,

however friction and resulting turbulence are not considered. This is not an issue for calculating the mentioned responses of the surrounding water since they are in general dominated by waves and pressure differences for a body such as a ship. The wave forces generated by the spar will not be modelled in frequency domain calculations. This is because these forces are friction and turbulence dominated. These effects will be treated and added to the time domain calculations. Figure 9 shows the synergy between both frequency and time domain calculations in the current study.

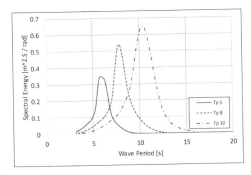

Figure 8. Wave spectrum for different peak periods and significant wave height "Hs" 1.5 [m].

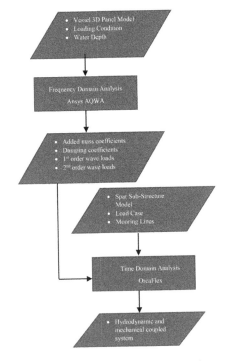

Figure 9. Frequency domain and time domain synergy in the current study.

376

4 RESULTS AND DISCUSSION

4.1 *Frequency domain results*

The geometry of the vessel in water depth 105m has been modelled as a mesh of panels the frequency domain analysis was performed considering frequency range from 0.05 to 1.70 [rad/s] with step of 0.05 [rad/s]. All values obtained are for zero speed, which is a realistic assumption for installation operation connected to the moored substructure.

Figure 10 shows the response amplitude operator (RAO) for heave, roll, and pitch for the float-over vessel. Results for heave and roll obtained presented for beam sea condition, however, for pitch results presented for head sea condition. Moreover, wave forces around Z axis and moments around X and Y axis.

4.2 *Time domain results*

Time domain simulations are performed to capture the non-linear hydrodynamic load effects and non-linear interaction effects between the vessel and the floating spar. Moreover, a time domain analysis gives the response statistics without making assumptions regarding the response distribution. Vessel and spar substructure responses are introduced in Figure 11 before using the motion compensated gripper clamp as a mechanical connection between the vessel and the spar. For the motion compensation clamps, clamping forces are determined for different sea states "Hs = 1.5 [m]

and range of wave peak periods from 6 to 10 [sec]". Since the free-floating substructure is also subject to motion, the relative motions between the vessel and the substructure may further complicate the mating operation.

For that reason, the relative motions between the vessel and spar substructure in Z—direction is investigated and introduced in Table 4.

In the first approach, only two degrees of freedom will be compensated in x and y directions. The maximum interface forces and moments are presented for the gripper clamp in Figure 12.

The maximum force in x and y directions at Hs of 1.5 [m] and peak periods range from 6 to 10 [sec] are 750 and 1400 [kN] which are reasonable values to compensate by a gripper clamp. The resulting interface moments at the connection are 1800 and 900 [kN.m] in x and y directions respectively. The highest interface force is found in beam sea condition due to the sway motion of the spar substructure and the vessel. Another large difference is for the force in the transverse direction between 90 and 270 degrees. This is when the wave heading reversed, the motions remain the same and the phases are reversed. The vessel and spar motions are now different than what was already presented in Figure 11 due to the mechanical coupling of the gripper clamp. The new vessel and spar maximum motions for the coupled system are presented in Figure 13.

The second approach where the gripper clamp will compensate for three degrees of freedom in x, y and z directions. For this concept, the maximum interface forces and moments are presented for the

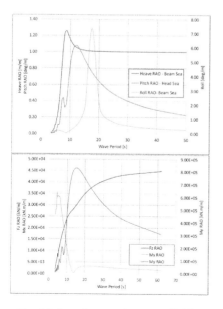

Figure 10. Motions, forces and moments RAOs around X, Y, Z axis.

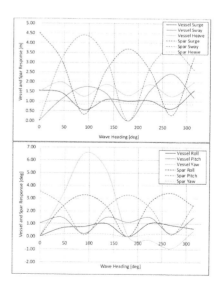

Figure 11. Maximum vessel and spar motions in sea state Hs = 1.5 [m] and different peak periods.

Table 4. Maximum relative motion in Z-direction.

Heading [deg]	Maximum relative motion [m]
0	1.5
45	2.2
90	1.5
135	1.8
180	1.5
225	2.1
270	1.5
315	2.0

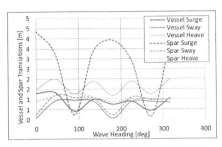

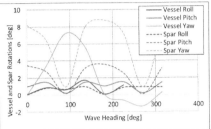

Figure 13. Maximum vessel and spar motions in case of gripper clamp compensation in x and y directions.

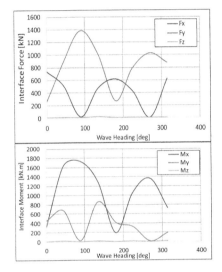

Figure 12. Maximum interface forces and moments in the gripper clamp for x, y motion compensation.

gripper clamp in Figure 14. The interface forces in x, y and z directions were increased compared to the first concept where only x and y motions were compensated. The maximum force in z direction around 10000 [kN] where the forces in x and y directions are 750 and 5500 [kN]. The maximum vessel motions for vessel and spar are altered due to the mechanical connection in x, y and z directions compared to the first concept. The new maximum vessel and spar motions are presented in Figure 15.

4.3 Operability analysis

For marine operations, which are sensitive to weather, it is useful to evaluate the probability of acceptable weather conditions. For this study, it is also important to show the influences of different factors affecting on the operational probability for a typical site. The North Sea scatter area is chosen

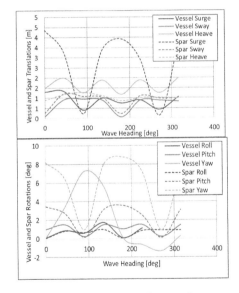

Figure 14. Maximum interface forces and moments in the gripper clamp for x, y and z motion compensation.

for the operability analysis. Figure 16 shows the scatter area for the North Sea. The operability is the percentage of time that an operation or process can statistically be expected to work satisfactorily under the conditions to be expected at the given location. From operational simulations of the skidding operations, 0.75 [m/sec^2] most probable extreme value for accelerations is a realistic assumption to restrict the operation.

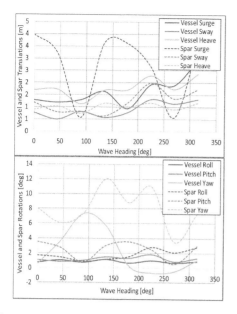

Figure 15. Maximum vessel and spar motions in case of gripper clamp compensation for x, y and z directions.

Figure 16. North Sea scatter area.

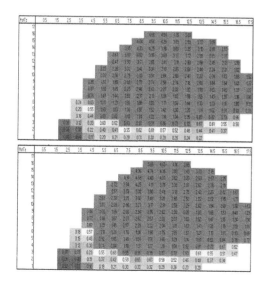

Figure 17. Operability results for wave headings 180 and 195 [deg].

Table 5. Operability percentage per wave heading.

Heading [deg]	Operability (%)
150	50.4
165	65.6
180	74.8
195	68.0
210	59.2

This criterion used to assess the vessel operability and uptime at the North Sea scatter areas. The Operability estimate shows a good operability percentage from 50 to 75% in the North Sea area depends on the wave heading. Figure 17 shows results of vertical acceleration level at the skidding system for heading +/− 180 degrees. Some detailed results are presented for different wave headings in table 5.

5 CONCLUSIONS AND RECOMMENDATIONS

The current study investigated the dynamic analysis during installation of floating wind turbine by float-over method particularly focusing on the feasibility of using different motions compensation gripper concepts during the operation. Two main concepts were investigated by connecting the vessel to the floating spar in x, y and z translations. The resulting forces and moments at the connection found reasonable values to compensate. This will lead to minimizing the relative motion between both bodies which will lead to an efficient mating operation.

In addition, the operability estimate shows a higher percentage from 50 to 75% at the North Sea depending on the wave heading. However, the skidding operation is considered a fast operation and require a relatively short weather window compared to heavy lifting operations. The risk of weather downtime during the installation can be reduced, resulting in lower installation costs.

The mating operation is not investigated in the current paper and needs to be studied to get insight about the impact force and its effect on the connection between the turbine and spar substructure. Radiation damping resulting from spar structure is not considered in the current analysis. Radiation damping could be an important factor affecting the response in short waves but can be neglected in long waves where the vessel response is dominant. For these large diameters of structures, comparison between wave forces from Morison equation and strip method might be required to give a better understanding about the effect of both methods in the overall response.

Shielding effects from the vessel can play an important factor in the overall method efficiency and could lead to increase the operability. Current and wind forces are excluded from this study and need to be further checked.

The operability study considers only the acceleration levels during skidding operations. It is important to include the other operational limits from individual activities and combine them to establish the operational limits of the complete operation. The methodology should hence be extended to consider different activities, sequence, and continuity.

REFERENCES

A RJM Lloyd, 1998. Ship Behavior in Rough Weather, Ellis Horwood, Chichester, Sussex, United Kingdom.

Aarset K, Sarkar A & Karunakaran D. 2011.Lessons learnt from lifting operations and towing of heavy structures in North Sea. In: Offshore Technology Conference, May 2–5, Houston, Texas, USA;

ANSYS, 2016. The AQWA Reference Manual—Version 16.2.

Bagbanci H., Karmakar D. & Guedes Soares C., 2015. Comparison of spar and semisubmersible floater concepts of offshore wind turbines using long-term analysis, Journal of Offshore Mechanics and Arctic Engineering, Vol.137, 061601-1 - 061601-10.

DNV, 2014a. Recommended practice DNV-RP-H103, modelling and analysis of marine operations; February

DNV, 2014b. Recommended practice DNV-RP-C205, environmental conditions and environmental loads; April

DNV, 2014c. Offshore Standard DNV-OS-J101, Design of onshore wind turbine structures;

EWEA, 2015. The European offshore wind industry—key trends and statistics Report, The European Wind Energy Association (http://www.ewea.org/).

EWEA, 2013. The European offshore wind industry—Deep water, the next step for offshore wind energy; (http://www.ewea.org/).

Faltinsen, O. M., 1990. Sea Loads on Ships and Ocean Structures. Cambridge University Press;

G. A. M. van Kuik, J. Peinke, R. Nijssen, D. Lekou, J. Mann, J. N. Sørensen, C. Ferreira1, J. W. van Wingerden1, D. Schlipf, P. Gebraad, H. Polinder1, A. Abrahamsen, G. J. W. van Bussel1, J. D. Sørensen, P. Tavner, C. L. Bottasso1, M. Muskulus, D. Matha, H. J. Lindeboom, S. Degraer, O. Kramer, S. Lehnhoff, M. Sonnenschein, P. E. Sørensen, R. W. Künnekel, P. E. Morthorst, & K. Skytte, Long-term research challenges References 87 in wind energy research agenda by the European academy of wind energy. Wind Energy Science 1, 1–39; 2016.

Guedes Soares C., Bhattacharjee J. & Karmakar D., 2014. Overview and prospects for development of wave and offshore wind energy, Brodogradnja, 65; 87–109.

Herman, S. A., Offshore wind farms—analysis of transport and installation costs, report no. ECN-I-02-002. Tech. rep., Energy research Centre of the Netherlands; 2002.

http://www.principlepowerinc.com/, accessed: Nov 2016.

Journée, J.M.J. & Massie, W.W., Offshore Hydromechanics Book, Delft University of Technology, First Edition, January 2001.

Kang, H.Y & Kim, M.H, Safety assessment of caisson transport on a floating dock by frequency—and time-domain calculations, Ocean Systems Engineering, Vol. 4, No. 2; 2014.

Karmakar D., Bagbanci H. & Guedes Soares C., 2016. Long-term extreme load prediction of spar and semisubmersible floating wind turbines using the environmental contour method, Journal of Offshore Mechanics and Arctic Engineering, 138:021601-1 - 021601-9.

Lee, D.H. & Choi, H.S., 1998, The motion behavior of shuttle tanker connected to a turret-moored FPSO, Proceedings of Third International Conference on Hydrodynamics 173–178.

Li, L., Gao, Z. & Moan, T., 2013a. Numerical simulations for installation of offshore wind turbine monopiles using floating vessels. In: Proceedings of the 32nd International Conference on Ocean, Onshore and Arctic Engineering, Nantes, France, June 9–14;

Li, L., Gao, Z. & Moan, T., 2015a.Comparative study of lifting operations of offshore wind turbine monopile and jacket substructures considering shielding effects. In: The 25th International Onshore and Polar Engineering Conference, Hawaii, USA, June 21–26;

Li, L., Gao, Z., & Moan, T., 2016a.Analysis of lifting operation of a monopile considering vessel shielding effects in short-crested waves. In: The 26th International Offshore and Polar Engineering Conference, Rhodes, Greece, June 26-July 2;

Mone, C. & Stehly, T., 2014. Cost of wind energy review report; (http://www.nrel.gov).

Npower Renewable, Capital grant scheme for the North Hoyle onshore wind farm annual report: July 2005-june 2006. Tech. rep., Npower Renewables Limited, Essen, Germany.

Orcaflex, 2016. The Orcaflex Reference Manual—Version 10.1.

Principle Power, 2011. WindFloat.

Stempinski, F. & Wenzel, S., Modelling Installation and Construction of Offshore Wind Farms, Proceedings of the ASME 2014 33rd International Conference on Ocean, Offshore and Arctic Engineering, June 8–13, 2014, San Francisco, California, USA.

Uzunoglu U., Karmakar D. & Guedes Soares C., 2016; Floating Offshore Wind Platforms, L. Castro-Santos & V. Diaz-Casas (Eds.), Floating Offshore Wind Farms. Springer International Publishing Switzerland; pp. 53–76.

Van den Boom H, Dekker J. & Dallinga R. Computer analysis of heavy lift operations, 22nd Offshore Technology Conference, Houston; 1990.

Wang, S. & Li, X., Dynamic analysis of three adjacent bodies in twin-barge floatover installation, Ocean Systems Engineering, Vol. 4, No. 1; 2014.

Maritime Transportation and Harvesting of Sea Resources – Guedes Soares & Teixeira (Eds)
© 2018 Taylor & Francis Group, London, ISBN 978-0-8153-7993-5

An investigation on a Mathieu-type instability in irregular seas: The case of a tension leg platform

J.C. Pérez
COPPE/UFRJ, Rio de Janeiro, Brazil

M.A.S. Neves
LabOceano, Rio de Janeiro, Brazil

ABSTRACT: The proposed paper is a continuation of numerical studies carried out on undesirable resonant phenomena observed during experimental tests of a TLWP platform coupled with a scaled model of a FPSO. The results from regular sea states showed that the system presents unstable behavior for a range of wave periods and heights in the TLWP's yaw mode. These phenomena were studied by Rivera et al. (2012), which showed that the unstable behavior were caused by Mathieu-type internal excitations of the TLWP-FPSO system induced by external (direct) excitations in the sway mode. Test was also performed in irregular seas. A point of interest here is: how tuned resonant responses observed in regular conditions would behave under irregular excitations. To verify this dynamic aspect, a nonlinearly coupled numerical model is developed capable of reproducing the platform responses for a given sea state. Various numerical analyses are performed for a given sea state. The main goal is the numerical verification of whether the platform (under random excitation) shall display the complex nonlinear coupling and change of attractors, complexities previously observed in the regular seas environment.

1 INTRODUCTION

The conceptual design of conventional Tension Leg Platforms (TLPs) has been developed with the aim of reducing vertical motions to a minimum while at the same time keeping horizontal translational motions circumscribed to a quite limited area; practical experience has confirmed the intrinsically stable characteristics of this type of offshore platform, Brewer (1975), Denise and Heaf (1979). However, the inclusion of additional mooring lines to the unit or the proximity of another body may alter the dynamical behaviour of the system; under these circumstances large unstable motions may develop. Essentially, this is the focus of the present investigation.

Model experimental results obtained from recently conducted tests with a TLP (in fact a Tension Leg Well-head Platform) connected to a nearby positioned FPSO (NMRI, Japan) surprisingly revealed, at a given range of wave exciting periods, the onset of quite large oscillatory yaw motions, whereas the FPSO remained rather stable when the TLWP was directly excited in sway, as shown in the NMRI Report, see Maeda et al. (2008). Clearly, in such wave incidence the TLWP's yaw mode is not directly excited. Previous numerical investigation conducted by the authors demonstrated that these large yaw motions are the result of internal excitations of the two-body system induced by sway motions, thus characterizing the occurrence of Mathieu-type parametric resonance, see Rivera et al. (2012).

The main objective here is to investigate whether the fine tuning required for sustained parametric amplification in regular waves can be encountered in a random frequency dependent scenario and could eventually prevail, thus contributing to endangered situations to the designed system in irregular seas.

Figure 1 shows a plan view of the test arrangement. The TLWP, at the left in the figure, is expected to remain always in close proximity to the large VLCC. According to the basin arrangement, waves are coming from the left.

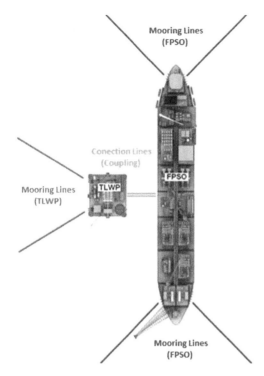

Mooring Lines (FPSO)

Conection Lines (Coupling)

Mooring Lines (TLWP)

FPSO

TLWP

Mooring Lines (FPSO)

Figure 1. Two-body arrangement showing the mooring and connection set-up during experiments.

2 MATHEMATICAL MODEL

The two-body system is defined by a 12-DOF non-linear model expressed as: where the displacement vector is:

$$(M + A)\ddot{s} + B_L\dot{s} + B_{NL}(\dot{s}) + c_r(s,\zeta) = c_{ext}(\zeta,\dot{\zeta},\ddot{\zeta}) \quad (1)$$

where the displacement vector is:

$$s(t) = [x_1 \ y_1 \ z_1 \ \phi_1 \ \theta_1 \ \psi_1 \ x \ y \ z \ \phi \ \theta \ \psi]^T.$$

In Eq. (1) M is a 12×12 matrix which describes hull inertia characteristics, A is also a 12×12 matrix, whose elements represent hydrodynamic generalized added masses, B_L and $\mathbf{B_{NL}}(\dot{s})$ describe the hydrodynamic reactions dependent on the FPSO and platform velocities (damping). The latter incorporates non-linear terms due to the great influence of viscosity on tendons, pontoons and columns of the TLWP and roll motion of the FPSO. $C_r(s,\zeta)$ is a 12×1 vector which describes non-linear restoring forces and moments dependent on the relative motions between the platform and the mooring lines. It also considers the influence of the wave passage effect on the tendons

tension. On the right hand side of Eq. (1), the generalized vector $c_{ext}(\zeta,\dot{\zeta},\ddot{\zeta})$ represents linear wave external excitation, composed of the Froude-Krilov and diffraction wave external forcing terms, dependent on wave heading, excitation frequency ω_w, wave amplitude A_w and time t.Taking into account the prevailing physics and the particular transversal wave excitation considered, without loss of generality the 12-DOF problem will be restricted to the following 7-DOF coupled problem: x_1 (surge), y_1 (sway), z_1 (heave) and ϕ_1 (roll) for the FPSO and x (surge), y (sway) and ψ (yaw) for the TLWP.

3 HYDRODYNAMIC COEFFICIENTS AND NON LINEAR DAMPING

Matrices **A** and **B** are computed by means of WAMIT computer code. The same applies to $c_{ext}(\zeta,\dot{\zeta},\ddot{\zeta})$. Potential damping has been considered for the following motions: a) horizontal motions of TLWP; b) horizontal motions of FPSO; viscous damping has been considered for the roll motion of FPSO. Drag quadratic terms in Morison formula have been used as an approximation for calculating the non-linear damping of the TLWP, being applied to columns, tendons and pontoons. These actions are computed by taking into account the instantaneous velocities of the body, and disregarding the influence of fluid velocities, Rivera et al. (2012). For some specific wave periods the TLWP responded with large angular displacements in yaw. For this reason a quadratic damping model was considered for this platform motion mode.

4 REGULAR SEAS ANALYSIS

A brief summary is given below of the experimental and numerical simulations. Table 1 shows the range of tested wave amplitudes and periods. More details may be obtained from Rivera et al. (2012)

Following the methodology proposed in Neves and Rodríguez (2007) for deriving limits of stability for coupled differential equations with time-dependent coefficients, Figure 2 is presented incorporating the dots defined according to Table 1 for the experimentally tested conditions. Yaw amplifications took place in conditions

Table 1. Tested conditions.

T(s)	7	9	10	11	12	13	14	15	17.2	20	25
)		1.5	2	2.0	2.2	2,4	2.6	2.8	3.0	3.4	4	5

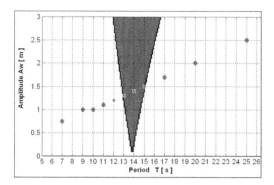

Figure 2. Dots representing experimental points are superimposed to yaw's Ince-Strut diagram.

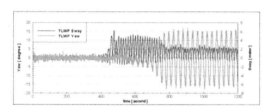

Figure 3. TLWP's sway and yaw time series obtained experiments in regular waves, T = 14 sec, H = 2.8 m.

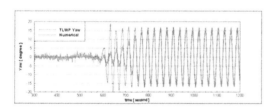

Figure 4. Comparison of numerical and experimental yaw coupled motions, period T = 14 sec, wave height H = 2.8 m.

inside the stability region, for wave periods around 14 seconds.

Figure 3 shows experimental time-series of TLWP's sway and yaw motions. Two aspects deserve attention here: firstly, the intense yaw resonant amplitudes of around 15 degrees; secondly, the strong coupling between sway and yaw. It is observed that as large yaw motions develop (at double period, a feature characteristic of parametric resonance), there is a complementary reduction in the sway motion (taking place at wave period), suggesting an interesting strong nonlinear coupling between the two modes. Figure 4 shows the good agreement of the mathematical model with the large experimental resonant yaw time-series.

5 IRREGULAR SEAS ANALYSIS

A JONSWAP sea spectrum $S_\zeta(\omega)$ is considered, with significant height and peak period defined as H sig = 7,8 m and T_ζ = 15.56 sec, respectively. In general the Taylor expansion model proposed by Rodríguez et al. (2016) for parametric rolling of ships is now adopted (and adapted) to model the TLWP-FPSO system behaviour in irregular seas. Accordingly, externally and internally exciting spectra are obtained in frequency domain as, for example:

$$S_{fw}(\omega) = \left[RAO_{fw}(\omega)\right]^2 * S_\zeta(\omega) \quad (2)$$

where: $RAO_{fw}(\omega)$: amplitude of force transfer function due to external wave action at a given mode. $s_{fw}(\omega)$: spectrum of wave force excitation at a given mode.

For internal actions due to wave passage, which are proportional to wave amplitude squared, frequency decomposition is accomplished by means of a corresponding quadratic transfer function (QTF) multiplied by wave spectrum, as exemplified below for the TLWP's surge mode:

$$S_{X_{\zeta x}}(\omega) = QTF_{X_{\zeta x}}(\omega) * S_\zeta(\omega) \quad (3)$$

where: $RAO_{X_{\zeta x}}$: amplitude operator of response due to wave passage along tendons in surge; $s_{X_{\zeta x}}(\omega)$: spectrum of amplitude operator of responses due to wave passage in surge;

After obtaining the spectra for all internal and external excitations, use is made of Fourier analysis for generating the corresponding group of time-series $\gamma(t)$, which may be obtained in general using the well-known spectral decomposition algorithm:

$$\gamma(t) = \sum_{n=1}^{N} \gamma_{a_n} * \cos\left(\omega t + \varepsilon_n + \alpha_n\right) \quad (4)$$

where:

$$\gamma_{a_n} = \sqrt{2 S_\gamma(\omega).\delta\omega}$$

γ_{a_n}: amplitudes of forces and moments (internal and external)
ε_n: represent phase difference between excitations and waves;
α_n: independent random variables with a uniform probability distribution in interval [0;2π].

After obtaining all the time series, Eq. (1) may be numerically integrated (here a 4th order Runge-Kutta algorithm is used). It is noted that due to the strong nonlinear character of the 12-DOF Eq. (1), depending on the random phasing, distinct time series will be obtained. A set of 960 sea spectra has been generated and some different typical cases of coupled responses have been observed. Figures 5, 6 and 7 are typical samples of such cases. In each case sway and yaw time series are shown together, describing different categories of coupling effects. The vertical segmented lines at 600 seconds indicate the beginning of the time-series considered for the purpose of spectral computations:

Figure 5 is a case in which no yaw parametric amplification is developed (!), sway motion remains bounded. Figure 6 is a case in which large yaw parametric resonance develops. In Figure 7 it is possible to observe the peculiar alternating of energy transfer between the sway and yaw modes. Large sway motions induce yaw motion amplification; when yaw motion develops large oscillations

levels, sway is reduced. This essentially corresponds to the exchange of energy that had been observed in regular seas, as seen and noted in Figure 3.

Figure 8 shows the sway and yaw response spectra corresponding to the time-series depicted in Figure 7 (case 3). The small red arrow indicates the frequency at which maximum energy is registered in both sway and yaw modes. This corresponds to $\omega_{n12} = 0.202$ rad per sec, the TLWP's natural frequency in yaw, which is close to half the value of $\omega_\zeta = 0.4$ rad per sec, the wave frequency which corresponds to the spectral peak period. Effectively, this corresponds to the typical first Mathieu tuning ratio. It is also noticed that yaw motion is only excited at the natural frequency. On the other hand, sway motion is excited at both ω_{n12} and ω_ζ, thus confirming the coupled nature of the sway and yaw modes.

Figure 9 shows yaw's response spectra for three cases. It shows the small amount of energy

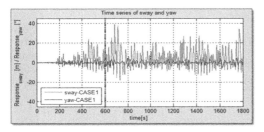

Figure 5. Sway and yaw time-series: Case of large sway, small yaw.

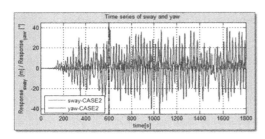

Figure 6. Sway and yaw time-series: Case of small sway, large yaw.

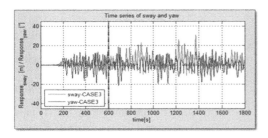

Figure 7. Sway and yaw time-series: Case of strong coupling.

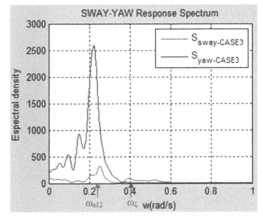

Figure 8. Sway and yaw spectra: Case of strong coupling.

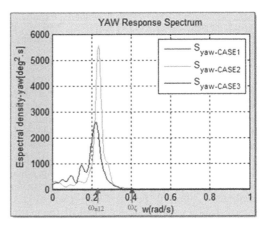

Figure 9. Yaw spectra for different yaw samples.

obtained in Case 1 and the energy gains in Cases 2 and 3. These distinct standards of response are indicative of the system's non-ergodicity due to the nonlinearities.

6 CONCLUSIONS

Tests on regular beam seas waves performed at NMRI surprisingly revealed high amplitude oscillatory yaw motions for a given range of wave periods. Rivera et al. (2012) investigated these phenomena with the aim of explaining the physical reasons for these amplifications. They elaborated a nonlinearly coupled mathematical model capable of reproducing with good accuracy even the large amplifications observed in the tests and showed that the large amplifications in yaw responses are due to coupling effects taking place between yaw and sway and revealed that these had to be enforced by the connections and mooring. The sway mode, after reaching some energy level, excites the yaw mode at its natural frequency, thus giving rise to parametric amplification. However, the limited test program conducted in irregular waves led the laboratory report to establish that the yaw parametric amplifications had no relevant tendency to occur in random seas. This motivated the Authors to perform a more comprehensive numerical investigation on the system's responses in irregular seas. In order to extend the knowledge on the system's dynamics, a numerical model has been developed to investigate the responses in irregular waves, based on the equations developed in Rivera et al. (2012) for regular waves as well as on the methodology developed by Rodríguez et al. (2016) for irregular seas analyses. Using the developed algorithm, 960 samples have been generated, therefore corresponding to 960 JONSWAP sea spectra generated seas, all with the same significant height of 7.8 m and peak period of 15.56 s. Phases required to generate the different time-series have been allowed to vary randomly. For each sea state, coupled motion equations have been integrated, resulting in time-series for each degree of freedom. As the main interest here resides in the TLWP dynamics, their sway and yaw results have been grouped, for the sake of analysis, into three distinct typical cases, describing distinct peculiar aspects. Case 1 confirms the results reported by NMRI, that is, the system shows small responses with no evidence of parametric resonance. Cases 2 and 3, differently, show large amplitudes for the

yaw motion, due to parametric resonance, but with clear distinct tendencies.

It has been observed that the irregular seas numerical algorithm is capable of reproducing the mechanisms of exchange of energy between the sway and yaw modes that had been observed in regular waves. The spectral decomposition shows that the yaw mode mainly oscillates at double the excitation period and close to its natural period. This gives the basis to establish that the TLWP's yaw dynamics in irregular seas corresponds to a case of parametric resonance, with possibility of reaching large amplitudes.

Finally, from the three distinct energetic levels observed in Figure 9, it is noted that due to the strong nonlinearities affecting the coupled system the resulting dynamics displays a clear non-ergodicity characteristic.

ACKNOWLEDGEMENTS

The present investigation is supported by CNPq within the STAB project (Non-linear Stability of Ships) coordinated by the second author. Authors also acknowledge financial support from CAPES, FAPERJ, PETROBRAS and LabOceano.

REFERENCES

Brewer, J.H. 1975. The Tension Leg Platform concept. *American Petroleum Institute*, Annual Meeting Papers, Division of Production, Dallas, Texas.

Denise, J-P. F. and Heaf, N.J. 1979. Comparison between linear and non-linear response of a proposed Tension Leg Production Platform. OTC 3555, In Proceedings of the 11th Offshore Technology Conference, Houston, Texas.

Maeda et al. 2008. TLWP-FPSO coupled model tests in NMRI, National Maritime Research Institute, Japan, Report OCN-07–03.

Neves, M.A.S. and Rodríguez, C.A. 2007. Influence of nonlinearities on the limits of stability of ships rolling in head seas. Ocean Engineering. Vol. 34, pp 1618–1630.

Rivera, L. et al. 2012. Study on unstable motions of a Tension Leg Platform in close proximity to a large FPSO. Proceedings STAB 2012 Conference, Athens, Greece.

Rodríguez, et al. 2016. A time efficient approach for nonlinear hydrostatic and Froude-Krylov forces for parametric roll assessment in irregulars seas. Ocean Engineering 120 (2016) pp 246–255.

Maritime Transportation and Harvesting of Sea Resources – Guedes Soares & Teixeira (Eds)
© *2018 Taylor & Francis Group, London, ISBN 978-0-8153-7993-5*

A new procedure of power prediction based on Van Oortmerssen regression

B. Xhaferaj
Department of Engineering and Marine Technology, University of Vlora "Ismail Qemali", Albania

A. Dukaj
Department of Maritime Science, University of Vlora "Ismail Qemali", Albania

ABSTRACT: Rapid prediction of resistance and power of ship is a very important task of ship design process, especially in early design stage. The software Ship Power V_1.0 is a parametric power prediction tools for resistance and power of ship. The Software is developed the University of Vlora "Ismail Qemali"—Albania. The actual version of the software was based on the well-known MARIN regression equation. These two procedures are the core procedure of the software. A new procedure which can be used for the evaluation of resistance and power of small ships, based on the model of regression Van Oortmerssen is developed and tested. The aim of the paper is to present the architecture of the new developed procedure, and the integration of this procedure with the actual version of the software. An example and verification of the procedure with other results available in literature is also presented.

1 INTRODUCTION

"Ship Power V_1.0" is a parametric prediction software of resistance and power of ship. The software is able to predict the resistance, power and propulsion parameters, based on the basic geometrical parameters of the ship, in a certain range of speeds defined by the user. The software enables the presentation of the results of calculation either in tabular or in graphic form. Moreover, the software enables the user to get these results both in paper and electronic format. The basic models on which the software is based are the models of regression Holtrop and Holtrop Mennen. Based on these two well-known models of the prediction of ship resistance has been developed the basic procedure of this software. To ensure the accuracy of the implementation of this procedure in the software "Ship Power V_1.0" are carried out comparisons of the respective results of calculations with the data available in the literature. Also the results of calculations of the software "Ship Power V_1.0" (according the Holtrop Mennen procedure) are also compared with the results of calculations of well-known software, which are used in the practice of ship design process. (Xhaferaj (2017)). The basic architecture and algorithm of the software "Ship Power V_1.0" are conceived to facilitate the implementation of other procedures. The Van-Oortmerssen procedure is another procedure developed in the "Ship Power V_1.0" software. This procedure is based on Van-Oortmersen regression, which can be used in initial design stages of tugboat and fishing vessels.(Van Oortmerssen (1971), Altosole M. et al. (2014), Tassetti A. et al. (2015)). The procedure is integrated with the basic procedure of "Ship Power V_1.0" in order to respect the format of presentation of results of calculation. Even in the case of the development of this procedure to ensure the best possible results, its algorithm is tested in every part, with data available in literature.

2 MATERIAL AND METHODS

Van Oortmerssen is another well-known regression equations, which is used for parametric evaluation of resistance of small ships, tugs, fishing vessels, stern trawlers and pilot boats etc (Molland (2011), Van Oortmerssen (1971)). This model is defined based on the multiple regression analysis of 970 model resistance data of fishing vessels and tugboats, carried out at MARIN towing tank. (Van Oortmerssen (1971), Molland (2011)). The model of regression Van Oortmerssen represents the specific residual resistance as function of Froude number and ship geometric parameters. Limits of application of the model of regression Van Oortmerssen are given in Table 1.

The mathematical expression of Van Oortmerssen regression is as follow: (Van Oortmerssen (1971), Molland (2011)).

Table 1. Limits of application of the model of regression Van Oortmerssen (Molland (2011)).

NR	Elements	Range of values
1.	Water line Length	$15 \div 75$ m
2.	Ratio L/B	$3.4 \div 6.2$
3.	Prismatic coefficient	$0.55 \div 0.70$
4.	Midship Coeffient	$0.76 \div 0.94$
5.	LCB position	$-4.4\%L \div +1.6\%L$
6.	half-angle of entrance i_e	$15° \div 35°$
7.	Froude Number (Fn)	$\div 0.5$

$$\frac{R_R}{\Delta} = C_1 e^{-mF_n^2/9} + C_2 e^{-mF_n^{-2}} + C_3 e^{-mF_n^2}$$
$$. \sin F_n^{-2} + C_4 e^{-mF_n^2} . \cos F_n^{-2} \quad (1)$$

where the coefficients in the above equation (1) can be calculated as in formulas (2), (3): (Van Oortmerssen (1971), Molland (2011)). Correction of the sign error in coefficient $d_{3,5}$, as indicated in (Molland (2011)) is taken is consideration for the development of the procedure in "Ship Power V_1.0".

$$m = 0.14347 C_P^{-2.1976} \quad (2)$$

$$c_i = (d_{i,0} + d_{i,1}LCB + d_{i,2}LCB^2 + d_{i,3}C_P +$$
$$+ d_{i,4}C_P^2 - d_{i,5}\left(\frac{L_D}{B}\right) + \left(\frac{L_D}{B}\right)^2 +$$
$$+ d_{i,7}C_{WL} + d_{i,8}C_{WL}^2 + d_{i,9}\left(\frac{B}{T}\right)$$
$$+ d_{i,10}\left(\frac{B}{T}\right)^2 + d_{i,11}C_M)10^{-3} \quad (3)$$

The wetted surface can be calculated as in the following formula: (Van Oortmerssen (1971), Molland (2011)).

$$S = 3.223\nabla^{2/3} + 0.5402 L_D \nabla^{1/3} \quad (4)$$

where S is the wetted surface in m² and ∇ is the displacement volume in m³.

The values of coefficients "di" can be defined according the data of reference (Molland (2011)). In formulas for the calculation of coefficients c_i values of C_{WL} are calculated based on the value of half angle of entrance of designed waterline, i_E in degrees ($C_{WL} = i_E.L/B$). According the model of regression Van Oortmersen for the calculations of Fr, Re, LCB, CP and CM is used the displacement length L_D. For the calculation of displacement length L_D, the following formula is used: (Van Oortmerssen (1971), Molland (2011)).

$$L_D = 0.5(L_{BP} + L_{WL}) \quad (5)$$

where L_D is the displacement length in m, L_{BP} is the length between perpendiculars in m, L_{WL} Length on loaded water line in m.

The model of regression Van Oortmerssen does not provide data regarding appendages resistance and the influence of bulb and transom.

Based on the model of regression Van Oortmerssen, the actual structure of the software, which is based on the models of regression Holtrop and Holtrop Mennen, the general expression for the calculation of total resistance is as following:

$$R_T = (R_F + R_R + R_{APP} + R_A + R_{AIR} + R_{ADD}) \quad (6)$$

R_F – Friction resistance calculated according to ITTC – 1957.
R_R – Residual Resistance.
R_{APP} – Appendages resistance
R_A – The model ship correlation resistance.
R_{AIR} – Air resistance is calculated as in (Lewis, (1988))
R_{ADD} – Additional resistance for meteorological conditions. In the current version of the program this resistance is simply calculated as a percentage addition of the resistance in calm seas.

Based on the specific residual resistance defined according the expression (1), calculation of residual resistance in this procedure of the software "Ship Power V_1.0" is carried out according the following formula:

$$R_R = \left(\frac{R_R}{\Delta}\right)(\nabla \rho g) \quad (7)$$

Calculations of resistance coefficients, power, propeller propulsion force, hull efficiency, and quasi propulsive efficiency are realized according the following formulas.

$$C_{XX} = \frac{R_{XX}}{0,5\rho SV^2} \quad (8)$$

$$T = \frac{R_t}{(1-t)} \quad (9)$$

$$\eta_H = \frac{(1-t)}{(1-w)} \quad (10)$$

$$\eta_D = \eta_o \eta_H \eta_R \quad (11)$$

$$P_E = R_t V \quad (12)$$

$$P_T = TV \quad (13)$$

$$P_D = \frac{P_E}{\eta_D} \quad (14)$$

In the above formulas, the propulsive coefficients are modelled using the formulas and

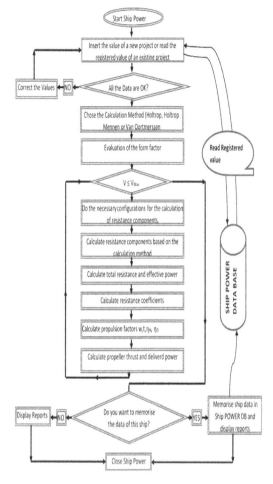

Figure 1. Ship Power general flow chart.

coefficients presented in (Van Oortmerssen (1971). The calculation of propulsive coefficient is done only for the service speed and the respective value of open water propeller efficiency declared by the user in the general graphic user interface of "Ship Power V_1.0".

In Figure 1 is presented the general flow chart of the new version of the software.

3 RESULTS AND DISCUSSIONS

3.1 Development of procedure in Ship Power V_1.0

The new procedure of the software "Ship Power V_1.0" is developed in order to harmonize the model of regression Van Oortmerssen with the actual structure of the program. Even in this procedure the program represents the reports of results

of calculations according the existing format and structure. The reports of calculation generated by Ship Power V_1.0 even for this procedure are:

- Report of general parameters of ship.
- Report of calculations of a-dimensional coefficients of resistance.
- Report of calculations of the components of resistance.
- Report of calculations of propulsion and power. Even in this procedure the software represent graphically the results of calculations. The graphical representation of Ship Power in this procedure is:
- Graph of resistance as function of ship speed.

Also this new version of the software has two main interfaces (Graphic User Interfaces, GUI) and five menus, which are activated and deactivated in function of the graphic interfaces in which we are working. The software operates in Albanian and English interfaces.

The first interface of Ship Power V_1.0 is the general data interface. Through this procedure the user can declare some general data of the ship, such as:

- Ship
- Type of ship
- Type of propulsion
- Propeller data
- General Ship data (Length between perpendiculars, length on waterline, length overall, beam, draft, depth)

In Figure 2 is presented the general interface of the software.

The general data user interface is managed by an algorithm, which signals in case of incorrect input of data, or when the data needed for calculations

Figure 2. General data interface of Ship Power, as implemented in Van Oortmerssen procedure.

are missing. In such a way the algorithm of this interface gives the user the possibility to correct the incorrect or missing data.

In Figure 3 is presented a typical message of error of Ship Power in case of an incorrect or missing data, as implemented in Van Oortmerssen procedure.

The Van Oortmerssen procedure of calculation, in software Ship Power is activated through the graphical general data interface, when the user decides to personalize the calculation method. By default the procedure of calculation is the procedure based on Holtrop and Holtrop Mennen regression analysis. In Figure 4 is presented a typical windows

message of "Power Ships" for choosing of calculation method.

The second graphical user interface of Ship Power V_1.0 is the graphic user interface of personalised data. Through this procedure the user can declare more detailed data for the ship, which are grouped according to the following blocks of this interface.

– Hull data (main dimensions; length on waterline, length between perpendiculars, beam, forward draft, aft draft; geometric coefficients, block coefficient, midship coefficient, water plane coefficient, prismatic coefficient, position of LCB, wetted surface, half angle of entrance).
– Water environment (water temperature, density, viscosity)
– Range of velocity (minimum speed value, maximum speed value, step of speed)
– Particularity of ship hull (stern ship hull sections, bulbous bow parameters, ship bow thrusters, transom area)
– Appendages modelling

In Figure 5 is presented the personalised graphic user interface of the software.

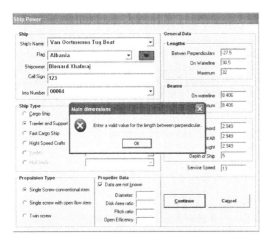

Figure 3. Typical message of error of Ship Power in case of an incorrect or missing data.

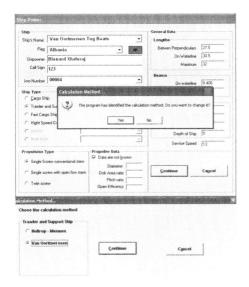

Figure 4. Typical windows message of "Power Ships" for choosing of calculation method.

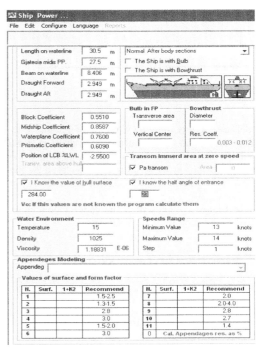

Figure 5. Personalised graphic user interface of the software.

3.2 Validation of the procedure

The validation of this procedure is realised using the results of calculations for a tug boat based on the model of regression Van Oortmerssen. (Helmore (2008)). Principal dimension of ship taken in consideration are presented in Table 2.

For the ship taken in consideration the results of calculation as in reference (Helmore (2008)) are presented in Table 3.

The procedure Van Oortmerssen developed in "Ship Power V_1.0" is executed based on the main dimension of the ship, as presented in Table 2. The results of calculation and the corresponding comparisons with results of reference (Helmore (2008)) are presented in Table 4.

Table 2. Principal dimension of Ship under consideration (Helmore (2008)).

Elements	Symbols	Values
Length on water line	L_{WL}	30.5 m
Length between perpendiculars	L_{BP}	27.5 m
Beam	B_{WL}	8.406 m
Draft on F.P	T_F	2.949 m
Draft on A.P	T_A	2.949 m
Displacement length	L_D	29 m
Displacement volume	∇	376 m^3
Wetted surface	S	284 m^2
Prismatic coefficient on L_{WL}	0,5790	
Half angle of WL entrance	23	deg
L_D/B	3.450	
B/T	2.850	
Position of LCB as% of L_D	−2.55	
Prismatic coefficient based on L_D, C_P	0.6090	
Midship coefficient	0.8587	
Block coefficient based on L_D, C_B		
Speed	13	Kn

Table 3. Results of calculation according Van Oortmerssen model as presented in reference (Helmore (2008)).

Nr	Elements	Value
1	Coefficient m	0,4267
2	Coefficient C_{WL}	79.35
3	Coefficient c_1	$2.096 * 10^{-3}$
4	Coefficient c_2	$248.3 * 10^{-3}$
5	Coefficient c_3	$−68.87 * 10^{-3}$
6	Coefficient c_4	$42.33 * 10^{-3}$
7	$f_1 = e^{-m/9Fn^2}$	0.7398
8	$f_2 = e^{-m/Fn^2}$	0.06634
9	$f_3 = e^{-m/Fn^2} \sin(1/Fn^2)$	0.004992
10	$f_4 = e^{-m/Fn^2} \cos(1/Fn^2)$	0.06615
11	R_R / V	$20.48 * 10^{-3}$
12	$R_R = (R_R / V) * (\rho V)$	R_R − 7893 kgf = 77.403 kN

Table 4. Comparison of results of calculation with those of reference (Helmore (2008)).

Nr	Elements	Value ship power	Differences in% with results of reference (Helmore (2008))
1	m	0.4266641	0.0084%
2	C_{WL}	79.348084	0.0024%
3	c_1	$2.09597*10^{-3}$	0.0014%
4	c_2	$248,28*10^{-3}$	0.0081%
5	c_3	$−6.8882*10^{-3}$	−0.0174%
6	c_4	$42.325*10^{-3}$	0.0118%
7	f_1	0.7397149	0.0115%
8	f_2	0.0663101	0.0451%
9	f_3	0.00506237	−1.4097%
10	f_4	0,0661165	0.0506%
11	R_R/V	$20.4639*10^{-3}$	0.0786%
12	R_R	77.264 kN	0.1796%

Table 5. Calculations of resistance and effective power in the range of velocity 3–13 knots.

V	R_R	R_T	P_E
3	27.52	1108.79	1.71
4	327.62	2185.43	4.50
5	1031.00	3860.17	9.93
6	1925.32	5916.52	18.26
7	2863.39	8204.06	29.54
8	4414.58	11289.53	46.46
9	7319.01	15910.80	73.66
10	16484.67	26973.96	138.75
11	20253.00	32818.77	185.70
12	36924.39	51744.15	319.41
13	77264.95	94514.85	632.04

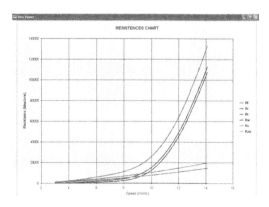

Figure 6. Graphical representation of resistance components according Ship Power, for the ship under consideration.

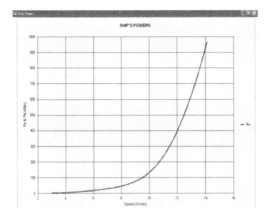

Figure 7. Graphical representation of power according Ship Power, for the ship under consideration.

From the comparison of the results is founded a satisfactory compliance of results of calculation according Van Oortmerssen procedure of software "Ship Power V_1.0" with results of calculations of reference (Helmore (2008)).

The results of calculation for resistance and effective power, for the ship under consideration in the range of velocity 3–13 knots, according Van Oortmerssen procedure of "Ship Power V_1.0" are presented in Table 5 and graphical representation, according Ship Power V_1.0, in Figures 6, 7.

4 CONCLUSIONS

Van Oortmerssen procedure is a new procedure developed in Ship Power. The procedure is developed in harmonisation and integration with the actual procedure, which is based on the model of regression of Holtrop and Holtrop Mennen. This procedure follows strictly the model of regression of Van Oortmerssen for the calculation of residual resistance, while for the calculations of other parameters the procedure follows the formulation of the existing model of the software.

Compared to the results available in literature, the results of calculation of this procedure, as presented in Table 4, are almost identical. These differences are mainly due to rounding and approximations.

Also the procedure Van Oortmerssen of the software Ship Power can be used as well as other similar procedure of other well-known software.

REFERENCES

Altosole M., Buglioni G., Figari M, (2014) Alternative propulsion technologies for fishing vessels: a case study. *International Review of Mechanical Engineering, Vol. 8, N. 2.*

Helmore, P. (2008). Update on van Oortmerssen's Resistance Prediction. *Proceedings of Pacific 2008 International Maritime Conference.*

Lewis E.V. (1988). *"Principles of Naval Architecture. Volume II.* Jersey City: The Society of Naval Architects & Marine Engineers.

Molland, A. (2011). *Ship Resistance and Propulsion— Practical Estimation of Ship Propulsion Power.* Cambridge Press.

Van Oortmerssen, G. (1971). A power prediction method and its application to small. *International Shipbuilding Progress,* 397–415.

Xhaferaj, B., Dukaj, A., & Lapa K. (2014). The architecture of software ship power V_1.0. *Annals of Costanta Maritime University,* 23–30.

Xhaferaj B., Dukaj, A. (2017). Results of calculations of Holtrop_Mennen procedure of "Ship Power V_1.0" versus other commercial software. *International Maritime Science Conference—IMSC 2017,* (pp. 283–289). Solin.

Hydrodynamics – foils

Maritime Transportation and Harvesting of Sea Resources – Guedes Soares & Teixeira (Eds)
© 2018 Taylor & Francis Group, London, ISBN 978-0-8153-7993-5

A BEM for calculating the effects of waves on WIG marine craft performance

K.A. Belibassakis
School of Naval Architecture and Marine Engineering, National Technical University of Athens, Greece

ABSTRACT: A novel method based on vortex elements is presented supporting the study of the effects of waves on WIG marine craft performance and especially on the aerodynamic characteristics. In particular, a modified Green's function is developed fulfilling the condition on a moving wavelike surface at small distance below the singularity. Then an asymptotic expansion is derived and applied, in conjunction with lifting surface theory, to the calculation of the aerodynamic characteristics of wings travelling at a constant speed and small height above the water. Also, the pressure distribution on the water surface is estimated, which is exploited for further calculations concerning the disturbance and propagation of waves on the water surface generated by the moving pressure distribution.

1 INTRODUCTION

Wing-In-Ground (WIG) marine crafts are frequently used as an alternative transportation means exploiting the proximity effects to improve aerodynamic loads and efficiency remaining airborne just above the water surface; see, e.g., Rozhdestvensky (2006), Yun et al. (2010). A WIG-craft operates at much higher speeds than ships and more efficiently than aircraft. Another distinct advantage of a WIG-craft is its ability to take off anywhere from the sea surface without the need for a runway. Due to its superb and unique features, WIG-crafts serve as a promising choice of fast, safe and efficient platform for future marine transportation systems.

The wing in ground (WIG) effect is a phenomenon where the lifting flow characteristics are modified essentially due to strong interaction of wing with nearby boundaries, i.e., when the aircraft is near to the ground which may refer to not only land, but also water, ice, snow and sand. The effects of the wing in ground effect can be beneficial since the presence of a boundary close to a wing results in increased pressure on the lower surface hence increasing the generated lift, Nebylov & Wilson (2002). As the efficiency of a wing is determined by the lift to drag ratio, it increases due to the ground effect, providing increased carriage capacity at moderate speeds.

Various types of WIG craft have been designed specifically to utilise the above benefits, and are not actually regarded as aircrafts. Although development of WIG craft has taken place over many decades, the technology has not progressed to the point where such craft can become a mainstream commercial success, due to early design inefficiencies and lack of funding for research and development in this area worldwide. Development and validation of simulation tools could further support the design and optimization of WIG craft, with possible applications not only focusing on civil practical utilization, but also military, like heavy lift carriage, special forces insertion, maritime strike, anti-submarine warfare and amphibious assault missions. Among other, operational advantages of WIGs are speed advantage over conventional sea borne vessels, invulnerability to torpedoes and mines, reduced detection range compared to aircraft, and very low acoustic signature.

In this work a novel method, based on vortex elements, is presented, supporting the study of the effects of waves on WIG marine craft performance and especially on the aerodynamic characteristics. In particular, a modified Green function is developed fulfilling the condition on a moving wavelike surface at small distance below the singularity. Then an asymptotic expansion is derived and applied, in conjunction with lifting surface theory, to the calculation of the aerodynamic characteristics of wings travelling at a constant speed and small height above the water. Also, the pressure distribution on the water surface is estimated, which is exploited for further calculations concerning the disturbance and propa-gation of waves on the water surface generated by the moving pressure distribution.

2 A SIMPLE TWO-DIMENSIONAL MODEL

In Howard & Yu (2007) an exact theory has been developed for linear irrotational wave motions over a corrugated bottom, under the assumption of simple harmonic time dependence. The corrugations used constitute a family characterized by arbitrary wavelength and amplitude parameters, and for small amplitude, the corrugations are nearly sinusoidal; see also Yu & Howard (2010). The precise form of the solutions is obtained by means of conformal map, transforming a strip above a flat bed and below a horizontal free-surface level to a strip above a simply periodic corrugated bed. In order to illustrate the present approach, we exploit in this work a modified version of the previous transformation mapping the halfspace above a horizontal boundary to a halfspace above a simply periodic boundary, defined as follows

$$kx = \xi + \varepsilon \cos(\xi) \exp(-\eta),\qquad(1a)$$

$$kz = \eta + \varepsilon \sin(\xi) \exp(-\eta),\qquad(1b)$$

where k denotes the wavenumber and ε the corresponding non-dimensional amplitude of the lower boundary corrugations. As an example, the transformation of the rectangle $-2\pi \le \xi \le 2\pi,\ 0 \le \eta \le 11$, for $\varepsilon = 0.3$, is illustrated in Fig. 1.

The above mapping permits the transformation of simple singularities on the plane (e.g., vortex, source and multipoles) over corrugated lower boundaries, supporting the study of field properties and the effect of waves on fundamental solutions of the Laplace equations. Similarly as in the study of ground effects on the lifting characteristics

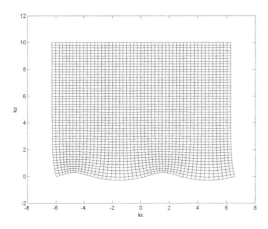

Figure 1. Transformation of a rectangle on the $\xi\eta$-plane, extended over two periodic cells, using Eq.(1), for $\varepsilon = 0.3$. Orthogonal curvilinear grid on xz-plane obtained by the mapping of lines $\xi = $ const and $\eta = $ const.

of WIG crafts operating over the sea, due to small value of the ratio of air to water density $\rho_a / \rho_w \ll 1$, in conjuction with the fact that the wave celerity is much smaller than the travel speed U of the WIG, permitting us to approximately neglect the effect of rate of free surface elevation deformation on the boundary condition, we consider as a first approximation that the air-water interface behaves like a hard boundary, where instantaneously a homogeneous Neumann condition applies. Following this assumption, in the sequel we consider a pair of vortex singularities with opposite strengths, identically satisfying Neumann boundary condition on the $\xi\eta$-plane as shown in Fig. 2(a), represented by the real part of the complex potential

$$F(u) = -\frac{i\Gamma}{2\pi}\left(\ln(u - u_v) - \ln(u - u_v^*)\right), u = \xi + i\eta, \quad(2)$$

where (ξ_v, η_v) denotes the position of a point vortex on the $\xi\eta$-plane with strength Γ, and $u_v^* = \xi_v - i\eta_v$ the complex conjugate of u_v.

Exploiting the conformal mapping, Eq. (1), and applying to the function $F(u)$, we easily obtain from the real part of the transformed function the potential generated by a point vortex singularity located at some level (x_v, z_v) above the lower wavelike boundary, where it fulfills the homogeneous Neumann boundary condition, and from the imaginary part the associated stream functions, as illustrated in Fig. 2(b). The specific example corresponds to a point vortex at $(\xi_v, \eta_v) = (0.5, 1.0)$ in the case of the transformation with parameters as in Fig. 1, and thus, also $(kx_v, kz_v) \approx (0.5, 1.0)$. We observe the discontinuity of the potential associated with the circulation and the perfect fulfillment of the no-entrance boundary condition on the surface of the wavelike lower boundary.

Subsequently, we proceed to derive, and validate against the above complete solution, an asymptotic expression in terms of the non-dimensional amplitude ε, measuring the slope of the wavelike lower boundary $z = \eta(x)$, corresponding to the instantaneous wave profile. Considering the latter to be a simple sinusoidal surface of the form

$$z = \eta(x) = 0.5H \sin(kx),\qquad(3)$$

where $H = 2A$ is the wave height, the solution of the potential is expanded in the form

$$\Phi = \Phi^{(1)} + \varepsilon\Phi^{(2)} + ...,\qquad(4)$$

which has to satisfy the no-entrance boundary condition

$$\partial_z\Phi + \partial_x\eta\partial_x\Phi = 0, \quad\text{on}\quad z = \eta(x).\qquad(5)$$

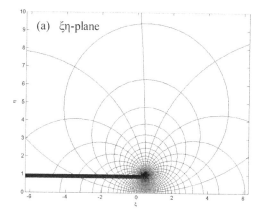

(a) ξη-plane

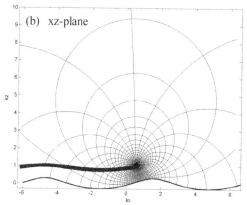

(b) xz-plane

Figure 2. A point vortex fulfilling Neumann boundary condition on the wavelike lower boundary. The discontinuity of the potential is evident on both the ξη and the xz planes by the thick line.

Under the assumption of small slope of the wavelike boundary, the above consition is linearized on the mean level $z = 0$, we use as a base solution $\Phi^{(1)}$ the field generated by the point vortex at (x_v, z_v) and its image (with opposite strength) below the flat surface at $(x_v, -z_v)$, as follows

$$\Phi^{(1)}(x,z) = \frac{\Gamma}{2\pi}\left(\tan^{-1}\left(\frac{z - z_v}{x - x_v}\right) - \tan^{-1}\left(\frac{z + z_v}{x - x_v}\right) \right),$$

(6)

which satisfies the Neumann condition

$$\partial_z \Phi^{(1)} = 0, \quad \text{on} \quad z = 0.$$

(7)

Keeping orders up to the second in the expansion, Eq. (4), the next term is found as a solution of the Laplace equation on the upper half xz-plane satisfying the inhomogeneous condition

$$\partial_z \Phi^{(2)} = -\partial_x \eta \partial_x \Phi^{(1)}, \quad \text{on} \quad z = 0.$$

(8)

Using the above Eqs.(3) and (6) we finally obtain the following solution for the second term in Eq.(4)

$$\varepsilon\Phi^{(2)} = -A\partial_x\Phi^{(1)}(x, z = 0)\cos(kx)e^{-kz},$$

(9)

and the complete field is easily calculated from Eq. (4) using Eqs. (6) and (9), from which also the components of velocity field are obtained by differentiation with respect to the spatial coordinates. The accuracy of the above asymptotic solution is illustrated in Fig. 3 in the case of the field generated by a point vortex over a wavy surface of Fig. 2(b). We observe that the flow details are very well represented in the whole domain, even for point singularities in close proximity of the lower wavy boundary.

Under the assumptions that the wave celerity is small in comparion with the travelling speed of the wing $(cU^{-1} \ll 1)$, and the rate of free surface elevation deformation on the boundary condition is small $(i.e., U^{-1}\partial_t\eta \ll 1)$, its effects may be neglected, and the above model can be approximately used for calculating the lifting flow characteristics above the moving surface $\eta(x;t) = 0.5H\sin(kx \pm \omega t)$ corresponding to harmonic propagating waves of frequency ω and wavenumber k. In the next section, an iterative scheme based on a three-dimensional lifting surface model will be described for the calculation of the effects of waves on WIGs performance. At the first step,

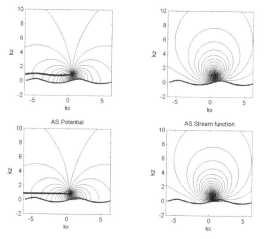

Figure 3. Comparison between the exact solution (upper sublots) and the asymptotic solution, Eqs. (4),(5),(9) (lower subplots) in the case of the point vortex above a wavy surface. Left column shows the potential and right column the stream function.

the strengths of the vortex elements on the wing and the vortex wake are calculated satisfying condition (6) on the flat surface $z = 0$. Then, the forcing term in the right-hand side of Eq. (8) is calculated and the solution carrying out the wave effect on the lifting flow is obtained from Eq. (9). Finally, the wave disturbance on the free surface, associated with an added wave resistance component on the WIG, can be calculated from the moving air pressure distribution generated by the travelling wings.

3 THREE-DIMENSIONAL LIFTING PROBLEM

We consider lifting surface(s) modeling the symmetrically arranged wing(s) of a WIG craft, flying at constant speed U and angle of attack α, at small height h above the water surface. The latter will be considered either calm or a wavy surface characterized by wavelength k and waveheight H in a constant and small water depth d strip; see Fig. 4.

The flow around the wing(s) is assumed to be inviscid and incompressible. Under the previous assumptions, the problem concerning the lifting flow around the wings generated from the combined motion of the wing with velocity $-U\mathbf{i}_x$ (in the inertial frame of reference) and the problem consists of the Laplace equation,

$$\nabla^2 \Phi(x,y,z;t) = 0, \tag{10}$$

where $\Phi(x,y,z;t)$ denotes the disturbance potential satisfying the solid (no-entrance) boundary condition on the wing surface,

$$\partial_n \Phi(x,z;t) = U \cdot n_x, \text{ on } S_b, \tag{11}$$

where n denotes the normal vector on the body surface and $\partial_n \Phi(x,z;t)$ the normal derivative of the potential. The above are considered in conjunction with the kinematical and the dynamical conditions on the free vortex sheets, and the Kutta condition along the trailing edge and the tip(s) of

the wing; see, e.g., Katz & Plotkin (1991). For small and moderate values of the angle of attack a, this problem is effectively treated in the framework of viscous-inviscid interaction techniques.

As before, for waves of small amplitude $\alpha \ll 1$ and $U^{-1}\partial_t \eta \ll 1$, a Neumann condition is considered to apply instantaneously on the air-water interface $\eta(x;t)$

$$\partial_z \Phi - \partial_x \eta \partial_x \Phi = 0 \text{ on } z = \eta(x;t), \tag{12a}$$

which in the case of simple harmonic waves is $\eta(x;t) = 0.5H \sin(kx \pm \omega t)$, with ω is the wave angular frequency.

The case of calm water corresponds to $H = 0$, in which case Eq. (12) simplifies in the form:

$$\partial_z \Phi = 0 \text{ on } z = 0. \tag{12b}$$

As each lifting component is steadily translated, the vorticity created in the boundary layers of its upper and lower surfaces is continuously shed into the vortex wake. Furthermore, this vorticity, which is subsequently convected away from the wing with the local velocity, is assumed to be concentrated into sheets of infinitesimal thickness $S_w(t)$, constituting the trailing vortex wakes. It is also assumed that the flow outside these vortical regions is irrotational. In particular, an unsteady vortex-lattice technique, coupled with a non-linear wake model, has been developed in Belibassakis & Politis (2001) and successfully applied to the problem of thrust production by oscillating wings. By using Green's theorem, the disturbance velocity field (due to the presence and the motion of lifting bodies) is represented by means of singularity distributions, as follows

$$\mathbf{v} = \nabla \Phi = \frac{1}{4\pi} \left\{ \int_{S_b(t)} \sigma \frac{\mathbf{r}}{r^3} dS + \int_{S_b(t) \cup S_w(t)} \gamma \times \frac{\mathbf{r}}{r^3} dS \right\}, \tag{13}$$

where γ is the surface vorticity, associated with the tangential discontinuity of the velocity field on the solid boundary (S_b) and on the vortex sheets (S_w), and stands for the source-sink distribution on the body. These singularities are represented by a discrete source-lattice and vortex-ring elements extended over their support. For attached flows the wake vorticity is shed from the separation lines, which in this case consist of the trailing edge and the tip(s) of the wing, at a rate determined by Kutta condition. The problem is solved by a time-marching technique permitting the modelling of continuous shedding and transport of vorticity in the wake, and the simultaneous deformation of the vortex sheets.

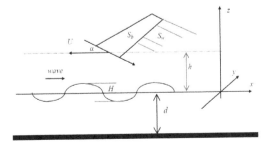

Figure 4.   Coordinate systems and notation.

3.1 Trailing vortex sheet and kutta conditions

Trailing vortex sheets are force-free material surfaces and the vorticity field (concentrated on these surfaces) is divergence-free. Shed vorticity is convected with the local velocities as calculated at the nodal points of each free vortex ring. Also, by this technique Kelvin's theorem,

$$\frac{D\Gamma}{Dt} = 0, \quad \text{where} \quad \Gamma = \oint_{c(t)} \mathbf{v} \cdot d\mathbf{l}, \tag{14}$$

holding for any closed material circuit $c(t)$, is automatically fulfilled. Bernoulli's theorem in unsteady flow (Katz & Plotkin 1991), states

$$p - p_\infty = -\rho_a \partial_t \Phi - \rho_a (\nabla\Phi + 2\mathbf{u})\nabla\Phi, \tag{15}$$

where $\mathbf{u}$ is the velocity due to the motion of the wing as observed from the body frame of reference (and in case of wings at steady translation as in Fig4 $\mathbf{u} = U\mathbf{i}_x$), and p_∞ is a reference pressure. Then, it can be shown that requirement (14) is practically equivalent to the satisfaction of the dynamical condition on the free-vortex sheet, that requires equal limit pressures at points at the upper and lower sides of S_w. Among various alternative approximate forms of the Kutta condition in unsteady flow, in the present work this condition is implemented by enforcing the vorticity generated at the trailing edge and the tips of the wing to shed with the local velocity. By applying Kelvin's theorem, Eq. (14), in conjunction with Bernoulli's theorem, one can derive that this postulate is asymptotically (for small time increments) equivalent to the requirement of equal limit pressures at points on the upper and lower surfaces of the wing (or the vortex wake), as these separation lines are approached.

3.2 Numerical example

As a demonstrative example we consider a wing of trapezoidal planform of semispan over root chord ratio $s/c = 3$, ratio of tip to root chord $c_T/c = 0.4$, leading edge swept angle 70 deg and linear chord distribution over the semispan, respectively. In this case the aspect ratio is significant $AR \approx 8.5$. The wing sections are symmetric of very small thickness. The wing is flying at small height $h/c = 2$ over the water surface at an angle of attack $\alpha = 5$ deg. In Fig. 5 a plotof the calculated solution at various instances is shown, with respect to the phase of an incoming propagating wave of period $TU/c = 4$ and height $H/c = 1$. The subplots of this figure illustrate the strengths of vortex ring elements both on the wing $S_b(t)$ and the wake surface $S_w(t)$. For simplicity in the calculations, the wake deformation effects are turned off and the shape

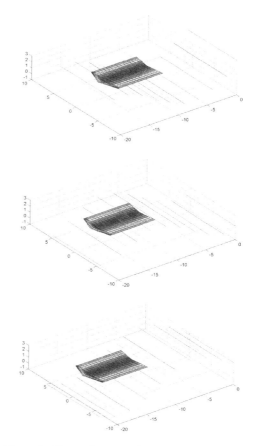

Figure 5. Vortex ring strength on body S_b and wake surface S_w for a trapezoidal wing travelling at constant speed and $\alpha = 5$ deg. Plot of at various instances, with respect to the incoming propagating wave.

of the trailing vortex sheets is calculated only by the translation velocities. In the results shown in Fig. 5 a mesh of 7 spanwise by 4 chordwise elements on S_b is used and the time step is selected so that the average size of wake elements on S_w to remain compatible with the corresponding size of wing elements on S_b. The effect of secondary air flow induced by the harmonic wave and represented by means Eq. (8) generates a harmonic oscillation of vorticity, circulation and lift characteristics as shown in Fig. 6, where the time history of circulation at various wing sections, near the root ($y/s = 0.071$), in the middle ($y/s = 0.5$) and near the tip of the wing ($y/s = 0.928$) is shown, as calculated by the present method starting from rest. For comparison, in this figure the value of the wing section circulation without any ground effect is plotted by using dashed lines, and the corresponding estimation for the wing flying above a flat horizontal boundary by using thin solid lines.

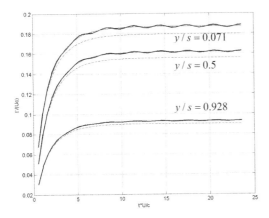

Figure 6. Circulation Γ/Uc at various spanwise positions on the trapezoidal wing travelling at constant speed and $\alpha = 5$ deg. Plot against non-dimensional time.

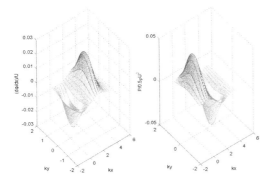

Figure 7. Distribution of (a) horizontal disturbance flow velocity $\partial_x \Phi/U$ and (b) corresponding pressure coefficient $(p - p_\infty)/0.5\rho_a U^2$ on $z = 0$ below the lifting surface.

The unsteadiness generated by the secondary airflow is clearly observed at the encounter (relative) frequency of oscillation, representing the effect of the wave on the lifting flow charateristics of the wing.

After the solution is obtained, forces and moments are calculated by integration of pressure from Eq. (15) on the wing surface S_b. Also, the latter Bernoulli equation is used to calculate from flow velocity the moving pressure applied on the air-water interface shown in Fig. 7, which will be used to calculate the wave disturbance generated by the travelling WIG-craft as described in the following section.

4 WAVE DISTURBANCE

The linearized system of Boussinesq equations (see, e.g., Peregrine 1967) is used in this work to calculate the wave disturbance generated by the moving applied air pressure $P(x,y,t) = p(x,y,z = 0;t)$ generated by the wing on the water surface. The system models free or forced wave propagation in water of small depth taking into account the effects of variable bathymetry $d(x,y)$, and consists of the continuity and momentum equations, as follows

$$\partial_t \eta + \partial_x (du) + \partial_y (dv) = 0, \tag{16}$$

$$\partial_t u + g\partial_x \eta = d^2 \left(\frac{1}{3} + b\right)\left(\partial_x^2 u_t + \partial_y^2 u_t\right)$$
$$+ dd_x \left(u_{tx} + v_{ty}\right) + bgd^2 \partial_x^3 \eta + \rho_w^{-1}\partial_x P, \tag{17a}$$

$$\partial_t v + g\partial_y \eta = d^2 \left(\frac{1}{3} + b\right)\left(\partial_x^2 v_t + \partial_y^2 v_t\right)$$
$$+ dd_y \left(u_{tx} + v_{ty}\right) + bgd^2 \partial_y^3 \eta + \rho_w^{-1}\partial_y P. \tag{17b}$$

The terms containing the b-coefficient have been introduced in order to enhance the dispersion characteristics of the above system; see, also Beji & Battjes (1994). It is easily seen for the simplified form of the system in constant depth and harmonic waves propagating in single direction that the dispersion relation is

$$\frac{c^2}{gd} = \frac{1 + b(kd)^2}{1 + (1/3 + b)(kd)^2}. \tag{18}$$

As shown in Fig. 8, a good choice is $b = 1/21$, bringing the dispersion characteristics of the system Eqs. (16), (17) very close to the analytical curve of linearized wave theory (i.e., $c^2/gd = \tanh(kd)/kd$) up to non-dimensional wavenumbers $kd = 2\pi$, which practically corresponds to deep water.

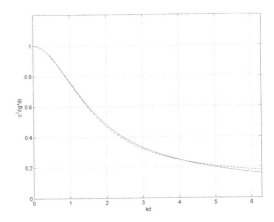

Figure 8. Dispersion characteristics of Boussinesq system, Eqs.(16),(17) with $b = 1/21$ (dashed line) against linearized wave theory (solid line).

Furthermore, the present Boussinesq system, Eqs. (16), (17), is modified near the ends of the domain to include an optimized PML absorbing layer model (Berenger 1994, Colino & Monk 1998). The latter model is used in order to truncate the horizontal computational domain with minimum backscattering and low contamination of the computational solution during the time-integration of the system.

As an example of application, the wave disturbance generated by the moving air pressure of the trapezoidal wing of Fig. 7b, flying at low height $h = 2c$ over the water, is presented in Fig. 9. In particular, only the wave disturbance is shown, as a result of the unsteady applied air pressure on the free surface, and its propagation in a water strip of constant depth of small depth $d = 3c$, corresponding to shallow water environment, as calculated by the enhanced Boussinesq system, Eqs. (16), (17).

The pattern in Fig. 9 presents an analogy with the Neumann-Kelvin wave pattern generated by surface ships travelling in waters of finite depth, which is treated by similar methods; see, e.g., Liu & Wu (2004), Bayraktar Ersan & Beji (2013) and the references cited there. We note here that although the results in the above example correspond to constant depth, variable bathymetry effects can be treated by the system, providing us also information and data concerning various propagation aspects including possible wash effects of WIG generated waves in coastal areas.

Moreover, the present method could offer valuable information that could be further exploited for the calculation of added-wave resistance effects on WIG crafts, due to generation and propagation of wave systems on the water surface. This could be based on energy methods estimating the added resistance from the rate of energy associated the above wave patterns. The effects of waves on both lift and added resistance characteristics of WIGs could be important, especially in cases of near sea-surface flying, and this subject will be examined in detail in future work.

5 CONCLUSIONS

A novel method is presented to investigate the effects of waves on WIG marine craft performance and especially on its aerodynamic characteristics. A modified Green function is developed fulfilling the condition on a moving wavelike surface at small distance below the singularity. Also, an asymptotic expansion is derived and applied, in conjunction with lifting surface theory, to the calculation of the flow characteristics of wings travelling at small height above the water, taking into account the wave free-surface. The aerodynamics of wings travelling at constant and small speed is efficiently calculated by means of BEM based on vortex elements, taking into account the wavy boundary. Subsequently, the applied air pressure distribution on the water surface is obtained, which is exploited for further calculations concerning the propagation of wave disturbances on the water surface generated by the moving pressure distribution. The present method takes into account the effect of the wavy free surface through the satisfaction of the corresponding boundary conditions, as well as the velocity component due to waves and currents on the formation of the incident flow. The effects of waves on both lift and added resistance characteristics of WIGs could be important, especially in cases of near sea-surface flying.

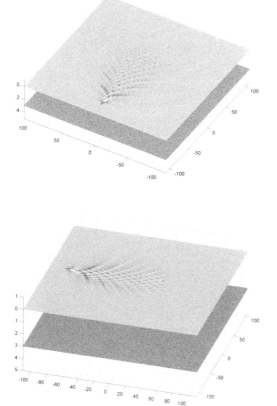

Figure 9. Wave disturbance generated from the moving air pressure applied on the water surface, as induced by the ground effect of the trapezoidal wing of Fig. 5 of chord $c = 1$ m travelling at a speed $U = 5$ m/s.

REFERENCES

Bayraktar Ersan D., Beji S. 2013. Numerical simulation of waves generated by a moving pressure field. *Ocean Engineering* 59, 231–239.

Beji S., Battjes J.A., 1994. Numerical simulation of non-linear wave propagation over a bar, *Coastal Engineering* 23, 1–16.

Belibassakis, K.A., Politis, G.K., Triantafyllou, M.S., 1997. Application of the Vortex Lattice Method to the propulsive performance of a pair of oscillating wing-tails, *Proc. 8th Inter. Conf on Computational Methods and Experimental Measurements, CMEM'97*, Rhodes, Greece.

Berenger J.P., 1994. A perfectly matched layer for the absorption of electromagnetic waves, *J Comput Physics* 114, 185–200.

Collino F, Monk P. 1998. Optimizing the perfectly-matched layer. *Comp Meth Appl Mech Eng* 164, 157–71.

Howard L., Yu J., 2007. Normal modes of a rectangular tank with corrugated bottom. *J. Fluid Mech.* 593, 209–234.

Katz, J. & Plotkin A., 1991. *Low-speed Aerodynamics*, Mc-Graw Hill, New York.

Liu P., Wu T. 2004. Waves generated by moving pressure disturbances in rectangular and trapezoidal channels *Journal of Hydraulic Research v* Vol. 42, No. 2, pp. 163–171.

Moran, J., 1984. *An Introduction to Theoretical and Computational Aerodynamics*, J. Wiley & Sons.

Nebylov A.V., Wilson P.A., 2002. *Controlled flight close to the sea*, WIT press.

Peregrine, D.H., 1967, Long waves on a beach, *Journal of Fluid Mechanics* 27, 815–827.

Rozhdestvensky K.V. 2006. Wing-in-ground effect vehicles, *Progress in Aerospace Sciences* 42 211–283.

Yu J., Howard L., 2010. On higher order Bragg resonance of water waves by bottom corrugations. *J. Fluid Mech.* 659, 484–504.

Yun L., Bliault A., Doo J., 2010. *WIG Craft and Ekranoplan*, Springer.

Maritime Transportation and Harvesting of Sea Resources – Guedes Soares & Teixeira (Eds)
© 2018 Taylor & Francis Group, London, ISBN 978-0-8153-7993-5

Estimation of planing forces in numerical and full scale experiment

A. Caramatescu, C.I. Mocanu & F.D. Pacuraru
University of Galati, Romania

G. Jagîte
Deepwater Technology Research Center, Bureau Veritas, Singapore

ABSTRACT: Numerical experiments involving Computational Fluid Dynamics (CFD hereafter) tools are becoming a widespread solution for analysis on hull panels exposed to hydrodynamic loadings. This paper presents a comparison between full scale experiment that was conducted to measure bottom loadings using strain gauges and the results of CFD numerical experiment. The availability of the robust commercial CFD software and high speed computing has led to increasing use of this computational tool to solve fluid engineering problems across industry and boat building is no exception. The results provide an interesting convergence of the calculations and the measurements, and are encouraging for the future use of this path in evaluation of new designs, determining the hydrodynamic loads and potential structural enhancements.

1   INTRODUCTION

CFD based tools are are often called to provide a solution to boat hulls design in the early design stages but this calls for powerful flow solvers able to accept complex geometries as well as interdependent physical phenomena, such as free surface, spray and dnamic planing forces. Accuracy depends on the attention to the physical modeling, particularly to the turbulent flows and to the delicate matter of numerical discretization. Moreover, the multiphase flow encountered at high Froude number and the development of the spray sheet increase the complexity of the numerical simulation. At the end, the representativeness of numerical simulations is increased by the validation with full scale measurements. The conclusions evaluate the accuracy of the computed loadings, the simulation parameters and improve the numerical modeling in order to converge the full scale experiment. At the same extent, the predicted loadings of the classification societies can be improved in order to provide a realistic estimation and a reasonable safety factor, thus leading to a better use of materials and an improved weight of the hull and its structural members.

In this case, a complex system of strain gauges, accelerometers and a two channel inclinometer was connected to the same data acquisition system in order to provide an accurate system of values that are compared with the numerical simulations results.

Figure 1 above describes the shape of the boat chosen for the study; it was selected so as to have the least sharp angles, despite they are specific for planning boats. Because the path of this project crosses through several fields of study, it was kept

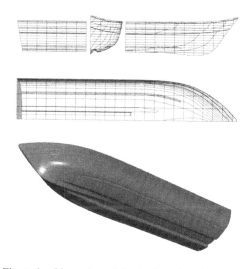

Figure 1.   Lines plan of the hull used in this project (top) and rendering of the 3d model (below).

in mind that the numerical modeling of the structure must represent as accurate as possible the full scale model.

2   EXPERIMENTAL FLUID DYNAMICS

2.1   *Experimental setup*

The full scale experiment was the starting point of this project, collecting data in open water conditions. In order to conduct the experiment a composite hull was laminated and while in the process of assembly a set of 10 strain gauges were installed on the inside of the shell and on the topside of some structural members as can be seen in Figure 2. Main characteristics of the boat are described in Table 1.

There were used also 2 accelerometers, an inclinometer and a GPS speed measurement device. Considering the real navigation conditions that were encountered in this experimental phase, accelerometers were used to separate the "calm water" measured data from the navigation in waves; nevertheless, high values of accelerations measured in this experiment will be used in a separate study that will determine the impact loads encountered during slamming. The inclinometer measured the rotation along the Ox and Oy axes in order to validate the trim angle resulting from CFD and to eliminate transversal inclined measurements resulting in viciated strain gauge recordings. Nevertheless, inclinometer readings are compared with the predictions of trim angle calculated with the empirical method proposed by Savitsky (1964), still one of the most used planning behaviour prediction method due to its relative low computational requirements.

All the measurement devices were connected to a twin Spider8 measurement bridge, receiving data on 14 channels with a rate of 50 readings per second, in blocks of 2040 values for each channel. The measurement bridges were installed in the bilge of the boat as it can be seen in Figure 3.

This study will focus on two values for navigation speed, 4 m/s and 5 m/s. These values were selected because they represent the transition between semi displacement and planning conditions, where large forces are predicted to be encountered as the hull

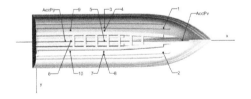

Figure 2. Location of each measurement device installed on the experimental setup. Strain gauges are 1 to 10, AccPv and AccPp are the accelerometers; the inclinometer shares the same position with AccPp.

Table 1. Main characteristics of the hull used in the project.

Length overall	5.65 m
Beam	1.78 m
Hull depth	0.70 m
Empty draught	0.27 m
Hull weight	220 kg
Loading capacity	550 kg
Installed power	40 HP

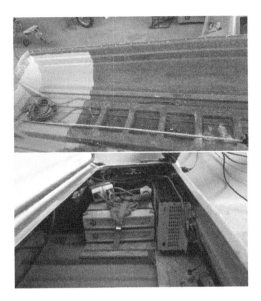

Figure 3. Installation of the measurement devices in the boat.

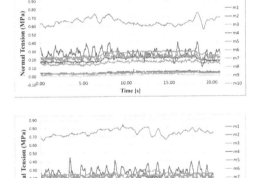

Figure 4. Strain gauges measurements for 4 m/s.

climbs on the own generated wave, a case similar to the hull on wave crest condition. The transition speed values were estimated using the Savitsky method and accurately confirmed by both CFD and EFD tests.

2.2 Recorded data

The variation of normal tension recorded using the strain gauges are represented graphically in Figure 4 for the 4 m/s trial and in Figure 5 for the 5 m/s trial.

The collected data shows similarities between the variations of the adjacent strain gauges, therefore the shape of the hull does not have tension concentrators that block the dispersion of the strain in a certain point. In the chart representing 4 m/s trial, the line representing a constantly higher value of the strain corresponds to the strain gauge number 5, located approximately amidships where there were expected to be higher pressure loads on the bottom, while the boat is still on the transition period. In the graphic representing the 5 m/s trial, there can be seen that the strain gauge # 5 keeps on recording the highest tension, but the values are smaller compared to the previous situation, due to the fact that the boat moved over the hump.

Figure 5 shows also an interesting phenomenon, detailed in Figure 6: the boat passes a small amplitude and large period wave travelling at the same direction with the boat but at a very small different angle. As the boat climbs over the wave the gauge # 5 begins to record higher values, as well as the trim angle keeps on increasing. Around the moment

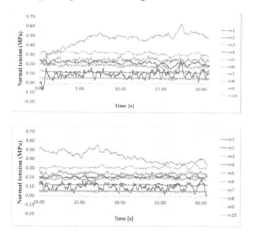

Figure 5.   Strain gauges measurements for 5 m/s.

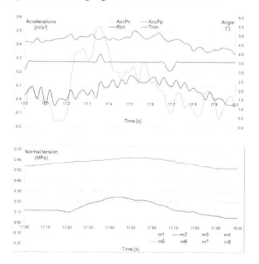

Figure 6.   An oblique wave event.

17.1 s of the measurement the bow of the boat falls over the crest of the wave for 1/10 of a second (shown by the measurement of the bow accelerometer); gauges begin to record the event in the order of its encounter: gauge #2 located near the bow accelerometer show an increase in the normal tension at the same time, 17.2 s and a peak at 17.4 s while the midship gauge #5 shows a delay of 0.1 s for the same event. The wave encounter was oblique because it produces a slight variation in the roll angle at the same time the peak acceleration is recorded, and most importantly it was recorder with a higher intensity by the gauge #2 located at starboard and at a lower intensity for the mirror gauge #1.

However, despite the interesting dynamics detailed in this chart excerpt, this is an event that was filtered from the processed data due to the fact that is was not conducted in calm water, as the CFD simulation. Data filtration was performed using center of gravity acceleration values: strain gauges measurements kept for the comparison must have been recorded while acceleration values were below an absolute threshold of 1/50 g or 0.2 m/s².

3   COMPUTATIONAL FLUID DYNAMICS

Considering that equilibrium position at high Froude numbers is significantly different from the hydrostatic position, one can expect large mesh deformation. In order to avoid this, a two-step preparation was used. First, the estimation of the final hull position (trim and sinkage) for the specific speed based on Savitsky method in order that to generate a better mesh for the hull close to the final position. Then the flow around the hull, sinkage and trim are solved iteratively until the solution converged for the boat that reaches the equilibrium position. The information targeted are the dynamic trim, sink and normal pressure on the wetted area.

The NUMECA/FineMarine commercial code is used to compute the flow solution on a multiblock, high-performance parallel computing fashion. The RANS solver is fully implicit, based on finite volume method to build the spatial discretization of the transport equations. The velocity field is obtained from the momentum conservation equations and the pressure field is extracted from the mass conservation constraint. In the case of turbulent flows, additional transport equations for modeled variables are discretized and solved using the same principles [Deng et al 2010, Duvigneau et al 2003]. The k-ω turbulence model is used for turbulence closure in this paper based on FineMarine recommendations for planning boat. Free-surface flow is simulated with a multi-phase flow approach. Incompressible and nonmiscible flow phases are modeled through the use of conservation equations for each volume fraction of phase/fluid [Queutey et al 2007].

For complex geometries such as a planning boat hull, the grid generation is a very complicated and time consuming task to accomplish. A cartezian mono-block unstructured H-H type grid of about 2.5 million cells has been generated to cover the entire computational domain along the bare hull.

Considering the narrow and long wake fields usually encountered in the planing hulls, the domain size has to be extended in order to capture the wake correctly. Since the wake development is particularly interesting for planning hull designers, a proper wake field capturing is the key of the computation. The domain covers one ship length upstream of the bow, half above, one and a half ship length out from the side and bottom of the hull, and four ship lengths downstream of the stern.

A noslip boundary condition was enforced at the hull surface, where the first grid point from the body surface was located at around $y+ = 30$ for turbulence modeling considerations. Details of the near-hull grids at the bow and stern center plane are shown in Figure 7. A refined grid distribution is used also on the free surface boundary layer and on the sharp edges where higher pressure values are expected to build. The refinement is cell-based: existing cells are subdivided into smaller cells to locally create a finer grid. Layers of high aspect ratio cells tangentially to the wall have been inserted in order to correctly resolve boundary layers. The technique is based on successive subdivisions of the cells connected to the walls, this contrasts with other techniques which insert layers by extrusion. This refinement technique has the advantage of robustness and speed. Besides, the inflation algorithm improves the mesh quality in viscous layers.

The geometry of the hull was reproduced using Rhinoceros CAD software then imported to CFD software.

Simulations were conducted based on a imposed constant speed of 4 respectively 5 m/s.

The previsualization of the the CFD results depicted in Figure 8 comes to confirm the experimental results previously recorded. Pressure field numerical values calculated on the shell were extracted in order to be used in the next step of the project, Finite Element Analysis.

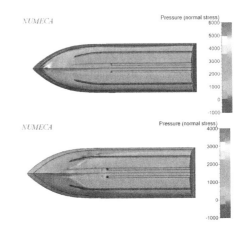

Figure 8. Pressure distribution [N/m²] for 4 m/s computation (top) and 5 m/s computation (below).

Pressure fields calculated in each of the simulations were divided into 16 strips parallel with the Ox axis, having an equal horizontal projected width of 100 mm, and individual values of pressure were extracted together with the cartesian position of each point. In order to reduce the time needed to introduce the pressure loadings into FEA software, a 6th order pressure equation dependent on the length was extracted (including the hydrostatic and the hydrodynamic components) for each strip, the example depicted in Figure 11 for the strip located at 600 mm aside the centerline and the speed of 4 m/s.

4 FINITE ELEMENT ANALYSIS

The numerical finite element analysis was conducted to calculate the effect of the pressure loads previously determined with CFD, using Femap FEA software. A linear analysis was run in order to determine the total deformations in the nodes located in the position of the strain gauges installed for the experiment.

Similar to the method used to divide the pressure field, the geometry of the model was divided into 16 strips and × depending variable pressure loads were applied, together with the own weight of the model and local loads of the engine and 2 persons participating in the experiment.

In order to accurately describe the materials used in the production of the hull, a sample covering 0.5 meters alongside the hull was laminated and shell samples were extracted to determine the material properties in the lab, using tensile testing method described by ASTM D 3039. Figure 9 shows the laminated hull section before samples were extracted, with the location of each test sample. In Figure 10 there can be observed several test samples tractioned to the breaking point.

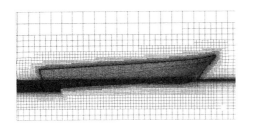

Figure 7. 2.5 million cells grid used in CFD computation.

The lab testing samples were laminated 4 months before the date of testing, in order to allow complete cure of the composite materials. Properties determined were Young modulus (5.685 Gpa) and Poisson's coefficient (0.25).

Both loading cases consisting in pressure loads for the 4 m/s simulation (Figure 12) and 5 m/s (Figure 13) simulation were run and results were collected in a data table and compared to the displacements recorded during the experiment.

The results show an accurate prediction of displacements calculated in FEA software, with a range of errors between –4.38 percent and 6.2 percent, as it can be observed in the Table 2 below.

Figure 11. Pressure distribution at Y = 600 mm section for 4 m/s simulation and the 6th order polynomial regression.

Figure 12. Total displacements for the loading case 4 m/s.

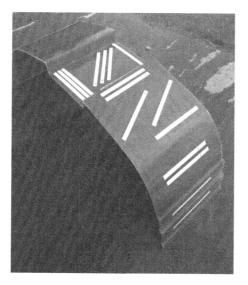

Figure 9. Hull excerpt before test material samples were extracted.

Figure 10. Mechanical properties testing in the lab and a few samples tested to the breaking point.

Figure 13. Total displacements for the loading case 5 m/s.

Table 2. Total displacements comparison between full scale experiment and numerical experiment.

Strain Gauge ID	FEM node location [mm]			Total displacements [mm]				Error [%]	
	X	Y	Z	4 m/s	5 m/s	4 m/s	5 m/s	4 m/s	5 m/s
m1	2156.31	399.76	9.16	11.34	4.45	11.72	4.54	3.38	1.91
m3	4053.12	302.02	27.88	7.60	2.08	7.30	2.18	–3.97	4.95
m4	2160.24	353.24	–6.66	11.23	4.38	11.75	4.20	4.63	–4.01
m9	1076.57	399.88	9.16	6.56	2.56	6.91	2.45	5.32	–4.38
m5	2176.67	0.00	0.00	11.23	4.33	11.93	4.57	6.20	5.63
m6	1092.08	0.00	0.00	6.74	2.65	7.13	2.79	5.90	5.45

5 CONCLUSIONS

The comparison of the values obtained for displacements in both numerical and full scale experimental studies show a reasonable convergence. This procedure uses a series of software solutions transferring the results of one simulation to the starting point of another. Validation of a numerical simulation procedure is a complex and laborious process that involves many experimental tests and many numerical simulations before a correct conclusion can be drawn. This paper is only one small but successful step of such a process.

A furthermore investigation needs to be performed for several speed values and results must be compared to see if the error remains in the same limits or the results become divergent.

With the development of the calculation capabilities, simulation software becomes a close partner for engineering development, but experimental studies must always be taken into consideration each time a new hull shape is considered or a new and different material is used.

REFERENCES

Azcueta R. 2003. Steady and unsteady RANSE simulations for planing RAFTS. *7th Conference on Fast Sea Transportation, FAST'03*, Abano Terme, Italy.

Battistin D., Iafrati A., 2003. A numerical model for hydrodynamic of laning surfaces. *Proc. 7th Int. Corif. Fast Sea Transportation FAST 2003*, Nanjing, 33–38.

Caponnetto M., 2001. Practical CFD simulations for planing hulls. *Proc. of Second International Euro Conference on High Performance Marine Vehicles*, Hamburg, 128–138.

Daniel, I.M., Composite Materials, Chapter 19 in Handbook on Experimental Mechanics, *SEM Inc.*, Ed. Albert Kobayashi, Pretice - Hall, 1987

Deng, G.B., Queutey, P., Visonneau, M., 2010. RANS Prediction of the KVLCC2 Tanker in Head Waves, *9th International Conference on Hydrodynamics*, October 11–15, Shanghai, China, pp.

Duvigneau R.; Visonneau M.; Deng G.B. 2003. On the role played by turbulence closures in hull shape optimization at model and full scale. *J. Marine Science and Technology*, 8(1), 1–25.

Faltinsen, O.M. 2005. Hydrodynamics of High-Speed Marine Vehicles, Cambridge University Press, Cambridge, UK.

Ghassemi, H., Ghiasi, M. 2008. A combined method for the hydrodynamic characteristics of planing craft. *Ocean Engineering 35*, 310–322.

Hadăr, A., 1997. Probleme locale la materiale compozite, Universitatea "POLITEHNICA" Bucureşti,

Judge,Q.C., 2013. Comparison Between Prediction and Experiment for Lift Force and Heel Moment for a Planing Hull, *Journal of Ship Production and Design, Volume 29, Number 1*, February 2013, pp. 36–46(11), Society of Naval Architects and Marine Engineers (SNAME).

Mocanu, I.C. 2005. Rezistenta Materialelor, editia a doua, Galati.

P. Queutey and M. Visonneau, 2007. An interface capturing method for free-surface hydrodynamic flows. *Computers & Fluids*, 36(9), 1481–1510.

Pendleton, R.L., Tutle, M.E., 1989. Manual of Experimental Methods for Mechanical Testing of Composites, *SEM*.

Queutey P.; Visonneau M. 2007. An interface capturing method for free-surface hydrodynamic flows. *Computers & Fluids, 36(9)*, 1481–1510.

R. Duvigneau, M. Visonneau, and G.B. Deng, 2003. On the role played by turbulence closures in hull shape optimization at model and full scale. *J. Marine Science and Technology, 8(1), 1–25.*

Santoro, N., Begovic, E., Bertorello, C., Bove, A., De Rosa, S., Franco, F., 2014. Experimental Study of the Hydrodynamic Loads on High Speed Planing Craft, *International Symposium on Dynamic Response and Failure of Composite Materials, DRaF2014.*

Savitsky, D. 1964. Hydrodynamic design of planing hulls. *Marine Technology 1 (No. 1)*, 71–95.

Savitsky, D., Brown W.P. 1976. Procedures for Hydrodynamic Evaluation of Planing Hulls in Smooth and Rough Water, *Marine Technology, vol. 13, No.4*, Oct.1976, pp 381–400.

Tripa, P., 2010. Metode experimentale pentru determinarea deformaţiilor şi tensiunilor mecanice, Editura MIRTON, Timişoara.

Volpi, S., Sadat-Hosseini, H., Diez, M., Kim, H.D., Stern, F., Thodal, R.S., Greeenestedt, J.L., 2015. Validation of High Fidelity CFD/FE FSI for Full-Scale High-Speed Planing Hull With Composite Bottom Panels Slamming, *VI International Conference on Computational Methods for Coupled Problems in Science and Engineering. COUPLED PROBLEMS.*

Wackers, J.; Visonneau, M. 2009. Adaptive grid refinement for ship flow computation, *Proceedings of ADMOS 2009*, Brussels, Belgium.

Yumin Su, Qingtong Chen Hailong Shen and Wei Lu 2012. Numerical Simulation of a Planing Vessel at High Speed, *J. Marine Sci. Appl.* 11: 178–183.

Zhao, R., Faltinsen, O.M., Haslum, H., 1997, A simplified non-linear analysis of a high-speed planing craft in calm water, *In Proc. FAST'97, ed. N. Baird, Vol. 1*, pp. 431–8.

Maritime Transportation and Harvesting of Sea Resources – Guedes Soares & Teixeira (Eds)
© 2018 Taylor & Francis Group, London, ISBN 978-0-8153-7993-5

Ground effect or grounding: Which happens when?

O.K. Kinaci, T. Cosgun, A. Yurtseven & N. Vardar
Yildiz Technical University, Istanbul, Turkey

ABSTRACT: A WIG craft is more efficient when compared to an airplane. The underlying reason of this phenomenon is known for more than half a century now: the existence of a ground beneath the craft enhances the lift. The ground pushes the body away. On the other hand, experienced captains know for sure that it is dangerous to draw very close to land due to the grounding risk in shallow waters. The ground proximity in this case works the other way around: it pulls the body to itself. These two distinct examples reveal the perplexing behavior of ground vicinity. This study distinguishes these two opposite behaviors of ground proximity and explains physically which happens when.

1 INTRODUCTION

Ground effect is usually known as "enhanced lift" by aerodynamicists although it may as well work negatively. WIG crafts benefit from ground vicinity to carry more load and spend less fuel but cases of extreme ground effect may result in crashing the vehicle. Researchers working in the field of ship hydrodynamics are familiar with negative ground effect which is also known as "grounding" or "ship squat". It may be concluded from these two examples that ground vicinity may enhance or reduce the vertical forces acting on a body.

There are many studies in the literature subjected the ground effect and large part of these works center upon the enhancement or decrement of lift and the pressure distribution on both surfaces of the wing to investigate the physics underlying this behavior. Luo et al. (2012) carried out an experimental work to observe the effect of ground proximity on a finite wing with a NACA0015 cross section. They have determined the areas where the ground has augmenting effect on lift coefficient for varying angels of attack. Jung et al. (2008) have experimentally studied ground effect with an asymmetrical wing with a NACA 6409 cross section. Besides researching many cases including ground vicinity, angle of attack and aspect ratio, they also reported that end plate has an enhancing effect on lift by reducing tip vortex, especially in low ground clearance. Ahmed et al. (2007) have investigated the effect of ground vicinity experimentally in a wind tunnel with moving ground simulation. They found that, influence of the ground proximity and the contribution to the lift coefficient is changing with the modification of the ground surface, due to the convergent-divergent passage between the wing and the ground and it's suction effect.

There are also numerical works on the subject ground effect. Hsiun et al. (1996) have investigated ground effect on 2D NACA4412 airfoil with various ground clearances and angles of attack using RANSE based numerical solver. Barber (2006) has studied Computational Fluid Dynamics (CFD) integration for wings operating on ground proximity by examining the influence of different boundary conditions and flow assumptions on the accuracy of numerical results. A number of researchers used more complicated numerical approaches like Detached Eddy Simulation (Mitchell, 2006; Nishino, 2008) to observe and visualize the flow in more detail. Also, some researchers have looked into different flow conditions including influence of compressibility (Doig, 2011), free surface effect (Zong, 2012) and swept wing (Chengjiong, 2010) via computational fluid dynamics. Other than viscous flow solvers, inviscid (Liang and Zong, 2011; Liang, 2013) and potential based (Philips and Hunsaker, 2013) plow solvers have also used by numerous researchers to investigate ground effect phenomenon.

Ground proximity is usually used in WIG crafts to enhance the lift and carry more load. However, the same ground proximity might be troublesome for ships due to the grounding risk. The former is usually named as wing-in-ground effect which is usually exploited by the aerodynamicists, while the latter is called squat in ship hydrodynamics. As it can be understood from these two distinct examples; ground proximity may be a phenomenon that pulls a body to the ground, or in some cases it may push it higher. The topic of this study is to scrutinize how to distinguish between these two cases and to identify the complex behavior of ground proximity mathematically. The mathematical derivations are validated with computational fluid dynamics approach and the borders of positive and negative ground effect are drawn with both methods.

2 GROUND EFFECT VERSUS SHIP SQUAT

2.1 *Unbounded fluid case*

Consider a wing in an unbounded fluid and assume that the flow is inviscid and irrotational. Therefore, the flow is potential and the Bernoulli equation is valid.

Using the Bernoulli equation to calculate the pressure at an arbitrary point on the wing,

$$\frac{1}{2}\rho V^2 + \rho gz + p_0 = p_a \qquad (1)$$

where ρ is the fluid density, V the local fluid velocity, g the gravitational acceleration, z the vertical coordinate of the arbitrary point, the atmospheric pressure and the local pressure. The total force on the wing can be calculated as:

$$F = \oint_S p_0 dS = \oint_S \left(p_a - \frac{1}{2}\rho V^2 - \rho gz \right) dS \qquad (2)$$

For this wing to fly, the total force should be greater than zero, $F > 0$. This means that the force created at the suction side (SS) should be higher than the force created at the pressure side (PS) of the wing.

$$\int \left(p_a - \frac{1}{2}\rho V^2 - \rho gz \right) dS_{SS} < \int \left(p_a - \frac{1}{2}\rho V^2 - \rho gz \right) dS_{PS} \qquad (3)$$

For this wing to land, the total force should be less than zero, $F < 0$. Therefore, the opposite of eqn. (3) should be valid.

2.2 *Ground proximity case for a slender wing*

WIG crafts benefit from the enhanced lift produced due to the proximity of the ground to the foil. For a wing to make use of the ground in its vicinity, eqn. (3) should hold. For slender wings, a flat plate approach may be used. Then,

$$S_{SS} = S_{PS} \text{ and } \int zdS_{SS} = \int zdS_{PS} \qquad (4)$$

Figure 1.   A wing in an unbounded fluid.

Using the conditions given in eqn. (4) and rearranging eqn. (3) will give the enhanced lift condition for a wing:

$$\text{Enhanced lift} \rightarrow \int_{S_{SS}} V^2 dS > \int_{S_{PS}} V^2 dS \qquad (5)$$

Eqn. (5) states that the velocities on the suction side should be higher than the pressure side. When the velocities are greater on the suction side, the pressure will be lower as compared to the pressure side. The airfoil will move towards the side with lower pressure which is the suction side in this case. This case usually happens when the wing has positive angle of attack, $AoA > 0$. This is discussed with an example in the next section. How the positive ground effect acts on the wing is shown in Figure 2.

In some cases, the wing may not be making use of the ground beneath it but the opposite happens: the wing is pulled by the ground and it loses altitude instead of flying higher. Such a case usually arises when the wing has negative angle of attack, $AoA < 0$. In that case the reverse of eqn. (6) holds:

$$\text{Reduced lift} \rightarrow \int_{S_{SS}} V^2 dS < \int_{S_{PS}} V^2 dS \qquad (6)$$

2.3 *Ship squat in shallow waters*

Ship squat in shallow waters happens when the total force on the hull is lower than zero, $F < 0$. A schematical representation is given in Figure 3.

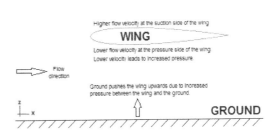

Figure 2.   Positive ground effect pushes the wing away from the ground.

Figure 3.   A ship in shallow water.

A conventional ship has no angle of attack and floats in between two media (positioned at the free surface between air and water) so eqns. (5) and (6) given previously are not relevant. A new approach should be adopted.

Using eqn. (2) and calculating the force over the entire hull surface, the condition for squat will be:

$$F = \int\left(P_a - \frac{1}{2}\rho_a V^2 - \rho_a gz\right)dS_a + \int\left(P_a - \frac{1}{2}\rho_w V^2 - \rho_w gz\right)$$
$$dS_w < 0 \tag{7}$$

Water is about 800 times denser than air; therefore, the density of air can be neglected, $\rho_a = 0$. With this assumption, the squat condition for a ship reduces to:

$$\int V^2 dS_w > -2g\int z dS_w \tag{8}$$

The left hand side of eqn. (8) represents a term related to the dynamic pressure on the hull while the right hand side represents a term related to the hydrostatic pressure. Eqn. (8) states that the *hydrostatic pressure has to be less than the dynamic pressure for a ship to squat in water*. Eqn. (8) also states that it would be enough to take into account only the wetted hull form to assess the risk of squat. Note that the right hand side of eqn. (8) always turns out to be positive as z denotes the sinkage and has always a negative sign.

3 POSITIVE GROUND EFFECT, GROUNDING AND NEGATIVE GROUND EFFECT ON A FLAT PLATE

The push or the pull of the ground is investigated on a simple example of a flat plate of chord c with an angle of attack of α in ground effect. The distance of the lumped vortex from the ground is h. The lumped vortex model of the flat plate is reproduced from (Katz and Plotkin, 2001) and schematically given in Figure 4. As it can be understood from the Figure, the model uses the mirror image of the flat plate to assess the ground effect.

This example is solved in (Katz and Plotkin, 2001) and in their book the circulation and the generated lift are respectively given as:

$$\Gamma = \pi V_\infty c \sin\alpha \left(\frac{1-(c/2h)\sin\alpha + c^2/16h^2}{1-(c/4h)\sin\alpha}\right) \tag{9}$$

$$L = \rho V_\infty \Gamma \left(1 - \frac{\Gamma}{4\pi V_\infty h}\right) \tag{10}$$

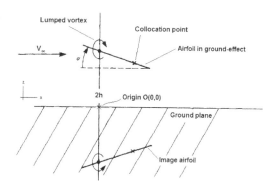

Figure 4. A flat plate in ground-effect. Figure reproduced from (Katz and Plotkin, 2001).

3.1 *Positive ground effect case*

The generalized version of the Kutta-Joukowsky theorem is given in eqn. (10) along with the circulation around the flat plate in ground proximity. Out of the ground proximity $h \to \infty$, which would mean that;

$$\Gamma_{OGE} = \lim_{h\to\infty}\Gamma = \pi V_\infty c \sin\alpha \tag{11}$$

$$L_{OGE} = \lim_{h\to\infty}L = \rho V_\infty \Gamma_{OGE} \tag{12}$$

where the subscript OGE denotes the initials of the "out of ground effect". Eqns. (11) and (12) are special versions of eqns. (9) and (10) where there is no ground boundary below the flat plate.

In order for the ground to create a positive lift, the following should be satisfied,

$$L_{OGE} < L \tag{13}$$

which means that the ground proximity creates an extra lift to support the flat plate. By letting,

$$a = \frac{1-(c/2h)\sin\alpha + c^2/16h^2}{1-(c/4h)\sin\alpha} \tag{14}$$

the expanded version of eqn. (13) becomes,

$$a\left(1 - \frac{ac\sin\alpha}{4h}\right) > 1 \tag{15}$$

3.2 *Grounding case*

Except for negative values of $\sin\alpha$, grounding can only occur during ground proximity and the grounding condition is,

$$L < 0 \tag{16}$$

An expanded version of this condition and substituting α,

$$\rho V_\infty^2 \pi a c \sin \alpha \left(1 - \frac{a c \sin \alpha}{4h} \right) < 0 \qquad (17)$$

3.3 Negative ground effect case

Negative ground effect occurs when neither of the above cases (positive ground effect or grounding) takes place. For the ground to create negative lift,

$$L_{OGE} > L \qquad (18)$$

should hold. Therefore, L_{OGE} is the upper limit of the lift in ground proximity for this case. The lower limit is 0; as when the lift is lower than 0, grounding takes place:

$$L > 0 \qquad (19)$$

Combining equations (18) and (19), we get the condition for the negative ground effect case:

$$L_{OGE} > L > 0 \qquad (20)$$

4 A BOUNDARY ELEMENT METHOD APPROACH

The case of a flat plate close to the ground is also investigated using a numerical approach implementing the boundary element method. To represent the flat plate a very thin two-dimensional NACA0001 wing is used. The forces produced on the flat plate arise from its image which is representing the ground as is the case given in Figure 4. The numerical scheme uses an iterative method to solve for the lift force acting on the plate. The flat plate is discretized into 100 panels with each panel having constant source and dipole distributions at their center. The formulation of the problem and the solution approach is extensively explained in (Kinaci, 2014).

The circulation Γ around the flat plate at each clearance h is calculated using the iterative boundary element method. Then using the equation,

$$L = \frac{1}{2} \rho c V^2 C_l \qquad (21)$$

and combining it with equation (10), the lift coefficient C_l of the flat plate becomes,

$$C_l = \frac{2\Gamma}{c V_\infty} \left(1 - \frac{\Gamma}{4 \pi V_\infty h} \right) \qquad (22)$$

Cases of the lift coefficient dictate whether the flat plate is in positive ground effect, negative

ground effect or grounding condition. When $C_l < 0$ the plate is in grounding condition and when $C_l > 0$, it may either be positive or negative ground effect condition. To understand which of these two cases hold for the flat plate, the lift coefficient of the plate should be calculated when it is out of the ground effect condition and the lift coefficient in this case is represented by $C_{l,OGE} > 0$. If, $C_l < C_{l,OGE}$, then the flat plate is in negative ground effect and in positive ground effect when the opposite holds. To summarize,

$$\begin{aligned} C_l < 0 &\rightarrow \text{Grounding} \\ C_l < C_{l,OGE} &\rightarrow \text{Negative ground effect} \\ C_l > C_{l,OGE} &\rightarrow \text{Positive ground effect} \end{aligned} \qquad (23)$$

A flat plate of unit length is subjected to an incoming uniform flow with unit velocity in a fluid having unit density. The angle of attack α of the flat plate and its clearance from the ground h are varied to specify in which condition the flat plate is. The results of the analytical approach explained in the previous section and the computational approach presented in this section are shown in Figure 5.

Both approaches reveal the intrinsic nature of a body in ground condition. All conditions (positive ground effect, negative ground effect and grounding) are present in both solutions although computational results suggest a wider range for the negative ground effect case. The difference in results is accounted to the assumptions made and numerical errors in the computational approach.

5 A FINITE VOLUME METHOD

A finite volume method approach using RANSE was used to solve for the turbulent flow around a NACA0015 foil in ground proximity. First, the experimental results of (Luo and Chen, 2012) were validated. Then, cases that generated positive and negative ground effects were picked and investigated in terms of the pressure distribution along the foil to have a better insight on what happens in the flow during ground proximity.

5.1 Computational approach

The flow around NACA0015 wing section with various attack angles (0 to 12) and ground clearances were studied via computational fluid dynamics. Computations were carried out for a single velocity at a Reynolds number of 1.872×10^5 Reynolds number was defined in this study as $Re = uL / v$, where u is the stream wise velocity, is the chord length and the v is the kinematic viscosity.

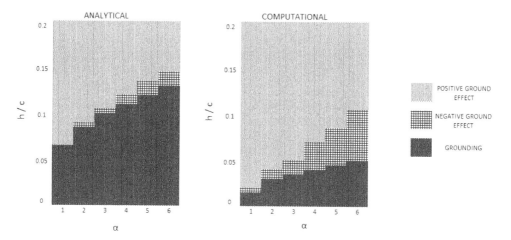

Figure 5.   The results of the analytical and computational approaches to specify the condition of the flat plate.

Figure 6.   The solution domain and the boundary conditions.

Solution domain, which is schematically shown in Figure 6, was created large enough to capture all of the pressure and velocity variations in the flow. Inlet boundary was placed 3 chord lengths (*c*) from the leading edge of the wing and outlet boundary was placed 7 chord lengths from the trailing edge. Top boundary is placed 3 chord lengths from the midline of the wing, while the position of the bottom of the domain is changing with the varying ground clearances. Spanwise dimension of the domain is one chord length, which is equal to the span dimension of the airfoil. Symmetry boundary condition was applied to the side walls of the domain and, together with the end-to-end placement of the wing, this condition satisfies 2D flow condition. Uniform velocity profile was imposed through the inlet and top boundaries along the x-axis. No-slip boundary condition was applied on the wing surfaces and bottom boundary.

For all cases investigated in this study, angle of attack (α) was measured as the angle between the midline of the wing and the bottom boundary of the domain. Ground clearances were measured between the trailing edge of the wing and bottom boundary and were denoted by *h*. Computations were repeated for varying ground clearances (h/c) from 0.1 to 0.75.

Computational grid was created using structural hexahedral elements. Detailed view of the

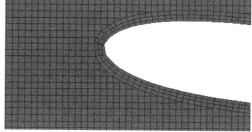

Figure 7.   The grid structure in the vicinity of the leading edge of the airfoil.

grid structure around the wing can be seen in Figure 7. A refined grid system was used in the near field of the wing to capture the strong velocity gradients. Also a second refinement on the grid was implemented on the leading and trailing edges. Prism layers were used on the wing surface and the bottom boundary of the domain to clearly identify the flow characteristics near the walls. Wall $y+$ on the wall boundaries were kept above the value of 30, for the proper use of wall function approach. Here, wall $y+$ was defined as $y+ = uy/v$, where u is the stream wise velocity, y is the wall distance of the first grid and v is the kinematic viscosity. Total number of elements in the fluid domain varied from $4.5*10^5$ to $7.15*10^5$, depending on the clearance of the wing from ground.

Unsteady Reynolds Averaged Navier Stokes Equations (U-RANS) was solved to simulate the flow field using finite volume discretization. To capture the effect of turbulence, realizable $k - \varepsilon$ turbulence model was used.

413

5.2 Numerical validation

Figure 8 shows the numerical versus experimental results of NACA0015 foil at different angle of attack conditions. The clearances are varied from 0.1 to 0.8 and it is of direct notice that numerically obtained lift coefficient values are somewhat higher than experiments (for 6 and 12 degrees cases). This is due to the two-dimensional flow approach which is adopted in numerical simulations. In the experiments of (Luo and Chen, 2012), the wing is not lying from end-to-end in the wind tunnel, which leads to flow escaping from the tips of the foil. Tip flow leads to three-dimensionality of the flow and considerably reduces the effective lift on the foil (Kinaci, 2016). The 12° angle of attack condition was not simulated well enough; the results are far off and the trend of the curve does not agree with the experiments. An angle of attack of 12° is very high which might lead to the instability of the foil. At high angle of attack conditions, vortices that might shed from the foil will stimulate high turbulent flow which is not possible to be simulated numerically with steady state approach.

For 0 angle of attack condition, the experimental lift coefficient values are higher. The numerical simulations have successfully captured the general trend of the lift coefficient curve. The numerical results also suggest a high grounding risk between $0 < h / c < 0.5$. This is parallel to the mathematical derivations given in previous section. The grounding case was defined by equation (8).

The experimental results at 6° angle of attack show a local minima at around $h / c = 0.15$. This is actually a negative ground effect condition, which is defined by equation (12) and was not captured numerically. Although there is still lift being produced by the foil, it is less than the out of ground effect case.

5.3 Investigation of pressure distributions for positive and negative ground effect

In this section the pressure distributions along the foil are investigated to reveal what happens when the foil experiences positive and negative ground effect. First, an ordinary case of 6° angle of attack case was investigated. Then, the 0° angle of attack case, which was the main focus in this study, was touched upon. It can be seen from Figure 8 that positive and negative ground effects exist simultaneously at this case. $h / c = 0.1$ case creates positive lift while a clearance of $h / c = 0.15$ leads to negative lift.

The pressure coefficients obtained for three different ground clearances for the 6° angle of attack case are given in Figure 9. The figure shows higher pressure at the pressure side as the foil gets closer to the ground which is parallel to expectations. Due to the higher pressure at the pressure side, the foil will tend to go up and clear itself from the ground.

The pressure at the lower side of the foil is lower in higher ground clearances which is a sign of lower lift generation. When Figure 9 is observed in the light of Figure 8, the reason of lower lift will be better understood. In Figure 8, it can be seen that lift coefficient decreases with respect to increasing ground clearance for 6° angle of attack. The reason of this is highlighted by Figure 9.

The case for 0° angle of attack condition is different than 6°. Figure 10 shows the change in pressure coefficient along the NACA0015 foil at different ground clearances. In that figure, it can be observed that for $h / c = 0.15$ case, the pressure

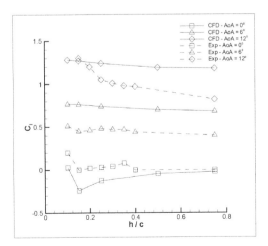

Figure 8. Lift coefficient values of NACA0015 foil at different AoA.

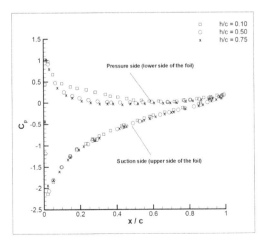

Figure 9. Pressure coefficients for NACA0015foil at 6° angle of attack at different ground clearances.

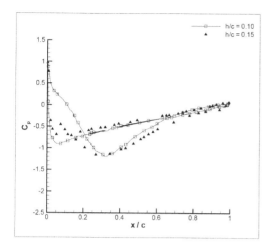

Figure 10. Pressure coefficients for NACA0015 foil at 0° angle of attack at different ground clearances.

is becoming lower at the lower side of the foil at around $x / c > 0.15$. Therefore, it can be said that 85% of the chord is generating negative lift while 15% is generating positive lift. The summation of all the pressures along the chord results in returning a negative lift, which is again highlighted by Figure 8. In Figure 8, it can be seen that the lift coefficient is negative at $h / c = 0.15$ which corresponds to a grounding risk defined by equation (8).

On the other hand, $h / c = 0.1$ case is generating positive lift as shown in Figure 8. The only part that is generating negative lift in this case is the portion of the foil between $0.2 < x / c < 0.65$ as given in Figure 10. The rest part of the foil covers for this portion and creates a positive lift for the foil in total.

6 DISCUSSION

The ground vicinity is a complicated phenomenon which needs careful examination. It does not always help a body to get off the ground and "float" in the fluid nor does it always lead to grounding as is the case in ship squat. The investigation of a simple flat plate reveals the intrinsic nature of a body in ground condition. Analytical and computational studies both reveal positive ground effect, negative ground effect and grounding conditions clearly.

7 CONCLUSIONS

Although the phrase "ground effect" usually evokes the idea of enhanced lift on a body, the actual ground effect may both work positively and negatively. It is not frequent to encounter the lift

reduction cases of the ground proximity because aerodynamicists usually want to benefit from the ground effect and enhance the lift on an airfoil. However, ship hydrodynamicists are aware of the grounding risk in shallow waters and therefore ground effect is closely investigated in ships that operate in rivers or coasts.

With these in mind, the different behaviors of ground proximity were investigated in this study. First; the cases of positive ground effect for a wing and grounding risk for a ship were clearly defined. Then; the analytical approach on a flat plate was explained and cases of positive ground effect, negative ground effect and grounding conditions were mathematically defined. A computational approach was also presented to observe if all these conditions could be captured numerically. Next; the forces on the flat plate were calculated using analytical and computational methods to reveal the perplexing behavior of the ground vicinity. A RANSE-based method was used to explain the underlying physics of the ground effect; from the perspective of the change in pressure in close proximity to the ground.

It was found out that different conditions that arise due to ground vicinity can be captured both analytically and computationally. The borders of each of these conditions were clearly drawn. Although the computational results differ from the analytical results, it was successful in capturing all the modes of a body experiencing ground condition.

ACKNOWLEDGEMENTS

This research has been supported by Yildiz Technical University Scientific Research Projects Coordination Department. Project No: 2013-10-01-KAP02.

REFERENCES

Ahmed, M.R., Takasaki, T. & Kohama, Y., 2007. Aerodynamics of a NACA4412 Airfoil in Ground Effect, AIAA Journal, 45(1), pp. 37–47.

Anthony, M.M., Scott, A.M., James, R.F. & Russell, M.C., 2006. Analysis of Delta-Wing Vortical Substructures Using Detached-Eddy Simulation, AIAA Journal, 44(5), pp. 964–972.

Barber T., 2006. Aerodynamic ground effect: a case study of the integration of CFD and experiments, Int. J. Vehicle Design, 40(4).

Chih-Min H. & Cha'o-Kuang C., 1996. Aerodynamic characteristics of a two-dimensional airfoil with ground effect, Journal of Aircraft, 33(2), pp. 386–392.

Doig, G., Barber, T.J. & Neely, A.J., 2011. The influence of compressibility on the aerodynamics of an inverted wing in ground effect, J. Fluids Eng, 133(6), 061102.

Jung, K.H., Chun, H.H. & Kim, H.J., 2008. Experimental investigation of wing-in-ground effect with a

NACA6409section. Journal of Marine Science and Technology, 13(4), pp.317–327.

Katz, A., Plotkin, J., 1991. Low speed aerodynamics – from wing theory to panel methods. 1st ed. Singapore: Mc-Graw Hill Inc.

Kinaci, O.K., 2014. An iterative boundary element method for a wing-in-ground effect. International Journal of Naval Architecture and Ocean Engineering, Vol. 6, No. 2, pp. 282–296.

Kinaci, O.K., Lakka, S., Sun, H., Bernitsas, M.M., 2016. Effect of tip-flow on vortex induced vibration of circular cylinders for $Re < 1.2*10^5$. Ocean Engineering, Vol. 117, pp. 130–142.

Liang, H & Zong, Z., 2011. A subsonic lifting surface theory for wing-in-ground effect. ActaMechanica, 219(3–4), pp.203–217.

Liang, H., Zhou, L., Zong, Z. & Sun, L., 2013. An analytical investigation of two-dimensional and three-dimensional biplanes operating in the vicinity of a free

surface. Journal of Marine Science and Technology, 18(1), pp.12–31.

Luo, S. C. & Chen, Y. S., 2012. Ground effect on flow past a wing with a NACA0015 cross-section. Experimental and Thermal Fluid Science, Vol. 40, pp. 18–28.

Nishino, T., Roberts, G.T. & Zhang, X., 2008. Unsteady RANS and detached-eddy simulations of flow around a circular cylinder in ground effect. Journal of Fluids and Structures, 24, pp. 18–33.

Phillips, W.F & Hunsaker, D.F., 2013. Lifting-line predictions for induced drag and lift in ground effect. Journal of Aircraft, 50(4), pp.1226–1233.

YING C., YANG W. & YANG Z., 2010. 3D numerical simulation on reverse forward swept wing-in-ground effect. Computer Aided Engineering, 2010–03.

Zong, Z., Liang, H. & Zhou, L., 2012. Lifting line theory for wing-in-ground effect in proximity to a free surface. Journal of Engineering Mathematics, 74, 143.

Maritime Transportation and Harvesting of Sea Resources – Guedes Soares & Teixeira (Eds)
© 2018 Taylor & Francis Group, London, ISBN 978-0-8153-7993-5

Effect of chordwise flexibility on flapping foil thruster performance by discrete vortex method

A.K. Priovolos, E.S. Filippas & K.A. Belibassakis
School of Naval Architecture and Marine Engineering, National Technical University of Athens, Greece

ABSTRACT: Recent research and development concerning flapping-wing propulsors has shown that such systems are able to achieve high levels of efficiency. Moreover, it has been demonstrated that chordwise flexibility further enhances their propulsive performance. Flapping foil biomimetic systems are also appropriate for other applications such as augmentation of ship overall propulsion by wave energy extraction and exploitation of wave and current renewable energy resources. In the present paper a hydroelastic model based on a Discrete Vortex Method (DVM) for the hydrodynamics, in conjunction with Kirchhoff plate theory equation for the flexural deflection, is used to study the effect of chordwise flexibility on the performance of flapping hydrofoil. The foil response is actuated by harmonic heaving motion and pitching about its leading edge. As a first approximation we assume that the thickness and transverse deflections are small compared to the chord length. Numerical results are presented concerning the thrust coefficient and the efficiency of the system over a range of design and operation parameters, including Strouhal number, heaving and pitching amplitudes, and flexural rigidity, indicating that chordwise flexibility can improve propulsive efficiency. The present method can serve as a useful tool for assessment and the preliminary design and control of such biomimetic systems for marine propulsion.

1 INTRODUCTION

It has been long assumed among the scientific and engineering community that animal propulsion mechanisms are very efficient, owing to millions of years of evolution. In this direction, many experiments, theoretical approaches and numerical models have been devised during the recent years, studying the efficiency and thrust production ability of flapping foil propulsors, which are inspired by aquatic animals' swimming mechanisms, performing a combination of heaving and pitching motion. The results show evidence that flapping foil thrusters are indeed able to achieve high efficiencies, simultaneously offering desirable levels of thrust forces, see Triantafyllou et al. (1998), Bose (1992), Belibassakis et al. (1997). Flapping foil configurations have been investigated as both main and auxiliary propulsion devices, either located below a ship's stern instead of a conventional propeller or placed under the ship's bow, taking advantage of the large motion amplitudes at forward stations due to the ship's response to sea waves (Belibassakis & Politis, 2013, Filippas & Belibassakis 2014)

A main component of fishes' swimming mechanisms is the flexural ability of their tails. This has attracted the researchers' interest, with works attempting to tackle the coupled hydroelastic problem of a flexible foil in unsteady flow. Wu (1961) considered a foil with prescribed chordwise

deformation. Katz & Weihs (1978) studied the coupled hydroelastic problem for slender unsteady propulsors for large motion amplitudes with small local angles of attack, assuming inviscid flow. Their results suggested that the efficiency of the propulsor is increased with increasing flexibility, although the thrust coefficient is decreased, thus rendering extreme flexibility of no practical value from an engineering point of view. Yamaguchi & Bose (1994) applied a vortex lattice method with nonlinear corrections to account for viscous and 3D effects for a flexible flapping foil. The open water efficiency of the flapping wing propulsor was up to 25% higher than that of a conventional screw propeller, although at their operation behind the ship and into its wake the efficiency of the flapping foil propulsor is 3–4% less than that of the screw propeller. Adding flexibility to the flapping foil, its efficiency behind the ship became 5% greater than that of the conventional screw propeller.

The efforts to tackle the problem have substantially increased in number in recent years. Alben (2008) proposed a potential theory coupled with a linear elastic sheet formulation to tackle the hydroelastic problem of a foil pitching with its pivot axis at the leading edge. Various peaks in thrust and efficiency were reported, corresponding to various non-dimensional rigidity parameters. De Sousa & Allen (2011) used a coupled fluid—structure solver for an elastic membrane in pitching motion.

They concluded that flexibility increases efficiency of a pitching wing, and that increasing its structural mass makes it more efficient. Barranyk et al. (2012) performed experiments with flat plates in flapping motion of varying rigid to flexible ratio and reported that increasing flexibility increased both thrust and efficiency in the parameter range tested. Prempraneerach & Triantafyllou (2013) performed experiments with chordwise flexible flapping foils in towing tanks. Propulsive efficiencies as high as 87%, up to 36% higher than those of rigid foils were recorded. Kancharala & Philen (2016) explored chordwise varying stiffness profiles, motivated by fish tails. They found that when stiffness is decreased towards the trailing edge the resulting force vector is better aligned with the motion direction, thus increasing thrust. Moreover, Paraz et al. (2016) studied the flexibility of heaving flat plates both experimentally and theoretically. They reported that the thrust displays peaks in motion frequency values coinciding with the resonance frequencies of the system comprising of the foil and the surrounding fluid. Their results also suggest that nonlinearities are certainly evident in the coupled system response. They devised an equation to model the elastic deformation based on standard Euler-Bernoulli beam models with nonlinear terms and coupled it with the unsteady flow with results from linear theory.

In the present work, a 2D flat plate is considered subject to flapping motion, pitching around its leading edge. It is flexible, in whole or in part, and is free to deform under inertia and reactive forces caused by its forced motion and hydrodynamic pressures formed around it. The foil is considered clamped at its leading edge, while its trailing edge acts as a free end, with the corresponding boundary conditions. A numerical model based Discrete Vortex Method (DVM) for the hydrodynamics, in conjunction with Kirchhoff plate theory is developed and applied to the hydrodynamic solution with the elastic deformations. Results are shown and compared with analytical theories, as well as numerical models found in literature and experiments.

1.1 Basic nomenclature

In this work a symmetric wing of infinite span travelling in a constant velocity and experiencing forced heaving and pitching motion around is considered.

The wing's thickness is assumed to be small compared to its chord length. The wing has finite chordwise flexural rigidity, therefore it passively deforms under the pressure variations generated along its chord. The surrounding fluid is assumed to be inviscid, and the resulting unsteady flow is irrational. These assumptions make the use of unsteady 2D thin foil theory acceptable. The foil

travels along the horizontal axis towards $-\infty$. Two reference frames are used; one that travels with the mean foil position, which is inertial, and a body-fixed one that is attached on the foil's leading edge.

The flow consists of two main components; a steady parallel flow $\mathbf{V}_\infty$ due to the constant travelling speed of the foil and an unsteady flow which, according to the assumptions stated above, can be represented by a scalar perturbation potential $\Phi(x,y;t)$. The unsteady velocities in the flow field are given by $\nabla\Phi(x,y;t)$. An unsteady background flow $\mathbf{V}_G$ can be included in the model provided that it is weakly rotational. A trailing vortex wake is formed behind the body due to the flow unsteadiness. If we define ∂D_B as the body surface and ∂D_W as the wake surface then $\partial D = \partial D_B \cup \partial D_W$. The positions of points $A \in \partial D_B$ are defined by $\mathbf{r}(s;t) = [x(s;t), y(s;t)]^T \in \partial D_B$, where s is a curvilinear coordinate. The pressure difference between the two sides of the foil's surface produces a transverse deflection, $\eta(\mathbf{r};t), \mathbf{r} \in \partial D_B$ defined on the body-fixed coordinate system. A sketch of the foil and the motion parameters is shown in Figure 1. The disturbance potential satisfies Laplace equation

$$\nabla^2 \Phi = 0, \tag{1}$$

and the no-entrance (Neumann) condition

$$\nabla\Phi \cdot \mathbf{n} = (\mathbf{V}_A - \mathbf{V}_\infty - \mathbf{V}_G) \cdot \mathbf{n}, \tag{2}$$

where $\mathbf{n}$ is a unit vector normal to the foil's surface and $\mathbf{V}_A = \partial_t \mathbf{r}(s,t)$ is the velocity of the surface points in the inertial reference frame. The unit normal vector and surface velocity are affected by both the forced flapping motion of the foil and its elastic deformation. The foil's surface and wake are considered as surfaces of potential discontinuity, leading to discontinuity of tangential velocities. We define the vorticity as the difference in tangential velocity

$$\gamma(s;t) = u_L(s;t) - u_U(s;t) = \nabla[\![\varphi]\!] \cdot \tau, \tag{3}$$

where the subscripts L, U correspond to the lower and upper sides of the surface, respectively, and τ is a unit vector tangent to the foil's surface. The difference of the trace potential between the lower and upper surfaces of the foil is denoted by $[\![\varphi]\!]$. The foil is represented as a zero thickness potential discontinuity surface with a distribution of discrete vortices

$$\Gamma_{Bi}(s;t) = \Gamma_B(s_i;t), i = 1\cdots N_B \in \mathbb{N}, \tag{4}$$

equivalent to a continuous distribution of bound vorticity in the sense that

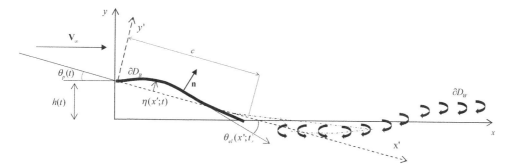

Figure 1. Foil and motion parameters sketch. The moving inertial and body fixed frames are denoted by x,y (solid lines) and x',y' (dashed lines), respectively. The foil deformation $\eta(x';t)$ is exaggerated for clarity. The pitch angle of the foil is decomposed in the forcing component $\theta_p(t)$ and the deformation component $\theta_{el}(x';t)$. The trailing wake is drawn in dash-dot line and wake point vortices are denoted by curved arrows. While only the structural response—induced camber is shown here, the foil may possess an arbitrary camber line.

$$\Gamma_{Bi} = \int_{s_i}^{s_{i+1}} \gamma_B(s;t)ds, i=1,...,N_B \qquad (5)$$

where $\gamma_B(s;t)$ is the distribution of bound vorticity as defined above in (3). Due to the vortex shedding from the trailing edge, the trailing wake has a vorticity distribution

$$\gamma_W(s;t), \mathbf{r}(s;t) \in \partial D_W \qquad (6)$$

This distribution is also approximated by an equivalent distribution of discrete vortices

$$\Gamma_W(s_j;t), \mathbf{r}(s_j;t) \in \partial D_W, j=1 \cdots N_W, \qquad (7)$$

defined in a similar fashion as in Eq.(5).

1.2 Solution of the lifting problem around the foil

Assuming a vorticity distribution on the surface of potential discontinuity ∂D the velocities induced on a point $\mathbf{r}(s;t) \in \partial D$ is given by the expression

$$\nabla \Phi \cdot \mathbf{n} = -\frac{1}{2\pi} \int_{\partial D_B} \frac{\gamma_B(\xi;t)}{|\mathbf{r}(\xi|s;t)|^2} \{\mathbf{k} \times \mathbf{r}(\xi|s;t)\} \cdot \mathbf{n} d\xi$$
$$-\frac{1}{2\pi} \int_{\partial D_W} \frac{\gamma_W(\xi;t)}{|\mathbf{r}(\xi|s;t)|^2} \{\mathbf{k} \times \mathbf{r}(\xi|s;t)\} \cdot \mathbf{n} d\xi. \qquad (8)$$

The first integral in Eq. (8) has to be interpreted in a Cauchy principal value sense, due to the fact that the kernel is singular. The relative positions of points on the surface are denoted by $\mathbf{r}(\xi|s;t)$. The unit vector $\mathbf{k}$ is normal to the x-y plane, according to the right-hand rule. Plugging (8) in the no-entrance boundary condition (2) and replacing the vorticity distributions by their equivalent discrete

vortices distributions we acquire a linear system to be solved for the intensities of the bound vortices. Defining a grid of point vortex points and a grid of collocation points on the foil's surface where the no-entrance boundary condition is imposed the above equation becomes (omitting the intermediate steps needed for its derivation)

$$\sum_{j=1}^{N_B} \Gamma_{Bj} \cdot w^B{}_{ij} = -\sum_{k=1}^{N_W} \Gamma_{Wk} w^W{}_{ik} + U_{Bi}, i=1\cdots N_B, \qquad (9)$$

which is solved at every time step t_n for the bound vortices' intensities. The second term in the RHS of (9) is the discrete equivalent of the RHS of (2). The velocities normal to the i – th foil collocation point induced from the j – th foil point vortex and the k – th wake point vortex are $w^B{}_{ij}$, $w^W{}_{ik}$. Each panel on the foil's surface contains a vortex point and a collocation point, in a proper distance between them, as discussed below in more detail. It is obvious from (5) that the bound vortices' intensities fully describe the lifting problem, as

$$\sum_{i=1}^{N_B} \Gamma_{Bi}(t_n) = \Gamma(t_n), \qquad (10)$$

that is, the sum of the discrete bound vortices gives the circulation around the foil at each time step. This method represents the flow as a superposition of bound and wake point vortices is a Discrete Vortex Method, which will be referred to as DVM in the course of this work (Katz & Plotkin, 1991).

1.3 Kutta condition—wake model

In the present work the trailing wake is assumed to be frozen, i.e its geometry is a byproduct of the foil kinematics only (Hess, 1972). The wake vorticity evolution is governed by the equation

$$\gamma_W(s;t) = \gamma_W(s_{TE}; t - t_{TE}),$$ (11)

where $t_{TE} = U^{-1}|\mathbf{r}(s|s_{TE};t)|$ and s_{TE} is the value of the curvilinear coordinate such that $\mathbf{r}(s_{TE}, t - t_{TE}) \in \partial D_W$ coincides with the foil's trailing edge. The quantity t_{TE} is the time duration for a material point on ∂D_W to travel from the trailing edge to its current position with the free-stream velocity in the inertial reference frame. Eq. (11) essentially states that the vorticity carried by points on the wake surface is the same as the vorticity that they had when they were shed at the trailing edge. This implies that the vorticity is a materially conserved quantity. This condition imposes continuity of the vorticity function in ∂D. It is also compliant with Kelvin's theorem, stating that the total circulation in the domain is conserved. In the present method Eq. (11) acts as a Kutta condition. In the fully linearized case (11) assumes the simpler form of a wavelike function $\gamma_W(x;t) = \gamma_W(x - Ut)$ which is the solution to a wave equation imposing zero pressure difference everywhere on the wake, resulting in zero pressure difference at the trailing edge (Newman, 1977).

Applying (11) to the DVM, the intensities and positions of wake vortices $\Gamma_{Wk}, 2 \le k \le N_W$ are known, since their intensity is equal to the shed vorticity at previous time steps integrated over the wake panel's length. However, the first wake vortex intensity is unknown at every time step. Invoking Kelvin's theorem, the first wake vortex has to compensate for the circulation change around the foil, thus its intensity is given by

$$\Gamma_{W1} = \Gamma(t_{n+1}) - \Gamma(t_n)$$ (12)

The first term in the RHS of the above relation is the circulation around the foil at the previous time step, hence known. The second term is the circulation around the foil at the current time step, currently unknown but equal to the sum of the unknown bound vortices in the current time step. This allows its incorporation in the LHS of the linear system. Transferring the unknown term to the LHS, the linear system to be solved for the bound vortices' intensities becomes

$$\sum_{j=1}^{N_B} \Gamma_{Bj} \cdot \left(w^B{}_{ij} - w^W{}_{i1}\right) = -\Gamma(t_{n-1}) \cdot w^W{}_{i1}$$
$$- \sum_{k=2}^{N_W} \Gamma_{Wk} \cdot w^W{}_{ik} + w_{Bi}.$$ (13)

Relation (13) is the linear system that gives the intensities of the bound vortices $\Gamma_{Bj}, j = 1,...,N_B$.

1.4 Pressure and forces calculation around body

The pressure difference coefficient on the foil is defined as $\Delta C_P = [\![C_P]\!] \triangleq (p_L - p_U)/0.5\rho U^2$ where p denotes the absolute pressure, U is the magnitude of the free stream velocity and ρ is the fluid density. The value of ΔC_P is calculated with Bernulli equation applied between points $\mathbf{r}_L(s;t)$ and $\mathbf{r}_U(s;t)$ on the foil's surface, where the subscripts L, U denote the lower and upper side, respectively. The expression used for its calculation is

$$\Delta C_P = \frac{2}{U^2}\left[-\frac{d[\![\varphi]\!]}{dt} + \nabla[\![\varphi]\!] \cdot (\mathbf{V}_A - \mathbf{V}_\infty - \mathbf{V}_G)\right],$$ (14)

It is important to note that (14) applies to time derivatives and gradients calculated by DVM in the body-fixed frame reference. Lift, thrust and foil moment coefficients are given by the following expressions, respectively,

$$C_L \triangleq \frac{L}{0.5\rho U^2 c} = \frac{1}{c}\int_{\partial D_B}(\Delta C_P \cdot \mathbf{n}) \cdot \hat{\mathbf{y}}ds,$$ (15)

$$C_T \triangleq \frac{T}{0.5\rho U^2 c} = -\frac{1}{c}\int_{\partial D_B}(\Delta C_P \cdot \mathbf{n}) \cdot \hat{\mathbf{x}}ds,$$ (16)

$$C_M \triangleq \frac{M}{0.5\rho U^2 c^2} = \frac{1}{c^2}\int_{\partial D_B}(\Delta C_P \cdot \mathbf{n}) \cdot \mathbf{r}(s|s^*;t)ds,$$ (17)

where $\mathbf{r}(s|s^*; t)$ is the vector pointing from the reference point (where moments are calculated) to the surface point $\mathbf{r}(s; t)$, and $\hat{\mathbf{x}}$ and $\hat{\mathbf{y}}$ are the orthonormal unit vectors of the inertial reference frame. Moreover, the input power coefficient is calculated by

$$C_P \triangleq \frac{P}{0.5\rho U^3 c} = -\frac{1}{Uc}\int_{\partial D_B}(\Delta C_P \cdot \mathbf{n}) \cdot \mathbf{V}_A(s;t)ds.$$ (18)

1.5 Solid model

The foil's deformation is modelled as that of a plate of large aspect ratio ignoring spanwise deformations. The deformations are assumed to be constrained on the $x' - y'$ plane. In the following relations, this body-fixed reference frame will be denoted as x, y for simplicity. Thus a two-dimensional model can be used. The equation that describes the plate's dynamic response is (e.g., Graf 1975).

$$m\partial_t^2\eta + C\partial_t\eta + D\partial_x^4\eta + \mu\partial_t\partial_x^4\eta + C_D\partial_t\eta|\partial_t\eta|$$
$$= f(x, \eta; t),$$ (19)

where $\eta(x;t)$ is the transverse deflection of the plate in its body-fixed system. This model complies with the Kirchhoff plate theory after ignoring strain in the spanwise direction. The first and third terms on the LHS of (19) are the inertia and elastic restoring forces, respectively. The second term is a linear damping term, and the fourth term is a viscoelastic damping term, according to the Kelvin-Voigt theory (Banks & Inman, 1989). The fifth term is a non-linear drag term, found to give results agreeing with experimental data for appropriate values of c_D (Paraz et al, 2015). The RHS term is the loading resulting from the coupling of the unsteady flow with the foil's dynamic response along with terms that manifest after changing the coordinate system from inertial to body-fixed. The flexural rigidity of the plate is $D = Et^3/12(1-v^2)$ where E is the Young's Modulus of the foil material, t is the foil thickness and v is the material's Poisson ratio. The foil is assumed to be of constant chordwise thickness, in accordance with the assumption of thin flat plate in the hydrodynamics part. However, the mass, stiffness and damping contribution are not constrained to be constant.

The viscoelastic coefficient μ and the non-linear drag coefficient c_D cannot be known on any physical basis (Adhikari, 2000). Their values are usually determined by experiments. The linear damping coefficient is calculated using Rayleigh's method (Chowdhury & Dasgupta, 2003).

The foil is clamped on the leading edge (or any other chordwise position) while the trailing edge is free. The boundary conditions therefore are

$$\eta|_{x=0}=0, \partial_x\eta|_{x=0}=0, \text{ clamped end,} \tag{20}$$

and

$$\partial_x\left(D\partial_x^2\eta\right)|_{x=c} +\mu\partial_t\left(\partial_x^3\eta\right)|_{x=c}=0,$$
$$D\partial_x^2\eta\,|_{x=c} +\mu\partial_t\left(\partial_x^2\eta\right)|_{x=c}=0, \text{ free end.} \tag{21}$$

The RHS of Eq.(19) is given by the expression

$$f(x,\eta,t)=\frac{1}{2}\rho U^2 \cdot \Delta C_P - m\left(\mathbf{a}_{rig}\cdot\mathbf{n}\right), \tag{22}$$

where $\mathbf{a}_{rig}$ is the acceleration of the foil's points in the inertia frame due to their rigid motion. The first term corresponds to the hydrodynamic pressure component while the second term arises due to the fact that the reference frame where (19) is applied is non-inertial.

1.6 Solution of coupled hydroelastic problem

The plate deformation model assumed allows no deflection in the chordwise direction, thus the position of collocation and point vortex points are always at the same chordwise position in the foil's body-fixed reference frame. The chord elements are distributed using the cosine spacing method, thus making the solution grid denser at the leading and trailing edge regions where the most significant variations in hydrodynamic data is expected.

Discretizing the foil into discrete panels, a point vortex point and a collocation point are placed at the positions 1/4 and 3/4 of its length, respectively. The significance of this configuration is extensively discussed in various works (James, 1971). As a result no collocation points are placed exactly on either the leading or the trailing edge.

Writing Eq. (19) as a first order system, after setting $\partial_t\eta=u$ and discretizing the equations, we have

$$\begin{bmatrix}0 & M \\ I & 0\end{bmatrix}\begin{bmatrix}\dot{\eta} \\ \dot{u}\end{bmatrix}+\begin{bmatrix}K & C \\ 0 & -I\end{bmatrix}\begin{bmatrix}\eta \\ u\end{bmatrix}=\begin{bmatrix}F \\ 0\end{bmatrix}. \tag{23}$$

The foil's transverse deflection is calculated at the collocation points of the hydrodynamic lifting problem. The deflections are given by the vector η and the corresponding response velocities by the vector $\mathbf{u}$. Utilizing a non-uniform finite difference scheme the spatial derivatives of the deflection are approximated by linear combinations of the deflection values in adjacent grid points. The elements of the matrices in (23) are $N_B \times N_B$ square matrices. Individually, $\mathbf{M}$ is the mass matrix, $\mathbf{K} = D\cdot\mathbf{FD}_4$ is the stiffness matrix, where $\mathbf{FD}_4$ is a matrix containing the 4th-derivative stencil in the collocation points grid, $\mathbf{C} = \mathbf{RD} + \mu\mathbf{FD}_4$ is the damping matrix containing both external linear and linear viscoelastic damping, where $\mathbf{RD}$ is the Rayleigh damping matrix, and $\mathbf{I}$ is the $N_B \times N_B$ identity matrix. The loading term $\mathbf{F}$ on the RHS of Eq. (23) is given by the expression

$$\mathbf{F} = \mathbf{f}(\eta,\mathbf{u};t) - \mathbf{C_D}\cdot\mathrm{diag}\left(|\mathbf{u}|\cdot\mathbf{u}^I\right), \tag{24}$$

where $\mathbf{C_D}$ is the non-linear damping matrix. The boundary conditions Eqs. (20), (21) are incorporated in the system of ODE's (23). This is a system of non-linear equations which is solved for the vector $\mathbf{w} = [\eta,\mathbf{u}]^I$ at each discrete time t_n based on the results of the previous time step t_{n-1}. The time integration of (23) is performed with an implicit Crank-Nicolson numerical method, coupled with an iteration scheme to treat the non-linear parts of the equations. The solution is implemented as follows:

i. Set $\mathbf{w}_1 = [\eta_1,\mathbf{u}_1]^I = 0$, solve the hydrodynamic problem with DVM (15) and calculate the pressure difference between the foil's surface upper and lower sides (14) for $n = 1$.

ii. At each time step, $n > 1$, the hydrodynamic problem is solved by initially setting $\mathbf{w}_n = \mathbf{w}_{n-1}$. The trailing wake point vortices are convected downstream by a distance equal to $U\Delta t$ towards the direction of the free stream velocity vector $\mathbf{V}_\infty$. After obtaining the bound vortices $\Gamma_{B_i}(t_n)$ and pressure difference coefficient ΔC_P an implicit time integration scheme is employed to approximate the solution, thus $\mathbf{w}_n^0$ is obtained. This initial approximation is used in the execution of an iterative Newton-Raphson scheme converging to a vector $\mathbf{w}_n^{m_n}$. Set $\mathbf{w}_n = \mathbf{w}_n^{m_n}$.

iii. When the time stepping algorithm is over, lift, thrust, moment and input power coefficients are calculated according to Eqs. (15)–(18).

2 RIGID CASE RESULTS

For validation purposes the present DVM algorithm is used to reproduce results and compare with unsteady hydrofoil theory and other methods, and against experimental data.

2.1 Flapping motion

The foil is subject to sinusoidal heaving and pitching motion according to the following relations

$$h(t) = h_0 \sin(\omega t), \quad \theta(t) = \theta_0 \sin(\omega t + \psi) \qquad (25)$$

where ψ is the phase difference between the two motions. It has been repeatedly reported in the literature that the value of ψ for optimum thrust production is around 90° (Schouveiler et al, 2005).

The most important parameters affecting the performance of the flapping foil are:

- Heaving amplitude to chord ratio, h_0/c
- Strouhal number $St = fA/U$, where $A = 2h_0$ and f is the flapping frequency.
- Reduced Frequency $k = \omega c/2U$
- Effective angle of attack amplitude $a_0 = \tan^{-1}(\pi St) - \theta_0$.
- Pivot axis chordwise position (non-dimensional) x_R.

2.2 Background gust velocity

In this case, the foil is subject to a parallel potential flow field and also encounters a slightly rotational background transverse sinusoidal gust velocity. The magnitude of this velocity is given by the expression

$$u_G(x;t) = V_G \cdot \cos(g_k x - \omega t), \quad g_k = \frac{2k}{c}, \qquad (26)$$

where k is defined as in the previous section and g_k is the wavelength. This implies that the sinusoidal gust is steady with respect to the free stream.

2.3 Numerical results for rigid foil

In Fig. 2 the DVM is applied to three cases: heaving motion, pitching motion and steady motion in a background sinusoidal gust for a range of reduced frequency values. The results for the lift coefficient $C_L = L/0.5\rho U^2 c$ are compared with analytic theory, expressed through Theodorsen function for heaving and pitching and Sears function for the gust problem (Newman, 1977), including added mass effects. The motion amplitudes are $h_0/c = 0.05$ for the heaving motion, $\theta_0 = 1^0$ for the pitching motion and the sinusoidal gust velocity amplitude is $V_G/U = 0.05$. It is evident that the present method is consistent with the linear theory in the range tested. As the reduced frequency approaches zero, all curves tend to reach the steady state result, $C_L/\alpha_0 = 2\pi$ where α_0 is the maximum angle of attack for each case; $\alpha_0 = \tan^{-1}(\pi St)$ for heaving motion, $\alpha_0 = \theta_0$ for pitching motion and $\alpha_0 = \tan^{-1}(V_G/U)$ for the background sinusoidal gust.

In Fig. 3, the results of present DVM are compared against the experimental results by Read et al. (2003). Systematic runs are performed for Strouhal numbers 0.20, 0.25, 0.30, 0.35, 0.40, 0.45, $h_0/c = 1$, $x_R = 1/3$ and $\psi = 90°$, with varying pitching motion amplitude. More specifically, results for the thrust coefficient $C_T = T/0.5\rho U^2 c$ are presented. The agreement is good for moderate to large effective angles of attack (shown in dotted contours), up to about 25° and Strouhal numbers up to 0.40, approximately. For large effective angles of attack (upper left region of Fig. 3) the experimental values of the thrust coefficient are larger than those of the DVM for $St > 0.3$ mainly due to the fact that the foil section used was a NACA 0012, possessing

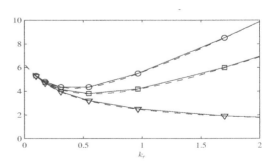

Figure 2. C_L/α_0 for pure heaving motion (squares), pure pitching motion (circles) and background sinusoidal gust (triangles), for small amplitudes. Comparison between DVM (solid line) and linear theory (dashed line).

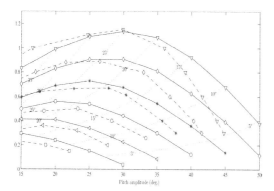

Figure 3. Comparison of thrust coefficient C_T (as function of Strouhal number St and pitching amplitude θ_0) against experimental data from Hover et al (2003). Solid lines: DVM, Dashed Lines: Experiments. Strouhal number is varied from 0.20 to 0.45.

considerably better hydrodynamic attributes that the flat plate considered in the DVM. On the other hand, the performance of the NACA foil in the experiments is significantly worse for small values of the effective angle of attack (lower right region of Fig. 3) as the Strouhal number increases. This has been attributed to the degraded angle of attack time signal, possessing many peaks in a period of the motion, as has been stated by Read et al. (2003).

3 FLEXIBLE FOIL RESULTS

Numerical simulations have been performed in order to study the effect of chordwise flexibility in flapping foils. For the validation of the present hydroelastic method the case of a heaving flexible plate studied by Paraz et al. (2016) is investigated. The body motion is enforced at the clamped leading edge while the trailing edge is free. The main purpose of this study is to validate the aforementioned authors' conclusion that the relative response amplitude of the trailing edge with respect to the amplitude of the harmonic motion is not constant for all excitation amplitudes. This is a manifestation of non-linearities in the coupled system.

In Fig. 4 systematic results are shown for three values of heaving amplitude $h_0/c = 0.035$, $h_0/c = 0.085$, $h_0/c = 0.12$ for direct comparison with the results of Paraz et al. (2015). The chord length is equal to 12 cm, the free stream velocity 0.0612 m/s and the flexural rigidity equal to 0.027 Nm. The mass distribution of the plate is 4.8 kg/m per meter span. The results are presented over a range of non-dimensional frequencies determined by the elastic and geometric parameters of the foil. In this setup, ω_0 is the corresponding

resonance frequency of the coupled system (foil & surrounding fluid), acquired through a linear analysis, Paraz et al. (2016). In general, the resonance frequency of the coupled system is lower than that of the plate in vacuum due to added mass effects. The nonlinear drag coefficient has been adjusted so that it is equivalent to the value given by the authors, $c_D = 12$ as has the viscoelastic damping parameter μ. The linear external damping has been adjusted using Rayleigh's method.

In Fig. 4a the relative amplitude of the trailing with respect to the input heaving motion is presented. The amplitude displays the first peak at a frequency close to the resonance frequency of linear analysis for all three of the heaving amplitudes. It is evident that as the excitation amplitude increases the response amplitude of the trailing edge decreases, and the peak frequency moves to the left. A second resonance frequency is included in the frequency range tested. The corresponding peak is lower than the first but spans across a larger range. Experimental results are shown for $h_0/c = 0.035$ in white-faced squares. The agreement is satisfactory both qualitatively and quantitatively.

In Fig. 4b the phase lag of the trailing edge response with respect to the leading edge amplitude is shown. The curves for the three cases essentially collapse into a single one. The phase lag at the first resonance is close to 90°, an expected result. In Fig. 4c the thrust coefficient variation is shown. The observation made by many authors (Dewey et al., 2013; Moore, 2014) that the thrust coefficient displays a peak at the structural resonant frequencies is clearly reproduced here. The thrust force is non-dimensionalized by a characteristic elastic force, and the resulting coefficient is further divided by a characteristic foil non-dimensional velocity to properly collapse the results; see Paraz et al (2016). The thrust curves have the same qualitative characteristic with those of the trailing edge response amplitude in Fig. 4a. The thrust coefficient as defined here is maximized for frequencies close to the resonant frequency of the linear analysis, while it displays less significant peak for the second resonant frequency. The curve for $h_0/c = 0.035$ corresponds to a case explicitly tested by the authors both experimentally and analytically. The agreement with the experimental results is only qualitative, since viscous resistive forces which are reported to be especially significant in larger frequencies are ignored in our model. The results of the theoretical model of Paraz et al. are shown in dashed line for this case. They correspond only to the reactive forces, i.e due to pressure difference. Resistive forces (i.e drag forces) are ignored. The thrust predicted by the current method displays good agreement with the reactive forces as predicted by Paraz et al. In Fig. 4d

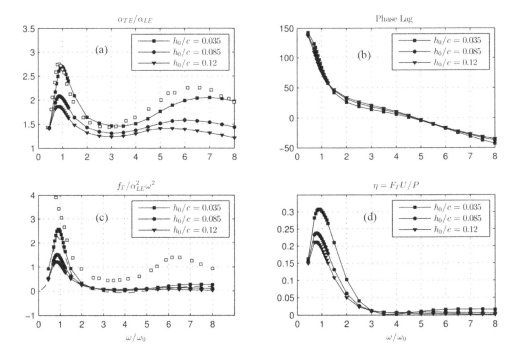

Figure 4. Numerical results for flat plate in heaving motion, clamped at its leading edge (Paraz et al, 2016). The excitation frequency is continuously varied from $0.4\omega_0$ to $8\omega_0$, where ω_0 is the first non-dimensional resonant frequency of the coupled system. Present method results ae shown by using lines and symbols. In subplot (c), the experimental data are shown by using squares, while theoretical results of Paraz et al. (2016) are indicated by dashed line.

the Froude efficiency is shown. It is maximized for all heaving amplitudes near the first resonant frequency and the corresponding efficiencies found with the hydroelastic coupling numerical model are: 0.31 for $h_0/c = 0.035$, 0.24 for $h_0/c = 0.085$ and 0.21 for $h_0/c = 0.12$. A peak is present for the second resonant frequency, however it is not interesting for propulsive applications since it is very small owing to the very large values attained by the Strouhal number: $St = 0.67$ for $h_0/c = 0.035$, $St = 1.25$ for $h_0/c = 0.085$ and $St = 1.94$ for $h_0/c = 0.12$.

The hydroelastic coupling numerical model is further used to study the effect of chordwise flexibility in a flapping foil, i.e a foil performing both heaving and pitching motions. The setup used is the one studied experimentally by Barranyk et al. (2012). Three cases are studied in total, each with different chordwise stiffness. The foils composed of two parts, a rigid forward one and a flexible rear. In each case the length ratio of these parts is varied: a) 100% rigid b) 50% rigid and c) 15% rigid.

The total chord length is equal to 0.2 m and the free stream velocity is equal to 0.22 m/s for all cases. According to the authors, the flexible parts were made of PDMS material which has density equal to 1200 kg/m³. The foil thickness is equal to 16 mm, so the surface mass density is equal to 19.2 kg/m per meter span. Its Young modulus, although not explicitly given, can be approximated by the natural frequencies of the structure in air as given by the authors, where added mass effects are negligible compared to those in water. We have found that its value is about $3.25 \cdot 10^6$ N/m².

The heaving amplitude is set to 0.08 m which results in the non-dimensional parameter $h_0/c = 0.4$ and the pitching motion amplitude is 8°. The phase lag between the two motions is $\psi = 90°$. The Strouhal number is varied between 0.25 and 0.45 for all three cases, where the experimental results of Barranyk et al. where purely thrust-producing. At lower Strouhal numbers they reported negative thrust and efficiencies which the present method is unable to predict. However the overall trends are correctly reproduced. In order to simulate the abruptly changing flexibility along the composite foil's chord, the clamped end is moved appropriately downstream while the forward part is considered absolutely rigid, acting only as a boundary as far as the plate's response solution is concerned. The results are presented in Fig. 5.

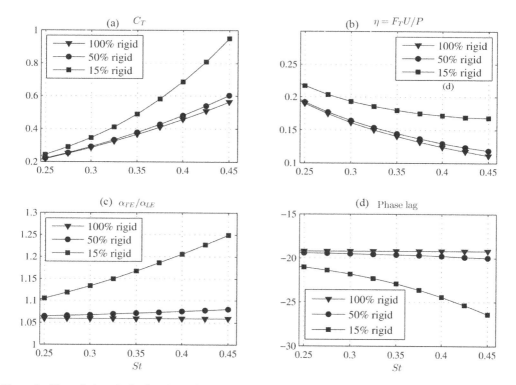

Figure 5. Numerical results for flat plate in flapping motion around its leading edge. The parameters are those of Barranyk et al. (2012): $c = 0.2$ m, $U = 0.22$ m/s, $h_0 = 0.08$ m, $\theta_0 = 8°$. The effective chordwise stiffness of the foil is controlled by modifying the ratio of the rigid to flexible parts of the composite foil.

In particular, in Fig. 5a the thrust coefficient is shown for the three cases. As the Strouhal number increases, the thrust coefficient of the 15% rigid foil is consistently larger than those of both the 100% rigid and 50% rigid foils. The thrust coefficient of the 50% rigid foil effectively displays improved thrust performance over the rigid one only for the largest Strouhal numbers tested.

In Fig. 5b the Froude efficiency is presented for the three foils. The effect of chordwise flexibility is dramatic in the propulsive efficiency; for the 100% and 50% rigid foils the efficiency reaches a minimum of around 0.11 while their derivative is negative, implying that it further drops to larger Strouhal numbers. On the contrary, the efficiency curve for the 15% rigid foil seems to have reached a plateau while its minimum value is around 0.17. This can be attributed to the fact that in subsequent Strouhal numbers the driving frequency approaches the natural frequency of the coupled system of the foil surrounded by the fluid, which according to the authors is 2.41 Hz for the 15% rigid foil. The maximum driving frequency of this study is 0.62 Hz, so it is expected that the foil that will be affected by the resonance phenomenon first is the 15% rigid one.

In Fig. 5c the relative amplitude of the trailing edge with respect to the leading edge is presented for the two flexible foils. The 50% rigid foil shows no significant trailing edge response, with the maximum value of the relative trailing edge response being 1.07. The 15% rigid foil displays a significant trailing edge response, the maximum value calculated being 1.25. These results are consistent with basic considerations in conventional propellers where the thrust of the propulsion device scales like the area swept by it. In Fig. 5d the phase lag of the trailing edge with respect to the leading edge motion forced by the heaving amplitude is presented for the two flexible foils. The phase lag of the 50% rigid foil is near the phase lag of the 100% rigid case, due to the combination of heaving and pitching motion. The phase lag curve of the 15% rigid case is substantially different, indicating the significant effects of flexibility in this case.

As a final remark we note that, although our results agree qualitatively well with experimental observations, better agreement can be obtained, if various parameters as 3D lifting flow and viscous effects, submergence of the thruster and wall effects (all present in the experimental configuration) are taken into account, and this is left to be considered in future extensions of the present method.

4 CONCLUSIONS

A numerical model has been developed and applied to the solution of the coupled problem concerning a flexible foil thruster in unsteady flow conditions. The foil is assumed to be thin and symmetric and thus it is represented by a flat plate of infinitesimal thickness. The flow is assumed irrotational, and a wake model is used to satisfy Kelvin's theorem. The foil material is assumed to be isotropic and its response is obtained on the basis of Kirchoff plate theory enhanced by appropriate viscoelastic and drag terms. The hydrodynamic and structural response problems are coupled mainly through the hydrodynamic pressure on the foil, and the problem is solved by a Discrete Vortex Method, in conjunction with an implicit scheme for the time integration of the system. It is demonstrated that foil flexibility has beneficial effects on the developed thrust and performance, in conformity with other methods and experimental observations. The present method can be extended to include 3D lifting flow and viscosity effects, as leading edge separation. Moreover, variable, finite thickness can be included in both the hydrodynamic and the structural response part. Finally, a free wake model can be used, which is expected to be important, especially at higher Strouhal numbers.

ACKNOWLEDGEMENTS

Support of E. Filippas by Alexander S. Onassis Public Benefit Foundation scholarship is acknowledged.

REFERENCES

Adhikari, S. 2000. Damping Models for Structural Vibration. PhD Dissertation. Trinity College, Cambridge.
Alben S. 2008. Optimal flexibility of flapping appendage in an inviscid fluid. *J. Fluid Mech.* Vol. 614: pp. 355–380.
Anderson, J.M., Streitlien, K., Barrett, D.S. & Triantafyllou, M.S. 1997. Oscillating foils of high propulsive efficiency, *J. Fluid Mech.*, Vol. 360, pp. 41–72.
Banks, H.T & Inman, D.J. 1989. On Damping Mechanisms in Beams, *NASA Contractor Report 181904, ICASE 89 – 64.*
Barranyk, O., Buckham, B.J. & Oshkai, P. 2012. On performance of an oscillating plate underwater propulsion system with variable chordwise flexibility at different depths of submergence, *J. Fluids and Struct.*, Vol. 28, pp. 152–166.
Belibassakis, K.A. & Politis, G.K. 2013. Hydrodynamic performance of flapping wings for augmenting ship propulsion in waves, *Ocean Engineering*, Vol. 72, pp. 227–240.
Belibassakis, K.A., Politis, G.K. & Triantafyllou, M.S. 1997. Application of the VLM to the propulsive performance of a pair of oscillating wing—tails, *Transactions on Modelling and Simulation*, Vol. 16, pp. 449–458.
Bose, N. 1992. A time—domain panel method for analysis of foils in unsteady motion as oscillating propulsors, *11th Australian Fluid Mechanics Conference, University of Tasmania, Hobart, Australia.*
Chowdhury, I & Dasgupta S. 2003. Computation of Rayleigh Damping Coefficients, *Electronic J. Geotechnical Eng.*
Dewey, P.A., Boschitsch, B.M., Moored, K.W., Stone, H.A. & Smits, A.J. 2013. Scaling laws for the thrust production of flexible pitching panels, *J. Fluid Mech.*, Vol. 732, pp. 29–46.
De Sousa, P.F. & Allen, J.J. 2011. Thrust efficiency of harmonically oscillating flexible flat plates, *J. Fluid Mech.*, Vol 674, pp. 43–66.
Filippas, E.S. & Belibassakis, K.A. 2014. Hydrodynamic analysis of flapping—foil thrusters operating beneath the free surface and in waves, *Engineering Analysis with Boundary Elements*, Vol. 41, pp. 47–59.
Graf K., 1975. *Wave motion in elastic solids*, Dover.
Hess, J.L. 1972. Calculation of potential flow about arbitrary three—dimensional lifting bodies, *Naval Air Systems Command, Report No. MDC J5679–01.*
Hover F.S., Read D.A., Triantafyllou M.S. 2002. Forces on oscillating foils for propulsion and maneuvering, *Journal of Fluids and Structures*, Vol. 17, pp. 163–183.
Hover, F.S., Haugsdal, O. & Triantafylloy, M.S. 2003. Effect of angle of attack profiles in flapping foil propulsion, *Journal of Fluids and Structures*, Vol. 19, pp. 37–47.
James, R.M. 1972. On the remarkable accuracy of the vortex lattice method, *Computer Methods in Applied Mechanics and Engineering*, Vol. 1, pp. 59–79.
Kancharala, A.K. & Philen, M.K. 2016. Optimal chordwise stiffness profiles of self—propelled flapping fins, *Bioinspiration & Biomimetics*, Vol. 11.
Katz J. & Weihs D. 1978. Hydrodynamic propulsion by large amplitude oscillation of an airfoil with chordwise flexibility, *J. Fluid Mech.*, Vol. 88, pp. 485–497.
Katz, J. & Plotkin, A. 1991. *Low—Speed Aerodynamics.* McGraw—Hill
Michelin, S. & Smith, S.L. 2009. Resonance and propulsion performance of a heaving flexible wing, *Physics of Fluids*, Vol. 21
Newman, J.N. 1977. *Marine Hydrodynamics.* MIT Press.
Paraz, F., Schouveiler, L. & Eloy, C. 2016. Thrust generation by a heaving flexible foil: Resonance, nonlinearities, and optimality, *Physics of Fluids*, Vol. 26 (Issue 1).
Prempraneerach, P., Hover, F.S., & Triantafyllou, M.S. 2013. The effect of chordwise flexibility on the thrust and efficiency of a flapping foil, *Proceedings of the 13th International Symposium on Unmanned Untethered Submersible Technology: Special session on bioengineering research related to autonomous underwater vehicles*, New Hampshire.
Schouveiler, L., Hover, F.S. & Triantafyllou, M.S. 2005. Performance of flapping foil propulsion, *Journal of Fluids and Structures*, Vol. 20, pp. 949–959.
Wu, T.Y. 1961. Swimming of a waving plate, *J. Fluid Mech.*, Vol. 10, pp. 321–344.

Hydrodynamics – resistance

Maritime Transportation and Harvesting of Sea Resources – Guedes Soares & Teixeira (Eds)
© 2018 Taylor & Francis Group, London, ISBN 978-0-8153-7993-5

Ship weather routing focusing on propulsion energy efficiency

N. Lamprinidis & K.A. Belibassakis
School of Naval Architecture and Marine Engineering, National Technical University of Athens, Greece

ABSTRACT: In this work a direct method for modelling ship weather routing with regards to energy efficiency is presented. Except of weather data, also bathymetry and coastline, ship added resistance, and the effect of the vertical stern motion of the propeller are taken into consideration. The ship route is represented by Fourier series in appropriate orthogonal curvilinear system on the surface of the earth, permitting the reformulation of the optimization problem with respect to the series coefficients. The optimization algorithm is validated in specific test problems from physics. Subsequently, the present method is implemented as a Matlab GUI, and as a first application the software tool is used for specific ships in conjunction with historical wave data in the Mediterranean Sea region. Numerical results illustrate the performance of the method concerning robustness, accuracy of results and speed of computations.

1 INTRODUCTION

No one can dispute the necessity and size of the shipping industry. Around 90% of the world trade is carried away by ships, while consuming 4 million barrels of oil per day. In addition, we are entering an era of greening of transportation, where energy efficiency is of outmost importance; see, e.g. IMO GloMEEP project (http://glomeep.imo.org/). With the use of ship routing optimization algorithms, energy and time can be significantly saved, as well as increase in safety at sea and travel comfort. Thus, environmentally friendly technical solutions with reduction of exhaust gases are requested, including methods for ship weather routing, taking into account hydrodynamic responses of specific ship, and its propulsion system characteristics for a particular voyage. To this aim, except of ship responses, also added resistance in waves and other factors are examined.

The optimal route can be considered in regards to safety and comfort (Maki, et al 2011, Kosmas et al 2012), maximum energy efficiency (Calvert et al 1991, Dewit et al 1990), minimum voyage time (Zhong et all 1992, Lunnon et al 1992) or the combinations of these factors (Padhy et al 2008, Hinnenthal et al 2010) under the encountered weather circumstances. Various methods have been developed, as e.g., calculus of variations and the modified isochrone method (Hagiwara et al 1987 & 1989), the isopone method (Spaans 1995, Klompstra et al 1992). In addition to above algorithms, many other approaches have also been employed, such as the iterative dynamic programming (Avgouleas, 2008), augmented Lagrange multiplier (Tsujimoto et al 2006) and genetic algorithms (Bekker et al 2006,

Vettor et al 2016). Recent progress has been presented by Waltheret al (2016) and Perera & Guedes Soares (2017).

In the present work a ship weather routing method is developed and tested through numerical simulations based on hindcast weather data in the Mediterranean Sea. The method is based on minimization of fuel consumption taking into account the influence of wave added resistance, in conjunction with the effects of vertical stern motion of the ship in waves on the overall propulsion system efficiency; see Belibassakis et al (2013). Also, seakeeping criteria are implemented as additional operability constraints. Finding the optimal route however, is not an easy task. An analytical solution under the framework of the rigorous mathematical approach (Calculus of Variations) is not possible in practice, due to the complexity of the problem and the sparseness of the data. The problem of ship weather routing is formulated with regards to energy saving criteria, is reduced to a finite variable optimization problem. The innovation on our part is the form of the representation of every possible path (which is highly accurate, robust and quickly calculated), with the conjunction of a Matlab-GUI implementation that combines hindcast weather, as well as geographical data and ship responses to calculate the optimal route.

2 OPTIMIZATION PROBLEM

In the following we focus on the cost associated with the increase of fuel consumption for a travelling ship due to waves and wind, and we present a relatively simple, yet effective alternative computational method to calculate the optimal ship weather route. At the present stage the method is developed without the consideration of additional limitations that could be imposed by the operability constraints of the ship, which will be studied and incorporated in future extensions.

2.1 *Downhill simplex method*

The downhill simplex method developed by Nelder and Mead (1965) is used for the solution of the optimization problem; see e.g., Press et al (1997, Chap.10). The method is very robust and requires only function evaluations, and not derivatives. It may be considered not very efficient in terms of the number of function evaluations that it requires. However, the downhill simplex method may frequently be the best method for a problem characterized by small computational burden. A simplex is the object consisting of $N+1$ points (or vertices) in N dimensions, and all the interconnecting line segments, polygonal faces, etc. In three dimensions it is a tetrahedron, not necessarily the regular one. Given the simplex starting points, the algorithm moves the points of the simplex where our function has smaller values, using reflections, expansions and contractions of the starting configuration converging to the local minimum; see Fig. 1.

2.2 *Optimal ship weather routing method*

For simplicity we consider below a simple trip in the sea or ocean region described by a continuous and partially smooth curve in geographical coordinates (latitude, longitude) joining two points: the departure (A) and the arrival point (B) on the map. The method has been extended to include complex ship routes that are defined by means of a set of control points in the geographical space. The control points play the role of intermediate arrival/departure points, and are used to define the corresponding parts (curves) defining the piecewise smooth ship route.

2.3 *The case of a single trip*

Let (φ, λ) be the latitude and longitude on the surface of the earth, respectively. The latitude $\varphi \in (-\pi/2, \pi/2)$, measured from the equator (positive north), and the longitude $\lambda \in (-\pi, \pi)$ is measured from the Greenwich meridian (increasing eastbound). Also, $A = (\varphi_1, \lambda_1)$ and $B = (\varphi_2, \lambda_2)$ are the coordinates of the departure and arrival point. Consider the rhumb line AB and an auxiliary orthogonal curvilinear coordinate system with x-axis along the loxodrome and y-axis in the normal direction, parameterized with respect to the physical distance; see Fig. 2. Using this coordinate system we consider the following representation of paths (by simple, smooth, non self intersecting curves ending at points A and B).

$$y(x) = \sum_{n=1}^{N} a_n \sin\left(\frac{n\pi x}{\ell}\right) \quad x \in [0, AB], \tag{1}$$

where $\ell = AB$ is the length of the loxodrome path joining the points A and B.

The above model could be extended in order to cover more complex routes (broken into separate paths) by adopting appropriate parametric representations which could be required if also constraints (like seakeeping criteria) are included in the formulation of the optimization problem (but this is left for future extensions). Next, let the total ship fuel energy consumption (for a given route $y(x)$) be given by the following functional:

$$J[y] = \int_{t_0}^{t_f(y)} S \frac{UR_{tot}(t; y)}{n_R n_H n_S n_{GB} n_o(t; y) n_W} dt, \tag{2}$$

where

$$R_{tot} = R_{calm} + R_{wave}(t; y) + R_{wind}(t; y), \tag{3}$$

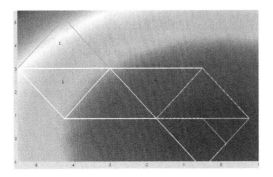

Figure 1. Application of simplex algorithm to the minimization of function $f(x, y) = x^2 + y^2$. First 7 iterations.

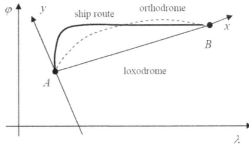

Figure 2. Coordinate systems on geographical space.

is the total resistance, U is the ships speed and S denotes the specific fuel consumption (sfc). The resistance components are R_{calm} which is the calm water resistance (constant for given ship hull and service speed U), R_{wave} which is the added wave resistance and R_{wind} which is the wind resistance. Both R_{wave} and R_{wind} depend on time, location and the relative orientation of the weather conditions with respect to the ship. Finally, n_R is the relative rotative efficiency, n_H the hull efficiency, n_o the open water propeller efficiency, n_s the shaft efficiency, n_{GB} the gear box efficiency. In the results that will be presented in the sequel all variables $U, n_R, n_H, n_S, n_{GB}, n_o$ have been considered constant, however, the present method could be extended to treat controllable variation of these variables along the ship route.

Limitation of the possible route with respect to shallow waters is considered by imposing a bathymetry penalty on the cost function increasing very rapidly whenever the route passes close to the coastline (in practice for water depths lower than 20 m). An important aspect of the method is that it can be extended to include also various equality and inequality constraints, like operability constraints, engine load, speed (and other restrictions), as well as safety constraints (e.g., probability of slamming, propeller emergence, water on deck which should remain below a specific threshold).

The above expression, Eq. (2), concerning the cost functional depends only on the discrete degrees of freedom associated with the coefficients of the present ship route representation Eq. (1), and the problem becomes one of multidimensional optimization with respect to them. As a starting guess of the solution the coefficients $\{a_n\}$ are set to zero, which corresponds to the loxodrome.

2.4 Added wave resistance

An important factor concerning ship operation in realistic sea-states, closely connected to ship dynamics, is added resistance in waves. This could also have a strong economical effect on ship operational and maintenance cost. There are several methods to obtain the added resistance in waves of a ship, and the validity of each method is dependent on the specific ship type. In Arribas (2007) several available methods are studied and validated against seakeeping tests for monohull models, focusing on head seas, that is, usually the most severe situation concerning added resistance. The analysis shows that radiated energy method (Gerritsma & Beukelman 1972) is a method leading to relatively good-quality results, although it could present numerical stability problem in short waves. In the present work we employ the radiated energy method, as extended by Loukakis & Sclavounos (1978) for the prediction of head-to-beam seas, in conjunction with strip

theory (Salvesen *et al* 1970), for the calculation of the added resistance and the vertical ship motion at the stern; see also Lewis (1989 Sec. 3.4). As concerns the added mass and damping coefficients of various ship sections, as well as the Froude-Krylov and diffraction forces, a low-order hybrid panel method, as described Belibassakis (2008), is used to calculate the involved 2D potentials. This method is based on domain decomposition in conjunction with boundary integral formulation based on simple source distribution for the representation of the wave potentials in the near field middle subdomain containing the section, and normal-mode expansions of the potentials in the two semi-infinite strips. The formulation is completed by means of matching conditions, ensuring continuity of all the potentials on the vertical boundaries separating the three subdomains. Although the present analysis is for the ship floating in deep water, the main advantage of the previous hybrid method is its applicability also to cases of limited water depth, permitting us to obtain seakeeping analysis results also in areas of finite depth, either constant or presenting variation along the transverse direction. This could be important in cases of ship operation in areas of limited waterdepths (as e.g. in some routes in the Baltic sea) and/or in channels, where the effects of water depth and its side variation (for ship routes parallel to coastline and bottom contours) on added resistance and ship motions could be important, similarly, as happens to be the case with calm water resistance, and especially for high-speed ships near critical conditions; see, e.g., Suzuki *et al* (2009). More details concerning the calculation of the seakeeping characteristics can been found in Belibassakis (2009). More recent results have been presented by Liu & Papanikolaou (2017).

2.5 Wind resistance

In this work a heuristic approximation of the wind resistance is used (Lewis 1989) as follows.

$$F = K\rho U_R^2 \left(A_L \sin^2(\theta) + A_T \cos^2(\theta)\right) / \cos(a - \theta), \quad (4)$$

where F is the wind resistance, ρ is the air density, U_R, θ is the value and angle (with respect to the ship) relative wind speed, a is the true wind angle, A_T is the transverse projected area and A_L is the longitudinal projected area of the ship. The constant K usually takes values between 0.5–0.65.

2.6 Description of solution method

Step 1. The starting time, as well as departure and arrival points are defined. Then, the geographical data and the weather data are loaded.

Step 2. Every possible path is approximated on the auxilliary orthogonal curvilinear system as a Fourier series of N terms by Eq.(1). The fuel energy functional is now dependent on N variables $\{a_n, n = 1,2,...N\}$. With the use of downhill simplex method, the a_n terms that minimize the merit function Eq. (2) are found.

Such a method however does not perform well for large N as mentioned before, both with regards to time and robustness. However, in the case of trips that are described by piecewise smooth curves in the geographical space, an hierarchy of the coefficients $a_n >> a_{n+m}$ is usually expected, and in such cases the following alternative algorithm can be used which accelerates substantially the solution:

Step 2a. Let $N_1 < N$ (in this work $N_1 = 4$). With the use of downhill simplex find $a_i^{(0)}$ $i = 1,...,N_1$ that minimize the cost functional,

Step 2b. Keeping the above solution concerning the first N_1 coefficients the optimization problem is solved in parameter subspaces of small dimension with respect to the rest of the coefficients $a_i^{(0)}$ $i = N_1 +1,..N$, and the above steps are iterated providing us with an $\ell -$ sequence of coefficients $\{a_i^{(\ell)}$ $i = 1,..N\}$ until convergence.

The above alternative algorithm offers good computational performance, since high dimensional optimisation is slow (complexity of the algorithm is exponential with regards to the dimension).

On the other hand, breaking the optimization into smaller dimensions allows one to achieve the required accuracy of the solution, as needed, very quickly.

3 BENCHMARK PROBLEMS

3.1 *The ray equations as an optimization problem*

In order to test the accuracy of the present method and its numerical efficiency, the application of the above algorithm, in conjunction with the present series representation is studied for the solution of ray equations in wave physics. The latter is a standard technique to treat high-frequency wave phenomena in stratified media as in the case of underwater acoustic and seismic wave propagation (see, e.g., Jensen et al 1994, Katsenelenbaum 1998) atmospheric acoustic (Salomons 2001). We mention that similar methods have been used to treat the optimal shiprouting problem (see, e.g., Papadakis & Perakis 1990). As an example we consider the case of rays emitted from a source at point A $(x_1, x_2) = (0, -300$ m$)$ in hydroacoustic environment, stratified along the x_2-direction, which is characterized by wave speed $c(x_1, x_2) = c_1 \cosh(c_2(x_2 - x_2(A)))$, where

$c_1 = 1510$ m/s and $c_2 = 0.0006$ m^{-1}. In this case an analytical solution is also available, as follows

$$x_2 - x_2(A) = c_1^{-1} \sinh^{-1}\left(\tan\theta_A \sin\left(c_2\left(x_1 - x_1(A)\right)\right)\right), \quad (5)$$

where θ_A is the slope of ray at the source point A. The calculated result by the system of ray differential equations is shown in the upper subplot of Fig. 4 for the bunch of rays emitted from the source-point at directions in the interval $(-50$ deg, 50 deg$)$. As an example the solution of the boundary value problem involving source point A and receiver point B located at $(x_1, x_2) = (4000$ m, -200 m$)$ is also plotted in lower subplot of Fig. 3, where it is compared with the solution by the present simplex algorithm using crosses, as obtained by keeping only 4 terms in the series (1).

3.2 *Calm water performance*

Another test, concerning the present method is the case of the ship travelling at constant speed in calm water, without any effects from waves. In this case, the resistance remains constant and the minimization of fuel consumption is equivalent to the shortest distance on the globe. Under the approximate assumption of spherical shape for the globe, this corresponds to the arc joining the departure and arrival points corresponding to the great circle (orthodrome).

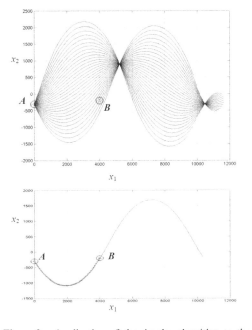

Figure 3.   Application of the simplex algorithm to the system of ray equations in a vertically stratified medium.

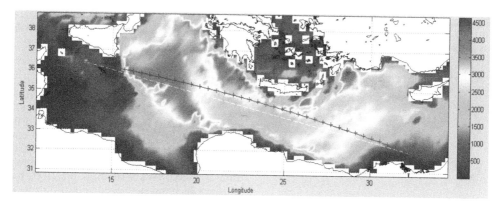

Figure 4. Orthodrome (continuous line), loxodrome (white dashed line) and present method solution (crosses) in the case of calm water. Bathymetric data (in metres) are shown by using colorplot.

The present method always verifies the above result in the absence of waves, as illustrated below (plotted in Figs. 4). The orthodrome (solid line) is used together with the route based on the loxodrome (dashed line), which also shown in Fig. 4. Both the orthodrome and the loxodrome are used as reference values in order to compare and evaluate the solution in the case of realistic weather routing solutions.

4 OPTIMAL SHIP WEATHER ROUTING

As an example, we will consider here the case of an AFRAMAX 105000 tn DWT tanker, for which various data are available concerning ship hull form, general arrangement and load conditions, as well as engine data and ship performance for various speeds, including information about shaft RPM, thrust and torque, environmental conditions etc.

The above AFRAMAX class tanker has length between perpendiculars $L_{BP} = 234$ m and 105000 tn DWT; see Fig. 5(a). The main dimensions and principal data of this ship are: $L = 238.5$ m (overall length), $B = 42$ m (max breadth), $T = 14.9$ m (scantling draft). In the full load condition ($T = 14.9$ m) the main hydrostatic data are displacement $D = 122770$ tn, $C_B = 0.82$, $KB = 7.76$ m. The center of gravity is located at $KG = 12$ m, $LCG = 6.8$ m, the metacentric height is $GM = 6.1$ m, and the radii of gyration about transverse and longitudinal axes passing through the center of gravity are estimated to be: $R_{yy} = 63$ m, $R_{xx} = 12$ m. For this ship information concerning bare-hull calm-water resistance was available as it was estimated from model tests at various drafts. Extrapolation of these data to full scale are shown in Fig. 6 concerning the full load condition ($T = 14.9$ m) and a second draft of the examined AFRAMAX tanker, $T = 13.6$ m. In

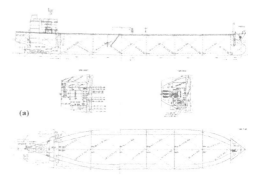

Figure 5. AFRAMAX 105000 tn DWT.

addition, data concerning the distribution of the axial wake $1 - w = \dfrac{U_A}{V}$ on the propeller plane is available. As observed a strong variation of incoming flow with azimuthal angle, especially around the top position ($\theta = 0$ deg). The global wake fraction has been estimated by self propulsion tests, and at $T = 14.9$ m draft is $1 - w = 0.645$.

The above wake data refers to calm water conditions and presents small variation for ship speeds in the range of interest (11–15 kn). The examined ship is equipped with one Diesel engine (MAN B&W 6S60MC type, MCR 15400 BHP @ 97 RPM), directly coupled to the propeller. The engine is capable of driving the ship (clean hull) at about 14.2 kn, in calm water and scantling draft loading condition ($T = 14.90$ m), with a sufficient power margin for real sea conditions. Based on the above propeller analysis and results, the behavior of the propulsion system based on the calm-water resistance for the two drafts $T = 7.2$ m (dotted line) and $T = 14.9$ m (continuous line) is calculated, and the results are plotted in Fig. 7. Numerical results concerning the AFRAMAX tanker are presented in Table 1 concerning the mean added resistance, for ship speeds corresponding to Froude numbers $F = 0.12–0.17$ in head waves ($\beta = 180°$). Similar results are available for various values of the direction of incident waves. Also, the effect of the vertical ship motion on the propeller efficiency can be incorporated in calculating engine power data and fuel consumption, as described in Belibassakis et al (2013) and is left to be inclded in future extensions.

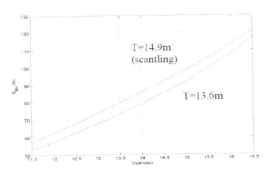

Figure 6.   Calm-water resistance vs. ship's speed at two drafts.

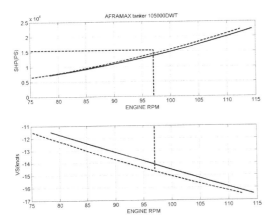

Figure 7.   Calm-water propulsion system performance of AFRAMAX tanker at two drafts (continuous line draft T = 14.9 m).

4.1   Metadata info

As concerns geographical data, the database General Bathymetric Chart of the Oceans (GEBCO www.gebco.net) is used, in conjunction with GMT for coastline data (Wessel, & Smith 1996). GEBCO is a high quality bathymetry database that covers the whole planet with resolution 30". The offshore wave data used in the present work are obtained using a 3-level SWAN (Booij et al., 1999) based scheme, which was developed in the framework of Thales CCSEAWAVS project. aimed to estimate the effects of climate change on sea level and wave climate of the Greek seas, the coastal vulnerability and the safety of coastal and marine structures (see Makris et al 2016). This simulation scheme uses past and future projections climatic wind fields (also produced in the context of CCSEAWAVS) for the estimation of wave characteristics with resolution 0.2 degrees in the Mediterranean basin (Level 1). These data provide boundary information for repeating the simulation using a finer mesh 0.05 degrees inside an Eastern Mediterranean subsection (Level 2). Details of the methodology are also described in Athanassoulis et al (2014).

The initial scope of SWAN was to compute random, short-crested wind-generated waves in coastal regions and inland waters. The model was later extended to allow simulation of waves in deep waters as well. As for the wind data, those were taken using the ICTP RegCM3 model. The dataset extends over the entire Mediterranean Sea, with spatial resolution of 25×25 km and temporal resolution of 6 hours. Although the resolution of the above data is low, it is used to demonstrate the present method, and the corresponding database will be enhanced in future work by using higher resolution forecasting models.

Table 1.   Mean added Resistance (% of calm water resistance at same ship speed at draft $T = 14.9$ m), in head waves ($\beta = 180°$) as function of sea condition and ship speed Vs (BF indicates Beaufort number).

				Vs(kn)				
BF	U_w(m/s)	H_s(m)	T_p(S)	11.5	12.5	13.5	14.5	15.5
2	2.0	0.2	2.24	1.31	1.10	0.94	0.57	0.50
3	4.4	0.6	3.87	3.93	3.31	2.46	1.71	1.49
4	6.7	1.0	5.00	6.54	5.51	3.78	2.85	2.49
5	9.3	2.0	7.07	12.80	9.53	7.08	5.70	4.97
6	11.8	3.0	8.66	17.38	13.39	10.38	8.55	6.99
7	15.3	4.0	10.00	40.87	32.99	26.76	21.90	18.03
8	19.0	5.5	11.73	129.30	109.37	92.12	78.39	67.38

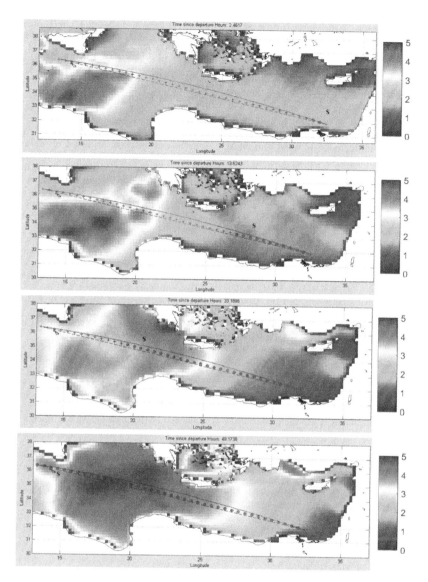

Figure 8. Calculation of optimal weather shiproute in the case of a storm in Eastern Mediterranean Sea. Sea severity (significant wave height, shown by means the colobar) is shown by using a colorplot. The subplots correspond to times near the beginning, in the middle and near the end of the first part of the trip, indicating also the position of the ship.

4.2 Numerical example

As an example, the calculated optimal shiproute, in the case of a storm in Mediterranean Sea, is presented in Fig. 8, concerning the **AFRAMAX** tanker for a trip from Port Said to for a trip from Port Seid to Marseille. In particular the first part of the route is presented (up to Sicily sea region). The ship departs on 1st of November, 12 am, which is an interesting test case since, from the simulated data, intense weather conditions are prevailing in the Eastern Mediterranean basin. The calculated results is shown by using symbols, and compared against the loxodrome (shown by using dashed lines) and the orthodrome (shown by using solid lines). In this particular example the calculated cost of optimal trip is found to 2% better compared to the orthodrome.

5 CONCLUSIONS

In this work a method for the ship weather routing problem with regards to energy efficiency is presented. The accuracy of the optimization algorithm has been validated against similar problems with analytical solutions. The algorithm is implemented in conjunction with geographical and simulated weather data as a Matlab GUI showing good numerical performance concerning energy consumption reduction. Future extension will include the consideration of operability and other criteria as additional constraints of the optimization problem, the enhancement of wind and wave databases, and the extension of the tool worldwide.

REFERENCES

Arribas, P.F. 2007, Some methods to obtain the added resistance of a ship advancing in waves, Ocean Engineering, 34, pp. 946–955.

Athanassoulis, G., Belibassakis, K., Gerostathis, T., Stefana kos, Ch., Spaan, G., 2003, WORLDWAVES: High quality coastal and offshore wave data within minutes for any global site, Proc. 22nd Int. Conference on Offshore Mechanics and Arctic Engineering, OMAE2003, Cancun, Mexico.

Athanassoulis, G.A., Belibassakis, K.A., Gerostathis, Th.P., Kapelonis, Z.G., 2014. Application of SWAN wave model for climatic simulation of sea condition at coastal areas of the Mediterranean. Proc. 6th Panhellenic Conf. Coastal Zones Manage. Athens, Greece.

Barstow, S., Mørk, G., Lønseth L, Schjølberg, P., Machado, U., Athanassoulis, G., Belibassakis, K., Gerostathis, T., Stefana kos, Ch., Spaan, G., 2003, WORLDWAVES: High quality coastal and offshore wave data within minutes for any global site, Proc. 22nd Int. Conference on Offshore Mechanics and Arctic Engineering, OMAE2003, Cancun, Mexico.

Barstow, S., Mørk, G., Lønseth L, Schjølberg, P., Machado, U., Athanassoulis, G., Belibassakis, K., Gerostathis, T., Stefanakos, Ch., Spaan, G., 2003, WORLDWAVES: Fusion of data from many sources in a user-friendly software package for timely calculation of wave statistics in global coastal waters, 13th Intern. Offshore and Polar Conference and Exhibition, ISOPE2003, Honolulu, Hawaii, USA.

Bekker JF, Schmid JP. 2006, Planning the safe transit of a ship through a mapped minefield. Journ. Opera-tions Research Society of South Africa, 22, pp. 1–18.

Belibassakis, K., 2009, Effects of wave-induced ship motion on propeller-hull interaction with application to fouling estimation and propulsion optimization. Proc. Int. Maritime Association Mediterranean Conference, IMAM 2009 Istanbul.

Belibassakis, K.A., 2008, A boundary element method for the hydrodynamic analysis of floating bodies in general bathymetry regions, Engineering Analysis with Boundary Elements, 32, pp. 796–810.

Belibassakis K.A., Politis, G.K., Gerostathis Th.P. 2013, Calculation of ship hydrodynamic propulsion in rough seas by non-linear BEM with application to reduction of energy losses in waves, 32th Int. Conference on Offshore Mechanics and Arctic Engineering (OMAE2013), June 9–14, 2013 – Nantes, France.

Calvert S, Deakins E, Motte R. 1991, A dynamic system for fuel optimization Trans—Ocean. Journal of Navigation 44, pp. 233–65.

Dewit C. 1988, Practical weather routeing of sail-assisted motor vessels. Journal of Navigation 41, p.134.

Dewit C. 1990, Proposal for low-cost ocean weather routeing. Journal of Navigation 43, pp. 428–39.

Gelfand I.M., Fomin S.V., 2000, Calculus of Variations, Dover Publications, Inc.

Gerritsma, J., Beukelman, W., 1972, Analysis of the resistance increase in waves of a fast cargo ship, Intern.Shipbuilding Progress 19(217), pp. 285–293.

Hagiwara H, Spaans JA. 1987. Practical weather routing of sail-assisted motor vessels. Journal of Navigation vol. 40, pp. 96–119.

Hagiwara H. 1989, Weather routing of (sail-assisted) motor vessels. Delft: Delft University of Technology.

Hinnenthal J, Clauss G. 2010, Robust pareto-optimum routing of ships utilising deterministic and ensemble weather forecasts. Ships and Offshore Structures; 5, pp. 105–14.

ITTC 1998, Report of the Resistance Committee, 22nd Inter. Towing Tank Conf.

ITTC 2002, Report of the Resistance Committee, 23rd Inter. Towing Tank Conf.

ITTC 2005, Report of the Resistance Committee, 24th Inter. Towing Tank Conf.

Jensen F., Kupperman W., Porter M., Schmidt H., 1994, Computational Ocean Acoustics. AIP Press.

Katsenelenbaum B.Z., Mercader del Rio L., Pereyaslavets M., Sorolla Ayza M., Thumm M., 1998, Theory of Nonuniform Waveguides, The Cross-Section Method. IEE London.

Klompstra MB, Olsde GJ, Van Brunschot PKGM, 1992, The isopone method in optimal control. Dynamics and Control vol.2, pp. 281–301.

Kosmas OT, Vlachos DS. 2012, Simulated annealing for optimal ship routing. Computers & Operations Research vol.39, pp. 576–81.

Lewis, E.V. (Ed), 1989, Principles of Naval Architecture, vol. 3, NJ: Society of Naval Architects & Marine Engin. (SNAME), New York.

Liu S. Papanikolaou A., 2017, On the prediction of the added resistance of large ships in representative seaways Journal Ships and Offshore Structures 12, 690–696.

Loukakis, T.A., Sclavounos, P., 1978, Some extensions of the classical approach to strip theory of ship motion, including the calculation of mean added forces and moments" Journal of Ship Research, 22 (1), pp. 1–19.

Lunnon RW, Marklow AD. 1992, Optimization of time saving in navigation through an area of variable flow. Journal of Navigation vol.45, pp. 384–99.

Maki A, Akimoto Y, Nagata Y, Kobayashi S, Kobayashi E, Shiotani S. 2011, A new weather-routing system that accounts for ship stability based on a real-coded genetic algorithm. Journal Marine Science and Technology vol.16, pp. 311–22.

Makris Ch., Galiatsatou P., et al, 2016, Climate change effects on the marine characteristics of the Aegean and Ionian Seas, Ocean Dynamics, vol. 66 (12), 1603–1635.

Muntean T., 2008, Ship propulsion train efficiency sensing, Wartsila Technical Journal, 02.

Nelder, J.A., Mead, R. 1965, Computer Journal, vol. 7, pp. 308–313.

Padhy CP, Sen D, Bhaskaran PK. 2008, Application of wave model for weather routing of ships in the North Indian Ocean. Natural Hazards vol.44, pp. 373–85.

Papadakis, N Perakis A., 1990, Deterministic Minimal Time Vessel Routing Operations Research, 426–438.

Perera L., Guedes Soares C., 2017, Weather routing and safe ship handling in the future of shipping Ocean Engineering vol. 130. 684–695.

Press W Teukolsky S, Vetterling W Flannery B, 1997, Numerical Recipes, Cambridge University Press.

Salomons E.M., 2001, Computational Atmospheric Acoustics, Kluwer Academic Publ,

Salvesen, N., Tuck, E.O., Faltinsen, O., 1970, Ship motions and sea loads, Transactions SNAME 78, pp 250–87.

Shao W, Zhou PL, Thong SK. 2012, Development of a novel forward dynamic programming method for weather routing. Journal of Marine Science and Techn. vol.17, pp. 239–51.

Spaans J.A. 1995, New developments in ship weather routing. Navigation vol. 169, pp. 95–106.

Suzuki K., Kai H., Hirai M., Tarafder Sh., 2009, Simulation of free surface flow of high speed ship in shallow or restricted water condition, Proc.10th Int. Conference on Fast Sea Transportation, FAST 2009, Athens, Greece.

Takashima K, Mezaoui B, Shoji R. 2009, On the fuel saving operation for coastal merchant ships using weather routing. International Journal on Marine Navigation and Safety of Sea Transportation vol.3, pp. 401–6.

Tsujimoto M, Tanizawa K. 2006, Development of a weather adaptive navigation system considering ship performance in actual seas. Proc. 25th int conf on offshore mechanics and arctic engineering. Hamburg, Germany.

Vettor, R., Tadros, M., Ventura, M., Guedes Soares, C. 2016, Route planning of a fishing vessel in coastal waters with fuel consumption restraint. In Guedes Soares, C. & Santos T.A., (Eds.). Maritime Technology and Engineering 3. London, UK: Taylor & Francis Group, pp. 167–173.

Walther L., Rizvanolli A. Mareike Wendebourg M. Jahn C., 2016, Weather Routing and Safe Ship Handling in the Future of Shipping International Journal of e-Navigation and Maritime Economy 4, 031–045.

Wessel, P., and W.H.F. Smith, 1996, A Global Self-consistent, Hierarchical, High-resolution Shoreline Database, J. Geophys. Res., vol.101.

Zhang LH, Zhang L, Peng RC, Li GX, Zou W. 2011, Determination of the shortest time route based on the composite influence of multidynamic elements. Marine Geodesy, vol. 34, pp. 108–18.

Maritime Transportation and Harvesting of Sea Resources – Guedes Soares & Teixeira (Eds)
© *2018 Taylor & Francis Group, London, ISBN 978-0-8153-7993-5*

Influence of the approximated mass characteristics of a ship on the added resistance in waves

I. Martić, N. Degiuli, P. Komazec & A. Farkas
Faculty of Mechanical Engineering and Naval Architecture, University of Zagreb, Zagreb, Croatia

ABSTRACT: Added resistance in waves is one of the main causes of the involuntary speed reduction of a ship. Total resistance of the ship sailing at actual seaway should be determined instead of considering the calm water resistance and taking the Sea margin into account. The exact ship mass characteristics are often unknown especially in the preliminary design phase. In order to calculate the ship response and loads acting on the ship hull in waves, the mass characteristics have to be approximated with sufficient accuracy. In this paper, the influence of the approximate determination of the ship's centre of gravity and gyration radii on the added resistance in waves is investigated for KCS and S175 container ships and KVLCC2 tanker for full load and ballast condition. Added resistance in waves is calculated based on the potential flow theory. The obtained results are compared with the available experimental data from the literature.

1 INTRODUCTION

Ship added resistance in waves is one of the main causes of involuntary speed reduction and an increase in the fuel consumption. Since it has a significant effect on CO_2 emission, it is important to predict an increase of the total resistance due to sailing in waves as accurate as possible rather than taking it into account as the Sea margin. In severe sea states, ship will experience an increase in resistance larger than the one usually taken into account as a percentage of the calm water resistance. Considering the increasing demands on the energy efficiency of ships adopted by International Maritime Organization (IMO), added resistance in waves should be known already in the preliminary design phase of a ship. However, in the preliminary design phase mass characteristics required for the added resistance calculation, i.e. position of the centre of gravity and radii of gyration, are often unknown parameters. Thus, it is necessary to estimate mass characteristics with sufficient accuracy in order to obtain reliable results regarding ship added resistance in waves.

Added resistance is considered as the longitudinal component of the time averaged part of the second-order wave force and is highly dependent on ship motions. It is a consequence of the interference of the ship wave system caused by the forward speed of the ship advancing through still water and waves caused by motions as a response to the incoming waves. Due to the oscillatory motions caused by incoming waves, energy is transmitted by the radiation waves away from the ship forming the radiation induced part of the added resistance which is characteristic for large ship motions. The diffraction induced added resistance on the other hand is characteristic for small ship motions in short incoming waves. Viscous part of added resistance is negligible (Arribas 2007). Along with approximate analytical methods and strip method, ship added resistance in waves can be calculated using numerical methods with sufficient accuracy (Duan & Li 2013). Even thou numerical methods based on the viscous flow theory are proven to be more accurate than the ones based on the potential flow theory, due to the required computational time as well as complexity of the simulation preparation, the advantage is often given to latter ones (Södig & Shigunov 2015). The accuracy of added resistance calculation using numerical methods based on viscous flow is highly dependent on the grid size. Using too coarse grid may lead to underestimation of the obtained results (Ozdemir & Barlas 2017). Also, considering the high non-linearity of the wave diffraction, an extensive computational effort is required due to numerical diffusion in short waves (el Moctar et al. 2015). The diffraction part of added resistance is usually under predicted when using numerical methods based on the potential flow theory, i.e. panel methods. However, it can be successfully taken into account by applying the correction of the results in short waves (Martić et al. 2016, Liu & Papanikolaou 2016).

During the experiments in towing tank, added resistance in waves cannot be measured directly.

Tests are conducted in calm water and in waves in order to derive an increase in resistance due to waves. In order to know the accuracy of the results measured during tests, it is necessary to conduct the uncertainty analysis of the experimental data. The uncertainty of derived added resistance in waves is greater than uncertainty of ship motions (Park et al. 2015). It can reach up to 16% in short wave region and up to 9% in moderate wave region mainly due to uncertainty in measurements of wave elevations and resistance in waves for KVLCC2. For very large wavelengths, the uncertainty can reach up to 60%. This was expected taking into account small absolute value of added resistance for very large wavelengths (Park et al. 2015). Experiments showed that added resistance is concentrated in fore part of the ship hull (Guo & Steen 2011), thus an optimization of the bow shape with the aim of reducing added resistance in waves could be beneficial. Bolbot & Papanikolau (2016) showed that reduced waterplane in the bow area and vertical stem can lead to the significant reduction of added resistance for KVLCC2 tanker in short waves.

Added resistance in waves is calculated generally for the full load condition of the ship. However, the draught of the ship changes within different operating conditions and ship spends 40–50% of its total voyage time in the ballast condition. In the ballast condition, the flow around the ship and the wave interference are different and velocity gradient as well as pressure differences are greater compared to full load condition (Luo et al. 2016). In the short wave region, where the diffraction part is dominant, added resistance is larger compared to the full load condition (Park et al. 2016). Change in the effective hull form in the bow area and larger relative motions in ballast condition lead to an increase in the diffraction part of the added resistance. Also, the position of the maximum added resistance in ballast condition moves towards higher frequencies (Park et al. 2016).

In this paper, influence of the approximated mass characteristics on ship added resistance in waves is investigated for three benchmark ships: S-175 and KCS containerships and very large crude carrier KVLCC2. Added resistance is determined utilizing the commercial software HydroSTAR v.7.3 (Bureau Veritas 2016), which uses 3D Green function panel method, based on the potential flow theory at regular head waves. Ship forward speed is taken into account through the encounter frequency. The obtained results are compared with the experimental data available in the literature. Added resistance is determined also for KVLCC2 ship in ballast condition and the obtained results are compared with the calculated results for full load condition.

2 NUMERICAL SETUP

2.1 3D panel method

Commercial software used for the calculation of added resistance in regular head waves uses 3D panel method based on potential flow theory (Bureau Veritas 2016). Motion of the incompressible, inviscid, and irrotational fluid is described using the potential Φ, which satisfies Laplace equation in the entire computational domain:

$$\Delta\Phi = 0 \tag{1}$$

The body surface is represented by quadrilateral flat panels with the distribution of sources and dipoles of constant strength on each panel. Linear equations for the unknown and constant source strengths are solved using Green function, which is the fundamental solution of the Laplace equation for a source of constant strength. In that way the velocity field is obtained. Green function is formulated in order to satisfy the boundary condition on the linearized free surface, boundary condition on the seabed and hull and radiation condition in the far field as follows:

$$-k\Phi + \frac{\partial\Phi}{\partial z} = 0 \quad \text{on} \quad z = 0 \tag{2}$$

$$\frac{\partial\Phi}{\partial n} = v_n \quad \text{on the wetted surface} \tag{3}$$

$$\frac{\partial\Phi}{\partial z} = 0 \quad \text{on the seabed} \tag{4}$$

$$\lim_{R\to\infty} \left[\sqrt{R}\left(\frac{\partial\Phi}{\partial R} - ik\Phi \right) \right] = 0 \quad \text{in far field} \tag{5}$$

where k is the wave number, n is the unit surface normal of boundary element and v_n is the collocation point velocity. Radiation condition, which requires that fluid velocity potential disappears in the infinite boundary $R \to \infty$ around the ship hull is automatically satisfied when fairly perfect fluid is considered.

The software provides the solution of the first-order motions and forces as well as the low-frequency second-order load. A fictitious force dependent on fluid velocity is added in momentum equation in order to include the energy dissipation due to fluid viscosity. Forces and moments acting on ship hull are evaluated by integrating the pressure obtained from known potentials in the centroids of panels, i.e. in the collocation points. First order potential is linearized about the mean wetted surface area of the hull and quadratic terms in the second order potential are obtained by weakly nonlinear expansion of the perturbation potential. Advance speed of the ship is taken into account through encounter frequency of the incoming waves as follows:

$$\omega_e = \omega - \frac{\omega^2}{g} v \cos \beta \qquad (6)$$

Second-order wave loads are calculated using so-called near-field formulation, i.e. the direct pressure integration, and added resistance in waves is determined through the quadratic transfer function (QTF) approximated by zeroth-term only. Therefore, the obtained force is a constant value at each given frequency of incoming wave.

2.2 Case study

2.2.1 Full load condition

Influence of the approximately determined position of the vertical centre of gravity (VCG) and gyration radii as the percentage of ship length (or breadth) is investigated for three benchmark ships: S-175 and KCS container ships and full form of very large crude carrier KVLCC2. Origin of the coordinate system is placed in the intersection of the aft perpendicular and mean free surface with positive x-axis directed towards the fore part of the ship, positive y-axis towards port side and positive z-axis upwards. Calculations of motions are performed with respect to the ship centre of gravity. Considered radii of gyration are considered about the axes through the ship centre of gravity: roll radius of gyration k_{xx} about x-axis, pitch radis of gyration k_{yy} about y-axis and yaw radius of gyration k_{zz} about z-axis.

Main particulars of ships used within this research are presented in Tables 1 and 2. Regular head waves are imposed on ships advancing with speeds corresponding to nominal Froude numbers for which the towing tank experiments were conducted. Hull forms are discretized up to the mean free surface by sufficient number of quadrilateral panels considering the recommendation that the

Table 1. Main particulars of S-175 and KCS container ships.

	S-175	KCS
Length, L_{PP} (m)	175.0	230.0
Breadth, B (m)	25.4	32.2
Draft, T (m)	9.50	10.8
Displacement volume, ∇ (m³)	24154.13	52030
Block coefficient, C_B	0.572	0.651
Position of LCG from $L_{PP}/2$, x_G (m)	−2.5475	−3.4
Position of VCG, KG (m)	9.55	7.28
Radius of gyration, k_{xx}/B	0.328	0.4
Radius of gyration, k_{yy}/L	0.24	0.25
Froude number, Fn	0.25	0.26

Table 2. Main particulars of KVLCC2 tanker.

	KVLCC2
Length, L_{PP} (m)	320.0
Breadth, B (m)	58.0
Draft, T (m)	20.8
Displacement volume, ∇ (m³)	312622
Block coefficient, C_B	0.8098
Position of LCG from $L_{PP}/2$, x_G (m)	11.1
Position of VCG, KG (m)	18.6
Radius of gyration, k_{xx}/B	0.4
Radius of gyration, k_{yy}/L	0.25
Froude number, Fn	0.142

Table 3. Number of panels in full load condition.

	S-175	KCS	KVLCC2
Hull surface	1259	1396	1931
Interior waterplane	560	878	1507

Figure 1. Panel model of the S-175 container ship.

Figure 2. Panel model of the KCS container ship.

length of the panel should be lower than 20% of the smallest wavelength regarding the required computational time. In order to eliminate irregular frequencies or shift them further in the frequency range, ship interior waterplane is discretized and boundary value problem is modified. Number of panels for ships in full load condition on the hull surface and the interior waterplane can be seen in Table 3. Panel models of S-175, KCS and KVLCC2 can be seen in Figures 1, 2 and 3 respectively.

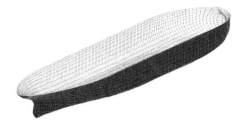

Figure 3. Panel model of the KVLCC2 tanker.

Added resistance in waves as the longitudinal component of wave force acting in the opposite direction of the ship speed is presented as the dimensionless value in dependence of the ratio of wavelength and ship length λ/L. λ/L ratio is determined with respect to ship length between perpendiculars or length of waterline depending on the available results regarding towing tank experiments. The numerically obtained results are compared with experimental data available in the literature (Fujii & Takahashi 1975, Simonsen et al. 2013, Guo & Steen 2011).

2.2.2 Ballast condition

Added resistance in waves is also investigated for KVLCC2 tanker in ballast condition. Main particulars different than the ones in full load condition can be seen in Table 4. Incoming wave characteristics as well as Froude number are taken the same as in calculations for full load condition. The results are compared with the available experimental data (Park et al. 2016). Number of panels used to discretize the hull form in ballast condition can be seen in Table 5.

The coefficient of added resistance in waves for both full load and ballast condition is obtained as follows:

$$C_{AW} = \frac{R_{AW}}{\rho g \zeta_a^2 B^2/L} \qquad (7)$$

where ζ_a is the incoming wave amplitude.

A correction of the results obtained for short wave region is taken into account for ballast condition. An increase in diffraction part of added resistance in ballast condition is expected (Park et al. 2016) as well as larger relative motions compared to full load condition. Thus, the obtained results are corrected based on the practical correction proposed by Tsujimoto et al. (2008). The correction takes into account the bluntness coefficient, effect of the draft and frequency and effect of the advance speed of ship. Correction due to wave diffraction R_{AWr} is added to the calculated added resistance force in order to correct

Table 4. Main particulars of KVLCC2 ship in ballast condition.

	KVLCC2
Draft of fore peak, T_F (m)	6.7
Draft of aft peak, T_A (m)	11.9
Displacement volume, ∇ (m³)	123910
Position of VCG, KG (m)	14.0

Table 5. Number of panels in ballast condition.

	KVLCC2
Hull surface	1848
Interior waterplane	1215

the obtained results for the diffraction effect that occurs especially in short waves. Total added resistance force is then obtained as:

$$R_{AW}{}^{\text{corrected}} = R_{AW} + R_{AWr} \qquad (8)$$

Correction due to wave diffraction is calculated by the following expression:

$$R_{AWr} = \frac{1}{2}\rho g \zeta_a^2 B B_f \alpha_d (1 + \alpha_U) \qquad (9)$$

where B_f is the bluntness coefficient of the non-shaded part of the waterline, α_d is the correction for draft and frequency and α_U is the correction for advance speed.

Correction factors in eq. (9) are obtained using the following expressions:

$$B_f = \frac{1}{B}\left\{\int_I \sin^2(\alpha + \beta_w)\sin\beta_w \mathrm{d}l \right.$$
$$\left. + \int_{II} \sin^2(\alpha - \beta_w)\sin\beta_w \mathrm{d}l\right\} \qquad (10)$$

$$\alpha_d = \frac{\pi^2 I_1^2(k_e d)}{\pi^2 I_1^2(k_e d) + K_1^2(k_e d)} \qquad (11)$$

$$k_e = k\left(1 + \frac{\omega U}{g}\cos\alpha\right)^2 \qquad (12)$$

$$1 + \alpha_U = 1 + C_U Fn \qquad (13)$$

where α is the wave heading, β_w is the slope of the line element along the waterline, I, II are the domains of the integration along the waterline, I_1 is the Bessel function of the first kind of first order, K_1 is the Bessel function of the second kind of first order, $k_e d$ is the non-dimensional frequency, U is the advance speed and C_U is the coefficient of the advance speed.

3 RESULTS

In comparison with the experimental results (Fujii & Takahashi 1975), the deviation of the numerically obtained results for S-175 containership can be noticed in the entire range of λ/L ratio, except at the lowest considered frequency, i.e. the highest value of λ/L ratio (Fig. 4).

The maximum relative deviation between the results is approximately 35%, when the results obtained at the highest considered frequencies are excluded. At these frequencies, values of added resistance is almost two times lower than the value obtained in the experiment. Also, the highest value of added resistance coefficient is slightly shifted towards higher frequencies, i.e. lower λ/L ratio, even thou the trend of the results is the same.

The relative deviation between the numerically obtained results and experimental data (Simonsen et al. 2013) is much smaller in the case of KCS container ship (Fig. 5). In the area of moderate λ/L ratios, relative deviation between the results is lower than 10%. As expected, in the short wave region, the maximum relative deviation between results is over 50%. On the other hand, the trend

of the results is the same and the maximum value for both measured and calculated data occurs at the same λ/L ratio.

The average relative deviation between the experimental results for KVLCC2 (Guo & Steen 2011) and the numerically obtained results is about 35% when the value at lowest considered λ/L is excluded. This value is underestimated by panel method for about 60% (Fig. 6). Even thou the trend of the results is the same, numerically obtained values underestimate measured data in the entire inspected range. It should be noticed, that in a very similar range of λ/L ratio compared to other two ships considered, values of the added resistance coefficient in short waves are significantly larger even thou Froude number is lower. In the short wave region, the diffraction part of added resistance is dominant. Blunt shape of the waterplane in the bow area leads to an increase in added resistance.

In Figures 7–9, the influence of the approximated position of the vertical centre of gravity is presented for S-175, KCS and KVLCC2 respectively. Only vertical position of centre of gravity is variated, while the longitudinal and transverse position are kept constant in order to avoid any angle of trim or heel. VCG is variated within 2 meters in the positive and negative direction. The range of the variation is determined using the

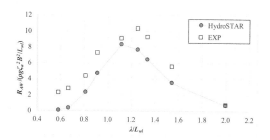

Figure 4. Comparison of the numerically obtained results with experimental data for S-175 at $Fn = 0.25$.

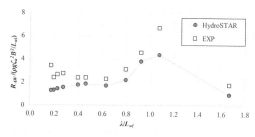

Figure 6. Comparison of the numerically obtained results with experimental data for KVLCC2 at $Fn = 0.142$.

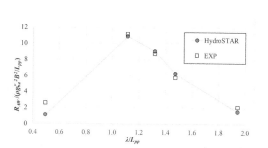

Figure 5. Comparison of the numerically obtained results with experimental data for KCS at $Fn = 0.26$.

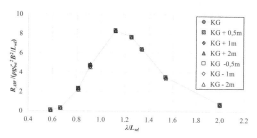

Figure 7. Influence of VCG approximation on coefficient of added resistance in waves for S-175.

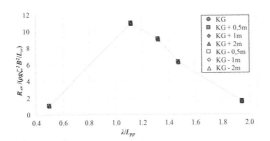

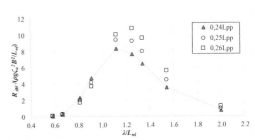

Figure 8. Influence of VCG approximation on coefficient of added resistance in waves KCS.

Figure 10. Influence of k_{yy} approximation on coefficient of added resistance in waves for S-175.

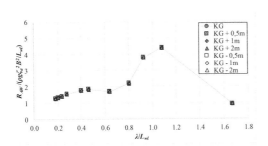

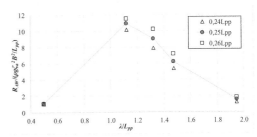

Figure 9. Influence of VCG approximation on coefficient of added resistance in waves for KVLCC2.

Figure 11. Influence of k_{yy} approximation on coefficient of added resistance in waves for KCS.

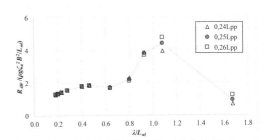

Figure 12. Influence of k_{yy} approximation on coefficient of added resistance in waves for KVLCC2.

formulae for approximation of the VCG based on the ship main dimensions and form coefficients, roll period as well as position of the centre of buoyancy (Schneekluth & Bertram 1998, Papanikolaou, 2014). Even thou motions are one of the main influence parameter of added resistance in waves, and are calculated with respect to position of the centre of gravity, the influence of changed VCG position seems to be negligible. On the other hand, the influence of the variation of pitch gyration radius is noticeable. The initial pitch radius of gyration for S-175 container ship is 24% L_{pp}. The usual range for approximation of the pitch gyration radius is 22–28% L_{pp}. Within this research, 24–26% L_{pp} variation is investigated for all benchmark ships. In the case of S-175, the influence on the results of added resistance coefficient can be noticed in the range of moderate λ/L ratios, where the difference between the initially obtained results and the ones re-calculated using 26% L_{pp} for pitch gyration radius is 40–50% (Fig 10). Despite the fact that maximum relative deviation is obtained in the case of longest wave or highest λ/L ratio, absolute value of added resistance coefficient for that point is significantly lower than the values obtained in the range of moderate λ/L ratios. Added resistance in waves due to very low relative motions in long waves significantly decreases.

A very similar trend can be noticed for KCS container ship (Fig. 11). However, relative deviations of these results are within 15% in the entire investigated range for both considered radius variations. Pitch gyration radius larger than 25% of L_{pp} leads to higher values and smaller radius leads to lower values of added resistance coefficient, as expected.

Relative deviations between the results obtained for KVLCC2 are insignificant except for λ/L ratio where maximum value of added resistance coefficient occur. Relative deviations of the added resistance coefficient obtained with gyration radius equal to 24% L_{pp} show underestimation for about 10% and with gyration radius of 26% L_{pp} overestimation for about 8% compared with initial pitch gyration radius equal to 25% L_{pp}. Differences between the results obtained for the longest considered wave are larger but the absolute value of added resistance coefficient is significantly smaller compared to the peak value.

For higher Froude numbers, for all three ships, the influence of the approximated pitch radius of gyration seems to be smaller compared to the values obtained for nominal Froude numbers unlike the approximated VCG position whose influence remains almost negligible.

The obtained results regarding the influence of approximated gyration radius for roll (k_{xx}), as expected, in the case of head waves was non-existent.

Added resistance in regular head waves obtained for KVLCC2 in ballast condition is compared to the available experimental data (Fig. 13). The trend is well captured, however, underestimation of the results is noticeable in the entire considered range of λ/L ratio. The deviation between the results was expected in short waves due to diffraction part of added resistance that is generally poorly captured by 3D panel method based on free surface Green function. However, in the region of moderate wavelengths, the results also show significant underestimation compared to experimental data. The results obtained using correction for the short wave region give much better agreement, even thou still underestimate them in comparison with the experimental results.

The uncertainty of towing tank experiments should not be neglected since it can be large, especially when second-order force is measured in very short waves. The noise introduced into the measured values has to be reduced in post-processing. The results show that lower frequency noise also exists possibly due to uncertainty in wave generation and disturbance caused by connection with carriage (Guo & Steen 2011). When it comes to towing a ship in ballast condition, these

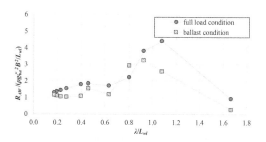

Figure 14. Comparison of added resistance coefficient for KVLCC2 in full load and ballast condition.

uncertainties might become even larger. With significant reduction in draft of the ship and larger ship motions, wetted surface area, as well as shape and area of the waterplane, effectively changes much more than in full load condition.

Considering larger ship motions in ballast condition and significant change in wetted surface area, panel code fail to perform pressure integration properly and provide reliable results.

In Figure 14 the comparison of numerically obtained results in full load and ballast condition for KVLCC2 without correction is shown. In the range of moderate wavelengths, i.e. outside the area of short waves where an increase in added resistance is expected, values of the added resistance coefficient in ballast condition are lower compared to values in full load condition. Also, the maximum value of added resistance is shifted towards higher frequencies, which coincide with the results that Park et al. 2016 obtained in their work. On the other hand, in the range of short wavelengths, added resistance coefficients are lower than the ones in full load condition. Considering the large deviation between the numerically and experimentally obtained results regarding added resistance in ballast condition, the comparison of the results in full load and ballast condition is not completely reliable.

4 CONCLUSIONS

In this paper, numerical investigation of added resistance in regular head waves was conducted using commercial software based on the potential flow theory. 3D panel method used in calculations is based on free surface Green function and the added resistance force is obtained by the direct pressure integration over wetted surface area. The obtained results are compared with the available experimental data. Very good agreement between the numerical and experimental results can be noticed for KCS containership, while

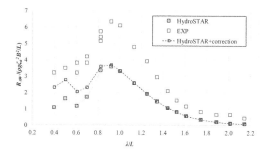

Figure 13. Comparison of the numerically obtained results and experimental data for KVLCC2 in ballast condition.

numerically obtained results for S-175 container ship and KVLCC2 tanker show significant under-estimation compared with experimental data. The influence of approximated ship mass character-istics, i.e. position of center of gravity and gyra-tion radii was investigated by varying position of vertical centre of gravity in positive and negative direction regarding the initial position and varying the gyration radii as percentage of ship length. It can be concluded that influence of approximated vertical centre of gravity is almost non-existent, as well as the influence of roll gyration radius, which was expected. On the other hand, it is quite important to determine pitch gyration radius of a ship as accurate as possible considering the sig-nificant effect that this parameter has on the added resistance.

When it comes to evaluation of the added resist-ance in waves for a ship in ballast condition, for KVLCC2 tanker, panel method shows significant underestimation in the entire investigated range of ratio of wavelength and ship length, even with cor-rection for short waves. Due to significant change in wetted surface area and effective hull shape in ballast condition, caused by larger ship motions, panel method failed to provide reliable results. However, the possible high uncertainty of towing tank measurements in ballast condition should be also taken into account.

REFERENCES

Arribas, F.P. 2007. Some methods to obtain the added resistance of a ship advancing in waves. *Ocean Engineering* 34: 946–955.

Bolbot, V. & Papanikolau, A. 2016. Parametric, multi-objective optimization of ship's bow for the added resistance in waves. *Ship Technology Research-Schiffstechnik* 63(3): 171–180.

Duan, W. & Li, C. 2013. Estimation of Added resistance for Large Blunt Ship in Waves. *Journal of Marine Science and Application* 12: 1–12.

Fujii, H. & Takahashi, T. 1975. Experimental Study on the resistance increase of a ship in regular oblique waves. *Proceedings of 14th ITTC, International Towing and Tank Conference, National Research Council of Canada*, 351–359.

Guo, B. & Steen, S. 2011. Evaluation of added resistance of KVLCC2 in short waves. *Journal of Hydrodynamics* 23(6): 709–722.

HYDROSTAR for Experts, v7.3, 2016. Bureau Veritas, Paris, France.

Liu, S. & Papanikolaou, A. 2016. Fast approach to the estimation of the added resistance of ships in head waves. *Ocean Engineering* 112: 211–225.

Luo, W., Guo, C., Wu, T., Dai, S. & Su, Y. 2016. Numeri-cal Simulation of Viscous Flow Field around Ships in Ballast. *Journal of Coastal Research* 32(4): 911–922.

Martić, I., Degiuli, N. & Ćatipović, I. 2016. Towards understanding the ship added resistance in waves. *Proceedings of the XXII. Symposium on the Theory and Practice of Shipbuilding In memoriam prof. Leopold Sorta*, 8th-10th October 2016., Seget Donji, Croatia.

el Moctar, O., Sigmund, S. & Schellin, T.E. 2015. Numer-ical and experimental analysis of added resistance of ships in waves. *Proceedings of the ASME 2015, 34th International Conference on Ocean, Offshore and Arc-tic Engineering OMAE2015*, May 31-June 5 2015, St. John's, Newfoundland, Canada.

Ozdemir, Y.H. & Barlas, B. 2017. Numerical study of ship motions and added resistance in regular incident waves of KVLCC2 model. *International Journal of Naval Architecture and Ocean Engineering* 9: 149–159.

Papanikolaou, A. 2014. Ship Design – Methodologies of Preliminary Design. Springer.

Park, D.-M., Lee, J. & Kim, Y. 2015. Uncertainty analy-sis for added resistance experiment of KVLCC2 ship. *Ocean Engineering* 95: 143–156.

Park, D.-M., Kim, Y., Seo, M.-G. & Lee J. 2016. Study on added resistance of a tanker in head waves at different drafts. *Ocean Engineering*, 111: 569–581.

Schkneekluth, H. & Bertram, V. 1998. Ship Design for Efficiency and Economy. Butterworth-Heinemann, Oxford.

Simonsen, C.D., Otzen, J.F., Joncquez, S. & Stern, F. 2013. EFD and CFD for KCS heaving and pitching in regular head waves. *Journal of Marine Science and Technology*, 18(4): 435–459.

Södig, H. & Shigunov, V. 2015. Added resistance of ships in waves. *Ship Technology Research-Schiffstechnik* 62(1): 2–13.

Tsujimoto, M., Shibata, K., Kuroda, M., Takagi, K. 2008. A Practical Correction Method for Added Resistance in Waves. *Journal of the Japan Society of Naval Architects and Ocean Engineers*, 8: 177–184.

Maritime Transportation and Harvesting of Sea Resources – Guedes Soares & Teixeira (Eds)
© 2018 Taylor & Francis Group, London, ISBN 978-0-8153-7993-5

Hydrodynamic and geometrical bow shape for energy efficient ship

M.A. Mosaad & H.M. Hassan
Department of Naval Architecture and Marine Engineering, Faculty of Engineering, University of Port Said, Egypt

ABSTRACT: Ship hydrodynamic studies are executed to improve hull form for minimizing the total ship resistance which reflects on capital and running cost by reducing fuel consumption. A good investigation reduction in total ship resistance is shown due to fitting bulbous bow. However, further research work on different bow shape is needed. This present work offers the effect of using new innovative X-bow shape to minimize the wave making resistance of DTMB545 model. The work involves a numerical simulated flow around hull model to estimate the wave making ship resistance using computational fluid dynamics, CFD. The numerical total ship resistance output data of DTMB5415 hull with sonar dome bow were validated with existing experimental model test. Geometrical modification on the DTMB5415 hull by adding X-bow on the fore portion is done. The work has shown significant reduction in wave making resistance due to using the X-bow for the range of Froude number from 0.3 to 0.45, if compared with the original hull of DTMB 5415 for the same range. Another work is carried out to study effect of change ranges of L/B, L/▼0.33 and CP through Maxsurf resistance software for DTMB5415 model with X-bow and shows the results of wave making resistance at different ranges.

1 INTRODUCTION

Energy Efficient Ship is a subject that has controlled the shipping world for the last years due to unsteady increase in fuel prices (WSC, 2008). Hull form modification is one of the hydrodynamical methods to reduce the power requirement by reducing the total resistance which will lead to savings in fuel consumption. Focusing on hull form modification of ship's bows which have great effect ship wave making resistance. Over the years, the ship hydrodynamic researches have created unique bow forms that reduce the ship wave making resistance.

Ship hull was designed with rack bow type which has a considerable high wave making resistances. Accordingly, many modifications were carried out on the fore portion of ship to minimize wave making resistance and create a new bow form shape called bulbous bow. Recently, there is bow form is generated called X-bow form which depends on wave cancellation theory. The effect of modifying bow shape used to test in towing tank model experiment. Today, Computational Fluid Dynamics (CFD) codes have become good and popular tools to do most of the same work in short time and low cost through geometrical 3d modelling creation.

Theoretical and experimental works were carried out for optimizing the geometrical shape of the bow.

Starting with the theoretical work attempted by Wigley (1936). It shows how far the effects of

bulb on reducing wave making resistance. The first experimental attempt to study the effect of fitting bulbous bow was carried out by Dillon & Lewis (1955). They found that there is a large reduction in wave making resistance in calm water if compared with that bulb in rough waves.

Johnson (1962) carried out experimental work on four different models. One has conventional bow and the three others have a bulbous bow with different frontal sectional area. The results have shown that decrease in required effective power by 20% for larger bulb area and 4% of effective power for the lower one. It is therefore, recommended that there is a good relation between bulb area and reduction in effective power.

Inui (1960) completed his theatrical work for determining the bulb size by using amplitude functions of regular waves from both the ship with conventional bow and ship with bulb. In the 1970's two design methods were created by Yim (1974) and Kracht (1978) for designing a bulbous bow. These are the most common design methods used in the design preliminary stage. The other literature review is related to numerical methods which carried out on different vessels types with different CFD codes. Fonfach & Guedes Soares (2010) showed simulation around series 60 model by using CFD-CFX Code. They found a good agreement between CFD-CFX Code and experimental work when used k-e turbulence model. The authors modified series 60 model to suit with two bulbs. They considered the bulbs as cylinder, in which the transversal area of the cylinder was estimated as 20% of the midship section area and the length was considered 0.04 Lpp for the first bulb and 0.06 Lpp for the second bulb. They concluded that, for low Froude number (i.e.0.22) the total resistance coefficient was reduce by 3.19% for the first bulb and 10.35% for the second one. At high Froude number (i.e.0.34), the total resistance coefficient was reduced by 8.25% for the first bulb and 9. 73% for the second one. Grzegorz Filip et al. (2013) carried out their study on the effect of bulbous bow on container ships. They used KCS model as original model for computational test and the numerical estimation was done by using CFD-OpenFOAM Code. They used three different bulbs with height 8 m, 9 m and 10 m permutations with three different bulb width 4 m, 5.4 m and 6.7 m. They generated nine new bulb geometry and retrofit on KCS model hull to do the numerical simulation. The result has shown that the bulb with 8 m height and 4 m width is the best one in power savings. The bulb with 10 m height and 6.7 m width have increased in total resistance by 25% if compare with the best bulb. Xie & Li (2012) used CFD-FLUENT Code for simulation flow around Deep Sea Trawler vessel. Validation results with existing experimental model test with length 4.67 m fitted with bulbous bow have been done. They concluded that reduction in total resistance due to fitted bulb is 4.31% at Froude number 0.326, if compared with original model.

Mosaad et al. (2017) illustrated that great reductions in wave making resistance due to using of X-bow for Series 60 and KCS models by using CFD-Fluent. They have shown that reduction in wave making resistance by 6% and 27% due to

using X-bow form for series 60 and KCS model respectively, at high Froude number if comparing with original models at same test case.

In this paper, CFD code is used to estimate the total resistance of DTMB5415 model which includes two main components; viscous resistance and wave making resistance. The change in wave making resistance with bow form was validated with experimental available data of DTMB5415 model. Modification has been made for DTMB5415 to be modelled with geometrical X-bow. Numerical estimation is calculated by commercial software ANSYS-FLUENT version 14.5.In addition, another work is performed to show the effect of some parameters such as L/B, L/$\nabla^{0.33}$ and C_p for wave making resistance of DTMB5415 model with X-bow. The methodology given in Figure 1 is used to simulate and evaluate new innovative X-bow form for DTMB5415 model and study the effect of change of some variables on wave making resistance of DTMB 5415 model with X-bow.

2 LOGIC OF WORK

The methodology of this paper is shown in Figure1. This methodology began with use an existing experimental work of DTMB5415 model and is validated with results of CFD-Fluent. Maxsurf is another tool used to predict total resistance of DTMB5415 hull with X-bow and is validated with result of CFD-Fluent to choose the optimum method and reduce consume of computer run time. Finally, the effect of change some parameters on DTMB5415 with X-bow hull resistance are illustrated.

3 MAIN CHARACTERISITICS AND EXPERIMENTAL DATA OF DTMB5415

The hull form selected in this study is DTMB 5415 Model which suitable for a preliminary design of navy combatant. DTMB 5415 model hull geometry includes both a sonar dome and transom stern. The main dimensions of original model of DTMB 5415 are shown in Table 1 and their perspective

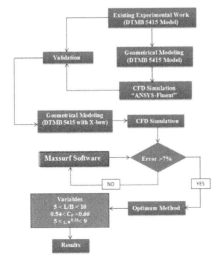

Figure 1. Logic of work.

Table 1. Main dimensions of DTMB 5415 Model.

DTMB5415	
L_{PP} (m)	**5.72**
B (m)	**0.76**
T (m)	**0.248**
S (m²)	**4.786**
▼ (m³)	**0. 549**
C_B (....)	**0.506**

Figure 2. Perspective View of DTMB5415 Model with sonar dome.

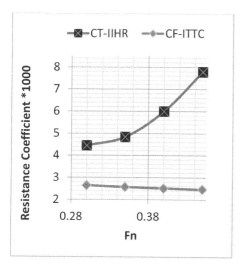

Figure 3. Experimental data of DTMB5415.

views of DTMB5415 model are shown in Figure 2. Experimental data of DTMB5415 is obtained from the Iowa Institute of Hydraulic Research (IIHR) (2001) and form factor (1+k) is estimated graphically to be 1.482. Figure 3 illustrates the experimental data of DTMB5415.

4 GEOMERTRICAL MODELING OF DTMB5415

The 3D solid model of original model of DTMB5415 geometry and modified hull of DTMB5415 to be with X-bow are created by using Rhinocerocs3d software. The X-bow design was recommended by Patent Cooperation Treaty (PCT) (2011). Modified hull model have the same principle dimensions of original hull with slightly different in wetted surface area and length over submerged (LOS). DTMB 5415 with Sonar dome Bow and X–bow are shown in Figure 4.

5 CFD SIMULATION SETUP OF DTMB5415 AND VALIDATION

CFD simulation is carried out using ANSYS-Fluent code to predict resistance components of DTMB5415 model with sonar dome bow in calm water.

Specification parameters of CFD simulation setup are shown in Table 2 and the pressure distribution contour at a maximum Froude number of the DTMB5415 is shown in Figure 5.

Comparison between experimental results and numerical CFD-FLUENT output at different Froude number for wave making resistance coefficient are illustrated in Figure 6.

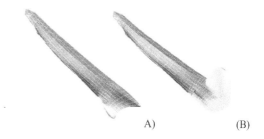

A)   (B)

Figure 4. DTMB 5415 Model with: A) Sonar dome Bow and B) X-Bow.

Table 2. parameters of CFD Simulation Setup.

Variables	Specification
Computational domain size	• Computational domain is taken as Box shape • Length of domain in front the hull is L_{pp} • Length of domain in behind the hull is 3 L_{pp} • Height of domain is L_{pp} • Width of domain is 3 L_{pp} • Hull at center line due to symmetry
Turbulence model	• Standard k-ω type
Mesh study	• Mesh type is unstructured Tetrahedral • cells number is 1.89 M • $y^+ \approx 30$ with 10 Layer of inflation
Boundary conditions	• Inlet is equal to normal speed • Outlet is represent the static pressure • Free surface elevation is taken steady • Hull, Side and Bottom is wall with no slip • Asymmetry plane condition at Y is zero • Degree of freedom of Rigid body is Y translation, Z rotation and X velocity is zero speed

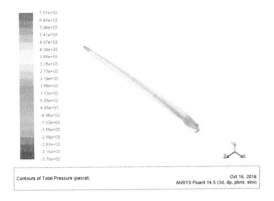

Figure 5. Pressure distribution along DTMB5415 model Sonar Dome at $F_n = 0.45$.

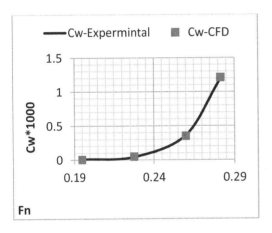

Figure 6. Validation of DTMB5415 Model.

Figure 6 shows that, the percentage error between CFD-FLUENT code and experimental data of DTMB5415 model with sonar dome for wave making resistance coefficient is not exceed 7% and this acceptable value for validation.

6 NUMERICAL SIMULATION OF DTMB5415 MODEL WITH X BOW

Numerical simulation was carried out on DTMB5415 model with X-bow. It has the same domain size, number of mesh, boundary conditions and turbulence model which are illustrated in Table 2.

Figure 7 shows the pressure distribution of DTMB5415 model with X-bow at Froude number is 0.45 and Figure 8 shows comparison between results of wave making resistance for DTMB5415 with sonar dome and with X-bow.

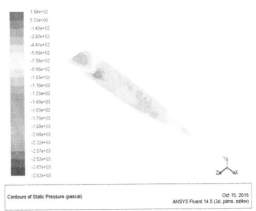

Figure 7. Pressure distribution along DTMB 5415 model with X-bow at $F_n = 0.45$.

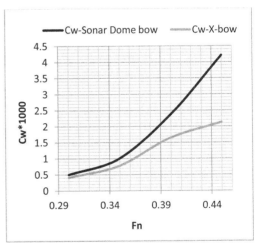

Figure 8. Wave making resistance coefficient of DTMB5415 hull with Bulbous and with X-bow.

It is obvious that there are reductions in wave making resistance at low Froude number about 18%. For intermediate Froude number, the reduction change from 22% to 30% by increase hull speed.

At high Froude number, the reduction of wave making is about 50%.

7 MAXSURF RESISTANCE SOFTWARE COMPUTATIONS

Maxsurf Resistance is naval architecture software that can predict hull resistance with different forms. Maxsurf Resistance software includes some empirical methods such as Holtro, Van oorsmeson and Fung's method to predict hull resistance components.

In this section, Maxsurf Resistance is used to predict resistance components of DTMB5415 with X-bow model at change the original ranges of length beam ratio (L/B), slenderness ratio (L/∇0.33) and prismatic coefficient (C_p) after that choose the optimum empirical method for estimation.

8 DATA VALIDATION OF MAXSURF AND CFD-FLUENT FOR DTMB5415

In this study, the choose of optimum empirical method for estimation depends on the accuracy of results if compare with CFD-results

Figure 9 shows the validation of results between three different empirical methods Holtrop, Van Oortmersse and fung's method with CFD-results.

Figure 9 shows that the best empirical method can be used for predicate the resistance coefficient components is Fung's method.

9 RESULTS AND DISCUSSIONS OF DTMB5415 MODEL WITH X-BOW

Fung's method is used to predicate wave making resistance coefficient for DTMB 5415 hull model at different ranges of L/B, L/∇0.33 and CP.

Figure 10 shows the wave making resistance coefficients of DTMB545 model with an X-bow at different ranges of length-beam ratio against Froude number. It is clear that at low range of L/B, there is an increase in wave making resistance by increase Froude number. At low beam length ratio, the reduction of wave making resistance between value of High Froude number (i.e 0.45) and Low Froude number (i.e 0.35) is 87% but the

reduction of the wave making resistance between value of high Froude number (i.e 0.45) and Low Froude number (i.e 0.35) at high beam length ratio is 81%.

Figure 11 shows the results of the wave making resistance coefficients of DTMB545 model with X-bow at different ranges of slenderness ratio against Froude number. The wave making resistance coefficient has increased dramatically with increase of slenderness ratio. For low and medium range of slenderness, there are an steady reduction in wave making resistance between value of high Froude number (i.e 0.45) and Low Froude number (i.e 0.35) ratio by 75% but at high slenderness ratio the reduction becomes 94%. Figure 12 illustrates the wave making resistance coefficients of DTMB545 model with X-bow at different ranges of prismatic coefficient against Froude number.

Increase of prismatic coefficient has no effect of wave making resistance along the range for low and medium.

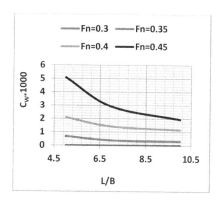

Figure 10. Wave making resistance coefficient of DTMB 5415 with X-bow at Different Ranges of L/B.

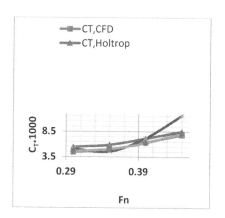

Figure 9. Total resistance coefficient of DTMB5415 by CFD-Fluent and three empirical methods.

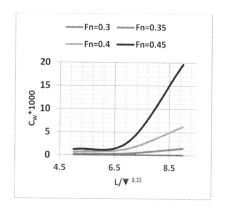

Figure 11. Wave making resistance coefficient of DTMB 5415 with X-bow at Different Ranges of L/∇1/3.

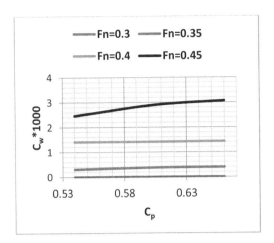

Fn=0.3 · · · · · · **Fn=0.35**

Fn=0.4 ——— **Fn=0.45**

Figure 12. Wave making resistance coefficient of DTMB 5415 with X-bow at Different Ranges of C_P.

Froude number but increase the value of wave making resistance for high Froude number by 16%.

10 CONCLUSIONS

An investigation was performed out on the resistance of a DTMB5415 model by CFD-Fluent software. There was good agreement between the experimental data and CFD for DTMB5415 model.

DTMB5415 model with X-bow illustrates a good reduction in wave making resistance at high Froude number by 49%, in contrast with the original hull. The optimum empirical method used to estimate the total resistance coefficient of DTMB 5415 model is Fung method. The increase of length beam ratio of DTMB 5415 with X- bow shows a good reduction in wave making resistance. The increase of the slenderness ratio of DTMB 5415 with X-bow, show that the wave making resistance coefficient is increased at low Froude number and increased significantly at high Froude number. The prismatic coefficient of DTMB 5415 with X- bow has no effect on ship resistance at low and medium Froude number.

REFERENCES

Dillon, E.S., Lewis, E.V. & Scott, E. (1955). Ships with Bulbous Bows in Smooth Water and in Waves. Trans. SNAME, 63, 726.

Doust, D.J. (1962). Optimised Trawler Forms. National Physical Laboratory.

Fonfach, J.M.A. & Guedes Soares, C. (2010). Improving the resistance of a series 60 vessel with a CFD code. In V European Conference on Computational Fluid Dynamics.

Grzegorz Filip, et al. (2013). Bulbous Bow Retrofit of a Container Ship Using an Open-Source Toolbox. Transactions of SNAME.

Inui, T., Takahei, T. & Kumano, M. (1960). Wave profile measurements on the wave-making characteristics of the bulbous bow. Journal of Zosen Kiokai, 1960(108), 39–51.

Kamsvag, G. (2011). Fore Ship Arrangement for a Vessel of Displacment Type. Patent Application Publication.

Kracht, A.M. (1978). Design of bulbous bows. SNAME Transactions, 86, 197–217.

Mosaad. M.A. et al. (2017). On the Design of X-bow for Ship Energy Efficiency. RINA.

Olivieri, A. et al. (2001). Towing Tank Experiments of resistance sinkage and trim, boundary layer, wake and free surface flow around a naval combatant 2340 model. The University of Iowa.

Wigley, W., (1936). The theory of the bulbous bow and its practical application. North East Coast Institution of Engineers and Shipbuilders.

World Shipping Council (WSC), (2008). Record fuel prices place stress on ocean shipping.

Yim, B. (1974). A simple design theory and method for bulbous bows of ships. Journal of Ship Research, 18(3).

Yonghe, L. et al. (2012). Study on Optimal Design of Bulbous Bow for Deep Sea Trawlers Based on Viscous Flow Theory. International Society of Offshore and Polar Engineers.

Maritime Transportation and Harvesting of Sea Resources – Guedes Soares & Teixeira (Eds)
© 2018 Taylor & Francis Group, London, ISBN 978-0-8153-7993-5

A ship weather-routing tool for route evaluation and selection: Influence of the wave spectrum

E. Spentza, G. Besio & A. Mazzino
Department of Civil, Chemical and Environmental Engineering, University of Genoa, Italy
CINFAI Consortium, Genova, Italy

T. Gaggero & D. Villa
Department of Naval Architecture, Electrical, Electronic and Telecommunications Engineering,
University of Genoa, Italy

ABSTRACT: A tool has been developed to assess safety and efficiency of shipping routes across the Mediterranean Sea. The tool is based on combining the strip-theory programme PDSTRIP with the WavewatchIII operational model run at DICCA, University of Genoa, to assess safety, efficiency and comfort parameters of vessels and assist in route selection and optimisation. In this paper, the focus is on analysing the effect of parametric versus full wave spectra on the response of the vessel and the consequent route selection. The influence of the wave spectrum is demonstrated initially for an example sea-state localised in Eastern Liguria, and subsequently for a route from Genoa, Italy to Tangier, Morocco, with three predetermined route variations. A typical passenger ferry is considered with an operational ship speed of 15 kn, corresponding to a Froude number of 0.22. It is shown that using standard parametric formulations for the wave spectrum versus the full (numerically forecasted) directional spectrum can affect the route evaluation and potentially lead to a different optimum route choice.

1 INTRODUCTION

The reduction of ship fuel consumption is becoming more and more important in the last years for both economic aspects and environmental protection reasons. These topics influence all the ship life aspects: at the design stage with, for example, the Energy Efficiency Design Index (EEDI) or during the operational activities with the corresponding Energy Efficiency Operational Indicator (EEOI). In 2013, the mandatory evaluation of the ship energy efficiency entered into force under Chapter 4 of the MARPOL Annex VI, which concern the adoption of the two previous indexes in conjunction with the Ship Energy Efficiency Management Plan (SEEMP) for all the new and existing ships. These activities are still in progress with successive updates of the regulations.

Both the EEOI and the SEEMP have an important impact on the ship operational activities in particular for the pre-existing ships. In this context, the most promising activities to improve the ship operational energy efficiency are related to the integration of numerical modelling tools with in-service monitoring systems such as weather routing systems (DNV-GL 2014, ABS 2014). The adoption of weather routing techniques for the

ship route plan also presents further benefits such as an increase of navigation safety and on board comfort depending on the ship type. For cargo ships for example the focus could be on the prevention of cargo loss while for passenger ships (as for ferries or cruise ships) the comfort aspects are predominant (considering that the safety constraints are commonly less restrictive) to plan the ship time schedule and the detailed route between two ports without neglecting the omnipresent constraint due to the fuel consumption.

In the last years, several tools able to provide weather routing information to the ship master have been generated (Wisniewski 2005, Marie 2009, Szlapczynska 2009, Vettor 2015, 2016). Such tools are mainly focused on the optimization aspects (how to select the best route) often using simplified models for the ship responses and sea state forecasts. To overcome these limitations, a preliminary activity has been developed at the University of Genova, in collaboration with the LaMMA Consortium (Coraddu et al. 2013) in the framework of the collaborative European project COSMEMOS (COSMEMOS 2012). This activity has been further developed within a joint collaboration at the University of Genoa between the DITEN and DICCA departments each one for its area of expertise: the DITEN for the naval aspects and the DICCA for the metocean data.

The present paper, after giving a brief description of the in-house developed tool, focuses on the influence of the accuracy of sea state data on the evaluation of the ship responses. Different types of wave spectral forms are investigated ranging from the full spectrum in terms of wave amplitude and direction, to different multimodal or unimodal parametric forms, calculated using commonly available forecasted parameters (such as significant wave height and peak period) and a standard Jonswap formulation. Compressing the weather data to a few parameters is commonly required for storage capacity reasons or for cost/efficiency of the data transmission on board. The impact of this simplification will be analysed to define the expected relative quality of the output and assess the importance of using the full spectra.

The evaluation of the influence of the different sea spectral inputs is carried out by comparing different weather conditions on a typical route in the Mediterranean Sea (from Genoa, Italy to Tangier, Morocco), in terms of ship motions, Motion Sickness Index (MSI) and total forward resistance (see Eq. (1)) and their effects on the best route selection.

2 DESCRIPTION OF THE TOOL

2.1 Overview

All weather routing tools use weather data, in terms of encounter waves and wind speed with associated directions, combining them with the ship response, typically the Response Amplitude Operation (RAO) or similar ship models, to generate useful information on the ship behaviour at sea. Obviously both the input data and the ship model should have comparable accuracy. Indeed, if high-resolution weather data are combined with a too simplified ship model, or vice versa, the expected output has the accuracy of the more simplified component. The here proposed tool adopts the state of art related to the ship motion evaluation and to the weather forecast model and brings together the detailed and accurate modelling of the weather conditions and the fast and efficient assessment of the behaviour of the analysed vessel. When combined through a three-dimensional spectral analysis these two aspects produce a number of desired outputs, such as roll motions, vertical bow acceleration, added resistance in waves, motion sickness index and so on.

The tool is modular and flexible so that the inputs can be easily substituted if different source data is required and the outputs can be easily modified to suit the user needs. The overall structure

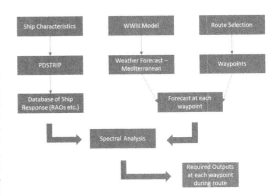

Figure 1. Flowchart of route analysis tool.

of the tool is shown in Figure 1, and the various components are described in more detail in the following sections.

2.2 Weather forecast

The weather database used in this application has been developed by the MetOcean group in the Civil, Chemical and Environmental Engineering Department at the University of Genova (DICCA). The database is based on the WaveWatch III (Tolman 2009) wave model run in operational mode to produce a 120-hour forecast updated every 12 hours. The model is run across the Mediterranean basin on a 10 km grid resolution and has been extensively validated and tested in the Mediterranean Sea (Mentaschi et al., 2013a, b; Mentaschi et al., 2015).

For the present application three different types of outputs are used from the WaveWatch III (WWIII) model, and the accuracy of each approach assessed. The three outputs, going from the more complex to the more simplified, are:

a. *full spectra: the whole spectrum is saved for 25 frequencies and 24 directions*
b. *partitioned spectra: only integral parameters relating to 3 spectral partitions (wind sea and two swell partitions) are considered*
c. *averaged parameter spectra: all the data are compressed by means of integrated parameters relating to the total sea state (significant wave height, peak period, peak direction).*

Options (b) and (c), obviously, involve reconstructing the full spectra using average sea state parameters and assuming a Jonswap spectrum for each partition with spectral and spreading parameters applicable to the Mediterranean Sea. This technique is widely adopted in the field of meteorology to compress the huge amount of data required to store all the frequencies and wave directions for a portion of sea commonly discretised with a fine regular grid. These type of data compressions are very useful in the weather forecast field, where, commonly, the global expected weather conditions are more important. On the contrary if the metocean data are used as input for a ship motion tool, all the wave components, with their respective directions, take on greater importance.

2.3 Hydrodynamic ship response

The hydrodynamic modelling of the vessel is performed using the Rankine-source strip theory programme PDStrip (Bertram et al. 2006). PDStrip is fast, allows the user to incorporate fins and append-ages to model viscous damping, and includes forward speed. The accuracy of PDStrip versus other panel methods and strip theory codes has been verified internally by the present authors and satisfactory results have been obtained. In addition, Gourlay et al (2015) have conducted extensive comparisons between AQWA, GL Rankine, MOSES, OCTOPUS, PDStrip and WAMIT, and verified against model tests for three cargo ships, demonstrating good performance of PDStrip versus the other codes and the model tests.

In the present application, PDStrip is used to calculate motion, velocity and acceleration Response Amplitude Operators (RAOs) and added resistance in waves at the centre of gravity of the vessel. The RAOs are then transposed to any reference point on the vessel to be used in the route assessment. In any case, the proposed tool is also able to overwrite these operators adopting data provided by means of other methods, such as panel codes or viscous solvers (Grasso et al 2010). This feature however can drastically affect the computational costs.

After properly combining the weather output with the hydrodynamic responses in the frequency domain it is possible to calculate any number of outputs which are of interest in the decision-making process (for example maximum roll angle, heave motion on the bridge, vertical acceleration of the bow, total forward resistance or seasickness indices in passenger cabins). The selected parameters can be easily modified on a case-by-case basis depending on the application e.g. passenger ferry, cargo ship, heavy lift vessel etc.

The total forward resistance is of interest in most practical cases. In this work, the total forward resistance (acting along the x-axis of the vessel) has been calculated as follows:

$$F_x = R_t + F_x^w + F_x^c + R_{aw} \qquad (1)$$

where R_t is the still water resistance calculated according to Holtrop (1984), F_x^w and F_x^c are the wind and current forces respectively, and R_{aw} is the added resistance in waves calculated from PDSTRIP. The wind and current forces are calculated using a standard drag force formulation with typical drag coefficients chosen for the hull and topsides. The wind speed and direction are taken from the WWIII model. The current speed and direction is not yet available within the weather database but will be incorporated in the near future. For this application, a zero current speed is assumed.

3 APPLICATION TO WAYPOINT

3.1 Example vessel

The selection of the vessel in this application is not a focal point, therefore the proposed test case is not a real ship currently in service. However, to avoid unphysical results a preliminary ship design has been performed to collect all the necessary data (length, beam, draft, etc.) and reproduce a realistic ship for the selected route. The hull of the example vessel is typical of ferries performing routine voyages in the Mediterranean Sea, and the body plan is shown in Figure 2. The vessel has a length between perpendiculars of 127.4 m, a breadth of 22.6 m, an operating draft of 5.9 m and a displacement of 10,370 tonnes. An operational forward speed of 15 kn is assumed. The peak of the pitch RAO in head seas is at approximately 10 secs, whilst the peak of the roll RAO in beam seas is at approximately 8.5–9.0 secs for zero forward speed.

The vessel has a calm water resistance of approximately 250 kN at 15 kn. Appendages have not been included in the still water resistance calculation (this can be easily added) but fins are incorporated in PDSTRIP for calculation of the hydrodynamic damping.

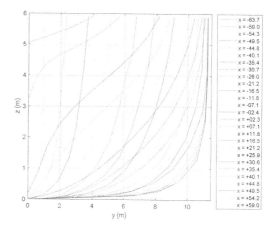

Figure 2. Body plan of example vessel.

3.2 Spectral inputs

Figures 3, 4 and 5 below show an example of options (a), (b) and (c) type of wave data input (see section 2.1) for a location offshore La Spezia on the 25th November 2016. In all spectral plots that follow the angular axis is in degrees with North at the top of the plot at 0 degrees, the wave direction convention is "Coming From", and the radial axis is the wave period in secs.

Option (a) arises directly from the WWIII model. Options (b) and (c) have been calculated using the following standard formulations:

$$S(f) = a \cdot H_s^2 \cdot T_p^{-4} \cdot f^{-5} \cdot \exp\left(-1.25\left(T_p \cdot f\right)^{-4}\right) \cdot \gamma^\beta \quad (2)$$

where

$$a = \frac{0.0624}{\left(0.230 + 0.0336\gamma - 0.185\left(1.9 + \gamma\right)^{-1}\right)}$$

$$\beta = \exp\left(-\frac{\left(T_p \cdot f - 1\right)^2}{2\sigma^2}\right)$$

$$\begin{matrix} \sigma = 0.07 \\ \sigma = 0.09 \end{matrix} \quad \text{for} \quad \begin{matrix} f \le f_p \\ f > f_p \end{matrix}$$

$$D(\theta) = \cos\left(\frac{\theta - \bar{\theta}}{2}\right)^{2s} \quad (3)$$

Equation (2) describes the Goda formulation of the Jonswap spectrum as given in ISO 19901-1:2015, where $S(f)$ is the wave energy spectrum in terms of frequency, H_s is the significant waveheight, T_p is the peak period, f_p is the peak frequency,

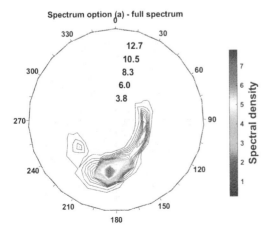

Figure 3. Example full spectrum from WWIII model. La Spezia location on 25th November 2016 02:00 hrs.

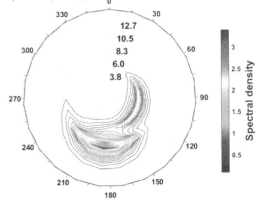

Figure 4. Reconstructed spectrum using three partitions. La Spezia location on 25th November 2016 02:00 hrs.

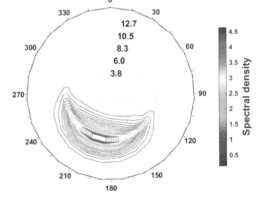

Figure 5. Reconstructed spectrum using averaged quantities (single partition). La Spezia location on 25th November 2016 02:00 hrs.

and γ is the peakedness parameter. Equation (3) describes a directional spreading function $D(\theta)$ (as per ISO 19901-1:2015) with θ being the mean direction, and s being the directional exponent.

The following available forecast parameters have been used in conjunction with these formulations:

- For option (b) – H_s, T_p and $\bar{\theta}$ for three partitions (wind sea and two swell partitions).
- For option (c) – H_s for the combined total sea-state, with T_p and θ_p corresponding to the primary wave system, as given by WWIII.

In addition, since the peakedness parameter and the directional spreading are not commonly found in available forecasts, standard values for the Mediterranean have been used: for option (b) a γ value of 2.2 for wind sea and 2.7 for swell has been chosen and a directional spreading exponent s of 6 for wind sea and 15 for swell; for option (c) a γ value of 2.5 and a directional spreading exponent s of 7.

A better comparison could be achieved by calculating the spreading exponent and the γ value from the full spectrum, and use those in the parametric formulations. However, in the majority of industry applications this information is assumed based on standard values for the area. Indeed, the most common input is a Jonswap spectrum with integrated parameters representing combined total sea condition. Therefore, it is important to include these formulations and assess the influence on the accuracy of the results compared to using the full model spectrum in as realistic a context as possible.

It should be noted that some slight differences exist between the total energy calculated from the full spectrum outputted by the model and the total average energy of the partitions in the forecast (for example see Fig. 14 in the following section). This is believed to be due to numerical accuracy when integrating the outputted spectrum (energy rounded to three significant figures) over frequency and direction during post-processing, compared to the accuracy in the WWIII model during operation.

3.3 Vessel responses

Figures 6 to 9 show corresponding example output parameters of interest and their variation with vessel heading. Vessel heading is expressed in degrees clockwise with respect to geographic North. Roll, pitch, vertical bow acceleration and added resistance in waves are compared. Pronounced differences and a slight shift in the heading producing

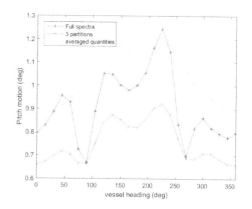

Figure 7. Comparison of pitch motion with vessel heading for the three spectral input types. La Spezia location on 25th November 2016 02:00 hrs.

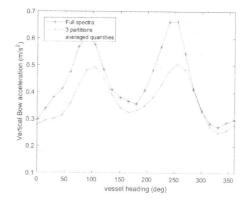

Figure 8. Comparison of vertical bow acceleration with vessel heading for the three spectral input types. La Spezia location on 25th November 2016 02:00 hrs.

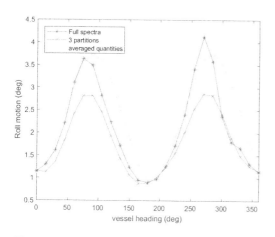

Figure 6. Comparison of roll motion with vessel heading for the three spectral input types. La Spezia location on 25th November 2016 02:00 hrs.

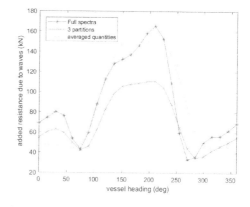

Figure 9. Comparison of added resistance in waves with vessel heading for the three spectral input types. La Spezia location on 25th November 2016 02:00 hrs.

457

the most severe responses are seen when using the averaged quantities due to the inability of this simplified spectral formulation to capture the multi-modal nature of the sea-state. Whereas using multiple partitions correctly captures the worst headings in terms of response, but the magnitude is in this case too low compared to using the full spectrum. The latter differences have been explored and are due to slight differences in the periods where most of the energy of the spectrum is located, in a range of periods where the corresponding RAOs have very steep gradients.

To further explore the differences between the three types of inputs a large number of individual points and time-steps have been considered and analysed in the same way as shown above. Overall it is seen that there are no consistent trends, where one of the three types of inputs is always lower or higher and the results show great sensitivity to small variations. These differences are explored further in the following section using as an example an entire route calculation.

4 APPLICATION TO A ROUTE

4.1 Route

The selected route for demonstration purposes is from Genoa, Italy to Tangier, Morocco (Fig. 10). In this application three predetermined variations of the route are considered, passing (1) west of the Balearics, (2) through the Balearics, and (3) east of the Balearics. The routes are extracted from the classical real ship routes performed between the two cities. The routes have been selected to be as much as possible of equal distance, so as to remove any bias in the results (it is expected that a ship Master would normally choose the shortest route). Route 1 is 887 nm, Route 2 is 899 nm and Route 3 is 903 nm. The same number of waypoints (31 in total) are maintained for ease of comparison. The last 10 waypoints of all 3 routes are the same. At a nominal speed of 15 kn arrival times are approxi-

mately 59–60 hours for all 3 routes. Three departure dates are chosen for testing and comparison: the 26th April 2017 and the 10th May 2017, both at 00:00 hrs, and the 28th June 2017, at 12:00 hrs. The results presented herein correspond to the departure date of the 26th April since for this departure date the vessel experiences the most severe responses and the results are of greater interest.

4.2 Full spectra versus parametric solutions

The Figures below show the results for roll (Fig. 11), pitch (Fig. 12) and vertical bow acceleration (Fig. 13) for Route 1 departing on the 26th April 2017. Figure 14 shows the significant wave

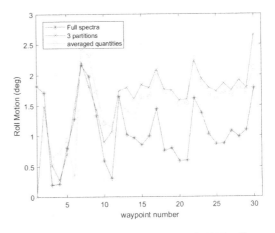

Figure 11. Route 1 departure 26th April 2017 – Comparison of roll motion during voyage calculated with the three weather input options.

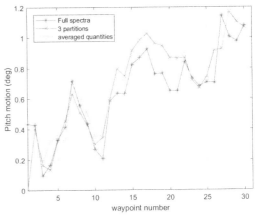

Figure 12. Route 1 departure 26th April 2017 – Comparison of pitch motion during voyage calculated with the three weather input options.

Figure 10. Chosen route variations from Genova to Tangier.

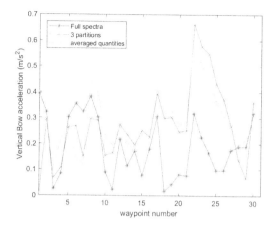

Figure 13. Route 1 departure 26th April 2017 – Comparison of vertical bow acceleration during voyage calculated with the three weather input options.

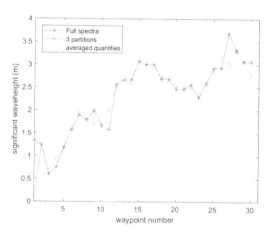

Figure 14. Route 1 departure 26th April 2017 – Comparison of significant wave height during voyage arising from the three weather input options.

height H_s along the voyage as calculated from the full spectra and as provided from the forecast and used in the parametric spectra with 3 partitions or a single partition (averaged quantities). Some minor differences are seen between the data sources, arising from numerical accuracy as discussed above. However, pronounced differences in the roll, pitch and vertical bow acceleration are seen particularly in the second half of the route which are not linked to differences in average H_s. To investigate these differences the spectra used in each option at waypoint 21 are plotted in Figures 15, 16, and 17. Waypoint 21 has been chosen since it shows large differences in all three of the responses being considered whilst no notable difference in H_s is present.

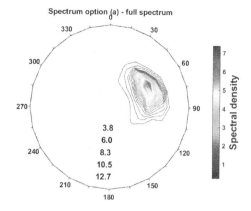

Figure 15. Route 1 departure 26th April 2017, Waypoint 21: full spectra from WWIII model.

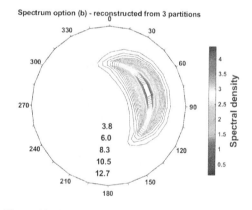

Figure 16. Route 1 departure 26th April 2017, Waypoint 21: reconstructed spectra using wind sea and two swell partitions.

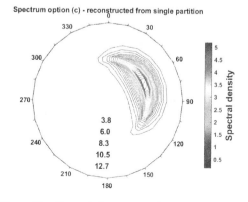

Figure 17. Route 1 departure 26th April 2017, Waypoint 21: reconstructed spectra using single average parameters.

459

Contrary to the example presented in Section 3, the sea-state in this case has only one predominant wave system. The spectra appear similar, with the main difference being in the shape, arising from the difference in peakedness parameter and directional spreading. It is clear that using a parametric spreading function with an average exponent in this case results in an overestimation of the vessel response compared to the full spectrum. Yet the directional spreading exponent is typically unavailable in most practical applications.

As discussed above for the single point comparisons, so far no consistent trend in the comparisons between the full spectra and the parametric formulations has been seen. Whether using a parametric formulation is conservative or unconservative in terms of responses appears to be varying on a case by case basis, depending as a minimum on the particular sea state, spectral shape, wave systems present, vessel heading, and vessel speed. However, if ever increasing efficiency in routing and route selection and assessment is to be achieved, a more refined weather input such as the full spectra is necessary to achieve the desired accuracy in response predictions.

The following sections proceed to make a comparison of the three route options using only the full spectra as input, and subsequently assess if the choice of route may vary if a more simplified input is used.

5 COMPARISON OF ROUTES

5.1 *Route comparison using full spectra*

Figures 18 to 21 show a comparison of the key outputs for the three route options calculated using

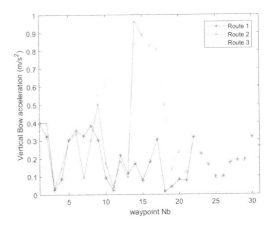

Figure 19. Comparison between the three routes, departure 26th April 2017: vertical bow acceleration.

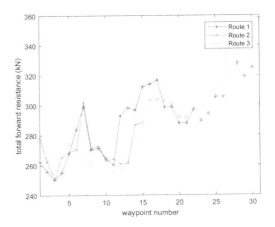

Figure 20. Comparison between the three routes, departure 26th April 2017: total forward resistance.

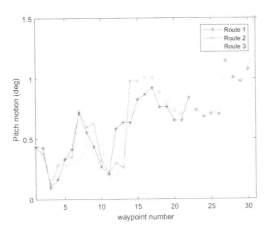

Figure 18. Comparison between the three routes, departure 26th April 2017: pitch motion.

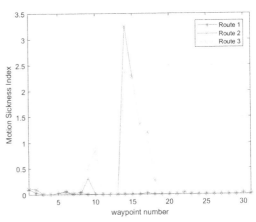

Figure 21. Comparison between the three routes, departure 26th April 2017: motion sickness index.

the full spectra for the 26th of April departure date. It can be seen that even though Route 1 experiences slightly higher Hs south of the Balearics it appears to be more favorable than Routes 2 and 3 in terms of roll motion, vertical bow acceleration and motion sickness index (MSI, calculated as per Cepowski 2012) and slightly less favorable in terms of total forward resistance. In a real weather-routing application, the final choice of the preferred route would largely depend on the particular vessel type and application. For example, a cruise ship may choose to minimize MSI, a heavy-lift vessel carrying cargo may choose to minimize accelerations and hence loading on the sea-fastenings, and a ferry performing a regular route may have a policy to minimise consumption (hence total forward resistance) subject to the MSI not exceeding a specific value.

5.2 Route comparison using different spectral inputs

Tables 1, 2 and 3 below show some comparisons of the effect of using the full spectra on the decision making versus Options (b) and (c) respectively.

Table 1. Mean and standard deviation of chosen output parameters for all three routes: full spectra, departure 26th April 2017.

Route	Route 1		Route 2		Route 3	
	Mean	Std	Mean	Std	Mean	Std
Heave (m)	0.28	0.12	0.37	0.21	0.33	0.18
Roll (deg)	1.10	0.51	1.60	1.14	1.35	0.95
Pitch (deg)	0.65	0.29	0.72	0.40	0.66	0.41
Acc_x (m/s^2)	0.62	0.29	0.59	0.33	0.54	0.36
Acc_y (m/s^2)	0.41	0.13	0.45	0.19	0.39	0.20
Acc_z (m/s^2)	0.20	0.11	0.26	0.15	0.23	0.13
Fx total (kN)	293	26.1	294	34.0	291	35.0

Table 2. Mean and standard deviation of chosen output parameters for all three routes: 3 partitions, departure 26th April 2017.

Route	Route 1		Route 2		Route 3	
	Mean	Std	Mean	Std	Mean	Std
Heave (m)	0.33	0.14	0.34	0.20	0.31	0.21
Roll (deg)	1.57	0.62	1.59	0.97	1.42	1.04
Pitch (deg)	0.69	0.31	0.61	0.39	0.54	0.40
Acc_x (m/s^2)	0.59	0.28	0.51	0.33	0.43	0.35
Acc_y (m/s^2)	0.41	0.15	0.40	0.23	0.33	0.23
Acc_z (m/s^2)	0.23	0.09	0.25	0.14	0.22	0.15
Fx total (kN)	290	21.2	285	27.6	280	28.0

Table 3. Mean and standard deviation of chosen output parameters for all three routes: averaged parameters, departure 26th April 2017.

Route	Route 1		Route 2		Route 3	
	Mean	Std	Mean	Std	Mean	Std
Heave (m)	0.29	0.11	0.34	0.17	0.35	0.15
Roll (deg)	1.24	0.44	1.57	0.90	1.55	0.85
Pitch (deg)	0.68	0.29	0.67	0.34	0.67	0.35
Acc_x (m/s^2)	0.63	0.29	0.57	0.32	0.51	0.36
Acc_y (m/s^2)	0.41	0.15	0.43	0.21	0.37	0.21
Acc_z (m/s^2)	0.19	0.08	0.24	0.12	0.25	0.11
Fx total (kN)	291	21.3	288	26.1	289	29.0

The reported accelerations are on the bow, at a position which is 40 m forward of amidships, on the centerline in y, and 15 m above keel. As an initial simplified means of comparing the routes the mean and standard deviation of each parameter of interest is considered along the route. This does not reflect the exact situation of using an optimisation algorithm to select the next waypoint in real time, however provides some valuable insight into the sensitivity of the calculation.

It is seen, for example, that if selecting a route to minimize roll using the full spectra, Route 1 may be selected whilst if using the three partitions Route 3 may be selected. Similarly, if minimizing forward resistance Route 3 would be selected if calculated using the full spectra or the three partitions, whilst Route 2 would be selected if using the averaged parameters. These results further point to the sensitivity of the calculation and the need to use as accurate input data as possible for optimized results.

6 CONCLUSIONS

A tool for calculating the vessel responses as part of a weather-routing system has been presented. Three types of spectral inputs, ranging from the complex to the simple, have been compared, and the accuracy of the results assessed in terms of the vessel performance and responses for an example vessel and a variety of weather conditions. Overall it is seen that not using the full spectrum, as is common practice in other weather-routing tools, can lead to differences and possibly also affect the final decision-making.

It should be noted that the tool presented herein is being integrated into a full optimization algorithm which will employ the full spectra in the decision-making. In the future, it will be possible to assess the influence of high quality weather data on the final chosen route in a more complete manner, compared to the preliminary comparisons presented in Section 5 above.

ACKNOWLEDGMENTS

The authors would like to thank the Italian flagship project RITMARE for the financial support and computational infrastructure.

REFERENCES

ABS (2014). Ship energy efficiency measures advisory. Status and guidance, American Bureau of Shipping, 2014.

Bertram, V., Veelo, B. & Soding, H., (2006). Program PDSTRIP: public domain strip method. software documentation.

Cepowski, T. (2012). The prediction of the Motion Sickness Incidence index at the initial design stage. Published in the Scientific Journal of the Maritime University of Szczecin, 31(103) p. 45–48.

Coraddu, A., Figari, M., Savio, S., Villa, D. & Orlandi, A. (2013) Integration of seakeeping and powering computational techniques with meteo-marine forecasting data for in-service ship energy assessment In C. Guede Soares and L. Pena eds. Developments in Maritime Transportation and Exploitation of Sea Resources IMAM2013, CRC Press Balkema, Taylor & Francis Group London, p. 93–101.

COSMEMOS—COoperative Satellite navigation for MEteomarine MOdelling and Services, (2012) European project FP7 3rd Call, Contract Number 287162.

DNV-GL, (2014). Energy management study 2014.

Gourlay, T., von Graefe, A., Shigunov, V. & Lataire, E. (2015). Comparison of AQWA, GL RANKINE, MOSES, OCTOPUS, PDSTRIP and WAMIT with model test results for cargo ship wave-induced motions in shallow water. Proceedings of the ASME 2015 34th International Conference on Ocean, Offshore and Arctic Engineering, May 31–June 5, 2015, St. John's, Newfoundland, Canada. OMAE2015-41691.

Grasso, A., Villa, D., Brizzolara, S. & Bruzzone, D., (2010). Nonlinear motions in head waves with a RANS and a potential code. Journal of Hydrodynamics, Ser. B, Volume 22, Issue 5, Supplement 1, 10–15 October 2010, p. 172–177.

Holtrop, J. (1984). A statistical re-analysis of resistance and propulsion data. International Shipbuilding Progress 28 (363), 272–276, Delft University Press.

ISO 19901-1:2015. Petroleum and natural gas industries – Specific requirements for offshore structures – Part 1: Metocean design and operating considerations.

Marie S. & Courteille E. (2009). Multi-Objective Optimization of Motor Vessel Route. Marine Navigation and Safety of Sea Transport. Balkema Book, Taylor & Francis Group. London.

Mentaschi L., Besio G., Cassola F. & Mazzino A. (2013) Problems in RMSE-based wave model validations. Ocean Modelling, 72, pp. 53–58.

Mentaschi L., Besio G., Cassola F. & Mazzino A., (2013) Developing and validating a forecast/hindcast system for the Mediterranean Sea. Journal of Coastal Research, SI 65, pp. 1551–1556.

Mentaschi L., Besio G., Cassola F. & Mazzino A. (2015) Performance evaluation of WavewatchIII in the Mediterranean Sea. Ocean Modelling, 90, pp. 82–94.

Orlandi, A., Capecchi, V., Rovai, L., Benedetti, R., Romanelli, S., Ortolani, A., Coraddu & Villa, D. (2015), Ship performances forecasting at the mediterranean scale: evaluation of the impact of meteocean forecasts on fuel savings for energy efficiency and weather routing, In Proceedings of 18th International Conference on Ships and Shipping Research NAV2015, June 24–26, Lecco, Italy, pp 538–549.

Orlandi, A., Pasi, F., Capecchi, V., Coraddu, A. and Villa, D., (2015) Powering and seakeeping forecasting for energy efficiency: Assessment of the fuel savings potential for weather routing by in-service data and ensemble prediction techniques Proceedings of 16th International Congress of the International Maritime Association of the Mediterranean IMAM 2015, 21–24 September 2015, Pula, Croatia.s

Tolman, H.L., (2009) User manual and system documentation of WAVEWATCHIII version 3.14. Technical Report NOAA/NWS/NCEP/MMAB.

Vettor, R. and Guedes Soares, C. (2015). Multi-objective evolutionary algorithm in ship route optimization. Guedes Soares, C. & Santos T.A. (Eds.). Maritime Technology and Engineering. London, UK: Taylor & Francis Group; pp. 865–876.

Vettor, R. & Guedes Soares, C. (2016) Development of a ship weather routing system. Ocean Engineering 123, 1–14.

Wisniewski B. & Chomski J. (2005). Evolutionary algorithms and methods of digraphs in the determination of ship time-optimal route, 2nd International Congress of Seas and Oceans, Szczecin-Świnoujście.

Hydrodynamics – seakeeping

Maritime Transportation and Harvesting of Sea Resources – Guedes Soares & Teixeira (Eds)
© 2018 Taylor & Francis Group, London, ISBN 978-0-8153-7993-5

Influences of a bow turret system on an FPSO's pitch RAO

T.S. Hallak

*Centre for Marine Technology and Ocean Engineering (CENTEC), Instituto Superior Técnico,
Universidade de Lisboa, Lisbon, Portugal*
Previously at Numerical Offshore Tank (TPN), Universidade de São Paulo, SP, Brazil

A.N. Simos

*Department of Naval Architecture and Ocean Engineering, Polytechnic School, Universidade
de São Paulo, São Paulo, Brazil*

ABSTRACT: The influences of mooring lines and risers on the first order responses of a floating system moored with catenary cables have not yet been systematically studied; likewise, it's not well understood when these influences may be seen primarily as a kind of added stiffness, added damping or added inertia, if not always a complex superposition of influences. The following paper presents several analyses and comparisons between methods in order to assess the actual influences of a bow turret in the pitch RAO of an FPSO system. Both frequency and time domain simulations were performed; and the results were compared with outputs from simplified algebraic formulations. Different sea depths and lines' parameters/configurations were considered in order to provide consistent comparisons and a coherent separation of influences into geometric stiffness, elastic stiffness, damping and inertia. The separation of influences gives important insights for further considerations of the turret influences on floating systems.

1 INTRODUCTION

Floating, Production, Storage & Offloading (FPSO)'s hulls were not initially designed to be moored in deep waters; moreover, since risers connect the vessel to the seabed and it's desirable to have the most profitable operability at the oil field, ship owners must choose a solution for the vessel's station keeping problem. An inventive solution for this problem is the turret mooring system; firstly designed in the 80s for an FSO system, it's still regarded as an advantageous solution, especially for FPSOs and Floating Liquefied Natural Gas (FLNG) systems. The turret system may also be referred as Single Point Mooring (SPM), since all mooring lines are connected to the vessel at the fairlead, which, for analytical purposes, may be seen as a single point at the vessel's hull. The risers, on their turn, are connected to the vessel through the inner part of the turret unity; this enables the vessel to weathervane around the fairlead without compromising the cables integrity.

In the latest years, several authors published studies on the turret system, e.g. (Kannah & Natarajan, 2006) and (Xie et al. 2015), who consolidated the idea that bow turrets are indeed the most effective in the purpose of reducing dynamic responses—when compared to other turret locations; (Howell et al. 2006), who performed a comparison between

SPM and Spread Mooring System (SMS), indicating several relative advantages of the turret system; and (Soares et al. 2005), who verified that turret moored FPSOs' RAOs are indeed linear in respect to wave amplitude for the vertical modes (surge, heave and pitch).

A turret mooring system, however, is designed basically in the purpose of minimizing vessel's offset from the stationary condition when drift forces are present due to ambient action (wind, current and waves), a problem governed by higher order effects. The influences of the turret system on the vessel's first order responses, on the other hand, still wait for a deep systematic study.

Moreover, the analyses presented in this paper had another motivation, i.e. the method addressed to (Simos et al. 2010, 2012), where a Bayesian inference model is developed in order to assess a sea directional wave spectrum based on the motions of a stationary vessel. Some of the model's most relevant inputs are the RAOs from a specific FPSO. Given that the method performs the spectral crossing in inverse order, which constitutes an underdetermined system, it was not known if discrepancies—at first look non-substantial—in the FPSO's RAOs were compromising the method, whereas systematic errors have been observed in the method's outputs. These discrepancies could have been a consequence of turret's influences.

Likewise, it's still not well understood what the attributes of these influences are. Even in a simplified linear model for the coupled dynamics, mooring lines and risers' influences may be composed as a sum of inertial, damping and stiffness influences—the latter may also be separated into elastic stiffness (due to line's stretching) and geometric stiffness (due to lines' buoyancy increments and decrements). The original objectives of the research were so to provide better understanding on the subject; and collaborate effectively with the inference model, indicating if the turret influences may be neglected or not.

Along the research lines, several conclusions were drawn, starting from quantitative conclusions for the specified turret-moored FPSO used by the inference model; and general qualitative conclusions for any bow-turret-moored system. As we'll see further on, the turret influences are relevant as long as cables' total suspended mass is large, thus only for deep waters' systems and substantial amount of cables; the research also points out that inertial influences are notably the most relevant.

2 METHODOLOGY

All analyses performed correspond to a single FPSO, whose main characteristics are presented in Table 1. Vessel's mooring systems' and risers' parameters presented in Table 2 correspond to two different scenarios: shallow waters (actual depth of the FPSO used by the inference method) and deep waters. In both of them, mooring lines are considered homogeneous. A third, hypothetical scenario, characterized by almost vertical cables, was also considered for the analyses. For this, all parameters

Table 1. FPSO's main dimensions.

Parameter	Description	Value
L_{PP}	Length between perpendiculars	320.0 m
L_{OA}	Total length	345.2 m
B	Moulded beam	58.0 m
D	Moulded depth	31.0 m
T	Intermediate draft	15.5 m
LCG*	Centre of gravity longitudinal position	163.0 m
KG**	Centre of gravity vertical position	15.5 m
Ltur*	Turret's fairlead longitudinal position	276.6 m
Δ	Vessel's displacement at op. draft	225,000 ton
Rgyr	Pitch's radius of gyration	24.0% Loa

*Counted from aft. perpendicular.
**Counted from baseline.

Table 2. Mooring Lines and risers' parameters.

Shallow waters (scenario 1)		Mooring	Risers
Parameter	Description	Values	
H	Sea depth	162.2 m	162.2 m
φ_W	Slope at fairlead	34.7°	72.6°
L	Total length	993 m	450 m
C_M	Added mass coeff.	2.25	1.00
C_D	Drag coeff.	2.16	1.20
w_0	Weight in air	1.58 kN/m	2.32 kN/m
EA	Axial stiffness	775 MN	588 MN
D	Diameter	0.10 m	0.38 m
μ	Friction coeff.	0.98	0.98

Deep waters (scenarios 2 & 3)		Mooring	Risers
Parameter	Description	Values	
H	Sea depth	1380 m	1380 m
φ_W*	Slope at fairlead	81.9°	83.0°
L*	Total length	2186 m	1890 m
C_M	Added mass coeff.	1.00	1.00
C_D	Drag coeff.	1.23	1.20
w_0	Weight in air	1.06 kN/m	0.66 kN/m
EA	Axial stiffness	241 MN	401 MN
D	Diameter	0.14 m	0.24 m
μ	Friction coeff.	0.98	0.98

*Different for scenario 3 (as specified in text).

are set equal to deep waters case with an exception, i.e. cables' slope at the fairlead are all set equal to $\varphi_W = 87.0°$, and cables' total length changing accordingly to maintain cables' total suspended mass (thus 2175 m of total length for mooring lines and 2036 m for risers). In this scenario, due to the almost vertical cables' profile, all damping and added mass influences are expected to be ineffective. For all time domain simulations, waves with 1.0 meter of amplitude were considered, which is in accordance with the actual mean sea state of scenario 1.

For scenario 3 simulations, the total suspended mass of mooring lines and risers was systematically increased to twice and four times the original value. This allowed the assessment of the marginal influences of cables' inertia in the vessel's first order response.

To validate all meshes, FPSO's RAOs were numerically evaluated by the low-order panel method by WAMIT®—frequency domain—but also on Ansys® AQWA's frequency and time domain simulation plug-ins. All curves matched. For time domain simulations with coupled dynamics, AQWA meshes all cables into 100 segments (maximum possible discretization), and then applies Morison's formula considering an embedded lumped mass method to simulate cables' dynamics.

In order to provide consistent analyses on the separated influences, i.e. geometric stiffness, elastic stiffness, damping and inertia, several approaches were taken. Software WAMIT was used changing systematically the values of external stiffness matrix; and comparing the results obtained for all scenarios. All values inputted within the matrix are in accordance with the algebraic formulation presented in the sub-section 2.1; afterwards, results were compared with stiffness parameters—defined in sub-section 2.2.

Several series of time domain simulations were then performed on Ansys AQWA. Pitch's RAOs were evaluated including coupled dynamics and then compared with hull's RAO for all scenarios; two extra series of simulations were performed for the almost vertical cables scenario incrementing cables' total mass, i.e. setting w as double and then four times the actual value. Afterwards, results were compared with damping and inertia parameters, defined in the sub-section 2.2; and also compared with hull's RAOs with different pitch's radii of gyration.

To define cables' configuration in a complete way, it's still necessary to specify the radial distribution of mooring lines and risers. Figures 1 (scenario 1) and 2 (scenario 2 and 3) present top pictures of the models simulated on AQWA; clearly, cables' configurations are symmetric in respect to vessel's symmetry plane.

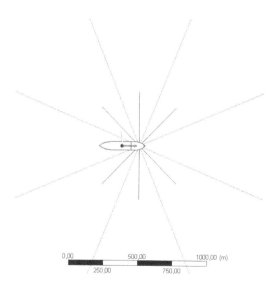

Figure 1. Cables' configuration from top—scenario 1. Darker lines represent risers; lighter lines represent mooring.

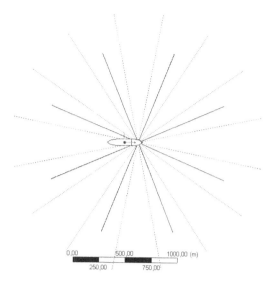

Figure 2. Cables' configuration from top—scenarios 2 and 3. Darker lines represent risers; lighter lines represent mooring.

It's important to highlight that the formulation addressed to (Aranha et al. 2001) for evaluating the dynamic tensions on cables was used in order to provide coherent analyses, especially on what concerns geometric stiffness effects. The algebraic formulation deduced in the—subsection 2.1 for the external stiffness matrix is strictly valid only in case of a quasi-static assumption for the cables' motions. Indeed, it was verified that dynamic tensions are far lower than the static ones in the shallow waters scenario; deep waters and almost-vertical lines, on the other hand, may reach substantial values of dynamic tension next to the seabed, thus the results outputted by WAMIT on these cases must be more carefully analyzed. Anyway, dynamic tension values never surpass the static ones.

2.1 Algebraic formulation for stiffness matrix

For the frequency domain simulations on WAMIT, an external stiffness matrix is inputted in order to account for the linear loads induced at the fairlead. It's assumed that cables' perform a quasi-static motion, so that the force acting at the fairlead balances the static condition (catenary profile) at all instants.

If F_1 is the horizontal component of this force pointing towards the anchor; and F_2 the vertical one, pointing downwards, it holds:

$$X_1 = L - X_2\sqrt{1 + \frac{2F_1}{wX_2}} + \frac{F_1}{w}\cosh^{-1}\left(1 + \frac{wX_2}{F_1}\right) \quad (1)$$

$$F_2 = wX_2\sqrt{1 + \frac{2F_1}{wX_2}} \qquad (2)$$

where X_1 = horizontal distance between fairlead and anchor; X_2 = vertical distance between fairlead and seabed; and w = weight in water. Stiffness coefficients are defined as:

$$K_{ij} = \frac{\partial F_i}{\partial X_j}\bigg|_{\substack{X_1 = X'_1 \\ X_2 = H}} \Rightarrow \begin{cases} K_{11} = K_H \\ K_{12} = F'_1/wH \\ K_{21} = K_H/K_V \\ K_{22} = K_V \end{cases} \qquad (3)$$

where:

$$K_H = w\left[-\frac{2}{K_V} + \cosh^{-1}\left(1 + \frac{wH'}{F'_1}\right)\right]^{-1} \qquad (4)$$

$$K_V = w\sqrt{1 + \frac{2F_1}{wH'}} \qquad (5)$$

And primes correspond to static values. The formulae above are valid in local coordinate system (vertical plane containing catenary profile). If one considers a radial set of identic cables, each with heading θ_i, symmetric in respect to vessel's symmetry plane, external stiffness matrix C^{ext} written in conventional vessel's fixed global coordinate system has non-zero elements as shown in formulae 6.

$$\begin{cases} C_{11}^{ext} = K_H \sum_{i=1}^{N} |\cos\theta_i| \\ C_{22}^{ext} = K_H \sum_{i=1}^{N} |\sin\theta_i| \\ C_{33}^{ext} = NK_V \\ C_{55}^{ext} = NK_V (L_{tur} - LCG)^2 \\ C_{66}^{ext} = K_H (L_{tur} - LCG)^2 \sum_{i=1}^{N} |\sin\theta_i| \\ C_{26}^{ext} = C_{62}^{ext} = K_H (L_{tur} - LCG)\sum_{i=1}^{N} |\sin\theta_i| \\ C_{35}^{ext} = C_{53}^{ext} = -NK_V (L_{tur} - LCG) \end{cases} \qquad (6)$$

2.2 Stiffness, damping and inertia parameters

Ratio $p_c = C_{55}^{ext}/C_{55}$ is taken as the stiffness parameter. It relates added stiffness on pitch and pitch's hydrostatic restoring coefficient

Parameter $p_b = B_{55}^{ext}/B_{55}$ is then defined as damping parameter. It relates added damping and potential damping based on maximum experienced

damping moment on pitch due to lines' motion. B_{55}^{ext} is defined as:

$$B_{55}^{ext} = \frac{\max(F_D)}{\omega_N \xi_2}(L_{tur} - LCG) \qquad (7)$$

where $\max(F_D)$ = maximum vertical drag force acting on the cables transferred to the vessel at the fairlead—estimated with Morison's formula; and ξ_2 = pitch's amplitude of motion at natural frequency ω_N. Due to drag's non-linearity, F_D and ξ_2 are evaluated in accordance with incident waves of 1.0 meter of amplitude.

FPSO's original radius of gyration in pitch is equal to 24.0% of L_{OA}, thus $I_{YY} = 1.207 \cdot 10^9$ ton m². Lastly, $p_a = M_{tur}(L_{tur}-LCG)^2/I_{YY}$ is defined as inertia parameter, relating added inertia and structural inertia. In the equation above, M_{tur} = total cables' suspended mass. It's important to highlight that p_a does not correspond quantitatively to the increment on pitch's radius of gyration.

3 RESULTS AND DISCUSSION

As presented in Figure 3, the addition of external stiffness matrix C^{ext} is clearly ineffective on modifying FPSO's pitch RAO for any scenario. Greatest modifications actually occur around 10 s and 12 s, where the modified transfer function differs from its original value at around 0.16%, 0.07% and 0.07% for scenarios 1, 2 and 3, respectively. Stiffness parameters on their turn were evaluated as

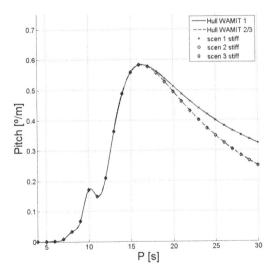

Figure 3. RAOs evaluated with different depths and stiffness matrices.

0.50%, 0.23% and 0.23%. Again, dynamic tensions evaluated for scenarios 2 and 3 reached significant values next to the seabed, but never surpassing static values. Thus, neglecting stiffness effects is reasonable for all scenarios. However, a preliminary relation between p_c and maximum transfer function's discrepancy cannot be reckoned, even though it looks linear at first sight.

Due to the quite small discrepancies, it's not visible in the figure, but all responses at periods below pitch's natural period follow the tendency of increments of response; and all responses above the same period are affected in the opposite way. Pitch's natural period is 11.2 s, and we may call it "inversion point" from now on.

As presented in Figure 4, when all effects are taken into account, responses are modified in the opposite way as experimented by only adding external stiffness. Thus, it can be said that other effects surpass stiffness effects; but pitch's natural period can still be regarded as an inversion point. Another observed tendency is a more pronounced decrement of response next to the inversion point. Highest increments are observed around 16 s: 0.83%, 2.3% and 2.5%; highest decrements around 9 s: 9.5% 17% and 18% – which are high values in percentage but not very significant for the FPSO's dynamics due to the period band; moreover, decrements of 0.92%, 7.6% and 7.6% were observed around the inversion point. Damping parameters are 1.8%, 2.4% and 0.36%; and inertia parameters 0.46%, 2.1% and 2.1% – thus scenarios 2 and 3 differ notably only on damping. Even elastic effects may be considered as equal, for cables' suspended length varies only slightly.

It's remarkable how scenarios 2 and 3 have quite similar response, even though cables' profiles are drastically different. Just as geometric stiffness, drag forces acting on the cables and transferred to the FPSO seem to be quite ineffective on changing FPSO's RAOs. Anyway responses in scenario 2 are slightly below scenario 3, this is in accordance with the fact that damping effects dissipate kinetic energy from body's motion.

As presented in Figure 5, changing systematically total cables' suspended mass in scenario 3 makes possible to finally find noteworthy modifications on FPSO's pitch RAO: highest increments of 3.1%, 5.1% and 8.9% around 15 s (transfer functions' peak); and decrements of 7.6%, 19% and 46% next to the inversion point, all effects of added inertia due to cables' motion. It's also clearer that, besides the significant drop at the inversion point, inertia effects modify pitch's RAO in the exact opposite way as stiffness effects. Inertia parameters are evaluated as 2.10%, 4.20% and 8.40%.

Rough estimates of pitch's radius of gyration variations based on cables' total suspended mass are now convenient. Originally, $R_{gyr} = 24.0\%$ of L_{OA}; and for the three simulated cases in scenario 3, R_{gyr} varies to 24.25%, 24.5% and 25.0% of L_{OA}—values estimated based on the assumption that there's a static force at the vessel's fairlead corresponding to cables' total suspended weight in water. A comparison between these values and the RAOs' variations indicates that the experienced effects are of strong non-linear nature.

A further inclusion of cables' hydrodynamic added mass on radii of gyration estimates ($C_M = 1$ for all cables) wouldn't take these values up to 26.0%

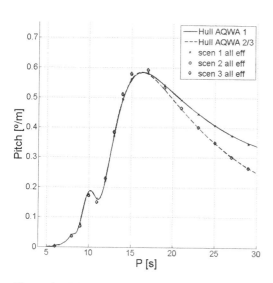

Figure 4. RAOs evaluated taking all effects into account.

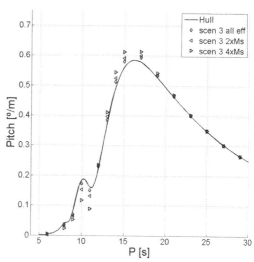

Figure 5. RAOs evaluated with different cables' total suspended mass (up to 7000 ton) – vertical lines case.

due to cables' profile. Figure 6 presents the results obtained for the case with heaviest cables (7000 ton of suspended mass in total) and compares it to pitch's RAOs as evaluated with different radii of gyration, up to 26.0%. Some points are out of the range considered for the radius of gyration, making clear that the inclusion of cables inertial effects cannot be seen simply as added moment of inertia, especially in low periods. This suggests that elasticity increments inertia effects, but validation is required at this point. To add more to that, significant drop at the inversion point cannot be regarded as added moment of inertia, since response at this point does not change as radius of gyration varies systematically (see Figure 6) – after all the inversion point corresponds to the situation where hull's moment of inertia balances pitch's hydrostatic restoration.

When to consider cables' influences on FPSO's pitch RAO is still not answered; and since FPSO's response also depends on the incident waves, other parameters related to the sea state must be known a-priori to answer this very question. To illustrate this, a brief evaluation of pitch's response spectra in scenario 3 simulated with heavy cables (7000 ton of suspended mass) and two different bow waves' conditions—one following a JONSWAP spectrum of 2.0 m of significant height and 8.0 s of peak period; the other following a JONSWAP spectrum of 2.0 m and 13.0 s—are presented in Figures 7 and 8. Even though non-linearities were primarily accounted for only on regular sea, the considered wave heights are correspondent with the regular sea simulations. Moreover, the first order

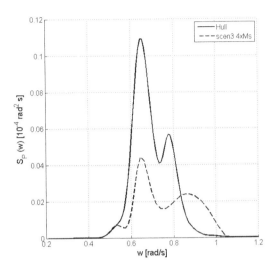

Figure 7. Pitch's response spectra—JONSWAP (2.0 m; 8.0 s).

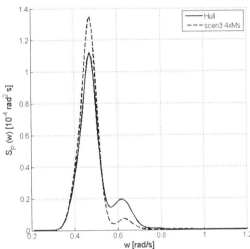

Figure 8. Pitch's response spectra—JONSWAP (2.0 m; 13.0 s).

responses are indeed expected to vary linearly with wave amplitudes in the cases studied.

Both spectra are compared with the spectra obtained when disregarding all cables' influences. Whilst in the low period case, significant amplitude of pitch response drops 24%, in the high period case it is increased by 1.8%. Anyway, in the latter case FPSO's pitch motion is also more energetic when compared with the prior case due to the period band—as it can be seen from y-axes' orders of magnitude. Hence, a strong dependency on the sea state (or period band under analysis) remains in order to account or not for cables' influences.

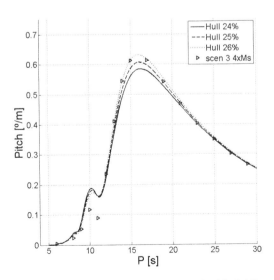

Figure 6. Heavy cables' RAO compared with RAOs from different pitch's radii of gyration.

4 CONCLUSIONS

As major conclusion of the study, turret system's inertial effects are notably the most relevant; and are of non-linear nature with respect to cables' total suspended mass. However, they are significant only in cases where cables' total suspended mass is large, thus only for deep waters systems and a big amount of cables (mooring lines and risers); which may rarely occur on offshore systems, for in these situations different kinds of mooring systems will be preferably employed.

The research thus contributed objectively to the mentioned inference method (see Introduction). Scenario 1's simulations, characterized by shallow waters, proved that no modifications on vessel's RAOs need to be accounted for.

Whilst geometric stiffness and damping proved to be insignificant in all cases studied, elasticity seems to increment inertia effects, but validation is still required at this point. Physically speaking, the loosening of the cables gives more freedom to the cables to oscillate, thus the inertial forces related to cables' motion may be experienced by the FPSO as higher, whereas the phase difference between cables' motion and fairlead's vertical motion may not change much. Neglecting damping and geometric stiffness' effects on the first order responses is a straightforward simplification, but the same should never be considered in a coupled analysis for the vessel's second order response, for in this situation it's known that both damping and geometric stiffness play important roles on the coupled dynamics.

The inclusion of effects so reshape pitch's RAO, usually slightly, according to the tendency of incrementing pitch's radius of gyration. There's not a single value of increment (e.g. from 24.0% to exactly 26.0% through the whole period band); moreover, next to the inversion point, this tendency is not followed, but a significant drop of response is experienced. This is the point where the greatest modification of response can be found; and may be a consequence of heave-pitch coupling, since it's next to this period that both heave and pitch's natural periods are situated; and pitch's response is considerably sensitive to the coupling.

Finally, sea state (or period band under analysis) has its contribution to the modifications of response. For instance, significant amplitude on pitch motion may drop substantially when sea's energetic band is found just before pitch's natural period; on the other hand, significant amplitude on pitch motion may also increase when sea's energetic band is found next to pitch's peak of response period.

REFERENCES

Aranha J.A.P., Pinto M.O. (2001). Dynamic Tension in Risers and Mooring Lines: an Algebraic Approximation for Harmonic Excitation. Applied Ocean Research 23, pp. 63–81.

Howell G.B., Duggal A.S., Heyl C, Ihonde O. (2006). Spread Moored or Turret Moored FPSOs for Deep Water Field Developments. Offshore West Africa 10.

Kannah T.R., Natarajan R. (2006). Effect of Turret Location on the Dynamic Behavior of an Internal Turret Moored FPSO System. Journal of Naval Architecture and Marine Engineering 2, pp. 23–27.

Simos A.N., Tannuri E.A., da Cruz J.J., Queiroz Filho A.N., Bispo I.B.S., Carvalho R.C.A. (2012). Development of an On-Board Wave Estimation System Based on the Motions of a Moored FPSO: Comissioning and Preliminary Validation. International Conference on Ocean, Offshore and Arctic Engineering 31, Proceedings. Rio de Janeiro, Brazil.

Simos A.N., Tannuri E.A., Sparano J.V., Matos V.L.F. (2010). Estimating Wave Spectra from the Motions of Moored Vessels: Experimental Validation. Applied Ocean Research 32, pp.191–208.

Soares C.G., Fonseca N., Pascoal R. (2005). Experimental and Numerical Study of Motions of a Turret Moored FPSO in Waves Journal of Offshore Mechanics and Arctic Engineering 127, pp. 197–204.

Xie Z.T., Yang J.M., Hu Z.Q., Zhao W.H., Zhao J.R. (2015). The Horizontal Stability of an FLNG with Different Turret Location. International Journal of Naval Architecture and Ocean Engineering 7, pp. 244–258.

Maritime Transportation and Harvesting of Sea Resources – Guedes Soares & Teixeira (Eds)
© 2018 Taylor & Francis Group, London, ISBN 978-0-8153-7993-5

Comparison of two approaches for prediction of wave induced loads in damaged ships

H. Jafaryeganeh & C. Guedes Soares
Centre for Marine Technology and Ocean Engineering (CENTEC), Instituto Superior Técnico, Universidade de Lisboa, Lisbon, Portugal

ABSTRACT: The present work compares the performance of the added-weight and lost buoyancy methods for simulating the damaged ship condition in steady state wave induced load calculations. The wave loads are evaluated by a three-dimensional linear panel method. Horizontal and vertical bending moment and shear force are calculated as the main induced wave loads to ship hull. In this study, a frigate model is adopted and the results obtained by each method are compared with reported results, which were obtained using the strip theory and model tests. In this case study, the response amplitude operator of the mentioned loads is evaluated for different wave heading angles. It is shown that each method has pros and cons for implementation and obtained results.

1 INTRODUCTION

Ship accidents cause loss of cargo and human life besides of environment pollution. Probable damage effects need to be considered in the initial design stage for reducing the costs after the accident and increasing the safety. One of the important damage effects is the change of induced load to the ship structure. Thus, proper design needs methods that are capable to adequate prediction of the loads to damaged ship.

Prediction of wave induced loads for damaged ships has been investigated in three phases of transient, progressive and steady state flooding. As described in Downes et al. (2007), these loads categorized in three groups of still water, wave and dynamic loads. Time-domain theoretical models are usually capable of predicting the induced loads in the transient and progressive flooding. In this regard, Santos & Guedes Soares (2008a, 2008b, 2009) presented a model to predict the behavior of a damaged ship subjected to progressive flooding in the wave conditions. The motion of the water inside the compartment was modeled by a system of non-linear hyperbolic equations. They also demonstrated that during the intermediate stages of flooding, floodwater distributions can be generated. Those may cause significantly higher shear forces or bending moments than the ones present in the final equilibrium damage condition. Additionally, the rough sea induces the effects like slamming (Wang, & Guedes Soares 2016) and green water on deck (Guedes Soares & Pascoal 2005). The current study is focused only on moderate sea

conditions and steady state of flooding for reliable estimation of wave loads.

If a damaged ship reaches equilibrium after the accident, the floating conditions become stable. Traditionally, two methods of added-weight and lost buoyancy have been used to determine the final stability condition (Lewis. 1989). Associated variation of weight or buoyancy distribution can predict still water loads for each damage scenario.

The assumptions of each of the above methods affect the prediction of induced wave loads too. The method of lost buoyancy requires a significant change of intact ship model for hydrodynamic simulation; while, the added mass method introduce more variation of the dynamic model.

Application of the added-weight method simplifies the damage problem, because, the flooded water effect in considered as mass for changing the stability condition like heel and trim. Thus, intact condition methodologies can be applied to predict wave induced loads.

The strip theory has been used to evaluate the wave induced loads for simulated models of damaged ship with the added-weight approach. Lee et al. (2014) used the linear strip theory to estimate the induced wave loads for a sample study of a frigate. The comparison was shown good agreement with the measured experimental data for their case study. Chen et al. (2003) applied a more complicated approach for prediction of dynamic global wave loads for a Ro-Ro ship. The responses were obtained by a nonlinear time-domain simulation method and compared with experimental measurement data. However, they did not provide more accurate result than two dimensional strip theory, in spite of using more complicated method.

In a different linear approach, Folso et al. (2008) have performed seakeeping computations on a damaged ship by the three-dimensional linear hydrodynamic code PRECAL. The added-weight concept was applied for the flooded water. The simulation resulted in the dynamic vertical bending moment on the hull girder. Jafaryeganeh et al. (2016) compared the two-dimensional strip theory and the three-dimensional panel method although, they only investigated the predicted vertical bending moment transfer function with experimental measurements for the intact ship.

A few studies applied the lost buoyancy simulations for the prediction of wave loads of damaged ship. In this regard, Parunov et al. (2015) compared the two methods of added-weight and lost buoyancy for assessment of vertical bending moment of damaged ship hull in head waves. Hydrostar was used as three-dimensional seakeeping software. The predicted result showed a larger value for added-weight and lower values for lost buoyancy in comparison with intact ship evaluation. They neglected the horizontal component of the loads in the prediction methods, as the effect of heel angle was ignored in their seakeeping model.

Furthermore, the effect of wave heading needs to be considered in the comparison of predicted loads by both added mass and lost buoyancy methods. As was studied in Chan et al. (2001), the vertical wave bending moment is usually highest in head sea, whereas oblique sea gives the highest horizontal bending moment. Also, the combination of vertical and horizontal bending moments needs to be considered in the prediction methods.

Regarding experimental studies, Begovic et al. (2011 and 2013) measured the motion response characteristics of frigate model in intact and damaged conditions in head, beam and quartering seas at zero speed. Later, Begovic et al. (2017) performed an experimental investigation of the hull girder loads on an intact for the same model and heading angels.

In the present study, two methods of added-weight and lost buoyancy are used to simulate the flooding condition in the steady state phase. The dynamic response of the damaged ship is predicted by linear three-dimensional panel method. The evaluated wave induced loads include the horizontal bending moment, vertical bending moment and shear forces in different headings. Finally, a damage scenario is considered for a frigate as a case study and both approaches are implemented. The predicted results for different wave headings are compared with experimental measurement and predicted strip theory results by Lee et al. (2014).

2 METHOD OF EVALUATION OF TRANSFER FUNCTIONS

The three-dimensional panel method is applied for the evaluation of vertical shear force, vertical bending moment and horizontal bending moment transfer function. The method was used for prediction of vertical bending moment of the intact ship in Jafaryeganeh et al. (2016). In this study, the formulation is adapted for load prediction in the damage models, due to account the wave induced load component in both methods. The ship speed is assumed to be zero for all the wave headings.

2.1 Dynamic shear force and bending moment evaluation

The dynamic shear force at a cross section is the difference between the inertia force and the sum of the external forces acting on the hull section (e.g. Salvesen et al. 1970).

$$V_i = I_i - P_i \qquad (1)$$

where, V_i is the generalized dynamic force or moments and P_i is the generalized external force and moment, and I_i is the inertia force and moment.

The external forces are the summation of the restoring force, exciting force and hydrodynamic response force or moments.

$$P_i = R_i + H_i \qquad (2)$$

where R_i is the restoring force and H_i is the sum of the exciting force and hydrodynamic responses, including the diffraction and radiation effects in potential fluid. So, the vertical shear force, vertical bending moment and horizontal bending moment can be calculated by:

$$V_3 = I_3 - (R_3 + H_3) \qquad (3)$$

$$V_5 = I_5 - (R_5 + H_5) \qquad (4)$$

$$V_6 = I_6 - (R_6 + H_6) \qquad (5)$$

2.2 Inertia forces and moments

The inertia force is the mass times the acceleration if the inertia force is expressed in the terms of the sectional inertial force, thus the inertia terms given by:

$$I_3 = \int m(\ddot{\eta}_3 - \xi\ddot{\eta}_5) d\xi \qquad (6)$$

$$I_5 = -\int m(\xi - x)(\ddot{\eta}_3 - \xi\ddot{\eta}_5)d\xi \tag{7}$$

$$I_6 = \int m(\xi - x)(\ddot{\eta}_2 + \xi\ddot{\eta}_6 - z_G\ddot{\eta}_4)d\xi \tag{8}$$

where m is the sectional mass per unit length and the $\ddot{\eta}_2, \ddot{\eta}_3, \ddot{\eta}_4, \ddot{\eta}_5, \ddot{\eta}_6$ are the accelerations in sway, heave, roll, pitch and yaw directions, respectively.

2.3 Restoring force and moments

The hydrostatic restoring force and moment are given by (e.g. Papanikolaou & Schellin, 1992):

$$R_3 = \iint_s n_3 p_s ds \tag{9}$$

$$R_5 = \iint_s (\xi - x)n_3 - z\, n_1)p_s ds + g\int mz_G\xi_5 d\xi \tag{10}$$

where $p_s = -\rho g(\eta_3 + \eta_4 y - \eta_5 x)$ is the increase of hydrostatic pressure at point of x, y, z on the hull surface due to body motion η_i. The restoring forces for every panel is evaluated and integrated along the hull surface.

2.4 Exciting force hydrodynamic responses evaluation

The boundary value problem is solved by using a source technique and applying Green's second identity, so the velocity potential can be evaluated for each panel. Consequently, the pressure is calculated on each panel (Lee, 1995).

Complex unsteady hydrodynamic pressure on the body boundary or in the fluid domain is related to the velocity potential by the linearized Bernoulli equation:

$$p = -\frac{\partial \varphi}{\partial t} \tag{11}$$

$$\varphi = \varphi_D + \varphi_R \tag{12}$$

where φ_D and φ_R are the diffraction and radiation velocity potential, respectively, given by equation 10 and 11.

$$\varphi_R = i\omega \sum_{j=1}^{6} \eta_j \varphi_j \tag{13}$$

where the constants η_j denote the complex amplitudes of the body oscillatory motion in its six rigid-body degrees of freedom, and φ_j the corresponding unit-amplitude radiation potentials.

$$\varphi_D = \varphi_0 + \varphi_S \tag{14}$$

φ_0 is velocity potential of the incident wave and the velocity potential φ_S represents the scattered

disturbance of the incident wave by the body fixed at its undisturbed position.

Thus, the evaluated pressure for every panel includes the incident wave, scattering and radiation effects. The following formulas present the force and moment due to the dynamic pressure of p on the point of x, y, z:

$$H_3 = \int n_3 p ds \tag{16}$$

$$H_5 = \int (n_3 x - n_1 z) p ds \tag{17}$$

$$H_6 = \int (n_2 x - n_1 y) p ds \tag{18}$$

where n_1, n_2, n_3 are the normal vectors to the panel center in the x, y, z direction respectively.

3 CASE STUDY

Table 1 presents the particulars of the frigate used in this study. Lee et al, (2014) evaluated the bending moment and shear force transfer functions experimentally and by a strip theory. Although they assumed four damage scenarios in three wave headings, only the experimental results of the largest damage scenario were compared with prediction results of the two-dimensional strip theory. Moreover, data are available for large and small waves, which is an advantage to evaluate the predicted transfer functions accuracy for both damaged simulation methods.

Figure 1 shows the three-dimensional hull model of the frigate used for panel discretization. The total weight is equal to 9034.24 Ton. Figure 3 shows the weight and hydrostatic pressure distributions along the ship length. An improved panel discretization has been performed to evaluate the equivalent pressure in each of the 142 sections that are considered from the aft to the fore,

Table 1. Main particulars of the frigate 5415.

Particular	Value	unit
L	153.3	m
L_{pp}	142.2.0	m
B_{wl}	19.082	m
B_{OA}	20.54	m
D	12.47	m
T	6.34	m
V	8042.4	m³
Δ	8635	(t,Kg)
C_B	0.505	
C_P	0.616	
C_M	0.815	

Figure 1. Hull model of the frigate 5415.

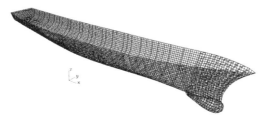

Figure 2. Panel discretization for hull frigate 5415.

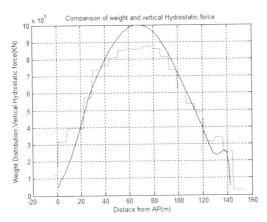

Figure 3. Weight and buoyancy distributions for intact ship.

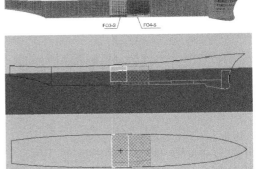

Figure 4. Schematic location of damaged compartments.

Table 2. Condition of intact and damage case.

	Intact	Damage	Unit
Draft @M.S	6.34	7.466	m
Trim by fwd	0	0.63	deg
Heel	0	1.11	deg

Table 3. Flooding summary of damage case study.

Item	Eff /Perm	Displ (MT)	LCB (m)	TCB (m)	VCB (m)
Hull	1	11,411.26	1.359f	0.096s	4.435
Mach2	0.85	−939.18	0.329a	0.115s	4.68
FuelOil3_s	0.99	−89.54	0.367a	2.862s	0.987
FuelOil4_s	0.99	−105.1	11.522f	2.549s	1.004
Mach3	0.85	−1,252.75	11.793f	0.113s	4.793

and the distance of 1 m. Figure 2 shows the panel discretization along the ship hull. The panels' sizes are considered below the recommended size range in the DNVGL rules. Also, the panel size ensures the convergence of intact ship results according to Jafaryeganeh et al. (2015).

In the current approach, the quadrilateral panels of the ship hull are confined between consecutive sections, i.e. there would not be any panel corner in two different sections, so, numerical integrations have more accuracy for each section.

3.1 *Damage scenario*

Figure 4 presents the location of damaged compartments, as is mentioned the damage scenario 2 with more detail in the Lee et al. (2014).

Table 2 shows the draft, trim and heel for intact and the damage case study. Table 3 presents the

flooding summary for the hull and four flooded tanks in the damage condition. The damage case is rather large compared to the length of the ship, besides unsymmetrical flooding of the tanks leads to heel for the damaged ship.

4 METHODS OF DAMAGED SHIP SIMULATION IN STEADY STATE OF FLOODING

Added-weight and lost buoyancy methods are applied for modeling steady-state flooding. These approaches result in the changes of the weight distribution and buoyancy respectively. The final equilibrium condition is the same for both methods. Though, the assumptions introduce variation in the wave load prediction.

4.1 Added-weight method

The ship is assumed to be undamaged in the added-weight method, but a completely filled tank is assumed at the location of the damaged compartment. Thus, the dynamic effect of free surface is not accounted for the induced water in the flooded compartment. This is equivalent to adding a rigid weight for shifting the ship center of gravity and displacement. The location of the added-weight is specified with the center of volume of the induced water in the damaged compartment.

The method can evaluate the new draft, trim and heel in the damaged ship. Additionally, the added-weight assumption causes a change of metacentric height, buoyancy and weight distribution and moment of inertia, relative to the intact ship. For each damage scenario, all the mentioned properties should be updated for providing the seakeeping model and predicting the still water and wave loads.

Figure 5 shows the panel discretization of added-weight approach for the mentioned damage case of the frigate 5415, the ship hull is modeled similar to the intact ship, while the condition of Table 2 is applied for added-weight simulation.

Figure 6 presents weight and buoyancy distribution for the added-weight approach. The changed hydrostatic properties of the damaged case are compared with the intact ship in Table 4.

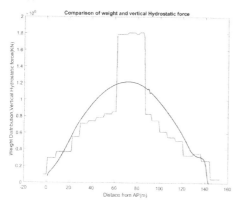

Figure 5. Panel discretization for the added-weight method.

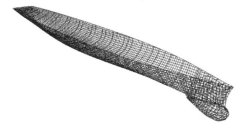

Figure 6. Weight and buoyancy distributions for added-weight approach.

Table 4. Hydrostatic properties of intact ship and damaged case with added-weight approach.

	Added weight	Intact	Unit
kxx	6.225359	6.932	m
kyy	31.33356	35.1234	m
kzz	31.17367	35.1234	m
KM	9.396	9.493	m
KG	7.021	7.555	m
GM	2.375	1.938	m
LCG	71.483	70.104	m-form A.P
VCG	5.897	6.283	m-form B.L
TCG	0.069 s	0	m-form C.L
Buoyancy	1119.2	9033.4	MT

4.2 Lost buoyancy

In the lost buoyancy method, the buoyancy variation is analyzed instead of the weight change. So, the center of gravity, weight distribution and moment of inertia remain the same. The total buoyant volume must remain constant since the weight of the ship is not changing.

The draft will increase and the ship will list and trim until the lost buoyant volume is regained. The final stability condition is the same as the added-weight approach, but the buoyancy distribution is updated according to the final equilibrium data.

A different three-dimensional model is provided with removing the damaged compartment space. Therefore, the inside space of the flooded compartment is simulated like the exterior domain of the fluid. The simplification allows measuring the horizontal pressure on the external boundary of undamaged compartments, which is part of the internal subdivision in the intact case.

Figure 7 and Figure 8 show the adopted model for the lost buoyancy approach in the assumed damage scenario. The model consists of all the wetted surfaces that contribute to the buoyancy provision and used for panel discretization. Figure 6 presents weight and buoyancy distribution for the lost buoyancy approach.

4.3 Weight and buoyancy distributions

The hydrostatic pressure is evaluated separately for the panels situated between two consecutive sections. Thus, the water height in every panel centroid was considered from the free surface and the vertical force is also calculated by the method described in section 2.

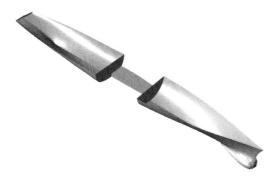

Figure 7. 3D view of lost buoyancy model for assumed damage scenario.

Figure 8. Side view of lost buoyancy model for assumed damage scenario.

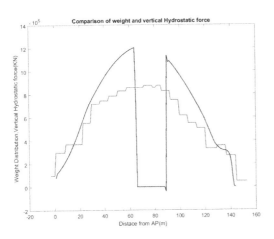

Figure 9. Weight and buoyancy distributions for lost buoyancy approach.

5 RESULTS AND DISCUSSION

The result of the simulated damage case obtained by the three-dimensional panel method is presented here for both added-weight and lost-buoyancy approaches. The response amplitude operators (RAOs) are compared with available experimental data and two-dimensional strip theory results, which are predicted with added-weight approach.

5.1 Comparison of dynamic vertical bending moment

Figure 10 to Figure 12 present the predicted results of the vertical bending moment for three wave head-

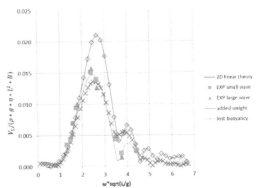

Figure 10. Dynamic vertical bending moment RAO of DS2 H5415 in head waves.

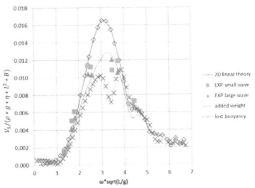

Figure 11. Dynamic vertical bending moment RAO of DS2 H5415 in stern quartering waves.

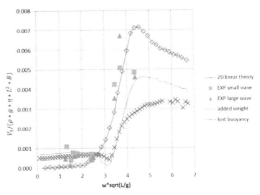

Figure 12. Dynamic vertical bending moment RAO of DS2 H5415 in beam waves.

ings. The maximum values of the vertical bending moment are observed in head seas. However, the minimum value of measured experimental data and predicted results are observed in beam seas.

In head waves, the added-weight results are slightly larger than the lost-buoyancy predictions in lower frequencies, while, the lost-buoyancy result are slightly larger at higher frequencies. Both simulations of damage have acceptable results for prediction the largest vertical bending moment in head sea.

Also, the predicted results of two-dimensional strip theory are larger than three-dimensional panel method in added-weight approach.

In stern quartering waves, the difference between added-weight and lost-buoyancy are more observable in predicting the largest value of vertical bending moment. The added-weight simulation predicts one maximum value for response amplitude operator (RAO), while the lost-buoyancy RAO shows two maximum values. However, the added-weight approach shows better agreement with the measured experimental data. Also, the results of two-dimensional linear theory overestimate the measured experimental data, while the three-dimensional panel method has better agreement with the same added-weight simulation.

In beam waves, the predicted results of both damage simulation methods have good agreement with measured experimental data in lower frequencies, while the predicted result underestimates the measured experimental data in higher frequencies. However, the difference between the prediction result and measured experimental data are significantly smaller in the added-weight simulation, especially in the two-dimensional strip theory.

5.2 Comparison of dynamic vertical shear force

Figure 13 to Figure 15 present the predicted and measured experimental results of vertical shear force for three wave headings. The maximum and minimum values are observed in the head and beam sea, respectively.

In head waves, a good agreement is observed between the predicted result of added-weight simulation and measured experimental data, although, a slightly overestimation can be seen in the maximum value. The lost-buoyancy simulation results underestimate the measured experimental data in lower frequencies, while, the prediction results of higher frequencies shows better agreement with the measured experimental data. Additionally, the three-dimensional panel method prediction with added-weight approach has better agreement with measured experimental data, relative to the two-dimensional strip theory.

In stern quartering sea, a significant difference is observed in predictions of the maximum value of vertical shear force by two methods of added-weight and lost-buoyancy. The correlation between the predicted results of the lost-buoyancy simulation

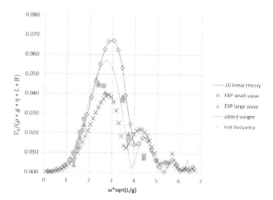

Figure 13. Dynamic vertical shear force RAO of DS2 H5415 in head waves.

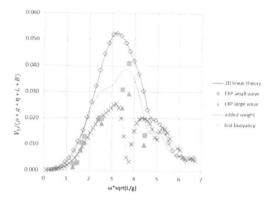

Figure 14. Dynamic vertical shear force RAO of DS2 H5415 in stern quartering waves.

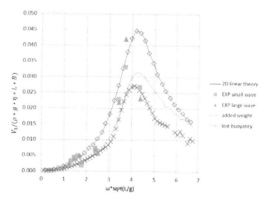

Figure 15. Dynamic vertical shear force RAO of DS2 H5415 in beam waves.

and the measured experimental data shows a good agreement in lower wave frequencies. The added-weight simulation can predict the maximum values of shear force with good accuracy. Moreover, the two-dimensional strip theory overestimates the measured experimental data with added-weight approach.

In beam seas, both methods of added-weight and lost-buoyancy are in good agreement with experimental measurements However, an underestimation is observed for the maximum values. The results of two-dimensional strip theory with added-weight approach have good prediction of the maximum vertical shear force value.

5.3 Comparison of dynamic horizontal bending moment

Figure 16 to Figure 18 present the predicted results of the horizontal bending moment for three wave headings. The maximum values are observed in the stern quartering sea. However, the minimum value of measured experimental data and predicted results are observed in the head sea.

In head waves, a significant difference is observed between the prediction results and meas-

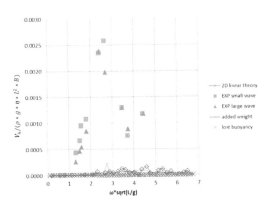

Figure 16. Dynamic horizontal bending moment RAO of DS2 H5415 in head waves.

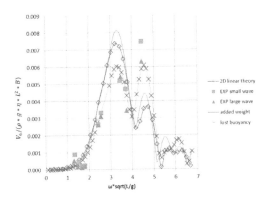

Figure 17. Dynamic horizontal bending moment RAO of DS2 H5415 in stern quartering waves.

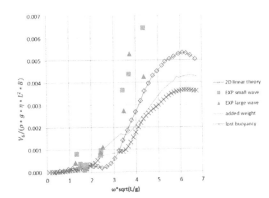

Figure 18. Dynamic horizontal bending moment RAO of DS2 H5415 in beam waves.

ured experimental data. It seems the differences are related to sloshing effects in the damaged compartment. Also, the uncertainty of the measured experimental data should be considered due to the model drift, as is mentioned in Lee et al. (2014).

In the stern quartering waves, the lost-buoyancy simulation results are in a good agreement with measured experimental data, especially in prediction of maximum values, while the added-weight approach shows an overestimation of the measured experimental data in lower wave frequencies, also, it underestimates the measured experimental data in higher wave frequencies. Moreover, the added-weight simulation does not have a significant difference in predicted results by two-dimensional strip theory and three-dimensional panel method.

In beam waves, the correlation between the predicted results and measured experimental data shows a significant difference, especially for the predicted maximum values. However, the added-weight and lost-buoyancy can predict the measured experimental data in lower wave frequencies with good agreement.

6 CONCLUSIONS

The damage condition is simulated with two methods of added-weight and lost-buoyancy in the steady state of flooding. A three-dimensional panel method is applied to the prediction of wave loads in both models for a damage scenario with a large opening. The predicted results are compared with experimental measurements and predicted results of two-dimensional strip theory by added-weight simulation.

The added-weight simulation has an advantage in simplification of the hydrodynamic model because the intact ship assumption applies with some modification in hydrostatic and mass properties. The predicted results of wave loads using the added-weight simulation indicates an adequate estimation of the maximum value of the vertical shear force and bending moment. However, the added-weight simulation does not provide an accurate prediction for the small values of wave loads, especially for the horizontal bending moment.

The lost-buoyancy simulation needs a more complicated hydrodynamic model, due to removing the flooded compartment from the hull model. The method provides an adequate prediction for maximum values of the dynamic horizontal bending moments, which is observed in the stern quartering sea. However, its prediction results underestimate the maximum values of measured experimental data for vertical shear force and bending moment.

The three-dimensional panel method indicates better agreement with experimental measurement relative to the two-dimensional strip theory when both methods applied the added-weight simulation.

Added-weight and lost-buoyancy simplify the damage problem for rapid evaluation of wave loads. However, other damage effects, like the sloshing of induced water in flooded compartment, needs to be considered in future studies.

ACKNOWLEDGEMENTS

The work has been conducted within the Strategic Research Plan of the Centre for Marine Technology and Ocean Engineering, which is financed by Portuguese Foundation for Science and Technology.

REFERENCES

Begovic, E; Day, A.H; Incecik, A. (2017): An experimental study of hull girder loads on an intact and damaged naval ship. *Ocean Eng.* 133:47–65.

Begovic, E; Mortola, G; Incecik, A; Day, A.H.(2013): Experimental assessment of intact and damaged ship motions in head, beam and quartering seas. *Ocean Eng.* 72:209–226.

Begovic, E; Incecik, A; Day, A.H.(2011): Experimental assessment of intact and damaged ship motions in head, beam and quartering seas. *In: HSMV Conference, Naples, Italy.*

Chan, HS; Atlar, M; & Incecik, A. (2003): Global wave loads on intact and damaged Ro-Ro ships in regular oblique waves. *J MarStruct.* 16:323–344.

Chan, HS; Incecik, A; & Atlar, M. (2001): Structural integrity of a damaged Ro-Ro vessel. *Proceedings of the 2nd International Conference on Collision and Grounding of Ships*, Technical University of Denmark, Lybgby. pp. 253–258.

Downes, J; Moore, C; Incecik, A; Stumpf, E; & McGregor J. (2007): A Method for the quantitative Assessment of Performance of Alternative Designs in the Accidental Condition, *10th International Symposium on Practical Design of Ships and Other Floating Structures, Houston, Texas.*

Folso, L; Rizzuto, E; & Pino, E. (2008): Wave induced global loads for a damaged vessel. *Ships Offshore Struct,* 3:4, 269–287.

Guedes Soares, C. & Pascoal, R (2005). Experimental Study of the Probability Distributions of Green Water on the Bow of Floating Production Platforms. *Journal of Offshore Mechanics and Arctic Engineering.* 127(3):234–242.

Jafaryeganeh, H; Rodrigues, J.M; & Guedes Soares, C. (2015): Influence of mesh refinement on the motions predicted by a panel code, *Maritime Technology and Engineering*, Guedes Soares, C. & Santos T.A. (Eds.), Taylor & Francis Group, London, UK, pp. 1029–1038.

Jafaryeganeh, H.; Teixeira, A.P; & Guedes Soares, C. (2016): Uncertainty on the bending moment transfer functions derived by a three-dimensional linear panel method, *Maritime Technology and Engineering*, Guedes Soares, C. & Santos T.A., (Eds.), Taylor & Francis Group, London, UK, pp. 295–302.

Lee, C.H. (1995): *WAMIT theory manual*, Dept. of Ocean Eng., MIT, Cambridge, MASS.

Lee, Y; Chan, H; Pu, Y; Incecik, A; & Dow, R. (2014): Global wave loads on a damaged ship. *Ship and offshore structure*, 7:3,237–268.

Lewis, E.V. Editor. (1989): Principles of Naval Architecture (2nd Rev.) Vol. 1 – *Society of Naval Architects and Marine Engineers*. ISBN 0-939773-00-7.

Papanikolaou, A; & Schellin, T.E. (1992): A three-dimensional panel method for motions and loads of ships with forward speed, *Ship Technology Research* 39: 147–156.

Parunov, J; Ćorak, M; & Gledić, I. (2015): Comparison of two practical methods for seakeeping assessment of damaged ships, *Analysis and Design of Marine Structures*, Guedes Soares C. & Shenoi RA (Eds), Taylor & Francis Group, London, UK, pp. 37–45.

Salvesen, N; Tuck, E.O.; & Faltinsen, O. (1970): Ship Motions and Sea Loads, *Trans. SNAME*, 78: 250–287.

Santos,T; & Guedes Soares,C. (2008a): Study of damaged ship motions taking into account flood water dynamics. *J. Mar. Sci. Technol*, 13 (3), 291–307.

Santos,T; & Guedes Soares,C. (2008b): Global loads due to progressive flooding in passenger Ro-Ro ships and tankers. *Ships Offshore Struct*, 3 (4), 289–303.

Santos, T; & Guedes Soares, C. (2009): Numerical assessment of factors affecting survivability of damaged Ro-Ro ships in waves. *Ocean Eng*, 36 (11), 797–809.

Wang, S. & Guedes Soares, C. 2016. Experimental and numerical study of the slamming load on the bow of a chemical tanker in irregular waves. *Ocean Eng.* 111:369–383.

Maritime Transportation and Harvesting of Sea Resources – Guedes Soares & Teixeira (Eds)
© 2018 Taylor & Francis Group, London, ISBN 978-0-8153-7993-5

Seakeeping performance of a Mediterranean fishing vessel

D. Obreja
Naval Architecture Faculty, University "Dunarea de Jos" of Galati, Galati, Romania

R. Nabergoj
NASDIS PDS, Isola, Slovenia

L. Crudu & L. Domnisoru
Naval Architecture Faculty, University "Dunarea de Jos" of Galati, Galati, Romania

ABSTRACT: To achieve good hydrodynamic performances several important factors and criteria have to be taken into consideration like hull forms, loading conditions and operational indexes in connection with environmental sources of excitation. To this purpose, during the initial design stages, the availability of enough powerful tools becomes mandatory. The analysis of dangerous situations and the corresponding risk must be carefully evaluated using a systematic investigation of the seakeeping performances. The paper presents the results of both numerical and experimental investigations carried out for a typical fishing vessel, operating in the Mediterranean Sea. An extensive experimental program, consisting in free running tests and semi-captive tests using the PMM equipment, was developed having as a target the validation of the theoretical models and computer codes which are practically used during the design process.

1 INTRODUCTION

A safe ship with optimum hydrodynamic performances constitutes the major concern in ship research and design activity (Santos Neves et al., 2003; Lee et al., 2005). Statistics reported in the literature reveal that fishing is one of the most dangerous occupations (Calisal et al., 1999). It has been recognized since many years that Mediterranean Sea constitutes a highly risky area for the hazardous operation of fishing vessels.

Appealing to the concept of global safety, the naval architect aims to obtain the optimum ship shape, which will be able to allow safe operation, in given environment conditions. Since the occurrence of dangerous situations may have undesirable consequences, critical situations should be carefully investigated. A deeper understanding of the complex hydrodynamic phenomenon is necessary. Possible major accidents can be avoided by imposing certain restriction within operation procedures. In this respect, it is very important for the designer to use powerful hydrodynamic tools in order to investigate ship safety, starting from the initial design stage.

In order to apply theoretical calculations, it is necessary to confirm the software validity by

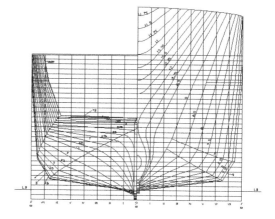

Figure 1. The body plan of the fishing vessel.

means of experimental tests. To this aim, the hydrodynamic characteristics of a double chine fishing vessel were experimentally investigated. The ship forms are shown in Figure 1, while the characteristics of the ship and experimental model are presented in Table 1. The fishing vessel sails at relatively high Froude numbers ($F_n = 0.4$).

Table 1. Main characteristic of ship and model at full loading.

Main characteristics	Full scale	Model scale (1/12)
Volumetric displacement, V	296.0 m³	0.171 m³
Length overall, L_{OA}	32.7 m	2.725 m
Length between perpendiculars, L_{BP}	25.0 m	2.083 m
Moulded breadth, B	8.0 m	0.667 m
Draught at fore perpendicular, T_F	2.42 m	0.202 m
Draught at aft perpendicular, T_A	2.74 m	0.228 m
Longitudinal centre of gravity, LCG	11.32 m	0.943 m
Vertical centre of gravity, KG	3.05 m	0.254 m
Transverse metacentre height, GM_T	0.65 m	0.054 m
Block coefficient, C_B	0.574	0.574
Waterline area coefficient, C_W	0.819	0.819
Natural roll period, T_ϕ	6.2 s	1.8 s
Roll radius of gyration, k_{xx}	2.46 m	0.205 m
Pitch radius of gyration, k_{yy}	6.78 m	0.565 m
Yaw radius of gyration, k_{zz}	6.9 m	0.575 m
Rudder area, A_R	2.88 m²	0.02 m²
Stock propeller diameter, D	1.8 m	0.15 m
Ship speed, U	12 kn	1.8 m/s

2 THEORETICAL MODEL

The researches on the ship hydrodynamic safety try to clear up the physical aspects of the complex mechanism of dynamic behaviour in waves and to determine the main hydrodynamic characteristics of the ship in typical dangerous situations. The real problem is difficult to be solved since it involves the consideration of non-stationary flow around a solid body, with the influence of the free surface of the viscous fluid.

Within the limits of the potential theory, the velocity potential has to fulfil the Laplace equation with boundaries and infinity conditions

$$\Delta\Phi = 0 \tag{1}$$

and, may be written under the form

$$\Phi = \Phi_U + \Phi_I + \Phi_D + \Phi_R \tag{2}$$

where, Φ_U is the velocity potential due to the running ship in still water at constant speed, Φ_I is the incident wave potential, Φ_D is the diffraction potential and Φ_R is the radiation potential.

The sum of the potentials $\Phi_I + \Phi_D$ describe the diffraction problem, generated by the interaction of the incident waves with the fixed ship hull.

The excitation hydrodynamic forces and moments F_e may be determined solving the diffraction problem.

The radiation problem is caused by ship forced motions, in initial still water. The radiation hydrodynamic forces and moments characterize the hydrodynamic properties of the body and have the following form

$$G = -A\ddot{\eta} - B\dot{\eta} - C\eta \tag{3}$$

where, A is the added mass matrix, B is the damping coefficients matrix, C is the matrix of restoring forces and η denotes ship's motions.

Taking into account the hypothesis of the linear hydrodynamic model, the ship motion equations with six degrees of freedom may be obtained, by using Newton's law

$$\frac{d}{dt}(M\dot{\eta}) = F_e + G \tag{4}$$

where, M is the body mass matrix. By substituting the Equation 3 in Equation 4, the general form of the motion equations is obtained

$$(M + A)\ddot{\eta} + B\dot{\eta} + C\eta = F_e \tag{5}$$

The Equation 5 can be described in the frequency domain, depending of the complex motions amplitudes η_k^A (Salvesen et al., 1970)

$$\sum_{k=1}^{6}\left\{-\omega^2\left[M_{jk} + A_{jk}(\omega)\right] + i\omega B_{jk}(\omega) + C_{jk}\right\}\eta_k^A = F_{ej}$$
$$j, k = 1, ..., 6 \tag{6}$$

3 ADDED MASSES AND DAMPING COEFFICIENTS

The experimental determination of hydrodynamic coefficients was performed on the basis of forced harmonic oscillation tests of the model in initial still water, at zero speed, by using a mechanical oscillator, sinus type, shown in Photo 1. The two columns can provide translation motions, both in phase and out of phase obtaining pure heave and pitch harmonic motions.

The radiation hydrodynamic forces and moments were measured by means of a six-component transducer. The interdependence matrix of six-component transducer is presented in Table 2. The maximum interdependence (7.6%) is obtained on the longitudinal force direction G_1, for pure roll moment loading G_4, but many other influences can be neglected.

Photo 1. General view of the PMM.

Table 2. Interdependence matrix of six-component transducer.

	Pure loading of transducer					
	G_1	G_2	G_3	G_4	G_5	G_6
G_1	1	−0.02339	0.00062	0.07636	0.00172	−0.03924
G_2	0.00297	1	−0.00489	0.01202	−0.00858	−0.00566
G_3	0.00105	0.00143	1	−0.00665	−0.00189	−0.01883
G_4	−0.00050	0.00168	0.00116	1	0.00091	−0.00029
G_5	−0.00436	−0.00102	−0.00061	0.03045	1	0.02458
G_6	−0.00226	0.00361	0.00338	−0.00659	0.00679	1

(leftmost vertical label: Interdependence)

The experimental tests have included forced harmonic oscillation at zero speed, for heave motion having amplitude $\zeta_3 = 0.01$ m and for pitch motions with amplitude $\zeta_5 = 1.15$ deg.

All experimental results have been obtained using a dedicated system, SAPDAT (Data acquisition and Processing System). The main modules of the system consist in: data acquisition on magnetic support, Analogue—Digital convertor, data visualization and inspection, data conversion from electrical output values to physical (degree, length, forces, moments, etc.) ones, data processing specific to the measuring device (like data correction based on interdependence matrix, etc.), FFT module, final experimental data bank (Bendat et al.,1980). Mention should be made that the system can be easily upgraded by introducing specific

applications based on the type of the measuring device or the required application.

The experimental hydrodynamic coefficients for heave and pitch motions are presented in Tables 3 and Table 4 and were calculated on the basis of general relations (Van Oortmerssen, 1976)

$$A_{jk} = -\frac{1}{\omega^2}\left(\frac{G_j \cos\varepsilon_{jk}}{\zeta_k} - C_{jk}\right) - M_{jk}$$
$$B_{jk} = \frac{G_j \sin\varepsilon_{jk}}{\omega\zeta_k}$$
(7)

where, G_j are the amplitudes of the radiation hydrodynamic forces and moments, ε_{jk} are the phase-difference between excitation and hydrodynamic response, ζ_k are the amplitudes of the harmonic forced motions and ω is the circular frequency of the imposed harmonic motion. The non-dimensional values are determined on the basis of the mass of the ship m, the length between perpendiculars, L_{BP}, and the gravity acceleration, g.

From theoretical point of view, the hydrodynamic coefficients have been calculated by using a computer code based on the source distribution method proposed by Frank (Frank, 1967). The evolutions of the theoretical and experimental non-dimensional hydrodynamic coefficients within non-dimensional circular frequency range are given in Figure 2. There is a satisfactory correlation between theoretical and experimental results.

Table 3. Non-dimensional added masses and damping coefficients for heave motions at zero speed. Experimental results.

$\omega \cdot (L/g)^{1/2}$	A_{33}^0/m	$B_{33}^0 \cdot (L/g)^{1/2}$
0.869	2.296	1.835
1.216	2.119	2.609
3.011	1.056	1.490
3.474	0.997	1.195
4.082	0.980	1.081
4.690	0.980	0.557

Table 4. Non-dimensional added masses and damping coefficients for pitch motions at zero speed. Experimental results.

$\omega \cdot (L/g)^{1/2}$	A_{55}^0/m	$B_{55}^0 \cdot (L/g)^{1/2}/m \cdot L^2$
0.869	0.116	0.108
1.737	0.096	0.154
3.011	0.076	0.112
3.474	0.073	0.093
4.082	0.065	0.083
4.690	0.059	0.060

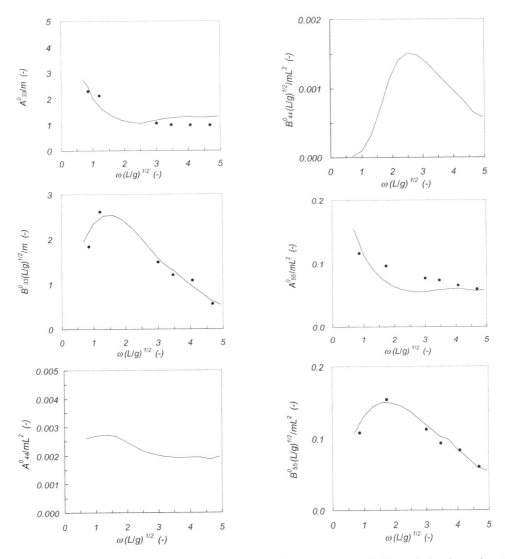

Figure 2. Non-dimensional added masses and damping coefficients at zero speed. Theoretical and experimental results.

4 SEAKEEPING PERFORMANCE

In order to evaluate the behaviour of the fishing vessel depending of the sea state and ship-wave incidence angle, a hybrid method was used. This methodology is based on the experimental response amplitudes operators RAO_η obtained in regular waves. For a given sea state, characterized from the energy point of view by the power spectral density $S_w(\omega_e)$, the response spectra of the ship's motions $S_\eta(\omega_e)$ can be calculated on the basis of encountering circular frequency ω_e with the expression

$$S_\eta(\omega_e) = RAO_\eta \cdot S_w(\omega_e) \qquad (8)$$

The spectra moments of zero order can be determined

$$m_0^{(\eta)} = \int_0^\infty S_\eta(\omega_e) d\omega_e \qquad (9)$$

as well as the root mean square values of the amplitudes of ship's motions on irregular waves

$$RMS_\eta = \sqrt{m_0^{(\eta)}} \qquad (10)$$

486

4.1 Free running and semi-captive model tests in regular waves

The experimental tests at design speed on regular following waves ($\alpha = 0°$) and head waves ($\alpha = 180°$) have been performed with a semi-captive model, with three freedom degrees (heave, roll and pitch motions). The experimental tests on regular beam waves ($\alpha = 90°$) were performed for the case of zero speed. For measuring the model motions, a special device with potentiometers system was used. The response amplitudes operators were determined for heave (RAO_z), roll (RAO_ϕ) and pitch (RAO_θ) motions

$$RAO_z = \left(\frac{\zeta_g}{\zeta_w}\right)^2 ; RAO_\phi = \left(\frac{\phi}{\zeta_w}\right)^2 ; RAO_\theta = \left(\frac{\theta}{\zeta_w}\right)^2$$

$$(11)$$

where, ζ_g, ϕ are θ the amplitudes of the heave, roll and pitch motions, respectively and ζ_w represents the amplitude of regular wave. Also, the transfer functions were determined for heave (Heave TRF), roll (Roll TRF) and pitch (Pitch TRF) motions

$$HeaveTRF = \frac{\zeta_g}{\zeta_w} ; RollTRF = \frac{\phi}{k\zeta_w} ; PitchTRF = \frac{\theta}{k\zeta_w}$$

$$(12)$$

where, k is the wave number.

The experimental values of the transfer functions for heave, roll and pitch motions, in longitudinal and beam waves, depending on the ratio between ship length and wave length L/λ are presented in Tables 5÷8.

From theoretical point of view, the response amplitude operators and transfer functions have been calculated using two computer codes, based on the source distribution method proposed by Frank and the conformal transformation by Lewis method (Bhattacharyya, 1978; Voitkunski, 1985). The diagrams of the numerical and experimental transfer functions of the ship motions on regular following and head waves at design speed, and on

Table 5. Transfer functions of heave and pitch motions in following waves ($\alpha = 0°$, $F_n = 0.40$). Experimental results.

L/λ	Heave TRF	Pitch TRF
0.250	0.90	0.865
0.404	0.648	0.666
0.556	0.447	0.484
0.769	0.173	0.221
0.855	0.10	0.145

Table 6. Transfer Functions of heave and pitch motions in head waves ($\alpha = 180°$, $F_n = 0.40$). Experimental results.

L/λ	Heave TRF	Pitch TRF
0.250	1.204	1.028
0.334	1.378	1.014
0.404	1.673	1.025
0.481	1.536	0.958
0.60	0.539	0.451

Table 7. Transfer function of heave motion in beam waves ($\alpha = 90°$, $F_n = 0$). Experimental results.

L/λ	Heave TRF	L/λ	Heave TRF
0.214	1.004	0.714	1.177
0.375	1.011	0.855	1.212
0.455	1.021	1.111	0.927

Table 8. Transfer function of roll motion in beam waves ($\alpha = 90°$, $F_n = 0$). Experimental results.

L/λ	Roll TRF	L/λ	Roll TRF
0.270	1.959	0.481	2.081
0.334	5.866	0.769	0.410
0.404	6.060	1.0	0.185
0.434	4.119	1.250	0.093

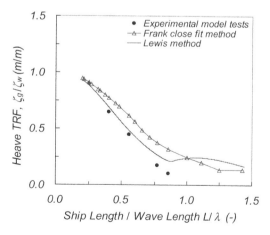

Figure 3. Transfer functions of heave motion in following waves ($\alpha = 0°$, $F_n = 0.40$). Comparison of theory and experiment.

regular beam waves at zero speed, depending on the ratio between ship length and wave length L/λ are presented in Figures 3÷8.

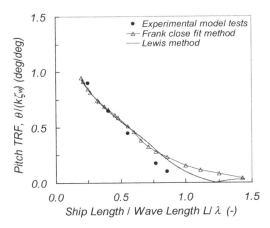

Figure 4. Transfer functions of pitch motion in following waves ($\alpha = 0°$, $F_n = 0.40$). Comparison of theory and experiment.

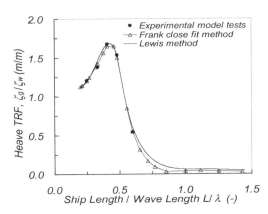

Figure 5. Transfer functions of heave motion in head waves ($\alpha = 180°$, $F_n = 0.40$). Comparison of theory and experiment.

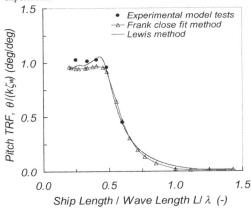

Figure 6. Transfer functions of pitch motion in head waves ($\alpha = 180°$, $F_n = 0.40$). Comparison of theory and experiment.

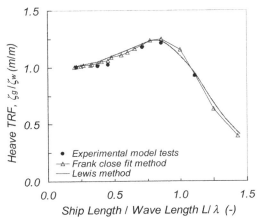

Figure 7. Transfer function of heave motion in beam waves ($\alpha = 90°$, $F_n = 0$). Comparison of theory and experiment.

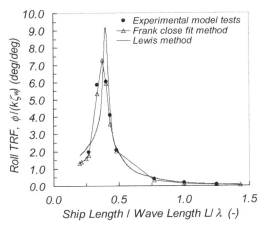

Figure 8. Transfer functions of roll motion in beam waves ($\alpha = 90°$, $F_n = 0$). Comparison of theory and experiment.

The correlation between theoretical and experimental results is satisfactory, but the computer codes did not offer adequate results on following regular waves, in the domain of encountering circular frequencies close to zero (Figs. 3 ÷ 4, $L/\lambda = 0.7 ÷ 0.9$). Also, a non-concordance regarding the transfer function of the roll motion on regular beam waves is remarked at resonance domain (Fig. 8, $L/\lambda = 0.3 ÷ 0.5$).

Also, the experimental tests have been performed at design speed with a self-propelled and remote controlled model, on regular stern quartering waves ($\alpha = 45°$) and bow quartering waves ($\alpha = 135°$). The device used for measuring the angular motions of the ship model consists in three

Table 9. Transfer functions of roll and pitch motions in stern quartering waves ($\alpha = 45°$, $F_n = 0.40$). Experimental results.

L/λ	Roll TRF	Pitch TRF
0.250	1.050	0.735
0.455	0.813	0.535
0.714	0.608	0.431
0.769	0.565	0.377
1.0	0.370	0.278
1.250	0.255	0.185

Table 10. Transfer functions of roll and pitch motions in bow quartering waves ($\alpha = 135°$, $F_n = 0.40$). Experimental results.

L/λ	Roll TRF	Pitch TRF
0.120	1.520	–
0.214	2.459	0.740
0.260	1.894	0.735
0.321	0.841	0.721
0.404	0.461	0.692
0.556	0.197	0.668
0.714	0.134	0.360
0.855	0.093	0.154

gyroscopes able to measure the heading angle, the roll motion and the pitch motion.

The experimental values of the transfer functions for roll and pitch motions, on regular oblique waves, at design speed, as a function on the L/λ ratio are presented in Tables 9÷10. Also, the comparison between theoretical and experimental diagrams of the transfer functions of the roll and pitch motions are presented in Figures 9÷12. The correlation between theoretical and experimental results is satisfactory.

4.2 Prediction of the ship motions on irregular waves

The ITTC (one parameter) spectrum was used for modelling the sea state. The power spectral density S_w was calculated using the significant wave height $H_{1/3}$ taken from the sea state tables for Mediterranean Sea, areas 26–27 (season December/February, all directions of wind) (Hogben et al., 1986). It was analyzed the ship behaviour on irregular waves for following sea states: moderate sea ($H_{1/3} = 2$ m), rough sea ($H_{1/3} = 3$ m) and very rough sea ($H_{1/3} = 4$ m).

The diagrams of RMS values for heave, roll and pitch motions are exemplified in Figures 13÷15, depending on the sea states and ship-wave incidence angles. The ship motions analysis on

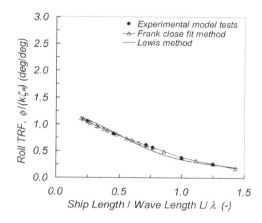

Figure 9. Transfer functions of roll motion in stern quartering waves ($\alpha = 45°$, $F_n = 0.40$). Comparison of theory and experiment.

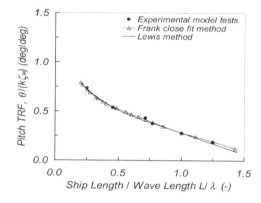

Figure 10. Transfer function of pitch motion in stern quartering waves ($\alpha = 45°$, $F_n = 0.40$). Comparison of theory and experiment.

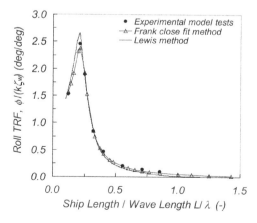

Figure 11. Transfer functions of roll motion in bow quartering waves ($\alpha = 135°$, $F_n = 0.40$). Comparison of theory and experiment.

489

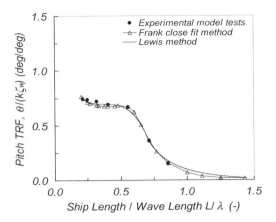

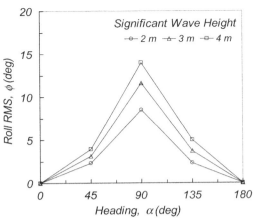

Figure 12. Transfer functions of pitch motion in bow quartering waves ($\alpha = 135°$, $F_n = 0.40$). Comparison of theory and experiment.

Figure 15. RMS of roll motion for different storm conditions.

Table 11. Comparison of heave RMS values in following waves ($\alpha = 0°$, Fn = 0.40).

| Significant height $H_{1/3}$ | Heave RMS, z (m) | | |
	Theoretical method	Hybrid method	Differences (%)
2 m	0.37	0.35	5.7
3 m	0.60	0.57	5.3
4 m	0.83	0.81	2.5

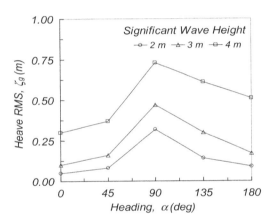

Figure 13. RMS of heave motion for different storm conditions.

Table 12. Comparison of pitch RMS values in following waves ($\alpha = 0°$, Fn = 0.40).

| Significant height $H_{1/3}$ | Pitch RMS, θ (deg) | | |
	Theoretical method	Hybrid method	Differences (%)
2 m	1.6	1.8	−11.1
3 m	2.3	2.6	−11.5
4 m	3.0	3.4	−11.8

Table 13. Comparison of roll RMS values in beam waves at zero speed ($\alpha = 90°$, Fn = 0).

| Significant height $H_{1/3}$ | Roll RMS, ϕ (deg) | | |
	Theoretical method	Hybrid method	Differences (%)
2 m	8.3	8.5	−2.4
3 m	11.3	11.6	−2.6
4 m	13.6	14.0	−2.9

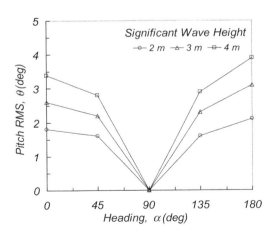

Figure 14. RMS of pitch motion for different storm conditions.

irregular waves at design speed shows important values of pitch angles on longitudinal waves and for heave displacement on beam waves. For the case of zero speed on beam waves, high RMS values for roll angle amplitude were obtained.

5 CONCLUSIONS

In order to determine the seakeeping performances of the fishing vessel (Fig. 1, Table 1) depending on the sea state and the ship-wave incidence angle, a hybrid method was used. This methodology is based on experimental RAO functions for model motions, obtained in regular waves. From theoretical point of view, the RAO functions were calculated using a computer code based on the source distribution method proposed by Frank.

Also, the study includes a comparison of the TRF transfer functions calculated by Frank method, by Lewis method and experimental results, proving a satisfactory correlation between the numerical simulations and model tests (Figs. 3÷12).

The RMS amplitudes of the ship motions on irregular waves were calculated, for given sea states (Figs. 13–15). The correlation between the theoretical and hybrid methods is satisfactory. The maximum differences are presented in Tables 11÷13.

It can be concluded that the computer codes for seakeeping provide reliable inputs to be used in the design process.

ACKNOWLEDGEMENTS

The authors wish to express their thanks to Mr. G. Messina, who supported the model experiments under contract financed by Ministero per le Politiche Agricole (MIPA, Italy), reference no. 4B01.

REFERENCES

Bendat, J.S. & Piersol, A.G. 1980. *Engineering applications of correlation and spectral analysis.* New York: Wiley—Interscience Publication.

Bhattacharyya, R. 1978. *Dynamics of marine vehicles.* New York: John Wiley & Sons Publication.

Calisal, S.M., Akinturk, A., Howard, D. & Mikkelsen, J. 1999. *Application of motion algorithms for safe fishing vessel designs.* International Shipbuilding Progress 46(447): 319–341.

Frank, W. 1967. *Oscillation of cylinders in or below the free surface of deep fluids.* Report No. 2375. Washington: Naval Ship Research and Development Center.

Hogben, N., Dacunha, N.M.C. & Olliver, G.F. 1986. *Global wave statistics.* London: British Maritime Technology.

Lee, S.K., Surendran, S. & Lee, G. 2005. *Roll performance of a small fishing vessel with live fish tank.* Ocean Engineering 32: 1873–1885.

Salvesen, N., Tuck, E.O. & Faltinsen, O. 1970. *Ship motion and sea loads.* New York: Transactions of SNAME. 78: 250–279.

Santos Neves, M.A., Perez, N. & Lorca, O. 2003. *Analysis of roll motion and stability of a fishing vessel in head seas.* Ocean Engineering 30: 2087–2104.

Van Oortmerssen, G. 1976. *The motion of a moored ship in waves.* Publication No.510. Wageningen: Netherlands Ship Model Basin.

Voitkunski, Y.I. 1985. *Ship theory handbook.* Sankt Petersburg: Sudostroenie.

Maritime Transportation and Harvesting of Sea Resources – Guedes Soares & Teixeira (Eds)
© 2018 Taylor & Francis Group, London, ISBN 978-0-8153-7993-5

Hydrodynamics performance of high speed multi-hulls in waves

G. Vernengo, C.M. Apollonio & D. Bruzzone
Department of Electric, Electronic and Telecommunication Engineering and Naval Architecture (DITEN),
University of Genova, Italy

L. Bonfiglio
MIT—Sea Grant College Program, Massachusetts Institute of Technology, Cambridge, MA, USA

S. Brizzolara
Aerospace and Ocean Engineering, Virginia Tech, Blacksburg, VA, USA

ABSTRACT: The proposed study aims at the understanding of the seakeeping responses of three multi-hull vessel configurations, namely catamarans, Small Waterplane Area Twin Hulls (SWATHs) and trimarans. Performance prediction has been obtained by using a first order 3D Boundary Element Method (BEM) based on a distribution of Rankine sources. Empirical models have been used to account for lifting effects of appendages, for viscous effects induced by the eddy generation and for hull friction. A validation of the method against available experimental data is proposed. The analysis then develops as comparison of the heave, roll and pitch responses among the three selected multi-hull configurations in terms of Response Amplitude Operators (RAOs).

1 INTRODUCTION

In the last decades multi-hull vessels have gained relevant success in many fields of the marine industry especially in those cases requiring significant performance at high speed, improved transverse stability during operations or large deck areas. There are several examples of high speed catamaran ferries as well as some catamarans and SWATHs used as wind farm support vessels (see among the other Papanikolaou et al. (1991), Jupp et al. (2014), Marsh (2015), Begovic et al. (2015) and Begovic et al. (2016)). A resonance-free SWATH have been designed by Yoshida et al. (2011) while the so-called 2nd generation SWATHs, exploiting canted double-struts configuration, have been proposed in Brizzolara et al. (2012) and, more recently, optimized by Vernengo and Brizzolara (2017). Recently, trimaran hulls have been proposed for civil and military applications. These examples refer to vessels that generally need to maintain high performance even when operating in rough sea environments; in fact for the ferries it is basically a matter of passenger comfort and safety; considering work-boats this requirement is more related to the safety during operations that is also relevant for navy vessels that meanwhile must ensure the higher possible level of precision while accomplish a specific mission.

In this context a fast and reliable prediction of ship motions is a crucial aspect both in the design phase, allowing the designers to achieve the best possible solution, and also during the operation of the vessels, providing for example useful information about possible changes of the routes. To this aim a fully 3D method, particularly suitable for multi-hulls, has been used to obtain ship responses to regular incoming waves.

The proposed work focuses on the prediction of the seakeeping responses of different types of multi-hull vessels operating at medium to high speed. Once the seakeeping prediction on the chosen multi-hulls has been validated against available experimental data, the analysis develops as comparison among the three selected hull configurations, i.e. catamaran, SWATH and trimaran. The comparative analysis aims at the identification of both the relative amplitude of the responses and the peculiar shapes of the RAOs, which, in some cases, could show double-peaked structures (see for instance Brizzolara et al. (2015)). Compared to traditional monohulls vessels, the literature on seakeeping of multi-hull is more limited, especially as regards lateral motion experiments. Some theoretical models have been developed for vertical plane motions (e.g. Fang et al. (1996) or Davis and Holloway (2003)) and fewer studies have analyzed the solution of lateral plane motions (see for instance Centeno et al. (2000), Davis et al. (2005), Katayama et al. (2008) and Grafton (2008)).

The numerical seakeeping analysis of vertical and lateral motions is performed in this study using a Boundary Element Method (BEM) based on a distribution of Rankine sources able to account for the effects of the ship speed on the motion responses (see for instance Bruzzone (2003)). The method, already used to predict the seakeeping responses of multihulls (see e.g. Vernengo et al. (2015) or Vernengo and Bruzzone (2016)),also includes some corrections for viscous and append-ages effects particularly relevant for lateral motions Apollonio et al. (2017).

2 SEAKEEPING ANALYSIS BY 3D RANKINE SOURCE BASED BOUNDARY ELEMENT METHOD

According to Bruzzone (2003), within the hypoth-esis of non viscous flow and irrotational motion, a first order linear 3D Boundary Element Method (BEM) is adopted to solve the radiation and dif-fraction problems. The ship responses to regu-lar incoming waves are then computed in the frequency domain. The rigid body motion in 6 Degrees of Freedom (DoF) is defined by the gen-eralized motion vector $\eta_k(\omega_e)$. Assuming small amplitude motions, the ODE system of equation of motion can be writtenin the frequency domain according to Eq. (1). M_{jk}, $A_{jk}(\omega_e)$, $B_{jk}(\omega_e)$, C_{jk}, $F_j^D(\omega_e)$ and $F_j^{FK}(\omega_e)$ are the mass, the added mass, the damping, the restoring coefficients, the diffraction and the Froude-Krylov force matrices, respectively.

$$\sum_{k=1}^{6} \{-\omega_e^2 [M_{jk} + A_{jk}(\omega_e)] + i\omega_e B_{jk}(\omega_e) + C_{jk}\} \eta_k(\omega_e) = F_j^D(\omega_e) + F_j^{FK}(\omega_e)$$

(1)

The solution is found in terms of a velocity potential Φ by imposing imposing the non-perme-ability condition on the hull surface and the com-bined kinematic and dynamic boundary condition on the free surface. The forward ship speed effect on the ship responses is included in the hull bound-ary condition through the m-terms and in the free surface boundary condition by additional speed terms. The total potential Φ is subdivided into a steady and a unsteady contribution, namely Φ_S and Φ_U, that are superimposed due to the linear-ity of the formulation. The latter is further defined as the summation of the incoming wave potential ϕ_I, the diffracted potential ϕ_D and the six radiation potentials ϕ_k, each one accounting for one of the 6 DoF of the ship following Eq. (2)

$$\Phi = \Phi_S + \Phi_U = $$
$$U_\infty x + \left[\phi_I + \phi_D + \sum_{k=1}^{6} \phi_k \eta_k\right] e^{i\omega_e t}$$

(2)

The BEM is formulated so that the radiation condition, imposing that no radiated waves propa-gate forward of the ship, is fulfilled for $\frac{\omega_e U_\infty}{g} > 0.25$.

Viscous corrections have been applied to improve the quality of the BEM prediction. Verti-cal motions (η_3, η_5) are corrected in case stabilizer fins are present following the method by Sclavou-nos and Borgen (2004). Their effect is accounted for also when lateral motions (η_2, η_4, η_6) are com-puted but, in addition, the viscous effects induced by eddy making and hull friction are considered by empirical formulation proposed by Schmitke (1978) by including the corresponding damping coefficients B_{44}^E and B_{44}^H, respectively. In the case of the trimaran, following Grafton (2008), the additional contribution to the roll damping due to the heave motion of the side hulls is accounted for. Such a new term is computed according to Eq. (3) where $B_{33}^{SH^*}(\omega)$ is the heave damping coefficient of a catamaran made of the two side hulls (i.e. exclud-ing the center hull of the trimaran).

$$B_{44}^{SH^*}(\omega) = \frac{s^2}{4} B_{33}^{SH}(\omega)$$

(3)

Since B_{44}^E and B_{44}^H are linear functions of η_4, a global iterative process is required to converge to the roll motion prediction (see e.g. Apollonio et al. (2017)).

3 VALIDATION OF MOTION PREDICTION FOR THE SELECTED MULTI-HULLS

Seakeeping response of three multi-hull configura-tions have been analyzed. Numerical predictions have been first compared to available experimental results on such a hulls, namely a catamaran from Van't Veer (1998), a SWATH from Kallio (1976) and a trimaran from Maki et al. (2009). The numer-ical analysis has then been extended to see the effect of the wave encounter angle μ and the ship speed by the Froude number F_N. Even keel condition has been considered in all the presented computations.

Figure 1 displays a perspective view of the three multi-hulls. Both the catamaran demi-hulls and the trimaran center and side hulls, Figure 1(a) and Fig-ure 1(b), respectively, have round-bilge shapes. The trimaran center hull has an inverse bow design while its outriggers and the catamaran demi-hulls have conventional bow shapes. The SWATH, Figure 1(c), presents a cylindrical torpedo shape without any humps and single symmetric strut configuration.

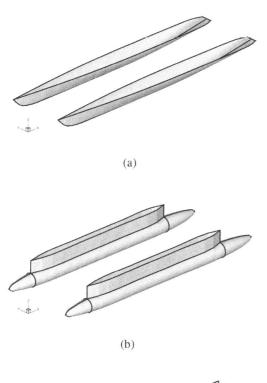

(a)

(b)

(c)

Figure 1. Perspective views of the three multi-hulls. Only the submerged part of the hulls are shown. Figure 1(a): catamaran. Figure 1(b): SWATH. Figure 1(c): trimaran.

Main particulars of the three hulls are reported in Table 1. $\frac{L_{WL}}{B}$ and $\frac{B}{T}$ ratios are comparable between the catamaran and the trimaran while they are very different for the SWATH. This is due to the smaller beam of the waterline of the latter vessel with respect to those of the other two more conventional hulls. Comparing the values of the waterline area coefficients C_{WP} of both the side and center hulls of the trimaran with that of the catamaran indicates that the former has narrower waterlines. The SWATH, on the other hand, shows

Table 1. Main characteristics of the selected multi-hulls.

	Catamaran (Demi-hull)	SWATH (Demi-hull)	Trimaran (Center hull)	Trimaran (Side hulls)
$L_{WL}[m]$	3.00	2.33	4.94	4.96
$\frac{L_{WL}}{B}$	12.50	23.76	16.76	16.79
$\frac{B}{T}$	1.61	0.27	1.54	1.54
$\frac{S}{\nabla^{\frac{2}{3}}}$	7.77	8.15	8.28	8.28
$\frac{L_{WL}}{\nabla^{\frac{1}{3}}}$	8.49	3.69	9.42	9.38
$\frac{s}{L_{WL}}$	0.23	0.44	–	0.17
C_{WP}	0.77	0.85	0.78	0.81
C_P	0.63	0.85	0.66	0.67

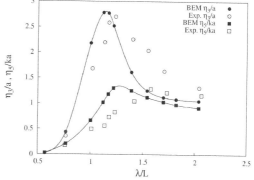

Figure 2. Comparison of computed and measured heave and pitch RAOs for the catamaran. $\mu = 225°$. $F_N = 0.60$.

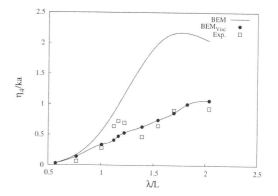

Figure 3. Comparison of computed and measured roll RAO for the catamaran. $\mu = 225°$. $F_N = 0.60$. BEM prediction with and w/o viscous corrections are shown.

a higher C_{WP} since most of its waterline breadth is constant.

Figure 2 and Figure 3 display the comparison between the numerical prediction and the experimental results for the catamaran in terms of vertical and lateral motions, respectively. Both responses have been computed at $F_N = 0.60$ for a wave encounter angle $\mu = 225°$ ($\mu = 180°$ corresponds to head waves). The hull is not equipped with any kind of appendages. Hence, viscous corrections have only been applied for roll prediction including eddy making and hull friction components. The computed heave and pitch RAOs generally well agree with the experimental measures. The two peaks are slightly anticipated and the decay of the heave response for high $\frac{\lambda}{L}$ is faster in the numerical response. Numerical roll prediction, including viscous corrections, is satisfactory even if the first peak of this response, which is seen in the experimental measures at $\frac{\lambda}{L} = 1.15$, is overdamped by the method.

Results of the validation carried out for the SWATH-6A highlight the relevant effect of the stabilizer passive fins mounted on the torpedo hull. In fact, both the heave and pitch responses ($\mu = 180°$, $F_N = 0.435$) and the roll RAO ($\mu = 90°$, $F_N = 0.384$), shown in Figure 4 and 5, respectively, found an agreement with the experimental measures only if their presence is accounted for. Both vertical and lateral motions are over-predicted by the BEM as is. This is partially due to the formulation of the method itself (as in fact it also happens for the catamaran) but here it is also caused by the particular hull type that has very small hydrostatic restoring forces and a significant component of viscous effects as highlighted e.g. by (Bonfiglio & Brizzolara 2004).

Predicted heave and pitch RAOs and the roll response are compared with the corresponding experimental measures in Figure 6 and Figure 7, respectively. The comparison is performed at $\mu = 90°$ with $F_N = 0.44$. A satisfactory prediction of

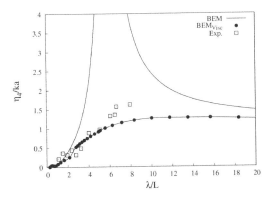

Figure 5. Comparison of computed and measured roll RAO for the SWATH-6A. $\mu = 90°$. $F_N = 0.384$. BEM prediction with and w/o passive stabilizer fins and viscous corrections are shown.

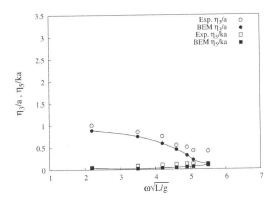

Figure 6. Comparison of computed and measured heave and pitch RAOs for the Trimaran. $\mu = 90°$. $F_N = 0.44$.

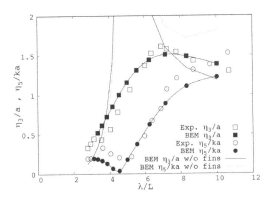

Figure 4. Comparison of computed and measured heave and pitch RAOs for the SWATH-6A. $\mu = 180°$. $F_N = 0.453$. BEM prediction with and w/o passive stabilizer fins are shown.

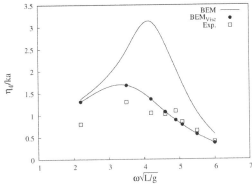

Figure 7. Comparison of computed and measured roll RAO for the Trimaran. $\mu = 90°$. $F_N = 0.44$. BEM prediction with and w/o viscous corrections are shown.

vertical motions is achieved. The pitch response, as expected, is very low. The trend of the computed heave RAO is consistent even if the absolute values are slightly reduced compared to the experiments, especially in the range of the higher tested frequencies. The agreement on the roll response is acceptable showing the greater errors at the lower frequencies. Such a difference in the roll prediction could be due to the fact that the viscous corrections have been originally formulated for monohulls and then adapted to account for multi-hull configurations.

4 SYSTEMATIC PERFORMANCE ANALYSIS OF THE THREE MULTI-HULLS

Once the method has been validated, the seakeeping responses of the three hulls have been systematically analyzed at Froude numbers in the range $F_N = (0.3; 0.7)$, for encounter angles of the incoming waves equal to $\mu = (90°; 135°; 180°)$, i.e. from beam to head waves. Numerical prediction of the roll RAO accounts for the viscous empirical corrections. The SWATH computations have been performed including the effects of the passive stabilizer fins on both the vertical and lateral motions. The tested conditions are listed in Table 2 together with the corresponding figure number.

Figure 8 and Figure 9 display the heave and pitch RAOs of the three hulls at $\mu = 180°$ and $\mu = 135°$, respectively. As F_N increases, the peaks of heave and pitch of the catamaran (see Figures 8(a) and 9(a)) rise in value and moves towards lower frequencies. Consistently, comparing the trends at the two wave encounter angles, namely $\mu = 180°$ and $\mu = 135°$, the peaks are higher for head waves with respect to bow waves. The η_3 and η_5 trends change in the case of the SWATH due to the effect generated by the passive stabilizer fins as seen from Figure 8(b) and Figure 9(b). The peak of the response at the higher F_N is lower with respect to that at the lower speed. According to the method used in this study to modify vertical motions (see Sclavounos

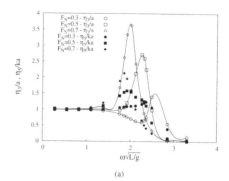

(a)

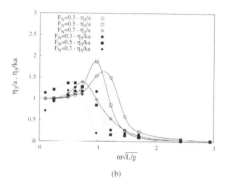

(b)

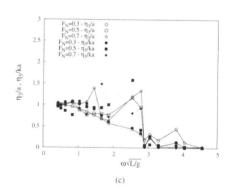

(c)

Figure 8. Heave and pitch RAOs at $\mu = 180°$ for $F_N = 0.30; 0.50; 0.70$. (a): Catamaran. (b): SWATH. (c): Trimaran.

Table 2. Analyzed conditions for the seakeeping study.

$F_N = 0.3; 05;$ 0.7	180°	135°	90°
Catamaran	η_3, η_5 (Fig. 8(a)) η_3, η_5, η_4	(Fig. 9(a)), 10(a))	η_4 (Fig. 11(a))
SWATH	η_3, η_5 (Fig. 8(b)) η_3, η_5, η_4	(Fig. 9(b)), 10(b))	η_4 (Fig. 11(b))
Trimaran	η_3, η_5 (Fig. 8(c)) η_3, η_5, η_4	(Fig. 9(c)), 10(c))	η_4 (Fig. 11(c))

and Borgen (2004)), being V_S the ship speed, the additional restoring and damping terms are proportional to V_S^2 and V_S (through the lift force). This means that, assuming $\frac{\delta C_L}{\delta \alpha} = 2\pi$, the effect of the foils increase with the ship speed. This could explain the particular trends of the SWATH heave and pitch RAOs. Heave and pitch responses of the trimaran, shown in Figure 8(c) and Figure 9(c) at $\mu = 180°$ and $\mu = 135°$, respectively, present more oscillatory trends. In particular considering the response at the higher F_N both η_3 and η_5 have

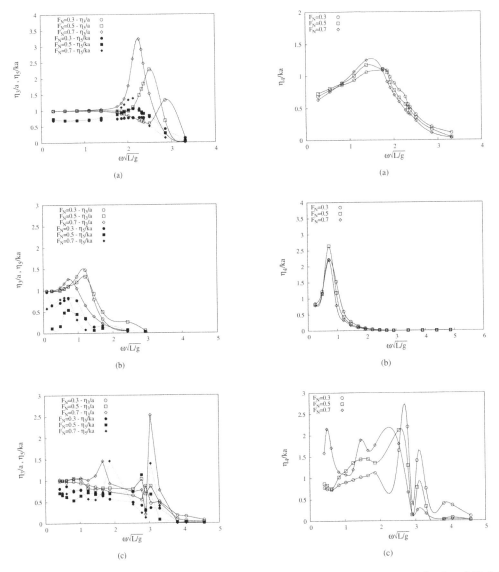

Figure 9. Heave and pitch RAOs at $\mu = 135°$ for $F_N = 0.30$; 0.50; 0.70. (a): Catamaran. (b): SWATH. (c): Trimaran.

Figure 10. Roll RAO at $\mu = 135°$ for $F_N = 0.30$; 0.50; 0.70. (a): Catamaran. (b): SWATH. (c): Trimaran.

two peaks at two distinct frequencies. Again, the peak frequencies move towards higher values as the wave encounter angle shifts from $\mu = 180°$ to $\mu = 135°$. Comparing the responses of the three hulls, in terms of vertical motions, it can be seen that the catamaran peaks occur at relatively high frequencies, in the range $\omega\sqrt{L/g} \cong [2.0; 2.5]$, while the SWATH has its maximum responses in the range $\omega\sqrt{L/g} \cong [0.8; 1.2]$. The double peaks of the trimaran RAOs happen for $\omega\sqrt{L/g} \cong [1.5; 2.0]$ and

$\omega\sqrt{L/g} \cong [2.6; 3.2]$. The catamaran at $F_N = 0.70$ at $\mu = 180°$ shows the maximum peak value, corresponding to $\frac{\eta_3}{a} = 3.62$. This value should be further checked in order to exclude the occurrence of resonant phenomena that can amplify the response and, eventually, to dampen it.

Figure 10 and Figure 11 present the roll RAOs at $\mu = 135°$ and $\mu = 90°$, respectively. The effect of the ship speed on such response is less marked, producing minimal shifts. Considering that the

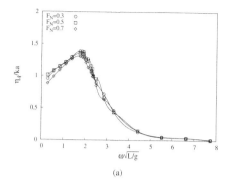

(a)

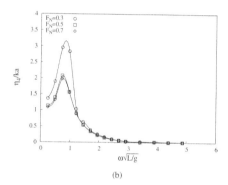

(b)

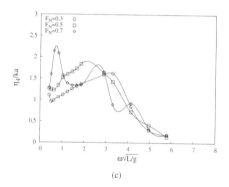

(c)

Figure 11.   Roll RAO at $\mu = 90°$ for $F_N = 0.30; 0.50; 0.70$. (a): Catamaran. (b): SWATH. (c): Trimaran.

SWATH response is largely dampen by the effect of the passive stabilizer fins, at $\mu = 135°$ the magnitude of the roll RAO decreases with the speed (see Figure 10(b)) for the same reason outlined above for the case of η_3 and η_5. The trimaran represents an exception since its roll response is more confused, showing multiple peaks especially at the higher speed. This unusual behavior could be associated to the particular configuration of the selected trimaran that is made of three comparable

hulls rather than on a center hull with outriggers. The asymptotic trends of the roll RAOs in the range of low frequencies are in both cases ($\mu = 135°$ and $\mu = 900°$) consistent with the theory.

5   CONCLUSIONS

Seakeeping responses of three different multihull configurations have been predicted by a first order Boundary Element Method. Corrections accounting for viscous effects (eddy making and hull friction), for motion interaction effects (roll damping induced by heave motion) and for lifting effects due to the presence of appendages have been included in the motion predictions by empirical formulations. The seakeeping analysis has been performed on three selected multi-hulls, namely a catamaran, a SWATH and a trimaran. The seakeeping performance prediction has been preliminary validated by comparing the numeric computations against experimental results in head and oblique waves available for the three hulls. Then a systematic analysis of the response of the selected hulls with respect to heave, roll and pitch has been carried out.

This study achieves two main results. First the seakeeping method, i.e. the BEM including all the mentioned corrections, has been validated on three significant hull types by comparison against experiments, leading to satisfying results both in terms of vertical and lateral motions prediction. Then the numeric analysis provided some insights into the seakeeping behaviors of these multi-hulls for highspeed applications. In particular, the trends of the responses of these different hull types have been computed showing for example the double peaked RAOs of the trimaran. In addition, the produced results allow to identify the ranges of frequencies in which the peaks occur for each of the three multi-hulls over a significant range of speeds.

REFERENCES

Apollonio, C., G. Vernengo, L. Bonfiglio, S. Brizzolara, & D. Bruzzone (2017). On the lateral motion prediction of high speed multi-hull vessels. In *The Twenty-seventh (2017) International Ocean and Polar Engineering Conference.* San Francisco, California, June 25–30, 2017. International Society of Offshore and Polar Engineers.

Begovic, E., C. Bertorello, A. Bove, & F. De Luca (2016). Exploitation of swath hull form concept for small workingpleasure craft. *Maritime Technology and Engineering III: Proceedings of the 3rd International Conference on Maritime Technology and Engineering—Guedes Soares & Santos (Eds), MARTECH 2016, 2016 Taylor & Francis Group, London.*

Begovic, E., C. Bertorello, & S. Mancini (2015). Hydrodynamic performances of small size swath craft. *Brodogradnja 66*(4), 1–22.

Bonfiglio, L. & S. Brizzolara (2004). Unsteady viscous flow with nonlinear free surface around oscillating swath ship sections. *WSEAS Transactions on Fluid Mechanics 9*, 49–57.

Brizzolara, S., T. Curtin, M. Bovio, & G. Vernengo (2012). Concept design and hydrodynamic optimization of an innovative swath usv by cfd methods. *Ocean Dynamics 62*(2), 227–237.

Brizzolara, S., G. Vernengo, L. Bonfiglio, & D. Bruzzone (2015). Comparative performance of optimum high speed swath and semi-swath in calm water and in waves. *Transactions - Society of Naval Architects and Marine Engineers 123*, 273–286.

Bruzzone, D. (2003). Application of a rankine source method to the evaluation of motions of high speed marine vehicles. In *Proceeding of the 8th International Marine Design Conference, Athens*, Volume 2, pp. 69–79.

Centeno, R., N. Fonseca, & C.G. Soares (2000). Prediction of motions of catamarans accounting for viscous e ects. *International shipbuilding progress 47*(451), 303–323.

Davis, M., N. Watson, & D. Holloway (2005). Measurement of response amplitude operators for an 86 m high-speed catamaran. *Journal of ship research 49*(2), 121–143.

Davis, M.R. & D.S. Holloway (2003). Motion and passenger discomfort on high speed catamarans in oblique seas. *International shipbuilding progress 50*(4), 333–370.

Fang, C.C., H. Chan, & A. Incecik (1996). Investigation of motions of catamarans in regular wavesi. *Ocean engineering 23*(1), 89–105.

Grafton, T.J. (2008). *The roll motion of trimaran ships*. University of London, University College London (United Kingdom).

Jupp, M., R. Sime, & E. Dudson (2014). Xss–a next generation windfarm support vessel. In *RINA Conference: Design & Operation of Wind Farm Support Vessels*, pp. 29–30.

Kallio, J.A. (1976). Seaworthiness characteristics of a 2900 ton small waterplane area twin hull (swath).

Technical Report SPD-620–03, David Taylor Naval Ship Research and Development Center.

Katayama, T., T. Taniguchi, & M. Kotaki (2008). A study on viscous effects of roll damping of a high-speed catamaran and a high-speed trimaran. In *Proc. 6th Osaka Colloq. Seakeeping and Stability of Ships*.

Maki, K., L. Doctors, R. Scher, W. Wilson, S. Rhee, A. Troesch, & R. Beck (2009). Conceptual design and hydrodynamic analysis of a high-speed sealift adjustable-length trimaran. *Trans.-Soc. Nav. Archit. Mar. Eng 116*, 16–39.

Marsh, G. (2015). Radical access solutions for distant offshore wind farms. *Renewable Energy Focus 16*(4), 81–83.

Papanikolaou, A., G. Zaraphonitis, & M. Androulakakis (1991). Preliminary design of a high-speed swath passenger/car ferry. *Marine Technology 28*(3), 129–141.

Schmitke, R.T. (1978). Ship sway, roll, and yaw motions in oblique seas. *SNAME Transaction 86*, 26–46.

Sclavounos, P.D. & H. Borgen (2004). Seakeeping analysis of a high-speed monohull with a motion-control bow hydrofoil. *Journal of ship research 48*(2), 77–117.

Van't Veer, R. (1998). Experimental results of motions and structural loads on the 372 catamaran model in head and oblique waves. *TU Delft report* (1130).

Vernengo, G. & S. Brizzolara (2017). Numerical investigation on the hydrodynamic performance of fast swaths with optimum canted struts arrangements. *Applied Ocean Research 63*, 76–89.

Vernengo, G., S. Brizzolara, D. Bruzzone, et al. (2015). Resistance and seakeeping optimization of a fast multihull passenger ferry. *International Journal of Offshore and Polar Engineering 25*(01), 26–34.

Vernengo, G. & D. Bruzzone (2016). Resistance and seakeeping numerical performance analyses of a semi-small waterplane area twin hull at medium to high speeds. *Journal of Marine Science and Application 15*(1), 1–7.

Yoshida, M., H. Kihara, H. Iwashita, & T. Kinoshita (2011). Seaworthiness of resonance-free swath with movable fins as an oceangoing fast ship. In *11th International Conference on Fast Sea Transportation FAST 2011*.

Hydrodynamics – CFD

Maritime Transportation and Harvesting of Sea Resources – Guedes Soares & Teixeira (Eds)
© 2018 Taylor & Francis Group, London, ISBN 978-0-8153-7993-5

Self-propulsion simulation of DARPA Suboff

A. Dogrul & S. Sezen
Yildiz Technical University, Istanbul, Turkey

C. Delen & S. Bal
Istanbul Technical University, Istanbul, Turkey

ABSTRACT: Hydrodynamic performance prediction of a propeller working behind a submerged body is a popular research field. For a submarine propeller, the propeller-hull interaction should be considered during the preliminary design stage. In this study, the resistance and propulsion analyses of the well-known benchmark DARPA Suboff with E1619 propeller have been done using Computational Fluid Dynamics (CFD) method. Self-propulsion of the submarine has been modeled with actuator disc based on body force method and with the propeller itself behind the submarine. The flow has been considered as 3-D, fully turbulent, incompressible and steady, thus the governing equations (RANSE) have been discretized with Finite Volume Method (FVM). Uncertainty analysis has also been carried out to determine the optimum cell number in terms of total resistance. The numerical results have been compared with the available experimental data. The applicability of CFD method on self-propulsion performance prediction of the underwater vehicles has been discussed.

1 INTRODUCTION

The technology of unmanned underwater vehicles, which provide the opportunity to work in hazardous areas, are improving rapidly especially for military and research purposes. The complexity of the work being done, the variability and the difficulty of environmental conditions is gaining importance in the design phase of the vehicle (Vaz et al. 2010). The interaction between propeller and the hull of underwater vehicle is therefore very important and should be determined precisely and reliably in the preliminary design stage of unmanned vehicles.

In the past, Zhang et al. have made a study involving the interaction between propeller and a submarine hull. The analyses have been made by taking the free surface effect into account. The results show a good agreement with the experiments (Zhang et al. 2014). Berger et al. have been focused on the propeller hull interaction with a coupled method. The numerical study has been carried out for the well-known benchmark case KRISO container ship (KCS). The velocity field gathered from the RANSE solver has been given as an input to the potential solver. After the newly calculated velocity field has been applied to the RANSE solver and the flow is solved by taking the propeller hull interaction into account. The numerical results of the model propeller have been compared with those of a fully RANSE computation. By employing the propeller model with the developed code, computation time has been decreased

drastically. Especially the thrust prediction has become quicker (Berger et al. 2011). Rijpkema et al. have studied the propeller-hull interaction by simulating the steady viscous flow around KCS hull with RANSE method and unsteady propeller flow with BEM. The numerical analyses have been carried out via a hybrid method. The coupled RANSE-BEM approach has given accurate results for thrust compared with the experimental data (Rijpkema et al. 2013). A comprehensive study has been made by Ozdemir et al. in order to predict resistance and wave profile of KCS numerically. The numerical method has been validated with the experimental data in terms of total ship resistance and wave profiles along the hull (Ozdemir et al. 2016). A numerical study has been conducted for resistance, propeller open water and self-propulsion performance prediction for KCS hull using a RANSE solver by Seo et al. (Seo et al. 2010). Local mesh refinements have been used in order to gain a convenient mesh structure. Sliding mesh technique has been chosen for propeller tests. Numerical results have been then compared with the existing available data. Villa et al. have made a numerical simulation of the flow around a ship with self-propulsion with RANSE solver. The coupled method solves the viscous flow around KCS hull with RANSE solver while the performance of KP505 propeller is calculated by an unsteady panel method (Villa et al. 2012). The propeller-ship interaction has been investigated by a commercial CFD program for DTC Post-Panamax Container Ship in Kinaci et al. The ship has been analyzed without taking free surface effect into account (Kinaci et al. 2013). In another study of Kinaci et al., CFD analyses have been carried out for resistance prediction of KRISO Container Ship. Experimental and numerical calculations have been performed for also a fully submerged body and a validation study has been made (Kinaci et al. 2016). The paper of Carrica et al. presents a method for self-propulsion calculation of surface ships. The method is based on controlling the propeller rotation speed (RPS) to find the self-propulsion point while reaching the target Froude number (Carrica et al. 2010). Chase has studied the self-propulsion problem of the well-known DARPA Suboff as a thesis work. A custom developed CFD solver has been employed for various advance coefficients. The effect of the turbulence has been observed via different turbulence models. The wake velocities have been compared with the experimental data for a constant advance coefficient (Chase 2012). A seven bladed INSEAN E1619 propeller has been studied in the presence of DARPA Suboff submarine model by Chase et al. The numerical analyses have been made by employing Delayed Detached Eddy Simulation (DDES) approach. The results have been compared with different turbulence models using four grids and three time steps for one advance coefficient. The results show that the present approach is applicable in self-propulsion performance prediction of submarines (Chase et al. 2013). The effect of bow and stern geometries on resistance of bared DARPA Suboff has been studied via CFD by Budak et al (Budak et al. 2016). A very recent study has been carried out by Delen et al. in order to predict the self-propulsion performance of DARPA Suboff bare hull with DTMB4119 model propeller for two different velocities (Delen et al. 2017).

In this study, resistance values and self-propulsion points of the DARPA Suboff bare form (AFF-1) have been computed. In resistance analyses, an uncertainty analysis has also been performed to identify the suitable mesh structure. Verification and validation has been made with the help of uncertainty assessment and available experimental data. The numerical results have been compared with the experiments. The suitable mesh structure has then been employed for the self-propulsion analysis. Single phase CFD analyses have been carried out for the bare hull. The flow has been considered as 3-D, fully turbulent, incompressible and steady. k-ε has been chosen as the turbulence model in the numerical calculations. Before calculation of self-propulsion point, open water results of the E1619 propeller have been computed. Two techniques have been used in self-propulsion analyses. First, a disc having the same diameter with

the actual propeller has been defined in the propeller plane. The self-propulsion analysis has then been performed with actual E1619 model propeller. Self-propulsion points for both techniques have been compared with each other for DARPA Suboff bare hull for velocity, V = 3.046 m/s. The applicability of the numerical method has been discussed via self-propulsion point.

2 THEORETICAL BACKGROUND

2.1 Numerical method

The governing equations are the continuity equation and the well-known RANSE equations for the unsteady, three-dimensional, incompressible flow. The continuity can be given as;

$$\frac{\partial U_i}{\partial x_i} = 0 \tag{1}$$

Velocity U can be decomposed as mean velocity and fluctuating velocity, respectively;

$$U_i = \overline{U}_i + u_i \tag{2}$$

While the momentum equations are expressed as;

$$\frac{\partial U}{\partial t} + \frac{\partial (U_i U_j)}{\partial x_j}$$
$$= -\frac{1}{\rho}\frac{\partial P}{\partial x_i} + \frac{\partial}{\partial x_j}\left[v\left(\frac{\partial U_i}{\partial x_j} + \frac{\partial U_j}{\partial x_i}\right)\right] - \frac{\partial \overline{u_i' u_j'}}{\partial x_j} \tag{3}$$

In this paper, since all simulations are run under steady state conditions, the first term in equation 3 is not taken into account. In momentum equations, U_i states the mean velocity while u_i' represents the fluctuation velocity components in the direction of the Cartesian coordinate. P expresses the mean pressure, ρ the density and v the kinematic viscosity.

The well-known k-ε turbulence model is employed in order to simulate the turbulent flow around the submarine precisely. This turbulence model is applicable when there are not high pressure changes along the hull and separation near the hull. In this case, k-ε turbulence model is used because the vessel is fully submerged and hence there are no free surface effects. During the analyses, Reynolds stress tensor is calculated as follow;

$$\overline{u_i' u_j'} = -v_t\left(\frac{\partial U_i}{\partial x_j} + \frac{\partial U_j}{\partial x_i}\right) + \frac{2}{3}\delta_{ij}k \tag{4}$$

Here, v_t is the eddy viscosity and expressed as an empirical constant ($C_\mu = 0.09$). k is the turbulent kinetic energy and ε is the turbulent dissipation rate. In addition to the continuity and momentum equations, two transport equations are solved for k and ε:

$$\frac{\partial k}{\partial t} + \frac{\partial (kU_j)}{\partial x_j} = \frac{\partial}{\partial x_j}\left[\left(v + \frac{v_t}{\sigma_k}\right)\frac{\partial k}{\partial x_j}\right] + P_k - \varepsilon \quad (5)$$

$$\frac{\partial \varepsilon}{\partial t} + \frac{\partial (kU_j)}{\partial x_j} = \frac{\partial}{\partial x_j}\left[\left(v + \frac{v_t}{\sigma_\varepsilon}\right)\frac{\partial \varepsilon}{\partial x_j}\right] + C_{\varepsilon 1}P_k\frac{\varepsilon}{k} - C_{\varepsilon 2}\frac{\varepsilon^2}{k}$$

$$(6)$$

$$P_k = -\overline{u_i'u_j'}\frac{\partial U_i}{\partial x_j} \quad (7)$$

where, $C_{\varepsilon 1} = 1.44$, $C_{\varepsilon 2} = 1.92$ turbulent Prandtl numbers for k and ε are $\sigma_k = 1.0$ and $\sigma_\varepsilon = 1.3$, respectively. Further explanations for the k-ε turbulence model may be found in (Wilcox 2006).

2.2 Uncertainty assessment

In this study, uncertainty analysis has been made via Grid Convergence Method as recommended in the ITTC procedure for CFD verification (ITTC 2011a). This method firstly was proposed by Roache and then improved with different studies (Roache 1998). The procedure implemented in this study has been explained below (Celik et al. 2008).

Let h_1, h_2 and h_3 are grid lengths and $h_1 < h_2 < h_3$. The refinement factors (r) are as follows:

$$r_{21} = \frac{h_2}{h_1} \quad r_{32} = \frac{h_3}{h_2} \quad (8)$$

Refinement factors should be greater than 1.3 in accordance with the experiments (Roache 1998). Grid lengths' refinement is selected as a value of $\sqrt{2}$. Because of the mesh algorithm, number of cells (N) has been taken into account during the calculation of the refinement factors. This choice is crucial in uncertainty analyses especially for unstructured mesh system. Therefore, these values have been differentiated.

$$r_{21} = \left(\frac{N_1}{N_2}\right)^{1/3} \quad r_{32} = \left(\frac{N_2}{N_3}\right)^{1/3} \quad (9)$$

The differences (ε) between generated cell numbers can be calculated below:

$$\varepsilon_{21} = X_2 - X_1 \quad \varepsilon_{32} = X_3 - X_2 \quad (10)$$

Here, X is the solution of the analysis.

At this point, convergence condition R can be examined.

$$R = \frac{\varepsilon_{21}}{\varepsilon_{32}} \quad (11)$$

In this study, R is calculated between 0 and 1 which means that the solution is converged monotonically. Detailed information about Grid Convergence Index (GCI) can be found in (Roache 1998).

3 COMPUTATIONAL METHOD

3.1 Grid structure and boundary conditions

A proper computational domain has been created in order to simulate the flow around the submarine model with/without self-propulsion. Three dimensional grids have been employed for modelling the flow region with fully hexahedral elements.

Figure 1 shows the computational domain with the assigned boundary conditions on the surfaces. The left side of the computational domain is defined as velocity inlet and the right side is defined as pressure outlet. The submarine surface is considered as no-slip wall. In addition, the side surfaces are also defined as symmetry. More detailed information on boundary conditions can be obtained from the theory guide of the commercial CFD software (Star CCM+ 2015).

The computational domain is divided into three dimensional finite volumes and discretized according to the finite volume method (FVM). The main dimensions of the computational domain are determined in accordance with the ITTC guidelines to properly determine the flow (ITTC 2011b). To create a computational domain, unstructured hexahedral elements are employed in the whole domain. The mesh refinements are also made in the bow, stern and the wake area of the form.

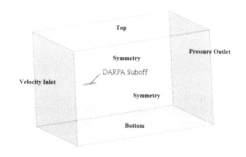

Figure 1. Computational domain and the boundary conditions.

505

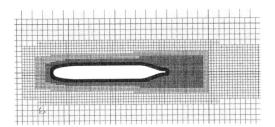

Figure 2. Unstructured mesh of the computational domain.

Unstructured mesh of the computational domain is given in Fig. 2. For propulsion analyses including open water and self-propelled cases; polyhedral mesh structure has been employed but also a fine mesh region has been created around the propeller disc in order to model the interaction of the hull and the propeller more accurately.

3.2 Solution strategy

k-ε turbulence model is used in the computational analyses because there are no high pressure gradients along the hull. In other words, the slenderness of the hull geometry makes the effect of boundary layer separations on the flow characteristics around the hull insignificant.

The pressure field is solved by using SIMPLE algorithm which is based on pressure-velocity coupling. SIMPLE is a commonly used algorithm for calculating pressure and velocity fields in an iterative manner. Especially for steady state analyses, it reduces the computational time rapidly (Versteeg et al. 2007). All the governing equations are discretized using a cell based finite volume method and the advection terms are discretized with a first-order upwind interpolation scheme.

Because of the flow characteristics, the analyses have been made via single phase assumption. Free surface effects are not taken into account in either resistance or propulsion analyses. Viscous effects near the ship are taken into account by modelling the boundary layer with an appropriate grid structure keeping y+ values of the hull in a reasonable range (30–300).

For self-propulsion analyses based on body force method, an actuator disc has been modeled using actuator disc theorem. Distribution of body forces has been applied in this actuator disc region. Here, actuator disc region represents an infinite-bladed propeller. So the characteristics of hydrodynamic performance of model propeller have been defined in the region. The actuator disc and the model propeller have the same diameter and identical distributions of elemental thrust along the radius (Krasilnikov 2013).

Also, self-propulsion analyses have been done using Moving Reference Frame (MRF) technique. Within this technique, a rotating region has been created behind the submarine including the model propeller itself. According to this technique, the governing equations are transformed into a rotating frame to get a steady-state solution (Moussa 2014).

4 VERIFICATION AND VALIDATION

In this section, five different grid sizes have been used as given in Table 1 for bare hull resistance analyses. In order to determine the uncertainty, three groups are created including three mesh cases. Different cell numbers have been used for modelling the computational domain for resistance analyses. After verification and validation, the optimum cell number has been chosen for the rest of the analyses including actuator disc and self-propulsion cases for different velocities.

Table 2 shows that the convergence condition R is between 0 and 1 as described in Section 2.2. As explained above, analysis groups are selected as 1-2-3, 2-3-4 and 3-4-5. The results of uncertainty analyses are shown below in Table 3.

Table 1. Total resistance via different cell numbers.

| | | Bare hull resistance | |
#	Name	Number of Cells	Resistance (N)
1	Finer	937.000	88.17
2	Fine	516.154	88.41
3	Medium	276.000	89.21
4	Coarse	175.659	90.93
5	Coarser	103.068	95.90

Table 2. Convergence conditions for uncertainty analyses.

Analysis set	R
1 2 3	0.300
2 3 4	0.465
3 4 5	0.346

Table 3. Uncertainty analyses for bare hull resistance.

| | Bare hull resistance |
Analysis set	% GCI$_{FINE}$
1 2 3	0.15
2 3 4	0.75
3 4 5	1.41

Table 4. Validation of the numerical method.

V_s (m/s)	$R_{T\text{-}EXP}$ (N)	$R_{T\text{-}CFD}$ (N)	Absolute relative difference (%)
3.046	87.4	88.41	1.15

Table 5. Main particulars of DARPA Suboff (AFF-1).

L_{OA} (m)	4.356
L_{BP} (m)	4.261
D_{max} (m)	0.508
S (m²)	5.980
∇ (m³)	0.717

The numerical results are also compared with the experimental results in order to validate the numerical method as given in Table 4. The numerical result is of the optimum mesh number. The uncertainty analyses have been made for total ship resistance analyses. Fine mesh algorithm has been chosen as the optimum one for resistance analyses. The finer mesh may not lead to more accurate results.

Attention has been paid in order to validate the numerical method with the experiments in addition to the uncertainty analysis of the method for verification and validation process. The optimum cell number has been determined and also validated with the available experimental data. The free stream velocity has been considered as 3.046 m/s for bare hull conditions during the uncertainty analyses. The number two (#2) mesh structure (fine) has been selected for the rest of the analyses.

Figure 3. 3-D view of DARPA Suboff bare hull.

Table 6. Main particulars of E1619 propeller.

	Open water	Self-propulsion
D (m)	0.485	0.262
P/D at 0.7R	1.15	1.15
Z	7	7
A_E/A_0	0.608	0.608

5 COMPUTATIONAL RESULTS AND DISCUSSIONS

This section focuses on the computational results of the flow simulation around DARPA Suboff geometry for bare hull condition using the optimum cell number determined with verification and validation process above. In addition, open water flow analyses for E1619 propeller have been performed. Following the resistance and open water analyses, self-propulsion tests have been done via body force propeller method. For this purpose, the open water propeller results have been coupled with RANSE solver by creating an actuator disc representing the finite-bladed model propeller. Then, the self-propulsion point has been determined for one velocity. This method has been discussed in terms of thrust, torque, wake fraction factor, thrust deduction factor, delivered power and thrust power. The advantageous and disadvantageous properties of these methods have been highlighted.

5.1 Geometrical dimensions

DARPA Suboff submarine model is a widely used benchmark form. Table 5 shows the main particulars of the model submarine. Also 3-D model of the submarine bare hull can be seen in Figure 3.

Figure 4. 3-D view of E1619 model propeller.

Table 6 on the other hand gives the main particulars of the model propeller E1619 used in the self-propulsion analyses. 3-D model of the model propeller can be found in Figure 4.

5.2 Computational results

A series of analyses has been conducted for prediction of total resistance of DARPA Suboff bare hull. One may see from Table 7 that the numerical method can calculate the submarine total resistance with an acceptable error when compared with the experiments.

Figure 5 shows the velocity contours around the aft body of the submarine. It can be seen that there is no separation near the hull surface.

Table 7. Comparison of the numerical and experimental results.

Rn * 10^6	V (m/s)	$R_{T\text{-}EXP}$ (N)	$R_{T\text{-}CFD}$ (N)	Absolute relative difference (%)
12.40	3.046	87.4	88.41	1.15
20.95	5.144	242.2	234.91	3.01
24.81	6.091	332.9	321.96	3.28
29.17	7.161	451.5	435.41	3.56
33.54	8.231	576.9	564.62	2.13
37.69	9.255	697	702.91	0.84

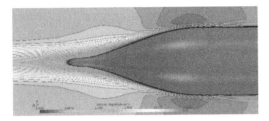

Figure 5. Velocity contours on the aft body of the submarine.

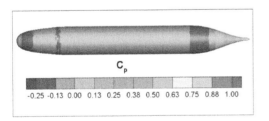

Figure 6. Non-dimensional pressure distribution on the submarine surface.

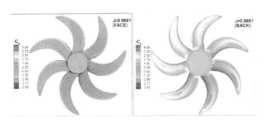

Figure 7. Non-dimensional pressure distribution on the propeller blades in open water condition.

Figure 6 shows the non-dimensional pressure distribution on the submarine surface. The pressure is higher on the aft and fore body as expected.

Figure 7 shows the non-dimensional pressure distribution on the propeller blades in open water

condition. Figure 8 presents the non-dimensional pressure distribution on the propeller blades behind the submarine.

Figure 9 shows the comparison of open water propeller characteristics for numerical and experimental methods. The results are quite satisfactory in a wide range of advance ratios.

Figures 10 and 11 show the computed self-propulsion point for DARPA Suboff bare hull at a constant velocity of 3.046 m/s.

It can be said that the actuator disc theory gives slightly higher self-propulsion point for the submarine when compared with the other technique having the actual propeller model behind the submarine. Later the nominal wake coefficient, thrust deduction factor, hull efficiency, relative rotative efficiency, open water propeller efficiency and propulsion efficiency have been determined using the following equations. Finally, effective power and delivered power have been calculated by both techniques.

The propulsion performance of a bare form is briefly described as below (ITTC 1978).

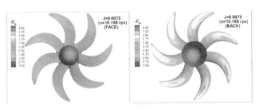

Figure 8. Non-dimensional pressure distribution on the propeller blades behind the submarine.

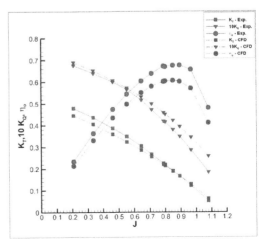

Figure 9. Comparison of thrust, torque and efficiency of E1619 propeller.

508

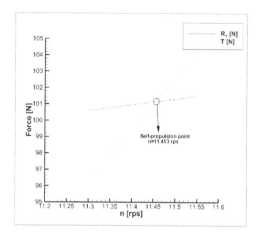

Figure 10. Self-propulsion point of DARPA Suboff with actuator disc theory.

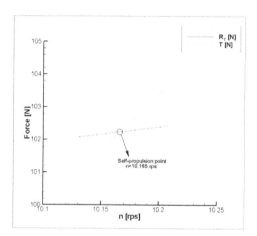

Figure 11. Self-propulsion point of DARPA Suboff in the presence of E1619 propeller.

The nominal wake coefficient here is calculated as follow:

$$w = 1 - \frac{V_A}{V_S} \qquad (12)$$

Here, V_A is the average flow velocity in the propeller plane, V_S is the incoming flow velocity towards the hull. V_A is calculated by

$$J = \frac{V_A}{n \cdot D} \qquad (13)$$

Here J is advance coefficient, D is propeller diameter and n is propeller rotation speed. The

thrust deduction factor on the other hand can be calculated by

$$t = 1 - \frac{R_T}{T} \qquad (14)$$

where, T is thrust force. The hull efficiency is expressed as the ratio of effective power to propeller thrust power. It can be expressed as follows:

$$\eta_H = \frac{1-t}{1-w} \qquad (15)$$

The relative rotation efficiency is expressed as the ratio of the open water propeller torque to the torque of the propeller working behind the hull.

$$\eta_R = \frac{Q_o}{Q} \qquad (16)$$

The open water propeller efficiency at V_A is calculated by the help of the following equation. This efficiency is calculated from open water data in momentum theory.

$$\eta_o = \frac{T \cdot V_A}{2\pi \cdot n \cdot Q_o} \qquad (17)$$

The propulsion efficiency is then expressed as follows:

$$\eta_D = \eta_H \cdot \eta_o \cdot \eta_R \qquad (18)$$

Effective power is the power required to pull the hull at constant speed (V_S).

$$P_E = R_T \cdot V_S \qquad (19)$$

The power delivered to the propeller is calculated as follow:

$$P_D = \frac{P_E}{\eta_D} \qquad (20)$$

Thrust loading factor in momentum theory can be calculated as below:

$$C_{T_h} = \frac{T}{\rho \cdot V_A^2 \cdot D^2 \cdot \dfrac{\pi}{8}} \qquad (21)$$

The ideal efficiency of propeller in momentum theory is expressed as below (Bertram 2012):

Table 8. The self-propulsion characteristics of DARPA Suboff with two methods.

	Actuator disc	Model propeller
V_S (m/s)	3.046	3.046
V_A (m/s)	2.690	2.169
J	0.896	0.815
n (rps)	11.453	10.165
R_T (N)	101.0708	102.1987
T (N)	101.0708	102.1987
Q (Nm)	5.5906	5.5357
Q_0 (Nm)	5.574	5.148
$w_{nominal}$	0.32	0.32
t	0.125	0.135
η_H	0.991	1.215
η_R	0.997	0.930
η_o	0.678	0.674
η_D	0.669	0.762
η_i	0.919	–
P_E (W)	269.297	269.297
P_D (W)	402.307	353.557

$$\eta_i = \frac{2}{1+\sqrt{1+C_{T_h}}} \qquad (22)$$

Table 8 shows the self-propulsion characteristics of DARPA Suboff with E1619 model propeller. During the calculations, thrust identity method has been used. By this method, non-dimensional thrust coefficient has been derived from the propeller thrust. Then, advance ratio of the propeller has been calculated using thrust coefficient in open water performance curve of the propeller. Finally, average inflow velocity coming to the propeller has been gained using advance ratio.

6 CONCLUSIONS

In this study, the bare hull resistance of DARPA Suboff form has been investigated using CFD based finite volume method. A verification and validation assessment has been done. After selecting the optimum mesh number, total resistance of the submarine has been calculated for different velocities and compared with the experimental results. In addition, open water flow analyses for E1619 model propeller have been carried out. Again, the results have been validated with available experimental data. Following the resistance and open water analyses of model propeller, self-propulsion analyses have been done using both the actuator disc theory and modeling the actual propeller itself behind the submarine. The self-propulsion characteristics of DARPA Suboff have then been determined for a constant velocity. It is found that CFD method is robust for prediction of self-propulsion point of underwater vehicles. Because prediction of self-propulsion point of an underwater vehicle is possible numerically without any need to complicated and expensive experimental methods. Moreover it is obtained that the actuator disc theory gives a slightly higher self-propulsion point while the total resistance is lower since there is no propeller behind the hull. Actuator disc theory has higher delivered power when compared with the model propeller technique. In the presence of the propeller, total resistance is higher, so the thrust deduction also increases. So one may say that the actuator disc theory can be applied safely in self-propulsion point estimation. Actuator disc theory may be suitable for prediction of self-propulsion characteristics such as delivered power.

The main highlight of this paper is that self-propulsion tests with CFD method using different techniques may be effective for preliminary design stage for obtaining fast results.

As a future work, self-propulsion calculations will be done using an average advance coefficient gained from thrust and torque identity methods.

NOMENCLATURE

U_i	mean velocity
u_i'	fluctuation velocity component
P	mean pressure
ρ	fluid density
k	turbulent kinetic energy
ε	turbulent dissipation rate
R_T	ship total resistance
L_{OA}	length overall
L_{BP}	length between perpendiculars
D_{max}	submarine maximum diameter
S	submarine wetted surface area
∇	submarine displacement
D	propeller diameter
P/D	propeller pitch ratio
Z	number of blades
A_E/A_0	propeller blade expansion ratio
w	nominal wake coefficient
V_A	the mean flow velocity at the propeller plane
V_S	incoming flow velocity
n	propeller rotation speed
J	advance ratio
t	thrust deduction factor
T	propeller thrust force
η_H	hull efficiency
η_R	relative rotation efficiency
η_o	open water propeller efficiency
η_D	propulsion efficiency
P_E	effective power

P_D delivered power
C_{T_h} thrust loading factor
η_i ideal efficiency of the propeller

ACKNOWLEDGEMENTS

We would like to thank Mr. Emre Kahramanoglu for his valuable support in uncertainty assessment throughout this study.

REFERENCES

Berger, S., Druckenbrod, M., Greve, M., Abdel-Maksoud, M. and Greitsch, L. 2011. An Efficient Method for the Investigation of Propeller Hull Interaction. *14th Numerical Towing Tank Symposium.*

Bertram, V. 2012. Practical Ship Hydrodynamics, 2nd edition. Elsevier.

Budak, G. and Beji, S. 2016. Computational Resistance Analyses of a Generic Submarine Hull Form and Its Geometric Variants. *J. Ocean Technology, 11, pp. 77–86.*

Carrica, P.M., Castro, A.M. and Stern, F. 2010. Self-propulsion computations using a speed controller and a discretized propeller with dynamic overset grids. *J. Mar. Sci. Technol., vol. 15, no. 4, pp. 316–330.*

CD-adapco 2015. *Star CCM+ Documentation.*

Celik, I.B., Ghia, U. and Roache, P.J. 2008. Procedure for estimation and reporting of uncertainty due to discretization in CFD applications. *J. Fluids Eng.-Trans. ASME, vol. 130, no. 7.*

Chase, N. 2012. Simulations of the DARPA Suboff submarine including self-propulsion with the E1619 propeller. *Master Thesis, University of Iowa.*

Chase, N. and Carrica, P.M. 2013. Submarine propeller computations and application to self-propulsion of DARPA Suboff. *Ocean Eng., vol. 60, pp. 68–80.*

Delen, C., Sezen, S., and Bal, S. 2017. Computational investigation of self-propulsion performance of DARPA Suboff vehicle. *Tamap Journal of Engineering, vol. 1, no. 4.*

ITTC 1978. Report of Performance Committee. *Proceedings of 15th ITTC, Hague.*

ITTC 2011a. 7.5-03-01-04 CFD, General CFD Verification. *ITTC - Recommended Procedures and Guidelines.*

ITTC 2011b. "Practical guidelines for ship CFD applications," Proceedings of 26th ITTC, Rio de Janeiro.

Kinaci, O.K., Kukner, A., and Bal, S. 2013. On Propeller Performance of DTC Post-Panamax Container Ship. *International Journal of Ocean System Engineering, 3(2), 77–89.*

Kinaci, O.K., Sukas, O.F. and Bal, S. 2016. Prediction of Wave Resistance by a RANSE based CFD Approach. *Proceedings of the Institution of Mechanical Engineers, Part M, Journal of Engineering for the Maritime Environment, Vol. 230, Issue 3, pp. 531–548.*

Krasilnikov, V. 2013. Self-Propulsion RANS Computations with a Single-Screw Container Ship. *Third International Symposium on Marine Propulsors, Tasmania, Australia.*

Moussa, K. 2014. Computational Modeling of Propeller Noise: NASA SR-7 A Propeller. *Master Thesis, University of Waterloo, Waterloo, Ontario, Canada.*

Ozdemir, Y.H., Cosgun, T., Dogrul, A., Barlas, B. 2016. A numerical application to predict the resistance and wave pattern of KRISO container ship. *Brodogradnja, vol. 67, no. 2, 47–65.*

Rijpkema, D., Starke, B. and Bosschers, J. 2013. Numerical simulation of propeller-hull interaction and determination of the effective wake field using a hybrid RANS-BEM approach. *Third International Symposium on Marine Propulsors, Launceston, Tasmania, Australia.*

Roache, P.J. 1998. Verification of Codes and Calculations. *AIAA J., vol. 36, no. 5, pp. 696–702.*

Seo, J.H., Seol, D.M., Lee, J.H., and Rhee, S.H. 2010. Flexible CFD meshing strategy for prediction of ship resistance and propulsion performance. *Int. J. Nav. Archit. Ocean Eng., vol. 2, no. 3, pp. 139–145.*

Vaz G., Toxopeus, S., and Holmes S. 2010. Calculation of Maneuvering Forces on Submarines using Two Viscous-Flow Solvers. *Proceedings of the ASME 2010 29th International Conference on Ocean, Offshore and Arctic Engineering Shanghai, China.*

Versteeg, H. and Malalasekera, W. 2007. An Introduction to Computational Fluid Dynamics: The Finite Volume Method, 2nd edition. *Harlow, England; New York: Prentice Hall.*

Villa, D. and Brizzolara, S. 2012. Ship Self Propulsion with different CFD methods: from actuator disc to viscous inviscid unsteady coupled solvers. *10th International Conference on Hydrodynamics, St. Petersburg, Russia.*

Wilcox, D.C. 2006. Turbulence Modeling for CFD. *3rd edition. La Cãnada, Calif.: D C W Industries.*

Zhang, N. and Zhang, S. 2014. Numerical simulation of hull/propeller interaction of submarine in submergence and near surface conditions, *J. Hydrodyn. Ser B, vol. 26, no. 1, pp. 50–56.*

Maritime Transportation and Harvesting of Sea Resources – Guedes Soares & Teixeira (Eds)
© 2018 Taylor & Francis Group, London, ISBN 978-0-8153-7993-5

CFD analysis of a fixed floating box-type structure under regular waves

J.F.M. Gadelho, S.C. Mohapatra & C. Guedes Soares
Centre for Marine Technology and Ocean Engineering (CENTEC), Instituto Superior Técnico, Universidade de Lisboa, Lisbon, Portugal

ABSTRACT: Numerical modelling of wave interaction with fixed floating structure has been performed based on CFD simulations using OpenFOAM which solves free surface Newtonian flows using Navier-Stokes equations coupled with a volume of fluid (VOF) method. The numerical results obtained for a fixed floating structure are compared with linearized analytical results based on linearized Boussinesq equations. The comparisons result on free surface elevations and vertical force acting on the floating structure. Further, the CFD results of vertical force are also compared with the experimental model results recently presented. The numerical results are in good agreement with the analytical and experimental model results. Hence, it is observed that the results show a good performance of the numerical CFD model for wave interaction with a fixed floating structure in 2D.

1 INTRODUCTION

There has been a significant progress in the development of various design techniques for the study of floating structure for different purposes such as floating platforms, wave energy devices, and floating breakwaters. Apart from the analysis of various environmental and design conditions, wave load analysis of floating structures are performed based on various analytical and numerical methods for better understanding the action of waves.

One of the approaches of wave-structure interaction analysis is based of CFD modeling due to its wider applicability in the field of ocean engineering and mathematical physics.

Different numerical methods have been adopted to study the wave effects on the floating structure in the time domain. For instance, Brorsen & Bundgaard (1990) developed a new approach to calculate the nonlinear drift forces on large bodies based on numerical time domain model. Koutandos et al. (2004) used the finite difference method to study the hydrodynamic behaviour of fixed and heave motion of floating breakwater based on Boussinesq type equations. Fuhrman et al. (2005) described a finite difference model based on higher order Boussinesq formulation for nonlinear wave-structure interactions having arbitrary piecewise-rectangular bottom-mounted (surface-piercing) structures. Bai & Teng (2013) used the boundary element method to perform a numerical simulation of the nonlinear wave interaction with surface piercing fixed and floating circular cylinders.

Hsiang & Li (2013) presented a recent numerical study on the point absorber wave energy conversion system using a Reynolds-averaged Navier–Stokes (RANS)-based Computational Fluid Dynamics (CFD) method. Lin (2006) developed a 3D multiple-layer σ-coordinate model to simulate surface wave interaction with various structural configurations and model validated with the Volume-Of Fluid Model. Palm et al. (2016) formulated a model of floating wave energy converters using CFD and validated against experimental measurements of a cylindrical buoy in regular waves. Paredes et al. (2016) investigated experimentally of mooring configurations floating wave energy converters to study the performance of the wave energy converters.

Park et al. (1999) investigated the non-linear wave motion characteristics and their interactions with a three-dimensional stationary body inside a numerical tank by a finite difference scheme using a modified marker-and-cell method associated with the Navier-Stokes equation in two-layer fluids. Rodriguez & Spinneken (2016) investigated experimentally the nonlinear loading and dynamic response of a heaving rectangular box in two-dimensions using a series of experimental tests in regular and irregular waves. Rodriguez et al. (2016) presented a numerical investigation to address the nonlinear heave response of a rectangular box.

Recently, attempts have been made to obtain the analytical solutions associated with Boussinesq-type equations over constant/variable water depths using different analytical methods. Using the hyperbolic tangent method and the first integral method, Mohapatra & Guedes Soares (2015a) compared the exact solutions of the coupled Boussinesq equations in shallow water. Further, Mohapatra et al. (2016) compared the analytical and numerical simulations of the solutions of the coupled Boussinesq equations between the mathematical and numerical spectral wave model MIKE 21 BW. Mohapatra et al. (2017) compared the long nonlinear internal solitary waves between the analytical and numerical simulations in shallow water. On the other hand, Mohapatra & Guedes Soares (2015b) studied the wave forces on a floating structure over flat bottom based on linearized Boussinesq formulations by applying eigenmode expansion method.

The objective of the present work is to analyse the free surface elevations after and before the structure and the vertical force acting on the fixed floating structure by comparing the simulation results of numerical CFD against analytical and experimental model results.

In order to achieve this aim, the governing equations for CFD numerical model are based on the Navier-Stokes equations whilst, the nonlinear Boussinesq equation model is based on the Euler's equations of motion. On the other hand, the fixed floating structure is modelled as floating box-type structure over flat bottom with finite width and draft. The numerical solution is solved based on OpenFOAM code whilst the linearized analytical solution is based on eigenmode expansion method. In order to analyse the effect of waves on the floating structure, the results of the free surface elevations and vertical wave force acting on the fixed floating structure have been performed against the linearized analytical solution.

It is observed that the free surface elevations of the numerical and the analytical model are in very good agreement. Further, the results on the vertical wave force are compared between the present numerical CFD and experimental model results recently presented by Rodríguez & Spinneken (2016). It is found that the results are good between the two models. Finally, the significant comparisons are discussed on the simulations of free surface elevations and vertical force between the numerical, analytical and experimental models.

2 BRIEF DESCRIPTION OF THE CFD MODEL

The Computational Fluid Dynamics (CFD) software used in this work is release version 2.4.0 of the open source CFD software OpenFOAM (Open Source Field Operation and Manipulation) running on Ubuntu 14.04 operating system. OpenFOAM has an extensive range of features to solve complex fluid flows and has a wide variety of applications including offshore and coastal engineering problems.

The library waves2Foam toolbox is used to generate and absorb free surface water waves (Jacobsen et al., 2012). Currently the method applies the relaxation zone technique (active sponge layers) and a large range of wave theories are supported and the relaxation zones can take arbitrary shapes. The main solver is waveFoam, which is based on the native solver interFoam and it is described as a fully viscous solver for two incompressible, isothermal immiscible fluids.

The governing equations are the Navier-Stokes equations for an incompressible laminar flow of a Newtonian fluid. In vector form, the equation is given by

$$\rho\left(\frac{\partial v}{\partial t} + v.\nabla v\right) = -\nabla p + \mu \nabla^2 v + \rho g, \tag{1}$$

where v is the velocity, p is the pressure, μ is the dynamic viscosity and ∇^2 is the Laplace operator. The continuity equation is of the form

$$\nabla v = 0. \tag{2}$$

The volume of fluid (VOF) method is used to track the free surface elevation. This method determines the fraction of each fluid that exists in each cell. The equation for the volume fraction is obtained as

$$\frac{\partial \alpha}{\partial t} + \nabla(\alpha U) = 0, \tag{3}$$

where U is the velocity field, α is the volume fraction of water in the cell and varies from 0 to 1, full of air to full of water, respectively.

The numerical setup geometry is a 2D wave flume with 45.0 m long and a water depth of 1.25 m, like the one used in the experimental setup presented by Rodriguez & Spinneken, (2016). The width is simulated with a unique cell with 0.01 m.

The chosen Cartesian coordinates, x-axis is parallel to the flume length, y-axis is parallel to the flume height and z-axis is parallel to the flume width. The origin in x-axis is located in the left boundary of the domain and origin in y-axis is located in the initial free surface elevation.

Figure 1 represents the numerical model geometry. The floating box has 0.5 m of length and 0.5 m of height and is placed at 29.0 m from the wave maker. The draft of the box is 0.25 m. Two numerical wave gauges are placed before and after the box, G1 and G2, respectively. They are used to evaluate the wave generation and the wave transmission.

The calibration of the numerical model to be used as a numerical 2D wave flume is well described in Gadelho et al. (2014) where it is demonstrated

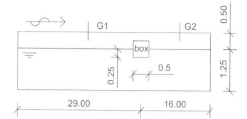

Figure 1. Numerical model geometry setup (box is 0.5 m × 0.5 m and draft is 0.25 m (not to scale)).

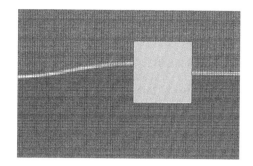

Figure 2. Representation of a section with the box used for case n° 5 at 35.5 s of simulation.

that the numerical model is able to generate and accurately model wave propagation and interaction with fixed floating structures. It can also describe the processes of wave transformation and energy exchange from low to high harmonics.

The use of analytical results based on the linearized Boussinesq-type equations to validate a CFD numerical model was successfully used by Gadelho et al. (2015). The analysis shows a good performance of the numerical model for proper generation and propagation of waves over a 2D wave flume and also accurately calculate the wave induced loads on bottom fixed rigid plates.

In the present work, a mesh sensitivity analysis was made for the given numerical setup. It consisted on raging the Δx (between 0.01 m and 0.02 m) and Δy (between 0.005 m and 0.01 m) for the same wave conditions. The differences observed on the free surface elevations and vertical force amplitudes were less than 1%.

In this way, the base mesh adopted for the full set of simulations is structured with hexahedron elements with Δx = 0.02 m and Δy = 0.01 m.

The mesh was created with blockMesh tool and the box was defined with SnappyHexMesh tool from a stereolithography file previous created by faceSetToSTL application.

Figure 2 is represents a section of the numerical model domain showing the fixed floating box. The snapshot represented is for case n° 5 at 35.5 s of simulation. It is visible the water in red, the air in blue and the mesh in dark colour.

The horizontal dimensions of the relaxation zones, used to absorb the reflected waves, were defined as one wave length λ in the inlet and the outlet, based on the highest wave length of the set of simulations. In this case the value of 8.0 m was used.

In the next section, the numerical CFD model associated with wave interaction with fixed box-type floating structure will be validated with the analytical model based on linearized Boussinesq-type equations and the results of free surface elevations before and after the box and the vertical force acting on the floating structure will be compared inn different cases.

3 BOUSSINESQ SOLUTION

The validation of the numerical model consists on comparing the time series of free surface elevation and vertical wave force are based on the CFD model and in the linearized Boussinesq type-equations for fixed floating structure. The numerical formulation of the wave interaction fixed floating structure is based on the CFD model (OpenFOAM code) whilst, the theoretical model is based on nonlinear Boussinesq formulation but linearized solutions with dispersive terms. The problem is analysed in two-dimensional Cartesian coordinate system (x, z), the x-axis being the horizontal direction and the z-axis is pointing upwards. A rectangular box-type rigid floating structure of width $2l$ with draft d is floating on the upper surface $-l < x < l$. The origin O is assumed to be the middle point of the rectangular rigid box. The domain of consideration is divided into three regions which are defined by $l < x < \perp$, $0 < z < h$ and $-l < x < -\perp$, $0 < z < h$ are referred as before and after structure and $-l < x < l$, $d < z < h$ is referred as structure covered region (as in Figure 1).

The velocity potential $\Phi(x, z, t)$ in the structure covered region is of the form $\Phi(x, z, t) = \mathrm{Re}\{\varphi(x, z) e^{-i\omega t}\}$. Therefore, potential flow theory is solved for the structure covered region, while Boussinesq equations are solved for before and after structure region.

The governing equation for fluid motion is the continuity equation, which can be expressed in non-dimensional form as:

$$\mu^2 u_x + w_z = 0 \text{ before and after the structure.} \quad (4)$$

Further, the governing equation for the fluid motion in the Euler's equations of motion for before and after structure regions can be expressed in non-dimensional form as:

$$\mu^2 u_t + \varepsilon \mu^2 u u_x + \varepsilon w u_z + \mu^2 p_x = 0, \quad (5)$$

515

$$\varepsilon w_t + \varepsilon^2 u w_x + \frac{\varepsilon^2}{\mu^2} w_z + \varepsilon p_z + 1 = 0, \qquad (6)$$

where the subscripts x, z, and t denote the partial derivatives w.r.t. x, z, and t respectively with ε and μ are the classical measures of nonlinearity and frequency dispersion defined by $\varepsilon = I_0/h_0$ and $\mu = h_0/\lambda_0$, where both ε and μ are assumed to be small. It may be mentioned that the non-dimensional variables are used for the formulation of the present model is same as in Mohapatra & Guedes Soares (2015a).

The boundary condition at the upper surface of the fluid is given by

$$p = p_{atm} = 0 \text{ at } z = \varepsilon\eta(x,t), \qquad (7)$$

where $\eta = \eta_1$ for before structure region and $\eta = \eta_3$ for after structure region, p_{atm} is the ambient atmospheric pressure which is taken as zero without loss of generality.

The kinematic boundary condition at the free surface is given by

$$w = \mu^2(\eta_t + \varepsilon u\eta_x) \text{ at } z = \varepsilon\eta. \qquad (8)$$

The rigid bottom boundary condition on $z = -h$ is given by

$$w = 0 \text{ at } z = -h, \qquad (9)$$

The pressure field equation for before and after the structure region is obtained by integration the vertical flow equation with respect to z and using (7) and (8) at the free surface as

$$p = \eta - \frac{z}{\varepsilon} + \frac{\partial}{\partial t}\int_z^{\varepsilon\eta} w\,dz + \varepsilon\frac{\partial}{\partial x}\int_z^{\varepsilon\eta} uw\,dz - \frac{\varepsilon}{\mu^2}w^2, \qquad (10)$$

where w is obtained by integrating (4) with respect to z,

$$w = -\mu^2\left[\frac{\partial}{\partial x}\int_{-h}^z u\,dz\right] \text{ at } z = -h. \qquad (11)$$

Integrating the continuity equation (4) from $-h$ to $\varepsilon\eta$ and applying the boundary condition as in Eqs. (8–9), and integrating Eq. (4–5) over the depth, applying Eqs. (7), (9–11), the one-dimensional nonlinear Boussinesq equation is obtained as

$$\eta_t + \hat{u}_x + \varepsilon(\eta\hat{u})_x = 0, \qquad (12)$$

$$\hat{u}_t + \eta_x + \varepsilon\hat{u}\hat{u}_x - \frac{1}{3}\mu^2\hat{u}_{txx} = 0. \qquad (13)$$

$$\hat{u} = \int_{-h}^{\eta} u\,dz, \qquad (14a)$$

with ψ is the dispersive (Boussinesq) term as

$$\psi = \frac{h^2}{3}\hat{u}_{xxt}. \qquad (14b)$$

Using eigenmode expansion method along with continuity of velocity and pressure at the vertical interfaces (structural edges), the free surface elevations and wave forces are same as in Mohapatra & Guedes Soares (2015b). However, there was no comparison with numerical modelling on the results of free surface elevations and wave forces. In the present paper, the results of free surface wave elevations and wave forces are compared with the CFD numerical model based on OpenFOAM code.

4 COMPARISON OF NUMERICAL AGAINST ANALYTICAL RESULTS

A comparison between the numerical CFD and linearized Boussinesq based model was made. The time histories of vertical force and free surface elevations before and after the box were compared. This comparison is made after several seconds, when the CFD model reaches a steady state, avoiding some transient effects.

A total of six cases were defined, varying the wave conditions. The list of the wave parameters, wave height H, wave period T and wave length λ, for each case is presented in Table 1.

The steepness $A_l k$ of the three first cases is 0.05 while in the remaining three is 0.1. A_l stands for wave amplitude and k is the wave number.

The first four cases were used to compare against the Boussinesq linear model results and cases n° 5 and n° 6 were used to compare with experimental results presented in the work of Rodriguez & Spinneken, (2016). All the results are presented for a rectangular box with width of 1.0 m.

The results of the vertical force and free surface elevations before and after the box for the case n° 1 are presented in Figures 3, 4 and 5.

Table 1. List of the wave parameters for each case.

Case n°	H (m)	T (s)	λ (m)
1	0.125	2.57	7.85
2	0.083	1.92	5.24
3	0.063	1.62	3.93
4	0.125	1.62	3.93
5	0.071	1.20	2.24
6	0.050	1.00	1.57

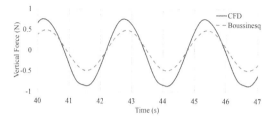

Figure 3. Comparison between the CFD and analytical model of the vertical force on the fixed box for case n° 1.

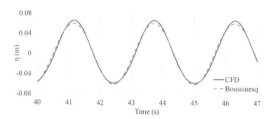

Figure 4. Comparison between the CFD and analytical model of free surface elevation before the box for case n° 1.

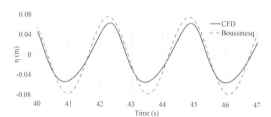

Figure 5. Comparison between the CFD and analytical model of free surface elevation after the box for case n° 1.

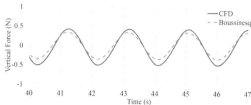

Figure 6. Comparison between the CFD and analytical model of the vertical force on the fixed box for case n° 2.

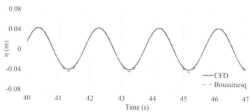

Figure 7. Comparison between the CFD and analytical model of free surface elevation before the box for Case n° 2.

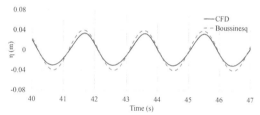

Figure 8. Comparison between the CFD and analytical model of free surface elevation after the box for case n° 2.

It is observed that the generated wave (Figure 4) is identical in both models, with a smaller crest described by the Boussinesq model, probably related to its linearized solution. But the vertical force amplitude (Figure 3) is higher for the CFD model than for the Boussinesq model while the transmitted wave amplitude (Figure 5) is higher in the Boussinesq model.

This suggests that for these wave parameters the non-linearity of the CFD model leads to an over prediction of the vertical force in the box, and as a consequence a smaller transmitted wave amplitude.

The results of the vertical force and free surface elevations before and after the box for the case n° 2 are presented in Figures 6, 7 and 8. It is observed that once again both models can generate almost the same wave profile (Figure 7). For the vertical force and the transmitted wave amplitudes the same behaviour is observed as in case n° 1, with the vertical force amplitude higher in the CFD

and the transmitted wave amplitude higher in the Boussinesq model. But the differences in amplitude are smaller.

The results of the vertical force and free surface elevations before and after the box for the case n° 3 are presented in Figures 9, 10 and 11.

For the case n° 3 the wave parameters are smaller than in the previous cases and therefore the second order effects of the wave theory are not that pronounced. This way the results for the free surfaces elevations before and after the box are almost identical, although, the vertical force is under predicted by the linearized Boussinesq model.

The results of the vertical force and free surface elevations before and after the box for case n° 4 are presented in Figures 12, 13 and 14.

The case n° 4 is important to study the influence of a higher steepness in the wave profile and vertical force on the box. The generated wave height is the same that in case n° 1, but the period was

517

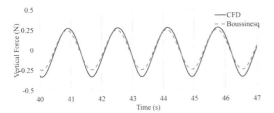

Figure 9. Comparison between the CFD and analytical model of the vertical force on the fixed box for case n° 3.

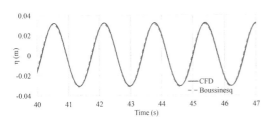

Figure 10. Comparison between the CFD and analytical model of free surface elevation before the box for case n° 3.

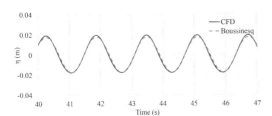

Figure 11. Comparison between the CFD and analytical model of free surface elevation after the box for case n° 3.

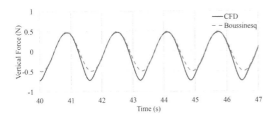

Figure 12. Comparison between the CFD and analytical model of vertical force on the fixed box for case n° 4.

changed to give a steepness of 0.1, instead of 0.05 of the first case.

It is observed that the wave profiles before and after the box are almost identical (Figures 13 and 14), but the vertical force is under predicted by the Boussinesq model in the negative part of the force time series. This is explained by the linear assumption of the Boussinesq model that cannot describe

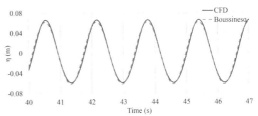

Figure 13. Comparison between the CFD and analytical model of free surface elevation before the box for case n° 4.

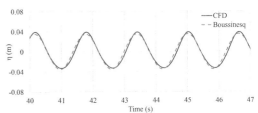

Figure 14. Comparison between the CFD and analytical model of free surface elevation after the box for case n° 4.

the non-linear phenomena of the force when there is an increase of the wave steepness.

In the next section, the results of vertical force obtained from the numerical CFD model in case n° 5 and 6 will be compared against experimental model results.

5 COMPARISON OF NUMERICAL AGAINST EXPERIMENTAL RESULTS

The CFD vertical forces are compared with the experimental results presented by Rodriguez & Spinneken, (2016). The experimental data was obtained in the Long Wave Flume located in the Imperial College London. The wave flume is 63.0 m, 2.8 m wide, with a water depth of 1.25 m. The box dimensions and positioning in the wave flume are the same that the ones used in the CFD model.

The generated waves for cases n° 5 and n° 6 are regular with the described parameters in Table 1.

The presented forces amplitudes are non-dimensional: $F(t)/\rho gAb$ where ρ is the water density, g is the gravity acceleration, and b is half of the box beam which takes the value of 0.25 m.

The comparison between the CFD and experimental model force time series on the box for case n° 5 is presented in Figure 15. Although, it is observed that the period and the shape of the force is well described by the numerical model but the amplitude of the force is lower in the experimental model.

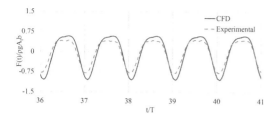

Figure 15. Comparison between the CFD and experimental model force time series on the box for Case n° 5.

Figure 16. Comparison between the CFD and experimental model force time series on the box for case n° 6.

Rodriguez & Spinneken (2016) also noticed that the magnitude of the experimental force is slightly lower than those predicted by the semi-analytical model they analysed. They decomposed the signal into first and second harmonics and say that these differences in the force amplitude can be associated with large local velocities in the fluid arising at higher frequencies.

The comparison between the CFD and experimental model force time series on the box for case n° 6 is presented in Figure 16. In this case it is observed that the shape of the force and the amplitude agrees very well. There is a little difference in the period between models that can be explained by some uncertainty related to the wave generation and wave transformation over the wave flume that may not be the same.

It should be mentioned that there are no free surface elevations time series to be compared against experimental model because of some uncertainty related to the wave generation and wave transformation over the wave flume that might be different between models.

6 CONCLUSIONS

This paper dealt with wave interaction with fixed box-type floating structure in 2D based on the CFD numerical model. The results of free surface elevations and vertical forces acting on the floating structure are compared with the linearized Boussinesq model. Further, the numerical results

of vertical force acting on the floating structure are compared with experimental results available in the literature. The comparison/validation of the numerical wave flume to work with a fixed floating structures based on the linearized Boussinesq theory demonstrated to be successful. The amplitudes of the free surface elevations (before and after the box) and the horizontal force on the fixed floating structure agreed well. The differences observed can be explained by the linearity assumptions of the Boussinesq model that cannot reproduce the peculiarities of the full non-linear CFD model employed.

Regarding the comparison between the CFD and the experimental model, for the cases presented, the CFD reproduced well the vertical force on the fixed-box. Although, because of lack of free surface elevations time records, there is some uncertainty related to the wave generation and wave transformation over the wave flume that might be different between models.

The results obtained demonstrate the potential of CFD modelling, to investigate the interaction between water waves and wave energy converters in a 2D wave flume. The present model may be compared against nonlinear analytical model of the same structural configuration in coming future.

ACKNOWLEDGEMENTS

This work was performed within the Project MIDWEST, Multi-fidelity Decision making tools for Wave Energy Systems, which is financed by the Portuguese Foundation for Science and Technology as part of the OCEANERANET program.

REFERENCES

Bai, W. & Teng, B. 2013. Simulation of second-order wave interaction with fixed and floating structures in time domain. *Ocean Engineering*, 74, 168–177.
Brorsen, M. & Bundgaard, H.I. 1990. Numerical model of the nonlinear interaction of waves and floating bodies. *In*: Edge, B.L. (ed.), *Coastal Engineering*. New York: American Society of Civil Engineers, pp. 3140–3151.
Fuhrman, D.R., Bingham, H.B. & Madsen, P.A. 2005. Nonlinear wave-structure interactions with a high-order Boussinesq model. *Coastal Engineering*, 52(8), 655–672.
Gadelho, J.F.M.; Lavrov, A., & Guedes Soares, C., 2014. Modelling the effect of obstacles on the 2D wave propagation with OpenFOAM, In: Guedes Soares, C. and López Peña, F., (eds.), *Developments in Maritime Transportation and Exploitation of Sea Resources*, London: Taylor and Francis Group, pp. 1057–1065.

Gadelho, J.F.M., Karmakar, D., Lavrov, A. & Guedes Soares, C. 2015. CFD analysis of a rigid bottom fixed submerged structure. *Renewable Energies Offshore*, Guedes Soares, C. (Ed.), Taylor & Francis Group, London, UK, pp. pp. 531–537.

Hsiang,Y, Y. & Li, Y 2013. Reynolds-averaged Navier–stokes simulation of the heave performance of a two-body floating-point absorber wave energy system. *Computers and Fluids*, 73:104–114.

Jacobsen, N.G., Fuhrman, D.R., & Fredsøe, J. 2012. A Wave Generation Toolbox for the Open-Source CFD Library: OpenFoam. International Journal of Numerical Methods in Fluids, 70(9): 1073–1088.

Koutandos, E.V., Karambas, Th.V. & Koutitas, C.G. 2004. Floating breakwater response to waves action using a Boussinesq model coupled with a 2DV elliptic solver. *Journal of Waterway, Port, Coastal, and Ocean Engineering*, 130 (5), 243–255.

Lin, P. 2006. A multiple-layer σ-coordinate model for simulation of wave-structure interaction. *Computers & Fluids*, 35:147–167.

Mohapatra, S.C. & Guedes Soares, C. 2015a. Comparing solutions of the coupled Boussinesq equations in shallow water. *In*: Guedes Soares, C. and Santos, T.A., (eds.), *Maritime Technology and Engineering*. London: Taylor & Francis Group, pp. 947–954.

Mohapatra, S.C. & Guedes Soares, C. 2015b. Wave forces on a floating structure over flat bottom based on Boussinesq formulation. *In*: Guedes Soares, C. (ed.), *Renewable Energies Offshore*. London: Taylor & Francis Group, pp. 335–342.

Mohapatra, S.C., Fonseca, R.B. & Guedes Soares, C. 2016. A comparison between analytical and numerical simulations of solutions of the coupled Boussinesq equations. *In*: Guedes Soares, C. and Santos, T.A. (eds.), *Maritime Technology and Engineering* 3. London: Taylor & Francis Group, pp. 1175–1180.

Mohapatra, S.C., Fonseca, R.B. & Guedes Soares, C. 2017. Comparison of Analytical and Numerical Simulations of Long Nonlinear Internal Waves in Shallow Water. *Journal of Coastal Research* DOI: 10.2112/JCOASTRES-D-16–00193.1.

Palm, J., Eskilsson, C., Paredes, G.M. & Bergdahl, L. 2016. Coupled mooring analysis for floating wave energy converters using CFD: Formulation and validation. *International Journal of Marine Energy*, 16:83–99.

Paredes, G.M., Palm, J., Eskilsson, C., Bergdahl, L. & Taveira-Pinto, F. 2016. Experimental investigation of mooring configurations for wave energy converters. *International Journal of Marine Energy*, 15:56–67.

Park, J-C., Kim, M-H. & Miyata, H. 1999. Fully nonlinear free surface simulations by a 3D viscous numerical wave tank. *International Journal for Numerical Methods in Fluids*, 29:685–703.

Rodriguez, M. & Spinneken, J. 2016. A laboratory study on the loading and motion of a heaving box. *Journal of Fluids and Structures*, 64:107–126.

Rodriguez, M. Spinneken, J. & Swan, C. 2016. Nonlinear loading of a two-dimensional heaving box. *Journal of Fluids and Structures*, 60:80–96.

Maritime Transportation and Harvesting of Sea Resources – Guedes Soares & Teixeira (Eds)
© 2018 Taylor & Francis Group, London, ISBN 978-0-8153-7993-5

Simulation dependency on degrees of freedom in RaNS solvers for predicting ship resistance

H. Islam
OSE, KAIST, Daejeon, South Korea
Centre for Marine Technology and Ocean Engineering (CENTEC), Instituto Superior Técnico, Universidade de Lisboa, Lisbon, Portugal

H. Akimoto
OSE, KAIST, Daejeon, South Korea

ABSTRACT: At present days, CFD simulation is the mostly used tool at the design stage to investigate ship resistance in calm water and in waves. The common practice is to simulate ships with restricted degrees of freedom, instead of considering all the motions. However, in actual sea, ships maneuver with all degrees of freedom, and the impact of ignoring some motions in the simulation is not well discussed in existing literature. This paper investigates the influence of degrees of freedom in ship resistance simulation, using an in-house RaNS solver. The paper compares resistance and motion prediction results from simulations performed with different degrees of freedom, both in calm water and in waves. The paper concludes that the influence of some motions are higher in simulation comparing to others, and for resistance simulation, considering heave and pitch motion is sufficient.

1 INTRODUCTION

Computational Fluid Dynamic (CFD) simulations have become a common tool for ship builders and researchers to investigate ship hydrodynamic properties. In the case of ship resistance prediction, most common practice in experimental facilities is to perform experiments with just two Degrees of Freedom (DOF) of motion, particularly heave and pitch. This is natural for resistance prediction in calm water and in waves since heave and pitch motion have the most significant influence on encountered resistance. Similarly, in the case of CFD, 2DOF is also preferred as it allows easy comparison to experimental data and application of symmetry condition, i.e., only half hull is simulated. Half hull simulation significantly reduces computational time and helps in keeping the conditions relatively simple. However, during the actual voyage, a ship moves with all 6 DOFs, thus, present practices for added resistance prediction may not be justified.

A ship in seaway moves with all 6 DOFs; heave, pitch, surge, roll, sway and yaw motion. Researchers like Hosseini et al. (2013), have already shown using experiments and CFD data that surge motion has limited impact on added resistance, thus it may not be essential to consider surge motion for added resistance prediction. However, very limited discussion is available on the impact of keeping other motions free, especially in case of

prediction using RaNS solvers. The impact of roll, sway, and yaw on added resistance prediction is not well described by existing literature.

Thus, in order to develop a better understanding on the matter, a series of simulations with different degrees of freedom were performed both in calm water and in waves, using an in-house RaNS solver, SHIP_Motion. Considering the computational resource requirement for full hull simulations and actual seaway condition for tanker ships, only short wave cases were simulated for this study. This consideration reduced both the required number of cases to be simulated and the necessary mesh resolution for simulation. In order to maintain confidence in results, simulation results were first validated for cases with 2DOF, and then simulations were performed with higher DOF, keeping the same mesh resolution and simulation conditions.

2 COMPUTATIONAL METHOD

2.1 *Solver's mathematical model*

The mathematical model of the used RaNS solver, SHIP_Motion, has been elaborately discussed by Kim et al. (2015), Orihara (2003) and Akimoto (2002) in their respective works. Thus, only a brief overview of the solver is given in this paper.

The governing equation for the mathematical model is the Reynolds averaged Navier-Stokes

(RaNS) equation and continuity equation. Two sets of coordinate system, body fixed and earth fixed, are used. The spatial discretization is by Finite Volume Method (FVM). 3rd order upwind differencing is used for advection, whereas, discretization in space is by 2nd order central difference scheme. Definition of physical values is in a staggered manner. Free surface capturing is by Marker density method, where, 3rd order upwind scheme performs space differencing and 2nd order Adams-Bashforth solves time differencing. Two type of turbulence models are incorporated, Baldwin-Lomax and Dynamic sub-grid scale model. Wall function is used to reduce dependency on mesh for capturing boundary layer properties. Parallel processing is done by the shared memory model of OpenMP.

A Marker and Cell (MAC) type pressure solution algorithm is employed. The pressure is obtained by solving the Poisson equations using the Successive Over Relaxation (SOR) method and velocity components are gained by correcting the velocity predictor with the implicitly evaluated pressure. In the overlapping grid system, inner domain moves according to floating body's equation of motion and outer domain represent free surface. Grid points located at the overlapping region exchange information through interpolation to update both the domains at every time step.

In this research, an overset structured single block mesh system was used. The coarse rectangular outer mesh with high resolution around the free surface was used to capture the free surface deformation. The fine O-H type inner mesh around the hull surface was used for capturing the flow properties around the hull surface. For both the domains, the orientation of x-axis was from bow to sternward direction, y-axis was positive towards starboard and z-axis was upward positive. Figure 1 shows the fine inner mesh (a), coarse outer mesh (b) and their combined arrangement (c) respectively. Image (a) shows a high resolution mesh in z direction at and around the free surface, and a high resolution in y direction near hull form. Image (b) shows a uniform mesh in x-y plane for the outer mesh and a gradual meshing in z direction with a hight resolution the near free surface. The ship deck was slightly extended for head wave simulations to avoid green water conditions.

2.2 Ship model and mesh resolution

The model ship used in this research was the KRISO Very Large Crude Carrier 2 (KVLCC2). Table 1 provides the specifications of the KVLCC2 model following Tokyo 2015 Workshop (2015) and Figure 2 shows its side view and body planes. The image of the body planes is magnified

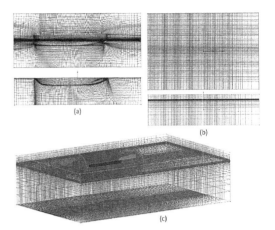

Figure 1. Mesh arrangement used for simulation, (a) inner mesh domain (front and bottom view), (b) outer mesh domain (top and front view) and (c) combined mesh domains (isometric view).

Table 1. Specifications of the oil tanker ship model KVLCC2.

Specification	Unit	KVLCC2 ship (full scale)
Length between perpendicular	Lpp (m)	320.0
Breadth	B (m)	58.0
Depth	D (m)	30.0
Draft	T (m)	20.8
Wetted surface area	S (m²)	27194.0
Displacement volume	V (m³)	312622
LCB from mid-ship	LCB (m)	11.136
Kyy	Kyy (m)	0.25 Lpp

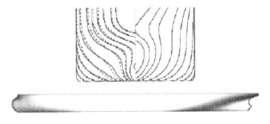

Figure 2. Body planes and side view (upto water line) of the KVLCC2 ship model.

for better ligebility. All the simulations were performed in non-dimensional scale, with complete hull form incluling freeboard.

For running simulation with higher degrees of freedom, the full hull of the ship was simulated. The inner mesh or inner domain was generated for half hull, and a symmetry was created by the solver

Table 2. Mesh configuration and simulation condition.

Condition	Settings
Fr number	0.142
Rn number	4.6×10^6
Wave amplitude, A/L	0.005
Wave direction	Head wave
Inner mesh resolution (half domain)	$197 \times 25 \times 109$
Inner mesh dimension (half domain)	$1.8 \text{ L} \times 0.35 \text{ L}$
Outer mesh resolution (full domain)	$190 \times 91 \times 64$
Outer mesh dimension (full domain)	$3.8 \text{ L} \times 2 \text{ L} \times 1 \text{ L}$

for full hull simulation. In the case of outer domain, however, a full mesh domain was generated instead of creating a symmetry. The mesh configuration and simulation conditions are shown in Table 2.

2.3 Computational resource

The OpenMP memory sharing model used in the solver limits its use to multi cores of only one node, not a cluster of nodes. Each simulation in this paper was performed in a single node of Intel(R) Xeon(R) CPU with 8 cores, clock speed 2.27 GHz and 8 GB of physical memory. The standard non-dimensional time step used was 1.5×10^{-4} and for simulating each non-dimensional time for added resistance, the required physical time was about 75 minutes per case. All the simulations were run up to 8 non-dimensional times for attaining stable results.

3 RESULTS

To investigate simulation dependency on DOF in resistance prediction, first, calm water simulation were performed for four different free movement cases. Next, head wave simulations were performed for same DOF cases. For both cases, same mesh resolution was used. The verification and validation of the solver have previously been studied by Islam et al. (2015), thus, it wasn't discussed in this paper.

3.1 Calm water resistance prediction with different DOFs

The calm water results for simulations with provided configuration (Table 2) at different degrees of freedom is shown in Figure 3. The result shows that there is a gradual decrease in resistance with higher degrees of freedom. This is because higher degrees of freedom implies lower restricting conditions and more freedom for the ship to move to low resistance position. The exception is observed at 6DOF, which is due to the high yaw motion the ship encounters because of the inherent poor

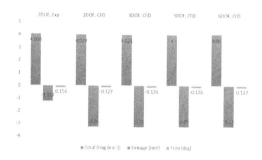

Figure 3. Calm water simulation results for KVLCC2 with different DOF.

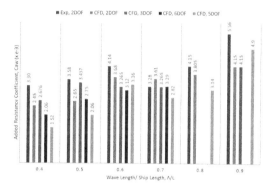

Figure 4. Added resistance coefficient in head waves for KVLCC2 with different degrees of freedom of motion.

directional stability of tanker models. Overall, the relative difference in the results is not significant, in fact, it is less than 1%. However, in the case of large container ships traveling from one end of the world to the other, a difference of 1% in fuel consumption can prove to be huge. The experimental data used here for comparison was from MOERI reported by Kim et al. (2013).

3.2 Added resistance prediction in head waves with different DOFs

Simulations in head waves were performed with different degrees of freedom (DOF), to check the added resistance dependency on DOF. The added resistance coefficient prediction for different degrees of freedom in short wave length cases is shown in Figure 4. The results show that the relative difference among added resistance coefficient prediction for different DOFs is comparatively small. For short wave length cases, the difference is very small; for $\lambda/L = 0.8$, a higher deviation is observed, but the difference again minimizes at

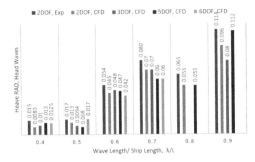

Figure 5. Heave RAO in head waves for KVLCC2 with different degrees of freedom.

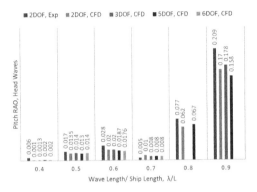

Figure 6. Pitch RAO in head waves for KVLCC2 with different degrees of freedom.

$\lambda/L = 0.9$. Thus, the influence of DOF is not very significant in simulating head wave cases.

The heave and pitch RAOs are also shown in Figure 5 and Figure 6. Heave and pitch RAOs show very little variation with higher degrees of freedom. This is because, at short wave lengths, ship's motions are very limited, thus, significant differences are not expected. However, the simulation cases here have been limited to short wave length cases and the trend might be different for long wave lengths. Thus, longer wave length cases are still subjected to investigation.

The linear and rotational motion response of the ship with 5DOF and 6DOF in head waves at $\lambda/L = 0.7$ and $A/L = 0.005$, is shown in Figure 7. The figures show that the yaw motion keeps on increasing with time in case of simulation with 6DOF and eventually will lose the initial heading direction if correction is not applied. Thus, it is recommended that, if rudder action is not possible, simulation should be performed with 5DOF (restricted yaw motion), to keep the reliability of the results.

Pressure contours for simulation with 6DOF and 5DOF at $\lambda/L = 0.7$ and $A/L = 0.005$, are provided in Figure 8 and Figure 9, respectively. Pressure distribution on ship hull also shows minimum variation for higher degrees of freedom.

Overall, minor difference is observed for resistance prediction, with respect to freedom of motion applied. This is because, in case of resistance prediction, ship hull shape and heaving and

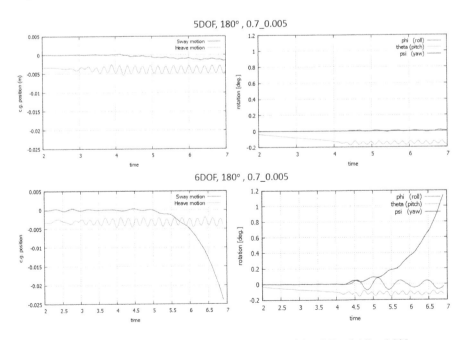

Figure 7. Linear and rotational response of KVLCC2 in head waves at $\lambda/L = 0.7$ and $A/L = 0.005$.

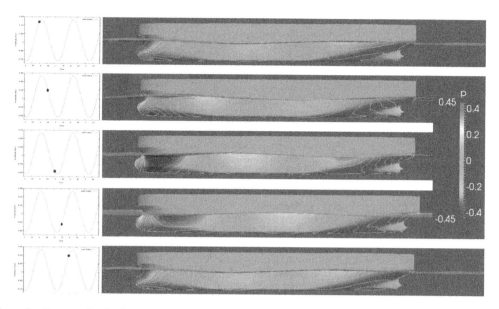

Figure 8.   Pressure distribution on hull surface of **KVLCC2** for head waves at $\lambda/L = 0.6$ and $A/L = 0.005$ with 6DOF.

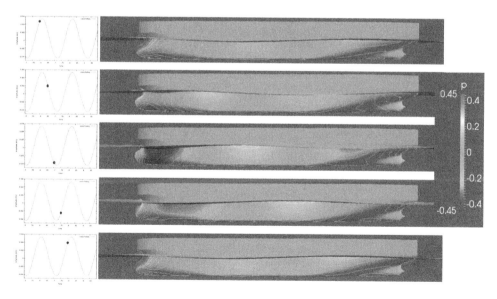

Figure 9.   Pressure distribution on hull surface of **KVLCC2** for head waves at $\lambda/L = 0.6$ and $A/L = 0.005$ with 5DOF.

pitching motion plays the most significant role. Roll motion is more significant for habitability checking and yaw motion is important for predicting rudder force. Thus, roll, sway, surge, and yaw adds minor changes to drag resistance. However, if these relatively small changes are scaled up to full-size ship, the interpretation might become considerable. Besides, different designs come with different features on stability and damping, which will not be fully revealed if motions are restricted during simulation. Still, considering the computational resource consumption for full hull simulation and the relative difference in prediction, it is recommended that full hull simulation with five or six degrees of freedom should be performed for finalizing a design, however, in case of the development stage, half hull simulation with 2DOF should prove enough to serve the purpose.

4 CONCLUSIONS

Simulations were performed for a tanker ship model at different degrees of freedom using a RaNS solver at both calm water and in head waves. Although initial simulation results for 2DOF show discrepancy with experimental data, simulation accuracy is enough for performing comparative study. By analyzing the results, the paper concludes that, except heave and pitch, other motions have limited influence in resistance prediction, both in calm water and in waves. However, while investigating habitability, other motions become important as well. Besides, the paper only presents results for head wave cases, whereas, in the actual voyage, ships mostly pass through oblique waves. Thus, further investigation is necessary for determining the influence of DOF on ships in oblique waves. Furthermore, similar study should also be performed with other RaNS solvers to ensure that the effects observed for different DOF of motion are not solver dependent.

REFERENCES

Akimoto, H. and. Miyata, H. 2002. "Finite-volume simulation method to predict the performance of sailing boat," Journal of Marine Science and Technology, vol. 7, pp. 31–42.

Islam, H. and Akimoto, H. 2015. "Prediction of ship resistance in head waves using RaNS based solver". 11th International Conference on Mechanical Engineering (ICME 2015), BUET, Dhaka.

Kim,H, H. Akimoto and Islam, H. 2015. "Estimation of the hydrodynamic derivatives by RaNS simulation of planar motion mechanism test," Ocean Engineering, vol. 108, pp. 129–139.

Kim, J., Park,I-R., Kim, K.-S., Kim, Y.-C., Kim, Y. Sik and Van., S.-H., 2013. "Numerical Towing Tank Application to the Prediction of Added Resistance Performance of KVLCC2 in Regular Waves," in Proceedings of the Twenty-third (2013) International Offshore and Polar Engineering (ISOPE) Anchorage, Alaska, USA.

Miyata,H., Akimoto, H. and Hiroshima, F. 1997. "CFD performance prediction simulation for hull-form design of sailing boats," Journal of Marine Science and Technology, vol. 2, pp. 257–267.

Orihara, H. and Miyata, H.. 2003. "Evaluation of added resistance in regular incident waves by computational fluid dynamics motion simulation using an overlapping grid system," Journal of Marine Science and Technology, vol. 8, pp. 47–60.

Sadat-Hosseini,H., Wu, P.-C, Carrica, P.M., Kim, H., Toda, Y. and Stern, F. 2013. "CFD verification and validation of added resistance and motions of KVLCC2 with fixed and free surge in short and long head Waves," Ocean Engineering, vol. 59, pp. 240–273.

"Tokyo 2015 Workshop," JBC, 2015. [Online]. Available: http://www.t2015.nmri.go.jp/jbc_gc.html. [Accessed 2016].

Maritime Transportation and Harvesting of Sea Resources – Guedes Soares & Teixeira (Eds)
© 2018 Taylor & Francis Group, London, ISBN 978-0-8153-7993-5

Prediction of ship resistance in head waves using OpenFOAM

H. Islam & C. Guedes Soares

Centre for Marine Technology and Ocean Engineering (CENTEC), Instituto Superior Técnico,
Universidade de Lisboa, Lisbon, Portugal

ABSTRACT: This paper discusses the estimation of resistance in calm water and in waves for a ferry ship model using the open source toolkit, OpenFOAM. In the calm water test, resistance, sinkage and trim values were predicted together with mesh dependency analysis and compared with experimental data. For resistance in head sea, short wave cases were simulated with 5 degrees of freedom (DOF) of motion (yaw restricted) and compared with experimental data. Overall the results capture well the trend and predicts the values with reasonable accuracy. The paper concludes that the solver still has some shortcomings regarding wave simulation with free motions.

1 INTRODUCTION

Prediction of ship's resistance and motion characteristics are essential at an early phase of ship design to understand its design efficiency. Although, model tests are popular for such predictions, in recent years, Computational Fluid Dynamics (CFD) has gained high popularity for such predictions because of its efficiency and economics. CFD is more efficient as it allows exploration of new engineering design frontiers in an economical way. Furthermore, over the years, the accuracy of CFD prediction has improved significantly. Thus, more and more research is being done on how to predict the sea-keeping and maneuvering capabilities of a ship in the design phase using CFD.

Prediction of ship hydrodynamic characteristics in the design phase is nothing new for the ship building industry. According to reports of 20th– 24th IITC, the first symposium on Naval Maneuverability was held in Washington DC in 1960. However, the first Numerical prediction for maneuverability using the boundary layer method was introduced in 1980. After that, various RANS models were developed throughout the 1980s. The 1994 Tokyo workshop brought a breakthrough in free surface calculation using RANS. For The Gothenburg 2000 workshop (Larsson et. al., 2003), three new ship models- KCS, KVLCC and DTMB 5415 were introduced with experimental results for validation. The self-propulsion system included CFD models were also introduced in this workshop.

In The SIMMAN 2008 workshop (SIMMAN, 2008), benchmarking was conducted for both Experimental Fluid Dynamics (EFD) based and CFD based methods. The Gothenburg 2010 workshop (Larson et. al., 2011) discussed global

and local flow variables, grid dependency, and turbulence modeling. However, practical use of CFD for ship maneuverability and design optimization started in early 2000 as shown by Percival et al. (2001), Peri et al. (2001) and Campana et al. (2006). Resistance prediction including sinkage and trim motion using CFD was first proposed by Yang et al. (2011). Since then, resistance and motion prediction using CFD has developed a lot and many researchers have contributed to the field by providing EFD and CFD results for different types of hull forms.

Over the years, application of CFD in ship hydrodynamics has gone through and is still going through continuous improvement. This paper aims at performing calm water and head wave simulation cases for a ferry ship using the open source CFD toolkit, OpenFOAM. The target of the research is to investigate the capabilities of the solver in simulating calm water and wave resistance with free ship motion.

2 SOLUTION PROCESS

2.1 *Solver*

OpenFOAM is an open source library that numerically solves a wide range of problems in fluid dynamics from laminar to turbulent flows, with single and multi-phases. It contains an extensive range of solvers to perform different types of CFD simulations. The solver has several packages to perform multiphase turbulent flow simulation for floating objects. The solver has been elaborately described by Jasak (2009).

The module used to perform ship hydrodynamic simulations in OpenFOAM simulates incompressible, two-phase flow. The solver follows earth and body fixed Cartesian coordinate system, with z axis upward positive. The governing equation is based on continuity and Reynolds averaged Navier-Stokes equation. The volume of Fluid (VOF) method is used to model fluid as one continuum of mixed properties. Finite Volume Method (FVM) is used to discretize the governing equations. Pressure-velocity coupling is obtained through PISO algorithm. OpenFOAM incorporates three different turbulence models, k-ε, k-ω and SST k-ω. Turbulence is discretized using a 2nd order upwind difference. For the present paper, SST k-ω model was used.

2.2 Turbulence parameters

Turbulence was modeled with the Reynolds-averaged stress (RAS) SST $k-\omega$ two equation model. The parameters were calculated as following:

$$I = 0.16 \cdot \text{Re}^{-1/8} \tag{1}$$

$$k\left(\frac{m^2}{s^2}\right) = \frac{3}{2}(u \cdot I)^2 \tag{2}$$

$$\delta(m) = \frac{L}{\sqrt{\text{Re}}} \tag{3}$$

$$l(m) = 0.4 \cdot \delta \tag{4}$$

$$C\mu = 0.09 \tag{5}$$

$$\omega\left(\frac{1}{s}\right) = \frac{\sqrt{k}}{C_\mu^{1/4} \cdot l} \tag{6}$$

$$v_T\left(\frac{m^2}{s}\right) = \sqrt{\frac{3}{2}} \cdot u \cdot I \cdot l \tag{7}$$

where, I is the turbulence intensity; k is the turbulent kinetic energy per unit mass; δ is the height of the boundary layer; l is the turbulence length scale; Cμ is an empirical constant; ω is the turbulence specific dissipation rate; and v_T is the turbulent kinetic eddy viscosity.

The density and the kinematic viscosity of the fluids used in simulations are reported in Table 1.

2.3 Boundary conditions

The control volume represented a deep water condition, so the two lateral sides and the bottom

Table 1. Physical properties.

	Water	Air
ρ, kg/m^3	998.8	1
v, m^2/s	1.337e-06	1.48e-05

were symmetry plane type faces; no additional information was required for this kind of boundary condition. Inlet, outlet, and atmosphere were patch faces with specific boundary condition for each one, and hull had a wall type boundary. For the calm water simulation cases, boundary conditions are as shown in Table 2, and for the head wave simulation cases, boundary conditions are as shown in Table 3.

In Tables 2 and 3, FV is fixedValue, specified by the user, OPMV is outletPhaseMeanVelocity, PIOV is pressureInletOutletVelocity that applies zero-gradient for outflow, whilst inflow velocity is the patch-face normal component of the internal cell value and MWV is movingWallVelocity. FFP is fixedFluxPressure that adjusts the pressure gradient such that the flux on the boundary is that one specified by the velocity boundary condition; ZG is zeroGradient; TP is totalPressure, calculated as static pressure reference plus the dynamic component due to velocity. IO is inletOutlet that provides a zero-gradient outflow condition for a fixed value inflow. kqRWF is the wallFunction for the turbulence kinetic energy, nutkWF is rough wall function for kinetic eddy viscosity and omegaWF is the wall function for frequency.

2.4 Ship model

The Ropax is a mono-hull ferry ship with wave piercing bow, low block coefficient, and low draft. It is fitted with skeg, two four-bladed counter-rotating (inwards) propellers and two Becker flap

Table 2. Boundary conditions used in calm water simulation.

	Inlet	Outlet	Atmosphere	Hull
U	FV	OPMV	PIOV	MWV
p_rgh	FFP	ZG	TP	FFP
$\alpha.water$	FV	VHFR	IO	ZG
k	FV	IO	IO	kqRWF
nut	FV	ZG	ZG	nutkRWF
$omega$	FV	IO	IO	omegaWF

Table 3. Boundary conditions used in head wave simulation.

	Inlet	Outlet	Atmosphere	Hull
U	waveVelocity	ZG	PIOV	FV
p_rgh	FFP	ZG	TP	FFP
$\alpha.water$	waveAlpha	ZG	IO	ZG
k	FV	ZG	IO	kqRWF
nut	FV	ZG	ZG	nutkWF
$omega$	FV	ZG	IO	omegaWF

spade rudders (Labanti et. al., 2016). For the resistance computation, only bare hull and skeg were considered. The ship's main particulars are shown in Table 4 and its hull form is shown in Figure 1.

2.5 *Meshing*

The domain size (blockMesh) for simulations was set following ITTC (2011) guidelines (ITTC Report, 2014); the inlet was placed roughly two ship length windward the bow, the outlet four ship length downstream the stern, each lateral boundary was two ship lengths away from the ship's symmetry plane, the depth or bottom of domain was set at one ship length and the atmosphere was at half ship length from free surface.

The hull form was integrated to the blockMesh by using snappyHexMesh utility, which created a "body fitted" hexahedral mesh around the hull surface from the specified STL file. The domain area near the free surface was refined multiple times using toposet and refinement, then snappyHexMesh was applied. Some refinement was also performed in snappyHexMesh. A configuration of the mesh used in calm water simulation is shown in Figure 2, however, the mesh topology used for calm water and wave simulation were different.

For calm water simulation, high mesh resolution is required near the free surface and ship hull surface. Thus, dense mesh is applied in the near the

free surface line throughout the domain and multiple refinements are done near the hull surface to properly capture the friction and pressure on hull surface, as shown in Figure 2.

In case of added resistance calculation, sufficient mesh resolution is required at the bow and stern part to properly capture the radiated waves, and at the entire hull form near the water line to capture the incident waves. Also high resolution is needed throughout the domain near the free surface to properly capture the incident waves. The mesh used for the wave simulations is shown in Figure 3.

2.6 *Computational resource*

The simulations were performed in a single node of Intel(R) Core i5 CPU with 4 cores, clock speed 2.27 GHz and 8 GB of physical memory. The time step used was adjustable runtime with minimum step size being 0.0001second, and for simulating

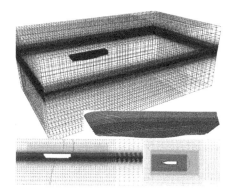

Figure 2. Mesh configuration for calm water simulation.

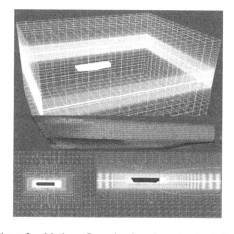

Figure 3. Mesh distribution for head wave simulations.

Table 4. Specifications of a Ropax ship.

Specification		Ropax ship (full scale)	Ropax ship (model scale)
Length between perpendicular	Lpp (m)	90	3.600
Breadth	B (m)	17.82	0.713
Depth	D (m)	14.8	0.592
Draft	T (m)	4	0.168
Wetted surface area	S (m²)	1743	2.788
Displacement volume	V (m³)	3725	0.238
LCB from mid-ship (fwd+)	xG (m)	−3.5	−0.14
Kyy	GML (m)	170	10.152

Figure 1. Body plan of Ropax model.

each case with stable output, the required physical time was about 40 hours per case for wave simulations, and about 8 hours for calm water simulation. All the simulations were run up to 100 seconds (simulation time) for attaining stable results.

3 RESULTS

3.1 Calm water resistance

Calm water resistance prediction is the estimation of drag force encountered by a ship while moving forward in calm water. Ship's drag resistance is summation of frictional resistance and pressure resistance. Frictional resistance arises from the hull surface friction and pressure resistance is mainly the wave making resistance encountered by ship during its forward motion.

The Ropax model was simulated in calm water using interDyMFoam solver, with free heave and pitch motion, and all other motions restricted. For investigating the grid dependency of the solver, three mesh resolutions were used maintaining a refinement ratio of $\sqrt{2}$. All simulations were performed with full hull (both starboard and port side). The mesh resolutions used were of 0.735 million (fine), 0.365 million (mid) and 0.321 million (coarse), for Froude numbers of 0.242, 0.200 and 0.087. The grid based uncertainty

analysis performed for Ropax model in Open-FOAM is shown in Table 5 The grid dependency analysis was performed by following the procedure prescribed by Celik et. al. (2008). Conventionally, the mesh resolution used for the study is too small. However, the ropax model has a very low draft, thus low surface area under water, and the simulated model size is also small, which allow the application of low mesh resolution.

For the analysis, the grid refinement is performed in all x, y and z axis, maintaining same refinement ratio. The order of accuracy, p and Grid Convergence Index (GCI) are predicted using the following equations,

$$p = \frac{1}{\ln(r_{21})} \left| \ln\left| \varepsilon_{32} \middle/ \varepsilon_{21} \right| + \ln\left(r_{21}^p - s \middle/ r_{32}^p - s \right) \right| \qquad (8)$$

$$GCI_{fine}^{21} = 1.25 e_a^{21}/(r_{21}^p - 1) \qquad (9)$$

As can be seen from the table, for Fr = 0.087 and Fr = 0.2, the analysis show monotonous convergence for resistance prediction, and oscillatory convergence for sinkage and trim prediction. For the Fr = 0.242, resistance and trim shows oscillatory convergence, and sinkage shows monotonous convergence. For proper sinkage and trim prediction, it is essential to have grid spacing in z direction in the same order as sinkage and

Table 5. The uncertainty analysis performed for Ropax model in OpenFOAM based on the mesh dependency.

		Total resistance (N)			Sinkage (mm)			Trim (deg)		
Fr. number		*0.087*	*0.200*	*0.242*	*0.087*	*0.200*	*0.242*	*0.087*	*0.200*	*0.242*
Output	$\varnothing_1$ (fine)	2.100	8.000	13.030	0.390	3.000	6.200	0.205	0.290	0.340
values	$\varnothing_2$ (mid)	2.200	8.120	12.540	0.380	3.130	6.170	0.210	0.300	0.350
	$\varnothing_3$ (coarse)	2.320	8.450	13.270	1.570	1.960	5.000	0.190	0.270	0.320
Refinement	$r_{12} = h_2/h_1$	1.385	1.385	1.385	1.385	1.385	1.385	1.385	1.385	1.385
ratio	$r_{32} = h_3/h_2$	1.300	1.300	1.300	1.300	1.300	1.300	1.300	1.300	1.300
Difference of	$\varepsilon_{21} = \varnothing_2 - \varnothing_1$	0.100	0.120	−0.490	−0.010	0.130	−0.030	0.005	0.010	0.010
estimation	$\varepsilon_{32} = \varnothing_3 - \varnothing_2$	0.120	0.330	0.730	1.190	−1.170	−1.170	−0.020	−0.030	−0.030
Convergence	$\varepsilon_{21/\varepsilon 32}$	0.833	0.364	−0.671	−0.008	−0.111	0.026	−0.250	−0.333	−0.333
	$s = 1.\ \mathrm{sgn}$ $(\varepsilon_{21}/\varepsilon_{32})$	1.000	1.000	−1.000	−1.000	−1.000	1.000	−1.000	−1.000	−1.000
Order of accuracy	p	1.400	3.820	1.100	18.200	8.210	9.270	5.060	3.960	3.000
Extrapolated	$\varnothing_{ext}^{21}$	1.927	7.951	14.167	0.390	2.990	6.202	0.204	0.286	0.334
values	$\varnothing_{ext}^{32}$	1.930	7.929	10.358	0.370	3.284	6.283	0.217	0.316	0.375
Approximate	e_a^{21}	0.048	0.015	−0.038	−0.026	0.043	−0.005	0.024	0.034	0.029
relative error	e_a^{32}	0.055	0.041	0.058	3.132	−0.374	−0.190	−0.095	−0.100	−0.086
Extrapolated	e_{ext}^{21}	0.090	0.006	−0.080	0.000	0.003	−0.0002	0.006	0.013	0.018
relative error	e_{ext}^{32}	0.140	0.024	0.211	0.027	−0.047	−0.018	−0.033	−0.052	−0.067
Grid	GCI_{fine}^{21}	0.103	0.0076	−0.109	−0.0001	0.004	−0.0003	0.007	0.016	0.022
convergence index (GCI)	GCI_{coarse}^{32}	0.1536	0.0295	0.218	0.033	−0.061	−0.023	−0.043	−0.068	−0.089

trim values. However, in case of coarse grid, due to relatively large spacing in z direction, the sinkage or trim values are either under or over predicted. Thus, oscillatory convergence is observed. Overall, all the results show a low level of uncertainty, except for the resistance prediction at Froude 0.087 and 0.242. The relative errors are also high for these cases.

The order of accuracy for all the predictions are also within acceptable range. Because of the low draft and sharp bow of the Ropax model, it is particularly difficult to simulate, and problems occur in the flow field at high Froude numbers, as reported by Labanti et al (2016) and Ciortan et al. (2007). Thus, convergence is difficult to attain, as is validation.

For validating the simulations, results were compared with experimental data already used by Labanti et. al. (2016). The comparison with experimental data is shown in Table 6. Unfortunately, experimental data was available only for two ship speeds, and sinkage and trim data were not available. The solution shows significant deviation for Fr = 0.242 case. However, even with significantly higher mesh resolution, the results could not be improved. Since, no other CFD or EFD data is available for the ship in same conditions at the moment, it is difficult to comment on the reliability of EFD data. Furthermore, the simulations were performed with free heave and pitch motion, as for experiments, detail is not available in this regard.

To further illustrate the results, the pressure distribution on free surface and on hull surface for the two Froude numbers is shown in Figure 4.

3.2 Resistance in head waves

Added resistance is mainly the additional resistance a ship faces while making its way through the waves. This resistance part is created by the loss of energy to both, radiated waves caused by ship motion, and diffraction of incident waves on ship hull. However, energy distribution among these two components are dependent on the ratio of incident wave length to ship length (λ/L).

The calm water resistance predicted for the present mesh using interDyMFoam solver is 1.23 (N) for Fr = 0.087, and 8.68 (N) for Fr = 0.242. The

Table 6. Comparison of Ropax simulation results with experimental data.

	Total calm water resistance	
Fr. Number	0.087	0.242
Exp. Data	2.21 (N)	16.879 (N)
CFD data	2.1 (N)	13.03 (N)

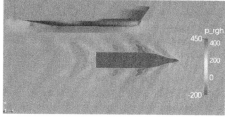

Figure 4. Pressure distribution on free surface and hull surface for Ropax model in Fr = 0.087 and Fr = 0.242, respectively.

Table 7. Head wave simulation conditions for Ropax model.

Ship model	Ropax
Froude number	0.087 (5 knots), 0.242 (14 knots)
Ship speed (in knots)	5, 14
Reynolds number	1.7×10^6, 4.76×10^6
Degrees of freedom	5DOF (yaw restricted)
Mesh resolution	1.34 million
Mesh dimension	5 L x 3 L x 2 L
Wave periods (s)	0.630, 0.629, 0.631, 0.721, 0.839, 1.006.
Wave heights (m)	0.015, 0.014, 0.011, 0.028, 0.057, 0.095.

*L = Ship length.

low prediction can be contributed to the lack of sufficient mesh resolution near the hull surface. In order to ensure smooth wave propagation, overall resolution of the domain had to be kept high. However, for accommodating ship motion using the grid deforming technique of OpenFOAM, high resolution around the hull surface could not be applied, which resulted in loss of accuracy. Because of insufficient resolution near hull area, both viscous and pressure resistance were under predicted. The simulation conditions for head wave cases are described in Table 7.

For performing the head wave simulations, waveFoam and waveDyMFoam solvers were used. Both of them are extended solvers of OpenFOAM, which are also available under open license. waveFoam performs simulation with all motions restricted, whereas, waveDyMFoam allows simulation with 6 DOF motion.

In order to check the simulation dependency on degrees of freedom, first limited cases were simulated using waveFoam, and compared with experimental data gained from the SHOPERA project. Figure 5 shows head wave simulation results using waveFoam solver for Fr = 0.087 (5 knots speed) and Fr = 0.242 (14 knots speed). As can be seen from the figure, CFD results follow the trend of experimental results closely for short wave periods and wave heights, where ship motion is not significant. However, with increase in wave height, ship motions become pronounced and so does radiation force, thus deviation with experiment increases.

Next, simulations were performed with waveDyMFoam solver, with 5 DOF of motion. The yaw motion was restricted in this case since the solver doesn't provide a yaw correction option. Furthermore, in experiments, a soft captive system was used for the ship model, with allowed it free motion in all DOF, with sufficient damping. Since similar restraining model could not be applied in the solver, comparison of motion data is not shown in the paper. The simulations were kept limited to short wave period and wave height cases, as the solver fails to accommodate pronounced motion of ship at higher wave periods and wave amplitudes. The predicted results are shown in Figure 6.

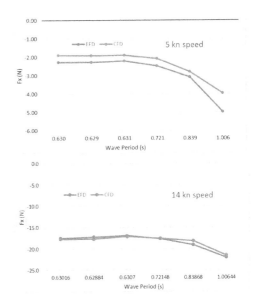

Figure 6. Resistance prediction in head waves for Ropax model using waveDyMFoam solver (Fr = 0.087 and Fr = 0.242, respectively).

As can be seen from the figure, the results show good agreement with experimental data, especially at Fr = 0.242 (14 knots speed) cases, where the average deviation with experimental data is less than 2.5%. For the Fr = 0.087 cases, although the overall trend is well followed, the deviation is significant. This is mostly because of the insufficiency of mesh resolution applied near hull surface. In high speed cases, viscous and pressure force generated are more pronounced and are relatively easy to capture with even a coarse mesh. However, for low speed cases, the change in forces are minor, thus higher mesh resolution is required near the hull surface to properly capture the changes in force.

OpenFOAM uses a mesh deformation technique for accommodating the motions of the ship. Thus, if motions become pronounced, smaller cells (high resolution mesh near hull surface) fail quicker under deformation and the simulation stops. Thus, mesh resolution near the hull surface had to be kept coarse to ensure continuation of simulation. As a result, accuracy was compromised and the resistance was under predicted.

To further illustrate the results, pressure distribution on free surface and hull form for wave period 0.84 s and wave height 0.057 m are shown in Figure 7 for Fr = 0.087 and Fr = 0.242, respectively.

Overall, waveDyMFoam results show significant improvement over waveFoam results, highlighting the significance of DOF of motion in wave simulations. Thus, waveDyMFoam should be preferred

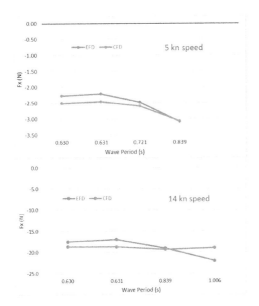

Figure 5. Resistance prediction in head waves for Ropax model using waveFoam solver (Fr = 0.087 and Fr = 0.242, respectively).

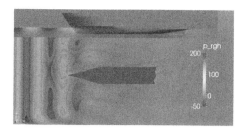

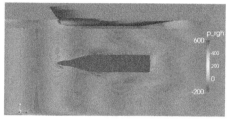

Figure 7. Pressure distribution on free surface and hull form for wave period 0.84 s and wave height 0.057 m are shown in for Fr = 0.087 and Fr = 0.242, respectively.

for performing wave simulations, provided that the mesh can accommodate the motions involved in the simulation.

4 CONCLUSIONS

In this paper, calm water and head wave simulation results for a ferry ship model are presented. In calm water simulation, three different mesh resolution were used to perform grid uncertainty analysis, then results were compared with experimental data for validation. The uncertainty analysis show oscillatory convergence for some resistance and motion cases, which can be contributed to the relatively low mesh resolution used for the cases of analysis. Furthermore, the analysis doesn't take the time step effect into consideration, which might also be a reason for observing oscillatory convergence. In case of validation study, the results show good agreement for low ship speed cases, however, deviation increases with increased speed.

The head wave simulations were performed with a higher mesh resolution and results were compared with experimental data. The results show good accuracy for high Froude number cases (Fr = 0.242), however, performs rather poorly for low Froude cases.

The waveDyMFoam solver is not fully equipped to perform simulations in waves with well pronounced ship motions. Thus, for short wave period and wave height cases, waveDyMFoam can be preferred, provided that the cases are at relatively high Froude number (speed). However, in some cases, waveFoam solver can be a better alternative, which, although restricts ship motions, allows application of sufficient mesh resolution around hull surface and improves resistance prediction significantly.

ACKNOWLEDGEMENTS

This work was performed within the scope of the Strategic Research Plan of the Centre for Marine Technology and Ocean Engineering (CENTEC), which is financed by the Portuguese Foundation for Science and Technology (Fundação para a Ciência e Tecnologia—FCT).

REFERENCES

Campana, E.F., Peri, D., Tahara, Y., Stern, F. (2006), Shape optimization in ship hydrodynamics using computational fluid dynamics. *Computer methods in applied mechanics and engineering 196:* 634–651.

Celik I. B., U. Ghia, P. J. Roache, C. J. Freitas, H. Coleman and P. E. Raad. (2008), Uncertainty due to discritization in CFD Journal of Fluids Engineering, *Transactions of the ASME, 2008.*

Ciortan, C.,Wanderley, J., Guedes Soares, C. (2007), Turbulent free-surface flow around a Wigley hull using the slightly compressible flow formulation. *Ocean Engineering 34:*1383–1392.

ITTC-Report, (2014), "Seakeeping Committee: Final Report and Recommendations to the 27th ITTC," 2014.

Jasak, H., (2009), OpenFOAM: Open Source CFD in research and industry. *International Journal of Naval Architecture and Ocean Engineering, 11(2), pp.89–94.*

L. Larsson, F. Stern and V. Bertram, (2003), Benchmarking of Computational Fluid Dynamics for Ship Flows: The Gothenburg 2000 Workshop. *Journal of Ship Research,* vol. 47, pp. 63–81, 2003.

L. Larsson, F. Stern and M. Visonneau, (2011), CFD in Ship Hydrodynamics- Results of the Gothenburg 2010 Workshop, in *MARINE 2011, IV International Conference on Computational Methods in Marine Engineering, Computational Methods in Applied Sciences,* 2011.

Labanti J., Islam H. and C. Guedes Soares (2016), CFD assessment of Ropax hull resistance with various initial drafts and trim angles. *Maritime Technology and Engineering 3 – Guedes Soares & Santos (Eds), ISBN 978-1-138-03000-8.*

Percival, S., Hendrix, D., Noblesse, F. (2001), Hydrodynamic optimization of ship hull forms. *Applied Ocean Research 23:* 337–355.

Peri, M., Rossetti, M., Campana, E.F. (2001), Design optimization of ship hulls via cfd techniques. *Journal of ship research 45:* 140–149.

SIMMAN (2008), Summary of proceedings of SIMMAN 2008 Workshop, 2008.

Yang, C., Huang, F., Wang, L. (2011), Numerical simulations of highly nonlinear steady and unsteady free surface flows. *Journal of Hydrodynamics 23(6):* 683–696.

Maritime Transportation and Harvesting of Sea Resources – Guedes Soares & Teixeira (Eds)
© 2018 Taylor & Francis Group, London, ISBN 978-0-8153-7993-5

Supporting development of the smart ship technology by CFD simulation of ship behavior in close to real operational conditions

K. Niklas

The Faculty of Ocean Engineering and Ship Technology, Gdansk University of Technology, Gdansk, Poland

ABSTRACT: The shipping industry is at the milestone of technological development—autonomous ships. Involving smart technologies accelerates development of green and safe shipping. Also IMO regulations, especially EEDI enhanced application of new technologies. Meeting economic and ecological requirements future ships need to be designed with respect to increased reliability and operational performance. Design process of ships may be improved by utilizing advanced computer simulations. Especially application of the CFD enables optimization and simulation driven design. Recent research project 'SmartPS' includes advanced seakeeping analysis into development of the smart ship technology. In the article the CFD simulation of real scale ship with self-propulsion is presented. The goal is to demonstrate possible modelling of a case study ship behavior in close to real operational conditions. Results concerning energy efficiency, comfort and safety are presented giving demonstration of computer simulation as a very effective tool for determination of important data for smart ship technology.

1 INTRODUCTION

Maritime transport is the backbone of globalized world enabling international trade in a huge scale. The total world's seaborne trade is estimated to exceed 10 billion tons per year (UNCTAD, 2015). No one deny that ships are the most cost efficient and environmental friendly mean of transport for international trade over long distances. According to the analysis presented in (International Maritime Organization (IMO), 2009a) the transportation by rail emits about 4 times more CO_2 and by roads about 8 times more CO_2 per ton-km of cargo—see Fig. 1. Although there is no alternative mean of transport, more and more attention is being paid to environmental protection. Similarly to other sectors the shipping industry is under great pressure to reduce emission of the Greenhouse Gasses. The toughest regulations concern toxic substances, especially NO_x and SO_x. Globally the shipping industry emits about 2.5% of the CO_2 (Smith et al., 2014). The rise of environmental footprint from the shipping industry is estimated between 50–250% by the year 2050 depending on the global economy development.

The IMO introduced several regulations within the convention MARPOL, annex VI (IMO, 2011) limiting air pollution from ships. The quality of fuel has risen dramatically by introducing the Tier limits and very soon the Tier III will cut allowed emissions of NO_x and SO_x by 70% in comparison with the Tier II level.

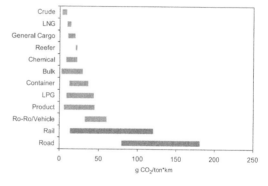

Figure 1. Comparison of different transport modes concerning the CO_2 emission per ton-km of cargo (International Maritime Organization (IMO), 2009b).

Also the UE limits allowable Sulphur content in fuel for ships to the level of 0.5% since the year 2020 (European Parliament and Council of the European Union, 2012). However the most important development in the shipping industry can results from Energy Efficiency Design Index (EEDI) (IMO, 2012a, 2012b, 2012c). The formulation of index defines the allowable CO_2 emissions per volume of cargo and ship's velocity. In upcoming years the unprecedented rise of ships energy efficiency by 30% is expected. Meeting the required efficiency index can be realized by applying technological and/or operational improvements.

Considering next 5 to 10 years the greatest reduction of environmental footprint can be reached by rising operational energy efficiency of existing ships. Both for the new and existing ships the significant increase of the energy efficiency can be realized by application of the smart ship technology supported by CFD simulations. This approach is presented in the next section.

2 SMART SHIP TECHNOLOGY DEVELOPEMENT SUPPORTED BY CFD

The technology of smart ships is at stage of rapid development (Kanegaonkar, 1999), (Murphy, 2000), (Liu & Mu, 2010), (Yan, Venayagamoorthy, & Corzine, 2011), (Lee, Shin, Lee, Kim, & Kim, 2013), (Bertram, 2016), (Ang, Goh, & Li, 2016), (Geertsma, Negenborn, Visser, & Hopman, 2017). The presence of autonomous shipping is now a matter of years, not decades. The development towards smart, green and safe shipping can be significantly supported by the application of advanced CFD simulations. During last years the maturity of numerical tools and accessibility of High Performance Computing enabled performing ship behavior simulations with increased accuracy. One of the greatest feature is possibility of calculation a ship and offshore structure response to different operational loads including rough sea conditions (Atluri, Magee, & Lambrakos, 2009; Darvishzadeh et al., 2016; Guo, Steen, & Deng, 2012; Niklas, 2017; Peder Kavli, Oguz, & Tezdogan, 2017; Rajendran, Fonseca, & Guedes Soares, 2016; Seo, Park, & Koo, 2017; Tezdogan et al., 2015).

Many of the computer simulations directly aim at rising of the energy efficiency by application of the CFD simulations(Haase et al., 2016; Peder Kavli et al., 2017; Seo et al., 2017; Tezdogan et al., 2015). Increased number of analyses were performed for full scale ship model (unsteady RANS CFD simulations). Direct goal of a study is most often accurate prediction of added resistance and ship motions in head sea. In seakeeping analysis wave loads play crucial role, but most often are simplified to regular wave model (i.e. Stokes 5th order). Different common applications of the CFD simulations in marine hydrodynamics are presented in Fig. 2.

At present the CFD technology features are utilized in a very limited scale. Most often the commercial use CFD software is limited to the determination of ship resistance on calm water conditions and in a model scale. Basically the whole design process is performed with an assumption of calm water conditions and the effect of waves is included by simplified analytical formulas for the estimation of so called 'added resistance'. Also

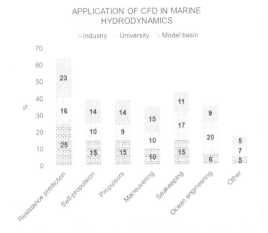

Figure 2. Applications of CFD in marine hydrodynamics (ITTC, 2011).

ship trials are being performed in a calm water condition to proof contracted speed of a ship. Whereas real ships in an operation are subjected to waves and wind almost all the time. The interesting question is if a new design optimized for calm water conditions will perform as expected in waves? From the author practical experience results an argument that especially seakeeping performance of small and medium size vessels needs to be addressed more carefully. Especially for many types of vessel operating very often in rough sea states the seakeeping analysis is needed and it's accuracy can be significantly raised by application of the CFD simulations. The direct benefits of extended analysis in close to the real conditions is in example increased power efficiency, comfort and safety.

For some of the ship types the seakeeping analysis seems to be obvious, but rarely performed in a real life. The overall performance of ships like the Anchor Handling Tug/Supply Vessel (AHTS/OSV), Wind Farm Support Vessels (WFSV), ferries, fishing vessels, research vessels, military vessels, future autonomous vessels strongly depends on proper prediction of a ship response to waves. Especially at sides like The North Sea and The Baltic Sea where dominates moderate and rough sea state. The influence of wave and wind loads on the behavior of any ship is significant and can be analyzed by advanced CFD simulations of real scale ship and close to the real loading conditions. This is also important that the CFD simulations can be applied during a design process of a new ship, as well as for optimizing performance of any of the existing ships. Rising operational energy efficiency of existing ships is even in a greater importance considering need of reducing environmental footprint from shipping industry. Considering smart ship technology for the propulsion system the data of ship response to environmental loads are very important. For the purpose of this article the selected data are being considered. Regarding propulsion efficiency the ship resistance including waves was analyzed. For measureable physical quantity of comfort the vertical acceleration of selected point at the wheelhouse was selected. As sample result considering safety the minimum distance from propeller tip to free surface of water was pointed. In the following section the sufficient CFD model of full scale ship including self-propulsion is presented. The following section demonstrates selected capabilities of modelling a case study ship behavior in a close to the real operational condition.

3 ADVANCED CFD SIMULATION OF A CASE STUDY SHIP

3.1 Aim & scope of the simulation

The simulation was performed for the demonstration of selected capabilities of the commercially available CFD RANS tools used for the determination of ship response to close to the real operational condition. For the purpose full scale and self-propelled numerical model of a case study ship was built. Presented approach uses the advanced computer simulation for the determination of selected data of the ship response:

- ship resistance on waves,
- vertical acceleration of selected point at location of an officer on duty (wheelhouse),
- distance from propeller tip to the free surface of water.

The calculated ship resistance for assumed loading condition is representative result from the energy efficiency point of view during ship's operation. Accurate numerical modelling of waves was included in the analysis. Second derived result is vertical acceleration and refers to a comfort criteria. The third calculated value is distance between propeller tip and free surface of water and refers to safety criteria (risk of ventilation, propulsion damage). All the results were calculated for one environmental condition of wind force 5 B and head wave at the Baltic sea side. Similar analysis can be performed for different loading conditions providing in example operational limits of the ship regarding efficiency, comfort and safety criteria.

3.2 Numerical model of a case study ship

The numerical model was built in STAR-CCM+ software. The implicit unsteady time integration was used. Free surface and waves were modelled

with the use of 'Volume of Fluid Waves' approach. The domain with standard boundary conditions is presented in Fig. 3. At the outlet boundary also numerical damping was applied. The rigid body ship motion was modelled by the 'Dynamic Fluid Body Interaction' approach including heave and trim movements. The rotating propeller was modelled as a rigid body which directly transfers thrust force to the hull. Rotational speed was equal 206 rpm with a propeller pitch of 15.2 degree.

The mesh consisted of over 11 million cells (5 million for rotating region and 6 million for static region). The sample surface mesh of the ship model is presented in Fig. 4.

For the case study analysis the research vessel 'Nawigator XXI' (IMO: 9161247) was selected. The overall length of the hull is equal LOA = 60.30 m, breath B = 10.47 m and drought T = 3.20 m. The displacement of the ship is equal D = 1126 m³. The assumed COG position and inertia moments were calculated from stability book for a common operational state.

The analyzed environmental loads condition was referring to the wind force 5 in the Beaufort scale. The irregular wave model was used with JON-SWAP spectrum type (Det Norske Veritas, 2014). Significant wave height was equal $H_{1/3}$ = 1.2 m, peak wave period was 4.4 s and the peak shape parameter was equal 3.3.

3.3 Results

The presented results are demonstrating selected features of CFD modelling of the ship response to

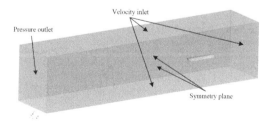

Figure 3. The fluid domain with boundary conditions.

Figure 4. The surface mesh representation of the ship model.

the loading scenario. The sample wave pattern of free surface of water is presented in Fig. 5. The calculated average ship resistance was equal 18.2 kN. The vertical acceleration of selected point of the vessel at the wheelhouse (location of officer on duty) is presented in Fig. 6. Maximum acceleration was equal 1.2 m/s² (0.12 g), which is high value considering the maximum acceptable criterion for commercial ships of 0.15 g (MARINTEC, 1987) and for naval vessels of 0.2 g. The operation of the ship would be very difficult in analyzed conditions.

The last presented data is position of the propeller tip with relation to a free surface of water—see Fig. 7. The results points that for analyzed loading

Figure 5. Sample wave pattern of free surface of water.

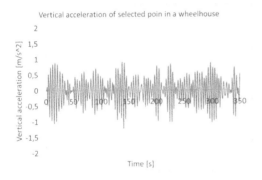

Figure 6. Vertical acceleration of selected point at location of officer on duty in the wheelhouse. Selected timespan of 350 s.

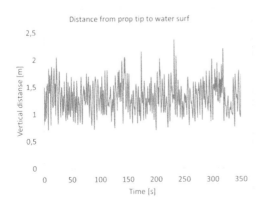

Figure 7. Distance from propeller tip to the free surface of water. Selected timespan of 350 s.

working conditions of the propeller are acceptable. The calculated propeller position was at least 0.7 m below the free water surface.

4 CONCLUSIONS

The ships play important role in globalized trade of goods and other human's activities like fishing, research, oil and gas extraction. For years the shipping industry has got the impression of 'conservative' and applying the technological development with an inertia of time. Today this seems to be no longer valid. The demand of reducing environmental impact from ships introduced significant changes in the industry. The new regulation of EEDI brings major technological and operational development. The need of rising energy efficiency is accelerating development of the smart ship technology. Moreover the technology of autonomous ships is coming with a big steps.

The technological development rising the energy efficiency, comfort and safety of the ships can be significantly supported by the advanced CFD simulations. With the use of commercially available software any ship can be modelled in a real scale, including fluid body interaction with waves. The effect of propulsion can be modelled in simplified manner i.e. by disk model, or by rigid body representation. Although computing power and time is the limitation. The accurate determination of ship response to the environmental loads is very important for the new design, as well as for the improvement of the performance of existing ships. The application of CFD for modelling of ship behavior in close to the real operational conditions can bring the following benefits:

- rising the ship energy efficiency by accurate prediction and optimization of ship behavior. Detail investigation of many design aspects like hydrodynamic flow, resistance, hull and propulsion efficiency, ship movements;
- rising comfort onboard a ship by optimizing ship movements, especially by lowering accelerations;
- rising safety by analysis of ship movements on waves (propeller ventilation, human factor, cargo),

Many other aspects can be mentioned here depending on the specific goal of the analysis.

In this article the CFD simulation of the case study vessel was presented for assumed loading condition. The ship was modelled in a natural scale with a propeller and rudder. The selected results were presented to demonstrate sample aspects of energy efficiency, comfort and safety. These and much more data determined by computer simulation can be utilized in the development of smart ship technology for standard and autonomous ships. In fact the analyzed case study vessel from practical experience was recognized as having low seakeeping abilities which was confirmed by the numerical simulation. The ship behavior on waves in practice significantly limits the operability and performance. Detail investigation is planned for the ship with a purpose of determination of the operational limits driven by energy efficiency, comfort and safety criteria.

From design point of view it is also worth to notice that by using the CFD technology the development of ships can be performed in revolutionary way. Especially in the field of concept evaluation of innovative hull form and propulsion.

The huge potential of CFD tools can be utilized for design of autonomous ships. Significantly greener ships with excluded all the requirements resulting from a human presence onboard. For the purpose of lowering the operational risk of autonomous ships their detailed response to several loading conditions is needed. Significantly extended analysis of ship behavior in different environmental conditions including rough sea states is important.

ACKNOWLEDGEMENTS

This research was supported by the National Centre for Research and Development within the 'SmartPS' project ERA-NET MARTECII/SmartPS/4/2016.

The research was supported by the Academic Computer Centre in Gdansk (CI TASK). All support is highly appreciated by the author.

REFERENCES

Ang, J.H., Goh, C., & Li, Y. (2016). Smart design for ships in a smart product through-life and industry 4.0 environment. In *2016 IEEE Congress on Evolutionary Computation, CEC 2016* (pp. 5301–5308). https://doi.org/10.1109/CEC.2016.7748364

Atluri, S., Magee, A., & Lambrakos, K. (2009). CFD as a design tool for hydrodynamic loading on Offshore structures. *Proceedings of the ASME 2009 28th International Conference on Ocean, Offshore and Arctic Engineering*, 1–9. https://doi.org/10.1115/OMAE2009-79502

Bertram, V. (2016). Autonomous Ship Technology – Smart for Sure, Unmanned Maybe. *Smart Ship Technology*, 26–27.

Darvishzadeh, T., Sari, A., Sarkar, A., Gudmestad, O. T., Cerik, B. C., Shin, H. K., ... Banister, K. (2016). CFD Applications in Offshore Engineering. *Marine Structures*, 46, 1–29. https://doi.org/10.4043/25930-MS.

Det Norske Veritas. (2014). Recommended Practice DNV-RP-C205—Environmental Conditions and Environmental Loads.

European Parliament and Council of the European Union. (2012). Directive 2012/33/EU of the European Parliament and of the Council. Retrieved from http://eur-lex.europa.eu/legal-content/EN/TXT/?uri=CELEX:32012L0033

Geertsma, R.D., Negenborn, R.R., Visser, K., & Hopman, J.J. (2017). Design and control of hybrid power and propulsion systems for smart ships: A review of developments. *Applied Energy*, 194, 30–54. https://doi.org/10.1016/j.apenergy.2017.02.060

Guo, B.J., Steen, S., & Deng, G.B. (2012). Seakeeping prediction of KVLCC2 in head waves with RANS. *Applied Ocean Research*, 35, 56–67. https://doi.org/10.1016/j.apor.2011.12.003

Haase, M., Zurcher, K., Davidson, G., Binns, J.R., Thomas, G., & Bose, N. (2016). Novel CFD-based full-scale resistance prediction for large medium-speed catamarans. *Ocean Engineering*, 111, 198–208. https://doi.org/10.1016/j.oceaneng.2015.10.018

IMO. (2011). MARPOL Annex VI, Chapter 4.

IMO. (2012a). MEPC.212(93) – 2012 Guidelines on the method of calculation of the attained Energy Efficiency Design Index (EEDI) for new ships.

IMO. (2012b). MEPC.214(93) – 2012 Guidelines on survey and certification of the Energy Efficiency Design Index (EEDI).

IMO. (2012c). MEPC.215(93) – Guidelines for calculation of reference lines for use with the Energy Efficiency Design Index (EEDI).

International Maritime Organization (IMO). (2009a). Second IMO GHG Study 2009. Retrieved from http://www.imo.org/en/OurWork/Environment/PollutionPrevention/AirPollution/Documents/SecondIMO-GHGStudy2009.pdf

International Maritime Organization (IMO). (2009b). Second IMO GHG Study 2009.

ITTC. (2011). The Specialist Committee on Computational Fluid Dynamics—Final Report and Recommendations to the 26th ITTC. Retrieved from http://ittc.info/media/5528/09.pdf

Kanegaonkar, H.B. (1999). Smart Technology Applications In Offshore Structural Systems: Status And Needs. Retrieved from http://www.onepetro.org/conference-paper/ISOPE-I-99-036

Lee, K.J., Shin, D., Lee, J.P., Kim, T.J., & Kim, H.J. (2013). Experimental investigation on the hybrid smart green ship. In *Lecture Notes in Computer Science (including subseries Lecture Notes in Artificial Intelligence and Lecture Notes in Bioinformatics)* (Vol. 8103 LNAI, pp. 338–344). https://doi.org/10.1007/978-3-642-40849-6-33

Liu, H., & Mu, L. (2010). Construction of integrated smart power system for future ship. In *2010 International Conference on Power System Technology: Technological Innovations Making Power Grid Smarter, POWERCON2010*. https://doi.org/10.1109/POWERCON.2010.5666659

MARINTEC. (1987). The Nordic Cooperative Project: Seakeeping Performance of Ship, Assessment of Ship Performace in a Seaway.

Murphy, B.P. (Brian P. 1963-. (2000). Machinery monitoring technology design methodology for determining the information and sensors required for reduced manning of ships. Retrieved from http://dspace.mit.edu/handle/1721.1/88345

NATO. (1997). NATO STANAG 4154, Common Procedures for Seakeeping in the Ship Design Process.

Niklas, K. (2017). Strength Analysis of a Large-Size Supporting Structure for an Offshore Wind Turbine. *Polish Maritime Research*, *24*(s1). https://doi.org/10.1515/pomr-2017-0034

Peder Kavli, H., Oguz, E., & Tezdogan, T. (2017). A comparative study on the design of an environmentally friendly RoPax ferry using CFD. *Ocean Engineering*, *137*(March), 22–37. https://doi.org/10.1016/j.oceaneng.2017.03.043

Rajendran, S., Fonseca, N., & Guedes Soares, C. (2016). Prediction of vertical responses of a container ship in abnormal waves. *Ocean Engineering*, *119*, 165–180. https://doi.org/10.1016/j.oceaneng.2016.03.043

Seo, S., Park, S., & Koo, B.Y. (2017). Effect of wave periods on added resistance and motions of a ship in head sea simulations. *Ocean Engineering*, *137*(October 2016), 309–327. https://doi.org/10.1016/j.oceaneng.2017.04.009

Smith, T.W.P., Jalkanen, J.P., Anderson, B.A., Corbett, J.J., Faber, J., Hanayama, S., … Hoen, M., A. (2014). *Third IMO Greenhouse Gas Study 2014*. *International Maritime Organization (IMO)*. https://doi.org/10.1007/s10584-013-0912-3

Tezdogan, T., Demirel, Y.K., Kellett, P., Khorasanchi, M., Incecik, A., & Turan, O. (2015). Full-scale unsteady RANS CFD simulations of ship behaviour and performance in head seas due to slow steaming. *Ocean Engineering*, *97*, 186–206. https://doi.org/10.1016/j.oceaneng.2015.01.011

UNCTAD. (2015). *Unctad Handbook of Statistic*. *United Nations Publication*. https://doi.org/ISBN 978-92-1-012077-7\re-ISBN 978-92-1-056876-0\rISSN 1992-8408

Yan, C., Venayagamoorthy, G.K., & Corzine, K.A. (2011). Optimal location and sizing of energy storage modules for a smart electric ship power system. In *IEEE SSCI 2011—Symposium Series on Computational Intelligence—CIASG 2011: 2011 IEEE Symposium on Computational Intelligence Applications in Smart Grid* (pp. 123–130). https://doi.org/10.1109/CIASG.2011.5953336

Maritime Transportation and Harvesting of Sea Resources – Guedes Soares & Teixeira (Eds)
© *2018 Taylor & Francis Group, London, ISBN 978-0-8153-7993-5*

Modification of traditional catamaran to reduce total resistance: Configuration of centerbulb

S. Samuel & D.-J. Kim
Department of Naval Architecture and Marine Systems Engineering, Pukyong National University, Republic of Korea

M. Iqbal
Department of Naval Architecture, Diponegoro University, Republic of Indonesia

A. Bahatmaka & A.R. Prabowo
Interdisciplinary Program of Marine Design Convergence, Pukyong National University, Republic of Korea

ABSTRACT: Traditional fishermen in Cilacap modify a monohull fishing vessel to be a catamaran. With the same draft, the capacity increases two times while the resistance increases almost four times than original monohull due to low S/L. The wave interference between hulls caused by low S/L increases the wave resistance leading to fuel consumption problem. This paper offers one of the solutions to overcome this problem by placing a centerbulb. The position of centerbulb is configured to get the lowest total resistance. The total resistance was analyzed using empirical formula and solved by a computational fluid dynamics (CFD) software. The results show that centerbulb configuration reduces the total resistance up to 25.76%. This is good news for the fishermen in Cilacap to modify their catamaran by adding centerbulb.

1 INTRODUCTION

Basically, catamaran ship has better resistance compared to monohull ship with the same displacement (capacity) (Setyawan et al. 2010). In Cilacap Region, Central Java—Indonesia, the fishermen modify the monohull vessel into catamaran without considering the S/L ratio and the capacity. With the same draft, the capacity increases two times while the resistance increases almost four times than original monohull due to low S/L. The wave interference between hulls caused by low S/L increases the wave resistance leading to fuel consumption problem (Samuel et al. 2015).

The component of total resistance comprises of normal stress and tangential stress. Normal stress relates to wave making resistance and viscous stress. Tangential stress relates with a viscosity of the fluid. But (Molland et al. 1996) simplify the component of resistance consists of viscous resistance and wave resistance. For catamaran ship, each of component has interference. Therefore, it is reasonable that the total resistance of catamaran in Cilacap increases as a result of wave interference between the hulls (τ) at S/L of 0.2. The value is far higher than at S/L 0.3 and 0.4.

The investigation of the effects of the separation distance between the resistance components was also carried out by (Millward 1992) and (Molland et al. 2004). Jamaluddin *et al.* modify the empirical Molland formula for calculating the viscous resistance of catamaran based on an experimental and numerical study (Jamaluddin et al. 2012). The modified formula considered S/L parameter to calculate the viscous resistance. They also purposed the empirical formula to calculate wave interference between hulls (τ) as used in this research. Meanwhile, number of computational investigations of the wave resistance of multi-hulls have been reported recently in the literature (Zhong & Xiao-ping 2011; Zotti 2007; Larsson et al. 1997).

A centerbulb can be described as a volume between two demihull of a catamaran. It aims to reduce the wave interference resistance by creating a secondary wave interaction (Zotti 2007; Danışman 2014). Other researchers have also been done by (Danışman 2014; Danisman et al. 2001) and his research group who analyzed the resistance with a centerbulb, which its geometry was optimized by artificial neural network (ANN). An efficient centerbulb geometry was obtained from ANN coupled with a computational flow solver based on the low—Froude number theory used a source-panel method following Dawson's algorithm (Dawson 1977). In this study, a CFD—Tdyn was used to determine resistance as suggested by previous researchers (Samuel et al. 2015; Yousefia et al. 2013; Iqbal & Utama 2014).

This research focuses on calculating the total resistance with configurations of centerbulb position to reduce the total of ship resistance based on Jamaluddin's equation and slender body method according to ITTC standard procedure (ITTC 2002).

2 METHODOLOGY

2.1 *Component of resistance for catamaran ship*

There are various methods of studying wake-wash phenomena. (Varyani 2006) describes an example of a full-scale study that is very costly. On the other hand, (Abdul Ghani & Wilson 2009) investigated the phenomena through laboratory experiment. Whereas others, such (Zaghi et al. 2010; Zaghi et al. 2011) used both numerical and experimental methods.

A ship moving on the water with any particular speed experiences resistance on its wetted surface area opposing its movement. Its resistance caused by some factors, i.e. its velocity, wetted surface area, and hull form.

Upon its physical process, that ship resistance on the free surface water consists of two main components, which are normal stress and tangential stress. Normal stress influences the wave resistance and viscous tension. Meanwhile, tangential stress influences the friction resistance.

By substituting the equation for acceleration, tangential stress, and normal stress will result from the equation of fluid dynamic at the moving fluid. This equation is so-called Navier-Stokes's equation (eq.1). That equation is used for an algorithm (Tdyn 2014a).

$$\rho\left[\frac{\partial u}{\partial t} + (u.\nabla)u\right] + \nabla p - \nabla.(\mu\nabla u) = \rho f \qquad (1)$$

where u is the velocity vector u; p is the pressure; ρ is fluid density; μ is the dynamic viscosity fluid coefficient; f is volume acceleration.

(Insel & Molland 1991) formulated the resistance into two components there are viscous resistance and wave resistance. A numeral factor used to determine the coefficient of total resistance (CT_{CAT}) from hull form factor and wave fluid is described by equation 2.

$$CT_{CAT} = (1 + \beta K) CF + \tau CW \qquad (2)$$

where CTCAT is the total resistance coefficient; (1+ βK) is the hull form coefficient; CF is the frictional

resistance coefficient; τ is the wave interference factor and CW is the wave resistance coefficient.

Jamaluddin et al. modified hull form coefficient by considering S/L effect. Than, equation 2 can be written as equation 3 according to slenderness ratio: L/V1/3 = 6–9, L/B = 6–12, B/T = 1–3 and CB = 0.33–0.45.

$$(1 + \beta k) = 3.03 (L/V^{1/3})^{-0.40} + 0.016 (S/L)^{-0.65} \qquad (3)$$

The value of τ is wave interference factor which caused by fluid flow around the catamaran at a certain distance (S) of ship length (L). Details of the derivation of those formulas are given in (Jamaluddin et al. 2012):

$$\tau = 0.068(S/L)^{-1.38}, \text{ at fr} = 0.19 \qquad (4)$$

$$\tau = 0.359(S/L)^{-0.87}, \text{ at fr} = 0.28 \qquad (5)$$

$$\tau = 0.574(S/L)^{-0.33}, \text{ at fr} = 0.37 \qquad (6)$$

$$\tau = 0.790(S/L)^{-0.14}, \text{ at fr} = 0.47 \qquad (7)$$

$$\tau = 0.504(S/L)^{-0.31}, \text{ at fr} = 0.56 \qquad (8)$$

$$\tau = 0.501(S/L)^{-0.18}, \text{ at fr} = 0.65 \qquad (9)$$

Wave resistance (RW) consists of ideal fluid components (invicid), while viscous resistance (RV) consists of friction drag and viscous pressure. Then total resistance can be calculated using equation 10.

$$RT = \frac{1}{2} \rho \, CT \, (WSA) \, V^2 \qquad (10)$$

where RT is the total resistance; CT is the total resistance coefficient; WSA is wetted surface area; V is speed of ship.

The subject of this study is a catamaran fishing vessel of the Cilacap native (Figure 1), MV Laganbar that scaled 1: 10 and S/L = 0.2 (see Table 1).

The geometry used to make the centerbulb is a 3D ellipsoid. Figure 2 describes the geometry of solid centerbulb made by (Danışman 2014).

The equation based on a mathematical function lies on the cartesian coordinate system in equation 11.

$$\frac{x^2}{a} + \frac{y^2}{b} + \frac{z^2}{c} = 1 \qquad (11)$$

The results of an experimental study reported by (Danışman 2014) was able to reduce up to 15% wave resistance and 13% of residual resistance.

The design of centerbulb used in this research was based on the relation of ship principal dimension to centerbulb principal dimension. The

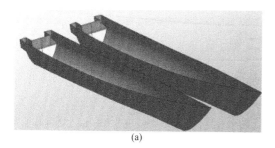

(a)

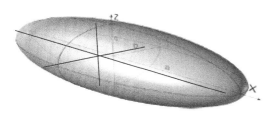

(b)

Figure 1. Hull form original (a) and 3D model of cata-maran vessel (b).

Table 1. Principal dimension of MV Laganbar.

Dimension	Full scale	Ship model
LWL (m)	8.720	0.872
B demihull (m)	1.013	0.101
BOA (m)	2.881	0.288
T (m)	0.500	0.050
WSA (m²)	23.760	0.237
Volume (m³)	4.650	0.004
Displacement (Ton)	4.768	0.004

Figure 2. The geometry of an ellipsoid centerbulb (Danışman 2014).

parameters used were the length of water line (LWL), beam over all (B demihull), and draught (T) as tabulated in Table 2. The configurations of centerbulb positions are fixed in the center line and certain longitudinal and height position are given in Figure 3.

Table 2. Parameters that used to determine geometry of centerbulb.

Dimension	Danisman m	MV. Laganbar m
LWL	25,25	8,72
B demihull	2,6	1,1
T	1,4	0,5
a	1,41	0,5
b	0,62	2,6
c	0,34	0,12

Figure 3. Configuration of centerbulb.

Table 3. Centerbulb coordinates.

Model	a (m)	b (m)	c (m)	x (m)	y (m)	z (m)
1	0.50	0.26	0.12	8.10	0	0.06
2	0.50	0.26	0.12	8.10	0	0.28
3	0.50	0.26	0.12	8.10	0	0.50
4	0.50	0.26	0.12	4.50	0	0.06
5	0.50	0.26	0.12	4.50	0	0.28
6	0.50	0.26	0.12	4.50	0	0.50
7	0.50	0.26	0.12	0.90	0	0.06
8	0.50	0.26	0.12	0.90	0	0.28
9	0.50	0.26	0.12	0.90	0	0.50

2.2 Centerbulb positions

The centerbulb positions that used for this analysis consist of 3 longitudinal positions according to stations position with the x/L = 0.9; x/L = 0.5; x/L = 0.1 along with the x-axis. Then, there are three variation vertical positions according to water line level for each longitudinal position along in z-axis, which is Z/T = 0.12; Z/T = 0.56; Z/T = 1. Therefore, there are no positions on y-axis because it lies in the center line of the ship. Therefore, there are 9 variant models to this analysis given in Figure 4 and Table 3.

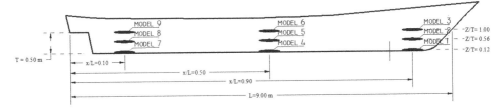

Figure 4. Centerbulb positions.

3 RESULTS AND DISCUSSION

3.1 *Simulation set-up*

The simulation program used in this study was a commercial package of CFD-Tdyn software (Tdyn 2014a; Tdyn 2014b). This software is capable to compute the resistances of a ship by potential flow method, boundary layer method and Reynolds Average Navier–Stokes equations (RANS). The domain is 3 dimensions (Ω) and the calculation interval is time (0,t) in equation 1. The spatial discretization of the Navier-Stokes equations has been conducted by means of the finite element method, while for time discretization, an iterative algorithm that can be considered as an implicit two-step "Fractional Step Method" (Kleinstreuer 1997). Using the standard Galerkin method to discretize the incompressible Navier-stokes equations leads to numerical in stabilities.

The two-equation model for turbulent flows with integration to the wall is expressed in terms of a k-ω model formulation. The k-ω shear-stress-transport (SST) model combines several desirable elements of standard k-ε and k-ω models. The two major features of this model are a zonal weighting of model coefficients and a limitation on the growth of the eddy viscosity in rapidly strained flows. The zonal modeling uses the k-ω model near solid walls and a standard k-ε model near boundary layer edges and in free-shear layers. This switching is achieved with a blending function of the model coefficients. The SST model also modifies the eddy viscosity prediction, improving the prediction of flows with strong adverse pressure gradients and separation. The SST model has also been used in many other studies (Bardina et al. 1997; Menter 1993; Menter 1994).

The final result is total resistance with the force unit shown in the post processor sub menu. The results from Tdyn is combined with equation 12 to determine the total force of this software.

$$u = u_c \text{ in } \Gamma_D \times (0,t)$$
$$\rho = \rho_c \text{ in } \Gamma_P \times (0,t)$$
$$n.\sigma_1 g_1 = 0, n.u = u_m \text{ in } \Gamma_M \times (0,T) \qquad (12)$$
$$u(x,0) = u_0(x) \text{ in } '\Omega_D \times (0)$$
$$\rho(x,0) = \rho_0(x) \text{ in } '\Omega_D \times (0)$$
$$\Gamma = \partial\Omega$$

Table 4. Mesh size of boundary condition.

Boundary condition	Mesh size
Ship	0.010
Free surface	0.050
Wall, inlet, outlet, bottom	0.100
Centerbulb	0.005

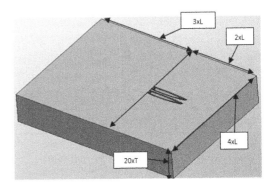

Figure 5. The dimensions of the computational domain for simulation resistance.

where $'\Omega$ is the domain boundary surface; n is the vector unit; g_1 is the tangential boundary surface vector u_c is the velocity field in Γ_D; Γ_D is the boundary surface in velocity field; Γ_P is the boundary surface in the pressure field; Γ_M is the boundary surface in free surface field.

The total force was calculated in every element with the Finite Element Method (FEM) around the boundary layer area (Figure 6).

The boundary layer was built on rectangular form with the dimensions 5L $\times$ 2 L $\times$ 20T. The mesh size for each boundary condition is shown in Table 4 and Figure 5.

3.2 *The results of simulation*

The total resistance of the original model of catamaran was calculated with CFD and empirical equation (Eq. 2–9). The results of resistance calculation from CFD were compared to results of

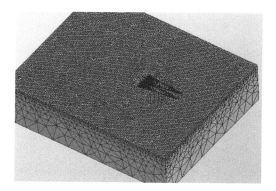

Figure 6. Computational grid.

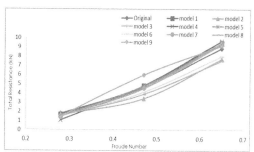

Figure 7. Results of total resistance.

Table 5. Resistance validation.

Fr	V (m/s)	$RT_{empiric}$ (kN)	RT_{CFD} (kN)	Difference (%)
0.28	2.62	1.81	1.76	2.84
0.47	4.40	4.82	4.58	4.98
0.65	6.09	8.84	8.80	0.50

Table 6. Total resistance comparative of original and the best model.

Fr	V (m/s)	$RT_{original}$ (kN)	$RT_{model 2}$ (kN)	Difference (%)
0.28	2.62	1.76	1.67	5.11
0.47	4.40	4.58	3.40	25.76
0.65	6.09	8.80	7.70	12.50

Table 7. Total resistance.

Model	Fr 0.28 (kN)	Fr 0.47 (kN)	Fr 0.65 (kN)
Original	1.76	4.58	8.80
1	1.65	4.77	9.45
2	1.67	3.40	7.70
3	1.47	4.48	9.30
4	1.08	4.80	9.67
5	1.56	4.76	9.40
6	1.52	4.10	7.99
7	1.47	4.58	9.46
8	1.52	3.88	7.50
9	1.50	5.96	9.16

resistance from empirical equation. The comparations were used for make sure that the setup and mesh size of CFD was valid and can be used to calculated resistance of catamaran with centerbulb. The difference between CFD & empirical result was quite small (below 5%) as shown in Table 5. The results of CFD calculation for catamaran with centerbulb configuration are shown in Table 6.

The results of Model 2, 6 and 8 are below the original model which are better because of lower the total resistance. Meanwhile, the other model which are other models has a higher value of total resistance compared to the initial model. Model 2 is selected as the best model in this research because has the lowest resistance than another model at an operational speed of catamaran (Figure 7 and Table 7).

The comparisons of original and the best model are shown in Table 6. The best model can reduce up to 25.76% of total resistance at Fr 0.47. This is a good news for fishermen to modify their catamaran by adding centerbulb.

4 CONCLUSIONS

The configurations of centerbulb position on catamaran to reduce the total resistance have been done. The research that uses CFD software Tdyn is good enough to develop other research. The result of CFD software is close to the empirical equation with the different result below 5%. Based on this research, the use of centerbulb in catamaran can reduce the total resistance up to 25.76% at Fr 0.47 where the centerbulb position at the fore and in the center between baseline and draft.

ACKNOWLEDGEMENTS

This research is fully supported by LPDP (Lembaga Pengelola Dana Pendidikan) and Diponegoro University.

REFERENCES

Abdul Ghani, M. & Wilson, P., 2009. Experimental analysis of catamaran forms with bulbous bows operating in shallow water. *International Shipbuilding Progress Journals*, 56(1), pp. 29–57.

Bardina, J.E., Huang, P.G. & Coakley, T.J., 1997. *Turbulence modeling validation, testing, and development*, California: Ames research center.

Danisman, D.B. et al., 2001. An optimasi Study for the bow form of high speed displacement catamaran. *Marine technology*, 3(2), pp.116–121.

Danışman, D.B., 2014. Reduction of demi-hull wave interference resistance in fast displacement catamarans utilizing an optimized centerbulb concept. *Ocean Engineering*, 91, pp.227–234.

Dawson, C.W., 1977. A practical computer method for solving ship–wave problems. In *Proceedings of the 2nd International conference on numerical ship hydrodynamics*. Berkeley.

Insel, M. & Molland, A.F., 1991. An investigation into the resistance components of high speed displacement catamarans. *The Royal Institution of Naval Architects*, 134, pp.1–11.

Iqbal, M. & Utama, I.K.A.P., 2014. An Investigation into the effect of water depth on the resistance components of trimaran configuration. In *Proc. The 9th International Conference on Marine Technology*. Surabaya.

ITTC, 2002. Report of the Resistance Commetee. In *Proceedings of the 23rd International Towing Tank Conference*. Vinece, Italy: INSEAN.

Jamaluddin, A. et al., 2012. Experimental and numerical study of the resistance component interactions of catamarans. *Proceedings of the Institution of Mechanical Engineers, Part M: Journal of Engineering for the Maritime Environment*, 227(1), pp.51–60.

Kleinstreuer, C., 1997. *Engineering Fluid Dynamics, An Interdisciplinary Systems Approach*, Cambridge, UK: Cambridge University Press.

Larsson, L., Janson, C.. & Brun, F., 1997. A numerical investigation of trimaran configurations. In *Proceedings of Fast '97*. Sydney.

Menter, F.R., 1994. Two-equation eddy-viscosity turbulence models for engineering applications. *AIAA Journal*, 32(8), pp.1598–1605.

Menter, F.R., 1993. Zonal two equation k-ω turbulence models for flows. *AIAA Journal*, pp.93–2906.

Millward, A., 1992. The effect of hull separation and restricted water depth on catamaran resistance. *Transactions of Royal Institute of Naval Architects*, 134, pp.341–349.

Molland, A.F. et al., 2004. Resistance and wash wave measurements on a series of high speed displacement monohull and catamaran forms in shallow water. *The International Journal of Maritime Engineering*, 146(a2), pp.19–38.

Molland, A.F., Wellicome, J.F. & Couser, P.R., 1996. Resistance experiments on a systematic series of high-speed displacement catamaran forms: variations of length-displacement ratio and breadth-draugh ratio. *Transaction RINA*, 138A.

Samuel, Iqbal, M. & Utama, I.K.A.P., 2015. An investigation into the resistance components of coverting a traditional monohull fishing vessel into catamaran form. *International Journal of Technology*, 6(3), pp.432–441.

Setyawan, D. et al., 2010. Development of catamaran fishing vessel. *The Journal for Technology and Science*, 21(4), pp.2–9.

Tdyn, 2014a. Tdyn. In *Tdyn Theory Manual*. Barcelona, Spain, pp. 1–33. Available at: http://www.compassis.com/downloads/Manuals/TdynTheory.pdf.

Tdyn, 2014b. Tdyn CFD+HT. In *Turbulence Handbook*. Barcelona, Spain, pp. 1–39. Available at: http://www.compassis.com/downloads/Manuals/TdynTurbulenceHB.pdf.

Varyani, K., 2006. Full scale study of the wash of high speed craft. *Ocean Engineering Journal*, 33, pp.705–722.

Yousefia, R., Shafaghata, R. & Shakerib, M., 2013. Hydrodynamic analysis techniques for high-speed planing hulls. *Appl. Ocean Res*, 42, pp.105–113.

Zaghi, S., Broglia, R. & Mascio, A., 2010. Experimental and Numerical Investigations on Fast Catamarans Interference Effects. *Journal of Hydrodynamics*, 22(5), pp.545–549.

Zaghi, S., Broglia, R. & Di Mascio, A., 2011. Analysis of the interference effects for high-speed catamarans by model tests and numerical simulations. *Ocean Engineering*, 38(17–18), pp.2110–2122.

Zhong, W. & Xiao-ping, L.U., 2011. Numerical simulation of wave resistance of trimarans by nonlinear wave making theory with sinking and trim being taken into account. *Journal of Hydrodynamics*, 23(2), pp.224–233.

Zotti, I., 2007. Medium speed catamaran with large central bulbs: experimental investigation on resistance and vertical motions. In *Proceedings of ICMRT'07*. Ischia, Naples, Italy, pp. 167–174.

Maritime Transportation and Harvesting of Sea Resources – Guedes Soares & Teixeira (Eds)
© 2018 Taylor & Francis Group, London, ISBN 978-0-8153-7993-5

Cavitation analyses of DTMB 4119 propeller with LSM and URANS approach

S. Sezen
Yildiz Technical University, Istanbul, Turkey

S. Bal
Istanbul Technical University, Istanbul, Turkey

ABSTRACT: Cavitation is a very important physical phenomenon to be investigated for hydrodynamic performance of marine propellers. This paper focuses on the numerical analysis of a cavitating propeller (DTMB 4119) in open water conditions. The flow around the propeller has been simulated for different cavitation numbers with both a potential based code and unsteady Reynolds-Averaged Navier Stokes (RANS) solver. The RANS solver is based on finite volume method while the potential code is using the lifting surface method. The Lifting Surface Method (LSM) models the three-dimensional unsteady cavitating flow around a propeller by representing the blade and wake as a discrete set of vortices and sources. For RANS analyses, the flow is considered as 3-D, fully turbulent, incompressible. Cavitation phenomenon has been modeled by using Volume of Fluid (VOF) technique to represent the two phases of liquid and vapor. Cavitation effect on the hydrodynamic performance of marine propeller has been investigated.

1 INTRODUCTION

The cavitation on propellers is one of the major problems encountered commonly in marine vessels. It affects the hydrodynamic performance of the marine propellers significantly. The practical importance of numerical simulations to predict the propeller cavitation pattern has been increased very much in recent years for marine industry. When cavitation occurs, the water liquid changes its phase from liquid into vapor at certain flow region where local pressure is very low because of the high local velocities. The high velocity gradients cause great challenge for prediction of cavitation numerically. Prediction techniques for cavitation on marine propellers have been classified as discrete bubble dynamics, interface tracking and interface capturing methods by 26th ITTC CFD committee (26th ITTC 2011).

Marine propellers can display different forms of cavitation such as propeller-tip vortex cavitation, propeller hub cavitation, bubble cavitation and sheet cavitation. The sheet type of propeller cavitation has also been simulated using Boundary Element Method (BEM) approaches (Kinnas et al. 2005a, Kinnas 1998b, Kinnas 1989c) in the past. In addition, viscous flow solvers such as Reynolds-Averaged Navier Stokes (RANS), Large Eddy Simulation (LES) have been widely used in very recent years (Morgut et al. 2015, Turunen

et al. 2014 and Salvatore et al. 2009). One of the main aims of a marine propeller design is to avoid from cavitation or reduce the effect of cavitation. In the past, Kerwin and Lee have studied on the prediction of steady and unsteady marine propeller hydrodynamic performance using numerical lifting-surface theory. The main aim of the study was developing a practical tool for prediction of hydrodynamic performance of propeller both for steady and unsteady flow problems (Kerwin and Lee 1978). On the other hand, Mishima and Kinnas have developed a numerical nonlinear optimization algorithm for automated and systematic design of propeller blades under cavitation conditions. First, the non-cavitating propeller blade geometries have been designed and directly compared with the LSM design approach. Later, the developed optimization algorithm has been applied to design propeller blade geometries under non uniform flow conditions. The main aim of the study was to obtain maximum propeller efficiency for given cavity conditions Mishima and Kinnas (1997). Kuiper and Jessup have developed a design method for non-cavitating propeller under non-uniform flow conditions. The blade sections have been specially designed for obtaining the systematic propeller series. The validation study has been done with model test results (Kuiper and Jessup 1993). Lloyd et al. have investigated Potsdam Propeller Test Case (PPTC) in oblique flow under non-cavitating and cavitating conditions using computational fluid dynamics code (ReFRESCO). All analyses have been done in an unsteady manner. Three different unstructured grids have been implemented in numerical simulations to compare the propeller hydrodynamic performance. The pressure pulses have been predicted inside the cavitation tunnel measurements. The results were in a good agreement with the experimental ones (Lloyd et al. 2015). Perali et al. have studied the prediction of pressure pulses of E779A propeller in a cavitation tunnel. The main aim of the study was to compare the viscous flow codes and potential flow tools in terms of propeller pressure pulses. The experimental results have been compared with RANS and BEM calculations (Perali et al. 2016). Salvatore et al. have studied propeller hydrodynamic performance at model scale via inviscid/viscous flow coupled methodology. The proposed methodology has been based on nonlinear boundary integral formulation. Unsteady sheet cavitation of a marine propeller has been investigated (Salvatore et al. 2003). Baltazar and Campos have developed an iteratively coupled solution method based on potential flow theory for the propeller cavitation. The numerical calculations have been examined for MARIN S-Propeller and INSEAN E779A propeller (Baltazar and Campos 2010). Bensow et al. have used different computational tools for propeller cavitation analysis. It has been stated that potential flow codes can demonstrate the cavitation effects properly on propeller hydrodynamic performance. Several computational tools such as LES, Euler methods have also been developed to capture the different cavitation dynamics for marine propellers (Bensow et al. 2012). Gaggero et al. have investigated the cavitation characteristics of the CP CLT propeller using in-house panel code, a commercial RANS solver (Star CCM+) and an open source RANS code (OpenFOAM). Steady and unsteady cavity extensions have been investigated with different numerical tools and the differences have been presented in terms of capturing the cavitation dynamics on propeller blades. The results have been validated with the available experimental data which have been conducted in cavitation tunnel of the University of Genoa (Gaggero et al. 2012). Sulaiman et al. have studied on cavitation flows around two different propellers using a viscous CFD approach. Analyses have been conducted to observe the cavitation patterns on the selected radii. Various parameters such as angle of attack, sea water viscosity, vapor pressure and lift-drag vectors of each blade sections have been investigated during the analyses. The results have been discussed within the theory (Sulaiman et al. 2014). Subhas et al. have studied the open water characteristics of propeller under cavitating and non-cavitating conditions using a commercial computational fluid dynamics code (Fluent

6.3). The hydrodynamic performance of the propeller has been compared for both cavitating and non-cavitating conditions at a fixed advance ratio. The results have indicated that the propeller thrust coefficient can decrease while the propeller torque coefficient can increase in comparison with cavitating and non-cavitating conditions at a constant advance ratio (Subhas et al. 2012). Dymarski has investigated open water propeller characteristics and cavitation on propeller blades using SOLAGA program based on travelling bubble model. Open water characteristics of conventional and skewed propellers have been calculated (Dymarski 2008). Fujiyama et al. have examined the hydrodynamic performance and cavitation patterns of the propeller using SC/Tetra CFD code based on finite volume method (FVM). Numerical simulations have been performed both under non-cavitating and cavitating conditions. The results showed that SC/Tetra RANS-based code is an effective tool to obtain the cavity dynamics on marine propellers (Fujiyama et al. 2011). Kınacı et al. have studied on the propeller ship interaction using commercial CFD program for DTC-Post Panamax Container Ship. The ship has been analyzed without taking free surface effect into account. The propeller hydrodynamic performance has also been calculated using potential based code and RANS solver (Kınacı et al. 2013). Bauer and Maksoud (2012) have tried to predict the operational loads induced by marine propellers. The numerical method has been based on potential theory. The three dimensional first order panel method has been implemented. The cavitation has been adopted in the program in order to include the influence of sheet cavitation effects on hydrodynamic performance of the propeller (Bauer and Maksoud 2012). Shin et al. have investigated the cavitating flow around the conventional and highly-skewed propellers in the behind-hull condition. The simulation has been done using in-house RANS solver based on the homogeneous equilibrium modeling. The hydrodynamic results have been validated with experimental data under cavitating and non-cavitating conditions. The cavity patterns have been compared with the cavitation tunnel test results (Shin et al. 2011). Brizzorala et al. have investigated the five different propeller hydrodynamic performance using a commercial RANSE solver and in-house panel code. The numerical results have been compared each other not only non-dimensional thrust and torque values but also the friction coefficient and pressure distributions on propeller blades for over a wide range of advance ratios (Brizzorala et al. 2008).

In this paper, numerical analyses have been conducted for prediction of hydrodynamic performance of the well-known benchmark DTMB 4119 model propeller with both a potential based code and RANS solver. The flow around the propeller has been simulated in a steady manner under non-cavitating conditions with both methods. The hydrodynamic performance characteristics of DTMB 4119 model propeller have been validated with the experimental results. Later, the hydrodynamic performance of the model propeller has been investigated for two different cavitation numbers using a potential based code and unsteady RANS solver under uniform flow conditions. The non-dimensional thrust, torque coefficients, propeller efficiency and cavity patterns on the propeller blades have been compared by both methods.

2 MATHEMATICAL MODEL

2.1 *Unsteady RANS approach*

For the numerical analyses, the governing equations are the continuity equation and the well-known RANS equations for the unsteady, three-dimensional, incompressible flow. The continuity can be given as;

$$\frac{\partial U_i}{\partial x_i} = 0 \tag{1}$$

While the momentum equations are expressed as;

$$\frac{\partial U}{\partial t} + \frac{\partial (U_i U_j)}{\partial x_j} = -\frac{1}{\rho}\frac{\partial P}{\partial x_i} + \frac{\partial}{\partial x_j}\left[\nu\left(\frac{\partial U_i}{\partial x_j} + \frac{\partial U_j}{\partial x_i}\right)\right] \\ -\frac{\partial \overline{u_i' u_j'}}{\partial x_j} \tag{2}$$

In momentum equations, U_i states the mean velocity while u_i' states the fluctuation velocity components in the direction of the Cartesian coordinate. P expresses the mean pressure, ρ the density and ν the kinematic viscosity.

Shear Stress Transport (SST) k-ω model has been used for turbulence modeling. In propeller hydrodynamics, both k-ε and k-ω turbulence models can be used. The k-ω turbulence model is composed of two equations as linear eddy viscosity equation and RANS equation. The SST k-ω turbulence model is a hybrid model combining k-ω and k-ε models. Specific dissipation rate has been calculated as the dissipation rate per unit turbulent kinetic energy:

$$\omega = \frac{\varepsilon}{k} \tag{3}$$

Detailed information about the turbulence model has been found in (Wilcox 2006).

2.2 *Lifting surface method*

A lifting surface method has been applied to calculate the propulsive performance due to the propeller, similar to the one given in (Bal 2011a and Bal 2011b). The lifting surface method (propeller analysis) models the three-dimensional unsteady cavitating flow around a propeller by representing the blade and wake as a discrete set of vortices and sources, which are conveniently located on the blade mean camber surface and wake surface. In particular, the three components of the discretization are as follows:

i. A vortex lattice on the blade mean camber surface and wake surface to represent the blade loading and trailing vorticity in the wake.
ii. A source lattice on the blade mean camber surface to represent blade thickness.
iii. A source lattice throughout the cavity extent to represent cavity thickness.

The sources representing blade thickness are line sources along the span-wise direction. The strengths of the line sources are given in terms of derivatives of the thickness in the chord-wise direction and are independent of time. The unknown bound vortices on the blade and the unknown cavity sources are determined by applying the kinematic boundary condition and the dynamic boundary condition.

In this method, a discretized version of the kinematic boundary condition can be employed as:

$$\sum_\Gamma \Gamma \vec{v}_\Gamma . \vec{n}_m = -\vec{v}_{in} . \vec{n}_m - \sum_{Q_B} Q_B \vec{v}_Q . \vec{n}_m - \sum_{Q_C} Q_C \vec{v}_Q . \vec{n}_m \tag{4}$$

where $\vec{v}_\Gamma$ the velocity vector is induced by each unit strength vortex element, $\vec{v}_Q$ is the velocity vector induced by each unit strength source element, and $\vec{n}_m$ is the unit vector normal to the mean camber line or trailing wake surface. Q_B and Q_C represent the magnitude of the line sources that model the blade thickness and cavity source strengths, respectively. The kinematic boundary condition must be satisfied at certain control points located on the blade mean camber surface. The kinematic boundary condition requires that the sum of the influences for all of the vortices' sources and the inflow normal to a particular control point on the blade is equal to zero. Another way to explain this is that the kinematic boundary condition requires the flow to be only tangential to the surface. Other assumptions employed throughout the method include:

i. The cavity thickness varies linearly across panels in the chord-wise direction and is piecewise constant across panels in the span-wise direction.
ii. There are no span-wise flow effects in the cavity closure condition.

iii. Viscous force is calculated by applying a uniform frictional drag coefficient.

The details of the method can be found in (Kerwin, Handler 2010 and Bal, Güner 2009).

On the other hand the non-dimensional cavitation number has been calculated via equation given below;

$$\sigma = \frac{P_{atm} + \rho gh - P_v}{0.5\rho(nD)^2} \qquad (5)$$

Here, n is propeller rotating speed, D is propeller diameter, ρ is fluid density and h is operation depth.

3 VALIDATION AND NUMERICAL RESULTS

3.1 *Non-cavitating DTMB 4119 propeller*

DTMB 4119 model propeller has been chosen for investigation of hydrodynamic performance under non-cavitating and cavitating conditions. Geometrical properties of the model propeller have been given in Table 1. 3-D view of the DTMB4119 model propeller has been shown in Figure 1.

For CFD analyses, the flow is considered as 3-D, fully turbulent, incompressible and steady for non-cavitating conditions. The turbulence model has been chosen as SST k-ω turbulence model. The pressure field has been solved by using SIMPLE algorithm which is based on pressure-velocity coupling. The reference axis has been chosen as moving reference frame (MRF) for steady analyses (Sezen 2016). The computational domain consists of rotating and static regions as given in Figure 2. The inlet and outlet surfaces have been defined as velocity inlet and pressure outlet, respectively. The common surfaces between the static and rotating regions have been defined as interfaces. The surfaces far from the propeller have been considered as symmetry planes in order to force the normal component of the velocity as zero on these surfaces. The propeller blades and shaft have been defined as non-slip wall for ensuring the no-slip condition. Detailed information about the boundary conditions applied in the analyses can be found in the theory guide of the commercial CFD solver (Star CCM+ 2015).

Unstructured mesh has been utilized to discretize the computational domain by three dimensional finite volumes in compliance with the finite volume method. Mesh refinements around the propeller blades have been made to calculate the pressure and velocity fields precisely. The unstructured mesh on the propeller surface has been given in Figure 3.

Table 1. Main particulars of DTMB 4119 propeller.

DTMB 4119 model propeller	
D (m)	0.3048
Z	3
Skew (°)	0
Rake (°)	0
Blade section	NACA66 a = 0.8
Rotation direction	Right

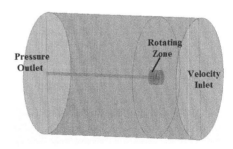

Figure 2. Boundary conditions of computational domain.

Figure 1. 3-D model of DTMB 4119.

Figure 3. Unstructured mesh on propeller surface.

The flow around the model propeller has also been investigated using lifting surface method (LSM) based on potential theory. The hydrodynamic performance of the propeller has been calculated for different advance ratios under non-cavitating and cavitating conditions. The panel distributions on the propeller have been given in Figure 4. The wake region behind the propeller has also been modeled.

The hydrodynamic performance of DTMB 4119 propeller has been calculated using both methods for different advance ratios under non-cavitating conditions. The results have been validated with the experimental ones in terms of non-dimensional thrust, torque and propeller efficiency.

As can be seen from Figure 5, the numerical results (LSM and RANS) are in good agreement

with the experimental results. RANS approach gives better results for open water propeller efficiency when compared with LSM code.

3.2 Cavitating DTMB 4119 propeller

The hydrodynamic performance of DTMB 4119 propeller has been analyzed at two different cavitation numbers using both methods in open water conditions. Viscous calculations of cavitation have been carried out with Volume of Fluid approach solving RANS equations. Rigid body motion (RBM) model has been chosen to designate the reference axis. RBM model is used to capture the non-homogeneity of the incoming flow.

Cavitation number which is given in eqn. (5) has been calculated by considering the operation depth of the propeller (h) is 5 meters while the saturation pressure (P_V) is taken as 2000.7 Pa. Atmospheric pressure has been taken as 101325 Pa. The operating conditions have been briefly described in Table 2.

The hydrodynamic performance of the propeller has been calculated for two different advance ratios under non-cavitating and cavitating conditions. Table 3 summarizes the results for the performance of the propeller in cavitating and non-cavitating conditions using both numerical methods.

As it can be seen in Table 3, cavitation numbers for these conditions are $\sigma = 3.3$ and $\sigma = 2.5$, respectively. There are high loadings on the propellers due to small advance coefficients. The non-dimensional thrust and torque values computed by LSM have been calculated less than those of RANS solver.

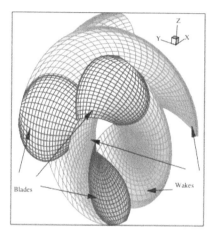

Figure 4. Perspective view of DTMB 4119 propeller and its wake.

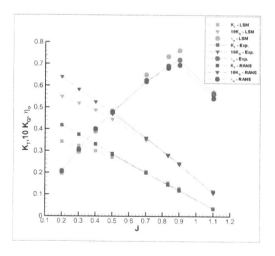

Figure 5. Comparison of thrust, torque and efficiency of DTMB 4119 propeller.

Table 2. Operating conditions of cavitating DTMB 4119 propeller.

Case	J	σ	V_A (m/sn)	n (rpm)
1	0.20	2.5	2.196	2146
2	0.23	3.3	2.196	1850

Table 3. Comparison of hydrodynamic results for non-cavitating and cavitating conditions.

Method	Cavity	J	σ	K_T	$10*K_Q$	η
RANS	No	0.20	∞	0.421	0.640	0.209
LSM	No	0.20	∞	0.343	0.550	0.198
RANS	No	0.23	∞	0.409	0.621	0.241
LSM	No	0.23	∞	0.340	0.532	0.233
URANS	Yes	0.20	2.5	0.410	0.666	0.195
LSM	Yes	0.20	2.5	0.350	0.570	0.195
URANS	Yes	0.23	3.3	0.408	0.649	0.230
LSM	Yes	0.23	3.3	0.340	0.540	0.230

The circulation distributions computed by LSM on the propeller blades have also been given in Figure 6 and Figure 7 for two different advance ratios under cavitating and non-cavitating conditions. As expected, the blade loading is increasing with a decrease in cavitation number.

The cavitation patterns on the propeller blades have been shown using both methods for two different advance ratios in Figure 8 and Figure 9, respectively. The cavity lengths and area of the RANS solver has been found relatively lower than those of LSM method.

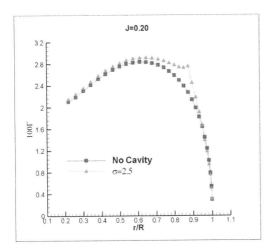

Figure 6. Circulation distributions on the propeller blades for cavitating and non-cavitating conditions at J = 0.20.

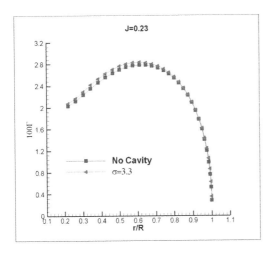

Figure 7. Circulation distributions on the propeller blades for cavitating and non-cavitating conditions at J = 0.23.

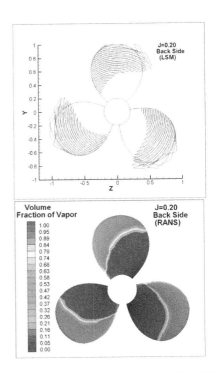

Figure 8. Comparison of cavity pattern with LSM (top) and RANS (bottom) at J = 0.20, σ = 2.5.

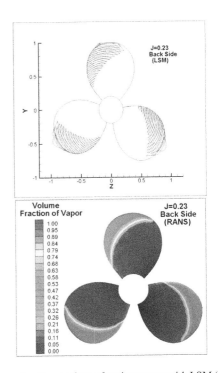

Figure 9. Comparison of cavity pattern with LSM (top) and RANS (bottom) at J = 0.23, σ = 3.3.

From Figures 8 and 9, it is clear that the cavity lengths and area on the blades have been found to be extended with an increase in blade loading. It is noted that the super-cavity has not been shown in RANS solver though it has been given in LSM method (see Figure 8). In Table 3, the thrust, torque and efficiency values of the propeller have been investigated with two methods under cavitating conditions for two different advance ratios. It is found that an increase in cavity area and lengths causes a slight decrease in efficiency for these conditions.

4 CONCLUSIONS

In this study, the hydrodynamic performance of the DTMB 4119 propeller has been investigated with a potential based code and RANS solver. Firstly, the flow around the propeller has been simulated under non-cavitating conditions with both methods. The results have been compared with the available experimental data. The hydro-dynamic performance of the propeller has then been simulated with both methods under cavitating conditions for two different advance ratios. The coefficients of thrust, torque and propeller efficiency have been compared with two methods under cavitating and non-cavitating conditions. In addition, the cavity patterns of the propeller have been compared with both methods. The results show that the potential based code (LSM) has predicted the propeller cavity lengths and area slightly higher than those of RANS solver. When the loading on the blade increases, the differences between the results of LSM and RANS methods may be slightly larger. It should also be mentioned that although the results of RANS method under higher loading conditions are promising, LSM method works very quick and efficient. Although these two CFD methods are quite different, potential based methods are cost efficient and also faster than RANS methods under cavitating and non-cavitating conditions.

NOMENCLATURE

U_i	mean velocity
u_i'	fluctuation velocity component
P	mean pressure
$\bar{v}_T$	velocity vector by unit strength vortex element
$\bar{v}_Q$	velocity vector by unit strength source element
Q_B	line source for blade thickness source strength
Q_C	line source for cavity source strength
$\bar{n}_m$	unit vector normal
Γ	circulation
Z	number of blades

σ	cavitation number
P_{atm}	atmospheric pressure
h	operating depth of the propeller
ρ	fluid density
g	gravitational acceleration
P_v	saturation pressure
n	propeller rotation speed
D	propeller diameter
J	advance ratio
V_A	the mean flow velocity at the propeller plane

ACKNOWLEDGEMENTS

We would like to thank Asst. Prof. Ali Dogrul and Mr. Ahmet Yurtseven from Yıldız Technical University for their valuable support throughout this study.

REFERENCES

Bal, S. 2011. A Method for Optimum Cavitating Ship Propellers. *Turkish Journal of Engineering and Environmental Sciences, Vol. 35, pp. 139–158.*

Bal, S. 2011. Practical Technique for Improvement of Open Water Propeller Performance. *Proceedings of the Institution of Mechanical Engineers, Part M, Journal of Engineering for the Maritime Environment, Vol. 225, Issue 4, pp: 375–386.*

Bal, S. Güner, M. 2009. Performance Analysis of Podded Propulsors. *Ocean Engineering, Vol. 36, pp: 556–563.*

Baltazar, J. and Campos, J.A.C. 2010. An iteratively coupled solution of the cavitating flow on marine propellers using BEM. *9th International Conference on Hydrodynamics, October 11–15, Shanghai, China.*

Bauer, M. and Maksoud, A.M. 2012. A 3-D Potential Based Boundary Element Method for Modelling and Simulation of Marine Propeller Flows. *IFAC Proceedings Volumes, Vol. 45, Issue 2, Pages 1179–1184.*

Bensow, R.E, Bark, G., Lu, N.X., Eskilsson, C. and Vesting, F. 2012. Computational Tools for Propeller Cavitation Analysis. *15th Numerical Towing Tank Symposium, October.*

Brizzolara, S., Villa, D. and Gaggero, S. 2008. A Systematic Comparison between RANS and Panel Methods for Propeller Analysis. *Proc. of 8th International Conference on Hydrodynamics, Nantes, France, Sep. 30–Oct. 3.*

Dymarski, P. 2008. Computations of the propeller open water characteristics using the SOLAGA computer program. Prediction of the cavitation phenomenon. *Archives of Civil and Mechanical Engineering, Vol. 8, Issue 1, Pages 15–25.*

Fujiyama, K., Kim, C.H. and Hitomi, D. 2011. Performance and Cavitation Evaluation of Marine Propeller using Numerical Simulations. *Second International Symposium on Marine Propulsors SMP'11, Hamburg, Germany, June.*

Gaggero, S., Viviani, M., Villa, D., Bertetta, D., Vaccaro, C. and Brizzolara, S. 2012. Numerical and Experimental Analysis of a CLT Propeller Cavitation

Behavior. *Proceedings of the 8th International Symposium on Cavitation, CAV2012, August 14–16, Singapore.*

ITTC 2011. Proceedings of 26th International Towing Tank Conference (ITTC). *The Specialist Committee on Computational Fluid Dynamics.*

Kerwin, J.E. and Hadler, J.B. 2010. The Principles of Naval Architecture Series-Propulsion. *Published by The Society of Naval Architects and Marine Engineers, New Jersey.*

Kerwin, J.E. and Lee, C.S. 1978. Prediction of Steady and Unsteady Marine Propeller Performance by Numerical Lifting Surface Theory. *Trans. SNAME, 86, 218–253.*

Kinaci, O.K., Kukner, A. and Bal, S. 2013. On Propeller Performance of DTC Post-Panamax Container Ship. *International Journal of Ocean System Engineering, Vol. 3, No:2, pp: 77–89.*

Kinnas, S.A., 1998. The prediction of unsteady sheet cavitation. *In: Proceedings of the Third International Symposium on Cavitation. pp. 16–36.*

Kinnas, S.A., Fine, N.E. 1989. Theoretical prediction of the midchord and face unsteady propeller sheet cavitation. *In: Proceedings of the Fifth International Conference on Numerical Ship Hydrodynamics. Hiroshima, Japan, pp. 685–700.*

Kinnas, S.A., Lee, H.S., Gu, H., and Deng, Y. 2005. Prediction of performance and design via optimization of ducted propellers subject to non-axisymmetric inflows. *Trans. SNAME, 113, 99–121.*

Kuiper, G. and Jessup, S.D. 1993. A Propeller Design Method for Unsteady Conditions. *Trans. SNAME, 101, 247–273.*

Lloyd, T., Vaz, G., Rijpkema, D. and Schuiling, B. 2015. The Potsdam Propeller Test Case in oblique flow: prediction of propeller performance, cavitation patterns and pressure pulses. *Second International Workshop on Cavitating Propeller Performance, Austin, Texas, 4th June.*

Mishima, S. and Kinnas, S.A. 1997. Application of a Numerical Optimization Technique to the Design of Cavitating Propellers in Non-Uniform Flow. *Journal of Ship Research, 41, 93–107.*

Morgut, M., Jost, D., Nobile, E. and Škerlavaj, A. 2015. Numerical investigations of a cavitating propeller in non-uniform inflow. *Fourth International Symposium on Marine Propulsors smp'15, Austix, Texas, USA, June.*

Perali, P., Vaz, G. and Lloyd, T. 2016. Comparison of URANS and BEM-BEM for propeller pressure pulse prediction: E779 A propeller in a cavitation tunnel. *NuTTS, France, September.*

Salvatore, F., Streckwall, H. and Terwisga, T. 2009. Propeller Cavitation Modelling by CFD-Results from the Virtue 2008 Rome Workshop. *First International Symposium on Marine Propulsors smp'09, Trondheim, Norway, June.*

Salvatore, F., Testa, C. and Greco. L. 2003. A viscous/inviscid coupled formulation for unsteady sheet cavitation modelling of marine propellers. *5th International Symposium on Cavitation (CAV2003), Osaka, Japan, November 1–4.*

Sezen, S. 2016. Numerical Investigation of Ship Propeller Hydro-Acoustics Performance. *MSc Thesis, Istanbul Technical University, Istanbul.*

Shin, K.W., Andersen, P. and Mikkelsen, R. 2011. Cavitation Simulation on Conventional and Highly-Skewed Propellers in the Behind-Hull Condition. *Second International Symposium on Marine Propulsors smp'11, Hamburg, Germany, June.*

Star CCM+ Documentation 2015, *CD-adapco.*

Subhas, S., Saji, V.F., Ramakrishna, S. and Das, H.N. 2012. CFD Analysis of a Propeller Flow and Cavitation. *International Journal of Computer Applications (0975–8887), Vol.55- No.16, October.*

Sulaiman, O.O, Nick, W.B.W. and Saharuddin, A.H. 2014. CFD Simulation for Cavitation of Propeller Blade. *Global Journal for Information Technology and Computer Science, Vol.1, Issue.1, January.*

Turunen, T., Siikonen, T., Lundberg, J. and Bensow, R. 2014. Open-Water Computations of a Marine Propeller Using OpenFoam. *11th World Congress on Computational Mechanics, July.*

Wilcox, D.C. 2006. Turbulence Modeling for CFD. *3rd edition. La Cãnada, Calif.: D C W Industries.*

Hydrodynamics – manoeuvring

Maritime Transportation and Harvesting of Sea Resources – Guedes Soares & Teixeira (Eds)
© 2018 Taylor & Francis Group, London, ISBN 978-0-8153-7993-5

Innovative systems improving safety of manoeuvring and berthing operations of inland vessels

T. Abramowicz-Gerigk & Z. Burciu
Department of Ship Operation, Gdynia Maritime University, Gdynia, Poland

ABSTRACT: The paper presents two aspects of safety of inland vessels manoeuvring operations. The most important problem is vessel manoeuvrability in confined waters, mainly narrow rivers and channels in close proximity to other manoeuvring or moored vessels and recreational boats. The second aspect is safety of berthing in shallow water conditions including scouring effects of propeller jets. Safety of berthing in shallow waters was investigated to determine the necessary lateral forces and yawing moments for the safe performance of berthing operations, determination of scoring, determination of propeller jet loads induced on the quay wall and monitoring system for propeller jet induced loads are presented.

1 INTRODUCTION

The waterborne inland transport in Poland has big reserves and its development is best placed in the frames of UE sustainable development policy with respect to the EC guidelines supporting sustainable transport system and its evolution to less energy consuming more ecological and safer forms.

The condition of waterways mainly due to many years insufficient or neglected river training and maintenance allows adapting the waterways to navigation in limited range. A big part of inland waterways became wildlife areas and they are now placed in Natura 2000 network. The development of waterborne inland transport is therefore closely related with the development of ecological means of transport adapted to the difficult navigational conditions.

From another side the introduction of regular navigation can decrease the amount of required maintenance works, especially dredging and winter ice breaking works necessary for flood protection. A new generation of inland vessels—new designs of river vessels, barges, push trains along with traffic management systems and automatic vessel operation systems improving energy efficiency, require a careful consideration of their possible negative impact on the river environment.

The European policy of innovative solutions in waterborne inland transport comprises the promotion of the energy efficiency—development of intermodal transport of goods and energy-efficient means of transport (Abramowicz-Gerigk & Burciu, 2016).

Article 8 of the Trans-European Transport Network (TEN-T) guidelines recommends appropriate ratings related to the Birds and Habitats Directives for evaluation of the environment impact on new projects (Burciu & Gasior).

The directions of innovative technological development and interventions included in Article 7.3 of the guidelines are as follows:

– Striving to create favourable conditions for the transfer of shipping from roads to rail, especially on the distances over 300 km
– Promotion of the ecological means of transport powered by the alternative clean energy sources, resulting with the reduction of air pollution
– Introduction of a new generation of inland vessels, including the energy efficient inland containerships with the low carbon dioxide footprint

2 INFLUENCE OF REGULAR RIVER NAVIGATION ON THE NATURAL ENVIRONMENT

The ships movement can be the reason of hydrodynamic alterations, turbidity increase, resuspension of sediments and emissions of toxic substances. The waves generated by ships, drawdown, return currents and propeller jets can be the reason of riverbed and bank erosion, uprooting plants, disturbing the benthic fauna and flora and spawning area for fish.

The acceleration of morphologic process can influence lowering the water quality and makes more difficult water treatment to a desired drinking quality (PIANC, 2016). Therefore the maintenance and development of inland waterways has to combine both the navigational and ecological demands (Abramowicz-Gerigk & Burciu, 2016).

The main factors influencing interactions of river navigation with nature are dependent on vessel design and operational characteristics (Gerig, 2004; Gerigk, 2014). The following characteristics should be considered:

– Ship main particulars and hull form
– Ship propulsion characteristics—type and powering, propeller jets characteristics
– Ship velocity relative to water velocity and water depth to ship draft ratio, river profile—water depth Froude number

Depending on the river profile, the flow field generated by a ship has different transport patterns. Water depth is the main factor determining the critical speed (Kulczyk & Wereszko, 2006).

The depth Froude number determines whether a ship navigates in shallow, transitional or deep water, affecting the amplitude and period of the generated waves, drawdown and ship squat (Althage, 2010. The influence of ship movement on the riverbed scouring may be neglected at water depth to draft ratios above 2. The limiting values of scouring velocities in dependence of bed material are from 0.1 m/s for fine sand to 4 m/s for boulders (0.7 m grain size). The over bed velocities can reach 160%–180% of the ship speed equal to 2.44 m/s to 4 m/s, exceeding the scouring limits.

In practice, the limiting ship speeds in shallow waterways are determined by ship squat and ship resistance to motion ratios.

Propeller jet streams and ship-generated waves have great influence on the waterway. The deteriorating effect is bigger under higher propeller loads, especially during manoeuvres.

The prediction and modelling of the scour should be based on the time period of ship-riverbed interactions (PIANC, 2008).

Wake wash and waves generated by vessels increase shear stress acting on the river bed and banks. The amplitude of the waves depends mainly on the vessel velocity and small fast crafts produce higher waves with shorter wave periods than the large slow ships and barges.

The drawdown is the reason of the induced flows in side channels and backwaters as well as the hydraulic impact on the slopes and banks dangerous for their stability. The generated waves can break upon the bank, the wave is following the drawdown exceeding the still water level prior to ship passage, blurring the shore.

3 DESIGN OBJECTIVES OF INNOVATIVE WATERBORNE INLAND TRANSPORT MEANS

The last innovative solutions of transport means for Polish waterways related to the ecological conditions comprise of several new concepts. The new concept of INBAT (Odra) push train developed within the frames of a collaborative research project INBAT (Innovative Barge Trains for Effective Transport on Shallow Waters) leaded by Wroclaw University of Technology, included the pusher boat and barge has been adapted for the shallow water operation with variable navigational conditions of Odra and Elba rivers. The concept of pusher boat included triple propellers enabling the push train operation in the ranges 0.6 m to 1.2 m ship draft. The special tool in the form of computer program have been developed for the optimisation of operational parameters of propulsion system as well as analysis of the influence of push train navigation on the river and numerical prediction of propulsion prognosis. The main particulars of INBAT (Odra) universal barge are presented in Table 1.

The capacity of the barge is 21 TEU in 1 layer, the minimum permissible draft is 0.6 m, the operation of the push train can be profitable at drafts greater than T = 0.8 m.

The studies on the minimised resistance and influence on the waterway, included:

– Drawdown
– Return current forming,
– Design and optimisation of a bow shape.

The vessels with a slender bow, longer entrance length generate lower waves.

The present research problems related to the waterborne inland transport in the Lower Vistula region are mainly related to the difficulties in transport of oversized loads and export of containers via A1 motorway and railway from Port of Gdansk (6000 containers per day). The parameters of the presently operated barges originally designed for Odra Waterway are not adapted to the navigational conditions of the Lower Vistula waterway mainly due to the small beam.

The expertise carried out within European project INVAPO (Burciu & Gasior) confirmed the possibilities of waterborne transport of containers on the route Gdansk-Tczew-Solec Kujawski. The transport of containers on barges in 3 layers independent on their mass is possible in relation Gdansk-Tczew, transport of two container layers is possible in relation to Solec Kujawski.

Table 1. Main particulars of the proposed universal barge INBAT (Odra) hull.

Parameter	Value [m]
Length	48.75
Breadth	9
Draft	1.7
Depth	2.2

The possibility of container transport in three layers on board is dependent on the mass of containers but also on occurrence of high navigational water.

The transport of containers in the Gulf of Gdansk, mainly the export of the empty containers from DCT (Deepwater Container Terminal in Gdansk) to the depot in Gdynia could solve the bottleneck problems in Gdansk and Gdynia Ports. It is related now to the development of a new generation of coastal barges or ships similar to LASH (Lighter Aboard Ships) systems and appropriate classification rules.

4 INNOVATIVE SOLUTIONS IN NEW DESIGNS OF INLAND VESSELS

4.1 *New design of small inland passenger catamaran—investigations on wave system and resistance*

The influence of passenger inland vessel hull form on hydrodynamic resistance and analysis of possible ship draught reduction due to the hull form modification without change of ship displacement, length and breadth has been investigated.

The change of both hull breadths, non-symmetric internal sides of hulls and modification of the bottom in the aft part of the hull were tested using CFD methods (Abramowicz-Gerigk et al. 2016a, 2016b). The catamarans with asymmetric hulls were investigated by Broglia et al. (2011), Array & Cheng (1995), Kaklis & Papanikolaou (1992). The efficiency problems were presented by Damiano et al. (2003), resistance problems by Dubrovsky et al. (2002) and Haase et al. (2012), stability problems were analysed by Krata (2006) and Krishna & Krishnannkuty (2009). High speed catamaran manoeuvring capabilities were investigated by Soares et al. (1999).

The modern tween hull passenger inland catamaran, built by Zegluga Mazurska (Masuria Shipping Ltd.) is presented in Figure 1.

On the basis of numerical simulation the ship resistance at the design speed was calculated and compared for each of the four tested hull forms (Figs. 2–4).

The CFD simulations of the flow field for both C1 and C2 models showed the significant lowering of water level in the aft area of the hull, making impossible the proper action of ship propeller. For this reason the model C4 with reduced draft equal to 0,65 m and slope of bottom in the aft part was adopted (Figs. 4, 5).

The lower draft of inland ship increases her area of operation but also influences ship resistance and significant height of generated waves. This has a direct impact on environment in form of side and bottom erosion and emission of harm substances.

Figure 1. Tween hull passenger inland catamaran (www. zeglugamazurska.com.pl).

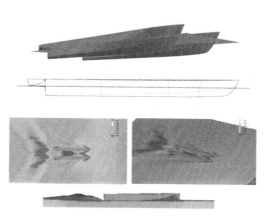

Figure 2. Hull form and wave system generated by model C1 at speed 5.8 m/s (Abramowicz-Gerigk et al. 2016).

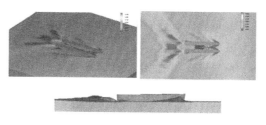

Figure 3. Wave system generated by model C2 with reduced draft 0.6 m and widened hulls of 20%, at speed 5.8 m/s (Abramowicz-Gerigk et al. 2016).

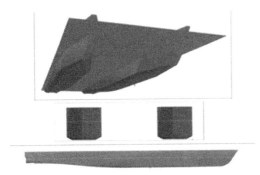

Figure 4. Hull form of model C4 (Abramowicz-Gerigk et al., 2016).

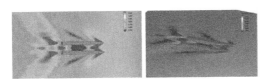

Figure 5. Wave system generated by model C4 at speed 5.8 m/s (Abramowicz-Gerigk et al., 2016).

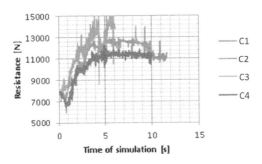

Figure 6. Results of CFD simulation—calm water resistance of ship models at speed 10 km/h (Abramowicz-Gerigk et al. 2016) for C1, C2, C3 and C4 models.

The effect of lower draft is increased leeway, much smaller for catamarans the single hull ships.

4.2 New design of push train integrated control system

The general assumptions related to minimisation of the possible harmful impact on river environment, are exploitation of modern technologies of energy storage and green energy production. The synergy of application potential of particular elements of the used technologies comprises the hull form optimisation, low emission drives—hybrid electric drives used as calm electric propulsion on arrival or departure form ports, LNG instead of fuel oil and use of electric charging in ports.

The Vistula river layout is full of bends therefore the idea of integrated control system with flexible connection between the pusher and barge in the push train has been developed (Fig. 7).

The push train performance in narrow river bends is presented in Figure 8. The proposed system comprised of a hydraulic control system of flexible connection between pusher and barge, twin rotors in the bow part of the barge and twin rudders, tween propellers propulsion system.

The different configurations of the bow part of the barge were tested:

– Conventional single hull
– Twin bulbous bows
– Twin bows with rotors

The CFD calculation of the push train resistance for different hull configuration in the bow part of the barge at 10 km/h speed and water depth 5 m are presented in Figure 9.

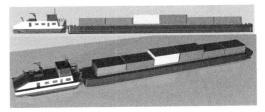

Figure 7. The push train with innovative flexible connection designed for operation on Lower Vistula River.

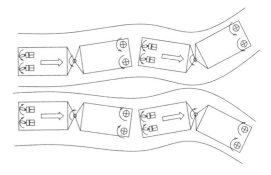

Figure 8. Performance in narrow river bends of a push train with flexible connection system between the pusher and barge, designed for operation on Lower Vistula River.

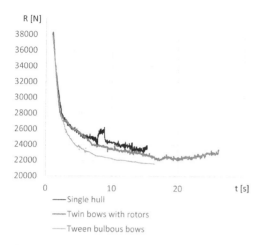

Figure 9. CFD calculation of push train resistance for different hull configuration in the bow part, speed 10 km/h, water depth 5 m.

5 INNOVATIVE SOLUTION OF A BERTHING MONITORING SYSTEM

Determination of lateral forces and yawing moment necessary for the safe performance of berthing operations of self-berthing vessels is based on the recognition of forces generated on the hull by propellers, rudders, thrusters including interaction effects (Abramowicz-Gerigk, 2008) and weather forces.

The empirical approach allows for the determination of a potential bed disturbances from a propeller jet in the second jet zone where the flow interacting with the bed is dependent on the axial distance to the propeller x, the radial distance to the propeller axis r, the initial efflux velocity V_0, and propeller diameter (de Jong, 2014):

$$U(x,r) = A \cdot V_0 \cdot \left(\frac{D_0}{x}\right)^a \cdot \exp\left(-\frac{b \cdot r^2}{x^2}\right)$$

where: A, a and b are coefficients determined through experiments depending on the vessel and local conditions (de Jong, 2014): $A = 0.1$ to 2.8; $a = 0.3$ to 1.6 and $b = 15.4$ to 22.2 (specific values are available for lateral walls and for vessels with twin propellers in BAW (2011).

The latest result of field tests carried out on inland waterways presented by Symonds et al. (2017) show the best agreement with the measured data for $A = 0.6$, $a = 0.7$ and $b = 15.4$ for the forward throttle bursts and $A = 1.2$, $a = 1.1$ and $b = 3.0$ for the reverse throttle bursts.

Monitoring of the propeller jets induced loads on the quay wall and bed along the quay can be obtained using measuring system installed on the quay wall (Fig. 10).

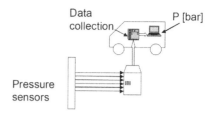

Figure 10. Measurement of the propeller jets induced pressure field on the quay wall using mobile laboratory for the data collection.

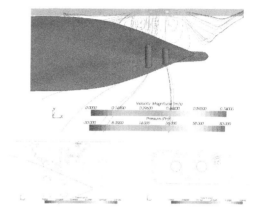

Figure 11. Example of CFD calculation of the flow field generated by bow thrusters during the first phase of ship unberthing.

The collected data of the pressure field induced by ship propellers and thrusters allows to determine the pressure and velocity fileds over the bed using CFD model (Tech. Report, 2009). The example of CFD calculations is presented in Figure 11.

The new approach to the development of the monitoring system is to collect and transmit the data online and present them on the website using Internet. This will be a subject of a recently started project "Development of a prototype of a system for monitoring the loads on berths and bed protection in the area of ship berthing along with the implementation of the final product on the market by Enamor Ltd. company from Gdynia" within "Smart Specialisations of Pomerania Region—offshore technology, ports and logistics" European program.

6 CONCLUSIONS

The innovative systems improving safety of manoeuvring and berthing operations of inland vessels proposed in the paper are concentrated on new design issues with respect to inland vessels construction. The innovative solutions for the push train will be developed and implemented on the market within the frames of the project "Research, design and construction of a prototype of an innovative push train: pusher with hybrid propulsion with innovative accompanying boat with changing draft and fuel cells power supply, innovative pushed barge for heavy oversized goods transport with an anti-heeling system for Ro-RO loading operations, and with bow thruster (with dimensions matched Przegalina lock) within "Smart Specialisations of Pomerania Region—offshore technology, ports and logistics" European program. The system for propeller induced loads monitoring during berthing operations is the most important for shallow water marinas.

REFERENCES

Abramowicz-Gerigk T. 2008. Experimental study on the hydrodynamic forces induced by a twin-propeller ferry during berthing. Ocean Engineering. Elsevier. Volume 35, Issues 3–4, p. 323–332.

Abramowicz-Gerigk, T. & Burciu Z.: The influence of regular river navigation in special protection areas of Natura 2000 network. Journal of Kones No. 3, 2016.

Abramowicz-Gerigk, T., Burciu, Z. & Jachowski, J. 2016a. Investigations of passenger inland vessel hull form influence on her operational characteristics—hull resistance. Scientific Journals of Gdynia Maritime University, No. 97/2016, in Polish.

Abramowicz-Gerigk, T., Burciu, Z., Jachowski, J. 2016b. Investigations of passenger inland vessel hull form influence on her operational characteristics—leeway. Scientific Journals of Gdynia Maritime University, No. 97/2016, 16–24, in Polish.

Althage, J. 2010. Ship-Induced Waves and Sediment Transport in Göta River, Sweden. Master's Thesis 2010. Water Resources Engineering, Department of Building and Environmental Technology, Faculty of Engineering, Lund University. 2010.

Arai, M., Cheng, L.Y. 1995. Numerical study of water impact loads on catamarans with asymmetric hulls fast 3rd Intl Conf on Fast Sea Transportation, 25–27 Sept 1995; Lubeck-Travemunde, Germany. Procs. Ed by C.F.L. Kruppa. Publ by Schiffbautechnische Gesellschaft, Germany. Vol 1, p 221.

BAW. 2011. Principles for the design of bank and bottom protection for inland waterways (GBB), BAW Code of Practise, Issue 2010. Karlsruhe, March 2011.

Broglia, R., Zaghi, S., Di Mascio, A. 2011. Numerical simulation of interference effects for a high-speed catamaran. Journal of Marine Science and Technology 16 (3), 254–269, 2011.

Burciu Z., Gasior A.: The analysis of the possibility of revitalization of lower Vistula based on INWAPO European Project. Zeszyty Naukowe Akademii Morskiej w Gdyni Nr 92/2015.

Damiano C., Lazzara S., Mancuso A., Virzimariotti G., 2003 "Study on the Efficiency of an Innovative Hull", Nav 2003, International Conference on ship and shipping research, Palermo, June 24–27.

Damiano C., Stroligo M., Virzi'Mariotti G., Zotti I., 2009, Theoretical and Experimental Comparison among the Resistance Components of a Fast Catamaran in Different Configurations; NAV 2009, 16th International Conference of Ship and Shipping Research, Vol I, ISBN 978 – 88- 904394-0-7, Messina (Italy), 2009, pp 31–39.

de Jong, J. 2014, Numerical modelling of bow thrusters at open quay structures", TU Delft MSc Thesis, January 2014.

Dubrovsky V., Lyakhovitsky A. Multi Hull Ships. Fair Lawn, NJ, Backbone Publishing Co. 2002,

Gerigk M. 2004. On a risk-based method for safety assessment of a ship in critical conditions at the preliminary design stage. Polish Maritime Research, No. 1 (39), Vol. 11, 8–13.

Gerigk, M. 2015. Modeling of event trees for the rapid scenario development. Safety and Reliability: Methodology and Applications. Nowakowski et al. (Eds), London: Taylor & Francis Group, 2015, 275–280.

Haase M., Davidson G., Friezer S., Binns J., Thomas G., Bose N. 2012. On the Macro Hydrodynamic Design of Highly Efficient Medium-Speed Catamarans with Minimum Resistance. International Journal of Maritime Engineering 154(A3), 2012. 131–141.

Kaklis P., Papanikolaou A., 1992 The Wave Resistance of a Catamaran with Non-Symmetric Thin Demihulls, Proc 1st Larsson L., Baba E., 1996 Ship resistance and flow computations, Advances in marine Hydrodynamics, M. Ohkusu (ed.), Comp. Mech. Publ.

Krata, P. 2006. Formal requirements with respect to stability of catamarans. Journals of Gdynia Maritime University, No. 18/2006, 69–88.

Krishna B., Krishnankutty P. Experimental and numerical study on trim and sinkage of a high speed catamaran vessel in shallow waterways. International Shipbuilding Progress, vol. 56, no. 3–4, pp. 159–176, 2009.

Kulczyk, J., Werszko, R. 2006. Influence of ship motion on waterway backward current velocity. Polish Maritime Research, Special Issue 2006/S2.

PIANC, 2016. Values of inland waterways. 2016. PIANC Report No. 139. Inland Navigation Commission. www.pianc.org.

PIANC, 2008. Considerations to reduce environmental impacts of vessels navigation. PIANC Report N° 99, Inland Navigation Commission, 2008. www.pianc.org.

Soares, G.C., Sutulo S., Francisco R., Santos F., Moreira L., 1999, Full scale measurements of the manoeuvring capabilities of a catamaran. In: Proc. Hydrodynamics of High-Speed Craft, RINA, London, UK, November, 1999, 24–25.

Soares, Guedes C., Teixeira, A.P. 2001. Risk assessment in maritime transportation. Reliability engineering and System Safety 74, 299–309.

Symonds A., Britton G, Donald J., Loehr H. 2017. Predicting propeller wash and bed disturbance by recreational vessels at marina. Technical articles dedicated to Australia, Host Country of PIANC's AGA 2017.

Technical report No. RH-2009/T-027. 2009. CTO S.A. CFD calculations of interactions during harbour manoeuvres of vessel Kolobrzeg. The research project conducted at Gdynia Maritime University N N509293635 sponsored by Polish Ministry of Science and Higher Education: Safety of berthing of ships in the Motorway of the Sea transportation system. Gdynia 2009.

http://www.zeglugamazurska.com.pl.

Maritime Transportation and Harvesting of Sea Resources – Guedes Soares & Teixeira (Eds)
© 2018 Taylor & Francis Group, London, ISBN 978-0-8153-7993-5

Experimental and numerical simulations of zig-zag manoeuvres of a self-running ship model

M.A. Hinostroza, H.T. Xu & C. Guedes Soares

Centre of Marine Technology and Ocean Engineering (CENTEC), Instituto Superior Técnico, Universidade de Lisboa, Lisbon, Portugal

ABSTRACT: Experimental and numerical simulations of manoeuvres of a self-running ship model are presented. The experimental manoeuvring tests were carried out for a self-running ship model in an outdoor pool, following the ITTC recommended procedures. The self-running ship model is 2.5 m length and it is equipped with motions sensor, positioning GPS, propulsion DC motors, wi-fi shore communication, and anemometer. Numerical simulations of ship manoeuvres were carried out based in a 3DOF nonlinear model the same ship model. The manoeuvres considered were 20°/20° and 10°/10° zigzag. Comparisons between the experimental results and numerical simulations are also presented.

1 INTRODUCTION

In recent years there has been a growing interest in the development of advanced autonomous vehicles for operations at sea. In this scenario, Autonomous surface vehicles allows fast and optimal access to otherwise unreachable regions and can simplify the task of acquiring hydrographic data fast and cost-effectively without placing human lives at risk.

According to Caccia (2006), there are three main groups of unmanned marine craft: (1) marine weapons (probably, historically the oldest but rather specific kind not considered here); (2) Autonomous Vehicles (AV) subdivided into underwater (AUV) and surface (ASV; USV) vehicles; (3) self-running or Free-Running scaled Models (FRM) aimed at studying manoeuvring and/or sea keeping qualities of corresponding full-scale objects.

In the USA, the Massachusetts Institute of Technology (MIT) have developed a family of ASVs which include a 1:7 scale fishing trawler type vessel, ARTEMIS (1993), the catamaran models, ACES (1997) and AutoCat (1999), and the SCOUT vessels (2004). Of all these prototypes, the kayak type, SCOUT vessels have successfully implemented COLREGs at a basic level for head-on situations whilst maintaining wireless communication (Benjamin and Curcio, 2004). A USV platform adapted from a SEADOO Challenger 2000, originating from the Space and Naval Warfare Systems Center (SSC) San Diego has also executed obstacle avoidance in accordance with the Rules of the Road during trials and is discussed in a later section

(Larson et al., 2006). The key to the development of ASV depends on advances in the underpinning technology which determines their capabilities. One popular development goal involves collaboration with one or more ASVs to create a relay network with wireless communication. In that sense, a series of model tests has been developed and tested with success using self-running scaled models. Moreira et al., (2008) used a model of the tanker "Esso Osaka", which was instrumented for autonomous operation and different guidance and control approaches were implemented and tested (Moreira et al., 2007), including the automatic performance of some standard manoeuvring tests. Also, typically ASVs are controlled by a rather sophisticated steering and guidance system, often based on artificial intelligence algorithms where e.g. a collision-avoidance option may be implemented for a self-propelled chemical tanker ship model, (Perera et al., 2013, 2015).

The present work is an extension of the work of Ferrari et al. (2015), where the goal was to study the experimental performance of manoeuvres for a self-running ship model. In this paper the objective is validate a mathematical model for path prediction in manoeuvring simulations with ship models, (Sutulo et al. 2002). Experimental and numerical simulations of zigzag manoeuvres are performed for a scaled ship model, of a chemical tanker.

The simulations were based in the 3DOF formulation proposed by Abkowitz (1980). The experimental tests were carried out in an outdoor pool, following the ITTC recommended procedures. A scaled model of a chemical tanker with a length of 2.5 m is equipped with motion sensors, positioning

GPS, propulsion DC motors, wi-fi communication and anemometer. Finally in the last chapters a comparison between numerical simulation and experimental results are presented and a good agreement was found.

The rest of the paper is organised as follows. The mathematical formulation used for the numerical simulations is given in Sect. 2.

Numerical simulations of zigzag manoeuvres for a "Esso Osaka" ship model using the hydro-dynamic parameters presented in (Moreira et al., 2007) is presented in Sect. 3, while the experimental setup and the experimental results of a zigzag manoeuvres in a pool with are discussed in Sect. 4. The comparison between numerical and experimental results of zigzag manoeuvres for a ship model is presented in Sect. 5, and Section 6 is the conclusion.

2 MATHEMATICAL FORMULATION

2.1 *Coordinate frames and notation*

When analysing the motion of marine vehicles in 6 DOF it is convenient to define two coordinate frames. The moving coordinate frame $X_0Y_0Z_0$ is conveniently fixed to the vehicle and is called the body-fixed reference frame. The origin 0 of the body-fixed frame is usually chosen to coincide with the centre of gravity (CG) when CG is in the principal plane of symmetry or at any other convenient point if this is not the case.

For marine vehicles, the body axes X_0, Y_0 and Z_0 coincide with the principal axes of inertia, and are usually defined as:

X_0—longitudinal axis (directed from aft to fore);
Y_0—transverse axis (directed to starboard);
Z_0—normal axis (directed from top to bottom).

The motion of the body-fixed frame is described relative to an inertial reference frame. For marine vehicles it is usually assumed that the accelerations of a point on the surface of the Earth can be neglected. Indeed, this is a good approximation since the motion of the Earth hardly affects low speed marine vehicles. As a result, an earth-fixed reference frame XYZ can be considered to be inertial. This suggests that the position and orientation of the vehicle should be described relative to the inertial reference frame while the linear and angular velocities of the vehicle should be expressed in the body-fixed coordinate system. The different quantities are defined according to the notation proposed by SNAME (1950), as indicated in Table 1. Based on this notation, the general motion of a marine vehicle in 6 DOF can be described by the following vectors:

Table 1. Notation used for marine vehicles.

DOF	Motion/rotation	τ	v	η
1	In x-direction (surge)	X	u	x
2	In y-direction(Sway)	Y	v	y
3	In z-direction(heave)	Z	w	z
4	About x-axis(roll)	K	p	ϕ
5	About y-axis (pitch)	M	q	θ
6	About z-axis(yaw)	N	r	ψ

$$\eta = [\eta_1^T, \eta_2^T]^T, \eta_1 = [x, y, z]^T, \eta_2 = [\phi, \theta, \psi]^T$$
$$v = [v_1^T, v_2^T]^T, v_1 = [u, v, w]^T, v_2 = [p, q, r]^T$$
$$\tau = [\tau_1^T, \tau_2^T]^T, \tau_1 = [X, Y, Z]^T v_2 = [K, M, N]^T$$

Here η denotes the position and orientation vector with coordinates in the earth-fixed frame, v denotes the linear and angular velocity vector with coordinates in the body-fixed frame and τ is used to describe the forces and moments acting on the vehicle in the body-fixed frame.

2.2 *Nonlinear manoeuvring mathematical model of marine surface ship*

A ship in a seaway has 6 degrees of freedom (DOF) to move freely in the space, as illustrated in Figure 1, in order to simplify the problem of manoeuvring modelling, some assumptions need to be adopted. The heave, roll and pitch motion are not important in the manoeuvring problem, as the ship moves in the horizontal plane. The coordinate frames of surface ship in 3 DOF are presented in Fig. 2. In estimation theory, a mathematical model of surface marine ship is needed to describe the dynamics of the system. The Abkowitz (1980) model will be modified in order to make the modelling more flexible and realistic physically. In this study, the current effect is considered as the main external excitation, because the ship model has a small above water structure.

As presented in Figure 2, u_c is the current's magnitude, α is the current's direction, ψ is the ship's heading angle, u is the forward component of velocity over ground, and v is the transverse component of velocity, the relative forward velocity and transverse velocity are given by

$$u_r = u - u_c \cos(\psi - \alpha)$$
$$v_r = v + u_c \sin(\psi - \alpha) \tag{1}$$

The time derivatives of u and v are given:

$$\dot{u} = \dot{u}_r - u_c r \sin(\psi - \alpha)$$
$$\dot{v} = \dot{v}_r - u_c r \cos(\psi - \alpha) \tag{2}$$

where the accelerations of the motion in 3 degree of freedom (surge, sway and yaw) are given by

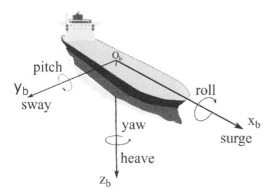

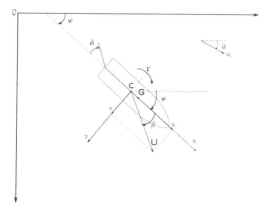

Figure 1. Motion of marine surface vehicle in 6 degree of freedom.

Figure 2. Coordinate frames for marine surface vehicle.

$$(m - X_{\dot{u}_r})\dot{u}_r - mv_r r - mx_G r^2 = f_1$$
$$(m - Y_{\dot{v}_r})\dot{v}_r + (mx_G - Y_{\dot{r}})\dot{r} + mu_r r = f_2 \quad (3)$$
$$(mx_G - N_{\dot{v}_r})\dot{v} + (I_z - N_{\dot{r}})\dot{r} + mx_G u_r r = f_3$$

where m the mass of the ship is, $X_{\dot{u}_r}, Y_{\dot{v}_r}, Y_{\dot{r}}, N_{\dot{v}_r}, N_{\dot{r}}$ are the added mass and moment, respectively. The dimensionless forces are defined as multi-variety third-order regression polynomials depending on the non-dimensional velocities.

$$f_1' = \kappa_{T_{f_1}} T_{f_1}' u_r'^2 + \kappa_{T_{f_2}} T_{f_2}' n' u_r' + \kappa_{T_{f_3}} T_{f_3}' n'^2 - \kappa_{C_R} C_R' + \kappa_{X_{v_r v_r}} + X_{v_r^2}' v_r'^2 + \kappa_{X_{ee}} X_{e^2}' e^2 + \kappa_{X_{rr}} X_{r^2}' r'^2 + \kappa_{X_{v_r r}} X_{v_r r}' v_r' r' + \kappa_{v_r^2 r^2} X_{v_r^2 r^2}' v_r'^2 r'^2$$

$$(4)$$

$$f_2' = \left\{ \kappa_{Y_{v_r}} Y_{v_r}' v_r' + \kappa_{Y_\delta} Y_\delta' (c - c_0) v_r' \right\} + \left\{ \kappa_{Y_r} Y_r' r' - \kappa_{Y_\delta} \frac{Y_\delta'}{2} (c - c_0) r' \right\} + \kappa_{Y_\delta} Y_\delta' \delta + \kappa_{Y_{rrv_r}} Y_{r^2 v_r}' r'^2 v_r' + \kappa_{Y_{eee}} Y_{e^3}' e^3 + \kappa_{Y_0} Y_0'$$

$$(5)$$

$$f_3' = \kappa_{N_0} N_0' + \left\{ \kappa_{N_{v_r}} N_{v_r}' v_r' - \kappa_{N_\delta} N_\delta' (c - c_0) v_r' \right\} + \kappa_{N_\delta} N_\delta' \delta + \kappa_{N_{rrv_r}} N_{r^2 v_r}' r'^2 v_r' + \kappa_{N_{eee}} N_{e^3}' e^3 + \left\{ \kappa_{N_r} N_r' r' + \frac{1}{2} \kappa_{N_\delta} N_\delta' (c - c_0) r' \right\}$$

$$(6)$$

Detailed information about the formulation can be found in Moreira et al. (2007).

3 NUMERICAL SIMULATIONS

This section presents numerical simulations of zig-zag manoeuvres for the "Esso Osaka" ship model based on the nonlinear 3DOF mathematical model of Abkowitz (1980). The vehicle main characteristics are listed in Table 2.

Table 3 presents the non-dimensional hydro-dynamics coefficients of the "Esso Osaka" ship model, these values were obtained from Moreira et al. (2007).

3.1 Zigzag manoeuvre

This type of manoeuvre is also known as the Z-Manoeuvre or the Kempf Manoeuvre. In some situations, a ship is required to change its course or heading. Sometimes in rough seas or in cases of directional errors, the ship may be required to change its direction more rapidly within a limited span of time. So, the ability to zigzag manoeuvre should be an inherent property in the manoeuvring characteristics of a vessel, (Sutulo and Guedes Soares 2011).

The procedure is as follows: with zero rudder, achieve steady speed for one minute, then deflect the rudder to 20°, and hold until the vessel turns 20°, after, deflect the rudder to -20°, and hold until the vessel turns to –20° with respect to the starting heading and repeat.

Each zigzag manoeuvre is defined with two constant parameters: the heading deviation and

Table 2. "Esso Osaka" model particulars.

"Esso Osaka" model	
Overall Length(mm)	3430
Length between perp.	3250
Breadth(mm)	530
Draught (estimated at the tests) (mm)	217
Displacement (estimated at trials) (Kg)	319.4
Rudder area (m^2)	0.0120
Propeller area (m^2)	0.0065
Scaling coefficient	100

Table 3. The base values of hydrodynamic derivative.

Coefficient	Value	Coefficient	Value
$(m - Y_{\dot{v}})'$	0.0352	Y_0'	$1.90 * e^{-6}$
$(I_z - N_{\dot{r}})'$	0.00222	Y_{v_r}'	−0.0261
$(m - X_{\dot{u}_r})'$	0.0116	Y_δ'	0.00508
η_1'	$-0.962 * e^{-5}$	Y_r'	0.00365
η_2'	$-0.446 * e^{-5}$	Y_{rrv_r}'	−0.0450
η_3'	$0.0309 * e^{-5}$	Y_{eee}'	−0.00185
C_R'	0.00226	N_0'	−0.00028
$X_{v_r^2}'$	−0.006	N_{v_r}'	−0.0105
X_{ee}'	−0.00224	N_δ'	−0.00283
X_{rr}'	0.00515	N_r'	−0.00480
$(X_{v_r r} + m)'$	0.0266	N_{rrv_r}'	0.00611
X_{rrvv}'	−0.00715	N_{eee}'	0.00116

the rudder angle deviation. In general, these parameters can be assigned arbitrary values but, mostly, either both are taken equal to 10 degrees (one speaks then about a 10°/10° zigzag), or to 20 degrees (20°/20° zigzag).

$$\delta_c(r, \psi) = \delta_z sign(\psi_z sign(r) - \psi) \quad (7)$$

where δ_c is the rudder command, δ_z and ψ_z are the zigzag parameters.

Overshoot angles (usually the first and second overshoots are considered) are the main numerical measures of a ship's dynamic qualities in the zigzag manoeuvre. For a 10°/10° zigzag, for instance, the first overshoot normally varies from 5 to 20 degrees and the second one from 5 to 35 degrees. However, these overshoot angles can reach 70–80 degrees for especially bad vessels. In the 20°/20° zigzag the overshoot angles lie mostly within the 10–30 degrees range.

3.2 Numerical simulations

Figure 3 presents the 20°/20° zigzag manoeuvre for the "Esso Osaka" ship model. From this plot several important characteristics of the yaw response can be established. The response time (time to reach a given heading) is 19 s, the yaw overshoot (amount the vessel exceeds ±20 deg.) when the rudder has turned the other way) is around 10 deg.,

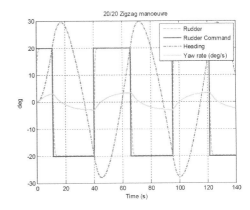

Figure 3. 20°/20° Zigzag manoeuvre for a ship model.

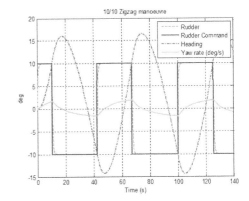

Figure 4. 10°/10° Zigzag manoeuvre for a ship model.

and the total period for the 20 deg. oscillations is around 60 sec.

Figure 4 presents the 10°/10° zigzag manoeuvre for the "Esso Osaka" ship model, and from this manoeuvre several important characteristics of the yaw response can be established. The response time (time to reach a given heading) is 18 s, the yaw overshoot (amount the vessel exceeds ±10 deg) when the rudder has turned the other way) is 6 deg. and the total period for the 20 deg. oscillations is around 60 sec.

From the plots of 20°/20° and 10°/10° zigzag manoeuvres it is clear to see that the response time and the total period are similar for both cases, but the yaw over shoot angles are different ~5 deg. This difference is because the yaw moment is different and proportional to ship rudder deflection.

4 EXPERIMENTAL SET-UP

In this section, the experimental set-up is presented, the self-running ship model particularities, software, hardware, place of tests and results are described. The ASV is a scale model of 2.5 meters, self-propelled and equipped with navigation and positioning equipment. The tests were conducted in a large pool.

4.1 Chemical tanker self-running ship model

The vessel model used in this study is presented in Fig. 5. It is a scaled model from a chemical tanker built in single skin glass reinforced polyester with plywood framings and has autonomous control of the navigation and platform, which can be divided into two components: hardware structure and software architecture. Its design speed is 0.98 m/s. Its main particulars are given in Table 4.

4.2 Hardware structure

The hardware structure consists of all the sensors and actuators that are used in the ASV real-time navigation and control platform. The hardware is further divided into two units of command and monitoring (CMU) and of communications and control (CCU) as described in (Perera et al., 2015).

The main objective of the CMU is to facilitate manual and autonomous control of the ASV that provides a human machine interface (HMI). As presented in Fig. 6, the CMU consists mainly of several instruments: Laptop, GPS unit, industrial

Figure 5. Free-running ship model with various sensors.

Table 4. Main dimensions of the model.

Chemical tanker model	
Overall Length (mm)	2587.5
Length between perp.	2450
Breadth (mm)	426.2
Draught (estimated at the tests) (mm)	102
Propeller diameter (mm)	82.2
Design speed (m/s)	0.984
Scaling coefficient	65.7

Wi-Fi unit, Compact-RIO, main AC power supply unit, DC power supply unit and anemometer.

A laptop is used in a HMI that is connected through industrial wi-fi unit for communication with the CCU. The laptop works as a data display unit as well as an automatic and manual control unit for the ASV. A GPS unit is used in the CMU for position measurements of the ASV. The unit consists of GNSS (Global navigation satellite system) antenna to absorb the electromagnetic signals transmitted by the GNSS satellites into RF signals. However, the complete GPS system consists of two units of: base station and rover station. Both GPS units are used to improve the position accuracy of the ASV that is around the accuracy of ± 20 (cm).

The anemometer is used to measure the relative wind direction and wind speed at the location of tests, the data from the sensor is acquired using a C-RIO using an analogue input module. All units in the CMU are powered by the shore based main AC unit that is also complemented by a NI DC power supply unit.

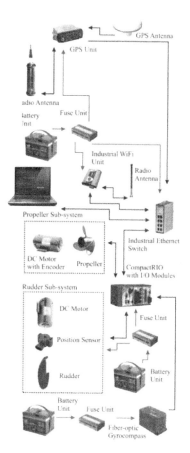

Figure 6. Sensors and equipment in communication and control unit.

4.3 Software structure

The software architecture is mainly programmed by LABVIEW software, based in the real-time acquisition program used in Hinostroza et al. (2016a; 2016b). The software architecture consists of several program loops: FPGA loop, real-time loop and TCP/IP loop. The FPGA loop is associated with collecting data from the sensors (i.e. GPS, IMS and fiber-optic gyrocompass units) and controlling the actuations of propeller and rudder sub-systems that have been programmed under a reconfigurable FPGA platform, where LABVIEW provides the VDHL software codes.

4.4 Place of tests

Model tests on the ASV were conducted at a pool shown in Figure 7. The weather was sunny and dry, but some wind was constantly present, changing its speed (approximately in the interval of 1–2 m/s) and direction as time passed.

The pond was certainly deep enough to neglect any shallow-water effects. The pool has a maximum length of 50 m and average breadth of 20 m.

4.5 Presentation of results

In this section data recorded from zigzag manoeuvre tests of the chemical tanker model is presented. Three different tests were performed, zigzag 20°/20°,

Figure 7. Location of the pool for the free-running model tests.

Table 5. Zigzag data recorded.

Data	Rudder angle	Data characteristics	Wind conditions
Z1	20°/20°	All equipment working, good data.	Moderate ≈ 2 [m/s]
Z2	25°/25°	Corrupted data, **not presented.**	Too windy ≈ 6 m/s
Z3	30°/30°	All equipment working, good data	Moderate ≈ 2 [m/s]

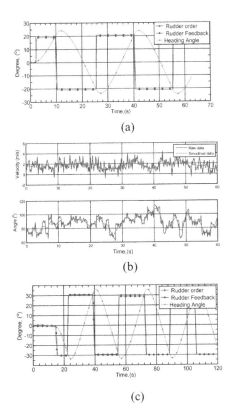

Figure 8. Experimental results from model scale tests, (a) 20°/20° zigzag manoeuvre (b) wind conditions during the tests, (c) 30°/30° zigzag manoeuvre.

25°/25° and 30°/30°. The tests have in average of 60 sec. duration, with all equipment working well and with moderate wind conditions, 2 m/s, details are presented in Table 5.

Figure 8a and 8c present results of 20°/20° and 30°/30° zigzag manoeuvre. Figure 8b presents the time history of wind speed and wind direction during the tests, about 2 m/s and the direction is about 100 deg.

5 COMPARISON BETWEEN NUMERICAL AND EXPERIMENTAL RESULTS

In this section a comparison between simulations and experimental results of 20°/20° zigzag manoeuvres for the chemical tanker ship model are presented. The hydrodynamic coefficients of the model ship were obtained by Xu et al., (2017), who applied system identification algorithm to experimental data, and the mathematical model considered is the 3DOF Abkowitz (1980) model.

5.1 Hydrodynamic coefficients of chemical tanker ship model

The hydrodynamic coefficients of the chemical tanker model are obtained from a system identification technique and are presented in Table 6. The system identification method considers different loss functions that are firstly defined considering the effect of noise and the empirical errors, because the "true" values of hydrodynamic coefficient of ship mode was not known. A good loss function will increase the accuracy of the identified results and help us to approach the "true" values. Then a global optimization algorithm, based in genetic algorithms, is introduced, as explained in detail in (Xu & Guedes Soares, 2016, Xu et al., 2017).

In order to estimate the hydrodynamic coefficients of equations (4–6), Adjustment coefficients are introduced, which will randomly change within the interval [0.5, 1.5] in the identification process.

$$Coef_{new} = \kappa_{Coef} * Coef_0 \qquad (8)$$

where, $Coef_0$ is the base values of hydrodynamic derivative of the similar ship model, which are given in Xu and Guedes Soares, (2016). The non-dimensional hydrodynamic coefficients of the similar ship model, "Esso Osaka" were presented in Table 3.

Table 6. The adjusted coefficients of the hydrodynamic parameters of nonlinear manoeuvring mathematical model.

Subscript	L^2-norm	Subscript	L^2-norm
η_1'	0.9036	Y_δ'	0.5092
η_2'	0.9439	Y_r'	0.5256
η_3'	0.5073	Y_{rrv_r}'	0.5019
C_R'	0.9968	Y_{eee}'	1.1637
$X_{v_r^2}'$	0.5052	N_0'	0.9962
X_{ee}'	0.5477	N_{v_r}'	1.2113
X_{rr}'	1.3335	N_δ'	0.5431
$X_{v_r r}' + m'$	0.5008	N_r'	0.5097
X_{rrvv}'	0.5028	N_{rrv_r}'	0.5143
Y_0'	0.9609	N_{eee}'	1.0670
Y_{v_r}'	0.5066		

In order to measure the difference or distance between target systems output, $y_i (i = 1 \cdots N)$, and the mathematical model output, $\hat{y}_i (i = 1 \cdots N)$, the loss function will need to be defined. It usually has an important effect on the precision and generalization performance of the desired mathematical model. With the assumed mathematical model, which can be obtained by a mechanical analysis or a priori considerations, the parameter identification consists in seeking values that would minimize the loss function. Some popular loss functions can be defined. The L^2-norm can be defined as follow:

$$F^2(y, \hat{y}) = \|y - \hat{y}\|_{L^2} = \frac{1}{N}\sqrt{\sum_{i=1}^{i=N}(y_i - \hat{y}_i)^2} \qquad (9)$$

5.2 Comparison between numerical simulations and experimental results

Figure 9 presents the comparison between numerical simulations and experimental results from 20°/20° zigzag manoeuvre for the chemical tanker ship model, and from the plot a good agreement can be found, Moreover the small differences are mainly due to wind disturbance.

Figure 10 presents the plots of 30°/30° zigzag manoeuvre for the chemical tanker ship model, which shows a good agreement between numerical simulations and experimental tests.

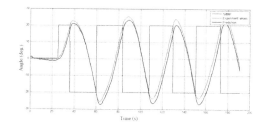

Figure 9. Comparison between experimental and numerical simulation 20°/20° Zigzag manoeuvre.

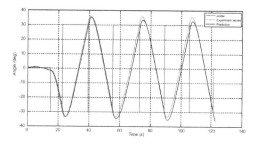

Figure 10. Comparison between experimental and numerical simulation 30°/30° Zigzag manoeuvre.

From the numerical simulations of 20°/20° zigzag manoeuvres for the "Esso Osaka" and the chemical tanker ship models presented in figures 3 and 20 respectively, is clear to see that the chemical tanker has a faster time of response, 15 sec., less yaw overshoot, 3 deg., and smaller period of oscillation, 47.5 sec., than the "Esso Osaka" model. These results are consistent because the differences in ship length and inertias.

6 CONCLUSIONS

The zigzag manoeuvres fit very well to the theoretical expected results and in general, the measured trajectories present a satisfactory agreement with the expected ones.

Experimental results presented in this report demonstrate that a self-running guided model of tanker ship was properly instrumented and brought to working condition. All equipment which included the powering, rudder gear, sensors, and all control and data acquisition software has been functioning properly during preliminary manoeuvring tests in open air swimming pool.

ACKNOWLEDGEMENTS

This work was performed within the Strategic Research Plan of the Centre for Marine Technology and Ocean Engineering (CENTEC), which is financed by Portuguese Foundation for Science and Technology (Fundação para a Ciência e Tecnologia-FCT). The experiments with the vessel model were not possible without the collaboration of Eng. José Costa, Coordinator of the "Piscina Oceanica", Oeiras, Portugal, who allowed the realization of the manoeuvring tests in their installations. The authors are grateful to FP Santos and R Ramachandran for their help during the execution of the model tests.

REFERENCES

Abkowitz, M.A. (1980). Measurement of hydrodynamic characteristics from ship maneuvring trials by system identification. SNAME Transactions, *88*, 283–318.

Benjamin, M.R., & Curcio, J.A. (2004, June). COLREGS-based navigation of autonomous marine vehicles. Autonomous Underwater Vehicles, *2004 IEEE/ OES* (pp. 32–39). IEEE.

Caccia, M., 2006, "Autonomous surface crafts: prototypes and basic research issues", in Proc. 14th Mediterranean Conference on Control and Automation, Ancona, Italy, pp. 1–6.

Ferrari, V., Perera, L.P., Santos, F.P., Hinostroza, M.A., Sutulo, S., & Guedes Soares, C. (2015). Initial experimental tests of a research-oriented self-running ship model. Guedes Soares, C. & Santos T.A. (Eds.) Maritime Technology and Engineering, Taylor & Francis Group, London, UK, 913–918.

Hinostroza, M.A., & Guedes Soares, C. (2016a). Nonparametric estimation of directional wave spectra using two hyperparameters. Guedes Soares, C. & Santos T.A., (Eds.) Maritime Technology and Engineering, 3, London, UK: Taylor & Francis Group; 287–293.

Hinostroza, M.A., & Guedes Soares, C. (2016b). Parametric Estimation of the Directional Wave Spectrum from Ship Motions. International Journal of Maritime Engineering, *158*, A121-A130.

Larson, J., Bruch, M., & Ebken, J. (2006). Autonomous navigation and obstacle avoidance for unmanned surface vehicles. Space and naval warfare systems center Sand Diego CA.

Moreira, L., Fossen, T.I., and Guedes Soares, C. 2007, "Path Following Control System for a Tanker Ship Model". Ocean Engineering. 34:2074–2085.

Moreira, L., Santos, F.J., Mocanu, A., Liberato, M., Pascoal, R., and Guedes Soares, C., 2008, " Instrumentation used in guidance, control and navigation of a ship model", 8th Portuguese Conference on Automatic Control, Vila Real, Portugal, pp. 530–535.

Perera, L.P., Ferrari, V., Santos, F.P., Hinostroza, M.A., and Guedes Soares, C. (2015). Experimental Evaluations on Ship Autonomous Navigation and Collision Avoidance by Intelligent Guidance. IEEE Journal of Oceanic Engineering, *40*(2), 374–387.

Perera, L.P., Ferrari, V., Santos, F.P., Hinostroza, M.A.,and Guedes Soares, C., 2013, "Experimental results on collisions avoidance of autonomous ship manoeuvres", Proceedings of the 32nd International Conference on Ocean, Offshore and Arctic Engineering (OMAE 2013), Nantes, France, paper OMAE 2013–11265.

SNAME, (1950). The Society of Naval Architects and Marine Engineers, Nomenclature for Treating the Motion of a Submerged Body Through a Fluid, Technical and Research Bulletin No. 1–5, 1950.

Sutulo, S. and Guedes Soares, C., (2011), Mathematical models for simulation of manoeuvring performance of ships. Guedes Soares, C. Garbatov Y. Fonseca N. & Teixeira A.P., (Eds.). Marine Technology and Engineering. London, UK: Taylor & Francis Group; pp. 661–698.

Sutulo, S.; Moreira, L., and Guedes Soares, C., (2002), Mathematical Models for Ship Path Prediction in Manoeuvring Simulation Systems . Ocean Engineering; 29(1):1–19.

Xu, H., & Guedes Soares, C. (2016). Vector field path following for surface marine vessel and parameter identification based on LS-SVM. Ocean Engineering, 113, 151–161.

Xu, H.,.Hinostroza M.A, & Guedes Soares, C. (2017). Identification of hydrodynamic coefficients of ship nonlinear manoeuvring mathematical model with free running model tests. C. Guedes Soares & A.P. Teixeira (Eds). Maritime Transportation and Harvesting of Sea Resources, UK: Taylor and Francis.

Maritime Transportation and Harvesting of Sea Resources – Guedes Soares & Teixeira (Eds)
© *2018 Taylor & Francis Group, London, ISBN 978-0-8153-7993-5*

Comparative simulation of definitive manoeuvres of the KVLCC2 benchmark ship using different empiric mathematical models

S. Sutulo & C. Guedes Soares

Centre for Marine Technology and Ocean Engineering (CENTEC), Instituto Superior Técnico, Universidade de Lisboa, Lisbon, Portugal

ABSTRACT: The in-house manoeuvring simulation code MANIST was augmented with the empiric mathematical model developed by Pershitz and under his supervision in the years 1960–1985. Although the model had been initially developed on the basis of a series of light cruisers and was known to provide good predictions for naval combatants, it was later extended to more full-bodied merchant vessels. Results of computations of the behaviour of the KVLCC2 ship in standards manoeuvres (turning circles, Dieudonné spiral, zigzags) as well as time histories of various kinematical components were obtained with the model mentioned above and also with the well-known Inoue and Kijima models. Comparisons with some independent published data were also carried out. The results demonstrate a considerable scatter of the predictions while some sub-models combinations showed better quality of predictions than the others.

1 INTRODUCTION

In spite of substantial progress achieved in application of the methods of Computational Fluid Dynamics (CFD) to hydrodynamics of ships in curvilinear manoeuvring motion, interest to fast and simple empiric methods for predicting manoeuvring properties of ships does not seem to fade. While the CFD methods are most consistent and promising, at present level of the hardware development they are prohibitively computationally heavy and practically unusable for manoeuvring predictions of ships at early design stages. The same is true for methods based on physical modelling.

The so-called empiric methods using databases of previously conducted systematic model tests or CFD computations are very fast and easy to use even when the hull shape is not defined in all details. However, these methods suffer from increased uncertainties and possible substantial errors in predicted manoeuvring qualities (Quadvlieg et al. 2015). Credibility of predictions can be somewhat increased when several independent methods are used which has some resemblance with the data assimilation method.

Practically all existing empirical methods are of modular type i.e. based on the decomposition of total hydrodynamic loads into those applied separately to the hull, propeller, and rudder (or other steering device but the term "rudder" is often used in the generalized sense).

Currently, the most popular and widely used group of empirical methods is associated with the Japanese Manoeuvring Modelling Group (MMG) and includes the procedures developed by Inoue et al (1981), Matsunaga (1993), Kijima (2003), Yasukawa & Yoshimura (2015).

Much less known to the international community is the empiric method developed by Pershitz and under his supervision at the Krylov Ship Research Institute (St. Petersburg, Russia) on the basis of multiple tests with scale physical models carried out on two rotating-arm facilities and using powered models. The method was primarily developed on the basis of captive-model tests performed with hull shapes typical for pre-war light cruisers and it was oriented to prediction of the turning ability of surface displacement naval combatants. This naval heritage can still be traced and will be commented further in course of description of the method. Obviously, in course of years many additional tests with fuller hull forms typical for merchant vessels were carried out and analysed. As result, nowadays the method is positioned as applicable to any type of surface displacement ships. In course of decades, the method was used as standard and de facto official in the Russian shipbuilding industry. Besides routine applications at the ship design centres, it was widely used in teaching and research including development of implicit manoeuvring standards embedded in the Rules of the Russian Maritime Register of Shipping (RS 2017).

First versions of the method and its fragments were mainly described in 50 s in internal reports of the Krylov Institute but then it was published in the first edition of the reference book (Voytkunsky et al. 1960). Improved and extended versions were later published in the 2nd and 3rd editions of the same book (Voytkunsky et al. 1973) and (Voytkunsky 1985). Also, interesting comments on the history of the method and some considerations behind its development are presented in (Pershitz 1983).

After the first relatively complete publication on an MMG method (Inoue et al. 1981) it appeared that the interest to the Pershitz method has diminished even among Russian specialists. This happened because the Pershitz method suffers from several internal inconsistencies many of which had been traced and commented by Sobolev (1976). These inconsistencies will be also commented below, in course of description of the method. However, Pershitz himself admitted that his method was exploiting mutual cancellation of some evident inaccuracies (Pershitz 1983).

An unbiased look to the situation with the empiric methods for ship manoeuvring indicates, however, that no one of the existing methods is able to provide accurate and reliable predictions in all cases of interest. In view of this truth, the authors of the present contribution judged reasonable implementation of the rather aged Pershitz method as an additional option in their multi-model offline manoeuvring simulation code MAN-IST (Sutulo & Guedes Soares 2005). In any case, this increases the potential of the code, especially in application to naval combatants, and bearing in mind that in spite of evident and sometimes crude imperfections the method accumulates many years of experimental research and a great deal of engineering perspicacity and intuition.

As the Pershitz method has never been described in English language literature, its concise commented description will be presented in the next section following the last published official version (Voytkunsky 1985). Unfortunately, it was not possible to give there a full exposure of the method allowing its reproduction because of too cumbersome approximations of hull hydrodynamic forces and from this viewpoint the method is less portable than any method of the MMG group. At least partly this is historically conditioned: the method had been devised much earlier than its counterparts and was almost exclusively oriented to manual calculation of the steady turning motion. The hydrodynamic coefficients of the hull depending on geometric parameters were to be estimated from rather complex nomograms which were much later approximated with sets of multivariate, multistage polynomials applied

piecewise. These approximations (mainly devised by Dr Melkozerova) facilitate greatly computer implementations of the method. It is clear, however, that unlike most other methods mainly based on direct regressions over experimental data, the Pershitz method uses de facto two-stage approximations not connected directly with experimental responses.

In the following sections a concise commented description of the method is presented and some results of its application for simulating turning and zigzag manoeuvres of the KVLCC2 benchmark ship. Also, the same manoeuvres were simulated using the Inoue and Kijima models. The authors did not focus too much on comparisons with other results available for this vessel but rather tried to demonstrate relative performance of the Pershitz method.

2 DESCRIPTION OF THE PERSHITZ MANOEUVRING MATHEMATICAL MODEL

2.1 Equations of motion

The Pershitz model is 3DOF and is only describing the ship's motion in the horizontal plane. The kinematic equations are standard and the dynamic equations are written in the central body axes $Gxyz$ with the z-axis pointing downwards:

$$
\begin{aligned}
(m + \mu_{11})\dot{u} - mvr &= X_H + X_P + X_R, \\
(m + \mu_{22})\dot{v} + mur &= Y_H + Y_R, \\
(I_{zz} + \mu_{66})\dot{r} &= N_H + N_R,
\end{aligned}
\tag{1}
$$

where m is the mass of the ship, I_{zz} is its moment of inertia; μ_{ij} are the ship added masses; u is the velocity of surge, v —the velocity of sway, r is the angular velocity of yaw; X, Y are the hydrodynamic forces of surge and sway, N is the yaw moment, and the subscripts H, P, R stand for the hull, propeller and rudder respectively.

The hull hydrodynamic forces are represented as:

$$
\begin{aligned}
X_H &= \mu_{22}vr - R_T(V), \\
Y_H &= -\mu_{11}ur + Y_H' \frac{\rho V^2}{2} LTC_L, \\
N_H &= N_H' \frac{\rho V^2}{2} L^2 TC_L,
\end{aligned}
\tag{2}
$$

where $V = \sqrt{u^2 + v^2}$ is the current speed of the ship, $R_T(V)$ is its total resistance in straight run but with the actual speed; Y_H', N_H' are the hull sway force and yaw moment coefficients, L and T are the ship length and draught, and C_L is the effective centerplane area coefficient calculated through a special procedure described below. It can be already noticed that the reference area is here slightly different from the more commonly used LT.

No special procedure is envisaged for the propeller force X_P while the rudder contributions are represented as

$$X_R = -C_{XR} \frac{\rho V^2}{2} A_{RE},$$

$$Y_R = -C_{YR}^{\alpha} \alpha_R \frac{\rho V^2}{2} A_{RE}, \qquad (3)$$

$$N_R = x_R Y_R,$$

where C_{XR} is the rudder drag coefficient, C_{YR}^{α} is the rudder lift coefficient gradient, A_{RE} is the effective rudder area, x_R is the rudder abscissa (relative to the midship plane which somewhat contradicts the central axes assumption), and α_R is the rudder attack angle defined as:

$$\alpha_R = \delta_R - \kappa_E \left(\beta - x_R' r' \right), \qquad (4)$$

where δ_R is the rudder deflection angle, κ_E is the straightening factor, β is the ship drift angle, $x_R' = x_R / L$, and $r' = rL/V$ is the ship usual dimensionless rate of yaw.

The description above already reveals some evident inconsistencies of the Pershitz method. In particular: the rudder drag and lift are assumed to be identical with the rudder surge and sway force respectively neglecting evident necessary transformations. The rudder speed with respect to water is identified with the ship speed although it can be very different from it. The same straightening factor is applied to the sidewash caused by the ship drift and its yawing although it is generally acknowledged that these contributions must be handled separately. Cavitation on the rudder is neglected even when high speed ships are considered.

In the following subsections procedures for estimating hull and rudder hydrodynamic coefficients as well as some auxiliary parameters will be outlined.

2.2 Hull coefficients

While the hull surge force contribution is based on the resistance curve, see eq. (2), the sway and yaw coefficients are calculated as

$$Y_H' = Y_\beta' \beta + Y_{\beta\beta}' \beta |\beta|,$$

$$N_H' = N_\beta' \beta + N_r' r', \qquad (5)$$

where the coefficients Y_β', $Y_{\beta\beta}'$, N_β', N_r' traditionally called "manoeuvring derivatives" depend on the

hull geometric characteristics and on the dynamic trim.

Comparing to other modular empiric methods the hull force model (5) looks extremely simple. The sway force is assumed to be independent of the yaw motion as experiments typically show relatively small absolute values of the coefficient Y_r' which may take different signs and certainly this contribution is much inferior to that by the inertial term mur in (1). Although experimental data show some nonlinearities for the yaw moment, they are typically much weaker than those for the sway force and are neglected.

As result, the structural model (5) is of minimum possible complexity which was, however, regarded as sufficient for capturing all quantitative peculiarities of manoeuvring models of surface displacement ships which are always substantially nonlinear objects. The fact that the drift angle is used instead of the more typical dimensionless velocity of sway is not very important for moderate manoeuvres.

The structure (5) is only valid for manoeuvres realizable at any speed with normal steering devices. Extension of the model (5) to arbitrary manoeuvres was carried out by Tumashik (Voytkunsky 1985) but this extended form is not presented and used here.

The "manoeuvring derivatives" in (5) depend on the following geometric parameters of the ship hull: the prismatic coefficient C_P, on the dimensionless abscissa of the centre of mass with respect to the midship plane $x_G' = x_G / L$, on the hull aspect ratio $k_H = T/L$ and on the hull centerplane area coefficient C_L. The latter is in general calculated as

$$C_L = 1 - \frac{3}{20 - i_{UV}} \cdot \frac{A_{CO}}{LT} + \frac{0.054}{k_H} (\tan\theta_1 + \tan\theta_2), \quad (6)$$

where i_{UV} is the not necessarily integer number of the theoretical section (all numbered from 0 at the fore perpendicular to 20 at the aft perpendicular) corresponding to the transition between U- and V-shaped sections, A_{CO} is the centerplane stern cutoff area, θ_1 is the static trim angle of the ship which is positive by the stern and include also the design trim, θ_2 is the dynamic trim depending on x_G' and on the ship Froude number $Fn = V / \sqrt{gL}$. The dynamic trim is assumed to be zero at $Fn \le 0.34$ and is estimated using a set of piecewise-defined 4th-order polynomials for the Froude number values up to 0.6. As long as all "manoeuvring derivatives" in (5) depend on C_L, they all depend on the Froude number exceeding 0.34 which betrays the primary "naval" purpose of the method. This dependence is only shown as indirect i.e. through the dynamic trim although direct influence of high Froude numbers is also present.

The Froude number affecting the hull forces according to the method description must correspond to the approach phase i.e. remains constant during the manoeuvre. More logical, however, would be to use the current Froude number and in this study the both variants were implemented.

Definition of the cutoff area A_{CO} is not straightforward and special instructions given in (Voytkunsky 1985) must be followed. For instance, the design trim must be ignored and the projection of the propeller shaft bossing, if present, must be excluded unlike shaft struts and any other appendages which should be ignored. The skegs are treated as part of the hull i.e. their presence just reduces the cutoff area.

Further, the first two terms in (6) must be substituted with 0.975 for the bulbous stern and with 0.962 for the stern with the thick skeg and the parameters A_{CO} and i_{UV} are not considered in those cases.

Of all "manoeuvring derivatives" only the yaw damping coefficient is approximated with the unique formula:

$$N_r' = -\left(0.739 + 8.7k_H\right)\left(1.611C_L^2 - 2.873C_L + 1.33\right) \quad (7)$$

while approximations for the remaining "derivatives" are much more complicated and should be consulted in the primary source: their full implementation required around 150 lines of code.

In general, concerning the hull forces the Pershitz method is characterized, besides the Froude-number dependence for fast vessels, by a rather thorough handling of the influence of the shape and arrangement of the afterbody. From this viewpoint, the Pershitz method was almost 40 years ahead of other empiric methods.

2.3 Estimation of the rudder force coefficients

The rudder lift gradient is represented as

$$C_{YR}^\alpha = kC_{YR0}^\alpha, \quad (8)$$

where the correction factor $k = 1.0$ for an all-movable (spade) rudder, $k = 1.3$ for a rudder behind a rudder post, and $k = 0.9$ for a steering nozzle.

For all rudders except that with the rudder post the base gradient C_{YR0}^α is estimated using the classic Prandtl formula:

$$C_{YR0}^\alpha = \frac{2\pi C_R k_R}{2 + k_R}, \quad (9)$$

where k_R is the aspect ratio of the rudder and the correction factor $C_R = 1$ for all-movable rudders and $C_R = 7/6 - k_R/3$ for horn rudders.

A special formula is reserved for rudders behind the rudder post:

$$C_{YR0}^\alpha = \frac{2\pi k_R}{2 + k_R}\sqrt{b_M'} + 1.4\frac{k_R^2 C_{TA0}}{k_R^2 + 0.47}, \quad (10)$$

where $b_M' = b_M/b_R$ is the mean chord of the blade non-dimensionalized by the mean chord of the whole rudder b_R and

$$C_{TA0} = \frac{8T_0}{\rho\pi D_P^2 V_0^2} \quad (11)$$

is the propeller loading coefficient in approach phase with D_P being the propeller diameter and T_0 is the propeller thrust corresponding to the propulsion point at the velocity V_0.

The Pershitz method provides also a formula for the lift gradient of a steering nozzle but this case is not discussed here.

The effective rudder area present in (3) is defined as

$$A_{RE} = A_{M0} + A_{MP}\left(1 + C_{TA0}\right), \quad (12)$$

where A_{M0} is the area of the part of the rudder blade outside the propeller slipstream and A_{MP} is the area of the remaining part of the blade.

The straightening coefficient

$$\kappa_E = \kappa_H \kappa_P, \quad (13)$$

where the hull coefficient κ_H varies from 0.3 to 1.0 depending on the stern type and on the rudder type and arrangement, see (Voytkunsky 1985) for detailed recommendations.

For all types of the rudder except for that behind the rudder post the propeller factor is:

$$\kappa_P = \frac{A_{R0} + A_{RP}\sqrt{1 + C_{TA0}}}{A_{R0} + A_{RP}\left(1 + C_{TA0}\right)}, \quad (14)$$

where A_{R0} is the whole rudder area outside the slipstream and A_{RP} is the area inside the slipstream.

The method for calculating the rudder lift described above is more or less consistent physically but contains many simplifying assumptions commented in (Sobolev 1976). For instance, the assumed value of the slipstream velocity corresponds to the flow at infinity while the rudder area in the slipstream is estimated without account for the contraction of the jet.

On the other hand, the assumption on the linear dependence of the lift on the attack angle in combination with the Prandtl formula overestimating the lift gradient for small aspect ratios is another simplification commented in more detail in (Sutulo & Guedes Soares 2011).

Due to all imperfections combined, the method for the rudder force recommended for predictions of the turning ability, has never been positioned to provide reliable estimates for the true force on the rudder and a totally independent and more complicated method was recommended for this purpose (Voytkunsky 1985) but that alternative method is only applicable for the straight motion with zero drift angle. That method was also recommended by Pershitz for estimating the rudder drag coefficient as historically it was considered not essential for turning ability predictions performed with non-dimensionalized equations.

However, the information on the rudder drag provided by the reference book (Voytkunsky 1985) is very scarce and is limited by a couple of experimental plots for one all-movable and one horn rudder which could not be accepted for an all-purpose manoeuvring code. As the major contribution for the rudder drag comes from the induction resistance, the Prandtl formula as presented in (Crane et al. 1989) was used in the present study:

$$C_{XR}(\alpha_R) = 0.0065 + \frac{C_{YR}^2}{0.9\pi k_R}. \qquad (15)$$

Comparisons of estimates obtained with the formula (15) with the plots mentioned above shows some over-prediction for the induction part. However, as no better alternatives or reliable data were found and the accuracy of the published plots presented some doubts, the formula (15) was used as is.

3 NUMERICAL RESULTS AND DISCUSSION

3.1 Brief description of the vessel

The virtual vessel KVLCC2 has lately become a very popular benchmark ship or model and its description can be found in numerous publications, see e.g. (Quadvlieg et al. 2015) and (Yasukawa et al. 2015). The ship has the following main particulars: $L = 320$m, the breadth $B = 58$m, $T = 20.8$m, $m = 320825$t, the block coefficient $C_B = 0.81$ (however, it was assumed that $C_P = 0.80$ as this was the limiting value in Pershitz' nomograms), the rudder area with the horn $A_R = 136.7$m^2, and $i_{UV} = 13.5$.

3.2 Trajectories in turning manoeuvre

All simulations were performed with the helm 35 deg starboard. The used propeller mathematical model was the same in all cases and is described in (Sutulo & Guedes Soares 2011, 2015). The propeller rotation frequency was kept constant. The trajectories obtained with various methods with the simulation time 2,000 s are shown in Figure 1. As it was difficult to determine the appropriate type of the stern, simulations were carried out for three variants: "normal" stern with the U–V transition (type 0), stern with bulb (type 1) and stern with the thick skeg (type 4). The results are rather different and the variant with the stern type 0 was considered as the most plausible and selected for the following analysis and simulations. As the initial Froude number for the vessel was only 0.142, it is obvious that the dynamic trim was absent and the high-Froude-number potential of the Pershitz method was not exploited.

The turning trajectories with the snapshots of characteristic positions of the vessel are also presented separately in Figures 2–4. It can be seen immediately that the Pershitz method and the MMG methods have performed very differently and the trajectory predicted by the former method looks rather weird containing some loop and with a part of the turning circle lying to the port side from the approach path. Such trajectories can be found sometimes in publications but do not look plausible and apparently the Pershitz method is not very suitable for this ship form. Also, the authors have never faced such trajectories in full-scale trials and this effect did not appear in many other KVLCC2 simulations (Quadvlieg et al. 2015). The both MMG methods were used here with the

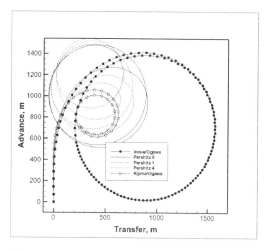

Figure 1. Trajectories in turning manoeuvre: various mathematical models.

rudder model suggested by Ogawa (1980) and look qualitatively similar although predicting very different degrees of the turning ability.

The computed values of the relative advance A_d/L and relative tactical diameter D_T/L are given in Table 1 where they are compared with simulated and experimental data from (Yusakawa et al. 2015) and with free-running models results presented at SIMMAN2008 (Stern & Agdrup 2009). Besides the hull–rudder models combinations mentioned above the data are also given for the Inoue and Kijima hull models combined with the rudder model used in (Sutulo & Guedes Soares 2015) and based on the model earlier proposed by Söding (1982).

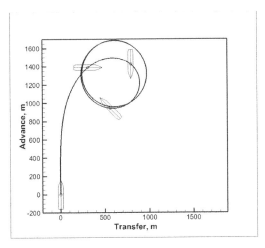

Figure 4. Trajectory with ship images for the Pershitz model.

Table 1. Values for relative advance and tactical diameter in 35 deg helm turning manoeuvre.

Model or source	A_d/L	D_T/L
Inoue/Ogawa	4.3	4.8
Inoue/Soeding	5.1	2.69
Kijima/Ogawa	3.13	1.84
Kijima/Soeding	3.99	3.05
Pershitz 0	3.6	2.5
Pershitz 1	3.36	2.07
Pershitz 4	3.12	1.68
(Yasukawa et al. 2015)	3.0–3.4	3.1–3.25
(Stern & Agdrup 2008)	2.5–3.0	3.0–3.3

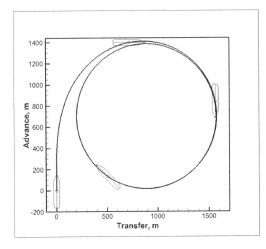

Figure 2. Trajectory with ship images for the Inoue model

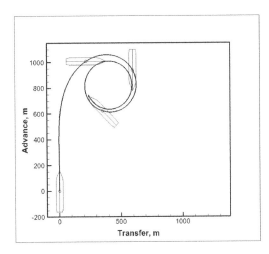

Figure 3. Trajectory with ship images for the Kijima model.

These data show relatively good prediction of the advance except for Inoue method which is also heavily overpredicting the tactical diameter. The other methods rather underpredict it except for the combination Kijima–Soeding which apparently hits well here although, as will be commented below, it is quite inacceptable in general.

3.3 Time histories in turning manoeuvre

Observation of time histories provides a somewhat deeper insight of peculiarities of various mathematical models and helps to explain some specifics of the trajectories. Responses in the ship speed and drift angle during the turning manoeuvre are presented in Figure 5 and those for the dimensionless yaw rate r' and the rudder attack angle α_R—in Figure 6.

Inspection of the time histories for the ship speed demonstrates considerable speed drop shown by all models which could be expected from a highly turnable vessel. This drop is smaller for

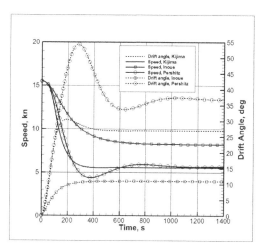

Figure 5. Time histories for the current ship speed and drift angle in turning manoeuvre.

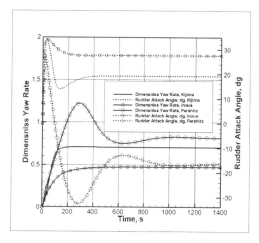

Figure 6. Time histories for the dimensionless yaw rate and for the rudder attack angle in turning manoeuvre.

the Inoue model which underpredicts the turning ability. The response for the Pershitz model exhibits a rather peculiar nonmonotonic behaviour. This corresponds to the specifics of the predicted trajectory but in general does not look credible. Regarding the simplistic nature of the method this is not very surprising.

Even larger anomalies are revealed from inspection of the plots for the drift angle. While it grows monotonically to a rather moderate value of 11 degrees for the Inoue model, it overshoots almost up to 55 degrees for the Pershitz model showing almost oscillatory behaviour. It must be noticed that at such large values of the drift angle the applicability of the equations (5) is no longer guaranteed

and this can be one of the reasons for the inappropriate overall behaviour of the model. The Kijima model also exhibits an overshoot although much less pronounced than in the previous case.

Inspection of the responses in the dimensionless turning rate and in the rudder attack angle reveals similar peculiarities including "wavy" behaviour of the dimensionless yaw rate predicted by the Pershitz model. The steady values for the rudder attack angle predicted by the Inoue and Kijima models are not very different although the initial transient looks rather different. The most peculiar is the behaviour of the attack angle predicted by the Pershitz model: the attack angle very soon changes the sign and remains negative in the steady turn. This is characteristic for directionally unstable ships and means that the rudder works as a stabilizer preventing the ship from falling into a tighter turn.

3.4 Spiral curves

The Dieudonné spiral manoeuvre was simulated for the model combinations presented in Figure 7 but also for the combination Kijima–Soeding. The latter case, however, is not shown as the manoeuvre failed: the ship turned out so unstable that was not able to change the direction of the turn.

However, other models combinations also resulted in very different steady turning diagrams. The Inoue hull model always resulted in a directionally stable ship and the application of the Soeding rudder model just reduced the predicted turning ability at larger rudder deflection angles. The Pershitz model lead to a moderately directionally unstable ship which agrees with the inversion of the rudder attack angle commented above. The most disappointing was the spiral curve obtained with the Kijima model: the resulting ship turned out prohibitively unstable. The most strange was amplification of this instability with the alternative Soeding rudder model while the same submodel change rather improved the inherent directional stability when applied to the Inoue hull model. The internal causes of this phenomenon still remain unclear but it is already obvious that the spiral curve is very informative and is able to effectively discriminate between ships with very similar turning ability at large helms.

3.5 Zigzag manoeuvres

The standard 10°–10° and 20°–20° were simulated for various combinations of the hull and rudder models and some results are presented in Figures 8 and 9. To avoid congestion, the rudder angle output being not very informative is only shown for one model. The response for the Kijima model in the 10°–10° zigzag is not shown in Figure 8 as this

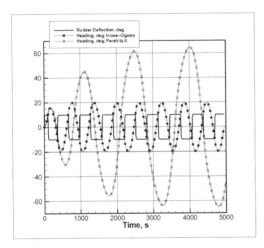

Figure 7. Time histories for the current ship speed and drift angle in turning manoeuvre.

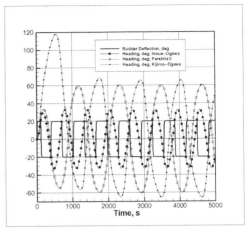

Figure 9. Time histories for the 20–20 zigzag manoeuvre.

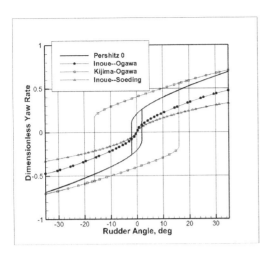

Figure 8. Time histories for the 10–10 zigzag manoeuvre.

manoeuvre failed which, however, could be expected regarding the spiral curve for this model shown in Figure 7. The 20°–20° zigzag turned feasible for the same model although with an almost 100 degrees first overshoot. The unstable Pershitz model has also produced large second and following overshoots. The first overshoot in 10–10 zigzag was 5.5 deg as predicted by the Inoue model and 6.4 deg according to the Pershitz model which corresponds to the data by Yasukawa et al. (2015): from 5.6 to 8.2 degrees. The values provided by earlier free-running models tests vary from 7 to 9 degrees (Stern & Agdrup 2009).

However, Figure 8 also demonstrates that when unstable models are involved, the disagreement in the overshoot angle may become much higher at the third and further overshoots.

4 CONCLUSIONS

The main result of the present study is the implementation of the empirical manoeuvring model developed by Pershitz. The model was tested on the benchmark ship KVLCC2 and the responses were compared with those obtained with more popular empirical models and with some independent published data. The discussion of the output data carried out in the preceding section makes possible drawing the following conclusions:

1. The scatter of results obtained with various empirical methods is considerable which corresponds to earlier observations (Quadvlieg et al. 2015).
2. The original Pershitz model did not perform well with the full-bodied vessel. As was already mentioned, one of the causes could be too large values of the drift angle exceeding normal limits for the approximation (5). This makes desirable implementation of the extended model developed by Tumashik and applicable to arbitrary manoeuvres. It may result in better behaviour of this model even with moderate manoeuvres performed by full ships. On the other hand, it will be reasonable to test and validate all the models with slimmer ship forms which had been laid in the foundation of the Pershitz method.
3. The performed comparisons have clearly shown that manoeuvring models must be validated using, first of all, the time responses for various phase coordinates although inspection of the trajectories can also be helpful as an auxiliary tool. Consideration of such convenient and popular numerical measures as the advance, tactical diameter, zigzag overshoots

etc. is absolutely insufficient: these measures can be predicted well while the overall modelled dynamic properties of the ship are not even close to reality.

4. Unsatisfactory observed performance of the Kijima model is somewhat disappointing but the authors cannot state that its implementation was absolutely free of flaws. In particular, the parameters of the stern shape used by this model were determined rather crudely and the corresponding sensitivity of the model must be explored. Also, implementation of the latest "standard" MMG model (Yasukawa & Yoshimura 2015) is planned

5. All empirical manoeuvring models in general contain a number of not so well defined parameters which may be varied for tuning that can be performed either manually or using some system identification procedure (Sutulo & Guedes Soares 2014). It was confirmed during the debugging and verification process that these models are robust in the sense that even large variations of many parameters do not lead to dramatic changes in the responses. This is exactly the property making the inverse problem of the parametric identification ill-posed.

In general, this study confirmed once more that creation of a reliable and accurate empirical method for simulating manoeuvring motion of ships is an extremely difficult task and such methods must be applied wisely and carefully.

ACKNOWLEDGEMENTS

The study was performed within the project PTDC/EMSTRA/5628/2014 "Manoeuvring and moored ships in ports—physical and numerical modelling", funded by the Portuguese Foundation for Science and Technology.

REFRERENCES

Crane, C. L., Eda, H., Landsburg, A.C. 1989. Controllability. In E. V. Lewis (ed.), *Principles of Naval Architecture*, Volume 3: 191–422. Jersey City, NJ: SNAME.

Inoue, S., Hirano, M, Kijima, K., Takashina, J. 1981. A practical calculation method of ship maneuvering motion. *International Shipbuilding Progress* 28(325): 207–222.

Kijima, K. 2003. Some studies on the prediction for ship manoeuvrability. In: *International Conference on Marine Simulation and Ship Maneuverability MAR-SIM'03, 25 28 August 2003, Kanazawa, Japan*, pp. KN–3–1–KN–3–10.

Matsunaga, M. 1993. Method of predicting ship manoeuvrability in deep and shallow waters as a function of loading condition. *NK Technical Bulletin:* 51–59.

Ogawa, A., Hasegawa, K., Yoshimura, Y. 1980. Mathematical modeling of the ship's maneuvering, *Nihon Zosen Gakkaishi (Techno Marine),* 616: 565–576 (in Japanese).

Pershitz, R.Y. 1983. *Manoeuvrability and ship steering (in Russian)*. Leningrad: Sudostroyeniye.

Quadvlieg, F., Simonsen, C., Otzen, J., Stern, F., 2015. Maneuvering simulation of a KVLCC2 tanker in irregular seas, MARSIM 2015: *International Conference on Ship Manoeuvrability and Maritime Simulation, Newcastle University, United Kingdom, 8–11 September 2015*, Paper 4–2–3, 16p.

RS 2017. *Rules for classification and construction of seagoing ships. Part III: Equipment, arrangements and outfit*, St. Petersburg: Russian Maritime Register of Shipping.

Sobolev, G.V. 1976. *Ship manoeuvrability and automated-Navigation (in Russian)*. Leningrad: Sudostroyeniye.

Söding, H. 1982. Prediction of ship steering capabilities, *Schiffstechnik* 29: 3–29.

Stern, F. & Agdrup, K. (Eds.) 2009. *SIMMAN2008: Workshop on verification and valdation of ship manoeuvring simulation methods, Copenhagen, 14–16 April 2008*, Force Technology Publ.

Sutulo, S. & Guedes Soares, C. 2005. An object-oriented manoeuvring simulation code for surface displacement ships. In C. Guedes Soares, Y. Garbatov, and N. Fonseca (Eds.), *Maritime Transportation and Exploitation of Ocean and Coastal Resources*, pp. 287–294. London: Taylor & Francis Group.

Sutulo, S. & Guedes Soares, C. 2011. Mathematical Models for Simulation of Manoeuvring Performance of Ships, In: C. Guedes Soares et al. (eds.), *Marine Technology and Engineering:* 661–698. London, Taylor & Francis Group.

Sutulo, S., Guedes Soares, C. 2014. An algorithm for offline identification of ship manoeuvring mathematical models from free-running tests, *Ocean Engineering.* 79: 10–25.

Sutulo, S. & Guedes Soares, C. 2015. Development of a core mathematical model for arbitrary maneuvers of a shuttle tanker, *Applied Ocean Research* 51: 293–308

Voytkunsky, Y.I., Pershitz, R.Y., Titov, I.A. 1960. *Handbook on ship hydrodynamics (in Russian)*, Leningrad, Sudpromgiz

Voytkunsky, Y.I., Pershitz, R.Y., Titov, I.A. 1973. *Handbook on ship hydrodynamics: propulsion and manoeuvring (in Russian)*, Leningrad, Sudostroyeniye

Voytkunsky, Y.I. (ed.) 1985. *Handbook on ship hydrodynamics (in Russian), Vol.3*, Leningrad, Sudostroyeniye.

Yasukawa, H., Hirata, N., Yonemasu, I., Terada, D., Matsuda, A. 2015. Maneuvering simulation of a KVLCC2 tanker in irregular seas, *MARSIM 2015: International Conference on Ship Manoeuvrability and Maritime Simulation, Newcastle University, United Kingdom, 8–11 September 2015*, Paper 2–2–3, 16p.

Yasukawa, H., Yoshimura, Y. 2015. Introduction of MMG standard method for ship maneuvering predictions, *J. Mar. Sci. Techno.* 20: 37–52

Ship structures

Maritime Transportation and Harvesting of Sea Resources – Guedes Soares & Teixeira (Eds)
© 2018 Taylor & Francis Group, London, ISBN 978-0-8153-7993-5

Assessment of the residual strength of ships after damage

M. Attia
Alexandria Shipyard, Alexandria, Egypt

A. Zayed, H. Leheta & M. Abd Elnaby
Department of Naval Architecture and Marine Engineering, Faculty of Engineering, Alexandria University, Alexandria, Egypt

ABSTRACT: This paper aims to investigate the residual ultimate strength capacity of ship hull after encountering damage. Both ship collision and grounding were considered with different levels of severity. The extent of damage was defined according to the specification of the classification society rules along with other codes related to the maritime field. The effect of corrosion degradation on the residual strength of ship hull after damage was investigated through the ship's life. As the reduction of computational time is an essential demand in certain situations, such as during the emergency response to ship damage, progressive collapse methods were used to assess the ultimate strength capacity of the ship hull. In order to further optimize the computational time, a correction factor was developed in order to assess the ultimate strength of the corroded ship hull based on the as-built one. This avoids the necessity to remodel the corroded ship hull. This correction factor was analytically verified and high accuracy was obtained. It was shown that there are many types of damage that must be considered in the assessment of residual strength of ship hull. Aging is an important factor and must be considered in the assessment of ship strength after damage.

1 INTRODUCTION

The ship as an asset sails in severe environment and may exhibit many types of damage that lead to environmental catastrophes. This issues the importance of assessment of the residual strength of ships after damage. In case of oil tanker damage, a fast and quick action must be taken to save the ship and protect the environment from the consequences of such accident. This action depends on the efficient assessment of the residual strength of the damaged section. When the ultimate strength assessment is addressed, several methods can be used, such as ISUM, FEA and Smith method. In order to optimize the damage assessment process, the effect of corrosion degradation on the ultimate strength must be considered, but in a quick and easy way to reduce the computational time.

As a first attempt to assess the hull girder ultimate strength, the ship is idealized as free-free end beam and the stress distribution due to external forces is calculated at the section of maximum load. By comparing this stress with the yield strength of the material, a judgment on the ultimate strength capacity can be obtained. By adjusting the values of the first and second moments of area, the damage state can be expressed, and the bending and shear stresses can be calculated.

However, this method has limitations in modelling the true behaviour of the damaged vessel's structure. The failure is based on the yield strength or an allowable elastic strength of the material and not the ultimate strength. Similarly the method does not allow the calculation of failure of the structure due to any local collapse mechanism.

Caldwell (1965) idealised the cross section and showed that it is composed of stiffened panels with equivalent thicknesses. Then, he calculated the fully plastic bending moment of the cross-section considering the influence of buckling.

The method employed by Caldwell (1965), to determine the ultimate strength, assumes that the material within the structure that is in tension has fully yielded and all the material that is under compression has reached its ultimate buckling strength. As more material is allowed to reach its limit state than will occur in an actual vessel, the resulting ultimate bending moment for the section will be greater than the actual value (Huilong et al. 2008), leading to an un-conservative assessment of the ultimate longitudinal bending strength capability.

To take into account the more realistic scenario where the structural material between the deck and keel is unlikely to reach its limit state, Caldwell (1965) method was further developed by Qi & Cui (2006), adopting a method developed by Paik &

Mansour (1995). This method has many weaknesses for application with damage structure as the influence of damage won't be accounted for.

ISUM is another method which is considered as an alternative to the progressive collapse method. This method is based on minimizing the degrees of freedom which are used in the finite element analysis. As it can be considered as a FEM method, modeling the structure of the vessel is from some elements. By combining all these elements, the cross section can be modeled. Despite the developments of ISUM over the years since its conception, Paik & Thayamballi (2003) stated that the number of users of the method in commercial application remains limited.

In the progressive collapse method proposed by Smith (1977), the cross-section is divided into small elements composed of stiffener(s) and attached plating. At the beginning, the average stress–average strain relationships of individual elements under axial load are derived considering the influences of yielding and buckling. Then, a progressive collapse analysis is performed assuming that a plane cross-section remains plane and each element behaves according to its average stress–average strain relationships.

The FEM can be a powerful method to perform progressive collapse analysis on a hull girder. In 1983, ABS group (Chen et al. 1983) was the first to apply the FEM to this collapse analysis. They developed special elements such as orthotropic plate element representing stiffened plate, and introduced the yielding condition in terms of sectional forces to reduce the number of degrees of freedom of the calculated model. The analyses were performed on 1+1/2 holds model. DNV group also performed this kind of progressive collapse analysis by the FEM employing specially developed elements (Valsgaard et al. 1991). The analyses were performed on 1/2+1/2 holds model and one frame space model. Generally speaking, hull girder is too huge to perform progressive collapse analysis by the ordinary FEM, and some simplified methods are required.

Based on the progressive collapse method of Smith (1977), the incremental iterative method was proposed (IACS 2014). In this method the midship section is discretized to certain elements where every element follows a predetermined load end shortening curve described by empirical equation. The ultimate bending moment capacity for a transverse section is defined as the maximum value of the curve of bending moment versus the curvature.

Table 1 is a comparison between the different methods extracted from the review with their cost, speed and accuracy of the outputs.

In the present study, as the computational time is an important factor, the progressive collapse

Table 1. Calculation methods: Advantages/disadvantages.

Method	Accuracy	speed	Cost
Full scale experiment	Very high	Very slow	Very high
Model scale experiment	Medium/ High	Very slow	High
Finite element method	Medium/ High	Slow	Medium
Analytical methods	Medium/ High	Fast	Low

method has been selected to assess the ultimate hull girder strength in damage condition.

2 ULTIMATE HULL GIRDER STRENGTH BY INCREMENTAL ITERATIVE METHOD

The incremental iterative method was proposed by IACS. It is based on the progressive collapse method proposed by Smith (1977). In this method the midship section is discretized to certain elements as plate element, stiffened plate element, and hard corner element. Every element follows a predetermined load end shortening curve. These curves are described in IACS (2014) by empirical equations. The ultimate bending moment capacity for a transverse section is defined as the maximum value of the curve of bending moment M versus the curvature χ (Figure 1). The M-χ curve is obtained by means of an incremental-iterative approach.

In each step of the incremental procedure a curvature χ_i is imposed and the bending moment M_i which acts on the hull transverse section is calculated. The value of χ_i relevant to any step is obtained by adding a curvature increment $\Delta\chi$ to the curvature relevant to the previous step χ_{i-1}. This curvature increment corresponds to an increment in the rotation angle of the hull girder transverse section around its horizontal neutral axis. This rotation increment induces axial strains ε in each hull structural element, whose value depends on the position of the element. In hogging condition, the structural elements above the neutral axis are lengthened, while the elements below the neutral axis are shortened and vice-versa in sagging condition. The stress σ induced in each structural element by the strain ε is obtained from the load-end shortening curve σ-ε of the element, which takes into account the behaviour of the element in the non-linear elasto-plastic domain. The stress σ is selected as the lowest among the values obtained from each of the considered load-end shortening curves σ-ε.

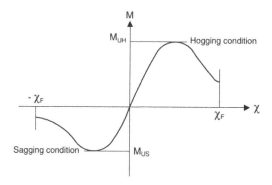

Figure 1. Bending moment versus curvature χ (IACS 2014).

The distribution of the stresses induced in all the elements composing the hull transverse section induces, for each step, a variation of the neutral axis position, since the relationship σ-ε is non-linear. The new position of the neutral axis relevant to the considered step is obtained based on an iterative process, where the equilibrium among the stresses acting in all hull elements should be verified. Once the position of the neutral axis is known and the relevant stress distribution through the elements of the structural section is obtained, the bending moment M_i about the new neutral axis corresponding to a curvature χ_i, can be obtained by summing the moment contribution of each element.

The procedure is to be repeated for each step, until the value of the imposed curvature permits the calculation of the peak of the M-χ curve.

According to IACS (2014), and Wang et al. (2008), when calculating ultimate hull girder longitudinal strength using incremental iterative method the following should be considered:-

1. The method for calculating the ultimate hull girder capacity is to identify the critical failure modes of all main longitudinal structural elements.
2. Structures compressed beyond their buckling limit have reduced load carrying capacity. All relevant failure modes for individual structural elements, such as plate buckling, torsional stiffener buckling, stiffener web buckling, lateral or global stiffener buckling and their interactions, are to be considered in order to identify the weakest inter-frame failure mode.
3. Only vertical bending is considered. The effect of shear force, torsional loading, horizontal bending moment and lateral pressure are neglected.

In applying the procedure described above the following assumptions are generally made:

- The ultimate strength is calculated at hull transverse sections between two adjacent reinforced rings.
- The hull girder transverse section remains plane during each curvature increment.
- The hull material has an elasto-plastic behaviour.
- The hull girder transverse section is divided into a set of elements, which are considered to act independently. These elements are:
- Transversely framed plate panels and/or ordinary stiffeners with attached plating,
- Hard corners, constituted by plates crossing.

3 RESIDUAL HULL GIRDER ULTIMATE STRENGTH

When a ship exhibits damage, some members lose their ability to sustain loads; this loss of ability leads to decrease in the strength capacity of the overall section of the ship at which damage occurred. The remaining amount of this strength is known as ship hull girder residual strength which is taken into consideration in the design of subject vessel. The ship must have a considerable residual strength to sustain the applied loads.

Simply talking, the hull girder ultimate residual strength is the ultimate strength of ship after damage excluding damaged members from contribution in strength or section modulus (Fujikubo et al. 2012, Makouei et al. 2015, Luís et al. 2009). The method of calculation is the same as the calculation of ship hull girder ultimate strength (Makouei et al. 2014). In the present study, the calculation of ultimate strength is made using Mars 2000 software that belongs to Bureau Veritas. A scenario of damage will be assumed using maritime rules and regulations considering both collision and grounding. Vessel specifications and midship section are as shown in Table 2 and Figure 2, respectively:-

The collision scenarios shown in Figure 3 and Table 3 are described with damage at deck beginning from 6% of ship breadth, which is around the ABS and CSR rule values that equal 2 m and 2.0125 m, respectively. The damage was increased gradually to 18% of the ship breadth, which is around the MARPOL, IBC and IGC value that

Table 2. Vessel specifications.

Length B. P.	176.0 m
Breadth (MLD.)	32.20 m
Depth (MLD.)	18.20 m
Draft (DESIGN)	11.00 m
Draft (Scantling)	12.60 m
Deadweight at design draft	42500 t
Deadweight at scantling draft	50500 t

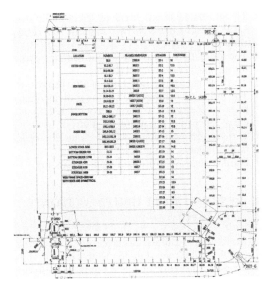

Figure 2.   Midship section of the considered vessel.

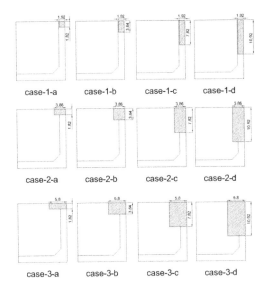

Figure 3.   Collision damage cases.

equal 6.44 m. The side damage was increased gradually from 10% to 60% of the ship depth which is the value specified by IACS (2014).

Hull girder ultimate strength after damage is calculated for various failure modes and the worst one, which is the Beam Column failure mode, was selected as the loading limit. The moment-curvature curve for this failure mode is shown in Figure 4–6 for each damage case, respectively.

Table 3.   Description of collision damage scenarios.

	Deck Damage 6%			
Case-1	Side damage	a		10%
		b		20%
		c		40%
		d		60%
	Deck Damage 12%			
Case-2	Side damage	a		10%
		b		20%
		c		40%
		d		60%
	Deck Damage 18%			
Case-3	Side damage	a		10%
		b		20%
		c		40%
		d		60%

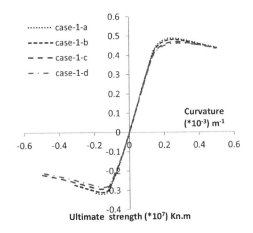

Figure 4.   Hull girder residual ultimate strength—Case-1.

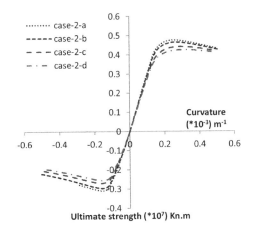

Figure 5.   Hull girder residual ultimate strength—Case-2.

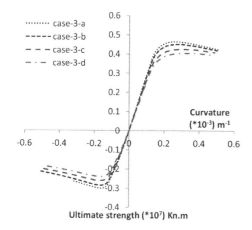

Figure 6. Hull girder residual ultimate strength—Case-3.

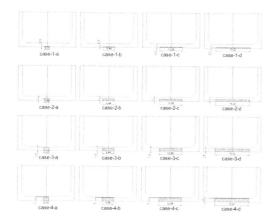

Figure 7. Grounding cases.

Table 4. Description of grounding damage cases.

		Girder damage	Bottom damage	
Case-1		0.5 m	a	10%
			b	20%
			c	40%
			d	60%
Case-2		1.0 m	a	10%
			b	20%
			c	40%
			d	60%
Case-3		1.5 m	a	10%
			b	20%
			c	40%
			d	60%
Case-4		2 m	a	10%
			b	20%
			c	40%
			d	60%

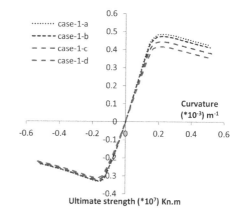

Figure 8. Hull girder residual ultimate strength—Case-1-grounding.

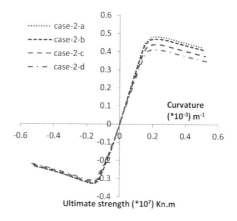

Figure 9. Hull girder residual ultimate strength—Case-2-grounding.

Moment here is only the vertical bending moment which is the prominent one.

The same procedure is repeated to study the effect of bottom damage on ship strength. 4 cases were considered where each case involves four levels a, b, c and d as shown in Figure 7 and Table 4.

These cases are according to the classification societies requirements IACS (2014) and ABS (1995) for residual strength assessment. The damage width is 60% of the considered ship breadth and its height is 2 m (IACS 2014), which is also compatible with MARPOL, IBC, IGC requirements of damage size, as shown in Table 4. Damage size was increased gradually to those requirements. For each level the moment-curvature relationship and the ultimate strength were obtained for various failure modes and the worst one, which is the beam column failure mode, was selected as the loading limit as indicated in Figures 8–11.

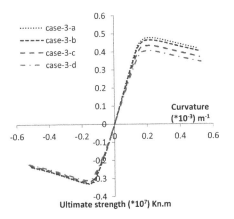

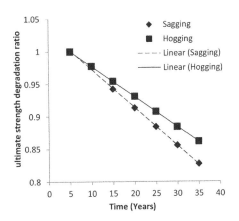

Figure 10. Hull girder residual ultimate strength—Case-3-grounding.

Figure 12. Ultimate strength degradation rate.

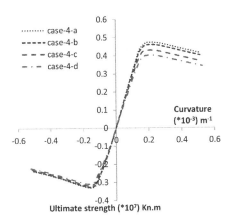

Figure 11. Hull girder residual ultimate strength—Case-4-grounding.

4 CONSIDERATIONS OF CORROSION EFFECT

4.1 Ultimate strength degradation model

Corrosion is one of the most important factors that degrades the ship hull and endanger its safety. There are many models describing the corrosion rate (Guedes Soares & Garbatov 1998,1999, Guedes Soares et al. 2008,2009,2011, Melchers 2003, Paik et al. 2003, Qin & Cui 2003, Yamamoto & Ikegami 1998). Wang et al. (2003) made a survey on approximately 140 single hull oil tankers and over 110,000 thickness measurements were obtained. Mean value, standard deviation and maximum value of corrosion rate for various structural members in oil tanker were estimated.

In the present study, constant corrosion rates, equal to the mean values predicted by Wang et al.

(2003), are considered for the different structural components. This accounts the corrosion thickness reduction as a linear function of time. Based on this consideration, the ultimate strength through the ship service life was determined. Five years are considered as the effective life of the protective coating where there is no corrosion. After the coating effective life, corrosion occurs and ultimate strength starts to decrease. In the present study, corrosion cases were developed over a life of approximately 35 service years, where after every 5 years the ultimate strength was calculated as shown in Figure 12. A formula (Eqn 1), representing the ratio between the ultimate strength of the corroded and the as-built hull, was obtained which can be considered as a tool to predict the ultimate strength after damage at any time through the ship service life.

Degradation ratio =
$$= -0.0046 \times time + 1.0236 \quad \text{(Hogging)} \ (R^2 = 1)$$
$$= -0.0059 \times time + 1.0314 \quad \text{(Sagging)} \ (R^2 = 0.999)$$
(1)

4.2 Verification of ultimate strength degradation model

In order to verify the previous model, the ultimate strength of damaged corroded models at different lives were evaluated and compared with the values obtained from damaged non-corroded models after applying the ultimate strength degradation formula. Damaged corroded cases are shown in Figures 13–16 where the ultimate strength was calculated according to corrosion data obtained from ABS (Wang et al. 2003). The error was ranging from 0.01% to 0.72%. In case-1 verification was done after 5 and 20 years. In case-2 verification was done after 10 and 25 year. In case-3 verification

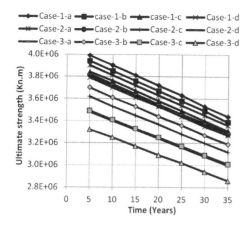

Figure 13. Strength degradation in hogging [Collision].

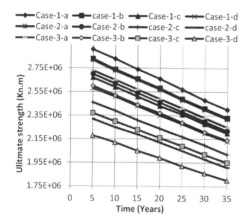

Figure 14. Strength degradation in sagging [Collision].

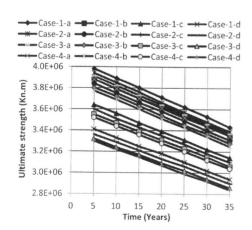

Figure 15. Strength degradation in hogging [Grounding].

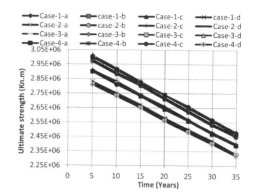

Figure 16. Strength degradation in sagging [Grounding].

was done after 15 and 30 years. The error range was 0.19%–0.56% in case-1, 0.01%–0.7% in case-2, and 0.05%–0.72% in case-3. The error is calculated as follows:-

$$\text{Error} = |M_{DN} - M_{DC}|/M_{DC} \times 100 \qquad (2)$$

where, M_{DC} is the ultimate strength of the damaged corroded models, and M_{DN} is the ultimate strength of the damaged non-corroded models after applying the ultimate strength degradation formula.

From Figure 13, 14 it can be shown that in the case of damage due to collision, sagging conditions are the worst as the ultimate strength capacities are smaller than those in hogging conditions. It should be noted that other damage cases than those suggested by IACS may be critical and need assessment depending on the loading conditions upon structural failure.

Case-1-a,b, case-2-a,b and case-3-a,b can be considered as minor damages as they are in the far corner of the outer shell and are relatively small. Also, depending on the loading condition during damage, they may have some marginal freeboard. However, when the damage penetrates the inner hull, the situation becomes worse and load redistribution may be needed in order to control the situation.

Figure 15 and 16 show the damage cases due to grounding. From the figures it can be noted that the rate of strength reduction is higher in the case of upward penetration of damage than in the case of transverse extension. If the upward penetration of damage reached the inner bottom, the situation will be more serious.

5 CONCLUSIONS

Based on this study, the following can be concluded:-

- Progressive collapse method can be considered as an alternative to finite element method when

589

- efficient calculation is a demand. It is a reliable tool for predicting the strength of ship hull.
- There are many types of damage that must be considered in the calculation of the residual strength, integrity and tightness of ship hull.
- There are many factors that must be considered in the calculation of ultimate strength of ship hull. Aging is highly important and should be predicted carefully.
- Downward breaching of the ship side shell appears to be the most critical damage.
- There are certain damage cases, other than what was suggested by IACS, that must be considered as their corresponding residual strength can be critical.
- The ultimate sagging bending moment capacity of the ship hull is impaired more by collision damage while the ultimate hogging bending moment capacity is impaired more by grounding.
- Immediate and quick assessment of the residual strength after damage is important as the reliability of the calculated cases based on IACS do not assure survival in all situations.

REFERENCES

ABS 1995. Guide for assessing hull-girder residual strength for tankers. New York: American Bureau of Shipping.

Caldwell, J.B. 1965. Ultimate longitudinal strength. RINA 107:411–30.

Chen, K.Y., Kutt, L.M., Piaszczyk, C.M. & Bieniek, M.P. 1983. Ultimate strength of ship structures. SNAME Transactions, 91:149–68.

Fujikubo, M., Zubair, M.A., Takemura, K., Iijima, K. & Oka, S. 2012. Residual Hull Girder Strength of Asymmetrically Damaged Ships. Journal of the Japan Society of Naval Architects and Ocean Engineers, 16, 131–140.

Guedes Soares, C. & Garbatov, Y. 1998. Non-linear time dependent model of corrosion for the reliability assessment of maintained structural component. In: Lydersen, S., Hansen, G. & Sandtorv, H. (eds.) Safety and Reliability. Balkema.

Guedes Soares, C. & Garbatov, Y. 1999. Reliability of Maintained, Corrosion Protected Plates Subjected to Non-Linear Corrosion and Compressive Loads. Marine Structures, 12, 425–445.

Guedes Soares, C., Garbatov, Y. & Zayed, A. 2011. Effect of environmental factors on steel plate corrosion under marine immersion conditions. Corrosion Engineering, Science and Technology, 46, 524–541.

Guedes Soares, C., Garbatov, Y., Zayed, A. & WANG, G. 2008. Corrosion wastage model for ship crude oil tanks. Corrosion Science, 50, 3095–3106.

Guedes Soares, C., Garbatov, Y., Zayed, A. & Wang, G. 2009. Influence of environmental factors on corrosion of ship structures in marine atmosphere. Corrosion Science, 51, 2014–2026.

Huilong, R., Chenfeng, L., Guoqing, F. & Hui, L. Calculation Method of the Residual Capability of Damaged Warships. In: 27th International Conference on Offshore Mechanics and Arctic Engineering, 2008 Estoril, Portugal. ASME.

IACS 2014. Common Structure Rules for Bulk-carrier and Oil-tankers. IACS.

Luís, R.M., Teixeira, A.P. & Guedes Soares, C. 2009. Longitudinal strength reliability of a tanker hull accidentally grounded. Structural Safety, 31, 224–233.

Makouei, S.H., Teixeira, A.P. & Guedes Soares, C. 2014. An approach to estimate the ship longitudinal strength using numerical databases of stress-strain curves of stiffened panels. In: Guedes Soares, C. & López Peña, F. (eds.) Developments in Maritime Transportation and Exploitation of Sea Resources. London: Taylor & Francis Group.

Makouei, S.H., Teixeira, A.P. & Guedes Soares, C. 2015. A study on the progressive collapse behaviour of a damaged hull girder. In: Guedes Soares, C. & Santos, T.A. (eds.) Maritime Technology and Engineering. London: Taylor & Francis Group.

Melchers, R.E. 2003. Probabilistic Models for Corrosion in Structural Reliability Assessment—Part 2: Models Based on Mechanics. Journal of Offshore Mechanics and Arctic Engineering, 125, 272–280.

Paik, J.K. & Mansour, A. 1995. A simple formulation for predicting the ultimate strength of ships. Marine Science and Technology, 52–62.

Paik, J.K. & Thayamballi, A.K. 2003. A concise introduction to the idealized structural unit method for nonlinear analysis of large plated structures and its application. Thin-Walled Structures, 329–355.

Paik, J.K., Lee, J.M., Hwang, J.S. & Park, Y.L. 2003. A Time-Dependent Corrosion Wastage Model for the Structures of Single and Double Hull Tankers and FSOs and FPSOs. Marine Technology, 40, 201–217.

Qi, E. & Cui, W. 2006. Analytical method for ultimate strength calculations of intact and damaged ship hulls. Ships and Offshore Structures 153–164.

Qin, S. & Cui, W. 2003. Effect of Corrosion Models on the Time-Dependent Reliability of Steel Plated Elements. Marine Structures, 16, 15–34.

Smith, C.S. Influence of local compressive failure on ultimate longitudinal strength of a ship's hull. In: Proceedings of the International Symposium on Practical Design in Shipbuilding (PRADS), October 1977 Tokyo. 73–79.

Valsgaard, S., Lorgensen, L., Boe, A.A. & Thorkildsen, H. Ultimate hull girder strength margins and present class requirements. In: Proceeding of the SNAME Symposium on Marine Structural Inspection, Maintenance and Monitoring, 1991 Arlington, USA.B.1–19.

Wang, G., Spencer, J. & Sun, H. Assessment of Corrosion Risks to Aging Ships Using an Experience Database. In: Proceeding of 22nd International Conference on Offshore Mechanics and Arctic Engineering, Paper OMAE2003–37299, June 2003 Cancun, Mexico. ASME.

Wang, X., Sun, H., Yao, T., Fujikubo, M. & Basu, R. Methodologies on hull girder ultimate strength assessment of FPSOs. In: 27th International Conference on Offshore Mechanics and Arctic Engineering, 2008 Estoril, Portugal. ASME.

Yamamoto, N. & Ikegami, K. 1998. A Study on the Degradation of Coating and Corrosion of Ship's Hull Based on the Probabilistic Approach. Journal of Offshore Mechanics and Arctic Engineering, ASME, 120, 121–128.

Maritime Transportation and Harvesting of Sea Resources – Guedes Soares & Teixeira (Eds)
© *2018 Taylor & Francis Group, London, ISBN 978-0-8153-7993-5*

Experimental investigation on explosive welded joints for shipbuilding applications

P. Corigliano, V. Crupi, E. Guglielmino & A.M. Sili
Department of Engineering, University of Messina, Messina, Italy

ABSTRACT: Nowadays dissimilar metals can be joint by means of the explosion welding technique, which can be used in the naval field giving a good compromise between weldability and mechanical properties. In the present study, the mechanical behavior of a commercial bar for shipbuilding applications was investigated. The bar is made of ASTM A516 structural steel, clad by explosion welding with AA5086 aluminum alloy and provided with an intermediate layer of pure aluminum. Two full-field techniques were applied during the bending tests: Digital Image Correlation and Infrared Thermography. The Digital Image Correlation technique allowed the analysis of the displacement and strain patterns of the different metals, and Infrared Thermography was used to detect the superficial temperature of the specimen. Micro-hardness tests were also performed in order to evaluate the properties of the transition zones.

1 INTRODUCTION

Material selections for ship or offshore structures involve complex considerations of cost, weight, producibility, marine corrosion, fatigue, fire resistance, vibration and sound damping. Unfortunately, a single structural material, which can satisfy all of these requirements, doesn't exist. Thus different metals are generally used throughout the structure; each is selected to be appropriate for a specific component.

Aluminium/steel welded joints have been widely adopted in different industries, but find their main application in the shipbuilding industry (Ayob 2010). One of the most performing techniques is the explosion welding (Crossland 1971), which is a solid-state process to join dissimilar materials. The metals surface to be joined are brought in close contact at extremely high velocity through the use of chemical explosives detonated against them. Impact velocity and impact angle are the two most critical parameters to determine the weld strength and the interface morphology (Botros et al 1980). However, it is difficult to directly achieve explosive welding of aluminum alloy to steel, so a thin intermediate plate is generally inserted (Han et al 2003, Li et al 2015).

The literature on the recent developments in explosive welding was reviewed by Findik (Findik 2011).

A shipbuilding application of bimetallic joints for the connection of aluminium superstructure to steel deck is shown in (Young, Banker 2004). The application of such transition joints on marine structures requires determining its properties. An extensive battery of tests was conducted in the past (McKenney, Banker 1971) to evaluate their corrosion resistance and mechanical properties.

Experimental investigations of explosive welding of aluminium to steel were reported in literature (Acarer, Demir 2008, Raghukandan 2003).

The aim of this research activity is to apply full field techniques in order to investigate the influence of the different materials on the mechanical strength of the exploded welded joints during three and four bending loading. The investigated aluminium/steel welded joints is used for shipbuilding applications. Two full-field techniques were applied during the bending tests: Digital Image Correlation (DIC) and Infrared Thermography (IR).

Some of the authors have already applied full-field and non destructive techniques for the analysis of different materials used for marine structures: aluminum T-shaped welded joints under high cycle fatigue loading (Crupi et al 2007), steel T-shaped welded joints under low cycle fatigue loading (Corigliano et a. 2015), Iroko wood under static loading (Bucci et al 2017, Corigliano et al 2017

2 MATERIALS AND METHODS

Static bending tests were carried out on explosive welded specimens using a servo-hydraulic load machine (INSTRON 8854). The explosive welded specimens were produced by TRICLAD for shipbuilding applications and consist of ASTM A516 Gr55 structural steel, clad by explosion welding with AA5086 aluminum alloy and provided with an intermediate layer of AA1050 commercial pure aluminum (Fig. 1). The interface aluminum/ aluminum alloy is wavy, as a consequence of the great plasticity of both the two metals. The waves formation produces an increase of the interfacial area with an improved bonding efficiency. Straight or wavy interfaces can be formed by explosion, according to metals involved and to process parameters: increasing in explosive loading increases the impact energy of flyers causing the transition from straight to wavy form.

Three points and four points bending tests were performed at a displacement rate of 4 mm/min. The experimental setup is shown in figure 2. The support span L was set equal to 300 mm in three points bending, while in four bending tests, the spans were respectively 450 mm for the lower one and 150 mm for the upper support span. DIC and IR cameras have been used during all tests. The DIC technique is a full-field non-contact measurements method, which allows the detection of displacement and strain fields, and can be used to monitor areas that are hard to analyze using traditional tools as strain gages. Two cameras with a resolution of 4000 × 3000 pixels, with a focal length of 50 mm, were used for the application of DIC technique. The system accuracy for the strain measurement is up to 0.01%, while the highest acquisition frequency is 58 Hz at the highest resolution. The specimens were coated with a black-white speckle pattern and ARAMIS 3D 12M system was used to analyze the strain pattern of the specimen surface. The specimens were coated with black paint and an IR camera was placed on the opposite side of the specimen with respect to the DIC equipment. The temperature increment of the specimen surface was detected during the static bending test tests by a micro-bolometric IR camera (model FLIR Systems SC640) with a resolution of 640 × 480 pixels and a measurement accuracy of ±2°C.

3 HARDNESS MEASUREMENTS

The hardness of the three metals of the welded joints was measured far from the interfaces, obtaining: 175 HV for ASTM A516 gr 55, 47 HV for AA 1050 interlayer, 109 HV for AA 5086. Microhardness profile recorded across the interface shows an increase of values due to the deformation hardening of metals caused by the explosion, above all at the steel side (Figs. 3–4).

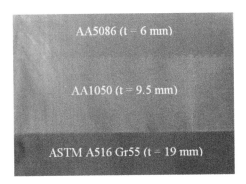

Figure 1.   TRICLAD explosive welded joints.

Figure 2.   Experimental setup.

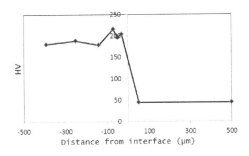

Figure 3.   ASTM A516-AA1050 microhardness profile.

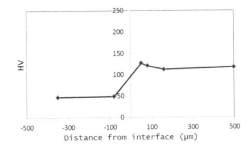

Figure 4.   AA1050-AA5086 microhardness profile.

4 THREE POINTS BENDING

Figure 5 shows the curves of the bending moment versus the imposed displacement, obtained by three point bending tests.

The definition of the neutral axes includes a first step to impose the equilibrium, in this specific case along the longitudinal axis. Referring to figure 6 ($H_1 = 19$ mm, $H_2 = 9.5$ mm, $H_3 = 6$ mm, $E_1 = 205$ GPa, $E_2 = E_3 = 71000$ GPa) and using the Hooke's law ($\sigma = E\varepsilon$, and $\varepsilon = -Y/R$, with R being the curvature radius), the position of the neutral axis Y_G was calculated with Equation 3 obtaining $Y_G = 13.3$ mm.

$$\int_{A_1} \sigma_{x1} dA + \int_{A_2} \sigma_{x2} dA + \sigma_{x3} dA = 0 \qquad (1)$$

$$\int_{A_1} E_1 \frac{Y}{R} dA - \int_{A_2} E_2 \frac{Y}{R} dA - \int_{A_3} E_3 \frac{Y}{R} dA = 0 \qquad (2)$$

then the value of the neutral axis is given by:

In some tests, the load was applied on the aluminium side, in order to have a tensile stress on the steel side.

Some characteristic points called "Stage points", as shown in figure 7, were introduced in the DIC images to evaluate the behavior of the different zones along the surface of the welded specimen during the bending tests: stage points 0, 1, 2 were positioned in the tensile (lower) side of the specimen. The DIC analysis does not take into account very close points to the edges, so two layers thick about 2.1 mm and located respectively at the Al alloy and steel external side are excluded from the analysis. Figure 7(a) allows the evaluation of the neutral axis position of the specimen during bending, i.e. the neutral axis is found where the strain is equal to zero which resulted at a coordinate of 19 mm of the section thickness, this means 21.1 mm from the Al alloy external side and 13.4 mm from the steel external side. This result is in good agreement with the theoretical one calculated imposing the equilibrium and the different

$$Y_G = \frac{E_1 H_1^2 + 2E_1 H_1 H_2 + E_2 H_2^2 + 2E_3 H_1 H_3 + 2E_3 H_3 H_2 + E_3 H_3^2}{2(E_1 H_1 + E_2 H_2 + E_3 H_3)} \qquad (3)$$

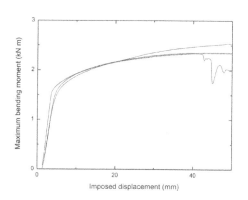

Figure 5. Three point bending curves.

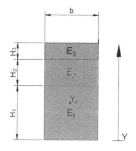

Figure 6. Scheme used for neutral axis calculation.

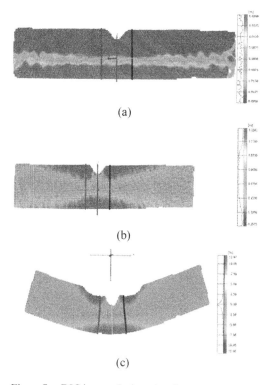

(a)

(b)

(c)

Figure 7. DIC images during a bending test.

Young moduli of the layers. Figure 8 shows the pattern of the longitudinal strain along the thickness for the stage points 0 and 2.

In other bending tests, the load was applied on the steel side, in order to have a tensile stress on the aluminium side.

Stage points 0, 1, 2 were positioned in the tensile side (lower side) of the specimen, while stage points 3, 4, 5 were positioned in correspondence of the upper side (distance between two stage points was equal to 10 mm). Figures 9(a) and 9(b) report a contour map of the longitudinal strain at yielding and right before the end of the test: it

is clearly visible that the aluminium side is much more strained with respect to the steel side. In fact, Figure 10 shows that stage points belonging to the aluminium alloy side have a higher level of strain during the whole test, this behavior is also confirmed by Figure 11 which plots the force vs. strain curve indicating that in correspondence of the maximum force steel reaches strain values close to 11.7%, which are almost double for the aluminium alloy.

Figure 12 shows the maximum bending moment vs. strain curve, where the strain was measured in

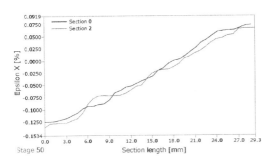

Figure 8.   Longitudinal strain along the thickness.

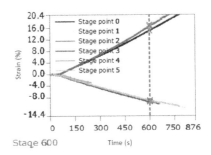

Figure 10.   Longitudinal strain at stage points.

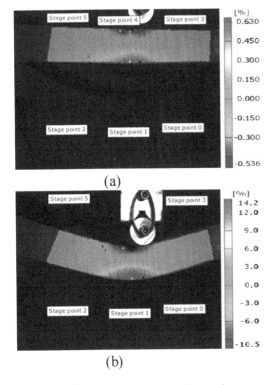

(a)

(b)

Figure 9.   DIC images of the longitudinal strain.

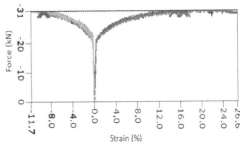

Figure 11.   Force vs. strain curve.

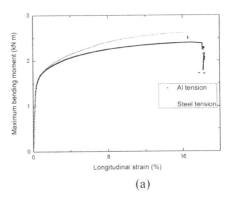

(a)

Figure 12.   (*Continued*)

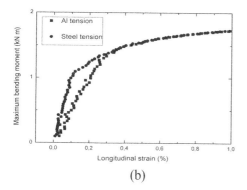

(b)

Figure 12.   Maximum bending moment vs. strain curve.

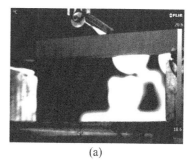

(a)

(b)

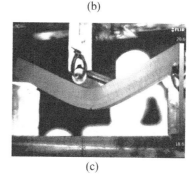

(c)

Figure 13.   IR images.

correspondence of the side in tension during two different experimental tests: in the first one the load was applied on the aluminium side and the in the second one on the steel side. In particular, the results show that, due to the higher elastic moduli, the steel layer deforms less than the others in the elastic phase, while there is a perfect correspondence of the curve at the beginning of the plastic phase, but a high difference is depicted in the plastic phase.

5   IR ANALYSIS

It is well known that a metallic specimen changes its temperature when it is subjected to stresses. The temperature change is produced by two main effects: elastoplasticity and thermoelasticity (Audenino et al 2004).

The surface temperature variations of specimens, undergoing bending loading, were recorded by an infrared camera. Figure 13 shows the IR images of the specimen during a bending test.

Figure 14 shows the temperature increment during the bending test. The temperature evolution on the specimen surface, detected by means of an infrared camera, is characterized by three phases: an initial approximately linear decrease due to the thermoelastic effect (phase I), then the temperature deviates from linearity until a minimum (phase II) and a very high further temperature increment until the failure (phase III).

The temperature evolution is similar to the thermal response of composite materials during static tests as reported in literature (Vergani et al 2014, Crupi et al 2015). There are many studies in literature on the thermal response of composites during static tests for the prediction of the fatigue limit as reported in (Vergani et al 2014), but few studies about the thermal response of steels during static tests, only two papers (Corigliano et al 2015b,

Corigliano et al 2016,) about the application to welded joints, and none to explosive welding, as far as the authors are aware. A similar trend of Figure 14 was detected by some of the authors (Corigliano et al 2015b, Corigliano et al 2016) during tensile tests carried out on steel welded joints.

Figure 15 shows the crack propagation in a specimen after a three point bending test. The crack starts from the external side of the Al alloy and does not propagate through the pure Al layer, following the preferable path, which is constituted by the interlayer formed from the explosive welding, which is more fragile, as confirmed by hardness measurements (Fig. 4). It is to be mentioned that only one of the six static tests ended with the specimen failure, because the other tests were stopped by the imposed limit of the deflection.

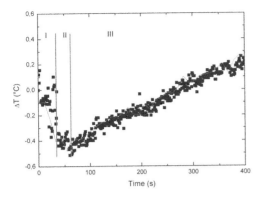

Figure 14. Temperature increment during the test.

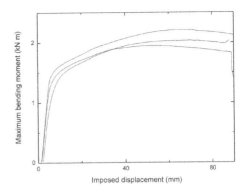

Figure 16. Four point bending curves.

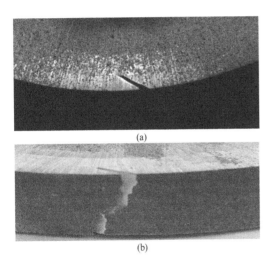

(a)

(b)

Figure 15. Crack propagation.

6 FOUR POINTS BENDING

Figure 16 shows the bending moment—imposed displacement: the measured maximum bending moment in the four points bending tests resulted lower than in the three points bending tests.

In some tests, the load was applied on the aluminium side, in order to have a tensile stress on the steel side.

Figure 17 plots a contour map of the longitudinal strain obtained by the four points bending tests, respectively at 21 and 26 kN.

Figure 18 plots the strain along the sections defining the position of the neutral axis also in four bending tests.

Figure 19 reports the force-strain curves for the different stage points (0, 1, 2 in correspondence of the steel and 3,4,5 in correspondence of the

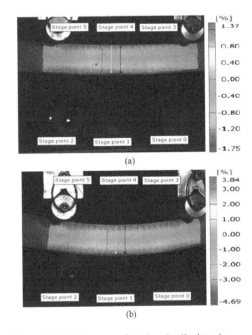

(a)

(b)

Figure 17. DIC images of the longitudinal strain.

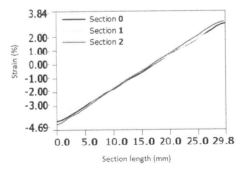

Figure 18. Longitudinal strain along the thickness.

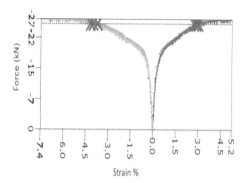

Figure 19. Force vs. strain curve.

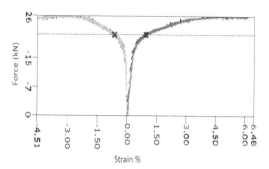

Figure 21. Force vs. strain curve.

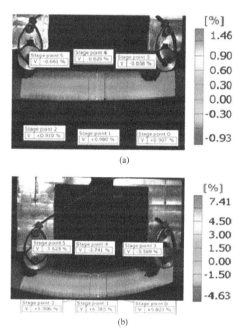

(a)

(b)

Figure 20. DIC images of the longitudinal strain.

aluminium alloy side) showing that the aluminium alloy has always a higher value of strain than steel. The red dots indicate the test load at which figure 17(b) was evaluated.

In other bending tests, the load was applied on the steel side, in order to have a tensile stress on the aluminium side.

Figure 20 plots a contour map of the longitudinal strain obtained by the four points bending tests, respectively at 21 and 26 kN. Also when the load is applied on the steel side, the aluminium side shows higher strain values.

Figure 21 reports the force-strain curves for the different stage points.

7 CONCLUSIONS

In this study, full field techniques were applied in order to investigate the influence of the different materials on the mechanical strength of the exploded welded joints during three and four bending loading.

An AA5086 aluminum alloy plate and an ASTM A516 Gr55 structural steel plate were cladded by an explosive welding method using an AA1050 commercial pure aluminum as an interlayer plate.

The investigated aluminium/steel explosion welded joints is used for shipbuilding applications.

Micro-hardness tests were performed on the explosive welded joints in order to evaluate the properties of the transition zones.

Two full-field techniques were applied during the bending tests: DIC and IR.

The DIC analysis allowed the determination of the position of the neutral axis during three points and four points bending tests. The aluminium alloy has a higher value of the measured strain.

DIC results for different load configurations have shown that, due to the higher elastic moduli, the steel layer deforms less than the aluminium alloy one in the elastic phase, while there is a perfect correspondence of the two external layers deformations at the beginning of the plastic phase, but a high difference is depicted in the plastic phase

The maximum bending moment in four bending tests has a lower value with respect to three points bending tests.

IR results have shown that the temperature evolution on the specimen surface, is characterized by three phases: an initial approximately linear decrease due to the thermoelastic effect (phase I), then the temperature deviates from linearity until a minimum (phase II) and a very high further temperature increment until the failure (phase III).

The crack started on the Al alloy side and propagated through the thickness towards the pure Al interlayer. The crack changed its propagation direction in correspondence of the Al-steel interface causing a sort of delamination of the transition joint.

ACKNOWLEDGEMENTS

The research reported in this paper was conducted with the facilities of the Research Project "CERISI" ("Research and Innovation Centre of Excellence for Structure and Infrastructure of large dimensions"), funded by the PON (National Operative Programme) 2007–2013. This study is part of the research activities of the Research Project PRIN (Announcement 2015) CLEBJOINT, project funded by the Italian Ministry of Scientific and Technological Research.

REFERENCES

Acarer, M. & Demir, B. 2008. An investigation of mechanical and metallurgical properties of explosive welded aluminium–dual phase steel; *Mater Lett* 62(25): 4158.

Audenino, A.L., Crupi, V. & Zanetti, E.M. 2004. Thermoelastic and elastoplastic effects measured by means of a standard thermocamera; *Experimental Techniques* 28(2): 23–28.

Ayob, F. 2010. Joining of dissimilar materials by diffusion bonding/diffusion welding for ship application; *Mar Front*, 1: 69–73.

Botros, K.K. & Groves, T.K. 1980. Characteristics of the wavy interface and the mechanism of its formation in high velocity impact welding; *J Appl Phys*, 51: 3715–3721.

Bucci, V., Corigliano, P., Crupi, V., Epasto, G., Guglielmino, E. & Marinò, A. 2017. Experimental investigation on Iroko wood used in shipbuilding. *P.I. Mech. Eng. C-J. Mec.* 231: 128–139.

Corigliano, P., Crupi, V., Fricke, W., Friedrich, N. & Guglielmino E. 2015. Experimental and numerical analysis of fillet-welded joints under low-cycle fatigue loading by means of full-field techniques; Special Issue "Fatigue Design and Analysis in Transportation Engineering", *P I MechEng C-J Mech* 229: 1337–1348.

Corigliano, P., Crupi, V., Epasto, E. Guglielmino, Maugeri, N. & Marinò, A. 2017. Experimental and theoretical analyses of Iroko wood laminates. *Compos. Pt. B—Eng.* 112: 251–264.

Corigliano, P., Crupi, V., Epasto, G., Guglielmino, E. & Risitano, G. 2015. Fatigue assessment by thermal analysis during tensile tests on steel; *Procedia Eng* 109: 210–218.

Corigliano, P., Crupi, V., Epasto, G., Guglielmino, E. & Risitano, G. 2016. Fatigue life prediction of high strength steel welded joints by Energy Approach; *Procedia Structural Integrity* 2: 2156–2163.

Crossland, B. 1971. The development of explosive welding and its application in engineering; *Metals Mater*: 401–402.

Crupi, V., Guglielmino, E., Risitano, A. & Taylor, D. 2007. Different methods for fatigue assessment of T welded joints used in ship structures; *J. Ship Res* 51(2): 150–159.

Crupi, V., Guglielmino, E., Risitano, G. & Tavilla, F. 2015. Experimental analyses of SFRP material under static and fatigue loading by means of thermographic and DIC techniques; *Composites Part B: Engineering* 77: 268–277.

Findik, F. 2011. Recent developments in explosive welding; *Materials and Design* 32: 1081–1093.

Han, J.H., Ahn, J.P. & Shin, M.C. 2003. Effect of interlayer thickness on shear deformation behaviour of AA5083 aluminium alloy/SS41steel plates manufactured by explosive welding; *J Mater Sci*: 38:13.

Li, X., Ma, H. & Shen, Z. 2015. Research on explosive welding of aluminum alloy to steel with dovetail grooves; *Materials and Design* 87: 815–824.

McKenney, C.R. & Banker, J. 1971. Explosion—bonded metals for marine structural applications; Marine Technology. *Society of Naval Architects and Marine Engineers*: 285–292.

Raghukandan, K. 2003. Analysis of the explosive cladding of Cu–low carbon steel plates; *J Mater Process Technol* 139: 573–7.

Vergani, L., Colombo, C. & Libonati, F. 2014. A review of thermographic techniques for damage investigation in composites; *Frat Integ Strut* 27: 1–12.

Young, G.A. & Banker, J.G. 2004. Explosion welded, bi-metallic solutions to dissimilar metal joining; 13th Offshore Symposium, Texas Section of the Society of Naval Architects and Marine Engineers, Houston, Texas.

Maritime Transportation and Harvesting of Sea Resources – Guedes Soares & Teixeira (Eds)
© 2018 Taylor & Francis Group, London, ISBN 978-0-8153-7993-5

Analytical treatment of welding distortion effects on fatigue in thin panels: Part I—closed-form solutions and implications

P. Dong, S. Xing & W. Zhou
Department of Naval Architecture and Marine Engineering, University of Michigan, Michigan, USA

ABSTRACT: In lightweight shipboard structures, welding-induced distortions can be characterized as two characteristic types: buckling distortions and angular distortions. There have been some recent studies showing that these distortion types can introduce complications in fatigue test data analysis, which may call for an adequate consideration of nonlinear geometry effects on stress concentration at weld locations. In this study, both linear and nonlinear treatments of the two types distortions are presented for calculating stress concentration factor for a stiffened panel subjected to remote cyclic tension. The final closed-form solutions are presented for both types of distortions with validations by means of finite element solutions considering nonlinear geometry effects. Their use in understanding existing empirical based distortion tolerance criteria is then examined in light of the present analytical developments.

1 INTRODUCTION

It has been well established that welding induced distortions have been a major issue in construction of lightweight ship structures (Dong, 2005; Jung, et al, 2007, and Yang & Dong, 2012).

In addition to developing effective mitigation techniques for reducing such distortions, existing distortion tolerances such as those given Class societies such as by ABS (2007) which were based on data for thick-section structures, some which dated back many decades ago (e.g., MIL-STD-1698,) may need to be revisited for its applicability in lightweight structures. As discussed in by Hunag et al (2004), in lightweight steel shipboard structures, plate thicknesses of equal or less than 5 mm have become increasingly dominant and post a major challenge in accuracy control in ship construction processes.

Effects of weld residual stresses on structural integrity have been extensively addressed in the context of fitness for service in the literature, as recently summarized by Dong & Brust (2000) and Dong et al (2014, 2017), Song & Dong (2017), and Dong (2005, 2008). However, how welding-induced distortions impact structural integrity of lightweight structures has not been adequately addressed. Some of the early publications on this subject are scarce, e.g., by Antoniou (1980) and Carlsen & Czujko (1978), and focused on limited experimental observations on certain specific type of distortions in ship construction environment.

These studies had a narrow focus on effects of some observed welding distortions on structural

buckling strength under compressive loading with plate thicknesses being above 10 mm and did not address how distortions affect fatigue behaviors in welded structures.

Recognizing the prevalence of distortions in thin plate structures, researchers have recently reported a series of investigations into fatigue behaviors observed in thin plate structures, including those by Lillemäe et al (2012) on distortion effects on fatigue strength in butt-welded joints, Xing et al (2016) on fatigue failure mode transition behaviors in thin plate cruciform joints, and Xing et al, (2016) and Liu & Dong (2014) on quantitative fillet weld sizing criteria for preventing weld throat cracking in load-carrying fillet welds. In the latter studies, the importance of proper treatment of joint misalignments in thin plate structures is demonstrated by means of a series of new analytical stress concentration factor solutions.

Both these experimental and finite element studies have showed that fatigue behaviors in thin plate structures tend to show a great deal of scatter, much more so than thick plate structural joints. To clarify some of the dominant effects of welding-induced distortions on fatigue, a fundamental approach is needed for separating distortion characteristics that are intrinsic to certain lightweight structural forms, which can be quantitatively related to fatigue performance, from those distortion features that may exhibit a great deal of variability. This way, the latter effects on fatigue may be deemed as being secondary, or being negligible for understanding basic interactions between distortions and effects on fatigue.

Along this line of thinking, the authors present an analytical treatment of distortion effects on fatigue behavior in welded thin plate structures in this paper. At first, two types of welding induced distortions, i.e., buckling type and angular type, are introduced based on a series of recent studies and idealized for analytical treatment in this paper.

For each type of distortions, the resulting stress concentration factors caused by remote loading are formulated and solved analytically by considering linear and nonlinear geometry effects. These solutions are then applied for interpreting the underlying basis upon which some of the existing distortion tolerance criteria were developed, such as MI-STD-1689 in Part I, and for understanding fatigue test data presented by Lillemäe et al (2012) in Part II (see Dong, et al, 2017), which contain complex welding induced distortions. The results show that these fatigue test results on thin plate butt-welded specimens which exhibit complex distortion features can be correlated reasonably well with the 2007 master S-N curve scatter band which contain large scale fatigue test data with plate thicknesses ranging from 5 mm up to over 100 mm.

2 DISTORTION-INDUCED SECONDARY BENDING STRESS

2.1 Two distinct distortion types

As discussed in Dong (2005, 2008) and Yang and Dong (2012), welding-induced distortions in thin plate structures tend to exhibit buckling distortions due to structural instability triggered by compressive residual stress distributions, often accompanied by angular distortions. Both distortion types are described in Fig. 1, in which Fig. 1a shows a LIDAR scan image of a 16 feet by 20 feet stiffened panel which clearly shows a checker-board pattern, particularly on the lower half of the panel.

A stiffened panel exhibiting angular distortion is illustrated in Fig. 1b. The corresponding deformation profiles for both cases are illustrated in Figs. 1c and 1d, respectively.

As such, buckling distortion profile can be characterized as a sinusoidal wave form in which rotation at stiffener locations is unrestricted (see Fig. 1c) while angular distortion profile shows no rotation at stiffener locations. In both cases, stiffener spacing is represented by s and out of plane distortion peak value by δ_0.

2.2 Linear analysis

2.2.1 Idealization and solutions

As a first order approximation, an equivalent beam bending problem for angular distortion profile (see Fig. 2) can be idealized as shown in Fig. 2b, which is statically indeterminate and can be readily solved by potential energy method.

Similarly, an equivalent beam bending problem for angular distortion profile (see Fig. 3) can be idealized as shown in Fig. 3b which is statically determinate and can be solved directly using beam theory.

For the boundary conditions given in Fig. 2 corresponding to angular distortion, a statically equivalent moment at fillet weld location as result of the peak out-of-plane distortion δ_0 can be found as:

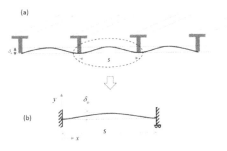

Figure 2. An equivalent beam bending problem definition—angular distortion type.

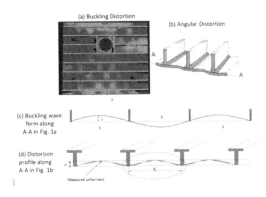

Figure 1. Two major distortion in thin plate structures.

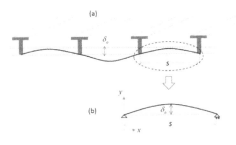

Figure 3. An equivalent beam bending problem definition—buckling distortion type.

$$M_0 = \frac{24\delta_o EI}{s^2} \qquad (1)$$

$$\sigma_b(T) = \frac{12\delta(T)Et}{s^2} \qquad (3)$$

where E and I are Young's modulus of material and bending moment of inertia. Similarly, the maximum statically equivalent moment at the beam mid span corresponding to buckling distortion type shown in Fig. 3 can be determined as:

$$M_0 = \frac{12\delta_o EI}{s^2} \qquad (2)$$

which is exactly one half of that given in Eq. (1) corresponding to the angular distortion conditions in Fig. 2.

From this point on, the bending moment given in Eq. (1) at fillet weld corresponding to angular distortion will be considered, unless specifically stated. Note that reaction moment is zero at two fillet weld locations for the case of buckling distortions shown in Fig. 3. For more complex boundary conditions under general misalignment conditions, solutions can be found in a recent publication by Xing & Dong (2016).

2.2.2 Implications on fatigue behavior

Based on the linear analysis results described in the previous section, as far as fatigue behaviors at fillet weld locations are concerned, one only needs to consider the angular distortion case (Fig. 2). Consider that a stiffened panel containing such angular distortions is subjected to a lateral load P(T) which varies as a function of time (T).

As shown in Fig. 4, with the presence of P(T) <0, beam deflection $\delta(T)$ increases above δ_0 (see Fig. 4d). When P(T) becomes positive, i.e., P(T)>0, beam deflection decreases. As suggested in Eq. (1), the corresponding bending moment at fillet weld location alternates over time in a linear manner with $\delta(t)$ as illustrated in Fig. 4e. As a result, the presence of the out-of-plane angular distortion generates additional bending stress in addition to alternating membrane stress P(T)/t, assuming unit beam depth into the paper. The corresponding bending induced stress becomes:

Note that $\delta(T)$ has a complex nonlinear relationship with initial distortion δ_0 and applied load level P(T), which will be explicitly addressed in a following section. However, with a simple analysis procedure demonstrated above, the results can still be used to make some useful interpretation of some of the existing distortion tolerance standards such as MIL-STD-1689 and implications on urgent needs for improved tolerance criteria for applications in modern lightweight structure construction.

Fig. 5 is taken directly from MIL-STD-1698, which specifies allowable distortions in terms of stiffener spacing (s) and plate thickness (t). Attempts in finding its basis and background has not been successful to this day. One relevant publication by Evan (1954) stated, in commenting on rationales behind MIL-STD-1689, states that "if initial unfairness is beyond some limiting value, at surface, the bending stress may easily be three times the mid-thickness stress".

This comment suggested that Fig. 5 was somehow based on a secondary bending stress limit caused by welding-induced distortions or "initial unfairness".

By using Eq. (3) and considering a reference position in Fig. 5 (e.g., t = 3/8″ and s = 22″, as indicated by the solid dot in Fig. 6), a set of distortion tolerance criteria can then be generated as shown by solid lines in Fig. 6 and compared with dashed

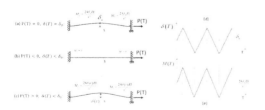

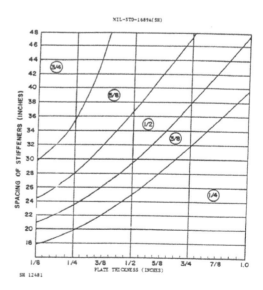

Figure 4. Interpretation of beam strengthening effects as a function of remote displacement controlled loading P(T).

Figure 5. Permissible distortions in steel welded structures according to MIL-STD-1689.

lines representing the original acceptance criteria (given in Fig. 5).

There exists a general agreement in Fig. 6 between the two sets of distortion tolerance criteria, suggesting that imposing a constant bending stress limit seems capable of reproducing, to a large extent, the distortion criteria given in MIL-STD-1689, given the empirical nature of Fig. 4.

To examine how both MIL-STD-1689 and the newly developed criteria correlate with one another for various plate thicknesses compare, Fig. 7 summarizes the comparison between criteria given in MIL-STD-1689 and a constant bending stress based criteria developed in this study. The constant bending stress level caused by distortion is determined based on a reference position defined in Fig. 4 at $t = 3/8''$ and $s = 22''$. It is interesting to note that as plate thickness increases from $t = 1/8''$ to $t = 1/2''$, the differences between the two

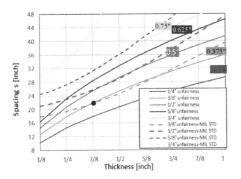

Figure 6. Comparison of distortion tolerance criteria between MIL-STD-1689 and those based on bending stress limit from Eq. 3.

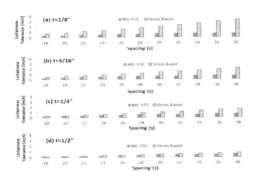

Figure 7. Comparison of distortion tolerances for different plate thicknesses (t) as a function stiffener spacing (s) with constant bending stress corresponding to $t = 3/8''$, $s = 22''$ in MIL-STD-1689.

sets of criteria decrease and become negligible at $t = 1/2''$. This implies that distortion tolerance criteria in MIL-STD-1689 can be interpreted as being based on secondary bending stresses caused by distortions to a large extent and more appropriate for thicker welded plates. The latter point is obvious since decades ago shipbuilding industry mostly dealt with thicker plates than today.

2.3 Nonlinear analysis

2.3.1 Formulation and solutions
Consider the beam bending problem with angular distortion defined in Fig. 2b or with buckling distortion defined in Fig. 3b, when a horizontal load (P) is applied (as illustrated in Fig. 4), the governing differential equation for beam defection v and corresponding general solution can be written as, if $P > 0$,

$$EIv^{(4)} - Pv'' = 0, \quad v = A\sinh \lambda x + B\cosh \lambda x + Cx + D,$$
$$\lambda^2 = \frac{P}{EI} \tag{4}$$

and if $P < 0$,

$$EIv^{(4)} - Pv'' = 0, \quad v = A\sin \lambda x + B\cos \lambda x + Cx + D,$$
$$\lambda^2 = \frac{-P}{EI} \tag{5}$$

The corresponding specific solution can be expressed for the angular distortion case in Fig. 2b becomes, in terms of bending stress generated at fillet weld location (i.e., $x = 0$):

$$\Delta\sigma_{b,\max}(0) =$$
$$\begin{cases} \left| -\frac{12E\delta_0 t}{s^2}\left(\frac{4}{\lambda L} \cdot \frac{\cosh\frac{\lambda s}{2} - 1}{\sinh\frac{\lambda s}{2}} - 1 \right)\right|, & P > 0, \lambda = \sqrt{\frac{P}{EI}} \\ \left| \frac{12E\delta_0 t}{s^2}\left(\frac{4}{\lambda s} \cdot \frac{\cos\frac{\lambda s}{2} - 1}{\sin\frac{\lambda s}{2}} + 1 \right)\right|, & P < 0, \lambda = \sqrt{\frac{-P}{EI}} \end{cases} \tag{6}$$

and for the buckling distortion case in Fig. 3b, at $x = s/2$ (i.e., beam mid-span),

$$\Delta\sigma_{b,\max}\left(\frac{s}{2}\right) =$$
$$\begin{cases} \left| \frac{6E\delta_0 t}{s^2}\left(\frac{2}{\lambda s}\tanh\frac{\lambda s}{2} - 1 \right)\right|, & P > 0, \lambda = \sqrt{\frac{P}{EI}} \\ \left| \frac{6E\delta_0 t}{s^2}\left(\frac{2}{\lambda s}\tan\frac{\lambda s}{2} - 1 \right)\right|, & P < 0, \lambda = \sqrt{\frac{-P}{EI}} \end{cases} \tag{7}$$

It can be shown that despite the existence of initial distortion δ_0 in both cases (Figs. 2b and 3b), critical compressive load for causing buckling remains the same as classical buckling solutions for the same beam configurations without considering any imperfections, i.e.,

$$P_{crit} = \frac{4\pi^2 EI}{s^2} \qquad (8)$$

for the case of angular distortions and,

$$P_{crit} = \frac{\pi^2 EI}{s^2} \qquad (9)$$

for the case of buckling distortions.

2.3.2 Bending stress from angular distortions

Consider an illustrative case with t = 1/4″ (6.35 mm), s = 24″ (609.6 mm), δ_0 = 5 mm, and applied axial load (P) varies over time from initial tension to peak and then compression, as shown in Fig. 8a. The corresponding beam deflection δ at first reduces from δ_0 = 5 mm to about δ = 3.7 mm, indicating straightening, then increases to a peak value of about δ = 7.3 mm, as shown in Fig. 8b. The resulting secondary stress as a function of time-varying load (P) is shown in Fig. 9. Due to straightening effects of axial load, when P reaches its peak tension, the bending stress, e.g., at $y = t/2$ is much less than that corresponding peak compression.

It is also interesting to note that if linear solution in Eq. (3) is used with deflection δ being calculated from Eq. (4), the bending stress is over-estimated somewhat, but still provides a reasonable estimation of the overall bending stress behavior. As a result, the comments made regarding MIL-STD-1689 in Sec. 2.2.2 can be therefore justified. Again, when axial force P becomes compressive, the maximum bending stress develops. As far as fatigue damage for fillet welds is concerned for this case, the resulting bending stress range is about 150 MPa shown in Fig. 8.

2.3.3 Bending stresses at beam mid-span

Often, there exists a butt-seam weld in between two stiffeners, say at x = s/2. The question then becomes how both types of distortions discussed in Sec. 2.1 contribute to secondary bending stresses at such a butt seam weld under remote horizontal loading in stiffened panels. The analytical solutions for both types of distortions discussed in Sec. 2.3.1 can be used by setting x = s/2. The corresponding bending stresses after being normalized by remote stress (i.e., P/t) are plotted in Fig. 10 as a function of remote axial load P normalized by their respective buckling loads (see Eqs. 8 and 9). As shown in Fig. 9, buckling distortions generate a much higher secondary

bending stress at beam mid-span than angular distortions, in fact, exactly twice as much, even though distortion magnitudes in both cases are the same, i.e., δ_0 = 5 mm. The nonlinear dependency of the stress concentration factor (SCF) on remote load P is clearly shown for both cases in Fig. 10, with a rapid increase in SCF when P becomes compressive and approaches its buckling limit. With computed SCF values, fatigue test data can be correlated in terms of master S-N curve as discussed in Dong & Hong (2004) and Xing et al (2016).

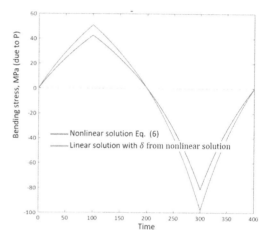

Figure 8. Secondary bending stresses at wel location and y = t/2 generated by axial load P interactiong with initial distortion – δ_0 angular distortion.

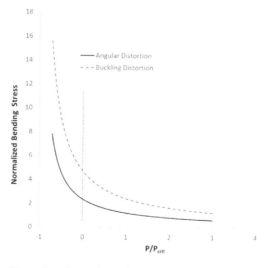

Figure 9. Comparison of normalized bending stresses ar beam mid span (Y = L/2) between angular and buckling distortion cases.

3 CONCLUDING REMARKS

Two distortion types in welded thin panels are generalized based on detailed observations of experimental and computational modeling results to date. These are angular distortions and buckling distortions that are prevalent in lightweight shipboard structures today. To quantitatively examine their effects on fatigue under remote fatigue loading conditions in service, the two types of distortions are idealized into two sets of compound beam problems containing pre-existing curvatures and being subjected remote tension and compression. Both linear and nonlinear treatments are considered for deriving analytical closed-form solutions of secondary bending stresses. With linear treatment, existing empirical-based distortion tolerance criteria, e.g., MIL-STD-1689, have been quantitatively examined using these solutions. It was found that existing distortion tolerance criteria are very conservative for applications in lightweight shipboard panels with plate thickness equal and less 5 mm. This finding is then further confirmed by nonlinear analysis results presented in the same study. Furthermore, nonlinear analysis performed in this study provides a set of closed form solutions for secondary bending stress concentration developed at both fillet welds and beam mid-span, which can be used for fatigue evaluation of panel structures dominated by distortion types examined in this study.

REFERENCES

ABS Guide, "Shipbuilding and repair quality standard for hull structures during construction", March 2007

Antoniou, A.C. (1980). On the maximum deflection of plating in newly built ships. *Journal of Ship Research*, 24(1).

Carlsen, C.A., & Czujko, J. (1978). The specification of post-welding distortion tolerances for stiffened plates in compression. *Structural Engineer*, 56.

Dong, P. (2005). Residual stresses and distortions in welded structures: a perspective for engineering applications. *Science and Technology of Welding and Joining*, 10(4), 389–398.

Dong, P. (2008). Length scale of secondary stresses in fracture and fatigue. *International Journal of Pressure Vessels and Piping*, 85(3), 128–143.

Dong, P., & Brust, F.W. (2000). Welding residual stresses and effects on fracture in pressure vessel and piping components: a millennium review and beyond. *Journal of Pressure Vessel Technology*, 122(3), 329–338.

Dong, P., & Hong, J.K. (2004). The master SN curve approach to fatigue evaluation of offshore and marine structures. In *Proceedings of the 23rd International Conference on Offshore Mechanics and Arctic Engineering*, Vol. 2, pp. 847–855.

Dong, P., Song, S., Zhang, J., & Kim, M.H. (2014). On residual stress prescriptions for fitness for service assessment of pipe girth welds. *International Journal of Pressure Vessels and Piping*, 123, 19–29.

Dong, P., Zhou, Q., and Xing, S. (2017). Analytical treatment of welding distortion effects on fatigue in thin panels: Part I—closed-form solutions and implications, Maritime Transportation and Harvesting of Sea Resources, Guedes Soares & Teixeira (Ed), Taylor and Franciso, London (in press).

Evan, H.R., (1954). Full scale ship structural experiments and the effect of unfair plating in tension, *ASNE Journal*, Feb, 1954, 55–78

Huang, T. et al, (2004). Fabrication and engineering technology for lightweight ship structures, part 1: distortions and residual stresses in panel fabrication. *Journal of Ship Production*, 20(1), 43–59.

Jung, G., et al. (2007). Numerical prediction of buckling in ship panel structures. *Journal of Ship Production*, 23(3), 171–179.

Lillemäe, I., et al (2012). Fatigue strength of welded butt joints in thin and slender specimens. *International Journal of Fatigue*, 44, 98–106.

Lu, H., Dong, P., & Boppudi, S. (2015). Strength analysis of fillet welds under longitudinal and transverse shear conditions. *Marine Structures*, 43, 87–106.

MIL-STD-1689, (1967) Fairness tolerance criteria, US Navy

Song, S., & Dong, P. (2017). Residual stresses at weld repairs and effects of repair geometry. *Science and technology of welding and Joining*, 22, 4, 265–277.

Xing, S., & Dong, P. (2016). An analytical SCF solution method for joint misalignments and application in fatigue test data interpretation. *Marine Structures*, 50, 143–161.

Xing, S., Dong, P., & Threstha, A. (2016). Analysis of fatigue failure mode transition in load-carrying fillet-welded connections. *Marine Structures*, 46, 102–126.

Xing, S., Dong, P., & Wang, P. (2017). A Quantitative Weld Sizing Criterion for Fatigue Design of Load-Carrying Fillet-Welded Connections. *International Journal of Fatigue*.

Yang, Y.P., & Dong, P. (2012). Buckling distortions and mitigation techniques for thin-section structures. *Journal of materials engineering and performance*, 21(2), 153–160.

Maritime Transportation and Harvesting of Sea Resources – Guedes Soares & Teixeira (Eds)
© 2018 Taylor & Francis Group, London, ISBN 978-0-8153-7993-5

Analytical treatment of welding distortion effects on fatigue in thin panels: Part II—applications in test data analysis

P. Dong, W. Zhou & S. Xing

Department of Naval Architecture and Marine Engineering, University of Michigan, Michigan, USA

ABSTRACT: As a sequel to Part I of the same study (see Dong, Xing, and Zhou, 2017), this paper (Part II) focuses on applications of their analytical solutions developed in Part I in interpretation of some of the recent test data available in literature. These fatigue tests involved thin plate butt-welded specimens containing complex angular distortions as well as axial misalignments. In extending the analytical solutions developed in Part I, both test load conditions and angular distortions in butt seam welded specimens are analytically treated. In this process, the angular misalignment equation given by BS 7910 incorporating nonlinear geometry effects is recovered by the present solution scheme. By introducing a superposition scheme incorporating local and global distortions as well as axial misalignments reported by Lillemäe et al. (2012), their fatigue test data are shown not only correlating among themselves in a reasonable narrow band, but also falling into the 2007 ASME master S-N curve scatter band in a satisfactory manner.

1 INTRODUCTION

It has been well established that welding induced distortions have been a major issue in construction of lightweight ship structures (Dong, 2005; Jung, et al, 2007, and Yang & Dong, 2012). In addition to developing effective mitigation techniques for reducing such distortions, existing distortion tolerances such as those given Class societies such as by ABS (2007) which were based on data for thick-section structures, some of which dated back many decades ago (e.g., MIL-STD-1698,) may need to be revisited for applications in lightweight structures today. As discussed in Huang et al (2004), in lightweight steel shipboard structures, plate thicknesses of equal or less than 5 mm have become increasingly dominant, posing a major challenge in accuracy control in ship construction processes.

Effects of weld residual stresses on structural integrity have been extensively addressed in the context of fitness for service in the literature, as recently summarized by Dong & Brust (2000) and Dong et al (2014, 2017), Song & Dong (2017), and Dong (2005,2008). However, how welding-induced distortions impact structural integrity of lightweight structures have not been adequately addressed. Some of the early publications on this subject are scarce, e.g., by Antoniou (1980) and Carlsen & Czujko (1978), and focused on limited experimental observations on certain specific types of distortions in ship construction environment. These studies had a narrow focus on effects of some observed welding distortions on structural buckling strength under compressive loading with plate thicknesses being above 10 mm and did not address how distortions impact fatigue behaviors in thin plate structures. Recognizing the prevalence of distortions in thin plate structures, an increasing number of researchers have recently reported their of investigations into fatigue behaviors observed in thin plate welded specimens containing various forms of distortions. These studies include those by Lillemäe et al (2012) on distortion effects on fatigue strength in butt-welded plate specimens, Xing, et al (2016) on fatigue failure mode transition behaviors in thin plate cruciform joints and Lu et al (2015) and Xing et al (2016) on quantitative fillet weld sizing criteria for preventing weld throat cracking in load-carrying fillet welds. In the latter studies, the importance of proper treatment of joint misalignments in thin plate structures is demonstrated by means of a series of new analytical stress concentration factor solutions. Both these experimental and finite element studies have showed that fatigue behaviors in thin plate structures tend to show a great deal of scatter, much more so than thick plate structures joint. To clarify some of the dominant effects of welding-induced distortions on fatigue, a fundamental approach is needed for separating distortion characteristics that are intrinsic to certain lightweight structural forms, which can be quantitatively related to fatigue performance from those distortion features that may exhibit a great deal of variability, whose effects on fatigue may be only secondary.

As a sequel to Part I of the same study (see Dong, et al, 2017), this paper (Part II) focuses on applications of their analytical solutions in interpretation of some of the recent test data available in literature (e.g., by Lillemäe et al, 2012). These fatigue tests involved thin plate butt-welded specimens containing complex angular distortions as well as axial misalignments. In extending the analytical solutions developed in Part I, both test load conditions and angular distortions in butt seam welded specimens are analytically treated. In this process, the angular misalignment equation given by BS 7910 incorporating nonlinear geometry effects is recovered by the present solution scheme. By introducing a superposition scheme incorporating local and global distortions as well as axial misalignments reported by Lillemäe et al (2012), the test data can be satisfactorily correlated with the 2007 ASME master S-N curve scatter band which contain large scale fatigue test data with plate thicknesses ranging from 5 mm up to over 100 mm.

2 TWO DISTINCT DISTORTION TYPES

As discussed in Dong (2005, 2008) and Yang and Dong (2012), welding-induced distortions in thin plate structures tend to exhibit buckling distortions due to structural instability triggered by compressive residual stress distributions, often also accompanied by angular distortions.

Both distortion types are described in Fig. 1, in which Fig. 1a shows a LIDAR scan image of a 16 feet by 20 feet stiffened panel, showing a checkerboard pattern in out of plane distortions, particularly on the lower half of the panel. A stiffened panel exhibiting angular distortion is illustrated in Fig. 1b. The corresponding deformation profiles for both cases are illustrated in Figs. 1c and 1d, respectively. As such, buckling distortion profile can be characterized as a sinusoidal wave form in which rotation at stiffener location is unrestricted (see Fig. 1c) while angular distortion profile shows

no rotation at stiffener locations. In both cases, stiffener spacing is represented by s and out of plane distortion peak value by δ.

3 DISTORTION AND LOAD INTERACTIONS

3.1 *Idealization, formulation, and solution*

As a first order approximation, pre-existing distortion and its interaction with a lateral load can be idealized as a compound beam bending problem shown in Fig. 2 for angular type of distortions and in Fig. 3 for buckling type of distortions.

In Figs. 2b and 3b, P represents axial force remotely applied with respect to the beam axis which may vary over time during service.

Consider the beam bending problem with angular distortion defined in Fig. 2b or with buckling distortion defined in Fig. 3b, when a horizontal load (P) is applied, the governing differential equation for beam deflection v and corresponding general solution can be written as, if $P > 0$,

$$EIv^{(4)} - Pv'' = 0, \; v = A \sinh \lambda x + B \cosh \lambda x + Cx + D,$$
$$\lambda^2 = \frac{P}{EI} \tag{1}$$

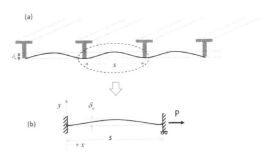

Figure 2. An equivalent beam bending problem definition—angular distortion type.

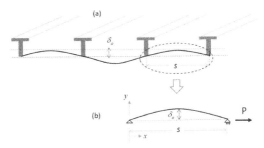

Figure 3. An equivalent beam bending problem definition—buckling distortion type.

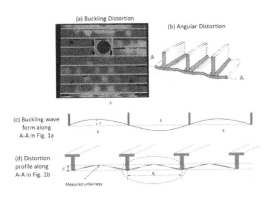

Figure 1. Two major distortion types in thin plate structures.

and if $P < 0$,

$$EIv^{(4)} - Pv'' = 0, \quad v = A\sin\lambda x + B\cos\lambda x + Cx + D,$$
$$\lambda^2 = \frac{-P}{EI} \tag{2}$$

The corresponding specific solution can be expressed for the angular distortion case in Fig. 2b becomes, in terms of bending stress generated at fillet weld location (e.g., $x = 0$):

$$\Delta\sigma_{b,\max}(0) =$$
$$\begin{cases} \left| -\dfrac{12E\delta_0 t}{s^2}\left(\dfrac{4}{\lambda s}\cdot\dfrac{\cosh\dfrac{\lambda s}{2}-1}{\sinh\dfrac{\lambda s}{2}}-1\right)\right|, & P>0, \lambda=\sqrt{\dfrac{P}{EI}} \\[4mm] \left|\dfrac{12E\delta_0 t}{s^2}\left(\dfrac{4}{\lambda s}\cdot\dfrac{\cos\dfrac{\lambda s}{2}-1}{\sin\dfrac{\lambda s}{2}}+1\right)\right|, & P<0, \lambda=\sqrt{\dfrac{-P}{EI}} \end{cases} \tag{3}$$

for the buckling distortion case in Fig. 3b, at $x = s/2$ (i.e., at beam mid-span),

$$\Delta\sigma_{b,\max}\left(\frac{s}{2}\right) =$$
$$\begin{cases} \left|\dfrac{6E\delta_0 t}{s^2}\left(\dfrac{2}{\lambda s}\tanh\dfrac{\lambda s}{2}-1\right)\right|, & P>0, \lambda=\sqrt{\dfrac{P}{EI}} \\[4mm] \left|\dfrac{6E\delta_0 t}{s^2}\left(\dfrac{2}{\lambda s}\tan\dfrac{\lambda s}{2}-1\right)\right|, & P<0, \lambda=\sqrt{\dfrac{-P}{EI}} \end{cases} \tag{4}$$

Additional derivation details in arriving at Eqs. (3) and (4) can be found in Part I of the same study by Dong, et al (2017). It can be shown that despite the existence of initial distortion δ_0 in both cases (Figs. 2b and 3b), critical compressive load for causing buckling remains the same as the classical buckling solutions for the same beam configurations, i.e.,

$$P_{crit} = \frac{4\pi^2 EI}{s^2} \tag{5}$$

for the case of angular distortion and,

$$P_{crit} = \frac{\pi^2 EI}{s^2} \tag{6}$$

for the case of buckling distortion.

3.2 Treatment of angular distortions in butt welded specimens

Some interesting fatigue tests on thin plate welded specimens were reported by Lillemäe et al (2012),

which were shown difficult to interpret with either finite element based hot spot stress or notch stress method. These are butt-seam welded plate specimens with distortions characterized as shown in Fig. 4 and allowed to rotate freely during cycle axial loading during fatigue testing.

In addition to axial misalignment e, angular distortions are further divided into local angular distortion α_L and global angular distortion α_G, as illustrated in Fig. 4. In concucting fatigue testing, the grips at specimen ends used a special pin connections to ensure free rotations during fatigue loading.

3.2.1 Global angular distortion

To facilitate analytical treatment of such conditions, a statically equivalent representation for a beam with an angular distortion of α_G and pin supports at both ends subjected to an axial tension F is considered as shown in Fig. 5.

Due to symmetry at weld location, only one half of the specimen needs to be considered as shown in Fig. 5b, which can be rotated to align beam axis with x axis.

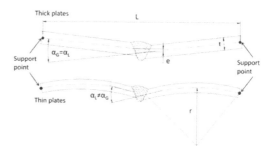

Figure 4. Distortion definitions associated with butt-welded thin plate specimens tested with cyclic axial tension with prin grips (taken from Lillemäe et al, 2012).

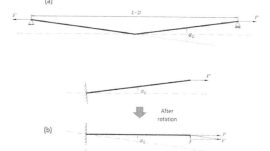

Figure 5. Treatment of global angular distortion in butt-welded plate specimen tested under pin-support conditions.

Eq. (1) is directly applicable here. After imposing appropriate boundary conditions corresponding to pin supports, the final beam defection equation becomes:

$$v = -\frac{F}{\lambda P}\sinh \lambda x + \frac{F}{\lambda P}\frac{\sinh \lambda l}{\cosh \lambda l}\cosh \lambda x + \frac{F}{P}x$$
$$- \frac{F}{\lambda P}\frac{\sinh \lambda l}{\cosh \lambda l} \tag{7}$$

Then, it can be shown that at $x = 0$,

$$\frac{\sigma_{b,nonlinear}}{\sigma_{b,linear}} = \frac{-\frac{F}{\lambda}\frac{\sinh \lambda l}{\cosh \lambda l}}{-Fl} = \frac{\tanh \lambda l}{\lambda l}$$

or,

$$\sigma_{b,nonlinear} = \sigma_{b,linear} \cdot \frac{\tanh \lambda l}{\lambda l} \tag{8}$$

where $\sigma_{b,linear}$ represents bending stress generated by F in Fig. 5 without considering geometric nonlinearity. Note that

$$\lambda = \sqrt{P/EI} = 2\sqrt{3P/Et^3} = (2/t)\sqrt{3\sigma_m/E}, \text{ where,}$$
$$\sigma_m = P/t. \text{ Now let,}$$
$$\beta = \lambda l = \lambda(L/2) = (L/t)\sqrt{3\sigma_m/E}, \quad \text{and}$$
$$b = \tanh(\beta)/\beta.$$

The normalized bending stress at $x = 0$ can then be written in the form of stress concentration factor, k:

$$k = 1 + \left(\frac{6}{4} \times \frac{L}{t} \times \alpha_G\right) \times b \tag{9}$$

where, the bracketed term in Eq. (9) represents the normalized bending stress without nonlinear geometry effects, β is a dimensionless function of slender ratio L/t. It should be pointed out that Eq. (9) is exactly the same as the one given in BS 7910 (2013).

3.2.2 Local angular distortion

To facilitate the use of Eq. (9) for treating local angular distortion α_L illustrated in Fig. 4, an equivalent local angular distortion α'_L is introduced in Fig. 6, which can be expressed as below:

$$\alpha'_L = \alpha_L - \alpha_G \tag{10}$$

which can be directly inserted into Eq. (9) for calculating stress concentration contributed by both global and local angular distortions. Note that in the present approach, actual curvature contributing to α_L is ignored here for simplicity. As a part

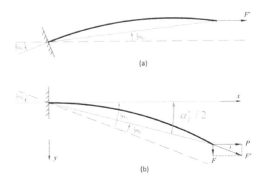

Figure 6. An equivalent local angular distortion definition.

of this study, an analytical solution considering the actual curvature effects is presented in Zhou and Dong (2017).

4 APPLICATION IN TEST DATA ANALYSIS

The fatigue test results on thin-plate butt welded specimens (3 mm in thickness) were reported by Lillemäe et al (2012), who also performed detailed finite element analysis (FEA) of these specimens with measured distortions. With geometric nonlinearity considerations, they evaluated the feasibility of using either surface extrapolated hot spot stress or local notch stress method recommended by IIW (Hobbacher, 2009). The results are directly taken from Lillemäe et al (2012) and shown in Fig. 7, indicating that either method provided a satisfactory correlation of the same test data. This is in view of the fact that the test data spread within a factor of 10 in life without exhibiting any definable slope, in terms of either hot spot stress (Fig. 7a) or local notch stress (Fig. 7b).

With the developments presented in Sec. 3, stress concentration due to misalignments (e) involved in the same specimen is calculated as k_e by means of an analytical solution given in Xing and Dong (2016), while stress concentration factor resulted from the measured global angular distortion (α_G) and local angular distortion α_L is calculated as k by means Eq. (9) with α_G being replaced by α'_L thorough Eq. (10). The results are summarized in Fig. 8. In Fig. 8, the equivalent structural stress range is defined as according to the 2007 ASME Div 2 Code as:

$$\Delta S_s = \frac{\Delta \sigma_s}{t^{\frac{2-m}{2m}} I(r)^{\frac{1}{m}}} \tag{11}$$

where $\Delta \sigma_s$ in the current case is calculated as $(k + k_e) \times \Delta \sigma_n$ in which $\Delta \sigma_n$ is nominal stress

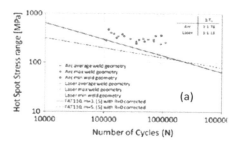
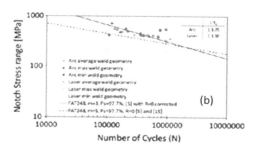

Figure 7. Test data correlation using nonlinear geometry FEA calculated stresses (taken from Lillemäe et al (2012): (a) IIW's surface extraspolated hot spot stress method; (b) IIW's local notch stress method.

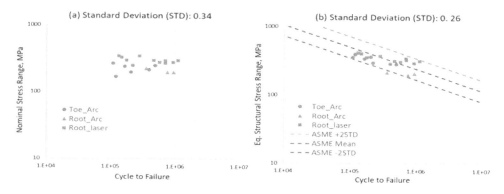

Figure 8. Test data correlation using nonlinear geometry FEA calculated stresses (taken from Lillemäe et al (2012): (a) IIW's surface extraspolated hot spot stress method; (b) IIW's local notch stress method.

range as reported by Lillemäe et al (2012). Note that m is 3.6 as given in 2007 ASME Div 2 Code and $I(r)^{1/m}$ is a dimensionless function bending stress ratio $r = |\Delta\sigma_b|/(|\Delta\sigma_m| + |\Delta\sigma_b|)$, also given in ASME Div 2 Code (see Dong and Hong (2004) and Dong et al (2010).

In Fig. 8, nominal stress range based data plot is shown in Fig. 8a while equivalent structural stress range (according to Eq. 11) based lot is shown in Fig. 8b. Note that the ASME's master S-N curve scatter band (based on *mean* ± 2 × STD) is also given as dashed lines in Fig. 8b. It can be seen that in Fig. 8b, the thin plate butt welded specimen test data not only correlate reasonably well among the test data themselves by forming a well-defined scatter band with a standard deviation of 0.26, but also with the 2007 ASME's master S-N curve scatter band which represents about 1000 large scale fatigue tests with plate thickness varies from 5 mm up to over 100 mm.

5 CONCLUDING REMARKS

This paper (Part II) focuses on application of a set of analytical solutions developed in Part I by the

same authors for treating effects of welding induced distortions on fatigue behaviors in welded thin plate structures reported by Lillemäe et al (2012).

These fatigue tests involved thin plate butt-welded specimens containing complex angular distortions as well as axial misalignments. In extending the analytical solutions developed in Part I, both test loading conditions and angular distortions (local and global) in butt seam welded specimens are analytically treated in deriving stress concentration factors at weld location in butt-seam welded specimens. As far as global angular distortions are concerned, the angular misalignment equation given by BS 7910 considering nonlinear geometry effects is recovered exactly by the present solution.

By incorporating stress concentration factors from local and global distortions as well as axial misalignments into the equivalent structural stress range parameter given by the 2007 ASME Div 2 Code, the test data by Lillemäe et al (2012) can be correlated reasonably well among themselves in forming a narrow band with a clearly defined slope.

Furthermore, all these test data points but one fall within the 2007 ASME master S-N curve scatter band defined as *mean* ± 2 × STD, which is

609

derived based on about 1000 large-scale fatigue tests with plate thicknesses ranging from 5 mm up to over 100 mm.

REFERENCES

ABS Guide, "shipbuilding and repair quality standard for hull structures during construction", March 2007.

Antoniou, A.C. (1980). On the maximum deflection of plating in newly built ships. Journal of Ship Research, 24(1).

ASME Codes and Standards, (2007). Div 2 Sec VIII Part 5.5.5.

BS 7910:2013, Guide to methods for assessing the acceptability of flaws in metallic structures, BSI Standards Publication.

Carlsen, C.A., & Czujko, J. (1978). The specification of post-welding distortion tolerances for stiffened plates in compression. Structural Engineer, 56.

Dong, P. (2005). Residual stresses and distortions in welded structures: a perspective for engineering applications. Science and Technology of Welding and Joining, 10(4), 389–398.

Dong, P. (2008). Length scale of secondary stresses in fracture and fatigue. International Journal of Pressure Vessels and Piping, 85(3), 128–143.

Dong, P., & Brust, F.W. (2000). Welding residual stresses and effects on fracture in pressure vessel and piping components: a millennium review and beyond. Journal of Pressure Vessel Technology, 122(3), 329–338.

Dong, P., & Hong, J.K. (2004). The master SN curve approach to fatigue evaluation of offshore and marine structures. In Proceedings of the 23rd Inter.Conference on Offshore Mechanics and Arctic Engineering, Vol. 2, pp. 847–855.

Dong, P., Hong, J.K., Osage, D.A., Dewees, D.J., & Prager, M. (2010). The master SN curve method: an implementation for fatigue evaluation of welded components in the ASME B&PV CODE, Section VIII, Division 2 and API579–1/ASME FFS–1. Welding Research Council Bulletin, 523.

Dong, P., Song, S., Zhang, J., & Kim, M.H. (2014). On residual stress prescriptions for fitness for service assessment of pipe girth welds. International Journal of Pressure Vessels and Piping, 123, 19–29. applications. Science and Technology of Welding and Joining, 10(4), 389–398.

Dong, P., Xing, S., and Zhou, Q., (2017). Analytical treatment of welding distortion effects on fatigue in thin panels: Part I—closed-form solutions and implications, Maritime Transportation and Harvesting of Sea Resources, Guedes Soares & Teixeira (Eds.), Taylor and Francis, London, (in press).

Dong, P., Zhou, W, and Xing, X. (2017), Analytical treatment of welding distortion effects on fatigue in thin panels: part i—closed form solutions and implications, in preparation.

Evan, H.R., (1954). Full scale ship structural experiments and the effect of unfair plating in tension, ASNE J, Feb, 1954, 55–78.

Hobbacher A. (2009). IIW recommendations for fatigue design of welded joints and components, WRC Bulletin 520, New York: The Welding Research Council.

Huang, T. et al, (2004). Fabrication and engineering technology for lightweight ship structures, part 1: distortions and residual stresses in panel fabrication. Journal of Ship Production, 20(1), 43–59.

Jung, G., et al. (2007). Numerical prediction of buckling in ship panel structures. Journal of Ship Production, 23(3), 171–179.

Lillemäe, I., et al (2012). Fatigue strength of welded butt joints in thin and slender specimens. International Journal of Fatigue, 44, 98–106.

Lu, H., Dong, P., & Boppudi, S. (2015). Strength analysis of fillet welds under longitudinal and trans-verse shear conditions. Marine Structures, 43, 87–106.

MIL-STD-1689, (1967) Fairness tolerance criteria, US Navy.

Song, S., & Dong, P. (2017). Residual stresses at weld repairs and effects of repair geometry. Science and technology of welding and Joining, 22, 4, 265–277.

Xing, S., & Dong, P. (2016). An analytical SCF solution method for joint misalignments and application in fatigue test data interpretation. Marine Structures, 50, 143–161.

Xing, S., Dong, P., & Threstha, A. (2016). Analysis of fatigue failure mode transition in load-carrying fillet-welded connections. Marine Structures, 46, 102–126.

Xing, S., Dong, P., & Wang, P. (2017). A Quantitative Weld Sizing Criterion for Fatigue Design of Load-Carrying Fillet-Welded Connections. International Journal of Fatigue.

Yang, Y.P., & Dong, P. (2012). Buckling distortions and mitigation techniques for thin-section structures. Journal of materials engineering and performance, 21(2), 153–160.

Zhou, W. and Dong, P., (2017). A set of analytical solutions of stress concentration factors for various distortion modes in welded thin plates, under preparation.

Maritime Transportation and Harvesting of Sea Resources – Guedes Soares & Teixeira (Eds)
© 2018 Taylor & Francis Group, London, ISBN 978-0-8153-7993-5

Fatigue reliability of butt-welded joints based on spectral fatigue damage assessment

Y. Garbatov & Y. Dong
Centre of Marine Technology and Ocean Engineering (CENTEC), Instituto Superior Técnico, Universidade de Lisboa, Lisbon, Portugal

J. Rörup
DNV GL—Maritime, Hamburg, Germany

S. Vhanmane
Indian Register of Shipping, Mumbai, India

R. Villavicencio
LLOYD'S REGISTER, Southampton, UK

ABSTRACT: This study deals with the fatigue reliability analysis of a butt-welded joint in a bulk carrier. The formulation for the assessment of the reliability is built on the S-N approach and FORM technique. The study uses the initial fatigue analysis related to the closed form approach and the ones based on the long-term response and spectral approach damage assessment and accounts for the existing uncertainties originating from fatigue phenomenon, load and stress calculations and fabrication. Sensitivity with respect to the random variables involved in the reliability limit state function are analyzed and the partial safety factors that can be used in the design are defined and determined.

1 INTRODUCTION

For ship structures, which are subjected to wave-induced loads during the service life, the fatigue strength assessment needs to be performed. The assessment based on the S-N approach is usually used on the design stage. In this respect, one of the most advanced approaches for marine structures is the spectral one.

It assumes that the sea elevation is stationary, relatively narrow-banded, Gaussian random process over a short period of time 1~3 hours, where the distribution of the wave energy over different frequencies is expressed by a wave spectrum. It is also assumed that the sea states are described by a single peaked spectrum, which is well modelled by the Pierson and Moskowitz form (Pierson and Moskowitz, 1964) or the JONSWAP form (Hasselmann, 1973). The long-term sea-state is modelled by a large amount of discrete stationary short-term sea states. The scatter diagram of the North Atlantic is usually adopted in ship design, based on the argument that it has the most extreme sea states of all ocean-going areas.

If the ship response to a wave excitation is a linear one, the total response in a seaway is described by a superposition of the responses to all regular wave components that constitute the irregular sea, which can be performed in a frequency domain analysis. The transfer functions generally referred to as Response Amplitude Operators, or RAOs. The RAO represents the response of the ship's structure to excitation by a wave of a unit height, and it is derived over the full range of (encounter) frequencies that will be experienced. Different approaches, such as the 2D strip theory, 3D diffraction theory and even close form solutions are developed to define the RAOs.

To evaluate the fatigue damage, the nominal stress, hot spot stress or the notch stress combined with a particular S-N curve are used (Hobbacher, 2014). Current classification rules are based on the hot spot stress approach, which needs a detailed finite element analysis (Niemi et al., 2004). The estimated local stresses at discontinuities strongly depend on the structural idealization, the element type and the mesh density (Fricke et al., 2008).

The stress spectrum of a selected fatigue prone location in a short-term sea state can be obtained by combining the wave spectrum, RAOs and the stress analysis, leading to a stress range distribution, which is the Rayleigh distributed if the narrow-banded Gaussian process assumption is fulfilled. Fatigue damage in the short-term sea state can be evaluated by means of a close form expression. The linear superposition of short-term damages over all sea states, heading angles and loading conditions gives the total fatigue damage for the structural detail.

The reliability based methods have been gaining increasing attention because it can assess the level of safety in the structures. The evaluated fatigue life may be subjected to considerable uncertainty due to the load calculation, fatigue phenomenon, fabrication and stress calculation.

Garbatov et al. (2004) presented a reliability assessment based on a complete stochastic spectral fatigue analysis of a FPSO that was converted from an oil tanker after a 20 year-operation, employing the FORM technique. Fatigue strength and reliability assessment of complex double hull oil tanker structures, based on different local stress approaches, was performed by Garbatov (2016) accounting for the uncertainties originated from load, nominal stresses, hot spot stress calculations, weld quality estimations and misalignments and fatigue S-N parameters including the correlation between load cases and the coating life and corrosion degradation. In this study, a fatigue reliability assessment of a butt-welded joint located on the deck of a structure of a bulk carrier was carried out.

The load estimated of transverse butt welds in the upper hull of a bulk carrier, as reported by Rörup et al. (2017), are evaluated to identify the existing uncertainties. The reliability analysis also accounts the uncertainties in the damage at failure, S-N curve, stress calculation and the misalignment of the welded joint, employing the FORM technique. The hot spot stress approach and a linear form S-N curve considering the thickness effect are adopted. Only the vertical bending moment is considered and the mean stress effect is ignored in the analysis.

2 FATIGUE DAMAGE EVALUATION

The fatigue damage of ship structural details of the service life of $T_d = 25$ years is calculated based on Miner's rule (Miner, 1945). The design S-N curve is chosen based on the assessment approach and structural details, according to DNV-GL (2015):

$$\log N = \log a - m \log \Delta \sigma \qquad (1)$$

where N is the predicted number of cycles to failure under a given stress range $\Delta \sigma$, m is the slope of the S-N curve and $\log a$ is the intercept of the $\log N$-axis of the S-N curve. To consider the thickness

effect, a correction on $\log a$ is adopted, depending on the thickness of the detail.

If the fatigue stress process is stationary, Gaussian and narrow band, the fatigue damage can be calculated by:

$$D = (n/a)\left(2\sqrt{2}\sigma\right)^m \Gamma\left(\frac{m}{2}+1\right) \qquad (2)$$

where n is the total number of cycles, σ is the root mean square of the process and $\Gamma(\cdot)$ is the Gamma function. This formula is used to calculate the fatigue damage of the short-term condition.

The fatigue damage, during the service life, is the summation of the damages from all short-term conditions and becomes:

$$D = \frac{T_d}{a} \Gamma\left(1 + \frac{m}{2}\right) \sum_{\chi=1}^{n_\chi} r_\chi \sum_{\substack{j=1 \\ i=1}}^{\substack{all\ seastates \\ all\ headings}} p_{ij} f_{0,ij\chi} \lambda_{ij\chi} \left(2\sqrt{2}\sigma_{ij\chi}\right)^m \quad (3)$$

where T_d is the service life of the ship, n_χ is the number of loading conditions, r_χ is the fraction of time in the loading condition χ, p_{ij} is the probability of occurrence of each sea state i combined with the heading angle j, $f_{0,ij\chi}$ is the zero-up crossing frequency and $\lambda_{ij\chi}$ is the correction factor that represents the difference between the damage caused by the wide-banded process and the damage based on narrow-banded assumption Wirsching (1980).

Deterministic spectral fatigue damage can be obtained based on the above general procedures. Details with respect to the determination of the parameters in Eq. (3) can be found for an example in DNV-GL (2015). Comparing the evaluated fatigue damage and the damage at failure, Δ, which is usually assumed to be 1, one can verify the fatigue strength of the details. The major difference between the probabilistic analysis and the deterministic one lies in treating some of the parameters as random variables instead of deterministic constants.

3 STOCHASTIC VARIABLES

1.1 S-N curve parameter

The uncertainty of the S-N curve is observed as a scatter band of the fatigue data. Even for smooth specimens, their fatigue lives distribute in a range under the same constant amplitude loading. The welded specimens, which include more uncertainties in the local geometry, residual stress, microstructure and misalignment, results in a more severe uncertainty. It is usually assumed that the slope parameter of the S-N curve is deterministic and the intercept parameter is log-normally distributed.

In the deterministic approach, employing the hot spot stress, the S-N curve of FAT 90 is used, which represents 97.6% of survival probability. It is determined by downward shifting the mean S-N curve by two times of the standard deviation:

$$\log a = \log a_o - 2s \qquad (4)$$

where a_o is the constant relating to the mean S-N curve and s is the standard deviation, which is 0.2 according to DNV-GL (2015).

1.2 Damage at failure

According to the Miner's rule, fatigue failure occurs when $D \geq 1$. Many factors such as the temperature, chemical environments, mean stress, loading rate, surface condition, etc. affect the fatigue. It is not surprising that the simple Miner's rule would fail to provide an accurate description of a complex phenomenon. It has been demonstrated experimentally, by many studies that the critical value of the cumulative fatigue damage at failure is not always close to 1, but in fact is treated as a random variable whose median values are close to 1 and the coefficient of variation is relatively large (Wirsching, 1980).

1.3 Load calculation

The actual root mean square of the fatigue stress in a short-term condition can be presented as (Wirsching, 1980):

$$\sigma_{ij\chi} = B\sigma_{o,ij\chi} \qquad (5)$$

where $\sigma_{o,ij\chi}$ is the best estimate of the root mean square of the fatigue stress and B is a random variable, which quantify the uncertainty in the stress analysis procedures (Folsø et al., 2002).

For the butt welded joint, the random variable B can be split into several components, including B_l modelling the uncertainty in the load calculation, B_σ modelling the uncertainty in the stress calculation and B_m modelling the uncertainty due to the lateral distortion effect.

The randomness of the zero-up crossing rate and the correction factor can be included by generalizing the B_l, which originally represents the uncertainty of the root mean square. The generalized random variable $G_l = B_l{}^m$ represents the uncertainty of the fatigue damage due to the uncertainty of the load calculation.

Since different approaches are applied to calculate the hydrodynamic loads in the benchmark

study as reported by Rörup et al. (2017), a sample of the random variable G_l can be obtained. It assumes that the stress load ratio and other parameters used in the fatigue damage calculation are the same for all the approaches. It also assumes that the best estimator of fatigue damage is the median. Thus, for this study, the realization of the random variable G_l is the ratio between the calculated damage value and the median. Statistical analysis can then be carried out on the sample and the distribution of the B_l can be obtained. B_l is assumed to be log-normally distributed.

In the round robin study reported by Rörup et al. (2017), four studies among total ten analyses employed advanced methods considering the ship speed. The variation in the results may originate from different software, settings, mesh conditions, weight distributions that one adopted. The median of the four analyses is used to determine G_l and the four analyses are used to evaluate B_l. Since the sample size is too small, the 95% confidence intervals of the median and CoV are calculated as well. The results are listed in Table 1.

It seems that the first four participants used a similar method and settings and the variation in the load used in the fatigue damage assessment are not significant. Comparing the results from participant 5 and 6, close results are obtained by the same method. In the following reliability analysis, the median of B_l is assumed to be 1 and the CoV of B_l is assumed to be 0.033, employing the upper bound of the confidence interval.

1.4 Stress calculation

The random variable B_s is used to quantify the uncertainty in the stress calculation. The round robin study of local stresses of ship structural details performed by Fricke et al. (2008) was with

Table 1. Evaluation of B_l.

Participant*	Hydro code	G_l	Median B_l	CoV B_l
1	3D	0.988	1	0.0088
2	3D	1.012	[0.986,	[0.005,
3	3D	0.969	1.014]	0.033]
4	3D	1.029		
5	2D	1.613	–	–
	3D	1.506		
	Close-form	1.127		
6	2D	1.596		
	Close-form	1.189		
7	3D (zero speed)	1.332		

*: see reference (Rörup et al., 2017)

the objective in quantifying the uncertainties related to the modelling and stress evaluation and identifying sources of scatter. The element type and properties of the finite elements, mesh density chosen and the type of extrapolation are the main sources. Typical coefficients of variation to be expected from local stress analyses are in the order of 5%.

1.5 Lateral distortion

The effect of axial misalignment less than 0.1t is taken into account implicitly in the S-N curve (DNV, 2014). The uncertainty of the S-N curve includes the uncertainty of the axial misalignment. While, the lateral distortion commonly accompanied with the axial misalignment, e in welded joints is not considered. A magnification factor due to the lateral distortion can be used to correct the fatigue stress (DNV-GL, 2015):

$$B_m = 1 + \frac{\lambda}{4}\alpha\frac{s}{t} = 1 + \frac{\lambda y}{t} \tag{6}$$

where λ is related to the boundary conditions ($\lambda = 3$ for fixed ends and $\lambda = 6$ for pinned end), s is the distance between the frames, α is the angular misalignment, which can be approximated by 4y/s, y is the lateral distortion and t is the plate thickness, see Figure 1.

The parameter y can be treated as a random variable. Its characteristic value corresponds to a confidence level of 95% is assumed to be 6 mm, which is the tolerance limit for the fabrication imperfection according to DNV (2014). A truncated normal distribution is assumed because the y is an absolute value. Its probability density function is in the form of:

$$f_Y(y) = \begin{cases} \dfrac{\sqrt{2}}{\sigma\sqrt{\pi}}\exp\left(-\dfrac{y^2}{2\sigma^2}\right) & y \geq 0 \\ 0 & y < 0 \end{cases} \tag{7}$$

The statistical descriptors of random variables and deterministic constants are listed in Table 2. The joint No.1 in hold 5 close to the hatch with a thickness of 34 mm (Rörup et al., 2017) is analyzed. A global FE analysis with a coarse mesh is used to determine the stress load ratio.

a. Mean and standard deviation of the corresponding normal distribution.

b. $\phi = \Gamma\left(1 + \dfrac{m}{2}\right)\displaystyle\sum_{\chi=1}^{n_\chi} r_\chi \sum_{\substack{j=1 \\ i=1}}^{\substack{all\ seastates \\ all\ headings}} p_{ij} f_{0,ij\chi}\lambda_{ij\chi}\left(2\sqrt{2}\sigma_{o,ij\chi}\right)^m$

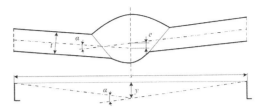

Figure 1. Lateral distortion and axial misalignment.

Table 2. Statistical descriptors of random variables and deterministic constants.

Parameter	Distribution	Median (Mean)	CoV (S. D.)
Δ	Log-normal	1	0.3
a	Log-normal	2.911E+12	0.486
B_l	Log-normal	1	0.033
B_s	Log-normal	1	0.05
y	Half-normal	(0)[a]	(6, 1.96)[a]
φ^b	–	858.8	
m	–	3	
λ	–	3	
t	–	34	
T_d	–	$0.85 \times 3600 \times 24 \times 365 \times 25$	

4 RELIABILITY ASSESSMENT

The first step in the reliability analysis using FORM is to identify a set of basic random variables, which influence the limit state. The limit state function can be formulated in terms of the set of random variables as:

$$g(X) = g(X_1, X_2, ..., X_n) \tag{8}$$

This function defines a failure surface when:

$$g(x) = g(x_1, x_2, ..., x_n) = 0 \tag{9}$$

This surface divides the space of n basic variables in a safe region when g(x)>0 and a failure region when g(x)<0. The failure probability can be written as:

$$P_f = P[g(x) \leq 0] = \int_{g(x) \leq 0} f_X(x)dx \tag{10}$$

where fx($\cdot$) is the joint probability density function of the n random variables and P_f denotes probability of failure.

In practice, the reliability cannot be evaluated based on the n-dimensional joint probability density function, which is usually not available. The FORM technique provides a way to evaluate the reliability efficiently and accurately (Hasofer and Lind, 1974, Rackwitz and Filessler, 1978).

The limit state function for fatigue failure can be written as:

$$g(x)=\ln\left(\frac{\Delta a}{B_I^m B_s^m \left(1+\frac{\lambda y}{t}\right)^m \phi}\right)-\ln(T_d) \qquad (11)$$

Through the FORM, a reliability index, β can be obtained, which is related to the probability of failure by:

$$P_f=\Phi^{-1}(-\beta) \qquad (12)$$

The importance of the contribution of each variable to the uncertainty of the limit state function can be assessed by the sensitive factors, which are determined by:

$$\alpha_i=-\frac{1}{\sqrt{\sum_{i=1}^{n}\left(\frac{\partial g(x)}{\partial x_i}\right)^2}}\frac{\partial g(x)}{\partial x_i} \qquad (13)$$

A positive sensitivity factor indicates that the increasing of the variable is benefit to the structure reliability and vice versa. Figure 2 shows the sensitivities of the failure function with respect to the changes in the random variables. It is visible that the uncertainty in the S-N curve and in the lateral distortion have almost the same importance and they are the most important variables. The additional uncertainty in the load calculation may be reduced if a similar hydrodynamic analysis method is employed.

The reliability index can be improved by reducing the stresses, which may be achieved by increasing the midship section modulus. A magnification

factor of the modulus, C is introduced. Assuming the plate thickness is not changed with the modulus, the limit state function then becomes:

$$g(x)=\ln\left(\frac{\Delta a C^m}{B_I^m B_s^m \left(1+\frac{\lambda y}{t}\right)^m \phi}\right)-\ln(T_d) \qquad (14)$$

Figure 3 illustrates the reliability index as a function of the magnification factor C. The fatigue reliability, after a 25- year service life.

An alternative way to improve the fatigue reliability index is to use a stricter tolerance limit for the lateral distortion in the fabrication, which can change its distribution parameter. It can be seen from Figure 4 that the reliability index increases from 1.5 to 2.15 by restricting the lateral distortion from 6 mm to 3 mm.

Partial safety factors for a target reliability can be assessed based on the design points and the characteristic values using the following design equation:

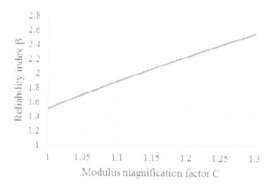

Figure 3. Reliability index as a function of modulus magnification factor.

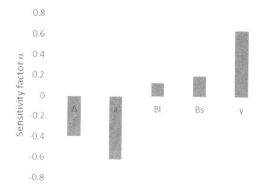

Figure 2. Sensitivity factors.

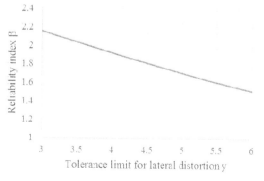

Figure 4. Reliability index as a function of tolerance limit for lateral distortion.

Table 3. Characteristic values and partial safety factors.

Random Variable	–	Δ	a	Bl	Bs	y
Xc (quantile)	–	1(50%)	1.159E12 (2.3%)	1(50%)	1(50%)	6(95%)
Parameter	C	γ_Δ	γ_a	γ_{Bl}	γ_{Bs}	γ_y
β = 1.5	1	1.19	0.61	1.01	1.02	0.71
β = 2	1.13	1.26	0.70	1.01	1.02	0.84
β = 2.5	1.28	1.34	0.81	1.01	1.03	0.97

$$G\left(x_c, \gamma\right) =$$

$$\ln\left(\frac{\Delta_c a_c C^m}{\gamma_\Delta \gamma_a \gamma_{B_l}^m \gamma_{B_s}^m B_{lc}^m B_{sc}^m \left(1 + \frac{\lambda \gamma_y y_c}{t}\right)^m \phi}\right) - \ln\left(T_d\right) \quad (15)$$

where variables with subscript c represent the characteristic values and γ is the partial safety factor. The partial safety factors corresponding to the reliability index of 1.5, 2 and 2.5 are listed in Table 3.

5 CONCLUSIONS

The fatigue reliability of a butt-welded joint located on the deck of a bulk carrier is assessed based on the S-N approach and FORM technique. The uncertainties originating from the fatigue phenomenon, load calculation, stress calculation and fabrication are considered. The spectral fatigue damage assessment serves as the basis in the formulation of the limit state function. Sensitivity factors and partial safety factors for the target reliability are determined.

The following conclusions may be established:

The additional uncertainty in the load calculation may be reduced if a similar hydrodynamic analysis method is employed.

The detail of interest has a fatigue reliability index of 1.5. An appropriated way to improve the fatigue reliability index is to increase the midship section modulus or adapt a stricter limit tolerance with respect to lateral distortion.

REFERENCES

DnV-Gl 2015. Fatigue Assessment of Ship Structures. *Class Guideline 0129*. Hovik: Det Norske Veritas.

DnV 2014. Fatigue Assessment of Ship Structures. *Classification Notes 30.7*. Hovik: Det Norske Veritas.

Folsø, R., Otto, S. & Parmentier, G. 2002. Reliability-based Calibration of Fatigue Design Guidelines for Ship Structures. *Marine Structures*, 15, 627–651.

Fricke, W., Bollero, A., Chirica, I., Garbatov, Y., Jancart, F., Kahl, A., Remes, H., Rizzo, C.M., Von Selle, H., Urban, A. & Wei, L. 2008. Round robin study on structural hot-spot and effective notch stress analysis. *Ships and Offshore Structures*, 3, 335–345.

Garbatov, Y. 2016. Fatigue strength assessment of ship structures accounting for a coating life and corrosion degradation. *International Journal of Structural Integrity*, 7, 305–322.

Garbatov, Y., Teixeira, A.P. & Guedes Soares, C. Fatigue reliability assessment of converted FPSO. Proceedings of the Offshore Mechanics and Arctic Engineering Specialty Conference on Integrity of FPSO System, 2004 Houston, USA. ASME, New York, paper OMAE-FPSO'04–0035.

Hasofer, A.M. & Lind, N.C. 1974. An exact and invariant first-order reliability format. *Journal of Engineering Mechanics Division*, 100, 111–121.

Hasselmann, K. 1973. Measurements of Wind-Wave Growth and Swell Decay During the Joint North Sea Wave Project (JONSWAP). *Deutchen Hydrographischen Institut*. Reihe.

Hobbacher, A. 2014. *Recommendations for fatigue design of welded joints and components*, Welding Research Council Shaker Heights, OH.

Miner, M.A. 1945. Cumulative Damage in Fatigue. *Journal of Applied Mechanics -Transactions of the ASME*, 12, A159-A164.

Niemi, E., Fricke, W. & Maddox, S. 2004. Structural Stress Approach to Fatigue Analysis of Welded Components - Designer's Guide. *IIW Doc. XIII-1819-00/XV-1090–01*.

Pierson, W.J. & Moskowitz, L. 1964. A Proposed Spectral Form for Fully Developed Wind Seas Based on Similarity Theory of S.A. Kitaigorodskij. *Journal of Geographical Research*, 69, 5181–5190.

Rackwitz, R. & Filessler, B. 1978. Structural Reliability under Combined Random Load Sequences. *Computers & Structures*, 9, 489–494.

Rörup, J., Garbatov, Y., Dong, Y., Uzunoglu, E., Parmentier, G., Andoniu, A., Quéméner, Y., Chen, K.-C., Vhanmane, S., Negi, A., Parihar, Y., Villavicencio, R. & Yue, J. 2017. Round robin study on spectral fatigue assessment of butt-welded joints. *Proceedings of the International Maritime Association of Mediterranean Congress, IMAM*. Lisboa, Portugal: Taylor & Francis.

Wirsching, P.H. 1980. Fatigue under wide band random Stress. *Journal of Structural Division*, 98, 1593–1606.

Maritime Transportation and Harvesting of Sea Resources – Guedes Soares & Teixeira (Eds)
© 2018 Taylor & Francis Group, London, ISBN 978-0-8153-7993-5

Assessment of distortion and residual stresses in butt welded plates made of different steels

M. Hashemzadeh, Y. Garbatov & C. Guedes Soares
Centre for Marine Technology and Ocean Engineering (CENTEC), Instituto Superior Técnico,
Universidade de Lisboa, Lisbon, Portugal

ABSTRACT: The objective of this work is to analyse the welding induced temperature, distortion and residual stresses of butt-welded plates made of different steels. Two types of steels, ASTM A36 and ST37 (S235) are considered in the butt-welded plates of A36-A36, ST37-ST37 and A36-ST37 steels. A two-step non-linear finite element analysis is employed to perform the thermo-mechanical analysis. The thermal analysis estimates the temperature distribution around the simulated weld, employing the mechanical properties that vary as a function of the applied temperature and the type of material that the welded plates are made. The thermal and mechanical analyses are performed separately. Both analyses are based on the finite element method using the commercial software ANSYS. The boundary conditions for the thermal analysis are considered to be conduction, convection and radiation and in the case of the mechanical analysis free boundary conditions are applied and only the two ends of the weld are fixed to restrain the structures. Several case studies are investigated and conclusions are derived.

1 INTRODUCTION

Welding is the most reliable and fastest way of metal joining, but in the other hand it is very sensitive to several variables and welding conditions and still needs to be analysed comprehensively. Due to the fast heating and cooling processes in welding, the microstructure of welded metal changes and causes distortions and residual stresses. These post weld effects have an impact on the strength of structures and because of that it is needed to be predicted and measured.

In this regard, Deng (2009) investigated the effects of the solid-state phase transformation on the welding residual stresses and distortions in the low carbon and medium carbon steels based on the sequentially coupled thermal, metallurgical, mechanical 3-D finite element models using the commercial software ABAQUS.

The welding simulation and improvement of the structural responses was analysed in several studies. Bachorski et al. (1999), Chang (1966) and Long et al. (2009) evaluated the post weld structural characteristics by analysing the thermos-mechanical properties of butt welded plates. In fact, the importance of their study is in defining of the temperature dependent material properties of the welded plate and their significance. In some studies, the experimental investigations were considered to calibrate the results estimated by FEM.

The thermal history and residual deformation of a double-sided fillet welding were studied numerically and experimentally by Biswas et al. (2009). Michaleris and DeBiccari (1997) compared numerical calculations with the experimental data from the welded small and large-scale specimens to validate the scalability of the results.

In complex structures, dissimilar material welding may happen. In this regard, several researches were performing studies to investigate the welding response. Attarha and Sattari-Far (2011), investigated the impact of dissimilar materials on the transient temperature distribution. They concluded that the peak temperature distribution vs. the distance in the weld pieces that the temperature decreasing in a nonlinear nature.

Nivas et al. (2017) studied the dissimilar material of welding between ASTM SA508 low alloy steel and 304 LN austenitic stainless steel. They discussed the microstructure of material.

Hu et al. (2012) developed a model to predict the heat and mass transfer in welding stainless steel and nickel and concluded that the mass transfer is highest during the initial stage of the weld pool formation and thereafter decreases with time.

In the present study, following the previous studies reported by Hashemzadeh et al. (2015a, 2015b), a numerical analysis is performed to investigate the impact of the dissimilar material properties of welded steel plates on the transient temperature distribution, distortions and residual stress of butt

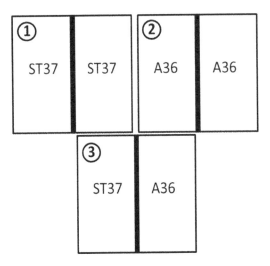

Figure 1. Case studies of similar and dissimilar material welding.

weld with a single pass of welding. In this regard, two similar materials welding of ASTM A36 and ST37 (S235) steel plates and one dissimilar material welding of A36-ST37 are modelled numerically. Figure 1 shows the cases studied in the present work.

2 FINITE ELEMENT ANALYSIS

The finite element analysis is performed using a transient thermo-mechanical analysis, which estimates the temperature distribution around the simulated weld, based on the mechanical properties that vary as a function of the applied temperature.

Considering that the mechanical response of the welded structure has negligible effects on the thermal analysis, indirect, decoupled, thermo-mechanical analysis is carried out as the thermal and mechanical analysis are performed separately. Figure 2 shows the decoupled thermo-mechanical analysis.

Finite element analysis employs commercial software ANSYS (2009). The finite element model is generated by eight-node three-dimensional brick thermal elements, Solid 70, which is switched automatically to Solid 180 for the mechanical part of the finite element calculations.

A refined finite element mesh is generated in the vicinity of the weld path, while a rough element size is used for the rest of the model, away from the welding path.

To simulate the arc welding, the moving heat source is assumed to be a transient process. During

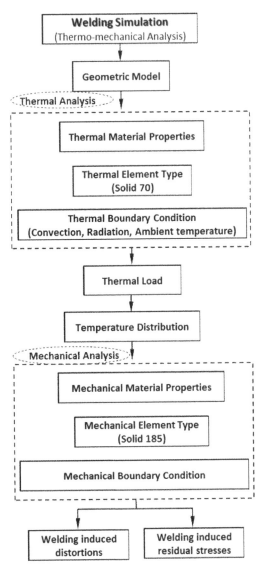

Figure 2. Transient thermo-mechanical analysis.

the movement of the heat source, the heat energy is kept as constant; however, the centre of the heat source is changed as a function of time. The surface of the butt—welded plates is subjected to heat convection, while the weld path is in a contact with the heat source and it is subjected to heat flux instead.

The finite element analysis is carried out on welded thin plates of a breadth × length × thickness of $160 \times 210 \times 4$ mm, employing the double ellipsoid heat input model as is introduced by Goldak et al. (1984). The mesh configuration near the weld is shown in Figure 3. The element size close to the

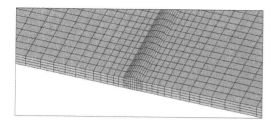

Figure 3. Finite element model.

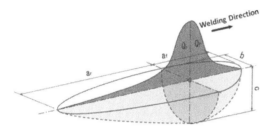

Figure 4. Double-ellipsoidal distributed heat source.

welding are 1 mm, 1 mm, and 4 mm in the x, y and z directions respectively.

The double ellipsoidal heat source model is shown in Figure 4.

For the arc welding processes, the deposition of the heat may be characterized as a distribution of heat flux on the weld surface. Assuming that the heat from the welding arc is applied at any given instant of time, t to a point within the first semi-ellipsoid, y = 0, which moves with a constant velocity, v along the z-axis, the rate of the internal heat generation by the welding torch is modelled by a double ellipsoidal power density distribution at instant and it is described by the following equation (Hashemzadeh et al. 2017):

$$q_{sub} = \frac{6\sqrt{3}Q_{net}}{b.c.\pi} \begin{cases} \dfrac{f_f}{a_f}\exp\left(-\dfrac{3(z-vt)^2}{a_f^2} - \dfrac{3y^2}{b^2} - \dfrac{3x^2}{c^2}\right) \\[2ex] \dfrac{f_r}{a_r}\exp\left(-\dfrac{3(z-vt)^2}{a_r^2} - \dfrac{3y^2}{b^2} - \dfrac{3x^2}{c^2}\right) \end{cases}$$

where f_f is the heat input ratio of the front part, 0.6 in the case before the torch passes the analysed region, f_r is the heat input proportion in the rear part, 1.4 in the case after the torch passes the analysed region, $f_f + f_r = 2$ and $Q = \eta UI/v$ is the welding heat input per unit weld length in W s/mm.

The welding process is characterized by the welding current of 190 A, voltage of 28 V and welding speed of 6 mms⁻¹. The convection and radiation coefficients of the external areas are

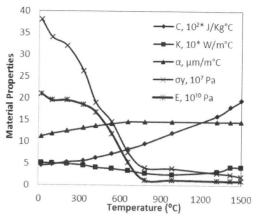

Figure 5. Temperature-dependent material properties, ASTM A36 carbon steel, Chang and Teng (2004).

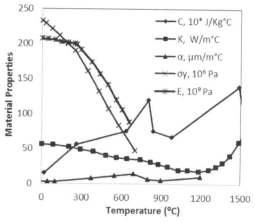

Figure 6. Temperature-dependent material properties, ST37 steel (S235), Barsoum and Barsoum (2009).

considered as $h_f = 15$ Wm⁻² and $\varepsilon = 0.9$ and the environmental temperature is 25°C. The boundary conditions are imposed in two ends of the weld line to let the edges to move free. Figure 5 and 6 show the temperature dependent material properties of ASTM A36 and ST37 respectively. It is noted that, in present study the material is considered to be elastic perfectly plastic.

3 RESULTS AND DISCUSSION

A 3D-transient thermo-mechanical finite element model using ANSYS software is developed. The temperature dependent material properties are applied for A36 and ST37 steel. Welding simulations of plates made of A36-A36, ST37-ST37 and A36-ST37 are performed and the results of thermal and mechanical analyses are presented in this section.

3.1 Thermal analysis

The temperature distributions due to the distance of the three cases are presented in Figure 7. In fact, this figure shows how the temperature varies as a function of the distance from the centre of the heat source.

Figure 7 shows that the maximum temperature occurs in the case of ST37 steel and it may be explained by the fact that the specific heat of the implemented type of steel is lower in a comparison to the A36 steel. It has to be noted that the specific heat is the amount of heat per unit mass required to raise the temperature by one degree Celsius. Also, from Figure 7 can be seen that the cooling rate of A36-A36 steel welded plates is higher than the ones of other cases.

Figure 8 and 9 show the temperature distribution with respect to the time for all cases on a 4 mm and 6 mm distance from the centre of the weld. It is noteworthy to point out that the dissimilar material of welded plates (case three, A36-ST37) changes the maximum temperature and cooling rate of the A36 steel. It can be seen that in right side, where the A36 steel is located that the maxi-

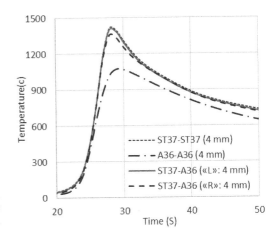

Figure 8. Temperature distribution at mid-section by 4 mm distance from welding.

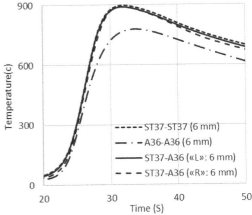

Figure 9. Temperature distribution at mid-section by 6 mm distance from welding.

mum temperature is equal to ST 37 steel, however, the cooling rate is higher than the one of A36 steel, but lower than the one of ST37 steel.

3.2 Mechanical analysis

The mechanical response results are presented in two different sections of the mid-section and weld line. Figure 10 shows the distortion distribution for the mid-section. It seems that the distortion is higher in the case of ST37 steel. The higher temperature leads to a higher distortion. In the case of dissimilar material in the welding, it is detected that the distortion distribution is between two distributions of the welding of plates of similar materials.

In the case of the transverse and longitudinal residual stresses as a result of welding (see

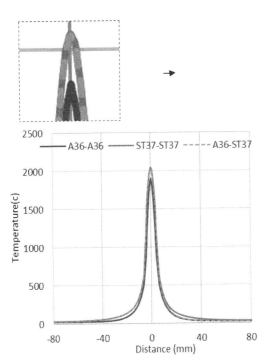

Figure 7. Temperature distribution from center of heat source.

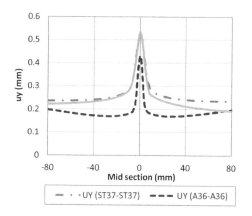

Figure 10. Vertical distortion distribution of welding at mid-section.

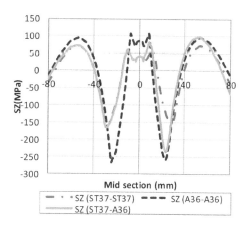

Figure 12. Longitudinal residual stress distribution of welding at mid-section.

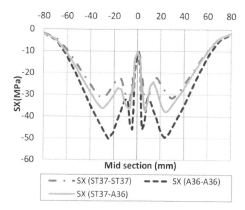

Figure 11. Transverse residual stress distribution of welding at mid-section.

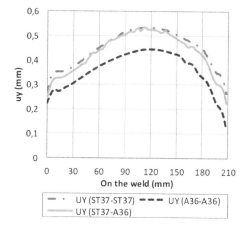

Figure 13. Vertical distortion distribution of welding at weld line.

Figure 11 and 12), it is found that the higher level of stress is created in A36 steel during the butt welding, applying the same welding condition. Considering the material properties and crystallization of ASTM A36 steel and simultaneously applying the faster cooling rate, results in higher induced residual stresses of as a consequence of welding.

In the case of the transverse residual stress, as shown in Figure 11, it can be observed that the stress distribution of dissimilar material of welding is symmetric and close to the distribution occurred for ST37 steel.

However, in the case of the longitudinal residual stress, as shown in Figure 12 it can be seen that the stress distribution of the welded plates of dissimilar material of welding is asymmetric and each material has the same trend of the longitudinal residual stress of the similar material welding on that side.

It seems that the longitudinal residual stress is more sensitive to material properties.

In the section of the weld line, which can be seen from Figure 13, it is detected that A36 steel has a lower distortion compared to other cases. Also, the distortion distribution of dissimilar material of welding is occurring between the distributions of A36 an ST37.

It can be seen from Figure 14 that the transverse residual stress has no significant variations for different materials on the weld line. However, the effect of the material properties of steel is remark-

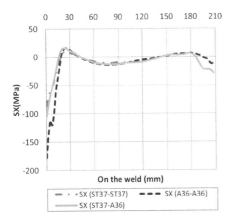

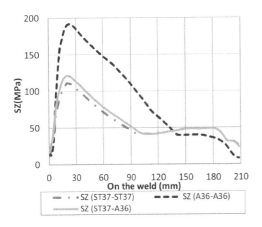

Figure 14. Transverse residual stress distribution of welding at weld line.

Figure 15. Longitudinal residual stress distribution of welding at weld line.

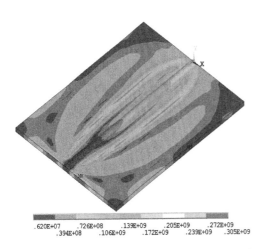

.620E+07   .726E+08   .139E+09   .205E+09   .272E+09
   .394E+08   .106E+09   .172E+09   .239E+09   .305E+09

Figure 16. Von-Mises stress distribution for A36-A36.

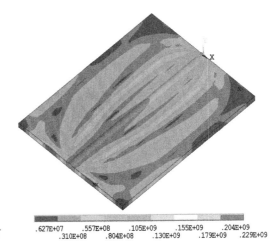

.627E+07   .557E+08   .105E+09   .155E+09   .204E+09
   .310E+08   .804E+08   .130E+09   .179E+09   .229E+09

Figure 17. Von-Mises stress distribution for ST37-ST37.

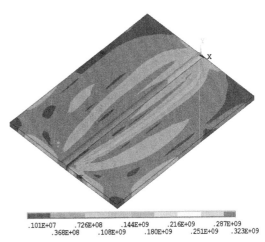

.101E+07   .726E+08   .144E+09   .216E+09   .287E+09
   .368E+08   .108E+09   .180E+09   .251E+09   .323E+09

Figure 18. Von-Mises stress distribution for A36 (right)-ST37 (left).

able on the longitudinal residual stress as can be seen in Figure 15.

Figure 16, 17 and 18 show the Von-Mises stress distribution of the cases A36-A36, ST37-ST37 and A36-ST37, respectively.

It seems that A36 steel material causes a higher residual stress induced by welding process considering the same welding conditions.

4 CONCLUSIONS

Two types of steels, ASTM A36 and ST35 are considered in the butt-welding of plates made of A36-A36, ST37-ST37 and A36-ST37 steels.

The transient thermo-mechanical analysis was performed to investigate the impact of dissimilar material on temperature distribution, distortion and residual stress induced by welding and following conclusions were derived:

- The maximum temperature occurs in ST37 steel and it may be explained by the fact that the specific heat is lower compared to A36 steel.
- The higher temperature leads to a higher distortion and in the case of dissimilar material of welding; it seems that the distortion distribution is between two distributions of similar materials.
- In the case of transverse and longitudinal residual stress of welding, it can be concluded that a more stress is created in A36 steel. The reason may be explained by the fact that more resistance to move can cause a higher stress. The transverse residual stress distribution of dissimilar material of welding is symmetric and close to the residual distribution of ST37 steel.
- In the case of longitudinal residual stress, it is found that the stress distribution of dissimilar material of welding is asymmetrical. Also, the trend of the longitudinal stress distribution on the left side (ST37) and right side (A36) is close to the longitudinal stress distribution of the welded plate of the same material of ST37 and A36 respectively, leading to the conclusion that welded plates of different steel material do not have a significant effect on the longitudinal stress distribution. It is also observed that longitudinal residual stress is more sensitive to the material properties.

ACKNOWLEDGEMENTS

The first author has been funded by the Portuguese Foundation for Science and Technology (Fundação para a Ciência e Tecnologia—FCT) under contract SFRH/BD/97682/2014.

REFERENCES

Ansys (2009). *Online Manuals*, Release 11.

Attarha, M. J. & Sattari-Far, I. (2011), Study on welding temperature distribution in thin welded plates through experimental measurements and finite element simulation. *Journal of Materials Processing Technology,* 211, 688–694.

Bachorski, A., Painter, M. J., Smailes, A. J. & Wahab, M. A. (1999), Finite-element prediction of distortion during gas metal arc welding using the shrinkage volume approach. *Journal of Materials Processing Technology,* 92–93, 405–409.

Barsoum, Z. & Barsoum, I. (2009), Residual stress effects on fatigue life of welded structures using LEFM. *Engineering Failure Analysis,* 16, 449–467.

Biswas, P., Mahpatra, M. & Mandal, N. (2009), Numerical and Experimental Study on Prediction of Thermal History and Residual Deformation of Double-sided Fillet Welding. *Journal Engineering Manufacture,* 223, 1–10.

Chang, P. H. & Teng, T. L. (2004), Numerical and experimental investigations on the residual stresses of the butt-welded joints. *Computational Materials Science,* 29, 511–522.

Chang, S. 1966. Experimental Investigation of Rigid Flat Bottom Body Slamming. *Journal of Ship research,* 1–17.

Deng, D. (2009), FEM prediction of welding residual stress and distortion in carbon steel considering phase transformation effects. *Materials & Design,* 30, 359–366.

Goldak, J., Chakravarti, A. & Bibby, M. 1984. A new finite element model for welding heat sources. *Metallurgical Transactions B,* 15, 299–305.

Hashemzadeh, M., Garbatov, Y. & Guedes Soares, C. (2015a). Analysis of Butt-weld Induced Distortion Accounting for the Welding Sequences and Weld Toe Geometry. *In:* Guedes Soares, C. & Santos, T. A. (eds.) *Marine Technology and Engineering.* London, UK: Taylor & Francis Group.

Hashemzadeh, M., Garbatov, Y. & Guedes Soares, C. (2015b), Numerical Investigation of the Thermal Fields due to the Welding Sequences of Butt-welds. *In:* Guedes Soares, C. & Santos, T. A. (eds.) *Marine Technology and Engineering.* London, UK: Taylor & Francis Group.

Hashemzadeh, M., Garbatov, Y. & Guedes Soares, C. (2017), Analytically based equations for distortion and residual stress estimations of thin butt-welded plates. *Engineering Structures,* 137, 115–124.

Hu, Y., He, X., Yu, G., Ge, Z., Zheng, C. & Ning, W. (2012), Heat and mass transfer in laser dissimilar welding of stainless steel and nickel. *Applied Surface Science,* 258, 5914–5922.

Long, H., Gery, D., Carlier, A. & Maropoulos, P. G. (2009), Prediction of welding distortion in butt joint of thin plates. *Materials and Design,* 30, 4126–4135.

Michaleris, P. & Debiccari, A. (1997), Prediction of Welding Distortion. *Welding Journal,* 76, S172-S181.

Nivas, R., Singh, P. K., Das, G., Das, S. K., Kumar, S., Mahato, B., Sivaprasad, K. & Ghosh, M. (2017), A comparative study on microstructure and mechanical properties near interface for dissimilar materials during conventional V-groove and narrow gap welding. *Journal of Manufacturing Processes,* 25, 274–283.

Maritime Transportation and Harvesting of Sea Resources – Guedes Soares & Teixeira (Eds)
© 2018 Taylor & Francis Group, London, ISBN 978-0-8153-7993-5

Approximation of maximum weld induced residual stress magnitude by the use of Meyer Hardness

P.R.M. Lindström

Faculty of Technology, Kalmar Maritime Academy, Linnaeus University, Sweden

ABSTRACT: Weld joints play a very important role in assessment of structural integrity of steel structures. The weld joint region is the location of weld induced residual stress and strain fields (WRS). For the time being there is a lack of engineering methods to be used for approximations of the maximum WRS magnitudes in the way of a weld joint proposed to be produced with a specific Welding Procedure Specification (WPS). Or screening of residual stress measurement results obtained by various measuring methods. This report describes how one can proceed to establish best estimate material data by the use of the Meyer's hardness. The Meyer's hardness can also be used to determine the maximum physically possible WRS magnitude

1 INTRODUCTION

Weld joints play a very important role in assessment of steel structures' structural integrity. Where the ultimate mechanical properties of a welded joint depends upon an intricate relationship between several contributing factors, see Figure 1.

Steel plate materials are constituted of polycrystalline iron-carbon alloys. Where the thermal and mechanical properties of a steel depend on the actual alloy's chemical composition; and the actual grain sizes and grain geometries. The steel material's chemical composition is controlled by the steel mill's steel making and casting process; and the grain size and geometry is controlled by the rolling or forging process, that is a transient thermomechanical process; imposing kinematic strain hardening of the plate material (Hosford 2012). Two different types of hardening models (isotropic and kinematic) and the effect of those on the yield surfaces are described in Figure 2; the effect is further elaborated by Dieter (1989), Hosford (2012) and Bower (2010).

A butt weld joint is constructed of two plate materials and the weld metal. Where the weld joint's mechanical properties are influenced by the interaction of the Heat Affected Zone (HAZ) (Svensson 1993) and the Weld Residual Stress Affected Zone (WRAZ) (Lindström 2015). Where the Weld Induced Residual Stresses and Strains (WRS) magnitude and distribution can be modelled and visualized by the use Computational Welding Mechanics (CWM) (Goldak & Akhlaghi 2005 & Lingren 2007) and/or measured by Residual Stress

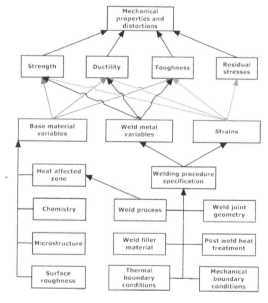

Figure 1. The relation between various variables and the ultimate mechanical properties of a welded joint.

and Distortion measurement techniques (Schajer 2013).

According to Bower (2010), selecting the correct material model is the most critical part of Finite Element (FE) modelling, as the use of an inaccurate material data and/or the wrong material model will always invalidate the Finite Element Analyses (FEA) results. The present author argue that use

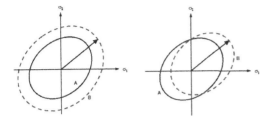

Figure 2. Illustration of different strain hardening models on von Mises yield surfaces: Isotropic hardening (left) Kinematic hardening (right).

of authentic material data is equal important as the use of the correct material model when the FEA objective is to understand and apply first-order physical principles (U.S. Department of energy 1999).

Constitutive thermal and mechanical material data for the ship steel plate material quality AH36 presented in Figures 3–6 has been compiled by the present author. The yield strength values were derived from a high-temperature test dataset of a St 52-3 steel plate (Ivarsson 1991). From (EN 12952-3:2011), the author derived the room temperature density (7850 kg/m3 at 20°C), Young's modulus, secant thermal expansion, thermal conductivity, and specific heat capacity values of up to 600°C. Material data values above 600°C were calculated by thermodynamic modelling and calculation (JMatPro) of a normalised AH36 ship steel plate's chemical composition (SSAB Oxelösund 2010). The solidus and liquidus threshold temperatures – 1490°C and 1530°C—were estimated with the help of an Iron-Carbon phase diagram (Welsh 2006). The tangential thermal expansion values were derived from the material's secant thermal expansion values.

The mechanical material modelling is optimised for Computational Welding Mechanics (CWM) and Elastic Plastic Fracture Mechanics (EPFM) by transient thermo-mechanical staggered coupled FEA; for numerical solution time reasons utilises tangential thermal expansion (βt) instead of secant thermal expansion (βs) (Shapiro 2008), see Figure 7.

The thermal material data for the AH36 steel plate material was benchmarked against experimental thermal data and JMatPro calculation data of an identical material quality presented in a recently published doctoral thesis (Bhatti 2015) and were find to be consistent with values here presented.

Furthermore, the thesis of Bhatti was found to be a good source of high-temperature test data of structural steels with room temperature yield strengths of 370, 400, 420, 540, and 790 MPa.

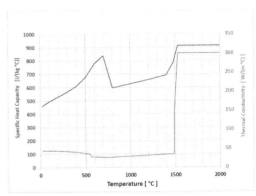

Figure 3. Specific Heat and Thermal Conductivity.

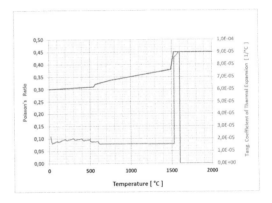

Figure 4. Poisson's ratio and Tangent Thermal Expansion.

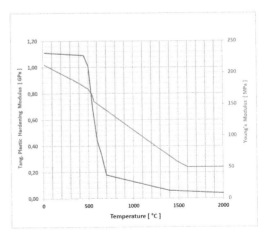

Figure 5. Tangent plastic hardening and Young's modulus.

It was also found that Bhatti assumes that the yield strength of the weld filler material is identical to the base material's yield strength, an approach

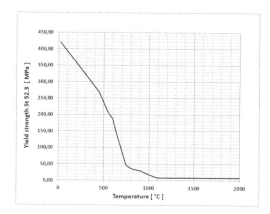

Figure 6. Yield strength.

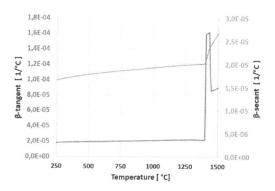

Figure 7. Illustration of the tangential (β_t) and secant thermal expansion's (β_s) temperature dependency for stainless steel 316 L.

also used in two Finite Element Analyses (FEA) described by Barsoum (2008). Both Bhatti and Barsoum consistently assume the tangent plastic hardening modulus (h) to be in the range 1/200 – 3/100 of the Young's modulus at the corresponding temperature, namely 1/200; 1/100; 1/84; and 3/100. According to Bhatti et al. (2015), this approach was used to avoid convergence issues during numerical analysis. The mechanical material modelling by Bhatti, and partly Barsoum, was driven by a desire to identify an empirical and simplified Computational Welding Mechanics (CWM) methodology by the use of regression analyses.

Nevertheless, the tangent plastic hardening modulus (h) is a function of the true tensile strength (s_{uts}); true yield stress (s_{yield}); and the true plastic strain (e_{plas}), see Equation 1–4 (Hilding 2012 & BS7910:2013). Therefore, the mechanical constitutive modelling and CWM-analyses by Bhatti and Barsoum are not in line with the present author's objective to understand and apply realistic material data and first-order physical principles.

$$h = \frac{s_{uts} - s_{yield}}{e_{plas}} \qquad (1)$$

$$s = \sigma \cdot (1 + \varepsilon) \qquad (2)$$

$$e = ln \cdot (1 + \varepsilon) \qquad (3)$$

$$e_{plas} = e_{uts} - \left(\frac{s_{yield}}{E} \right) \qquad (4)$$

Furthermore, in a report sponsored by the Swedish Radiation Safety Authority (SSM) (Dillström et al. 2009), it is stated that the stainless steel material data for the tube and weld metal was taken from Brickstad et al. (1998). The material models used for the CWM-analysis of the WRS and the EPFM FE-analyses were all based on the yield strength 230 MPa for both the base and weld filler material. This suggests that a typical Specified Minimum Yield Stress (SMYS) value for the austenitic stainless-steel grade 316 L was used for the simulations, despite the lack of justifications for doing so. Indeed, by Brickstad et al., which the report claims to follow, the weld metal yield strength is explicitly stated to be 460 MPa. The report also did not use the tangent plastic hardening modulus values stated Brickstad et al. The target plastic hardening modulus used by Dillström et al. was about ten times higher than the values used by Lindström (2015).

In another report sponsored by the Swedish Radiation Safety Authority (SSM) (Mullins et al. 2009) and published the same year as the report of Dillström et al., it is stated that the common material data source for the austenitic stainless-steel grade 316 for CWM-analyses is the standard ASME SA240.

The present author argue that standard SMYS values should not be used, as less than one of hundred rolled plates would have an yield stress ≤ SMYS; and just a small fraction of such one of hundred plates should pass through the classification societies' control out on the market. Therefore, the present author suggest that best estimate material data should be used instead of SMYS, that also is known as the characteristic yield stress value (σ_{yield}).

2 WELD JOINT MATERIAL PROPERTIES

A first glance of DNV GL Material Rules (2017) is giving an impression of that there are just a few ship steel plate qualities i.e. steel plates with the characteristic yield strength values (σ_{yield}): 235 MPa; 265 MPa; 315 MPa; 355 MPa; 390 MPa. Where the characteristic yield stress value is presented in the form of engineering stress (σ). That is fairly different compared to the true stress (s) when plastic strain are at hand i.e. when ε > 0.2%, see Figure 8.

Furthermore, as an option to the toughness transition temperature variation a ship steel can also have guaranteed through thickness tensile properties (so called Z-plates) and brittle crack arrestability (so called BCA-plates).

In addition to that, a specific ship steel plate quality can be produced with various conditions of delivery and heat treatments. For example, can a 15 mm thick 355 MPa steel plate be manufactured by the use of five different processes: As Rolled (AR); Normalised Rolled (NR); Normalised (N); Thermo-Mechanical Rolled (TM); and Quenched and Tempered (QT).

Where each delivery condition can have its own unique chemical composition, grain size and geometry; and accumulated thermo-mechanical history.

Compiling all known variables in one table, presents the fact that a '355 MPa ship steel plate' in the reality is one material quality that can have about 50 different unique constitutive data sets, see Table 1.

Therefore, marine structures in the reality are constructed from several different material qualities. Where each material's mechanical properties are changing (increasing and/or decreasing) over the structures entire lifetime times as a function of the imposed thermal and mechanical strains.

For example, a butt weld joint is constituted of its weld metal and two base materials with its WRAZ and HAZ, that is constituted of five sub-zones: coarse grained zone; fine grained zone; intercritically heated zone; and tempered zone. Where the two base materials rarely are identical to each other as they most likely will possess different

Table 1. Steel plate delivery condition.

Plate designation	AR	NR	N	TM	QT
VL 36 A	x	x	x	x	x
VL 36 A Z25	x	x	x	x	x
VL 36 A Z35	x	x	x	x	x
VL 36 A BCA	x	x	x	x	x
VL 36B	x	x	x	x	x
VL 36B Z25	x	x	x	x	x
VL 36D Z35	x	x	x	x	x
VL 36D BCA	x	x	x	x	x
VL 36E	x	x	x	x	x
VL 36E Z25	x	x	x	x	x
VL 36E Z35	x	x	x	x	x
VL 36E BCA	x	x	x	x	x

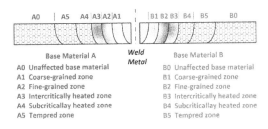

Base Material A
A0 Unaffected base material
A1 Coarse-grained zone
A2 Fine-grained zone
A3 Intercritically heated zone
A4 Subcriticallay heated zone
A5 Tempred zone

Weld Metal

Base Material B
B0 Unaffected base material
B1 Coarse-grained zone
B2 Fine-grained zone
B3 Intercritically heated zone
B4 Subcriticallay heated zone
B5 Tempred zone

Figure 9. Illustration of the 13 different thermal, mechanical and thermo-mechanical material properties that exists in a singel pass weld joint of one VL 36 A (TM) plate and one VL 36EZ35 (N) plate.

physical constitution i.e. chemical compositions, grain sizes, grain geometries, microstructures and accumulated thermo-mechanical histories. This is exemplified in Figure 9 where the HAZ of a single pass but weld joint of one VL 36 A (TM) plate and one VL 36EZ35 (N) plate. In Figure 9 on can see that the weld joint is constituted of at least thirteen different materials (Svensson 1993), each one with its own unique thermal, mechanical and thermo-mechanical constitutive data.)

3 MEYER HARDNESS TESTING

The present author has used the Meyer hardness as a qualitative tool for assessment of the accumulated plastic work hardening (yield surface) in conjunction with manufacturing, alteration and repair of power boilers, recovery, boilers pressure vessels and nuclear components. But also in situations where one is in the need of best estimate mechanical data, for example by a series of hardness indentions with progressively increasing loads a plastic hardening curve can be established.

The ultimate tensile strength (σuts) can be approximate by the use of the Brinell hardness

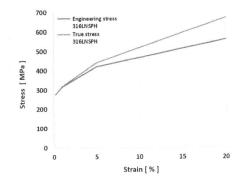

Figure 8. The diagram illustrates the deviation between the engineering stress and the true stress as a function of the engineering strain.

(BHN) and Equation 5; and the yield strength (σyield) can be approximated by means of Equation 6–9, the Meyer's hardness index (n´); Hardness Vickers (HV) (Dieter, 1989). For fully annealed steels the Meyer's hardness index (n´) is about 2.5; and about 2 for fully strain hardened steels; p = pressure (Pa); q = metal's initial resistance to penetration; and d = diameter of indentation (m).

$$\sigma_{uts} = 3.4 \cdot BHN \tag{5}$$

$$R_{p0.2\%} = g \frac{HV}{3} \cdot (0.1)^{n´-2} \tag{6}$$

$$p = q \cdot d^n \tag{7}$$

$$n´ = 2 + \frac{log\left(\dfrac{R_{p0.2\%}}{g \cdot \dfrac{HV}{3}}\right)}{log\,0.1} \tag{8}$$

$$n´ = \frac{log\left(\dfrac{p}{q}\right)}{log\,d} \tag{9}$$

4 APPROXIMATION OF MAX POSSIBLE WRS

Research projects aiming to measure and quantify residual stresses by the use of hardness testing have been reported (Abbate et al. 1993 & Schroeder et al. 1995). However, such a method is not possible because residual stresses are constituted of elastic strain (Radaj 1992 & 2003 and elastic strain does not significantly affect the measurable hardness value of a material (Ma et al. 2012).

Given this understanding, it should not be physically possible for a metallic material to withstand an effective elastic stress magnitude exceeding the material's flow stress (Saxena 1998). The author proposes that hardness value profiles should be used to determine the highest possible WRS-magnitude that can be experienced in the unaffected base material of a specific weld joint configuration.

At CWM analyses of the IIW RSDP Phase II benchmark problem. The presenting author carried out feasibility test welding to identify the yield strength value of the plate material and weld metal by the use of hardness test results. The feasibility test welding was carried out by manual GTAW of an ER316 LSi welding consumable with the following chemical composition in weight percent: C < 0.01; Si 0.9; Mn 1.8; P 0.015; S 0.010; Cr 18.4; Ni 12.2; Mo 2.6; Cu 0.05; N 0.05; and FN 10 (ESAB Lot No. PV3503724641). For this test, two weld passes were constructed with the weld heat inputs 1.8 kJ/mm (Root) and 3.2 kJ/mm (Run- 2). The weld was made in a rectangular weld groove (5 × 5 mm) that had been machined in the longitudinal centre-line of a 300-mm long, 200-mm wide and 30-mm thick 316 L austenitic stainless steel plate.

Particulars of the test plate are: ASME IIA ED.2010+AD11 SA 240/M—UNS S31603 (316 L); EN 10028–7:2007 + AD2000 W2 + ADW10 – X2CrNiMo17-12-2 (1.4404); 1050°C – 1 min/mm Air solution annealed; manufactured by ArcelorMittal, Marchienne-au-Pont, Belgium, Doc. No. 2013-117720; Ref. No. 9415 rev 00/SH NOR2013/0077; Heat No 40261.

The plate's yield strength is 283 MPa and its chemical composition in weight percent is: C 0.022; Si 0.288; Mn 1.855; P 0.0357; S 0.021; Cr 16.997; Ni 10.022; Mo 2.010; and N 0.0540.

A sample was extracted from the plate's mid-section for the sake of hardness testing, macroscopic, and optical microscopy examination. The sample was polished (Grit 2500) and etched with V2 A-Beis at 55°Gr.

The hardness testing was done along 3 lines as illustrated in Figure 10, and the hardness test values along the three lines are presented in Figures 11–12.

Using the yield strength 283 MPa from the base material's material certificate and the hardness values for unaffected base material presented in Figure 11 and Figure 12, the Meyer's law exponent n´ = 2,267 was calculated by Equation 8. From the literature, it is understood that n´ should be approximately 2.5 for a fully annealed metal and about 2 for a fully strain-hardened metals. By the use of the Meyer's law exponent value, 2,267 and the hardness values in Figure 11 and Figure 12, the weld metal's yield strength was approximated to about 334 MPa.

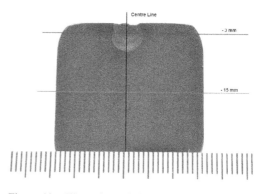

Figure 10. Illustration of the three hardness testing lines' location on the macro section.

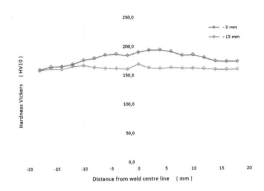

Figure 11.   Hardness test values along the lines – 3 mm and – 15 mm on both sides of the weld centreline.

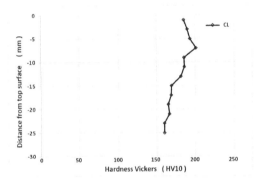

Figure 12.   Hardness test values along the weld centreline from the top surface in the downward direction.

Given this understanding, it should not be physically possible for a metallic material to withstand an effective elastic stress magnitude exceeding the material's flow stress, the hardness test result values in Figures 11–12, were used to approximate the maximum possible WRS-magnitude of the weld test specimen by the use of Equation 6 and the Meyer's law exponent, $n' = 2,267$, see Figure 13.

The approximated yield strength profiles in Figure 13 were plotted in diagrams together with the results of the present author's fully 3D and 2D CWM-analyses.

In Figures 14–15, one can see that the fully 3D CWM-analysis predicts the maximum WRS-magnitude to be less than the unaffected base material's yield strength. One can also see that the 2D CWM-analysis, on a distance of about 3 mm from the fusion lines, overestimates the maximum WRS-magnitude at about 100 MPa, as compared with the unaffected base material's yield strength.

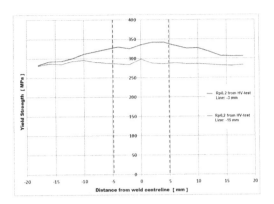

Figure 13.   the red line indicates the unaffected base material's approximated yield strength 3 mm below the top surface. The green line indicates the base material's approximated yield strength in the mid-thickness of the plate.

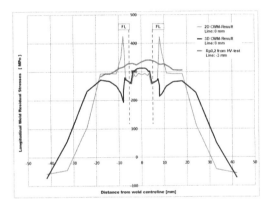

Figure 14.   Approximated yield strength versus fully 3D and 2D CWM-results in the longitudinal (zz) direction, line: 0 mm.

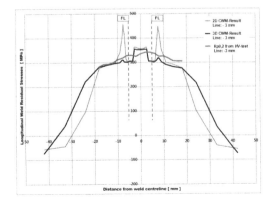

Figure 15.   Approximated yield strength versus fully 3D and 2D CWM-results in the longitudinal (zz) direction, line: –3 mm.

5 CONCLUSIONS

Thus, the work deviates somewhat from traditional and theoretical examinations of weld properties and stress and strain fields. The perspective of the work is more that of a welding engineer rather than someone working on the solid mechanics side.

Nonetheless, best estimate material data can be established by the use of the Meyer's hardness.

The Meyer's hardness can also be used to determine the maximum physically possible WRS magnitude.

REFERENCES

Abbate, A., Frankel, J, Scholz, W.; 1993, Measurement and Theory of the Dependence of Hardness on Residual Stresses, Technical Report: ARCCB-TR-93022, US Army Armament Research and Development and Engineering Center, Close combat armaments center Benét Laboratories, Watervliet, New York, USA.

Barsoum, Z., 2008, Residual stress analysis and fatigue assessment of welded steel structures, Doctoral thesis, ISSN 1651-7660, Kungliga tekniska högskolan, Stockholm, Sweden.

Bhatti, A.A., 2015, Computational weld mechanics: Towards simplified and cost-effective FE Simulations, Doctoral thesis, ISSN 1651-7660, ISBN 978-91-7595-626-8, Kungliga tekniska högskolan, Stockholm, Sweden.

Bhatti, A.C., Barsoum, Z.A.B, Murakawa, H.C., Barsoum, I., 2015, Influence of thermo-mechanical material properties of different steel grades on welding residual stresses and angular distortion, Materials and Design 65 (2015), pp. 878–889.

Bower, A.F., 2010, Applied Mechanics of Solids, ISBN 978-1-4398-0247-2, CRC Press Taylor & Francis Group, Boca Raton, Florida, USA, pp. 115–124.

Bower, A.F., 2010, Applied Mechanics of Solids, ISBN 978-1-4398-0247-2, CRC Press Taylor & Francis Group, Boca Raton, Florida, USA, pp. 2–12.

Brickstad, B., Josefson, B.L., 1998, A parametric study of residual stresses in multi-pass butt-welded stainless steel pipes, International Journal of Pressure Vessels and Piping, Vol. 75, pp. 11–25

BS7910:2013, "Guide to methods for assessing the acceptability of flaws in metallic structures", ISBN 9780580743344, BSI Standards Limited, UK, pp. 40.

Dieter, G.E. 1989, Mechanical Metallurgy—SI Metric Edition, 3 Revised Ed., ISBN 9780071004060, MacGraw-Hill Book Company, London, UK, pp. 81–85.

Dieter, G.E. 1989, Mechanical Metallurgy—SI Metric Edition, 3 Revised Ed., ISBN 9780071004060, MacGraw-Hill Book Company, London, UK, pp. 326–332

Dillström, P., Andersson, M., Sattari-Far, I., Zang, W., 2009, Analysis strategy for fracture assessment of defects in ductile materials, SSM Report 2009:27, ISSN: 2000-0456, Swedish Radiation Safety Authority, Stockholm, Sweden, pp. 31–41.

EN 12952-3:2011, Water-tube boilers and auxiliary installations—Part 3: Design and calculation for pressure parts of the boiler, European Committee for Standardization, pp. 153

Goldak, J., Akhlaghi, M., 2005, Computational Welding Mechanics, ISBN 9780387232874, Springer-Verlag New York Inc., USA

Hilding, D., 2012, Dynamore Nordic AB, Brigadgatan 14, Linköping, Sweden, private communication

Hosford, W.F., 2012, Iron and steel, ISBN 978-1-107-01798-6, Cambridge University Press, Cambridge, UK.

Hosford, W.F., 2012, Iron and steel, ISBN 978-1-107-01798-6, Cambridge University Press, Cambridge, UK, pp. 186–187.

Ivarsson, B., 1991, "Elevated temperature tensile properties of candidate steels for fire and blast walls and for cable ladders", Avesta Report No. RE91047, Outokumpo Stainless AB, Sweden.

JMatPro www.sentesoftware.co.uk.

Lindgren, L-E, 2007, Computational Welding Mechanics—thermomechanical and microstructural simulations, ISBN 978-1-84569-221-6, Woodhead publishing Ltd., Cambridge, UK.

Lindström, P.R.M., 2015, Improved CWM platform for modelling welding procedures and their effects on structural behavior, Doctoral Thesis, ISBN 978-91-87531-07-1, University West, Trollhättan, Sweden.

Ma, Z.S., Zhou, Y.C., Long, S.G., Lu, C., 2012, Residual stress effect on hardness and yield strength of Ni thin film, Surface & Coatings Technology, Volume 207, 25 August 2012, pp. 305–309.

Mullins, J., Gunnars, J., 2009, Influence of Hardening Model on Weld Residual Stress Distribution, Report 2009:16, ISSN: 2000-0456, Swedish Radiation Safety Authority, Stockholm, Sweden.

Radaj, D., 1992, Heat effects of welding—temperature field—residual stress—distortion, ISBN 0-540-54820-3, Springer-Verlag, Berlin, W. Germany, pp. 200–204.

Radaj, D., 2003, Welding Residual Stresses and Distortion: Calculation and Measurement, ISBN 3-87155-791-9, Woodhead Publishing Ltd, Cambridge, UK, pp. 5–12.

Saxena, A., 1998, Nonlinear Fracture Mechanics for Engineers, ISBN 0-8493-9496-1, CRC Press Inc., USA, pp. 54.

Schajer, G.S., 2013, Practical Residual Stress Measurement Methods, ISBN 9781118402818, John Wiley & Sons Ltd., Chichester, UK.

Schroeder, S.C.; Frankel, J.; Abbate, A., 1995, The Relationship between Residual Stress and Hardness and the Onset of Plastic Deformation, Technical Report: ARCCB-TR-95029, US Army Armament Research and Development and Engineering Center, Close combat armaments center Benét Laboratories, Watervliet, New York, US.

Shapiro, A., 2008, Mysteries behind the Coefficient of Thermal Expansion (CTE) Revealed, 10 International LS-Dyna Users Conference, 8–10 June 2008, Dearborn, Michigan USA.

Ship steel plate AH36, 2010, Normalised, Charge no. 093006, Steel plate identification no. 093006-259508, SSAB Oxelösund AB, Oxelösund, Sweden.

Svensson, L.-E., 1993,Control of Microstructure and Properties in Steel Arc Welds, ISBN 9780849382215, CRC Press Inc, USA, pp. 83–84.

U.S. Department of energy, office of basic energy sciences, 1999, Washington DC., USA Computational materials science: A scientific revolution about to materializeSSI Materials white paper, Materials component strategic simulation initiative.

Welsch, G.E., 2006, Iron-Carbon phase diagram, 5M0106 FN00454 Rev. 1, Buehler Ltd., 41 Waukegan Road, Lake Bluff, Illinois, USA.

Maritime Transportation and Harvesting of Sea Resources – Guedes Soares & Teixeira (Eds)
© 2018 Taylor & Francis Group, London, ISBN 978-0-8153-7993-5

Nonlinear FEA of weld residual stress influence on the crack driving force

P.R.M. Lindström
Faculty of Technology, Kalmar Maritime Academy, Linnaeus University, Sweden

Erling Østby
A DNV GL Materials Laboratory, Høvik, Norway

ABSTRACT: Weld joints play a very important role in the assessment of structural integrity of steel structures. The likelihood of defects is significantly higher in the weld joint compared to the unaffected base material. Here the Computational Welding Mechanics platform developed by Lindström (2015) has been used to model representative residual stress fields by the use of the IIW RSDP Phase II Initiative documentation (Janosch 2001). Where the CWM-platform includes an option for tracking the welding's influences on the evolution of the base and weld material's tensile properties. A detectable defect is introduced in the weld metal and the crack driving force in terms of the CTOD is extracted from the FEA. The loading applied covers both globally elastic and fully plastic conditions. The analysis allows evaluation of the influence of both residual stresses and changes in the tensile properties on the crack driving force as a function of applied global load.

1 INTRODUCTION

Fracture assessment of the integrity of welded structures is very important at various stages over the entire lifetime of a structure. Assessment may be necessary at any point, from the initial design to controlled decommissioning, including various phases of construction, in-service inspections, failure, and remaining lifetime analyses (Koçak 2006). These assessments are conducted in order to verify that the equipment is appropriate for its intended operation, with respect to safety, environment, and commercial availability. Weldments are the most likely location for defects or cracks. As well, metallurgical conditions are highly complex and inhomogeneous, with possible local deterioration of the fracture toughness as compared to the virgin base material (Taylor et al. 2003). These aspects represent major challenges when assessing the criticality of possible weld defects with respect to brittle or ductile fracture. Here, linear elastic fracture mechanics (LEFM) are used to solve brittle fracture problems, defined as having almost zero plastic deformation at the time of failure. Elastic plastic fracture mechanics (EPFM) are used to solve ductile fracture problems, or plastic deformation at the failure mode. The principles of LEFM and EPFM are described by Anderson (2005) and Saxena (1998), respectively. One main issue in this respect is how to include welding residual stresses and strains (WRS) in an appropriate manner in the fracture mechanics assessment.

The main focus of the present authors is the influence of residual stresses on the crack driving force (CDF), expressed through the CTOD (Crack Tip Opening Displacement), in a GTMA weld in a 30 mm thick stainless steel plate. This case has been subject to residual stress measurement, thus, validation of numerical schemes to determine residual stresses is possible. Prior to outlining the procedures to simulate the WRS, the main principles for fracture mechanics assessments are presented. The welding induce residual stresses and strains are obtained from fully 3D Computational Welding Mechanics (CWM) simulations. The simulations are shown to compare well with experimental measurements. In the interest of computations efficiency, it is further demonstrated that sufficiently accurate results may be obtained using a symmetry approach which is substantially less expensive to perform. Different scenarios to introduce the defect in the weld are considered, both addressing assumed room temperature formation of the defect and defects introduce at higher temperatures (in this case akin to a hot ductility dip crack). FE simulations are carried out to quantify the effect of the different assumptions and parameters on the resulting CDF.

2 FRACTURE MECHANICS ASSESSMENT

2.1 *Background*

A material's resistance to fracture is called fracture toughness, and can be obtained by experimental testing in which the elastic fracture toughness parameter, K_{mat}, is typically quantified by the use of a compact tension test specimen, see Figure 1. The material's resistance to elastic plastic fracture (J_{mat} or $CTOD_{mat}$) is typically quantified by the use of a single-edge notch bend (SENB) test specimen, see Figure 2. However, elastic fracture toughness material data (K_{mat}) can be used for EPFM analysis if one is utilizing the failure assessment diagram (FAD) approach (Anderson 2005).

In a FAD, the plastic collapse ratio (Lr) is indicated by the x-axis and the brittle fracture ratio (K_r) is indicated by the y-axis, see Figure 3. Here, Lr is the ratio between the reference stress (σ_{ref}) and the yield strength (σ_{yield}), while K_r is the ratio between the stress intensity (K_I) and the material fracture toughness (K_{mat}).

Furthermore, use of the FAD approach is not limited to situations where elastic fracture material toughness data is available. Using the

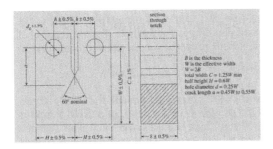

Figure 1.   Compact tension test specimen used for testing of Kmat (Demaid 2004).

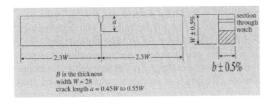

Figure 2.   Single-edge notch bend (SENB) test specimen (Demaid 2004).

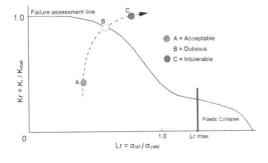

Figure 3.   Illustration of a failure assessment diagram (FAD).

equations 1–4 one can convert an elastic plastic fracture toughness dataset, e.g., $CTOD_{mat}$ to K_{mat} (Lindström 2015).

Typical J_{mat} values for austenitic stainless steel of the 316 L type when subjected to a GTAW weld process are presented in Table 1. (Mills 2009 & 1997). Also included are calculated $CTOD_{mat}$ and K_{Jmat}-values, which have been approximated by Equation 1 and Equation 3.

$$K_{Jmat} = \sqrt{\frac{E \cdot J_{mat}}{(1-v^2)}} \quad (1)$$

$$K_{CTODmat} = \sqrt{\frac{m \cdot \sigma_{yield} \cdot E \cdot CTOD_{mat}}{(1-v^2)}} \quad (2)$$

$$CTOD_{Jmat} = \sqrt{\frac{J_{mat}}{(m \cdot \sigma_{yield})}} \quad (3)$$

$$m = 1.517 \cdot \left(\frac{\sigma_{yield}}{\sigma_{uts}}\right)^{-\frac{797}{2500}} @ 0.3 < \frac{\sigma_{yield}}{\sigma_{uts}} < 0.98 \quad (4)$$

However, if one wants to avoid unit conversion artefacts similar to those in Table 1, one should use a FAD that is based on the available elastic plastic fracture toughness dataset, see Figure 4. However, if one intends to perform an EPFM-analysis and possesses the elastic plastic fracture toughness data, e.g. $CTOD_{mat}$, one can use the aforementioned FAD approach or the Crack Driving Force (CDF) approach (Saxena 1998 & Koçak et al. 2008).

The authors consider the use of the CDF approach in combination with the crack driving force parameter, $CTOD_{CDF}$, (Wells 1961) a better option as compared with the FAD-approach, as it is more straight-forward and eliminates the risk of unit conversion artefacts. In a CDF-diagram, the plastic collapse ratio (Lr) is indicated on the x-axis between the reference stress (σ_{ref}) and the yield strength (σ_{yield}). On the y-axis, the crack driving force, in this case $CDOD_{CDF}$, is stated. All load cases with a $CTOD_{CDF}$ value below the material's

Table 1.   Typical fracture resistance values for 316 L GTAW weld.

Condition/ Type	Temperature (°C)	J_{mat} (kJ/m²)	$CTOD_{Jmat}$ (mm)	K_{Jmat} (MNm⁻³/²)
316 L	20	672	1,09	377
(base	125		1,13	368
metal)	400	421	0,77	273
	550		0,81	263
ER361 L/	20	492	0,80	323
ER316 LSi	125		0,82	315
(weld	400	293	0,53	228
metal)	550		0,56	219

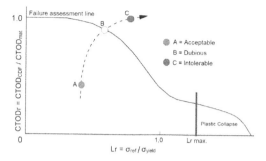

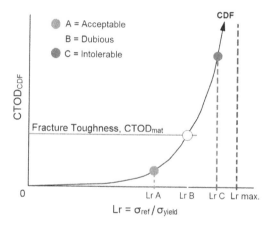

Figure 4. Illustration of a CTODmat based failure assessment diagram (FAD).

Figure 5. Illustration of a Crack Driving Force (CDF) diagram.

fracture toughness value (CTOD$_{mat}$) is thought stable and acceptable, see Figure 5.

2.2 Residual stresses and strains

One particular issue that has received much attention in the field of fracture mechanics is the importance of welding-induced residual stresses and strains (Radaj 2003 & Fuchs et al. 2000). Recently published general fracture assessment procedures utilizing the FAD approach prescribe that if specific knowledge regarding the residual stress conditions is unavailable, residual stresses with a magnitude equal to the yield strength of the material in which the flaw is located should be accounted for in the analysis (BS7910:2013), an approach that is in line with (R6 Procedure, Rev. 4 & API 579-1/ ASME FFS-1, 2007). There is no general consensus on how to modify the treatment of residual stresses in engineering fracture assessments when one is using the CDF approach. However, the interaction between WRS and the CTOD$_{CDF}$ has been studied by the use of 2D FEA (Fu et al. 2014 & Liu et al. 2008). The authors take the CTOD$_{CDF}$

analysis methodology further through the use of the fully 3D CWM methodology, where the accumulated thermal and mechanical influences from the first weld pass to the final fracture assessment are brought forward in one and the same meshed geometry, along with the mechanical material models' accumulated yield surfaces and element formulation.

3 THE CASE CONSIDERED

3.1 The weld join configuration

In this specific case, the weld joint represents steel mill's correction of a surface defect of a 30-mm thick NV 316 L stainless steel plate intended for use on the lower part of a corrugated longitudinal bulkhead of a 27 000 DWT chemical tanker with Finnish-Swedish Ice Class. If this weld were to fail, the consequences could be catastrophic, see Figure 6 (Ship Structure Committee 1998). Other examples of commercial application of this type of problem (i.e., repair welding of centreline cracks) include in-service repair welding of ship and offshore structures (IACS No. 47 2013 & Lindström & Ulfvarson 2002), structural weld overlay (SWOL) of nuclear components (Rishel et al. 2007 & Marlette et al. 2010), and cladded tubes and plates (BS7910:2013).

The interaction between all processes during a welding operation and leading up to the final residual stress field is very intricate (Totten 2002). In this study, only the thermo-mechanical contribution of the boundary condition and the actual arc welding process are taken into account, while, for example, local thermomechanical plastic deformation and strain hardening of the surface material due to grinding and chip hammering are not considered. Both the thermomechanical properties of the material and the time of formation of the defects

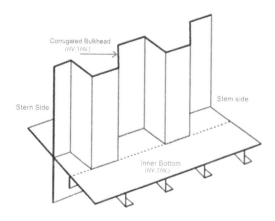

Figure 6. Illustration of a corrugated bulkhead.

will likely have a significant impact on the influence of the residual stresses on the crack driving force.

Here, the influence of residual stresses on the crack driving force ($CTOD_{CDF}$) for a hot crack in ER316 LSi weld metal is studied. The scope of this study includes both an investigation of the general effect of welding-induced residual stresses (WRS) on the crack driving force and a comparison to cases where welding only affects the material properties or where the effects of welding are disregarded. Furthermore, different assumptions about the effect of the formation of the defect—from a trivial case where the defect is introduced after the simulation of the welding process to a more realistic scenario where the defect is formed at elevated temperatures during the welding process—are considered part of the study. In this study, a hot crack of the ductility dip crack (DDC) type will be examined. DDC is caused by a solid-state grain boundary embrittlement at approximate 50% of the solidus temperature (Saluja & Moeed 2012).

As for weld geometry, the IIW numerical and experimental Residual Stress and Distortion Prediction (RSDP) Round Robin Initiative, Phase II (Janosch 2001) has been chosen. This two-pass butt weld forms a virtual CTOD test coupon for the present study. Fully 3D CWM methodology is applied to calculate the stress and strain history during welding to model the influence of materials parameters and the sequence of defect formation. The calculated welding residual stresses for this CTOD test plate were validated (Lindström 2015) against experimental and numerical results in the IIW RSDP final report (Wohlfahrt et al. 2012). A crack size has been selected to represent the limit crack size, which is the minimum detectable crack size in the weld centreline of an austenitic stainless steel weld joint using ordinary NDT-techniques (BS7910:2013 & Crutzen et al. 1999 & Hogberg 2014).

In the present case (i.e., GTAW repair welding of a 316 L plate by a 316 LSi consumable), the risk of fracture should be fairly low as the material quality is very resistant to fracture. The delta ferrite content in austenitic stainless steel 135 weld metals depends on the chromium-nickel ratio, and a 316 LSi stainless steel weld metal (20% Cr and 10% Ni) should have about 5–10 FN (Folkhard 2013). That said, it is not unrealistic to consider the possibility that an Ice Classed Chemical tanker will be subjected to an asymmetric loading operation at –40°C during its design lifetime. The main purpose of this study is to explore the feasibility of the CWM platform to analyse the influences of WRS on the criticality of potential weld defects, addressing both the WRS and the material hardening of the base and weld filler material.

3.2 Fully 3D CWM procedures

The fully 3D transient thermo-mechanical staggered coupled FEM-weld simulation of the crack initiation and propagation was carried out using symmetry conditions along the weld centreline, see Figure 7, although the layout of the welding experiments was asymmetric (Lindström 2015).

The run-on plate was assumed to be connected to the main plates through a tack weld at the top and reverse side surfaces of the weld centreline with thermal and mechanical contact conditions between the run-on plate and the main plate's right-end surface. Essential particulars of the CWM analysis are presented in Table 1 and Table 2.

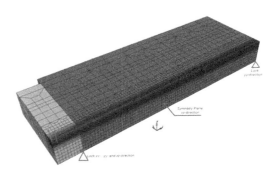

Figure 7. Fully 3D CWM-model and boundary condition used.

Table 2. Weld process data used at the 3D CWM Analysis.

Base material	316LNSPH	Weld joint geometry
Consumable	ER 316LSi	
Weld heat power	837 W	
Wlding speed	41 mm/min	
Groove area	30.0 mm²	
Root pass area	12.2 mm²	
Second pass area	17.8 mm²	
Preheat and inter pass temp.	20°C	
Material hardening formulation	100% Kinematic	
Presumed grain size of base material	20 μm	Smallest element size used (μm): 250 × 375 × 1000
Presumed grain size HAZ	50 μm	
Presumed grain size of weld metal	50 μm	
Goldak weld heat source config.		Goldak weld heat source nomenclature
Forward fraction f_f	0,500	
Aftward fraction f_a	1,500	
Weld pool radius a	2,357 mm	
Weld pool radius b	2,357 mm	
Weld pool radius c_1	1,000 mm	
Weld pool radius c_2	3,000 mm	

Figure 8. 3D mesh with its CWM-results (left); complete weld test coupon in equilibrium (center); formatted test plate brought into equilibrium with the run-on plate removed (right).

Table 3. Essential particulars of the 3D CWM analysis.

Element data		Degrees of freedom	
Mech. element type	ELFORM-2	Therm. DOF	157190
Number of elements	178946	Mech. DOF	934872
Number of nodes	157190	Total DOF	1092062
Temp. activated functions		Start temp.	Stop temp.
Residual stress release activation		1030°C	1100°C
Therm. Weld filler material activation		1407°C	1433°C
Mech. Weld filler material activation		1500°C	1510°C
Implicit solver settings		Therm. solver	Mech. solver
Time stepping		0,017 s	0,0595 s.
Absolute convergence tolerance		$1,0 \cdot 10^{-10}$	$1,0 \cdot 10^{-20}$
Convergence tolerance		$1,0 \cdot 10^{-4}$	–
Relative convergence tolerance		$1,0 \cdot 10^{-4}$	–
Max. temp. change in each time step		100°C	–
Displacement convergence tolerance		–	$1,0 \cdot 10^{-3}$
Energy convergence tolerance		–	$1,0 \cdot 10^{-2}$
Residual force convergence tolerance		–	$1,0 \cdot 10^{-10}$
Line search convergence tolerance		–	$9,0 \cdot 10^{-1}$

At completion of the second pass, the mesh of the symmetry model and its CWM-results (see Figure 8—left) were integrated into a complete weld test coupon in two steps. First, the mesh, along with the copied results, were mirrored and merged onto a complete weld test coupon simply supported on three points and brought into equilibrium by thermo-mechanical coupled FEA, see Figure 8—center. Second, the run-on plate was discarded, and the final formatted test plate was brought into equilibrium once again by thermomechanical coupled FEA, with a simple support of three points, see Figure 8—right.

3.3 Symmetry CWM versus fully 3D CWM

Before the crack-generating simulations begun, the results of the 3D symmetry approach were validated toward the fully 3D CWM-results (Lindström 2015). In see Figure 9 and see Figure 10, on the right-hand side of the weld centreline, one can see that the CWM results match one another very well. On the left-hand side, there is an increasing deviation between the two result curves as they move farther away from the weld centreline. At a distance of about 38 mm from the weld centreline, the results of the symmetry approach transition from compressive to tensile residual stresses. In see Figure 11, there is an almost perfect match between the two result curves, and in see Figure 12, one can see that there is a minor difference between the two curves' geometry and magnitude in the peak tensile and compressive stress regions.

All in all, the authors conclude that the CWM-results correspond fairly well to one another and to the WRS-measurement results presented in (Wohlfahrt et al. 2012). Therefore, the symmetry CWM-method is considered to be precise enough for the purpose of generating trustworthy WRS-fields to be used in computational $CTOD_{CDF}$ analyses.

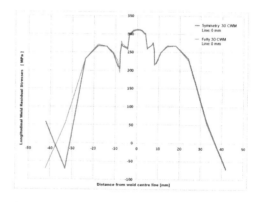

Figure 9. 3D symmetry CWM versus fully 3D CWM-results in the longitudinal (zz) direction; line: 0 mm.

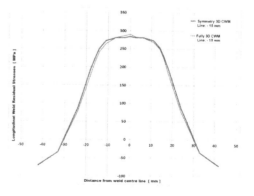

Figure 10. 3D symmetry CWM versus fully 3D CWM-results in the longitudinal direction; line: –3 mm.

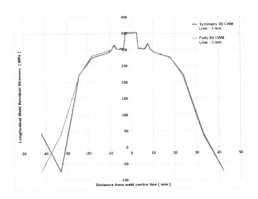

Figure 11. 3D symmetry CWM versus fully 3D CWM-results in the longitudinal direction; line: –15 mm.

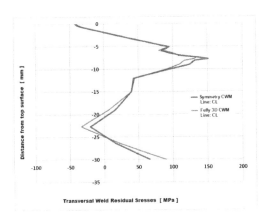

Figure 12. 3D symmetry CWM versus fully 3D CWM-results in the transversal (xx) direction; Line: CL.

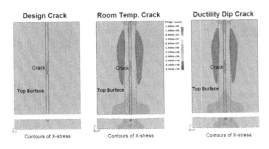

Figure 13. Illustration of the crack location and the Weld Residual Stress Affected Zone (WRAZ) of the three different CTODCDF test plates (The fringe plot level is Pa).

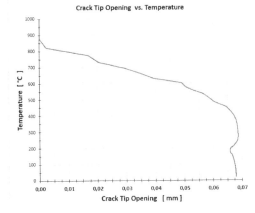

Figure 14. Diagram showing the formation of the Crack Tip Opening (CTO) as a function of the weld metal's cool-down temperature.

3.4 Procedures for introducing cracks in FE model

Three different numerical $CTOD_{CDF}$ test plates, all with a 35-mm long and 5.5-mm deep weld centreline crack, were generated based on the mesh of the CWM-test plate. The first plate represents the theoretical situation one might encounter in a design drawing, i.e., a perfect geometry with nil WRS distribution and magnitude. The second plate represents a situation in which the crack is formed in room temperature at the completion of the weld metal's solidification process. The last plate represents a ductility dip crack (DDC) that initiates and propagates during the welding process, see Figure 13.

The Design Crack $CTOD_{CDF}$ plate, without any WRS-field values or backstresses, was created by the use of the CWM-plate's original and ideal design geometry.

The Room-Temperature $CTOD_{CDF}$ plate, with its associated WRS-fields, was created by forming the crack geometry in conjunction with the merge of the weld centreline nodes in step 1, as described above.

The Ductility Dip Crack $CTOD_{CDF}$ plate, with its associated WRS-fields, was created by resimulation of the entire second weld pass. Here, the weld centreline's boundary condition, near the crack, was released in 12 sequential time steps (193–244 s) during the CWM-analysis. In Figure 14, one can see how the crack tip opening (CTO) of the DDC opened up during the cooling-down period.

3.5 Basis for $CTOD_{CDF}$ analyses

Prior to the main CDF simulation activity, a number of analyses were carried out verifying that the mesh density, as well as the element formulation used during the fully 3D CWM simulation, were also suitable for the purpose of $CTOD_{CDF}$ analyses. The numerical $CTOD_{CDF}$ analyses were performed by the use of thermo-mechanical staggered coupled FEA, subjecting the test plates to

Figure 15. Illustration of the boundary conditions used during the CTOD$_{CDF}$ simulations.

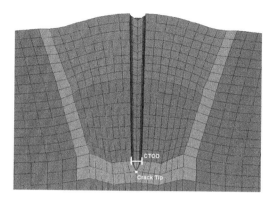

Figure 16. Illustration of the crack tip node and the location of the nodes used to measure the CTOD$_{CDF}$.

Figure 17. Contours of base material temperature (°C) in the quarter symmetry planes of a CTOD$_{CDF}$ plate.

transversal strain (ε_{xx}) with the mechanical boundary conditions shown in Figure 15.

The CTOD$_{CDF}$ value is calculated using the displacement between the two nodes just above the crack tip node (see Figure 16) as a function of the actual mean stress or strain.

During the verification activity, it was noted that the material temperature in the region of the crack tip increased by less than 2°C at the actual strain rate of 2 ms^{-1}, see Figure 17. This temperature increase is too small to significantly affect the simulation result; therefore, thermal boundary conditions were not used during the CTOD$_{CDF}$ analyses.

4 CRACK DRIVING FORCE RESULTS

4.1 Design crack CDF results

The crack driving force of the design crack plate was calculated for two alternative material and

Table 4. Particulars of the design crack plate's CDF-analysis.

Particulars	Alternative 1	Alternative 2
Base material yield strength	220 MPa	275 MPa
Weld metal yield strength	220 MPa	323 MPa

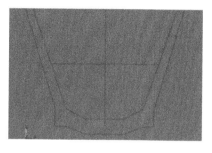

Figure 18. Mid-cross-section crack geometry of the design crack plate prior to CDF analysis.

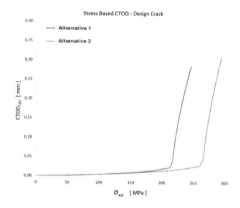

Figure 19. CTOD as function of applied mean stress—design crack.

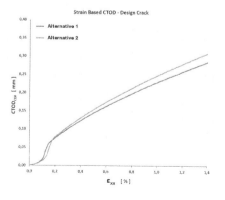

Figure 20. CTOD as function of applied strain—design crack.

WRS combinations; the particulars of the two alternatives are presented in Table 4.

The crack's mid-cross-section geometry prior to the CTOD$_{CDF}$ analysis is presented in Figure 18. In Figure 19 and Figure 20, one can see that the yield strength of the base and weld material affects the CTOD$_{CDF}$ value.

4.2 Room temperature CDF results

The CDF of the room-temperature crack plate was calculated with four alternative material and WRS combinations; the particulars of the four alternatives are presented in Table 5.

The mid-cross-section geometry of the crack prior to the CDF-analysis is presented in Figure 21. One can see that the crack is closed and that its upper part is subjected to compressive stresses while its lower part—the crack tip region—is surrounded by a tensile stress field.

The result of the room-temperature crack plate's CDF-analyses are presented in Figure 22 and Figure 23. In Figure 22, one can see that the weld-induced residual stresses increase the CDF-magnitude by almost 100% in the elastic region (100–250 MPa mean stress). Thereafter, the WRS contribution rapidly trends down to zero, at a mean stress of about 260 MPa. Furthermore, one can observe that the material hardening tends to increase the CDF in the mean stress interval of 200–260 MPa.

In Figure 23, one can see clearly that the accumulated yield surface of the materials drives the CDF magnitude at strain values above 0.15%, while the CDF is driven by the WRS at strain

Table 5. Particulars of the room-temperature and ductility dip crack plates.

Particulars	Alternative 1	Alternative 2	Alternative 3	Alternative 4
Accumulated WRS from CWM-analysis	Yes	Yes	No	No
Base material yield strength	275 MPa	275 MPa	275 MPa	275 MPa
Accumulated kinematic yield surface from CWM-analysis	Yes	No	Yes	No
Weld metal yield strength	323 MPa	323 MPa	323 MPa	323 MPa
Accumulated kinematic yield surface from CWM-analysis	Yes	No	Yes	No

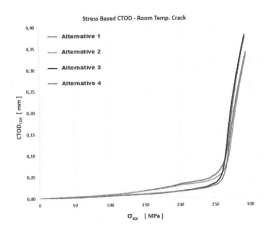

Figure 22. CTOD as function of applied stress—room temperature crack geometry.

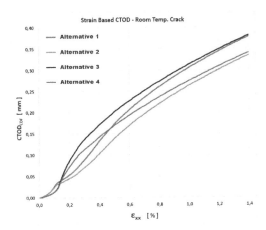

Figure 23 CTOD as function of applied strain—room temperature crack geometry.

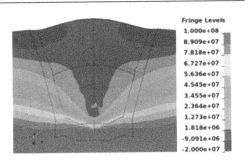

Figure 21. Mid-cross-section crack geometry of the room-temperature crack with its associated contours of transversal stress prior to CTOD$_{CDF}$ analysis. The fringe plot unit is Pa.

values less than 0.1%. This indicates that the WRS-magnitudes should have a major influence in the case of LEFM applications (e.g., fatigue) and EPFM analysis of materials with fracture toughness below their typical values, see Table 5.

4.3 Ductility dip crack CDF results

The CDF of the ductility dip crack plate was calculated with exactly the same material and WRS combinations as the room-temperature crack plate, see Table 5. In Figure 24, the crack's mid-cross-section geometry is presented prior to the CDF-analysis. The crack mouth opening (CMO) is closed and subjected to compressive stresses. One can see that the crack opens up about 1 mm below the top surface, and that the crack tip region is enclosed by a tensile stress field.

The CDF-analysis results of the ductility dip crack plate are presented in Figure 25–26, and the diagrams are very similar to the diagrams of the room temperature crack plate results in Figure 22–23. One can see in Figure 25 that the

CDF increased by about 100% in the elastic region (100–250 MPa) due to the influences of the WRS. At a mean stress of about 260 MPa, the WRS contribution to the CDF disappeared, as it did in the room-temperatures diagrams. One can also see that the materials' hardening tends to increase the CDF in the mean stress interval of 200–260 MPa.

Furthermore, in Figure 26, it is apparent that the CDF at strain values above 0.15% is driven by the hardening of the materials, and that the WRS drives CDF at strain values of less than 0.1%.

4.4 Room temperature crack vs. ductility dip crack

In order to test if there is a difference in the crack driving force between a room-temperature crack and a ductility dip crack, the Alternative 1 results for the two crack types were plotted in the same diagrams, see Figure 27 and Figure 28. In the two

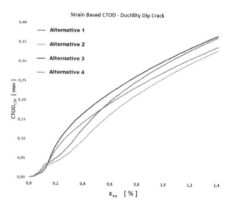

Figure 26. CTOD as function of applied strain—ductility dip crack geometry.

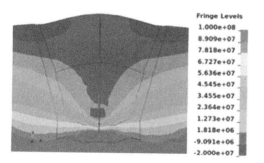

Figure 24. Mid-cross-section crack geometry of the ductility dip crack with its associated contours of transversal stress prior CTOD_{CDF} analysis (the fringe plot unit is Pa).

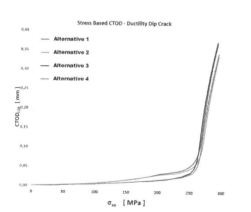

Figure 25. CTOD as function of applied mean stress—ductility dip crack geometry.

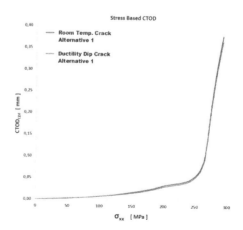

Figure 27. CTOD as function of applied means stress—room temperature vs. ductility dip crack geometry.

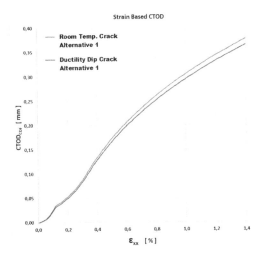

Figure 28.   CTOD as function of applied strain—room temperature vs. ductility dip crack geometry.

diagrams, one can clearly see that there is not a significant difference in the crack driving force between the two crack type's geometry.

5   INFLUENCE OF THE WELD RESIDUAL AFFECTED ZONE

The crack driving force of the three crack types were plotted in the same diagram in order to illustrate the Weld Residual Stress Affected Zone's (WRAZ) influence on the crack driving force, see Figure 29.

In Figure 29, one can see, in the elastic stress range 50–220 MPa, that the WRS increases the crack driving force with about 100%, thereafter the WRS influences is disappeared at the tensile stress 240 MPa. Therefore, one can conclude that in this specific case will the plate's LEFM and fatigue resistance be reduced due to the existence of a WRAZ around the weld repaired area.

Regarding the increased CDF-values in the vicinity of the WRAZ it can be discussed if it originates from an increased energy release rate at the crack tip or if it is a combination of enhanced crack energy and a boosted elastic stress resulting from the alternated flow stress of WRAZ affected material.

6   DISCUSSION

First-order selective reduced integrated element formulation has been used for the FEM-simulations, because higher-order element formulations deteriorate in accuracy more rapidly as compared to first order elements in large deformations (Bower 2010 & ABAQUS 2015 & Belytschko et al. 2000).

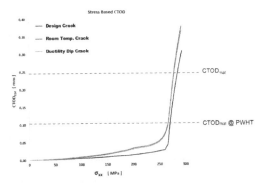

Figure 29.   The diagram shows the crack driving forces, expressed in CTOD, of the design, room-temperature and the ductility dip crack geometry as the function of the applied strain. The red dotted lines indicates the minimum CTODmat acceptance values according to ISO 19902:2007.

Where the smallest element size used was 250 × 375 × 1000 μm. These were selected with the goal of performing CDF-calculations with consistent WRS-fields and geometrical mesh and element formulations.

For fracture mechanics FEA, it is desirable to use the smallest acceptable element size in the vicinity of the crack tip in order to capture the stress singularity effect (LEFM) (Bower 2010) or the CDF (EPFM), respectively. For CWM analyses with first-order element formulation for both the thermal and mechanical simulations, it is also desirable to use the smallest acceptable size near the weld joint in order to avoid inconsistent stress and strain results in areas with steep temperature gradients (Oddy et al. 1990).

A metal material is built up by grains bonded to each other, and where the various grains have different mechanical properties in different directions, depending on their chemical composition, size, and thermo-mechanical history. As well, the undesirable impurities between the grains have various constitutive properties (Svensson 1993). Therefore, the smallest acceptable element size is limited by the origin of the mechanical material data and the grain size. Metallic material models constructed on the basis of industrial standardized mechanical test results will have validity for element volumes containing no less than 125 grains (5 × 5 × 5 grains) (Lemaitre et al. 1994).

The crack driving force effects revealed and expressed through the CTOD can be divvied into two different parts: globally elastic response regime and globally plastic response regime. The elastic response regime for the ductility dip crack rises up to about 250 MPa mean stress and 0.16% strain, respectively. The plastic response regime achieves levels above those values, see Figure 25 and Figure 26.

In the first regime (globally elastic response), the results of the finite element analyses demonstrate that accounting for weld residual stresses will modify the applied crack driving force. In the second regime (globally plastic response), an effect on the crack driving force believed to be associated with the modification of the tensile properties was observed. In this regime, the material flow stress was modelled by the yield surface of the kinematic material hardening model.

The use of a linear tangent plastic hardening modulus was believed to cause ambiguous crack driving force values at strains above 0.2–0.3%. Therefore, the role of the plastic hardening and accurate nonlinear plastic hardening modulus modelling are of increasing importance as the strain rises into higher magnitudes. In Figure 25 and Figure 26 one can see how the CTOD curves 1 and 3 (that have identical flow stress values) are approaching each other with increased strain in the same way as the CTOD curves 2 and 4 (that also have identical flow stress values) are approaching each other. This indicates that the influences and importance of the WRS gradually decrease at higher strain levels.

Thus, two different effects on the crack driving force are seen in the simulation results. The first effect is due to the presence of a residual stress field, whereas the second is due to modifications of the tensile properties from the finite element weld simulation.

7 CONCLUSIONS

The computational welding mechanics method applied has demonstrated its capability to yield results showing the effects of the arc-weld process on the applied crack driving force expressed through the CTOD. The effects revealed can be divided into two different parts: globally elastic response regime and globally plastic response regime.

In the globally elastic response regime, the weld residual stresses will modify the applied crack driving force. During the ceases studied, the crack driving force was augmented due to the presence of the residual stresses. The increase in driving force is apparently a function of the applied global load and does not act as an additive term.

In the globally plastic response regime, an effect on the crack driving force was observed and is believed to be associated with the modification of the tensile properties. The use of a linear tangent plastic hardening modulus is believed to yield inaccurate crack driving force result at large applied strains.

Therefore, finite element weld simulations, intended for consistent elastic plastic fracture mechanics analyses, should be performed by the use of a nonlinear plastic hardening modulus.

For smaller loads, the former effect dominates, whereas for large applied strains, the latter effect governs the response. A parametric study varying both the arc-weld process parameters and the defect positions should promote a better understanding of this phenomenon's full nature.

REFERENCES

Abaqus 6.14, 2015, Theory Guide, Chapter 3.2.1, Simulia, Dassault Systèmes.

Anderson, T.L., 2005, Fracture Mechanics: Fundamentals and Applications, 3rd Ed., ISBN 780849316562 CRC Press Inc., USA, pp. 410–432.

API 579-1/ASME FFS-1, 2007, Fitness-For-Service, 2nd Ed., standard by American Petroleum Institute and ASME, ASME International, New York, USA.

Belytschko, T., Liu, W.K., and Moran, B., 2000, Non-linear Finite Elements for Continua and Structures, ISBN 9780471987741, John Wiley & Sons Ltd, UK, pp. 453–456.

Bower, A.F., 2010, Applied Mechanics of Solids, ISBN 978-1-4398-0247-2, CRC Press Taylor & Francis Group, Boca Raton, Florida, USA, pp. 428–432.

Bower, A.F., 2010, Applied Mechanics of Solids, ISBN 978-1-4398-0247-2, CRC Press Taylor & Francis Group, Boca Raton, Florida, USA, pp. 567–588.

BS7910:2013, "Guide to methods for assessing the acceptability of flaws in metallic structures", ISBN 9780580743344, BSI Standards Limited, pp. 418–428.

BS7910:2013, "Guide to methods for assessing the acceptability of flaws in metallic structures", ISBN 9780580743344, BSI Standards Limited, UK, pp. 370–372.

BS7910:2013, "Guide to methods for assessing the acceptability of flaws in metallic structures", ISBN 9780580743344, BSI Standards Limited, UK, pp. 47–48.

Crutzen, S., Frank, F., Fabbri, L., Lemaitre, P., Schneider, C., Visser, W., 1999. SINTAP Task 3.4 final report—Compilation of NDE effectiveness data. Final report for EC Brite-Euram Project number BE95-1426, March 1999, pp. 56–57.

Demaid, A., 2004, Fail-safe, 2nd ed., ISBN 0 7492 5900 0, Fracture Training Associates, Sheffield, UK, pp. 46.

Folkhard, E., 2013, Welding Metallurgy of Stainless Steels, Softcover reprint of the original 1st ed. 1988, ISBN 978-3709189672, Springer-Verlag Wien NewYork, pp. 80–94.

Fu, G., Lourenco, M.I., Duan, M.; Estefen, S.F., 2014, Effect of boundary conditions on residual stress and distortion in T-joint welds, Journal of Constructional Steel Research 102 (2014), pp. 121–135.

Fuchs, H.O., Fatemi, A., Stephens, R.I., Stephens, R.R., 2000, Metal Fatigue in Engineering 2nd Ed., ISBN 9780471510598, John Wiley & Sons Inc. pp. 261–265.

Hogberg, K., 2014, Swedish Qualification Centre AB, Sweden, Private Communication.

IACS No.47 Shipbuilding and Repair Quality Standard, 1996, Rev.7 2013, Part B—Repair Quality Standard for Existing Ships, International Association of Classification Societies Ltd., London, UK, pp. 17.

ISO 19902:2007/A12013, 22013, Petroleum and natural gas industries -Fixed steel offshore structures, Chapter

20.2.2.5.3 CTOD Fracture toughness requirements, International Organization for Standardization, Geneva, Switzerland.

Janosch, J.J, IIW Doc. X/XV-RSDP-60-01, 2001, IIW Doc. X/XIII/XV-RSDP-68-02/XV-1121-02/XIII-1904-02, 2002, IIW Doc. X/XV-RSDP-68-03, 2003, IIW Round Robin Updated Results for Residual Stress and Distortion Prediction, IIW Doc. X/XIII/XVRSDP-97-04.

Koçak, M., 2006, FITNET Final Technical Report, Version: 27 Nov. 06, GTC1-2001-43049, European Fitnessfor-service Network, pp. 10–14.

Koçak, M., Webster, S., Janosch, J.J., Ainsworth, R.A., Koers, R., 2008, FITNET Fitness-for-Service Procedure, Volume 1, ISBN 978-3-940923-00-4, pp. 6-8-6-14.

Lemaitre, J., Chaboche, J.L., 1994, Mechanics of Solid Materials, ISBN 9780521477581, Cambridge University Press, UK, pp. 69–71.

Lindström, P.R.M., 2015, Improved CWM platform for modelling welding procedures and their effects on structural behavior, Doctoral Thesis, ISBN 978-91-87531-07-1, University West, Trollhättan, Sweden.

Lindström, P.R.M., 2015, Improved CWM platform for modelling welding procedures and their effects on structural behavior, Doctoral Thesis, ISBN 978-91-87531-07-1, University West, Trollhättan, Sweden, pp. 24.

Lindström, P.R.M., 2015, Improved CWM platform for modelling welding procedures and their effects on structural behavior, Doctoral Thesis, ISBN 978-91-87531-07-1, University West, Trollhättan, Sweden, pp. 65.

Lindström, P.R.M., 2015, Improved CWM platform for modelling welding procedures and their effects on structural behavior, Doctoral Thesis, ISBN 978-91-87531-07-1, University West, Trollhättan, Sweden, pp. 83–116.

Lindström, P.R.M., Ulfvarson, A., 2002, "Weld Repair of Shell Plates During Seagoing Operations", Proceedings of OMAE'02, 21st International Conference on Offshore Mechanics and Artic Engineering, June 23–28, 2002, Oslo, Norway, OMAE2002–28583.

Liu, J., Zhang, Z.L., B. Nyhus, B., 2008, Residual stress induced crack tip constraint, Engineering Fracture Mechanics; Volume 75, Issue 14, ISSN 0013-7944, pp. 4151–4166.

Marlette, S., et al., 2010, "Simulation and measurement of through-wall residual stresses in a structural weld overlaid pressurizer nozzle" PVP2010-25736, Proceedings of the ASME 2010 Pressure Vessels & Piping Division/K-PVP Conference, July 18–22, 2010, Bellevue, Washington, USA.

Mills, W.J., 1997 "Fracture Toughness of Austenitic Stainless Steels and Their Welds", ASM Handbook—Volume 19: Fatigue and Fracture, 1997, ISBN 0-87170-385-8, ASM International, Ohio, USA, pp 1860.

Mills, W.J., 2009, Fracture Toughness of Austenitic Stainless Steels and Their Welds, ASM Handbook, ISBN 978-1-61503-005-7, ASM International, Vol. 19, pp. 733–756.

Oddy, A.S., McDill, J.M.J., Goldak, J.A., 1990, "Consistent Strain Fields in 3D Finite Element Analysis of Welds", Transactions of the ASME Journal of Pressure Vessel Technology, AUGUST 1990, Vol. 112/311.

R6 Procedure, Rev. 4, Assessment of the integrity of structures containing defects, EDF Energy Nuclear Generation Ltd., Assessment Technology, Barnwood, UK.

Radaj, D., 2003, Welding Residual Stresses and Distortion: Calculation and Measurement, ISBN 3-87155-791-9, Woodhead Publishing Ltd, Cambridge, UK, pp. 5–12.

Rishel, R., Lenz, H., Turley, G., Newton, B., 2007, "NDT with the structural weld overlay program: Recent field experience and lessons learned", Proceedings of the International Conference Nuclear Energy for New Europe, Portorož, Slovenia, Sept. 10–13, 2007.

Saluja, R.; Moeed, K.M., 2012, The Emphasis of phase transformation and alloying constituents on hot cracking susceptibility of 304 L and 316 L stainless steel welds, International Journal of Engineering Science and Technology, Vol. 4 No.05 May 2012, ISSN: 0975-5462, pp 2206–2216.

Saxena, A., 1998, Nonlinear Fracture Mechanics for Engineers, ISBN 0-8493-9496-1, CRC Press Inc., USA, pp. 81–105.

Saxena, A., 1998, Nonlinear Fracture Mechanics for Engineers, ISBN 0-8493-9496-1, CRC Press Inc., USA.

Ship Structure Committee, 1998, MSC CARLA Complete Hull Failure in a Lengthened Container Vessel, Case Study IX, US Coast Guard, Washington DC, USA.

Svensson, L.-E., 1993, Control of Microstructure and Properties in Steel Arc Welds, ISBN 9780849382215, CRC Press Inc, USA, pp. 55–162.

Taylor, N., Kocak, M., Webster, S., Janosch, J.J., Ainsworth, R.A., Koers, R., 2003, FITNET Final Report for Work Package 2 "State of the art and Strategy", Version 1, FITNET TR3-03, European Fitnessfor-service Network, pp. 29–30.

Totten, G.E., 2002, Handbook of Residual Stress and Deformation of Steel, ISBN 9780871707291, ASM International, Ohio, USA, pp. 11–26.

Wells, A.A., 1961, "Unstable Crack Propagation in Metals: Cleavage and Fast Fracture", Proceedings of the Crack Propagation Symposium, Vol.1, Paper84, Cranfield, UK.

Wohlfahrt, H., Nitscke-Pagel, Th., Dilger, K., Siegel, D., Brand, M., Sakkiettibutra, J. and Loose, T., 2012, Residual Stress Calculations and Measurements—Review and Assessment of the IIW Round Robin Results, Welding in the World, 09/10 2012, Vol. 56, pp. 120–140.

Maritime Transportation and Harvesting of Sea Resources – Guedes Soares & Teixeira (Eds)
© 2018 Taylor & Francis Group, London, ISBN 978-0-8153-7993-5

Analysis of structural crashworthiness on a non-ice class tanker during stranding accounting for the sailing routes

A.R. Prabowo, A. Bahatmaka & J.H. Cho
Interdisciplinary Program of Marine Convergence Design, Pukyong National University, Pusan, Korea

J.M. Sohn & D.M. Bae
Department Naval Architecture and Marine Systems Engineering, Pukyong National University, Pusan, Korea

S. Samuel
Department Naval Architecture, Diponegoro University, Semarang, Indonesia

B. Cao
China Shipbuilding Industry Corporation Economic Research Center, Beijing, China

ABSTRACT: Certain ships are built with capability to survive in voyage through the North Sea Route. However, non-ice class ship especially tankers should be given adequate attention accounting for its behavior when accidental loads are experienced, e.g. stranding. Aim of this paper is to estimate structural behavior of a non-ice class tanker during stranding. Failure and damage extent of the ship were assessed to provide reference data regarding countermeasure and casualty estimations for further study. In this study, a chemical tanker was modelled to be the target ship in a series of stranding scenario model. Condition of the structural damage and tendency of crashworthiness criteria, namely: rupture energy, crushing force and structural acceleration were observed. The results indicated that the lateral impact was capable to produce higher extent of damage on the bottom structure than the frontal impact. Finally, recommendation of the ice class material's application on marine structures is summarized.

1 INTRODUCTION

In the recent years, expedition to the polar territory through the Arctic Ocean has been conducted for various reasons, including researches of the North Pole's species, oil and gas explorations and export-import activities. This route is regarded as a good prospect to distribute commodities from Asia to Europe using a water transportation mode, namely ship. High rate of production and demand for raw material (crude oil, liquefied natural gas, etc.) and daily product (shoes, shirt, car etc.) on these two continents, as well as requirement of industry and society to reduce fuel consumption and operational cost will push the related parties to conduct faster distributions, which an alternative route to reach the designated destinations in Asia and Europe is urgently required. This situation becomes a reasonable consideration that the voyage route through the Arctic Ocean and the North Sea will be main shipping route for northern countries, including Korea, Japan and China for Asia continent as well as Russia, Norway, Netherlands in Europe. Even though this solution is judged available and good

enough, challenge raises to ensure ship and cargo can reach destination safely. During its operation, ship may be subjected by several lethal impact phenomena which can delay the distribution or even cause catastrophic incidents, e.g. grounding accident of the Exxon Valdez in Alaska and the Costa Concordia in Italy and also high profile collision case of the Titanic during its sailing on the Atlantic Ocean, US. Possibility of other impact scenarios on a ship prior entering the Polar region and during sailing at the Arctic in form of stranding is exist. Water territory where the assumed stranding takes place, presents a variety of the sea ground materials which may have contact with bottom structure of a ship.

This study addresses its focus to observe a series of stranding scenario during a ship conducts voyage through the North Sea. The scenarios are fundamentally composed based on the possible material on the sea ground which hard rock on the warm sea territory and grounded ice in the Polar region will be considered in analysis. Sea ground topology is defined as the indenter, and several penetration positions on the ship's bottom structure are applied to assess structural crashworthiness of the structure against various scenario models of stranding.

2 FUNDAMENTAL CONCEPT OF STRANDING PHENOMENON

In maritime accident, stranding is defined as a contact between an object on sea ground so called indenter and a bottom part of ship or vessel as classified as target ship. The indenter penetrates the target ship in stranding and the process stops before the indenter separates from the target ship. In this case, the indenter is in struck position on the target ship. Stranding is classified as a high nonlinear phenomenon which several crushing events (i.e. plate tearing, girder crushing, bottom raking etc.) take place during a ship experiences stranding. A brief literature review is presented in this section in order to provide solid basis in understanding stranding analysis and phenomenon on marine structures.

2.1 Structural estimation by FE approach

As rapid development in terms of computational methodology and virtual technology, finite element method has become a popular method which can be deployed to assess various phenomena in science and engineering. The main concept of this method is discretizing a large model into a finite numbers of smaller individual entity which contain a series of nodal point and connecting element. Deformation of a structural model is defined as two assumptions, the displaced distance of the nodal points from their initial position in both axial and rotational directions, and the assumed deformation field for the elements. Formulation of the finite element approach is presented by direct equilibrium formulation as shown in Equation 1. The most common method to solve dynamic state of a structure under spontaneous impact load, such as crash (Klinger & Bohraous 2014), collision (Prabowo et al. 2016a, b) and grounding (Abu-Bakar & Dow 2013) in this study is referred by the explicit method. This method is known for its ingenious which simply put everything except inertia on the p(t), solve the acceleration and integrate the acceleration twice in order to claim velocity and displacement as a function of time.

$$m\ddot{v} + c\dot{v} + kv = p(t) \qquad (1)$$

where m = the mass matrix; c = the damping constant; k = the spring stiffness; $\ddot{v}$ = the acceleration; $\dot{v}$ = the velocity; v = the displacement; and $p(t)$ = the global external loads.

The finite element approach is applied in pioneer works and combined with other methods in order to validate configuration in the numerical model and cross-check the results. In impact engineering, Alsos & Amdahl (2009a) conducted a laboratory tests and re-conduct it using the FE approach (Alsos et al. 2009b). A series of this research was re-analyzed as a benchmark particulars to assess rebounding phenomenon (Prabowo et al. 2017a). In a collaboration research, the empirical formulations of Minorsky (1958) and Woisin (1979) were used to compare numerical result of ship collision (Bae et al. 2016a) and furthermore is continued to striking bow-side structure collision and effects of material properties to structural behavior after side impact (Prabowo et al. 2017b, c). Analytical expression of Stronge (2004) is used to re-conduct analysis to proposed new analytical model for iceberg-ship collision by Liu & Amdahl (2010). Application of the finite element for ice-structure interaction case was also considered by Bae et al. (2016b). Based on these pioneer studies, the finite element approach is judged powerful enough to analyze impact phenomena which it is expected suitable for stranding calculation.

2.2 Analytical basis of the hard ground model

The analytical concept for grounding model is proposed by Simonsen (1997a) who adopted so called the *Energy Method*. According to this method, when external loads are applied to a deformable target, the work rate of these loads must be equal to the incremental energy that is stored elastically or dissipated in the structure. If a rigid-plastic interaction is assumed, there will be no elastic energy can be stored and the work rate of the external loads equals with the rate of energy dissipated by plastic deformation, fracture and frictional effects on the surface of the involved entities.

In terms of global deformation kinematics, since the bottom structure elements are attached to the shell or the inner bottom plating, then it is convenient to use the deformation of the plating. It is required that all attached elements to the plating, consistently follow the plating during the deformation process. The illustration of the deformation process is shown in his doctoral thesis (Simonsen 1997b) and the longitudinal and transverse gap opening consecutively, are presented in Equations 2 and 3.

$$u_0 = B_{de}\sqrt{(1-\cos\alpha)^2 \sin^2\theta + (1-\cos\theta)^2 \sin^2\alpha} \qquad (2)$$

$$v_0 = B_{de}(1/\cos\alpha - 1) \qquad (3)$$

where u_0 = the longitudinal gap opening; v_0 = the transverse gap opening; B_{de} = the half width of the deforming zone; α = the angle from horizontal to the flaps; and θ = the plate split angle.

3 BENCHMARK VERIFICATION

The benchmark study was conducted based on the penetration experiment by rigid indenter of Alsos & Amdahl (2009a). Origins of this study presented an idealization of ship structure subjected to impact load. It is considered suitable for verification of various impact analyses, including stranding. The target ship is modelled using three elements, namely plate, stiffener and hollow frame. In contact between the ship structure and rock so called *indenter*, penetration is set to occur approximately until 0.25 m. In this study, benchmark is conducted by finite element method which ANSYS LS-DYNA (ANSYS, 2017) is considered to solve penetration model. The ship structure is denoted as the *stiffened plate*. Material properties for the stiffened plate are presented in Table 1. The indenter is augmented by a constant velocity 0.6 m/s to move to center of the plate.

The meshing size is determined to be 10 and 15 mm for the elements of the stiffened plate. Results in term of structural response will be compared with the experiment results. Further analysis is conducted with implementation of the failure criterion based on Peschmann & Kulzep (2000). Mathematical expression for this criterion is presented in Equation 4. Confirmations regarding the crushing force and other structural response in form of the rupture energy are summarized to observe the plate behavior.

$$\varepsilon_f(l_e) = \varepsilon_g + \alpha \cdot \frac{t}{l_e} \tag{4}$$

where ε_f = the failure strain; ε_g = the uniform strain; l_e = the discretised element length; t = the plate thickness; and $\alpha = \varepsilon_m \cdot (x_e/t)$. The detail for ε_f, ε_g and α constants in this works considers the presented value based on measurements in Lehmann & Peschmann (2002) and Ehlers et al. (2008).

The results based on a series of finite element simulation presented a good correlation with the

experimental study. As shown in Figure 1, the crushing force of element length l_e = 15 mm produced the most similar with the experiment in terms of magnitude and peak-force displacement. In the analysis, it was obtained that element length l_e = 10 mm produced earlier failure which made the force of this element size could not provide good similarity.

Further assessment on failure criterion was concluded that larger size of l_e would produce higher force magnitude. The results per l_e were summarized and it was obtained that l_e became less effective during 35 and 45 mm were used in analysis. The similarity was observed between the l_e = 10 mm and l_e = 25 mm applied by the Peschmann criterion. The force was found perpendicularly equal with the rupture energy (Figure 2) which was satisfy fundamental relation of energy-

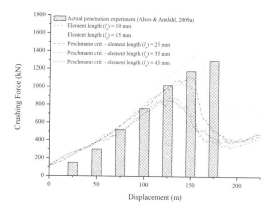

Figure 1. The crushing force of the actual experiment and a series of finite element simulation.

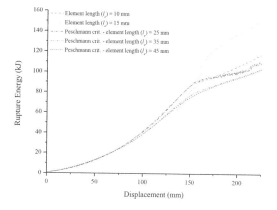

Figure 2. The rupture energy by the numerical simulation. Tendency of the energy is similar with the experienced stress by the stiffened plate during penetration.

Table 1. The properties of steel for target components (Alsos & Amdahl, 2009a).

Material type	Component	n (–)	ε_f (–)	σ_Y (MPa)
A	Plate	0.22	0.35	260
B	Stiffener	0.225	0.35	340
C	Hollow frame	0.18	0.28	390

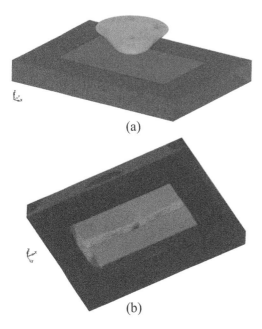

(a)

(b)

Figure 3. Idealization of the penetration experiment: (a) set-up model and (b) stress contour on bottom view.

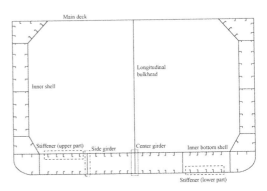

Figure 4. Midship section of the double hull tanker. The boxes highlight detail members on the bottom structure.

force expression. The energy results were obtained from numerical analysis since it was not provided in reference (Alsos & Amdahl 2009a). The set-up and extent of damage on the stiffened plate using $l_e = 15$ mm are presented in Figure 3. The von Mises contour indicated that the initial failure would be experienced by the stiffener. The overall results suggest that the present methodology in conducting the finite element simulation has successfully produced similarity result with the actual experiment. The configuration and setting of the benchmark will be applied further in composing stranding scenarios.

4 ENGINEERING MODEL AND ANALYSIS PREPARATION

The ship model is built as an idealization of a 17,000 DWT non-ice class chemical tanker with principal particulars: 144 m in length, 22.6 m in width and 12 in depth with average length for one cargo tank is 13.95 m. To avoid interference with boundaries, the ship model is focused on the double bottom structure which spans for both starboard and portside. The midship section and full-scale double bottom structures are shown in Figures 4 and 5 consecutively. Stranding analysis is performed by numerical calculation using FE codes ANSYS LS-DYNA (ANSYS 2017) to pro-

duce results for various penetration scenarios. The bottom structure is augmented by the plastic-kinematic model with properties of a medium carbon steel material with density $\rho = 7850$ kg.m^{-3}, Young's modulus $E_x = 210000$ MPa; Poisson's ratio $n = 0.3$ and yield stress $\sigma_Y = 440$ MPa are used to fulfil the characteristic of this material model.

The ship structure is assumed using shell element and to restrict hourglass phenomenon in analysis, the fully integrated Belytschko-Tsay is applied on the ship model. The boundaries of the structures is implemented on the end of the model which axial and vertical displacements are restrained and fixation for rotational movement during contact with rigid rock and grounded ice is applied. The mentioned contact is accounted by the general automatic contact option. Since contact of two entities takes place, frictional constant is applied based on the coulomb coefficient in the range 0.2–0.4. In the following simulations that are conducted in the following sections, the standard value of 0.3 is adopted for rock-stranding scenario while as influenced by nature of ice, a lower friction value 0.05 based on an actual experiment (Cho et al. 2011) is applied for stranding to a grounded ice. The mesh size should be adequately small so that detail of deformation contour can be captured but large enough to obtain the calculation results in a reasonable time process. Ratio of the element-length-to-thickness (ELT) of the discretized model 5–10 is considered in analysis configuration.

The applied material of the rock takes the *plagioclase mineral* (Christensen 1996), a mineral rock which can be found in oceanic ground. The *plagioclase mineral* has mechanical properties as follows: density $\rho = 2690$ kg.m^{-3}; Poisson's ratio $n = 0.296$; assumed modulus $E_x = 67450$ MPa and compressional and shear wave velocity ratio $V_p/V_s = 1.859$. In other hand, the grounded ice is augmented with properties according to Jones (2006) which are

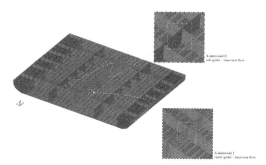

Figure 5. The finite element model of the bottom structure. The circles show details of the intersections for the stranding model.

Table 2. Testing results of the tensile test (Min et al. 2011).

Testing condition	σ_Y (MPa)	σ_U (MPa)	n (–)	K (MPa)
Room temp.	385.05	552.06	0.20	929.54
−30°C	425.27	599.13	0.21	1021.28
−50°C	448.01	596.26	0.23	1048.78

defined: $\rho = 960$ kg.m^{-3}; Poisson's ratio $n = 0.27$ and Young's modulus $E_x = 1000$ MPa. Both entities are modelled using rigid material. However, especially for ice, isotropic material is also used to build crushable geometry. The fracture on the bottom structure is defined using failure criterion of Peschmann & Kulzep (2000) which has been described in previous section. The bottom structure is denoted as the *target structure* while oceanic rock and grounded ice are denoted as the *indenter* in stranding simulation which is classified to assess the ship structure in various extreme conditions.

The first analysis is conducted to observe the structure under stranding with impacting the indenters to four different locations, namely the center girder, side girder, x-intersection I and x-intersection II as indicated in Figures 4 and 5. These locations are classified into two penetration preference. The penetration 1, represents frontal (head-to-head) contact as the target structure is approaching the indenter prior stranding process. In this scenario, the indenter is placed in front of the target structure on the *x-axis* (longitudinal direction) according to the Cartesian coordinate system. Contact is determined happen between the center and side girders with the indenter during this penetration model. Impact model of the penetration 1 is assumed based on general stranding-grounding model (Sormunen et al. 2016 and Prabowo et al. 2017d). In the penetration 2, the indenter penetrates the target structure on the *z-axis* (vertical direction) from below part of the structure. Target points are determined in the x-intersection I (center girder—transverse floor) and x-intersection II (side girder—transverse floor). This penetration is possibly taking place during a ship experiences heavy slamming or get drifted by wind and tide (Simonsen & Hansen 2000) in shallow water territory or rocky strait during rough sea or bad weather.

The second analysis is addressed to study behavior of the ship structure at low temperature as

can be experienced during its sailing through the North Sea passage. The strategy is adopted based on previous work of Yamada (2006) which the ship material is experimented by the mechanical tensile test. Since condition of polar region is considered, the test is conducted under several conditions (Table 2), namely room temperature, −30°C (approx. 243 K) and −50°C (approx. 223 K). The testing data refers to previous testing results by cooperation of Hyundai, Mokpo National University and University of Ulsan in South Korea (Min et al. 2011).

5 SIMULATION RESULTS

The calculated stranding scenarios are presented in this section. Discussion is addressed to assess structural crashworthiness in stranding per target component and response of the bottom structure during stranding at low temperature.

5.1 *Stranding damage on various targets*

Several components of the double bottom are determined to be a target during stranding. Each target is expected to produce different resistance level and increment tendency during the ship impacts the indenter. The presented results in Figure 6 is stranding case before the ship enters the polar region, which rock-structure interaction is considered in this part. The rupture energy is taken as part of the crashworthiness criteria that describes an energy level that is used to plastically deform in contact between two entities. As further displacement of the ship, the indenter penetrated until certain depth of the double bottom structure. In this situation, destroyed material occurs after the plastic deformation was experienced by the ship.

The head-to-head impact produced stable increment along penetration which in same time also explicitly described that the center girder successfully produced higher resistance level than the side girder. Direct contribution of the components' thickness is expected as the main cause of this phenomenon. The stable increment also indicated

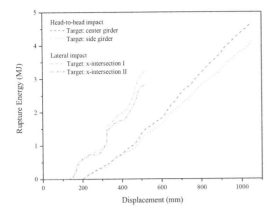

Figure 6. The rupture energy of the stranding scenarios to various target components. The energy of the lateral impact ended earlier than the head-to-head impact since velocity during for these scenarios were determined different with the head-to-head velocity $v_{head} = 7$ m/s and lateral velocity $v_{lat} = 3.5$ m/s.

Figure 7. The damage extent after stranding occurred on the center girder. Stress contour was found well distributed by the bottom stiffener.

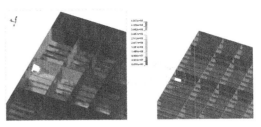

Figure 8. The damage extent after stranding occurred on the side girder. Distribution of the stress were observed focus on the connection between the bottom plating and side girder.

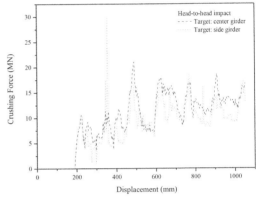

Figure 9. Tendency of the crushing force for the head-to-head impact. The side girder produced lower level as it was weaker than the center girder during stranding.

that during stranding, the indenter penetrated the double bottom structure which the structural components near the targets (center and side girders) had similar preference which in this case was longitudinal structure. As presented in Figures 7 and 8 for targets on the center and side girders, the girders, stiffeners and bottom plating were arranged in longitudinal position and only the solid floor was found as the transverse component, which made the energy were produced in stable increment along the indenter's penetration.

This tendency was also found satisfy the energy-force correlation where the behavior of the crushing force (Figure 9) showed stable tendency. Distinction between force levels illustrated the experienced stress of the structure in impact. Further comparison in term of the damage extent (Figures 7 and 8) concluded that narrower space of the bottom stiffener (center girder: 0.38 m) provided better resistance than wider space of the stiffener near the side girder (0.68 m). Tearing on

the bottom plating during impact to the center girder (0.958 m) was observed narrower than on the side girder (1.943 m) which stress was concentrated on the T-connection between the bottom plating and side girder.

Other stranding model was calculated in form of lateral impact. In this case ship was assumed approach the seabed rock from *z-axis*. It can be concluded based on results in Figure 6 that even though the penetration ended earlier than the head-to-head scenarios, amount of the energy during penetration was remarkable. If penetration was continued as far as displacement in the previous scenarios, the energy was expected 50–70% larger than the head-to-head stranding. This result was verified with the tendency of the crushing force (Figure 10) which was found reach higher level than the head-to-head stranding. This tendency could be affected by the movement direction of the ship when it was approaching the indenter. The downward movement made weight of the ship also contributed that initiated larger impact.

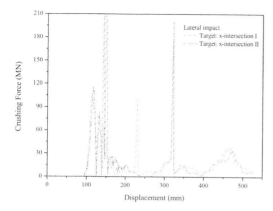

Figure 10. The crushing force of the lateral impact. One cause of the high level force was expected from movement direction of the target ship. In this situation, body weight was concentrated during the ship slammed the indenter in z-axis.

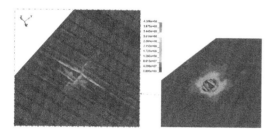

Figure 11. The extent of damage of the lateral stranding with the target on the x-intersection I. In this location, the intersection consisted the transverse floor and center girder.

Besides the movement effect, this phenomenon is also influenced directly by the structural topology on the targets. Both targets in the lateral stranding were full-solid x-intersection between the transverse floor and girders which in order to destroy these targets would need higher energy than T-connection in the head-to-head scenario. This topology pattern is very different with the previous scenarios that after the transverse floor was breached, only longitudinal components resisted the penetration.

The extent of damage is presented in Figures 11 and 12 for different targets in the lateral stranding. Damage of the x-intersection I was found smaller as the bottom stiffener acted as main resistor which prevented the indenter to directly penetrate the intersection. High stress contour was found in four points on the stiffener (Figure 11). Different tendency was shown during a stranding occurred on the x-intersection II which in this scenario, the

Figure 12. Damage of the lateral stranding with the target on the x-intersection II. Distribution of the stress contour on the intersection was found farther than the intersection I.

indenter directly destroyed the intersection. The concentrated stress (Figure 12) was experienced by the transverse floor and side girder after impact.

5.2 Crashworthiness criteria during ice-structure interaction

After discussion of structural response during rock-structure interaction, behavior of the non-ice class ship is observed in terms of the crashworthiness criteria during stranding at low temperature region, e.g. Arctic Ocean. In this region, ice-structure interaction has high possibility to occur, including stranding with part of a grounded ice (ICEX 1979 and Riska 2006). Accounting for the structural crashworthiness, the first observed criterion is the rupture energy (Figure 13). The produced energy by the crushable ice during stranding indicated more fluctuation when ice and structure contacted each other in stranding. This behavior occurred as two deformable bodies (ice and structure) deformed in same time.

During impact to the center girder, a peak point happened after the displacement of the ship passed 400 mm which in this moment, due to its thickness t was larger than the side girder, the ice was penetrated by the girder during both entities collided each other. After the ice was penetrated the energy reduced and further deformation was dominated by the bottom structure. This phenomenon was not experienced by ice during stranding to the side girder. Component strength on the bottom structure was concluded will provide various deformation processes during stranding with ice. Confirmation of the characteristic of ice material is presented with comparison to rigid ice. The rigid body did not experience any deformation during impact of two entities as its geometry was constant along the process. Since it had constant geometry, larger structural responses were expected than the crushable model. The results of stranding to the rigid ice is observed has more stable increment as

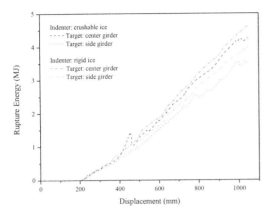

Figure 13. The rupture energy in stranding with ice-structure interaction was considered. The rigid ice was analyzed to verify result's characteristic of the crushable ice.

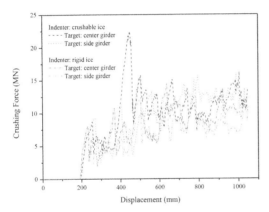

Figure 14. Characteristic of the crushing force for each indenter type. High fluctuation form was observed occur on the crushable ice stranding.

deformation process only occurs on the ship structure. Furthermore, the energy shows that the produced magnitude is higher but not in significant level, which disparities of the rigid and crushable ices are in range 6–13%. During the time constraint is determined as an absolute parameter, analysis using the rigid ice can be considered as a reasonable option.

The second criterion is addressed to the crushing force (Figure 14) in stranding. The energy behavior of the crushable ice in contact with the center girder was successfully validated by the peak fluctuation after passing 400 mm displacement. The results also confirms that in moment of two deformable entities made contact, higher fluctuation was likely taking place for the crashworthiness criteria.

In other hand, during the rigid ice was deployed, fluctuation of the force was found similar. It also indicated that the crushing process of the girders on the bottom structure would occur in similar form if the indenter is very hard, i.e. rigid-grounded ice and seabed rock. These indenters are considered suitable for assessment of critical state of a marine structure during impact which valid for both routes (North Sea and South Sea passages) as the indenters represent most-likely encountered indenter for each route.

5.3 Assessment of structural crashworthiness: Non-ice and ice class ships

Structural crashworthiness is considered for three criteria to focus the observation, i.e. energy, force and acceleration. In term of the rupture energy (Figure 15), it was obtained that the ice class ship had significantly higher resistance during impact with ice in room temperature condition. An approximation difference was calculated 25%. During the stranding scenarios were deployed at the most extreme temperatures which considered condition of the Arctic Ocean in winter (The Arctic 2017), the ice class ship provided the best resistance against impact. In temperature −50°C or approximately equal with 223 K, the rupture energy almost reached 7 MJ which 1.5 times better than the non-ice class ship in experiencing stranding scenario.

Tendency of the rupture energy is also found match with characteristic of the crushing force and structural acceleration in Figures 16 and 17 consecutively. These results indicated that the experienced fluctuation by the bottom structure using ice class material in stranding was higher than

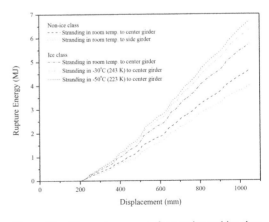

Figure 15. The rupture energy for non-ice and ice class ships during ice-structure interaction. The ice class ship successfully produced the best resistance capability in encountering ice-stranding scenario.

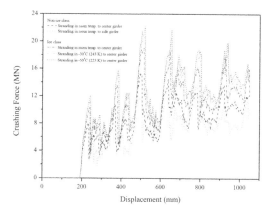

Figure 16. Behavior of the crushing force for non-ice and ice class ships during ice-structure interaction.

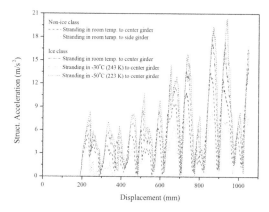

Figure 17. Fluctuation of the experienced acceleration by the ship structure using non-ice and ice class regulations during ice-structure interaction.

the non-ice class ship, which explicitly described the capability increment of the structure after ice class regulation was applied. If the initial velocity is applied or in other words, the ship detects the obstacle and reduce its speed, the damage extent of the ice-class ship will be lower as it is implemented by stronger material. In other scenario, possibility ice rebounding during ice-structure collision, can be expected and observed (Prabowo et al. 2017a). A serious consideration to implement regulation of the ice class requirement (ABS 2005 and IACS 2016) is encouraged based on the presented results in this subsection.

Better safety has been proven that the ice-class ship can provide better safety for ship, crew and cargo during the North Sea passage is considered as an alternative routes from Asia to reach Europe (and vice versa). Implementation of the regulation

is also expected can reduce possibility of terrible chain reaction, e.g. human life loss, oil spill species extinction and environmental damage.

6 CONCLUSIONS

This present work aimed to assess structural crashworthiness during a non-ice class tanker experiences stranding in both South and North Sea passages. The numerical method which was validated using a laboratory test reference was conducted to perform a series of stranding analysis.

The results indicated that the lateral impact was capable to produce higher extent of damage on the bottom structure. This phenomenon may occur during the ship encounters high sea level and storm in a voyage. Main support of navigational instrument to detect the topology of seabed should be considered. In discussion regarding non-ice and ice class ships, the structural capability to resist stranding impact of the non-ice ship was overwhelmed by the ice class. Even though a ship is not embedded by specific requirements of ice class, implementation of ice class material on the ship hull below maximum draught line is recommended to increase structural strength against under water impact. Finally, to ensure safety for all related parties in ship voyage, further regulations to oblige requirement of polar region on non-ice class marine structures which operate on and through North Sea are seriously acknowledged.

REFERENCES

AbuBakar, A. & Dow, R.S. 2013. Simulation of ship grounding damage using the finite element method. *International Journal of Solids and Structures* 50: 623–636.

ABS. 2004. *Guidance notes on ice class*. Texas: American Bureau of Shipping.

Alsos, H.S. & Amdahl, J. 2009a. On the resistance to penetration of stiffened plates, Part I - Experiments. *International Journal of Impact Engineering* 36: 799–807.

Alsos, H.S., Amdahl, J. & Hopperstad, O.S. 2009b. On the resistance to penetration of stiffened plates, Part II – Numerical analysis. *International Journal of Impact Engineering* 36: 875–887.

ANSYS. 2017. *ANSYS LS-DYNA User's Guide*. Pennsylvania: ANSYS. Inc.

Bae, D.M., Prabowo, A.R., Cao, B., Zakki, A.F. & Haryadi, G.D. 2016a. Study on collision between two ships using selected parameters in collision simulation. *Journal of Marine Science and Application* 15: 63–72.

Bae, D.M., Prabowo, A.R., Cao, B., Sohn, J.M., Zakki, A.F. & Wang, Q. 2016b. Numerical simulation for the collision between side structure and level ice in event of side impact scenario. *Latin American Journal of Solids and Structures* 13: 2991–3004.

Cho, S.R., Chun, E.J., Yoo, C.S., Jeong, S.Y. & Lee, C.J. 2011. The measuring methodology of friction coefficient between ice and ship hull. *Journal of the Society of Naval Architects of Korea* 48: 363–367 (in Korean).

Christensen, N.I. 1996. Poisson's ratio and crustal seismology. *Journal of Geophysical Research* 101: 3139–3156.

Ehlers, S., Broekhuijsen, J., Alsos, H.S., Biehl, F. & Tabri, K. 2008. Simulating the collision response of ship side structures: A failure criteria benchmark study. *International Shipbuilding Progress* 55: 127–144.

IACS. 2016. *Requirements concerning polar class*. London: International Association of Classification Societies.

ICEX. 1979. Ice and climate experiment. Report of Science and Application Working Group. Maryland: Goddard Space Flight Center.

Klinger, C. & Bohraus, S. 2014. 1992 Northeim train crash – A root cause analysis. *Engineering Failure Analysis* 43: 171–185.

Lehmann, E. & Peschmann, J. 2002. Energy absorption by the steel structure of ships in the event of collisions. *Marine Structures* 15: 429–441.

Liu, Z. & Amdahl, J. 2010. A new formulation of the impact mechanics of ship collisions and its application to a ship-iceberg collision. *Marine Structures* 23: 360–384.

Min, D.K., Shim, C.K., Shin, D.W. & Cho, S.R. 2011. On the mechanical properties at low temperatures for steels of ice-class vessels. *Journal of the Society of Naval Architects of Korea* 48: 171–177 (in Korean).

Minorsky, V.U. 1958. An analysis of ship collision with reference to protection of nuclear power ship. *Journal of Ship Research* 3: 1–4.

Peschmann, J. & Kulzep, A. 2000. Final Report for BMBF Life-Cycle Design, Part D2A: Side Collision of Skin Ship. Hamburg: Technical University of Hamburg.

Prabowo, A.R., Bae, D.M., Sohn, J.M. & Zakki, A.F. 2016a. Evaluating the parameter influence in the event of a ship collision based on the finite element method approach. *International Journal of Technology* 4: 592–602.

Prabowo, A.R., Bae, D.M., Sohn, J.M. & Zakki, A.F. 2016b. Energy behavior on side structure in event of ship collision subjected to external parameters. *Heliyon* 2: e00192.

Prabowo, A.R., Bae, D.M., Sohn, J.M., Zakki, A.F., Cao, B. & Cho, J.H. 2017a. Effects of the rebounding of a striking ship on structural crashworthiness during ship-ship collision. *Thin-Walled Structures* 115: 225–239.

Prabowo, A.R., Bae, D.M., Sohn, J.M., Zakki, A.F., Cao, B. & Wang, Q. 2017b. Analysis of structural behavior during collision event accounting for bow and side structure interaction. *Theoretical & Applied Mechanics Letters* 7: 6–12.

Prabowo, A.R., Bae, D.M., Sohn, J.M., Zakki, A.F. & Cao, B. 2017c. Rapid prediction of damage on a struck ship accounting for side impact scenario models. *Open Engineering* 7: 91–99.

Prabowo, A.R., Cao, B., Bae, D.M., Bae, S.Y., Zakki, A.F. & Sohn, J.M. 2017d. Structural analysis of the double bottom structure during ship grounding by finite element approach. *Latin American Journal of Solids and Structures* 14: 1106–1123.

Riska, K. 2006. *Ship-ice interaction in ship design: Theory and practice*. Trondheim: Norwegian University of Science and Technology.

Simonsen, B.C. 1997a. Ship grounding on rock – I. Theory. *Marine Structures* 10: 519–562.

Simonsen, B.C. 1997b. *Mechanics of Ship Grounding*. Lyngby: Technical University of Denmark.

Simonsen, B.C. & Hansen, P.F. 2000. Theoretical and statistical analysis of ship grounding accidents. *Journal of Offshore Mechanics and Arctic Engineering* 122: 200–207.

Sormunen, O., Castrén, A., Romanoff, J. & Kujala, P. 2016. Estimating sea bottom shapes for grounding damage calculations, *Marine Structures* 45: 86–109.

Stronge, W.J. 2004. *Impact Mechanics*. Cambridge: Cambridge University Press.

The Arctic. 2017. *Climate change*. [arctic.ru/climate/], accessed in the 27 April 2017.

Woisin, G. 1979. Design against collision. *Schiff & Hafen* 31: 1059–1069.

Yamada, Y. 2006. Bulbous buffer bows: A measure to reduce oil spill in tanker collision. Lyngby: Technical University of Denmark.

Maritime Transportation and Harvesting of Sea Resources – Guedes Soares & Teixeira (Eds)
© 2018 Taylor & Francis Group, London, ISBN 978-0-8153-7993-5

The measurement of weld surface geometry

M. Randić
Croatian Register of Shipping, Branch office Rijeka, Rijeka, Croatia

D. Pavletić & G. Turkalj
Faculty of Engineering, University of Rijeka, Rijeka, Croatia

ABSTRACT: The weld surface geometry greatly influences the fatigue life of welded joints. Expressions given in relevant literature used for the calculation of fatigue stress concentration factors, usually take into account five parameters of weld surface geometry, i.e., the thickness of the welded material, weld toe radius, weld toe angle, weld face reinforcement and weld width. Thus, to determine the fatigue stress concentration factor for selected welds, it is necessary to measure the weld surface geometry. There are several methods of measuring it. A non-contact 3D measurement method based on structured light projection and relevant measurement software is presented. The adopted approach is discussed as measurements are significantly influenced by the operator, consequently also the resulting stress concentration factors.

1 INTRODUCTION

A higher stress concentration ocurrs at the point of a sudden change in geometry of the weld, Perović (2011), these points being highly sensitive to the initiation of surface cracks, Pilkey & Pilkey (2007). Therefore, the precise determination of the geometry of the weld is of utmost importance in order to have the precise calculation of the stress concentration factor. In this way it is possible to identify points of a higher stress concentration on the weld surface, that is, points of potential initiation of surface cracks during the exploitation of the weld.

The toe radius and the weld toe angle have the greatest impact on stress concentration, Fujisaki et al. (1990), Ohta et al. (1990), so that a special attention will be devoted to the measurement of these geometry variables. Besides these two, there are other three geometrical variables that affect stress concentration, i.e., the thickness of the base material, Berge (1985), the weld width, Kim et al. (1996), and the reinforcement height, Fricke (2003).

The analysis of weld geometrical variables can be divided into scanning methods and evaluation. Figure 1 presents both the scanning methods and evaluation of the geometries of the weld.

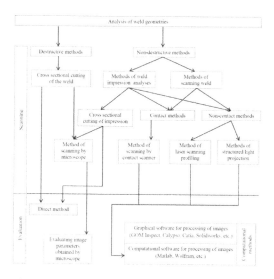

Figure 1. Analysis of the geometry of the weld.

2 SCANNING METHODS OF THE WELD GEOMETRY

The scanning of the geometry of the weld can be divided into two methods:

- Destructive methods and
- Non-destructive methods.

2.1 Destructive methods

These methods are applicable for research purposes only as the weld on which geometrical variables are measured has to be cut and thus it can no longer be used, this also being the greatest drawback of this method. During cutting the shape of the weld should not be changed locally so that due to this the cutting is performed by procedures of cold cutting, such as cutting with the saw, that is, without introducing the heat that changes the shape of the weld surface.

The measurement of geometries can be performed at the cross section of the weld only, so that using this method it is not possible to follow the changes of geometrical variables along the welded joint. After the cutting, the geometrical variables of the weld are measured by the direct method, as described in sub-paragraph 3.1, or, the cross section can be scanned by a measuring microscope and the geometrical variables are measured on the obtained scan.

2.2   Non-destructive methods

Unlike destructive methods and depending on the equipment used for testing, non-destructive methods can be used on welded joints that are exploited, as the testing does not change the characteristics of the weld. Non-destructive methods of measuring geometrical variables of the welded joint can be divided in:

a. Methods of weld impression analyses (WIA) and
b. Methods of scanning the weld surface.

2.2.1   Method of weld impression analyses

This is a simple method that does not require expensive equipment while the obtained data are quite correct, Harati et al. (2014).

The application of this method begins with cleaning of the welded joint, onto which two-component putty for making the impression of the weld surface is applied. After a period of setting, the putty is removed from the weld and thus the impression is made, ready for the analysis. The advantage of this method is that the impression of the welded joint can be made in the area that otherwise cannot be reached with instruments for non-contact scanning of welded joints, also that the impression of the weld can be analysed in laboratory conditions.

2.2.2   Methods of scanning weld surfaces

According to Groš et al. (2012), methods of scanning weld surfaces can be divided into two groups:

a. Contact methods and
b. Non-contact methods.

a. *Contact methods*
Contact methods of scanning the weld are those that use the touch for assessing the weld. The best known is the method that employs the contact scanner.

It is based on the physical contact of the scanner probe that is attached to the coordinate measuring machine. The measuring takes a long time while

specific measuring equipment is needed so that it is rarely used for the scanning of welded joints.

b. *Non-contact methods*
The best known are the following two non-contact methods, Harati et al. (2014):

– Laser Scanning Profiling (LSP) and
– Structured Light Projection (SLP).

2.2.3   Laser scanning profiling

The Laser Scanning Profiling method uses the principle of optical triangulation. A laser line is projected on the weld surface, the reflected light going back to the sensor. On the basis of information gathered in this way, a three-dimensional model of the weld is constructed, suitable for further processing by one of the computational methods for measuring geometrical variables, Harati et al. (2014).

2.2.4   Structured light projection

The Structured Light Projection method is a technique of projection of known light pattern on the object. The structured light can be projected with the Liquid Crystal Display (LCD) or any other stable source of light. As the light is projected on the object, images of the object are captured by the camera. By using the triangulation method, the three-dimensional shape of the weld analysed is derived from the images, suitable for further computational processing. In order to obtain the necessary precise measurements, the object is captured from more angles (usually 8 of them), Harati et al. (2014).

The structured light can function in visible or non-visible light spectre. During the work in the visible spectre, the interference of the structured light with the surrounding light may occur, which can be avoided by applying the source of structured light that functions in the non-visible light spectre. According to the length of the band of light projected by its source, there are three types of non-visible structured light, Fofi et al. (2004):

– Infrared Structured Light (IRSL);
– Imperceptible Structured Light (ISL) and
– Filtered Structured Light (FSL).

The advantage of this method is a great speed and precision of caption. Instead of scanning one point only when using laser, the whole area is scanned at once with this method.

3   METHODS OF MEASURING THE GEOMETRIES OF THE WELD

The geometries of the weld can be measured with two methods:

a. Direct method and
b. Computational methods for the analysis of the surface.

3.1 *Direct method of evaluation*

This method is the simplest and fastest so that it can easily be applied during the everyday work. According to Lindgren and Stenberg (2011), the results obtained by this method are unreliable and subjective as they depend on the operator who assesses the measurements. This especially regards the weld toe radius and the weld toe angle.

The thickness of the base material, the weld width and the reinforcement height are measured by the moving gauge, the weld toe angle by the protractor while the weld toe radius by the feeler gauge that contains a set of blades, where the end of every blade has a different predefined radius.

3.2 *Computational methods for surface analysis*

A few computational methods have been developed for the processing of the image of the weld by measuring weld geometries. They can be divided into graphical and computational methods, i.e., software for image processing.

3.1.1 *Graphical methods for image processing*

Graphical methods are based on the software for processing the image of the weld. In such software, a three-dimensional representation of the weld is intersected in areas where the surface analysis is to be conducted. Measurements of geometrical features of the weld surface are conducted on the obtained diagonal cross section. The way of measuring depends on the software used. Among the software more frequently used for this purpose are *GOM Inspect*, *Calypso*, *Catia* and *Solidworks*.

3.1.2 *Computational methods for image processing*

Computational methods are conducted by the application of corresponding software intended for the calculation of geometrical features of the weld surface on the basis of the data obtained by scanning the weld, i.e., the three-dimensional representation of the weld surface, Stenberg et al. (2012). Among the software that can be used for this purpose there are, for example, *Matlab* and *Wolfram Mathematica*.

4 AN EXAMPLE OF SCANNING AND MEASURING WELD GEOMETRIES

4.1 *Scanning of samples*

In the chosen example, the scanning of weld samples was carried out by the scanner for three-dimensional scanning in the Centre for Advanced Computing and Modelling of the University in Rijeka, Figure 2.

The *ATOS II Triple Scan* system was used for scanning, consisting of:

– Projector;
– Two cameras and
– Control unit.

A base stand for gauge, a rotating table for the sample measured and a computer were added to the above.

The projector that is built in the centre of the sensor head is used for emitting narrowband blue light that enables precise measurements and digitalization to be carried out independently of environmental lighting conditions. The projected blue light is reflected from the object that is measured into two cameras with the resolution of 5×10^6 pixel each. The minimum distance between the points on the sample measured is 0.02 mm, according to the instructions of Acquisition Basic (2016).

4.2 *Measurement of weld geometries*

GOM Inspect computational software was used to process the geometries of the weld. It is software that was developed for the analysis of data obtained by three-dimensional scanning. STL data of the scanned samples are imported into software, with the X-axis placed along the weld with the starting point in the place where welding started.

Weld geometries were measured on the sample in three bands. The first band was determined in the area of 20 to 30 mm after the start of welding, the second band was set in the centre of the sample from 70 to 80 mm from the start of welding, while the third band in the area from 120 to 130 mm from the start of welding. As during welding the automatic reading of welding parameters was made (the strength of welding electric current, the power of welding electric current and the wire feed speed), it was established that welding parameters

Figure 2.   Scanning of samples by 3D scanner.

became stabilised before the initial band in which weld geometries are measured.

Each band is 10 mm wide, while the distance between cross sections is 1 mm. This is how in each section 11 cross sections were obtained where weld geometries were measured. Figure 3 presents the sample generated in the *GOM Inspect* with bands in which cross section geometries were measured, while Figure 4 presents the actual sample with marked bands. Figure 5 presents the transverse appearance of a cross section; in this case it is the cross section in the distance of 20 mm from the start of welding.

Each cross section was geometrically processed separately, on each one the measuring of the following geometries made:

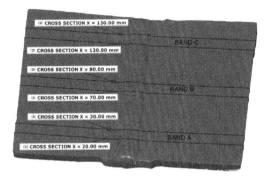

Figure 3. Sample with marked bands generated in *GOM Inspect*.

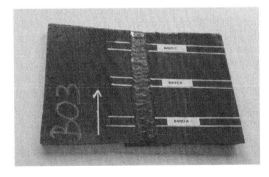

Figure 4. Sample with marked bands.

Figure 5. Graphical representation of cross section X = 20.00 mm.

– Toe radius;
– Weld toe angle;
– Reinforcement height and
– Weld width.

4.3 *Toe radius*

The toe radius was measured for each cross section for the left and the right side. According to Cerit et al. (2010), the greatest stress concentration occurs in the toe radius, while according to Radaj et al. (2006), it is the geometrical variable that affects the stress concentration most, so that a comparison was made of the three methods by which the toe radius was obtained, i.e., the following:

a. Direct measuring of the toe radius with the feeler gauge for radius measuring;
b. Graphical reading of the toe radius by the *GOM Inspect* software and
c. Calculation of the weld toe radius by *Wolfram Mathematica* software.

4.3.1 *Direct measuring of the toe radius with the radius-measuring gauge*

The sample was marked with bands and cross sections, measuring with the moving gauge from the welding start. At the point of each cross section, the radius was measured with the feeler gauge Figure 6.

The feeler gauge used for measuring had the smallest radius of 0.5 mm, the radii made bigger at each 0.125 mm. The obtained values are shown on diagrams in Figure 8. Figure 6 presents direct measuring on the sample, for the cross section of 20 mm to the left.

4.3.2 *Graphical evaluation of the toe radius with GOM Inspect*

At each cross section a point was established in which the surface of the base material changes into the weld reinforcement height. For the left side of the weld, this point has the mark "POINT LEFT 02".

Figure 6. Direct measuring on the sample, for the cross section X = 20.00 mm to the left.

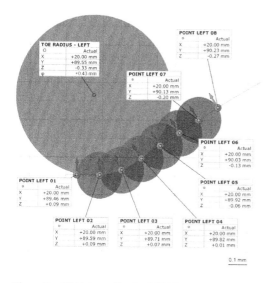

TOE RADIUS - LEFT
	Actual
O	
X	+20.00 mm
Y	+89.55 mm
Z	-0.33 mm
φ	+0.43 mm

POINT LEFT 08
	Actual
X	+20.00 mm
Y	+90.23 mm
Z	-0.27 mm

POINT LEFT 07
	Actual
X	+20.00 mm
Y	+90.13 mm
Z	-0.20 mm

POINT LEFT 06
	Actual
X	+20.00 mm
Y	+90.03 mm
Z	-0.13 mm

POINT LEFT 01
	Actual
X	+20.00 mm
Y	+89.46 mm
Z	+0.09 mm

POINT LEFT 05
	Actual
X	+20.00 mm
Y	+89.92 mm
Z	-0.06 mm

POINT LEFT 02
	Actual
X	+20.00 mm
Y	+89.59 mm
Z	+0.09 mm

POINT LEFT 03
	Actual
X	+20.00 mm
Y	+89.71 mm
Z	+0.07 mm

POINT LEFT 04
	Actual
X	+20.00 mm
Y	+89.82 mm
Z	+0.01 mm

0.1 mm

Figure 7.   Weld toe radius in *GOM Inspect*.

On all cross sections for all the samples this point is marked at the point where the weld reinforcement height begins.

The toe radius can be defined as the radius of a circle that passes through three points on the weld surface. In that case the value of the toe radius depends on the distance between the points through which the circle passes. In this paper we chose the distance between points of 0.125 mm, following the research of Lawrence & Mazumdar (1981), where it was established that the biggest stress concentration occurs in the very area of 0.125 mm along the weld.

As the weld surface is of an irregular shape, there are no two cross sections with the same shape of the surface, so that for each cross section n points were evaluated, where the number of n points for each section is not necessarily the same.

While measuring the weld toe radius using the *GOM Inspect* software package, circles were constructed through each of the three marked points

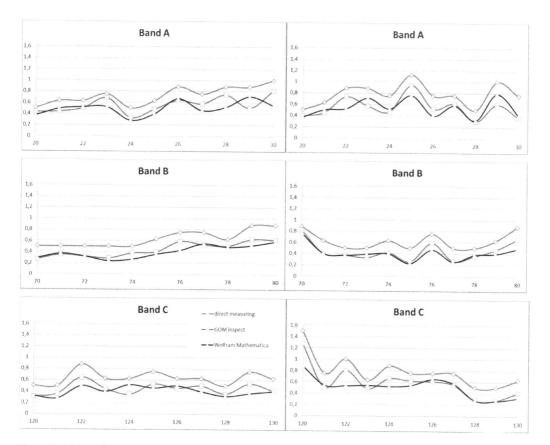

Figure 8.   Comparison of three methods of measuring weld toe radius.

on the weld surface. As the stress concentration is the biggest in the area of the smallest toe radius, the circle with the smallest radius is evaluated in further analysis. In the case represented on Figure 7, these are the "POINT LEFT 02", "POINT LEFT 03" and "POINT LEFT 04", while the circle that passes through these three points is marked with the red colour.

4.3.3 Calculation of the weld toe radius by Wolfram Mathematica software

The basis for the calculation of the weld toe radius is the equation of the circle with the centre in point (a,b), of the radius r, which passes through the point (x,y), Equation 1,

$$(x-a)^2 + (y-b)^2 = r^2 \tag{1}$$

where a,b = points of circle centre; r = circle radius; x,y = circle points.

The circle points (x,y), taken, for instance, from the *GOM Inspect* software, are entered into the *Wolfram Mathematica* software where the circle radius is determined computationally, i.e., the weld toe radius. In the *GOM Inspect*, n points were evaluated for each cross section, so that Equation 1 can be written in the form of equation system of n points, Expression 2.

$$(x_1 - a_1)^2 + (y_1 - b_1)^2 = r_1^2$$
$$\vdots \qquad \vdots \qquad \vdots \tag{2}$$
$$(x_n - a_n)^2 + (y_n - b_n)^2 = r_n^2$$

An algorith that calculates the radius for each combination of three points was made in the *Wolfram Mathematica* software. For n points that were marked in the *GOM Inspect*, the algorithm calculates (n-2)² toe radius. As the stress concentration is found in the area of the smallest radius, as the final solution the algorithm gives the radius that is the smallest of all the calculated ones.

4.3.4 Comparison of results

Three methods by which the weld toe radius can be measured have been presented. The first method consists in the direct assessment of the radius by a feeler gauge, while for the other two methods it is common that the coordinates of the weld surface points are determined by the *GOM Inspect* software, while each of the methods calculates the smallest radii in a different way.

The graphs in Figure 8 represent the comparison of results obtained by the three methods of measuring the toe radius. From these graphs it is possible to see that direct measuring gives the highest values of the toe radius so that it is possible

to conclude that for practical purposes the direct measuring is accurate enough, taking into account that the results obtained by direct measuring are 50% bigger than the radius values obtained by the other two methods.

4.4 Weld toe angle

The weld toe angle is the angle between the line of the base material and the tangent at the point of inflection on the weld surface.

This angle was measured with the *GOM Inspect* software. It was measured with the help of points that were marked on the weld surface, as it is explained in 4.3. The area of inflection is the point in which the curve of the weld surface changes from the concave shape into the convex shape. This point was determined by drawing circles through the three points that are marked on the weld surface. The Figure 9 presents the circle that passes through points 05, 06 and 07, where the weld surface has a concave shape. The circle that passed through the following neighbouring points 06, 07 and 08, has a convex shape. This means that the point of inflection is between points 06 and 07. On the Figure 9, the point of inflection is marked at the half of the distance between these two points, "POINT OF INFLECTION—LEFT".

The line that passes through this point, being also the tangent on the weld surface curve, is marked as the "LINE OF WELD REINFORCEMENT HEIGHT—LEFT". The angle between the "LINE OF BASE MATERIAL—LEFT" and the "LINE OF WELD REINFORCEMENT HEIGHT—LEFT" is the "WELD TOE ANGLE—LEFT".

Theoretically speaking, this is the biggest angle that weld reinforcement height acquired for each cross section.

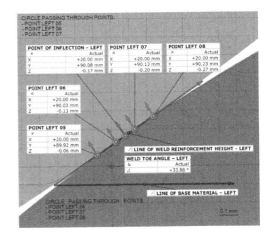

Figure 9. Graphical representation of weld toe angle.

660

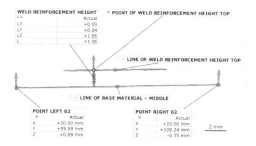

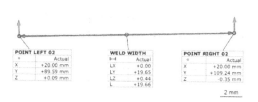

Figure 10. Graphical representation of weld reinforcement.

Figure 11. Graphical presentation of the weld width.

4.5 Weld toe reinforcement height

Weld toe reinforcement height is the distance between the line of the base material in the area of the weld toe and the line parallel to it that passes through the highest point of weld toe angle. The line of base material can be defined as the line that connects points in which the toe angle begins from the left and the right side, these being "POINT LEFT 02" and "POINT RIGHT 02". On Figure 10 this line is marked with "LINE OF BASE MATERIAL—MIDDLE". It is passing through the highest point, which is at the farthest distance from the "LINE OF BASE MATERIAL—MIDDLE". The distance of these two lines is the height of the toe angle. In Figure 10 this geometrical variable is marked as "TOE ANGLE REINFORCEMENT HEIGHT". For the cross section of 20 mm left it amounts to 1,56 mm.

4.6 Weld width

The weld width is the distance measured between the points "POINT LEFT 02" and "POINT RIGHT 02". Figure 11 presents the weld width on the cross section of 20.00 mm.

5 CONCLUSIONS

The accurate measurement of weld geometries is of great significance for the precise determination of geometrical factor of stress concentration, thus also the critical area in which surface cracks might occur, especially so with regard to the weld toe radius. This is why it is necessary to pay a special attention to the measuring of the toe radius.

The three methods that were used to measure the toe radius give results that can be qualitatively compared, remarking that the results obtained by the direct analysis are around 50% bigger than the results obtained by the other two methods. The results obtained by the *GOM Inspect* software are influenced by the distance between the points of the weld, with the help of which the toe radius is calculated. This research adopted the distance of 0.125 mm, as this is the width of the area of the greatest stress concentration.

It can be concluded that by the direct method, data that are precise enough for everyday use can be obtained, i.e., in the areas where the scanning of the weld is not possible due to unavailability. Scanning, followed by the measuring of the toe radius by the *GOM Inspect,* is possible in the conditions where the weld is available and where there are good conditions, especially with regard to workshops and laboratory conditions.

ACKNOWLEDGEMENTS

This work was supported by the Croatian Science Foundation—project 8722 and the University of Rijeka (contract no. 13.09.1.1.05).

REFERENCES

Acquisition Basic 2016, GOM Software, Braun-schweig, Njemačka.

Berge, S. (1985): "On the Effect of Plate Thickness in Fatigue of Welds", Engineering Fracture Mechanics, Vol. 21, no. 2, pages 423–435.

Cerit, M., Kokumer, O., Genel, K. 2010: "Stress Concentration Effects of Undercut Defect and Reinforcement Metal in Butt Welded Joint", Engineering Failure Analysis 17, pages 571–578.

Fofi, D., Sliwa, T., Voisin, Y. 2004: "A Comparative Survey on Invisible Structured Light", *Machine Vision Applications in Industrial Inspection* XII, 90

Fricke, W.F. 2003: Fatigue analysis of welded joints: state of development, *Marine Structures 16*, pages 185–200.

Fujisaki, W., Noda, N., Tanaka, H., Nisitani, H. 1990: "Effects of Reinforcement Geometry and Welding Condition on Stress Concentration Factor of Butt Welded Joint", pages 1533–1538. (na japanskom).

Groš, J., Medić, S., Brozović, M. 2012: Metode trodimenzionalnog optičkog mjerenja i kontrole geometrije oblika, *Zbornik Veleučilišta u Karlovcu, godina II, Broj 1*, pages 43–48. (na hrvatskom).

Harati, E., Svensson, L.-E., Karlsson, L. 2014: The Measurement of Weld Toe Radius Using Three

Non-destructive Techniques, *6th International Swedish Production Symposium*.

Kim, I.S., Kwon, W.H., Park, C.E. 1996: "The Effect of Welding Process Parameters on Weld Bead Width in GMAW Processes", Journal of KWS, Vol. 14, No. 4, pages 204–213.

Lawrence, F.V., Ho, N., Mazumdar, P.K. 1981: Pre-dicting the Fatigue Resistance of Welds, *Ann. Rev. Mater. Sci., Vol. 11*, page 401–425.

Lindgren, E., Stenberg, T. 2011: "Quality Inspection and Fatigue Assessment of Welded Structures", Stockholm, Švedska.

Ohta, A., Mawari, T., Suzuki, N 1990.: "Evalution of Effect of Plate Thickness on Fatigue Strength of Butt Welded Joints by a Test Maintaining Maximum Stress at Yield Strength", Engineering Fracture Mechanics, Vol. 37, no. 5, pages 987–993.

Perović, D.Z. 2011: "The Weld Profile Effect on Stress Concentration Factors in Weldments", 15th International Research/Expert Conference "Trends in the Development of Machinery and Associated Technology", Prague.

Pilkey, W.D., Pilkey, D.F. 2007: "Peterson's Stress Concentration Factors", John Wiley & sons, Inc, treće izdanje, New Jersey.

Radaj, D., Sonsino, C.M., Fricke, W. 2006: "*Fatigue Assessment of Welded Joints by Local Approaches*", Woodhead Publishing Ltd., Cambridge.

Stenberg, T., Lindgren, E., Barsoum, Z. 2012: "Development of an Algorithm for Quality Inspection of Welded Structures", Proceedings of the Institution of Mechanical Engineers, Part B: Journal of Engineering Manufacture, Vol. 226, pages 1033–1041.

Maritime Transportation and Harvesting of Sea Resources – Guedes Soares & Teixeira (Eds)
© 2018 Taylor & Francis Group, London, ISBN 978-0-8153-7993-5

Round robin study on spectral fatigue assessment of butt-welded joints

J. Rörup
DNV GL—Maritime, Hamburg, Germany

Y. Garbatov, Y. Dong & E. Uzunoglu
CENTEC, Universidade de Lisboa, Portugal

G. Parmentier & A. Andoniu
Bureau Veritas Marine, Paris, France

Y. Quéméner & K.-C. Chen
CR Classification Society, Taipei, Taiwan

S. Vhanmane, A. Negi & Y. Parihar
Indian Register of Shipping, Mumbai, India

R. Villavicencio & V. Parsoya
Lloyd's Register, GTC Southampton, UK

L. Peng & J. Yue
Wuhan University of Technology, Wuhan, China

ABSTRACT: The objective of this study is to perform a round robin study on spectral fatigue damage assessment of butt-welded joints and to identify the existing uncertainties and challenges in the commonly used approaches. Various methods are employed to estimate the fatigue damage of butt-welded joints located in the deck and shell structure of a bulk carrier. Hydrodynamic loads are calculated based on strip and panel theory and closed form hydrodynamic loads. The finite element method and beam theory are both feasible to calculate the stress responses and the analysis procedures for each method are introduced and the corresponding results are compared and conclusion are derived.

1 INTRODUCTION

The most accurate way of fatigue life estimation in practice today is the spectral approach, which simulates the time history of ship structural response in waves by a linear superposition. The basis for developing loads in the spectral analysis method is the development of transfer functions generally referred to as Response Amplitude Operators, or RAO's. RAO represents the response of the ship's structure to excitation by a wave of unit height, and it is derived over the full range of (encounter) frequencies that will be experienced. RAO's express the amplitude and phase relationship between the wave load and the response.

The frequency response of the load components is determined by hydrodynamic analysis. Different numerical hydrodynamics methods have been developed.

First methods were based on a 2D-theory and used a velocity potential with linear assumptions only. To compensate the insufficient solution of strip theory in long waves and of slender hulls in short waves a unified theory was developed (Maruo 1970, Newman & Sclavounos 1980, Kashiwagi, 1995). This 2D-seakeeping analysis yields good results, but cannot completely represent the 3D problem of a ship in waves.

Consequently, 3D boundary element methods were developed, using either a wave Green's function or a Rankine source in frequency or time domain.

Green function methods discretize the wetted hull surface into many small surface elements (panels). For each panel, a Green function defines the velocity potential. For numerical scheme and related problems with the Green's function, refer to Inglis & Price (1982), Iwashita & Ohkusu (1992) and Takaki, Iwashita & Lin (1992).

Only the Rankine method includes the potential for steady flow. In addition, more complicated boundary conditions on the free surface and the hull are considered. However, the free surface surrounding the hull as well as the hull itself must be discretized by panels. This method originally was applied successfully to the wave-ship interaction problem by Nakos (1989) Sclavounos & Nakos (1990). A comprehensive overview of various Rankin singularity methods for seakeeping is documented by Bertram & Yasukawa (1996).

For fatigue investigations, the ship response to the relevant wave excitation is linear and accordingly the response in a seaway is described by a superposition of the responses to all regular wave components that constitute the irregular sea, which can be performed in a frequency domain analysis. Given the linearity, the response is described by a stationary and ergodic, but not necessarily narrow-banded Gaussian process.

The stress transfer functions (RAOs) were evaluated directly from structural analysis with finite element method (FEM) or by simple beam models for selected fatigue locations. Stress range is normally expressed in terms of probability density functions for different short-term intervals corresponding to the individual cells of the wave scatter diagram. The linear addition of short term damages sustained over all sea states gives the total fatigue damage for the structural detail. Fatigue damage is calculated on the Palmgren-Miner approach and accumulated over operational service life by accounting for all sea states encountered.

This paper presents a round robin study on fatigue damages for transverse butt welds in the upper hull of a bulk carrier. Considering an axial misalignment of 10% in plate thicknesses a 'FAT80' was employed by all participants for the nominal stress approach. Thickness effect was considered from all participants in the same way, here exists a common understanding by all applied class rules. Mean stress effects were considered individually according to the applied classification rules. For the hydrodynamic analysis and statistical approach, all participants use following unified input parameters:

– wave lengths λ, 25 to 1125 m, Δλ = 50 m
– wave directions Θ, 0 to 360°, ΔΘ = 30°
– 75% of service speed, ~11 knots
– directional spreading (cos, n = 2)
– wave spectrum: Pierson-Moskowitz form
– long term scatter diagram: World-wide
– service life: part of life at sea: 0.85*25 years
– uniform weighting factors for heading angles

The round robin study was carried out to identify existing challenges in commonly used fatigue spectral approaches as part of the "Fatigue and Fracture" committee at ISSC 2018. Seven institutes participated in the round robin study with nine different approaches. Ship data of the investigated Suez-max bulk carrier, a global FE-model and nodal forces representing the mass distributions of four loading conditions were offered to the participants.

The participants were requested to assess fatigue damage for transverse butt welds at two x-positions in the mid ship area, for each at three different locations in the upper hull. Beside the fatigue damages, a comparison of the RAO's and long-term values for vertical bending moment was carried out, too.

2 SHIP DATA

A bulk carrier design with a double hull side structure was chosen as target ship for the round robin study. Principal dimensions of the bulk carrier are given in Table 1. The fatigue life of transverse butt welds in the upper hull of a bulk carrier is investigated. Two frame locations are chosen, see Figure 1. One in hold 6, the heavy ballast hold and the other in hold 5, which is a loaded hold in alternate condition. Fatigue damage will be computed for each x-position at three different locations, see Figure 2.

Hull girder cross section properties at the two investigated x-locations, based on the gross and net50 scantlings, were distributed to the participants.

2.1 Loading conditions

The spectral fatigue assessment was conducted for four loading conditions, namely:

Table 1. Principal dimensions of the bulk carrier.

Length	~285 m
Breadth	~47 m
Depth	~25 m
Draft	~17 m

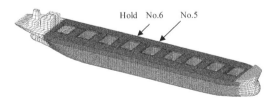

Figure 1. Finite element model of the bulk carrier.

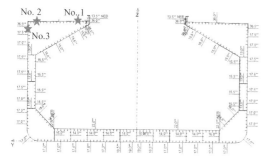

Figure 2.   Cross section of hold 5 at x/L = 0.55.

Table 2.   Still water bending moments.

| | SWBM in 10^6 [kNm] | | |
Condition	at x/L = 0.45	at x/L = 0.55	Fraction of time
Homogenous	−1.316	−1.530	0.25
Alternate	3.272	2.057	0.25
Normal Ball.	2.969	2.803	0.20
Heavy Ball.	−1.660	0.365	0.30

- *Homogeneous* full load departure condition, all cargo holds filled with cargo of density 0.9 t/m³. All wing and double bottom water ballast tanks empty.
- *Alternate departure* condition loaded with a cargo of density 3 t/m³ in hold 1, 3, 5, 7 and 9. All wing and double bottom water ballast tanks empty.
- *Normal Ballast* departure condition, double wing and bottom tanks filled with 70,000 t water ballast. All cargo holds empty.
- *Heavy Ballast* departure condition, all, double wing and bottom tanks filled with ~75,000 t water ballast. Cargo hold No.6 filled with ~22,000 t ballast water.

Drafts, centre of gravity, GM-values, Radii of Gyration, displacements and block coefficients of the investigated loading conditions were distributed to the participants.

Table 2 shows the still water bending moments at the two relevant x-positions and considered the fraction of time in operation for all loading conditions.

3   PARTICIPANTS AND THEIR METHODS

Seven institutes participated in the round robin study with nine different approaches in total. For each participant, the applied hydrodynamic software is shortly introduced. For the homogeneous loading condition, each member demonstrates his hydrodynamic results for vertical bending moment by a RAO diagram. Furthermore, the way of stress determination and if relevant, the mass application in the global FE-model is explained.

3.1   *Method of DNV GL*

With DNV GL software ShipLoad all masses were applied directly on the structural model and resulting nodal forces in the FE model reproduced the total mass of the ship, including the mass of structure, equipment, engine etc. Tank liquid and cargo masses are reproduced by nodal forces, too, considering the hydrostatic pressure distribution on all tank and cargo hold boundaries. With this mass distribution in the global FE-model the hydrodynamic calculations are conducted and are used later for stress determination, too. The hydrodynamic potentials are determined by a 3-D Rankine code. The fatigue approach was carried out according to DNV GL (2015a, b).

3.2   *Method of CENTEC*

The 2-D linear strip theory is employed to estimate the hydrodynamic loads among which only

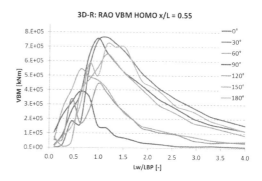

Figure 3.   RAO's for VBM from DNV GL.

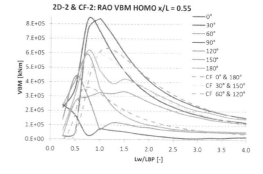

Figure 4.   RAO's for VBM from CENTEC.

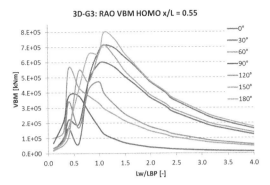

Figure 5. RAO's for VBM from BV.

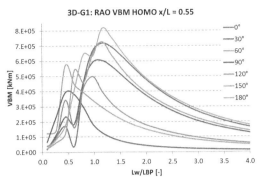

Figure 6. RAO's for VBM from CR.

the wave-induced vertical bending moment is taken into account. In this case, the global finite element method is used to determine the stress load ratios and static stresses at the locations of interest.

As an alternative to the 2-D linear strip theory a parametric approach, developed by Jensen & Mansour (2002) for specific vessel types and loads, is also employed here. The stress load ratios and static stresses are calculated based on the beam theory.

3.3 Method of Bureau Veritas (BV)

The spectral fatigue assessment has been carried out according to BV's guidelines NI611, Bureau Veritas (2016) based on stress RAOs obtained using the global FE model of the ship. The lightship weight was modeled by means of nodal masses derived from the available nodal forces. Liquid ballast was taken into account by directly modeling the ballast tanks for which the radiation problems are solved in order to obtain pressures on the tank boundaries. Finally, dry cargo is modeled by means of nodal masses placed at the indicated center of gravity and inertia loads are distributed on the boundaries of the cargo hold.

BV's software Homer v2.1.4 was used to perform the coupled hydro-structure analysis. The hydrodynamic coefficients for both the hull and ballast tanks are calculated using the 3D-Green function potential flow code Hydrostar v7.3. Pressures are then computed at the FE mesh points and integrated over the FE mesh which ensures an implicit balance between the inertia loads and hydrodynamic and hydrostatic forces. Finally, structural analysis is performed in order to determine the stress RAOs at the locations of interest.

3.4 Method of CR classification society

The coupled hydro-structure analyses were conducted using Homer 1.2 from Bureau Veritas (BV). The finite element model of the ship structure included nodal masses to reproduce the dry cargo and water ballast inertia loads that were distributed to the hold/tank boundaries using NX Nastran's rigid body elements (RBE3).

The seakeeping analyses were carried out using Hydrostar 7.25 (BV), a 3-D-Green-Function potential flow code. The hydrodynamic pressures were then transferred on the structural mesh elements, and the ship motion accelerations were applied to balance the model. The finite element analyses were then conducted for each combination of wave frequency and heading by NX Nastran, and the RAOs of stress were extracted at the considered hot spots. The conversion to fatigue hot spot stress RAO was conducted accordingly to the CSR (IACS, 2015).

3.5 Method of Indian Register of Shipping (IRS)

Fatigue assessment is based on two methods (2-D strip and closed form) to compute nominal stress. Consideration is given to global loads (vertical and horizontal bending moments). However, the closed form method considers only vertical bending moment (Jensen & Mansour, 2002).

The beam theory is employed to obtain the nominal stresses for the given welded joints. Spectral analysis is performed accounting the appropriate mean stress effect, thickness factor (reference thickness is assumed as 25 mm) and rain flow correction factor (Wirching & Light, 1980). A suitable bi-linear S-N curve is selected for given structural details.

3.6 Method of Lloyds Register (LR)

The Lloyd's Register (LR) software WaveLoad-FD is used for the hydrodynamic analysis, which uses the Green function for computing the hydrodynamic potentials. In the hydrodynamic analysis, the mass distribution is determined with the LR software ShipRight where the cargo and ballast

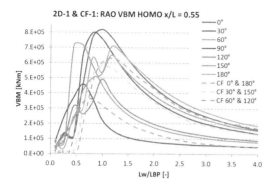

Figure 7. RAO's for VBM from IRS.

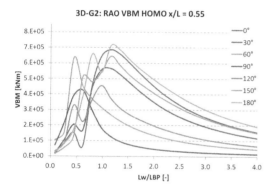

Figure 8. RAO's for VBM from LR.

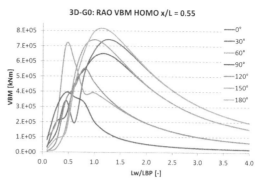

Figure 9. RAO's for VBM from WHUT.

Table 3. Summary of methods.

	Hydro code	Stresses from	Fatigue rules
DNVGL	3D- Rankine	Global FE	DNVGL
CR	3D-Green-func.	Global FE	CSR
LR	3D-Green-func.	Cargo Hold FE	LR
BV	3D-Green-func.	Global FE	BV
IRS	2D-Strip	Beam, V&H BM	IRS
	Closed Form	Beam, VBM	IRS
CENTEC	2D-Strip	Global FE, VBM	DNVGL
	Closed Form	Beam, VBM	DNVGL
WHUT	3D-Green-func.	Global FE	DNV

tanks, as well as other compartments, are automatically detected and then the respective loading is assigned as per the loading manual.

The fatigue assessment uses the three-hold finite element model (holds 4, 5 and 6) and follows LR procedures (LR 2016a). The loading of the model and the combination of the stresses is carried out using the LR software ShipRight (LR 2016b). ShipRight adopts a unit load approach to estimate the total stress response by combining the results of discrete unit load cases and the applied loads.

The unit cases include hull girder global loadings, external hydrodynamic wave pressure loads, and internal solid cargo/water ballast inertia pressure loads. All these loads are further computed for any loading condition and sea state resulting from the hydrodynamic analysis and scatter diagram. The distribution and magnitude of the internal inertia pressure loads are determined by simplified expressions for each ship motion.

3.7 Method of Wuhan University (WHUT)

Hydro model was built according to the hull of the bulk carrier. Hydrodynamic loads were determined

based on the Green function and wave induced forces for a specified set of wave frequencies and heading angles were obtained by the DNVGL software SESAM. Wave loads were mapped from the panels to the hull elements in the global model, which indicated that all loads effects would be involved in SFA.

The FE analysis was based on the Classification Note No. 30.7 from DNV. With the Palmgren-Miner approach, the accumulated fatigue damage of six hot spots in each loading condition was calculated. It should be noted that only the responses in zero speed were considered in the analysis.

3.8 Summary of methods

Table 3 summarizes the applied methods from all participants.

4 COMPARISON OF RESULTS

4.1 Response amplitude operators

The vertical bending moment represents the primary load component for the investigated details. By example, the RAO's for the homogenous loading condition at x/L = 0.55 are shown for each participant in Figures 3 to 9. The maximum values in the head or following seas are in the range of 710,000–840,000 kNm/m. An exception exists for the closed form solution in Figs. 4 and 7 where the VBM is only of 610,000 kNm/m. For all the loading conditions, the closed form predictions represent either the lower or the upper bound of the VBM values produced by all the participants (see also long-term values in Figs. 10 to 13).

For direct investigated RAO's the differences in aft-oblique sea are much more present than in other heading angles. Particularly for a heading angle of 60°, occurs here in some cases a pitch resonance and results in a high VBM. For 90°–150° heading angle, this phenomenon does not exist and the results scatter less than in aft-oblique sea.

4.2 Long-term values

The long-term values of the vertical bending moment were assessed through spectral analysis by

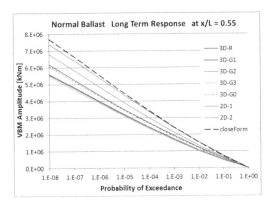

Figure 12. Long-term values for VBM in Normal Ballast.

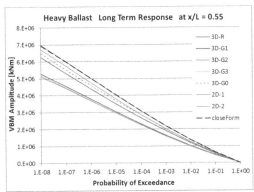

Figure 13. Long-term values for VBM in Heavy Ballast.

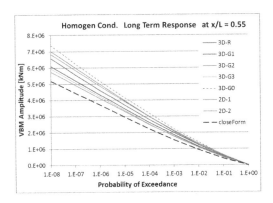

Figure 10. Long-term values for VBM in Homogen Condition.

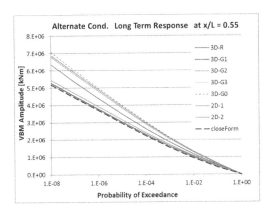

Figure 11. Long-term values for VBM in Alternate Condition.

all participants using their own investigated RAOs or from closed form and following settings:

- Pierson-Moskowitz wave spectrum,
- Angular spreading of wave energy by cos^2-funct.,
- Equal heading probability, and
- World-wide wave scatter diagram

Figures 10 to 13 present the long-term values of VBM for each loading condition. By trend, the long-term values of VBM produced by 2D-seakeeping analyses are higher than those obtained by 3D methods. For the long-term values of all participants (without closed form) exist a maximum deviation of about 30% for all conditions and probability levels.

In Table 4 the fatigue interesting long-term values of the 10^{-2} probability level are listed for all methods and loading conditions.

Application of the hydrodynamic closed form solution results always in an imprecise forecast. For all loading conditions, the closed form solutions

Table 4. 10-2 long-term value of **VBM** at x/L = 0.55.

10^6 [kNm]	Homogen loaded	Alternate loaded	Normal Ballast	Heavy Ballast
3D-R	1.244	1.156	1.222	1.188
3D-G1	1.143	1.018	1.069	1.016
3D-G2	1.095	1.059	1.052	0.997
3D-G3	1.058	0.951	1.057	0.988
2D-1	1.319	1.270	1.342	1.250
CF-1	0.934	0.934	1.526	1.293
2D-2	1.265	1.330	1.542	1.363
CF-2	0.990	0.990	1.560	1.367
3D-G0	1.407	1.309	1.192	1.275

Table 5. Correction factors for mean stress effect.

	Homogen loaded	Alternate loaded	Normal Ballast	Heavy Ballast
3D-R	0.810	0.969	1.000	0.911
3D-G1	0.612	0.976	1.000	0.918
3D-G2	1.000	1.000	1.000	1.000
3D-G3	1.000	1.000	1.000	1.000
2D-1	0.839	1.000	1.000	0.927
CF-1	0.807	1.000	1.000	0.928
2D-2	0.781	0.941	0.976	0.894
CF-2	0.751	0.984	0.969	0.867
3D-G0	0.893	0.963	0.994	0.903

either represent the lower or upper bound in the scatter of all results.

4.3 Fatigue damage

For this benchmark study, the fatigue life of transverse butt-weld connections amidship in the upper hull were assessed using the nominal stress approach. Considering an axial misalignment of 10% in plate thicknesses 'FAT80' was applied for the fatigue investigation from all participants.

The fatigue damages on the Palmgren-Miner approach were directly calculated through a spectral analysis using the nominal stress RAOs on which were applied correction factors that include the effect of corrosion, mean stress and thickness effect. These factors were determined according to the applied class rules. Clear differences exist for the mean stress. By example Table 5 shows for the location No.2 at x/L = 0.55 the applied mean stress factor for all four loading conditions.

Figure 14 shows for the location No.2 at x/L = 0.55 the fatigue damages for the different loading conditions and the combined one. Although for a single loading condition obvious differences exist due to hydrodynamic inputs and mean stress effects, the combined damages are on a comparable level for all approaches. In case of an approach with RAOs from the closed form, the Normal Ballast conditions contribute much more than the other loading conditions, particularly the two loaded conditions have a much lower contribution.

The combined fatigue damages for all investigated locations are shown in Figure 15. The differences in fatigue damages for the combined case are reduced in comparison with single loading conditions. The approaches CF-1, 2D-2 and CF-2 that consider only the vertical bending moment for the stress assessment, indicate always the location P1 as the most critical for both cross sections because P1 is at the largest vertical distance to the ship section

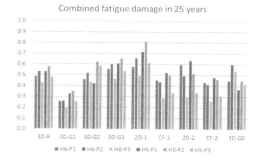

Figure 14. Fatigue damages for location No.2 at x/L = 0.55.

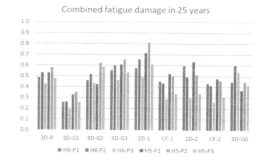

Figure 15. Combined fatigue damages for all locations.

neutral axis. All other approaches that employ a direct stress assessment from the FE model loaded accordingly to the hydrodynamic analyses results, indicate always the location P2 as the most critical. The direct stress assessment approach is more precise because it allows for considering further hull girder load components and possible local bending effects.

5 CONCLUSIONS

The work presented here performed a round robin study on the spectral fatigue damage assessment of butt-welded joints employing various methods to estimate the fatigue damage of butt-welded joints located in the deck and shell structure of a bulk carrier. Hydrodynamic loads were calculated based on strip and panel theory and closed form hydrodynamic loads. The finite element method and beam theory demonstrated to be feasible in estimating the ship hull stress responses.

Since wave-induced loads are the most significant contributing factor to fatigue, the outcomes of the ship motion and load analysis will determine the fatigue damage of the butt-welded joints. All participants used the same inputs for their hydrodynamic analyses. However, the methods to compute the hydrodynamic potentials differ among them. By trend the long-term values of VBM are higher from 2D-seakeeping analysis than from 3D-methods. The application of the hydrodynamic closed form solution results in an imprecise forecast, as it either represents the lower or the upper bound in the scatter of all results and for all loading conditions.

It has been noticed that the horizontal bending and torsion have a moderate impact with following effect on the fatigue results: The assessment methods that only consider the vertical bending moment effect predict that the location P1 close to the hatch coaming is the most critical. On the other hand, when considered further hull girder load components and possible local bending effects, the other approaches indicate that the location P2 is the most critical.

Although all loading conditions exhibit obvious differences in the fatigue damage, mainly due to the different hydrodynamic loads and mean stress effects, the combined damage is comparable at a certain level.

REFERENCES

Bertram V. and Yasukawa H. (1996). Rankine Source Methods for Seakeeping Problems. Jahrbuch Schiffbautechnische Gesellschaft, Vol. 90, pp. 411–425.

Bureau Veritas. (2016). NI 611 Guideline for Fatigue Assessment of Steel Ships and Offshore Units. BV.

DNV GL (2015a). DNVGL-RU-9111:2015-7, "Rules for Classification, Part 3 Hull, Chapter 9 Fatigue", DNV GL, Oslo.

DNV GL (2015b). Class Guideline DNVGL-CG-0129: "Fatigue Assessment of Ship Structures", DNV GL, Oslo.

IACS (2015). "Common Structural Rules for Bulk Carriers and Oil Tankers", International Association of Classification Societies, London.

Inglis R.B. & Price, W.G. (1981). A Three-Dimensional Ship Motion Theory; Comparison between Theoretical Predictions and Experimental Data of the Hydrodynamic Coefficients with Forward Speed. Trans. RINA, 124.

Iwashita H. & Ohkusu M. (1992). Green Function Method for Ship Motions at Free Speed, Ship Technology Research, 32, 2.

Jensen JJ. & Mansour AE. (2002). Estimation of ship long-term wave-induced bending moment using closed-form expressions. Royal Institution of Naval Architects Transactions Part A International Journal of Maritime Engineering.

Kashiwagi M. (1995). Prediction of Surge and its Effect on Added Resistance by Means of the Enhanced Unified Theory. Trans. West-Japan Society of Naval Architects, 89, pp. 77–89.

LR (2016a). ShipRight Design and Construction: Fatigue Design Assessment – Application and Notations (Notice 1 and Notice 2). Lloyd's Register Group Limited, London, UK.

LR (2016b). ShipRight Design and Construction: Fatigue Design Assessment – Level 3 Procedure Guidance on Direct Calculations (Notice 1). Lloyd's Register Group Limited, London, UK.

Maruo H. (1970). An improvement of the Slender Body Theory for Oscillating Ships with zero Forward Speed. 19. Bulletin of the Faculty of Engineering, Yokohama National University, pp. 45–66.

Nakos D. (1989). Free surface methods for unsteady forward speed flows, Proc. 4th Int. workshop on water wave and floating bodies, Norway.

Newman J.N. & Sclavounos P. (1980). The unified theory of ship motions. In: Proc. of the 28th Symposium of Naval Hydodynamics, Tokyo.

Sclavounos P. & Nakos D. (1990). Ship motions by Three-dimensional Rankine panel method. 18 th Symposium on Naval Hydrodynamics.

Takai M., Iwashita H. & Xin Lin. (1992). Forces on a ship with forward speed in waves, Proc. Int. Ocean Space and resource utilization seminar and 29th ocean engineering research workshop, Ulsan, Korea.

Wirsching PH & Light MC (1980). Fatigue under wide band random stresses. Journal of the Structural Division. 1980; 106:1593–607.

Maritime Transportation and Harvesting of Sea Resources – Guedes Soares & Teixeira (Eds)
© 2018 Taylor & Francis Group, London, ISBN 978-0-8153-7993-5

Ultimate bending capacity of multi-bay tubular reinforced structures

S. Saad-Eldeen
Centre for Marine Technology and Ocean Engineering (CENTEC), Instituto Superior Técnico, Universidade de Lisboa, Lisbon, Portugal (On leave from the Naval Architecture and Marine Engineering Department, Faculty of Engineering, Port Said University, Port Fouad, Egypt)

Y. Garbatov & C. Guedes Soares
Centre for Marine Technology and Ocean Engineering (CENTEC), Instituto Superior Técnico, Universidade de Lisboa, Lisbon, Portugal

ABSTRACT: The aim of the present analysis is to investigate the effect of different geometrical parameters and imperfections on the ultimate bending moment capacity of multi-bay tubular reinforced structures, subjected to four-point bending, through conducting a series of nonlinear finite element analysis. Three affecting parameters are considered: initial imperfections, shell thickness and local circumferential imperfections. For the first parameter, two cases are studied; the model only with local imperfections or with combined local and global imperfections. The second parameter, deals with six cases with varying shell thicknesses and for the third parameter, different numbers of imperfection half-waves around the cylindrical shell circumferences are considered. The resulting structural responses are presented by moment-curvature relationships, and post-collapse deformed shapes are analysed and several conclusions are derived.

1 INTRODUCTION

Tubular structures, modelled as cylindrical shells, either stiffened or unstiffened are used in many structural applications, such offshore structures. Those structures are subjected to different loading conditions according to their locations, i.e. in the case of submerged structures as a submarine, the cylindrical shells are subjected to external water pressure. For the offshore wind turbine supporting structures, the cylindrical shells are subjected to bending loads and for fixed offshore structures as jacket platform, they are also subjected to axial compressive loading, due to structural weight and impact load as a result of the wave impact or any interaction with other floating objects.

For the external pressure, the concept of using ring stiffeners to increase the strength of cylindrical shells was introduced by Von Mises (1933) who identified that the buckling strength of cylinders vary as a function of the unsupported length.

Kirkpatrick & Holmes (1989) tested thin cylindrical shells under axial compression beyond the critical buckling load and performed finite element simulations of the shell including the pre-test measured imperfections in the shell geometry and asymmetry in the axial load have been carried out.

Kim et al. (2003) investigated numerically the effects of the initial imperfection and the variation of hull thickness on the collapse pressure of stiffened cylindrical shells.

Song et al. (2004) presented a systematic numerical investigation of the nonlinear load carrying behaviour and imperfection sensitivity of cylindrical shells subjected to non-uniform axial load. It was demonstrated that the weld depressions may be considered as the most detrimental imperfection form for cylindrical shells under a partial axial compression.

Houliara & Karamanos (2006) examined the non-linear response of long pressurized thin-walled elastic tubes. The effects of both internal and external pressure, initial curvature, as well as radius-to-thickness ratio were investigated. It was concluded that the response is governed by the strong interaction of the cross-sectional ovalization, which characterizes the non-linear pre-buckling path, and bifurcation instability, in the form of uniform wrinkles along the tube.

Guo et al. (2013) performed a series of bending tests on thin-walled circular tubes with different diameter-to-thickness ratios, to examine the influence of the section slenderness on the inelastic and elastic bending properties. It was observed that the specimens with small diameter-to-thickness ratios failed by extensive plasticity on the central part of the tube. But for a higher diameter-to-thickness ratio, the local buckling phenomena became more pronounced.

Grandez & Netto (2015) studied the effect of structural geometrical parameters as stiffener spacing and thickness of shell plating of reinforced cylindrical shells under external pressure. It was concluded that the geometric radial imperfections between the frames are the most detrimental to the collapse pressure and the importance of the geometric radial imperfection shape is related to the maximum imperfection amplitude.

The aim of the present study is to analyse the effect of different geometrical parameters as initial imperfection combinations, shell thickness and number of half-waves around the circumference of the cylindrical shell on the bending capacity of multi-bays tubular reinforced structures.

2 CONFIGURATION OF TUBULAR STRUCTURE

The analysed structure is defined as a cylindrical shell consisting of five bays, divided by full depth internal reinforcements made of circular plates. The use of a multiple—bay model instead of a single one allows having a realistic structural response by avoiding the effect of the boundary conditions to the centrally located bay, due to the eccentricity of the load and the interference between adjacent bays.

The total length of the model is 5 meters, divided into five bays of length L_1 of 1 meter, as shown in Figure 1. The model is subjected to a four-point bending moment and the load is acting exactly at the two intermediate plates and the boundary conditions at the two ends of the cylindrical shell are simply supported. The shell thickness, t varies from 5 mm to 10 mm, with a constant cylindrical shell radius, R of 500 mm, resulting in a ratio of R/t of 1:2.

The load-carrying capacity of the cylindrical shell is strongly influenced by the shape, distribution and amplitude of geometric initial imperfections. The initial geometrical imperfections of the cylindrical shell are simulated by employing a sinusoidal form, in both axial and circumferential direction of the cylinder as:

$$\omega_{local} = \omega_0 \sin \frac{k_1 \pi z}{L} \cos \left(a \cos \left(\frac{x}{R} \right) \frac{k_2}{2} \right)$$

$$\omega_{Global} = \omega_1 \sin \frac{k_3 \pi z}{L} \tag{1}$$

where ω_0 and ω_1 are the local and global imperfection amplitude, L is the total length of the cylinder, k_1 is the local number of half waves along the cylinder length, k_2 is the number of half waves in the circumferential direction, k_3 is the number of half waves of the global imperfection along the cylinder length, $z \in [0, L]$ and $x \in [-R, R]$.

Figure 2 and Figure show both local and global initial imperfections along the cylinder length as well as the imperfection around the circumference whereas Figure 3 represents the combination between local and global imperfection along the length. The characteristics of the initial imperfections shown in Figure, are as follows, $\omega_0 = 20$ mm, $\omega_1 = 50$ mm, k_1, k_2 and k_3 are 15, 8 and 1, respectively.

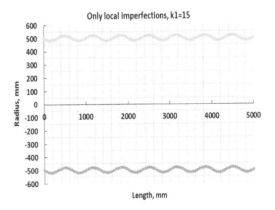

Figure 2. Local initial imperfections along the length.

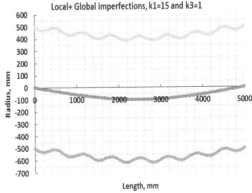

Figure 3. Local and global initial imperfections along the length.

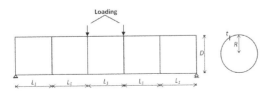

Figure 1. Cylindrical shell under four-point bending.

3 FE MODELLING AND ANALYSIS

In order to estimate the capacity of the cylindrical shell subjected to four-point bending, the nonlinear finite element commercial code ANSYS (2009) has been used, solving the geometric and material nonlinearities.

The shell 181 element, which is suitable for analysing thin to moderately-thick shell structures has been used to generate the entire cylindrical shell model. It is a four-node element with six degrees of freedom at each node. The shell 181 element is well-suited for a linear, large rotation, and/or large strain nonlinear applications.

The cylindrical shell material is mild steel with a yield stress and Young's modulus of 235 MPa and 206 GPa, respectively, modelled by an elastic-perfectly plastic material stress-strain curve.

The implemented boundary conditions for the two ends of the model are simply supported as shown in Figure 5. The load is applied as a nodal force, for a set of nodes located at the intermediate bay reinforcements as shown in Figure 5, simulating the condition of four-point bending, producing compression at the upper shell and tension on the lower shell. Both local and global initial imperfections, as presented in Figure 4, have been generated by ANSYS (2009) by changing the nodal locations without inducing any additional stresses.

The current analysis investigates the effect of different geometrical parameters on the ultimate bending moment capacity of a cylindrical shell, subjected to four-point bending. Three affecting parameters are considered as the initial imperfection combinations, shell thickness and local circumferential imperfection, as given in Table 1.

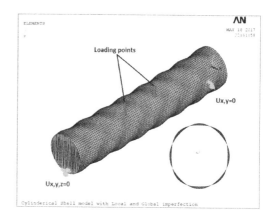

Figure 5. Finite element model of cylindrical shell.

Table 1. Geometrical properties.

	Imperfection	t,	ω_0,	ω_1,	k_1, k_2, k_3
Effect of imperfection combinations	Local		5 mm, 20 mm, 0 mm, 15, 8,0		
	Combined		5 mm, 20 mm, 50 mm, 15, 8,1		
Shell thickness	Combined		5 mm, 20 mm, 50 mm, 15, 8,1		
			6 mm, 20 mm, 50 mm, 15, 8,1		
			7 mm, 20 mm, 50 mm, 15, 8,1		
			8 mm, 20 mm, 50 mm, 15, 8,1		
			9 mm, 20 mm, 50 mm, 15, 8,1		
			10 mm, 20 mm, 50 mm, 15, 8,1		
Local circumferential imperfections	Combined		5 mm, 20 mm, 50 mm, 15, 2,1		
			5 mm, 20 mm, 50 mm, 15, 3,1		
			5 mm, 20 mm, 50 mm, 15, 4,1		
			5 mm, 20 mm, 50 mm, 15, 5,1		
			5 mm, 20 mm, 50 mm, 15, 6,1		
			5 mm, 20 mm, 50 mm, 15, 7,1		
			5 mm, 20 mm, 50 mm, 15, 8,1		

For the first parameter, two cases are studied; the model with only local imperfections or with combined local and global imperfections.

The second parameter is related to the shell thickness, where six cases are studied. For the third parameter, different numbers of imperfection half-waves around the cylindrical shell circumference are considered.

For that purpose, an element size of 50 mm is considered as the most appropriate element size for the current model, resulting in a mesh density of 10,756 elements.

3.1 Effect of imperfections

Two types of initial imperfections are applied, the first one is the local imperfections around the cylinder circumference and along the cylinder length,

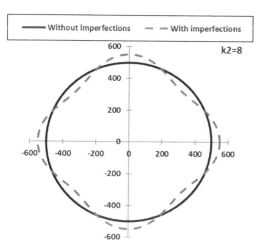

Figure 4. Local initial imperfections along the circumference.

where the second one is the global imperfection along the cylinder length. For the local imperfections, see Figure 2 and Figure 4, the amplitude ω is 20 mm with the number of half waves along the cylinder length k1 of 15 and with the number of half waves around the circumference, k2 = 8.

For the combined imperfections, the global imperfection along the length with an amplitude ω_1 of 50 mm with the number of half waves along the cylinder length k3 = 1, as shown in Figure 3 is considered, in addition to the local ones in both along the circumference and the length of cylinder. The moment-curvature relationships for the two initial imperfection combinations are shown in Figure 6 and the ultimate bending moment capacities, UBM are given in Table 2. Figure 6 shows that both curves are coinciding with each other in all loading regimes, with a slightly higher ultimate bending capacity of the one with the local imperfections than the one with combined imperfections of 0.23%, see Table 2.

Therefore, it may be concluded that for the considered imperfection amplitudes and the number of half waves, both combinations give almost the same ultimate bending moment capacity and the model with combined imperfections is more appropriate to be used. The Von Mises stress distributions for both imperfection combinations are shown in Figure 7. It is visible that both combi-

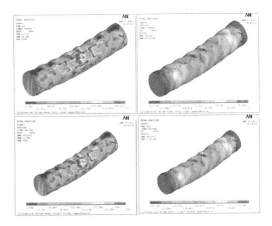

Figure 7. Von Mises stress (left) and vertical displacement (right).

nations react in the same way along the cylinder length, especially in the central bay. The same behaviour is also observed from the vertical deformations, where a slight difference is registered.

3.2 Shell thickness

The effect of different shell thicknesses on the ultimate bending moment capacity of the cylindrical model is investigated here. The shell thickness varies from 5 mm to 10 mm with an interval of 1 mm, as given in Table 1, taking into consideration the combined initial imperfections mode. The structural response represented by the moment-curvature relationship is presented in Figure 8. As may be seen, with increasing the shell thickness, the ultimate bending capacity increases, in addition to a visible increase of the flexural rigidity. Up to 7 mm, the global behaviour of cylinder is similar in all loading regimes, which is the contrary of the thicknesses bigger than 7 mm, where the structure reaches its ultimate capacity without post-collapse discharge. This gives an indication that the shell thickness is one of the most important controlling parameter in both local and global response of the cylindrical shell model.

Based on the ultimate bending capacity of each shell thickness given in Table 3, a relationship between cylinder radius/shell thickness ratio R/t, and the ultimate bending moment is shown in Figure 9. It is seen that by decreasing the R/t ratio, the ultimate bending moment increases nonlinearly.

The Von Mises stress distributions for the shell thicknesses of 5 mm and 9 mm are shown in Figure 10 and Figure 11. It is clear that the spread of the highly-stressed locations in the central and the adjacent bays for a 9 mm shell thickness is big-

Table 2. Effect of different imperfection combinations.

	UBM, N.mm	Difference %
With local imperfections	6.41E + 08	
With combined imperfections	6.39E + 08	0.23%

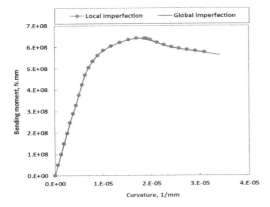

Figure 6. Moment-curvature relationship.

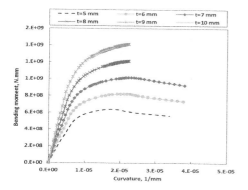

Figure 8.   Moment-curvature relationship.

Table 3.   Effect of different shell thicknesses.

t, mm	R, mm	R/t	UBM, N.mm
5	500.00	100.00	6.39E + 08
6	500.00	83.33	8.21E + 08
7	500.00	71.43	1.01E + 09
8	500.00	62.50	1.21E + 09
9	500.00	55.56	1.41E + 09
10	500.00	50.00	1.37E + 09

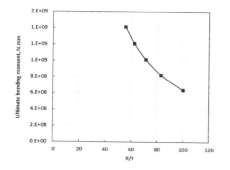

Figure 9. Relationship between ultimate bending moment and R/t.

ger than the one of 5 mm. Also, the middle of the adjacent bays to the central one is highly stressed, where the ones for 5 mm are less stressed. This confirms the domination of the shell thickness on the global response and the contribution of the adjacent bays to the central one with respect to both global and local responses as in Figure 11.

On the contrary, for a shell thickness of 5 mm, (Figure 10), the central bay is highly stressed and the adjacent bays are less stressed, especially in the middle part, which confirms the domination of

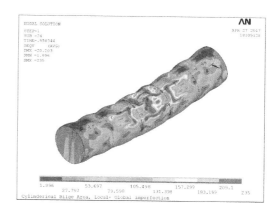

Figure 10.   Von Mises stress distributions.

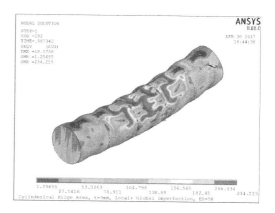

Figure 11.   Vertical deformation.

the initial imperfections on the structural response rather than the shell thickness.

3.3   Local circumferential imperfections

The effect of different initial imperfection half-wave numbers around the cylinder circumferential, k_2, on the bending capacity is analysed here. The k_2 varies from 2 to 8 with an interval of 1 and the shapes of the cylinder with and without initial imperfections for different k_2 values are presented in Figure 12. The results of the moment-curvature relationships for different k_2 for a shell thickness of 5 mm and combined local and global imperfections are given in Figure 13 to Figure 16.

For even k_2 numbers of 2, 4, 6 and 8, the moment-curvature relationships are presented in Figure 13, which shows that for lower k_2, the behaviour of the cylinder becomes stiff with a higher bending capacity. With increasing k_2 from 2 to 4 and 6, the ultimate bending capacity decreases, but the flexural rigidity is the same in the elastic zone, with a deviation in the buckling and post-buckling regimes, especially for k_2 of 6, for which a discharge of the capacity occurs in the post-collapse regime. For higher k_2 of 8, the behaviour of the cylindrical shell is different in all loading regimes, represented by a higher reduction of the flexural rigidity, deviating from the other k_2 values. Also, the post-collapse regime is much softer with respect to other configurations.

The reason for that may be because both k_2 of 2 and 4 create a shape of imperfection that resists the applied bending and forces the cylindrical shell not to deform and prevents the occurrence of earlier buckling, see Figure 12, up-middle, left, which is the contrary to the ones for k_2 of 6 and 8.

The Von Mises stress distributions at the ultimate load for k_2 of 4 and 6 are shown in Figure 14 (up), in which the deformed shape in the central bay is different, where for k_2 of 4, the imperfection creates a pre-deformed shape in the upper shell and out of plan pre-deformation in the two side shells, which results in excessive downward deformation of the upper shell with increasing the bending loading. On the other hand, for k_2 of 6, the initial imperfection shape, Figure 12 (left, down),

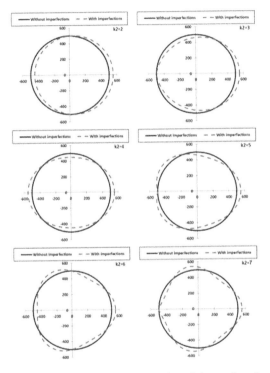

Figure 12. Imperfection as a function of the number of half-waves around the circumference of cylinder.

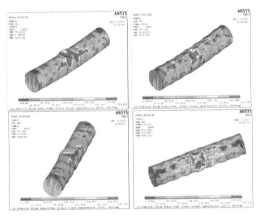

Figure 14. Von Mises stress distributions for different k_2.

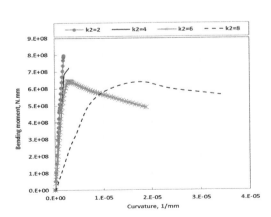

Figure 13. Moment-curvature relationship.

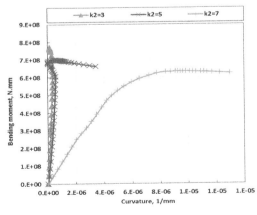

Figure 15. Moment-curvature relationship.

creates a pre-deformed shape at the upper-right quarter of the cylindrical shells, which forces the shell to deform asymmetrically.

For the odd k_2 values, the moment-curvature relationships are presented in Figure 15. The general observation is that with increasing the k_2 values, the ultimate bending capacity decreases and for higher values i.e. $k_2 = 7$, the behaviour of the cylindrical shell is much soft than other odd values 2 and 5. For $k_2 = 5$, it is visible that at the post-buckling regime, the curvature of the cylindrical shell decreases with increasing the bending moment, after that the curvature increases until reaching the ultimate bending capacity, then the post-collapse occurs.

This may be explained by an initial shape of the imperfection at the bottom shell, where the peak of the imperfection is not exactly at the middle, (see Figure 12, middle, right), therefore, after reaching the buckling, more resistance of the cylindrical shell occurs, which results in a change of the deformed shape and then the imperfection dominates the final collapse mode, represented by an increase of the curvature.

For $k_2 = 3$, from the beginning, the initial imperfection forced the cylindrical shell to buckle near the right-side, which creates an in-plane curvature that decreases as the bending load increases. Therefore, the case with $k_2 = 3$ shows the higher bending capacity than k_2 of 5 and 7, respectively as given in Table 4 and shown in Figure 15.

The Von Mises stress distributions for k_2 of 3 and 7 are presented in Figure 14 (up), where the initial out of plane deformation of the right-side shell propagates with increasing bending load, excessive in-plane deformation occurs at the left side.

This creates locations with plastic deformation, which forces the cylindrical shell to change the post-buckling deformation shape, represented by a decrease of the curvature, as shown in Figure 15. In the case when $k_2 = 7$, the initial imperfection shapes becomes more irregular, which facilitate the occurrence of earlier buckling and the spread of locations whit high stresses, gives the structures more freedom to deform softly.

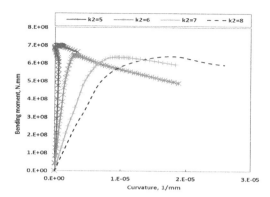

Figure 16.   Moment-curvature relationship.

From k_2 of 5 to 8 the response of the cylindrical shell behaves in a soft way with a gradual reduction of both flexural rigidity and ultimate bending capacity. This may be explained by the complex initial imperfection shapes in Figure 12, which create locations from which the buckling may initiate and then the imperfections dominate the global response of the cylindrical shell as the number of half waves k_2 increases.

4   CONCLUSIONS

The ultimate bending capacity of a multi-bay tubular reinforced structure has been investigated numerically, through varying different geometrical parameters.

Based on the performed analysis, it was observed that for a thickness bigger than 7 mm, the structure reaches its ultimate capacity without a post-collapse discharge, even with the presence of a combined initial imperfection. This gives an indication that the higher shell thickness is one of the most important controlling parameter in both local and global response of the cylindrical shell.

With decreasing the R/t ratio, the ultimate bending moment increases nonlinearly.

For lower even numbers of half waves around the imperfection circumference, the response of the structure is stiff, without a post-collapse regime. For higher even numbers, the behaviour of the cylindrical shell is different in all loading regimes, represented by a higher reduction of the flexural rigidity, with a much softer post-collapse regime.

The number of half waves around the circumference has a significant effect on both flexural rigidity, ultimate load carrying capacity as well as the final deformation shape, which in some cases in the key parameter.

Table 4.   Ultimate bending moment as a function of k_2.

	k_2	UBM, N.mm
Even	2	7.96E + 08
	4	7.23E + 08
	6	6.43E + 08
	8	6.39E + 08
Odd	3	7.71E + 08
	5	7.00E + 08
	7	6.34E + 08

ACKNOWLEDGEMENTS

The first author has been funded by the Portuguese Foundation for Science and Technology (Fundação para a Ciência e Tecnologia—FCT) under contract SFRH/BPD/84823/2012.

REFERENCES

Ansys 2009. *Advanced Analysis Techniques Guide,* Southpointe, 275 Technology Drive, Canonsburg, PA 15317, Ansys, Inc.

Grandez, E.V. & Netto, T.A. 2015. Influence of geometric parameters on the structural strength of reinforced cylindrical shells under external hydrostatic pressure. *In:* Guedes Soares, C. & Santos, T. a. R. (eds.) *Maritime Technology and Engineering.* Taylor & Francis Group, London, UK, 475–484.

Guo, L., Yang, S. & Jiao, H. 2013. Behavior of thinwalled circular hollow section tubes subjected to bending. *Thin-Walled Structures,* 73, 281–289.

Houliara, S. & Karamanos, S.A. 2006. Buckling and post-buckling of long pressurized elastic thin-walled tubes under in-plane bending. *International Journal of Non-Linear Mechanics,* 41, 491–511.

Kim, K., Kim, U. & Park, J. 2003. A Study on Effects of Initial Deflection on Ultimate Strength of Ringstiffened Cylindrical Structure under External Hydrostatic Pressure. *The Thirteenth (2003) International Offshore and Polar Engineering Conference.* Honolulu, Hawaii, USA: The International Society of Offshore and Polar Engineers.

Kirkpatrick, S.W. & Holmes, B.S. 1989. Axial buckling of a thin cylindrical shell: experiments and calculations. *The American Society of Mechanical Engineers, Computational Experiments,* 176, 67–74.

Song, C.Y., Teng, J.G. & Rotter, J.M. 2004. Imperfection sensitivity of thin elastic cylindrical shells subject to partial axial compression. *International Journal of Solids and Structures,* 41, 7155–7180.

Von Mises, R. 1933. The Critical External Pressure of Cylindrical Tubes Under Uniform Radial and Axial Load," Experimental Model Basin". Berlin.

Maritime Transportation and Harvesting of Sea Resources – Guedes Soares & Teixeira (Eds)
© 2018 Taylor & Francis Group, London, ISBN 978-0-8153-7993-5

Fatigue life prediction method of fairleads installed on a FPSO

J. Yue & P. Chen
Wuhan University of Technology, Wuhan, China

Y. Liu
The University of Tokyo School of Engineering, Tokyo, Japan

W. Mao
Chalmers University of Technology, Goteborg, Sweden

ABSTRACT: In this paper, fatigue strength analysis is carried out for the fairleads of a FPSO served in the South Sea in China. Hydrodynamic performance of the FPSO coupling with its mooring system is firstly obtained, and then the short-term history of cable tension can be obtained by wave scatter diagram. The cable tension is regarded as the dynamic loading acted on the fairleads; and the corresponding history of stress range in the fairleads can be obtained by Finite Element Analysis (FEA). At last, fatigue accumulating damage in the fairleads is calculated based on the rain-flow counting method and S-N curves. The estimation framework based on the ship motion and mooring tension coupling analysis in time domain proposed in this work can provide a fast fatigue life prediction of the fairleads when the ship is operated in different conditions.

1 INTRODUCTION

In recent years, with the increase of exploration reserves of marine oil and gas fields, the oil industry gradually developed from shallow water to deep sea. In process of oil exploitation in deep water, the fixed platform, which is suitable for shallow sea, is unable to meet the requirements of deep water oil production platform because of some problems such as high cost, design difficulty and low efficiency. Fairlead is an important part of mooring system, which can change the direction of the mooring line and maintain mooring state of the ship or platform. The damage of the fairlead easily changes mooring state of floating structure which can lead to great accident, especially in the case of typhoon or storms.

The load applied on fairlead mainly come from mooring lines. Floating structure in ocean is offered by wind, current and wave, which can force motion on the structure. Most of this motion is limited by mooring line and part of load of mooring lines is transformed to fairlead. These loads not only cause fatigue failure on mooring line but also on fairlead, which is one of locations that fatigue always happen.

For FPSO, scholars and organizations take most attention on the floating motion and cable tension. For the moored FPSO, traditional separated approach of calculation floating motion

and dynamic response of moorings is inaccurate for floating structures operating in deep waters (Ormberg and Larsen, 1989). To better evaluate the dynamic motion and cable tension of FPSO in deep water, it is necessary to use the dynamic coupling analysis method. Based on the dynamic coupling analysis method, by using the asynchronous coupling algorithm, Ma & Duan (2014) developed the dynamic coupled program between platform and mooring. In order to check the code, they used the software package *AQWA* to analyze the single point moored FPSO and compared the results with program, which proved the program numerical results had good agreement. Up to now, for FPSO or other Floating, the combination method of numerical analysis and pool test is regarded as the most effective way to solve the characteristics of mooring systems. Based on it, Munipalli *et al.* (2007) studied the instability of FPSO single point system. The results show that the effect that the influence of nonlinear drift force acting on the hull structure is very obvious.

In addition to floating motion and cable tension, scholars and organizations will also focus on the fatigue problem of FPSO systems. During their lifetime, FPSO systems are subjected to sea wave induced random loads that may lead to fatigue failure and several studies have dealt with the fatigue problem of FPSOs (Zhao *et al.*, 2001; Jiang *et al.*, 2003; Sun and Yong, 2003; Sun and Soares, 2003).

Spanos et al. (2003) focused on the fatigue life estimation of pipelines conveying fluid located on the deck of FPSOs, since their repair or replacement will imply shutdown of the production. After that, the fatigue damage assessment of topside facilities became an important issue. Wu et al. (2016) studied the fatigue problem of soft yoke mooring system based on monitoring technique, which was a type of the single point mooring system. The soft yoke mooring system established the mooring functions via the multi-dynamic mechanism of thirteen hinge joints and the damage failure of hinge joints would cause great financial loss.

The fairlead can be seen as the bearing roller, and the function is similar to hinge joints, so the fatigue life of the fairlead is important to the mooring system. However, studies of the fairlead installed on a FPSO single point mooring system is less. In 2010, fairlead which was installed on a FPSO was found to be damaged (McKeown et al, 2010), this is believed to be the first time that a fairlead has been changed-out in situ on any floating installation in the North Sea. "Position mooring" contains regulations of the material steel grade, roller diameter and designing load in mooring conditions.it requires that fairlead roller need choose high strength steel, and working life is at least 20 years, but there are no direct calculation guidelines for fairlead design and check (DNV GL OS-E301, 2015). Therefore, the accuracy of fatigue life of the fairlead is important to ensure the safety of mooring systems.

In this paper, a fast fatigue life prediction method of the fairlead is proposed. Firstly, the cable tension of the FPSO in the South China Sea can be obtained by time domain coupling method under the environmental condition with maximum wind, waves, and associated currents based on a 100-year recurrence interval, which was considered as the external loads of fairlead. Then, based on FEA, the structural strength of the fairlead can be conducted. Then, the relationship between cable tension and fairlead stress can be established, which is used to obtain the fairlead stress time history. Finally, based on the rain-flow counting method and S-N curves, the fatigue life of the fairlead can be predicted.

2 COMPUTATIONAL THEORY

2.1 Coupling theory in time domain

Floating structure moored in the ocean is affected by the ocean environmental load which contains wave loads, wind loads and current loads. The loads from wind and current are estimated using model tests from Oil Companies International Marine Forum curves (OCIMF, 1994) advantage of the formula is that the wind and the current force coefficients are from model tests and has relatively good prediction accuracy. According to the three-dimensional potential flow theory (Newman, 1997), hydrodynamic parameters, wave load transfer function, additional mass and amplitude response operator etc can be calculated, so the time domain wave loads can be obtained by Cummins impulse theory (Cummins, 1962). In addition to the effect of ocean environmental loads, the floating body will be affected by the restoring force provided by the mooring system. In deep water, mooring lines become more flexible, the dynamic effect is more pronounced under the motions of platform. Based on the finite element method (Ma, 2009), by solving mooring cable dynamic differential equation, the mooring force can be obtained and the dynamic effect can be considered.

Therefore, on the basis of known hydrodynamic coefficients, wave loads, wind loads and mooring restoring forces, the time domain coupling equations of mooring system (Garrett et al., 2002) can be expressed as follows:

$$\left[M + A(\infty) \right] \ddot{x}(t) + \int_0^t h(t - \tau) \dot{x}(\tau) d\tau + Kx(t) = F \quad (1)$$

where, M is the mass matrix; is the additional mass matrix with infinite frequency; K is the hydrostatic restoring matrix; $\int_0^t h(t - \tau) \dot{x}(\tau) d\tau$ is the delay function; F is the force vector, including wave force, air force and mooring restraint.

2.2 Fatigue analysis

The fatigue design is based on use of S-N curves, which are obtained from fatigue tests. The design S-N curves which follows are based on the mean-minus-two-standard-deviation curves for relevant experimental data. The S-N curves are thus associated with a 97.7% probability of survival. The basic design S-N curve is given as:

$$\lg N = \lg A - m \lg S \quad (2)$$

where, N is the predicted number of cycles to failure for stress range S; S is the stress range with unit MPa; m is the negative inverse slope of S-N curve; lg A is the intercept of log N-axis by S-N curve.

The fatigue life may be calculated based on the S-N fatigue approach under the assumption of linear cumulative damage (Palmgren-Miner rule).

According to the Palmgren-Miner rule (Singh and Ahmad, 2015), it is considered that the accumulated fatigue damage D of the structure under the action of the constant amplitude alternating stress is the sum of the damage degree D_i in the each stress range. Following the Miner criterion, the damage degree D_i on a certain level of stress range is equal to the ratio between the actual number n_i of cycle in the stress range and the required cycle number N_i of structure under the action of stress reaching failure. If the stress range levels have k levels, there is:

$$D = \sum_{i=1}^{k} D_i = \sum_{i=1}^{k} \frac{n_i}{N_i} \tag{3}$$

So the fatigue life T is:

$$T = \frac{1}{D} \tag{4}$$

where, D is the accumulated fatigue damage; k is the number of stress range levels; D_i is the fatigue damage degree; n_i is the number of stress cycles; N_i is the number of cycles to failure at constant stress range; T is the fatigue life.

3 TIME DOMAIN COUPLING ANALYSIS OF THE FPSO

3.1 Parameters of FPSO

The structure of FPSO is single deck, double bottom and side, no bulbous bow, no forecastle, and no propulsion. 12 point mooring positions are selected in the bow of the ship. The ship works in the South China Sea where the depth of water is 1000 meters and requirement of the fatigue life is 20 years. Main parameters are shown in Table 1.

3.2 FPSO single point mooring system

The FPSO owns internal turret mooring system and the mooring radius is 2200 m. The mooring system includes three groups mooring lines, each group is consisted of four lines and the angle between groups is 120 degree. The mooring line is divided into three parts, the part close to the hull

Table 1. Principal dimension.

Project	Data	Units
Total length	268	m
Waterline length	262.5	m
Molded breadth	48	m
Molded depth	26	m
Designed load draft	16.2	m
Design draft displacement	188843	t
Design draft weight	150025	t

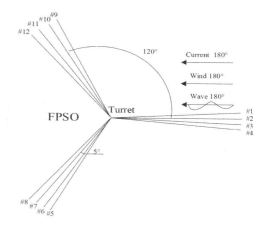

Figure 1. Mooring arrangement diagram.

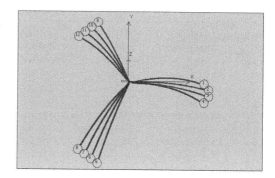

Figure 2. Single point mooring system model.

Table 2. Mooring cable parameters.

Position	Ingredients	Length m	Diameter mm	Breaking strength N	Wet weight kg/m
Upper	Steel	200	146	1.89E+07	370
Middle	Polyester	2040	270	1.82E+07	10.8
Distal	Steel	220	146	1.89E+07	370

and seabed are using the steel chain, intermediate part use polyester line and the angle between mooring lines is 5 degree at the same group (Fig. 1, Fig. 2, Table 2).

3.3 Environment condition

According to the ABS specification (ABS, 2014), the design environmental condition for a single point system design is defined as the environmental condition with maximum wind, waves, and associated currents based on a 100-year recurrence interval. The FPSO operates in the South China Sea and the environment condition is selected (Table 3).

Table 3. Environment condition.

Environment	Attribute	
Wave	Wave spectrum	JONSWAP
	Spectrum peak factor γ	2.4
	Significant wave height Hs	13.3 m
	Peak period Tp	15 s
	Direction	180°
	Initial frequency	0.1 rad/s
	Terminal frequency	1.5 rad/s
Wind	Steady wind	45 m/s
	Direction	180°
Current	Water surface velocity	1.5 m/s
	Direction	180°

Table 4. The values of FPSO cable tension.

Type	Average	Minimum	Maximum
#1	2.582E+06	3.241E+06	3.800E+06
#2	8.488E+06	7.808E+06	9.059E+06
#3	8.488E+06	7.808E+06	9.060E+06
#4	3.241E+06	2.581E+06	3.800E+06
#5	7.245E+06	6.948E+06	7.521E+06
#6	5.373E+06	5.026E+06	5.709E+06
#7	4.516E+06	4.122E+06	4.912E+06
#8	4.695E+06	4.255E+06	5.147E+06
#9	7.245E+06	6.948E+06	7.523E+06
#10	5.373E+06	5.026E+06	5.712E+06
#11	4.516E+06	4.122E+06	4.914E+06
#12	4.695E+06	4.256E+06	5.150E+06

Table 5. The values of FPSO motion.

Project	Average	Minimum	Maximum
Surge (m)	−82.19	−96.75	−64.97
Sway (m)	0	−0.025	0.028
Heave (m)	−3.70	−8.84	1.20
Roll (deg)	0	−0.0568	0.051
Pitch (deg)	0.896	−4.812	6.421
Yaw (deg)	0	−0.0276	0.0247

3.4 *Results analysis*

Under the environment condition, values of the FPSO cable tension and motion response are shown in Table 4 and Table 5.

Under the condition of the first wave, Surge is the major motion of the FPSO. In Table 4, the maximum Surge value of the FPSO is 96.75 m, which is 9.675% of water depth. The maximum tension of mooring lines is an important parameter to the FPSO mooring system. In Table 5, the maximum cable tension is 9.059E+06 N, the cables are #2 and #3, and the safely factor is 2.01. According to the API specifi-

cation (API, 2005), maximum surge value cannot exceed 10% of water depth under working status, and the minimum mooring line dynamic safety factor of 1.67 when the numerical analysis were used. So the results show that the design of the FPSO mooring system is safe and reliable.

4 STRUCTUAL STRENGTH ANALYSIS OF THE FAIRLEAD

4.1 *Finite element model*

The height and length of rotary fairlead is 3.87 m and 3.68 m respectively and its thickness is 30 mm / 40 mm. The contact part of the fairlead and the mooring cable is a roller which is connected with the circular shaft, and the force is transmitted to the fairlead. The schematic diagram of the mooring system fairlead is shown in Figure 3. In order to simplify the calculation, the calculation model ignored the roller, and the load will be directly applied to the shaft. The finite element model is shown in Figure 4. When the fairlead is connected with the FPSO contact part, the load is directly loaded on the circular shaft.

4.2 *Results analysis*

In the process of changing the direction of mooring cable, the fairlead bears the pressure from mooring cable. Considering the extreme design condition of FPSO, maximum tension of #2 mooring cable was selected as external load, so the strength of fairlead is analyzed. According to the DNV GL specification (DNV GL OS-E301, 2015), the mooring cables angle of the fairlead for combined chain is from 120 degree to 150 degree, and the most unfavorable direction should be considered during strength check. In Figure 5, #2

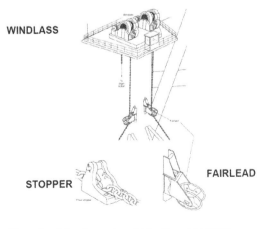

Figure 3. Mooring system fairlead's arrangement.

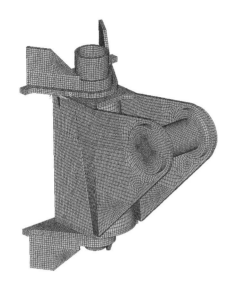

Figure 4. Fairlead finite element model.

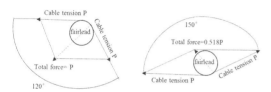

Figure 5. The force diagram of fairlead.

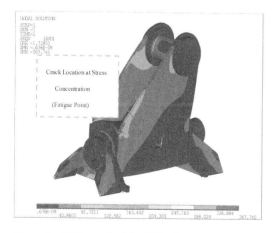

Figure 6. FEA result of the fairlead.

cable tension is supposed as P, the resultant force on fairlead is P when the angle is 120 degree, and the resultant force on fairlead is 0.518P when the angle is 150 degree, so the angle between fairlead has a great influence on the load.

Based on safety consideration, the angle between fairlead is thought to be 120 degree. The fairlead

stress is obtained by FEA through software *Ansys* and the result is shown in Figure 6, the maximum stress is 367.745 MPa. The material of the fairlead is low alloy high strength steel Q420, and the yield strength is 420 MPa. The FEA indicates that the structural strength of fairlead meets the requirements. The maximum stress point occurs at the intersection of the side plate and the front axle, which is selected as the fatigue point.

5 FAST PREDICTION METHOD OF THE FAIRLEAD FATIGUE LIFE

5.1 Calculation illustrates

In evaluation of fatigue life, long-term state of wave can be considered as composed of several short-term sea state sequences. Each short-term sea state can be described by parameters as wave height, period, spectrum, velocity, wind speed and so on, and each short-term sea state has corresponding probability of occurrence.

The FPSO long term operations in the South China Sea S4 area (Table 6).

Based on it, the long-term sea state is discretized into 64 kinds of short-term sea states (Table 7). For example, when H_s belongs to 0–0.5 m, the average value 0.25 m is taken as the representative value of it and the probability of occurrence can be obtained by the T_z. The wave spectrum of each short-term sea state is simulated by JONSWAP spectrum and the spectral wind factor γ is 3.3. The numerical simulation adopts steady wind and current. The direction of wind, wave and current is 180 degree.

Based on the coupling calculation method of time domain, the time histories of #2 cable tension above 64 kinds of short-term sea states can be calculated. Based on FEA, the relationship between #2 cable tension and the fairlead stress can be established, the relationship is shown in Figure 7.

Therefore, by time domain coupled analysis and FEA, time histories of fairlead stress on the condition of 64 kinds of short-term sea states can be calculated quickly.

5.2 Time domain fatigue analysis

Typical #2 cable tension time history and fairlead stress time history are shown in Figure 8 and Figure 9 for sea states 1 and 64 respectively. The black line represents the #2 cable tension time history and the red line represents the fairlead stress history. They have the same trend and only values are different.

Compared the Figure 8 with the Figure 9, the variation range of the stress amplitude (or the cable tension amplitude) is large when the environment condition is harsh, and the fatigue damage will be high.

Table 6. Wave dispersion pattern of S4 sea area.

| H_S | Average Zero-crossing Period Tz(s) | | | | | | | |
	<3.54	5	6	7	8	9	10	11	Total
0–0.5	1	10	29	29	15	5	1		90
0.5–0.85		8	31	37	21	8	2	1	107
0.85–1.25		7	34	50	33	13	4	1	142
1.25–1.85		5	37	72	57	25	8	2	206
1.85–2.5	1	19	55	57	30	10	2		174
2.5–3.25			7	32	46	31	12	3	1	132
3.25–4			1	11	24	2	10	3	1	52
4–5			1	2	12	17	10	3	1	45
5–6				2	6	6	2	1	17
6–7.5						2	3	2	7
Total	1	31	159	288	267	139	66	19	4	974

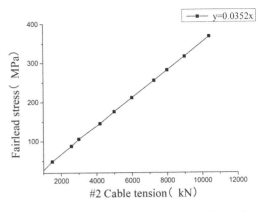

Figure 7. The relationship between #2 cable tension and fairlead stress.

Table 7. 64 kinds of short-term sea states.

Sea State	Hs m	Tz S	P %	Sea State	Hs m	Tz S	P %
1	0.25	3.5	0.1	33	2.175	8	3.08
2	0.25	4	1.03	34	2.175	9	1.03
3	0.25	5	2.98	35	2.175	10	0.21
4	0.25	6	2.98	36	2.875	5	0.72
5	0.25	7	1.54	37	2.875	6	3.29
6	0.25	8	0.51	38	2.875	7	4.73
7	0.25	9	0.1	39	2.875	8	3.19
8	0.675	4	0.82	40	2.875	9	1.23
9	0.675	5	3.19	41	2.875	10	0.31
10	0.675	6	3.81	42	2.875	11	0.1
11	0.675	7	2.16	43	3.625	5	0.1
12	0.675	8	0.82	44	3.625	6	1.13
13	0.675	9	0.21	45	3.625	7	2.47
14	0.675	10	0.1	46	3.625	8	0.21
15	1.05	4	0.72	47	3.625	9	1.03
16	1.05	5	3.49	48	3.625	10	0.31
17	1.05	6	5.14	49	3.625	11	0.1
18	1.05	7	3.39	50	4.5	5	0.1
19	1.05	8	1.34	51	4.5	6	0.21
20	1.05	9	0.41	52	4.5	7	1.23
21	1.05	10	0.1	53	4.5	8	1.75
22	1.55	4	0.51	54	4.5	9	1.03
23	1.55	5	3.8	55	4.5	10	0.31
24	1.55	6	7.4	56	4.5	11	0.1
25	1.55	7	5.86	57	5.5	7	0.21
26	1.55	8	2.57	58	5.5	8	0.62
27	1.55	9	0.82	59	5.5	9	0.62
28	1.55	10	0.21	60	5.5	10	0.21
29	2.175	4	0.1	61	5.5	11	0.1
30	2.175	5	1.95	62	6.75	8	0.21
31	2.175	6	5.65	63	6.75	9	0.31
32	2.175	7	5.86	64	6.75	10	0.21

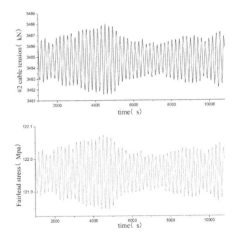

Figure 8. #2 Cable tension time history and fairlead stress time history for sea state-1.

When time histories of fairlead stress is obtained, based on the rain-flow counting method, the mean stress, the variation of stress amplitude and number of cycles can be obtained, which is

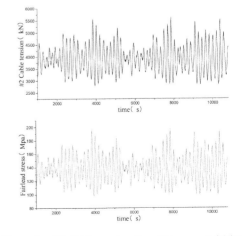

Figure 9. #2 Cable tension time history and fairlead stress time history for sea state-64.

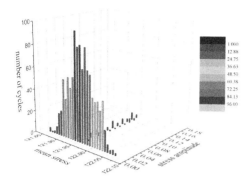

Figure 10. The rain flow histogram for sea state-1.

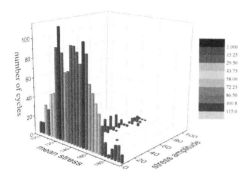

Figure 11. The rain flow histogram for sea state-64.

called the rain flow histogram or the fatigue load spectrum. The rain flow histogram for sea state 1 and 64 is shown in the Figures 10 and 11. Form the 3D histogram, number of cycle is added for a particular stress range with different mean stress.

In Figure 10, the maximum stress amplitude is 0.161 MPa, for the number of 3 cycles and mean stress of 121.955 MPa. The maximum number of cycles is 96, for the stress amplitude of 0.007 MPa and mean stress of 121.94 MPa.

In Figure 11, the maximum stress amplitude is 102.934 MPa, for the number of 2 cycles and mean stress of 142.328 MPa, the number of 1 cycle and mean stress of 146.01 MPa, the number of 1 cycle and mean stress of 149.69 MPa. The maximum number of cycles is 115, for the stress amplitude of 4.474 MPa and mean stress of 120.24 MPa.

For fairleads, when the high stress amplitude of cycles is occurred during their lifetime, the fatigue damage will be very large, although the number of cycles is very small.

5.3 Stress concentration factor and Goodman's relation

On the basis of DNV GL specification (DNV GL RP-C203,2016), the joint classification and corre-

sponding S-N curves takes into account the local stress concentrations created by the joints themselves and by the weld profile. The fatigue design stress can therefore be regarded as the nominal stress, adjacent to the weld under consideration. However, if the joint is situated in a region of stress concentration resulting from the gross shape of the structure, this must be taken into account.

According to Figure 6, the fatigue point of the fairlead is situated in the region of stress concentration, so the stress concentration factor (SCF) should be considered when fatigue analysis is performed. Thus the local stress is derived as:

$$\sigma_{local} = SCF \sigma_{nominal} \tag{5}$$

where, σ_{local} is the fatigue design stress; $\sigma_{nominal}$ is obtained by the FEA to the fairlead.

The σ_{local} shall be used together with the relevant S-N curves D through G. The D-curve is the reference curve (DNV GL RP-C203, 2016).

The crack location can be seen as the web-stiffened cruciform joint. According to the DNV GL specification (DNV GL CG-0129, 2015), the SCF is 1.12 for this paper.

Since the stress cycle are not completely reversed, so the effect of the mean stress value should be accounted for. For that purpose, the Goodman's relation to estimate mean stress effect on fatigue life is used. The Goodman's relation is derived as:

$$\frac{S_a}{S_{NF}} + \frac{S_m}{S_u} = 1 \tag{6}$$

where, S_a is the stress amplitude; S_{NF} is the effective stress at failure for a lifetime of NF cycles; S_m is the mean stress and S_u is the ultimate stress.

5.4 Result analysis

Based on the fatigue load spectrum and S-N curve, the fatigue damage of the fairlead under 64 kinds of short-term sea states can be calculated, and the result is shown in Table 8. The fatigue damage in the Table 8 is obtained based on the time domain coupling analysis for 3 hours, and there are 2920 numbers of 3 hours in a year. Therefore, the total fatigue damage D_p is 2920 times of the calculation results of Table 8.

According to the Table 8, most of the fatigue damage is caused by sea states 51 to 64, and the most damaging fatigue sea state is sea state 64 with Hs = 6.75 m and Tp = 10 s. The total fatigue damage D of the fairlead within one year is 0.054, and the fatigue life of the fairlead is 18.5.

Based on the DNV-GL specification (DNV GL OS-E301, 2015), the fatigue life of the fairlead is at least 20 years. Therefore, if the fairlead installed on

Table 8. Fatigue damage of the fairlead.

Sea state	Di	P %	DP	Sea state	Di	P %	DP
1	4.88E-14	0.1	4.88E-17	33	2.00E-07	3.08	6.16E-09
2	7.26E-14	1.03	7.48E-16	34	1.14E-07	1.03	1.18E-09
3	5.75E-13	2.98	1.71E-14	35	2.06E-07	0.21	4.33E-10
4	1.04E-12	2.98	3.09E-14	36	2.87E-07	0.72	2.07E-09
5	4.18E-13	1.54	6.44E-15	37	6.03E-07	3.29	1.98E-08
6	1.32E-12	0.51	6.72E-15	38	4.31E-07	4.73	2.04E-08
7	1.22E-12	0.1	1.22E-15	39	1.08E-06	3.19	3.43E-08
8	8.61E-11	0.82	7.06E-13	40	6.68E-07	1.23	8.21E-09
9	7.13E-11	3.19	2.27E-12	41	1.07E-06	0.31	3.31E-09
10	5.69E-11	3.81	2.17E-12	42	2.54E-06	0.1	2.54E-09
11	7.00E-11	2.16	1.51E-12	43	1.54E-06	0.1	1.54E-09
12	4.72E-10	0.82	3.87E-12	44	3.17E-06	1.13	3.59E-08
13	2.09E-10	0.21	4.40E-13	45	2.22E-06	2.47	5.47E-08
14	5.77E-10	0.1	5.77E-13	46	6.30E-06	0.21	1.32E-08
15	8.80E-10	0.72	6.34E-12	47	5.32E-06	1.03	5.48E-08
16	1.68E-09	3.49	5.85E-11	48	5.23E-06	0.31	1.62E-08
17	6.24E-10	5.14	3.21E-11	49	1.60E-05	0.1	1.60E-08
18	1.21E-09	3.39	4.10E-11	50	2.26E-06	0.1	2.26E-09
19	3.58E-09	1.34	4.80E-11	51	2.11E-05	0.21	4.43E-08
20	2.13E-09	0.41	8.72E-12	52	1.77E-05	1.23	2.17E-07
21	6.55E-09	0.1	6.55E-12	53	5.04E-05	1.75	8.82E-07
22	9.23E-09	0.51	4.71E-11	54	2.48E-05	1.03	2.55E-07
23	1.44E-08	3.8	5.46E-10	55	2.92E-05	0.31	9.07E-08
24	9.92E-09	7.4	7.34E-10	56	8.62E-05	0.1	8.62E-08
25	8.33E-09	5.86	4.88E-10	57	1.23E-04	0.21	2.58E-07
26	4.06E-08	2.57	1.04E-09	58	3.29E-04	0.62	2.04E-06
27	1.63E-08	0.82	1.34E-10	59	1.60E-04	0.62	9.92E-07
28	3.97E-08	0.21	8.34E-11	60	1.26E-04	0.21	2.65E-07
29	6.13E-08	0.1	6.13E-11	61	5.20E-04	0.1	5.20E-07
30	7.22E-08	1.95	1.41E-09	62	7.46E-04	0.21	1.57E-06
31	9.50E-08	5.65	5.37E-09	63	1.77E-03	0.31	5.47E-06
32	9.05E-08	5.86	5.30E-09	64	2.62E-03	0.21	5.51E-06

the FPSO, the fairlead may be fatigue failure during FPSO's lifetime. But it needs to be noted that the life prediction of the fairlead is conducted based on the smaller angle. And then, for FPSO, the direction of the wind, wave and current are the same of 180 degree, which are not like this in most cases.

6 CONCLUSIONS

In this paper, a fast fatigue life prediction methodology of the fairlead is proposed. By this methodology, when the mooring system and operating area of floating are defined, it can be used to rapidly predict the fatigue life of the fairlead, which provides a reference for mooring system designers to select fairleads. The conclusions are as follows:

1. For the FPSO mooring system, it's safe and reliable under the environmental condition with maximum wind, waves, and associated currents based on a 100-year recurrence interval. The

maximum Surge value of the FPSO is 96.75 m, which is 9.675% of water depth. The maximum cable tension is 9.059E+06 N, the cables are #2 and #3, and the safely factor is 2.01.

2. Under the environmental condition of a 100-year recurrence interval, the structural strength of the fairlead meets the requirements, the maximum stress of the fairlead is 367.745 MPa and the yield strength is 420 MPa, which the angle between cables on the fairlead is smaller. After that, the fatigue point will be confirmed, which is situated in a region of stress concentration.

3. Most of the fatigue damage of the fairlead is caused by sea states 51 to 64, and the most damaging fatigue sea state is sea state 64 with $Hs = 6.75$ m and $Tp = 10$ s. The total fatigue damage D of the fairlead within one year is 0.054, and the fatigue life of the fairlead is 18.5.

ACKNOWLEDGEMENTS

This research is sponsored by the Excellent Dissertation Cultivation Funds of Wuhan University of Technology (Grant No. 2016-YS-020) and the National Natural Science Foundation of China (Grant No.51679177), these financial supports are gratefully acknowledge.

REFERENCES

ABS, Rules for Building and Classing Single Point Moorings. 2014.

API, Recommended Practice for Design and Analysis of Station Keeping Systems for Floating Structures. API RP 2SK, 2005.

Cummins W E, The impuse response function and ship motions[R]. Department of the NAVY David Taylor Model Basin, report 1661, 1962.

DNV GL, Offshore Standard-Position Mooring. DNV GL OS-E301, 2015.

DNV GL, Fatigue design of offshore steel structures. DNV GL RP-C203, 2016.

DNV GL, Fatigue assessment of ship structures. DNV GL CG-0129, 2015.

Garrett D L, Gordon R B, Chappell J F, Mooring and riser-induced damping in fatigue sea states [C]. Proceedings of the 21st International Conference on offshore Mechanics and Arctic Engineering, Oslo, Norway, June 23–28, 2002: 793~799.

Jiang, H., Gu, Y., and Hu, Z., 2003, "Fatigue Assessment of FPSO," J. of Ship Mech., 7, pp. 70–80 (in Chinese).

Ma, S., and Duan, W. Y., 2014, "Dynamic Coupled Analysis of the Floating Platform Using the Asynchronous Algorithm," J. Mar. Sci. Appl., 13(1), pp. 85–91.

Ma, G., 2009, "Dynamic Research of Deepwater Mooring Line and Riser Based on Elastic Rod Theory," M.S. thesis, Harbin Engineering University, Harbin, China.

McKeown, R., Bisset, A., and McKeown, S. J., Offshore Replacement of a Damaged FPSO Fairlead. Society of Petroleum Engineers, 2011.

Munipalli J, Pistani F, Thiagarajan K P, et al., Weathervaning instabilities of a FPSO in regular waves and consequence on response amplitude operators[C]. Proceedings of 26th International Conference on Offshore Mechanics and Arctic Engineering. San Diego, USA, June 10–15, 2007: OMAE 2007–29359.

Newman J.N., Marine Hydrodynamics[M]. Massachusetts: The Massachusetts Institute of Technology, 1977.

OCIMF, 1994, Prediction of Wind and Current Loads on VLCCs, 2nd ed., Oil Companies International Marine Forum, London.

Ormberg, H., and Larsen, K., 1998, "Coupled Analysis of Floater Motion and Mooring Dynamics for a Turret-Moored Ship," Appl. Ocean Res., 20(1–2), pp.55–67.

Singh M, Ahmad S, Fatigue Life Calculation of Deep Water Composite Production Risers by Rain Flow Cycle Counting Method[C]. ASME 2015, International Conference on Ocean, Offshore and Arctic Engineering. 2015:102–112.

Spanos, P., Wang, J., Peng, B., and Song, S., 2003, "A Procedure for Stochastic Fatigue Analysis of Equipment on Floating Production/Storage/Offloading Structures(FPSO)," *Computational Stochastic Mechanics*, edited by P. D. Spanos and G. Deodatis, Millpress, Rotterdam, pp. 591–598.

Sun, H., and Yong, B., 2003, "Time-Variant Reliability Assessment of FPSO Hull Girders," Mar. Struct., 16, pp. 219–253.

Sun, H., and Soares, C. G., 2003, "Reliability-Based Structural Design of Ship-Type FPSO Units," ASME J. Offshore Mech. Arct. Eng., 125, pp. 108–113.

Wu W, Lv B, Li W, et al, Fatigue Life Analysis Method of Upper-Hinge Joints of FPSO SYMS Based on Real-Time Proto-Type Monitoring Technique[C]. ASME 2016, International Conference on Ocean, Offshore and Arctic Engineering. 2016:V003T02 A027.

Zhao, C., Bai, Y., and Shin, Y., 2001, "Extreme Response and Fatigue Damages For FPSO Structural Analysis," Proc. Int. Offshore and Polar Eng. Conf., Vol. 1, pp. 301–308.

Author index

Printed and bound by CPI Group (UK) Ltd, Croydon, CR0 4YY

17/10/2024

01775696-0004